ENCYCLOPEDIA OF VIROLOGY

THIRD EDITION

EDITORS-IN-CHIEF

Dr BRIAN W J MAHY
and
Dr MARC H V VAN REGENMORTEL

Amsterdam • Boston • Heidelberg • London • New York • Oxford
Paris • San Diego • San Francisco • Singapore • Sydney • Tokyo
Academic Press is an imprint of Elsevier

Academic Press is an imprint of Elsevier
Linacre House, Jordan Hill, Oxford, OX2 8DP, UK
525 B Street, Suite 1900, San Diego, CA 92101-4495, USA

Copyright © 2008 Elsevier Inc. All rights reserved

The following articles are US government works in the public domain and are not subject to copyright:
Bovine Viral Diarrhea Virus, Coxsackieviruses, Prions of Yeast and Fungi, Human Respiratory Syncytial Virus,
Fish Rhabdoviruses, Varicella-Zoster Virus: General Features, Viruses and Bioterrorism, Bean Common Mosaic
Virus and Bean Common Mosaic Necrosis Virus, Metaviruses, Crimean-Congo Hemorrhagic Fever Virus and Other
Nairoviruses, AIDS: Global Epidemiology, Papaya Ringspot Virus, Transcriptional Regulation in Bacteriophage.

Nepovirus, Canadian Crown Copyright 2008

No part of this publication may be reproduced, stored in a retrieval system or transmitted in any form or by
any means electronic, mechanical, photocopying, recording or otherwise without the prior written
permission of the publisher

Permissions may be sought directly from Elsevier's Science & Technology Rights Department in Oxford,
UK: phone (+44) (0) 1865 843830; fax (+44) (0) 1865 853333; email: permissions@elsevier.com. Alternatively
you can submit your request online by visiting the Elsevier web site at (http://elsevier.com/locate/permission),
and selecting *Obtaining permission to use Elsevier material*

Notice
No responsibility is assumed by the publisher for any injury and/or damage to persons or property as a
matter of products liability, negligence or otherwise, or from any use or operation of any methods, products,
instructions or ideas contained in the material herein. Because of rapid advances in the medical sciences,
in particular, independent verification of diagnoses and drug dosages should be made

British Library Cataloguing in Publication Data
A catalogue record for this book is available from the British Library

Library of Congress Catalog Number: 200892260

ISBN: 978-0-12-373935-3

For information on all Elsevier publications
visit our website at books.elsevier.com

PRINTED AND BOUND IN SLOVENIA
08 09 10 11   10 9 8 7 6 5 4 3 2 1

Working together to grow
libraries in developing countries

www.elsevier.com | www.bookaid.org | www.sabre.org

ELSEVIER   BOOK AID International   Sabre Foundation

Encyclopedia of VIROLOGY

Third Edition

EDITORS-IN-CHIEF

Brian W J Mahy MA PhD ScD DSc
Senior Scientific Advisor,
Division of Emerging Infections and Surveillance Services,
Centers for Disease Control and Prevention,
Atlanta GA, USA

Marc H V Van Regenmortel PhD
Emeritus Director at the CNRS,
French National Center for Scientific Research,
Biotechnology School of the University of Strasbourg,
Illkirch, France

ASSOCIATE EDITORS

Dennis H Bamford, Ph.D.
Department of Biological and Environmental Sciences
and Institute of Biotechnology, Biocenter 2,
P.O. Box 56 (Viikinkaari 5),
00014 University of Helsinki,
Finland

Charles Calisher, B.S., M.S., Ph.D.
Arthropod-borne and Infectious Diseases Laboratory
Department of Microbiology, Immunology and Pathology
College of Veterinary Medicine and Biomedical Sciences
Colorado State University
Fort Collins
CO 80523
USA

Andrew J Davison, M.A., Ph.D.
MRC Virology Unit
Institute of Virology
University of Glasgow
Church Street
Glasgow G11 5JR
UK

Claude Fauquet
ILTAB/Donald Danforth Plant Science Center
975 North Warson Road
St. Louis, MO 63132

Said Ghabrial, B.S., M.S., Ph.D.
Plant Pathology Department
University of Kentucky
201F Plant Science Building
1405 Veterans Drive
Lexington
KY 4050546-0312
USA

Eric Hunter, B.Sc., Ph.D.
Department of Pathology and Laboratory Medicine, and
Emory Vaccine Center
Emory University
954 Gatewood Road NE
Atlanta Georgia 30329
USA

Robert A. Lamb, Ph.D., Sc.D.
Department of Biochemistry,
Molecular Biology and Cell Biology
Howard Hughes Medical Institute
Northwestern University
2205 Tech Dr.
Evanston
IL 60208-3500
USA

Olivier Le Gall
IPV, UMR GDPP, IBVM,
INRA Bordeaux-Aquitaine, BP 81,
F-33883 Villenave d'Ornon Cedex
FRANCE

Vincent Racaniello, Ph.D.
Department of Microbiology
Columbia University
New York, NY 10032
USA

David A. Theilmann, Ph.D., B.Sc., M.Sc
Pacific Agri-Food Research Centre
Agriculture and Agri-Food Canada
Box 5000, 4200 Highway 97
Summerland
BC V0H 1Z0
Canada

H. Josef Vetten, Ph.D.
Julius Kuehn Institute, Federal Research Centre for
Cultivated Plants (JKI)
Messeweg 11-12
38104 Braunschweig
Germany

Peter J Walker, B.Sc., Ph.D.
CSIRO Livestock Industries
Australian Animal Health Laboratory (AAHL)
Private Bag 24
Geelong
VIC 3220
Australia

PREFACE

This third edition of the *Encyclopedia of Virology* is being published nine years after the second edition, a period which has seen enormous growth both in our understanding of virology and in our recognition of the viruses themselves, many of which were unknown when the second edition was prepared. Considering viruses affecting human hosts alone, the worldwide epidemic of severe acute respiratory syndrome (SARS), caused by a previously unknown coronavirus, led to the discovery of other human coronaviruses such as HKU1 and NL63. As many as seven chapters are devoted to the AIDS epidemic and to human immunodeficiency viruses. In addition, the development of new molecular technologies led to the discovery of viruses with no obvious disease associations, such as torque-teno virus (one of the most ubiquitous viruses in the human population), human bocavirus, human metapneumovirus, and three new human polyomaviruses.

Other new developments of importance to human virology have included the introduction of a virulent strain of West Nile virus from Israel to North America in 1999. Since that time the virus has become established in mosquito, bird and horse populations throughout the USA, the Caribbean and Mexico as well as the southern regions of Canada.

As in the two previous editions, we have tried to include information about all known species of virus infecting bacteria, fungi, invertebrates, plants and vertebrates, as well as descriptions of related topics in virology such as antiviral drug development, cell- and antibody-mediated immunity, vaccine development, electron microscopy and molecular methods for virus characterization and identification. Many chapters are devoted to the considerable economic importance of virus diseases of cereals, legumes, vegetable crops, fruit trees and ornamentals, and new approaches to control these diseases are reviewed.

General issues such as the origin, evolution and phylogeny of viruses are also discussed as well as the history of the different groups of viruses.

To cover all these subjects and new developments, we have had to increase the size of the Encyclopedia from three to five volumes.

Throughout this work we have relied upon the 8th Report of the International Committee on Taxonomy of Viruses published in 2005, which lists more than 6000 viruses classified into some 2000 virus species distributed among more than 390 different genera and families. In recent years the criteria for placing viruses in different taxa have shifted away from traditional serological methods and increasingly rely upon molecular techniques, particularly the nucleotide sequence of the virus genome. This has changed many of the previous groupings of viruses, and is reflected in this third edition.

Needless to say, a work of this magnitude has involved many expert scientists, who have given generously of their time to bring it to fruition. We extend our grateful thanks to all contributors and associate editors for their excellent and timely contributions.

Brian W J Mahy
Marc H V van Regenmortel

HOW TO USE THE ENCYCLOPEDIA

Structure of the Encyclopedia

The major topics discussed in detail in the text are presented in alphabetical order (see the Alphabetical Contents list which appears in all five volumes).

Finding Specific Information

Information on specific viruses, virus diseases and other matters can be located by consulting the General Index at the end of Volume 5.

Taxonomic Groups of Viruses

For locating detailed information on the major taxonomic groups of viruses, namely virus genera, families and orders, the Taxonomic Index in Volume 5 (page...) should be consulted.

Further Reading sections

The articles do not feature bibliographic citations within the body of the article text itself. The articles are intended to be a first introduction to the topic, or a 'refresher', readable from beginning to end without referring the reader outside of the encyclopedia itself. Bibliographic references to external literature are grouped at the end of each article in a Further Reading section, containing review articles, 'seminal' primary articles and book chapters. These point users to the next level of information for any given topic.

Cross referencing between articles

The "See also" section at the end of each article directs the reader to other entries on related topics. For example. The entry *Lassa, Junin, Machupo and Guanarito Viruses* includes the following cross-references:

See also: Lymphocytic Choriomeningitis Virus: General Features.

CONTRIBUTORS

S T Abedon
The Ohio State University, Mansfield, OH, USA

G P Accotto
Istituto di Virologia Vegetale CNR, Torino, Italy

H-W Ackermann
Laval University, Quebec, QC, Canada

G Adam
Universität Hamburg, Hamburg, Germany

M J Adams
Rothamsted Research, Harpenden, UK

C Adams
University of Duisburg–Essen, Essen, Germany

E Adderson
St. Jude Children's Research Hospital, Memphis, TN, USA

S Adhya
National Institutes of Health, Bethesda, MD, USA

C L Afonso
Southeast Poultry Research Laboratory, Athens, GA, USA

P Ahlquist
University of Wisconsin – Madison, Madison, WI, USA

G M Air
University of Oklahoma Health Sciences Center, Oklahoma City, OK, USA

D J Alcendor
Johns Hopkins School of Medicine, Baltimore, MD, USA

J W Almond
sanofi pasteur, Lyon, France

I Amin
National Institute for Biotechnology and Genetic Engineering, Faisalabad, Pakistan

J Angel
Pontificia Universidad Javeriana, Bogota, Republic of Colombia

C Apetrei
Tulane National Primate Research Center, Covington, LA, USA

B M Arif
Great Lakes Forestry Centre, Sault Ste. Marie, ON, Canada

H Attoui
Faculté de Médecine de Marseilles, Etablissement Français Du Sang, Marseilles, France

H Attoui
Université de la Méditerranée, Marseille, France

H Attoui
Institute for Animal Health, Pirbright, UK

L Aurelian
University of Maryland School of Medicine, Baltimore, MD, USA

L A Babiuk
University of Alberta, Edmonton, AB, Canada

S Babiuk
National Centre for Foreign Animal Disease, Winnipeg, MB, Canada

A G Bader
The Scripps Research Institute, La Jolla, CA, USA

S C Baker
Loyola University of Chicago, Maywood, IL, USA

T S Baker
University of California, San Diego, La Jolla, CA, USA

J K H Bamford
University of Jyväskylä, Jyväskylä, Finland

Y Bao
National Institutes of Health, Bethesda, MD, USA

M Bar-Joseph
The Volcani Center, Bet Dagan, Israel

H Barker
Scottish Crop Research Institute, Dundee, UK

A D T Barrett
University of Texas Medical Branch, Galveston, TX, USA

J W Barrett
The University of Western Ontario, London, ON, Canada

T Barrett
Institute for Animal Health, Pirbright, UK

R Bartenschlager
University of Heidelberg, Heidelberg, Germany

N W Bartlett
Imperial College London, London, UK

S Basak
University of California, San Diego, CA, USA

C F Basler
Mount Sinai School of Medicine, New York, NY, USA

T Basta
Institut Pasteur, Paris, France

D Baxby
University of Liverpool, Liverpool, UK

P Beard
Imperial College London, London, UK

M N Becker
University of Florida, Gainesville, FL, USA

J J Becnel
Agriculture Research Service, Gainesville, FL, USA

K L Beemon
Johns Hopkins University, Baltimore, MD, USA

E D Belay
Centers for Disease Control and Prevention, Atlanta, GA, USA

M Benkő
Veterinary Medical Research Institute, Hungarian Academy of Sciences, Budapest, Hungary

M Bennett
University of Liverpool, Liverpool, UK

M Bergoin
Université Montpellier II, Montpellier, France

H U Bernard
University of California, Irvine, Irvine, CA, USA

K I Berns
University of Florida College of Medicine, Gainesville, FL, USA

P Biagini
Etablissement Français du Sang Alpes-Méditerranée, Marseilles, France

P D Bieniasz
Aaron Diamond AIDS Research Center, The Rockefeller University, New York, NY, USA

Y Bigot
University of Tours, Tours, France

C Billinis
University of Thessaly, Karditsa, Greece

R F Bishop
Murdoch Childrens Research Institute Royal Children's Hospital, Melbourne, VIC, Australia

B A Blacklaws
University of Cambridge, Cambridge, UK

C D Blair
Colorado State University, Fort Collins, CO, USA

S Blanc
INRA–CIRAD–AgroM, Montpellier, France

R Blawid
Institute of Plant Diseases and Plant Protection, Hannover, Germany

G W Blissard
Boyce Thompson Institute at Cornell University, Ithaca, NY, USA

S Blomqvist
National Public Health Institute (KTL), Helsinki, Finland

J F Bol
Leiden University, Leiden, The Netherlands

J-R Bonami
CNRS, Montpellier, France

L Bos
Wageningen University and Research Centre (WUR), Wageningen, The Netherlands

H R Bose Jr.
University of Texas at Austin, Austin, TX, USA

H Bourhy
Institut Pasteur, Paris, France

P R Bowser
Cornell University, Ithaca, NY, USA

D B Boyle
CSIRO Livestock Industries, Geelong, VIC, Australia

C Bragard
Université Catholique de Louvain, Leuven, Belgium

J N Bragg
University of California, Berkeley, Berkeley, CA, USA

R W Briddon
National Institute for Biotechnology and Genetic Engineering, Faisalabad, Pakistan

M A Brinton
Georgia State University, Atlanta, GA, USA

P Britton
Institute for Animal Health, Compton, UK

J K Brown
The University of Arizona, Tucson, AZ, USA

K S Brown
University of Manitoba, Winnipeg, MB, Canada

J Bruenn
State University of New York, Buffalo, NY, USA

C P D Brussaard
Royal Netherlands Institute for Sea Research, Texel, The Netherlands

J J Bugert
Wales College of Medicine, Heath Park, Cardiff, UK

J J Bujarski
Northern Illinois University, DeKalb, IL, USA and Polish Academy of Sciences, Poznan, Poland

R M Buller
Saint Louis University School of Medicine, St. Louis, MO, USA

J P Burand
University of Massachusetts at Amherst, Amherst, MA, USA

J Burgyan
Agricultural Biotechnology Center, Godollo, Hungary

F J Burt
University of the Free State, Bloemfontein, South Africa

S J Butcher
University of Helsinki, Helsinki, Finland

J S Butel
Baylor College of Medicine, Houston, TX, USA

M I Butler
University of Otago, Dunedin, New Zealand

S Bühler
University of Heidelberg, Heidelberg, Germany

P Caciagli
Istituto di Virologia Vegetale – CNR, Turin, Italy

C H Calisher
Colorado State University, Fort Collins, CO, USA

T Candresse
UMR GDPP, Centre INRA de Bordeaux, Villenave d'Ornon, France

A J Cann
University of Leicester, Leicester, UK

C Caranta
INRA, Montfavet, France

G Carlile
CSIRO Livestock Industries, Geelong, VIC, Australia

J P Carr
University of Cambridge, Cambridge, UK

R Carrion, Jr.
Southwest Foundation for Biomedical Research, San Antonio, TX, USA

J W Casey
Cornell University, Ithaca, NY, USA

R N Casey
Cornell University, Ithaca, NY, USA

S Casjens
University of Utah School of Medicine, Salt Lake City, UT, USA

R Cattaneo
Mayo Clinic College of Medicine, Rochester, MN, USA

D Cavanagh
Institute for Animal Health, Compton, UK

A Chahroudi
University of Pennsylvania School of Medicine, Philadelphia, PA, USA

S Chakraborty
Jawaharlal Nehru University, New Delhi, India

T J Chambers
Saint Louis University School of Medicine, St. Louis, MO, USA

Y Chang
University of Pittsburgh Cancer Institute, Pittsburgh, PA, USA

J T Chang
Baylor College of Medicine, Houston, TX, USA

D Chapman
Institute for Animal Health, Pirbright, UK

D Chattopadhyay
University of Calcutta, Kolkata, India

M Chen
University of Arizona, Tucson, AZ, USA

J E Cherwa
University of Arizona, Tucson, AZ, USA

V G Chinchar
University of Mississippi Medical Center, Jackson, MS, USA

A V Chintakuntlawar
University of Oklahoma Health Sciences Center, Oklahoma City, OK, USA

W Chiu
Baylor College of Medicine, Houston, TX, USA

J Chodosh
University of Oklahoma Health Sciences Center, Oklahoma City, OK, USA

I-R Choi
International Rice Research Institute, Los Baños, The Philippines

P D Christian
National Institute of Biological Standards and Control, South Mimms, UK

M G Ciufolini
Istituto Superiore di Sanità, Rome, Italy

P Clarke
University of Colorado Health Sciences, Denver, CO, USA

J-M Claverie
Université de la Méditerranée, Marseille, France

J R Clayton
Johns Hopkins University Schools of Public Health and Medicine, Baltimore, MD, USA

R J Clem
Kansas State University, Manhattan, KS, USA

C J Clements
The Macfarlane Burnet Institute for Medical Research and Public Health Ltd., Melbourne, VIC, Australia

L L Coffey,
Institut Pasteur, Paris, France

J I Cohen
National Institutes of Health, Bethesda, MD, USA

J Collinge
University College London, London, UK

P L Collins
National Institute of Allergy and Infectious Diseases, Bethesda, MD, USA

A Collins
University of Wisconsin School of Medicine and Public Health, Madison, WI, USA

D Contamine
Université Versailles St-Quentin, CNRS, Versailles, France

K M Coombs
University of Manitoba, Winnipeg, MB, Canada

J A Cowley
CSIRO Livestock Industries, Brisbane, QLD, Australia

J K Craigo
University of Pittsburgh School of Medicine, Pittsburgh, PA, USA

M St. J Crane
CSIRO Livestock Industries, Geelong, VIC, Australia

J E Crowe, Jr.
Vanderbilt University Medical Center, Nashville, TN, USA

H Czosnek
The Hebrew University of Jerusalem, Rehovot, Israel

T Dalmay
University of East Anglia, Norwich, UK

B H Dannevig
National Veterinary Institute, Oslo, Norway

C J D'Arcy
University of Illinois at Urbana-Champaign, Urbana, IL, USA

A J Davison
MRC Virology Unit, Glasgow, UK

W O Dawson
University of Florida, Lake Alfred, FL, USA

L A Day
The Public Health Research Institute, Newark, NJ, USA

J C de la Torre
The Scripps Research Institute, La Jolla, CA, USA

X de Lamballerie
Faculté de Médecine de Marseille, Marseilles, France

M de Vega
Universidad Autónoma, Madrid, Spain

P Delfosse
Centre de Recherche Public-Gabriel Lippmann, Belvaux, Luxembourg

B Delmas
INRA, Jouy-en-Josas, France

M Deng
University of California, Berkeley, CA, USA

J DeRisi
University of California, San Francisco, San Francisco, CA, USA

C Desbiez
Institut National de la Recherche Agronomique (INRA), Station de Pathologie Végétale, Montfavet, France

R C Desrosiers
New England Primate Research Center, Southborough, MA, USA

A K Dhar
Advanced BioNutrition Corp, Columbia, MD, USA

R G Dietzgen
The University of Queensland, St. Lucia, QLD, Australia

S P Dinesh-Kumar
Yale University, New Haven, CT, USA

L K Dixon
Institute for Animal Health, Pirbright, UK

C Dogimont
INRA, Montfavet, France

A Domanska
University of Helsinki, Helsinki, Finland

L L Domier
USDA–ARS, Urbana, IL, USA

L L Domier
USDA-ARS, Urbana-Champaign, IL, USA

A Dotzauer
University of Bremen, Bremen, Germany

T W Dreher
Oregon State University, Corvallis, OR, USA

S Dreschers
University of Duisburg–Essen, Essen, Germany

R L Duda
University of Pittsburgh, Pittsburgh, PA, USA

J P Dudley
The University of Texas at Austin, Austin, TX, USA

W P Duprex
The Queen's University of Belfast, Belfast, UK

R E Dutch
University of Kentucky, Lexington, KY, USA

B M Dutia
University of Edinburgh, Edinburgh, UK

M L Dyall-Smith
The University of Melbourne, Parkville, VIC, Australia

J East
University of Texas Medical Branch – Galveston, Galveston, TX, USA

A J Easton
University of Warwick, Coventry, UK

K C Eastwell
Washington State University – IAREC, Prosser, WA, USA

B T Eaton
Australian Animal Health Laboratory, Geelong, VIC, Australia

H Edskes
National Institutes of Health, Bethesda, MD, USA

B Ehlers
Robert Koch-Institut, Berlin, Germany

R M Elliott
University of St. Andrews, St. Andrews, UK

A Engel
National Institutes of Health, Bethesda, MD, USA and
D Kryndushkin
National Institutes of Health, Bethesda, MD, USA

J Engelmann
INRES, University of Bonn, Bonn, Germany

L Enjuanes
CNB, CSIC, Madrid, Spain

A Ensser
Virologisches Institut, Universitätsklinikum, Erlangen, Germany

M Erlandson
Agriculture & Agri-Food Canada, Saskatoon, SK, Canada

K J Ertel
University of California, Irvine, CA, USA

R Esteban
Instituto de Microbiología Bioquímica CSIC/University de Salamanca, Salamanca, Spain

R Esteban
Instituto de Microbiología Bioquímica CSIC/University of Salamanca, Salamanca, Spain

J L Van Etten
University of Nebraska–Lincoln, Lincoln, NE, USA

D J Evans
University of Warwick, Coventry, UK

Ø Evensen
Norwegian School of Veterinary Science, Oslo, Norway

D Falzarano
University of Manitoba, Winnipeg, MB, Canada

B A Fane
University of Arizona, Tucson, AZ, USA

R-X. Fang
Chinese Academy of Sciences, Beijing, People's Republic of China

D Fargette
IRD, Montpellier, France

A Fath-Goodin
University of Kentucky, Lexington, KY, USA

C M Fauquet
Danforth Plant Science Center, St. Louis, MO, USA

B A Federici
University of California, Riverside, CA, USA

H Feldmann
National Microbiology Laboratory, Public Health Agency of Canada, Winnipeg, MB, Canada

H Feldmann
Public Health Agency of Canada, Winnipeg, MB, Canada

F Fenner
Australian National University, Canberra, ACT, Australia

S A Ferreira
University of Hawaii at Manoa, Honolulu, HI, USA

H J Field
University of Cambridge, Cambridge, UK

K Fischer
University of California, San Francisco, San Francisco, CA, USA

J A Fishman
Massachusetts General Hospital, Boston, MA, USA

B Fleckenstein
University of Erlangen – Nürnberg, Erlangen, Germany

R Flores
Instituto de Biología Molecular y Celular de Plantas (UPV-CSIC), Valencia, Spain

T R Flotte
University of Florida College of Medicine, Gainesville, FL, USA

P Forterre
Institut Pasteur, Paris, France

M A Franco
Pontificia Universidad Javeriana, Bogota, Republic of Colombia

T K Frey
Georgia State University, Atlanta, GA, USA

M Fuchs
Cornell University, Geneva, NY, USA

S Fuentes
International Potato Center (CIP), Lima, Peru

T Fujimura
Instituto de Microbiología Bioquímica CSIC/University of Salamanca, Salamanca, Spain

R S Fujinami
University of Utah School of Medicine, Salt Lake City, UT, USA

T Fukuhara
Tokyo University of Agriculture and Technology, Fuchu, Japan

D Gallitelli
Università degli Studi and Istituto di Virologia Vegetale del CNR, Bari, Italy

F García-Arenal
Universidad Politécnica de Madrid, Madrid, Spain

J A García
Centro Nacional de Biotecnología (CNB), CSIC, Madrid, Spain

R A Garrett
Copenhagen University, Copenhagen, Denmark

S Gaumer
Université Versailles St-Quentin, CNRS, Versailles, France

R J Geijskes
Queensland University of Technology, Brisbane, QLD, Australia

T W Geisbert
National Emerging Infectious Diseases Laboratories, Boston, MA, USA

E Gellermann
Hannover Medical School, Hannover, Germany

A Gessain
Pasteur Institute, CNRS URA 3015, Paris, France

S A Ghabrial
University of Kentucky, Lexington, KY, USA

W Gibson
Johns Hopkins University School of Medicine, Baltimore, MD, USA

M Glasa
Slovak Academy of Sciences, Bratislava, Slovakia

Y Gleba
Icon Genetics GmbH, Weinbergweg, Germany

U A Gompels
University of London, London, UK

D Gonsalves
USDA, Pacific Basin Agricultural Research Center, Hilo, HI, USA

M M Goodin
University of Kentucky, Lexington, KY, USA

T J D Goodwin
University of Otago, Dunedin, New Zealand

A E Gorbalenya
Leiden University Medical Center, Leiden, The Netherlands

E A Gould
University of Reading, Reading, UK

A Grakoui
Emory University School of Medicine, Atlanta, GA, USA

M-A Grandbastien
INRA, Versailles, France

R Grassmann
University of Erlangen – Nürnberg, Erlangen, Germany

M Gravell
National Institutes of Health, Bethesda, MD, USA

M V Graves
University of Massachusetts–Lowell, Lowell, MA, USA

K Y Green
National Institutes of Health, Bethesda, MD, USA

H B Greenberg
Stanford University School of Medicine and Veterans Affairs Palo Alto Health Care System, Palo Alto, CA, USA

B M Greenberg
Johns Hopkins School of Medicine, Baltimore, MD, USA

I Greiser-Wilke
School of Veterinary Medicine, Hanover, Germany

D E Griffin
Johns Hopkins Bloomberg School of Public Health, Baltimore, MD, USA

T S Gritsun
University of Reading, Reading, UK

R J de Groot
Utrecht University, Utrecht, The Netherlands

A J Gubala
CSIRO Livestock Industries, Geelong, VIC, Australia

D J Gubler
John A. Burns School of Medicine, Honolulu, HI, USA

A-L Haenni
Institut Jacques Monod, Paris, France

D Haig
Nottingham University, Nottingham, UK

F J Haines
Oxford Brookes University, Oxford, UK

J Hamacher
INRES, University of Bonn, Bonn, Germany

J Hammond
USDA-ARS, Beltsville, MD, USA

R M Harding
Queensland University of Technology, Brisbane, QLD, Australia

J M Hardwick
Johns Hopkins University Schools of Public Health and Medicine, Baltimore, MD, USA

D Hariri
INRA – Département Santé des Plantes et Environnement, Versailles, France

B Harrach
Veterinary Medical Research Institute, Budapest, Hungary

P A Harries
Samuel Roberts Noble Foundation, Inc., Ardmore, OK, USA

L E Harrington
University of Alabama at Birmingham, Birmingham, AL, USA

T J Harrison
University College London, London, UK

T Hatziioannou
Aaron Diamond AIDS Research Center, The Rockefeller University, New York, NY, USA

J Hay
The State University of New York, Buffalo, NY, USA

G S Hayward
Johns Hopkins School of Medicine, Baltimore, MD, USA

E Hébrard
IRD, Montpellier, France

R W Hendrix
University of Pittsburgh, Pittsburgh, PA, USA

L E Hensley
USAMRIID, Fort Detrick, MD, USA

M de las Heras
University of Glasgow Veterinary School, Glasgow, UK

S Hertzler
University of Illinois at Chicago, Chicago, IL, USA

F van Heuverswyn
University of Montpellier 1, Montpellier, France

J Hilliard
Georgia State University, Atlanta, GA, USA

B I Hillman
Rutgers University, New Brunswick, NJ, USA

S Hilton
University of Warwick, Warwick, UK

D M Hinton
National Institutes of Health, Bethesda, MD, USA

A Hinz
UMR 5233 UJF-EMBL-CNRS, Grenoble, France

A E Hoet
The Ohio State University, Columbus, OH, USA

S A Hogenhout
The John Innes Centre, Norwich, UK

T Hohn
Basel university, Institute of Botany, Basel, Switzerland

J S Hong
Seoul Women's University, Seoul, South Korea

M C Horzinek
Utrecht University, Utrecht, The Netherlands

T Hovi
National Public Health Institute (KTL), Helsinki, Finland

A M Huger
Institute for Biological Control, Darmstadt, Germany

L E Hughes
University of St. Andrews, St. Andrews, UK

R Hull
John Innes Centre, Colney, UK

E Hunter
Emory University Vaccine Center, Atlanta, GA, USA

A D Hyatt
Australian Animal Health Laboratory, Geelong, VIC, Australia

T Hyypiä
University of Turku, Turku, Finland

T Iwanami
National Institute of Fruit Tree Science, Tsukuba, Japan

A O Jackson
University of California, Berkeley, CA, USA

P Jardine
University of Minnesota, Minneapolis, MN, USA

J A Jehle
DLR Rheinpfalz, Neustadt, Germany

A R Jilbert
Institute of Medical and Veterinary Science, Adelaide, SA, Australia

P John
Indian Agricultural Research Institute, New Delhi, India

J E Johnson
The Scripps Research Institute, La Jolla, CA, USA

R T Johnson
Johns Hopkins School of Medicine, Baltimore, MD, USA

W E Johnson
New England Primate Research Center, Southborough, MA, USA

S L Johnston
Imperial College London, London, UK

A T Jones
Scottish Crop Research Institute, Dundee, UK

R Jordan
USDA-ARS, Beltsville, MD, USA

Y Kapustin
National Institutes of Health, Bethesda, MD, USA

P Karayiannis
Imperial College London, London, UK

P Kazmierczak
University of California, Davis, CA, USA

K M Keene
Colorado State University, Fort Collins, CO, USA

C Kerlan
Institut National de la Recherche Agronomique (INRA), Le Rheu, France

K Khalili
Temple University School of Medicine, Philadelphia, PA, USA

P H Kilmarx
Centers for Disease Control and Prevention, Atlanta, GA, USA

L A King
Oxford Brookes University, Oxford, UK

P D Kirkland
Elizabeth Macarthur Agricultural Institute, Menangle, NSW, Australia

C D Kirkwood
Murdoch Childrens Research Institute Royal Children's Hospital, Melbourne, VIC, Australia

R P Kitching
Canadian Food Inspection Agency, Winnipeg, MB, Canada

P J Klasse
Cornell University, New York, NY, USA

N R Klatt
University of Pennsylvania School of Medicine, Philadelphia, PA, USA

R G Kleespies
Institute for Biological Control, Darmstadt, Germany

D F Klessig
Boyce Thompson Institute for Plant Research, Ithaca, NY, USA

W B Klimstra
Louisiana State University Health Sciences Center at Shreveport, Shreveport, LA, USA

V Klimyuk
Icon Genetics GmbH, Weinbergweg, Germany

N Knowles
Institute for Animal Health, Pirbright, UK

R Koenig
Biologische Bundesanstalt für Land- und Forstwirtschaft, Brunswick, Germany

R Koenig
Institut für Pflanzenvirologie, Mikrobiologie und biologische Sicherheit, Brunswick, Germany

G Konaté
INERA, Ouagadougou, Burkina Faso

C N Kotton
Massachusetts General Hospital, Boston, MA, USA

L D Kramer
Wadsworth Center, New York State Department of Health, Albany, NY, USA

P J Krell
University of Guelph, Guelph, ON, Canada

J Kreuze
International Potato Center (CIP), Lima, Peru

M J Kuehnert
Centers for Disease Control and Prevention, Atlanta, GA, USA

R J Kuhn
Purdue University, West Lafayette, IN, USA

G Kurath
Western Fisheries Research Center, Seattle, WA, USA

I Kusters
sanofi pasteur, Lyon, France

I V Kuzmin
Centers for Disease Control and Prevention, Atlanta, GA, USA

M E Laird
New England Primate Research Center, Southborough, MA, USA

R A Lamb
Howard Hughes Medical Institute at Northwestern University, Evanston, IL, USA

P F Lambert
University of Wisconsin School of Medicine and Public Health, Madison, WI, USA

A S Lang
Memorial University of Newfoundland, St. John's, NL, Canada

H D Lapierre
INRA – Département Santé des Plantes et Environnement, Versailles, France

G Lawrence
The Children's Hospital at Westmead, Westmead, NSW, Australia and
University of Sydney, Westmead, NSW, Australia

H Lecoq
Institut National de la Recherche Agronomique (INRA), Station de Pathologie Végétale, Montfavet, France

B Y Lee
Seoul Women's University, Seoul, South Korea

E J Lefkowitz
University of Alabama at Birmingham, Birmingham, AL, USA

J P Legg
International Institute of Tropical Agriculture, Dar es Salaam, Tanzania,
UK and
Natural Resources Institute, Chatham Maritime, UK

P Leinikki
National Public Health Institute, Helsinki, Finland

J Lenard
University of Medicine and Dentistry of New Jersey (UMDNJ), Piscataway, NJ, USA

J C Leong
University of Hawaii at Manoa, Honolulu, HI, USA

K N Leppard
University of Warwick, Coventry, UK

A Lescoute
Université Louis Pasteur, Strasbourg, France

D-E Lesemann
Biologische Bundesanstalt für Land- und Forstwirtschaft, Brunswick, Germany

J-H Leu
National Taiwan University, Taipei, Republic of China

H L Levin
National Institutes of Health, Bethesda, MD, USA

D J Lewandowski
The Ohio State University, Columbus, OH, USA

H-S Lim
University of California, Berkeley, Berkeley, CA, USA

M D A Lindsay
Western Australian Department of Health, Mount Claremont, WA, Australia

R Ling
University of Warwick, Coventry, UK

M L Linial
Fred Hutchinson Cancer Research Center, Seattle, WA, USA

D C Liotta
Emory University, Atlanta, GA, USA

W Ian Lipkin
Columbia University, New York, NY, USA

H L Lipton
University of Illinois at Chicago, Chicago, IL, USA

A S Liss
University of Texas at Austin, Austin, TX, USA

J J López-Moya
Instituto de Biología Molecular de Barcelona (IBMB), CSIC, Barcelona, Spain

G Loebenstein
Agricultural Research Organization, Bet Dagan, Israel

C-F Lo
National Taiwan University, Taipei, Republic of China

S A Lommel
North Carolina State University, Raleigh, NC, USA

G P Lomonossoff
John Innes Centre, Norwich, UK

M Luo
University of Alabama at Birmingham, Birmingham, AL, USA

S A MacFarlane
Scottish Crop Research Institute, Dundee, UK

J S Mackenzie
Curtin University of Technology, Shenton Park, WA, Australia

R Mahieux
Pasteur Institute, CNRS URA 3015, Paris, France

B W J Mahy
Centers for Disease Control and Prevention, Atlanta, GA, USA

E Maiss
Institute of Plant Diseases and Plant Protection, Hannover, Germany

E O Major
National Institutes of Health, Bethesda, MD, USA

V G Malathi
Indian Agricultural Research Institute, New Delhi, India

A Mankertz
Robert Koch-Institut, Berlin, Germany

S Mansoor
National Institute for Biotechnology and Genetic Engineering, Faisalabad, Pakistan

A A Marfin
Centers for Disease Control and Prevention, Atlanta, GA, USA

S Marillonnet
Icon Genetics GmbH, Weinbergweg, Germany

G P Martelli
Università degli Studi and Istituto di Virologia vegetale CNR, Bari, Italy

M Marthas
University of California, Davis, Davis, CA, USA

D P Martin
University of Cape Town, Cape Town, South Africa

P A Marx
Tulane University, Covington, LA, USA

W S Mason
Fox Chase Cancer Center, Philadelphia, PA, USA

T D Mastro
Centers for Disease Control and Prevention, Atlanta, GA, USA

A A McBride
National Institutes of Health, Bethesda, MD, USA

L McCann
National Institutes of Health, Bethesda, MD, USA

M McChesney
University of California, Davis, Davis, CA, USA

J B McCormick
University of Texas, School of Public Health, Brownsville, TX, USA

G McFadden
University of Florida, Gainesville, FL, USA

G McFadden
The University of Western Ontario, London, ON, Canada

D B McGavern
The Scripps Research Institute, La Jolla, CA, USA

A L McNees
Baylor College of Medicine, Houston, TX, USA

M Meier
Tallinn University of Technology, Tallinn, Estonia

P S Mellor
Institute for Animal Health, Woking, UK

X J Meng
Virginia Polytechnic Institute and State University, Blacksburg, VA, USA

A A Mercer
University of Otago, Dunedin, New Zealand

P P C Mertens
Institute for Animal Health, Woking, UK

T C Mettenleiter
Friedrich-Loeffler-Institut, Greifswald-Insel Riems, Germany

H Meyer
Bundeswehr Institute of Microbiology, Munich, Germany

R F Meyer
Centers for Disease Control and Prevention, Atlanta, GA, USA

P de Micco
Etablissement Français du Sang Alpes-Méditerranée, Marseilles, France

B R Miller
Centers for Disease Control and Prevention (CDC), Fort Collins, CO, USA

C J Miller
University of California, Davis, Davis, CA, USA

R G Milne
Istituto di Virologia Vegetale CNR, Torino, Italy

P D Minor
NIBSC, Potters Bar, UK

S Mjaaland
Norwegian School of Veterinary Science, Oslo, Norway

E S Mocarski
Emory University School of Medicine, Atlanta, GA, USA

E S Mocarski, Jr.
Emory University School of Medicine, Emory, GA, USA

V Moennig
School of Veterinary Medicine, Hanover, Germany

P Moffett
Boyce Thompson Institute for Plant Research, Ithaca, NY, USA

T P Monath
Kleiner Perkins Caufield and Byers, Menlo Park, CA, USA

R C Montelaro
University of Pittsburgh School of Medicine, Pittsburgh, PA, USA

P S Moore
University of Pittsburgh Cancer Institute, Pittsburgh, PA, USA

F J Morales
International Center for Tropical Agriculture, Cali, Colombia

H Moriyama
Tokyo University of Agriculture and Technology, Fuchu, Japan

T J Morris
University of Nebraska, Lincoln, NE, USA

S A Morse
Centers for Disease Control and Prevention, Atlanta, GA, USA

L Moser
University of Wisconsin – Madison, Madison, WI, USA

B Moury
INRA – Station de Pathologie Végétale, Montfavet, France

J W Moyer
North Carolina State University, Raleigh, NC, USA

R W Moyer
University of Florida, Gainesville, FL, USA

E Muller
CIRAD/UMR BGPI, Montpellier, France

F A Murphy
University of Texas Medical Branch, Galveston, TX, USA

A Müllbacher
Australian National University, Canberra, ACT, Australia

K Nagasaki
Fisheries Research Agency, Hiroshima, Japan

T Nakayashiki
National Institutes of Health, Bethesda, MD, USA

A A Nash
University of Edinburgh, Edinburgh, UK

N Nathanson
University of Pennsylvania, Philadelphia, PA, USA

C K Navaratnarajah
Purdue University, West Lafayette, IN, USA

M S Nawaz-ul-Rehman
Danforth Plant Science Center, St. Louis, MO, USA

J C Neil
University of Glasgow, Glasgow, UK

R S Nelson
Samuel Roberts Noble Foundation, Inc., Ardmore, OK, USA

P Nettleton
Moredun Research Institute, Edinburgh, UK

A W Neuman
Emory University, Atlanta, GA, USA

A R Neurath
Virotech, New York, NY, USA

M L Nibert
Harvard Medical School, Boston, MA, USA

L Nicoletti
Istituto Superiore di Sanità, Rome, Italy

N Noah
London School of Hygiene and Tropical Medicine, London, UK

D L Nuss
University of Maryland Biotechnology Institute, Rockville, MD, USA

M S Oberste
Centers for Disease Control and Prevention, Atlanta, GA, USA

W A O'Brien
University of Texas Medical Branch – Galveston, Galveston, TX, USA

D J O'Callaghan
Louisiana State University Health Sciences Center, Shreveport, LA, USA

W F Ochoa
University of California, San Diego, La Jolla, CA, USA

M R Odom
University of Alabama at Birmingham, Birmingham, AL, USA

M M van Oers
Wageningen University, Wageningen, The Netherlands

M B A Oldstone
The Scripps Research Institute, La Jolla, CA, USA

G Olinger
USAMRIID, Fort Detrick, MD, USA

K E Olson
Colorado State University, Fort Collins, CO, USA

A Olspert
Tallinn University of Technology, Tallinn, Estonia

G Orth
Institut Pasteur, Paris, France

J E Osorio
University of Wisconsin, Madison, WI, USA

N Osterrieder
Cornell University, Ithaca, NY, USA

S A Overman
University of Missouri – Kansas City, Kansas City, MO, USA

R A Owens
Beltsville Agricultural Research Center, Beltsville, MD, USA

M S Padmanabhan
Yale University, New Haven, CT, USA

S Paessler
University of Texas Medical Branch, Galveston, TX, USA

P Palese
Mount Sinai School of Medicine, New York, NY, USA

M A Pallansch
Centers for Disease Control and Prevention, Atlanta, GA, USA

M Palmarini
University of Glasgow Veterinary School, Glasgow, UK

P Palukaitis
Scottish Crop Research Institute, Invergowrie, Dundee, UK

I Pandrea
Tulane National Primate Research Center, Covington, LA, USA

O Papadopoulos
Aristotle University, Thessaloniki, Greece

H R Pappu
Washington State University, Pullman, WA, USA

S Parker
Saint Louis University School of Medicine, St. Louis, MO, USA

C R Parrish
Cornell University, Ithaca, NY, USA

R F Pass
University of Alabama School of Medicine, Birmingham, AL, USA

J L Patterson
Southwest Foundation for Biomedical Research, San Antonio, TX, USA

T A Paul
Cornell University, Ithaca, NY, USA

A E Peaston
The Jackson Laboratory, Bar Harbor, ME, USA

M Peeters
University of Montpellier 1, Montpellier, France

J S M Peiris
The University of Hong Kong, Hong Kong, People's Republic of China

P J Peters
Centers for Disease Control and Prevention, Atlanta, GA, USA

M Pfeffer
Bundeswehr Institute of Microbiology, Munich, Germany

H Pfister
University of Köln, Cologne, Germany

O Planz
Federal Research Institute for Animal Health, Tuebingen, Gemany

L L M Poon
The University of Hong Kong, Hong Kong, People's Republic of China

M M Poranen
University of Helsinki, Helsinki, Finland

K Porter
The University of Melbourne, Parkville, VIC, Australia

A Portner
St. Jude Children's Research Hospital, Memphis, TN, USA

R D Possee
NERC Institute of Virology and Environmental Microbiology, Oxford, UK

R T M Poulter
University of Otago, Dunedin, New Zealand

A M Powers
Centers for Disease Control and Prevention, Fort Collins, CO, USA

D Prangishvili
Institut Pasteur, Paris, France

C M Preston
Medical Research Council Virology Unit, Glasgow, UK

S L Quackenbush
Colorado State University, Fort Collins, CO, USA

F Qu
University of Nebraska, Lincoln, NE, USA

B C Ramirez
CNRS, Paris, France

A Rapose
University of Texas Medical Branch – Galveston, Galveston, TX, USA

D V R Reddy
Hyderabad, India

A J Redwood
The University of Western Australia, Crawley, WA, Australia

M Regner
Australian National University, Canberra, ACT, Australia

W K Reisen
University of California, Davis, CA, USA

T Renault
IFREMER, La Tremblade, France

P A Revill
Victorian Infectious Diseases Reference Laboratory, Melbourne, VIC, Australia

A Rezaian
University of Adelaide, Adelaide, SA, Australia

J F Ridpath
USDA, Ames, IA, USA

B K Rima
The Queen's University of Belfast, Belfast, UK

E Rimstad
Norwegian School of Veterinary Science, Oslo, Norway

F J Rixon
MRC Virology Unit, Glasgow, UK

Y-T Ro
Konkuk University, Seoul, South Korea

C M Robinson
University of Oklahoma Health Sciences Center, Oklahoma City, OK, USA

G F Rohrmann
Oregon State University, Corvallis, OR, USA

M Roivainen
National Public Health Institute (KTL), Helsinki, Finland

L Roux
University of Geneva Medical School, Geneva, Switzerland

J Rovnak
Colorado State University, Fort Collins, CO, USA

D J Rowlands
University of Leeds, Leeds, UK

P Roy
London School of Hygiene and Tropical Medicine, London, UK

L Rubino
Istituto di Virologia Vegetale del CNR, Bari, Italy

R W H Ruigrok
CNRS, Grenoble, France

C E Rupprecht
Centers for Disease Control and Prevention, Atlanta, GA, USA

R J Russell
University of St. Andrews, St. Andrews, UK

B E Russ
The University of Melbourne, Parkville, VIC, Australia

W T Ruyechan
The State University of New York, Buffalo, NY, USA

E Ryabov
University of Warwick, Warwick, UK

M D Ryan
University of St. Andrews, St. Andrews, UK

E P Rybicki
University of Cape Town, Cape Town, South Africa

K D Ryman
Louisiana State University Health Sciences Center at Shreveport, Shreveport, LA, USA

K D Ryman
Louisiana State University Health Sciences Center, Shreveport, LA, USA

K H Ryu
Seoul Women's University, Seoul, South Korea

M Safak
Temple University School of Medicine, Philadelphia, PA, USA

M Salas
Universidad Autónoma, Madrid, Spain

S K Samal
University of Maryland, College Park, MD, USA

J T Sample
The Pennsylvania State University College of Medicine, Hershey, PA, USA

C E Sample
The Pennsylvania State University College of Medicine, Hershey, PA, USA

R M Sandri-Goldin
University of California, Irvine, Irvine, CA, USA

H Sanfaçon
Pacific Agri-Food Research Centre, Summerland, BC, Canada

R Sanjuán
Instituto de Biología Molecular y Cellular de Plantas, CSIC-UPV, Valencia, Spain

N Santi
Norwegian School of Veterinary Science, Oslo, Norway

C Sarmiento
Tallinn University of Technology, Tallinn, Estonia

T Sasaya
National Agricultural Research Center, Ibaraki, Japan

Q J Sattentau
University of Oxford, Oxford, UK

C Savolainen-Kopra
National Public Health Institute (KTL), Helsinki, Finland

B Schaffhausen
Tufts University School of Medicine, Boston, MA, USA

K Scheets
Oklahoma State University, Stillwater, OK, USA

M J Schmitt
University of the Saarland, Saarbrücken, Germany

A Schneemann
The Scripps Research Institute, La Jolla, CA, USA

G Schoehn
CNRS, Grenoble, France

J E Schoelz
University of Missouri, Columbia, MO, USA

L B Schonberger
Centers for Disease Control and Prevention, Atlanta, GA, USA

U Schubert
Klinikum der Universität Erlangen-Nürnberg, Erlangen, Germany

D A Schultz
Johns Hopkins University School of Medicine, Baltimore, MD, USA

S Schultz-Cherry
University of Wisconsin – Madison, Madison, WI, USA

T F Schulz
Hannover Medical School, Hannover, Germany

P D Scotti
Waiatarua, New Zealand

B L Semler
University of California, Irvine, CA, USA

J M Sharp
Veterinary Laboratories Agency, Penicuik, UK

M L Shaw
Mount Sinai School of Medicine, New York, NY, USA

G R Shellam
The University of Western Australia,
Crawley, WA, Australia

D N Shepherd
University of Cape Town, Cape Town, South Africa

N C Sheppard
University of Oxford, Oxford, UK

F Shewmaker
National Institutes of Health, Bethesda, MD, USA

P A Signoret
Montpellier SupAgro, Montpellier, France

A Silaghi
University of Manitoba, Winnipeg, MB, Canada

G Silvestri
University of Pennsylvania, Philadelphia, PA, USA

T L Sit
North Carolina State University, Raleigh, NC, USA

N Sittidilokratna
Centex Shrimp and Center for Genetic Engineering and
Biotechnology, Bangkok, Thailand

M A Skinner
Imperial College London, London, UK

D W Smith
PathWest Laboratory Medicine WA, Nedlands, WA,
Australia

G L Smith
Imperial College London, London, UK

L M Smith
The University of Western Australia,
Crawley, WA, Australia

E J Snijder
Leiden University Medical Center, Leiden, The
Netherlands

M Sova
University of Texas Medical Branch – Galveston,
Galveston, TX, USA

J A Speir
The Scripps Research Institute, La Jolla, CA, USA

T E Spencer
Texas A&M University, College Station, TX, USA

P Sreenivasulu
Sri Venkateswara University, Tirupati, India

J Stanley
John Innes Centre, Colney, UK

K M Stedman
Portland State University, Portland, OR, USA

D Stephan
Institute of Plant Diseases and Plant Protection,
Hannover, Germany

C C M M Stijger
Wageningen University and Research Centre, Naaldwijk,
The Netherlands

L Stitz
Federal Research Institute for Animal Health, Tuebingen,
Gemany

P G Stockley
University of Leeds, Leeds, UK

M R Strand
University of Georgia, Athens, GA, USA

M J Studdert
The University of Melbourne, Parkville, VIC, Australia

C A Suttle
University of British Columbia, Vancouver, BC,
Canada

N Suzuki
Okayama University, Okayama, Japan

J Y Suzuki
USDA, Pacific Basin Agricultural Research Center, Hilo,
HI, USA

R Swanepoel
National Institute for Communicable Diseases,
Sandringham, South Africa

S J Symes
The University of Melbourne, Parkville, VIC, Australia

G Szittya
Agricultural Biotechnology Center, Godollo, Hungary

M Taliansky
Scottish Crop Research Institute, Dundee, UK

P Tattersall
Yale University Medical School, New Haven, CT, USA

T Tatusova
National Institutes of Health, Bethesda, MD, USA

S Tavantzis
University of Maine, Orono, ME, USA

J M Taylor
Fox Chase Cancer Center, Philadelphia, PA, USA

D A Theilmann
Agriculture and Agri-Food Canada, Summerland, BC,
Canada

F C Thomas Allnutt
National Science Foundation, Arlington, VA, USA

G J Thomas Jr.
University of Missouri – Kansas City, Kansas City, MO, USA

J E Thomas
Department of Primary Industries and Fisheries, Indooroopilly, QLD, Australia

H C Thomas
Imperial College London, London, UK

A N Thorburn
The University of Melbourne, Parkville, VIC, Australia

P Tijssen
Université du Québec, Laval, QC, Canada

S A Tolin
Virginia Polytechnic Institute and State University, Blacksburg, VA, USA

L Torrance
Scottish Crop Research Institute, Invergowrie, UK

S Trapp
Cornell University, Ithaca, NY, USA

S Tripathi
USDA, Pacific Basin Agricultural Research Center, Hilo, HI, USA

E Truve
Tallinn University of Technology, Tallinn, Estonia

J-M Tsai
National Taiwan University, Taipei, Republic of China

M Tsompana
North Carolina State University, Raleigh, NC, USA

R Tuma
University of Helsinki, Helsinki, Finland

A S Turnell
The University of Birmingham, Birmingham, UK

K L Tyler
University of Colorado Health Sciences, Denver, CO, USA

A Uchiyama
Cornell University, Ithaca, NY, USA

C Upton
University of Victoria, Victoria, BC, Canada

A Urisman
University of California, San Francisco, San Francisco, CA, USA

J K Uyemoto
University of California, Davis, CA, USA

A Vaheri
University of Helsinki, Helsinki, Finland

R Vainionpää
University of Turku, Turku, Finland

A M Vaira
Istituto di Virologia Vegetale, CNR, Turin, Italy

N K Van Alfen
University of California, Davis, CA, USA

R A A Van der Vlugt
Wageningen University and Research Centre, Wageningen, The Netherlands

M H V Van Regenmortel
CNRS, Illkirch, France

P A Venter
The Scripps Research Institute, La Jolla, CA, USA

J Verchot-Lubicz
Oklahoma State University, Stillwater, OK, USA

R A Vere Hodge
Vere Hodge Antivirals Ltd., Reigate, UK

H J Vetten
Federal Research Centre for Agriculture and Forestry (BBA), Brunswick, Germany

L P Villarreal
University of California, Irvine, Irvine, CA, USA

J M Vlak
Wageningen University, Wageningen, The Netherlands

P K Vogt
The Scripps Research Institute, La Jolla, CA, USA

L E Volkman
University of California, Berkeley, Berkeley, CA, USA

J Votteler
Klinikum der Universität Erlangen-Nürnberg, Erlangen, Germany

D F Voytas
Iowa State University, Ames, IA, USA

J D F Wadsworth
University College London, London, UK

E K Wagner
University of California, Irvine, Irvine, CA, USA

P J Walker
CSIRO Australian Animal Health Laboratory, Geelong, VIC, Australia

A L Wang
University of California, San Francisco, CA, USA

X Wang
University of Wisconsin – Madison, Madison, WI, USA

C C Wang
University of California, San Francisco, CA, USA

L-F Wang
Australian Animal Health Laboratory, Geelong, VIC, Australia

R Warrier
Purdue University, West Lafayette, IN, USA

S C Weaver
University of Texas Medical Branch, Galveston, TX, USA

B A Webb
University of Kentucky, Lexington, KY, USA

F Weber
University of Freiburg, Freiburg, Germany

R P Weir
Berrimah Research Farm, Darwin, NT, Australia

R A Weisberg
National Institutes of Health, Bethesda, MD, USA

W Weissenhorn
UMR 5233 UJF-EMBL-CNRS, Grenoble, France

R M Welsh
University of Massachusetts Medical School, Worcester, MA, USA

J T West
University of Oklahoma Health Sciences Center, Oklahoma City, OK, USA

E Westhof
Université Louis Pasteur, Strasbourg, France

S P J Whelan
Harvard Medical School, Boston, MA, USA

R L White
Texas A&M University, College Station, TX, USA

C A Whitehouse
United States Army Medical Research Institute of Infectious Diseases, Frederick, MD, USA

R B Wickner
National Institutes of Health, Bethesda, MD, USA

R G Will
Western General Hospital, Edinburgh, UK

T Williams
Instituto de Ecología A.C., Xalapa, Mexico

K Willoughby
Moredun Research Institute, Edinburgh, UK

S Winter
Deutsche Sammlung für Mikroorganismen und Zellkulturen, Brunswick, Germany

J Winton
Western Fisheries Research Center, Seattle, WA, USA

J K Yamamoto
University of Florida, Gainesville, FL, USA

M Yoshida
University of Tokyo, Chiba, Japan

N Yoshikawa
Iwate University, Ueda, Japan

L S Young
University of Birmingham, Birmingham, UK

R F Young, III
Texas A&M University, College Station, TX, USA

T M Yuill
University of Wisconsin, Madison, WI, USA

A J Zajac
University of Alabama at Birmingham, Birmingham, AL, USA

S K Zavriev
Shemyakin and Ovchinnikov Institute of Bioorganic Chemistry, Russian Academy of Sciences, Moscow, Russia

J Ziebuhr
The Queen's University of Belfast, Belfast, UK

E I Zuniga
The Scripps Research Institute, La Jolla, CA, USA

CONTENTS

Editors-in-Chief	v
Associate Editors	vii
Preface	ix
How to Use the Encyclopedia	xi
Contributors	xiii

VOLUME 1

A

Adenoviruses: General Features	B Harrach	1
Adenoviruses: Malignant Transformation and Oncology	A S Turnell	9
Adenoviruses: Molecular Biology	K N Leppard	17
Adenoviruses: Pathogenesis	M Benkő	24
African Cassava Mosaic Disease	J P Legg	30
African Horse Sickness Viruses	P S Mellor and P P C Mertens	37
African Swine Fever Virus	L K Dixon and D Chapman	43
AIDS: Disease Manifestation	A Rapose, J East, M Sova and W A O'Brien	51
AIDS: Global Epidemiology	P J Peters, P H Kilmarx and T D Mastro	58
AIDS: Vaccine Development	N C Sheppard and Q J Sattentau	69
Akabane Virus	P S Mellor and P D Kirkland	76
Alfalfa Mosaic Virus	J F Bol	81
Algal Viruses	K Nagasaki and C P D Brussaard	87
Allexivirus	S K Zavriev	96
Alphacryptovirus and *Betacryptovirus*	R Blawid, D Stephan and E Maiss	98
Anellovirus	P Biagini and P de Micco	104
Animal Rhabdoviruses	H Bourhy, A J Gubala, R P Weir and D B Boyle	111
Antigen Presentation	E I Zuniga, D B McGavern and M B A Oldstone	121
Antigenic Variation	G M Air and J T West	127
Antigenicity and Immunogenicity of Viral Proteins	M H V Van Regenmortel	137

Antiviral Agents H J Field and R A Vere Hodge	142
Apoptosis and Virus Infection J R Clayton and J M Hardwick	154
Aquareoviruses M St J Crane and G Carlile	163
Arboviruses B R Miller	170
Arteriviruses M A Brinton and E J Snijder	176
Ascoviruses B A Federici and Y Bigot	186
Assembly of Viruses: Enveloped Particles C K Navaratnarajah, R Warrier and R J Kuhn	193
Assembly of Viruses: Nonenveloped Particles M Luo	200
Astroviruses L Moser and S Schultz-Cherry	204

B

Baculoviruses: Molecular Biology of Granuloviruses S Hilton	211
Baculoviruses: Molecular Biology of Mosquito Baculoviruses J J Becnel and C L Afonso	219
Baculoviruses: Molecular Biology of Sawfly Baculoviruses B M Arif	225
Baculoviruses: Apoptosis Inhibitors R J Clem	231
Baculoviruses: Expression Vector F J Haines, R D Possee and L A King	237
Baculoviruses: General Features P J Krell	247
Baculoviruses: Molecular Biology of Nucleopolyhedroviruses D A Theilmann and G W Blissard	254
Baculoviruses: Pathogenesis L E Volkman	265
Banana Bunchy Top Virus J E Thomas	272
Barley Yellow Dwarf Viruses L L Domier	279
Barnaviruses P A Revill	286
Bean Common Mosaic Virus and Bean Common Mosaic Necrosis Virus R Jordan and J Hammond	288
Bean Golden Mosaic Virus F J Morales	295
Beet Curly Top Virus J Stanley	301
Benyvirus R Koenig	308
Beta ssDNA Satellites R W Briddon and S Mansoor	314
Birnaviruses B Delmas	321
Bluetongue Viruses P Roy	328
Border Disease Virus P Nettleton and K Willoughby	335
Bornaviruses L Stitz, O Planz and W Ian Lipkin	341
Bovine and Feline Immunodeficiency Viruses J K Yamamoto	347
Bovine Ephemeral Fever Virus P J Walker	354
Bovine Herpesviruses M J Studdert	362
Bovine Spongiform Encephalopathy R G Will	368
Bovine Viral Diarrhea Virus J F Ridpath	374
Brome Mosaic Virus X Wang and P Ahlquist	381
Bromoviruses J J Bujarski	386

Bunyaviruses: General Features R M Elliott	390
Bunyaviruses: Unassigned C H Calisher	399

C

Cacao Swollen Shoot Virus E Muller	403
Caliciviruses M J Studdert and S J Symes	410
Capillovirus, *Foveavirus*, *Trichovirus*, *Vitivirus* N Yoshikawa	419
Capripoxviruses R P Kitching	427
Capsid Assembly: Bacterial Virus Structure and Assembly S Casjens	432
Cardioviruses C Billinis and O Papadopoulos	440
Carlavirus K H Ryu and B Y Lee	448
Carmovirus F Qu and T J Morris	453
Caulimoviruses: General Features J E Schoelz	457
Caulimoviruses: Molecular Biology T Hohn	464
Central Nervous System Viral Diseases R T Johnson and B M Greenberg	469
Cereal Viruses: Maize/Corn P A Signoret	475
Cereal Viruses: Rice F Morales	482
Cereal Viruses: Wheat and Barley H D Lapierre and D Hariri	490
Chandipura Virus S Basak and D Chattopadhyay	497
Chrysoviruses S A Ghabrial	503
Circoviruses A Mankertz	513
Citrus Tristeza Virus M Bar-Joseph and W O Dawson	520
Classical Swine Fever Virus V Moennig and I Greiser-Wilke	525
Coltiviruses H Attoui and X de Lamballerie	533
Common Cold Viruses S Dreschers and C Adams	541
Coronaviruses: General Features D Cavanagh and P Britton	549
Coronaviruses: Molecular Biology S C Baker	554
Cotton Leaf Curl Disease S Mansoor, I Amin and R W Briddon	563
Cowpea Mosaic Virus G P Lomonossoff	569
Cowpox Virus M Bennett, G L Smith and D Baxby	574
Coxsackieviruses M S Oberste and M A Pallansch	580
Crenarchaeal Viruses: Morphotypes and Genomes D Prangishvili, T Basta and R A Garrett	587
Crimean–Congo Hemorrhagic Fever Virus and Other Nairoviruses C A Whitehouse	596
Cryo-Electron Microscopy W Chiu, J T Chang and F J Rixon	603
Cucumber Mosaic Virus F García-Arenal and P Palukaitis	614
Cytokines and Chemokines D E Griffin	620
Cytomegaloviruses: Murine and Other Nonprimate Cytomegaloviruses A J Redwood, L M Smith and G R Shellam	624
Cytomegaloviruses: Simian Cytomegaloviruses D J Alcendor and G S Hayward	634

VOLUME 2

D

Defective-Interfering Viruses L Roux	1
Dengue Viruses D J Gubler	5
Diagnostic Techniques: Microarrays K Fischer, A Urisman and J DeRisi	14
Diagnostic Techniques: Plant Viruses R Koenig, D-E Lesemann, G Adam and S Winter	18
Diagnostic Techniques: Serological and Molecular Approaches R Vainionpää and P Leinikki	29
Dicistroviruses P D Christian and P D Scotti	37
Disease Surveillance N Noah	44
DNA Vaccines S Babiuk and L A Babiuk	51

E

Ebolavirus K S Brown, A Silaghi and H Feldmann	57
Echoviruses T Hyypiä	65
Ecology of Viruses Infecting Bacteria S T Abedon	71
Electron Microscopy of Viruses G Schoehn and R W H Ruigrok	78
Emerging and Reemerging Virus Diseases of Plants G P Martelli and D Gallitelli	86
Emerging and Reemerging Virus Diseases of Vertebrates B W J Mahy	93
Emerging Geminiviruses C M Fauquet and M S Nawaz-ul-Rehman	97
Endogenous Retroviruses W E Johnson	105
Endornavirus T Fukuhara and H Moriyama	109
Enteric Viruses R F Bishop and C D Kirkwood	116
Enteroviruses of Animals L E Hughes and M D Ryan	123
Enteroviruses: Human Enteroviruses Numbered 68 and Beyond T Hovi, S Blomqvist, C Savolainen-Kopra and M Roivainen	130
Entomopoxviruses M N Becker and R W Moyer	136
Epidemiology of Human and Animal Viral Diseases F A Murphy	140
Epstein–Barr Virus: General Features L S Young	148
Epstein–Barr Virus: Molecular Biology J T Sample and C E Sample	157
Equine Infectious Anemia Virus J K Craigo and R C Montelaro	167
Evolution of Viruses L P Villarreal	174

F

Feline Leukemia and Sarcoma Viruses J C Neil	185
Filamentous ssDNA Bacterial Viruses S A Overman and G J Thomas Jr.	190
Filoviruses G Olinger, T W Geisbert and L E Hensley	198
Fish and Amphibian Herpesviruses A J Davison	205
Fish Retroviruses T A Paul, R N Casey, P R Bowser, J W Casey, J Rovnak and S L Quackenbush	212

Fish Rhabdoviruses *G Kurath and J Winton*	221
Fish Viruses *J C Leong*	227
Flaviviruses of Veterinary Importance *R Swanepoel and F J Burt*	234
Flaviviruses: General Features *T J Chambers*	241
Flexiviruses *M J Adams*	253
Foamy Viruses *M L Linial*	259
Foot and Mouth Disease Viruses *D J Rowlands*	265
Fowlpox Virus and Other Avipoxviruses *M A Skinner*	274
Fungal Viruses *S A Ghabrial and N Suzuki*	284
Furovirus *R Koenig*	291
Fuselloviruses of Archaea *K M Stedman*	296

G

Gene Therapy: Use of Viruses as Vectors *K I Berns and T R Flotte*	301
Genome Packaging in Bacterial Viruses *P Jardine*	306
Giardiaviruses *A L Wang and C C Wang*	312

H

Hantaviruses *A Vaheri*	317
Henipaviruses *B T Eaton and L-F Wang*	321
Hepadnaviruses of Birds *A R Jilbert and W S Mason*	327
Hepadnaviruses: General Features *T J Harrison*	335
Hepatitis A Virus *A Dotzauer*	343
Hepatitis B Virus: General Features *P Karayiannis and H C Thomas*	350
Hepatitis B Virus: Molecular Biology *T J Harrison*	360
Hepatitis C Virus *R Bartenschlager and S Bühler*	367
Hepatitis Delta Virus *J M Taylor*	375
Hepatitis E Virus *X J Meng*	377
Herpes Simplex Viruses: General Features *L Aurelian*	383
Herpes Simplex Viruses: Molecular Biology *E K Wagner and R M Sandri-Goldin*	397
Herpesviruses of Birds *S Trapp and N Osterrieder*	405
Herpesviruses of Horses *D J O'Callaghan and N Osterrieder*	411
Herpesviruses: Discovery *B Ehlers*	420
Herpesviruses: General Features *A J Davison*	430
Herpesviruses: Latency *C M Preston*	436
History of Virology: Bacteriophages *H-W Ackermann*	442
History of Virology: Plant Viruses *R Hull*	450
History of Virology: Vertebrate Viruses *F J Fenner*	455

Hordeivirus J N Bragg, H-S Lim and A O Jackson	459
Host Resistance to Retroviruses T Hatziioannou and P D Bieniasz	467
Human Cytomegalovirus: General Features E S Mocarski Jr. and R F Pass	474
Human Cytomegalovirus: Molecular Biology W Gibson	485
Human Eye Infections J Chodosh, A V Chintakuntlawar and C M Robinson	491
Human Herpesviruses 6 and 7 U A Gompels	498
Human Immunodeficiency Viruses: Antiretroviral Agents A W Neuman and D C Liotta	505
Human Immunodeficiency Viruses: Molecular Biology J Votteler and U Schubert	517
Human Immunodeficiency Viruses: Origin F van Heuverswyn and M Peeters	525
Human Immunodeficiency Viruses: Pathogenesis N R Klatt, A Chahroudi and G Silvestri	534
Human Respiratory Syncytial Virus P L Collins	542
Human Respiratory Viruses J E Crowe Jr.	551
Human T-Cell Leukemia Viruses: General Features M Yoshida	558
Human T-Cell Leukemia Viruses: Human Disease R Mahieux and A Gessain	564
Hypovirulence N K Van Alfen and P Kazmierczak	574
Hypoviruses D L Nuss	580

VOLUME 3

I

Icosahedral dsDNA Bacterial Viruses with an Internal Membrane J K H Bamford and S J Butcher	1
Icosahedral Enveloped dsRNA Bacterial Viruses R Tuma	6
Icosahedral ssDNA Bacterial Viruses B A Fane, M Chen, J E Cherwa and A Uchiyama	13
Icosahedral ssRNA Bacterial Viruses P G Stockley	21
Icosahedral Tailed dsDNA Bacterial Viruses R L Duda	30
Idaeovirus A T Jones and H Barker	37
Iflavirus M M van Oers	42
Ilarvirus K C Eastwell	46
Immune Response to Viruses: Antibody-Mediated Immunity A R Neurath	56
Immune Response to Viruses: Cell-Mediated Immunity A J Zajac and L E Harrington	70
Immunopathology M B A Oldstone and R S Fujinami	78
Infectious Pancreatic Necrosis Virus Ø Evensen and N Santi	83
Infectious Salmon Anemia Virus B H Dannevig, S Mjaaland and E Rimstad	89
Influenza R A Lamb	95
Innate Immunity: Defeating C F Basler	104
Innate Immunity: Introduction F Weber	111
Inoviruses L A Day	117
Insect Pest Control by Viruses M Erlandson	125
Insect Reoviruses P P C Mertens and H Attoui	133
Insect Viruses: Nonoccluded J P Burand	144

Interfering RNAs	K E Olson, K M Keene and C D Blair	148
Iridoviruses of Vertebrates	A D Hyatt and V G Chinchar	155
Iridoviruses of Invertebrates	T Williams and A D Hyatt	161
Iridoviruses: General Features	V G Chinchar and A D Hyatt	167

J

Jaagsiekte Sheep Retrovirus	J M Sharp, M de las Heras, T E Spencer and M Palmarini	175
Japanese Encephalitis Virus	A D T Barrett	182

K

Kaposi's Sarcoma-Associated Herpesvirus: General Features	Y Chang and P S Moore	189
Kaposi's Sarcoma-Associated Herpesvirus: Molecular Biology	E Gellermann and T F Schulz	195

L

Lassa, Junin, Machupo and Guanarito Viruses	J B McCormick	203
Legume Viruses	L Bos	212
Leishmaniaviruses	R Carrion Jr, Y-T Ro and J L Patterson	220
Leporipoviruses and Suipoxviruses	G McFadden	225
Luteoviruses	L L Domier and C J D'Arcy	231
Lymphocytic Choriomeningitis Virus: General Features	R M Welsh	238
Lymphocytic Choriomeningitis Virus: Molecular Biology	J C de la Torre	243
Lysis of the Host by Bacteriophage	R F Young III and R L White	248

M

Machlomovirus	K Scheets	259
Maize Streak Virus	D P Martin, D N Shepherd and E P Rybicki	263
Marburg Virus	D Falzarano and H Feldmann	272
Marnaviruses	A S Lang and C A Suttle	280
Measles Virus	R Cattaneo and M McChesney	285
Membrane Fusion	A Hinz and W Weissenhorn	292
Metaviruses	H L Levin	301
Mimivirus	J-M Claverie	311
Molluscum Contagiosum Virus	J J Bugert	319
Mononegavirales	A J Easton and R Ling	324
Mouse Mammary Tumor Virus	J P Dudley	334
Mousepox and Rabbitpox Viruses	M Regner, F Fenner and A Müllbacher	342
Movement of Viruses in Plants	P A Harries and R S Nelson	348

Mumps Virus	B K Rima and W P Duprex	356
Mungbean Yellow Mosaic Viruses	V G Malathi and P John	364
Murine Gammaherpesvirus 68	A A Nash and B M Dutia	372
Mycoreoviruses	B I Hillman	378

N

Nanoviruses	H J Vetten	385
Narnaviruses	R Esteban and T Fujimura	392
Nature of Viruses	M H V Van Regenmortel	398
Necrovirus	L Rubino and G P Martelli	403
Nepovirus	H Sanfaçon	405
Neutralization of Infectivity	P J Klasse	413
Nidovirales	L Enjuanes, A E Gorbalenya, R J de Groot, J A Cowley, J Ziebuhr and E J Snijder	419
Nodaviruses	P A Venter and A Schneemann	430
Noroviruses and Sapoviruses	K Y Green	438

O

Ophiovirus	A M Vaira and R G Milne	447
Orbiviruses	P P C Mertens, H Attoui and P S Mellor	454
Organ Transplantation, Risks	C N Kotton, M J Kuehnert and J A Fishman	466
Origin of Viruses	P Forterre	472
Orthobunyaviruses	C H Calisher	479
Orthomyxoviruses: Molecular Biology	M L Shaw and P Palese	483
Orthomyxoviruses: Structure of Antigens	R J Russell	489
Oryctes Rhinoceros Virus	J M Vlak, A M Huger, J A Jehle and R G Kleespies	495
Ourmiavirus	G P Accotto and R G Milne	500

VOLUME 4

P

Papaya Ringspot Virus	D Gonsalves, J Y Suzuki, S Tripathi and S A Ferreira	1
Papillomaviruses: General Features of Human Viruses	G Orth	8
Papillomaviruses: Molecular Biology of Human Viruses	P F Lambert and A Collins	18
Papillomaviruses of Animals	A A McBride	26
Papillomaviruses: General Features	H U Bernard	34
Paramyxoviruses of Animals	S K Samal	40
Parainfluenza Viruses of Humans	E Adderson and A Portner	47
Paramyxoviruses	R E Dutch	52
Parapoxviruses	D Haig and A A Mercer	57

Partitiviruses of Fungi *S Tavantzis*	63
Partitiviruses: General Features *S A Ghabrial, W F Ochoa, T S Baker and M L Nibert*	68
Parvoviruses of Arthropods *M Bergoin and P Tijssen*	76
Parvoviruses of Vertebrates *C R Parrish*	85
Parvoviruses: General Features *P Tattersall*	90
Pecluvirus *D V R Reddy, C Bragard, P Sreenivasulu and P Delfosse*	97
Pepino Mosaic Virus *R A A Van der Vlugt and C C M M Stijger*	103
Persistent and Latent Viral Infection *E S Mocarski and A Grakoui*	108
Phycodnaviruses *J L Van Etten and M V Graves*	116
Phylogeny of Viruses *A E Gorbalenya*	125
Picornaviruses: Molecular Biology *B L Semler and K J Ertel*	129
Plant Antiviral Defense: Gene Silencing Pathway *G Szittya, T Dalmay and J Burgyan*	141
Plant Reoviruses *R J Geijskes and R M Harding*	149
Plant Resistance to Viruses: Engineered Resistance *M Fuchs*	156
Plant Resistance to Viruses: Geminiviruses *J K Brown*	164
Plant Resistance to Viruses: Natural Resistance Associated with Dominant Genes *P Moffett and D F Klessig*	170
Plant Resistance to Viruses: Natural Resistance Associated with Recessive Genes *C Caranta and C Dogimont*	177
Plant Rhabdoviruses *A O Jackson, R G Dietzgen, R-X Fang, M M Goodin, S A Hogenhout, M Deng and J N Bragg*	187
Plant Virus Diseases: Economic Aspects *G Loebenstein*	197
Plant Virus Diseases: Fruit Trees and Grapevine *G P Martelli and J K Uyemoto*	201
Plant Virus Diseases: Ornamental Plants *J Engelmann and J Hamacher*	207
Plant Virus Vectors (Gene Expression Systems) *Y Gleba, S Marillonnet and V Klimyuk*	229
Plum Pox Virus *M Glasa and T Candresse*	238
Poliomyelitis *P D Minor*	243
Polydnaviruses: Abrogation of Invertebrate Immune Systems *M R Strand*	250
Polydnaviruses: General Features *A Fath-Goodin and B A Webb*	257
Polyomaviruses of Humans *M Safak and K Khalili*	262
Polyomaviruses of Mice *B Schaffhausen*	271
Polyomaviruses *M Gravell and E O Major*	277
Pomovirus *L Torrance*	283
Potato Virus Y *C Kerlan and B Moury*	288
Potato Viruses *C Kerlan*	302
Potexvirus *K H Ryu and J S Hong*	310
Potyviruses *J J López-Moya and J A García*	314
Poxviruses *G L Smith, P Beard and M A Skinner*	325
Prions of Vertebrates *J D F Wadsworth and J Collinge*	331
Prions of Yeast and Fungi *R B Wickner, H Edskes, T Nakayashiki, F Shewmaker, L McCann, A Engel and D Kryndushkin*	338

Pseudorabies Virus *T C Mettenleiter*	342
Pseudoviruses *D F Voytas*	352

Q

Quasispecies *R Sanjuán*	359

R

Rabies Virus *I V Kuzmin and C E Rupprecht*	367
Recombination *J J Bujarski*	374
Reoviruses: General Features *P Clarke and K L Tyler*	382
Reoviruses: Molecular Biology *K M Coombs*	390
Replication of Bacterial Viruses *M Salas and M de Vega*	399
Replication of Viruses *A J Cann*	406
Reticuloendotheliosis Viruses *A S Liss and H R Bose Jr.*	412
Retrotransposons of Fungi *T J D Goodwin, M I Butler and R T M Poulter*	419
Retrotransposons of Plants *M-A Grandbastien*	428
Retrotransposons of Vertebrates *A E Peaston*	436
Retroviral Oncogenes *P K Vogt and A G Bader*	445
Retroviruses of Insects *G F Rohrmann*	451
Retroviruses of Birds *K L Beemon*	455
Retroviruses: General Features *E Hunter*	459
Rhinoviruses *N W Bartlett and S L Johnston*	467
Ribozymes *E Westhof and A Lescoute*	475
Rice Tungro Disease *R Hull*	481
Rice Yellow Mottle Virus *E Hébrard and D Fargette*	485
Rift Valley Fever and Other Phleboviruses *L Nicoletti and M G Ciufolini*	490
Rinderpest and Distemper Viruses *T Barrett*	497
Rotaviruses *J Angel, M A Franco and H B Greenberg*	507
Rubella Virus *T K Frey*	514

S

Sadwavirus *T Iwanami*	523
Satellite Nucleic Acids and Viruses *P Palukaitis, A Rezaian and F García-Arenal*	526
Seadornaviruses *H Attoui and P P C Mertens*	535
Sequiviruses *I-R Choi*	546
Severe Acute Respiratory Syndrome (SARS) *J S M Peiris and L L M Poon*	552
Shellfish Viruses *T Renault*	560
Shrimp Viruses *J-R Bonami*	567

Sigma Rhabdoviruses	*D Contamine and S Gaumer*	576
Simian Alphaherpesviruses	*J Hilliard*	581
Simian Gammaherpesviruses	*A Ensser*	585
Simian Immunodeficiency Virus: Animal Models of Disease	*C J Miller and M Marthas*	594
Simian Immunodeficiency Virus: General Features	*M E Laird and R C Desrosiers*	603
Simian Immunodeficiency Virus: Natural Infection	*I Pandrea, G Silvestri and C Apetrei*	611
Simian Retrovirus D	*P A Marx*	623
Simian Virus 40	*A L McNees and J S Butel*	630
Smallpox and Monkeypox Viruses	*S Parker, D A Schultz, H Meyer and R M Buller*	639
Sobemovirus	*M Meier, A Olspert, C Sarmiento and E Truve*	644
St. Louis Encephalitis	*W K Reisen*	652
Sweetpotato Viruses	*J Kreuze and S Fuentes*	659

VOLUME 5

T

Taura Syndrome Virus	*A K Dhar and F C T Allnutt*	1
Taxonomy, Classification and Nomenclature of Viruses	*C M Fauquet*	9
Tenuivirus	*B C Ramirez*	24
Tetraviruses	*J A Speir and J E Johnson*	27
Theiler's Virus	*H L Lipton, S Hertzler and N Knowles*	37
Tick-Borne Encephalitis Viruses	*T S Gritsun and E A Gould*	45
Tobacco Mosaic Virus	*M H V Van Regenmortel*	54
Tobacco Viruses	*S A Tolin*	60
Tobamovirus	*D J Lewandowski*	68
Tobravirus	*S A MacFarlane*	72
Togaviruses Causing Encephalitis	*S Paessler and M Pfeffer*	76
Togaviruses Causing Rash and Fever	*D W Smith, J S Mackenzie and M D A Lindsay*	83
Togaviruses Not Associated with Human Disease	*L L Coffey,*	91
Togaviruses: Alphaviruses	*A M Powers*	96
Togaviruses: Equine Encephalitic Viruses	*D E Griffin*	101
Togaviruses: General Features	*S C Weaver, W B Klimstra and K D Ryman*	107
Togaviruses: Molecular Biology	*K D Ryman, W B Klimstra and S C Weaver*	116
Tomato Leaf Curl Viruses from India	*S Chakraborty*	124
Tomato Spotted Wilt Virus	*H R Pappu*	133
Tomato Yellow Leaf Curl Virus	*H Czosnek*	138
Tombusviruses	*S A Lommel and T L Sit*	145
Torovirus	*A E Hoet and M C Horzinek*	151
Tospovirus	*M Tsompana and J W Moyer*	157
Totiviruses	*S A Ghabrial*	163

Transcriptional Regulation in Bacteriophage   R A Weisberg, D M Hinton and S Adhya   174
Transmissible Spongiform Encephalopathies   E D Belay and L B Schonberger   186
Tumor Viruses: Human   R Grassmann, B Fleckenstein and H Pfister   193
Tymoviruses   A-L Haenni and T W Dreher   199

U

Umbravirus   M Taliansky and E Ryabov   209
Ustilago Maydis Viruses   J Bruenn   214

V

Vaccine Production in Plants   E P Rybicki   221
Vaccine Safety   C J Clements and G Lawrence   226
Vaccine Strategies   I Kusters and J W Almond   235
Vaccinia Virus   G L Smith   243
Varicella-Zoster Virus: General Features   J I Cohen   250
Varicella-Zoster Virus: Molecular Biology   W T Ruyechan and J Hay   256
Varicosavirus   T Sasaya   263
Vector Transmission of Animal Viruses   W K Reisen   268
Vector Transmission of Plant Viruses   S Blanc   274
Vegetable Viruses   P Caciagli   282
Vesicular Stomatitis Virus   S P J Whelan   291
Viral Killer Toxins   M J Schmitt   299
Viral Membranes   J Lenard   308
Viral Pathogenesis   N Nathanson   314
Viral Receptors   D J Evans   319
Viral Suppressors of Gene Silencing   J Verchot-Lubicz and J P Carr   325
Viroids   R Flores and R A Owens   332
Virus Classification by Pairwise Sequence Comparison (PASC)   Y Bao, Y Kapustin and T Tatusova   342
Virus Databases   E J Lefkowitz, M R Odom and C Upton   348
Virus Entry to Bacterial Cells   M M Poranen and A Domanska   365
Virus Evolution: Bacterial Viruses   R W Hendrix   370
Virus-Induced Gene Silencing (VIGS)   M S Padmanabhan and S P Dinesh-Kumar   375
Virus Particle Structure: Nonenveloped Viruses   J A Speir and J E Johnson   380
Virus Particle Structure: Principles   J E Johnson and J A Speir   393
Virus Species   M H V van Regenmortel   401
Viruses and Bioterrorism   R F Meyer and S A Morse   406
Viruses Infecting Euryarchaea   K Porter, B E Russ, A N Thorburn and M L Dyall-Smith   411
Visna-Maedi Viruses   B A Blacklaws   423

W

Watermelon Mosaic Virus and Zucchini Yellow Mosaic Virus   *H Lecoq and C Desbiez*   433

West Nile Virus   *L D Kramer*   440

White Spot Syndrome Virus   *J-H Leu, J-M Tsai and C-F Lo*   450

Y

Yatapoxviruses   *J W Barrett and G McFadden*   461

Yeast L-A Virus   *R B Wickner, T Fujimura and R Esteban*   465

Yellow Fever Virus   *A A Marfin and T P Monath*   469

Yellow Head Virus   *P J Walker and N Sittidilokratna*   476

Z

Zoonoses   *J E Osorio and T M Yuill*   485

Taxonomic Index   497

Subject Index   499

Defective-Interfering Viruses

L Roux, University of Geneva Medical School, Geneva, Switzerland

© 2008 Elsevier Ltd. All rights reserved.

This article is reproduced from the previous edition, volume 1, pp 371–375, © 1999, Elsevier Ltd.

History

In 1943, Henle and Henle reported the decreased infectivity for mice of influenza virus stocks obtained after a series of undiluted passages in embryonated chicken eggs. In the early 1950s, von Magnus showed that such undiluted passages generate incomplete virus particles capable of limiting the growth of infectious virus (hence exhibiting interference). This first characterization was soon followed by similar reports by Mims, on the one hand, and Cooper and Bellet, on the other, dealing respectively with Rift Valley fever virus and vesicular stomatitis virus (VSV). In the late 1950s, Cooper and Bellet went so far as to assign interference to sedimentable particles, but failed to identify them as antigenically related to VSV. From the mid-1960s on, the characterization of other positive- and negative-stranded RNA virus defective particles continued. In 1970, a review by A. Huang and D. Baltimore set the basic definition of defective interfering (DI) particles and emphasized their widespread occurrence. Since then, DI particles have been described for almost all the known DNA and RNA viruses, including plant and even fungal viruses.

Structure

DI particles have the same protein composition as their homologous nondefective 'parents', often called St. However, they differ from the St particles in the primary structure of their genome. As emphasized later, DI genomes lack part of the genetic information. They may or may not serve as coding sequences. However, they always conserve the *cis*-sequences needed for replication initiation (origins of replication), sometimes present in more than one copy, and sequences involved in encapsidation. Foreign sequences can also be inserted. DI particles can sometimes be separated from St particles on the basis of size, when the size of the particle closely corresponds to the size of the genome (for instance rhabdovirus), or on the basis of particle density differences (changes in nucleic acid to protein ratios). Often, however, only viral stocks enriched in DI particles are available owing to the size heterogeneity of the virus particles.

Generation of DI Genomes

DI DNAs very likely arise from various recombinational events not necessarily linked to genome replication, and which result in deletion, tandem duplication, insertion of host DNA, and polymerization of small monomer sequences. DI RNAs have been proposed to arise almost exclusively during genome replication by a mechanism of 'leaping polymerase' consisting of polymerase stop/fall-off or slippage/reinitiation events (**Figure 1**). In this model the replicase complex moves with the nascent RNA still attached to it. Depending on where reinitiation takes place, and on the number and the direction of the leaps, the resulting molecules can be of the copy-back type, with more or less intramolecular inverted complementary sequences (a), of the internal deletion type (b), and of the duplication type (c). Multisteps (b) or (c) and combinations of steps (b) and (c) can, moreover, lead to various mosaic types. Insertion of host RNA is also observed, especially in plant DI RNA. The frequency at which the polymerase leaps and resumes its synthesis is unknown. The probability for this exercise to be successful in producing a viable DI genome has been estimated for VSV to range in the order of 10^{-7}–10^{-8} per genome replication.

Defectiveness

The DI genomes contain interrupted or rearranged open reading frames. They partly or completely lack the full

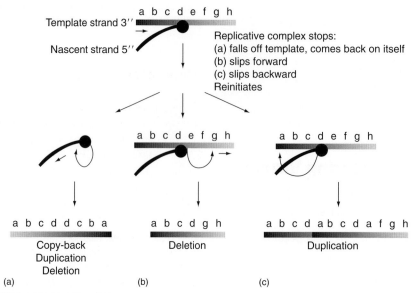
Figure 1 Defective RNA genome generation.

coding capacity of the viral genome. They are therefore defective, and depend for their replication and for their propagation (formation of virus particles) on the functions provided by the homologous standard virus (helper). Co-infection of cells with DI and standard particles is therefore essential for DI particle multiplication. Consequently, low-multiplicity infections, and particularly plaque purification, represent conditions which decrease, and potentially eliminate, DI particles from a viral preparation.

Interference

As stated earlier, the generation of a defective genome is likely to represent a rare event. This event would never be seen unless it was successfully amplified. During this amplification step the defective genome is preferentially replicated over the nondefective genome. This ability to replicate efficiently at the expense of the nondefective genome is called interference. The mechanisms of interference are not completely understood. They obviously change depending on the specificity of the viruses, and appear to be also affected by the host cell types. In general, interference involves an early step in genome replication, and can be pictured as a competition for limiting replication 'factors' (viral replicase, encapsidation proteins, host cell factors). Reiterated origins of replication or encapsidation sites on DI DNAs, presence of higher affinity sites for the replicase or for the encapsidation on both positive and negative polarity DI RNAs, and shorter length of the replicating units, higher availability for replication of molecules not involved in transcription, have been shown or postulated as taking part in the interference mechanism.

Defective Interfering versus Defective Viruses

Based on the outcome of experimental co-infections of defective with nondefective viruses, a distinction has been made between defective interfering or defective noninterfering particles, according to the ability of the defective viruses to selectively restrict nondefective virus replication. This distinction may not apply during the first events following the generation of a defective genome. As this is bound to be a rare event, an interference mechanism has to be invoked any time this defective genome is amplified to the point it can be detected, or become predominant.

Cyclic Variations of Defective Interference

The dependence of DI genome replication on functions provided by the nondefective genome on the one hand, and the interference exhibited by DI genome on the other hand, result in out-of-phase cyclic variations of both DI and St genome replication. As illustrated in **Figure 2**, efficient St genome replication must precede extensive DI genome replication. This in turn establishes conditions of high interference which results in inhibition of St genome replication. Decrease of helper function availability leads to DI genome replication dampening, and therefore to release from interference, allowing efficient St genome replication to resume. These cyclic variations have been observed in serial passages of St and DI viruses in cell culture, as well as in persistent infections. The periodicity of a complete cycle is generally a matter of days or of a few serial passages.

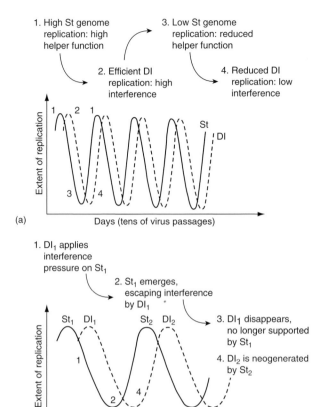

Figure 2 Cyclic variation of DI replication in (a) days, and (b) months.

the other hand, will yield in each series different DI particles or different sets of DI particles.

Biological Effects

DI particles have been shown to modulate the course of an infection. In cell culture, attenuation of the cytopathic effects is the most frequently described, and DI particles can promote cell survival and establishment of persistent infections. As far as negative-stranded RNA viruses are concerned, copy-back DI RNAs appear to prevent induction of apoptosis through a mechanism which has still to be unravelled, in which a certain category of small leader RNAs may participate. A possible role of the matrix M protein, the concentration of which is decreased to the point where viral assembly and budding at the cell surface is highly diminished in the presence of DI genomes, has also been considered. DI ability selectively allows emergence of St virus (St_2) which escapes interference (**Figure 2(b)**). St_2, resistant to interference, is selectively amplified over St_1, still sensitive to interference, and soon becomes predominant. It loses its ability to support DI_1, which therefore disappears. St_2 will generate its own DI_2, which in turn will favor emergence of a new St variant. Thus, DI viruses serve as mutational drivers favoring virus evolution, through cycles of high and low interference whose periodicity is this time measured in months or in hundreds of viral passages (compare **Figures 2(a)** and **2(b)**).

Assay for DI Particles

DI particles can be detected by physical separation on velocity or density gradients when applicable (see the section titled 'Structure'). The presence of subgenomic nucleic acids in viral stocks or in infected cells (distinct from viral messengers) can also be diagnostic. The ability to decrease the infectivity of a viral stock or to protect infected cells from the lytic infectious virus (see the section titled, 'Biological effects') are used in various biological assays to estimate quantitatively and qualitatively the DI particle composition of a viral stock. These assays, although appropriate to characterize DI particle preparations, are generally not sensitive enough to exclude, when negative, the presence of DI particles in a viral preparation. The test that still remains the most dependable to assess presence or absence of DI particles consists of multiple independent serial undiluted passages. It is based on the observation that a viral stock contaminated with an undetectable amount of DI particles will, on subsequent independent serial passages, promote in each series amplification of the same contaminating DI particles. A DI particle-free stock, on

DI Particles in Experimental Animals

DI particles are generated and amplified in the whole animal, as well as in cell culture. They can change the symptoms of viral infection from rapid death to slow, progressing paralysis. They can sometimes fully protect the animals from an otherwise lethal infection. Interference is likely to be involved in this modulation of symptoms, but other phenomena like increased interferon induction and immune response modulation are reported.

DI Particles in Natural Infections

Involvement of DI particles in natural infections is still poorly documented. This is partly because the experimental results supporting the strong potential for infection modulation of DI particles have not been fully recognized, as detection of DI particles in natural infections is not straightforward. Unpredictable cyclic variations in DI replication, efficiency of DI replication changing with the types of infected tissues, and potent interfering ability associated with poor DI particle replication are all factors which undoubtedly make DI particle detection difficult *in vivo*. Last but not least, virus isolation, which is used to characterize the infectious agent associated with a disease, often represents conditions (low multiplicity of infection) known to impair DI particle replication drastically.

Nevertheless, association of a chicken influenza virus strain, efficiently producing DI particles, with an epidemic of low morbidity and low mortality, and conversely, a high-mortality epidemic associated with a strain free of DI particles, have been reported. Murine and feline leukemia virus strains causing immune deficiency syndromes are shown to contain predominantly replication-defective viral genomes before onset and during the development of the disease. The pathogenicity of some bovine and swine pestiviruses has clearly been associated with presence of DI RNAs in the animals. For the bovine viral diarrhea virus (BVDV), a pestivirus of the same family as hepatitis C virus, the presence of a particular DI RNA can turn noncytopathic virus into a fatal infectious agent. In plants, at least three examples of DI RNAs are described to be involved in infection modulation. Interestingly, depending on the types of viruses, DI RNAs can either attenuate or exacerbate the symptoms.

DI RNAs are identified in stool and blood samples of humans suffering from hepatitis A virus, an infection known to be rather moderate and prolonged. DI particles are identified in measles virus-attenuated vaccine preparations which have been, and are being, widely and successfully used (raising the question of DI particle participation in vaccine attenuation). Measles viruses defective in viral assembly are currently found associated with human subacute sclerosing panencephalitis (SSPE). The brain cells of SSPE patients were, moreover, shown to harbor many species of measles virus copy-back DI RNAs. Direct amplification of a portion of the HIV *tat* gene from infected patients demonstrates that about a third of the sequences correspond to defective *tat* function. Moreover, human immunodeficiency virus (HIV)-1 sequences isolated from a cohort of six blood or blood product recipients infected with one donor all contained a similar deletion in the *nef* gene. Remarkably, all the patients harboring this deleted viral genome remained free of HIV-related diseases 12–16 years after infection, suggesting that this defective species of HIV genome is responsible for this decreased pathology. Epstein–Barr virus (EBV) replicative infections developing in human epithelial lesions involve a deleted rearranged form of EBV DNA (het DNA). This het DNA is associated, in experimental infections, with disruption of latency and persistent productive infections. Specific identification of viral hepatitis B (HBV) genomes containing an interrupted precore antigen (HBeAg) coding sequence in patients dying from fulminant hepatitis suggests that such defective genomes may be responsible for the exacerbation of the disease. This contrasts with more recent data reporting experimental evidence for the existence of DI-like viruses in HBV human chronic carriers; fluctuations between these naturally occurring core internal deletion variants and helper HBV in three chronic carriers were reminiscent of the cycling phenomenon in other DI viral systems.

Future Perspectives

Defective interfering particles are ubiquitous in the realm of animal and plant viruses. In experimental conditions they appear as necessary companions of their nondefective homologs. Capable of affecting the extent of viral growth, the course of viral infections, and serving as selective pressure to drive mutational changes, they can be seen as natural regulators of virus evolution. The demonstration of their participation in natural infections, and of their ability to affect the course of diseases, constitutes a challenge for the years to come. As pointed out by the few examples listed earlier, their direct detection in infected tissues will certainly be needed to assert their involvement in natural infections. The availability of sensitive detection techniques (like polymerase chain reactions), allowing direct observation of viral genomes without the possible distortion of virus isolation, bears great hope. More than giving increased insights into the physiopathology of viral infections, in times where the modifications of the viral genomes represent an imperative step in the generation of viral recombinant vaccines or of appropriate vectors for gene targeting, DI viral genomes represent natural versions of defective genomes that can serve as model tools for creation of more adapted vectors.

See also: Orthobunyaviruses; Orthomyxoviruses: Molecular Biology; Orthomyxoviruses: Structure of antigens.

Further Reading

Barrett ADT and Dimmock NJ (1986) Defective interfering viruses and infection of animals. *Current Topics in Microbiology and Immunology* 128: 55.

Perrault J (1981) Origin and replication of defective interfering particles. *Current Topics in Microbiology and Immunology* 93: 151.

Roux L, Simon AE, and Holland JJ (1991) Effects of defective interfering viruses on virus replication and pathogenesis. In: Shatkin A (ed.) *Advances in Virus Research,* vol. 40, 181pp. New York: Academic Press.

Dengue Viruses

D J Gubler, John A. Burns School of Medicine, Honolulu, HI, USA

© 2008 Elsevier Ltd. All rights reserved.

Glossary

Arbovirus A virus transmitted to vertebrates by hematophagous (blood-feeding) arthropods.
Endemic A disease constantly present in a human population.
Extrinsic incubation The time between infection and becoming infectious.
Hyperendemicity The co-circulation of multiple dengue virus serotypes in the same population.
Vertical transmission Transmission of a virus from a female arthropod to her progeny.

History

Dengue fever is a very old disease; the earliest record of a dengue-like illness found to date is in a Chinese encyclopedia of disease symptoms and remedies, first published during the Chin Dynasty (AD 265–420) and formally edited in AD 610 (Tang Dynasty) and again in AD 992 during the Northern Sung Dynasty. There are reports of epidemics of dengue-like illnesses in the French West Indies in 1635 and in Panama in 1699. By the late 1700s, the disease had a worldwide distribution in the tropics, with epidemics of a clinically compatible disease occurring in 1779 in Batavia (Jakarta), Indonesia and Cairo, Egypt, and in 1780 in Philadelphia, Pennsylvania, USA. From the late 1700s to World War II, repeated epidemics of dengue-like illness occurred in most tropical and subtropical regions of the world at 10- to 30-year intervals. There is no documentation, however, that dengue viruses were responsible for all these epidemics because diagnosis was based only on clinical reports. Clinical descriptions of some early epidemics were compatible with chikungunya virus infection, which has a transmission cycle similar to that of the dengue viruses. It is likely that epidemic chikungunya did occur, but recent data show that the dengue viruses, not chikungunya virus, were responsible for the majority of epidemics in the past 50 years.

The virus etiology of dengue fever was not documented until 1943–44, when Japanese and American scientists simultaneously isolated the viruses from soldiers in the Pacific and Asian theaters during World War II. Albert Sabin isolated dengue viruses from soldiers who became ill in Calcutta (India), New Guinea, and Hawaii. The viruses from India, Hawaii, and one strain from New Guinea were antigenically similar, whereas three others from New Guinea were different. These viruses were called dengue 1 (DENV-1) and dengue 2 (DENV-2) and designated as prototype viruses (DENV-1, Hawaii, and DENV-2, New Guinea C). The Japanese virus, isolated by Susumu Hotta, was later shown to be DENV-1. Two more serotypes, called dengue 3 (DENV-3) and dengue 4 (DENV-4), were subsequently isolated by William McD. Hammon and his colleagues from children with hemorrhagic disease during an epidemic in Manila, Philippines, in 1956. Although thousands of dengue viruses have been isolated from different parts of the world since that time, all fit antigenically into the four serotype classification.

Many early workers suspected that dengue viruses were transmitted by mosquitoes, but actual transmission was first documented by H. Graham in 1903. In 1906, T. L. Bancroft demonstrated transmission by *Aedes aegypti*, later known to be the principal urban mosquito vector of dengue viruses. Subsequent studies in the Philippines, Indonesia, Japan, and the Pacific showed that *Aedes albopictus* and *Aedes polynesiensis* also were efficient secondary vectors for dengue viruses.

During and following World War II, *Ae. aegypti* greatly expanded its distribution in Asia, becoming the dominant day-biting mosquito in most Asian cities. Multiple dengue virus serotypes were also disseminated widely at that time. A dramatic increase in urbanization in the postwar years created ideal conditions for increased transmission of urban mosquito-borne diseases. These changes, plus an increased movement of people within and among countries of the region via airplane, resulted in increased movement of dengue viruses between population centers, increased frequency of epidemic activity, the development of hyperendemicity (co-circulation of multiple serotypes), and the emergence of epidemic dengue hemorrhagic fever/dengue shock syndrome (DHF/DSS) in many countries of Southeast Asia during the 1960s. By 1975, DHF/DSS was a leading cause of hospitalization and death among children in the region. During the 1980s and 1990s, epidemic DHF/DSS continued to expand geographically in Asia. In the 1970s DHF/DSS moved into the Pacific Islands after an absence of 25 years. In the Americas, where *Ae. aegypti* had been eradicated from many countries as a result of efforts to control yellow fever, increased epidemic dengue fever closely followed the reinfestation of countries by this mosquito in the 1970s, 1980s, and 1990s.

With the development of hyperendemicity, DHF/DSS has emerged as a global public health problem in the past 25 years. In 2007, dengue fever is the most important arbovirus disease of humans, with more than 2.5 billion people living in areas at risk for dengue in a belt around the tropics of the world (**Figure 1**). An estimated 100 million dengue infections and 500 thousand cases of DHF/DSS occur each year. The average case–fatality rate for DHF/DSS is 5%.

Taxonomy and Classification

Dengue viruses belong to the family *Flaviviridae*, genus *Flavivirus*. There are four serotypes: DENV-1, DENV-2, DENV-3, and DENV-4. They belong to a larger, heterogeneous group of viruses called arboviruses. This is an ecological classification, one which implies that transmission between vertebrate hosts, including humans, is dependent on hematophagous arthropod vectors.

As are other flaviviruses, dengue viruses are comprised of a single-stranded RNA genome surrounded by an icosahedral nucleocapsid. The latter is covered by a lipid envelope, which is derived from the host cell membrane from which the virus buds. The complete virion is about 50 nm in diameter. The mature virion contains three structural proteins as follows: the nucleocapsid core protein (C), a membrane associated protein (M), and the envelope protein (E). Functional domains responsible for virus neutralization, hemagglutination, fusion, and interaction with virus receptors are associated with the E protein. Epitope mapping has demonstrated three to four major antigenic sites.

Antigenically, the four dengue viruses make up a unique complex within the genus *Flavivirus*. Although the four dengue serotypes are antigenically distinct, there is evidence that serologic subcomplexes may exist within the group. For example, a close genetic relationship has been demonstrated between DENV-1 and DENV-3 and between DENV-2 and DENV-4. The sizes of the genomic open reading frames of DENV-1, DENV-2, DENV-3, and DENV-4 are 3392 to 3396, 3391, 3390, and 3387 amino acids, respectively, the shortest among the mosquito-borne flaviviruses. An amino acid sequence positional homology of 63–68% is observed among the DENV serotypes compared to 44–51% between DENVs and other flaviviruses such as yellow fever and West Nile.

There are 53 flaviviruses recognized by the International Committee on Taxonomy of Viruses (this classification considers dengue 1 virus with four serotypes instead of four distinct viruses), including the genus prototype, yellow fever virus, and Japanese encephalitis, Murray Valley encephalitis, St. Louis encephalitis, West Nile, Zika and other viruses, all of which are transmitted by mosquitoes. Another group of flaviviruses are tick-borne and include tick-borne encephalitis, Omsk hemorrhagic fever, and Kyasanur Forest disease viruses. A small number of flaviviruses have no known arthropod vector, and three have been isolated only from insects.

Geographic and Seasonal Distribution

Dengue viruses have a worldwide distribution in the tropics (**Figure 1**). The viruses are endemic in most urban centers of the tropics, with transmission occurring year-round, and epidemics occurring every 3–6 years. It is well documented, however, that dengue viruses are maintained during interepidemic periods in most tropical areas and, although the risk of infection is lower than during epidemic periods, it is still substantial to unsuspecting visitors.

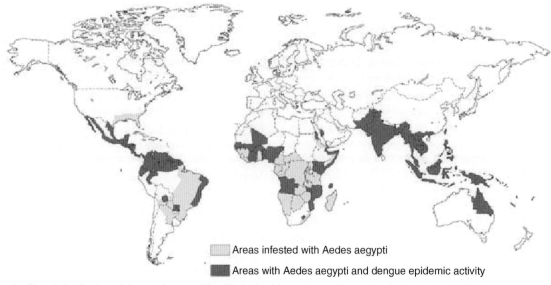

Figure 1 Global distribution of dengue fever and its principal epidemic mosquito vector, *Aedes aegypti, 2007*.

Peak transmission of dengue viruses usually is associated with periods of higher rainfall in most endemic countries. Factors influencing seasonal transmission patterns of dengue viruses are not well understood, but obviously include mosquito density, which may increase during the rainy season, especially in those areas where water level in larval habitats is dependent on rainfall. In areas where water in storage containers is not influenced by rainfall, however, other factors such as higher humidity and moderate ambient temperatures associated with the rainy season increase survival of infected mosquitoes, thus increasing the chances of secondary transmission to other persons. Virus strain and serotype, and herd immunity also influence transmission dynamics.

Host Range and Virus Propagation

There are only three known natural hosts for dengue viruses: *Aedes* mosquitoes, humans, and lower primates. Viremia in humans may last 2–12 days (average, 4–5 days) with titers ranging from undetectable to more than 10^8 mosquito infectious doses $50 (MID_{50}) \, ml^{-1}$. Experimental evidence shows that several species of lower primates (chimpanzees, gibbons, and macaques) become infected and develop viremia titers high enough to infect mosquitoes, but do not develop illness. Viremia levels in lower primates are more transient, often lasting only 1–2 days if detectable, with titers seldom reaching $10^6 \, MID_{50} \, ml^{-1}$.

Dengue viruses are known to cause clinical illness and disease only in humans. Baby mice, which are used for the isolation and assay of many other arboviruses, may show no signs of illness after intracerebral inoculation with most unpassaged strains of dengue viruses. Experimentally, however, some strains can be adapted to produce illness and death in baby mice. SCID, $Rag^{2-1-8c-1}$ and AG129 mice have been used for pathogenesis studies.

Mosquitoes only of species of the genus *Aedes* appear to be natural hosts for dengue viruses. Species of the subgenus *Stegomyia* are the most important vectors in terms of human transmission, and include *Ae. (S.) aegypti*, the principal urban vector worldwide, *Ae. (S.) albopictus* (Asia, the Pacific, Americas, Africa, and Europe), *Ae. (S.) scutellaris* spp. (Pacific), and *Ae. (S.) africanus*, and *Ae. (S.) luteocephalus* (Africa). It is uncertain what role *Ae. Albopictus* plays in transmission in areas where it has been recently introduced. Species of the subgenera *Finlaya* (Asia) and *Diceromyia* (Africa) appear to be important mosquito hosts involved in forest maintenance cycles of these viruses. Two other species, *Ochlerotatus (=Aedes) (Gymnometopa) mediovittatus* (Caribbean) and *Oc. (=Aedes) (Protomacleaya) triseriatus* (North America), have been shown to be excellent experimental hosts of dengue viruses.

Low passage or unpassaged dengue viruses can be propagated with consistent results only in laboratory-reared mosquitoes and in mosquito cell lines. Mosquito species most commonly used for *in vivo* propagation include *Ae. aegypti, Ae. Albopictus,* and *Toxorhynchites* spp., all of which can be reared with ease in the laboratory. Only three mosquito cell lines show high susceptibility to dengue viruses: C6/36 from *Ae.albopictus*, AP-61 from *Ae. Pseudoscutellaris,* and TRA-284 from *Tx. amboinensis.*

Dengue viruses can also be propagated in baby mice (see above) and in several vertebrate cell lines. These all have lower susceptibility to infection than do mosquito cells however, and dengue viruses must be adapted to each system by serial passage before consistent results can be obtained. Mammalian cell lines commonly used include LLC-MK2 and VERO (monkey kidney), BHK-21 (baby hamster kidney), FRhL (fetal rhesus lung), and PDK (primary dog kidney).

Genetics

Laboratory and epidemiologic studies have documented that genetic variants of dengue viruses occur in nature. DENV-3 isolated during epidemics in Puerto Rico in 1963 and 1977, and in Tahiti in 1965 and 1969, are antigenically and biologically very similar to each other, but very different from Asian strains of the same serotype. Similar antigenic differences were observed between Caribbean and Asian strains of DENV-4.

Oligonucleotide fingerprinting, restriction enzymes, primer extension sequencing, and nucleotide sequence comparison all have been used to study genetic variation among dengue viruses. In general, viruses circulating in the same geographic region during the same general time frame show genetic homogeneity, while differing from viruses of the same serotype from other regions. Because there is no good animal model for dengue however, it is not well understood how genetic variation influences phenotypic expression, in terms of clinical presentation or epidemic potential.

With increased transmission worldwide, dengue viruses have increased in diversity in recent years, most likely influencing both virulence and epidemic potential. This is supported by epidemiologic and virologic studies conducted in areas with sequential, but contrasting disease severity and transmission dynamics, such as Sri Lanka prior to and after the first DHF epidemic there in 1989. Other evidence includes the striking difference in replication characteristics between the South Pacific/American and the Southeast Asian subtypes of DENV-3 and DENV-2.

The number of genetic subtypes identified in each serotype varies with the method used, but more viruses have now been studied by partial nucleotide sequencing. Based on sequencing a 600 bp region of the envelope protein, which correlates well with sequencing the entire envelope protein, there are two distinct genotypes of DENV-1, four of DENV-2, four of DENV-3, and three of DENV-4. Of considerable epidemiologic interest is that there are currently two genotypes of DENV-2 circulating in the American region. One, designated American, has been in the Americas since at least 1952, whereas the second, designated Southeast Asia, was isolated for the first time from dengue patients in Jamaica in 1982 shortly after the 1981 Cuban epidemic of DHF/DSS. Analysis by restriction enzymes and primer extension sequencing has shown that the Jamaican virus is nearly identical to strains of DENV-2 isolated in Vietnam in 1987, suggesting that the DENV-2 virus causing the Cuban epidemic was introduced from Vietnam. This conclusion is supported by the fact that many Cubans were working in Vietnam on various aid projects during the late 1970s and early 1980s. The Southeast Asia genotype has subsequently become the predominant and most wide-spread strain of DENV-2 virus in the American region.

Although there have been increasingly frequent reports of intraserotypic recombination events among the dengue viruses, the data suggest that this has not been important in their evolution. However, with increased occurrence of the cocirculation of multiple serotypes in an area (hyperendemicity), there have been increased reports of concurrent infections with two serotypes. This will increase the probability that interserotypic recombination might occur.

Evolution

The two principal theories of flavivirus evolution are: (1) that the tick-borne and mosquito-borne flaviviruses evolved from viruses with no known vector and (2) that the tick-borne and no known vector viruses evolved from a common ancestor and the mosquito-borne viruses arose separately. The origin of dengue viruses is unknown.

Their natural history, however, suggests a long association with mosquitoes, possibly prior to becoming adapted to lower primates and humans. Biologically, dengue viruses are highly adapted to their mosquito hosts, being maintained by vertical transmission (from female mosquito to her offspring) in those species responsible for the forest maintenance cycle, with periodic amplification in lower primates (**Figure 2**). Such forest cycles have been documented in Southeast Asia and Africa. At some point in the past, probably with the clearing of the forests and development of human settlements in Asia, these viruses moved out of the jungle and into a rural environment where they were, and still are, transmitted to humans by peri-domestic mosquitoes such as *Ae. albopictus*. Migration of people ultimately moved the viruses into the cities of the tropics where they became 'urbanized' and transmitted by the highly domesticated, urban *Ae. aegypti* mosquito, which had been spread around the world via sailing ships and increased commerce.

Because of the rather slow rate of change (genetic drift) of the dengue virus genome, viruses isolated over long periods of time in the same geographic region still show striking homogenicity. The greatest genetic difference between dengue virus strains was observed between DENV-2 and DENV-3 isolated from forest mosquitoes in Africa and Asia, respectively, and viruses of the same serotype isolated from humans or mosquitoes in nearby urban areas. This would suggest that there is little gene flow between the forest and urban cycles. On the other hand, both laboratory and field evidence suggest that significant genetic changes that influence epidemic potential do occur in nature (see above).

Serologic Relationships and Variability

Dengue viruses share common morphology, genomic structure, and antigenic determinants with 52 other flaviviruses.

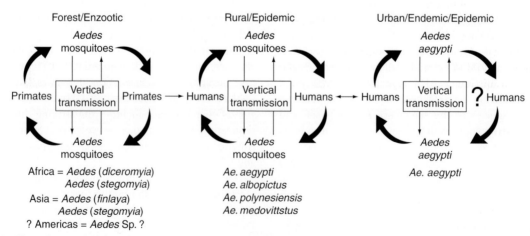

Figure 2 Natural transmission and maintenance cycles of dengue viruses.

Serologic tests most frequently used to determine antigenic relationships have included the hemagglutination-inhibition (HI), complement fixation (CF), and the plaque reduction neutralization (PRNT) tests. Because all flaviviruses share common antigenic determinants, identification of individual family members using these tests is difficult. The dengue viruses make up one antigenic complex within the family *Flaviviridae*. They share complex-specific antigenic determinants on both structural and nonstructural proteins. Serotypes within the dengue virus complex are most accurately and easily identified with an indirect immunofluorescent antibody (IFA) assay using serotype-specific monoclonal antibodies which react with epitopes on the structural protein. They can also be readily identified using polymerase chain reaction (PCR).

Both antigenic and biologic variation among dengue viruses have been documented. As noted above, DENV-3 viruses isolated in the Caribbean and the South Pacific in the 1960s were found to be antigenically distinct from the prototype and Asian strains of DENV-3 using PRNT. They were also biologically unique in that they did not grow as well in baby mice and mosquitoes as did Asian strains. DENV-4 viruses isolated in the Caribbean after the introduction of this serotype into that region in 1981 were antigenically distinct from DENV-4 viruses from Asia.

Field and epidemiologic evidence for natural strain variation among dengue viruses is more circumstantial. When DENV-2 was introduced into the South and Central Pacific islands in 1971 after an absence of more than 25 years, epidemics occurred on numerous islands. Marked variation was observed in disease severity, viremia levels, and epidemic potential in epidemics on the various islands. This variation was observed with both DENV-1 and DENV-2 in the Pacific and with DENV-3 in Indonesia. Some DENV strains appeared naturally attenuated, causing mild illness with low viremia levels of short duration, whereas others caused explosive epidemics with severe hemorrhagic disease and high viremia levels. Factors that could influence epidemic transmission and disease severity, other than differences in the virus strain, were ruled out as a cause of this variation. Recent studies in the American region have shown that variation among strains of DENV-2 and DENV-4 has influenced epidemic transmission.

Epidemiology

Dengue viruses occur in nature in three basic maintenance cycles (**Figure 2**). The primitive forest cycle involves canopy-dwelling mosquitoes and lower primates. A rural cycle, primarily in Asia and the Pacific, involves peridomestic mosquitoes (*Ae. albopictus* and *Ae. scutellaris* Spp.) and humans. The urban cycle, which is the most important epidemiologically and in regard to public health and economic impact, involves the highly domesticated *Ae. aegypti* mosquito and humans. The viruses are maintained in most large urban centers of the tropics, with epidemics occurring at periodic intervals of 3–6 years.

A combination of increased urbanization in the tropics, changing life styles, and lack of effective mosquito control has made most tropical cities highly permissive for transmission of dengue viruses by *Ae. aegypti*. Increased air travel by humans provides the ideal mechanism to transport dengue viruses between population centers. As a result, in the past 25 years there has been a dramatic global increase in the movement of dengue viruses within and between regions, resulting in increased epidemic activity, development of hyperendemicity, and the geographic spread and increased incidence of the severe and fatal form of disease, DHF/DSS. Once observed only in Southeast Asia, epidemic DHF/DSS has spread to west Asia, the Peoples Republic of China, the Pacific islands, and the Americas in the past 25 years.

Factors responsible for the emergence and spread of the severe form of disease, DHF/DSS, are not fully understood. The changing disease pattern described above provides support for both principal hypotheses regarding the pathogenesis of DHF/DSS, secondary infection, and virus virulence. Thus, increased movement of viruses between population centers results in increased transmission and the development of hyperendemicity, which then increases the probability of a secondary infection, of a virulent virus emerging via genetic change or being imported from another area.

Increased transmission of multiple dengue serotypes raises the iceberg further out of the water, and increases the probability that severe disease will occur, regardless of whether the underlying cause is due to increased virulence, immune enhancement, or, more likely, a combination of both (**Figure 3**).

Dengue is primarily an urban disease. Most major epidemics of DHF/DSS occur in tropical urban centers where

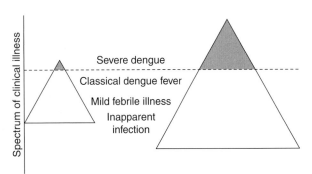

Figure 3 The iceberg concept of dengue/dengue hemorrhagic fever. The severe form of disease represents only the tip that protrudes from the water. As the incidence of infection increases, so too does the severe and fatal form of disease.

large and crowded human populations live in intimate contact with the principal mosquito vector, *Ae. aegypti*. This mosquito is a highly domesticated, day-biting species that lives and breeds in and around the home. High mosquito densities often occur in tropical cities because of water-storage practices and the accumulation of domestic trash. Primary larval habitats for *Ae. aegypti* include a variety of domestic water-storage containers such as clay jars and pots, metal drums, cement cisterns, and many other artificial containers found in the domestic environment that collect and hold rain water. The latter include, but are not limited to flower vases and pots, used automobile tires, buckets, bottles, cans, old machinery, etc.

Transmission and Tissue Tropism

Most dengue virus transmission is by the bite of an infective mosquito vector. Any of the four serotypes may cause high levels of viremia in humans ($\geq 10^8$ MID$_{50}$ ml^{-1}) that lasts an average of 4–5 days (range, 2–12 days). If a competent mosquito vector takes a blood meal from a person during this viremic phase, virus is ingested with the blood meal and infects the cells of the mosquito mesenteron. After 8–12 days, depending on ambient temperature, the virus, and the mosquito, the virus will disseminate and infect other tissues, including the mosquito salivary glands. When the mosquito takes a subsequent blood meal, virus is injected into the person along with the salivary fluids. Dengue virus infection has no apparent effect on the mosquito, which is infected for life.

Ae. aegypti is a highly competent epidemic vector of dengue viruses. It lives in close association with humans because of its preference to lay eggs in artificial water-holding containers in the domestic environment, and to rest inside houses and feed on humans rather than other vertebrates. It has a nearly undetectable bite and is very restless in the sense that the slightest movement will make it interrupt feeding and fly away. It is not uncommon, therefore, to have a single mosquito bite several persons in the same room or general vicinity over a short period of time. If the mosquito is infective, all of the persons bitten may become infected.

In addition to transmitting the virus to humans or lower primates, the female mosquito may also transmit the virus vertically to her offspring through her eggs. Although the implications of vertical transmission are not fully understood, it is thought to be an important mechanism in the natural maintenance cycles of dengue viruses, especially in rural and forest settings.

The primary site of replication of dengue viruses after injection into humans by the feeding mosquito is believed to be dendritic cells. Other tissues from which these viruses have been isolated include phagocytic monocytes, liver, lungs, kidneys, lymph nodes, stomach, intestine, and brain, but it is not known to what extent the virus replicates in these tissues. Pathological changes similar to those observed in yellow fever, with focal central necrosis, have been observed in the liver of some patients who died of dengue virus infection. There is some evidence that the viruses also replicate in endothelial cells and possibly in bone marrow cells. Encephalopathy has been documented in dengue infection but whether dengue viruses cross the blood–brain barrier and replicate in the central nervous system is still open to question. Dengue viruses have been transmitted by blood transfusion and organ transplantation.

Pathogenicity

There is still some controversy about the pathogenesis of DHF/DSS. Evidence suggests that at least two pathogenetic mechanisms are associated with severe dengue infection. Classical DHF/DSS is characterized by a vascular leak syndrome which, if not corrected, may rapidly lead to hypovolemia, shock, and death. The underlying pathogenetic mechanism for this syndrome is thought to be an immune enhancement phenomenon whereby the infecting virus complexes with non-neutralizing dengue antibody, thus enhancing infection of mononuclear phagocytes. The latter produce vasoactive mediators, which are responsible for increased vascular permeability. Loss of plasma from the vascular compartment may range from mild and transient to severe and prolonged, the latter often resulting in irreversible shock and death. Although classical DSS is most commonly associated with secondary dengue infections, it has also been documented in primary infections, which suggest that subneutralizing levels of homologous antibody or other immune factors may also cause immune enhancement.

In vitro studies have shown that not all dengue viruses can be enhanced and that there are qualitative differences in the enhancing ability of antibody to dengue viruses. This raises the question as to whether dengue virus strains vary in their ability to stimulate production of enhancing antibody, whether this is associated with virulence, and, if so, how this relates to the immune enhancement hypothesis. Because an animal model is not available, no good data exist that demonstrate variation in virulence among dengue viruses. However, an accumulating body of both experimental and field data suggest the dengue viruses, like most other animal viruses, vary in their virulence and in their epidemic potential. When DENV-1 and DENV-2 viruses were introduced into the Pacific in the early 1970s after an absence of more than 25 years, some islands experienced explosive epidemics, with patients having high viremia levels and severe and fatal hemorrhagic disease, whereas other islands with similar ecology experienced only sporadic or silent transmission, with low or undetectable viremias and mild illnesses. Virus strain variation was the only logical explanation for these differences. Recent laboratory evidence suggests that a major Cuban epidemic of DHF/DSS in 1981, the first of its kind in the American Region, was caused by a DENV-2 introduced from Vietnam, which was genetically distinct from the original American DENV-2. Data from this epidemic support both the immune enhancement and virus virulence hypotheses, which are not mutually exclusive. The most consistent feature associated with the emergence of DHF/DSS in an area is the development of hyperendemicity. This increases the probability of secondary infection, which is thought to be associated with DHF/DSS. However, hyperendemicity is also associated with increased transmission and movement of viruses between population centers, which increases the probability of genetic change and introduction of virus strains that have greater epidemic potential or virulence (**Figure 4**).

Patients infected with dengue viruses may also experience severe and uncontrolled bleeding, usually from the upper gastrointestinal (GI) tract. This severe hemorrhagic disease may be associated with multiple organ failure, and is more difficult to manage than classical DHF/DSS. The underlying pathogenetic mechanism for this type of bleeding is clearly different from that of the vascular leak syndrome, and involves disseminated intravascular coagulation and thrombocytopenia.

A third type of severe and fatal dengue infection, which may or may not involve overt hemorrhagic disease, is encephalopathy. Although many patients with this syndrome present clinically as viral encephalitis, conclusive evidence that dengue viruses infect the central nervous system has not yet been documented. Available data suggest that neurologic symptoms may be secondary to cerebral hemorrhage, edema, or other indirect effects of dengue virus infection.

Clinical Features of Infection

Dengue infection causes a spectrum of illness in humans ranging from clinically inapparent to severe and fatal hemorrhagic disease with the latter representing only the tip of the iceberg (**Figure 3**). The incubation period may be as short as 3 days and as long as 14 days, but most often is 4–7 days. The majority of patients present with mild, nonspecific febrile illness, or with classical dengue fever. The latter is generally observed in older children and adults, and is characterized by sudden onset of fever, frontal headache, retroocular pain, and myalgias. Rash, joint pains, nausea and vomiting, and lymphadenopathy are common. The acute illness, which lasts for 3–7 days, is usually benign and self-limiting, but it can be very debilitating, and convalescence may be prolonged for several weeks.

The hemorrhagic form of disease, DHF/DSS, is most commonly observed in children under the age of 15 years, but it also occurs in adults in areas of lower endemicity. It is characterized by acute onset of fever and a variety of nonspecific signs and symptoms that may last 2–7 days. During this stage of illness, DHF/DSS is difficult to distinguish from many other viral, bacterial, and protozoal infections. In children, upper respiratory symptoms caused by concurrent infection with other viruses or bacteria are not uncommon. The differential diagnosis should include other hemorrhagic fevers, hepatitis, leptospirosis, typhoid, malaria, measles, influenza, etc.

The critical stage in DHF/DSS occurs when the fever subsides to or below normal. At that time, the patient's condition may deteriorate rapidly with signs of circulatory failure, neurologic manifestations, shock and death if proper management is not implemented. Skin hemorrhages such as petechiae, easy bruising, bleeding at the sites of venepuncture, and purpura/ecchymoses are the most common hemorrhagic manifestations; GI hemorrhage may occur, usually after, but in some cases before, onset of shock.

The World Health Organization (WHO) has defined strict criteria for diagnosis of DHF/DSS, with four major clinical manifestations: high fever, hemorrhagic manifestations, hemoconcentration, and circulatory failure. WHO has classified DHF/DSS into four grades according to severity of illness: grades I and II represent the milder form of DHF and grades III and IV represent the more severe form, DSS (**Table 1**). Thrombocytopenia and hemoconcentration are constant features. However, there is some disagreement with the WHO case definition in that some patients may present with severe and uncontrollable upper GI bleeding with shock and death in the absence of

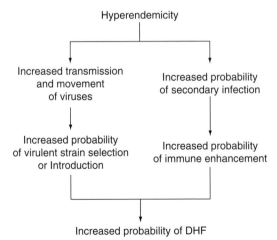

Figure 4 The cocirculation of multiple virus serotypes in a community (hyperendemicity) is the most important risk factor for the occurrence of dengue hemorrhagic fever and is compatible with the two principal hypotheses of pathogenesis, immune enhancement and virus strain variation.

Table 1 World Health Organization classification of dengue hemorrhagic fever

Grade I	Fever accompanied by nonspecific constitutional symptoms, with a positive tourniquet test and/or scattered petechiae as the only hemorrhagic manifestation
Grade II	The same as grade I, but with spontaneous hemorrhagic manifestations
Grade III	Circulatory failure manifested by rapid, weak pulse, narrowing of pulse pressure (20 mmHg or less), or hypotension
Grade IV	Profound shock with undetectable pulse and blood pressure

From Anonymous (1997) *Dengue Hemorrhagic Fever: Diagnosis, Treatment and Control*. Geneva: World Health Organization.

hemoconcentration or other evidence of the vascular leak syndrome. These patients by the WHO criteria cannot be categorized as having DHF/DSS. In addition, hepatomegaly may not be a constant feature in all epidemics of DHF/DSS. The WHO is currently reevaluating the case definitions for dengue fever, DHF and DSS.

Dengue virus infection is associated with a variety of neurologic and psychiatric disorders, including headache, dizziness, hysteria, and depression. In addition, some patients present with clinical symptoms of viral encephalitis, but as noted above, there is no conclusive evidence that CNS infection occurs.

Treatment for DHF/DSS is symptomatic, and the prognosis of the disease depends on early recognition, initiation of corrective fluid replacement, and management of shock. Definitive diagnosis can only be made in the laboratory by serologic and/or virologic methods.

Pathology and Histopathology

The pathology of dengue virus infection is not well understood because systematic postmortem studies have not been done on patients representing all types of clinical presentations. The major pathophysiologic abnormality in classical DHF/DSS is an increase in vascular permeability which leads to leakage of plasma. Patients may have serous effusions in the pleural and abdominal cavities and a variable amount of hemorrhaging in most major organs. Studies have not revealed destructive inflammatory vascular lesions, but some swelling and occasional necrosis have been observed in endothelial cells.

Limited studies on patients with a fatal outcome have demonstrated focal necrosis of the hepatic cells, Councilman bodies, and hyaline necrosis of Kupffer cells in the liver. Changes in the kidney are suggestive of an immune complex type of glomerulonephritis. There is depression of bone marrow elements, which improve when the patient becomes afebrile. Biopsy studies of the skin rash have demonstrated perivascular edema with infiltration of lymphocytes and monocytes.

Immune Response

Persons infected with dengue viruses produce IgM and IgG antibodies, both of which appear 5–7 days after onset of illness in primary infections. The highest titers of IgM antibody are produced in primary dengue infections, but IgM antibody is also produced in secondary and tertiary infections. IgM antibody is transient and generally disappears 30–90 days after onset of illness in primary infections and after shorter periods in secondary and tertiary infections. IgG antibody, by contrast, persists for at least 60 years and probably for the life of the patient. In persons experiencing their first dengue or flavivirus infection, peak IgG titers are reached 14–21 days after onset of illness and seldom exceed 640–1280, although there are exceptions. In secondary infections, on the other hand, there is an immediate anamnestic IgG immune response to dengue complex-and/or flavivirus-specific antigenic determinants. In these patients, IgG antibody titers may exceed 20 480. Both IgM and IgG antibodies neutralize dengue viruses, and infection provides life-long immunity to that specific dengue virus serotype.

Both IgM and IgG antibodies to dengue viruses cross-react with other flavivirus antigens, including those of yellow fever, Japanese encephalitis, West Nile, and St. Louis encephalitis viruses. Cross-reactivity with viruses in the dengue complex is more extensive than with other flaviviruses, and makes interpretations of serologic results difficult. In patients with second and third flavivirus infections, original antigenic sin reactions are not uncommon. In geographic areas where several flaviviruses are endemic, therefore, definitive laboratory diagnosis can only be made by virus isolation or nucleic acid detection, and in patients with primary infection, by PRNT. Normally, a combination of laboratory (serologic and virologic), clinical, and epidemiologic data is used to make a diagnosis of dengue and other flavivirus infections.

Because IgG antibody persists for many years, its presence in a single serum sample is not diagnostic unless it occurs at high titer (≥ 1280 by HI and PRNT, ≥ 256 by CF or $\geq 163\,840$ by IgG ELISA), which is considered presumptive evidence of a recent infection. Lower IgG titers simply indicate that the person has had a previous infection at some time in the past. Paired serum samples are required to confirm a current infection by demonstrating a fourfold or greater rise in IgG antibody. The presence of detectable IgM antibody in a singe serum sample is considered to be diagnostic for a recent infection because this isotype does not persist for long periods. The diagnosis is considered presumptive, and not confirmatory for a current infection, however, because IgM antibody may persist for 90 or more days.

Extensive work done in recent years has demonstrated that immunopathogenetic mechanisms play a major role in the pathophysiology of severe dengue infection. Generally, CD4+ and CD8+ T-cell responses are directed against multiple viral proteins. After primary infection, memory T cells proliferate in response to multiple dengue serotypes with both specific and cross-reactivities. After secondary infection, the T cell proliferation is of low-affinity and may be more cross-reactive to the first infecting virus serotype than the second, a phenomenon analogous to original antigenic sin seen in the antibody response.

Dengue virus infection results in the production of a number of chemokines, including IL-8, RANTES, MIP-1α, and MIP-β, which attract T cells and other inflammatory cells. Some, such as IL-8, are associated with plural effusions and may be important in inducing increased vascular permeability in patients with dengue infection.

There is accumulating evidence that cell-mediated immunity also plays a role in terminating dengue infections. Recent work suggests that dendritic cells (DCs) are likely the initial site of virus replication. Dengue virus infection stimulates DC maturation and activation, and production of TNF-α and INF-α. The DCs migrate to T-cell-rich lyphoid organs where T cells are activated, stimulating memory responses and releasing cytokines and chemokines. Circulating virus in the blood activates more T and B cells. During a second dengue infection, T-cell responses are likely to be dominated by a subset of specific memory T cells, which produce IFN-δ and CD40L, resulting in greater DC activation, T cell stimulatory capacity, IL-12 release, increased secretion of TNF-α and IFNδ, as well as potential dysregulation of cytokine responses. As viremia is cleared, the cascade of events initiated by an early type 1 cytokine response may contribute to the pathogenesis of DHF. The cross-reactive T cells that have low affinity to the second infecting virus serotype (see above) may be ineffective in clearing viremia, thus permitting a higher virus load and more severe disease.

Prevention and Control of Dengue

The options available for prevention and control of DF/DHF are limited. Although currently not available, considerable progress has been made in recent years in development of a vaccine (for DF/DHF). Effective vaccination to prevent DF/DHF will likely require a tetravalent, live attenuated vaccine. Promising candidate attenuated vaccine viruses have been developed and have been evaluated in phase I and II trials.

Progress has also been made on developing second-generation, recombinant dengue vaccines by using cDNA infectious clone technology. At least three candidate chimeric vaccines have been constructed by inserting the PrM and E genes into the backbone of yellow fever 17D vaccine virus, into an attenuated DENV-2 (PDK-53) backbone and into an attenuated DENV-4 backbone. The yellow fever 17D chimeric vaccine is further along and appears promising after phase I safety trials. The other two, plus a subunit candidate vaccine, are still in the preclinical phase.

The development of other new technology vaccines is in its infancy. Despite the promising progress, it will likely be 5–7 or more years before an effective, safe, economical dengue vaccine is commercially available. Also very promising has been the rapid progress in developing antiviral drugs that can be used in treatment of dengue infection, and perhaps, even in prevention and control programs.

Currently, the only way to prevent dengue infection is to control the mosquito vector that transmits the virus. Unfortunately, our ability to control *Ae. aegypti* is limited. For more than 25 years, the recommended method of control was the use of ultralow volume (ULV) application of insecticides to kill adult mosquitoes. Field trials in Puerto Rico, Jamaica, and Venezuela, however, showed that this method was not effective in significantly reducing natural mosquito populations for any length of time. This supports epidemiologic observations that ULV has little or no impact on epidemic transmission of dengue viruses.

The only truly effective method of controlling *Ae. aegypti* is larval control, that is to eliminate or control the larval habitats where the mosquitioes lay their eggs. Most important larval habitats are found in the domestic environment, where most transmission occurs. To have sustainability of prevention and control programs, some responsibility for mosquito control should be transferred from government to citizen homeowners. For long-term sustainability, mosquito control programs must be community-based and integrated. Persons living in *Ae. aegypti* infested communities have to be educated to accept responsibility for their own health destiny by helping government agencies control the vector mosquitoes, and thus prevent epidemic DF/DHF/DSS.

Countries with endemic dengue should develop active, laboratory-based surveillance systems that can provide some degree of epidemic prediction. Finally, prevention of excess mortality associated with DHF/DSS can be achieved by educating physicians in endemic areas on clinical diagnosis and management of DHF/DSS. As demonstrated in countries such as Thailand, early recognition and proper management are the key to keeping DHF/DSS case–fatality rates low.

Future

Continued population growth and urbanization of the tropics, changing lifestyles, increased air travel and lack of effective mosquito control have been the most important factors responsible for the dramatic increased incidence and geographic expansion of DHF/DSS in the past

25 years. DF/DHF/DSS has become a global public health problem in the tropics and it is anticipated that this trend will continue unless something is done to reverse it. More effective integrated prevention and control strategies must be developed and implemented worldwide in the tropics. Ultimately, development of an economical, safe, and effective vaccine holds the greatest promise for sustainable prevention and control.

See also: Arboviruses; Diagnostic Techniques: Serological and Molecular approaches; Flaviviruses: General Features.

Further Reading

Anonymous (1997) In: *Dengue Haemorrhagic Fever: Diagnosis, Treatment and Control*, 58 p. Geneva: World Health Organization.

Calisher CH, Shope RE, Brandt W, *et al.* (1989) Antigenic relationships among viruses of the genus Flavivirus (family Flaviviridae). *Journal of General Virology* 70: 37–43.

Gubler DJ (1989) Aedes aegypti and Aedes aegypti-borne disease control in the 1990's: Top down or bottom up. *American Journal of Tropical Medicine and Hygiene* 40: 571.

Gubler DJ (1998) Dengue and dengue hemorrhagic fever. *Clinical Microbiology Reviews* 11: 480–496.

Gubler DJ (2002) Epidemic dengue/dengue hemorrhagic fever as a public health, social and economic problem in the 21st century. *Trends Microbiology* 10: 100–103.

Gubler DJ and Kuno G (eds.) (1997) *Dengue and Dengue Hemorrhagic Fever*. Wallingford, UK: CAB International.

Gubler DJ, Kuno G, and Markoff L (2007) Flaviviruses. In: Knipe D and Howley P (eds.) *Fields Virology*, 5th edn., pp. 1153–1252. Philadelphia: Lippincott Williams and Wilkins.

PAHO (1994) *Dengue and Dengue Hemorrhagic Fever in the Americas: Guidelines for Prevention and Control*, No. 548. Washington, DC: Publicaciones Cientificas.

Phuong CX, Nhan NT, Kneen R, *et al.* (2004) Clinical diagnosis and assessment of severity of confirmed dengue infections in Vietnamese children: Is the World Health Organization classification system helpful? *Americal Journal of Tropical Medicine Hygiene* 70: 172–179.

Pugachev KV, Guirakhoo F, and Monath TP (2005) New developments in flavivirus vaccines with special attention to yellow fever. *Current Opinion Infectious Diseases* 18: 387–394.

Rico-Hesse R (2003) Microevolution and virulence of dengue viruses. *Advances in Virusus Research* 59: 315–341.

Thomas SJ, Strickman D, and Vaughn DW (2003) Dengue epidemiology: Virus epidemiology, ecology, and emergence. *Advances in Virusus Research* 61: 235–289.

Thongeharoen P (ed.) (1993) *Monograph on Dengue/Dengue Hemorrhagic Fever*, No. 22. New Delhi: World Health Organization.

Diagnostic Techniques: Microarrays

K Fischer, A Urisman, and J DeRisi, University of California, San Francisco, San Francisco, CA, USA

© 2008 Published by Elsevier Ltd.

Motivation for New Diagnostic Methods

Numerous human diseases exist for which viral etiologies are suspected, yet specific causal agents are not known. Among these are up to 20% of cases of acute hepatic failure, up to 35% of cases of acute aseptic meningitis and acute encephalitis, up to 50% of cases of acute respiratory infections, and numerous other conditions. In addition, infectious agents may be involved in the pathogenesis of a number of chronic conditions, most notably disorders such as chronic inflammation, autoimmune and degenerative conditions, as well as some forms of cancer. While it is unlikely that viruses cause all of these diseases, identifying causative agents in even a modest number of disorders will have profound implications for understanding, diagnosis, and treatment of these conditions.

New approaches to viral diagnostics and discovery are needed to overcome the shortcomings of existing methods. These methods include viral culture, electron microscopy, serology, specific polymer chain reaction (PCR)-based methods, and techniques based on subtractive hybridization. While these methods have been critical for identifying many important human and nonhuman pathogens, each of these methods has intrinsic limitations. For example, many viruses are refractory to culture. Inspection by electron microscopy may prove difficult depending on the titer and morphological features of the virus. Serology- and PCR-based techniques are highly specific methods, but their specificity frequently renders them ineffective for the detection of variant or novel viral species. In the case of PCR, there exist dozens if not hundreds of variations on the method which may extend the scope of the assay, usually through multiplexing or by the use of degenerate primers, yet the number of possible targets that can be interrogated remains small relative to the number of known viral pathogens. Finally, subtractive hybridization techniques, while unbiased, are difficult to troubleshoot and essentially impossible to scale up for high throughput.

DNA Microarrays

While there exist many different forms of DNA microarrays, produced by both researchers and corporations, they fundamentally share the same properties. All microarrays exist as some form of solid substrate, typically glass or silicon, to which is bound different species of nucleic acid. Each microarray may contain 10^5–10^6 different species of DNA, arranged in a grid. Fluorescently labeled nucleic acid derived from any biological sample may be interrogated by hybridization to the microarray. In this manner, the abundance of thousands of different nucleic acid species may be simultaneously measured. The most common application of microarray technology is the measurement of relative gene expression, often for entire genomes. While gene expression profiling has enjoyed tremendous success over the last decade, it has long been realized that the microarray format would be amendable to the detection of exogenously derived nucleic acids for purpose of identifying the presence of pathogens in a background of host material. The general concepts and methodologies are similar to those required for expression profiling; however, there are key differences to consider for the design of the array and the mechanics of sample processing. Furthermore, most implementations of virus-detection microarrays strive only to determine the presence of viral sequences, rather than attempting to quantify the amount of virus in a particular sample. In practical terms, this allows more aggressive amplification strategies to be used during sample preparation.

Microarray Design for Viral Detection

Algorithms to guide the process of sequence selection for expression microarrays are well developed, and many of the design principles apply directly to microarrays for viral detection. The parameters shared in common include physical properties of the oligonucleotide itself, such as the propensity of the sequence to form hairpins, melting temperature, and sequence complexity. Beyond this, the design considerations for viral detection and expression profiling differ substantially. First, the majority of microarrays intended for viral detection are designed to specifically detect the products of a multiplex PCR reaction using specific primers. In this role, the array simply sorts the product of amplification. The design specifications for these types of arrays are straightforward and are primarily guided by the choice of flanking PCR primers for virus amplification. This configuration of array does not exploit the full potential of the microarray for panviral detection. However, the design parameters for a generalized virus detection chip that does not rely on specific multiplex PCR primers must take into account additional factors.

While it is essential that probes for expression arrays are unique with respect to target sequence (to prevent cross hybridization from other mRNA species), the same does not necessarily apply to viral probes. In fact, to maximize the probability of detecting any viral species from a known family, it is often desirable to choose sequences that are the most conserved among a group of viruses. For example, while there exists a large range of sequence diversity among human rhinoviruses, sequences in the 5' UTR are highly conserved, even among more distant picornaviruses. These sequences may thus serve as a type of universal 'hook' to capture both existing and new variants of these RNA viruses. In case of novel pathogens, such as the SARS coronavirus, the use of conserved sequences was a key determinant for successful detection by microarray. Clearly, unique species-specific or even genotype-specific oligonucleotides can further augment the discrimination of the microarray. A logical extension of this strategy takes into account features of viral taxonomy. Rather than choosing all conserved, or all unique sequences, one may attempt to cover, by design, each level of the taxonomic tree for each family. Thus, some sequences would be chosen to be species specific (terminal nodes on the tree), some would be genus specific, and so on. Various bioinformatic tools, including online databases, are currently available to assist in such design efforts. In all cases, it is critical to prescreen each choice for matches within the human genome to prevent inappropriate cross-hybridization to host material.

A more simple approach is to simply tile overlapping oligonucleotides spanning the entire genome of the viruses in question. This approach is appropriate for relatively small panels of viruses since each species will result in large numbers of sequences, depending on genome length. In general, this approach becomes impractical when extended beyond a few viral species.

After satisfying the basic design requirements, more sophisticated considerations may also contribute to choice of viral sequence for representation on a microarray. For example, to enhance detection of latent herpesviruses, it may be advantageous to overrepresent sequences specific for genes specific for latent phase expression, rather than those involved in lytic processes. In this case, it is assumed that the RNA rather than DNA will be analyzed, which highlights the importance of sample processing and amplification considerations.

Sample Processing and Amplification

The protocol by which nucleic acids are isolated from specimens, and the subsequent amplification of the material, if needed, is also important to consider, as this may also affect both microarray and experimental design. In general, isolation of total RNA is the preferred and more conservative route. While this may seem biased toward RNA viruses, all DNA viruses produce mRNA as part of their lifecycle, so this choice does not exclude them. However, if viral particles are collected, or host material is removed from the specimen by filtration or other size-selection techniques, it may be advisable to isolate total nucleic acid (both RNA and DNA) to maximize sensitivity. The origin of the sample also bears on this issue. When processing a relatively acellular material, such as cerebral spinal fluid, a total nucleic acid extraction would be appropriate, whereas in the case of a solid organ, such as liver or brain, an RNA extraction would avoid the unnecessary complexity brought by co-purifying massive quantities of the host genomic DNA.

After nucleic acid has been isolated, an amplification step is typically employed to generate sufficient quantities for successful microarray hybridization. In certain situations, where large amounts of primary material are available, the yield of nucleic may be such that an amplification step may be bypassed altogether, but these situations are the exception. The choice of an amplification strategy is closely linked to the design of the array itself. Several virus-detection microarrays have been designed to serve as detectors for the products of multiplex PCR amplification strategies, but in these cases, the broad spectrum and unbiased nature of the microarray is not realized.

For panviral microarrays, where no assumptions are made as to the probable identity of the target, a general, randomized amplification strategy is required. Numerous random amplification strategies exist, but all begin with a priming step using a randomized oligo of various lengths, ranging from 6 to 15 bp. At this point, PCR adapters may be added, either by ligation, or through priming via a common sequence linked to the random primer. Alternatively, various RNA polymerase promoter sequences may be appended which then allow linear amplification. For all these methods, contamination with previously isolated material is a critical concern and good laboratory practices and appropriate controls, such as nontemplated amplification reactions, must be an integral part of the protocol.

The overall complexity of the sample, with respect to the viral species, is a critical factor for the success of random amplification strategies. As previously noted, the ideal samples are those that contain relatively low amounts of host cellular material and high titers of virus. For many biological specimens, it may be possible to reduce the complexity by filtration, centrifugation, or by pretreatment of the raw material with various nucleases. In the latter case, free host material such as genomic DNA and ribosomal RNA may be degraded, yet viral-packaged nucleic acid will escape destruction. Such 'preprocessing' can significantly enhance the signal to noise of the final microarray assay, although these steps add both cost and complexity to the overall protocol. Potentially, the use of novel microfluidic techniques for particle size discrimination may serve as a rapid and reproducible way to deterministically reduce complexity in biological samples.

Bioinformatics

Analysis of hybridization results depends greatly on the design of the experiment and of the DNA microarray in particular. Microarrays that are narrowly targeted, for the purposes of distinguishing between strains of a particular virus species, for example, human influenza or smallpox, are subjected to different analysis techniques than are multigenus, multifamily, or even panviral arrays. The utility of DNA microarrays in all these fields is demonstrable when investigators use appropriate analysis techniques.

In strain and species typing applications there are usually ample controls available to researchers. By use of control hybridizations, characteristic species or strain hybridization patterns can be classified manually. Often these patterns are the result of iterative microarray design where features with the desired specificity are spatially clustered on the chip. These patterns can be used as templates for visual inspection of the experimental arrays where different classes of microarray features are enumerated and the input sample is thereby placed into one of the known groups or unclassified as ambiguous. Machine-learning techniques, such as probabilistic neural networks, are also sometimes used in these studies, but to date have not been show to be more accurate than simple enumerative methods.

Broader explorations of viral populations, for example, panviral studies, must face the problem that the number of control samples available is small when compared to the number of distinct viral genomes that may be encountered. Many viral targets are extremely mutable, introducing one or more mutations per virus genome for every cycle of replication. Such diversity quickly outstrips the resources of any effort to perform exhaustive control hybridizations.

In panviral studies standard methods such as hierarchical clustering and several particular bioinformatic approaches can be employed. First the microarray should be designed to include conserved regions of the target viruses to minimize the chance that a divergent but related virus will escape detection. Second, estimates of the hybridization patterns expected from the possible targets for which sequence data is available can be generated using biophysical models of hybridization. Experimental hybridization patterns can then be compared to the model profiles using a correlation metric. If a virus present in a hybridization shares homology with one of the sequences used to estimate the hybridization profiles, this similarity will be reported as a significant correlation between the profile and the experimentally observed pattern. In the last method, the intensity history of the microarray's features can be compared to an experimental hybridization. Extraordinarily strong signal at a particular feature can indicate *bona fide* virus, especially when signal at taxonomically related features is similarly elevated. During the design stage of the experiment, it is important to consider that in multifamily, metagenomic studies it is critical to have negative controls to characterize the background of microarray designed with broad specificity, while in studies with narrower focus a significant number of positive controls are needed for each class of virus being considered.

Real-World Applications: Research and Clinical

To date, only a few studies have examined the use of viral-detection microarrays using actual prospectively collected patient samples, as opposed to viruses cultured in the lab or previously characterized retrospective samples. As of the time of this writing, no large-scale study has been published comparing the use of virus microarrays (using a random amplification) to traditional laboratory medicine diagnostics, such as commercial DFA kits and PCR assays. Preliminary data and studies using modest numbers of clinical samples suggest that a microarray-based viral diagnosis outperforms conventional DFA assays, both in terms of sensitivity and specificity. In comparison to a PCR assay using specific primer pairs, microarray assays are likely to have comparable specificity, yet the sensitivity is anticipated to be somewhat lower, depending on amplification strategy. However, microarray-based assay coupled to a random amplification protocol provides much broader detection capabilities, often including essentially every known viral pathogen. When considering the potential of DNA microarrays for clinical diagnostics, it may be the case that the microarray will serve a complementary role to specific PCR assays, especially when the latter fails to yield positive results.

For clinical cases where conventional diagnostic assays have failed, the use of a panviral microarray assay may allow identification of new, or unanticipated pathogens. In a recent case report, Chiu and colleagues reported the diagnosis of a previously healthy 28-year-old woman suffering from a severe respiratory tract infection of unknown etiology. Extensive panels of diagnostics for bacteria, fungi, and viruses failed to reveal any positive results, and all viral cultures remained negative. DNA microarray analysis of RNA isolated from endotracheal aspirate revealed the presence of human parainfluenza virus-4, a virus that is not normally included on standard DFA panels or PCR tests. Viral sequence was recovered directly from the patient sample confirming the identity of the virus, and it was further shown that the patient seroconverted during the time of the illness. While it is generally believed that human parainfluenza virus-4 causes only mild, self-limiting infections, this data taken together suggests that the spectrum of disease may extend to respiratory failure in an otherwise healthy adult. This example demonstrates an attractive feature of the microarray assay, namely, the power to detect the unexpected.

In addition to the detection of previously known pathogens, DNA microarrays have also been effective for the detection of novel viral species. In the case of SARS, total nucleic acid from a supernatant from an infected vero cell culture revealed a coronavirus signature consisting of oligonucleotides originating from avian infectious bronchitis, human and bovine coronaviruses, and, interestingly, several astroviruses. At first glance, the hybridization signal from astrovirus-derived oligonucleotides would seem to be aberrant. In fact, this is expected, since several astroviruses and coronaviruses share conserved sequences at the 3′ end of their genomes. These particular sequences were represented on the microarray since the panviral design algorithm purposely selected conserved sequences within and among viral families.

The same principles applied to a separate study in which a novel xenotropic gamma retrovirus was detected in prostate tumor biopsies of men with a mutant variant of the *RNASEL* gene. Integration sites and full-length genomes were subsequently cloned and the virus was demonstrated to be replication competent, thus validating the microarray result. Again, the broad-spectrum nature of the DNA microarray was critical to the success of the project, since there were no preconceptions that such a virus might be a candidate, given that no xenotropic gamma retrovirus had been previously observed in a human subject.

Limitations

Several important limitations of using microarrays for viral detection and discovery should not be overlooked. The most important limitation of the approach is its reliance on known viral sequences. Although most of the novel viruses discovered in the last decade share homology with previously known viruses, viruses lacking even short regions of homology cannot be detected by any hybridization-based method. In the case of profoundly divergent viruses, more brute-force approaches, such as shotgun sequencing, are likely to be applicable. Other existing and likely surmountable limitations of the microarray-based methods are their cost, the need for specialized equipment, and access to computational resources. It is likely that these limitations will become less pronounced as streamlined versions of the technology become available through academic and commercial efforts. It may also be the case that the utility of DNA microarrays may be surpassed by next-generation, massively parallel shotgun sequencing technologies, which would permit cheap, fast, and unbiased analysis of clinical samples.

Scientific and Public Health Implications

Currently, a substantial fraction of human disease, with a presumed viral cause, goes without a clinical diagnosis. This is especially true for common ailments, such as upper respiratory tract infections, where despite advances in PCR assays, the etiology of 30–60% of infections remains unidentified. Without considering the complexities of diagnostic regulatory approvals, an unbiased DNA microarray approach to the detection of viral pathogens should substantially increase the number of successful diagnoses, and as a consequence, may lead to improved therapeutics and supportive care. It should be noted that use of a virus microarray extends beyond the analysis of clinical samples. The wide scope of detection and the power to discover new pathogens has broad application to agriculture, veterinary medicine, ecology, and environmental genomics. While it is impossible to accurately predict the number of undiscovered viral pathogens remaining on this planet, tools such as virus-detection DNA microarrays permit rapid inroads into this fascinating and important aspect of virology.

See also: Diagnostic Techniques: Plant Viruses; Diagnostic Techniques: Serological and Molecular approaches.

Further Reading

Chiu CY, Rouskin S, Koshy A, et al. (2006) Microarray detection of human parainfluenzavirus 4 infection associated with respiratory failure in an immunocompetent adult. *Clinical Infectious Diseases* 43: e71–e76.

Lin B, Wang Z, Vora GJ, et al. (2006) Broad-spectrum respiratory tract pathogen identification using resequencing DNA microarrays. *Genome Research* 16: 527–535.

Lin FM, Huang HD, Chang YC, et al. (2006) Database to dynamically aid probe design for virus identification. *IEEE Transactions on Information Technology in Biomedicine* 10: 705–713.

Urisman A, Fischer KF, Chiu CY, et al. (2005) E-Predict: A computational strategy for species identification based on observed DNA microarray hybridization patterns. *Genome Biology* 6: R78.

Urisman A, Molinaro RJ, Fischer N, et al. (2006) Identification of a novel gammaretrovirus in prostate tumors of patients homozygous for R462Q RNASEL variant. *PLoS Pathogen* 2: e25.

Wang D, Coscoy L, Zylberberg M, et al. (2002) Microarray-based detection and genotyping of viral pathogens. *Proceedings of the National Academy of Sciences, USA* 99: 15687–15692.

Wang D, Urisman A, Liu YT, et al. (2003) Viral discovery and sequence recovery using DNA microarrays. *PLoS Biology* 1: e2.

Wang Z, Daum LT, Vora GJ, et al. (2006) Identifying influenza viruses with resequencing microarrays. *Emerging Infectious Diseases* 12: 638–646.

Diagnostic Techniques: Plant Viruses

R Koenig and **D-E Lesemann**, Biologische Bundesanstalt für Land- und Forstwirtschaft, Brunswick, Germany
G Adam, Universität Hamburg, Hamburg, Germany
S Winter, Deutsche Sammlung für Mikroorganismen und Zellkulturen, Brunswick, Germany

© 2008 Elsevier Ltd. All rights reserved.

Glossary

Cylindrical inclusions (CI) CI are induced in infected cells by all viruses belonging to the family *Potyviridae*. The term pinwheel inclusion is also widely used, because it describes the typical appearance of the CI in cross sections. CI are composed of one virus-encoded 66–75 kDa nonstructural protein which aggregates to monolayer sheets forming a complicated structure in which several curved sheets are attached to a central tubule. Details of the CI architecture are specific for the virus species and can serve as an additional identification feature of the virus.

Immunosorbent techniques (e.g., enzyme-linked immunosorbent assay, immunosorbent electron microscopy) Techniques in which a viral antigen is trapped on a solid matrix by means of specific antibodies that are bound to the matrix by adsorpiton.

Plant virus vectors Plant viruses can be transmitted specifically by various vector organisms, for example, aphids, mites, nematodes, or plasmodiophorid protozoans (*Olpidium* sp., *Polymyxa* sp.).

Viroplasm Cytoplasmic inclusions induced by members of the *Caulimoviridae*, *Rhabdoviridae*, *Reoviridae*, and *Bunyaviridae* are about the size of nuclei and are not bound by any membrane. The viroplasm consists of amorphous and/or fibrillar material and may or may not enclose immature or mature virus particles. It is generally assumed that viroplasms are the site of virus synthesis.

Introduction

Diagnostic techniques for plant viruses are indispensable for at least three major sets of applications:

1. the identification and classification of viruses associated with 'new' plant diseases (primary diagnosis);
2. the routine detection of known viruses, for example, in
   - indexing programs designed to provide healthy plant propagation material (seeds, tubers, grafting material, etc.);
   - breeding programs designed to select virus-resistant plants; and
   - epidemiological investigations designed to monitor the spread of viruses into new areas and in quarantine tests designed to prevent this spread;
3. investigations on the functional, pathogenic, and structural properties of a virus, for example, on
   - its spread within a plant or a vector;
   - changes produced in the ultrastructure of infected cells;
   - intracellular localization of the virions and of nonstructural viral proteins;
   - the time course of the expression of various parts of the viral genome; and
   - the surface structure of virus particles, for example, the accessibility or nonaccessibility to antibodies of various parts of coat protein in the virions.

The primary diagnosis usually requires the application of a number of different techniques, whereas for routine diagnosis only one well-adapted technique is employed. For studying additional properties of a virus, in order to understand its pathogenic effects, one or several techniques may be necessary. The most important diagnostic techniques for plant viruses are based on their morphological, serological, molecular, and pathogenic properties, quite often on a combination of them.

Techniques Used for the Identification and Classification of Viruses Associated with New Plant Diseases (Primary Diagnosis)

Inspection of Symptoms

The first step in attempts to identify the causal agent of a new plant disease is usually a careful inspection of the type of symptoms present in the natural host. **Figure 1** shows some typical virus symptoms. Mosaics of green and yellow (chlorotic) tissue areas (**Figure 1(a)**) or of green and brownish (necrotic) ones (**Figure 1(c)**) or irregular yellow spots (mottling) (**Figure 1(b)**) may be seen on leaves, flowers, or fruits. Flowers may show irregular discolorations (flower break) (**Figure 1(d)**). More or less severe growth reduction (stunting) and/or malformations of either the whole plants or of individual parts, such as flowers and tubers, are also often observed.

Adsorption Electron Microscopy

A quick means to detect the presence of viruses – preferentially in symptom-showing parts of a plant – is the electron microscopical examination of a crude sap extract. In adsorption electron microscopy (**Figure 2(a)**), the virus particles are adsorbed to support films on electron microscope grids and are negatively stained, for example with uranyl acetate. Rod-shaped, filamentous, or isometric particles are readily distinguished from normal plant constituents. More than 30 different morphological types of viruses (examples in **Figure 3**) have been recognized. Particle morphology may serve as a guide for further identification, for example, by means of serology or molecular analyses (see below). Unfortunately, many viruses with very different pathogenic and other properties may be morphologically indistinguishable. Another severe limitation of this test is its rather low sensitivity. Thus, viruses occurring in low concentrations may not be detected. These drawbacks may be overcome in the immunoelectron microscopical tests described in the following sections.

Immunosorbent Electron Microscopy

In immunosorbent electron microscopy (ISEM) (**Figures 2(b), 4(a),** and **4(b)**), the electron microscope grids carrying a support film are coated with antibodies specific for viruses having the same morphology as the ones found in the electron microscopical adsorption test or with antibodies to viruses being suspected to be present in the plant material under investigation. Over a thousand times more virus particles can be detected on such antibody-coated grids than on uncoated grids or on grids coated with antibodies to an unrelated virus. This ISEM test is a very versatile technique. Its sensitivity and the broadness of cross-reactivities detected can be modulated by the length of time during which the virus-containing fluids are in contact with the antibody-coated grids. When the incubation time is short (15 min or less), mainly the homologous and very closely related viruses are trapped. However, after incubation overnight, the efficiency of virus trapping increases and more distantly related viruses are also detected. Very small amounts of antisera are needed (1 µl of crude antiserum can support 200 ISEM tests) and the quality of the antisera need not necessarily be high, since antibodies to normal host constituents do not interfere with virus-specific reactions and low-titered antisera can also be used. Since ISEM tests are labor intensive – they are not suitable for the routine detection of viruses in large numbers of samples.

Figure 1 Typical symptoms induced by plant viruses: (a) mosaic symptoms on abutilon, (b) mottling symptoms on zucchini fruits, (c) chlorotic and necrotic stripe mosaic on barley leaves, (d) flower break on lilies.

Immunoelectron Microscopical Decoration Tests

In the immunoelectron microscopical decoration test (**Figures 2(c), 4(c),** and **4(d)**), the reaction between free antibodies and viruses which have been immobilized on electron microscope support films either by adsorption or serological binding can be visualized as antibody coating of particles ('decoration'). At low antiserum dilutions (e.g., 1:50), homologous antisera produce a dense coating. Reactions with heterologous antisera may be recognized by a much weaker coating efficiency. The closeness of serological relationships may be assessed by comparing the dilution end points (titers) of antisera with homologous and heterologous viruses. The more refined immunogold-labeling technique uses antibodies that are labeled with colloidal gold particles having diameters between 5 and 20 nm. The binding of small amounts of antibodies to individual epitopes, for example, at the extremities of rod-shaped or filamentous virus particles (**Figure 4(d)**), can be visualized by means of this technique.

The immunoelectron microscope decoration tests are the most reliable means for verifying reactions between

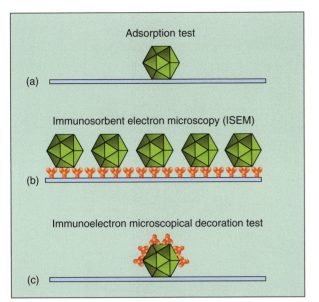

Figure 2 Schematic representation of various electron microscopical tests. For detailed explanations, see text. Adapted with permission from Koenig R and Lesemann D-E (2001) Plant virus identification. In: *Encyclopedia of Life Sciences*. Copyright 2001, © John Wiley & Sons Limited.

viruses and antibodies. Mixed infections by morphologically indistinguishable viruses are readily detected (**Figure 4(c)**). The presence of antibodies can be recognized even on a single virus particle. Thus, viruses occurring at low concentrations, as often happens in field samples, can nevertheless be analyzed directly. The detectability of viruses occurring at low concentration may be increased when the decoration test is preceded by an ISEM step. Antisera even with high host plant reactivities can be used successfully, because the specific antibody attachment to virus particles is directly visualized. The technique is not suited for detecting small antigenic differences between viruses and since it is labor intensive and cannot be automated, it is not suitable for routinely testing large numbers of samples.

Observations on Host Plant Reactions

The transmission of new viruses to experimental host plants is another helpful method to identify and characterize suspected 'new' viruses. Commonly used indicator and propagation hosts are *Chenopodium* species, for example, *C. quinoa* (**Figure 5**), and *Nicotiana* species, for example, *Nicotiana benthamiana*. These plant species are susceptible to a great number of plant viruses, but by far not to all of them. Host range studies may be helpful in finding a suitable host for propagating a new virus and for differentiating it from other similar viruses. In experimental hosts, viruses may reach higher concentrations than in their natural hosts and they may often be purified more easily from them, because natural hosts frequently contain tannins, glycosides, slimy substances, and other inhibitory compounds. Virus purification is necessary for the production of specific antisera which are needed for many diagnostic procedures, such as the above-mentioned immunoelectron microscopical techniques and many others, some of which will be described in the following sections.

Nucleotide Sequence Analyses

Nucleotide sequence analyses are the most reliable means to identify a virus either as a 'new' one or as a 'new' strain of a known virus. These analyses are more time-consuming than the above-described immunoelectron microscope tests. The latter can provide preliminary information concerning the putative genus or family membership of a new virus within 1–2 days. This preliminary information may serve as a valuable basis for further molecular studies, because it allows the design of broad specificity primers for cDNA synthesis and amplification of genome portions by means of the polymerase chain reaction (PCR). Such primers will be derived from highly conserved genome areas in the genera or families suggested by electron microscope studies.

Genome sequences of viruses with circular DNAs can be amplified directly by means of isothermic rolling-circle amplification (RCA) techniques using random primers and the DNA polymerase of *Bacillus subtilis* bacteriophage Phi29. This enzyme in addition to polymerase activity possesses strand-displacement activity, thus allowing circular DNA to be replicated to a nearly unlimited extent. Since random primers can be used, no previous sequence information is necessary. Virus variants or viruses in mixed infections can be recognized after cleavage with restriction endonucleases.

For RNA viruses, reverse transcription of the genome into cDNA is necessary. The PCR products obtained with broad specificity primers can often be sequenced directly without the need of time-consuming and costly cloning procedures. Several commercial companies offer this service. The sequences are compared with those of other viruses in the respective taxonomic group that can be obtained from gene banks, for example, 'Pubmed' or the 'Descriptions of Plant Viruses'. The latter as well as the 'ICTVdB Index of Viruses' offer additional information on previously described viruses.

If no information on the possible genus or family membership of a new virus is available from initial electron microscope studies, the double-stranded RNA (dsRNA) found as a replication intermediate in plants infected with RNA viruses may serve as a template for cDNA synthesis. Reverse transcription may be initiated by a 'universal' primer containing a sequence of about 20 defined nucleotides on its $5'$ end and a $3'$ tail consisting of six variable nucleotides that may bind randomly to various internal sites in the denatured dsRNA. The cDNAs

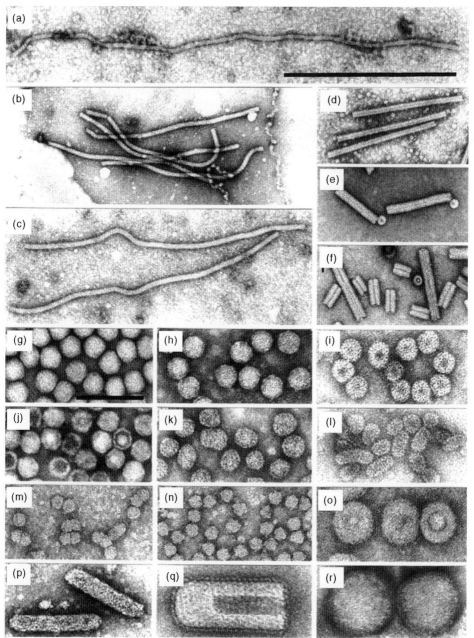

Figure 3 Various morphological types of plant viruses. Filamentous and rod-shaped particles: (a) beet yellows virus (genus *Closterovirus*), (b) potato virus X (genus *Potexvirus*), (c) potato virus Y (genus *Potyvirus*), (d) tobacco mosaic virus (genus *Tobamovirus*), (e): barley stripe mosaic virus (genus *Hordeivirus*), (f) tobacco rattle virus (genus *Tobravirus*). Isometric particles: (g) maize chlorotic mottle virus (genus *Machlomovirus*) , (h) barley yellow dwarf virus (genus *Luteovirus*), (i) cucumber mosaic virus (genus *Cucumovirus*), (j) arabis mosaic virus (genus *Nepovirus*). Irregular isometric to bacilliform particles: (k) apple mosaic virus (genus *Ilarvirus*), (l) alfalfa mosaic virus (genus *Alfamovirus*). Geminate or isometric 18 nm diameter particles: (m) tomato yellow leaf curl virus (genus *Begomovirus*), (n) faba bean necrotic yellows virus (genus *Nanovirus*). Isometric or bacilliform particles: (o) cauliflower mosaic virus (genus *Caulimovirus*), (p) cacao swollen shoot virus (genus *Badnavirus*) . Bullet-shaped or spherical particles: (q) Laelia red leaf spot-type virus (genus *Rhabdovirus*), (r) tomato spotted wilt virus (genus *Tospovirus*). The magnification bar shown in (a) equals 500 nm; the same magnification is also used in (b) to (f). The magnification bar shown in (g) equals 100 nm; the same magnification is also used in (h) to (r). Adapted with permission from Koenig R and Lesemann D-E (2001) Plant virus identification. In: *Encyclopedia of Life Sciences*. Copyright 2001, © John Wiley & Sons Limited.

obtained are rendered double-stranded and are amplified by means of PCR using a second primer that contains only the sequence of the 20 defined nucleotides forming the 5′ end of the 'universal' primer. The many different PCR products obtained will form a smear after electrophoresis in an agarose gel. The more slow-moving fraction of this smear is eluted from the gel. It contains the larger-sized PCR products that are cloned into a suitable

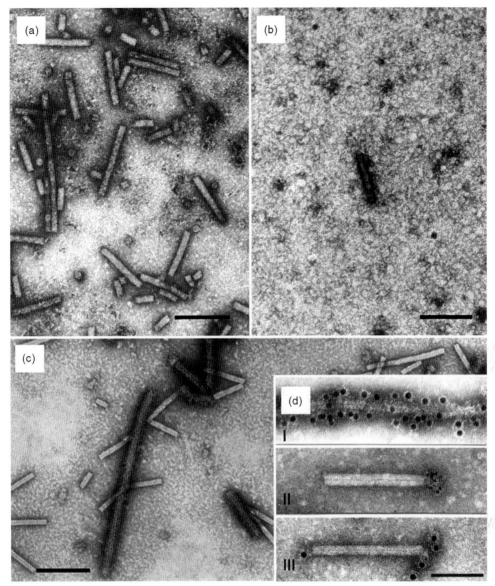

Figure 4 Immunoelectron microscopical analyses of virus-containing fluids: (a) Many virus particles are trapped within 15 min on a grid which had been coated with the homologous antiserum. (b) Only one particle is adsorbed nonspecifically from the same extract as in (a) on a grid coated with serum from a nonimmunized rabbit. (c) Detection of a mixed infection with two morphologically indistinguishable viruses: only those virus particles are decorated for which the antiserum is specific. (d) Binding of monoclonal antibodies to virus particles visualized by immunogold labeling (10 nm gold particles in the upper and lower parts and 5 nm gold particles in the middle part of the figure, respectively). In the upper part of the figure, the antibodies are bound along the entire surface of the particle, and in the middle and lower parts of the figure, on one or both extremities of the particles, respectively. Scale = 200 nm (a–c); 100 nm (d). Adapted with permission from Koenig R and Lesemann D-E (2001) Plant virus identification. In: *Encyclopedia of Life Sciences*. Copyright 2001, © John Wiley & Sons Limited.

vector. Nucleotide sequences are determined with clones containing differently sized inserts. Gaps between individual portions of a sequence may be bridged up by means of PCRs using suitably designed primers on each of these sequences. Complete sequences may be obtained by means of various RACE techniques (rapid amplification of cDNA ends). Since genome recombinations are frequently found with plant viruses, it is important to determine the full-length or almost-full-length nucleotide sequences for newly described viruses in order to allow comparisons to be made along the entire length of the genome.

Techniques That are Especially Useful for the Routine Detection of Plant Viruses

Most of the techniques that are presently used for the routine detection of plant viruses are either based on the use of enzyme-labeled antibodies or on the amplification of portions of the viral genome by means of the PCR or other nucleic acid amplification techniques (see below). Tests based on symptom formation on indicator plants to which the viruses are transmitted either mechanically or by grafting have been used extensively as diagnostic tools in earlier times, but they are increasingly replaced by molecular or serological techniques which provide results within 1–2 days rather than after weeks or months. In addition, these latter techniques do not require extensive greenhouse or outdoor space, are more specific, do not require the propagation of unwanted pathogens, and – very importantly – can be readily automated.

Enzyme-Linked Immunosorbent Assay

Enzyme-linked immunosorbent assay (ELISA), especially its double antibody sandwich form (DAS ELISA), is presently probably the most widely used technique for the routine detection of plant viruses. The principle of DAS ELISA is outlined in **Figure 6**. Virus particles present even in low concentrations in crude sap extracts of diseased plants are trapped by specific antibodies that have been immobilized on the surface of multiwell polystyrene plates.

The trapped particles are detected by means of enzyme-labeled antibodies. The enzyme attached to these antibodies is able to convert an unstained substrate into a brightly colored reaction product. Alkaline phophatase is used most commonly for antibody labeling in plant virus work. One molecule of alkaline phosphatase attached to an antibody molecule can dephosphorylate large numbers of molecules of the colorless p-nitrophenylphosphate to yield the bright yellow p-nitrophenol. Both the initial trapping step and the dye-production step make the test more than a 1000-fold more sensitive than the previously used precipitin tests which relied on the detection of a visible precipitate formed as a result of the antigen/antibody reaction. Triple antibody sandwich ELISAs are used especially in work with monoclonal antibodies (MAbs), because the latter often lose their activity when labeled directly with an enzyme. The binding of mouse or rat MAbs is detected by means of enzyme-labeled antibodies from rabbits specific for mouse or rat immunoglobulins, respectively. This requires an additional assay step, but often results in increase in sensitivity of virus detection and a broadening of the range of serologically related viruses which can be detected.

Tissue Print Immunoblotting

In the tissue print immunoblotting assay (**Figure 7**), the surfaces of freshly cut virus-infected plant organs, for example, of leaves or roots, are firmly pressed on a nitrocellulose or nylon membrane. Viruses contained in the sap become adsorbed to this membrane and are detected by means of antibodies labeled with an enzyme that is able to convert a colorless soluble substrate into a colored insoluble product. For alkaline phospatase, 5′-bromo-4-chloro-3-indolylphosphate (BCIP) is commonly used as a substrate. Together with 4-nitrotetrazoliumchloride blue (NTB), its dephosphorylation products form an intense blue deposition of diformazan. The binding of unlabeled antibodies produced in a rabbit can be detected by means of enzyme-labeled antibodies produced against rabbit immunogloblins in a different animal species, such as goats. Small portions of a leaf are sufficient for the routine detection of a virus. The test also allows studies of the distribution of a virus in large organs, such as sugar beet tap roots.

PCR and Other Nucleic Acid Amplification Techniques

PCR techniques using specific primers have become very popular for the routine detection of plant viruses. They are in general more sensitive than serological techniques and hence can be used even for the detection of many viruses in woody plants, for example, in fruit trees, for which ELISA is usually not sensitive enough.

With RNA viruses, transcription of the RNA into cDNA prior to PCR is necessary (reverse transcription PCR, RT-PCR). Viral RNAs can be extracted from

Figure 5 Chlorotic local lesions produced by a plant virus (beet necrotic yellow vein virus) on a leaf of the test plant *Chenopodium quinoa*.

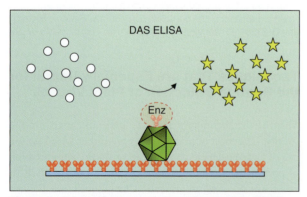

Figure 6 Schematic representation of the principle of DAS ELISA. For a detailed explanation, see text.

infected tissues by means of various commercially available kits which often are based on the selective binding of RNAs to silica gel-containing membranes and microspin technology. Alternatively, virus particles can be trapped from plant sap by means of specific antibodies which have been immobilized, as in ELISA, on a polystyrene surface, for example, in an Eppendorf tube. In this immunocapture RT-PCR (IC-RT-PCR), the nucleic acid is released from the virus by heat treatment and reverse-transcribed (**Figure 8**). Fragments of the generated cDNA are amplified and converted to dsDNA by means of a heat-stable DNA polymerase in the presence of deoxynucleotide triphosphates (dNTPs) and specific upstream and downstream primers flanking the sequence. The formation of correctly sized PCR products is checked by means of electrophoresis in agarose gels which are stained with ethidium bromide.

As mentioned earlier, viruses with circular DNA can readily be detected by means of the isothermic RCA technique using the DNA polymerase of *B. subtilis* bacteriophage Phi29. RT-LAMP (loop-mediated isothermal amplification) might be a promising isothermic alternative to PCR that can be used for RNA viruses.

In multiplex PCRs, several viruses can be detected simultaneously using pairs of primers specific for different viruses. The primers are designed in a way that each virus in a mixture yields a differently sized PCR product. The sensitivity of conventional PCRs can be further increased by means of a second amplification step using nested primers that anneal to internal sequences in the first PCR product which may be present in a concentration too low for direct detection.

The technical simplicity of PCR has made it a routine procedure in many laboratories, including virus-diagnostic laboratories. However, despite the ease of the method, careful interpretation of the results and positive, internal, and negative controls are extremely important in PCR analyses. The reason is that samples may easily become contaminated during working procedures by minute amounts of cDNAs or PCR products that may be present, for example, in aerosols in laboratories. UV irradiation is a useful means to decontaminate working areas.

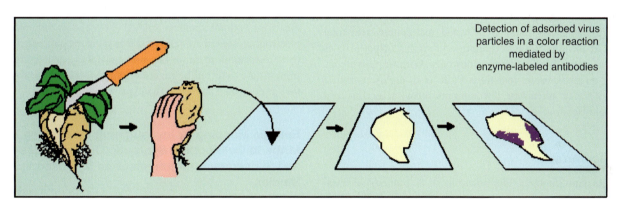

Figure 7 Schematic representation of the tissue print immunoblotting assay – for details, see text. Adapted with permission from Koenig R and Lesemann D-E (2001) Plant virus identification. In: *Encyclopedia of Life Sciences*. Copyright 2001, © John Wiley & Sons Limited.

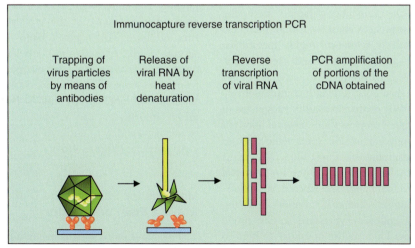

Figure 8 Schematic representation of the principle of IC-RT-PCR – for details, see text. Adapted with permission from Koenig R and Lesemann D-E (2001) Plant virus identification. In: *Encyclopedia of Life Sciences*. Copyright 2001, © John Wiley & Sons Limited.

Real-Time PCR

Real-time PCR combines the amplification and identification of target sequences in one complex reaction. Between the upstream and downstream amplification, primers most commonly a TaqMan probe is annealed specifically to the target sequence. This probe carries a fluorophore quencher dye attached to its 3′ end and a fluorpophore reporter dye attached to its 5′ end. During PCR, the polymerase replicates the template to which the probe is bound. Due to the 5′ exonuclease activity of the enzyme, the probe annealed to the target sequence is then cleaved and the reporter is thus released from the proximity of the quencher. This leads in each cycle to an increase in the fluorescence of the reporter dye that will be proportional to the number of amplicons that have been created. Because the increase in fluorescence signal is detected only if the target sequence is complementary to the probe, nonspecific amplification is not detected.

Real-time PCR is an attractive alternative to conventional PCR because of its rapidity, sensitivity, reproducibility, and the reduced risk of carryover contamination. In addition, it is cost effective when employed in high-throughput diagnostic work where it can be conveniently automated. As other PCR-based techniques, real-time PCR permits the differentiation of strains or pathotypes that differ in a few nucleotides only. With the possibility of quantification, real-time PCR is a highly versatile molecular technique for routine virus detection and identification.

Microarrays

The simultaneous identification of a number of viral nucleotide sequences, theoretically comprising all possible virus targets in a given sample, is possible by means of microarray analysis, a technique originally designed for gene expression studies with large numbers of nucleic acid sequences. With the availability of rapidly increasing viral sequence data, sequences conserved in all members of a genus or, alternatively, sequences highly specific for one or a few isolates can be defined and used in such hybridization analysis. PCR products or nowadays rather oligonucleotides less than 50 nt in length complementary to informative conserved or discriminatory sequences in virus genomes are printed on glass slides. Several hundred to a few thousand oligonucleotide spots which act as capture probes are compiled on a so-called DNA chip. The target sequences, for example, cDNA for the total RNA of a virus-infected plant, are labeled with fluorescent dyes, for example, cyanine 3/cyanine 5, during reverse transcription and are hybridized to the probe sequences on the slide. Their ability or failure to bind to the probes is subsequently analyzed using specific laser scanners and image analysis software.

Microarray analysis permits the parallel detection of a large number of pathogen sequences in one plant sample and can be useful, for example, in quarantine assays for identifying the 'array' of pathogen sequences present in one sample. The technique is commonly used for the detection and genotyping of human viruses, for example, HIV and hepatitis virus C. Its application in plant virology, however, is still at the level of proof of principle and its use for high-throughput testing is still limited by the high cost of chip production and target labeling. Also, preamplification steps may be necessary to enrich target sequences present in low concentrations.

Techniques That are Especially Useful for Studying Functional, Physical, and Structural Properties of Plant Viruses

Western Blotting

The correct sizes and the time course of the expression of structural and nonstructural proteins of a virus can be determined by means of Western blotting. In this technique, the proteins present in a plant extract are first separated by means of sodium dodecyl sulfate (SDS) polyacrylamide electrophoresis and are then transferred electrophoretically to a nitrocellulose or nylon membrane where the viral proteins are detected by means of enzyme-labeled antibodies as described above for tissue print immunoassay.

Epitope Mapping by Investigating the Binding of MAbs to Overlapping Synthetic Oligopeptides and Subsequent Immunoelectron Microscopical Studies

Epitopes, that is, binding sites of viral coat proteins recognized by antibodies, can be identified by studying the ability or inability of MAbs to bind to a series of overlapping synthetic oligopeptides corresponding to the amino acid sequence of the viral protein. The binding sites of those MAbs which have reacted with one or several of the oligopeptides are subsequently checked on the native virus particles by means of immunoelectron microscopical decoration tests using the sensitive immunogold-labeling technique (see above). By this means, information can be obtained on the accessibility or nonaccessibility of antibody-binding sites in the intact virus particles (**Figure 4(d)**).

Methods for Studying the Distribution of a Virus within a Plant or a Vector

The distribution of a virus inside a plant or a vector can be investigated macroscopically or by means of a light microscope using the above-mentioned tissue print immunoassay or by means of antibodies labeled with fluorescent dyes. Alternatively, the coding sequences for the green fluorescent protein (GFP) or other fluorescent tags can be attached to the coding sequences for structural or nonstructural viral proteins, provided that infectious cDNA clones of a virus are available. Investigations on the movement of a tagged virus or its tagged nonstructural proteins, such as of movement proteins, between different cell compartments are best performed by means of confocal laser scan microscopy (CLSM), which produces blur-free high-resolution images of thick specimens at various depths as well as computer-generated three-dimensional (3-D) reconstructions.

Studies on the Ultrastructure of Infected Host Cells by Means of Conventional Electron Microscopy

Alterations of the fine structure in infected host cells are studied by means of conventional electron microscopy in ultrathin sections that are obtained from plant tissues after chemical fixation, dehydration, and embedding in resin-like polymers such as epoxides or methacrylates (Epon, Araldite, Spurr's resin, LR White, or LR Gold). Cryosectioning after rapid freezing procedures is used less frequently. Cytological alterations are easily detected in ultrathin sections from well-selected tissue samples. The preparation of the sections, however, is time consuming and requires special equipment and expertise. Also, not all plant viruses induce characteristic cellular alterations.

Cytological alterations appearing as unusual inclusions may be recognized in infected cells also by means of light microscopy, especially after appropriate staining. However, details of their fine structure and the complete spectrum of cellular alterations can be analyzed only in ultrathin sections. Tissues from plant organs showing macroscopically visible symptoms like chlorosis and necrosis may show pathological alterations of various cell organelles, for example, chloroplasts, mitochondria, peroxisomes, nuclei or membranes of the endoplasmic reticulum. These alterations may be indicative of the damaging physiological effects induced by the infection in incompletely adapted host plants, rather than of virus-specific effects on cytology. However, virus-specific alterations may also be recognized in infected cells. They are most clearly seen in cells of hosts which do not produce pronounced macroscopically visible symptoms (latent infections).

Characteristic inclusions may be formed by accumulated virus particles (**Figure 9(a)**), nonstructural viral proteins (e.g., in the typical cylindrical inclusions induced by potyviruses; **Figure 9(b)**), proliferated host membrane elements, or they may be composed of complex

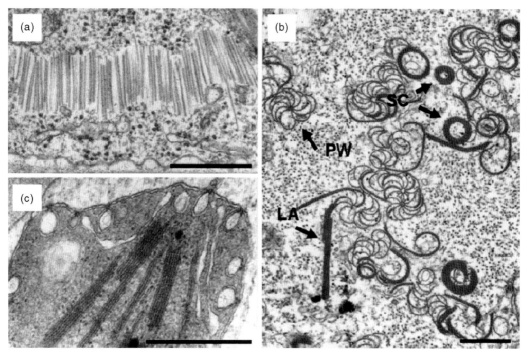

Figure 9 Electron-microscopically visible cytological alterations due to virus infections. (a) Plate-like cytoplasmic aggregate formed by the particles of a rod-shaped virus (cross section) – the highly regular *in vivo* arrangement with particle ends in register has been disturbed during fixation. Scale = 500 nm. (b) Ultrathin cross sections of pinwheels (PWs), scrolls (SCs), and laminated aggregates (LAs) formed by a nonstructural protein of a potyvirus. (c) Flask-shaped vesicles associated with the peripheral membrane of a chloroplast typical for tymovirus infections; the fibrillar content of the vesicles suggests the presence of dsRNA. Scale = 500 nm. Adapted with permission from Koenig R and Lesemann D-E (2001) Plant virus identification. In: *Encyclopedia of Life Sciences*. Copyright 2001, © John Wiley & Sons Limited.

aggregates of one or more of these components with host cell constituents. The particular structure and composition of such inclusions may be characteristic for a virus. Members of the Sindbis-like supergroup of positive-stranded single-strand RNA (ssRNA) viruses may induce virus-specific or genus-specific membrane-associated vesicles (**Figure 9(c)**), whereas members of the picorna virus-like supergroup may induce accumulations of free vesicles. Both types of vesicles contain dsRNA and are believed to be associated with viral RNA replication. Viruses with genomic nucleic acids other than positive-stranded ssRNA may induce viroplasm-like inclusions presumably involved in virus replication.

The fine structure of virus particle aggregates is usually determined by the morphology of the viruses (see **Figure 9(a)**) and is, thus, mostly specific for a virus genus. Occasionally, however, the particle arrangement may be specific for a certain virus strain. Inclusions composed of nonstructural viral proteins may exhibit very diverse and sometimes virus-specific shapes and fine structures (see **Figure 9(b)**). Proliferated membrane inclusions and membrane-associated vesicles are also species or genus specific (see **Figure 9(c)**).

Outlook: New and Neglected Techniques with Promise for Plant Virus Work

New technical developments that are promising for plant virus work include the use of liquid arrays and Padlock probes. A useful technique for RNA quantification that has been widely neglected by plant virologists is the RNase protection assay.

Liquid Arrays

Instead of applying antibodies to different viruses as an array to a planar surface, differently colored plastic beads are coated with different antibodies and serve as a basis for a miniaturized DAS ELISA in which the second antibody is labeled with a green fluorescent dye. In ELISA plates, mixtures of such beads carrying different antibodies allow the simultaneous testing of up to 100 different antigens. The assay is demultiplexed by means of a flow-through fluorescence reader that identifies by means of a laser beam on the basis of the color of a bead the antibody that has been used for coating the bead. Subsequently, a second laser beam shows how much fluorescent antibody has been bound to a specific type of bead because of the presence of the antigen in the testing material.

Padlock Probes

Padlock probes are DNA molecules whose 3′ and 5′ ends are complementary to adjacent parts of the target sequence (**Figure 10**). They become circularized in a ligation reaction when they fit the target sequence completely. In addition, they contain forward and backward primer-binding regions (P1 and P2) which can be used for amplification using unlabeled (**Figure 10(c)**) or labeled primers (**Figure 10(b)**). The probes also contain a ZipCode region which can be used either for hybridizing

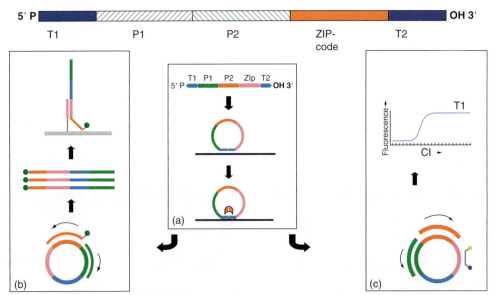

Figure 10 The Padlock probe, structure, and application for nucleic acid-based detection. The top drawing shows the construction of a Padlock probe. It consists of the two target-specific regions T1 and T2, the primer-binding regions P1 and P2, and the ZipCode, which serves as identification sequence. The probe is hybridized to the target and in case of a complete fit it is circularized (a). Subsequently the probe can serve either as a template for PCR amplification using fluorescent-labeled primers (b). In this case, the labeled amplicons are hybridized via the ZipCode to oligonucleotide arrays for differential detection (b). Alternatively, the circularized probe can be used for quantitative PCR with a TaqMan probe directed against the ZipCode (c). Courtesy of C. D. Schoen and P. J. M. Bonants, Plant Research International BV, Wageningen, The Netherlands.

the labeled Padlock amplicons to oligonucleotide arrays (**Figure 10(b)**) or as binding region for a TaqMan probe or a molecular beacon in quantitative real-time PCR (**Figure 10(c)**).

RNase Protection Assay

These tests offer a simple and cheap method to quantify RNA without PCR. They are especially useful when several RNAs have to be quantified simultaneously. Discrete-sized complementary RNAs are labeled, for example, by radioactive isotopes, and are hybridized in solution with the target RNAs to yield labeled dsRNAs. After RNase digestion of the unhybridized probe and target RNAs, the labeled dsRNAs are separated electrophoretically in a denaturing polyacrylamide gel and are visualized and quantified according to the label used.

See also: Diagnostic Techniques: Microarrays; Electron Microscopy of Viruses.

Further Reading

Abdullahi I, Koerbler M, Stachewicz H, and Winter S (2005) The 18S rDNA sequence of *Synchytrium endobioticum* and its utility in microarrays for the simultaneous detection of fungal and viral pathogens of potato. *Applied Microbiology and Biotechnology* 68: 368–375.

Baulcombe DC, Chapman SN, and Santa Cruz S (1995) Jellyfish green fluorescent protein as a reporter for virus infections. *Plant Journal* 7: 1045–1053.

Boonham N, Fisher T, and Mumford RA (2005) Investigating the specificity of real time PCR assays using synthetic oligonucleotides. *Journal of Virological Methods* 130: 30–35.

Boonham N, Walsh K, Smith P, *et al.* (2003) Detection of potato viruses using microarray technology: Towards a generic method for plant viral disease diagnosis. *Journal of Virological Methods* 108: 181–187.

Bystrika D, Lenz O, Mraz I, *et al.* (2005) Oligonucleotide-based microarray: A new improvement in microarray detection of plant viruses. *Journal of Virological Methods* 128: 176–182.

Commandeur U, Koenig R, Manteuffel R, *et al.* (1993) Location, size and complexity of epitopes on the coat protein of beet necrotic yellow vein virus studied by means of synthetic overlapping peptides. *Virology* 198: 282–287.

Francki RIB, Milne RG, and Hatta T (1985) *Atlas of Plant Viruses* Vols. I and II. Boca Raton, FL: CRC Press.

Froussard P (1992) A random-PCR method (rPCR) to construct whole cDNA library from low amounts of RNA. *Nucleic Acids Research* 20: 2900.

Haible D, Kober S, and Jeske H (2006) Rolling circle amplification revolutionizes diagnosis and genomics of geminiviruses. *Archives of Virology* 135: 9–16.

Koenig R and Lesemann D-E (2001) Plant virus identification. *Encyclopedia of Life Sciences.* London: Nature Publishing Group.

Rott ME and Jelkmann W (2001) Characterization and detection of several filamentous viruses of cherry: Adaptation of an alternative cloning method (DOP-PCR), and modification of an RNA extraction profile. *European Journal of Plant Pathology* 107: 411–420.

Tzanetakis IE, Keller KE, and Martin RR (2005) The use of reverse transcriptase for efficient first- and second-strand cDNA synthesis from single- and double-stranded RNA templates. *Journal of Virological Methods* 124: 73–77.

van Regenmortel MHV, Bishop DHL, Fauquet CM, Mayo MA, Maniloff J, and Calisher CH (1997) Guidelines for the demarcation of virus species. *Archives of Virology* 142: 1505–1518.

Varga A and James D (2006) Use of reverse transcription loop-mediated isothermal amplification for the detection of plum pox virus. *Journal of Virological Methods* 138: 184–190.

Webster CG, Wylie SJ, and Jones MGK (2004) Diagnosis of plant viral pathogens. *Current Science* 86: 1604–1607.

Relevant Websites

http://www.pri.wur.nl – Characterization identification and detection, Plant Research International – Research, Wageningen UR.

http://www.csl.gov.uk – Diagnostic Protocols, Special Interest, Central Science Laboratory, York.

https://www.invitrogen.com – D-LUX Assays, Invitrogen.

http://www.whatman.com – FTA Cards (Product of Whatman).

http://www.ncbi.nlm.nih.gov – ICTVdB Index of Viruses; PubMed, NCBI.

http://www.pcrstation.com – Inverse PCR, PCR Station.

http://www.brandeis.edu – LATE-PCR, Wangh Laboratory of Brandeis University.

http://www.luminexcorp.com/technology – Liquid Arrays Technology, Luminex.

http://www.ambion.com – Nuclease Protection Assays: The Basics, Ambion, Inc.

http://www.dpvweb.net – Plant Virus and Viroid Sequences (compiled 11 Jul. 2007), Descriptions of Plant Viruses.

http://pathmicro.med.sc.edu – Real Time PCR Tutorial (updated 7 Dec. 2006), Microbiology and Immunology On-line, University of South Carolina.

http://aptamer.icmb.utexas.edu – The Ellington Lab Aptamer Database.

Diagnostic Techniques: Serological and Molecular Approaches

R Vainionpää, University of Turku, Turku, Finland
P Leinikki, National Public Health Institute, Helsinki, Finland

© 2008 Elsevier Ltd. All rights reserved.

Glossary

EIA Enzyme immunoassays are methods used to estimate virus-specific IgG and IgM antibodies or virus antigens by enzyme-labeled conjugates.

PCR By the polymerase chain reaction (PCR) and with specific primers, DNA sequences can be multiplied.

RT-PCR For RNA, nucleic acid has to be transcribed with reverse transcriptase (RT) enzyme to complementary DNA prior to PCR.

Introduction

Specific virus diagnostics can be used to determine the etiology of acute viral infection or the reactivation of a latent infection. Two approaches can be used: demonstration of a specific antibody response or of the presence of the virus itself. Serological methods are used for measuring the antibody response while the presence of virus can be demonstrated by cultivation or demonstration of specific antigens or gene sequences. For the latter, molecular diagnostic methods have become more and more widely applied.

In this article, we briefly describe the principles of the most important serological methods and molecular applications that are used to provide information about the viral etiology of the clinical condition presumed to be caused by a viral infection.

The diagram of the course of acute virus infection (**Figure 1**) indicates the optimal methods for viral diagnosis. Following transmission, the virus starts to multiply and after an incubation period clinical symptoms appear with simultaneous shedding of infectious virus. Virus-specific antibodies appear somewhat later (from some days to weeks, called a window period). When the virus-specific antibody production reaches the level of detection, at first immunoglobulin M(IgM) antibodies and some days later immunoglobulin G(IgG) antibodies appear, and the amount of infectious virus starts to decrease. If this is the first encounter with this particular virus, that is, a primary immune response, IgG antibody levels can stay at a relatively low level, whereas in a later contact with the same antigen, that is, in secondary response, IgG levels increase rapidly and reach high levels while IgM response may not be detectable at all. Antibodies are usually investigated from serum samples taken at acute and convalescent phase of the infection. In selected cases other materials such as cerebrospinal fluid and other body fluids can also be analyzed.

The presence of infectious virus or viral structural components can be investigated directly from various clinical specimens either by virus isolation, nucleic acid detection assays, or antigen detection assays. In order to reach the best diagnosis for each patient, it is important to select the most suitable method using the right sample collected at the right time.

Principles of Serological Assays

During most primary infections IgM antibody levels peak at 7–10 days after the onset of illness and then start to decline, disappearing after some weeks or months. An IgM response is usually not detected in reactivated infections or reinfections. The production of IgG antibodies starts a few days after IgM response and these antibodies often persist throughout life.

Serological diagnosis is usually based on either the demonstration of the presence of specific IgM antibodies or a significant increase in the levels of specific IgG antibodies between two consecutive samples taken 7–10 days apart. The antigen for the test can be either viable or inactivated virus or some of its components prepared by virological or molecular methods. Isotype-specific markers or physical separation are used to demonstrate the isotype of the reacting antibody. In some cases, even IgG subclass specificities are determined although they have limited value in diagnostic work.

During the early phase of acute infection the specific avidity of IgG antibodies is usually low but it increases

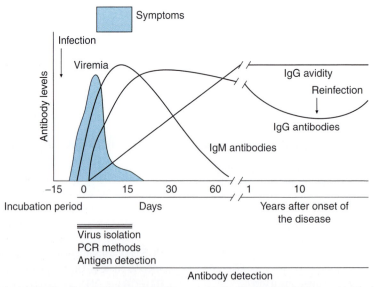

Figure 1 The course of virus infection. The shedding of infectious virus after incubation period and typical antibody response. Recommended diagnostic laboratory methods have been marked.

during the maturation of the response. Diagnostic applications of the measurement of the avidity of IgG antibodies against specific antigens have been developed to help distinguish serological responses due to acute infections from those of chronic or past infections.

Serological assays are useful for many purposes. In primary infections they often provide information about the etiology even after the acute stage when infectious virus or its components can no longer be demonstrated in the samples. They are widely used for screening of blood products for the risk of certain chronic infections, evaluation of the immune status, and need for prophylactic treatments in connection with certain organ transplantations. They are also widely used for epidemiological studies, determination of vaccine-induced immunity, and other similar public health purposes.

Serological assays have their limitations. In some infections the antibody response is not strong enough or the limited specificity of the antigens used in the assay does not allow unambiguous interpretation of the results. In infections of newborns the presence of maternal antibodies may render the demonstration of the response in the baby impossible. In immunocompromised patients the serological response is often too weak to allow the demonstration of specific responses. In these cases other virological methods should be considered.

Other clinical specimens than sera can be used for antibody assays. IgM and IgG antibody determinations from cerebrospinal fluid are used for diagnosis of virus infections in the central nervous system although new molecular methods are increasingly replacing them. Recently, increasing attention has been given to the use of noninvasive sample materials such as saliva or urine. They are becoming important for public health purposes but their value for diagnosing individual patients is still limited.

Principles of the Most Common Serological Tests

Neutralizing Antibody Assay

Antibodies that decrease the infectious capacity of the virus are called neutralizing antibodies. They are produced during acute infection and often persist during the entire lifetime. They are also useful as an indication of immunity. Both IgM and IgG antibodies participate in the neutralization.

In the assay, known amounts of infectious virus are mixed with the serum sample and incubated for a short period after which the residual infectivity is measured using cell cultures or test animals. This infectivity is then compared with the infectivity of the original virus and the neutralizing capacity is calculated from this result. Today, neutralizing antibody assays are often done by plaque reduction assays with better accuracy but with somewhat more complex technical requirements.

Neutralizing antibody assay is specific and sensitive, but time-consuming and laborious, and therefore it is not widely used in routine diagnostic services.

Hemagglutination Inhibition Test

Many viruses bind to hemagglutinin molecules found at the surface of red blood cells of various animal species and this can cause aggregation of red cells in suitable conditions. Prevention of this aggregation, called hemagglutination inhibition, by specific antiviral antibodies in the patient's serum has been widely used for diagnostic purposes. The test, known as hemagglutination inhibition test, has important diagnostic and public health applications in certain infections, most notably in influenza where antibodies measured by this test show additional specificity compared to other tests and therefore provide more detailed information about the immunity and past infections of individuals. However, for the diagnosis of individual patients, the assay is no longer widely used and is replaced by more modern immunoassays.

In the test, a virus preparation with a predetermined hemagglutinating capacity is mixed with the serum sample and after proper incubation the residual hemagglutination capacity is measured. Both IgM and IgG antibodies are able to inhibit hemagglutination.

Complement Fixation Test

The complement fixation test (CFT) is a classical laboratory diagnostic test, which is still used for determination of virus antibodies in patient sera or cerebrospinal fluid samples during an acute infection. The test mainly measures IgG antibodies.

The test is based on the capacity of complement, a group of heat-labile proteins present in the plasma of most warm-blooded animals to bind to antigen–antibody complexes. When the complexes are present on the surface of red blood cells, complement causes their lysis which can be visualized by a suitable experimental setup.

In the actual test, the complement in the patient's serum is first destroyed by heating; the serum is then mixed with appropriate viral antigen and after incubation; when the antigen–antibody complexes are formed, exogenous complement (usually from fresh guinea pig serum) is added. This complement then binds to the complexes and having been 'fixed', it is then no longer able to cause lysis of added indicator red cells. Usually, sheep red cells coated with antisheep red cell antibodies are used as indicator to measure the presence of any residual complement. The effect is measured by a suitable test protocol. Serial dilutions of the patient serum are used and the highest dilution where the serum can still prevent complement activity in the indicator system is taken as the CFT titer of the sample. The tests are usually carried out on microtiter plates and the results are observed by eye.

CFT is still used for diagnosis of acute virus infection. It measures certain types of antibodies which occur only during the acute phase of the infection. Therefore, CFT is not suitable for investigation of immune status. The assay procedure is quite complex, because the test is dependent on several biological variables, which have to be standardized by pretesting. The method is less sensitive than many other immunoassays. In addition, the method is very labor intensive and is not amenable to automation. The use of CFT in virus diagnostics is increasingly replaced by modern immunoassays.

Immunoassays

In immunoassays, antibodies binding to specific immobilized antigens can directly be observed using bound antigens and proper indicators such as labeled anti-immunoglobulin antibodies. The antigens can be immobilized to plastic microtiter plates, glass slides, filter papers or any similar material. Different immunoassays are nowadays widely used to measure virus-specific IgM and IgG antibodies. The most recent formats of immunoassays make it possible to detect simultaneously both antigens and antibodies decreasing significantly the window period between infection and immune response. Numerous commercial kits with high specificity and sensitivity are available. Automation has made immunoassay techniques more rapid, accurate, and easier to perform.

In the basic format of solid-phase immunoassays, virus-infected cells, cell lysates, purified or semipurified, recombinant viral antigens or synthetic peptides are immobilized to a solid phase, usually plastic microtiter wells or glass slides. Patient's serum is incubated with the antigen and the bound antibody, after washing steps, is visualized using labeled anti-immunoglobulin antibodies ('conjugate') (**Figure 2(a)**). If the label used is an enzyme, the test is called enzyme immunoassay (EIA) or enzyme-linked immunosorbent assay (ELISA) and the bound antibody is detected by an enzyme-dependent color reaction. If a fluorescent label is used, the method is called immunofluorescent test (IFT). The enzyme labels most commonly used are horseradish peroxidase (HRP) and alkaline phosphatase (AP). In HRP-EIA the color-forming system consists of ortho-phenyldiamine (OPD) as a chromogen and hydrogen peroxidase (H_2O_2) as a substrate. If the HRP-conjugate is bound to antibody–antigen complexes, the colorless chromogen becomes yellow and color intensity is measured with a photometer at a wavelength of 490–492 nm. The intensity of the color is proportional to the amount of bound conjugate and to the amount of specific antibodies in a patient serum sample. If the serum contains no specific antibodies, the conjugate is not bound and no color reaction occurs. By using either anti-IgG or anti-IgM conjugates it is possible to determine separately immunoglobulin subclasses.

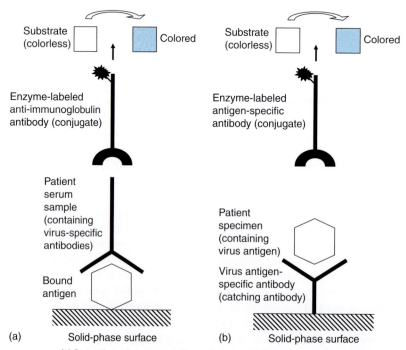

Figure 2 Enzyme immunoassay. (a) Detection of virus-specific antibodies. (b) Detection of virus antigens.

The specificity and sensitivity of these immunoassays are high. The sensitivity can be improved further, by using an additional incubation step where IgM antibodies are first enriched ('captured') in the sample by using anti-IgM immunoglobulin. Modifications to improve assay specificity by various methods of antigen handling and by using monoclonal antibodies or synthetic peptides have been developed.

Immunofluorescent tests were used in the past for measuring virus-specific antibodies, but are now replaced by EIA techniques. The principle of the method is similar to EIAs. In IFT infected cells are placed on a glass slide and bound antibodies are detected by fluorescein-labeled anti-immunoglobulin antibodies. The glass slides are examined under a fluorescence microscope. The method is specific and sensitive, but quite labor intensive and reading the test demands considerable experience.

Immunoblotting

In some infections (e.g., that caused by human immunodeficiency virus (HIV)), antibodies against certain components of the virus are more informative than other less-specific antibodies and they are detected by immunoblotting assays. Different virus antigens, prepared by gel diffusion or other techniques, are absorbed as discrete bands on a solid strip of cellulose or similar material and the strip is incubated with the patient's serum. Antibodies present in the serum bind to specific antigens and are detected using an HRP-conjugate and nitroblue tetrazolium as the precipitating color chromogen. The color reaction is observed and compared to positive and negative control samples assayed on separate strips.

Lateral-Flow and Latex Tests

A technique known as lateral-flow technology has also been used to identify antibodies or antigens. These tests involve application of serum or other samples directly on a strip of suitable material such as cellulose, where the antibodies are diffused laterally and eventually reach a site in the strip where appropriate antigen has been applied and chemically fixed. Specific antibodies become bound to the site while nonreacting antibodies diffuse out from the area. The presence of antibodies is visualized using labeled conjugates.

Although such tests are not quantitative, they are valuable for infections where the presence of specific antibodies is indicative, such as HIV infection. Performance of the test is often very simple and the result is available in a few minutes or a few hours, making such tests suitable for bed-side screening. In more advanced tests, several different antibodies can be detected by a single assay and the test conditions can be modified further so that antigens can also be detected. Many such tests have become commercially available in recent years.

For some applications, coated latex particles have replaced strips with fixed antigen as the solid phase. Binding of specific antibodies can be visualized with chromogenic or otherwise labeled indicator antibodies or a positive reaction can be detected by agglutination of the latex particles.

Point-of-Care Tests

Point-of-care tests (POC tests) are becoming increasingly common in clinical practice. Most of them are based on easy-to-use lateral-flow or latex particle technology and are able to give the result in a few minutes. POC tests are nowadays available for antibody screening of an increasing number of virus infections (HIV, hepatitis C virus (HCV), varicella-zoster virus (VZV), cytomegalovirus (CMV), Epstein–Barr virus (EBV)). Some authorities still question the validity of POC tests for clinical use although there is considerable evidence that many of the commercially available kits give reliable results.

Detection of Viral Antigens

The presence of viral antigens in clinical specimens, such as nasopharyngeal aspirates, fecal specimens, vesicle fluids, tissue specimens, as well as serum samples can be demonstrated by antigen detection assays.

In immunofluorescence tests, cells from a clinical specimen are fixed on a glass slide and viral antigens present in the cells are detected by fluorescein-labeled virus-specific antibodies. More reliable results can be obtained using enzyme immunoassay or time-resolved fluoroimmunoassay (TR-FIA). Europium-labeled monoclonal antibodies can be used as a conjugate. Solubilized antigens in clinical specimens are first captured using specific monoclonal antibodies bound to a solid phase, and are then detected with enzyme- or europium-labeled virus-specific antibodies (**Figure 2(b)**).

Antigen detection methods are especially recommended in the case of virus reactivation, for example, for herpes simplex and varicella zoster virus diagnosis where the serological response can be very weak. Antigen detection assays are also widely used in respiratory tract infections like influenza and respiratory syncytial virus infections. A simple test for the demonstration of rotavirus and adenovirus antigens in children with gastroenteritis is also available.

Nucleic Acid Detection Assays

Direct demonstration of viral nucleic acids in clinical samples is an increasingly used technique for virus diagnosis. Using the polymerase chain reaction (PCR) with specific primers, viral sequences can be rapidly multiplied and identified. These techniques are largely replacing classical virus isolation. They are rapid to perform and in many cases more sensitive than virus isolation or antigen detection methods making earlier diagnosis possible. They have proved particularly valuable for the diagnosis of viruses that cannot be cultivated such as papillomaviruses, parvoviruses, and hepatitis viruses. Semiquantitative and quantitative applications have been developed allowing monitoring of viral load during antiviral treatment. These tests cannot distinguish between viable and replication-incompetent virus, warranting caution in the interpretation of the results in certain cases. Also, sensitivity to cross-over contamination in the laboratory has caused some problems in clinical laboratory settings.

The specificity of these tests is based on the extent of pair-matching sequences between the viral nucleic acids and the primers. Extremely high sensitivity is typical for PCR methods; 1–10 copies of viral nucleic acid can be detected in a few hours. PCR methods are available for both RNA and DNA viruses. For RNA viruses viral nucleic acid has to be transcribed with reverse transcriptase (RT) enzyme to complementary DNA (RT-PCR).

Viral nucleic acid is extracted from the sample material and amplified in three successive steps. The double-stranded DNA is first heat-denaturated and separated into single strands. The specific target fragment of DNA strand is then amplified (**Figure 3**) by pairs of target-specific oligonucleotide primers, each of which hybridize to one strand of double-stranded DNA. The hybridized primers act as an origin for heat-stable polymerase enzyme and a complementary strand is synthesized via sequential addition of deoxyribonucleotides. After annealing of the primers, extension of the DNA fragment will start. These cycles are repeated 35–40 times, each cycle resulting in an exponentially increasing numbers of copies.

After the amplification is completed, the products can be detected by several methods. Agarose gel electrophoresis combined with ethidium bromide staining of the products is a classical method (**Figure 4**). The size

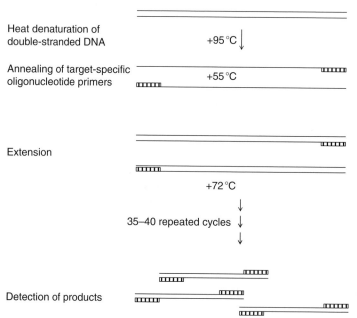

Figure 3 Polymerase chain reaction.

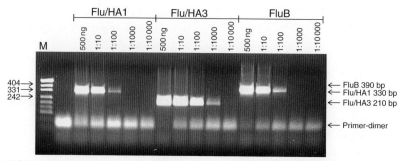

Figure 4 Detection of PCR products (amplicons) by an agarose gel electrophoresis after ethidium bromide staining.

of the amplified product is compared to control amplicons and other standards in the same gel. Various hybridization assays, based on labeled complementary oligonucleotides (probes), are also used to improve the sensitivity and specificity of the detection.

The amplified fragments can also be sequenced giving additional information about the virus. Comparison of the sequences with known virus sequences allows identification of species, strains, or subtypes that may be important for public health or medical purposes. Sequencing after RT-PCR is also the current method-of-choice for investigating the emergence of antiviral drug resistance among HIV-infected patients.

Real-time PCR instruments monitor accumulation of amplicons by measuring the fluorescence continuously in each cycle of the reaction. The earlier the amplification product becomes detectable over the background, the higher is the amount of virus in the sample (**Figure 5**). One application, based on the use of melting temperatures, allows simultaneous detection and analysis of several different nucleic acids. It also allows testing for more than one virus from the same sample (**Figure 6**).

The PCR assays are extremely sensitive and can therefore be influenced by inhibitors of the polymerase enzyme that are sometimes present in clinical samples. Internal controls can be included into reaction mixtures. Nucleases present in samples or in reagents can also cause false negative results by degrading viral nucleic acids. Furthermore, amplicons may also cause product carry-over and false positive results. Extremely high care has to be applied in handling the clinical specimens, the reagents, as well as the reaction products.

One of the great advantages of the PCR technology is its potential to detect new emerging viruses. By using primers from related viruses or so-called generic primers important information regarding the new virus can be obtained for further development of more specific tests. A good example is the severe acute respiratory syndrome (SARS) virus, for which specific diagnostic tests became available soon after the taxonomic position of the virus became known. The technology also allows safe handling and transport of virus samples, since extraction buffers added to the samples inactivate virus infectivity.

Future Perspectives

Driven by public health, scientific and commercial interests, new diagnostic tests for the laboratory diagnosis of viral infections are continuously being developed. The main area for development will probably be new molecular detection methods, where automation will provide rapid, well-standardized, and easy-to-use technology.

Use of multianalyte methods is becoming a practical reality and they might significantly change diagnostics of infectious diseases in future. They provide an opportunity to screen simultaneously for a wide range of viruses increasing the rapidity of the diagnostic procedure. A single microarray test (microchip) can contain thousands of virus-specific oligonucleotide probes spotted on a glass slide. Several kit applications for detecting viral nucleic acids and antigens or virus-specific antibodies already exist. Microarrays based on random PCR amplification can be used to detect a variety of viruses belonging to different families. Screening of some other infection markers can be also included in the same test format. Microarrays are not widely used for clinical purposes because of limited sensitivity and the difficulties of developing analytical instruments suitable for diagnostic laboratories. Another line of

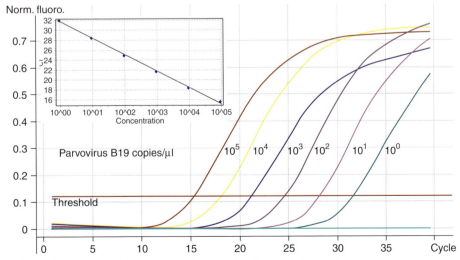

Figure 5 Quantitative real-time PCR with fluorescent-labeled probes for parvovirus B19.

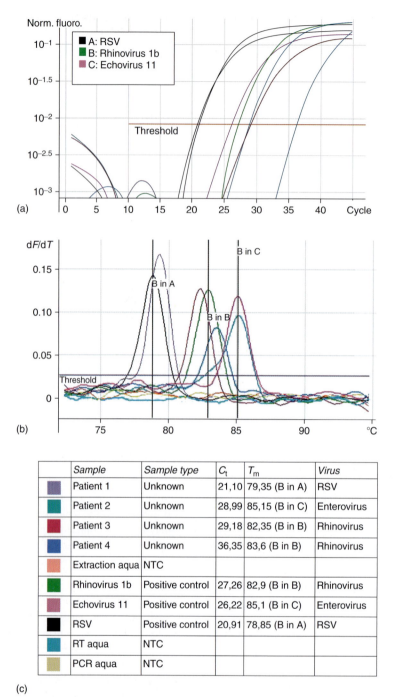

Figure 6 RT-PCR with real-time detection (a) and with melting curve analysis (b) for the detection of respiratory syncytial virus (RSV), rhinovirus, and enterovirus in respiratory secretions (c). C_t is a threshold cycle number, T_m is a melting temperature, and NTC is a nontemplate control. Unpublished results by Waris M, Tevaluoto T, and Österback R.

development is the increasing number of POC tests, which may form an important part of future diagnostic testing of infectious diseases.

See also: Antigenicity and Immunogenicity of Viral Proteins; Diagnostic Techniques: Microarrays; Immune Response to viruses: Antibody-Mediated Immunity.

Further Reading

Halonen P, Meurman O, Lövgren T, *et al.* (1983) Detection of viral antigens by time-resolved fluoroimmunoassay. In: Cooper M, *et al.* (eds.) *Current Topics in Microbiology and Immunology*, vol. 104, pp. 133–146. Heidelberg: Springer.

Hedman K, Lappalainen M, Söderlund M, and Hedman L (1993) Avidity of IgG in serodiagnosis of infectious diseases. *Review of Medical Microbiology* 4: 123–129.

Hukkanen V and Vuorinen T (2002) Herpesviruses and enteroviruses in infections of the central nervous system: A study using time-resolved fluorometry PCR. *Journal of Clinical Virology* 25: S87–S94.

Jeffery K and Pillay D (2004) Diagnostic approaches. In: Zuckerman A et al. (eds.) *Principle and Practise of Clinical Virology*, pp. 1–21. Chichester, UK: Wiley.

Wang D, Coscoy L, Zylberberg M, et al. (2002) Microarray-based detection and genotyping of viral pathogens. *Proceedings of the National Academy of Sciences, USA* 99: 15687–15692.

Dicistroviruses

P D Christian, National Institute of Biological Standards and Control, South Mimms, UK
P D Scotti, Waiatarua, New Zealand

© 2008 Elsevier Ltd. All rights reserved.

Glossary

Cohort A group of similar individuals.
Dipteran A member of the insect order Diptera: true flies.
Hemipteran A member of the insect order Hemiptera: true bugs (including aphids).
Hymenopteran A member of the insect order Hymenoptera: wasps and bees.
Intergenic region Region between the two open reading frames in the dicistrovirus genome.
Lepidopteran A member of the insect order Lepidoptera: moths and butterflies.
Orthopteran A member of the insect order Orthoptera: crickets and grasshoppers.
Penaeid A shrimp from the family Penaeidae.
Picorna-like Viruses that are ostensibly like members of the *Picornaviridae*; but the term is generally used to refer to any small (*c.* 30 nm in diameter) icosahedral viruses with single-stranded RNA genomes.
Polyprotein A protein that is cleaved after synthesis to produce a number of smaller functional proteins.
Vertical transmission Transmission of virus directly from an infected mother to her offspring.
VPg A virally encoded protein covalently linked to the 5′ end of the viral genome.

Introduction

After the early 1960s, it became apparent that invertebrates, as well as vertebrates and plants, played host to a number of small (<40 nm in diameter) icosahedral viruses with RNA genomes. The initial descriptions of many of these viruses involved little more than their physical and biochemical characteristics such as diameter, density, and S-value. By the 1970s, many new invertebrate small RNA viruses were being isolated and described and even minimal characterization made it clear that new families of viruses were emerging from this assemblage of small RNA-containing viruses of invertebrates, along with apparent members of existing virus families. Two major families of viruses recognized in the late 1970s and early 1980s were the *Tetraviridae* and *Nodaviridae*. Any of the other viruses were simply considered to be 'invertebrate picornaviruses' or 'small RNA viruses of insects'.

The properties of many of the yet-unclassified viruses were found to be very similar to those of the mammalian picornaviruses. In particular, the size of the virions (*c.* 30 nm), the composition of the capsids (three major proteins of around 30 kDa), and single-stranded, positive-sense RNA genomes all suggested these were invertebrate picornaviruses and this was very much the prevalent feeling – until 1998. At this point, the first full genome sequence of one such invertebrate virus, drosophila C virus (DCV), was published and surprisingly revealed a genome organization strikingly different from the picornaviruses, and indeed quite different from any other viruses known at that time. During the next several years, the genomes of a number of insect small RNA-containing viruses were sequenced and it became clear that two organizational paradigms existed. The first group became the *Dicistroviridae* while the second has become the (currently) unassigned genus *Iflavirus*.

Taxonomy and Classification

The family *Dicistroviridae* currently comprises 12 species, most of them in the only genus recognized so far, *Cripavirus* (**Table 1**). There are a number of other potential candidates for the family but these have yet to be accepted as species by the International Committee on Taxonomy of Viruses. For the purposes of this article, we will limit our discussion to only those species shown in **Table 1**.

Biophysical Properties

Dicistroviruses appear roughly spherical under the electron microscope in negative stained preparations with particle diameters of approximately 30 nm and no envelope (**Figure 1**). The mature virions contain three major structural proteins, VP1, VP2, and VP3 of between 28 and 37 kDa, although taura syndrome virus (TSV) does appear to have one larger structural protein of 56 kDa. In many mature virion preparations, one or more minor structural components can be present which are larger than the major capsid proteins and are presumed to be the precursor(s) of structural proteins. In some viruses, a fourth smaller structural protein (VP4) – of between 4.5 and 9 kDa – is also present in the virion. A summary of the biophysical properties of dicistrovirus virions and the size and composition of the genome is given in **Table 2**.

The virions exhibit icosahedral, pseudo $T = 3$ symmetry and are composed of 60 protomers. Each of the protomers is composed of a single molecule of each of the structural proteins VP2, VP3, and VP1. The protomers are arranged so that the molecules of VP1 are set around the fivefold axes (**Figure 2**), with VP4 lying inside the virion below the molecules of VP1. Where the molecules of VP1 come together the surface of the virion shows a slightly raised crown – similar to that on the surface of poliovirus and other picornaviruses (**Figure 3**). In contrast to poliovirus, the surface of cricket paralysis virus (CrPV) does not show the characteristic deep canyon around the fivefold axes – which in the case of the former is where the receptor-binding site is known to be. The mature virions have a buoyant density in neutral CsCl of between 1.34 and 1.39 g cm^{-3} and sedimentation coefficients that range between 153S and 167S. For those viruses where physicochemical stability has been assessed, for example, CrPV, the virions are stable at pH 3.0 and are resistant to treatment with detergents and organic solvents such as ether and chloroform.

The virions contain a single molecule of linear, positive-sense, single-stranded RNA (ssRNA) of approximately 9000–10 000 nt in size. Structural studies have not revealed any ordered structure to the RNA within the virion. Terminal modifications to the RNA include a covalently linked protein at the 5′ end of the genome (referred to as the VPg) and a polyA tract at the 3′ end.

Organization of the Dicistrovirus Genome

The single-stranded genomes possess a 5′ untranslated region (5′ UTR) of 500–800 nt followed by two open reading frames (ORFs) of c. 5500 and 2600 nt. The ORFs are separated by an untranslated region of ~190 nt, commonly referred to as the intergenic region (IGR) (**Figure 4**).

Table 1 Members of the virus family *Dicistroviridae*. Isolate and vernacular names are shown in brackets. Accession number for whole genome sequences are also given. A recently suggested taxonomy that places ABPV, KBV, SINV-1, and TSV as unassigned species in the family is followed

Genus	Species (isolate name)	Accession number	Abbreviation
Cripavirus	*Cricket paralysis virus* [type species] (cricket paralysis virus)	[AUF218039]	CrPV
	Aphid lethal paralysis virus (aphid lethal paralysis virus)	[AUF536531]	ALPV
	Black queen-cell virus (black queen-cell virus)	[AUF183905]	BQCV
	Drosophila C virus (drosophila C virus)	[AUF014388]	DCV
	Himetobi P virus (himetobi P virus)	[AUB017037]	HiPV
	Plautia stali intestine virus (plautia stali intestine virus)	[AUB006531]	PSIV
	Rhopalosiphum padi virus (rhopalosiphum padi virus)	[AUF022937]	RhPV
	Triatoma virus (triatoma virus)	[AUF178440]	TrV
Unassigned species in the family	*Acute bee paralysis virus* (acute bee paralysis virus)	[AUF150629]	ABPV
	Kashmir bee virus (Kashmir bee virus)	[AUY452696]	KBV
	Solenopsis invicta virus-1 (solenopsis invicta virus-1)	[AUY634314]	SINV-1
	Taura syndrome virus (taura syndrome virus)	[AUF277675]	TSV

The virion proteins (VPs) have been shown, by direct sequence analysis, to be encoded by an ORF proximal to the 3′ end, while the more 5′ ORF encodes protein(s) that have sequence motifs lying in the order (5′–3′) Hel-Pro-Rep, a feature common among a large number of other positive-sense RNA viruses such as picornaviruses, comoviruses, sobemoviruses, caliciviruses, sequiviruses, and potyviruses.

Where present, the sequence coding for the small virion protein VP4 is in the ORF encoding the structural proteins, between the region coding for the capsid proteins VP2 and VP3 (as it is also in the iflaviruses). VP4 is cleaved from VP3 during the maturation of the virion. The VPg is encoded in the nonstructural protein-encoding region, and in most dicistroviruses there are multiple copies of the VPg coding sequence which show some degree of heterogeneity.

Virus Replication and Genome Expression

The mechanism of virus entry into susceptible cells is unknown. Initiation of protein synthesis coincides with the shutdown or downregulation of host cell protein synthesis. During infection, a large number of precursor proteins are produced which are then cleaved to produce an array of smaller polypeptides. In the case of CrPV (one of the few dicistroviruses for which a permissive cell culture system exists), the structural proteins are produced in supramolar excess relative to the nonstructural proteins. Like many positive-sense RNA viruses, no subgenomic RNAs (sgRNAs) are produced during the infection cycle.

The absence of sgRNAs indicates that the translation of ORF2 could be initiated by an internal ribosome entry site (IRES) similar to the mechanism known from picornaviruses. Experimental studies did indeed demonstrate this and that both the 5′ UTR and IGR of several dicistroviruses including CrPV, plautia stali intestine virus (PSIV), and rhopalosiphum padi virus (RhPV) can act as IRES elements to direct initiation of translation in either *in vitro* translation systems or in cultured invertebrate cells.

However, perhaps the most unique feature of the replicative strategy of dicistroviruses is the fact that translation of the structural proteins from IGR–IRES element does not require the presence of a methionine codon. Computer modeling has shown that the IGRs of all dicistroviruses have predictable stem–loop structures. While there are two basic types among the dicistroviruses (**Figure 5**), these are fundamentally the same with six stem–loops – the most 3′ of which forms a pseudoknot with the codons involved in translation initiation. The initiation of translation is thought to be mediated by the codon which forms the part of the pseudoknot with stem loop VI, immediately upstream of the triplet which encodes the first amino acid of the mature VP2 (**Table 3**). In most cases, the initiation codon is CCU (Pro) while the first codon of

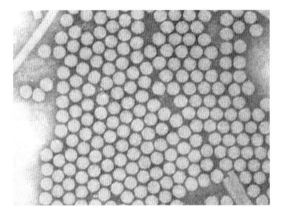

Figure 1 Negative stained electron micrograph of isometric particles CrPV showing rod-shaped particles of tobacco mosaic virus (diameter 18 nm) in the upper left- and lower right-hand corners. Electron micrograph supplied courtesy of Carl Reinganum.

Table 2 Summary of some biophysical properties of dicistroviruses

Virus	Molecular weight of major capsid proteins (kDa)[a]	Buoyant density in CsCl (g ml^{-1})	Particle diameter (nm)
Acute bee paralysis virus	35, 33, 24, 9	1.34	30
Aphid lethal paralysis virus	34, 32, 31 (41)	1.34	27
Black queen-cell virus	34, 32, 29, 6	1.34	30
Cricket paralysis virus	35, 34, 30 (43)	1.37	27
Drosophila C virus	31, 30, 28, 9 (37)	1.34	27
Himetobi P virus	37, 33, 28	1.35	29
Kashmir bee virus	41, 37, 25, 6	1.37	30
Plautia stali intestine virus	33, 30, 26, 5	n.d.	30
Rhopalosiphum padi virus	31, 30, 28 (41)	1.37	27
Solenopsis invicta virus-1	n.d.	n.d.	31
Taura syndrome virus	55, 40, 24 (58)	1.34	31
Triatoma virus	39, 37, 33	1.39	30

[a]Minor virion components are shown in brackets. These are presumed to be precursors of VP4–VP3.

VP2 encodes an alanine residue (**Table 3**). It has been shown experimentally that the glutamine at the 5′ end of PSIV VP2 can be replaced with any other amino acid to produce a mature protein in an *in vitro* translation system.

Where cell culture systems are available, pulse-chase studies have shown that translation of dicistrovirus genomes results in the production of polyproteins that are then cleaved to produce the structural and nonstructural proteins. For the virion proteins of some dicistroviruses, the cleavage sites between proteins in the structural polypeptide have been experimentally determined and are shown in **Table 3**. These sites show some degree of conservation and point toward the involvement of cysteine proteases. While the viruses themselves encode cysteine protease-like peptides in the nonstructural region there is also some evidence, for CrPV at least, that host-cell-encoded proteases may also be involved in the processing of viral polypeptides.

Host Range

To date, all members of the *Dicistroviridae* have been isolated from invertebrates, generally from a single species, or at most three closely related species. The hymenopteran viruses, acute bee paralysis virus (ABPV) and black queen-cell virus (BQCV), are known only from honeybees (*Apis mellifera*) but Kashmir bee virus (KBV) has also been isolated from the Asiatic hive bee *Apis dorsata*. It should also be noted that ABPV and KBV have been identified in the parasitic mite *Varroa destructor*, although there is no clear evidence that the virus is capable of replicating in this host. Solenopsis invicta virus-1 (SINV-1) is only known to infect the ant species, *Selonopsis invicta*.

Dicistroviruses found in homopterans and hemipterans (which includes aphids and true bugs, respectively) have slightly broader host ranges. RhPV has been isolated from laboratory and field populations of the aphids *Rhopalosiphum padi*, *R. maidis*, *R. rufiabdominalis*, *Schizaphis graminum*, *Diuraphis noxia*, and *Metapolophium dirrhodum*. Aphid lethal paralysis virus (ALPV) has only been isolated from aphids, but in this case from only three species: *R. padi*, *M. dirrhodum*, and *Sitobian avenae*. Himetobi P virus (HiPV) and PSIV are found in true bugs rather than aphids with HiPV having been isolated from the leafhoppers *Laodelphax striatellus*, *Sogatella furcifera*, and *Nilaparvata lugens*, and PSIV from the brown-winged green bug, *Plautia stali*. Triatoma virus is also a virus of hemipterans and has been found in the hematophagous triatomine bug, *Triatoma infestan*, which is also a vector of the protozoan agent that causes Chagas' disease in South America.

DCV is the only dicistrovirus with a host range restricted to dipterans and has been isolated from *Drosophila melanogaster* and the sibling species *D. simulans*. TSV is a virus of penaeid shrimps and has been isolated from a number of species including *Litopenaeus vannamei*, *L. stylirostris*, *Metapenaeus ensis*, and *Penaeus monodon*. The majority of the dicistroviruses have relatively restricted host ranges and, at most, have only been isolated from insects of a single order. CrPV is the striking exception. Originally isolated from the field crickets *Teleogryllus*

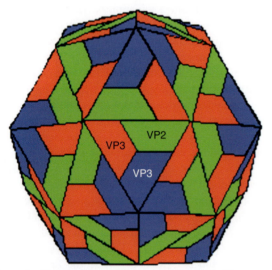

Figure 2 Diagram showing the surface packing of the coat proteins (VP1, VP2, VP3) of CrPV. Reproduced with permission from Fauquet CM, Mayo MA, Maniloff J, Desselberger U, and Ball LA (eds.) (2005) *Virus Taxonomy: Eighth Report of the International Committee on Taxonomy of Viruses*. San Diego, CA: Elsevier Academic Press.

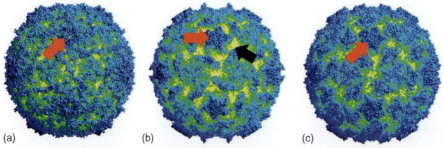

Figure 3 The surface structure of (a) CrPV, (b) poliovirus, and (c) rhinovirus 14. The red arrows show the raised crown around the fivefold axes and in poliovirus the black arrow indicates the deep canyon around this raised crown. Courtesy of John Tate.

oceanicus and *T. commodus*, CrPV has subsequently been isolated from a further 20 species belonging to five taxonomic families: *Orthoptera*, *Hymenoptera*, *Lepidoptera*, *Hemiptera*, and *Diptera*. Interestingly, while a number of other RNA-containing viruses are known from lepidopterans (moths and butterflies), that is, the iflaviruses and tetraviruses, as well as a number of uncharacterized viruses, CrPV is the only dicistrovirus isolated from lepidopterans – in fact from ten lepidopteran species.

All of the above records refer to the natural host range of the viruses. To a certain extent, studies on the experimental host range of dicistroviruses are limited and those that have been carried out have not substantially extended the known host ranges. Again the exceptions are CrPV and DCV. CrPV replicates in a number of established insect cell lines including those from *Drosophila*, the hemipteran *Agallia constricta*, and the lepidopterans *Pieris rapae*, *Plutella xylostella*, *Spodoptera ornithogalli*, and *Trichoplusia ni*. In addition to insect cell lines, CrPV has also been found to replicate readily in larvae of the greater waxmoth, *Galleria mellonella*. This is an easy insect to rear and maintain and virus yields can be very high. Apart from CrPV, the only other dicistrovirus shown to replicate in cultured cells is DCV, which multiplies in several *Drosophila* cell lines (some DCV isolates also replicate in the greater waxmoth). Reports that TSV can replicate in some mammalian cell lines have never been substantiated and may simply be attributable to the production of a cytopathic effect in the absence of virus replication.

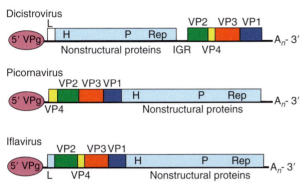

Figure 4 Diagrammatic representation of the genomic arrangement of the dicistroviruses, picornaviruses, and iflaviruses. The helicase (H), protease (P), and replicase (Rep) domains of the nonstructural proteins are indicated.

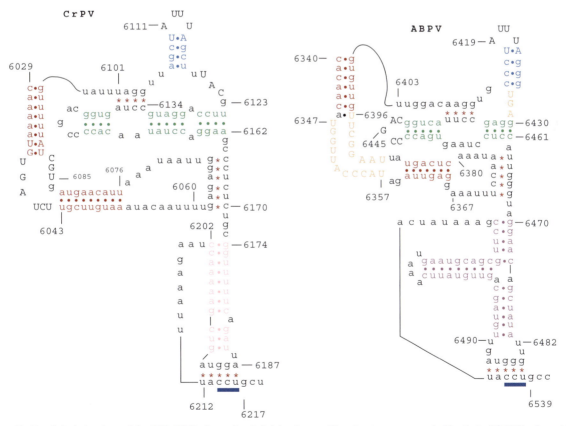

Figure 5 Predicted structure of the IGR–IRES elements of dicistroviruses. The structure represented by that of CrPV is shared by most dicistroviruses with the exception of ABPV, KBV, SINV-1, and TSV that have a structure similar to that represented by ABPV. The stem loop VI pseudoknot stuctures are shown in pink for CrPV and in purple for ABVP. The first translated codon is underlined in blue. Adapted with permission from Nobuhiko Nakashima.

Table 3 Some properties of the translation and processing of the structural polyprotein of the dicistroviruses. Only cleavage sites that have been empirically deduced from sequencing of the virion proteins are shown

Virus	First residue[a]	Initiation codon	Cleavage VP2/VP4	Cleavage VP4/VP3	Cleavage VP3/VP1
Cricket paralysis virus	A	CCU	IYAQ/AASE	LFGF/SKPT	n.d.
Acute bee paralysis virus	n.d.	CCU	VTMQ/INSK	IFGW/SKPR	ASMQ/INLA
Aphid lethal paralysis virus	A	CCU	n.d.	n.d.	n.d.
Black queen-cell virus	A	CCU	MLAQ/AGLK	LFGF/SKPL	MVAG/SNSG
Drosophila C virus	A	CCU	n.d.	MLGF/SKPT	IVAQ/VMGE
Himetobi P virus	A	CUA	AREQ/VNLN	APGF/KKPD	STAQ/EQAN
Kashmir bee virus	n.d.	CCU	n.d.	n.d.	n.d.
Plautia stali intestine virus	Q	CUU	LILQ/SGET	AFGF/SKPQ	LTLQ/SGDT
Rhopalosiphum padi virus	A	CCU	n.d.	THGW/SKPL	SIAQ/VGTD
Solenopsis invicta virus-1	n.d.	CCU	n.d.	n.d.	n.d.
Taura syndrome virus	A	CCU	n.d.	MFGF/SKDR	PSTH/AGLD
Triatoma virus	n.d.	CUC	n.d.	ALGF/SKPL	PIAQ/VGFA

[a]The first residue of the mature VP2.

Pathology and Transmission

The names of several of the dicistroviruses imply that virus infection can produce a noticeable disease symptom, that is, CrPV, ABPV, ALPV, and BQCV (potential queens die as propupae or pupae in hive cells in which the walls turn black). However, the majority of virus–host interactions produce no noticeable disease. Although BQCV kills some queens in their cells, most infected larvae, pupae, and workers appear to be completely unaffected.

Nevertheless, while disease symptoms may not be evident, in most of the cases studied, dicistrovirus infections reduce the life span of the infected individual. Or, more precisely, the presence of a dicistrovirus in a cohort (group of similar individuals) of insects coincides with a reduced life span relative to an uninfected cohort. This point is explicitly made since no studies have yet been undertaken where an invertebrate host has been nonlethally sampled for virus through its life span. In many other virus–host systems, for example, plants, mammals, or even fish and amphibians, it is possible to sample for virus without killing the host so the progress of infection can be measured at the level of the individual. With invertebrates, this has not been done and would generally be very difficult or impossible. The experimental approach has therefore been to introduce the virus into an uninfected cohort and compare the effects to a control cohort.

Despite these practical limitations, a number of studies have demonstrated the routes of virus transmission. BQCV has been detected in the eggs, larvae, and offspring of queens that were found to be infected – indicating that the virus is vertically transmitted. Similar findings have been made with ALPV which can be vertically transmitted in the aphid host, *R. padi*; this infection subsequently results in reduced longevity and fecundity. It has also been shown that ALPV RNA can be detected in the developing embryos inside infected females.

DCV is also vertically transmitted although virus is not present in the cytoplasm of the egg but is associated with the chorion on the egg surface. Infection presumably occurs when emerging larvae ingest the virus since they do not become infected if the chorion is removed. A similar phenomenon is found with CrPV and *T. oceanicus* where surface sterilization/dechorionation of eggs with dilute hypochlorite blocks the transmission of the virus to emerging nymphs.

Evidence of horizontal transmission is in many respects even more difficult to obtain than data on vertical transmission. However, experimental feeding studies indicate that dicistroviruses can be transmitted horizontally. In the case of triatoma virus (TrV) and its host *T. infestans*, infected insects excrete virus in their feces and since triatomine bugs are coprophagous and TrV is infectious *per os*, the virus can be readily spread. With DCV, it has been found that when uninfected males are placed with uninfected females, these males become infected (and vice versa). In fact, even if uninfected female flies are placed on media on which infected males have been allowed to feed for several hours, the females become infected. Such manipulations with other virus and invertebrate host systems are not as easy as those involving *Drosophila*, as it is difficult to obtain colonies free of viruses and the insects are not as easy to manipulate in the laboratory.

Evidence of vectors playing a role in dicistrovirus transmission is limited to honeybee viruses and their parasitic mite, *Varroa destructor*. It has been known for some time that the prevalence of a number of honeybee viruses increases in the presence of varroa mites. Initially, it was thought that the stress caused by varroa infestation induced the viruses to replicate. However, recently it has been shown that ABPV and KBV can be detected in mites implicating them more strongly in transmission of the viruses. In contrast, BQCV, although present at high frequency in adult honeybees, has not been detected in mites, suggesting that they play little or no role in transmission.

Geographic and Strain Variation

Some dicistroviruses infect specific hosts with a wide geographic distribution or have been found to infect a number of species spread over a large geographic range. Several studies have looked at strain variation between geographic isolates of dicistroviruses using a variety of techniques. CrPV and DCV possess a range of biological and serological characteristics that show differentiation between geographical isolates. With CrPV, two major serogroups have been identified that separate Australian and New Zealand isolates from North American isolates. With DCV, isolates from different localities varied both in their pathogenicity and virus yield after injection into virus-free flies. Additional genetic studies on CrPV and DCV have utilized ribonuclease T1 fingerprinting and subsequently polymerase chain reaction restriction endonuclease analysis. Estimates of the maximum nucleotide divergence between isolates of CrPV and DCV were about 10%. In the case of CrPV, it was found that that the North American isolates were quite distinct from the antipodean isolates, a finding which reflects the serological data.

More recently, a number of molecular studies with the honeybee dicistroviruses and TSVs have been undertaken to determine the levels of genetic variation between isolates of these viruses. In most instances, slightly different regions of the virion protein coding regions (usually regions of VP3) have been used, which makes direct comparison between studies quite difficult. Nevertheless, ABPV isolates show levels of nucleotide identity between 90% and 100%, but isolates from the same geographic region are more closely related than isolates from more distant locations. One such study has revealed that viruses isolated from different central Europe regions were more similar to each other than to isolates from North America or from the UK. Similar patterns are also evident for BQCV and KBV with nucleotide identity within species at 90–100%. To put these values in perspective, the region used for the KBV studies is $c.$ 75% identical to the same region from ABPV.

For TSV, the situation is slightly different, and it has been found that levels of nucleotide identity between isolates from North and Central America and Asia range from 95% to 100%. These lower levels of diversity may indicate that TSV has rapidly spread into many of the regions and hosts where it is now found – a hypothesis that is supported to some extent by the rapid emergence of the disease over the last two decades. There is also some evidence that there are serological differences between isolates.

Relationships within the Family

The taxonomy of the dicistroviruses currently recognizes only one genus, *Cripavirus*, into which most of the species are placed. However, comparisons of the coding sequences of the dicistroviruses reveal that they are only distantly related. For instance, the amino acid identity between the structural polyproteins of different species ranges from 19% to 66% while the amino acid similarity ranges from 37% to 80%. Using these data, the phenogram presented as **Figure 6** shows the pattern of relationships and reveals only two closely related pairs of species: CrPV and DCV (which also share a serological relationship) and ABPV and KBV. All other species are only distantly related. While the genus *Cripavirus* is currently proposed to include all but ABPV, KBV, SINV-1, and TSV, this genus will probably be divided and new genera established to accept the currently unassigned species listed above.

Similarity with Other Taxa

The dicistroviruses share various properties with a number of other positive-sense ssRNA virus genomes. Historically, these viruses have been collectively grouped as the picornavirus superfamily which has been considered to include the virus families *Comoviridae*, *Picornaviridae*, *Potyviridae*, *Sequiviridae*, *Caliciviridae*, and more recently the unassigned genera *Iflavirus*, *Sadwavirus*, and *Cheravirus* and the recently created family *Marnaviridae*. In all of the groups mentioned above, the gene order for the nonstructural proteins is the same, viz. Hel-Pro-Rep (as in **Figure 4**). When compared to other positive-sense viruses, these genes appear to be more closely related to each other

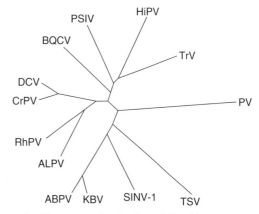

Figure 6 Phenogram showing the relationships among members of the family *Dicistroviridae*, constructed from the amino acid identity of the structural proteins encoded in ORF2. The phenogram was constructed using the neighbor-joining algorithm of the MEGA software. The sequence from Poliovirus type 3 [L23844] was used as an outgroup for the analysis. Branch lengths are drawn approximately to scale.

than to genes from other viruses, for example, coronaviruses, toroviruses, and tetraviruses. However, there is a subset of the above taxa which shares a larger set of properties: for example, isometric virus particles with pseudo $T = 3$ symmetry (formed of three units of eight-stranded β-barrel); the absence of sgRNAs during replication; and the presence of a 3–4 kDa VPg. On the basis of these properties, the grouping would exclude the *Caliciviridae* (sgRNAs, large VPg, and true $T = 3$ symmetry) and the *Potyviridae* (large VPg, rod-shaped particles with helical symmetry).

There are also a large number of other isolates of invertebrate picorna-like viruses that share some properties with dicistroviruses, for example, single-stranded postive-sense RNA genomes, isometric virions of around 30 nm in diameter, and up to three virion proteins of around 30 kDa. Further studies will undoubtedly reveal additional structural and organizational paradigms among the yet uncharacterized picorna-like viruses of invertebrates besides increasing the number of known dicistroviruses.

See also: Iflavirus; Insect Pest Control by Viruses; Picornaviruses: Molecular Biology; Taura Syndrome Virus.

Further Reading

Bailey L and Ball BV (1991) *Honey Bee Pathology,* 2nd edn. Sidcup: Harcourt Brace Jovanovich.

Fauquet CM, Mayo MA, Maniloff J, Desselberger U, and Ball LA (eds.) (2005) *Virus Taxonomy: Eighth Report of the International Committee on Taxonomy of Viruses.* San Diego, CA: Elsevier Academic Press.

Miller LK and Ball LA (eds.) (1998) *The Insect Viruses.* New York: Plenum Press.

Disease Surveillance

N Noah, London School of Hygiene and Tropical Medicine, London, UK

© 2008 Elsevier Ltd. All rights reserved.

Glossary

Case–fatality rate Number of persons dying of an infection divided by total number of persons with the disease. Thus, CFR of 5% means that 5 of 100 persons with the infection died.

Epidemic and outbreak Most infectious disease epidemiologists do not distinguish between these terms and use them interchangeably. Generally, an outbreak is defined as a localized increase in cases, whereas an epidemic is more widespread, perhaps affecting a whole country.

GUM clinics The term GenitoUrinary Medicine clinics is often used for special clinics for sexually transmitted infections (STIs).

Incidence The rate of new infections (number by population) in a given time – for example, 5 cases of influenza per 1000 per year. Good for short-term infections.

Outbreak When an infection occurs at a frequency higher than expected for that time or place. It is basically an increased incidence which is usually unexpected. Two linked cases is also theoretically an outbreak.

Pandemic This term should be restricted to infections affecting many countries, though it does not have to be worldwide. SARS spared many countries and indeed some continents, but it was a pandemic. AIDS is undoubtedly a pandemic.

Prevalence The number of infections at any one time in a given population, expressed as a rate. Good for chronic infections and serological studies (e.g., prevalence of varicella antibody at age 15 in a given population is 95%).

Introduction

Surveillance is undoubtedly an essential – indeed critical – ingredient of any disease control program. It is used to monitor the impact of an infection, the effect of an intervention or health promotion strategy, health policy, planning, and delivery. Surveillance is the ongoing and systematic collection of routine data which are then analyzed, interpreted, and acted upon. It is essentially a practical process, which nevertheless can be useful in other ways. Its main purpose is to analyze time trends – but these can include not simply fluctuations in overall numbers, but also changes in age and sex distributions, geographical locations, and even possibly, in some of the more sophisticated established surveillance systems, at-risk groups (such as particular social, ethnic and occupational groups). Surveillance is essential for evaluating the impact of an intervention, such as mass vaccination, on a population. It can be a fairly sensitive system for the detection of outbreaks earlier than they would have been recognized otherwise.

The word ongoing in the description of surveillance helps to distinguish it from a survey, which is usually finite, tends to focus more on one or more groups of persons, and involves a questionnaire. Nevertheless, surveillance does not have to continue forever – when it is no longer useful, it should be stopped.

To be systematic is another important ingredient of surveillance. If reporting centers are not consistent in what they report, nor regular, the data they send in will be uninterpretable, and probably useless. Defining what needs to be reported and agreeing on the criteria for making a reportable diagnosis are necessary to make sense of the data.

The collection of routine data is another important characteristic of surveillance, especially with surveillance of laboratory infections. Generally, testing of samples is done for diagnosis, not primarily for surveillance. As laboratory testing is expensive, making further use of the results by contributing to surveillance makes for more efficient use of information, and the contribution to surveillance itself can often justify further testing. Typing echoviruses or coxsackie viruses can seldom be justified on clinical grounds alone, but the surveillance of serotypes can provide valuable information on the epidemiology of these viruses, their clinical characteristics, seasonality, and age and sex distributions. Echovirus 4 for example tends to be rare, fairly localized even within a country, but, when it occurs, it has a high aseptic meningitis rate, and causes a short and sharp autumn outbreak. Echovirus 9 on the other hand is far more common, more widespread geographically when it causes an epidemic, more benign, with a macular rash, and with meningitis not an especially common manifestation.

Although surveillance is essentially a practical exercise, this article attempts to show that surveillance can also be useful in giving us clues about an infection, whether it be its natural history, etiology, severity, or outcome.

Collection of Data: Sources of Data in Surveillance of Viruses

Death Certification

In most developed and middle-income countries, deaths are certified primarily for legal reasons, but have proved to be an important data source to use for surveillance. Clearly, they tend to be useful mainly for serious infections, and may suffer from inaccuracies, but remain a useful basic data source. Diseases with a high mortality rate and a short duration of illness (the viral hemorrhagic fevers for example) will obviously be better represented by death certification, than those with a low mortality. HIV/AIDS, before the era of HAART, had a high mortality but long periods of symptomless and symptomatic infection, so that death certification data needed careful interpretation. Laboratory reporting systems can be another useful source of information on mortality from infection.

Notifications

Surveillance systems such as statutory notification tend to be based mainly on clinical features. These can be very useful for common diseases with distinctive clinical syndromes, such as measles and mumps. It is important however not to make a disease notifiable unless there is a good reason for it: 'good' reasons include a mass vaccination program (when surveillance is virtually mandatory), any other mass control program, serious diseases for which contact tracing, mass or close contact prophylaxis, or investigation into source, is necessary. Serious less common viral infections such as poliomyelitis are notifiable in most countries. This is because contact tracing and preventive measures can be taken. Broader clinical diagnoses, such as aseptic (viral) meningitis, may on the face of it be less useful to notify, as it is usually impossible at the bedside to distinguish such causes of it as, for example, the coxsackie B viruses, echoviruses, and mumps. Nevertheless, notification of aseptic meningitis can be useful because a rapid rise in notified cases may need to be investigated. Moreover, the timing of any epidemic may give a clue to etiology – mumps meningitis tends to increase in spring and is usually accompanied by a concurrent outbreak of clinical mumps, while enterovirus epidemics are more likely to occur in autumn.

Other Clinical Sources of Data

Specific general practitioner (GP) surveillance systems are useful to provide data of epidemiological value for infections that are not notifiable, such as the common cold or chickenpox (in UK). They are of course clinically based, but are nevertheless useful, and often surprisingly accurate, possibly because those GPs who subscribe to a surveillance system are motivated to do so. GP surveillance systems are often sentinel-based, that is, based on a sample of GPs in a country, region, or area. Thus, they are good for common infections which to make notifiable would possibly be wasteful, for example, chickenpox. Moreover each sentinel would normally provide complete reporting. In the English system, GPs provide data on the base populations of their practices, so that rates of infection can be provided as a routine, a feature that is almost unique among surveillance systems. GP surveillance systems also tend to be good for timeliness and completeness.

Laboratory Data

It is useful to think of laboratory data as being of qualitative rather than quantitative value as they add quality and detail to disease surveillance. Thus, in the example already used for aseptic meningitis, a precise diagnosis of mumps, echovirus, or coxsackie type is particularly useful in clustering and outbreaks. Indeed, as the enteroviruses exhibit a strong late summer/autumn seasonal pattern with each enterovirus returning to its baseline in winter, when numbers of a particular type continue to be reported at a level higher than the winter baseline, the return of this virus to cause another epidemic in the following summer can usually be safely predicted. Laboratory data are also essential in qualifying food poisoning and gastroenteritis. Separating viral from bacterial causes is a useful first step, as their management tends to be very different. Norovirus gastroenteritis can be food-borne, but also spreads very easily from person-to-person because an extremely small dose is necessary for infection to occur, and the virus being fairly resistant to the environment will survive for some time. Management therefore must concentrate on hygiene. With most bacterial causes of food poisoning, especially the salmonellas, management often depends on the removal of the offending food. Laboratory surveillance is particularly essential for unraveling the mass of respiratory viral infections that inflict humans – respiratory syncytial virus, adenoviruses, parainfluenza viruses, rhinoviruses, etc. Indeed, it is particularly useful for influenza – not only separating it from influenza-like illness, but also in identifying influenza A and B, and if A, the subtype and variant.

Surveillance of Outbreaks

Surveillance of outbreaks (as opposed to individual infections) can be revealing, and important to allow public health measures. The existence of the noroviruses was suspected in the UK many years before the organisms were identified. This was because some outbreaks which did not fit the characteristics of known infections but had characteristics of their own had occurred.

Hospital Admissions

These can be useful for certain more serious infections, such as hepatitis and encephalitis. They can be unwieldy, generally lack detail, and often are published a year or more after the events have occurred.

Serological Surveillance

Serological surveillance has become increasingly important, and is a useful tool in assessing the immunity of a population, though it can also be used to identify vulnerable individuals. The immunity of a population can be vaccine-induced, and serological surveillance is a valuable adjunct to the methods available to monitor a mass immunization program. Vulnerable age groups can be identified, and booster doses of the vaccine introduced. An example of the importance of serological surveillance in determining public health policy is included below (analysis by person).

Surveillance of Viruses in Nonhuman and Environmental Sources

To build a picture of an infection, animals, birds, and the environment have been placed under surveillance. Rabies in foxes and other wildlife, influenza in birds, pigs, and other animals, are examples of important and fairly successful surveillance systems.

Other Sources of Data

Records of sickness absence, absence from school, calls to an emergency room, if available, can provide speedy information that something has happened, but tend to be nonspecific. Surveillance of antiviral resistance will become important.

Surveillance of HIV/AIDS

The association of HIV/AIDS with stigma makes surveillance of this infection especially difficult. The uniqueness and seriousness of this infection warrants a separate section. It is an example of the importance of tailoring surveillance to a specific serious infection if it becomes necessary to do so.

In the UK and some other countries with data protection acts, HIV infection, as with other STIs, is not notifiable. Special confidential surveillance systems through clinicians and GUM clinics, as well as laboratories, are in place. These are especially important for assessing risk factors. Inclusion of risk factors is essential for targeted intervention – for example, the proportions and rates of new diagnoses attributed to men who have sex with men, heterosexual sex, mother-to-infant, blood transfusion, IVDUs, and other needlestick injury.

Laboratory reporting is essential. Death certification is useful, though it has been shown that men who have sex with men, and probably those with other risk factors, are under-represented. In the UK, matching reported cases with death certificates is very important, as it allows for detection of deaths due to AIDS (such as pneumonia) as well as deaths associated with AIDS, and which are seen now in HIV-infected individuals – these include liver and cardiovascular disease, overdoses, and malignancies.

A surveillance system, based on unlinked anonymous testing of samples of blood taken routinely from certain at-risk population or occupational groups has been shown to provide valuable information on HIV infection in these populations. Specific screening systems for blood donors, military recruits, commercial sex workers, and family planning/termination of pregnancy clinics are also useful if these populations are to be targeted. Behavioral surveillance should also be seriously considered, to assist in identifying future trends, healthcare planning, as well as for specific health promotion efforts.

Attributes of Surveillance Systems

Completeness

Incompleteness is an almost universal drawback of most notification systems. They should never be dismissed for this reason alone. Statutory notification systems can be essential for surveillance and control. For common infections completeness may not be worth striving for, because notifications (assuming consistency in reporting) will generally provide information on trends, as well as fairly accurate information on age, sex and seasonal distributions, and possibly on place. The effect of mass vaccination programs can also be monitored fairly closely with statutory notification systems, as with measles (**Figure 1**) and acute paralytic poliomyelitis (**Figure 2**) in the UK. When a mass vaccination or other universal control program reduces the incidence of an infection to low levels, completeness becomes much more essential. For serious infections also, such as SARS or Lassa fever, for which contact tracing or other control measure is necessary, completeness is essential.

In active surveillance, reporters make negative returns if they have had no cases during the reporting period, to ensure completeness. In some countries, enhanced surveillance has been used to assess more accurately the true incidence of an infection – regions or districts are chosen to report all cases of a particular infection or infections. It is a hybrid of active and sentinel surveillance.

Timeliness

Timeliness is important for infections for which urgent public health measures have to be undertaken, such as poliomyelitis and viral hemorrhagic fever, as well as any outbreak. In some instances, infections, not normally urgent, can become so as an elimination program progresses. In a country with elimination of measles as its goal, a case of indigenous or imported measles needs to be dealt with urgently, as it may lead to an outbreak if not controlled immediately. Laboratory data are often not timely, and hospital data generally even less so, but can make up in accuracy what they lose in timeliness.

Accuracy

Accuracy is clearly important, though some minor degrees of inaccuracy can be tolerated in some common infections. Clinical data are most liable to have some inaccuracies, though even laboratory data can be inaccurate. Case definitions and quality control systems can be useful to improve accuracy.

Representativeness

For surveillance to provide an accurate picture of the impact of a particular infection, representativeness is essential. It is perhaps the most important quality for any surveillance system. Having a wide coverage of reporting clinicians and laboratories, or a well-chosen sample of sentinel sites, is necessary for the data collected to be representative of an infection in a country. Sometimes it may be necessary to assess data from various sources, such as notifications/GPs (clinical), hospital, laboratory, and death certificates.

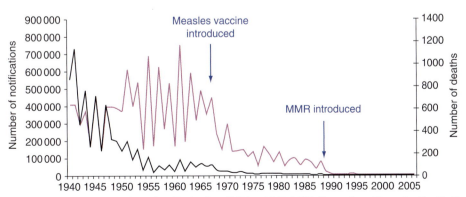

Figure 1 Measles notifications: cases and deaths, England and Wales 1940–2006. Reproduced from the Health Protection Agency (www.hpa.org.uk).

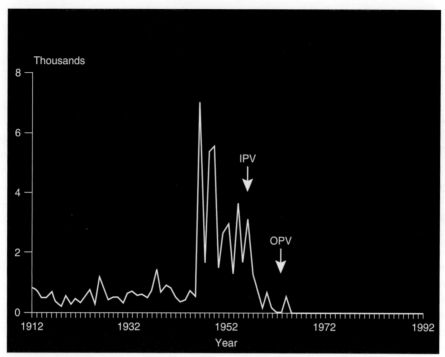

Figure 2 Notifications of acute poliomyelitis in England and Wales 1912–93. Reproduced from the Health Protection Agency (www.hpa.org.uk).

Consistency

Consistency is another crucial basic attribute of any surveillance system. Reporters must know what to report (case definition) and how often. Otherwise it will not be possible to interpret trends.

Analysis of Data

Time

The three basic analyses by time, place, and person should be routine. Computer programs have made analyses of data quicker but somewhat less flexible. There is no standard period for analysis by time. Depending on what the surveillance intends to show, yearly, quarterly, monthly, four-weekly, or weekly time intervals can be used. In surveillance, time intervals shorter than this are rarely used. Monthly intervals have the disadvantage of having unequal number of days in each month, and are difficult to use when reporting is weekly; for seasonal trends four-weekly periods are better but cannot be divided into quarterly periods. In viral surveillance, four-weekly rather than weekly intervals tend to be most useful in showing seasonal changes. There is more likely to be more variation ('noise') in weekly intervals, making for less smooth changes. For secular trends, quarterly or annual intervals are generally used. Analyzing by time can reveal regular changes in the periodicity of viruses, enabling some of them to be predicted.

A basic knowledge of seasonal and secular patterns makes it easier to detect changes that signify a possible epidemic, and to differentiate these from a random variation. It is important to remember when analyzing laboratory data that there is often an interval, which can be 2 weeks or more, between date of onset and date of reporting.

Person

Analysis by age and sex is another basic analysis in surveillance. It can identify those most affected, and vulnerable groups. Changes in age distributions may provide important clues about a changing viral infection, and the effect of mass interventions on the age distribution of an infection can be monitored. Changes in the age distribution of measles in 1994 in the UK signified that an epidemic in older children was imminent, and the vaccine schedule was changed to include an extra booster injection (MR) to children aged 5–16 years. This averted the outbreak and the booster dose became a permanent feature of the routine immunization schedule in the UK. Indeed the changes in age distribution following mass vaccination could be considered an epidemiological side effect of mass vaccination. Requests for occupational groups and travel histories should be selective. For poliomyelitis, SARS, dengue, and the viral hemorrhagic fevers, travel histories are required. Occupational group may be useful for norovirus, and hepatitis types A, B, or C. Specific risk factors may be worthwhile for HIV, hepatitis B and C.

Place

Analysis by place can pinpoint local outbreaks. Some echoviruses (e.g., echovirus type 4) can cause rare short local outbreaks, other types (e.g., types 9 and 11) are more common and more widespread. Food-borne outbreaks of hepatitis A are generally picked up locally through routine surveillance, but sometimes more extensive outbreaks caused by a more widely distributed foodstuff, including shellfish or frozen soft fruit, may be identified.

Interpretation

Collection and analysis are generally routine functions; skill is required in interpretation of the data. No statistic is perfect, and surveillance data, like all data, must be interpreted with caution. One must take into account the origins of the data – clinical, laboratory, or hospital. Not only must the reliability, or otherwise, of the data be evaluated, but what the data signify in the natural history of the infection must also be recognized.

Every viral infection has its stages and these must be recognized before surveillance data can be sensibly interpreted. At what stage are the data being collected important to understanding and interpretation? Using influenza or hepatitis A as examples, and a defined population (**Table 1**), only a proportion of persons in the defined population will be infected with the virus. They can only be comprehensively detected by screening, and serological surveillance will identify these persons, or assess population immunity (B1 in **Table 1**). In HIV/AIDS surveillance, unlinked anonymous testing of samples of blood taken routinely at say, an antenatal clinic, can give vital information on the prevalence of HIV infection, since in this infection, presence of antibody denotes infection, not immunity. A smaller proportion will be ill (B2), but only some of these will visit a doctor (B3). Surveillance systems based on GP consultation rates have now been recognized as an important addition to the spectrum of a disease, and many countries have excellent systems. Of those patients that do visit their family doctor, only some will be admitted to hospital (B4). Finally, only some will die (B5).

For laboratory data, the stages are slightly different (**Table 2**), but still important to understanding what the reported data mean. As before, only a proportion of persons will be infected (B1), a smaller proportion will be ill (B2), and a smaller proportion still visit their doctors (B3). Not all doctors will send specimens to a laboratory (B4), and only a proportion of these specimens (B5), depending on accuracy of the identification process, the method of transport, the fragility of the organism, and the swabbing or other sampling technique, will be positive. Finally, depending on the level of consistency of reporting, only some of these will be reported (B6). It is important to recognize these stages in interpreting surveillance data.

It is important moreover to recognize the biases that will inevitably occur between these stages. Collection of data in routine surveillance is not normally a scientific process as one has to rely on readily available data – data obtained mostly for other reasons, such as to make a definitive diagnosis. Only the most severe cases die, and death certification thus provides, at best, a limited view of any disease. Similarly, only certain types and severity of cases will be admitted to hospital (some admissions are for social reasons for example) or even visit their family doctor. Certain age, sex, and perhaps social or occupational groups are more likely to seek medical help, be investigated and be reported. In laboratory data, more severe cases, or children, are perhaps much more likely to be investigated in detail.

These shortcomings of surveillance data do not make them useless – but their strengths and limitations must be recognized.

Feedback

If interpretation is turning statistics into information, feedback is getting the information across to those that matter, and those that need to know, so that action – the objective of surveillance – can be taken. Without feedback, surveillance is pointless. Feedback is most likely to be informative if undertaken by those most closely involved in the surveillance cycle, and who understand the significance of the data they are receiving.

Feedback should be aimed at contributors and those in public health. Contributors will then be aware of which viruses are circulating and this will help them to know what to look for in their own tests (e.g., what echovirus or

Table 1 Stages of a viral infection

Clinical
A. Uninfected
B. Infected
  1. Asymptomatic
  2. Symptomatic unreported
  3. Symptomatic, sees a doctor
  4. Symptomatic, admitted to hospital
  5. Symptomatic, dies/survives

Table 2 Stages in laboratory diagnosis

A. Uninfected
B. Infected
  1. Asymptomatic
  2. Symptomatic, unreported
  3. Symptomatic, sees a doctor
  4. Symptomatic, specimens submitted
  5. Symptomatic, specimens positive
  6. Symptomatic, specimens reported to surveillance system

adenovirus types are in circulation). Moreover, routine surveillance will undoubtedly uncover outbreaks of infection, which will need further investigation and control at local, national, and even international level. An interesting side effect of a flourishing microbiological/feedback surveillance system is that it often stimulates better quality control within reporting laboratories.

Regular feedback is helpful, not only to contributors, but also to those who can act for the public health. Generally, the periodicity of feedback should reflect the frequency of reporting – weekly feedback for weekly reports for example. Regular topic-based reviews are important.

Evaluation of Surveillance

A surveillance system is like a country's train system. Once the rail lines have been built, the goods that will be carried along those lines can be changed according to need. Similarly, in a surveillance network, once the lines of communication have been laid down, the data being reported can be changed according to what is most important at the time (though probably not too frequently). Nevertheless, surveillance systems should ideally be frequently evaluated for usefulness, as well as for accuracy, efficiency, and effectiveness.

They should also be sufficiently flexible, so that 'new' or emerging infections can be included in an emergency or when the need arises. The successful implementation of international surveillance for SARS was instrumental in controlling it. Emergency surveillance was also essential following the tsunami of 2004, and is also necessary for the successful management of other disasters following earthquakes, hurricanes, and floods.

Surveillance systems should be evaluated before they are set up, and again at regular intervals thereafter. Before implementing a surveillance system, is there an adequate public health and administrative infrastructure in place to take action? Are the data to be collected representative and sufficiently timely for the specific infection? Are they useful, and is action being taken on the information? If not, is the feedback inadequate?

Global and International Surveillance

The ease of modern travel, the distribution of goods (especially foodstuffs) across increasingly wide parts of the world, and the uncontrollable spread of birds and other wildlife across boundaries has made global and international surveillance essential for outbreak and infection control. Surveillance of influenza now requires the expertise of many professionals – epidemiologists, virologists, and vaccinologists, clinicians, statistical modelers, veterinarians, managers, and planners – in many different countries so that information can be exchanged, and attempts made on a global basis, to prevent the next pandemic. Only recently, a new variant of chikungunya virus jumped from Kenya, where it seems to have started, to islands in the Indian Ocean and hence to India. It has now, in 2007, even reached Italy. In Reunion alone, it affected 265 000 people, an astoundingly high incidence of 34%, and an estimated case-fatality rate of 1/1000. In India 1.3 million persons are thought to have been affected (so far, to February 2007). Two species of mosquito have been involved, *Aedes aegypti* and *A. albopictus*. Epidemics of dengue and West Nile virus have also spread widely recently. AIDS/HIV was destined to become a global problem almost from the time of its first discovery. On a more positive note, SARS was contained through the use of international surveillance; and surveillance was the backbone of the smallpox eradication program.

International surveillance can also be used for the detection of international outbreaks of food poisoning caused by the distribution of foodstuffs across a wide number of countries. An outbreak of hepatitis A in England was caused by frozen raspberries grown and frozen in another country and another outbreak of hepatitis A, this time in Czechoslovakia (before it became separate republics) was caused by strawberries used to make ice cream; the strawberries had been imported from another Eastern European country. There are now well-established trans-European surveillance systems for salmonella infections and legionnaires' disease.

The need for surveillance will never diminish or disappear. Surveillance systems will only improve, become increasingly sophisticated, and become increasingly relied upon and used. Control of infection will not be possible without it.

See also: Bovine Spongiform Encephalopathy; Central Nervous System Viral Diseases.

Further Reading

Charrel RM, de Lamballerie X, and Raoult D (2007) Chikungunya outbreaks – The globalization of vectorborne diseases. *New England Journal of Medicine* 356: 769–771.

Chin J (ed.) (2002) *Control of Communicable Diseases Manual,* 17th edn. Washington, DC: APHA.

Chorba TL (2001) Disease surveillance. In: Thomas JC and Weber DJ (eds.) *Epidemiologic Methods for the Study of Infectious Diseases*, ch. 7. Oxford, UK: Oxford University Press.

Communicable Disease Report(1994) National measles and rubella immunisation campaign. *Communicable Disease Report Weekly* 4(31): 146–150.

Heymann D and Rodier GR (1998) Global surveillance of communicable diseases. *Emerging Infectious Diseases* 4: 362–365.

Noah N (2006) *Controlling Communicable Disease*, chs. 1–4, pp. 14–19. Maidenhead, UK: Open University Press.

UNAIDS (2003) Introduction to second generation HIV surveillance. http://www.data.unaids.org/Publications/IRC-pub03/2nd_generation_en.ppt (accessed September 2007).

Thomas MEM, Noah ND, and Tillett HE (1974) Recurrent gastroenteritis in a preparatory school caused by *Shigella sonnei* and another agent. *Lancet* 1: 978–981.

Relevant Websites

http://www.cdc.gov – CDC (Centers for Disease Control and Prevention).

http://www.hpa.org.uk – Communicable Disease Report Weekly, Communicable Disease Report, Health Protection Agency (HPA).

http://www.hpa.org.uk – Enter-net(Gastrointestinal), International, Health Protection Agency.

http://www.who.int – Epidemiologic Surveillance, WHO.

http://www.who.int – Epidemiology, WHO.

http://www.eiss.org – European Influenza Surveillance Scheme (EISS).

http://www.eurosurveillance.org – Eurosurveillance is Freely Available Weekly and Monthly, European Centre for Disease Prevention and Control (ECDC).

http://www.hpa.org.uk – Health Protection Agency (HPA).

http://www.hpa.org.uk – HIV and Sexually Transmitted Infections, Health Protection Agency.

http://www.who.int – National Influenza Pandemic Plans, WHO.

http://www.hpa.org.uk – Response, Co-ordination and Specialist Support for Outbreaks and Incidents, Centre for Infections, HPA.

http://www.who.int – Surveillance of Noncommunicable Disease Risk Factors, WHO.

http://www.EWGLI.org – The European Working Group for Legionella Infections, EWGLI.

DNA Vaccines

S Babiuk, National Centre for Foreign Animal Disease, Winnipeg, MB, Canada
L A Babiuk, University of Alberta, Edmonton, AB, Canada

© 2008 Elsevier Ltd. All rights reserved.

Glossary

Electroporation The perturbution of cell membranes by very weak electrical impulses to allow the uptake of plasmid DNA.

Th1 immune response The quality of the immune response represented by a more active cell-mediated immune response with the concomitant secretion of specific cytobines such as interferon gamma.

Th2 immune response The quality of the immune response represented by a larger humoral antibody response, with the secretion of specific cytobines such as interleukin-4.

Introduction

Although the era of vaccinology has begun over 200 years ago, some of the early advances continue to form the foundation of our most successful vaccines. For example, even though it is over 200 years since Jenner introduced the concept of vaccination, using a live virus vaccine to protect people from smallpox, the concept of vaccinating with live agents continues. Indeed, some of the most effective vaccines are live-attenuated or live-heterologous vaccines. These vaccines have been successful in eliminating smallpox, or dramatically reducing the economic impact of diseases such as measles, polio, and many animal viral infections such as rinderpest. The reason for this success is that as the agent replicates, it resembles a natural infection, thereby stimulating the appropriate immune responses, including humoral and cell-mediated immunity. Furthermore, the persistence of the infecting agent, over an extended period of time, also provides long-term memory. Approximately 100 years after Jenner, Pasteur introduced a new approach to vaccination by killing the infectious agent and introducing the killed agent into the body as a nonreplicating vaccine. Similar to the original live vaccines, killed vaccines continue to be used for many infectious agents and are an integral component of our current armamentarium for disease control. Indeed, in some cases, live vaccines, which, in many cases, are more effective than killed vaccines, can be used initially to reduce the level of circulating agent in the environment, followed by killed vaccines to 'mop up' the remaining agent. The best example of such an approach is the use of live polio virus vaccine to provide broad immunity and reduce the viral load in the environment or country of interest and then use killed polio virus vaccine as a final step in disease eradication. The reason for such a strategy is that live vaccines can revert to virulence and may have certain risks associated with their use, which are worth taking when the disease burden is high but less so when the disease burden is low.

A century after Pasteur, we embarked on a new era of vaccination, primarily driven by advances in molecular biology and biotechnology, as well as a much deeper understanding of the host immune responses and antigens of each pathogen that are involved in providing protection from infection. These advances are the underpinning of the era of vaccination employing genetic engineering. This resulted in the development of vaccines against hepatitis B and human papillomavirus, which not only prevent infection but also reduce the development of tumors. These vaccines were genetically engineered to contain all of the critical epitopes involved in inducing protective immunity. However, they do not contain any nucleic acid, thereby making them relatively safe and they are considered to be killed vaccines. Unfortunately, they require adjuvants and generally do not induce the broadest range of immune responses often seen with live vaccines.

Since production and purification of genetically engineered subunit vaccines are generally very expensive, require strong adjuvants, and generally induce a skewed systemic Th2 response, the quest continues for the development of more effective vaccines resembling live vaccines. To combine the benefits of a subunit vaccine such as safety with the broad immune responses produced by live vaccines, the concept of introducing the gene encoding putative protective proteins in a plasmid into the host was proposed. Indeed, this was heralded as a 'third generation of vaccinology', genetic vaccination, or DNA vaccination. The major advantage of such an approach, especially for viral infections, is that the antigen is expressed endogenously and the antigens are, therefore, processed in a manner resembling a viral infection. As a result, the antigen is presented by both major histocompatibility complex (MHC) I and MHC II pathways, inducing a more balanced immune response. This article describes the progress made to date in the area of DNA vaccination, including a description of the DNA vaccines that are already licensed as well as some of the challenges faced by this technology. Furthermore, the opportunities to overcome these challenges will also be discussed.

Concept of DNA Vaccines

The major reason for continued interest in DNA-based vaccines is their simplicity in concept, ease of production, potential to develop a broad range of immune responses, as well as their perceived safety and ability to induce immunity in neonates in the absence or presence of maternal antibodies. With regards to simplicity, a DNA vaccine is comprised of a plasmid containing various regulatory elements to ensure efficient production of the plasmid in bacterial systems, such as an origin of replication and a selectable marker as well as an expression cassette containing the gene of interest under a eukaryotic promoter usually human cytomegalovirus for efficient expression of the gene inserted into mammalian cells. Since the general features of a plasmid are identical for all vaccines, the single platform makes DNA vaccines very attractive from the prospective of manufacturing. Thus, the only difference between different vaccines would be the gene insert. Thus, if a company establishes a process for manufacturing one plasmid-based vaccine, they can use the same process for production and purification of a variety of different vaccines.

Recent advances in our understanding of pathogenesis, comparative biology, molecular biology, bioinformatics, and immunology make identification of putative protective antigens to most pathogens relatively easy. Second, identifying and isolating the gene encoding a protective antigen, combined with gene sequencing to ensure the cloned gene contains the correct sequence and is correctly positioned in the plasmid, is relatively straightforward. Furthermore, the gene sequence can also be modified to optimize the codon biases for expression in the host cell of interest. Once the plasmid is constructed, the production of the plasmid is relatively simple, using well-established fermentation processes in *Escherichia coli*. Since *E. coli* grows to very high densities, the quantity of plasmid produced in this manner is very high. Combining fermentation with well-established downstream processing currently developed for plasmid purification makes this process very attractive to commercial companies. Indeed, processes for plasmid purification are considered easier to perform than protein purification used in subunit vaccine production. Currently, a number of companies have established the downstream processing steps for purifying high-quality plasmids suitable for DNA vaccine production. These processes can easily be transferred to developing countries to provide secure supplies of vaccine for the developing world, should that be necessary. Indeed, all of the processes including fermentation, plasmid purification, and quality control can occur in a time frame as short as 1 month. This is a significant advantage over current subunit vaccine protein production.

The process of inserting a gene into the plasmid is routine and once the process is developed for one vaccine, the same process can be used for inserting any gene of interest. The development of a new vaccine would not require additional skills, except knowledge of the gene of interest. As a result, extensive trials and procedures required to demonstrate the absence of reactogenic components often required for subunit or conventional vaccines is reduced to a minimum. This not only reduces the concerns for safety, but also reduces the time for reaching the market.

Safety Issues with DNA-Based Vaccines

DNA vaccines are considered to be safe since they can be highly purified to remove extraneous materials and are noninfectious. The infectious nature is a common concern with live-attenuated vaccines where reversion to virulence has occurred. Preliminary trials in animals and humans demonstrated low inherent toxicity and since the plasmids are often administered in physiological saline, in liposome-based formulations, or various other carriers which have already been tested and used over extended periods in humans, these vaccines are considered relatively safe. Initially, there was concern regarding the potential of the introduced plasmids to integrate into somatic or germline cells. However, studies have suggested that the probability of this occurring is 3 orders of magnitude lower than spontaneous mutations. Finally, there was concern that introduction of large amounts of DNA might lead to the generation of anti-DNA antibodies. Fortunately, a number of studies designed to test this possibility have demonstrated that this is likely to be a rare event if it would occur at all. Thus, it is currently widely believed that DNA-based vaccines are relatively safe.

Attractiveness of DNA-Based Vaccines

One desirable feature of any vaccine is the ability to protect individuals against a variety of disease agents following single administration of a vaccine. Thus, multicomponent vaccines are gaining popularity, not only because of the breadth of protection they induce, but also because of improved compliance since individuals do not need to return for numerous vaccinations. One limitation of combining multiple conventional vaccines is the possibility of interference between various components in the vaccine. With DNA vaccines this is also a possibility, but it should be possible to identify which component is interfering in induction of immunity. For example, in the study of a vaccine containing nine different plasmids encoding nine different malarial antigens, reduced immune responses to various components were observed compared to responses induced by individual plasmids. Similarly, we observed interference between plasmids encoding the genes for bovine herpesvirus-1 gD, parainfluenza-3 HA, and influenza virus HA. We also showed that co-administration of BHV-1 gD reduced the immune responses to BHV-1 gB. Since in these last two instances, the gD gene was responsible for the interference, it will be necessary to re-engineer this gene to remove the interference. In contrast, the use of genes encoding the four different dengue serotypes did not result in any interference. Indeed, combining the four serotypes resulted in higher antibody levels against dengue-4 than if animals were immunized with a monovalent dengue-4 plasmid-based vaccine. Similarly, broadening of the immune response was seen with multiple DNA vaccine components of various HIV proteins. Thus, it appears that each plasmid combination may need to be investigated independently for maximal benefit.

In addition to combining plasmids encoding different genes of interest, it is also possible to introduce two different genes encoding two different proteins from either the same or different pathogens in a single plasmid, thus further reducing the cost of production since one vaccine would protect against two different agents. This can be achieved by generating a plasmid which contains two gene expression cassettes or by simply linking the genes from two pathogens and inserting them into the plasmid. In this case, the two genes would be expressed as a single protein from a single promoter. Since the conformational changes imposed by linking two proteins together may alter their immunogenicity and the stability of the plasmid may be compromised as a result of its size, a number of plasmids have been constructed with each gene driven by its own promoter or by insertion of an encephalomyocarditis virus (EMCV) internal ribosome entry site (IRES) to translate the second gene. In some of these cases, the level of expression of the second gene was not as effective as that of the first gene, whereas in other instances there was no reduction in gene expression of either gene. Using such manipulations, it should be possible to broaden the protection to induce immunity to a number of different diseases simultaneously.

Neonatal Immunization

Since many diseases occur during the first few weeks or months of life, it is critical to induce immunity from vaccines as early as possible. Unfortunately, most vaccines given to neonates are less effective than in adults. There are many reasons for this. In the case of live vaccines, the presence of maternally derived antibodies generally limit the degree of replication of the vaccine, resulting in poor immune responses due to reduced antigenic mass produced *in vivo*. In the case of killed vaccines where there is less interference by maternally derived antibodies, the immune response is generally skewed to a Th2 bias response, thought to be due to the immaturity of the immune system. Using DNA vaccines, it has been shown that not only was it possible to induce immunity in neonates, but immunity could even be induced *in utero*. Thus, sheep immunized *in utero* during the third trimester were born fully immune. More importantly, the animals developed long-term memory. Indeed, the induction of memory by DNA vaccines is a very attractive feature of these vaccines since immune memory and duration of immunity are desirable for many vaccines. Numerous other studies have also shown the induction of long-term memory of T cells and B cells following DNA vaccination. This has been shown in a variety of models including pigs, sheep, and cytomologous monkeys. Memory could be further increased by incorporating plasmids encoding for IL-12 or IL-15 with the DNA vaccine.

It is also interesting that DNA vaccines appear to be able to induce immune responses in the presence of high levels of antibody. Thus, the fact that DNA vaccines can induce immunity in neonates in the presence or absence of maternal antibodies makes this approach to vaccination extremely attractive, especially for diseases such as herpes simplex virus-2, human immunodeficiency virus (HIV), hepatitis B, group B strep, and chlamydia which often infect children during birth or shortly thereafter.

Limitations and Clinical Applications of DNA Vaccines

Although there are over 1000 publications demonstrating the effective induction of immunity using DNA vaccines in mice which protect mice from subsequent infection, DNA vaccines have not been as successful in larger animals or humans. Possibly the greatest challenge to adopting DNA vaccination as a routine in large animals and humans is the poor efficiency of transfection leading to suboptimal induction of immunity. This has limited the introduction of vaccines into the market. However, recently two DNA vaccines were licensed for commercial use. The first was a vaccine to immunize fish against infectious hematopoietic necrosis virus, followed shortly after with one for West Nile virus in horses. Although there have been numerous clinical trials in humans, the greatest success has been when DNA vaccines were used to prime the individual, followed by booster with a recombinant or subunit vaccine. Indeed, the DNA prime, followed by protein boost is currently the method of choice for induction of the immune responses using DNA vaccines in humans. Thus, it is our contention that if better delivery mechanisms could be introduced to enhance the transfection efficiency in large animals and humans, DNA vaccines have the potential to become a critical component in our armamentarium against infectious diseases.

Vaccine Delivery

Once the plasmids are purified, they are introduced into the host (animals or humans) and the DNA is taken up by specific cells where the gene is expressed and protein is produced. Unfortunately, only a very small percentage of the DNA is internalized, which is one of the drawbacks of DNA vaccination. Even for the DNA molecules that are internalized, not all are transported to the nucleus where they express their encoded antigen. However, the few molecules that enter the nucleus are transcribed and translated to produce protein, making the animal cells a bioreactor. Thus, the vaccine antigen is produced in the animal. Even though the quantity of protein produced by the transfected cell is very low, the protein is presented to the immune system and induces a broad spectrum of immune responses including antibody- and MHC-I-restricted CTL responses. Clearly, the magnitude of the immune response as well as the quality of the immune response can be influenced by the route of delivery; intradermal delivery is generally better than intramuscular since the numbers of antigen-presenting cells are more numerous in the skin than in muscle. Indeed, muscle cells are not generally capable of presenting antigen to the immune system, although immunity occurs as a result of antigen release from myocytes leading to cross-priming. The quality of the immune response can also be modulated by the background of the plasmid. A number of groups have modified the plasmid background to incorporate sufficient and appropriate CpG motifs to stimulate the innate immune response. By stimulating the innate immune response, it is possible to recruit the appropriate cells to the site of antigen expression and create a cytokine microenvironment conducive to developing an adaptive immune response. To improve the cytokine microenvironment conducive to induction of immunity, a number of investigators have co-administered DNA vaccines with plasmids expressing cytokines, defensins, or other immune stimulator molecules to recruit dendritric cells to the vaccination site. Other constructs have contained both the antigen and the cytokine/chemokines within the single plasmid. In most cases, the immune response was enhanced by the co-administration of cytokines such as granulocyte-macrophage colony-stimulating factor (GMCSF) and interleukins IL-2, IL-12, and IL-15. In addition, the cytokine–chemokine combination plasmids could bias the immune response. For example, macrophage inflammatory protein-1 (MIP-1)β biased the immune response of Th2-like response, whereas MIP-2 and MIP-1α favored a Th1 like response. Currently, a number of studies are ongoing using IL-12 and IL-15 for enhancing immune response to HIV vaccines.

Since intracellular uptake of DNA is not a very efficient process, numerous investigators have employed a variety of methods to enhance DNA uptake into cells. One very efficient delivery system is encapsulation of DNA in liposomes or polylactide-L-glycolide (PLG) microparticles. DNA delivery by these microparticles serves two purposes. First, the encapsulated DNA is protected from degradation, and, second, the liposomes or microparticles facilitate uptake of the DNA into antigen-presenting cells. Indeed, it is even possible to deliver the microparticles orally to induce mucosal immunity as well as systemic immunity. Since encapsulation of DNA in PLG is thought to cause DNA damage due to the process of encapsulation, other investigators have attached the DNA to the surface of PLG microspheres. This process has resulted in excellent immune responses regardless of whether the PLG microparticles were given mucosally or delivered systemically, suggesting the microspheres play a crucial role in DNA uptake in the cells.

Improving uptake of DNA is thought to be critical for enhancement of immunity, and since the response is better when more cells are transfected, significant efforts have been focused on enhancing DNA uptake by various physical methods. The earliest approach to enhancing cell uptake was the use of a gene gun. In this instance, the DNA is coated on gold particles and is propelled into the skin. Some of these particles directly penetrate cells, whereas others are taken up by antigen-presenting cells. Although this approach has been shown to be effective, the quantity of DNA that can be coated onto gold beads is generally low, thereby reducing the effectiveness of such an approach.

The limitation of the quantity of DNA that can be taken up by cells by a gene gun can be overcome by jet injection of DNA. Depending on the pressure used to deliver the DNA, the DNA can be deposited either in the epidermis, subcutaneously, or even intramuscularly. However, an even more efficient method to deliver DNA into cells is the use of electroporation. Various companies have developed devices to both inject and electroporate the DNA simultaneously. Electroporation does not only enhance DNA uptake, but it also causes localized inflammation and induces the various mediators of innate immunity, resulting in enhanced immune responses. These studies have demonstrated the importance of modifying the local environment at the site of injection since the mere enhancing of expression does not always increase the immune responses. These studies demonstrate that one requires all of the elements of the immune response to be present at the site, as well as the specific antigen for the immunity to be induced.

Future Prospects

DNA vaccines represent an exciting addition to an already impressive track record of vaccines, which have saved millions of lives and improved the quality of life of almost every individual on the planet. Furthermore, vaccines have added significant economic benefits to society. Unfortunately, individuals still suffer from many infections. The reasons for this are varied, including the cost of vaccines, politics, as well as distribution issues. One of the major challenges to distribution is the need for a cold chain. If this could be overcome, the vaccines could be distributed to even remote regions of the world. DNA vaccines offer an advantage in this regard. Furthermore, it should be possible to develop these vaccines more economically, ensuring that the poor in remote regions could be vaccinated. Unfortunately, for this dream to become a reality, it will be necessary to improve the transfection efficiency of DNA vaccines to ensure that sufficient antigen is produced to induce an immune response. Currently, improvements in DNA vaccines often have been marginal. What is required is a 10–50-fold improvement to make the vaccines economical in most veterinary species and for humans in the developing world. However, the fact that two different DNA vaccines have already been licensed provides hope that within the next decade we will see a number of new DNA-based vaccines licensed and shown to be effective in reducing both morbidity and mortality. Whether these vaccines will be comprised solely of DNA-based vaccines or will be comprised of DNA-based vaccines which will prime the immune response, followed by boosting with conventional or recombinant vaccines, remains to be determined. Indeed, we will probably see both types of vaccine configurations in the future.

See also: AIDS: Vaccine Development; Immune Response to viruses: Antibody-Mediated Immunity; Immune Response to viruses: Cell-Mediated Immunity; Vaccine Strategies.

Further Reading

Babiuk S, Babiuk LA, and van Drunen Littel-van den Hurk S (2006) DNA vaccination: A simple concept with challenges regarding implementation. *International Reviews of Immunology* 25: 51–81.

Gerdts V, Snider M, Brownlie R, Babiuk LA, and Griebel P (2002) Oral DNA vaccination in utero induces mucosal immunity and immune memory in the neonate. *Journal of Immunology* 168: 1877–1885.

O'Hagan DT, Singh M, and Ulmer JB (2004) Microparticles for delivery of DNA vaccines. *Immunological Reviews* 199: 191–200.

Ulmer JB, Wahren B, and Liu MA (2006) DNA vaccines: Recent technological and clinical advances. *Discovery Medicine* 6: 109–112.

van Drunen Littel-van den, Hurk S, Babiuk SL, and Babiuk LA (2004) Strategies for improved formulation and delivery of DNA vaccines to veterinary target species. *Immunological Reviews* 199: 113–125.

Ebolavirus

K S Brown and A Silaghi, University of Manitoba, Winnipeg, MB, Canada
H Feldmann, National Microbiology Laboratory, Public Health Agency of Canada, Winnipeg, MB, Canada

© 2008 Elsevier Ltd. All rights reserved.

History

Ebola virus (EBOV) first emerged as the causative agent of two major outbreaks of viral hemorrhagic fever (VHF) occurring almost simultaneously along the Ebola River in Democratic Republic of Congo (DRC, formerly Zaire) and Sudan in 1976 (see **Table 1**). Over 500 cases were reported, with case fatality rates (CFRs) of 88% and 53%, respectively. It was later recognized that these two outbreaks were caused by two distinct species of EBOV (*Zaire ebolavirus* and *Sudan ebolavirus*). In 1989, a novel virus, Reston ebolavirus (REBOV), was isolated from naturally infected cynomolgus macaques (*Macaca fascicularis*) imported from the Philippines into the United States. All shipments except one were traced to a single supplier in the Philippines; however, the actual origin of the virus and mode of contamination for the facility have never been ascertained. While pathogenic for naturally and experimentally infected monkeys, limited data indicate that REBOV may not be pathogenic for humans as animal caretakers were infected without producing clinical symptoms. In 1994, the first case of Ebola hemorrhagic fever (EHF) occurred in western Africa in the Tai Forest Reserve in Côte d'Ivoire (Ivory Coast). An ecologist was infected by performing a necropsy on a dead chimpanzee whose troop had lost several members to infection with Côte d'Ivoire ebolavirus (CIEBOV). A single seroconversion was later documented, suggesting another nonfatal human case in nearby Liberia. Zaire ebolavirus (ZEBOV) reemerged in Kikwit, DRC, in 1995, causing a large EHF outbreak with 81% CFR. Sudan ebolavirus (SEBOV) reemerged in 2000–01 in the Gulu District in northern Uganda. There were over 425 cases (53% CFR), making it the largest EHF epidemic documented so far. Starting in 1994, an endemic focus of ZEBOV activity became obvious in the northern boarder region of Gabon and the Republic of Congo (RC) with multiple small EHF outbreaks over the past decade. Most infections there have been associated with the hunting and handling of animal carcasses (mainly great apes). In addition, ZEBOV has virtually decimated the chimpanzee and gorilla populations in those areas. At least three laboratory exposures to EBOV have occurred; one in Russia (2004) was fatal.

Taxonomy and Classification

Filoviruses are classified in the order *Mononegavirales*, a large group of enveloped viruses containing nonsegmented, negative-sense (NNS) RNA genomes. The family *Filoviridae* is separated into two distinct genera, *Marburgvirus* and *Ebolavirus*. The genus *Ebolavirus* is subdivided into four species – *Zaire ebolavirus*, *Sudan ebolavirus*, *Côte d'Ivoire ebolavirus*, and *Reston ebolavirus*. Filoviruses are classified as maximum containment (biosafety level 4 (BSL-4)) agents as well as category A pathogens based on their generally high mortality rate, person-to-person transmission, potential aerosol infectivity, and absence of vaccines and chemotherapy.

Biological and Physical Properties of Virion

EBOV particles are pleomorphic, appearing as U-shaped, 6-shaped, circular forms, or as long filamentous, sometimes branched forms varying greatly in length (up to 14 000 nm), but have a uniform diameter of ~80 nm (see **Figures 1(a)** and **1(b)**). EBOV virions purified by ratezonal gradient centrifugation are bacilliform in outline and show an average length associated with peak infectivity of 970–1200 nm. Except for the differences in length, EBOVs seem to be very similar in morphology. Virions contain a helical ribonucleoprotein complex RNP or nucleocapsid roughly 50 nm in diameter bearing cross-striations with a periodicity of approximately 5 nm, and a dark, central axial space 20 nm in diameter running

Table 1 Ebola hemorrhagic fever (EHF) outbreaks from 1976 to 2005

Ebola species	Year	Outbreak location (country)	Place of origin	Human cases (% case fatality rate)
Zaire ebolavirus	1976	Yambuku (DRC)	DRC	318 (88)
	1977	Tandala (DRC)	DRC	1 (100)
	1994	Ogooue-Invindo province (Gabon)	Gabon	51 (60)
	1995	Kikwit (DRC)	DRC	315 (79)
	1996	Mayibout (Gabon)	Gabon	37 (57)
	1996	Booue (Gabon); Johannesburg (South Africa)	Gabon	61 (74)
	2001–02	Ogooue-Invindo province (Gabon); Cuvette region (RC)	Gabon?[a]	124 (79)
	2002–03	Cuvette region (RC); Ogooue-Invindo province (Gabon)	RC?[a]	143 (90)
	2003	Mboma and Mbandza (RC)	RC	35 (83)
	2005	Etoumbi and Mbomo in Cuvette region (RC)	RC	12 (75)
	2007	Kampungu (DRC)	DRC	Ongoing
Sudan ebolavirus	1976	Nzara, Maridi, Tembura, Juba (Sudan)	Sudan	284 (53)
	1979	Nzara, Yambio (Sudan)	Sudan	34 (65)
	2000–01	Gulu District in Mbarrara, Masindi (Uganda)	Uganda	425 (53)
	2004	Yambio Country (Sudan)	Sudan	17 (41)
Côte d'Ivoire ebolavirus	1994	Tai Forest (Ivory Coast)	Ivory Coast	1 (0)
	1995	Liberia (Liberia)	Liberia?[a]	1 (0)
Reston ebolavirus	1989	Reston, Virginia (also Pennsylvania and Texas) (USA)	Philippines[b]	4 (0)[c]
	1992	Siena (Italy)	Philippines	0[d]
	1996	Alice, Texas (USA)	Philippines	0[d]

[a]Place of origin unconfirmed.
[b]Reston virus has only been traced to a single monkey-breeding facility in the city of Calamba, the Philippines, which was depopulated in 1996 and is no longer in operation.
[c]Mortality in monkeys was estimated at 82%.
[d]Only monkeys were infected in these outbreaks; no reported human cases.
DRC, Democratic Republic of the Congo; RC, Republic of the Congo.

the length of the particle. The RNP complex is composed of the genomic RNA and the RNA-dependent RNA polymerase (L), nucleoprotein (NP), and virion proteins 35 and 30 (VP35 and VP30). A lipoprotein unit-membrane envelope derived from the host cell plasma membrane surrounds it. Spikes approximately 7–10 nm in length, spaced apart at ~10 nm intervals, are visible on the virion surface and are formed by the viral glycoprotein (GP). Virus particles have a molecular weight of approximately $3-6 \times 10^8$ Da and a density in potassium tartrate of 1.14 g cm^{-3}. Virus infectivity is quite stable at room temperature. Inactivation can be performed by ultraviolet (UV) light and γ-irradiation, 1% formalin, β-propiolactone, and brief exposure to phenolic disinfectants and lipid solvents, like deoxycholate and ether.

Properties of Genome

The EBOV genome consists of a molecule of linear, nonsegmented, negative-stranded RNA which is noninfectious, not polyadenylated, and complementary to viral-specific messenger RNA. The genome amounts to c. 1.1% of the total virion. EBOV genomes are ~19 kbp in length and fairly rich in adenosine and uridine residues. Genomes show a linear gene arrangement in the order 3′ leader–NP–VP35–VP40–GP–VP30–VP24–L–5′ trailer (see **Figure 1(b)**). All genes are flanked at their 3′ and 5′ ends by highly conserved transcriptional start (3′-CUnCnUn-UAAUU-5′) and termination signal sequences (3′-UAAUUCUUUUU-5), respectively, all of which contain the pentamer 3′-UAAUU-5′. Most genes are separated by intergenic sequences variable in length and nucleotide composition. A feature of all EBOV genomes is the fact that some intergenic regions overlap by the conserved pentamer (UAAUU) sequence. ZEBOV and SEBOV show three such overlaps within the intergenic sequences of VP35/VP40, GP/VP30, and VP24/L, whereas REBOV shows only two between VP35/VP40 and VP24/L. Extragenic leader and trailer sequences are present at the 3′ and 5′ genome ends. These sequences are complementary at their very extremities, showing the potential to form

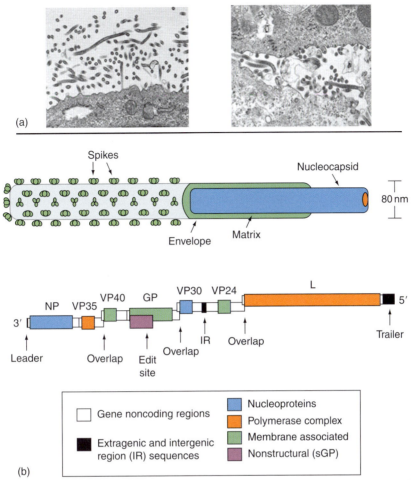

Figure 1 Particle morphology. (a) Transmission electron microscopy of Ebola virus particle. Both graphs show images of Vero E6 cells infected with Zaire ebolavirus. (b) Ebola virus particle structure and genome organization. The upper part provides a scheme of the virus particle separated into four components. The inner core consists of the nucleocapsid which is a structure formed by the single-stranded, negative-sense RNA genome associated with the two nucleoproteins (NP, VP30) and the polymerase complex (L and VP35). The matrix is built by the viral matrix protein VP40 and a minor component VP24. The envelope is derived from the infected cell during assembly/budding. The spikes consist of a homotrimer of the glycoprotein (GP).

stem–loop structures. Phylogenetic analyses on the basis of the GP gene of the different Ebola species show a 37–41% difference in their amino acid and nucleotide sequences. Analysis within one species shows remarkable genetic stability between strains (the variation in nucleotide sequences has been shown to be <7% and even <2% among distinct ZEBOV strains), unexpected for an RNA virus, but highly indicative that these viruses have reached a high degree of fitness to fill their respective niches.

Properties of Viral Proteins

Virions contain seven structural proteins with presumed identical functions for the different viruses (see **Table 2**). The electrophoretic mobility patterns of these proteins are characteristic for each species.

RNP Complex and Matrix Proteins

Four proteins are associated with the viral RNP complex: NP, L, VP30, and VP35 (see **Figure 1(b)**). These proteins are involved in the transcription and replication of the genome. NP and VP30 represent the major and minor nucleoproteins, respectively. They interact strongly with the genomic RNA molecule, and are both phosphoproteins. VP30 is also a zinc-binding protein that behaves as a transcriptional activator. The L and VP35 proteins form the polymerase complex. The L protein, like other L proteins of NNS RNA viruses, represents the RNA-dependent RNA polymerase. Motifs linked to RNA (template) binding, phosphodiester bonding (catalytic site), and ribonucleotide triphosphate binding have all been described. VP35 appears to behave in a mode similar to that of the phosphoproteins found in other NNS RNA viruses, acting as a cofactor that affects the mode of RNA synthesis (transcription vs

Table 2 Ebola virus proteins: functions and localization

Gene order[a]	Ebola virus proteins	Protein function (localization)
1	Nucleoprotein (NP)	Major nucleoprotein, RNA genome encapsidation (component of RNP complex)
2	Virion protein 35 (VP35)	Polymerase complex cofactor, type I interferon antagonist (component of RNP complex)
3	Virion protein 40 (VP40)	Matrix protein, virion assembly, and budding (membrane associated)
4 (primary)	Secreted glycoprotein (sGP) (nonstructural)	May have immunomodulatory role (secreted)
4 (secondary)	Glycoprotein (GP)	Receptor binding and membrane fusion (membrane associated)
5	Virion protein 30 (VP30)	Minor nucleoprotein, RNA binding, transcriptional activator (component of RNP complex)
6	Virion protein 24 (VP24)	Minor matrix protein, virion assembly, type I interferon antagonist (membrane associated)
7	Polymerase (L)	RNA-dependent RNA polymerase, enzymatic portion of polymerase complex (component of RNP complex)

[a]Gene order refers to 3′–5′ gene arrangement as shown in **Figure 1(b)**.

replication). VP35 is also known to be a type I interferon (IFN) antagonist; it blocks IFN-α/β expression by inhibiting IFN regulatory factor 3 (IRF-3). VP40 functions as the viral matrix protein and represents the most abundant protein in the virion. It plays a number of roles in viral infection related to assembly and budding of the viral particles. The production of VP40 by itself is sufficient to initiate the budding process for the production of virus-like particles (VLPs); however, the production of these particles is greatly enhanced by the addition of GP and NP. Recent studies have shown that EBOV hijacks the cellular protein machinery in order to mediate assembly and budding from the cellular membranes. These functions have been associated with overlapping late domain sequences (PTAP and PPEY) found in VP40. Late domains interact with cellular proteins such as Tsg101 and Nedd4 and affect a late step in the budding process. VP24 is thought to be a secondary (minor) matrix protein that, unlike VP40, is only incorporated into virions in small amounts. It has an affinity for the plasma membrane and perinuclear region of infected cells. The precise role of VP24 in replication is unclear; however, VP24 is a known type I IFN antagonist. Unlike VP35, VP24 blocks the translocation of phosphorylated STAT1 into the nucleus, subverting the antiviral response. More recently, VP24 has been associated with adaptation in rodent hosts.

Glycoproteins

The structural GP is a type I transmembrane protein inserted into the membrane as a trimer (see **Figure 1(b)**). It functions in viral entry, influences pathogenesis, and acts as the major viral antigen. The GP gene shows two open reading frames (ORFs) encoding for the precursors of GP (pre-GP) and a nonstructural secreted glycoprotein (pre-sGP), which is the primary product of this gene. Translation of pre-GP can only be achieved through mRNA editing, where one adenosine residue is added at a seven-uridine-stretch template sequence, resulting in a frameshift of the primary ORF. GP is cytotoxic when expressed at higher levels, leading to the hypothesis that mRNA editing may be evolutionarily related to the control of overexpression of this protein. The N-terminal ∼300 amino acids of the pre-GP are identical to those of pre-sGP, but the C termini of each protein are unique. pre-GP is translocated to the endoplasmic reticulum (ER) by an N-terminal signal sequence, and anchored by an extremely short membrane-spanning sequence at the C-terminus. The protein is glycosylated in the ER and Golgi apparatus with both N-linked and O-linked glycans. Most of the O-linked glycans are located in a mucin-like region (rich in theronine, serine, and proline residues) located in the middle of the protein. pre-GP is then cleaved by a subtilisin/kexin-like convertase such as furin, leading to the formation of disulfide-bonded $GP_{1,2}$. The smaller C-terminal cleavage fragment GP_2 contains the transmembrane region that anchors GP to the membrane. Interestingly, proteolytic cleavage of pre-GP is not required for infectivity or for virulence, indicating that the uncleaved precursor can mediate receptor binding and fusion. The production of pre-sGP differentiates EBOV from Marburg viruses, which do not express this soluble protein. pre-sGP is also translocated into the ER, modified in the secretory (e.g., oligomerization, glycosylation) pathway, and cleaved by furin near the C-terminus to release a short peptide termed delta peptide. No biological properties have been attributed to delta peptide. Recent matrix-assisted laser desorption/ionization time-of-flight mass spectrometry (MALDI-TOF MS) analysis suggests that sGP forms a homodimer in a parallel orientation, held together by disulfide bonds between the most N-terminal and C-terminal cysteine residues. sGP circulates in the blood of acutely infected humans. Its exact function is still unknown; however, an interaction with the cellular and humoral host immune responses has been postulated. In contrast to GP which mediates endothelial cell (EC) activation and decreases EC barrier functions, sGP seems to have an anti-inflammatory role.

Viral Life Cycle

The viral life cycle consists of several events: binding/entry, uncoating, transcription, translation, genome replication, and packaging/budding (see **Figure 2**). EBOVs have selective tropism primarily for monocytes, macrophages, and dendritic cells (DCs), although other cell types such as fibroblasts, hepatocytes, and ECs can be infected. In contrast, lymphocytes generally do not support EBOV replication, which is believed to be due to the lack of expression of the viral receptor(s). Although several molecules have been identified as possible viral receptors in various *in vitro* systems, it is not clear if any of the identified molecules are actually used *in vivo* or the degree of requirement for infection and disease. Folate receptor alpha was the first identified possible receptor but entry was also shown to occur in the absence of the molecule. C-type lectins such as DC-SIGN, DC-SIGNR, and hMGL were shown to be able to enhance binding but are not required for infection. Expression of members of the Tyro3 family converted the poorly susceptible Jurkat T cells into susceptible cells for particles pseudotyped with ZEBOV GP and enhanced ZEBOV infection. Receptor binding results in endocytosis into endosomes, possibly via clathrin-coated pits and caveolae although some studies suggest that they may not be necessary. Acidification of the endosomes is necessary for fusion of the viral and endosomal membranes, which is mediated by a region in GP_2. Proteolysis by cathepsins B and L in the acidic endosomes might be essential for infectivity.

Transcription and genome replication seems to follow the general principles for *Mononegavirales*. There seems to

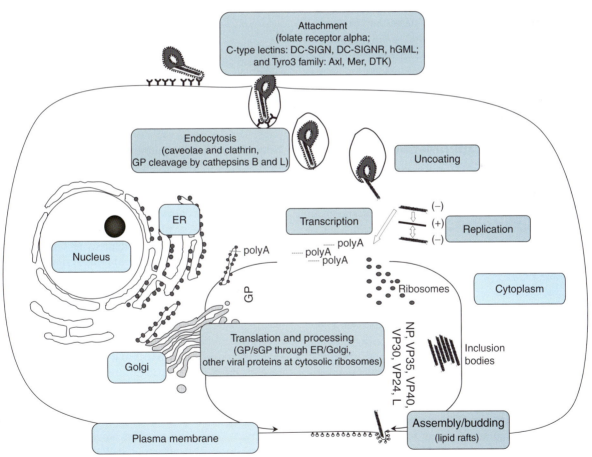

Figure 2 Ebola virus life cycle. The virus attaches to specific receptor(s) on the plasma membrane which leads to endocytosis via caveolae and clathrin-mediated pathways. The viral GP is cleaved in the endosome by cathepsins L and B. After uncoating, transcription and replication take place in the cytoplasm. All proteins except the GPs are translated on free ribosomes in the cytoplasm, while the glycoproteins are produced and modified in the ER and Golgi. Nucleocapsids formed in inclusion bodies interact with the matrix protein VP40 at the plasma membrane. Assembly and particle budding occurs at lipid rafts in the plasma membrane.

be a gradual decrease in mRNA levels from 3′ to 5′ end of the genome. The encapsidated viral genome serves as the template for transcription, which in the case of EBOV requires the L proteins, VP35 and VP30. mRNA transcripts are monocistronic, capped, polyadenylated, and contain long noncoding regions at their 3′ and/or 5′ ends. Stem–loop structures at the 5′ ends of mRNA transcripts may affect transcript stability, ribosome binding, and translation. The switch from transcription to replication seems to be triggered by an accumulation of viral proteins, especially NP, in the cytoplasm. Large amounts of NP are localized in inclusion bodies to which other viral proteins are recruited. These inclusion bodies may function as sites of RNP complex formation.

Membrane/lipid rafts have been identified as platforms for the assembly of virions; GP trimers conveyed to the surface membrane have an affinity for these rafts that is associated with palmitoylation of the membrane-spanning anchor sequence. RNP complexes interact with VP40, which is deposited at the plasma membrane via the late retrograde endosomal pathway. Abolition of the late domains in VP40 blocks particle formation, but only partially affects viral replication, suggesting that other domains or proteins must be involved. VP24 is also associated with the plasma membrane and enhances particle formation, but a specific role in particle maturation has not been identified. GP trimers seem to interact with VP40 and/or VP24 to finalize the budding process.

Host Range and Experimental Models

Natural Hosts and Geographic Distribution

EBOVs typically infect humans and nonhuman primates (NHPs); ZEBOV and SEBOV appear to be the main causes of lethal infections in humans, and thus are of primary public health concern. EBOVs appear to be indigenous to the tropical rain forest regions of central Africa (with the exception of REBOV), as indicated by the geographic locations of known outbreaks and seroepidemiological studies. The discovery of REBOV in the Philippines suggests the presence of a filovirus in Asia. Many species have been discussed as possible natural hosts; however, no nonhuman vertebrate hosts or arthropod vectors have yet been definitely identified. Epidemiological data have suggested monkeys as a potential reservoir of filoviruses; however, the high pathogenicity of EBOV for NHPs does not generally support such a concept. Similarities in biological properties to other viral hemorrhagic fever agents, such as 'Old World' arenaviruses, favor a chronic infection of an animal that regulates survival of the viruses in nature. More recently, viral RNA and virus-specific antibodies could be detected in African fruit bat species, suggesting that bats might be a reservoir for EBOV.

Experimental Models

Experimental hosts include monkeys (specifically rhesus and cynomolgus macaques), for which infection with ZEBOV is usually 100% lethal, guinea pigs (which show febrile responses 4–10 days after inoculation, but not uniform lethality), and newborn and immunocompromised mice when inoculated with wild-type virus. The resistance of the adult rodent models to EBOV infection has led to the production of guinea pig- and mouse-adapted ZEBOV strains (GPA-ZEBOV and MA-ZEBOV), produced through the serial passage of the virus through progressively older animals. These adapted strains are able to present uniform lethality in their respective hosts. MA-ZEBOV demonstrates reduced virulence in NHPs, whereas GPA-ZEBOV remains uniformly lethal in this model. Infected mice do not exhibit strong coagulation abnormalities (a hallmark of EBOV infection in humans and NHPs) and also slightly differ in other clinical symptoms. Mice, however, represent a good screening model for studies of antiviral and host immune responses. Infected guinea pigs do present coagulation defects that more closely resemble a human or an NHP infection. For growth in cell culture, primary monkey kidney cells and monkey kidney cell lines (e.g., Vero) are often used, but EBOV can replicate in many mammalian cell types including human ECs and monocytes/macrophages.

Clinical Features

Nonspecific, flu-like symptoms such as fever, chills, and malaise appear abruptly in infected individuals after an incubation period that ranges from 2 to 21 days, but on average lasts 4–10 days. Subsequently, multisystemic symptoms such as prostration, anorexia, vomiting, chest pain, and shortness of breath develop. Macropapular rash associated with varying degrees of erythema may also occur and is a valuable differential diagnostic feature. At the peak of the disease, vascular dysfunction signs appear ranging from petechiae, echymoses, and uncontrolled bleeding at venipuncture sites, to mucosal bleeding and diffuse coagulopathy. Massive blood loss is atypical, although it may happen in the gastrointestinal tract, and is not sufficient to lead to death. Fatal cases develop shock, multiorgan failure, and coma with death occurring between days 6 and 16. Survivors can have multiple sequelae such as hepatitis, myelitis, ocular disease, myalgia, asthenia, and psychosis. The mortality and severity of symptoms are viral species dependent, with ZEBOV causing 60–90% and SEBOV 50–60% lethality. Viremia in fatal cases can reach peak levels of 10^9 genomes ml^{-1}, while survivors have peak levels of about 10^7 genomes ml^{-1}. Viral antigen can be found systemically, although it is most abundant in the spleen and liver.

Diagnosis of EHF is based on the detection of virus-specific antibodies, virus particles, or particle components. The procedures are the same as for Marburg hemorrhagic fevers.

Pathogenicity

Most of the information available regarding pathogenicity is derived from ZEBOV infections in humans and animal models. Death seems to be the result of systemic shock due to vascular dysfunction, which is caused by a complex interaction of the immune system with vascular physiology. Three major processes – increase in vascular permeability, disseminated intravascular coagulation, and impaired protective immunological responses – are the main events that lead to shock and death. The first two processes are the product of a series of events starting with infection of primary target cells (monocytes/macrophages (Mø) and DCs) and later leading to activation and decrease of barrier function of ECs, while the third is still being actively investigated (see **Figure 3**).

Infected Mø are strongly activated, secreting pro-inflammatory molecules such as interleukin 6(IL-6), tumor necrosis factor alpha (TNF)-α, IL-1β, nitric oxide (NO), and spread the infection systemically. In contrast, infection of DCs results in impaired activation, with no upregulation of co-stimulatory molecules such as major histocompatibility complexes (MHCs) I and II, CD80, CD86, and CD40. Early interaction with primary target cells is independent of virus replication and current models suggest that GP in the repetitive context of a particle is required for activation, possibly by binding and cross-linking cellular receptors. Activation of ECs by mediators such TNF-α and NO is believed to be the main cause for the decrease in EC barrier function. TNF-α, NO, and pro-inflammatory cytokines increase vascular permeability and EC surface adhesion molecule expression, which are

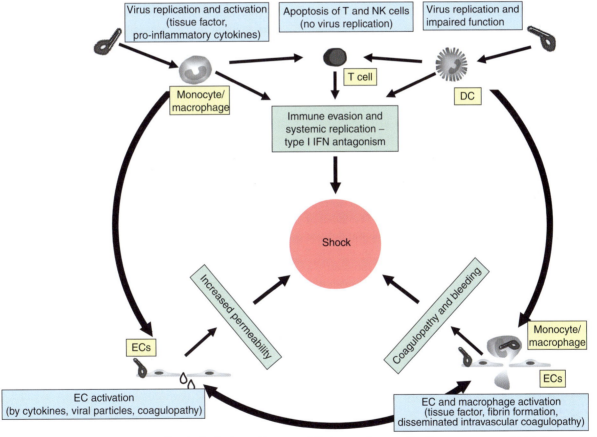

Figure 3 Ebola virus pathogenesis. Shock, the final event in severe/lethal cases, is caused by three processes, which influence each other: systemic viral replication and immune evasion, increase in vascular permeability, and coagulopathy. Infection of primary target cells such as monocytes/macrophages and DCs results in systemic spread of the virus and differential activation. Monocytes/macrophages are activated to produce pro-inflammatory cytokines and tissue factor (TF), while DC activation is impaired, leading to poor protective immune responses. Type I interferon responses are also inhibited by virus-encoded inhibitors (VP35 and VP24). Despite no infection of T- and natural killer (NK) cells, there is extensive apoptosis in those cell types. ECs are activated by pro-inflammatory cytokines and virus, which leads to increased permeability. TF expression induces coagulopathy, which is also able to increase inflammation.

necessary in extravasation of immune cells in inflamed tissue. Although these pro-inflammatory cytokines are an integral part of a normal, localized immune response by attracting and activating immune cells to the site of infection, in the context of a systemic and extensive response, they have a negative effect by inducing shock. ECs also serve as target cells and infection results in activation as indicated by upregulation of adhesion molecule expression followed by cytolysis. However, *in vivo* EC destruction can only be observed at the end stage of the disease.

Infection of MØ also results in the expression of tissue factor (TF), which can impair the anticoagulant–protein-C pathway by downmodulating thrombomodulin. TF is believed to be an important protein in the development of disseminated intravascular coagulopathy (DIC), as inhibition of TF delays death in NHPs and even protects 33% of the infected animals. TNF-α can also induce TF expression in ECs, which would further amplify coagulopathy and DIC. Coagulopathy is known to enhance inflammation, and this could lead to a vicious cycle where inflammation enhances coagulopathy, which in turn would amplify inflammation.

Rapid systemic viral replication is an integral part of the pathogenesis, as infection and even binding of particles results in not only MØ activation but also DC impairment. ZEBOV infection in humans and NHPs results in rapid massive viremia and high titers in various organs, especially spleen and liver. This correlates with an impaired innate and adaptive immune response. However, high viremia and lethality are only seen in guinea pigs and mice after host adaption or infection of various immune-deficient mice, suggesting that evasion of immune responses plays a pivotal role in pathogenesis. Mice lacking the IFN-α/β receptor or STAT1, the main signaling molecule in response to type I IFN, are extremely susceptible to ZEBOV. Treatment of NHPs with IFN-α did not protect the animals but delayed death, suggesting that type I IFN may be an important innate molecule in delaying viral replication. Interestingly, MA-ZEBOV is less sensitive to IFN-α treatment in murine macrophages compared to wild-type ZEBOV, and analysis of genomic mutations in GA- and MA-ZEBOV indicated that mutations in NP and VP24 are sufficient for a lethal phenotype. VP24 inhibits responses to interferon *in vitro*, but it is not clear yet if the mutations in the adapted strains are required for type I IFN evasion or other functions. ZEBOV is also able to inhibit IFN gene induction through VP35, and deletion of the region in VP35 involved in IFN antagonism results in a highly attenuated virus. Furthermore, comparative *in vitro* gene microarray analysis demonstrated a correlation between cytotoxicity and IFN antagonism, which was strongest with ZEBOV and weakest with REBOV.

Adaptive immune responses are also impaired during infection. As mentioned above, upregulation of co-stimulatory molecules is inhibited in DCs, which reduces activation of T cells. There is also a dramatic drop in number of T and natural killer (NK) cells despite the fact that lymphocytes do not support ZEBOV replication. It is believed that lymphopenia is caused by 'bystander' apoptosis, although the mechanism is not well understood. Activation of T cells could also be inhibited by treatment *in vitro* with a 17-amino-acid peptide present in GP_2, which has homology to a known immunosuppressive peptide in the retroviral Gag protein; yet there is no evidence that the complete ZEBOV GP has the same capabilities.

Treatment

The current treatment of EHF is strictly supportive, involving fluid and electrolyte replenishment and pain reduction. Due to the remote location of the outbreaks and limited resources available in the affected regions, treatment options have not been tested in patients. Several experimental treatment strategies have been successful in the rodent models, but failed in the NHP model, which is considered the most accurate in modeling human disease.

Therapeutic antibodies are still considered a short-term solution despite varying success in animal models and humans. Convalescent serum was used in a limited number of patients during the Kikwit 1995 ZEBOV outbreak but the success is a matter of dispute. Passive immunization with hyperimmune horse serum resulted in protection of hamadryl baboons, whereas it only delayed death in cynomolgus macaques. Monoclonal antibody treatment is successful in rodent models but has failed in preliminary NHP studies.

Currently the most feasible and promising approach relates to interference with coagulation using the recombinant nematode anticoagulant protein c2 (rNAPc2). Administration of the drug, which is already in clincial trials for other applications, as late as 24 h post infection resulted in 33% survival in the rhesus macaque model. Even more potent seems to be post-exposure treatment with a recombinant vesicular stomatitis virus (VSV) expressing the ZEBOV GP, which resulted in 50% protection when given 30 min post infection, but the mechanism of post-exposure protection is not yet understood. It is expected that approval of this attenuated replication-component vector will be difficult.

The recent advances in the understanding of EBOV pathogenesis and replication will open new avenues for intervention therapy. Novel antiviral strategies such as viral gene silencing through specific siRNA, cathepsin inhibition, and functional domain interference with small peptides showed promise in tissue culture and partially also in rodent models. Strategies targeting host responses are important alternative options and include anticytokine therapy and modulation of coagulation pathways. It should be noted that more classical approaches such as ribavirin treatment, which has been successfully used to treat other VHFs, are not indicated for EHF.

Vaccines

There are no approved vaccines against EBOV. Similar to therapeutic approaches, many initial vaccination strategies were successful in rodents but failed in NHPs. Nevertheless, there are several promising experimental strategies. Nonreplicating adenoviruses expressing the ZEBOV GP and NP were able to induce sterile immunity in cynomolgus macaques within 28 days or more post vaccination, either alone in a single shot approach or in combination with DNA priming and adenovirus boost. Although DNA priming is not necessary for protection of NHPs, it may be required to overcome the problem of preexisting immunity to human adenoviruses. DNA vaccination alone against EBOV is already in Phase I clinical trial, where it was shown to be safe and effective in inducing humoral and cellular immune responses.

Recombinant VSV expressing ZEBOV GP also induced sterile immunity when given 28 days prior to challenge of cynomolgus macaques, but was also able to protect 50% of rhesus macaques when administered 30 min post challenge. These results, together with the fact that prior immunity against the vector is extremely low, and the only target of neutralizing antibodies, VSV G protein, is removed from the vaccine vector, indicate that this platform may be successful in humans if licensing for the replication-competent vector can be achieved. Human parainfluenza virus-based vectors and VLPs are successful in protecting rodents, and preliminary results suggest that they even show efficacy in NHPs. Safety testing of vaccine vectors and the establishment of immune correlates is a priority for all these and future vaccine platforms.

Note: Investigation of an ongoing hemorrhagic fever outbreak in southwestern Uganda revealed what appears to be an additional distinct species of Ebola virus associated with this outbreak. Preliminary genome sequence analysis suggests the most closely related Ebola virus species would be the *Côte d'Ivoire ebolavirus*. Initial outbreak investigations suggest that infection with this newly discovered virus is associated with a lower case fatality than infections with Zaire ebolavirus (60–90%) or Sudan ebolavirus (50–60%) (T. G. Ksiazek, Centers for Disease Control and Prevention, Atlanta, GA, United States, personal communication).

Acknowledgments

The Public Health Agency of Canada (PHAC), Canadian Institutes of Health Research (CIHR), and CBRNE (Chemical, Biological, Radiological & Nuclear) Research and Technology Initiative (CRTI), Canada, supported work on filoviruses at the National Microbiology Laboratory of the Public Health Agency of Canada.

See also: Marburg Virus.

Further Reading

Feldmann H, Geisbert TW, Jahrling PB, et al. (2005) Filoviridae. In: Fauquet CM, Mayo MA, Maniloff J, Desselberger U, and Ball LA (eds.) *Virus Taxonomy: Eighth Report of the International Committee on Taxonomy of Viruses*, pp. 645–653. San Diego, CA: Elsevier Academic Press.

Feldmann H, Jones SM, Klenk HD, and Schnittler HJ (2003) Ebola virus: From discovery to vaccine. *Nature Reviews Immunology* 3: 677–685.

Feldmann H, Jones SM, Schnittler HJ, and Geisbert TW (2005) Therapy and prophylaxis of Ebola virus infections. *Current Opinion in Investigational Drugs* 6: 823–830.

Geisbert TW and Hensley LE (2004) Ebola virus: New insights into disease aetiopathology and possible therapeutic interventions. *Expert Reviews in Molecular Medicine* 6: 1–24.

Geisbert TW and Jahrling PB (2004) Exotic emerging viral diseases: Progress and challenges. *Nature Medicine* 10(supplement 12): S110–S121.

Hensley LE, Jones SM, Feldmann H, Jahrling PB, and Geisbert TW (2005) Ebola and Marburg viruses: Pathogenesis and development of countermeasures. *Current Molecular Medicine* 5: 761–772.

Hirsch M (ed.) (1999) *Journal of Infectious Diseases* 179(supplementum 1): S1–S288.

Jasenosky LD and Kawaoka Y (2004) Filovirus budding. *Virus Research* 106: 181–188.

Paragas J and Geibert TW (2006) Development of treatment strategies to combat Ebola and Marburg viruses. *Expert Reviews of Anti-Infective Therapy* 4: 67–76.

Pattyn SR (ed.) (1978) *Ebola Virus Hemorrhagic Fever*. Amsterdam: Elsevier/North-Holland Biomedical Press.

Reed DS and Mohamadzadeh M (2007) Status and challenges of filovirus vaccines. *Vaccine* 25: 1923–1934.

Sanchez A, Geisbert TW, and Feldmann H (2007) Marburg and Ebola viruses. In: Knipe DM, Howley PM Griffin DE, et al. (eds.) *Fields Virology*, 5th edn., pp. 1409–1448. Philadelphia: Kluwer/Lippincott Williams and Wilkins.

Walsh PD, Abernethy KA, Bermejo M, et al. (2003) Catastrophic ape decline in western equatorial Africa. *Nature* 422: 611–614.

Zaki SR and Goldsmith CS (1999) Pathologic features of filovirus infections in humans. *Current Topics in Microbiology and Immunology* 235: 97–116.

Echoviruses

T Hyypiä, University of Turku, Turku, Finland

© 2008 Elsevier Ltd. All rights reserved.

History and Classification

Echoviruses belong to the species *Human enterovirus B* (HEV-B), in the genus *Enterovirus* of the family *Picornaviridae*. Original classification of picornaviruses was based mainly on physicochemical properties and pathogenesis in experimental animals. At the beginning of the twentieth century, poliomyelitis was transmitted to monkeys using a filterable agent from clinical patients and 40 years later, new related human picornaviruses (coxsackieviruses), that characteristically caused disease in newborn mice, were isolated. Based on the disease signs observed in mice they were classified into A and B subgroups. After introduction of tissue cultures for virus propagation, a number of new viruses were isolated that exhibited similar physicochemical characteristics (resistance to organic solvents and low pH) with polio- and coxsackieviruses, but grew exclusively in cell culture. Since the disease association of these viruses was at that time unclear, they were grouped among ECHO viruses, which stands for enteric (isolated mainly from stool samples), cytopathogenic (cytopathic effect observed in tissue culture), human (no disease in monkeys or in newborn mice), orphan (disease association not confirmed). There are 28 distinct echovirus serotypes (EV1–33); EV1 and EV8 represent the same serotype, serotype 10 was identified to be a reovirus, serotype 28 was reclassified as a rhinovirus, and serotypes 22 and 23 were reclassified as members of a separate picornavirus genus, *Parechoviridae* (**Table 1**). Polioviruses, coxsackieviruses, and echoviruses formed human enteroviruses. Later, new enterovirus serotypes were not classified into the subgroups but were given successive numbers in the order of their identification (enteroviruses 68–71).

When sequence analysis became available, it was possible to determine the genetic relatedness of individual virus serotypes and strains. It was shown that echoviruses form a genetically rather coherent cluster and they are closely related to certain other enteroviruses. Molecular criteria replaced the previous enterovirus subgroup division in classification and currently echoviruses belong to the HEV-B species together with coxsackie B virus 1–6, coxsackievirus A9, and enterovirus 69 serotypes. More recently, identification of new enterovirus types has been exclusively based on comparison of the VP1 gene sequences, which correlate well with the previous serotype division. Currently, there are approximately 100 distinct enterovirus types, 63 of which have been assigned to serotypes. Further studies of the new members of the genus will show how many of them share biological properties with the classical echovirus serotypes.

Molecular Characteristics and Replication

Virus Particle

In general, echoviruses share the basic properties of other representatives of the enterovirus genus, where polioviruses are considered as the type members. Enteroviruses are composed of a protein capsid consisting of 60 copies of each of the four structural proteins (VP1–4) and enclosing a single-stranded infectious RNA genome, approximately 7400 nt in length. The diameter of the icosahedral virus particle is around 30 nm and the RNA genome represents about 30% of the molecular mass of the virion. VP1–3 are exposed on the surface of the virus particle and they all share the eight-stranded antiparallel β-barrel structure. These capsid proteins are responsible for the recognition of the cellular receptors and they contain the neutralizing B-cell epitopes, located in the variable loops connecting the β-strands and in the C-termini of the proteins. VP4 is located inside the virion and it contains a myristic acid, covalently linked to the N-terminus. Five VP1 molecules form a star-shaped structure around the fivefold symmetry axis that is surrounded by a surface depression (a canyon), while VP2 and VP3 molecules alternate around the threefold axis. The structural subunits are a protomer, composed of one copy of each of the capsid proteins and a pentamer including five protomers. The three-dimensional structure of at least two echovirus serotypes (EV1 and EV11) have been determined by X-ray crystallography and some additional serotypes have also been studied by cryoelectron microscopy. In general, the detailed architecture of these echoviruses closely resembles that of other enteroviruses. Structural, immunological, and other biological

Table 1 Summary of the molecular and clinical characteristics of echoviruses

Family	*Picornaviridae*
Genus	*Enterovirus*
Species	*Human enterovirus B*
Serotypes	1–7, 9, 11–21, 24–27, 29–33
Virus capsid	Nonenveloped, icosahedral, contains 60 copies of each of the capsid proteins VP1–4
Genome	Single-stranded, infectious RNA, approximately 7400 nt
Viral proteins	Cap-independent translation utilizing the internal ribosome entry site; translated as a polyprotein (2200 amino acids); the capsid proteins are at the N-terminus and the C-terminal part of the polyprotein contains the nonstructural proteins (2A, 2B, 2C, 3A, 3B, 3C, and 3D)
Receptors	
Most echoviruses	Decay-accelerating factor (CD55)
Echovirus 1	α2β1 integrin
Echovirus 9 (Barty)	αV integrins
Entry route	Endocytosis
Replication	Viral RNA synthesis takes place on cytoplasmic membranes
Clinical manifestations	Meningitis, encephalitis, generalized infections of newborns, myocarditis, rashes, respiratory infections, severe infections in immunocompromised patients
Diagnosis	RT-PCR, virus isolation, serology

reasons for the existence of almost 30 distinct echovirus serotypes in contrast to three polioviruses are currently not known.

The organization of the echovirus genome is analogous to other enteroviruses and the functions of the proteins have been mainly predicted on the basis of the findings reported for polioviruses. The capsid proteins are located at the N-terminal part (P1 region) of the polyprotein encoded by the viral RNA and the nonstructural proteins, necessary for replication, are at the C-terminal part (P2 and P3 regions). The 5′ noncoding region (NCR) contains around 750 nt, and it includes the structured elements important in translation (internal ribosome entry site, IRES) and genome replication. A small viral protein (3B, also known as VPg) is covalently linked to the 5′ terminus of the genome. The length of the 3′ NCR, also containing predicted secondary structures, is approximately 100 nt and it is followed by a poly(A) tract. The variation in the lengths of all echovirus genomes is in the range of 100 nt. The maximum difference between the 5′ NCR sequences is less than 25%; the capsid protein genes are rather variable (around 60–80% identity between the amino acid sequences of VP1), and less diversity is found at the nonstructural region (80–100% amino acid identity).

Replication Cycle

In most picornaviruses, the interaction with a specific cellular receptor takes place through structures located on the bottom of the canyon with the terminal domains of immunoglobulin (Ig) superfamily molecules. This interaction leads to conformational changes which give rise to further uncoating and release of the genome into the cytoplasm. Instead, many echoviruses attach to decay-accelerating factor (DAF, CD55) that belongs to a family of regulatory proteins protecting the cells from autologous complement-mediated lysis. DAF contains four short consensus repeat domains and is anchored to the surface of the plasma membrane. DAF-binding echoviruses interact mainly with the consensus repeat domain 3 but may also recognize binding sites in domains 2 or 4. Interaction of DAF with echoviruses does not lead to similar alterations in the conformation of the virus particle that are observed after interaction of the Ig superfamily member receptors. Structural studies have revealed that the binding site of DAF in echovirus 7 is located in the VP2 protein close to the twofold symmetry axis in contrast to the binding site of the Ig superfamily molecules in the canyon. DAF is also recognized by certain coxsackie A and B viruses but the binding mechanisms are different and the conformational changes in these viruses are induced by subsequent interaction with members of the Ig superfamily molecules. Therefore, it is thought that additional cellular factors are also needed for successful internalization and uncoating of the DAF-binding echoviruses. Candidate molecules include $\beta 2$-microglobulin, complement control protein CD59, and heparan sulfate.

Sequence analysis of the mouse-pathogenic echovirus 9 Barty strain revealed that there was a 10-amino-acid-long insert to the C-terminal end of the VP1 protein when compared to the Hill strain of the same serotype which does not cause disease in mice. This insert contains an arginine-glycine-aspartic acid (RGD) motif, known to interact with αV integrins, and this sequence has also been found in clinical isolates. Interestingly, coxsackievirus A9, another mouse-pathogenic member of the HEV-B species, contains a functional RGD motif in a similar position and recognizes αV integrins.

Echovirus 1 is the only picornavirus that is known to interact with $\alpha 2\beta 1$ integrin (a collagen receptor) on the cell surface. The virus binds to the I (inserted) domain in the $\alpha 2$ subunit which is also recognized by collagen but there are remarkable differences between these interactions. The binding site has been localized to the outer wall of the canyon. As in the case of DAF interaction with other echoviruses, binding of the $\alpha 2$I-domain to EV1 does not bring about the conformational changes needed for uncoating *in vitro*. Attachment of the virus to the integrin gives rise to clustering of the receptor molecules which is known to be able to induce signaling events which may play an important role in the subsequent internalization processes.

Entry of echoviruses has been studied in serotypes 1 and 11. The integrin interaction of EV1 is followed by endocytosis of the virus–receptor complex into cytoplasmic structures which are rich in caveolin-1. This internalization process is initiated in lipid rafts and, subsequently, the virus entry can occur either in caveolae or through a faster route into caveosomes, which are preexisting intracellular organelles. How the EV1 genome is released from the caveosomes to the cytoplasm, is currently not known. Echovirus 11 appears to use similar mechanisms during entry. In the plasma membrane, DAF is present in lipid rafts, which are cholesterol-rich domains, and it has been shown that cholesterol depletion inhibits DAF-mediated echovirus infection. The virus can also be copurified with components found in lipid rafts.

When released to the cytoplasm, the picornavirus genome is infectious and it can act directly as an mRNA encoding a large polyprotein that is subsequently processed to the mature viral polypeptides. Based on similarity echovirus nonstructural proteins with those of poliovirus, viral replication cycles are evidently very similar. The proteolytic cleavages take place as a cascade and some of the intermediate products (2BC, 3AB, and 3CD) exhibit also specific activities. 2A is a protease responsible for the cleavage between the VP1 and 2A proteins and, in addition, it causes shut-off of host-cell protein synthesis by cleaving a cellular factor needed for cap-dependent translation. Proteases 3C and 3CD are responsible for other proteolytic processing events except the cleavage between VP4 and VP2 which may be autocatalytic.

In the infected cells, the viral nonstructural proteins associate with host-cell membrane compartments and modify these to establish a specialized complex for viral RNA synthesis. In these replication complexes, containing almost all the nonstructural proteins with multiple interactions with each other and viral RNA, the genome is copied through a complementary template to new genomic strands. 3D is an RNA-dependent RNA polymerase responsible for the synthesis of the positive and negative strands. Due to the deficient proof-reading activity of the viral RNA-dependent RNA polymerases, errors are frequently generated during the replication and in picornaviruses every new genome contains approximately one mutation. 3B (VPg), found at the 5' termini of the RNA strands during replication but not during translation, plays a role in the priming of the RNA synthesis. The role of the 3AB protein is thought to be anchoring of VPg to the membranes for priming. The exact role of the 2B protein in the replication cycle is not known. It increases membrane permeability that may facilitate release of the mature viruses from the cells and it also interferes with cellular protein secretion. The 2C protein contains three motifs found in NTPases/helicases. Moreover, 2C and some other nonstructural proteins are involved in rearrangements of intracellular vesicles during formation of the replication complexes. In the assembly of infectious virions, the protomers form pentamers which can associate to form empty capsids. It is not clear whether the viral genome can be inserted into these particles or if they represent a reservoir of the structural units, finally assembling around viral RNA. The mature viruses are released as a result of the lysis of the cells.

Viral replication has dramatic effects in the host cell, including inhibition of cellular protein synthesis and secretion, interference with RNA synthesis and transport, and activation of other processes that together induce the cytopathic effect. Both apoptotic and anti-apoptotic effects, caused by enterovirus replication in cells, have been reported and also more detailed alterations can be detected. In EV1-infected cells, approximately 2% of the studied genes were induced whereas around 0.5% were downregulated. The activated genes included immediate-early response genes as well as genes involved in apoptotic pathways and cell growth regulation. Viral macromolecule synthesis was also observed to activate protein kinases known to control expression of several cellular genes. Further studies on the events during the infection may reveal cellular activities that are necessary for the infection or play a crucial role in the pathogenesis and they can become new potential targets of chemotherapy.

Clinical Manifestations

In general, the disease pattern caused by echoviruses is highly variable and most of the infections are evidently subclinical like those caused by other human enteroviruses. No specific disease condition can be directly associated with a certain echovirus and different serotypes cause identical clinical illnesses, while the same serotype can have a different clinical outcome in patients during the same epidemic.

Members of HEV-B species are among the most common causes of meningitis that may occur as sporadic cases as well as large epidemics. Typically, the symptoms and signs include fever, headache, nausea, stiffness of the neck, and may be associated with respiratory illness, rash, or myalgia. The manifestations may also exhibit signs of meningoencephalitis. Paralysis is a rare consequence of echovirus infection although some cases have been reported.

In neonates, a severe generalized infection, often clinically indistinguishable from bacterial sepsis, can include meningitis or meningoencephalitis, myocarditis, and hepatitis. A similar syndrome is caused by echoviruses and coxsackie B viruses. The transmission occurs transplacentally or soon after delivery from the mother but can also originate from other infected infants. Immunocompromised patients may develop serious chronic echovirus infections, like meningoencephalitis, occasionally with fatal outcome. These complications occur in individuals with B-cell deficiencies and gammaglobulin has been used for the prevention and treatment of the infections.

Echoviruses are associated with mild respiratory infections (e.g., common cold, bronchiolitis, and herpangina) that cannot be clinically distinguished from illnesses caused by other viral respiratory pathogens. Entroviruses cause maculopapular rashes with similar manifestations caused by other infections. Roseola-like skin manifestations can also occur during enterovirus infections and the 'Boston exanthem', caused by echovirus 16, was the first one of these infections recognized. Although echoviruses and other enteroviruses replicate in the gastrointestinal tract and nausea as well as diarrhea can be associated with the infection, they are not considered as major causative agents of acute viral gastroenteritis. Echoviruses have been isolated from individual cases of conjunctivitis but their role in the disease compared to enteroviruses causing epidemics (coxsackievirus A24 and enterovirus 70) is not important. Epidemics of uveitis, a severe eye infection, caused by echoviruses 11 and 19, have been described in Russia.

Myocarditis is often associated with coxsackie B viruses but other enteroviruses are evidently also responsible for this disease where specific diagnosis is usually difficult. Although neonatal myocarditis can be fatal, the manifestations of the disease in older age groups are often nonspecific, and the cardiac involvement is usually suspected on the basis of ECG findings and severe symptoms are uncommon. There is also evidence that enteroviruses could have an etiological role in the development of dilated cardiomyopathy.

Members of the HEV-B species have also been associated with the development of type 1 diabetes. This is based largely on seroepidemiological studies but, for instance, echoviruses have also been isolated from prediabetic and diabetic individuals and enterovirus genomes have been detected in the blood of patients with recent onset of diabetes. However, the role of enteroviruses in the process leading to type 1 diabetes remains unknown today. Echoviruses and other enteroviruses have also been suspected to have a role in amyotrophic lateral sclerosis, a chronic progressive neurological disorder, but further studies are required to clarify this association. Involvement of viruses, including enteroviruses, in the chronic fatigue syndrome has also been suspected, but this disease is still poorly characterized and the pathogenesis is not understood. It is known that echoviruses can cause chronic infections in cell culture models and availability of new animal models will further clarify their pathogenesis in the future. Sensitive and specific molecular methods will also cast light in the occurrence of echoviruses in diseases where the etiology is presently unknown.

Based on the observations about the protective effects of the inactivated poliovirus vaccine and appearance of severe chronic enterovirus infections in agammaglobulinemic patients, it is thought that protective immunity in enterovirus infections is mainly antibody mediated. It seems that the response is largely serotype specific and systemic infections may result in life-long immunity. Cell-mediated immunity against enterovirus infections is still rather poorly understood but there is evidence that cross-reactive epitopes between virus serotypes may exist.

Laboratory Diagnosis

Echoviruses can be detected from the cerebrospinal fluid (CSF) and respiratory tract samples during the acute phase of infection, but they may be excreted in stool samples for weeks or even months after the infection. Identification of enteroviruses has been classically based on isolation in cell culture followed by neutralization typing using antiserum pools. Although the advantage of the methods is that it gives information about the serotypes, the problem is the time and labor required. In spite of accelerated protocols, based on immunological staining of the infected cells, the isolation method is of limited use when specific diagnosis is required at the acute phase of the illness.

Introduction of the reverse transcription-polymerase chain reaction (RT-PCR) methods for the molecular detection of human picornaviruses revolutionized rapid diagnosis of the enterovirus infections, in particular those affecting the central nervous system. By this means, it is possible to obtain the results in a few hours which has an important impact in the differential diagnosis and treatment of the illness. Currently used RT-PCR methods take advantage of the conserved regions in the $5'$ noncoding region of the genome and do not therefore allow further typing of the detected enteroviruses. However, it has been shown that sequences of the VP1 gene correlate well with the current enterovirus serotypes which makes it possible to obtain more detailed information about the individual virus types. Often amplification of the VP1 sequences, utilizing less well-conserved primers, is not possible directly from the clinical material and requires a cell culture enrichment step. In the future, multiplex RT-PCR and microarray methods may provide us with new tools where the high sensitivity of the molecular methods can be combined with an informative collection of data about the pathogen.

There are no highly sensitive and specific standardized serological assays available for enteroviruses. Neutralization assays are serotype specific, but due to the large number of different viruses, the usage of this method is restricted to epidemiological studies. In-house methods, based on enzyme immonoassays or complement fixation tests, are widely used in diagnostic laboratories. The antigens are usually heat-treated to increase cross-reactivity and synthetic peptides containing conserved sequences are also in use. The presence of enterovirus-specific IgM antibodies or an increase in antibody titers between samples collected at the acute and convalescent phase of the illness are considered as a sign of acute infection.

Epidemiology

Transmission of echoviruses occurs through the respiratory and gastrointestinal routes and the epithelial tissue at these sites is the primary location of replication. This can be followed by viremia and infection of secondary target organs. It is possible that respiratory transmission may predominate in areas with improved hygiene. Several nosocomial outbreaks have been described in neonatal units. In addition to direct contact with infected individuals, the infection can be obtained from contaminated water and food. There is no evidence of animal reservoir for these viruses.

In many studies, echoviruses have been the most commonly isolated enteroviruses and there is an extensive number of reports of outbreaks from different geographical areas during several decades in the literature. Most frequently, the echoviruses have been isolated from children of less than 5 years of age that can be understood for immunological reasons and because of the increased risk of transmission. The occurrence of the infections in this age group does not necessarily correlate with the appearance of disease, since, for instance, meningitis may be detected more frequently in older children. Approximately half of the isolations originate from infections of the central nervous system and isolation from respiratory and gastrointestinal infections are more rarely used. One of the reasons for this is most probably that in mild infections specific diagnostic procedures are more rare. EV4, EV6, EV7, EV9, EV11, and EV30 have been among the predominating serotypes.

Enterovirus infections occur typically during the late summer and early autumn months in temperate climates while elsewhere, the viruses often appear to circulate through the year. The infections are more common in males and in individuals with lower socioeconomic status. Due to the high number of serotypes and the population immunity, there are considerable differences between the epidemiological periods, and variation from small local outbreaks to widespread epidemics is seen. There is also extensive fluctuation in the occurrence of the serotypes during the years and, for instance, echovirus 30 that has been responsible for large meningitis epidemics worldwide may sometimes represent almost half of the isolations but be absent during another epidemic season.

The epidemiological studies during the twentieth century were based on virus isolation and subsequent typing which provided information about the circulating serotypes. Since the RT-PCR assays only can give information about the presence or absence of enteroviruses in clinical samples, this epidemiological surveillance is significantly changing its character. However, introduction of molecular epidemiological analysis, based on the comparison of genomic sequences, has given new information about the circulation and evolution of the virus strains. The approach has been used in studies of, for example, echovirus 30 and the investigations have shown that there has been one predominant genotype with several distinct lineages circulating in Europe and in Northern America since 1978. These strains are clearly different from those isolated some decades ago and some genetic lineages seem to have completely disappeared during this time.

It has become clear that recombination is a frequent phenomenon in the evolution of enteroviruses. It has been mostly found to occur among the members of the species and in particular in the genome region encoding nonstructural proteins. Since the virus isolates are identified primarily on the basis of the serotype, based on the capsid region sequences, this heterogeneity of the genomes has not previously been observed in the epidemiological studies. Therefore, future studies may give a new picture about the circulating and reemerging genetic lineages. Although it is clear that the nonstructural gene region is evolving rapidly, the rules for the appearance of genetically mosaic strains which become predominant by their superior fitness and other properties remain still unclear.

The importance of these events in the generation of new pathogenic strains would be highly important to understand and molecular epidemiological approaches and studies on picornavirus evolution in general will hopefully be able to cast new light on these questions.

Future Perspectives

There are currently no vaccines available against other human picornaviruses except polioviruses and hepatitis A virus. The number of echovirus and other enterovirus serotypes makes vaccine development particularly challenging, since the protection from infection is primarily thought to be mediated by serotype-specific neutralizing antibodies. The most promising chemotherapeutic agent developed against clinical picornavirus infections so far has been pleconaril, a capsid-stabilizing small molecule. It has been successfully used in severe enterovirus infections but due to side effects observed in a treatment study against common cold it is not in general use. Other molecules, based on a similar principle, as well as inhibitors of the virus-specific proteases have also been developed, and they are likely to become tools for the treatment of enterovirus infections in the near future. Since echoviruses, with a few exceptions, do not replicate in experimental animals, testing of antiviral drugs is problematic. However, there are promising results showing that the infection can be successfully transmitted to transgenic mice expressing human receptors for an echovirus.

An interesting application in cancer therapy is the use of naturally occurring or modified viruses. For instance, adenovirus-based systems have been used for this purpose. There are also enteroviruses that have been reported to have such an effect. An example is EV1 shown to exhibit tropism for human ovarian cancer cells.

As illustrated by these few examples, increasing knowledge of echoviruses and their replication is likely to provide us with new tools for the treatment of these infections in the future. New disease associations may also be found by using the improved diagnostic methods. Development of an enterovirus vaccine is a particularly important and demanding goal. Understanding of the cell tropism and detailed multiplication mechanisms of these viruses in the body may also provide us with new tools for the development of tools against clinical illnesses.

Acknowledgments

The original studies carried out in our laboratory were supported by grants from the Academy of Finland and the Sigrid Juselius Foundation.

See also: Coxsackieviruses; Enteric Viruses; Enteroviruses: Human Enteroviruses Numbered 68 and Beyond; Enteroviruses of Animals.

Further Reading

Minor PD and Muir PA (2004) Enteroviruses. In: Zuckerman AJ, Banatwala JE, Pattison JR, Griffiths PD, and Schoub BD (eds.) *Principles and Practice of Clinical Virology*, pp. 467–489. New York: Wiley.

Pallansch MA and Roos RP (2001) Enteroviruses: Polioviruses, coxsackievirus, echoviruses, and newer enteroviruses. In: Knipe DM and Howley PM (eds.) *Fields Virology*, 4th edn., pp. 723–775. Baltimore, MD: Lippincott Williams and Wilkins.

Racaniello VR (2001) *Picornaviridae*:The viruses and their replication. In: Knipe DM and Howley PM (eds.) *Fields Virology*, 4th edn., pp. 685–722. Lippincott Wiliams & Wilkins.

Rotbart H (2002) Enteroviruses. In: Richman DD, Whitley RJ, and Hayden FG (eds.) *Clinical Virology*, pp. 971–994. Washington, DC: ASM Press.

Semler B and Wimmer E (eds.) (2002) *Molecular Biology of Picornaviruses*. Washington, DC: ASM Press.

Ecology of Viruses Infecting Bacteria

S T Abedon, The Ohio State University, Mansfield, OH, USA

© 2008 Elsevier Ltd. All rights reserved.

Glossary

Adsorption Phage diffusion and attachment to a bacterium.

Burst Release of phage from a bacterium (usually upon lysis). Burst size is the number of phage released per bacterium.

Chronic release The continuous release of phage progeny from infected bacteria in a manner that does not involve bacterial death or infection loss.

Community The species (more than one) found within a given environment.

Ecosystem The sum of the biotic (living community) and abiotic (nonliving) components of a single environment, such as a lake or a field.

Epifluorescent microscopy Microscopic observation of light reemitted from specimen-bound dyes.

Extracellular search The diffusion-mediated movement of a phage virion from the point of release from an infected bacterium through the irreversible attachment of a phage to a susceptible bacterium. The extracellular search for new bacteria to infect may be thwarted by virion decay.

Flow cytometry Technique involving detection of suspended particles (such as bacteria) as they flow through a laser beam.

Latent period The time beginning with phage adsorption to a bacterium and ending with phage release from a bacterium.

Metagenomic sequencing The isolation and sequencing of genomic nucleic acid from whole communities including bacteriophage communities such as those found within a sample volume of seawater.

Population A geographically associated group of interbreeding, potentially interbreeding, or clonally related individuals. A localized grouping of individuals of a single species.

Pseudolysogeny A latent phage infection in which penetration of the phage genome into the bacterial cytoplasm has occurred but subsequent steps towards either productive infection or lysogeny are delayed. Pseudolysogeny is thought to occur primarily when phage infect bacteria that are in a significantly starved state.

Pulsed-field gel electrophoresis A technique for gel separation of large DNA molecules such as the whole genomes of bacteriophage.

Single-step growth A technique involving adsorption of phage to bacteria that is followed by a determination of the timing of phage-progeny release from the now-infected bacteria as well as the total number of phage released per bacterium. The word 'single' (or 'one') is used to indicate that the released phages are prevented from infecting (or adsorbing to) subsequent bacteria within the experimental vessel.

Spatial structure Environmental impediments to mixing, diffusion, and organism motility such that the present spatial position of an organism serves as a good predictor of future spatial position. In the microbiology laboratory spatial structure is often imposed upon cultures via the addition of agar to growth media.

Introduction

Bacteriophages are viruses of bacteria. An enormous number of bacteriophages are found the world over, on the order of 10^{30} or more. By way of comparison, 10^{30} is approximately the weight of the Sun, in pounds. This chapter presents an overview of the interaction between bacterial viruses and their environments: phage ecology. Phage ecology is a process- or mechanism-centered exploration of how bacteriophages exist in the wild and of their impact there.

Phage Existence within Environments

Phage populations have been documented in numerous environments. With the availability of modern molecular, microscopic, and flow-cytometric techniques, it is relatively easy to obtain direct counts or to molecularly characterize uncultured phage virions obtained from environmental samples. A variety of techniques may also be employed to study viral infection/production (aka, proliferation) within environments, though none of these methods are ideal. Estimations of lysogeny in environments are possible but similarly are not ideal.

Determination of phage numbers by viable (typically plaque; **Figure 1**) count originally dominated phage environmental endeavors. Direct microscopic counts, however, are generally much higher than viable counts, for example, by as much as 10 000-fold. For isolation of viable phage from environmental samples, selective enrichment techniques are often employed. Representative phage (more generally, virus) numbers by direct counts range from 1×10^4 to 2×10^7 per ml seawater. Assuming virion degradation between sampling and enumeration in the above estimations, these numbers are estimated to be as high as $(3-100) \times 10^6$ per ml. Sediment and terrestrial counts can be even higher than phage numbers in open water.

Especially using epifluorescence microscopy, direct counts are accomplished more easily than is either phenotypically characterizing environmental phage or demonstrating phage environmental impact on specific bacteria. Partially making up for these deficiencies, the various molecular approaches – including pulsed-field gel electrophoresis and metagenomic sequencing of phage environmental populations – allow characterization of phage environmental diversity, especially at the level of genotype. While electron microscopy of environmental samples does allow some phenotypic characterization, particularly of virion morphology, none of these approaches are substitutes for phenotypic analysis of individual phage in at least 'phage' pure cultures. In most cases such characterization is not possible because the majority of presumptive phage hosts cannot be cultured in the laboratory.

Phage Ecology

The science of ecology may be differentiated into a number of more-or-less distinct subdivisions. These include organismal, population, community, ecosystem, and landscape ecologies. Also included, especially under the heading of organismal ecology, are physiological and evolutionary ecologies. Though important to an understanding of ecology in general, in most cases little effort is made within the phage ecology literature to distinguish among these various categories, plus there exist numerous overlaps among these categories. Indeed, phage biologists (including this author) often are trained first as microbiologists (or molecular geneticists, etc.) and only later or much less so as ecologists. As a consequence, phage publications relevant to our understanding of phage ecology typically will not describe their contents from a classically ecological perspective. A major goal of this entry, therefore, will be to expose readers to phage ecological thinking as organized within a more general ecological framework, presented in overview as follows.

Phage organismal ecology considers phage adaptations within environments, with emphasis on phage survival, reproduction, and dissemination. Phage physiological ecology considers the impact of environments on phenotypes, which for phage would be manifested either

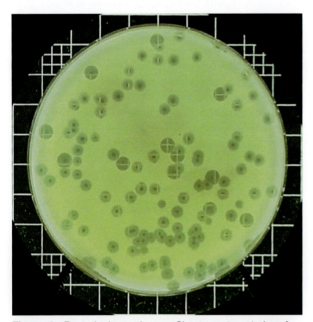

Figure 1 Bacteriophage plaques. Shown are two strains of phage RB69 as grown on the host bacterium *Escherichia coli* CR63. The larger plaques are made by a mutant, called sta5, which displays a significantly shorter latent period than the smaller-plaqued wild type. To accentuate the plaque-size differential, incubations were done at room temperature for 2 days.

through modifications in the physiology of bacterial hosts or in the environmental impact on virion properties. Phage evolutionary ecology considers the selective benefits of phage adaptations especially from an evolutionary perspective.

Phage population ecology considers phage intraspecific interactions, particularly in terms of how those interactions control phage population size, dispersion, and growth rates. Phage community ecology considers phage interactions with other species, especially with bacteria but also with other types of phages as well as with biotic components of the greater bacterial environments, for example, such as in terms of phage-encoded exotoxins which impact the eukaryotic hosts of bacteria. Phage ecosystem ecology considers phage impact on nutrient cycling and energy flow within ecosystems, which by necessity represents a much more abiotic emphasis than other approaches. Phage landscape ecology, which has been little considered at least explicitly within the phage literature, would consider the phage impact on interactions between distinct ecosystems, such as between a pond and a surrounding forest, a sewage treatment plant and downstream environments, or an animal (serving as an ecosystem itself) and the extra-animal environment. Note that these various means of looking at phage ecology build upon one another such that ecosystem ecology builds upon community ecology which builds upon population ecology which builds upon organismal ecology which, finally, is understood best in light of a strong molecular appreciation of phage biology.

Phage Organismal Ecology

Phage organismal ecology may be considered from the perspective of environmental impacts on phage 'growth' parameters, as well as the impact of these parameters on phage ecological success as measured in terms of reproduction, dissemination, and survival. 'Growth' parameters include things such as adsorption rates, infection timing, productive-infection versus lysogeny decisions, burst sizes, and virion decay rates. Since our understanding of phage growth parameters is derived from the laboratory characterization of phages, in this section the author describes this aspect of phage ecology employing a very much laboratory-centered emphasis.

The traditional means of studying most phage growth parameters are single-step growth experiments and phage adsorption determinations. We can fit these growth parameters into a generalized scheme of the phage life cycle, and then consider how variation in the assorted steps of this life cycle can impact phage ecological success. These steps, in order, include (**Figure 2**) (1) a diffusion-driven phage-virion extracellular search for bacteria to infect, (2) the phage adsorption process during which a relatively

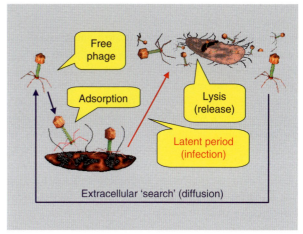

Figure 2 Lytic bacteriophage life history. Images © James A. Sullivan, used with permission.

inert phage-virion and adsorption-susceptible bacterium are together converted into a phage-infected bacterium, (3) the phage infection which can vary in its phenotypic expression from restricted to abortive to productive to pseudolysogenic (or 'carrier state') to lysogenic, and (4) some means of phage-virion release into the extracellular environment. The latter, depending on the phage, may or may not involve host-cell and infection death (i.e., via lysis versus chronic/continuous release, respectively).

Consideration of how variation in these assorted steps can impact phage ecological success is presented in the following sections. For example, we can expect rapid virion adsorption – particularly rapid phage attachment given encounter with a phage-susceptible bacterium – to be favored over less-rapid adsorption unless phenotypes conferring rapid adsorption interfere with phage survival or fecundity. For instance, T4-like phages may enhance their survival within extra-colonic environments by taking on a temporarily adsorption-incompetent but inactivation-resistant state. Another counter to the evolution of more-rapid phage adsorption/attachment may be phage growth within relatively spatially structured environments, if there exist tradeoffs between time spent disseminating versus time spent infecting bacteria. Note that these considerations of phage adsorption rates more generally reflect differences between phage properties 'as virions' versus 'as during bacterial infection', differences which are explored more fully in the following section.

Bacteria-Like versus Virion-Like Modes of Existence

The selective advantage associated with various phage infection and release strategies, similar to phage adsorption rates, will be dependent on environmental conditions, particularly in terms of bacteria numbers, bacteria physiology, and phage decay rates. In general, we can assume that conditions that favor bacterial growth and survival over phage population growth and survival will tend to favor more bacteria-like modes of phage replication/survival over more phage/virion-like modes; more precisely, we can expect life history biases toward phage existence infecting a bacterium rather than existence as a free virion. Thus, low bacterial densities, bacterial physiologies that poorly support significant phage fecundity, or virion-specific environmental antagonists (including phage-restricting bacteria) will tend to favor extended phage latent periods, once commitment to adsorption has occurred. These infecting states may include pseudolysogeny, lysogeny, or productive but longer-lasting infections, including infections by phage that release their progeny chronically rather than lytically. Alternatively, we can assume that conditions that favor phage population growth will tend to favor more phage-like modes of replication, that is, ones that emphasize the virion state over the infected state. This more virion-emphasizing state can include shorter, productive phage infection – especially lytic infections that favor earlier virion release over higher infection fecundity, but also more-rapid chronic release at the expense of longer-term infection survival.

The advantage of displaying a more phage-like existence under these latter conditions may be viewed from two perspectives. The first is that such conditions should allow for rapid phage-population expansion. Less intuitively, that expansion comes at the expense of bacteria survival, and therefore any phage which persists within a more bacteria-like (extended latent period) mode of infection may be more susceptible to attack by unrelated phages under conditions which are favorable toward phage population expansion (see, e.g., 'Kill the winner', below). Also favoring the virion over the infecting state, a temperate phage can increase its target number and reduce its genomic target size via lysogen induction (i.e., by changing from a single bacterial lysogen to multiple copies of phage). Such induction can occur following bacterial exposure to DNA damaging agents. In considering phage population growth and survival, we thus can envisage phage adaptation within real ecosystems as involving especially a compromise between conflicting tendencies toward greater emphasis on existence as an infection versus greater emphasis toward existence as a virion.

Phage Population Ecology

Considerations of phage population ecology may be differentiated into two categories: (1) the impact of phage adaptations and environmental conditions on phage population growth within well-mixed, fluid environments (such as within broth in the laboratory) and (2) the impact of phage adaptations and environmental conditions on phage population growth within spatially structured environments (such as within soft-agar overlays in the laboratory). These categories will be considerd in turn, emphasizing phage population growth in broth-like environments. Like phage organismal ecology, much of our understanding of phage population ecology derives from laboratory characterization of phage, but also from exercises in modeling phage population growth. We can also distinguish phage population considerations into those operating at low phage densities versus those operating at high phage densities. In both cases density refers to a single phage population since inter-(as opposed to intra-) species interactions are considered under the province of phage community rather than phage population ecology. These latter considerations are much simpler to envisage during phage broth-culture growth.

The growth of phage in broth can be considered to occur over at least four distinct steps. These are as follows:

1. Phage entrance into an environment. This entrance can be physical, in terms of phage movement from one place to another, it can be physiological in terms of the induction of an already-present lysogen, or it can be genetical in terms of phage host-range mutation (or recombination event) such that a previously phage-insusceptible bacterial population suddenly becomes susceptible to sympatric phage.
2. A period of phage-population exponential growth that spans from the point of the initial phage infection of a susceptible bacterium within an environment through to the point of transition whereby most susceptible bacteria have become phage infected.
3. The transition to and then span over which most of the phage-susceptible bacteria within a population are phage infected.
4. A postinfection period during which phage numbers greatly outnumber those of bacteria.

In many cases, each of these steps is at best a transient phenomenon for a given combination of phage, bacterium, and environment. To keep things simple, during this initial discussion assume the least complex of environments: one that is homogeneous, well-mixed, closed (e.g., not a continuous culture), and of finite volume. That is, the equivalent of a shaken or stirred broth-filled flask. Also for the sake of simplicity, assume that bacteria do not replicate nor change physiologically, other than by phage infection, over the course of phage population growth.

In general terms, once a phage virion has been released from an infected bacterium, it has entered into a period of what is described as an extracellular search for new bacteria to infect. The duration of this search will be a function of the product of the density of phage susceptible bacteria within a given environment – which in turn is a function of the range of bacterial types that a given phage may infect – and the phage adsorption constant, which will vary as a function of phage, bacterium, and environmental parameters. The likelihood that a given phage will successfully adsorb to and infect a bacterium is additionally a function of a phage's decay rate, with higher decay rates reducing the likelihood of successful virion infection. Decay can be a function of abiotic insults that result in virion capsid or nucleic acid damage, phage adsorption to abiotic materials, biotic insults including engulfment by eukaryotes as well as restriction (i.e., death or inactivation) mediated by phage-adsorbed bacteria, or even emigration out of an environment. To a phage entering a previously phage-free environment, the likelihood that phage population growth will occur therefore is a function of bacterial numbers, rates of phage adsorption per bacterium, and rates of phage decay.

If a successful adsorption occurs, the likelihood that phage population growth will continue beyond this first adsorption will additionally be a function of phage burst size, with each additional phage produced per infection essentially possessing an additional, nearly identical likelihood of successful bacterial adsorption. Extending this point, we can assume with either higher bacterial numbers, more-effective/rapid virion adsorption per bacterium, lower rates of decay, or higher initial phage numbers that there will be a greater likelihood that phage population growth will be initiated upon phage entrance into a new environment. The potential for a phage population to grow within a given environment can be described as its existence conditions.

Once phage population growth has begun, it seems unlikely that there would exist any intrinsic inhibitions on phage evolution toward faster population growth. This is not to say that there may not be blocks extrinsic to phage population growth, including various tradeoffs that could exist, such as potential conflicts between, on the one hand, the breadth of a phage's host range (more generally, niche) and, on the other hand, rates of phage population growth on any given host population (which would be a conflict between ecological generalization vs. specialization). However, both in terms of intraspecific and interspecific competition among phages, it is the phage that reaches a bacterium first that should have the greatest potential to 'claim' that bacterium for infection. Thus, a phage that grows its population fastest will have access to more bacteria sooner, and therefore enjoy among its progeny greater overall numbers of bacterial infections within a given environment. In general, we can expect faster overall population growth given greater phage burst sizes, shorter phage latent periods, or faster phage adsorption. Phage decay can also affect rates of phage population growth, and this occurs via reductions in effective phage burst size, where the qualifier 'effective' refers to the number of phage released from a given bacterial cell which go on to successfully infect a new bacterium.

Things change, within a given environment, as a phage population reaches the limits of its population growth. At this point the selective pressures on phage growth change from ones that are phage-density independent to ones which are phage-density dependent. These phage-density-dependent limits vary in their relevance, however, depending on bacterial densities along with phage fecundities. At one extreme, bacterial densities are so low, or phage fecundities (effective burst sizes) so small, that phage populations never reach densities of sufficient size to greatly affect bacterial populations. Under these conditions multiple phage adsorption to a given bacterium is of low likelihood and competition between phage for individual bacteria small. Selection here should be biased toward greater virion or infection durability, for example, rather than higher rates of phage population growth. We have this expectation because bacterial rarity presumably would put a premium on phage survival prior to bacterium encounter and also because phage potential to migrate to new bacteria-containing environments could be a function of longer-term phage survival, either as virions or as bacterial infections.

At the other extreme, bacterial densities as well as phage fecundities may be sufficiently high that a majority of phage-susceptible bacteria may become phage infected and therefore no longer available for further infection by the same phage type. Prior to this point we may expect selection to favor phage that display rapid population growth, even if such rapidity should come at the expense of effective burst size, either in terms of actual burst size or in terms of virion resistance to decay. Thus, at higher bacterial densities a phage could display faster population growth by shortening its latent period. Such shortening, however, is most easily accomplished at the expense of the overall duration of the period during which phage-progeny mature, resulting in a reduced phage burst size. Alternatively, one could envisage a loss of phage genes that otherwise could contribute to long-term virion survival should the presence of those genes come at the expense of effective phage burst size or further latent-period shortening. One possible example of such evolution could be a common loss of temperance among phage contaminating industrial dairy ferments.

Adaptation that results in declines in phage survival could be shortsighted within environments in which densities of new bacteria to infect are in rapid decline. Instead, we can envisage an advantage bestowed upon phage infections during 'final rounds' of infection – that is, at the point where most bacteria within a population have become phage infected – that would result from larger effective burst sizes. In this way more or more robust phages are produced to better assure phage survival until access to new bacterial populations occurs. Somewhat equivalently, a phage may accrue advantage by displaying a more bacteria-like lifestyle under conditions where bacterial densities are in decline, such as via temperate-phage reduction to lysogeny rather than display of a lytic infection.

Many of these same effects take place during phage population growth within spatially structured environments, such as within bacterial lawns immobilized within soft agar, that is, as phage plaques (**Figure 1**) or as within soil. A difference, though, is that in addition to temporal differentiation between density- and density-dependent selection, there are also spatial differences. Thus, phage multiplicity is higher toward the center of plaques versus toward their periphery. In other words, at a plaque's leading edge, selection acting upon phage characteristics is more density independent than toward the center of plaques. Thus, we can expect selection for faster phage population growth to occur at a plaque's periphery, whereas, by contrast, selection at a plaque's center should be biased more toward a greater effective burst size.

Phage Community Ecology

Phage community ecology includes issues of phage predation on bacteria as well as the establishment of mutualistic-like relationships such as those between prophage and bacteria, forming lysogens. The phage community ecology literature is more closely aligned with the ecological literature, which is to say that a good deal of phage community ecology has been done by individuals trained first as ecologists who have then taken to phage as model systems rather than by individuals trained first by studying more organismal or more molecular aspects of phage biology. As such, phage community ecology has a much more theoretical basis than either phage organismal or population biologies, including the employment of simulations of phage–bacterial interactions, especially within continuous cultures (chemostats). Phage community ecology has a much more environmental emphasis than these other ecologies, as workers consider, especially, the phage impact on bacterial populations. Phage community ecology can also involve consideration of interactions between different phage species or grazing of phage by protists.

We can distill three general governing principles of phage community ecology, especially with regard to phage impact on bacteria: (1) The greater a phage population density then, all else held constant, the greater the phage impact on a bacterial population, and the faster that phage populations reach these higher densities, then the sooner their impact. (2) The greater the density of phage-sensitive bacteria, then, potential changes in bacterial physiology aside, the greater the population size that phage can grow to and the sooner they will grow to that population size. (3) Bacteria can hide in various ways from phage – numerically, genetically, physiologically, spatially – but doing so can come at a cost in terms of, for example, bacterial growth rates.

Real world applications of these concepts include that of 'Kill the winner' where it is expected that populations of same or similar bacterial types, that grow to higher densities, will tend to become more susceptible to phage-mediated population crashes. These crashes result solely from bacteria having reached these higher densities. By culling especially more successful bacterial populations, phage may serve as mediators of a frequency-dependent selection among bacterial communities that results in greater bacterial diversity than environments could otherwise sustain.

A second and related real-world consideration is that bacteria within especially eutrophic (i.e., high-nutrient) environments, where phage virions can be especially abundant, may display physiologies that represent trade-offs between rates of bacteria growth and phage-free survival. That is, bacteria may be subject to physiological and other burdens that come with deploying protective measures. Protective measures are potentially diverse and certainly include things such as restriction-modification systems. Additionally, bacteria may take on more-protective modes of existence such as within extracellular polysaccharides or biofilms.

Bacteria may also display reduced versatility because of loss (temporarily or permanently; fully or partially) of membrane proteins involved, for example, in nutrient transport, proteins which could otherwise serve as phage receptors. That is, the more proteins (or other molecules) a bacterium displays on its surface, the more phage types that bacterium may be susceptible to. Because many of these mechanisms of phage resistance can result in reduced bacterial fitness, we should expect that the strength of selection for phage resistance should be directly proportional to phage density within a given environment. Phage, in turn, can display adaptations that serve to overcome various bacterium-mediated mechanisms of phage resistance. Bacteria can also evolve to mitigate the costs associated with displaying phage resistance.

A third real-world consequence of phage impact on bacteria is as a consequence of phage-mediated transduction of bacterial genetic material. Phage do this by two basic mechanisms: generalized transduction in which bacterial DNA is incorporated into consequently defective phage virions (with essentially equal probability of incorporation across the bacterial chromosome) and specialized transduction, which traditionally involves the incorporation of the bacterial genomic DNA that flanks prophage as a consequence of imprecise prophage excision upon induction. A third means, essentially an extension of the concept of specialization transduction, is via phage 'morons', which involves the incorporation via illegitimate recombination of bacterial genes into various locations within temperate phage, which are then passed

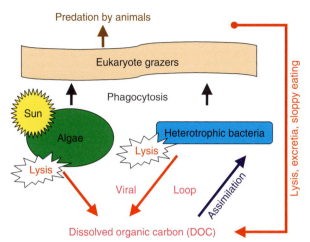

Figure 3 Microbial loop as short circuited by virus-(particularly phage-) induced lysis. The viral loop consists of generation of dissolved organic carbon via virus-induced lysis of especially heterotrophic bacteria and cyanobacteria (the latter here included under the heading, Algae), which is followed by assimilation of a fraction of that material into heterotrophic bacteria. The viral loop reduces the efficiencies by which photosynthetically fixed organic carbon is transferred to higher trophic levels, such as to aquatic animals, via the microbial loop.

on to bacteria via subsequent establishment of lysogeny. These genes, along with prophage as a whole, can impact bacterial phenotype including the fitness of lysogens, the latter at least some of the time for the better. These latter mechanisms of transduction serve as a means of transferring genetic material from bacterium to bacterium while also transferring genetic material from bacterium to phage.

Phage Ecosystem Ecology

Phage ecosystem ecology is more firmly rooted in considerations of phage environmental interactions than in the theory of phage community ecology. The primary emphasis of this field is on the impact of phage infection of prokaryotes on the movement of nutrients and energy from prokaryotes to higher trophic levels. That is, phage infection influences the acquisition of bacteria (including cyanobacteria) by bacteria-eating eukaryotes. The real-world significance of this emphasis derives from prokaryotes serving as the base of many, especially aquatic, ecosystems and the role that these bacteria play in carbon sequestration. That is, within aquatic environments most dissolved organic material becomes available to eukaryote grazers only once bacteria have assimilated it. With phage-induced bacterial lysis, however, a significant fraction, up to approximately 25% of the carbon entering these ecosystems photosynthetically, is converted to a dissolved organic state. This creates a 'viral loop' that shunts organic carbon away from what is known as the aquatic 'microbial loop', which otherwise moves carbon from prokaryotes to grazers (i.e., especially phagotrophic protists) to grazer-consuming animals (**Figure 3**). Another way of stating the significance of the phage impact on the survival of aquatic bacteria is that it is approximately equivalent to the impact of protist grazing. The impact of phage infection versus protist grazing, however, varies with habitat, with phage displaying a greater impact especially in more extreme, including low-oxygen environments.

See also: Epstein–Barr Virus: General Features; Virus Evolution: Bacterial Viruses.

Further Reading

Abedon ST (2006) Phage ecology. In: Calendar R and Abedon ST (eds.) *The Bacteriophages*, pp. 37–64. Oxford: Oxford University Press.

Abedon ST (2007) *Bacteriophage Ecology: Population Growth, Evolution, and Impact of Bacterial Viruses*. Cambridge, UK: Cambridge University Press.

Abedon ST and LeJeune JT (2005) Why bacteriophage encode exotoxins and other virulence factors. *Evolutionary Bioinformatics Online* 1: 97–110.

Abedon ST and Yin J (2008) Bacteriophage plaques: Theory and analysis. In: Clokie M and Kropinski A (eds.) *Bacteriophages: Methods and Protocols*, Totowa, NJ: Humana Press.

Ackermann H-W and DuBow MS (1987) *Viruses of Prokaryotes*. Boca Raton, FL: CRC Press.

Breitbart M, Rohwer F, and Abedon ST (2005) Phage ecology and bacterial pathogenesis. In: Waldor MK, Friedman DI, and Adhya SL (eds.) *Phages: Their Role in Bacterial Pathogenesis and Biotechnology*, pp. 66–91. Washington, DC: ASM Press.

Brüssow H and Kutter E (2005) Phage ecology. In: Kutter E and Sulakvelidze A (eds.) *Bacteriophages: Biology and Application*, pp. 129–164. Boca Raton, FL: CRC Press.

Carlson K (2005) Working with bacteriophages: Common techniques and methodological approaches. In: Kutter E and Sulakvelidze A (eds.) *Bacteriophages: Biology and Application*, pp. 437–494. Boca Raton, FL: CRC Press.

Chibani-Chennoufi S, Bruttin A, Dillmann ML, et al. (2004) Phage-host interaction: An ecological perspective. *Journal of Bacteriology* 186: 3677–3686.

Goyal SM, Gerba CP, and Bitton G (1987) *Phage Ecology*. Boca Raton, FL: CRC Press.

Lenski RE (1988) Dynamics of interactions between bacteria and virulent bacteriophage. *Advances in Microbial Ecology* 10: 1–44.

Levin BR and Lenski RE (1985) Bacteria and phage: A model system for the study of the ecology and co-evolution of hosts and parasites. In: Rollinson D and Anderson RM (eds.) *Ecology and Genetics of Host-Parasite Interactions*, pp. 227–242. London: Academic Press.

Paul JH and Kellogg CA (2000) Ecology of bacteriophages in nature. In: Hurst CJ (ed.) *Viral Ecology*, pp. 211–246. San Diego: Academic Press.

Weinbauer MG (2004) Ecology of Prokaryotic Viruses. *FEMS Microbiological Reviews* 28: 127–181.

Wommack KE and Colwell RR (2000) Virioplankton: Viruses in aquatic ecosystems. *Microbiology and Molecular Biology Reviews* 64: 69–114.

Relevant Website

http://www.cellsalive.com – Cells alive.

Electron Microscopy of Viruses

G Schoehn and R W H Ruigrok, CNRS, Grenoble, France

© 2008 Elsevier Ltd. All rights reserved.

Introduction

Most viruses are too small to be observed with light microscopy and many are too large or too irregular to crystallize. Therefore, electron microscopy (EM) is the method of choice for the direct visualization of viruses and viral proteins or subviral particles. However, electrons do not travel far in air and therefore the inside of the microscope has to be under very high vacuum (10^{-5} to 10^{-8} torr). This implies that biological samples have to be dehydrated and/or fixed. It is important that the biological sample is stabilized (or fixed) so that its ultrastructure remains as close as possible to that in the biologically active material when exposed to the vacuum.

The penetrating power of electrons in biological material is also very limited, which means that the specimens must either be very thin or must be sliced into thin sections (50–100 nm) to allow electrons to pass through. This is not a real problem for the imaging of viruses because the sizes of viruses are usually in this range but if we want to study viruses interacting with cells during infection or replication, sectioning is necessary.

Contrast in the transmission electron microscope basically depends on the atomic number of the atoms in the specimen; the higher the atomic number, the more electrons are scattered and the larger the contrast. Biological molecules are composed of atoms of very low atomic number (carbon, hydrogen, nitrogen, phosphorus, and sulfur). Staining methods using heavy metal salts or image analysis have to be used to enhance the contrast in electron micrographs.

In order to make this article easier to read we will illustrate the various EM-preparation techniques with images of intact adenovirus particles, individual capsid proteins or protein complexes from adenoviruses. Adenoviruses are double-stranded DNA viruses with an icosahedral capsid with a pseudo T number of 25 and a total protein mass of around 125 MDa. The major capsid components are the hexon, a trimeric protein with a basal hexagonal shape, and the penton that is a noncovalent complex between the pentameric penton base and the trimeric fiber protein. Two hundred and forty hexons form the 20 facets of the icosahedron, whereas the pentons form and project from the 12 vertices. The fiber binds to host cell receptors with its C-terminal knob domain whereas protein loops that extend from the penton base and that contain an Arg-Gly-Asp (RGD) sequence are required to bind to a secondary cell receptor to trigger endocytosis of the virus. The capsid also contains minor components, proteins IIIa, VI, VIII and IX that glue the hexons and pentons together at specific positions in the capsid. For a schematic drawing of the virus, see **Figure 1(a)**.

Techniques for Single Particle Imaging

Before any sample preparation takes place, one should be informed about the nature of the sample: its molecular weight, some information on its physical size, and the chemical nature of the components (proteins, nucleic

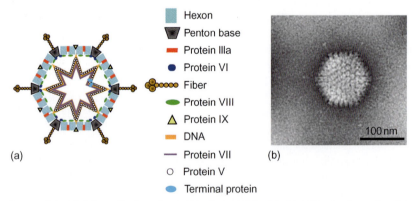

Figure 1 The adenovirus particle. (a) Schematic view of an adenovirus particle. A schematic view through a sliced adenovirus showing the different structural proteins of the virus. The adenovirus particle is an icosahedral particle with on each vertex a noncovalent complex between a trimeric fibre and a pentameric base. This complex is called the penton. (b) Electron micrograph of an adenovirus particle negatively stained with a 1% solution of ammonium molybdate pH 7.5. The particle is seen in the same orientation as in (a). The approximate molecular weight of an adenovirus is 150 MDa and the size of the particle is 1000 Å in diameter.

acids, lipid membranes). Further, it is important to verify by sodium dodecyl sulfate polyacrylamide gel electrophoresis (SDS PAGE) whether the protein(s) in the sample is (are) intact and not totally or partially degraded. Information on possible contamination of protein samples with nucleic acids can be obtained by measuring the optical density spectrum at 220–350 nm. Such a spectrum can also give information on whether a protein sample contains aggregated protein since a significant absorption at wavelengths above 300 nm is then observed. Such controls are important because the imaging techniques that are described here will always lead to a picture. However, one wants to know how the picture relates to the active biological sample.

Negative Staining

For negative staining the sample is placed on a support film, bathed in a solution of a heavy metal salt that is subsequently blotted, and the sample is then left to dry in air (**Figure 2**). The stain will dry around the molecules, and the area where the molecules happen to be will appear light and the space around it dark, that is, opaque for electrons. As support films one can use plastic films, commercially available or home-made, but these films tend to become hydrophobic after some time and need to be cleaned first with a plasma cleaner. Plastic films are mechanically resistant but are often rather thick and give a high background noise in the images. Thin carbon films give better results. These films are evaporated onto the surface of freshly cleaved mica by heating carbon at high temperatures under vacuum.

For sample preparation it is important to know the concentration of the sample. For small molecules like isolated capsomer proteins, a concentration between 0.05 and 0.1 mg ml^{-1} often gives the best results. When the concentration is too low the stain will not be spread out evenly on the support film and if the concentration is too high the molecules will be one on top of the other and the picture will be difficult to interpret. For intact virus particles such as adenoviruses and influenza viruses, a concentration of 1 mg ml^{-1} should be used. Using a drawn-out glass Pasteur pipette (the sample is taken up into the pipette by capillary force) 1–2 μl of the sample is touched at the carbon–mica interface of a small piece of precut carbon on mica (**Figure 2(a)**). The surface of the carbon on the outside is hydrophobic but at the interface the carbon is clean and hydrophilic and the sample is sucked in between the carbon and mica layers. Then the carbon film with the adsorbed sample is floated off in a pool of negative-stain (usually uranyl acetate (UA), ammonium molybdate, or sodium silicotungstate, SST) and a copper EM grid is placed on the top of the floating film. The grid plus carbon film is then picked up by placing a small piece of paper on the grid and, as the paper wets, it is lifted up taking grid and carbon film up from the pool of stain. In this way the edges of the carbon film that stick out over the rim of the copper grid do not fold back onto the bottom of the grid, creating carbon layers at both sides of the grid with an inclusion of stain. The paper-grid-carbon-sample sandwich is placed on filter paper to blot and air-dry for at least 5 min before insertion into the column of an electron microscope (**Figure 2(a)**).

Different heavy atom salts that are available commercially can be used for negative staining. However, chemical reactions between the salt and the specimen are possible. When the salt binds specifically to the sample molecules one may obtain positive staining rather than negative staining. With positive staining one does not obtain an image of a clear sample molecule on a dark background but a dark molecule on a clear background. It should also be taken into consideration that some salts for staining might have non-neutral pH values. Finally, stain molecules have defined sizes and the resolution of the image may be limited by the grain size of the stain molecules.

Uranyl acetate has a low pH (pH 4.4) and is known to cross-link and stabilize fragile macromolecular assemblies. This stain is especially useful for enveloped viruses since the binding of uranyl ions to the negatively charged lipid heads stabilizes the membranes. This stain does not stay close to the support film but will also stain the top of virus particles so that one obtains a superimposed image of the surface features that touch the support film plus the features on the opposite side of the virus particle. It is not such a good stain for thin objects that lie flat on the support film because the contrast is reduced by the stain lying on top of the object. In **Figure 2** the molecules in plate C (right) were stained with uranyl acetate. Note that uranyl acetate, because of its ability to chemically react with proteins, lipids, and nucleic acids, is known to cause positive staining as well as negative staining. One serious problem with this salt is that it is no longer available in some countries since even nonenriched uranium may be considered to be of military importance. Another problem is that it is slightly radioactive. Sodium silicotungstate is a chemically inert stain with a neutral pH. The stain stays very close to the support film and outlines only those features of a virus or protein that are in contact with the film and it is therefore often used to stain small proteins. The size of the stain molecule is rather large (i.e., 10 Å in diameter) and this feature may limit the resolution of image reconstructions. The stain is often sold as silicotungstic acid. The acid needs to be solubilized in water, neutralized with NaOH and the salt is precipitated with cold ethanol. In **Figure 2** plates (b), (c) (left), and (d) were stained with SST. This stain allows visualization (and sometimes determination of their oligomeric state) of proteins as small as 60 kDa (**Figure 2(d)**).

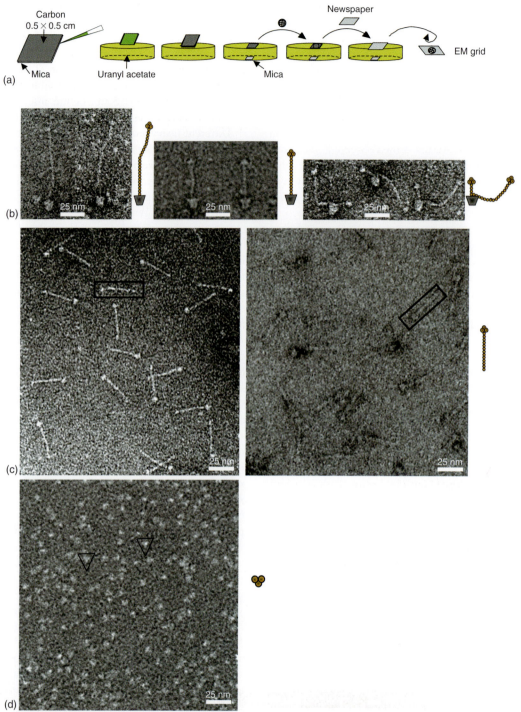

Figure 2 Negative staining technique and results. (a) The different steps of the negative staining method: the sample is adsorbed onto the carbon support film by touching a pipette with the sample to the mica and carbon interface. Then the sample/carbon is floated off the mica onto the surface of a solution of a heavy atom salt. An EM grid is placed on the surface of the carbon and then a piece of paper is used to pick up the grid with carbon film. The paper with grid, carbon, and sample is blotted on filter paper and the grid is then air-dried before observation in the microscope. (b) Electron micrographs of different adenovirus pentons negatively stained with sodium silicotungstate. From left to right one can see a bovine adenovirus 3 penton base (very long and bent fiber), a human adenovirus 41 penton with the long fiber, and an avian adenovirus penton (one base with two different fibers sticking out). The penton base is around 300 kDa is size, whereas the fiber is between 120 and 350 kDa. (c) Electron micrograph of an isolated adenovirus 2 fiber stained with sodium silicotungstate (left) and uranyl acetate (right). In each case, a single fiber is highlighted by a black rectangle. The molecular weight of one fiber is 190 kDa. (d) Electron micrograph of a recombinant canine adenovirus fiber-head expressed in *Escherichia coli*, stained with sodium silicotungstate. The molecular weight of the trimeric particle is around 60 kDa and it is sometimes possible to recognize the triangular shape of the particle (black triangles).

Phosphotungstic acid (PTA) can be used at pH values between 5 and 7. The staining characteristics are similar to those of SST although PTA often does not spread out well during air-drying and can remain in thick puddles. The stain in these puddles is so thick that protein structures cannot be visualized. This stain can be rather destabilizing for lipid membranes and enveloped viruses often show membrane blebs that are squeezed out of the regular lipid envelope during air-drying.

Ammonium molybdate has a neutral pH. During drying, the stain stays quite high on the support film. It is therefore especially useful for the staining of large particles (**Figure 1(b)**). This stain can be used for cryo-negative staining when very high stain concentrations (16%) are used. The preparation of the grids is as described below for regular cryo-microscopy. The advantage over regular cryo-microscopy is the much higher contrast. The disadvantage is that of all negative staining techniques: only the outside of the particles is visualized by negative staining without any information on the structures inside the virus particles.

Advantages and limitations

With the negative staining technique it is possible to visualize small proteins and sometimes determine their oligomeric state in a very rapid manner. The main limitation of this technique results from the air-drying step. If the sample is fragile, which is the case with enveloped viruses, some flattening can occur during drying and lipid blebs can be squeezed out. Finally, the molecular forces that act upon proteins, complexes, or viruses during the adsorption phase onto the support film can also induce conformational changes.

Cryo-Electron Microscopy

In order to overcome the adsorption step, stain artifacts and, most importantly, the drying step, Dr. Jacques Dubochet and collaborators developed a new sample preparation technique: the frozen-hydrated sample imaging or cryo-EM (**Figure 3**). For this technique the sample concentration must be much higher than that for negative staining because there is no direct support film that has a

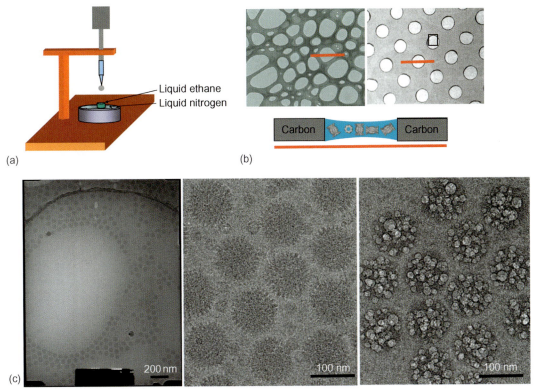

Figure 3 Cryo-electron microscopy (EM) technique and results. (a) Plunger used for rapid freezing of the sample. A grid covered with a holey carbon film is loaded with the sample just before blotting and plunging it very rapidly into liquid ethane cooled down by liquid nitrogen. (b) Examples of home-made holey carbon (left) and commercial quantifoil grid (right). The red bar represents a slice through a hole showing the sample in the frozen vitrified water (bottom). The rectangle on the quantifoil grid represents the portion of field indicated by the red bar above. (c) Cryo-EM images of adenovirus particles. The sample was prepared on a quantifoil film. The picture on the left part is an entire micrograph showing the repartition of the adenovirus particles according to the ice thickness (one part of the hole is totally empty due to the fact that the ice is too thin. The middle part shows a higher magnification of the cryo-EM micrograph. Details are visible on the virus particle; the smaller particles are adenovirus-associated viruses. On the right, the same adenovirus sample is shown after an 1 min exposure to the electron beam. Irradiation damage (bubbles on the picture) is clearly visible.

concentrating effect. The support film for the sample in solution is a carbon film with holes, either home-made (rather irregular, **Figure 3(b)**, left film) or commercially obtained (these films have very regular, evenly spaced holes, **Figure 3(b)**, right film). An aliquot of 2–4 μl of purified sample at a concentration of 0.5–1 mg ml^{-1} is pipetted onto a copper grid covered by the carbon film. Excess liquid is removed by brief blotting with a piece of blotting paper and the grid is then plunged into liquid ethane cooled by a liquid nitrogen bath (**Figure 3(a)**). The rate at which the temperature of the buffer in the holes is lowered is so high that there is no time for the formation of ice crystals; the water forms vitreous ice (also called amorphous ice). Because of the absence of ice crystals, the sample is not deformed but stays in its native state and the vitreous ice is translucent for electrons. The frozen grids have to be transferred into a cryo-holder and inserted into the microscope. The temperature of the sample has to stay very low (i.e., under $-160\,°C$) in order to prevent the vitreous ice to transform into hexagonal or cubic ice. If this irreversible conversion occurs the structure of the sample will be destroyed. All cryo-EM work has to be done at low temperature and the microscope has to be equipped with cryo-EM tools (cryo-holder, anti contaminator blades around the specimen in the microscope in order to avoid the deposition of water and other vapors onto the sample). The ranges of salt concentration and pH that can be used are quite large and include physiological conditions.

Advantages and limitations

Because no heavy atoms are added to the sample, the contrast in the images is very low (**Figure 3(c)**) (the contrast is partly generated by the difference in density between the frozen water and the protein, i.e., 1 g/ml vs. 1.3 g ml^{-1}). This low contrast makes the use of image analysis indispensable. The combination of low contrast and noise also limits the minimum size of the protein/macromolecular assemblies that can be studied by cryo-EM to about 200 kDa. Radiation damage is also a big problem in cryo-EM. As soon as the sample is exposed to the electron beam its starts to be destroyed (**Figure 3(c)**, right). However, minimal dose systems have been developed to limit the electron dose that hits the sample by exposing it only at the nominal magnification during the photographic or charged-coupled device (CCD) camera exposure time. Basically, the electron dose has to be less than 10 e^{-}Å^{-2} in order to preserve the structure of the sample.

There have been a number of recent technological advances that have made the use of this technique easier. The blotting time of the grid is one of the important steps in sample freezing, but is also one of the less reproducible ones. The thickness of the ice in the hole has to be as close as possible to the diameter of the sample. Automated procedures for producing reproducible cryo-EM grids have been developed and commercial devices for obtaining reproducible ice thickness are now available.

The introduction of support films with regularly spaced holes in parallel with automated freezing will make automated data acquisition possible.

Cryo-Electron Tomography

Electron tomography allows the study of the 3D organization of thin individual cell organelles and bacterial cells at nanometer resolutions without slicing them. This technique is also able to reconstruct the 3D structure of nonregular viruses. In contrast to all other EM reconstruction techniques, the samples are unique objects. For tomography, a tilt-series is taken from a single object (from $-70°$ to $+70°$). CCD cameras are used for the recording of many images of the same sample without serious radiation damage. The second step is to align all the images coming from the tilt series in order to generate a 3D model of the object. Tomography holds the potential to routinely enable reconstructions that will allow the fitting of intermediate-resolution EM-structures or high-resolution X-ray-structures into the features of the tomography reconstruction.

Metal Shadowing

The angular shadowing of protein samples consists of evaporation under vacuum of a thin film of metal at an angle over a dried sample on support film (**Figure 4(a)**). The sample is mixed with glycerol and then deposited on the surface of a mica sheet. The mica is then introduced into an evaporator and dried. The sample is then tilted for evaporation. Different kinds of metal can be evaporated onto the surface of the sample (titanium, gold, platinum-paladium alloy). Subsequently a carbon film is evaporated onto the mica, which is then floated off on a water bath, much like the methods used for negative staining above, and the metal comes off the mica with the carbon. Different metals will lead to different grain sizes. Differences in sample height will lead to differences in metal deposit and, hence in contrast (see **Figure 4(b)**). This technique is the only technique that can determine the handedness of a helix and distinguish, for example, between icosahedral 7d or 7l surface lattices. Rotational shadowing is based on the same principle as angular shadowing but the sample is rotated during evaporation (a fixed angle of evaporation is kept). This technique is especially helpful for the visualization of small proteins or nucleic acids (**Figure 4(b)**). Note that the dimensions of shadowed proteins are difficult to interpret because the metal shadow will add significantly to the thickness of the protein.

Preparation of Cells or Very Large Particles for EM Observation

One of the most interesting aspects of virus studies is the analysis of virus infection, trafficking, and particle

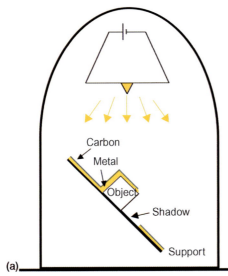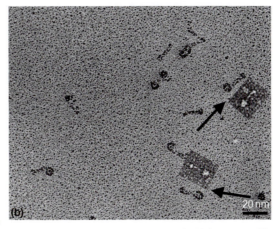

Figure 4 Principle of rotary shadowing. (a) Principle of shadowing. The shadowing process occurs under high vacuum. The specimen is in a tilted position. At the top, the metal to be evaporated is shown in yellow. It is heated until evaporation by a high-intensity current and will be deposited onto the sample. For rotating shadowing, the specimen is not fixed in a tilted position but rotated while keeping the same tilt angle. (b) Human adenovirus 3 penton base was rotatively shadowed using platinum. For comparison the same object negatively stained using sodium silicotungstate has been placed on the same figure (insets are arrowed).

formation in cells. Unfortunately, cells and tissues are generally too thick to be directly observed by EM and they first have to be sectioned.

Thin and Ultrathin Sectioning

Classical thin sectioning

The cells or tissues are first stabilized by chemical fixation (usually with aldehydes such as formaldehyde or gluteraldehyde) in order to immobilize the protein and nucleic acid components of the cell (**Figure 5**). Subsequently, fixation with osmium tetroxide will immobilize and stain the membranes of the cell. Note that all these fixatives are very toxic. Then the material is dehydrated in ethanol or acetone and embedded in plastic. Ultrathin sections (60 nm) cut with diamond knives using an ultramicrotome are floated on water, transferred to specimen support grids and examined in the transmission electron microscopy (TEM). Before observation, the sections are further contrasted with uranyl acetate and lead citrate. There are many different types of plastic resins. The hardest resins (epoxy) will best preserve the ultrastructure of the cells during sectioning but will not allow antibody labeling of sections because the antibodies have no or little access to their antigens. There exist soft or spongy resins (Lowicryl) that do permit antibody labeling but these resins are not hard enough to guarantee ultrastructure conservation.

Application: Antibody labeling

Successful immunocytochemistry is only possible when antibodies (or other affinity markers) are able to bind to their target ligands. For extracellular antigens this is not a problem, but for molecules inside cells the markers must be able to access the antigens. For EM it is important to use preparation methods that open the cells and allow introduction of antibodies without destroying the normal cellular organization. The antigens must also be preserved during the different sectioning steps.

The cryo-section technique is now one of the two most important techniques for subcellular immunocytochemistry. Specialized resin formulas have been developed specifically for use in immunocytochemistry. These include the Lowicryl resins, LR White and LR Gold, Unicryl, and MonoStep. They all have low viscosity and can be polymerized, at low temperatures, with ultraviolet light.

In practice, this method is used with electron opaque markers, which bind to the antibodies and show their location within the cell. In this way, subcellular antigens recognized by antibodies can be localized directly with TEM. Colloidal gold coupled to protein A (a protein from bacterial cell walls that binds to the Fc portion of antibodies) has been used extensively in recent years to localize antibodies on resin and frozen sections of biological materials (**Figure 5**). The ability to produce homogeneous populations of colloidal gold with different particle sizes enabled the use of these probes to co-localize different structures on the same section.

Cryo-sectioning

As for single particles, it is possible to freeze biological material fast enough to vitrify the water present inside the cells. Vitrification of water occurs when the freezing is so fast that ice crystals have no time to form. Vitrified biological material can be sectioned at low temperatures. Because no drying occurs and no salt has been added,

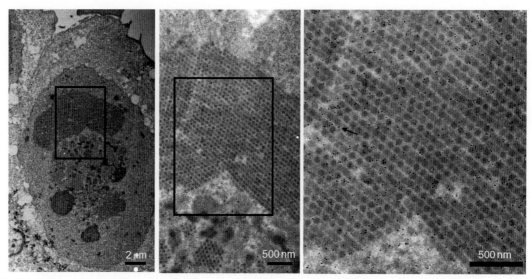

Figure 5 Thin sectioned, adenovirus infected cells. Going from left to right one can see an increase of magnification going from a general view of the infected nucleus to clearly recognizable particles. On the right, the infected cells have been labeled with anti-adenovirus-hexon (major capsid component) antibody coupled with colloidal gold (arrow). Cells were fixed with 2.5% glutaraldehyde in 100 mM HEPES buffer pH 7.4, post-fixed with 1% osmium tetraoxide and dehydrated with ethanol and then embedded in Epon prior to slicing. Grids were stained with saturated uranyl acetate in 50% ethanol and then with 1 M lead citrate.

cryo-sections better represent the native state of cells although the contrast in the sample is low. Presently, there are seven main rapid freezing methods possible:

1. Immersion freezing – the specimen is plunged into the cryogen.
2. Slam (or metal mirror) freezing – the specimen is impacted onto a polished metal (usually copper) surface cooled with liquid nitrogen or helium.
3. Cold block freezing – two cold, polished metal blocks attached to the jaws of a pair of pliers squeeze-freeze the specimen.
4. Spray freezing – a fine spray of sample in liquid suspension is shot into the cryogen (usually liquid propane).
5. Jet freezing – a jet of liquid cryogen is sprayed onto the specimen.
6. Excision freezing – a cold needle is plunged into the specimen, simultaneously freezing and dissecting the sample.
7. High-pressure freezing – freezing the specimen at very high pressure avoids the expansion of water that occurs during freezing.

CEMOVIS: Cryo-EM of vitreous sections

Dr. Jacques Dubochet in Lausanne has developed the cryo-EM of vitreous sections (CEMOVIS) technique. This technique which consists of cutting very thin sections of a vitrified sample requires good vitrification of the sample and good cutting. Progress in these two domains has been of crucial importance. The first one is high-pressure freezing, which made it possible to increase the thickness of a typical biological sample that can be vitrified from 10 to 100 μm. The second progress came from a better understanding of the cutting process which allowed optimization of the cutting. It is now possible to obtain reproducibly ribbons of homogeneous, 50 nm thick sections in which biological structures are remarkably preserved. It is also possible to observe for the first time, at a resolution approaching 2 nm, the ultrastructure of cells and tissues in their native state.

The next step using CEMOVIS is to combine it with electron tomography. Recording a large number of tilted images (typically 100) should allow the calculation of a 3D model of the specimen to a resolution approaching 4 nm. This is a real challenge for studying virus replication.

Image Analysis

As mentioned above, in tomography, about a hundred images of the same particle are combined in a computer in order to calculate a 3D reconstruction of a single particle. The resolution obtained is presently around 30 Å. However, computer analysis of images is also needed for cryo-EM images because the contrast in cryo-EM is too low. The images of thousands of particles can be combined to obtain a high-contrast model of the particle provided that all particles have the same composition and conformation. The resolution of models derived with this method can be as good as 7 Å for large protein complexes or intact virus particles. A large number of mathematical approaches are available to calculate such 3D reconstructions. Different approaches exist to calculate models of nonsymmetrical complexes, for helical structures, and for icosahedral virus capsid structures. These techniques requires classical EM photographs to be digitized by scanning the negatives or the images have to be directly recorded in the microscope using a CCD camera.

Figure 6 shows 3D reconstructions of adenovirus type 5 particles from negatively stained particles and from cryo-EM images. The two models are identical but the cryo-model has a higher resolution (around 10 Å compared to 35 Å for the negative stain model). Because the contrast in the negatively stained images is high, such a model from negatively stained particles can be used as starting model for the cryo-reconstruction. The major difference between the two models is that in the model derived from the stained particles only the outside of the virus capsid is visualized, whereas for the cryo-model all proteins inside and outside of the capsid are visible. The cryo-EM model of the capsid was combined with the atomic resolution data derived from the crystal structure of the two main adenovirus capsomers, the trimeric hexon

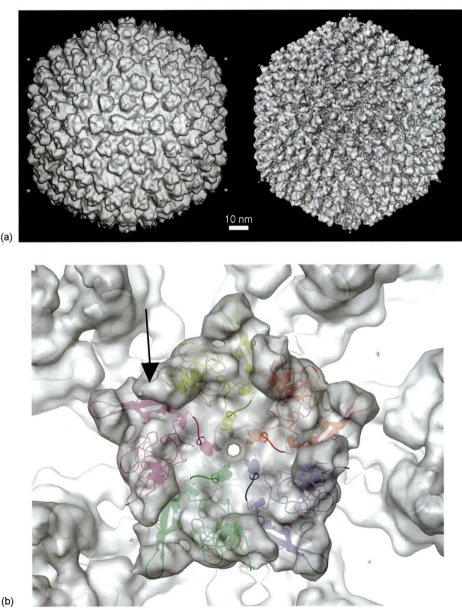

Figure 6 Image analysis and combination with other techniques (a) 3D reconstruction of adenovirus 5 particle after imaging in negative staining (left) and cryo-EM (right). Respectively 150 and 3000 individual negatively stained or cryo-EM images were combined to obtain 3D reconstructions at 35 Å (negative staining) and 10 Å (cryo-EM) resolution. (b) Combination of EM with X-ray crystallography. Fitting of the X-ray structure of the adenovirus penton base into the EM density. The arrow highlights a mobile part of the protein that is resolved in electron microscopy but not in X-ray crystallography.

and the pentameric penton base. By fitting the crystal structures into the EM density, a quasi-atomic model can be derived that describes the exact position of all capsomers and the interacting surfaces and protein loops between the capsomers. **Figure 6** shows details of the fit of the crystal structure of the penton base into a penton base from the EM model. A number of different algorithms are available to perform these fits, some of which take into account the possibility of flexibility between protein domains. The fit in **Figure 6** (arrow) shows density in the EM model that is not filled by density from the crystal structure. Some loops of the molecule are flexible and therefore invisible in the crystal structure, whereas the lower resolution of the EM model does allow imaging.

The Future

In our opinion, the future of EM in virology lies in an extended use of the combination of structural data as shown above. Fitting crystal structures of intact proteins or domains into protein complexes (viral complexes or complexes between viral and host proteins) will show how proteins interact in the infected cell. Although the EM models have a resolution of only around 10 Å, the information on the interface of the proteins in the complex is of near-atomic resolution. These techniques can be applied to regular viruses. For irregular viruses, such as influenza and paramyxoviruses, tomography will allow the construction of models with a resolution of around 40 Å that will show how the proteins in the virus particle interact. In particular, it will be interesting to see the packaging of the nucleocapsids in these viruses. At present we do not know how the DNA is packaged in adenovirus particles and it is possible that the icosahedral averaging techniques result in a loss of this information if the packing symmetry of the DNA is in fact not icosahedral. The same problem exists for capsid features that are not icosahedral such as unique capsid vertices that may be used for the packaging of DNA in adenoviruses and herpesviruses. Such problems may be overcome by using reconstruction techniques that are not based on icosahedral symmetry.

See also: Virus Particle Structure: Nonenveloped Viruses; Virus Particle Structure: Principles.

Further Reading

Al-Amoudi A, Dubochet J, and Studer D (2002) Amorphous solid water produced by cryosectioning of crystalline ice at 113 K. *Journal of Microscopy* 207: 146–153.

Al-Amoudi A, Chang JJ, Leforestier A, *et al.* (2004) Cryo-electron microscopy of vitreous sections. *EMBO Journal* 23: 3583–3588.

Chiu W, Baker ML, Jiang W, Dougherty M, and Schmid MF (2005) Electron cryomicroscopy of biological machines at subnanometer resolution. *Structure* 13: 363–372.

Dubochet J, Adrian M, Chang JJ, *et al.* (1988) Cryo-electron microscopy of vitrified specimens. *Quarterly Reviews of Biophysics* 21: 129–228.

Fabry CM, Rosa-Calatrava M, Conway JF, *et al.* (2005) A quasi-atomic model of human adenovirus type 5 capsid. *EMBO Journal* 24: 1645–1654.

Griffiths G, McDowall A, Back R, and Dubochet J (1984) On the preparation of cryosections for immunocytochemistry. *Journal of Ultrastructure Research* 89: 65–78.

Grunewald K and Cyrklaff M (2006) Structure of complex viruses and virus-infected cells by electron cryo tomography. *Current Opinion in Microbiology* 9(4): 437–442.

Harris JR (1997) *Royal Microscopy Society Handbook 35: Negative Staining and Cryoelectron Microscopy*. Oxford: Bios Scientific Publishers.

Sartori N, Richter K, and Dubochet J (1993) Vitrification depth can be increased more than 10-fold by high pressure freezing. *Journal of Microscopy* 172: 55–61.

Tokuyasu KT (1986) Application of cryoultramicrotomy to immunocytochemistry. *Journal of Microscopy* 143: 139–149.

Tokuyasu KT and Singer SJ (1976) Improved procedures for immunoferritin labeling of ultrathin frozen sections. *Journal of Cell Biology* 71(3): 894–906.

Wepf R, Amrein M, Burkli U, and Gross H (1991) Platinum/iridium/carbon: A high-resolution shadowing material for TEM, STM and SEM of biological macromolecular structures. *Journal of Microscopy* 163: 51–64.

Emerging and Reemerging Virus Diseases of Plants

G P Martelli and D Gallitelli, Università degli Studi and Istituto di Virologia Vegetale del CNR, Bari, Italy

© 2008 Elsevier Ltd. All rights reserved.

Introduction

A number of RNA plant viruses constitute new threats to economically relevant vegetable crops, grapevine, and citrus. These are listed among new disease-causing agents, and are therefore denoted 'emerging or re-emerging' being often new viruses or virus strains responsible for serious diseases. The emerging plant viruses and their known vectors are quickly expanding in new areas as a consequence of the increasing international trade and the large numbers of people traveling in many countries. Once introduced, the fitness of these viruses to new agroclimatic conditions increases, leading to the development of severe epiphytotics.

Vegetable Viruses

Pepino Mosaic Virus

Geographical distribution
Pepino mosaic virus (PepMV) was originally described in Peru on pepino (*Solanum muricatum*) and found to infect tomato and related wild species symptomlessly, in experimental trials. Since 1999, PepMV outbreaks have been reported almost simultaneously in many European countries (Austria, Bulgaria, Finland, France, Germany, Hungary, Italy, Netherlands, Norway, Poland, Slovakia, Spain, Sweden, Switzerland, Ukraine, and UK) where it is considered an emerging pathogen of tomato glasshouses. PepMV was also found in Canada, USA (Arizona, California, Colorado, Florida, Oklahoma, Texas), Ecuador, and Chile. After discovering the PepMV occurrence in tomato grown in Europe and North America, a survey carried out in central and southern Peru and Ecuador demonstrated that the virus was present in Peruvian tomato crops as well as in the following wild *Lycopersicon* species: *L. chilense*, *L. chmielewskii*, *L. parviflorum*, *L. peruvianum*, and in *L. pimpinellifolium* in Ecuador.

Disease symptoms and yield losses
In pepino, the virus causes yellow mosaic in young leaves, whereas affected tomato plants show a wide array of symptoms. These include stunting of the whole plant, bubbling of the leaf surface, interveinal chlorosis, mosaic and green striations on stem and sepals. Foliar symptoms resemble hormonal herbicide damage, while lower leaves show necrotic lesions that resemble damage caused by water that dripped onto the plant. Fruits from early infected tomato plants may show blotchy ripening and gold marbling. Ripe fruits develop yellow speckles and spots that make them unmarketable. Observations by Dutch scientists indicate that symptoms are more readily seen during autumn and winter months, and are masked during warmer months. Yield losses caused by PepMV are probably below 5%, although surveys in glasshouse-grown tomatoes reported crop reductions up to 40%. The disease spreads very rapidly and crop losses may be significant if early action is not taken to eliminate infected sources. Although Peruvian virus strains seem to be little virulent, new strains of higher virulence could arise through recombination. Because of the adverse effects on quality and yield, PepMV is becoming one of the most important tomato pathogens in Europe.

Causal agent and classification
PepMV (genus *Potexvirus*, family *Flexiviridae*) has helically constructed semirigid filamentous particles with a modal length of 508 nm and a diameter of 11 nm. Sequencing of a number of strains has shown that virions contain a single molecule of linear, positive-sense, single-stranded RNA (ssRNA) 6410–6425 nt in size. The genome organization is typical of the genus *Potexvirus*, with five open reading frames (ORFs): ORF1, encoding a putative replicase of 164 kDa; ORFs 2–4, coding for the triple gene block proteins 1–3 (TGBp) of 26 kDa, 14 kDa, and 9 kDa, respectively; and ORF5, encoding the 25 kDa coat protein (CP). Phylogenetic analyses carried out on replicase, TGBp1, and CP amino acid sequences revealed that PepMV is closely related to narcissus mosaic virus (NMV), scallion virus X (SVX), cymbidium mosaic virus (CymMV), and potato aucuba mosaic virus (PAMV) (**Figure 1**). Tomato isolates of PepMV can be distinguished from the original isolate from pepino on the basis of symptomatology, host range, and sequence data. On the other hand, tomato isolates from Europe appear very similar to each other sharing nucleotide sequence identities higher than 99% and in the range of 95–96% with PepMV from *L. peruvianum* from Peru. Nucleotide sequence comparisons between US and the European tomato isolates show only 79–82% identity. Gene-for-gene comparison between sequenced isolates suggests that TGBp1 and TGBp3 are more suitable than either the replicase or CP gene products for discriminating virus isolates.

Epidemiology
PepMV is transmitted by contact and readily spread mechanically by contaminated hands, tools, shoes, clothing, and plant-to-plant contact. The virus is thought to remain viable in dry plant material for as long as 3 months. Seed transmission or surface seed contamination in tomato is suspected but not demonstrated. In the UK, the virus is frequently found in imported fruits. Infection of other solanaceous crops such as eggplant, tobacco, and potato has only been observed in experimental trials, while infection in pepper has not been demonstrated. Cucumber can be artificially infected but the virus does not appear to spread systemically in the plant.

Alternative hosts that may serve as virus reservoirs were studied in Spain, where native plants with virus-like symptoms growing in or around tomato fields were collected and analyzed for the presence of PepMV. As many as 18 weed species were found to be infected.

Because of the high similarity of tomato virus isolates, a common origin seems likely although where this origin is located is still unclear. The extent of PepMV distribution in Peru, together with the fact that many of the wild Peruvian *Lycopersicon* populations sampled were isolated and had not been manipulated by man, led to the conclusion that this virus has been present in the region for a

Figure 1 (a) Strong symptoms of PepMV on tomato; (b) symptom of CYSDV on melon; (c) severe vein yellowing on cucumber infected by CVYV; (d) swelling at the graft union and death of the scion of a vine grafted on Kober 5BB, infected by *Grapevine leafroll-associated virus 2*; (e) a citrus grove affected by sudden death disease; and (f) a citrus plant killed by sudden death disease. (a) Courtesy of Dr. E. Moriones. (b) Courtesy of Dr. E. Moriones. (c) Courtesy of Dr. I. M. Cuadrado. (e, f) Courtesy of Dr. J. Bovè.

long time and that other factors might be involved in its spread. *Myzus persicae* does not transmit PepMV. In southern Spain a faster spread of PepMV was observed in bumble-bee pollinated greenhouses than in those with no pollinator insects.

Control

Recommended control strategies for PepMV focus on sanitation. Use of certified seed lots, complete removal (including roots) of infected plants, limited access to affected rows, and sanitation of clothing and tools are all critical. The increasing concern caused by this new disease has led the European Commission to authorize member countries to take measures to prevent the spread of PepMV within the European Union.

Criniviruses

Criniviruses are whitefly-transmitted RNA viruses that cause plant diseases with increasingly important yield impact, consequent to the sudden explosion of whitefly populations in temperate regions in the last 10–20 years. All these viruses induce yellowing symptoms in their hosts, are generally phloem-limited, nonmechanically transmissible, and have large ssRNA genomes. Cucurbit yellow stunting disorder virus (CYSDV), tomato infectious chlorosis virus (TICV), and tomato chlorosis virus (ToCV) are important emerging criniviruses.

Cucurbit Yellow Stunting Disorder Virus

Geographical distribution

CYSDV was first observed in the southeastern coast of Spain, on melon and cucumber grown under plastic. The virus was also found in the Canary Islands, Egypt, France, Israel, Jordan, Lebanon, Mexico, Morocco, Portugal, Saudi Arabia, Syria, Texas, Turkey, and the United Arab Emirates.

Disease symptoms and yield losses

Protected crops of cucumber and melon show severe yellowing that starts as interveinal mottle of older leaves to develop into complete yellowing of the leaf lamina, except for the veins, followed by rolling, brittleness, and stunting. On the whole, symptoms are virtually indistinguishable from those caused by beet pseudo yellows virus (BPYV). Zucchini can also be infected but symptoms have not been described. Disease incidence in protected crops can easily reach 100%. In Spain and many other countries, CYSDV is considered the prevailing virus of protected cucurbit crops.

Causal agent and classification
CYSDV (genus *Crinivirus*, family *Closteroviridae*) has a narrow host range limited to the family *Cucurbitaceae* and is confined to phloem tissues. Virions are flexuous filaments with lengths between 750 and 800 nm. Genome consists of two molecules of ssRNA of plus-sense polarity designated RNA-1 and -2. RNA-1 is 9123–9126 nt long and contains five ORFs with papain-like protease, methyltransferase, RNA helicase, and RNA-dependent RNA polymerase domains in the first two overlapping ORFs, a small 5 kDa hydrophobic protein, and two further downstream ORFs potentially encoding proteins 25 and 22 kDa in size, respectively. RNA-2 is 7976 nt long and contains the hallmark ORFs of the family *Closteroviridae*, encoding, in the order, a heat shock protein 70 (HSP70) homolog, a 59 kDa protein, the major (CP) and the minor (CPm) CP. In the 3′-terminal region, RNA-1 contains an ORF potentially encoding a protein of 25 kDa which has no homologs in any databases, and RNA-2 has an unusually long 59 noncoding region. Subgenomic RNAs were detected in CYSDV-infected plants, suggesting that they serve for the expression of internal ORFs. CYSDV can be divided into two divergent groups of isolates. One group is composed of isolates from Spain, Lebanon, Jordan, Turkey, and North America, the other of isolates from Saudi Arabia. Nucleotide identity between isolates of the same group is greater than 99%, whereas identity between groups is about 90%.

Epidemiology
Natural hosts of CYSDV are restricted to cucurbits: watermelon, melon, cucumber, and zucchini. *Cucurbita maxima* and *Lactuca sativa* are experimental host plants. The life cycle of CYSDV is dependent on its vector, the whitefly *Bemisia tabaci*, as viral outbreaks are associated with heavy infestations of whitefly biotypes A, B, and Q. Transmission of CYSDV by biotype B is more effcient than by biotype A, whereas biotype Q transmits as efficiently as biotype B. *Trialeurodes vaporariorum* was displaced *by B. tabaci* as the dominant whitefly along the southeastern coast of Spain when CYSDV took over BPYV as the agent of yellowing diseases of cucurbits. Acquisition periods of 18 h or more and inoculation periods of 24 h or more seem necessary for high transmission rates of CYSDV, which can persist for at least 9 days in the vector. The virus is not known to be seed-borne.

Control
Control of CYSDV consists in controlling *B. tabaci*, and on eliminating infection sources. Chemical control of *B. tabaci* has not been effective in preventing the spread and in reducing the incidence of the disease because the vector has a wide host range and quickly develops resistance to most of existing insecticides. Roguing infected plants and weeds that can act as hosts for the vector and removal of overwintering crops prior to the emergence of adult whiteflies may prove useful. This helps if applied over large areas and where there is no continuous cropping in glasshouses, which are the sites of whitefly survival and the source of virus spread throughout the year. Growing plants under physical barriers such as low-mesh tunnels may also have a positive effect. No resistant cultivars are currently available commercially but experimental evidence for delayed viral infection and decreased symptom severity in accessions of *Cucumis sativus* has recently been obtained.

Tomato Infectious Chlorosis Virus

Geographical distribution
TICV infections have been reported from California, France, Greece, Indonesia, Italy, Japan, North Carolina, Spain, and Taiwan. This virus may also be present in the Czech Republic but the record has not been confirmed.

Disease symptoms and yield losses
Symptomatic tomato plants in open field and greenhouses exhibit interveinal yellowing in older leaves, followed by generalized yellowing. Symptoms can be confused with nutritional disorders (i.e., magnesium deficiency), pesticide toxicity, or natural senescence as older leaves may also turn red. Necrosis and occasional upward rolling of the leaves have also been reported. On the whole, infected plants are less vigorous and with fruits that may show delayed ripening. There is no information on the effects of TICV on artichoke and lettuce crops or ornamental species known to be natural hosts. Disease incidence may vary from one or few plants to severe outbreaks, depending on the abundance of whitefly populations. In California and Greece, disease incidence between 80% and 100% was reported.

Causal agent and classification
The natural hosts of TICV (genus *Crinivirus*, family *Closterovidae*) include members of the families *Chenopodiaceae*, *Compositae*, *Ranuncolaceae*, and *Solanaceae* where the virus is confined to phloem tissue. Virions are flexuos filaments with lengths between 750 and 800 nm. Genome consists of two molecules of ssRNA of positive polarity, denoted RNA-1 and RNA-2, whose nucleotide sequence has partially been determined. Both RNAs contain the hallmark ORFs of the family *Closteroviridae*.

Epidemiology
TICV is spread by the whitefly *T. vaporariorum* but not by *B. tabaci* and it is not mechanically transmitted. Besides

tomato, natural infections have been detected in *Callistephus chinensis*, *Chenopodium album C. murale*, *Cynara cardunculus*, lettuce, *Physalis ixocarpa* (tomatillo), globe artichoke, *Nicotiana glauca*, *Petunia hybrida*, *Picris echioides*, *Ranunculus* sp., and *Zinnia elegans*.

Control

TICV has the potential to cause significant losses to tomato and other naturally susceptible crops if it becomes established. Control largely depends on the efficacy of treatments and strategies to limit populations of *T. vaporariorum*. Tomato seedlings for planting should come from disease-free stocks. The general management strategies outlined for CYSDV can also be adopted for TICV.

Tomato Chlorosis Virus

Geographical distribution

ToCV infections have been reported from the US (Colorado, Connecticut, Florida, Louisiana), Europe (France, Greece, Italy, Portugal, Spain), Morocco, Israel, Puerto Rico, South Africa, and Taiwan.

Disease symptoms and yield losses

Symptoms are very similar to those induced by TICV, that is, progressive yellowing of the whole plant. Apparently, there are no estimates of yield losses, although since its discovery, the virus represents a serious problem for tomato production in many parts of the world.

Causal agent and classification

ToCV (genus *Crinivirus*, family *Closteroviridae*) virions are flexuous filaments encapsidating two molecules of positive-sense ssRNA denoted RNA-1 and RNA-2, whose complete nucleotide sequence has been determined. RNA-1 consists of 8595 nt organized into four ORFs and encodes replication-associated proteins. RNA-2 is 8244 nt long and putatively encodes nine ORFs comprising in the order, the HSP70 homolog, a 59 kDa protein, CP, and CPm, which may be involved in determining the unique, broad vector transmissibility of the virus. Phylogenetically, ToCV is closely related to sweet potato chlorotic stunt virus (SPCSV) and CYSDV.

Epidemiology

ToCV is transmitted by *T. vaporariorum*, *B. tabaci* biotypes A and B, and *T. abutilonea*. Besides tomato, the natural host range includes pepper, *Datura stramonium*, *Physalis wrightii*, *Solanum nigrum*, and *Z. elegans*.

Control

ToCV has the potential to cause significant losses to tomato if it becomes established. Control depends on the efficacy of treatments and strategies to limit populations of its broad range of vectors. The management strategies outlined for CYSDV apply also for ToCV.

Cucumber Vein Yellowing Virus

Geographical distribution

Cucumber vein yellowing virus (CVYV) has been recorded form Greece, Israel, Jordan, Portugal, Spain, Sudan, and Turkey.

Disease symptoms and yield losses

In cucumber, CVYV causes pronounced vein clearing deformation of the leaves followed by a generalized chlorosis and necrosis of affected plants. Fruits show light to dark green mottling. Nonparthenocarpic cucumbers are symptomless carriers of CVYV while parthenocarpic cucumbers develop severe symptoms. Stunting was also observed in cucumber and melon and sudden death in protected melons in Portugal. In watermelon, symptoms are often light or not expressed, but splitting of the fruits has occasionally been observed. In zucchini, symptoms vary from chlorotic mottling to vein yellowing of the leaves, but symptomless infections have also been recorded. Symptoms severity may be increased by synergistic reactions with other viruses. Yield losses have not been quantified. CVYV could present a threat to cucurbits grown outdoors or under glasshouses, although its actual impact in the complex pathosystem affecting cucurbit crops in the Mediterranean and subtropical regions has not been determined.

Causal agent and classification

CVYV (genus *Ipomovirus*, family *Potyviridae*) has filamentous particles and is transmitted by *B. tabaci* but, unlike other whitefly-transmitted viruses, it can also be transmitted by mechanical inoculation. CVYV genome consists of a single molecule of plus-sense ssRNA of 9751 nt (ALM 23 isolate) containing most hallmarks of the genome of members of the family *Potyviridae*. The absence of a coding region for the helper component-proteinase seems to be a distinctive trait of CVYV. Two CVYV strains have been recognized from Israel and Jordan, that is, CVYV-Is and CVYV-Jor. The two strains induce similar vein-clearing symptoms in cucumber and melon, but CVYV-Jor infections in cucumber are more severe. CVYV is more closely related to *Sweet potato mild mottle virus* (SPMMV) than any other species in the family *Potyviridae*.

Epidemiology

CVYV naturally infects cucumber, melon, watermelon squash, and zucchini and several weed species are also natural hosts of the virus (e.g., *Ecballium elaterium*, *Convolvulus arvensis*, *Malva parviflora*, *Sonchus oleraceus*, *Sonchus asper*, and *Sonchus tenerrimus*). All experimental hosts belong to the family *Cucurbitaceae* and include *Cucurbita moschata*, *Cucurbita foetidissima*, and *Citrullus colocynthis*. CVYV is semipersistent in its white-fly vector (*B. tabaci*) which retains the virus for less than 6 h. Therefore, individuals moving to nonhost plants may not remain viruliferous long enough to transmit the virus. *Aphis gossypii* and *M. persicae* are not vectors. Whether CVYV is seedborne has not been determined.

Control

CVYV management is mainly based on the use of virus-free stock plants as well as of *B. tabaci* populations as already outlined for CYSDV. Several detection methods based on molecular hybridization and polymerase chain reaction (PCR) have been developed and used for screening *Cucurbitaceae* germplasm that can show some degree of resistance to CVYV.

Graft Incompatibility in Grapevines

Geographical Distribution

Cases of graft union disorders have been documented from Europe ('Kober 5BB incompatibility', 'Syrah decline'), California, New Zealand, Italy, Australia, and Chile ('Young vine decline'), and again California ('Roostock stem lesions', 'Necrotic union', and 'Stem necrosis'), but are likely to occur also in other viticultural countries.

Disease Symptoms and Yield Losses

The increased use of grapevine clonal material is revealing unprecedented and widespread conditions of generalized decline that develop dramatically in certain scion–rootstock combinations. Newly planted vines grow weakly, shoots are short, leaves are small sized, with margins more or less extensively rolled downwards, and the vegetation is stunted. The canopy shows autumn colors off season so that leaves turn reddish in red-berried varieties or yellow in white-berried varieties much earlier than normal. A prominent swelling forms at the scion/roostock junction ('Kober 5BB imcompatibility', 'Young vine decline') sometimes accompanied by necrosis at the graft union ('Necrotic union'), and variously extended necrotic lesions develop on the roostock stem ('Roostock stem lesion', 'Stem necrosis'). Severely affected vines decline and may die within 1 or 2 years. Syrah decline is a severe disease characterized by early reddening of the leaves and swellings with grooves and deep cracks at the graft union. Appearance of graft union disorders depends more on the rootstock than on the scion. For instance, European grape varieties grafted on tolerant rootstocks (e.g., Freedom, Harmony, Salt creek, 03916, 101-14) exhibit a green canopy and perform well, whereas varieties grafted on susceptible roostocks (e.g., Kober 5BB, 5C, 1103P, 3309) develop a discolored canopy, decline, and may die.

Causal Agents and Classification

An ordinary strain of grapevine leafroll-associated virus 2 (GLRaV-2) is consistently associated with Kober 5BB incompatibility in Europe, appears to be involved in California's young vine decline, and was detected in diseased Chilean and Argentinian grapes. GLRaV-2, a definitive member of the genus *Closterovirus* (family *Closteroviridae*), has flexuous filamentous particles *c.* 1600 nm long an RNA genome 15 528 nt in size made up of nine ORFs. A virus originally detected in cv. Redglobe in California called grapevine rootstock stem lesion-associated virus (GRSLaV) proved to be a molecular and biological variant of GLRaV-2 (GLRaV-2 RG). Other molecular variants of GLRaV-2 were reported from New Zealand, Chile, and Australia in association with young vine decline conditions. Based on the differential responses of a panel of 18 rootstocks, up to five different graft-transmissible agents inducing incompatibility were detected in California. Of these, only GLRaV-2 RG, the putative agent of roostock stem lesion, was identified. The agents of necrotic union and stem necrosis are unknown. Equally unkown is the agent of Syrah decline, although there is circumstantial evidence that grapevine rupestris stem pitting-associated virus (GRSPaV) may have a bearing on its aetiology. GRSPaV, a definitive member of the genus *Foveavirus* (family *Flexiviridae*), has filamentous particles *c.* 730 nm in length, a genome 8726 nt in size comprising five or six ORFs, and occurs in nature as a family of molecular variants.

Transmission

GLRaV-2 and GRSPaV have no known vectors, but GRSPaV is pollen- and seed-borne. Infected propagative material is the major means for dissemination of both viruses.

Control

Use of certified virus-free scionwood and rootstocks is recommended. Currently known graft incompatibility agents can be eliminated with reasonable efficiency by heat therapy, meristem tip culture, or a combination of the two. If scionwood is infected, the use of sensitive rootstocks is to be avoided.

Sudden Death of Citrus

Geographical Distribution

Citrus sudden death (CSD) has only been reported from the State of São Paulo in Brazil, where it has already killed about 1 million trees. It has the potentiality for spreading to other Brazilian States and neighboring countries.

Disease Symptoms and Yield Losses

CSD is a destructive disease first observed in 1999 in Brazil on sweet orange and mandarin trees grafted on Rangpur lime (*Citrus limonia*) or Volkamer lemon (*Citrus volkameriana*). Outward symptoms of infected trees and modifications of the bark anatomy at the bud union resemble very much those elicited by citrus tristeza, the major difference being that CSD develops on trees grafted on tristeza-resistant rootstocks. Symptoms of CSD are characterized by a generalized discoloration of the leaves which are pale green initially, then turn yellowish and abscise. With time, defoliation becomes more intense, the trees do not push new vegetation, and the root system decays. A rapid decline and death of the plant ensues, due also to phloem degeneration in the graft union region. The phloem of the susceptible rootstocks Rangpur lime and Volkamer lemon shows a characteristic yellow strain. Until their sudden collapse, infected trees bear a normal crop.

Causal Agent and Classification

All trees affected by CSD host a strain of CTV and, with 99.7% association, a spherical virus denoted citrus sudden death-associated virus (CSDaV). This virus is a tentative member of the genus *Marafivirus* (family *Tymoviridae*), has isometric particles about 30 mn in diameter, a ssRNA genome 6805 nt in size with a high cytosine content (37.4%), encompassing two ORFs. ORF1 codes for a large polyprotein with a predicted M_r of 240 kDa which is processed to yield the replication-associated proteins (methyltransferase, helicase, and RNA-dependent RNA polymerase), a papain-like protease, and two CP subunits 21 and 22 kDa in size, respectively. One of the CP subunits is produced from the cleavage of the C-terminus of the polyprotein, the other is translated from a subgenomic RNA. ORF2 codes for a putative protein 16 kDa in size, with some relationship with a movement protein of viruses belonging to the sister genus *Maculavirus* (family *Tymoviridae*). It has not been ultimately established whether CTV or CSDaV are the causal agents of CSD, but their consistent association with the disease suggests that both may have a bearing on its etiology.

Epidemiology

CSD is transmitted by grafting and can be disseminated with propagation material. Natural spreading is by aphids (*Toxoptera citricida*) with a temporal and spatial pattern similar to that of citrus tristeza. CSDaV was detected in *T. citricida* that had fed on infected trees and was transmitted experimentally to healthy plants. Since marafiviruses are not aphid-borne, it was hypothesized that CSDaV may use CTV as a helper virus.

Control

Production and distribution of healthy scion material to be grafted onto tolerant rootstocks (Cleopatra mandarin, Swingle citrumelo) can help restraining the spread of CSD in newly established groves. Grafting the affected trees above the graft union with seedlings of tolerant roostocks allows their recovery.

See also: Pepino Mosaic Virus; Plant Virus Diseases: Economic Aspects; Plant Virus Diseases: Fruit Trees and Grapevine; Virus Databases.

Further Reading

Boubals D (2000) Le dépérissement de la Syrah. Compte-rendu de la réunion du Groupe de Travail National. *Progrés Agricole et Viticole* 117: 137–141.

Célix A, López-Sesé A, Almarza N, Gómez-Guillamón ML, and Rodríguez-Cerezo E (1996) Characterization of cucurbit yellow stunting disorder virus, a *Bemisia tabaci*-transmitted closterovirus. *Phytopathology* 86: 1370–1376.

Duffus JE, Liu HY, and Wisler GC (1996) Tomato infectious chlorosis virus – A new clostero-like virus transmitted by *Trialeurodes vaporariorum*. *European Journal of Plant Pathology* 102: 219–226.

Greif C, Garau R, Boscia D, *et al.* (1995) The relationship of grapevine leafroll-associated virus 2 with a graft incompatibility condition of grapevines. *Phytopathologia Mediterranea* 34: 167–173.

Lecoq H, Desbiez C, Delécolle B, Cohen S, and Mansour A (2000) Cytological and molecular evidence that the whitefly-transmitted cucumber vein yellowing virus is a tentative member of the family *Potyviridae*. *Journal of General Virology* 81: 2289–2293.

Legin R and Walter B (1986) Etude de phénomènes d'incompatibilité au greffage chez la vigne. *Progrés Agricole et Viticole* 103: 279–283.

Maccheroni W, Allegria MC, Greggio CC, *et al.* (2005) Identification and genomic characterization of a new virus (*Tymoviridae* family) associated with citrus sudden death disease. *Journal of Virology* 79: 3028–3037.

Meng B and Gonsalves D (2003) Rupestris stem pitting associated virus of grapevines: Genome structure, genetic diversity, detection, and phylogenetic relationship to other plant viruses. *Current Topics in Virology* 3: 125–135.

Roman MP, Cambra M, Juares J, *et al.* (2004) Sudden death of citrus in Brazil: A graft-transmissible bud union disease. *Plant Disease* 88: 453–467.

Rubio L, Soong J, Kao J, and Falk BW (1999) Geographic distribution and molecular variation of isolates of three whitefly-borne closteroviruses of cucurbits: Lettuce infectious virus, cucurbit yellow stunting disorder virus, and beet pseudo-yellows virus. *Phytopathology* 89: 707–711.

Soler S, Prohens J, Diez MJ, and Nuez F (2002) Natural occurrence of *Pepino mosaic virus* in *Lycopersicon* species in Central and Southern Peru. *Journal of Phytopathology* 150: 49–53.

Uyemoto JK, Rowhani A, Luvisi D, and Krag R (2001) New closterovirus in 'Redglobe' grape causes decline of grafted plants. *California Agriculture* 55(4): 28–31.

Van der Vlugt RAA, Cuperus C, Vink J, Stijger ICMM, Lesemann DE, Verhoeven JTJ, and Roenhorst JW (2002) Identification and characterization of Pepino mosaic potexvirus in tomato. *Bulletin OEPP/EPPO Bulletin* 32: 503–508.

Verhoeven JTJ, van der Vlugt RAA, and Roenhorst JW (2003) High similarity between tomato isolates of *Pepino mosaic virus* suggests a common origin. *European Journal of Plant Pathology* 109: 419–425.

Wisler GC, Duffus JE, Liu H-Y, and Li RH (1998) Ecology and epidemiology of whitefly-transmitted closteroviruses. *Plant Disease* 82: 270–279.

Emerging and Reemerging Virus Diseases of Vertebrates

B W J Mahy, Centers for Disease Control and Prevention, Atlanta, GA, USA

© 2008 Elsevier Ltd. All rights reserved.

Glossary

Amplicon The product nucleic acid obtained from a polymerase chain reaction.
Bocavirus A genus of the family *Parvoviridae* containing bovine, canine, and human species.
Kaposi's sarcoma A skin tumor which frequently develops in young males infected with HIV.
Kawasaki disease A mucocutaneous lymph node syndrome that has features of a virus infection but so far no causative agent has been discovered. There are about 120 cases per 100 000 population in Japan, a sixfold higher incidence than in the USA.
picobirnavirus A virus containing a genome consisting of two segments of double-stranded RNA, 2.6 and 1.5 kb in length.
Sigmodontinae A subfamily of rodents in the family Muridae that contains over 500 species, confined to the American continent.
Vero cells A heteroploid cell line derived from the kidney of a normal African green monkey (*Cercopithecus aethiops*).

Introduction

It became apparent during the last two decades of the twentieth century that new infectious diseases were increasingly being recognized in the human and animal populations. This led to the establishment of a formal committee of the Institute of Medicine of the National Academy of Sciences, USA, who reported on their deliberations in 1992, in a report edited by Joshua Lederberg and Richard Shope. This was followed 10 years later by a second report, edited by Mark Smolinski, Margaret Hamburg, and Joshua Lederberg, which appeared in 2003. Among the factors they cited as contributing to emergence were microbial adaptation and change, human susceptibility to infection, climate and weather, changing ecosystems, economic development and land use, human demographics and behavior, technology and industry, international travel and commerce, breakdown of public health measures, poverty and social inequality, war and famine, lack of political will, and finally intent to harm.

Recognition of Emerging and Reemerging Virus Diseases

The advent of highly specific molecular techniques such as the polymerase chain reaction (PCR) in the early 1980s permitted the detection and grouping of viruses on the basis of genome nucleotide sequence analysis, and in several respects these techniques have replaced serological analyses for the characterization of viruses. Although it is still important to isolate viruses in cell culture for their complete characterization, it is now possible directly to detect viruses in diseased tissues by PCR, then, by sequencing the amplicon, to determine whether a new virus has emerged to cause the disease. In fact, many viruses which do not readily grow in cell culture can only be differentiated by sequence analysis. The papillomaviruses are an example. Their study was very difficult until the advent of sequence analysis, which now has revealed more than 100 types in humans, and many more in animals and birds. For differentiation, three virus genes (E6, E7, and L1) are sequenced, and if the combined sequence of these three genes differs by more than 10% from known papillomaviruses, the virus is considered to be a new type. Other viruses which have not been grown in cell culture include many caliciviruses, and the ubiquitous anelloviruses, such as torqueteno (TT) virus, which can be detected and sequenced in the blood of most humans and many other vertebrate species.

Hepatitis C virus was originally described as non-A, non-B hepatitis virus because of the severe disease it caused but the virus would not grow in cell culture, and eventually was detected in blood known to be infected with the virus by reverse transcription of the RNA present using random primers then expressing the resultant DNA in the bacteriophage lambda gt 11. Thousands of clones were screened using patient blood as a source of antibody before positive clones were detected, which then allowed the development of enzyme immunoassays that could detect the virus in blood and so were used to screen blood destined for transfusion, saving millions of lives worldwide. Once the complete genome of hepatitis C virus was sequenced, it became apparent that there are many different genotypes circulating in the world, with different pathogenic properties.

Nucleotide sequence analysis has also been extremely useful in tracing the origins of viruses. For example, when hantavirus pulmonary syndrome, caused by a bunyavirus of rodents, Sin Nombre virus, was initially detected in 1993 in the Four Corners region of Western USA, it was found that rodents inside a house where people had been infected carried a virus identical in sequence to virus isolated from human cases. However, rodents caught at various distances from the house had increasingly variable genome RNA sequences, providing strong evidence that these rodents, deer mice (*Peromyscus maniculatus*), were the source of the infection. Subsequently more than 30 other hantaviruses, some of which also cause hantavirus pulmonary syndrome in humans, were isolated from rodents throughout North and South America. Each new virus seems to be associated with a different genetic variant of rodent host, and all rodents that carry the virus belong to the subfamily Sigmodontinae, unique to the American continent.

A particularly powerful tool for the initial recognition of an emerging virus is the application of immunohistochemistry to diseased tissues. Provided a comprehensive collection of antibodies is available, the particular virus or related group of viruses can often be detected. For example, when Hendra virus first appeared in 1995 in Australia, causing the death of a horse trainer and 14 of his horses, antibody against the virus was sent to the Centers for Disease Control and Prevention (CDC). Then, in 1999, CDC was asked to investigate a newly emerged epidemic that had appeared in Malaysia, killing more the 100 people and causing disease in many pigs. Initially, it was suspected to be caused by a virus related to Japanese encephalitis virus, but a virus was isolated from the pigs that replicated in Vero cells, and reacted in an immunofluorescence test against the Hendra virus antiserum. This could subsequently be used on patient tissues to study the pathogenesis of the disease, and after comparison of the genome sequences of Hendra virus and Nipah virus they were found to be closely related and are now classified in the genus *Henipavirus* of the *Paramyxoviridae*.

Finally, molecular methods can be used to detect new, emerging viruses in the absence of disease in the host. In 2001, Allander and colleagues searched for RNA viruses in human respiratory secretions using random primer PCR, and discovered a hitherto unknown parvovirus with a sequence related to the bovine and canine parvoviruses, which are grouped together in the genus *Bocavirus*. The new virus was called human bocavirus, and many research groups worldwide have now confirmed the presence of the virus, particularly in pediatric samples, although it is still not certain how important this virus is in causing morbidity and mortality. Their method also amplified a human coronavirus from the respiratory samples, and when sequenced this turned out to be HKU1, a recently emerged coronavirus detected by scientists at Hong Kong University. It is possible that a systematic search of human samples using such molecular techniques might reveal more hitherto unknown human viruses.

Human Demographics and Behavior

In some cases, the emerging viruses themselves have contributed to other viruses emerging and reemerging in the population. This is especially true of human immunodeficiency virus (HIV), the cause of acquired immune deficiency syndrome (AIDS), which rapidly spread following its emergence in the early 1980s to infect more than 40 million people worldwide by the end of the twentieth century. Because of its severe effects on the immune system, the virus leads to numerous other infections in the HIV-infected population. For example, picobirnaviruses, that had been detected in fecal samples from chickens and rabbits, were difficult to detect in human fecal samples until a cohort of men with AIDS was examined, and in these humans picobirnavirus was detected for the first time. Some rare diseases have become common in persons with AIDS. For example, the human polyomavirus known as JC virus can cause the rare brain disorder known as progressive multifocal leukoencephalopathy (PML). Normally, the virus remains dormant in the kidney, but in HIV-infected individuals the HIV-encoded transactivator Tat acts as a transactivator of JCV leading to PML which progresses to death within 4 months after infection. Other important virus infections which emerge in AIDS patients are human herpesviruses (cytomegalovirus, herpes simplex viruses 1 and 2, varicella-zoster virus, and human herspesvirus 8, which causes Kaposi's sarcoma).

HIV is mainly spread through sexual activity between an infected and a noninfected person, and is most common in those who indulge in high-risk sexual behavior with multiple partners. It can also be transmitted by direct contact with infected blood, and is common in persons who indulge in intravenous drug use, particularly when needles, syringes, or equipment used to prepare drugs for injection are shared. It is therefore an example of a virus disease which is dependent on risky human behavior for its maintenance in the human population.

The ability of such new infections to spread in the population has been greatly enhanced by population growth and ease of movement as a result of rapid air travel. A dramatic recent example of this was the appearance of the coronavirus causing severe acute respiratory syndrome (SARS) in late 2002 which spread by air travel from a single infected Chinese physician who infected 12 persons in a Hong Kong hotel. These infected persons then traveled by air and spread the infection to more than 8000 individuals worldwide, 10% of whom died. The virus then apparently receded from the human population in July 2003. Only recently was it discovered that the SARS coronavirus has a natural reservoir in Chinese horseshoe bats (*Rhinolophus sinicus*). Some other species such as Himalayan palm civets and raccoon dogs from which the virus has been isolated may serve as amplification hosts.

Following the recognition of the human coronavirus SARS, research on coronaviruses intensified, and this led to the discovery in 2004 of two previously unrecognized human viruses, one found by Hong Kong University, called HKU1 virus, and another reported almost simultaneously from the Netherlands, called NL63, and from Yale University, called New Haven coronavirus. The latter viruses probably represent two isolates of the same virus species. They are clearly associated with lower respiratory tract infection in children, but initially it was claimed that New Haven coronavirus was also associated with Kawasaki disease in children. This intriguing claim was rapidly investigated and refuted by several different groups in Japan, Taiwan, and elsewhere, and the cause of Kawasaki disease, which has features resembling a virus infection, remains unknown.

Zoonotic Diseases

A majority of recent emerging virus diseases have been zoonoses (i.e., diseases transmitted from animals to humans under natural conditions). Some of the more important of these include HIV-1, which was transmitted to humans from chimpanzees in Central Africa around 1931, and HIV-2, transmitted from sooty mangabeys to humans in West Africa around 1940.

Other important recent examples are the viruses of the genus *Henipavirus*. Hendra virus was first recognized through a disease outbreak in some horse stables in Hendra, Queensland, Australia, when 14 horses and their trainer died from pulmonary disease with hemorrhagic manifestations in 1994. The reservoir of the virus was found to be in large fruit-eating bats (*Pteropus* spp.), and one year later a horse farmer 600 miles away in Mackay, Queensland, died of encephalitis from the same virus. Then, in 1999, a related virus was discovered in Malaysia following a major outbreak of respiratory disease in pigs and neurological disease in humans in their close contact. More than 100 humans died, and in a successful effort to control the disease 1.1 million pigs were slaughtered. The causative virus was isolated from a fatal human case that had lived in Nipah River Village, and so was named Nipah virus. Hendra and Nipah viruses are clearly members of the family *Paramyxoviridae*, but have been placed in a separate genus as their RNA genome is about 19 kb in length, larger than that of any other paramyxovirus.

Nipah virus, like Hendra virus, was found to have a reservoir in *Pteropus* bats, and has since been identified in fatal human disease outbreaks in India in 2003 and Bangladesh in 2004.

Other new viruses which apparently have a reservoir in fruit bats include Menangle virus, a new paramyxovirus which emerged in a commercial piggery near Sydney, Australia, to cause stillbirths and abortion in pigs. Menangle virus also caused disease in two workers in the piggery. A new virus related to Menangle virus emerged during an investigation of urine samples from *Pteropid* bats collected on Tioman island, off the coast of Malaysia, in 2001, and was named Tioman virus. During the same investigation, a new orthoreovirus was isolated from *Pteropus hypomelanus* in 1999 and called Pulau virus, and more recently a related orthoreovirus called Melaka virus was isolated from a human case of acute respiratory disease in Melaka, Malaysia. Serological studies of sera collected from human volunteers on Tioman island showed that 13% had antibodies against both Pulau and Malaka viruses.

Another important group of zoonotic diseases are rodent-borne, and caused by members of the genus *Hantavirus* of the family *Bunyaviridae*. These viruses first emerged during the Korean War of 1950–52, when thousands of UN troops developed a mysterious disease with fever, headache, hemorrhage, and renal failure with a fatality rate of 5–10%. It was more than a quarter of a century before the causative virus was isolated from field mice in Korea, and named Hantaan virus, the cause of hemorrhagic fever with renal syndrome (HFRS) in humans.

Then, in 1993, a new hantavirus emerged in the Four Corners region of Southwestern USA as the cause of a severe acute respiratory disease syndrome, with a fatality rate close to 40%, and named Sin Nombre virus. This virus was shown to be transmitted to humans by inhalation of virus present in the urine, feces, or saliva of deer mice (*Peromyscus maniculatus*). It seems likely that this disease had existed for many years, and was only recognized in 1993 because of a clustering of human cases as a result of a regional upsurge in the rodent population resulting from climatic conditions causing increased availability of rodent food. Fortunately, in most of these infections, humans appear to be a dead-end host, and transmission between humans does not occur except with the Andes virus in South America.

Rodent-borne viruses of the family *Arenaviridae* also cause a number of serious zoonotic diseases in humans. The 'Old World' arenaviruses such as Lassa fever virus have been known for some time, but still cause thousands of fatal hemorrhagic fever cases every year in West Africa. However, 'New World' arenaviruses such as Junin virus causing Argentinian hemorrhagic fever and Machupo virus causing Bolivian hemorrhagic fever have long been recognized in South America. Recently, new arenaviruses have emerged, probably as a result of deforestation, which results in rodents seeking shelter in human habitation, and brings them into closer contact with people. These viruses include Guanarito virus that causes Venezuelan hemorrhagic fever with 36% mortality rate from confirmed cases, and Sabia virus isolated in 1990 that causes Brazilian hemorrhagic fever with a high fatality rate, including two laboratory acquired cases.

Rabies is a zoonotic disease of great antiquity that has mainly been associated with carnivores, such as dogs. The virus is excreted in the saliva of infected animals, and following infection it moves through the nervous system to attack the brain, causing aggressive behavior which results in the animal biting humans and animals with which it comes into contact and thereby spreading the virus infection. Fortunately, due to early work by Louis Pasteur, a vaccine was developed that protects humans or other animals from infection, and can also be given immediately post exposure, and the domestic dog population in the developed world is vaccinated and does not pose a risk to humans. However, in some developing countries, it is not uncommon for a rabid dog to bite and infect more than 25 people before it can be put down, and worldwide there are still some 30 000 human rabies deaths per year. Using molecular sequencing techniques, it is now possible to distinguish the genotypes of rabies viruses associated with different species of host, as the virus has become adapted through frequent transmission between members of the same host species. In the USA, there are six recognized terrestrial animal genotypes, in raccoons in eastern states, skunks in north-central states, skunks in south-central states, coyotes in southern Texas, red foxes in Alaska, gray foxes in Arizona, and several genotypes associated with particular species of bat. In fact, most fatal cases of human rabies in the USA can now be traced to bats, which are often not detected when the person is bitten; so rabies is not suspected and vaccination is not undertaken until the disease has taken hold.

Ecological Factors Favoring Virus Emergence

Many important virus diseases are spread by arthropods, and exposure to new arthropods and the viruses they carry is critical to the emergence of new virus diseases. Dengue hemorrhagic fever is caused by dengue virus which is transmitted mainly by the Asian mosquito (*Aedes albopictus*), and dengue fever is one of the most rapidly emerging diseases in tropical regions of the world. There are four serotypes of dengue virus, and it seems that consecutive infections with two antigenic types can lead to the more serious disease of dengue hemorrhagic fever with shock syndrome, which, if untreated, can result in up to 50% mortality. Unfortunately, through the importation of vehicle tires containing water from Korea, the Asian mosquito was introduced into the USA, and is now present in several regions of the Southern states. It can act as a vector not only for dengue virus, but also for California encephalitis virus.

In Europe, the emergence of two important animal diseases has occurred through the movement of arthropod vectors into the Iberian Peninsula. African horse sickness virus causes a disease that can be fatal to horses, mules, and donkeys, and is transmitted by nocturnal biting flies of the genus *Culicoides*. These were introduced inadvertently into Spain, and the disease is now endemic around Madrid and regions to the south. African swine fever virus is transmitted by ticks of the genus *Ornithodorus* and it causes a fatal disease resembling classical swine fever in domestic pigs. It first emerged in Portugal and Spain in 1957, France in 1964, Italy in 1967, and Cuba in 1971. Through slaughter of infected animals, the disease was eradicated from Europe, except Sardinia, by 1995.

The most recent dramatic example of the movement of a virus vector is provided by West Nile virus, a flavivirus first isolated in Uganda in 1937. This virus uses birds as a reservoir host, and is transmitted from birds to humans and other vertebrates by mosquitoes. In 1999, cases of encephalitis in New York were found to have been caused by a strain of West Nile virus that was phylogenetically similar to a virus isolated from geese in Israel.

At the same time, many birds, especially corvids, began dying in New York State. Since the introduction in 1999, West Nile virus has become well established throughout the USA and moved north into Canada and south into the Caribbean and into Mexico. It is not known how the virus moved from Israel to the USA, but the most reasonable explanation is that it was carried in an infected mosquito or possibly an infected bird in the hold of an aircraft. Transmission by an infected human seems less likely since the titer of virus in human blood is usually too low for efficient mosquito transmission.

It is clear, nevertheless, that once it arrived in North America, West Nile virus found an extremely favorable environment with abundant avian and arthropod hosts that facilitated its spread throughout the American continent.

Prospects

The emergence of new viruses is likely to continue as viruses evolve and find new ecological niches in the human and animal population. It is noticeable that most newly recognized viruses have been RNA viruses, perhaps since RNA evolves at a faster rate than DNA, for which host cells have developed efficient proofreading enzymes. It will be important in the future to detect new viruses before they can emerge to cause disease in the population. The SARS epidemic provides an excellent example. Before the epidemic, only two human coronaviruses were known, human coronaviruses 229E and OC43. Despite the fact that serious coronavirus diseases were well known in other vertebrates, such as feline infectious peritonitis and avian infectious bronchitis virus, it was not until the SARS epidemic that research on human coronaviruses led to the discovery of three new human coronaviruses – SARS, HKU1, and NL63/New Haven.

There are other genera of viruses that cause serious disease in animals but have not been adequately investigated in humans. An example is the genus *Arterivirus*, which has members causing serious disease in horses and pigs, but has not been reported at all in humans. This could be a worthwhile area for future investigation.

Another critical factor in the future control of emerging viruses is better vector control. When mosquito control was conducted using DDT, dengue fever virus was virtually eliminated from the Americas in the 1970s, but environmental concerns led to the widespread banning of the use of DDT, so that since the 1980s there has been a considerable expansion of dengue fever in South America, with the appearance of dengue hemorrhagic fever there for the first time. There is a real need to improve mosquito control measures to control this disease. Although there are prospects for a dengue virus vaccine, this is so far not available.

Finally, one of the most important viruses that continue to emerge in different antigenic forms is influenza virus. The main reservoir of influenza viruses is in birds, and over the past century several pandemics of influenza have emerged, the most serious of which was in 1918. Pandemic strains usually arise by a process of antigenic shift, where one of the genes encoding the hemagglutinin and/or the neuraminidase of influenza virus is replaced by one from birds. New pandemics occurred in 1918 (H1N1 subtype), 1957 (H2N2 subtype), and 1968 (H3N2 subtype). Since 1968, there have been no new pandemics, but it is widely expected that another will occur. At the time of writing, there is worldwide concern that a highly pathogenic avian influenza virus (H5N1 subtype), which has caused some human infections and deaths in persons in close contact with infected birds, might mutate or recombine to generate a virus which would be highly transmissible in the human population. Plans are being developed in many countries and by the WHO to try to prepare for such an event by generating possible vaccines against such a virus and stockpiling antiviral drugs.

See also: Emerging and Reemerging Virus Diseases of Plants; Epidemiology of Human and Animal Viral Diseases.

Further Reading

Choo QL, Kuo G, Weiner A, *et al.* (1992) Identification of the major, parenteral non-A, non-B hepatitis agent (hepatitis C virus) using a recombinant cDNA approach. *Seminars on Liver Diseases* 12: 279–288.

Chua KB, Bellini WJ, Rota PA, *et al.* (2000) Nipah virus: A recently emergent deadly paramyxovirus. *Science* 288: 1432–1435.

De Villiers EM, Whitley C, and Gunst K (2005) Identification of new papillomavirus types. *Methods in Molecular Medicine* 119: 1–13.

Gratz NJ (2004) Critical review of the vector status of *Aedes albopictus*. *Medical and Veterinary Entomology* 18: 215–227.

Hayes EB and Gubler DJ (2006) West Nile virus: Epidemiology and clinical features of an emerging epidemic in the United States. *Annual Review of Medicine* 57: 181–194.

Hsu VP, Hossain MJ, Parashar UD, *et al.* (2004) Nipah virus encephalitis reemergence, Bangladesh. *Emerging Infectious Diseases* 10: 2082–2087.

Kahn JS (2006) The widening scope of coronaviruses. *Current Opinion in Pediatrics* 18: 42–47.

Korber B, Muldoon M, Theiler J, *et al.* (2000) Timing the ancestor of the HIV-1 pandemic strains. *Science* 288: 1789–1796.

Ksiazek TG, Erdman D, Goldsmith CS, *et al.* (2003) A novel coronavirus associated with severe acute respiratory syndrome. *New England Journal of Medicine* 348: 1953–1966.

Mahy BWJ and Brown CC (2000) Emerging zoonoses: Crossing the species barrier. *Revue Scientifique Et Technique Office International Des Epizooties* 19: 33–40.

Mahy BWJ and Murphy FA (2005) The emergence and re-emergence of viral diseases. In: Mahy BWJ and ter Meulen V (eds.) *Topley & Wilson's Microbiology & Microbial Infections*, 10th edn., pp. 1646–1689. London: Hodder-Arnold.

Mushahwar IK, Erker JC, Muerhoff AS, *et al.* (1999) Molecular and biophysical characterization of TT virus: Evidence for a new virus family infecting humans. *Proceedings of the National Academy of Sciences, USA* 96: 3177–3182.

Nichol ST, Spiropolou CF, Morzunov S, *et al.* (1993) Genetic identification of a hantavirus associated with an outbreak of acute respiratory illness. *Science* 262: 914–917.

Smolinski MS, Hamburg MA, and Lederberg J (eds.) (2003) *Microbial Threats to Health, Emergence, Detection and Response*, 367pp. 367pp. Washington, DC: The National Academies Press.

Stephenson I (2005) Are we ready for pandemic influenza H5N1? *Expert Review of Vaccines* 4: 151–155.

Emerging Geminiviruses

C M Fauquet and M S Nawaz-ul-Rehman, Danforth Plant Science Center, St. Louis, MO, USA

© 2008 Published by Elsevier Ltd.

Glossary

Geminivirus A class of plant viruses with a genome composed of single-stranded DNA encapsidated in geminate particles.

Pandemic An epidemic at the regional or worldwide level.

Satellite An infectious molecule that depends on a helper virus for crucial functions.

Introduction

Geminiviruses, which are circular single-stranded DNA (ssDNA) plant viruses, are divided into four different genera (*Curtovirus*, *Mastrevirus*, *Topocuvirus*, and *Begomovirus*). All geminiviruses are transmitted by insects; the most devastating begomoviruses are transmitted by whiteflies (*Bemisia tabaci*). In the last 20 years, we have witnessed in many parts of the world the emergence of newly discovered or previously recorded geminivirus diseases. These emergences are attributable to several factors, including an explosion of whitefly populations in many parts of the world, an increase in human activities, and a large development of global trade introducing new plant hosts, vectors, and viruses in different ecosystems. As there are too many geminivirus emerging diseases to be discussed here, we will restrict our description to the three most important diseases caused by begomoviruses that will illustrate the impact of the various factors influencing geminivirus emergences in the world.

The Tomato Yellow Leaf Curl Disease

Geographical Distribution

The tomato yellow leaf curl virus (TYLCV) was originally reported/isolated from Israel and was described in the 1960s, mostly infecting tomatoes and causing very severe symptoms and huge losses. In Spain and Italy, tomato production increased considerably in the 1980s, and closely related viruses were identified in these countries which were first mistakenly identified as new strains of TYLCVs, although the Israeli virus was subsequently identified in these countries. Furthermore, recombinants between local viruses and TYLCV were found and are now prevalent in several regions of both countries. TYLCV was later exported to several countries around the Mediterranean region and worldwide. In 1994, TYLCV was identified in the Caribbean islands and in 1997 was identified for the first time in the USA. Since then, the virus has been exported worldwide (**Figure 1**) and is present in the following countries: USA (California, Georgia, New Mexico, Arizona), Mexico, Puerto Rico, Dominican Republic, Cuba, Tunisia, Morocco, Egypt, Spain, Italy, France, Greece, Turkey, Sudan, India, China, Japan, and Australia.

Symptoms and Yield Losses

The symptoms induced by TYLCV are typically a leaf curling of the leaves with different levels of yellowing (**Figure 2**). However in a single field of infected tomato plants a variety of symptoms can be observed from green to purple leaf curling, thickening of the veins and the stems, and various levels of stunting. In some instances, the leaf surface is reduced a minimum, with large veins and thick lamina. When plants are infected at an early stage, the plant fail to produce fruits and will stay stunted. Losses are in general very high, and depending on the earliness of the infection and on the level of resistance of the variety can reach 100%. TYLCV is considered a major threat for farmers growing tomatoes, both for commercial or subsistence purposes.

Causal Agent with Classification

TYLCV belongs to the genus *Begomovirus*, in the family *Geminiviridae*. Begomoviruses are either monopartite and bipartite, meaning that their genome is composed of a single or two molecules of circular ssDNA and TYLCV is a monopartite virus. This feature does not explain the prevalence of TYLCV in the world, as there are many monopartite geminiviruses in the Old World that do not show the characteristics of a new virus emergence. Regarding the genome organization, TYLCV is a typical monopartite begomovirus coding for six open reading frames (ORFs). A total of five strains of TYLCV, called TYLCV-Iran, -Gezira, -mild, -Oman, and -Israel, have been identified, but it is mostly the Israeli strain that has, so far, spread all over the world.

Epidemiology

Begomoviruses are not seed transmitted, nor mechanically transmitted (with a few exception in lab conditions); they are only transmitted by the whitefly *Bemisia tabaci*. In the 1960s, TYLCV was found in local weeds in Israel, serving as virus reservoir, when tomatoes were planted in the field. In the 1980s in Spain, the year-round cultivation of tomatoes under plastic screen houses promoted the spread of the disease, but beans were also found as a host of the virus. Over the years, it was found that the privileged mode of spreading of the disease in the world was via international trade of tomato plantlets, but it was demonstrated that the green tissue attached to the fruits (sepals, stems) was a very efficient virus reservoir for local whiteflies. In several instances (Dominican Republic and Florida), it was observed that TYLCV became the prevalent virus, displacing the local geminiviruses previously infecting tomatoes. This could be due to the fact that TYLCV is very aggressive and could become prevalent because of a better fitness to the host. In addition to the trade impact, the explosion of the B-type population of whiteflies can explain the worldwide spread of TYLCV. This particular B-type population of whiteflies is very ubiquitous and infects more than 1000 different plant hosts, providing the insect with enormous resources for winter survival and local adaptation.

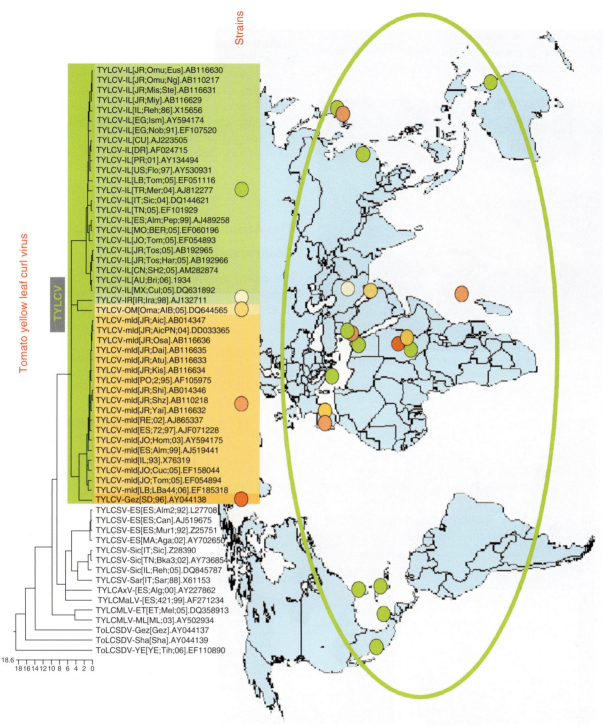

Figure 1 World map on which each of the 44 members representing five begomovirus strains of the TYLCV species have been indicated. The abbreviated name and accession numbers of the isolates are listed in phylogenetic tree in the upper part of the diagram. The phylogenetic tree has been built using their complete A component sequence. The Clustal V algorithm of the program MegAlign from DNAStar has been used and distances in percentage difference are indicated on the left. The tree shows a partition in six major clusters, one for each of the six designated viruses, TYLCV, tomato yellow leaf curl Sardinia virus (TYLCSV), tomato yellow leaf curl Axarquia virus (TYLCAxV), tomato yellow leaf curl Malaga virus (TYLCMalV), tomato yellow leaf curl Mali virus (TYLCMLV), and tomato leaf curl Sudan virus (ToLCSDV). These six species constitute the so-called TYLCV cluster of the Old World begomoviruses. The individual viruses composing the TYLCV cluster are positioned on the world map, as filled circles of various colors representing their pertaining to one of the six specific TYLCV strains, as indicated in the colored boxes at the bottom of the tree. On the world map, the individuals pertaining to the TYLCV species are shown with filled circles.

Control

Geminiviruses are in general very difficult to control. The most efficient way is the use of virus-resistant varieties when they are available, acceptable, and affordable. The cost of such varieties is limiting their use to industrial tomato plantations and therefore most of the tomato fields in poor farmers fields are extremely susceptible. Another limiting factor is the taste and shape of virus-resistant tomatoes that do not match local requirements and encourage the use of susceptible varieties. Controlling the insect, by chemical means, is practically impossible and very expensive. Agricultural practices, although efficient in certain conditions, have been of little help to control TYLCV in a practical manner.

Figure 2 Typical symptoms of TYLCV on tomato in a field in Jordan. Courtesy of Mohammed Abhary.

The African Cassava Pandemic

Geographical Distribution

The cassava geminivirus pandemic in Africa started in the early 1990s and is still going on today. It started in Uganda in 1994, invaded several countries in East Africa around Lake Victoria, before crossing the mountains in Rwanda-Burundi, to infect the entire Congo Basin and invade up to Gabon and the south of Cameroon (**Figure 3**). Cassava plants were already infected with African cassava mosaic

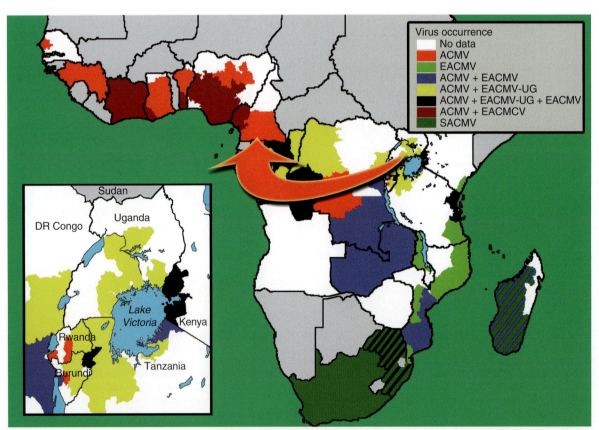

Figure 3 Distribution of cassava mosaic geminiviruses in Africa obtained from CMD surveys in Africa, 1998–2004. ACMV, African cassava mosaic virus; EACMV, East African cassava mosaic virus; EACMV-UG, East African cassava mosaic virus – Uganda; EACMCV, East African cassava mosaic Cameroon virus; SACMV, South African cassava mosaic virus.

virus (ACMV) and other viruses, but became superinfected with an Ugandan strain of East African cassava mosaic virus (EACMV-UG), which displaced other viruses.

Symptoms and Yield Losses

Symptoms due to the EACMV-UG are extreme reduction of the surface leaf, yellowing of the lamina, and downcurling of the leaves. The entire infected plants look yellow compared to healthy plants, with reduced size and extremely limited foliage. Furthermore, in many instances, plants are infected with both EACMV-UG and ACMV, and these two viruses synergize to cause a very dramatic syndrome, called 'candlestick syndrome'. In this instance, the stem is thick and turns purple, the leaves are even more reduced in size, with very thick veins and almost no lamina (**Figure 4(d)**). Cassava is normally propagated using the stem, but if the plants have been co-infected with ACMV and EACMV-UG, the stems cannot be used anymore and cassava cannot be propagated to the next season. This caused dramatic cassava shortages in Uganda between 1994 and 2000, estimated to losses amounting to 6 million tons per year, and famine and death for thousands of people.

Causal Agent with Classification

ACMV and EACMV-UG are both members of the genus *Begomovirus*, in the family *Geminiviridae*, similarly to TYLCV above. These viruses however are bipartite viruses, with a second DNA molecule coding for two movement proteins. Furthermore, EACMV-UG is a recombinant between ACMV and EACMV, with a fragment of 550 nt from the core part of the coat protein of ACMV integrated into the EACMV genome. Although there is no direct proof that this recombinant has any specific advantage over the nonrecombinant EACMV, it is a fact that the EACMV-UG strain is present in all the plants that show severe symptoms and is strictly correlated with the development of the pandemic across the African continent. In addition, these two viruses do synergize to cause a dramatic 'candlestick' symptom, preventing the cuttings from these infected plants to grow for the next generation. The explanation for this phenomenon resides in the fact that each virus codes for a very strong and differential gene silencing suppressor, and when both suppressors are present, the infected plant collapses from the resulting dual virus infection. This is the first time that such synergism between two geminiviruses has been found.

Epidemiology

ACMV and EACMV-UG are spread by the whitefly *Bemisia tabaci* and through the cassava cuttings, as there is no virus-clean material propagation system in place in Africa. The explosion of the cassava mosaic pandemic in Africa is currently explained by a conjunction of several mechanisms. (1) The synergism between the two viruses certainly played a key role, as it translates into a huge increase in viral DNA in the tip of the plants by 30–100 times, thereby enhancing the whitefly transmission capacity. (2) The recombinant fragment could provide an advantage to EACMV-UG, as it has been shown that the whitefly transmission epitopes are coded by this fragment and that there has been co-adaptation as far as the rate of transmission is concerned, between the local whitefly and viruses. (3) Finally, there has been a tremendous increase in whitefly populations adapted to cassava in Uganda, where the epidemic started. The conjunction of these three elements is believed to be at the triggering and maintenance of the cassava mosaic disease pandemic in Africa.

Control

Similar to the TYLCV, the only available method of control of the cassava mosaic disease is the use of virus-resistant cassava cultivars. Fortunately, such plants were made available to farmers in Uganda by the end of the 1990s, and the use of these plants allowed to restore the cassava production in Uganda. This however did not stop

Figure 4 Typical symptoms of cassava infected with ACMV (b) and EACMCV from Cameroon (c), and synergism between the two viruses expressing the 'candlestick' symptoms (d), compared to the control healthy cassava leaves (a).

the pandemic that is still going on in West Africa. Furthermore, many of the farmers in Uganda, and elsewhere, are now going back to their preferred cassava plant material that is not virus resistant, but highly appreciated for their organoleptic and processing qualities. Consequently, more virus-resistant cassava cultivars, acceptable to farmers, are needed.

The Cotton Leaf Curl Disease in the Indian Subcontinent

Geographical Distribution

The leaf curl disease of cotton (*Gossypium hirsutum* L.) was first observed in Multan, Pakistan, in 1967. The disease reappeared in 1987 in its epidemic form in most of the cotton-growing areas of Pakistan. In 1992, the disease was observed throughout the central and southern parts of Punjab in Pakistan. During 1997–98, the disease was established in the border areas of India (Rajasthan, Haryana, and Punjab), joining the southern Punjab of Pakistan. The most devastating virus known as cotton leaf curl Multan virus (CLCuMV) is attributed to huge losses throughout the Indian subcontinent. The estimated direction of movement for the disease is from the center of Pakistan to northern and southern districts of India. Nowadays, the disease can be observed in all primary or secondary cotton-growing areas of the Indian subcontinent (**Figure 5**).

Symptoms and Losses

Cotton leaf curl disease (CLCuD) is not a seed-borne disease, but is only transmitted through whiteflies. After 2–3 weeks of virus inoculation by the insects, vein thickening on young leaves can be observed. Later, upward or downward curling of leaves starts and a cup-shaped leaf enation appears at the underside of the leave (**Figure 6**). The infected plants remain stunted throughout their life cycle and may lose 100% of the total yield. Disease spread can be observed with the wind direction. During the last two decades, the disease caused heavy cotton yield losses which reached up to US$ 5 billion in Pakistan in 1992–97. Recently, the introduction of relatively resistant varieties has selected a more aggressive virus in 2002 in central Pakistan, called cotton leaf curl Burawalia virus (CLCuBuV), and this strain is now prevailing in all cotton-growing areas of Pakistan.

Causal Agent and Classification of Cotton Leaf Curl Disease

The begomoviruses associated with CLCuD are monopartite begomoviruses with an associated ssDNA satellite molecule known as DNA-β (beta satellite).

According to sequence analysis of the available sequences in gene bank, there are viruses belonging to seven different species involved in the etiology of CLCuD named as, CLCuMV, cotton leaf curl Kokhran virus (CLCuKV), cotton leaf curl Rajasthan virus (CLCuRV), cotton leaf curl Allahabad virus (CLCuAV), cotton leaf curl Burawalia virus (CLCuBuV), cotton leaf curl Bangalore virus (CLCuBV), and papaya leaf curl virus (PaLCuV). Cotton leaf curl Bangalore virus is only present in the Bangalore district of southern India, while viruses belonging to the other six species are widespread throughout the cotton-growing areas of the north of the Indian subcontinent.

The area of maximum diversity for CLCuD is in central parts of Pakistan, hosting viruses belonging to five species of cotton leaf curl viruses (CLCuVs) and one of PaLCuV.

CLCuD is in reality caused by a complex of viruses associated with virus satellites. A monopartite geminivirus, belonging to different species of viruses, is typically associated with a beta satellite (DNA-β) or a DNA-1 satellite. If DNA-β's have only been found associated with geminiviruses, DNA-1 satellites have originally been found associated to nanoviruses. Both satellite genomes are encapsidated by the geminivirus capsid protein, and thereby both are transmitted by whiteflies. Full-length clones of several CLCuVs DNA-A were unable to reproduce the symptoms in cotton and other different hosts such as tobacco. Infectious clones of DNA-1 were also not capable of inducing symptom development of CLCuVs in the inoculated plants.

With the discovery of DNA-β, it was proved that both DNA-A and DNA-β are necessary to induce typical symptoms of CLCuD, while DNA-1 is not necessary to induce the symptoms and is seldom found in association with the disease complex. At present, the origin of DNA-β is unknown but it can be transreplicated by diverse geminiviruses associated with the CLCuD complex. The DNA-β is only ±1300 nt long and codes only for one ORF called βC1. The gene βC1 codes for a strong suppressor of gene silencing, thereby providing a tremendous advantage to the cognate helper virus. DNA-1 satellite codes for a Rep protein that allows this satellite to be replicated on its own, but the satellite depends on the helper virus for all the other functions, such as gene-silencing suppression, movement, and transmission. DNA-β is dependent on DNA-A for replication, movement, and transmission. The genome organization of DNA-A is typical of any other begomovirus and has no similarity with its associated satellite molecules. Both DNA-β and DNA-1 are approximately half the size of DNA-A, that is, 1350 nt.

Epidemiology

Cultivated cotton or upland cotton (*G. hirsutum*) was introduced into the Indian subcontinent from southern

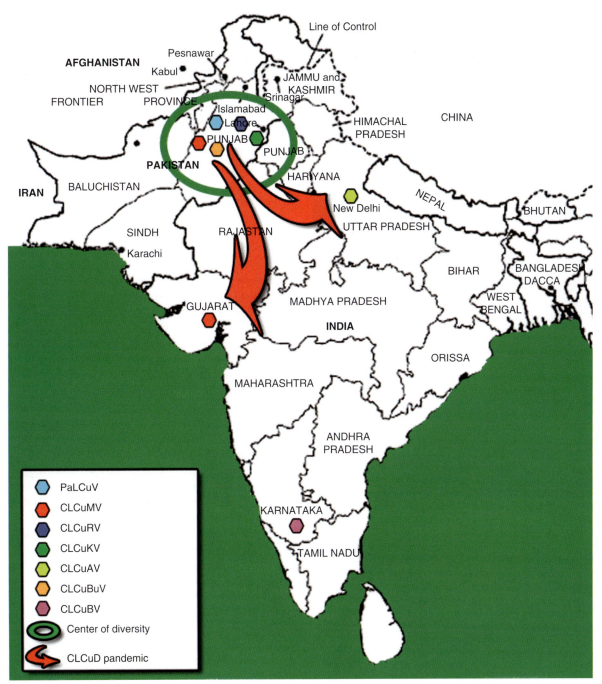

Figure 5 Localization of cotton leaf curl geminiviruses in the Indian subcontinent. PaLCuV, papaya leaf curl virus; CLCuMV, cotton leaf curl Multan virus; CLCuRV, cotton leaf curl Rajasthan virus; CLCuKV, cotton leaf curl Kokhran virus; CLCuAV, cotton leaf curl Allahabad virus; CLCuBuV, cotton leaf curl Burawalia virus; CLCuBV, cotton leaf curl Bangalore virus.

Mexico during the nineteenth century. In its native place, cotton is not associated with virus diseases as in the Indian subcontinent. CLCuD complex most probably evolved from some wild species of cotton growing over a long period of time or from other weed hosts present before the cultivation of upland cotton in this part of the world. In 1991, a new variety of cotton (S12) was introduced from Texas, because of its excellent fiber qualities; unfortunately, this new variety was hypersensitive to CLCuD and the disease propagated very quickly after this introduction. The CLCuD threat was greatly increased due to intensive cotton cultivation, monocropping pattern of agriculture, and overlapping seasons of other crop plants. CLCuVs can infect a number of diverse hosts in the

Figure 6 Downward leaf curl symptoms of cotton induced by the CLCuMV in association with the cotton leaf curl Multan beta satellite.

families Malvaceae and Solanaceae. The virus infection was not highly noticed at the early beginning of the epidemic, when perhaps it could have been possible to eradicate the disease. The disease is transmitted through whitefly in a persistent manner. The disease spread is directly correlated with the spread of the whitefly population, along the prevailing wind direction.

Control

Controlling the whitefly vector population is an important aspect to control CLCuD. Insecticides used to control whitefly play a vital role in the control of disease, but still cannot completely eliminate the disease. Over the years, breeders have selected several varieties of cotton with various levels of resistance, but either this level of resistance was unsatisfactory or has been broken by the appearance of new strains of the virus, such as the CLCuBuV from Burawalia. Recently RNA interference (RNAi) has been used and shown to be efficient, in model plants, to control the disease. However, there is still a need to do more work to control the disease in crop plants. In order to create a broad-spectrum resistance to control the disease complex, there is a need of integrated efforts of conventional and genetic engineering approaches.

Conclusion

The three examples of geminivirus emergence presented here provide examples of the importance of different factors that favor the emergence of geminiviruses under very different conditions and environments. In the case of TYLCV, it is a relatively simple case with a single monopartite geminivirus that is transported, through human-based international trade, to various places in the world, and where this very effective virus overcomes existing local viruses and prevails. Although we do not know the molecular and biological basis for the prevalence of TYLCV, it is clear that this virus has a better fitness in tomato and other hosts, compared to local geminiviruses, whether they be bipartite New World geminiviruses or monopartite Old World geminiviruses.

In the case of cassava mosaic geminiviruses, the situation is more complex as it involves at least two different geminiviruses, a new whitefly population, better adapted to cassava, as well as human participation to move the viruses, and probably the whiteflies, in new environments over chains of mountains, where the disease can again explode. It is therefore the apparent conjunction of various biological and human-based activities that promoted the pandemic in Africa. On the biological side, at least three elements can be related to the emergence: (1) the encountering of two geminiviruses with differential and combinatorial gene silencing suppressors that promoted synergism; (2) the occurrence of an apparent successful recombination between these two viruses; and (3) the adaptation of a new population of whiteflies to cassava. The pandemic has been able to travel eastward more than 3500 km in 12 years, and this would not have happened without human intervention. It remains to be seen if the pandemic will continue in West Africa or if it will fade away in this new ecological zone.

The CLCuD case is even a more complex situation where human intervention by introducing new cotton varieties, highly susceptible to local unknown viruses and satellites, as well as the absence of human intervention to stop the disease, triggered the CLCuD pandemic in the Indian subcontinent, costing billions of dollars to poor farmers and impacting Pakistan economy to about 30% for years. On the virological side, this pandemic revealed a new concept, that is, the crucial role of satellites like the DNA-β for inducing symptoms and finally for allowing the whole pandemic. It is the association of a single DNA-β to a variety of helper geminiviruses that is causing the disease and not a single monopartite geminivirus. It is now obvious that recombination of both DNA-β and its geminivirus helpers is an important evolutionary mechanism to create new diversity to overcome virus resistance, revealing the power of geminiviruses and their satellites to adapt to new ecological situations.

The common factor among these three examples is that human intervention, or lack of intervention, is providing viruses and their satellites opportunities to promote new disease emergences in combination with ecological changes, which are also linked to human impact. These dramatic pandemics, costing billions of dollars in losses and thousands of lives, offer however opportunities to the scientists to better understand the biological causes of these emerging diseases. Taking advantage of these global scale 'experiments', it may be possible to identify the relevant biological factors that could in future allow such diseases to be controlled.

See also: African Cassava Mosaic Disease; Beta ssDNA Satellites; Cotton Leaf Curl Disease; Nanoviruses; Plant Resistance to Viruses: Geminiviruses; Tomato Leaf Curl Viruses from India.

Further Reading

Briddon RW (2003) Cotton leaf curl disease, a multicomponent begomovirus complex. *Molecular Plant Pathology* 4: 427–434.

Czosnek H (ed.) (2007) *Tomato Yellow Leaf Curl Virus Disease*. New York: Springer.

Legg JP and Fauquet CM (2004) Cassava mosaic geminiviruses in Africa. *Plant Molecular Biology* 56: 585–599.

Mansoor S, Briddon RW, Zafar Y, and Stanley J (2003) Geminivirus disease complexes: An emerging threat. *Trends in Plant Science* 8: 128–134.

Polston JE and Anderson PK (1997) Emergence of whitefly transmitted geminiviruses in tomato in the Western Hemisphere. *Plant Disease* 81: 1358–1369.

Rojas MR, Hagen C, Lucas WJ, and Gilbertson RL (2005) Exploiting chinks in the plant's armor: Evolution and emergence of geminiviruses. *Annual Review of Phytopathology* 43: 361–394.

Endogenous Retroviruses

W E Johnson, New England Primate Research Center, Southborough, MA, USA

© 2008 Elsevier Ltd. All rights reserved.

Glossary

Metazoan A multicellular animal.
Monophyletic A taxonomic grouping comprised exclusively of an ancestor and its descendants.
Phylogenetic tree A diagram of the evolutionary relationships between entities descended from a common ancestor.
Proto-oncogene A gene whose normal function is to control cell growth or differentiation, but which can be converted into a cancer-causing gene (oncogene) by mutation or disregulation.

Introduction

At the heart of the retrovirus replication cycle is a stage in which the viral genome, having been converted from single-stranded RNA into double-stranded DNA by the viral reverse transcriptase, is inserted at random into a chromosome of the infected cell. The resulting DNA 'provirus' directs expression and assembly of progeny virions. The integrated provirus itself is contiguous with the rest of the host-cell chromosome, just like any other stretch of genomic DNA sequence. Consequently, if a cell harboring a provirus continues to divide and multiply, the provirus will be inherited by daughter cells during subsequent rounds of cell division like any other cellular gene. If retroviral integration occurs in a cell belonging to germline tissue of a host organism (e.g., in an oocyte or in an embryonic cell destined to develop into germline tissue), the resulting provirus constitutes an insertional mutation, and the new allele can potentially be passed on to the next generation like any other chromosomal locus. As a new allele, the proviral locus will be subject to the processes of random genetic drift and selection and, as a result, many such events may quickly be lost. But some will spread in the population and, given enough time, may eventually become fixed. In fact, this scenario has played out many millions of times over the course of metazoan evolution, such that the genomes of almost all animal species contain numerous sequences resembling proviruses. These elements are referred to as endogenous retroviruses (ERVs). While exogenous retroviruses spread from individual to individual by infection (horizontal transmission), endogenous retroviruses, as components of the germline, are inherited in the same fashion as other genes. The two modes are not mutually exclusive, and virions expressed from ERVs can give rise to spreading infections within the host and may also be transmitted to other individuals.

Structure

The typical retroviral provirus consists of *gag*, *pro*, *pol*, and *env* open reading frames, which encode the viral structural and nonstructural proteins, bookended by two identical regulatory regions called long terminal repeats (LTRs).

No chromosomal DNA is lost as a consequence of integration; however, a short stretch of host-cell DNA (typically around 4–6 bp) is duplicated, resulting in direct repeats (DRs) of cellular sequence flanking the provirus. Important *cis*-acting sequences for replication are located within the LTRs, and primer recognition sequences for reverse transcription are located at the junctions between the 5′ LTR and *gag* (first-strand synthesis), and between *env* and the 3′ LTR (second-strand synthesis). Retroviruses use a cellular transfer RNA molecule (tRNA) as the primer for the first step in reverse transcription; thus, the primer binding site (PBS) is complementary to the 3′ end of a cellular tRNA. The second primer is derived from a purine-rich stretch of the viral RNA itself (polypurine tract or PPT). The generic structure of a provirus is therefore:

DR-5′LTR-PBS-gag-pro-pol-env-PPT-3′LTR-DR

In general, ERVs can be recognized as genomic loci conforming to this basic arrangement, even though in many cases significant portions of the original provirus may be missing (**Figure 1**). There are no viral or cellular mechanisms for precise deletion of integrated proviruses, but ERV sequences can be lost or rearranged, in whole or in part, by the same general processes affecting all chromosomal DNA. For example, homologous recombination can occur between the two LTRs of an ERV, resulting in deletion of the intervening proviral sequences and leaving behind a single LTR (or solo-LTR). Solo-LTRs are abundant in animal genomes, and in many cases may significantly outnumber intact proviruses.

Discovery and Distribution

Even prior to the discovery of reverse transcriptase in 1970 (and confirmation that retroviruses replicate through a proviral intermediate), there was some phenotypic evidence for the existence of ERV loci. Then beginning in the early 1970s, the widespread existence of ERVs was quickly revealed using genomic DNA samples and nucleic acid hybridization techniques. Novel ERV sequences were cloned initially using probes derived from existing retroviruses, by accident during the cloning and characterization of unrelated genes, and later by using a variety of polymerase chain reaction (PCR)-based approaches. By far the richest vein of new ERV sequences has come from mining the human genome databases and subsequent whole-genome sequences of other organisms.

The distribution of ERV among modern species speaks of their age. ERV families or specific ERV loci are often shared among the genomes of related species, indicating that the integration events occurred in the genomes of ancestors shared by those species. For example, a large number of ERVs are located in identical positions in humans and closely related primates, such as chimpanzees, bonobos, and gorillas. Thus, these loci must have originated in a common ancestor of these species. This also provides a lower bound to the estimated time of integration based on the approximate time of divergence of these lineages. In the case of an ERV locus shared exclusively by humans and chimpanzees, for example, the provirus will be no less than about 5 million years old, whereas a locus shared by humans, apes, and Old World monkeys will be greater than 25 million years old.

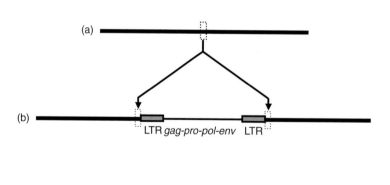

Figure 1 Structure of an ERV. Thick black line represents a segment of a host-cell chromosome. Thin line (labeled *gag-pro-pol-env*) and gray boxes (LTR) represent integrated viral sequences. Small boxes (dashed line) indicate short segment of chromosomal DNA duplicated (arrowheads) at the time of proviral integration. (a) Chromosomal segment prior to integration of a provirus. (b) Same chromosomal locus, but containing an integrated provirus. No chromosomal DNA is lost due to integration, and a short segment (~4–6 bp) is duplicated. When this structure occurs in the germline, it constitutes an ERV. At the time of integration, the LTRs of the ERV are identical, but these sequences will thereafter diverge due to the accumulation of random mutations; thus, the degree of divergence between the LTRs of an ancient ERV are a reflection of the age of the provirus. (c) Same locus, after recombination between the two LTRs, has resulted in loss of the intervening proviral sequence. Only a single copy of the LTR sequence remains, along with the flanking repeats.

In general, the wider spread an ERV-sequence family is among extant species, the more ancient that family is. More modern ERVs are found among a smaller subset of species or are even confined to a single species. (This is a generalization because viruses have the potential for horizontal, cross-species transmission; ERV sequences derived from a modern retrovirus could appear in the genomes of more than one lineage.) Modern ERVs are also more likely to be intact (and capable of expressing virus), and are often closely related to known exogenous retroviruses.

Origins

There are at least four possible scenarios to explain the presence of multicopy ERV families in the germline: (1) repeated, *de novo* infection of germline tissue by an exogenous retrovirus; (2) expression of an existing endogenous provirus with subsequent reinfections of germline tissue; (3) intracellular retro-transposition from an existing ERV; and (4) duplication of genomic DNA fragments already containing an integrated ERV (however, even in this case, the original ERV sequence must have arisen via one of the first three mechanisms). Although there is experimental proof-in-principle for all four processes, it is generally believed that scenarios involving extracellular replication and infection (i.e., the first two mechanisms) predominate in nature.

Whatever the mechanism, presence in the germline requires that integration, takes place either in cells of germline tissue or, because the provirus is maintained during cell division, in a cell lineage destined to develop into germline tissue. Once integrated, the selective forces that work on the proviral sequences will depend on a number of factors, including the state of the provirus at the time of integration. For example, did the viral genome suffer attenuating or debilitating mutations during reverse transcription? Was it intact and capable of expression at the time of provirus formation? Did integration occur in junk DNA, close to a gene, or in a gene? Was the provirus silenced after integration (e.g., by methylation), and in what developmental stages and in which tissues was the provirus expressed? In other words, whether or not a newly formed provirus will persist as an endogenous retrovirus is determined in part by its immediate effect on the survival of the infected cell and its ultimate effect on reproductive fitness of the host organism. ERVs that are strongly counterselected due to pathogenic or detrimental effects on the host are likely to be lost from the gene pool.

Although the bulk of integrations may occur in junk DNA, experimental evidence suggests that integration may occur preferentially into transcriptionally active areas, which increases the likelihood of insertion in and near genes. Insertional inactivation of critical genes can be immediately deleterious to the host, or may only be lethal in the homozygous state (in the latter case, an ERV-containing locus could persist for many generations as a minor allele). Integration can also lead to activation or aberrant regulation of genes near the site of integration. For example, activation of proto-oncogenes in the vicinity of integrated proviruses has been detected many times in experimental animal models of retroviral pathogenesis; it is possible that similar scenarios could also occur during formation of an ERV. Expression of the ERV can potentially be a source of pathogenic virus or can contribute to pathogenecity by recombination with other retroviruses.

ERV sequences may also have beneficial consequences for the host. Laboratory studies with both chickens and inbred mice have identified several ERV-derived genes that now function to reduce or prevent exogenous retroviral infection. ERV sequences also have the potential to provide new cellular functions during the course of evolution. The human Syncytin protein may be a striking example of the latter; this protein is in fact the envelope protein of an ancient endogenous provirus that appears to have been co-opted during evolution to function in human placental morphogenesis.

Stochastic factors will also influence the persistence of an ERV locus in the genome. Proviruses in the germline are subject to the same processes that affect all genomic sequences, including substitutions, deletions, insertions, recombination, gene conversion, random assortment, and genetic drift. Even in the absence of selective pressures, a newly formed ERV represents a minor allele and stands a good chance of being lost by chance. In the absence of positive selection, random accumulation of mutations over time will eventually degrade the viral open reading frames and the *cis*-acting elements required for expression. Thus, the interplay between random processes, as well as the nature and extent of selective forces acting on the ERV, will affect the way in which these sequences evolve during their residence in the germline.

Taxonomy and Classification

The term 'endogenous retrovirus' does not refer to a biological entity distinct from other retroviruses, but simply describes any DNA provirus, retroviral in origin, that has found its way into an organismal germline. This is true regardless of whether the provirus is still capable of expressing infectious virions. In fact, even highly degraded proviruses containing large deletions, insertions, and substitutions are often referred to as ERVs, as long as they are still clearly derived from a retrovirus.

Phylogenetic trees incorporating multiple genera of retroviruses are typically constructed using amino acid alignments corresponding to conserved domains in the reverse transcriptase gene. When representatives of multiple ERV families are included in such analyses, they do

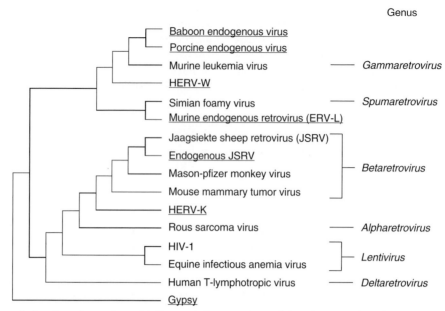

Figure 2 Phylogenetic tree based on amino acid alignment of reverse transcriptase from a variety of retroviruses. ERVs are underlined. Viruses representing genera in the family *Retroviridae* are indicated. Gypsy is an endogenous retroviral element from the invertebrate species *Drosophila melanogaster*. Tree was generated by maximum parsimony and represents one of five equally parsimonious trees. Note that ERV sequences are interleaved with exogenous sequences and do not represent a separate lineage.

not form a monophyletic cluster, but rather are scattered within and between the various genera of exogenous retroviruses (**Figure 2**). Moreover, modern ERVs may have very closely related, exogenous counterparts. Thus, ERVs do not constitute a separate taxonomic division from exogenous retroviruses (this refers to ERVs as a whole; any individual ERV locus or family may still represent a novel (and possibly extinct) retroviral genus).

Although there are exceptions, the vast majority of ERVs (particularly the ancient ERVs) is not closely related to known exogenous retroviruses, is no longer capable of expressing virus, and has no other associated biological or phenotypic properties to facilitate classification. The interleaving of ancient ERV sequences with extant retroviruses in phylogenetic trees, along with similarities to modern viruses in provirus organization, indicates that ancient ERVs are not ancestral stages in the evolutionary pathway leading to retroviruses (i.e., they are not protoretroviruses), but rather represent fully evolved (albeit extinct) sibling species to modern retroviruses.

There is as yet no officially recognized nomenclature for endogenous retroviruses, and a variety of naming conventions exist. In general, these represent myriad naming practices coined by various investigators at the time each novel ERV was discovered. For ERVs originally identified in the human genome, the practice has been adopted of using the acronym HERV (for human endogenous retrovirus) followed by a single letter that specifies the most likely tRNA primer (based on complementarity to the putative PBS sequence). Thus, HERV-K refers to endogenous proviruses with a PBS sequence complementary to a lysine tRNA. This convention has made the HERV literature much more accessible, but in practice it still leads to some ambiguity, because: (1) otherwise unrelated retroviruses may share the same or similar tRNAs; (2) many ERV loci may not contain a recognizable PBS element; and (3) in some cases, the PBS will be sufficiently degraded by substitutions that unambiguous prediction of tRNA specificity is not possible. Despite the name, the vast majority of HERV loci is not unique to humans but is shared with other, closely related primates (having first appeared in a common ancestor); thus, HERV loci should not be thought of as human specific, but rather as human orthologs of loci found in other primate species. Because some HERV families have hundreds or thousands of members, when a specific HERV locus is being described in reference to the chromosomal location, database accession number or cosmid clone is sometimes used for clarity (e.g., HERV-K6p22 refers to a provirus of the HERV-K family found at band 22 on the short arm of chromosome 6). However, this practice is recent and many of the original clone names are still in use.

Conclusion

Given the sheer number and ubiquity of ERVs among extant species, it is likely that retroviruses have made major contributions to both the content and structure of animal genomes for hundreds of millions of years. Retroviruses can give rise to new DNA content via proviral integration as well as by retro-transposition or transduction of nonviral sequences. Because many ERVs constitute multilocus families of closely related sequences, they can mediate intergenic recombination events leading to large-scale deletions, duplications, interchromosomal translocations, and inversions of chromosomal sequence. All of these processes are also mutagenic, and therefore provide sources of variation upon which selective forces may operate. ERVs can also recombine with exogenous retroviruses during infection, providing additional sources of viral variation and giving rise to retroviruses with altered or novel properties. In modern times, the advent of xenotransplantation (using appropriate animals as organ donors for human transplantation) has raised concerns that ERVs expressed in the donor organs could introduce new retroviruses into the human population. Finally, because of the vast archive of retroviral sequences present in animal genomes in the form of ERVs, the deep evolutionary past of the *Retroviridae* is amenable to scientific exploration in a way that other viral families are not.

See also: Bovine and Feline Immunodeficiency Viruses; Equine Infectious Anemia Virus; Evolution of Viruses; Feline Leukemia and Sarcoma Viruses; Fish Retroviruses; Foamy Viruses; Host Resistance to Retroviruses; Human Immunodeficiency Viruses: Molecular Biology; Human T-Cell Leukemia Viruses: General Features; Jaagsiekte Sheep Retrovirus; Mouse Mammary Tumor Virus; Organ Transplantation, Risks; Origin of Viruses; Reticuloendotheliosis Viruses; Retrotransposons of Vertebrates; Retroviral Oncogenes; Retroviruses: General Features; Retroviruses of Birds; Retroviruses of Insects; Simian Immunodeficiency Virus: Natural Infection; Simian Retrovirus D; Visna-Maedi Viruses.

Further Reading

Best S, Le Tissier PR, and Stoye JP (1997) Endogenous retroviruses and the evolution of resistance to retroviral infection. *Trends in Microbiology* 5: 313–318.

Boeke JD and Stoye JP (1997) Retrotransposons, endogenous retroviruses, and the evolution of retroelements. In: Coffin JM, Hughes SH, and Varmus H (eds.) *Retroviruses*, pp. 343–435. New York: Cold Spring Harbor Laboratory Press.

Gifford R and Tristem M (2003) The evolution, distribution and diversity of endogenous retroviruses. *Virus Genes* 26: 291–315.

Jern P, Sperber GO, and Blomberg J (2005) Use of endogenous retroviral sequences (ERVs) and structural markers for retroviral phylogenetic inference and taxonomy. *Retrovirology* 2: 50.

Lower R, Lower J, and Kurth R (1996) The viruses in all of us: Characteristics and biological significance of human endogenous retrovirus sequences. *Proceedings of the National Academy of Sciences, USA* 93: 5177–5184.

Mager DL and Freeman JD (1995) HERV-H endogenous retroviruses: Presence in the New World branch but amplification in the Old World primate lineage. *Virology* 213: 395–404.

Mi S, Lee X, Li X, *et al.* (2000) Syncytin is a captive retroviral envelope protein involved in human placental morphogenesis. *Nature* 403: 785–789.

Sverdlov ED (2000) Retroviruses and primate evolution. *BioEssays* 22: 161–171.

Turner G, Barbulescu M, Su M, Jensen-Seaman MI, Kidd KK, and Lenz J (2001) Insertional polymorphisms of full-length endogenous retroviruses in humans. *Current Biology* 11: 1531–1535.

Weiss RA (2006) The discovery of endogenous retroviruses. *Retrovirology* 3: 67.

Endornavirus

T Fukuhara and H Moriyama, Tokyo University of Agriculture and Technology, Fuchu, Japan

© 2008 Elsevier Ltd. All rights reserved.

Glossary

Cytoplasmic male sterility (CMS) Male sterility caused by a cytoplasmic factor.
Isogenic line Genetically identical pure-breeding group of individuals.
Mycovirus A virus that infects fungal species.
Replicon A genetic unit of replication.

Introduction

Double-stranded RNAs (dsRNAs) have frequently been identified in various healthy plants, from algae through to higher plants. These dsRNAs are not transcribed from the host genome DNAs, and they have important properties in common that differ from those of conventional viruses: (1) most of these dsRNAs have no obvious effect on the phenotype of their host plants; (2) they are present at a constant low concentration in the host plant; (3) they are efficiently transmitted to the next generation via seeds, but their horizontal transfer (infection) to other plants has never been proved. The size of these dsRNAs varies from 1.5 to 20 kbp. Smaller dsRNAs ($\sim$2.0 kbp) are often found with virus-like particles and some of these dsRNAs have already been classified as viruses in the genera *Alphacryptovirus* and *Betacryptovirus* of the family *Partitiviridae*. Partitiviruses have two unrelated linear dsRNA segments, each about 2.0 kbp in size. Larger dsRNA segments (more than 10 kbp) are unlikely to be associated with distinct virus-like particles, because no distinct virus-like particles have been detected in preparations obtained with various purification procedures. Thus, these large dsRNAs were previously referred to as RNA plasmids, enigmatic dsRNAs, or endogenous dsRNAs.

Large dsRNAs (Endornaviruses) in Plants

In 1966, a cytoplasmic male sterility (CMS) trait, termed '447', in broad bean (*Vicia faba*) was described, and in 1976, electron microscopy observations revealed that the cytoplasm of 447 male-sterile plants contained cytoplasmic spherical bodies (CSBs) that resembled vesicles of about 70 nm in diameter. These CSBs were found in all tissues of the 447 plants, and contained a high-molecular-weight dsRNA. An absolute correlation between the CMS trait and the presence of the CSB containing the dsRNA has been established.

In the 1990s, large dsRNAs of about 14 kbp in length were found in many strains of cultivated rice (*Oryza sativa* L.) in Japan (temperate *japonica* rice) and the Philippines (tropical *japonica* rice), but not in the *indica* subspecies of *O. sativa*. A similar dsRNA was also detected in one strain (W-1714) of wild rice (*O. rufipogon*), which is considered an ancestor of *O. sativa*. dsRNA-carrying plants and dsRNA-free plants were found for the Nipponbare cultivar of temperate *japonica* rice, but these two isogenic lines were indistinguishable in their phenotypes. Even rice breeders and farmers could not distinguish these two lines; in fact, dsRNA-carrying and dsRNA-free rice plants were found in the same rice paddy field at our university farm. If the dsRNA-carrying line had a lower harvest yield than the dsRNA-free line, rice breeders and/or farmers would have recognized this phenotype and discarded the lower-yielding line. Given that this did not occur, these two lines can be considered identical with respect to all phenotypes, including harvest yield. Electron microscopy observations revealed that all dsRNA molecules from rice plants are linear. Differential centrifugation and sucrose density-gradient centrifugation revealed that the dsRNA in rice is localized in the cytoplasm.

Large dsRNAs of about 14 kbp in length have also been found in other crops, but only three of these dsRNAs, from the 447 male-sterile strain of broad bean (*V. faba*), the Nipponbare cultivar of cultivated rice (*O. sativa*), and the W-1714 strain of wild rice (*O. rufipogon*), have been sequenced completely. All three dsRNAs encode a single long open reading frame (ORF), in which conserved motifs for RNA-dependent RNA polymerase (RdRp) and RNA helicase (Hel) are found (**Figure 1**). The large dsRNAs are RNA replicons that can

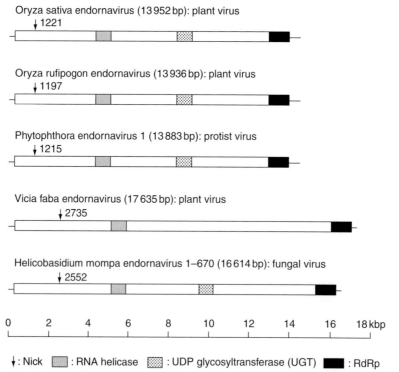

Figure 1 Comparison of the genome organizations of five endornaviruses: oryza sativa endornavirus (OsEV), oryza rufipogon endornavirus (OrEV), phytophthora endornavirus 1 (PEV1), vicia faba endornavirus (VfEV), and helicobasidium mompa endornavirus 1–670 (HmEV1–670).

replicate independently of their host genome by using their own RdRps, as will be discussed later. Phylogenetic analyses of the amino acid (aa) sequences of the RdRp and Hel motifs indicate that the three large dsRNAs that have been sequenced share a common ancestor with the alpha-like supergroup of positive-sense, single-stranded RNA (ssRNA) viruses, which includes many plant viruses, such as those of the genera *Tobamovirus* and *Cucumovirus*. Based on this information, in its eighth report, the International Committee on Taxonomy of Viruses (ICTV) accepted that these large plasmid-like dsRNA replicons are members of a new virus genus, *Endornavirus*. The dsRNAs found in cultivated rice, wild rice, and broad bean have been named oryza sativa endornavirus (OsEV), oryza rufipogon endornavirus (OrEV), and vicia faba endornavirus (VfEV), respectively (**Figure 1**).

Nonplant Endornaviruses

dsRNAs are found not only in plants, but also frequently in a variety of fungi; in fact, they have been recorded in all major fungal taxonomic groups, including various plant pathogens. The genomes of most mycoviruses consist of dsRNAs, and these fungal dsRNA viruses have so far been classified into five families: *Reoviridae*, *Totiviridae*, *Partitiviridae*, *Chrysoviridae*, and *Hypoviridae*. Members of only one of these families, the *Hypoviridae*, contain a single large dsRNA genome about 10–12 kbp in length, in contrast to members of the other families that contain one or multiple small dsRNA genomes about 2–6 kbp in length. However, many unclassified dsRNAs are still found in various fungi, some of which are larger than 10 kbp in length.

The AG3 isolate of the fungus *Rhizoctonia solani* is the major cause of rhizoctonia disease in potato, and contains two large dsRNAs, L1 (25 kbp) and L2 (23 kbp), in some strains of the fungus. The nucleotide sequence of the L2 dsRNA has been partially determined, and was found to be similar to the RdRp regions of the ORFs encoded by plant endornaviruses.

The violet root rot fungus, *Helicobasidium mompa*, occurs on various plants, and contains various-sized dsRNAs. A large dsRNA (L1 dsRNA) in the V670 strain of *H. mompa* has been identified as a hypovirulence factor, and has been sequenced completely. It encodes a single long ORF (5373 aa), in which conserved RdRp and Hel motifs are found (**Figure 1**). A BLAST search using the aa sequence of this ORF revealed significant sequence similarities to the dsRNAs of members of the genus *Endornavirus*. These results indicated that the large dsRNA in *H. mompa* belongs to the genus *Endornavirus*, and it was thus named helicobasidium mompa endornavirus 1–670 (HmEV1–670).

Plant pathogens of the genus *Phytophthora* have many of the same biological properties as fungi. However, on the basis of sequence similarities they have been classified together with diatoms and brown algae in a protist group known as the stramenopiles or chromista. The complete nucleotide sequence of the 13.9 kbp dsRNA isolated from a *Phytophthora* isolate from Douglas fir has also been published recently. The dsRNA encodes a single long ORF (4548 aa), whose genome organization is very similar to that of known endornaviruses (**Figure 1**). Phylogenetic analyses support the classification of this dsRNA (named *Phytophthora endornavirus* 1 (PEV1)) as a member of the genus *Endornavirus*.

Therefore, entire nucleotide sequences have been published for a total of five endornaviruses, three from plants (OsEV, OrEV, and VfEV), one from a fungus (HmEV1-670) and one from a protist (PEV1). Despite the diversity of their hosts, a comparison of the genome organizations of the five endornaviruses indicates that they share a unique structure (**Figure 1**): all five endornaviruses encode a single unusually long ORF, in which conserved motifs for RdRp and Hel are found in the same regions, and they contain a site-specific nick in the 5′ region of the coding strand.

Tentative Members of the Genus *Endornavirus*

In the 1990s, dsRNAs of about 14 kbp in size were found in many cultivars of barley (*Hordeum vulgare*), kidney bean (*Phaseolus vulgaris*), and bell pepper (*Capsicum annuum*). Barley and bell pepper plants hosting the dsRNAs are available commercially, so these plants must be healthy and common (not special) cultivars. These dsRNAs have been reported to be efficiently transmitted to the next generation. Therefore, these large dsRNAs with plasmid-like properties may be members of the genus *Endornavirus*. Unfortunately, only limited sequence information for these dsRNAs is available (except for an approximately 630-nt sequence of the dsRNA from kidney bean), so it is currently unclear whether they do in fact belong to the genus *Endornavirus*.

Partial sequences of the large dsRNAs from kidney bean (*P. vulgaris*, cv. Black Turtle) and bottle gourd (*Lagenaria siceraria*) have been reported recently, and a single ORF was found in dsRNAs from both species. Homologous sequence searches for these ORFs indicate that both ORFs contain sequences similar to the entire RdRp regions of endornaviruses, and that, in turn, the RdRp domains of endornaviruses are similar to the pfam00978 (RNA_dep_RNApol2) domain in the Conserved Domain Database. RdRps of ssRNA viruses of the alpha-like superfamily are classified as the pfam00978 domain. Molecular phylogenetic analyses using the aa sequences encoded by the RdRp regions (about 470 aa residues) of the kidney bean and bottle gourd dsRNAs as well as of the five endornaviruses and the 10 most diverse members of pfam00978 were carried out. The resulting phylogenetic tree shows that the kidney bean and bottle gourd dsRNAs form a monophyletic group with the known endornaviruses, a grouping that was supported by a high bootstrap probability (100%; **Figure 2(a)**), indicating that these two dsRNAs are members of the genus *Endornavirus*.

Furthermore, partial sequences of dsRNAs from barley (*H. vulgare*, cv. Kashima), melon (*Cucumis melo*), Malabar spinach (*Basella alba*), seagrass (*Zostera marina*), and strain V1369 of the violet root rot fungus (*H. mompa*) have also been recently reported. The sequences are short, but commonly contain core RdRp regions of about 80 aa residues. Phylogenetic analyses were carried out for these five dsRNA sequences, seven known endornaviruses (including kidney bean and bottle gourd dsRNAs), and 10 ssRNA viruses. Although the analyzed RdRp regions of the dsRNAs and viruses were short (~80 aa), the phylogenetic tree was essentially the same as the tree in **Figure 2(a)**, in which the five dsRNAs mentioned above formed a monophyletic group with the seven endornaviruses (**Figure 2(b)**). These results suggest that these five dsRNAs from four plants and one fungus are members of the genus *Endornavirus*. Therefore, there is a growing list of new dsRNA viruses that belong or very probably belong to the genus *Endornavirus*, distributed over the plant, protist, and fungal kingdoms (**Table 1**).

Proteins Encoded by Endornaviruses

All five endornaviruses that have been completely sequenced encode a single unusually long ORF, from 4500 to 5500 aa residues in length, which is one of the unique molecular features of endornaviruses (**Figure 1**). Conserved motifs encoding RdRp and Hel are found in the same regions of these long ORFs. The genes encoding RdRp and Hel are structurally similar to those encoded by the ssRNA genome of members of the alpha-like supergroup of RNA viruses. Thus, the endornavirus RdRp and Hel probably function in the replication of the dsRNA genome, just as their homologs encoded by ssRNA viruses function in the replication of the ssRNA genome. Indeed, viral RdRp activity has been detected in plants containing OsEV or VfEV (to be discussed later).

The enzyme uridine diphosphate (UDP) glycosyltransferase (UGT) is commonly found in eukaryotes. This enzyme catalyzes the addition of the glycosyl

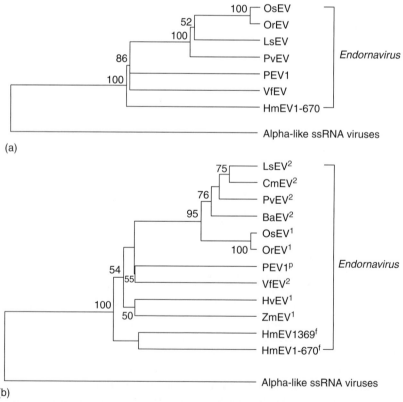

Figure 2 Phylogenetic positions of endornaviruses. (a) About 470 aa of the RdRp regions of seven endornaviruses and 10 alpha-like ssRNA viruses and (b) about 80 aa of the RdRp regions of 12 endornaviruses and 10 alpha-like ssRNA viruses were analyzed by using the ClustalX and MEGA2 (Molecular Evolutionary Genetics Analysis) programs, and the resulting neighbor-joining (NJ) trees are shown. A bootstrap test was performed with 100 resamplings. The hosts of endornaviruses are monocotyledonous plants (1), dicotyledonous plants (2), fungi (f), and a protist (p).

Table 1 Members in the genus *Endornavirus*

	Host	Virus	Abbreviation
Plant	Rice	*Oryza sativa* endornavirus	OsEV
	Wild rice	*Oryza rufipogon* endornavirus	OrEV
	Broad bean	*Vicia faba* endornavirus	VfEV
	Kidney bean	*Phaseolus vulgaris* endornavirus	PvEV
	Bottle gourd	*Lagenaria siceraria* endornavirus[a]	LaEV
	Barley	*Hordeum vulgare* endornavirus[a]	HvEV
	Malabar spinach	*Basella alba* endornavirus[a]	BaEV
	Melon	*Cucumis melo* endornavirus[a]	CmEV
	Seagrass	*Zostera marina* endornavirus[a]	ZmEV
Fungus	Violet root rot fungus	*Helicobasidium mompa* endornavirus 1–670	HmEV1–670
	Violet root rot fungus	*Helicobasidium mompa* endornavirus 1369[a]	HmEV1369
Protist	Phytophthora	*Phytophthora* endornavirus 1	PEV1

[a]Tentative member.

group from a UDP-sugar to a small hydrophobic molecule. UGTs have been identified in dsDNA viruses, including baculoviruses and nucleopolyhedroviruses from insects. In insect cells, the viral ecdysteroid UGT inactivates the molting hormones (ecdysteroids) of the host insect by sugar conjugation. Although a UGT gene has never been found in the genomes of the alpha-like supergroup of ssRNA viruses, it is found in strains of two hypoviruses (fungal dsRNA viruses), cryphonectria hypovirus 3 and 4 (CHV3 and CHV4). Amino acid sequences significantly similar to that of UGT are found in four of the five endornaviruses that have been sequenced in their entirety (**Figure 1**; they are not found in VfEV). Because the locations of the UGT motifs between the Hel and RdRp domains are conserved among the four endornaviruses, a common ancestor of the endornaviruses was likely to have also had a putative UGT gene. VfEV might have lost this gene during course of evolution, or a putative UGT gene encoded by VfEV might vary so much from other putative UGT genes that it was not identified as such. The functions of the putative UGT proteins in endornaviruses are unknown.

Because the single ORF of endornaviruses is very large (**Figure 1**), it very likely encodes several other proteins in addition to those already discussed. The ORF must encode a polyprotein, so it must also encode one or more proteinases to cleave the putative polyprotein into functional units (including RdRp, Hel, and UGT). A similar situation occurs in the potyviral genome, which contains a single large ORF encoding a polyprotein of more than 3000 aa residues; this also has proteinase activities which cleave the polyprotein into functional units. Furthermore, endornaviruses may encode an RNA-silencing suppressor, given that many other viruses contain this protein for protecting themselves against the host defense (RNA silencing) system.

Site-Specific Nick in the Coding Strand

All five endornaviruses that have been sequenced completely contain a site-specific nick in the 5′ region of the coding strand, which divides not only the coding strand but also the single long ORF (**Figure 1**). The biological implications of the nick remain unknown. However, it must affect at least two important steps in the life cycle of the endornavirus, because the divided coding strand can no longer be used as either a template for noncoding strand synthesis or an mRNA for translation of the putative polyprotein. The molecular mechanism that generates the nick is also unknown: an unknown endonuclease encoded by either the host genome or the endornavirus itself might cleave the coding strand site specifically. These two molecular features, the single long ORF and the site-specific nick, are unique, and have never been found in other known RNA viruses.

Inheritance (Vertical Transmission)

The transmission of endornaviruses relies essentially on cell division, and endornaviruses may survive in cooperation with their host cells. No horizontal spread has been observed in the field, and no potential vectors have been identified. OsEV-carrying plants and OsEV-free plants are found coexisting in the *japonica* rice cultivar Nipponbare, and these two isogenic lines cannot be distinguished on the basis of their appearance. Crossing experiments with OsEV-carrying and OsEV-free plants have indicated that the efficiency of OsEV transmission via pollen is more than 98%, and that of transmission via ova is 100%. Although OsEV is localized in the cytoplasm, it is transmissible to progeny plants via pollen as well as ova. Analyses of the F2 progeny plants of cv. Nipponbare indicated that the absence of OsEV is not associated with any particular gene(s) in the OsEV-free plants. The observed inheritance of OsEV is very different from that of other cytoplasmic genetic elements (e.g., chloroplasts and mitochondria), which are usually inherited only via egg cells. Thus, the propagation of OsEV, and probably other plant endornaviruses, may be entirely dependent on seed-mediated transmission. The high efficiency of transmission of OsEV via both pollen and ova is likely to be responsible for the wide distribution of OsEV in cultivars of *japonica* rice.

OsEV has been found in many cultivars of *japonica* rice (*O. sativa*) but not in any cultivars of *indica* rice (a subspecies of *O. sativa*). *Japonica* and *indica* are distinguishable on the basis of their phenotype (e.g., grain shape) and genotype (e.g., the distribution of transposons in their genomes). To determine the reason why OsEV is not found in *indica* rice cultivars, reciprocal crosses between an OsEV-carrier *japonica* variety (cv. Nipponbare) and an OsEV-free *indica* variety (cv. IR 26 or Kasalath) were performed. When cvs. IR 26 and Nipponbare were used, efficient transmission of OsEV via ova (93%) and pollen (89%) was observed. However, when the Kasalath and Nipponbare cultivars were used, OsEV transmission efficiency to F1 progeny was 68% via ova and 20% via pollen, and transmission to F2 progeny plants also followed a complicated, non-Mendelian inheritance pattern. These results suggest that OsEV is unstable in *indica* rice plants, which may lack one or more genes that are involved in the maintenance (replication) of OsEV.

Because OsEV is efficiently transmitted to progeny plants via pollen, reciprocal crosses between cultivated rice (*O. sativa*, cv. Nipponbare) and wild rice (*O. rufipogon*, W-1714) were performed to introduce two evolutionarily related endornaviruses (OsEV and OrEV) into F1 progeny plants. Unusual cytoplasmic inheritance of these endornaviruses was observed in F1 hybrids: the evolutionarily related endornaviruses were incompatible, and in some F1 plants the resident OsEV of an egg cell from cultivated rice was excluded by the incoming OrEV from a pollen cell from wild rice. The two coexisting endornaviruses in some F1 hybrids segregated away from each other in the F2 plants. However, the total number of the two coexisting endornaviruses in the F1 hybrid cells remained constant relative to the number of endornaviruses in the parent cells (approximately 100 copies/cell). The stringent regulation of endornavirus copy number may be responsible for their unusual inheritance pattern. This phenomenon is similar to the plasmid incompatibility that occurs with two evolutionarily related DNA plasmids in *Escherichia coli*.

Copy Number Regulation of OsEV

OsEV is found in every tissue as well as at every developmental stage in its host plants. Regardless of the host rice cultivar, the relative concentration of OsEV in comparable cell types is approximately constant. A comparison of the relative amounts of OsEV and rice DNA (430 Mb per haploid genome) showed that there are approximately 100 copies of OsEV per cell. No significant difference in OsEV concentration was found between seedlings, roots, mature leaves, and stems. Crossing experiments indicated that a mechanism for stringent regulation of endornavirus copy number may exist in host rice cells.

A 14-kb dsRNA has also been found in another ecospecies of cultivated rice, the tropical *japonica* rice cultivars grown in the Philippines, which are also classified as *O. sativa*. A partial nucleotide sequence from the RdRp region of the dsRNA from cv. Gendjah Gempel (tropical *japonica* rice) has been determined, revealing that this sequence has already diverged from that of the dsRNA from cv. Nipponbare (temperate *japonica* rice from Japan). The dsRNAs from cv. Nipponbare (temperate *japonica* rice) and cv. Gendjah Gempel (tropical *japonica* rice) are hereafter referred to as OsEV-N and OsEV-G, respectively. Although the three dsRNAs (OsEV-N, OsEV-G, and OrEV) in the genus *Oryza* have already diverged from one another at the nucleotide sequence level, the copy number of each dsRNA genome in their respective host cells is almost identical (100 copies per cell).

In contrast, the number of endornavirus copies in cultured cells was found to be about 10 times that in the cells of seedlings; a similarly high level of OsEV has been observed in each culture established from an independent temperate *japonica* seed (cv. Nipponbare). Although no changes in the copy number of OsEV were found in suspension cultures over the course of 3 years, the copy number returned to the original low level in rice plants regenerated from the cultured cells. An approximately tenfold increase of OsEV-G and OrEV copy numbers also occurred in suspension cultures of cv. Gendjah Gempel (tropical *japonica* rice) and W-1714 (wild rice) seeds, respectively. Thus, the copy numbers of OsEV-N, OsEV-G, and OrEV are dependent on the physiological conditions of the host plant, and may be maintained by a similar regulatory system in each rice genotype.

The amount of OsEV-N, OsEV-G, and OrEV also increases about 50-fold in the pollen grains of the host plants relative to the rest of the plant. There is no evidence for the horizontal transmission of these endornaviruses, so their propagation seems to depend on steady replication before every host cell division (not only mitosis but also meiosis) and efficient transmission to the next generation via egg and pollen. An unregulated increase in the amount of endornaviruses in cells could result in disease or death of the host plant, as seen with conventional viruses. Conversely, a decrease in the number of endornaviruses could cause the virus to disappear from a host plant (germ cells). The mating of host plants must be an opportunity for endornavirus propagation. An increase in the endornavirus copy number only in pollen grains must be a strategy to ensure their transmission to host progeny, despite their cytoplasmic localization. Unlike almost all other viruses, endornaviruses are not able to transmit horizontally, but they are capable of vertical transmission, which the majority of viruses cannot do.

Replication (RdRp Activity)

In 1990, the RdRp activity of endornaviruses was reported from a study of dsRNA in the 447 male sterile strain of broad bean (VfEV). Purified cytoplasmic membranous vesicles containing the dsRNA (VfEV) had RdRp activity, which required Mg^{2+} ions and the four nucleotide triphosphates, and was unaffected by inhibitors of cellular DNA-dependent RNA polymerases (e.g., alpha-amanitin and actinomycin D). Treatment of the vesicles with a nonionic detergent released the VfEV dsRNA together with the RdRp, which retained specific activity for synthesis of VfEV dsRNA.

RdRp activity can also be detected in the crude microsomal fraction of cultured rice cells that contain OsEV. Like VfEV RdRp activity, this RdRp activity was also highest in the presence of all four nucleotide triphosphates and Mg^{2+} ions, and was resistant to inhibitors of DNA-dependent RNA polymerases. This RdRp activity increased approximately 2.5-fold in the presence of 0.5% deoxycholate. Treatment of the purified microsomal fraction with proteinase K plus deoxycholate suggested that the RdRp enzyme complex, with its own 14-kb RNA template, is located in vesicles. So far, RdRp activity has been found only in two endornavirus-containing plants, broad bean containing VfEV and rice containing OsEV. Therefore, the available data concerning endornavirus replication are limited. There is a need to identify and characterize the endornavirus-and/or host-encoded proteins that are involved in endornavirus replication.

Molecular Evolution of Endornaviruses

Endornaviruses have been found in two ecospecies of cultivated rice (temperate *japonica* rice cultivars in Japan and tropical *japonica* rice cultivars in the Philippines), both of which are classified as *O. sativa*. They have also been found in one strain (W-1714) of wild rice (*O. rufipogon*, an ancestor of *O. sativa*). A comparison of the nucleotide and deduced amino acid sequences of the core regions of the RdRp domains of these three endornaviruses (OsEV-N, OsEV-G, and OrEV) indicates that the three endornaviruses are evolutionarily related and that OsEV-N is evolutionarily closer to OsEV-G than to OrEV. The relationships between these endornaviruses are consistent with the evolutionary relationships between the host plants. Therefore, the endornaviruses found in the genus *Oryza* are likely to have been transmitted vertically through seeds at least since the time when *O. sativa* and *O. rufipogon* diverged from each other several thousand years ago. These endornaviruses have probably evolved independently within their own host plants, which have always generated self-pollinated progeny.

Similar evolutionary divergence of endornaviruses within host plants has been reported for the bottle gourd (*L. siceraria*) endornaviruses. Large dsRNAs (*Lagenaria siceraria endornavirus*) are found in several cultivars of bottle gourd, including three Japanese cultivars and two Western cultivars. Northern blot analysis of these endornaviruses from five cultivars indicated that these endornaviruses have evolutionarily diverged from each other within their host plants. Therefore, because of their plasmid-like properties, namely vertical transmission only, endornaviruses in rice and bottle gourd are thought to have evolved within their host plants.

However, as shown in **Figure 2(b)**, the four endornaviruses in monocotyledonous plants (OsEV, OrEV, HvEV, and ZmEV) do not form a monophyletic group in the endornavirus phylogenetic tree. Similarly, the kidney bean endornavirus (PvEV) and the broad bean endornavirus (VfEV) do not cluster together, although their hosts both belong to the family Fabaceae Therefore, the phylogenetic relationships among these endornaviruses are not consistent with the phylogenetic relationships of their host plants. There are three possible explanations for these unexpected relationships between endornaviruses and their hosts. (1) Endornaviruses have speciated independently of their hosts. Several types of endornaviruses may have diverged before monocotyledonous and dicotyledonous plants diverged, and the families Poaceae (rice and barley) and Fabaceae (kidney and broad beans) now carry at least two independent lineages of endornaviruses. (2) The ancestors of endornaviruses were horizontally transmissible. (3) Some endornaviruses are horizontally transmissible, but horizontal transmissions occur only very rarely. The last two explanations may conceivably be possible, because the rare transmission events could have occurred during a sufficiently long time, over millions of years.

dsRNAs of various sizes are frequently found in plants, fungi, protozoans, and insects; in fact, dsRNAs (or dsRNA viruses) with plasmid-like properties can be considered to be widely distributed throughout all four eukaryotic kingdoms. Plasmid-like dsRNA replicons including endornaviruses and ssRNA viruses could have evolved from a common RNA replicon ancestor, in the process developing two different propagation strategies: ssRNA viruses became independent of the host cell (e.g., developed rapid replication of its genome RNA(s) and extracellular stability by encapsulation within a capsid protein), and dsRNA replicons developed a highly symbiotic relationship with the host cell (e.g., developed stringent copy-number regulation and efficient vertical transmission). dsRNA replicons have probably coevolved with their host cells to be symbiotic. Indeed, efficient vertical transmission and constant copy number are likely necessary to ensure the stability of the symbiotic relationship.

See also: Alphacryptovirus and Betacryptovirus; Hypoviruses; Totiviruses.

Further Reading

Brown GG and Finnegan PM (1989) RNA plasmids. *International Review of Cytology* 117: 1–56.

Fukuhara T (1999) Double-stranded RNA in rice. *Journal of Plant Research* 112: 131–138.

Gibbs MJ, Pfeiffer P, and Fukuhara T (2004) Genus *Endornavirus*. In: Fauquet CM, Mayo MA, Maniloff J, Desselberger U, and Ball LA (eds.) *Virus Taxonomy: Eighth Report of the International Committee on Taxonomy of Viruses*, pp. 603–605. San Diego, CA: Elsevier Academic Press.

Tavantzis SM (ed.) (2001) *dsRNA Genetic Elements: Concepts and Applications in Agriculture, Forestry, and Medicine*. Boca Raton, FL: CRC Press.

Enteric Viruses

R F Bishop and C D Kirkwood, Murdoch Childrens Research Institute Royal Children's Hospital, Melbourne, VIC, Australia

© 2008 Elsevier Ltd. All rights reserved.

Introduction

The intestinal tract, lined by replicating epithelial cells, bathed in nutrient fluids and maintained at optimal temperature provides an ideal milieu for growth of many viruses. 'Enteric viruses' represent a wide spectrum of viral genera that invade and replicate in the mucosa of the intestinal tract, and that can be grouped as follows:

- viruses causing localized inflammation at any level of the intestinal tract, predominantly in small intestinal mucosa, resulting in acute gastroenteritis, for example, rotaviruses, caliciviruses, adenoviruses, astroviruses;
- viruses that multiply at any level of the intestinal tract, causing few enteric symptoms prior to producing clinical disease at a distant site, for example, measles virus, reoviruses (in mice), enteroviruses (including polioviruses, coxsackieviruses, enteroviruses, hepatitis A and E); and
- viruses that spread to the intestinal tract during the later stages of systemic disease, generally in an immunocompromised host, for example, human immunodeficiency virus (HIV), cytomegalovirus.

This article focuses upon the first category of viruses that cause enteric disease associated with primary replication in the intestinal tract.

Viruses Associated with Acute Gastroenteritis

Acute gastroenteritis is one of the most common health problems worldwide. More than 700 million cases are estimated to occur annually in children less than 5 years of age, resulting in few deaths in developed countries, but more than 2 million deaths in developing countries.

Worldwide, a diverse group of viral, bacterial, and parasitic pathogens cause acute enteric symptoms including nausea, vomiting, abdominal pain, fever, and acute diarrhea. Infections with viral agents, unlike those with bacterial or parasitic pathogens, cannot be treated with antibiotics, and many cannot be prevented by improvements in quality of drinking water, food, or sanitation.

Until the early 1970s most viral agents causing gastroenteritis in humans were largely unknown. Studies using electron microscopy of intestinal contents resulted in the discovery of numerous viral enteropathogens now classified as caliciviruses, rotaviruses, astroviruses, or 'enteric' adenoviruses. Caliciviruses are now recognized as the most important cause worldwide of outbreaks of viral gastroenteritis in humans of all age groups. Rotaviruses are the single most important cause of life-threatening diarrhea in children <5 years old. Astroviruses and adenoviruses also cause severe diarrhea in children. **Table 1** lists the characteristics of the major viruses associated with acute gastroenteritis. Other viruses linked to gastroenteritis in humans include coronaviruses, toroviruses, picornaviruses, and picobirnaviruses. Understanding many features of these 'enteric viruses' has been based on parallel studies of related viruses infecting animals.

Caliciviruses

History

The family *Caliciviridae* contain small RNA viruses that cause enteric disease in a wide variety of hosts including cattle, pigs, rabbits, and humans. Infections in other hosts, for example, sea lions, cats, and primates, appear to cause predominantly systemic and respiratory symptoms. Caliciviruses are small nonenveloped viruses of 27–35 nm diameter (**Figure 1**) with a genome comprising

Table 1 Characteristics of major enteric viruses causing acute gastroenteritis in humans

Characteristic Family	Norovirus/ Sapovirus Caliciviridae	Rotavirus Reoviridae	Adenovirus Adenoviridae	Astrovirus Astroviridae
	Divided into genogroups, each with distinct genetic clusters	Six groups (A–F) Group A-multiple serotypes based on outer capsid proteins.	Six subgenera, more than 50 serotypes Enteric serotypes (Ad40–41)	Eight serotypes
Virion size (nm)	28–35	70	80	28
Capsid organization	Two structural proteins (orf2–56–62 kDa, orf3–22 kDa)	Two outer capsid proteins (VP7–38 kDa, VP4–88 kDa) Two inner protein layers (VP6–41 kDa, VP2–88 kDa)	Capsomer – composed of three proteins: hexon, penton, and fiber.	Precursor cleaved into several proteins (e.g., 20 kDa, 29 kDa and 31 kDa)
Nucleic acid	ssRNA (plus sense)	dsRNA	dsDNA	ssRNA (plus sense)
Genome organization	Three open reading frames (1, 2, and 3)	11 segments which encode specific protein.	Linear chromosome with multiple transcription/ translation units	2 open reading frames (1a, 1b, and 2)

ss, single-stranded; ds, double-stranded; kDa, kilodalton.

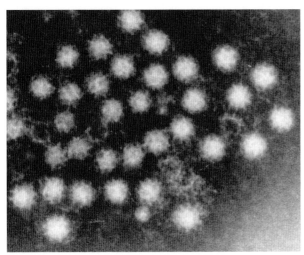

Figure 1 Electron micrograph of negatively stained calicivirus particles (NoV) in fecal extract.

a single-stranded positive-sense polyadenylated RNA genome of 7400–7700 nucleotides (nt). Typical calicivirus particles show cup-like hollows (calices) on the virus surface. Some caliciviruses have an indistinct appearance described as 'feathery'.

The 'Norwalk agent' (now classified as a calicivirus) was identified in 1972 by Kapikian and colleagues, using immune-electron microscopy (IEM) to search for the causative agent of an 1968 outbreak of gastroenteritis in humans in Norwalk, Ohio, USA. Many related viruses have been implicated as causes of gastroenteritis, and given names identifying the geographical location of the outbreak (e.g., Marin County, Snow Mountain, Hawaii, Sapporo).

Classification

Classification of caliciviruses was hampered for many years by the inability to culture these viruses, and study their genetic and protein structure. Caliciviruses causing enteric infections (in humans and other animals) are classified as belonging to the family *Caliciviridae*, which is divided into four genera. The genus *Norovirus* (NoV), and genus *Sapovirus* (SaV) cause human and animal infections. Other genera infect rabbits (*Lagovirus*) or sea-lions, cats, and primates (*Vesivirus*). The genome organization and reading frame usage differs between the four genera. For noroviruses, the genome contains three open reading frames (ORFs). The largest (ORF1) encodes a polyprotein which undergoes proteolytic cleavage to produce an NTPase, a 3c-like protease, and an RNA-dependent RNA polymerase (RdRp). ORF2 encodes the major capsid protein and ORF3 a putative minor capsid protein.

Noroviruses are subdivided into five genogroups (GI–GV). Genogroups GI, GII, and GIV infect humans, GIII infects cattle and GV infects mice. GI and GII contain at least 10 and 20 distinct genetic clusters, respectively. Sapoviruses are divided into genogroups (GI–GV), of which GI, GII, GIV, and GV infect humans. GI and GII are further divided into four and three genetic clusters.

The inability to culture human caliciviruses delayed the introduction of diagnostic tests, resulting in an under-appreciation of the significance of these agents for many years. Currently, over 20 different reverse transcription-polymerase chain reaction (RT-PCR) assays targeting regions on the RdRp gene and the capsid gene have been described and utilized in epidemiological studies. This large genetic diversity of human caliciviruses makes routine detection difficult.

Geographic and Seasonal Distribution

Human NoV are the leading causes of 'nonbacterial gastroenteritis' outbreaks in all age groups worldwide. Outbreaks frequently occur in communities such as nursing homes, hospitals, schools, and cruise ships. No consistent seasonal variation has been observed. Infection involves transmission via person-to-person contact or ingestion of contaminated food and water.

Epidemiological studies have identified caliciviruses in 60–95% of outbreaks in many countries. Estimates of disease burden in USA suggest that caliciviruses are responsible annually for 23 million illnesses and 50 000 hospitalizations. Strains of NoV GII cluster 4 (GII-4) have been the most common type identified worldwide in the past 5 years (2001–2005) in both adults and children. Prevalence rates of calicivirus (predominantly NoV) in young children admitted to hospital with acute gastroenteritis in many countries range from 3.5% to 20% annually. Strains of SaV, while playing a minor role overall, are more generally associated with childhood gastroenteritis than with disease in older children and adults. Caliciviruses may be important enteric pathogens in patients with hereditary or acquired immunodeficiency.

Genetics

The genera *Norovirus* and *Sapovirus* are genetically diverse, and multiple strains co-circulate in human populations. Individual dominant strains emerge every 2–5 years, and often have a global impact, such as the GII-4 strains identified in USA, Europe, Japan, and Australia in 1995/1996 and again in 2004. NoV recombinant strains, with polymerase and capsid genes derived from different ancestral clusters, have been identified in Thailand and Australia. Repeated attempts to adapt NoV and SaV to growth in cell culture have failed. Diagnostic techniques and analysis of antigenic variation rely predominantly on molecular biological techniques. Cloning and expression of the major viral capsid protein (VP1) in baculovirus expression systems has led to the formation of virus-like particles (VLPs) morphologically similar to native virus and their incorporation into enzyme immunoassay (EIA) assays.

Pathogenesis

Pathogenesis of NoV and SaV infection is poorly understood as a result of the long-standing inability to adapt these viruses to cell culture, and the absence of a small animal model. Symptomatic enteritis in human volunteers infected with 'Norwalk agent' showed changes in jejunal biopsies (mucosal inflammation, absorptive cell abnormalities, villus shortening, and crypt hypertrophy) that persisted for at least 4 days after remission of clinical symptoms and reverted to normal after 2 weeks. No identifiable viral particles were detected by electron microscopy in any affected intestinal tissue. The recent demonstration that human noroviruses can infect and replicate in a three-dimensional cell culture model of human intestinal epithelium, should improve our understanding of the pathogenesis, and antigenic diversity of this important group of enteric viruses. These studies will also be enhanced by discovery of a norovirus that infects mice, and that replicates after transfection of cultured kidney cells.

Immune Responses, Prevention and Control

Mechanisms of immunity to NoV are unclear. Infection results in formation of IgG and IgM serum antibody that are broadly reactive within, but not between, genogroups. The role of these antibodies in immune protection is unknown. Infected individuals can develop short-term immunity to homologous viruses but the molecular diversity of NoV circulating in communities makes it difficult to predict whether long-term immunity can develop. It is unclear why a proportion of exposed individuals remain uninfected during outbreaks. Recent studies suggest that histo-blood group antigens and the secretor status may be genetic susceptibility markers for infection. At present the major control strategies for prevention of human calicivirus infection rely on prevention of contamination of food and water supplies.

Rotaviruses

History

Virus particles, later classified in the genus *Rotavirus*, were first described in 1963 by Adams and Kraft as a cause of epidemic diarrhea in infant mice (EDIM). Similar particles (NCDV) were recognized in 1969 by Mebus and colleagues as a cause of severe diarrhea in newborn calves in Nebraska, USA. Neither virus was considered relevant as a causative agent of severe diarrhea in young children until 1973 when Bishop, Davidson, Holmes, and Ruck described a 'new virus' (later shown to be antigenically related to EDIM and NCDV) in duodenal biopsies and diarrheal feces from young children admitted to hospital in Melbourne, Australia with severe acute diarrhea. Named because of their wheel-like appearance in negatively stained extracts examined by electron microscopy (rota = Latin for wheel) rotaviruses have since become established as causes of severe acute diarrhea in the young of many mammalian and avian species worldwide. Rotavirus enteritis affects all children regardless of socioeconomic status, and results in over 600 000 deaths annually in young children in developing countries.

Classification

Rotaviruses are nonenveloped icosahedral viruses of 70 nm (**Figure 2**) diameter that belong to the genus *Rotavirus*

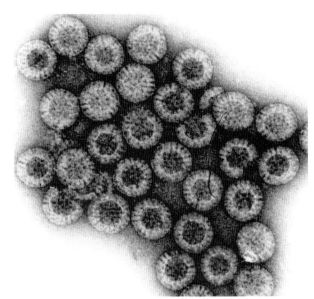

Figure 2 Electron micrograph of negatively stained triple-shelled rotavirus particles in fecal extract.

within the family *Reoviridae*. The double-stranded RNA genome is contained within a triple-layer of viral proteins (VPs) comprising a core (VP1, VP2, VP3), an inner capsid (VP6), and an outer capsid (VP4, VP7). Rotaviruses are classified into groups A to G based on serology of the VP6 protein. The majority of human and mammalian infections, due to group A viruses, are further classified into serotypes by antigenic differences on VP7 (G-serotypes) and into genotypes by genetic differences on VP4 (P-genotypes). To date there are at least 11 of 15 G-serotypes and 15 of 26 P-genotypes identified in humans. Groups B and C have been identified infrequently in humans. Groups D to G have been identified only in nonhuman mammalian or avian species.

Geographic and Seasonal Distribution

Five rotavirus serotypes G1P[8], G3P[8], G4P[8], G2P[4], and G9P[8] have been the most common serotypes causing severe human disease globally during the past 30 years. G1P[8] strains have been consistently present worldwide. Yearly winter epidemics of rotavirus disease are regularly observed in countries with temperate climates, whereas rotavirus disease is prevalent year-round in tropical climates lacking defined winter seasons.

Host Range

Group A rotaviruses infect humans and other mammals repeatedly throughout life. Most primary rotavirus infections in animals occur during the neonatal period. Most primary rotavirus infections in humans occur during the first 24 months of life. Children worldwide will experience at least one rotavirus infection by 5 years of age.

In general, group A rotaviruses are species specific. However, rotavirus strains with gene segments of feline, bovine, or porcine origin have been isolated from children, suggesting the occurrence of cross-species infection in nature. Cross-species infections can be established experimentally, and comprise one of the strategies for human vaccine development.

Genetics

The rotavirus genome consists of 11 segments of double-stranded RNA that can be separated by polyacrylamide gel electrophoresis, allowing epidemiological studies mapping the genetic diversity of strains within and between serotypes. Each gene segment encodes a separate protein, with the exception of gene segment 11 which encodes two proteins. Reassortment of genes between human strains and human and animal strains occurs *in vivo* and *in vitro*.

Pathogenesis

Rotaviruses are transmitted from person to person by the fecal–oral route, or via aerosols. Rotaviruses replicate in the cytoplasm of mature nonreplicating enterocytes lining the upper portions of the small intestinal villi, eventually causing cytolysis. Profuse watery diarrhea results from a combination of mechanisms including malabsorption secondary to loss of enterocytes responsible for absorption and digestion, activation of the enteric nervous system, and stimulation of intestinal secretion by the rotavirus nonstructural protein NSP4. Rotavirus antigenemia and viremia occur during the acute phase of severe primary rotavirus disease. As a result, complete rotavirus particles have been found in liver, lung, spleen, pancreas, thymus, and kidneys of experimental animals. It is not clear if the virus is replicating at these sites.

Clinical Features

Clinical symptoms in children are strongly influenced by age, with severe often life-threatening diarrhea occurring after primary infection in young children, and in aged people in nursing homes. Excretion of rotavirus particles in detectable numbers (by EIA, RT/PCR) continues for 5–10 days, and occasionally up to 50 days. Excretion can continue for months in immunodeficient children and animals. Reinfections occur throughout life, and are usually asymptomatic or associated with mild symptoms. Symptoms of primary infection require medical attention in 1:5 children, result in hospitalization in 1:65 children and death in 1:293 (almost all in young children in developing countries). Treatment is based upon replacement of fluid and electrolyte loss, usually achieved by oral administration of fluids containing glucose and electrolytes. Occasionally, delayed repair of the small intestinal mucosa is associated with disaccharide or monosaccharide malabsorption leading to malnutrition.

Immune Response/Prevention and Control

Primary rotavirus infection protects against severe symptomatic disease on reinfection, and is associated with humoral and cellular immune responses to individual rotavirus proteins. Neutralizing antibody to VP4 and VP7 outer capsid proteins contribute to protection, possibly by interfering with viral replication and limiting the extent of intestinal damage. The role of immune responses to other proteins is uncertain. Virus-specific cytotoxic T cells are not essential for protection.

The importance of rotavirus disease worldwide and its contribution to childhood mortality in developing countries has resulted in strong initiatives, supported by the World Health Organization, to develop live oral rotavirus vaccines to be administered to infants before 3 months of age. Two contrasting vaccines, a single attenuated G1P[8] human rotavirus and a pentavalent human–bovine G1-G4,P[8] reassortant vaccine, have been proved to be safe and effective in preventing severe rotavirus disease. Both have the potential to radically change global childhood mortality and morbidity.

Adenovirus

History and Classification

Adenoviruses were first detected in 1953 in cultured fragments of tonsillar and adenoidal tissue from children. They are nonenveloped icosahedral viruses approximately 80 nm in diameter (**Figure 3**). The genome is composed of double-stranded DNA. The family *Adenoviridae* comprises three genera: *Mastadenovirus* (mammalian) classified into subgenera A–F representing more than 50 serotypes, *Aviadenovirus* (birds), and a newly recognized genus *Atadenovirus* identified in sheep and reptiles. Most are readily cultivatable. Adenovirus infections occur worldwide in many mammalian species, are species specific, usually associated with disease in the respiratory, urinary, and ocular systems, and are frequently shed in feces in the absence of any gastrointestinal symptoms.

In 1975, Flewett and colleagues in Birmingham, UK, noticed the presence of large numbers of adenovirus particles in negatively stained extracts of diarrheal stools examined by EM. These proved difficult to culture, were designated 'enteric' adenoviruses (EAd), and are now classified as serotypes EAd40 and EAd41 within subgenus group F. Cultivation of EAd remains difficult. The most reliable growth has been achieved in human embryonic kidney cells (293 cells) immortalized by transfection with regions of Ad5.

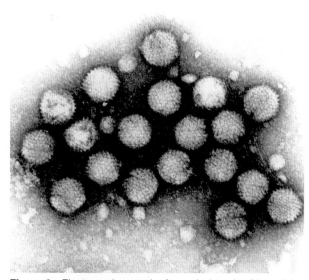

Figure 3 Electron micrograph of negatively stained 'enteric' adenovirus particles (showing characteristic hexagonal shape) in fecal extract. Courtesy of professor M. Studdert.

Geographic and Seasonal Distribution

EAd40 and EAd41 occur worldwide, causing severe acute enteritis in 5–15% of hospitalized young children. Outbreaks occur at unpredictable intervals year-round with no seasonal prevalence. Nosocomial epidemics occur in day-care nurseries and in hospital wards for children and adults. Group A adenoviruses (serotypes 12, 18, 31) have also been implicated in epidemics, usually in older age groups.

Genetics/Evolution

The adenovirus virion is composed of at least 10 different structural polypeptides and contains a linear 33–45 kbp DNA. Virus capsomers are arranged as hexons, the corners of which have antenna-like (fiber) projections presumed involved in cell attachment. The DNA genomes of groups A–F are genetically diverse, and differences can be illustrated by analysis using genome restriction endonucleases. The heterogenous genome in groups A–F makes recombination between subgenera unlikely, with exception of groups A and F which show a close evolutionary relationship.

Diagnoses of EAd infection rely on EIA that detects the hexon antigen common to groups A–F, followed by determination of restriction enzyme patterns and/or reactions in EIA incorporating neutralizing, monoclonal antibodies specific for Ed40 and 41.

Pathogenesis

EAd replicate within the epithelial cells of the small intestine. Group A adenoviruses have also been grown from mesenteric lymph nodes and appendices. The mechanisms causing diarrhea are not clear, but destruction of infected epithelial cells has a role.

Clinical Features

Adenovirus diarrhea is more common in infants <12 months old than in older children, and can be protracted with a mean duration of 12 days. Adenovirus diarrhea occurs in immunocompromised patients. Nonseasonal epidemics of EAd diarrhea occur in hospital wards, orphanages, and day-care nurseries. Occasional fatal cases have been reported in children. Evidence from animal models (with non-EAd) suggests that viremia occurs, and can lead to infection of other tissues. The natural history of disease and development of immunity is unknown.

Astroviruses

History and Classification

Astroviruses were first described in the UK in 1975 by Madeley and Cosgrove studying an outbreak of diarrhea in newborn babies in an obstetric hospital nursery. Astroviruses are small, round nonenveloped plus-stranded RNA viruses 28–30 nm diameter (**Figure 4**) occasionally exhibiting virions with a superficial star shape. They are members of the genus *Astrovirus* in the family *Astroviridae*. They have been detected in humans (children and adults) and a range of mammalian (sheep, cattle, pigs, dogs, cats, and mice) and avian (turkeys, ducks) species, usually associated with diarrhea. There are currently eight serotypes of human astrovirus, designated HAstV 1–8, based on reactivity with polyclonal antisera. HAstV 1 is most common worldwide.

Prevalence rates as a cause of diarrhea vary from 2% to 16% (hospital-based studies), and 5% to 17% (community-based studies). Most astrovirus infections have been recorded during colder months in temperate climates and year-round in tropical countries. A longitudinal study in Mayan children in a poor community in Mexico found a high prevalence (61%) of astrovirus infection in a birth cohort of 271 children followed for 3 years. Infection occurred primarily in infants <12 months old, and showed a high rate of asymptomatic infection and prolonged shedding (2–17 weeks) in many infants. Astrovirus infection has also been associated with persistent diarrhea (lasting for 14 days or more) in children in Bangladesh. Astroviruses are widespread in developed countries, causing outbreaks in day-care centers, hospitals, and nursing homes for the elderly. They are an important cause of enteritis in immunocompromised patients.

Genetics and Evolution

The HAst genome is a polyadenylated plus-stranded RNA molecule of approximately 7 kbp. The genome contains two open reading frames (ORFs), ORF1a and -1b, code for nonstructural proteins and ORF2 encodes for the capsid protein. Genetic diversity in all serotypes exists, but no association has been shown between serotypes and ability to cause severe gastroenteritis. Astrovirus diagnostic assays include commercially available EIA kits, electron microscopy and RT-PCR detection and genotyping of diarrheal feces.

Pathogenesis

Acute astrovirus infection induces a mild watery diarrhea in young children that lasts for 2–3 days and may be associated with vomiting, fever, and anorexia. The lack of a small animal model has hampered studies of the mechanism of astrovirus-induced diarrhea. Experimental models of astrovirus enteritis in turkeys and in gnotobiotic lambs show mild histopathological changes in the intestine (despite high mortality from severe osmotic diarrhea) together with viremia. Experimental astrovirus infection in calves is asymptomatic, with viral replication apparently targeted to M cells. It is possible that none of these animal models illuminate pathogenesis of HAstV infection in humans. Cultivation of HAstV was initially difficult, but can now regularly be achieved using a human colon cancer derived epithelial cell line (CaCo2 cells).

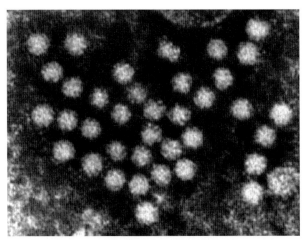

Figure 4 Electron micrograph of negatively stained astrovirus particles (showing star-shape) in fecal extract.

Coronaviruses/Toroviruses

Members of the genera *Coronavirus* and *Torovirus* are enveloped plus-strand single-stranded RNA viruses belonging to the family *Coronaviridae*. Electron microscopy shows them to be pleomorphic fringed particles 100–140 nm at maximum dimension (**Figure 5**). Coronaviruses and toroviruses can be distinguished by differences in peplomer structure and reaction in IEM using specific antisera.

Coronaviruses and toroviruses cause diarrhea, respiratory, and/or hepatic disease in many animal species, including cattle, mice, swine, cats, and dogs. In general, most of these viruses are species specific and disease is most severe in infant animals. Transmission is fecal–oral and due to virus lability may require close contact. Coronaviruses and toroviruses have been implicated in human diarrheal disease but there is still no consensus about their importance. Similar particles have been seen frequently in children without diarrhea, particularly in children in developing countries. Morphological similarities between these viruses and fragments of intestinal brush border make diagnosis difficult. Several studies have implicated coronaviruses as causative agents of necrotizing enterocolitis outbreaks in newborn babies.

Toroviruses were first described as a cause of diarrhea by Woode and colleagues in 1979 when Breda virus was identified in a severe outbreak of neonatal calf diarrhea in USA. Toroviruses are now also known to infect horses (Berne virus) and swine. Human infections were first described by Flewett *et al.* in Birmingham UK in 1984, but have rarely been reported since then.

The pathogenesis of diarrhea has been studied in animal models using infection with coronavirus (TGE) in piglets, and with Breda viruses in calves. Both replicate in epithelial cells of small intestine and descending colon causing diarrhea 24–72 h later. Breda virus also replicates in crypt cells.

Many animal coronaviruses can be propagated in cell culture. Isolation of human enteric coronaviruses is difficult and serological studies can be confounded by antibody resulting from repeated respiratory coronavirus infection. Enteric infection can be confirmed in feces by detection of viral RNA by RT-PCR, or IEM using antibodies to the viral envelope glycoproteins.

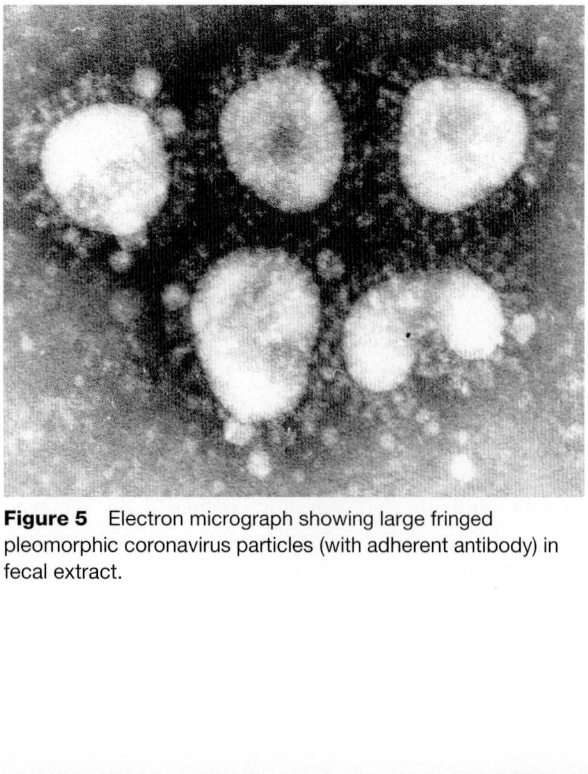

Picornaviruses

Picornaviruses are 24–30 nm featureless spherical particles containing single-stranded positive-sense RNA. They have been found in diarrheal feces from humans but their etiological role is often not clear. The first clear evidence implicating a picornavirus as an enteric pathogen identified Aichi virus, as a cause of oyster-associated epidemics of gastroenteritis in Japan in 1989. Aichi viruses are now classified as a new genus *Kobuvirus* within the family *Picornaviridae* (kobu = Japanese for knob). Isolation of Aichi virus in Vero cells has permitted development of EIA and RT-PCR assays based on nucleotide sequence data. Serological assays show seroconversion resulting from infection and a prevalence rate for antibody of 7.2% in Japanese children aged 7 months to 4 years and rising to 80% in adults by age 35.

The new genus *Parechovirus* within the family *Picornaviridae* contains at least one serotype (previously echovirus 22) that has been implicated as an enteric pathogen in humans.

Parvoviruses

Parvoviruses are small 22–26 nm single-stranded DNA viruses comprising a genus *Parvovirus* in the family *Parvoviridae*. Some animal parvoviruses have been clearly linked to enteritis including bovine, feline, mink, and canine strains. Canine parvovirus infection emerged after 1977. This lethal neonatal enteric infection, accompanied by viremia and widespread systemic infection, shows a pathogenesis distinct from most enteropathogenic viruses. The virus infects and destroys crypt epithelial cells resulting in flat mucosa with fused and stunted villi. Damage has been likened to that caused by radiation.

Other small viruses, resembling parvoviruses, have been seen by EM in diarrheal feces in humans. Evidence linking them to causation of disease is not convincing. They have often been present as dual infections with known enteric pathogens. In addition, their resemblance to some phages makes diagnosis uncertain.

Figure 5 Electron micrograph showing large fringed pleomorphic coronavirus particles (with adherent antibody) in fecal extract.

Picobirnaviruses

These are a group of currently unclassified small viruses detected in the feces of humans and animals without diarrhea. Picobirnaviruses are 35–41 nm particles with a bi- or tri-segmented dsRNA genome and have been detected in Europe, South America, and Australia. They are found significantly more often in patients with HIV-related diarrhea than those without diarrhea. Their role in gastroenteritis in healthy individuals remains unknown.

See also: Astroviruses; Birnaviruses; Enteroviruses: Human Enteroviruses Numbered 68 and Beyond.

Further Reading

Atmar RL and Estes MK (2001) Diagnosis of noncultivable gastroenteritis viruses, the human caliciviruses. *Clinical Microbiology Reviews* 14: 15–37.

Bresee JS, Widdowson M-A, Monroe SS, and Glass RI (2002) Foodborne viral gastroenteritis: Challenges and opportunities. *Clinical Infectious Diseases* 35: 748–753.

Chadwick D and Goode JA (2001) *Gastroenteritis Viruses. Novartis Foundation Symposium*, p. 238. New York: Wiley.

Chiba S, Estes MK, Nakata S and Calisher CH (eds.) (1996) *Viral gastroenteritis. Archives of Virology* (supplement 12): 119–128.

Duckmanton L, Luan B, Devenish J, Tellier R, and Petric M (1997) Characterisation of torovirus from human faecal specimens. *Virology* 239: 158–168.

Franco MA, Angel J, and Greenberg HB (2006) Immunity and correlates of protection for rotavirus vaccines. *Vaccine* 24: 2718–2731.

Glass RI (2006) New hope for defeating rotavirus. *Scientific American* 294: 33–39.

Hall GA (1987) Comparative pathology of infection by novel diarrhoea viruses. In: Bock G and Whelan J (eds.) *Novel Diarrhoea Viruses. Ciba Foundation Symposium,* vol. 128, pp. 192–217. Chichester: Wiley.

Hansman GS, Oka T, Katayami K, and Takeda N (2007) Human Sapovirus: Genetic diversity, recombination and classification. *Reviews in Medical Virology* 17: 133–141.

Mendez-Toss M, Griffin DD, Calva J, *et al.* (2004) Prevalence and genetic diversity of human astroviruses in Mexican children with symptomatic and asymptomatic infections. *Journal of Clinical Microbiology* 42: 151–157.

Ramig RF (1997) Genetics of the rotaviruses. *Annual Reviews of Microbiology* 51: 225–255.

Ramig RF (2004) Pathogenesis of intestinal and systemic rotavirus infection. *Journal of Virology* 78: 10213–10220.

Schwab KJ, Estes MK, and Atmar RL (2000) Norwalk and other human caliciviruses: Molecular characterization, epidemiology and pathogenesis. In: Cary JW, Linz JE and Bhatnagar D (eds.) *Microbial Foodborne Diseases: Mechanisms of Pathogenesis and Toxin Synthesis*, pp. 469–493. Lancaster, USA: Technomic Publishing Company.

Straub TM, Honer zu Bentrup K, Orosz-Coghlan P, *et al.* (2007) In vitro cell culture infectivity assay for human noroviruses. *Emerging Infectious Diseases* 13: 396–403.

Tiemessen CT and Kidd AH (1995) The subgroup F adenoviruses. *Journal of General Virology* 76: 481–497.

Yamashita T, Sugiyama M, Tsuzuki H, *et al.* (2000) Application of a reverse transcription-PCR for identification and differentiation of Aichi virus, a new member of the picornavirus family associated with gastroenteritis in humans. *Journal of Clinical Microbiology* 38: 2955–2961.

Enteroviruses of Animals

L E Hughes and M D Ryan, University of St. Andrews, St. Andrews, UK

© 2008 Elsevier Ltd. All rights reserved.

Introduction

In the 1950s, the use of monkey kidney cell cultures for the growth of poliovirus revealed the presence of simian viruses and further study showed some of these to have properties consistent with enteroviruses. In parallel, investigations of viruses infecting domestic animals also revealed the presence of enteroviruses (and enterovirus-like particles) in pigs and cattle. Enteroviruses have since been isolated from African buffalo, water buffalo, sheep, goat, deer, and impala and many have been shown to be related to bovine enterovirus isolates.

In the past, the classification of enteroviruses was primarily based upon the physico-chemical properties of virions, growth in tissue-cultured cell lines, and by serotyping. In practice, this can be difficult with some isolates being poorly recognized by the reference antisera, or, the occurrence of (misleading) cross-reactivities between serotypes. The expansion of the sequence database together with more sensitive cloning/sequencing techniques have facilitated the elucidation of the genome structures of many picornaviruses. Such analyses have replaced other techniques in the classification of picornaviruses, and this article discusses characteristics which are important for the classification of animal enteroviruses.

Animal Enteroviruses

Bovine Enteroviruses

Bovine enteroviruses (BEVs) are endemic in cattle in many regions of the world with infection typically asymptomatic and apparently healthy animals acting as carriers. There has, however, been some association of BEVs with diarrhoea and abortion. Studies in Spain have shown that BEV is widespread and variants co-circulate in cattle around the country, with the virus less prevalent in cattle from extensive farms (69%) than in cattle from intensive farms (94%). Analysis of samples collected from a farm in the USA indicated that the virus was present in the spring in 2–4-month-old calves but that the infection had probably been cleared by summer.

Recently, it has been shown that infectious virus particles are present in water from animal watering tanks, pastures, and streams/rivers in regions where BEVs are endemic, demonstrating the ability of these viruses to survive in aqueous environments. Interest in these viruses has widened over recent years with proposals that they can be utilized as markers for fecal contamination of the environment. BEVs have also been proposed as surrogate viruses for evaluating FMDV contamination of farms and the evaluation of extraction/detection techniques.

Porcine Enteroviruses

Historically, the more economically significant diseases of pigs caused by the porcine enteroviruses (PEVs) included neurological disorders (Teschen/Talfan disease), fertility disorders, dermal lesions, pneumonia, and diarrhoea. With the major reappraisal of the taxonomy of these viruses, these diseases are now associated with the teschoviruses (see below).

Swine Vesicular Disease Virus

Swine vesicular disease virus (SVDV) causes a highly contagious disease of pigs that spreads rapidly through contact with infected pigs and a contaminated environment. The disease is variable from mild or subclinical infections to lesions on the snout and feet that are indistinguishable from those caused by FMDV. For this reason, routine surveillance of SVDV is maintained in European countries.

SVDV was first described in Italy in 1966 and since then, numerous outbreaks have occurred throughout Europe and Asia. SVDV was largely eradicated from Europe during the 1970s and 1980s, but a new strain, possibly originating in the Far East, entered Europe during 1992. This strain spread to the Netherlands, Belgium, Portugal, Spain, and Italy. In 2003–04, clinical outbreaks were only reported in Portugal (leading to the slaughter of 2168 pigs) while subclinical infections are continuously detected in southern Italy. A long-term study of pigs infected with a recent Italian isolate showed the vRNA and virus could be detected long after the initial infection, and provided good preliminary evidence that this virus can establish persistent infections.

Simian Enteroviruses

Many simian viruses – including enteroviruses – were isolated from a range of nonhuman primate tissue-cultured cells used in biomedical research and vaccine development and from primates used for biomedical research. Twenty enterovirus serotypes were defined, with cross-reactivity in some cases. All these isolates were, however, distinct in that they did not show cross-reactivity with the (then) known human enteroviruses. Very little is known about the pathogenesis of these viruses. Simian enterovirus A (A-2 plaque virus) was isolated from a human, but caused viremia in tamarin monkeys 1–2 weeks post inoculation. Virus could not be detected in fecal samples.

'Animal' versus 'Human' Enteroviruses

Viruses of the genus *Enterovirus* of the family *Picornaviridae* primarily cause infections of the gastrointestinal tract, where large numbers of progeny virions are produced and shed in the feces. The virus particles are stable in a wide range of pH, temperature, and salinity conditions and may remain infective in the environment for long periods.

A number of viruses formerly thought to be enteroviruses are now known to be members of other genera. Presently, those animal viruses assigned to the enteroviruses include SVDV, BEV-1 and -2, and PEV-9 and -10.

Currently five human and three animal species comprise the genus *Enterovirus*. The separation of the genus into human and animal species is not as clear as in the past. Nucleotide sequence and serological data strongly suggest that some viruses have passed from man to domestic animals, or vice versa. For many years, SVDV was known to have a close relationship with human coxsackievirus B5 (CBV-5) and, indeed, this animal enterovirus is now classified in the species *Human enterovirus B* (HEV-B). The antigenic and molecular relationships between the two viruses suggest that CBV-5 crossed from humans to pigs – probably in the early 1950s – and has since adapted to the new host.

Another human enterovirus, type 70 (EV70; spp. *Human enterovirus D*) first appeared in humans in 1969 causing widespread outbreaks of acute hemorrhagic conjunctivitis. In a small proportion of cases (1:10 000–1:17 000), this generally benign ocular infection was accompanied by a disease with a presentation very similar to poliomyelitis. Interestingly, studies on animal sera collected prior to the human outbreaks showed the presence of anti-EV70 antibodies in cattle, sheep, swine, chickens, goats, dogs, and wild monkeys. EV70 disappeared from the human population in the 1980s, but the data suggest that this human disease was the result of a zoonosis.

Recent sequence analyses of simian enteroviruses has shown that some of these viruses are closely related to viruses within the species *Human enterovirus A*, while others are proposed as new members of *Human enterovirus B*. It is not surprising, therefore, that the present 'human' enterovirus classification scheme contains several animal viruses.

Genome Structure and Classification

The classification of picornaviruses has been in a state of flux for a number of years. Originally classification was based on serological, pathogenic and biophysical properties, but now genome structure is paramount. In some instances, such analyses confirmed previous classifications, while in others reclassification occurred – either by simply renaming, regrouping, or, more fundamentally, by the formation of new genera.

Entero- and Rhinovirus Genome Structures

Briefly, a viral protein (VPg), is covalently linked to the 5′ end of the genome and followed by a nontranslated region, the 5′ NTR (**Figure 1(a)**). A single, long, open reading frame (ORF) encodes a polyprotein of ∼2200 aa. The N-terminal domain of the polyprotein (P1) comprises the four capsid proteins, while the replicative proteins comprise the central domain (P2) and C-terminal domain (P3) of the polyprotein. The full-length translation product predicted by the single ORF is not observed within infected cells due to rapid, intramolecular, proteolysis by the two virus-encoded proteinases, 2A^{pro} and 3C^{pro} (**Figure 1(a)**). The P1, P2, and P3 'primary' cleavage products subsequently undergo proteolytic processing to yield the individual virus proteins. After the stop codon, there is a short 3′ NTR and a poly(A) tail (**Figure 1(a)**).

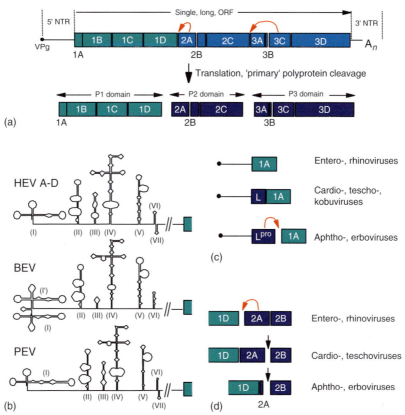

Figure 1 Enterovirus genome organization. The oligopeptide VPg (3B) is covalently attached to the 5′ terminus of vRNA. The long 5′ nontranslated region (5′ NTR) precedes the single, long, ORF (boxed area). The polyprotein comprises three domains: the capsid proteins precursor (P1; green boxed areas) and the nonstructural replication protein domains P2 and P3 (blue boxed areas). The 3′ portion of the genome comprises a short 3′ NTR and a poly(A) tail. The 'primary' (cotranslational) cleavages of the polyprotein mediated by the 2A and 3C proteinases are shown (red curved arrows) (a). The type 1 IRES RNA secondary structural features together with the individual domain features (I–VII) are shown for *Human enteroviruses A–D* (HEV A–D), bovine enterovirus (BEV), and porcine enterovirus (PEV) (b). In the case of enteroviruses, the N-terminal portion of the polyprotein is the capsid protein 1A (VP4), whereas in cardio- and teschoviruses, for example, the polyprotein starts with an L protein. Aphtho- and erbovirus polyproteins commence with an L proteinase (L^{pro}) which cleaves at its own C-terminus (c). Entero- and rhinovirus 2A proteins are proteinases (2A^{pro}) which cleave the polyprotein to form their own N-termini, whereas the 2A proteins of other genera (cardio-, tescho-, aphtho-, and erboviruses) mediate a cotranslational ribosome 'skipping' effect (producing an apparent prolyprotein cleavage), but these 2A proteins are not proteinases (d).

A few, but major, differences between the genomes of enteroviruses and other genera are now of prime importance in the classification of viruses within the *Picornaviridae*. Specifically, the RNA secondary structural features of the 5′ NTR (**Figure 1(b)**), the presence/absence of an L protein at the N-terminus of the polyprotein (**Figure 1(c)**), and the nature of the 2A region of the polyprotein (**Figure 1(d)**) are now taken to be the discriminative features. It has been shown, however, that comparison of VP1 (1D) sequences alone is sufficient to assign isolates to enterovirus species. Note that the high-level of recombination within noncoding regions and regions encoding nonstructural proteins (see below) means care should be taken when comparing sequences from these regions for taxonomic purposes.

The 5′ nontranslated region

The 5′ terminal ~100 nt of picornavirus vRNA folds into a cloverleaf-shaped structure (domain I; **Figure 1(b)**). A large proportion of the remainder of the 5′ NTR forms a complex secondary structure comprising multiple domains which functions as an internal ribosome entry site (IRES). This region confers a cap-independent mode of initiation of translation of the viral RNA: the ^{7me}G cap structure associated with eukaryotic mRNAs is replaced in picornaviruses by the oligopeptide 3B (or VPg) covalently linked to the 5′ terminus of vRNA. While the 5′ NTR has the same functions across the *Picornaviridae*, there are differences in the overall structure and this may be used to aid the classification of viruses. The entero- and rhinoviruses have a 'type 1' IRES (**Figure 1(b)**), while many of the other picornaviruses have a type 2 IRES.

The leader protein

A leader protein (L) is present at the N-terminus of the polyprotein in the aphtho-, erbo-, cardio-, tescho-, and kobuviruses (**Figure 1(c)**). In the cardioviruses, L has been shown to interact with Ran-GTPase disrupting nucleocytoplasmic trafficking of cellular proteins. In the aphtho- and erboviruses, however, L is a papain-like proteinase (L^{pro}). This proteinase cleaves at its own C-terminus and is involved in shut-off of host cell protein synthesis via cleavage of eIF4G. Expression of the erbovirus L protein in cells did not, however, lead to efficient inhibition of cap-dependent translation. In the entero- and rhinoviruses, no L protein is present: the N-terminus of the polyprotein is the capsid protein, VP4 (1A).

The 2A protein

Picornavirus 2A proteins are diverse in sequence and function (**Figure 1(d)**). The 2A proteins of entero- and rhinoviruses are very similar, and are structurally related to the small subclass of serine proteinases – although the active site nucleophile is not serine, but cysteine. They catalyze an intramolecular proteolytic cleavage at their own N-termini (*in cis*), separating the structural protein polyprotein domain (P1) from the nonstructural proteins (P2, P3; **Figure 1(a)**). Importantly, this virus-encoded proteinase also cleaves a host-cell translation factor (like L^{pro}), the initiation factor eIF4G, which brings about the 'shut-off' of host-cell (cap-dependent) mRNA translation.

The 2A regions of all other picornaviruses are quite different (**Figure 1(d)**). In the case of the aphtho-, cardio-, erbo-, and teschoviruses, the 2A protein is not a proteinase. This type of 2A may be identified by a C-terminal –NPG– motif. In these cases, the N-terminal residue of protein 2B is, invariably, proline. An apparent 'cleavage' of the nascent polyprotein occurs between the glycine and proline residues at the 2A/2B junction (-NPG ⇓ P-). This processing event is not, however, proteolytic but rather mediated by a translational effect termed ribosome 'skipping'.

Genome Structure of BEV

The genome organizations of several BEV isolates have been reported. BEV 5′ NTRs range from 812 to 822 nt and are unusual in that the 5′ cloverleaf-like secondary structure is duplicated (domains I and I′; **Figure 1(b)**). Furthermore, the putative domains III and VI of the IRES region differ in size and shape from other enteroviruses. The ORF varies from 6498 nt in BEV-2 to 6525 nt in BEV-1. The 3′ NTRs of BEV are 71–75 nt and although the predicted folding pattern (two stem–loop structures – a potential pseudoknot-like element) resembles other enteroviruses, no nucleotide sequence similarity exists.

Originally classified into several serotypes, only two serotypes (BEV-1 and BEV-2) are now recognized within the species *Bovine enterovirus*. A recent study has suggested, however, that the taxonomy of these viruses be revised. Sequence comparisons provided data that the bovine enteroviruses form two major clusters characterized by structural features and levels of sequence identity. These clusters have been designated BEV-A and BEV-B and appear to correlate with the two serotypes BEV-1 and BEV-2. It is also suggested that these clusters represent species, rather than serotypes. Further subgrouping within the clusters is proposed with BEV-A comprising two geno-/serotypes and BEV-B comprising three geno-/serotypes.

Genome Structure of PEV

Sequence analyses of what were formerly regarded as PEVs have shown that all but three now form the genus *Teschovirus*. Based upon the presence of an L protein and the nature of the 2A proteins, it is further proposed that one of these three, PEV-8 (together with some simian enteroviruses; see below) forms a new genus – *Sapelovirus*. The two remaining 'true' enteroviruses are PEVs 9 and 10. The 5′ NTR of PEV-9 is 809 nt (814 nt for PEV-10).

Both 5′ NTRs are predicted to adopt a secondary structure similar to that of the human enteroviruses. However, the 5′ cloverleaf has a unique insertion that enlarges domain I (**Figure 1(b)**). Interestingly, this domain has been previously shown to bind the 3CD proteinase in poliovirus and rhinovirus 14, forming a ribonucleoprotein complex required for initiation of positive-strand RNA synthesis. The long ORFs encode a polyprotein (2168 aa – PEV9; 2171 aa – PEV10) with a very similar organization to other enteroviruses (**Figure 1(a)**). The 3′ NTRs of both viruses are similar in sequence and length (72 nt – PEV9; 71 nt – PEV10), and predicted to form two stem–loop structures.

Genome Structure of SVDV

Complete genome sequences of the virulent J1′73 strain of SVDV and the avirulent H3′76 strain have been determined. Furthermore, partial sequences (5′ NTR, capsid protein VP1, protein 3BC) have been determined from a large number of isolates for epidemiological studies. The genome structure is very similar to that of the other enteroviruses (**Figure 1(a)**). Although only a single serotype, SVDV shows high genetic variability. The IRES is very similar to the human enterovirus type 1 IRES (**Figure 1(b)**; spp. HEV-B), although variability between isolates was observed in the 'spacer' region extending from the 3′ end of the IRES to the initiation codon. Indeed, several isolates had blocks of sequence (6–125 nt) deleted in this region.

A key determinant of virulence was mapped to a single residue within the 2A proteinase. Residue 20 of the avirulent H3′76 strain is isoleucine, the virulent J1′73 strain, arginine (this site is adjacent to His21, a component of the catalytic triad of 2A^{pro}). Pigs were infected with viruses rescued from an infectious copy of the virulent strain bearing mutations at this single site. Each mutant tested showed much reduced virulence correlating with the different efficiencies of the two forms of the proteinase to promote translation of the vRNA.

Genome Structure of Simian Enteroviruses

Formerly, the 20 simian picornaviruses were provisionally classified as enteroviruses. Analysis of partial (5′ NTR, VP1, and 3D) sequences showed that the simian viruses SV2, -16, -18, -42, -44, -45, and -49 are not enteroviruses, but comprise a new genus *Sapelovirus*. The complete genome sequence of SV2 confirmed that this virus was not an enterovirus since it does not possess a type 1 IRES and is thought, like PEV-8, to possess an L protein. The sequence of the 2A protein was quite different to other picornaviruses, although the presence of the conserved motif –GxCG– (also present in PEV-8) suggests that SV2/PEV8 2A proteins may be proteinases. If these viruses do possess an L protein and a 2A proteinase, then they represent an interesting link between the enteroviruses and other genera.

Analyses of the 5′ NTR and VP1 sequences showed that A13, SV19, -26, -35, -43, -46 were members of *Human enterovirus A*, SA5 a member of *Human enterovirus B*, and SV6, N125 plus N203 were related, but apparently form a new enterovirus species most closely related to *Human enterovirus A*. The complete genome sequence of simian enterovirus A (A-2 plaque virus) showed an organization characteristic of enteroviruses (**Figure 1(a)**), but along with SV4 and SV28 forms another new enterovirus species, this time most closely related to *Human enterovirus B*.

Virus Replication

Most of the receptors identified for picornaviruses belong to the immunoglobulin superfamily or the integrin receptor family. However, the receptors for the animal enteroviruses remain to be determined. Internalization occurs via endocytosis, and as the endosome undergoes acidification, changes in the virion structure lead to the release of the vRNA into the cytoplasm. The vRNA acts as an mRNA; the first step of replication is, therefore, translation of the single, long ORF. The autocatalytic processing by the virus-encoded proteinases 2A^{pro} and 3C^{pro} produces the three 'primary' processing products P1, P2, and P3 (**Figure 1(a)**) – the full-length (predicted) translation product is not observed. Subsequently further, 'secondary', proteolytic processing of these precursors occurs mediated by 3C^{pro}. In the case of poliovirus, it has been demonstrated that the P1 capsid protein precursor is processed not by 3C^{pro}, but by 3CD^{pro} – also a proteinase. The function of the vRNA soon switches from that of an mRNA, to that as a template for −ve strand synthesis. The −ve-strand RNA then serves as a template for the synthesis of +ve-strand RNA. A large excess (∼80-fold) of +ve-, over −ve-, sense RNA is observed. Although the RNA-dependent RNA polymerase (3D^{pol}) is an enzyme, overall the replication of vRNA is not 'enzymic' as such, since protein 3B is used to prime RNA synthesis and is covalently linked to the 5′ terminus of both +ve- and −ve-sense RNA.

A characteristic feature of picornavirus replication is the disappearance of the Golgi apparatus concomitant with the appearance of virus-induced vesicles. It is upon these structures that RNA replication occurs; proteins 2C and 3A are key players in this remodeling of the endomembrane system. An increase in the intracellular level of Ca^{2+} is also observed; protein 2B has been shown to function as a 'viroporin' releasing calcium ions stored in the endoplasmic reticulum.

Recombination occurs with high frequency, reported to be in the region of 10%. The polymerase complex, together with the nascent strand, may 'switch' template and complete the synthesis of the vRNA from another parent. Analyses of recombinants have shown that viable progeny arise largely from template switching within noncoding regions, or regions encoding nonstructural proteins.

The process by which vRNA is encapsidated is not clear. A feature of this process is the 'maturation' cleavage of VP0 (1AB) into VP4 and VP2 (1B) – again poorly understood. Replication is rapid with cell death and release of progeny virions occurring within $\sim$9 h for BEV and SVDV for most cell types.

Virion Structure and Properties

Particles have a buoyant density of 1.30–1.34 g ml^{-1} in cesium chloride gradients, are resistant to ether, and stable throughout a wide range of pH (3–10). The ability to tolerate low pH is seen as an adaptation to survive passage through the acidic conditions of the stomach. Under the electron microscope, particles are seen as icosahedrons, 25–30 nm in diameter (**Figure 2(a)**). Capsids are nonenveloped, comprising 60 copies of each of the capsid proteins VP4, VP2, VP3 (1C), and VP1. Capsid proteins are arranged into protomers and 60 of these structural units form an icosahedron (symmetry: pseudo $T = 3$; **Figure 2(b)**).

The atomic structures of both BEV and SVDV capsids have been determined. The folding pattern of BEV polypeptides VP1–3 is similar resulting in an eight-stranded antiparallel β-sheet structure. The VP4 protein is much smaller than the other capsid proteins, and lies across the inner surface of the capsid. The N-terminus of VP4 is close to the icosahedral fivefold axes of symmetry and the C-terminus close to the threefold axes. Furthermore, in all picornaviruses, the N-terminal residue of VP4 is covalently bonded to a myristic acid group giving the capsid five symmetry-related myristoyl moieties around the inner surface of the capsid, a channel running from the inner to outer surface at this point. It has been proposed that the myristic groups attached to the VP4 proteins may insert into the host membrane aiding entry.

A radial depth-cued image of the BEV particle is shown in **Figure 2(c)**. The topologies of the BEV major capsid proteins and the overall architecture of the virion are similar to those of related picornaviruses. Some differences were observed, however; the external loops are relatively truncated giving a comparatively 'smooth' appearance. A 'canyon' receptor binding region (running around the fivefold axes of symmetry) is observed in the structures of polio- and rhinoviruses. In BEV, this depression is partially filled by a five-residue extension of the G-H loop of VP3. The extended VP3 loop of BEV has been implicated in receptor binding, although the cell-surface receptor has not been identified. BEV is usually cultured in hamster kidney (BHK-21) cells but is known to be readily adaptable to grow in human cervical carcinoma (HeLa) cells to equivalent titer and can cause cytopathic effect in an extensive range of cell types *in vitro*. The BEV receptor is, therefore, thought to be a ubiquitous cell-surface glycoprotein. The crystal structure also revealed that the virus maintains a hydrophobic pocket within VP1, occupied by a specific 'pocket factor' which appears to be myristic acid. The pocket factor is thought to stabilize the capsid and it has been proposed that a kinetic equilibrium exists between occupied and unoccupied pocket states, with occupation inhibiting 'uncoating' of the vRNA. In purified BEV preparations, a small proportion of the precursor protein VP0 is detected and it is suggested that in a few protomers this precursor remains uncleaved. This maturation cleavage is completely absent in the genera *Parechovirus* and *Kobuvirus* – these viruses possessing only three structural proteins.

The crystal structures of two SVDV isolates, UK/27/72 and SPA/2/93, have been reported. These two structures are in agreement and are similar to those of other enteroviruses, with SVDV being most similar to coxsackievirus B3 (CBV-3). The major capsid proteins (VP1–3)

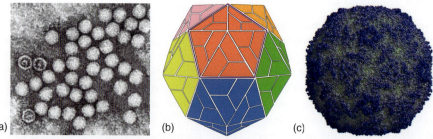

Figure 2 The structure of enterovirus particles. (a) A negatively stained electron microscopic image of enterovirus shows unenveloped particles $\sim$25 nm in diameter. (b) The icosahedral structure of the particle is shown. Five copies each of capsid proteins 1A–D form a pentamer, twelve of which form the complete particle. (c) A radially depth-cued image produced from the X-ray crystallographic data of the BEV particle showing surface projections (dark blue) – particularly notable at the fivefold axis of symmetry. The particle is smoother than the surface of other enteroviruses. Rasmol image courtesy of Dr. J.-Y. Sgro, University of Wisconsin, USA.

possess the conserved β-sheet structure with some notable differences. VP1, making up most of the outer surface area, is often the most variable, and while the β-sheet is conserved, the connecting loops vary in length between SVDV and CBV-3. Five copies of VP1 associate to form each of the fivefold vertices, and in SVDV it has been noted that an arginine residue from each of the five VP1s form a crown at the apex of the fivefold axes of symmetry. The hydrophobic pocket formed by VP1 is also present in SVDV and electron density clearly showed that this accommodates a fatty acid pocket factor. The hydrophobic pocket situated at the base of the canyon extends from a pore on the surface inward, until it is blocked by VP1 loops. In CBV-3, the pocket factor has been identified as palmitic acid. The dimensions of the pocket are similar in CBV-3 and SVDV, but in the case of SVDV the electron density suggests that the pocket factor is a longer molecule – similar to sphingosine.

The VP2 and VP3 proteins alternate around the threefold axes of symmetry. The VP2 of SVDV possesses a puff region composed of two sequential loops between sheets E and F, with the second loop being more exposed on the surface. This is the least conserved area between SVDV and CBV-3. Furthermore, in poliovirus, this area has been implicated in the binding of the virus to its cell-surface receptor. The C-terminus of the VP3 protein is external and forms a major surface protrusion termed the 'knob'; this structure is present in SVDV and CBV-3. VP3 is also important in the stability of the pentamer with fivefold neighboring VP3 proteins forming a β-cylinder.

VP4 is situated on the inner surface of the capsid and is the most conserved of the capsid proteins. The central region of this protein was found to be disordered with little secondary structure. VP4 begins close to the fivefold axes and snakes toward the nearest threefold axes, with the C-terminus of VP4 lying close to the N-terminus of VP2. The N-terminal glycine of VP4 is covalently attached to myristic acid and the myristoyl groups of adjacent VP4s group around the fivefold axes, under the VP3 β-cylinder. In general, the surface structure of SVDV is very similar to that of coxsackieviruses.

Rather than a continuous, circular, canyon in SVDV, there are five distinct depressions. The C-terminus of VP3, the first loop of the VP2 puff, and residues of the C-terminal VP1 loop form a ridge between these depressions. Further depressions are present on the twofold axes with the enclosing four walls composed of symmetrically related pairs of VP2 and VP3. SVDV, the six CBV serotypes (CBV-1–6), and many adenoviruses share a common receptor on human cells. The coxsackieadenovirus receptor (CAR) has two immunoglobulin-like extracellular domains, a transmembrane domain and a cytoplasmic domain. Cryoelectron microscopy has provided evidence that CAR binds into the canyon of CBV-3, mostly interacting with VP1 but with contributions from VP2 and VP3. CBV-5 (thought to have given rise to SVDV in pigs) uses CAR as a primary receptor and decay-accelerating factor (DAF; CD55) as a co-receptor. Recent isolates of SVDV have been shown to have lost the ability to bind human DAF but have not developed the ability to bind pig DAF, suggesting that SVDV may have adapted to use another co-receptor which is tailored to the new host.

Concluding Remarks

Understandably, in terms of research effort, for many years the animal enteroviruses have been the poor cousins of their human 'relatives'. The greatly expanded sequence database has led to a wholesale reappraisal of the taxonomy of these viruses. It is clear that within this group close relationships exist between animal and human viruses and that viruses have passed from man to animals, and vice versa. The high mutation rate in combination with high frequency of recombination means these viruses can rapidly adapt to new host species. Animal enteroviruses are, therefore, worthy of study not only for their intrinsic interest and economic impact but, through enzoonosis, they pose an ever-present threat to human health.

See also: Coxsackieviruses; Enteroviruses: Human Enteroviruses Numbered 68 and Beyond.

Further Reading

Hughes AL (2004) Phylogeny of the *Picornaviridae* and differential evolutionary divergence of picornavirus proteins. *Infections, Genetics and Evolution* 4: 143–152.

Hyypia T, Hovi T, Knowles NJ, and Stanway G (1997) Classification of enteroviruses based on molecular and biological properties. *Journal of General Virology* 78: 1–11.

Krumbholz A, Dauber M, Henke A, et al. (2002) Sequencing of porcine enterovirus groups II and III reveals unique features of both virus groups. *Journal of Virology* 76: 5813–5821.

Oberste MS, Maher K, Flemister MR, et al. (2000) Comparison of classic and molecular approaches for the identification of untypeable enteroviruses. *Journal of Clinical Microbiology* 38: 1170–1174.

Oberste MS, Maher K, and Pallansch MA (2002) Molecular phylogeny and proposed classification of the simian picornaviruses. *Journal of Virology* 76: 1244–1251.

Oberste MS, Maher K, and Pallansch MA (2003) Genomic evidence that simian virus 2 and six other simian picornaviruses represent a new genus in *Picornaviridae*. *Virology* 314: 283–293.

Pöyry T, Kinnunen L, Hovi T, and Hyypiä T (1999) Relationships between simian and human enteroviruses. *Journal of General Virology* 80: 635–638.

Zang G, Haydon DT, Knowles NJ, and McCauley JW (1999) Molecular evolution of swine vesicular disease virus. *Journal of General Virology* 80: 639–651.

Zell R, Dauber M, Krumbholz A, et al. (2001) Porcine teschoviruses comprise at least eleven distinct serotypes: Molecular and evolutionary aspects. *Journal of Virology* 75: 1620–1631.

Zell R, Krumbholz A, Dauber M, Hoey E, and Wutzler P (2006) Molecular-based reclassification of the bovine enteroviruses. *Journal of General Virology* 87: 375–385.

Enteroviruses: Human Enteroviruses Numbered 68 and Beyond

T Hovi, S Blomqvist, C Savolainen-Kopra, and M Roivainen, National Public Health Institute (KTL), Helsinki, Finland

© 2008 Elsevier Ltd. All rights reserved.

Glossary

Acute flaccid paralysis (AFP) A descriptive diagnosis used in the context of poliomyelitis eradication initiative (PEI), coordinated by the World Health Organization (WHO). A case of AFP is a patient with acute onset paralytic symptoms without known etiology. Within the PEI, an AFP patient is suspected of having poliomyelitis and should be tested for poliovirus excretion.

Acute hemorrhagic conjunctivitis (AHC) A rapidly developing severe inflammatory disease of the skin covering the eyeballs, including bleeding of the inflamed skin.

Hand-foot-and-mouth disease (HFMD) An acute febrile disease of children with flat or raised red spots, which may later develop to vesicles, in the mouth as well as on palms and soles of feet.

Phylogenetic typing Determination of genetic type of an unknown strain by sequence and phylogenetic analysis of partial genome sequence.

Prime strains Virus strains showing one-way cross-reactivity with prototype strains of a serotype and thus reflecting within-serotype antigenic variation. Antisera against a prototype strain do not neutralize a prime virus strain but an antiserum against a prime strain will neutralize the prototype strain.

Introduction

Historical Aspects

Original taxonomic subgrouping of human enterovirus (EV) isolates was based on host species range and pathogenicity in humans or experimental animals. Accordingly, the strains were classified as polioviruses (PVs), coxsackie viruses of subgroup A (CVA), or B (CVB), or echoviruses (E). Each subgroup was further divided in three or more serotypes according to results of cross-neutralization assays with hyperimmune antisera raised against individual strains. Toward the end of 1960s, after initial characterization of 3 PV serotypes, 24 CVA serotypes, 6 CVB serotypes, and 34 E serotypes, it was realized that antigenic relationships did not always correlate with other phenotypic features used in the taxonomic subgrouping. Subsequently, new serotypes were no more distributed in the established subgroups but instead given a running number from 68 onward. Within a few years, five new serotypes, EV68 through EV72, were identified. However, EV72 turned out to be very different from typical EVs and was later reclassified as hepatitis A virus in a genus of its own (*Hepatovirus*) in the family *Picornaviridae*.

Development of molecular techniques in virus strain characterization, notably phylogenetic analysis of (partial) genomic sequences, enabled novel approaches in virus taxonomy and also revived taxonomical interest in EV strains in the late 1990s. The official taxonomic subclassification of the genus *Enterovirus* now includes four phylogenetically distinct EV species with strains infecting humans, *Human enterovirus (HEV) A–D* (**Table 1**).

General Features of Human Enteroviruses

Human enteroviruses (HEVs) naturally share many basic biological properties with other picornaviruses, but may also show features typical of the species, serotype, or even strain in question. Although many of the features often referred to as typical of EVs have in reality been examined only for poliovirus type 1/Mahoney, we assume here that in principle they are also likely to be true for enterovirus types 68 and beyond. Exceptions to the 'rules', if known, will be mentioned in the subsequent serotype-specific articles.

EVs are typically acid stabile, nonenveloped, icosahedral particles composed of 60 structural subunits each comprising one copy of each of the four structural proteins, capsid protein 1–4 (VP1–VP4) surrounding the about 7500 nucleotide (nt) long RNA genome. Steps of the replication cycle follow the common pattern of viruses with messenger-sense single-stranded RNA genome.

Humans are the sole natural hosts of HEVs. Transmission of EV readily takes place through close direct contacts, for example, within a family, but also through indirect contacts such as by touching by fingers objects or surfaces contaminated with feces or respiratory excreta. Out of the large number of human enterovirus serotypes, only polioviruses can be prevented today by vaccines, developed already half a century ago.

Enterovirus 68 (HEV-D)

History

Human enterovirus 68 (EV68) was originally isolated from four children with pneumonia and bronchiolitis in

Table 1 Distribution of enterovirus (sero)types in the old and new taxonomic subclasses

Species	Classical subgroup of enteroviruses/sero(type) number				
	Poliovirus	Coxsackie-virus A	Coxsackie-virus B	Echovirus	New enterovirus
Human enterovirus A (N = 18)		CVA2–CVA8, CVA10 CVA12, CVA14, CVA16			EV71, EV76[a] EV89[a]–EV92[a]
Human enterovirus B (N = 57)		CVA9	CVB1–CVB6	E1–E7, E9 E11–E21 E24–E27 E29–E33	EV69, EV73[a]–EV75[a] EV77[a]–EV88[a] EV93[a], EV97[a], EV–98[a] EV100[a], EV101[a]
Human enterovirus C (N = 16)	PV1–PV3	CVA1, CVA11, CVA13 CVA17 CVA19–CVA22 CVA24			EV95[a], EV96[a] EV99[a], EV102[a]
Human enterovirus D (N = 3)					EV68, EV70, EV94[a]

[a]Novel types identified by the VP1 coding sequence and registered by the Picornavirus Study Group of ICTV.
PV, poliovirus; CVA, coxsackievirus subgroup A; CVB, coxsackievirus subgroup B; E, echovirus; EV, enterovirus. Explanation for lacking numbers: CVA15 and CVA18 have been reclassified as strains of CVA11 and CVA13, respectively; CVA23 was found to be same as echovirus 9; echovirus 8 is a strain of E1; echovirus 10 has been reclassified as reovirus type 1, echovirus 28 as rhinovirus type 1A; echovirus types 22 and 23 now belong to another genus, *Parechovirus*; the initially identified echovirus 34 is a strain of E33, enterovirus 72 was classifies as hepatitis A virus in the genus *Hepatovirus*.

California in 1962. One of these simultaneous isolates, recovered from an oropharyngeal swab of a 10-month-old female, was defined as the prototype strain Fermon. Since then isolations of EV68 have been rare. Human rhinovirus 87 (prototype strain Corn) was recently shown to both antigenically and genetically represent the same enterovirus serotype as EV68. Neutralization assays with antisera raised against EV68 Fermon and HRV87 Corn demonstrated one-way cross-reactivity between these virus strains indicating that strain Fermon could be considered a prime strain of the Corn.

Epidemiology

Most of the relatively few known EV68 isolates have been obtained from respiratory specimens, but occasional isolations from stool and sewage specimens have also been made. The transmission of EV68 is more plausible by the respiratory than by the fecal–oral route. EV68 may affect persons of different ages, although it has been most often detected in young children. Despite the infrequent isolation from clinical specimens, EV68 is probably a highly common virus since antibodies against EV68 are frequently found among Finnish people.

Cellular Interactions

Isolates of EV68 have been obtained from a variety of human and monkey cell lines. Besides, at least some of the currently circulating EV68 strains are able to induce cytopathic effect (CPE) in nonprimate mouse L cells. The host cell surface molecules involved in the initiation of EV68 infection are not known, but recent studies indicate that both the decay-accelerating factor (DAF) and a yet unknown sialic acid moiety may be required. Receptor specificity of HRV87 Corn was studied among other human rhinoviruses. The requirement for the presence of sialic acids on the HeLa cells for attachment of HRV87 distinguished it from other HRVs, the receptor of which was either LDL receptor or ICAM-1. On the other hand, pretreatment of HeLa Ohio cells with a monoclonal antibody to the SCR-3 epitope of DAF inhibited both the HRV87 and EV68 induced CPE.

Clinical Presentation and Pathogenesis

EV68 has a clear association with respiratory diseases. Most clinical isolates of EV68 have been obtained from respiratory specimens of patients with respiratory tract infection. Studies on strains EV68 Fermon, HRV87 Corn, and some recent clinical EV68 isolates have shown that these viruses share important biological features with human rhinoviruses, the most prominent causative agents of common colds. Unlike typical EVs, strains of EV68 lose their infectivity in acidic environment and besides, grow more effectively at +33 °C than at +37 °C. Clinical illness or symptoms described upon isolation of EV68 included pneumonia, bronchiolitis, upper respiratory tract infection, wheezing, asthma, emesis, respiratory distress, crackles, cough, and asthma exacerbation.

Enterovirus 69 (HEV-B)

Enterovirus 69, strain Toluca, was isolated in 1959 from a rectal swab of a healthy Mexican child. Later detections of EV69 among enterovirus isolates have remained very rare but the virus has been found in different parts of the world, mainly in patients with nontypical respiratory disease. Respiratory excreta and stool specimens may yield the virus demonstrating typical enteroviral CPE in monkey kidney and human tumor cell lines. Further details of the replication cycle have not been elucidated. Some antigenic cross-reactivity with echovirus 6 has been reported but according to phylogenetic analysis of the entire genome, EV69 is a typical member of the species *HEV-B* with its own distinct branch in a tree constructed on VP1 sequences and variable, region-dependent clustering with other HEV-B serotypes as regards the non-structural protein coding sequences.

Enterovirus 70 (HEV-D)

History

An emergence of a new type of viral conjunctivitis was observed in Acera, Ghana, in June 1969. The disease, locally first called 'Apollo 11 disease' because of the coincidental landing of Apollo 11 on the moon, was named as acute hemorrhagic conjunctivitis (AHC) and it spread rapidly over Africa and Asia. The causative agent was isolated in Japan, Singapore, and Morocco in 1971 and assigned as enterovirus 70 (EV70).

Phylogenetic and epidemiological analyses have confirmed that EV70 emerged from one focal place in August 1967 ± 15 months, which is about 2 years before the first recognition of the AHC pandemic. It has been speculated that EV70 may have originated from an animal or insect picornavirus. This hypothesis is supported by the wide *in vitro* host range of EV70 and detection of EV70 neutralizing antibodies from animal sera collected from Ghana and Senegal already around the estimated time of emergence of EV70.

Epidemiology

EV70 has caused two pandemics and several epidemics of AHC. The first pandemic began in Ghana and spread over Africa and Asia during 1969–1971. The largest number of cases occurred in crowded coastal areas of tropical countries. At the same time, outbreaks were also observed in some European countries. The second pandemic started in southern India in 1980 and spread extensively during 1980–1982, now also including the Western Hemisphere. The epidemic EV70 strains from Asia and America in 1980 and 1981 were shown to be genetically closely related and to differ significantly from the strains recovered during the first pandemic. Since the two large pandemics, which caused about 100 million cases of AHC, smaller outbreaks have been described at least in Singapore, China, Brazil, and Cuba in 1980s and in American Samoa, Israel, Japan, and India in 1990s.

EV70 is highly contagious and spreads fast under crowded and unhygienic conditions. The virus may be transferred by fomites as well as by direct inoculation by contaminated fingers. High levels of relative humidity prolong the survival of EV70. Transmission of EV70 occurs usually at school or home, but may also occur, for example, in eye clinics. During the epidemic, the highest number of cases is typically among school-aged children. Women in child-bearing age have been reported to be affected more frequently than men in the same age.

EV70 induces immune response, which provides only short-lived protection. Consequently, multiple epidemics have been seen in a given geographical region even within a 5-year period. Although EV70 typically causes clear-cut epidemics, the virus may continue endemic circulation in the affected population after the main outbreak. Sporadic cases of AHC or neurological diseases may occur over a period of some years.

Cellular Interactions and Pathogenesis

In humans, EV70 exhibits a restricted tropism to conjunctival and corneal cells, but like several other HEVs, it also has a propensity for infecting the central nervous system (CNS). *In vitro*, EV70 has the ability to infect a wide spectrum of mammalian cell lines, which is untypical for most HEVs.

EV70 attaches to the short consensus repeat domain 1 of DAF, on HeLa cells. However, additional sialylated binding factors on the cell surface are needed for productive infection. The DAF is also used as a receptor by several other EVs, but the requirement for the cell surface sialic acid is unique among HEVs. It has been speculated that this recognition of sialic acid attached to underlying glycans by a particular glycosidic linkage might contribute to the exceptional *in vitro* host range and tissue tropism of EV70.

Clinical Presentation and Pathology

EV70 is, together with a variant of CVA24, the leading causative agent of AHC. The typical site for EV70 entry is conjunctiva, where the virus may be transferred by fomites as well as by direct inoculation by contaminated fingers. The incubation period is about 24 h. The disease is usually localized to the eye and characteristically produces subconjunctival hemorrhage, which ranges from discrete petechiae to large patches of hemorrhage covering the bulbar conjunctiva. Symptoms include severe eye pain, photophobia, and blurred vision. Less specific symptoms like headache, nasal discharge, and sore throat may also occur. The cornea may be transiently involved in the form of epithelial keratitis. Asymptomatic EV70 infections are uncommon. Recovery is usually complete within less than 10 days with no significant ocular sequelae.

During the two EV70 pandemics, neurological manifestations were reported in about 1 out of 10 000 AHC patients. Male adults were more likely to be affected than women. The neurological symptoms resembling poliomyelitis developed as late as several weeks after the onset of AHC. Significant antibody titers against EV70 were demonstrated in the serum and more relevantly in the cerebrospinal fluid (CSF) of the patients.

Diagnosis

AHC caused by EV70 or other viruses are clinically indistinguishable. EV70 is most frequently isolated from conjunctival and oropharyngeal specimens, occasionally also from feces. Although several different cell lines may support EV70 replication, isolation of EV70 in cell culture is considered insensitive and today, EV70 infection is typically diagnosed by serological means or by reverse transcriptase-polymerase chain reaction (RT-PCR) methods. Demonstration of EV70 specific IgM antibodies in sera indicates a recent EV70 infection. The titers of IgM antibodies against EV70 begin to drop by the fifth week after onset of illness. In addition, neutralizing antibodies can be measured from sera, tears, or CSF. EV70 specific RNA by RT-PCR has been successfully detected from eyewash specimens, eye swabs, conjunctival swabs, and tears. Sequence analysis of (partial) VP1 is an useful tool for determination the origin of the outbreak.

Prospects for Prevention, Control, and Treatment

Currently no practical treatment method or effective vaccine is available for AHC caused by EV70. Preliminary studies aiming at development of the vaccine have been conducted with immunogenicity of the candidate materials tested in rabbits. Immunizations with UV-inactivated EV70 or bacterially produced EV70 VP1 protein or its nonoverlapping N and C terminal fragments were reported to elicit a classical humoral immune response including EV70-neutralizing antibodies in rabbit sera.

Disinfection and control of EV70 has been studied in experimental eye clinic conditions. When placed in eye drop solution and kept at room temperature, the infectivity of EV70 is sustained for more than 20 days. The most convenient and efficient disinfection procedures against EV70 is either drying the instruments and keeping them in a moisture-free environment or heating at 90 °C for 5 s.

Enterovirus 71 (HEV-A)

History

EV71 was initially isolated in 1969 from a case of fatal encephalitis in California. Since then it has caused several outbreaks in different parts of the world including Australia, Bulgaria, Hong Kong, Hungary, Japan, and Sweden. Clinical symptoms of infection have been various, sometimes associated with CNS involvement. In recent years Southeast Asia has suffered from large epidemics of EV71, manifested, for example, in Taiwan 1998 with 405 patients hospitalized and 78 fatalities.

Epidemiology and Genetic Characteristics

EV71 appears worldwide with both endemic and epidemic circulation. Most infections occur in children under 5 years. Infections are transmitted by feco–oral route, but respiratory droplet transmission is also possible. Epidemics occur year around with peaks in late summer and autumn, occasionally in 3-year cycles.

EV71 is a member of HEV-A species with a close genetic and antigenic relationship to CV-A16. Nucleotide and amino acid homologies are 77% and 89%, respectively, between these viruses. At least four distinct genogroups A, B, C, D, with subgroups B1–4 (5), C1–4, have been identified. The EV71 prototype strain, BrCr, is the sole member of genogroup A. Strains belonging to genogroups B and C have mainly caused the epidemics in the Asia-Pacific region. Large sets of strains worldwide have been subjected to molecular epidemiological and sequence analysis, but neurovirulence has not been connected to a single genotype and its genetic determinants have remained elusive.

Cellular Interactions and Pathogenesis

The receptor molecules involved in EV71 are not known. In the laboratory, EV71 strains replicating in Vero and RD or other primate cell lines produce CPE typical of EVs. Like other EVs, EV71 interferes with the cellular apoptosis program by inhibiting the activation during early steps of infection and then enhancing apoptosis in the later stages of the infection. It is not known if apoptosis has a significant role in the generation of symptoms *in vivo*. Immune response after infection is considered to be typical of enterovirus infection, involving both humoral and cellular responses with the humoral immunity likely to provide protection against reinfections. Animal models using either mice or monkeys have been developed for *in vivo* studies of EV infection. Alpha interferon seems to be an important component of the innate immunity limiting spread of the virus in the body. In an oral-infection model in newborn mice, the virus was found to spread from the intestines to several tissues including the spinal cord, eventually causing brain-stem encephalitis. The virus also spread to unprotected littermates and the dams. The model thus has several features similar to the human EV71 infection and is likely to be useful in future.

Clinical Presentation and Pathology

Enterovirus 71 is one of the two major HEV-A serotypes causing the relatively common hand-foot-and-mouth disease (HFMD) in children, the other being CVA16. In the western world, CVA16 has been dominating but in Southeast Asia, EV71 has caused severe outbreaks since the late 1990s with sometimes transmission of CVA16 coinciding. The usually self-limiting HFMD starts with fever followed by appearance of flat or raised red spots in the tongue, gums, and inside of the cheeks, as well as on palms, soles of feet, and occasionally buttocks. The spots may develop to blisters and in the mouth further turn to ulcers. The rash does not usually itch. Neurological complications, aseptic meningitis, encephalitis, and poliomyelitis-like polyneuritis are relatively common in EV71 but rare in CVA16 infections. In the large Southeast Asian epidemics during the last decade, meningo-encephalitis has occasionally been an alarmingly common complication with significant associated mortality especially in young children. Neurological disease may occasionally occur without recognizable rash. Pulmonary edema may contribute to the death and is considered to be secondary to dysregulation of blood circulation due to virus-induced damage in the brain stem resulting in dysregulation of glucose homeostasis, rather than putative infection of pulmonary tissues.

Diagnosis

EV71 infection has traditionally been confirmed in cell culture with virus isolation and serotyping using samples from throat swab or vesicular fluid or postmortem samples from the brainstem or spinal cord. Serological diagnosis with a high single titer or in a minority of cases a fourfold rise in antibody has also been used. New methods for rapid serotype-specific identification to replace the traditional tedious ones have been developed prompted by recent outbreaks. These include RT-PCR, real-time RT-PCR hybridization probe assay, DNA microchip array, and IgM-capture enzyme-linked immunosorbent assay (ELISA). Phylogenetic analysis of VP1 and VP4 capsid protein sequences have been used in tracing genetic origin of the strains.

Prospects for Prevention, Control, and Treatment

Severe outbreaks of EV71 in Southeast Asia have triggered studies aiming at development of vaccine and/or antiviral chemotherapy, but neither vaccine nor specific chemotherapy is available in the clinic. Initial studies in experimental animals have reported immunogenicity of Vero-cell-grown inactivated virions, baculovirus-expressed empty VP0-VP3-VP1 procapsids, and even transgenic tomatoes expressing the VP1 protein, but no clinical studies have been published by mid-2006. Spreading of the virus via both respiratory and feco–oral routes restricts the efficacy of attempts to control spreading of the virus in human populations. Pregnant women have been advised to avoid contacts with children suffering from HFMD although no data have been accumulated to suggest that maternal EV71 infection would be more dangerous to the fetus than infections in general. *In vitro* and animal model studies suggest that alpha interferon and lactoferrin may limit EV71 replication. In cell culture, siRNAs targeted to different genomic regions of EV71 were found to have antiviral activity.

The Newly Discovered HEV Types

History

Nontypable enterovirus isolates have been detected as long as enterovirus isolations have been carried out. Between 1970 and 2000, they were more or less ignored, and their existence perhaps interpreted by assuming that they might represent prime strains of known serotypes, rare serotypes not represented in the antiserum pools used, or mixtures of two EVs too difficult to sort out in the neutralization assays. By the end of 1990s, it became evident that phylogenetic analysis of VP1 coding sequences of conventionally serotyped enterovirus isolates results in defined clusters showing complete agreement with the phenotypic serotype. Subsequently, several groups have used this approach to identify nontypable isolates in historical collections of enterovirus strains, or among new isolates. The Picornavirus Study Group of ICTV has by mid-2006 listed almost 30 new types and the running number of the next new serotype is beyond 100. It is interesting that new enterovirus types have been described in all four HEV species and that most of them are in the species *HEV-B,* already traditionally the largest HEV species (**Table 1**).

For all the newly discovered enterovirus types, detection of perhaps only one or at the best, a handful of independent isolates have been reported. For most strains, only the source of origin and the VP1 sequence is known. Therefore, it is not possible to describe in further detail more than a few examples of these novel types.

New Types Forming a Subcluster within the Species HEV-A

Four of the new HEV-A serotypes have been studied in some detail. All but one of the altogether 19 isolates were derived from Bangladeshi children with acute flaccid paralysis (AFP) and collected in 1999–2002. The remaining isolate was from an adult gastroenteritis patient in France from 1991. It is not clear if these infections had any role in the generation of the recorded symptoms. Sequences of the VP1 coding region formed four serotype-like clusters close to but separated as a group from the classical HEV-A cluster as well as from the simian enterovirus isolates tentatively classified in the HEV-A species. Eleven strains, including the French one, were given type number 76 while two, four, and two strains segregated in the clusters labeled EV-89, EV-90, and EV-91, respectively. Sequences of the entire capsid coding region confirmed this clustering and monophyletic origin of each new type as well as that of the new subcluster within HEV-A. The observed monophyletic segregation of the new types from the classical HEV-A serotypes extended to the nonstructural protein coding part of the genomes. Hence, evidence for recombination between these new HEV-A types and the classical ones (or the simian ones) was not obtained. Analysis of larger numbers of independent sequences may be required before final conclusion about the matter but it is possible that this subcluster has diverged from the other HEV-A strains far enough so that functional combatibility of genomic parts joined in putative recombination events is too poor to produce viable progeny with sufficient fitness for enrichment and widespread transmission.

Selected New Types of the Species HEV-B

Epidemiological background and some genomic features of HEV types 73–75 and 77–78 have been published. While parts of these strains are derived from poliovirus surveillance, either from patients with AFP or their contacts, a significant proportion is from patients suffering from a variety of clinical entities typical of classical HEV-B infection, ranging from acute respiratory infection to aseptic meningitis. In phylogenetic analysis, these new types do not form a subcluster within HEV-B but are rather distributed in different parts of the multipronged tree of HEV-B.

HEV-94, A Proposed New Type in the Species HEV-D

During environmental polio surveillance in Egypt four nonpoliovirus strains were isolated from sewage specimens in L20 B cells, a recombinant murine cell line expressing human poliovirus receptor. The VP1 sequences of these isolates clustered together and showed less than 70% nucleotide and less than 80% amino acid sequence similarity to the closest established enterovirus serotype, EV70. This suggests that the isolates form a new enterovirus type, EV94, within the species *HEV-D*. In parallel with this finding AFP surveillance in the Democratic Republic of Congo revealed two nonserotypable virus strains whose VP1 sequences clustered very close to those of the above Egyptian isolates suggesting that they belong to the same serotype. Altogether, VP1 sequences of all these six EV94 isolates formed a monophyletic group within the species *HEV-D*. Enterovirus 94 cannot be considered as a new virus type although just recently discovered, since, according to the results of seroprevalence studies carried out in Finland, neutralizing antibodies to this virus strain were common already two decades ago.

Comparison of the whole genome sequence of the prototype strain of EV94 with those of EV68 and EV70 revealed no evidence for recombination between the prototype strains of these three HEV-D serotypes. In further characterization, the prototype strain turned out to be acid resistant suggesting that this virus strain, unlike EV68 and some strains of EV70, is a typical enteric pathogen capable of surviving the acidic stomach during the passage to the presumed intestinal replication sites.

The natural host range and the cellular receptor for EV94 are not yet known. Although two strains of EV94 were isolated from stool specimens of patients with AFP, it is not yet known whether these virus strains could be causative agents of neurological symptoms.

Future Prospects

It is likely that the number of human enterovirus (sero) types will still increase as more and more previously nontypable isolates will be characterized genetically. While genetic and serological identification of the HEV serotype usually is in good agreement, there are exceptions to this rule. Some serotypes contain strains showing wide genetic divergence in the VP1 coding region and sometimes strains of genetically relatively homogenous clusters show strong antigenic variation. On the other hand, it is unlikely that anybody would take the trouble and serotype all the new genetically identified types in the classical way using monotypic antisera. Serotyping of isolates is a tedious and expensive procedure and may in the future remain as a research tool only. For identification of clinical isolates, genetic tools are more rapid and useful, for example, in tracking the transmission routes.

See also: Coxsackieviruses; Enteroviruses of Animals; Epidemiology of Human and Animal Viral Diseases; Evolution of Viruses; Human Eye Infections; Picornaviruses: Molecular Biology; Poliomyelitis; Quasispecies; Rhinoviruses.

Further Reading

McMinn PC (2002) An overview of the evolution of enterovirus 71 and its clinical and public health significance. *FEMS Microbiology Reviews* 26: 91–107.

Nokhbeh MR, Hazra S, Alexander DA, et al. (2005) Enterovirus 70 binds to different glycoconjugates containing α-2, 3-linked sialic acid on different cell lines. *Journal of Virology* 79: 7087–7094.

Oberste MS, Maher K, Michele SM, Belliot G, Uddin M, and Pallansch MA (2005) Enteroviruses 76, 89, 90 and 91 represent a novel group within the species *Human enterovirus A*. *Journal of General Virology* 86: 445–451.

Oberste MS, Nix WA, Maher K, and Pallansch MA (2003) Improved molecular identification of enteroviruses by RT-PCR and amplicon sequencing. *Journal of Clinical Virology* 26: 375–377.

Pallansch MA and Roos RP (2001) Enteroviruses: Polioviruses, coxsackieviruses, echoviruses and newer enteroviruses. In: Knipe DM, Howley PM, Griffin DE, *et al.* (eds.) *Fields Virology*, 4th edn., pp. 723–775. Philadelphia, PA: Lippincott Williams and Wilkins.

Racaniello VR (2001) *Picornaviridae*: The viruses and their replication. In: Knipe DM, Howley PM, Griffin DE, *et al.* (eds.) *Fields Virology*, 5th edn., pp. 839–893. Philadelphia, PA: Lippincott Williams and Wilkins.

Smura T, Junttila N, Blomqvist S, *et al.* (2007) Enterovirus 94, a proposed new serotype in human enterovirus species D. *Journal of General Virology* 88: 849–858.

Relavant Website

http://www.picornaviridae.com – Picornavirus (refer to Enterovirus).

Entomopoxviruses

M N Becker and R W Moyer, University of Florida, Gainesville, FL, USA

© 2008 Elsevier Ltd. All rights reserved.

Introduction

The entomopoxviruses (EVs) comprise the *Entomopoxvirinae*, one of the two subfamilies of the family *Poxviridae*. EVs infect insect hosts and were first described in 1963. These viruses are found in a number of insect species but are particularly well characterized within the butterflies and moths. The other subfamily of poxviruses is the *Chordopoxvirinae*, which contains the well-known variola virus, the causative agent for smallpox, and vaccinia virus (VV), the vaccine strain for smallpox. The EVs possess a number of the same features as the chordopoxviruses (CVs); however, there are striking differences between the two subfamilies. The similarities include general virion morphology, a double-stranded DNA genome, and a cytoplasmic life cycle. The differences include the host range of the viruses, the presence of occluded virus in EVs, the composition and organization of the DNA genome, differences in gene regulation, and the optimal temperature for growth, which is 27 °C for EVs versus 37 °C for CVs. Although a large number of EVs have been identified, very little is known about most of them with the exception of the *Amsacta moorei* EV (AMEV), which is amenable to growth in culture.

Phylogeny

Within the subfamily *Chordopoxvirinae* are eight genera. The causative agent of smallpox (variola virus) and the virus used for vaccination against smallpox (VV) are members of the genus *Orthopoxvirus*. Other CVs that cause human disease are cowpox virus and monkeypox virus within this same genus and molluscum contagiosum virus in the genus *Molluscipoxvirus*.

Within the subfamily *Entomopoxvirinae* are three genera, recently renamed by the International Committee on Taxonomy of Viruses: *Alphaentomopoxvirus*, *Betaentomopoxvirus*, and *Gammaentomopoxvirus*. The genera are defined predominantly by the host range of the viruses and by the morphology of the virus particle. The genus *Alphaentomopoxvirus* contains members that infect beetles (coleopterans), and the type species is *Melolontha melolontha entomopoxvirus*. Little is known about the molecular biology of the members of this genus. The genus *Betaentomopoxvirus* contains the best-studied members of the EVs. Betaentomopoxviruses infect either butterflies and moths (Lepidoptera) or grasshoppers and locusts (Orthoptera). The type species is *Amsacta moorei entomopoxvirus*. The host of AMEV is the red hairy caterpillar from India, a member of the tiger moths. Members of the genus *Gammaentomopoxvirus* infect flies, mosquitoes, or midges (Dipterans), and the type species is *Chironomus luridus entomopoxvirus*. There are also a number of unclassified EVs. Included in this group is the well-studied virus from the grasshopper, *Melanoplus sanguinipes* (MSEV). Formerly classified as a betaentomopoxvirus owing to its host range, this virus has been reclassified based on genomic sequencing data and comparison to the genomic sequence of AMEV. Evolutionary trees based on the DNA polymerase sequence indicate not only that the EVs are distinct and evolutionarily separated from the CVs, but that AMEV and MSEV are also highly divergent from each other. There are other unclassified EVs that infect Hymenoptera (bees and wasps). The best studied of these is *Diachasmimorpha* EV, whose host is the parasitic wasp, *Diachasmimorpha longicaudata*.

Virion Structure

The general virion structure is similar to that of the orthopoxviruses in that it consists of an inner core of electron-dense material surrounded by one or two lateral bodies (**Figure 1**). Within the core are the double-stranded DNA genome and a number of proteins required for the cytoplasmic life cycle of the virus. The entire virion is enveloped by a membrane. **Table 1** indicates the differences in virion size and shape of the lateral bodies of virions from different genera.

One of the unique features of EVs in comparison to CVs is that in nature they exist in an occluded form (**Figure 1**). A variable number of virions are embedded within a paracrystalline protein matrix consisting almost exclusively of a single protein, spheroidin. This is a large protein of 100–115 kDa that is abundant at late times during infection. Spheroidins from different viruses infecting hosts of the same insect order have greater than 80% identity to each other; however, spheroidins from viruses that infect hosts from different orders, such as coleopterans versus lepidopterans, have little similarity. The baculoviruses are another family of double-stranded DNA insect viruses that also produce occluded virus. In baculoviruses, it is the polyhedrin protein that forms the occlusion bodies (OBs). Despite a similar function, spheroidin has little similarity to the baculovirus polyhedrin protein. A large number of cysteines are present with the spheroidin protein, leading to the hypothesis that the OB structure is held together by disulfide bonds.

The function of the OBs is to provide stability for the virus in the environment until ingested by another caterpillar. OBs contain a variable number of virion particles and only mature virions are occluded. OBs are alkaline sensitive, consistent with the high pH of the insect gut. Spheroidin is not required for virulence and the gene is nonessential in tissue culture. Nonoccluded virus is poorly infectious in caterpillars when ingested, but quite virulent when injected. OBs can be induced to form in the absence of virions by the expression of spheroidin protein in cell culture. An alkaline protease is associated with some preparations of EV OBs, and may be useful for dissolution of the OB in the insect gut. AMEV OBs are associated with an alkaline protease when isolated from insects, but not when the virus is grown in culture.

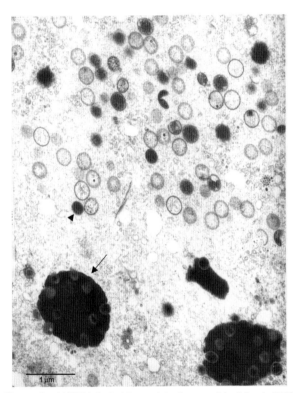

Figure 1 AMEV-infected *Lymantria dispar* cell (cell line Ld652). Different stages of virion development are seen, including mature virions indicated by an arrowhead and OBs indicated by an arrow.

Pathology

EVs infect primarily the larvae of insects rather than the adult insects. There are reports of infections of adult insects within laboratory settings, however. Most EVs have a restricted host range, although MSEV does infect several grasshopper species. Infection occurs primarily through an oral route with the ingestion of OBs. Once in the gut, the OBs are dissolved by the alkaline environment. The released virions attach to the midgut epithelium and then appear to fuse with the cell membrane. The virus then infects the fat body of the insect and this becomes the primary site of replication for the virus.

Table 1 Properties of EV virions

	Alphaentomopoxvirus	Betaentomopoxvirus	Gammaentomopoxvirus
Size	450 × 250 nm	350 × 250 nm	320 × 230 × 110 nm
Core	Unilateral concave core	Cylindrical	Biconcave
Lateral body	Single	Sleeve shaped	Two
Virion shape	Ovoid	Ovoid	Brick

EVs are reported to infect the hemocytes and AMEV does infect primary hemocytes in culture. Some researchers argue that the presence of virus in the hemocytes *in vivo* represents phagocytosis rather than infection. As the disease progresses within the insect, the hemolymph frequently turns white with OBs, and the larva itself can take on a whitish or white spotted appearance. The fat body disintegrates and the virus is disseminated throughout the body.

Infected larvae are lethargic and uncoordinated and exhibit a decrease in feeding. The duration of the instar phase is increased before progression to the next instar and few infected larvae continue through pupation. In some EV infections, this is thought to be due to the increased levels of juvenile hormone. There are reports of regurgitation and defecation of AMEV particles by infected *Estigmene acrea* larvae. The time to death of an infected insect is dependent both on the dose of virus received and the larval instar stage. Younger larvae appear to be more susceptible to lower doses of virus and die more quickly than older larvae. The lowest reported LD_{50} is 2.4 spheroids. *Melanoplus sanguinipes* larvae succumbing to an MSEV infection within less than 12 days are reported to lack OBs, although infections of greater than 14 days do possess OBs. The shorter time to death is the result of larger doses of virus.

Fusolin/Spindles

In addition to spheroidin, some, but not all, members of the alphaentomopoxviruses and many of the lepidopteran members of the betaentomopoxviruses produce another protein termed fusolin that forms crystalline spindles. Among the viruses known to produce spindles are *Melolontha melolontha* EV, *Heliothis armigera* EV, *Choristoneura fumiferana* EV, *Choristoneura biennis* EV, *Pseudaletia separata* EV, and *Anomala cuprea* EV (ACEV). Spindles do not contain virions, though spindles are occasionally found within the OBs of these viruses. Similar to spheroidin, fusolin is highly expressed at late times during infection and is a smaller protein of 38 kDa.

Spindles purified from EV infections enhance the infectivity of a number of baculoviruses. One proposed mechanism of this enhancement is through dissolution of the noncellular peritrophic membrane found within the insect gut. An alternate hypothesis is that the spindles increase the amount of baculovirus-to-cell fusion. Spindles from ACEV have also been demonstrated to enhance the infectivity of ACEV. Clearly, spindles are not required for pathogenicity as evidenced by their absence from most members of the gammaentomopoxviruses and the orthopteran betaentomopoxviruses. It should be noted that although AMEV is a lepidopteran virus it does not produce spindles or encode a gene for fusolin.

Genome Organization

Genome sizes of the EVs are larger than that of the best-studied CV, VV. The estimated sizes for EV genomes range from 200 to 390 kbp. Two EV genomes have been sequenced, those of AMEV and MSEV. The genome size of AMEV is 232 kbp and of MSEV is 236 kbp. Both genomes are highly A+T-rich, with an A+T content for each genome of approximately 82%, which is in line with DNA melting experiments for EV genomes. In comparison, VV has a genome that is 67% A+T. Both the MSEV and AMEV genomes contain inverted terminal repeats at the ends of the genome.

The organization of the AMEV and MSEV genomes is surprisingly unlike that of the orthopoxviruses. Within the orthopoxviruses there is a core of conserved genes flanked by variable genes involved in pathogenesis and host range. Although many of the same genes are present in EVs, the collinear core is not conserved in either MSEV or AMEV. In fact, it is the lack of conservation of genome organization between AMEV and MSEV, both originally classified as betaentomopoxviruses, that resulted in MSEV being removed from this genus and listed as an unclassified virus.

The available sequence data from the EVs has been included in recent bioinformatics analysis of all available poxvirus sequences. From this analysis, 49 gene families were identified that are conserved across all species of poxviruses. Homologs of the VV G3L gene have recently been identified in AMEV and MSEV, bringing the total of conserved gene families to 50. These 50 genes presumably comprise the minimal complement of genes required for poxvirus function. These conserved gene families include proteins necessary for DNA replication and transcription, polyadenylation of mRNA, and the major structural proteins found within the virion core. An analysis of CVs alone increases the number of conserved genes from 50 to 90. Clearly, the inclusion of the EVs in this type of analysis refines the minimal core of proteins required by a poxvirus.

Replication

Like other poxviruses, EVs complete their life cycle entirely within the cytoplasm of the cell. A number of enzymes required for DNA synthesis and for temporal regulation of RNA expression are encoded by EVs. These include DNA polymerase, RNA polymerase, poly(A) polymerase, topoisomerase, and a number of transcription factors. Replication within the insect host occurs primarily within the fat body. Many EVs then proceed to infect the hemocytes and other tissues. In cell culture, DNA replication occurs between 6 and 12 h post infection (hpi). At the optimum temperature of 27 °C, the life cycle takes approximately 18–24 h. By 9 hpi, host-cell protein synthesis is shut off. The sites of DNA replication and virion assembly are known as viroplasms or viral factories. Although many virions are packaged into OBs, some are not. Those virions that are not occluded proceed to bud through the cell membrane into the hemocoel of the insect or the cell culture medium *in vitro*. These nonoccluded virus particles are infectious and are thought to be responsible for virus spread with the organism.

RNA production occurs in a temporal fashion with early gene expression occurring from 0 to 6 hpi. Included in the early gene products are transcription factors required for intermediate and late gene production. No intermediate genes have been studied in the EVs, but homologs of several VV intermediate genes are present within both the AMEV and MSEV genomes. Late gene expression appears to follow DNA replication and begins at 9 hpi. The DNA regulatory elements that are present in VV to control this temporal gene expression are present in both AMEV and MSEV, although the functionality of these elements during an EV infection has not been explored. Among the conserved regulatory elements are the TTTTTNT termination sequence at the end of early genes, and the consensus late promoter TAAATG. Similar to the CVs, there is no RNA splicing of EV transcripts.

A unique discovery within the late genes of AMEV is the presence of discrete late transcripts. The late transcripts of VV lack specific 3′ termini, resulting in polydisperse 3′ ends. The rare exceptions to this are several genes that have been identified as having discrete 3′ RNA termini that are formed via cleavage of polydisperse transcripts. A number of major late transcripts of AMEV are discrete in length, unlike what is observed in VV. These include the transcripts for the structural proteins p4a and p4b, the superoxide dismutase, and RAP94, a component of the early RNA polymerase. The spheroidin transcript is also discrete.

Molecular Biology

Since the sequencing of two EV genomes, a number of interesting features of the EVs have come to light. One of these is the apparent absence of some members of the RNA polymerase complex. There are eight components of the RNA polymerase complex in VV and these are conserved among all sequenced CVs. Three of these proteins have no obvious homologs in AMEV and MSEV. Whether these subunits are not required in an insect cell environment or whether their roles are fulfilled by other nonhomologous proteins is not yet clear. Several transcription factor homologs are also missing from the EVs. Together, this might indicate that the EVs regulate transcription in a different fashion from their orthopoxvirus relatives.

In vitro transcription catalyzed by permeabilized AMEV virions requires different conditions from those needed for transcription from VV virions. Unlike VV reactions, which use low levels of detergent and reducing agent, AMEV transcription requires higher levels of reducing agent and the requirement for detergent is less stringent, as transcription can occur in the absence of detergent but not in the absence of reducing agent. AMEV reactions are optimal when an ATP-generating system is present. Such a system is not required for VV virion-mediated transcription. This indicates that the structure of the AMEV virion is different in its response to both reducing agents and detergents. Our attempts to isolate AMEV cores with methodology optimized for the preparation of VV cores has also indicated that key AMEV core proteins partition differently from their VV counterparts.

The transcripts of AMEV are polyadenylated at the 3′ termini, but the composition of the poly(A) polymerase complex appears to be different from that of VV. In VV, poly(A) polymerase is a heterodimer. AMEV has two homologs of the small VV subunit in addition to one homolog of the large subunit. This raises the question of whether both small subunits are involved in polyadenylation or whether there might be tissue or temporal specificity conferred on the complex by the different subunits. It is not clear how common two small poly(A) polymerase subunits are among EVs, since the only other fully sequenced EV (MSEV) only has one small subunit. Spheroidin and other transcripts from EVs have been shown to contain a nonencoded 5′ poly(A) head similar to that reported for a number of VV transcripts.

Protein gel analysis of ^{35}S-labeled virions indicates that AMEV contains 36–37 structural proteins and MSEV contains 39–45. The proteolytic processing of the large structural proteins that is detected in VV infections has not been found in EV infections. The enzymes thought to be responsible for this proteolytic cleavage in VV are the products of the I7L and G1L genes. These genes are conserved in AMEV and MSEV, indicating that protein processing may occur even though it is as yet undetected.

Both AMEV and MSEV encode an NAD^+-dependent DNA ligase. In contrast, VV encodes an ATP-dependent DNA ligase. The AMEV DNA ligase is capable of joining singly nicked DNA fragments. These are the first examples of an NAD^+-dependent DNA ligase outside of the eubacteria. The recently sequenced crocodile poxvirus has also been reported to encode an NAD^+-dependent DNA ligase.

Poxviruses are known for the wide variety of proteins that they encode to evade the host immunomodulatory response. VV and other orthopoxviruses encode serine protease inhibitors as well as chemokine binding molecules and decoy receptor molecules. Due to the host range of the

EVs, they encode other types of defense molecules. Among these are the *inhibitor of apoptosis* (*iap*) genes. The *iap* of AMEV has been well characterized and functionally inhibits apoptosis. A related AMEV gene that functions to inhibit apoptosis is a homolog of the baculovirus pan-caspase inhibitor, p35.

Another novel protein expressed by AMEV is a Cu–Zn superoxide dismutase (SOD). Although a number of the orthopoxviruses encode genes with homology to this class of SODs, neither the VV or myxoma virus proteins are functional in that capacity, although they are present within the virion. The SOD expressed by AMEV is functional as an SOD but is not essential for virus growth in culture. The deletion of the *sod* gene from AMEV appears to have no effect on the growth of the virus in gypsy moth larvae.

Summary

It is clear that this large subfamily of the family *Poxviridae* provides a wealth of possible information about the basic mechanisms of the poxvirus lifecycle. There appear to be a number of interesting variations on the molecular details which define this overall family of viruses. There are clear similarities to the vertebrate poxviruses in virion morphology, double-stranded DNA genome, cytoplasmic life cycle, and RNA expression. Yet the differences between the CVs and EVs are significant and represent an area of research that has not been fully explored. The data that have been obtained from genomic sequencing has been essential to identifying some of the different proteins that are present in the EVs, as well as identifying potentially missing homologs of VV proteins. It is important to note that there are large differences at the DNA level between the two sequenced EVs, indicating that there is probably a wide variety of unique features within the EVs as a group. As more sequence information becomes available, the diversity of this family of viruses may become more evident.

See also: Apoptosis and Virus Infection; Baculoviruses: Molecular Biology of Nucleopolyhedroviruses; Poxviruses; Vaccinia Virus.

Further Reading

Afonso CL, Tulman ER, Lu Z, et al. (1999) The genome of *Melanoplus sanguinipes* entomopoxvirus. *Journal of Virology* 73: 533–552.

Arif BM and Kurstak E (1991) The entomopoxviruses. In: Kurstak E (ed.) *Viruses of Invertebrates* pp. 179–195. New York: Marcel Dekker.

Bawden AL, Glassberg KJ, Diggans J, et al. (2000) Complete genomic sequence of the *Amsacta moorei* entomopoxvirus: Analysis and comparison with other poxviruses. *Virology* 274: 120–139.

Becker MN, Greenleaf WB, Ostrov DA, and Moyer RW (2004) *Amsacta moorei* entomopoxvirus expresses an active superoxide dismutase. *Journal of Virology* 78: 10265–10275.

Becker MN and Moyer RW (2007) Subfamily *Entomopoxvirinae*. In: Mercer A, Schmidt A, and Weber O (eds.) *Poxviruses*, pp. 251–269. Basel: Birkhäuser.

Gubser C, Hue S, Kellam P, and Smith GL (2004) Poxvirus genomes: A phylogenetic analysis. *Journal of General Virology* 85: 105–117.

Li QJ, Liston P, and Moyer RW (2005) Functional analysis of the *inhibitor of apoptosis* (*iap*) gene carried by the entomopoxvirus of *Amsacta moorei*. *Journal of Virology* 79: 2335–2345.

Li QJ, Liston P, Schokman N, Ho JM, and Moyer RW (2005) *Amsacta moorei* entomopoxvirus inhibitor of apoptosis suppresses cell death by binding grim and hid. *Journal of Virology* 79: 3684–3691.

Miller LK and Ball LA (1998) *The Insect Viruses*. New York: Plenum.

Moss B (2001) *Poxviridae*: The viruses and their replication. In: Knipe DM and Howley PM (eds.) *Fields Virology*, 4th edn., pp. 2849–2883. Philadelphia: Lippincott Williams & Wilkins.

Upton C, Slack S, Hunter AL, Ehlers A, and Roper RL (2003) Poxvirus orthologous clusters: Toward defining the minimum essential poxvirus genome. *Journal of Virology* 77: 7590–7600.

Winter J, Hall RL, and Moyer RW (1995) The effect of inhibitors on the growth of the entomopoxvirus from *Amsacta moorei* in *Lymantria dispar* (gypsy moth) cells. *Virology* 211: 462–473.

Epidemiology of Human and Animal Viral Diseases

F A Murphy, University of Texas Medical Branch, Galveston, TX, USA

© 2008 Elsevier Ltd. All rights reserved.

Glossary

Airborne transmission Transmission via large droplets and via very small droplet nuclei (aerosols) emitted from infected persons during coughing or sneezing or from environmental sources.

Case-control study An epidemiological study in which the risk factors of humans or animals with a disease are compared with those without the disease.

Cohort study Attempt to identify the cause of a disease by comparing exposed and nonexposed (control) populations in a prospective epidemiological study.

Common vehicle transmission Pertains to fecal contamination of food and water supplies.

Direct contact transmission Involves actual physical contact between an infected subject and a susceptible subject (e.g., kissing, biting, coitus).
Epidemic Major increase in disease incidence affecting either a large number of humans or animals or spreading over a large area.
Epidemiology The study of the determinants, dynamics, and distribution of diseases in populations.
Fomite An inanimate object that may be contaminated with virus and become the vehicle for transmission.
Herd immunity The immune status of a population that affects viral transmission rates. Often used in describing the elimination of a virus from a population when there are too few susceptible hosts remaining to sustain a transmission chain.
Horizontal transmission The transfer of infectious virus from one human or animal to another by any means other than vertical transmission.
Iatrogenic transmission Transmission via health care procedures, materials, and workers (e.g., physicians, nurses, dentists, veterinarians).
Incidence rate (or attack rate) A measure of the occurrence of infection or disease in a population over time – it refers to the proportion of a population contracting a particular disease during a specified period.
Mathematical model (epidemiological) A means to convey quantitative information about a host-virus interaction, such as an epidemic or an emerging disease episode, by the construction of a set of predictive mathematical algorithms.
Nosocomial transmission Pertains to infections acquired while a patient, human or animal, is in hospital.
Prevalence rate The ratio, at a particular point in time, of the number of cases currently present in the population divided by the number of subjects in the population at risk; it is a snapshot of the occurrence of infection or disease at a given time.
Species jumping (or host range extension) Referring to a virus that derives from an ancient reservoir life cycle in animals, but has subsequently established a new life cycle in humans or a different animal species and no longer uses, the original animal reservoir.
Transmission The process by which a pathogen is shed from one host and infects the next.
Vector-borne transmission Involves the bites of arthropod vectors (e.g., mosquitoes, ticks, sandflies).
Vertical or transplacental transmission Occurs from mother to fetus prior to or during parturition, either across the placenta, when the fetus passes through the birth canal, or via colostrum and milk.
Vertical transmission Transmission of virus from parent to progeny through the genome, sperm, or ovum or extracellularly (e.g., through colostrum or across the placenta).
Zoonosis Disease which is naturally transmitted to humans from an ongoing reservoir life cycle in animals or arthropods, without the permanent establishment of a new life cycle in humans.

Introduction

Viral disease epidemiology is the study of the determinants, dynamics, and distribution of viral diseases in populations. The risk of infection or disease in a population is determined by characteristics of the virus, the host, and the host population, as well as behavioral, environmental, and ecological factors that affect virus transmission from one host to another. Epidemiology attempts to meld these factors into a unified whole. The depiction of the interaction of factors favoring the emergence of a viral disease (**Figure 1**), called 'the convergence model', is taken from the US Institute of Medicine study, Microbial Threats to Health, Emergence, Detection and Response (National Academy Press, 2003). At the center is a box representing the convergence of factors leading to 'the black box', reflecting the reality that many unknown interactions are important virologically and epidemiologically.

The foundations of epidemiology predate the microbiological and virological sciences, starting with Hippocrates, the Greek physician and father of medicine, who in the fourth century BC made important epidemiologic observations on infectious diseases. John Snow is called the father of modern epidemiology because he developed excellent quantitative methods while studying the source of a cholera outbreak at the Broad Street pump in London in 1849. Snow was followed by William Farr, who in the 1870s advanced the use of vital statistics and clarified many of the principles of risk assessment and retrospective and prospective studies. Their vision is reflected in the fast-changing science of epidemiology which is now supported by advanced computer technology, sophisticated statistical methods, and very sensitive and specific diagnostic systems.

Assessment of Disease Occurrence and Outcome

By introducing quantitative measurements of disease trends, epidemiology has come to have a major role in improving our understanding of the overall nature of

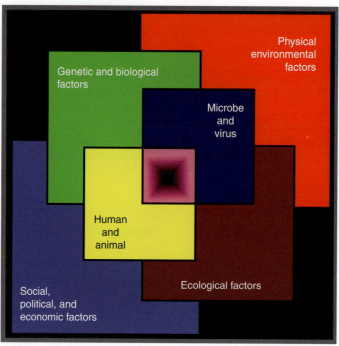

Figure 1 The Convergence Model. Factors contributing to the emergence of viral diseases. At the center is a box representing the convergence of factors leading to the emergence of a viral disease. The black center represents unknown interactions, 'the black box'. Reproduced from: Smolinski MS, Hamburgh MA, and Lederburg J (eds.) (2003) *Microbial Threats to Health, Emergence, Detection and Response*, **Figure 1** p. 3. Washington, DC: National Academies Press. Reprinted with permission from the National Academy Press, copyright (2003), National Academy of Sciences.

disease and in alerting and directing disease control activities. Epidemiology is also effective in (1) clarifying the role of particular viruses and viral variants as the cause of disease, (2) clarifying the interaction of viruses with environmental determinants of disease, (3) determining factors affecting host susceptibility, (4) unraveling modes of transmission, and (5) field testing of vaccines and antiviral drugs.

The comparison of disease experience between populations is expressed in the form of 'rates'. The terms 'incidence rate' and 'prevalence rate' are used to describe quantitatively the frequency of occurrence of infection or disease in populations. 'Incidence rate' (also called attack rate) is defined as the ratio of new cases occurring in a population to the size of the population during a specified period of time. Prevalence rate is the ratio of the total number of cases occurring in a population to the size of the population during a specified period of time. 'Seroprevalence rate' relates to the occurrence of antibody to a particular virus in a population. Because viral antibodies, especially neutralizing antibodies, often last a lifetime, seroprevalence rates usually represent cumulative experience with the virus. The term 'case–fatality rate' is used to indicate the percentage of subjects with a particular disease that die from the disease. All these rates may be affected by various attributes that distinguish one individual from another: age, sex, genetic constitution, immune status, pregnancy, nutritional status, and various behavioral and medical care and patient management parameters. The most widely applicable attribute is age, which may encompass immune status as well as various physiological variables.

A viral disease is characterized as 'endemic' when there are multiple or continuous chains of transmission resulting in continuous occurrence of disease in a population over a period of time. 'Epidemics' are peaks in disease incidence that exceed the endemic baseline or expected rate of disease. The size of the peak required to constitute an epidemic is arbitrary and is related to the background endemic rate and the anxiety that the disease arouses (e.g., a few cases of rabies is regarded as an epidemic, whereas a few cases of influenza is not). A 'pandemic' is a worldwide epidemic.

Epidemiologic Studies

A proper description of an outbreak of disease or an epidemic must include the parameters of 'person (or subjects in the case of animals), place, and time'. Such descriptive information is a necessary first step in describing the occurrence, distribution, course, threat, and anticipated action response to the initial recognition of a cluster of cases of disease. Much of the initial investigation called for rests in common sense, observational acuity, and an insightful 'index of suspicion'. Much of the initial investigation has been termed 'shoe-leather epidemiology'. The trigger for such initial investigation is most often an astute clinician (physician or veterinarian) or an astute pathologist.

Case-Control Studies and Cohort Studies

There are two basic analytic techniques used to investigate relationships between cause and effect and to evaluate risk factors of disease. These are the 'case-control study' and the 'cohort study'. In the case-control study, investigation starts after the disease has occurred – it is a retrospective study, going back in time to determine causative events. Although this kind of study does not require the creation of new data or records, it does require careful selection of the control group, matched to the test group so as to avoid bias. The retrospective case-control study lends itself to quick analysis and is relatively inexpensive to carry out. In the cohort study, the prospective study, investigation entails the gathering of new data to identify cause–effect relationships. This kind of study is expensive and does not lend itself to quick analysis as groups must be followed until disease is observed. However, when cohort studies are successful, proof of cause–effect relationship is often incontrovertible.

Molecular Epidemiologic Studies

The term 'molecular epidemiology' is used to denote the use of any of a large number of molecular biological methods in support of epidemiologic investigations. For example, with herpesviruses, restriction endonuclease mapping has provided a means of identification of unique viral genotypes – in an epidemiologic study recognized as the first based upon viral molecular characterization, the source of herpes simplex virus 1 causing disease in a hospital newborn nursery was traced to one persistently infected nurse rather than any of several other possible shedders. With rotaviruses and bluetongue viruses, polyacrylamide gel electrophoresis of the segmented viral RNA has been used epidemiologically, for example, to unravel outbreaks involving multiple viral variants. Panels of monoclonal antibodies have been used to distinguish virus variants for epidemiologic purposes; they have been particularly useful in elucidating host-range and geographic variants of rabies virus. Today, partial sequencing has become the most commonly used molecular epidemiologic methodology; partial sequencing of poliovirus isolates recovered from patients indicates whether they are wild type (even local or introduced wild type), attenuated vaccine type, or a vaccine type that has reacquired neurovirulence during human passage. Partial sequencing of foot-and-mouth disease viruses can offer the same kind of geographic information of virus movement as has proved so useful in polio control and eradication programs, but because of political sensitivities in some countries a robust international reference laboratory system has not been established that could provide the same kind of practical disease control information as has been the case with polio. Thus, with many human and animal viruses molecular epidemiologic studies are flourishing, but more such studies should lead to international reference laboratory systems to guide prevention and control actions. Such studies are developing rapidly today to deal with the threat of a human pandemic of avian influenza, but there are many more viral diseases, especially animal diseases, in need of this kind of development.

Etiologic Studies and the Proof of Causation

One of the landmarks in the history of infectious diseases was the development of the Henle-Koch postulates which established the evidence required to prove a causal relationship between a particular infectious agent and a particular disease. These simple postulates were originally drawn up for bacteria, but were revised in 1937 by Rivers and again in 1982 by Evans in attempts to accommodate the special problem of proving disease causation by viruses (**Table 1**). In many cases, virologists have had to rely on indirect evidence, 'guilt by association', with associations based on epidemiologic data and patterns of serologic positivity in populations. Today, many aspects of epidemiologic investigation play roles, especially in trying to distinguish an etiological, rather than coincidental or opportunistic relationship between a virus and a given disease.

For example, early in the investigation of human acquired immunodeficiency syndrome (AIDS), before its etiology was established, many kinds of viruses were being isolated from patients and many candidate etiologic agents were being advanced. Prediction that the etiologic agent would turn out to be a member of the family *Retroviridae* was based upon years of veterinary research on animal retroviruses and animal retroviral diseases. This prediction was based upon recognition of common biologic and pathogenetic characteristics of AIDS and animal retroviral diseases. This prediction guided many of the early experiments to find the etiologic agent of AIDS; later, after human immunodeficiency virus (HIV1) was discovered, its morphological similarity to equine infectious anemia virus, a prototypic member of the genus *Lentivirus*, family *Retroviridae*, was the key to unraveling confusion over the fact that the human virus killed host lymphocytes rather than transforming them as typical oncogenic retroviruses would do. Ever since, this essence of comparative medicine has been guiding HIV/AIDS research in many areas, including drug design,

Table 1 Criteria for disease causation: a unified concept appropriate for viruses as causative agents of disease, based on the Henle–Koch postulates, and modified by A. S. Evans

1. Prevalence of the disease is significantly higher in subjects exposed to the putative virus than in those not so exposed.
2. Incidence of the disease is significantly higher in subjects exposed to the putative virus than in those not so exposed (prospective studies).
3. Evidence of exposure to the putative virus is present more commonly in subjects with the disease than in those without the disease.
4. Temporally, the onset of disease follows exposure to the putative virus, always following an incubation period.
5. A regular pattern of clinical signs follows exposure to the putative virus, presenting a graded response, often from mild to severe.
6. A measurable host-immune response, such as an antibody response and/or a cell-mediated response, follows exposure to the putative virus. In those individuals lacking prior experience, the response appears regularly, and in those individuals with prior experience, the response is anamnestic.
7. Experimental reproduction of the disease follows deliberate exposure of animals to the putative virus, but nonexposed control animals remain disease free. Deliberate exposure may be in the laboratory or in the field, as with sentinel animals.
8. Elimination of the putative virus and/or its vector decreases the incidence of the disease.
9. Prevention or modification of infection, via immunization or drugs, decreases the incidence of the disease.
10. "The whole thing should make biologic and epidemiologic-sense."

diagnostics, and vaccine development. HIV/AIDS epidemiologic research has often been intertwined with research on the several simian immunodeficiency viruses (SIVs).

Seroepidemiologic Studies

Seroepidemiology is useful in public health and animal health investigations and in research to determine the prevalence or incidence of particular infections, to evaluate control and immunization programs, and to assess past history when a 'new' virus is discovered. When paired serum specimens are obtained from individuals several weeks apart, the initial appearance of antibody in the second specimen or a rise in antibody titer indicates recent infection. Similarly, the presence of specific immunoglobulin M (IgM) antibody in single serum samples, indicating recent infection, may be used in seroepidemiologic studies. Correlation of serologic tests with clinical observations makes it possible to determine the ratio of clinical to subclinical infections.

Sentinel Studies

Because of advanced diagnostic/serologic methods, sentinel studies can yield many valuable data in timely fashion about impending disease risks. For example, sentinel chicken flocks are set out for the early detection of the presence of arboviruses such as West Nile virus in the United States. These flocks are bled and tested weekly for the presence of virus or antiviral antibody; they provide an early warning of the levels of virus amplification that occur before epidemics.

Vaccine Trials

The immunogenicity, potency, safety, and efficacy of vaccines are first studied in laboratory animals, followed by small-scale closed trials, and finally in large-scale open trials. Such studies employ epidemiologic methods, rather like those of the cohort (prospective) study. In most cases, there is no alternative way to evaluate new vaccines, and the design of trials has now been developed so that they yield maximum information with minimum risk and acceptable cost.

Virus Transmission among Individuals

Viruses survive in nature only if they are able to be transmitted from one host to another, whether of the same or another species. Transmission cycles require virus entry into the body, replication, and shedding with subsequent spread to another host.

Virus Entry

Portals of virus entry into the body include the skin, respiratory tract, intestinal tract, oropharynx, urogenital tract, and conjunctiva. In some cases, viruses use a particular portal of entry because of particular environmental or host-behavior factors and in other cases because of specific viral ligands and host-cell receptors. In many cases, disruption of normal host-defense mechanisms leads to entry that might otherwise be thwarted; for example, papillomaviruses may enter the deep layers of the skin via abrasions, acid-labile coronaviruses may enter the intestine protected by the buffering capacity of milk, and influenza viruses may enter the lower respiratory tract because a drug has dampened cilial action of the respiratory epithelium.

Virus Shedding

The exit of virus from an infected host is just as important as entry in maintaining its transmission cycle. All portals used by viruses to gain entry are used for exit. The important elements in virus shedding are virus yield (from the standpoint of the virus, the more shedding the better) and timeliness of yield (again, the earlier the shedding the better). Viruses that cause persistent infections often employ remarkable means to avoid host inflammatory and immune responses so as to continue shedding. For example, the epidemiologically important shedding of herpes simplex viruses 1 and 2 that perpetuates the viruses in populations requires recrudescence of persistent ganglionic infection, centrifugal viral genomic transit to peripheral nerve endings, and productive infection of mucosal epithelium, all in the face of established host immunity.

Modes of Virus Transmission

Virus transmission may be 'horizontal' or 'vertical'. The vast majority of transmission is horizontal, that is, between individuals within the population at risk. Modes of horizontal transmission of viruses can be characterized as direct contact, indirect contact, common vehicle, airborne, vector-borne, iatrogenic, and nosocomial. Vertical or transplacental transmission occurs between the mother and her fetus or newborn. Some viruses are transmitted in nature via several modes, others exclusively via one mode (see **Table 2**).

'Direct contact transmission' involves actual physical contact between an infected subject and a susceptible subject (e.g., kissing, Epstein–Barr virus, the cause of mononucleosis, biting (e.g., rabies); coitus (sexually transmitted viral diseases)). Indirect contact transmission occurs via 'fomites', such as shared eating utensils, improperly sterilized surgical equipment, or improperly sterilized non-disposable syringes and needles.

'Common vehicle transmission' pertains to fecal contamination of food and water supplies (e.g., norovirus diarrhea). Common vehicle transmission commonly results in epidemic disease.

'Airborne transmission' typically results in respiratory infections (and less typically in intestinal infections), but these infections may also be transmitted by direct and

Table 2 Examples of human and animal virus transmission patterns

Infectious agent/disease	Mode of transmission	Portal of entry
Influenza virus/influenza	Contact/direct/indirect via droplets and droplet nuclei	Respiratory tract
Rhinoviruses/common cold	Contact/direct/indirect via droplets and droplet nuclei and fomites	Respiratory tract
Rubella virus/congenital rubella	Contact/direct/indirect via droplets and droplet nuclei	Respiratory tract
	Vertical/congenital	Transplacental
Rotaviruses/diarrhea	Contact/direct/indirect via fomites	Intestinal tract (oral)
Poliovirus/poliomyelitis	Contact/direct/indirect via fomites	Intestinal tract (oral)
Norovirus/diarrhea	Common vehicle/fecal contamination of water	Intestinal tract (oral)
Hepatitis A virus/hepatitis	Common vehicle/fecal contamination of food	Intestinal tract (oral)
Variant Creutzfeldt–Jakob disease prion/prion disease (spongiform encephalopathy)	Common vehicle/bovine spongiform encephalopathy prion contamination of beef or beef products	Intestinal tract (oral)
Herpes simplex virus/genital herpes	Contact/direct (sexual)	Genital tract
Human immunodeficiency virus 1/acquired immunodeficiency syndrome (AIDS)	Contact/direct (sexual), contact/direct (blood) Vertical/congenital	Genital tract, bloodstream, transplacental, at birth and via breast feeding
Rabies virus/rabies	Zoonotic/contact/direct (saliva)	Skin (bite wound)
Russian spring summer encephalitis virus/encephalitis	Zoonotic/arthropod-borne	Skin (tick bite)
Dengue viruses/dengue	Zoonotic/arthropod-borne	Skin (mosquito bite)
Sin Nombre and related viruses	Zoonotic/contact/direct (rodent urine, saliva and feces)	Respiratory tract
Lassa virus	Zoonotic/contact/direct (rodent urine, saliva and feces)	Respiratory tract and intestinal tract (oral)
Ebola and Marburg viruses	Zoonotic/reservoir host unknown; secondary cases contact/direct/nosocomial and iatrogenic	Index cases unknown, likely respiratory tract or skin and mucous membranes; secondary cases, contact and iatrogenic (injection)
Leukemia viruses/leukemias (proven only in animals)	Vertical/germ-line	Transmitted as genetic trait

indirect contact. Airborne transmission occurs via large droplets and via very small droplet nuclei (aerosols) emitted from infected persons during coughing or sneezing (e.g., influenza) or from environmental sources. Large droplets (>10 µm in diameter) settle quickly, but droplet nuclei evaporate forming dry particles (<5 µm in diameter) which remain suspended in the air for extended periods. Droplets may travel only a meter or so while droplet nuclei may travel over much longer distances.

'Vector-borne transmission' involves the bites of arthropod vectors (e.g., mosquitoes, ticks, and sandflies).

'Iatrogenic transmission' involves health care procedures, materials, and workers (e.g., physicians, nurses, dentists, and veterinarians).

'Nosocomial transmission' pertains to infections acquired while a patient, human or animal, is in hospital.

'Vertical or transplacental transmission' occurs from mother to fetus prior to or during parturition. Certain retroviruses are vertically transmitted in animals via the integration of viral DNA directly into the DNA of the germline of the fertilized egg. Other viruses are transmitted to the fetus across the placenta; yet others are transmitted when the fetus passes through the birth canal. Another vertical transmission route is via colostrum and milk. Vertical transmission of a virus may or may not be associated with 'congenital disease' (i.e., disease that is present at birth) which may be lethal (and the cause of abortion or stillbirth) or the cause of congenital abnormalities. The herpesviruses, especially cytomegaloviruses, and rubella virus cause important congenital diseases in humans, and pestiviruses, such as bovine viral diarrhea virus, in animals.

Common Patterns of Virus Transmission

Enteric infections are most often transmitted by direct contact and by fomites in a 'fecal–oral cycle' that may include fecal contamination of food and water supplies; diarrheic feces may also splash to give rise to aerosols (droplets and droplet nuclei). Respiratory infections are most often transmitted by the airborne route or by indirect contact via fomites in a 'respiratory cycle', that is, virus is shed in respiratory secretions and enters its next host through the nares during inhalation. The respiratory cycle is responsible for the most explosive patterns of epidemic disease in humans and all domestic animal species.

Perpetuation of Viruses in Nature

Perpetuation of a virus in nature depends upon the maintenance of serial infections, that is, a chain of transmission; the occurrence of disease is neither required nor necessarily advantageous.

Influence of the Clinical Status of the Host

Infection without recognizable disease is called 'subclinical' or 'clinically inapparent'. Overall, subclinical infections are much more common than those that result in disease. Their relative frequency accounts for the difficulty of tracing chains of transmission, even with the help of laboratory diagnostics. Although clinical cases may be somewhat more productive sources of virus than subclinical infections, because the latter do not restrict the movement of the infected host, they can be most important as sources of viral dissemination. In most acute infections, whether clinically apparent or not, virus is shed in highest titers during the late stages of the incubation period, before the influence of the host-immune response takes effect. Persistent infections, whether or not they are associated with episodes of clinical disease, also play an important role in the perpetuation of many viruses in nature. For example, prolonged virus shedding can reintroduce virus into a population of susceptibles all of which have been born since the last clinically apparent episode of infection. This is important in the survival of rubella virus in some isolated populations. Sometimes the persistence of infection, the production of disease, and the transmission of virus are dissociated; for example, togavirus and arenavirus infections may have little adverse effect on their reservoir hosts (arthropods, birds, and rodents), but transmission may be very efficient. On the other hand, the persistence of infection in the central nervous system, as with measles virus in subacute sclerosing panencephalitis (SSPE), is of no epidemiological significance, since no infectious virus is shed from this site.

Influence of Virulence of the Virus

The virulence of the infecting virus may directly affect the probability of its transmission. The classic example of this is rabbit myxomatosis. In Australia, mosquito-borne transmission of myxoma virus was found to be most effective when infected rabbits maintained highly infectious skin lesions for several days before death. Highly virulent strains of the virus were found to kill rabbits so quickly that transmission did not occur, and naturally attenuated strains were found to produce minimal lesions that healed quickly and did not permit transmission. Virus strains at either extreme of this virulence spectrum were found not to survive in nature, but virus strains of intermediate virulence have circulated for many years.

Influence of Host Population Immunity

With most viruses, endemic or epidemic transmission leads to a level of immunity in the host population that affects or even interrupts further transmission. The 'herd immunity' effect is countered in some cases by viral antigenic variation. For example, influenza viruses undergo genetic variations ('shift' and 'drift') such that persons immune to previously circulating virus strains are susceptible to new strains. Assessing these genetic changes is the main objective of laboratory-based surveillance programs, which in turn are the basis for decisions on the formulation of each year's influenza vaccine.

Influence of Population Size

It is self-evident that the long-term survival of a virus requires that it be continuously transmitted from one host to another. In general, for rapidly and efficiently transmitted viruses such as many respiratory viruses, local survival of the virus requires that the susceptible host population be very large. A virus may disappear from a population because it exhausts its potential supply of susceptible hosts as they acquire immunity to reinfection with the same virus. Depending on duration of immunity and the pattern of virus shedding, the 'critical population size' varies considerably with different viruses and with different host species. The most precise data on the importance of population size in acute nonpersistent infections come from studies of measles. Persistence of measles virus in a population depends upon a continuous supply of susceptible children. Analyses of the incidence of measles in large cities and in island communities have shown that a population of about half a million persons is needed to ensure a large enough annual input of new susceptible hosts, by birth or immigration, to maintain measles virus in the population. Because infection depends on respiratory transmission, the duration of epidemics of measles is correlated inversely with population density. If a population is dispersed over a large area, the rate of spread is reduced and the epidemic may last longer, so that the number of susceptible persons needed to maintain transmission chains is reduced. On the other hand, in such a situation a break in the transmission chain is much more likely. When a large proportion of the population is initially susceptible, the intensity of the epidemic builds up very quickly and attack rates are almost 100% ('virgin-soil epidemic'). On the other hand, when measles vaccination programs are implemented properly the virus disappears completely from the regional population.

Influence of Zoonotic Transmission Cycles

Because most viruses are host-restricted, most viral infections are maintained in nature within populations of the same or related species. However, there are a number of viruses that may have multiple hosts and spread naturally between several different species of vertebrate host, for example, rabies and eastern equine encephalitis viruses. The term 'zoonosis' is used to describe multiple-host infections that are transmissible from animals to man.

The zoonoses, whether involving domestic or wild animals or arthropods, usually represent important problems only under conditions where humans are engaged in activities involving close contact with animals or exposure to arthropods.

Influence of Arthropod Transmission Cycles

Many viral zoonoses are caused by arboviruses. Arboviruses have two classes of hosts, vertebrate and invertebrate. Over 500 arboviruses are known, of which about 100 cause disease in humans and 40 in domestic animals; some of these are transmitted by ticks, some by mosquitoes, and yet others by phlebotomine flies (sandflies) or *Culicoides* spp. (midges). Arthropod transmission may be 'mechanical', where the arthropod acts as a 'flying pin', or more commonly, 'biological', involving replication of the virus in the arthropod vector. The arthropod vector acquires virus by feeding on the blood of a viremic person or animal. Replication of the ingested virus, initially in the arthropod's gut, and its spread to the salivary glands takes several days; the interval varies with different viruses and is influenced by ambient temperature. Virions in the salivary secretions of the vector are injected into human or animal hosts during subsequent blood meals. Most arboviruses have localized natural habitats in which specific receptive arthropod and vertebrate hosts are involved in the viral life cycle. Vertebrate reservoir hosts are usually wild mammals or birds; humans are rarely involved in primary transmission cycles, although the exceptions to this generalization are important (e.g., Venezuelan equine encephalitis, yellow fever, and dengue viruses). Humans are in most cases infected incidentally, for example, by the geographic extension of a reservoir vertebrate host and/or a vector arthropod. Ecological changes produced by human activities disturb natural arbovirus life cycles and have been incriminated in the geographic spread or increased prevalence of arbovirus diseases.

Mathematical Modeling

From the time of William Farr, who studied epidemic disease problems in the 1870s, mathematicians have been interested in 'epidemic curves' and secular trends in the incidence of infectious diseases. With the development of computer-based mathematical modeling techniques, there has been a resurgence of interest in the population dynamics of infectious diseases. There has also been a resurgence in controversies surrounding the use of models; critics say 'for every model there is an equal and opposite model'. So, the proof of the value of models lies in their practical application, and in recent years there have been more and more successes. For example, when for counterterrorism reasons universal smallpox vaccination was being considered, models that showed that vaccine could be used effectively after rapid detection of a terrorism incident led to a decision to stockpile, but not widely use vaccine. As another example, when a foot-and-mouth disease epidemic raged in the United Kingdom in 2001, a model showed that only the most vigorous stamping-out campaign could get ahead of the movement of the virus across the country. The model, seeming eminently logical now, importantly provided the kind of veracity and political will needed to accelerate the stamping-out campaign.

Models may be used to determine (1) patterns of disease transmission, (2) critical population sizes to support the continuous transmission of viruses with short and long incubation periods, (3) the dynamics of endemicity of viruses that become persistent in their hosts, and (4) the variables in age-dependent viral pathogenicity. Computer modeling also provides useful insights into the effectiveness of disease control programs. Much attention has been given to modeling the future of the AIDS epidemic in the United States and the rest of the world. Such models usually start with historical data on the introduction of the etiologic virus, HIV1, proceed to the present stage of the epidemic where the disease has become well established in many countries and in fewer countries subject to prevention and treatment strategies, and then proceed to project its course into the future. During the first 10 years of the AIDS epidemic in the United States, African countries, and then in Asian countries, most models underestimated developing trends; more recently models have become more accurately predictive – but in many places more and more sobering.

Implications for Disease Prevention

Knowledge of the epidemiology and modes of transmission of infectious diseases is critical to the development and implementation of prevention and control strategies. Data on incidence, prevalence, and mortality contribute directly to the establishment of priorities for prevention and control programs while knowledge of viral characteristics and modes of transmission are used in deciding prevention strategies focusing on vaccine development and delivery, environmental improvements, enhancement of nutritional status, improvement in personal hygiene, and behavioral changes.

See also: Disease Surveillance; Viral Pathogenesis; Zoonoses.

Further Reading

Evans AS and Kaslow RA (eds.) (1997) *Viral Infections of Humans. Epidemiology and Control,* 4th edn. New York: Plenum Medical Book Company.

Heymann DL (ed.) (2005) *Control of Communicable Diseases Manual,* 18th edn. Washington, DC: American Public Health Association Press.

Mandell GL, Bennett JE, and Dolin R (eds.) (2000) *Mandell, Douglas, and Bennett's Principles and Practice of Infectious Diseases,* 5th edn. New York: Churchill Livingstone.

Monto A (2005) The epidemiology of viral infections. In: Mahy BWJ and ter Meulen V (eds.) *Virology: Topley and Wilson's Microbiology and Microbial Infections,* vol. 1. London and Washington, DC: Hodder Arnold and ASM Press.

Murphy FA, Gibbs EPJ, Horzinek MC, and Studdert MJ (1999) *Veterinary Virology,* 3rd edn. New York: Academic Press.

Nathanson N (2006) Epidemiology. In: Knipe DM, Howley PM, Griffin DE, Lamb RA, and Martin MA (eds.) *Fields Virology,* 5th edn. Philadelphia: Lippincott Williams and Wilkins.

Nathanson N and Murphy FA (1997) Evolution of viral diseases. In: Nathanson N, Ahmed R, Griffin DE, *et al.* (eds.) *Viral Pathogenesis,* Philadelphia: Lippincott-Raven Press.

Nowak MA and May R (2001) *Virus Dynamics: Mathematical Principles of Immunology and Virology.* Oxford, UK: Oxford University Press.

Smolinski MS, Hamburg MA, and Lederberg J (2002) *Emerging Microbial Threats to Health in the 21st Century.* Institute of Medicine/National Academy of Sciences. Washington, DC: National Academy Press.

Smalinski MS, Hamburgh MA, and Lederburgh J (eds.) (2003) *Microbial Threads to Health, Emergence, Detection and Response,* 367pp. Washington: National Academies press.

Thrusfield MV and Bertola G (2005) *Veterinary Epidemiology,* 3rd edn. London: Blackwell Publishing.

Epstein–Barr Virus: General Features

L S Young, University of Birmingham, Birmingham, UK

© 2008 Elsevier Ltd. All rights reserved.

Taxonomy and Genome Structure

Epstein–Barr virus (EBV; species *Human herpesvirus 4*) is a member of the genus *Lymphocryptovirus*, which belongs to the lymphotropic subfamily *Gammaherpesvirinae* of the family *Herpesviridae*. EBV is closely related to the lymphocryptoviruses (LCVs) present in Old World non-human primates, including EBV-like viruses of chimpanzees and rhesus monkeys. These viruses share homologous sequences and genetic organization, and infect the B lymphocytes of their host species, resulting in the establishment of latent infection *in vivo* and transformation *in vitro*. A transforming, EBV-related virus has also been isolated from spontaneous B-cell lymphomas of common marmosets and is thus the first EBV-like virus to be identified in New World primates. The genome of this marmoset LCV revealed considerable divergence from the genomes of EBV and Old World primate EBV-related viruses, suggesting that this virus represents a more primitive predecessor of the LCVs infecting higher-order primates.

The EBV genome is composed of linear, double-stranded DNA, approximately 172 kbp in length. EBV has a series of 0.5 kbp terminal direct repeats (TRs) and internal repeat sequences (IRs) that divide the genome into short and long, largely unique sequence domains. EBV was the first herpesvirus to have its genome completely cloned and sequenced. Since the EBV genome was sequenced from an EBV DNA *Bam*HI fragment cloned library, open reading frames (ORFs), genes, and sites for transcription or RNA processing are frequently referenced to specific *Bam*HI fragments, from A–Z, in descending order of fragment size. The virus has the coding potential for around 80 proteins, not all of which have been identified or characterized.

History

In 1958, Denis Burkitt described a lymphoma that represented the most common tumor affecting children in certain parts of East Africa. The geographical distribution of this malignancy suggested that the development of Burkitt lymphoma (BL) might be due to an infectious agent possibly related to malaria. In 1964, the successful establishment of cell lines from explants of BL enabled Tony Epstein and Yvonne Barr to identify herpesvirus-type particles by electron microscopy within a subpopulation of tumor cells *in vitro* (**Figure 1**). Werner and Gertrude Henle subsequently demonstrated that BL-derived cell lines expressed antigens that were recognized not only by sera from patients with BL but also by sera from patients with infectious mononucleosis (IM). Similar seroepidemiological studies also suggested a link between infection with this virus (now called Epstein– Barr virus after its discoverers) and undifferentiated nasopharyngeal carcinoma (NPC), leading to the subsequent direct demonstration of EBV DNA in the tumor cells of NPC. The ability of EBV to immortalize B lymphocytes efficiently *in vitro* and to induce tumors in nonhuman primates established this virus as a putative oncogenic agent in humans. Over the last 20 years EBV has been implicated in a variety of other lymphoid and epithelial malignancies (**Table 1**).

Geographic and Seasonal Distribution

EBV is ubiquitous, being found as a widespread and largely asymptomatic infection in all human communities.

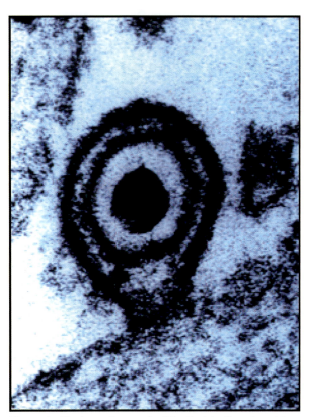

Figure 1 Electron micrograph of the Epstein–Barr virus virion.

Table 1 EBV-associated tumors

Tumor	Subtype	EBV genome +ve (%)
Burkitt's lymphoma	Endemic	100
	Sporadic	10–20
	AIDS	30–40
Nasopharyngeal carcinoma	Undifferentiated	100
	SCC	10–40
Hodgkin's disease	Mixed cellularity	>80
	Nodular sclerosing	30–40
	Lymphocyte predominant	<10
T-cell lymphoma	Nasal	100
	Others	10–40
Immunoblastic lymphoma	Transplant	100
	AIDS	>90

Primary infection often occurs early in life, particularly in tropical areas and in persons of lower socioeconomic class. Thus, in tropical Africa and New Guinea primary EBV infection is common in the first year of life whereas in Western communities infection is continually acquired throughout childhood and early adulthood. Although the EBV-associated malignancies BL and NPC exhibit an unusual geographic distribution, this appears not to be due to differences in EBV infection but due to additional cofactors.

Host Range and Virus Propagation

Humans are the natural host for EBV infection. EBV has a unique ability to transform resting B cells into permanent, latently infected, lymphoblastoid cell lines (LCLs) in which every cell carries multiple copies of circular, extrachromosomal, viral DNA (episomes) and produces a number of latent proteins, including six EBV-encoded nuclear antigens (EBNAs 1, 2, 3A, 3B, 3C, and -LP) and three latent membrane proteins (LMPs 1, 2A, and 2B); this form of infection is referred to as latency III. When peripheral blood lymphocytes from healthy EBV seropositives are placed in culture, the few EBV-infected B lymphocytes that are present regularly give rise to spontaneous outgrowth of EBV-transformed LCLs, provided that immune T lymphocytes are either removed or inhibited by the addition of cyclosporin A to the culture. LCLs can also be generated by direct infection of resting B lymphocytes with EBV derived from the throat washings of seropositive individuals or from producer B-cell lines. LCLs have provided an invaluable, albeit incomplete, model of the lymphomagenic potential of EBV.

Certain nonhuman primates, particularly the cotton-top tamarin, have been used as experimental hosts for EBV, but virus infection in these animals is associated with the induction of lymphomas. LCLs generated by EBV transformation of cotton-top tamarin B lymphocytes *in vitro* spontaneously produce large amounts of virus compared with their human counterparts and as such have been used as a source of EBV; for example, the prototype strain of EBV is B95.8, which is produced from a tamarin LCL originally immortalized with EBV from a patient with IM. However, the lack of a fully permissive system for propagating EBV *in vitro* has hampered our understanding of virus replication and prevented the generation of EBV mutants. Over recent years a number of systems have been developed for the generation of recombinant forms of EBV which rely on the manipulation of the entire virus genome within bacteria followed by rescue and propagation in human cells.

Genetics and Strain Variation

EBV isolates from different regions of the world or from patients with different virus-associated diseases are remarkably similar when their genomes are compared by restriction fragment length polymorphism analysis. However, variations in repeat regions of the EBV genome are observed among different EBV isolates. Analysis of the EBV genome in a number of BL-cell lines revealed gross deletions, some of which account for biological differences. For example, P3HR-1 virus, which is nontransforming, has a deletion of the EBNA2-encoding gene. Strain variation over the EBNA2-encoding (*Bam*HI WYH) region of the EBV genome permits all virus isolates to be classified as either 'type 1' (EBV-1, B95.8-like) or 'type 2' (EBV-2, Jijoye-like). This genomic variation results in the production of two antigenically distinct forms of the EBNA2 protein that share only 50% amino acid sequence homology. Similar allelic polymorphisms (with 50–80% sequence homology depending on the locus) related to the EBV type occur in a subset of latent genes, namely those encoding EBNA-LP, EBNA3A, EBNA3B, and EBNA3C. These differences have functional consequences as EBV-2 isolates are less efficient in *in vitro* B-lymphocyte transformation assays compared with EBV-1 isolates. A combination of virus isolation and seroepidemiological studies suggest that type 1 virus isolates are predominant (but not exclusively so) in many Western countries, whereas both types are widespread in equatorial Africa, New Guinea, and perhaps certain other regions.

In addition to this broad distinction between EBV types 1 and 2, there is also minor heterogeneity within each virus type. Individual strains have been identified on the basis of differences compared with B95.8, ranging from single-base mutations to extensive deletions. While infection with multiple strains of EBV was originally thought to be confined to immunologically compromised patients, more recent studies demonstrate that normal healthy seropositives can be infected with multiple EBV isolates and that their relative abundance and presence appears to vary over time. Coinfection of the host with multiple virus strains could have evolutionary benefit to EBV, enabling the generation of diversity by genetic recombination. Such intertypic recombination has been demonstrated in human immunodeficiency virus (HIV)-infected patients and appears to arise via recombination of multiple EBV strains during the intense EBV replication that occurs as a consequence of immunosuppression.

The possible contribution of EBV strain variation to virus-associated tumors remains contentious. Many studies have failed to establish an epidemiological association between EBV strains and disease and suggest that the specific EBV gene polymorphisms detected in virus-associated tumors occur with similar frequencies in EBV isolates from healthy virus carriers from the same geographic region. However, this does not exclude the possibility that variation in specific EBV genes is responsible for the distinct geographic distribution of virus-associated malignancies. In this regard, an LMP1 variant containing a 10-amino-acid deletion (residues 343–352) was originally identified in Chinese NPC biopsies and has oncogenic and other functional properties distinct from those of the B95.8 LMP1 gene. It is therefore likely that variation in LMP1 and other EBV genes can contribute to the risk of developing virus-associated tumors, but more biological studies using well-defined EBV variants are required.

Evolution

EBV, like other herpesviruses, has probably evolved with humans. The relatedness of the LCVs at both the genomic and protein levels does not correlate with the evolutionary relatedness of their Old World primate hosts. This implies that the selective pressures governing the evolution of these viruses are different from those responsible for the evolution of primate species. It has been suggested that the specific tissue tropism of LCVs may have constrained their evolutionary divergence. A comparison of marmoset LCV with the LCVs present in Old World primates provides some insight into LCV evolution approximately 35 million years before the appearance of EBV. The acquisition of accessory genes by Old World LCVs supports the contention that these viruses have coevolved with their hosts possibly as a means of maintaining their biological properties.

The evolutionary relationship between the type 1 and type 2 strains of EBV remains obscure. They may have evolved from a common progenitor virus or through recombination of the ancestral LCVs infecting Old World primates. The pronounced (but not exclusive) segregation of EBV-2 isolates within equatorial regions suggests that environmental factors (including immunological competence) may have influenced EBV evolution and may still be responsible for the effective competition between EBV-2 isolates and the ubiquitous EBV-1 family.

EBV Primary Infection and Persistence

EBV infects the majority of the world's adult population and, following primary infection, the individual remains a lifelong carrier of the virus (**Figure 2**). In underdeveloped countries, primary infection with EBV usually occurs during the first several months to few years of life and is often asymptomatic. However, in developed populations, primary infection is more frequently delayed until adolescence or adulthood, in many cases producing the characteristic clinical features of IM. EBV is orally transmitted, and infectious virus can be detected in oropharyngeal secretions from IM patients, from immunosuppressed patients, and at lower levels from healthy EBV seropositive individuals. These observations, together with the fact that EBV-transformed LCLs *in vitro* tend to be poor producers of the virus and B lymphocytes permissive of viral replication have not been demonstrated *in vivo*, suggest that EBV replicates and is shed at epithelial sites in the oropharynx and/or salivary glands. This is supported by the demonstration of replicating EBV in the differentiated epithelial cell layers of oral 'hairy' leukoplakia, a benign lesion of the tongue found in immunocompromised patients. However, it is also likely that EBV-infected B cells are reactivated within the local mucosal environment and that this contributes to virus shedding at oropharyngeal sites (**Figure 2**).

The inability to detect EBV routinely in normal epithelial cells and the demonstration that EBV can be completely eradicated by irradiation in bone marrow transplant recipients suggests that B lymphocytes are the main site of EBV persistence (**Figure 2(b)**). This is supported by the B lymphotropism of EBV which is mediated by the binding of the major viral envelope glycoprotein gp350 to the CR2 receptor on the surface of B cells. Virion penetration of the B-cell membrane requires further interactions between the EBV glycoprotein gp42 (which forms a ternary complex with gH and gL viral glycoproteins) and human leukocyte antigen (HLA) class II molecules. It appears that the presence or absence of HLA class II in virus-producing cells influences the tropism of EBV for B cells or epithelial cells by affecting the availability of gp42. Other CR2-independent pathways may be responsible for EBV infection of epithelial cells, including secretory component-mediated immunoglobulin (Ig) A transport, integrin interactions with polarized epithelium, and direct cell-to-cell contact, but these are relatively inefficient and of unknown relevance to EBV infection *in vivo*.

The B-cell tropism of EBV and its ability to establish a latent infection in these cells both *in vitro* and *in vivo* further supports the contention that B lymphocytes are the reservoir of lifelong persistent infection (**Figure 2(b)**). Detailed examination of EBV infection *in vivo* has shown that the virus persists in the IgD^-CD27^+ memory B-cell subset and that these cells have downregulated the expression of most, if not all, viral genes. The precise route of entry of EBV-infected B cells into the memory compartment remains a subject of much debate. This reservoir of infected cells is stably maintained thereafter, apparently subject to the same physiologic controls as the general mucosa-associated memory B-cell pool. EBV persistence within this B-cell population brings with it the possibility of fortuitous antigen-driven recruitment of infected cells into germinal centers (GCs), leading to progeny that either re-enter the circulating memory pool or differentiate to plasma cells that may migrate to mucosal sites. The different forms of latency that are manifest in virus-associated malignancies (discussed below) may represent latency programs that have evolved to accommodate such changes in host cell physiology (**Figure 2**). Thus GC transit appears to activate a latency program where only the genome maintenance protein EBNA1 is expressed (latency 0), while exit from GCs is possibly linked to the transient expression of the EBV-encoded latent membrane proteins, LMP1 and LMP2 (latency II). The ability of these proteins to mimic the key signals required for B cells to undergo a GC reaction, namely T-cell help via the CD40 pathway (constitutively provided by LMP1) and activation of the cognate B-cell receptor (augmented by LMP2A), supports this strategy whereby EBV exploits the physiological process of B-cell differentiation to access and maintain persistent

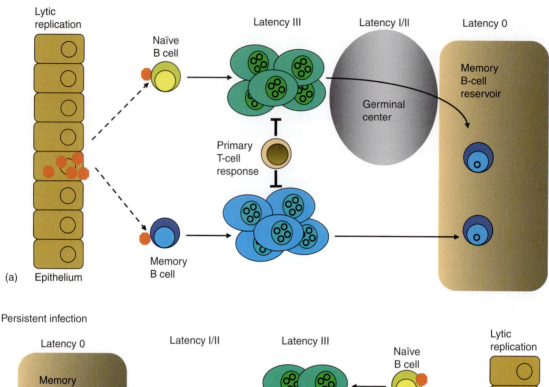

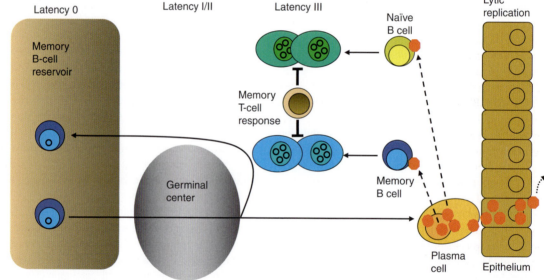

Figure 2 EBV primary infection and persistence. Diagram showing putative *in vivo* interactions between EBV and host cells. (a) Primary infection. EBV replicates in epithelial cells and spreads to lymphoid tissues as a latent growth-transforming (latency III) infection of B cells. Many infected B cells are removed by the emerging EBV-specific T-cell response. Some infected cells escape by downregulating EBV latent genes to establish a stable pool of resting virus-positive memory B cells. (b) Persistent infection. EBV-infected memory B cells become subject to the physiological controls governing memory B-cell migration and differentiation. Occasional recruitment into germinal center (GC) reactions resulting in activation of different EBV latency programs and re-entry into the memory cell reservoir or plasma cell differentiation with activation of virus lytic cycle. Infectious virions then initiate foci of EBV replication in epithelial cells and also new growth-transforming infection of naïve and/or memory B cells.

infection in the memory B-cell pool. The possible commitment of these cells to plasmacytoid differentiation may trigger entry into lytic virus replication, thereby providing a source of low-level virus shedding into the oropharynx. There may also be circumstances in which infected cells in the reservoir can reactivate back to proliferative infections similar to those of an EBV-transformed LCL.

Immune Response

EBV elicits both humoral and cell-mediated immune responses in infected hosts. Primary infection with EBV is associated with the rapid appearance of antibodies to replicative antigens such as viral capsid antigen (VCA), early antigen (EA), and membrane antigen (MA; gp350/220) with a later serological response to EBNA proteins. In IM, these responses are exaggerated and are accompanied by autoantibodies such as rheumatoid factor as well as a heterophile antibody response directed against antigens on the surface of sheep erythrocytes. These autoantibodies are the result of EBV-induced polyclonal B-cell activation. In the chronic asymptomatic virus carrier, antibodies to VCA, MA, and EBNAs are found, the titers of which remain remarkably stable. Of these antibodies, those against MA are particularly important as they have virus-neutralizing ability and can also mediate antibody-dependent cellular cytotoxicity. As discussed below, the levels of these EBV-specific antibodies are elevated in different EBV-associated diseases.

As with other persistent viruses, cell-mediated immunity plays an important role in controlling EBV infection. Primary EBV infection elicits a robust cellular immune response and the lymphocytosis observed in IM is a consequence of the hyperexpansion of cytotoxic CD8+ T cells with reactivities against both latent and lytic viral antigens. These reactivities are subsequently maintained at high level (up to 5% of the total circulating pool) in the CD8+ memory T-cell pool. An EBV-specific CD4+ T-cell response also contributes toward the control of EBV infection and, along with the CD8+ response, appears to be important in preventing the unlimited proliferation of EBV-infected B cells. Thus, impairment of the T-cell response, either by immunosuppressive drug therapy or by HIV infection, is responsible for the development of polyclonal lymphoproliferations that can progress to frank monoclonal non-Hodgkin's lymphomas (see below). These lesions can be controlled by adoptive therapy with EBV-specific T cells. The growth and survival of BL, NPC, and Hodgkin's lymphoma (HL) in immunocompetent individuals implies that the tumor cells can evade EBV-specific T-cell surveillance. This may be achieved by restricting EBV latent gene expression to those viral proteins not efficiently recognized by the T-cell responses (i.e., away from expression of the highly immunogenic EBNA3 family) and/or by downregulation of target cell molecules required for immune recognition such as major histocompatability complex (MHC) class I and the antigen-processing machinery.

EBV-Associated Diseases

Infectious Mononucleosis

Primary infection with EBV in childhood is usually asymptomatic but when delayed until adolescence or early adulthood can manifest clinically as IM, a self-limiting lymphoproliferative disease associated with the hyperexpansion of cytotoxic CD8$^+$ T cells that are reactive to both lytic and latent cycle viral antigens. These reactivities are subsequently maintained in the CD8$^+$ T-cell memory pool at high levels (up to 5% of circulating CD8$^+$ T cells), even in individuals with no history of IM. The incidence of IM is low in developing countries where asymptomatic primary infection predominantly occurs in childhood. In certain poorly defined situations, IM-like symptoms can persist, resulting in chronic active EBV infection associated with elevated antibody titers to virus lytic antigens but low titers to the EBNAs.

Lymphomas in Immunosuppressed Individuals

Patients with primary immunodeficiency diseases such as X-linked lymphoproliferative syndrome (XLP) and Wiscott–Aldrich syndrome are at increased risk of developing EBV-associated lymphomas. Because these tumors are extremely rare, little is known about the precise contribution of EBV and the associated pattern of viral gene expression in these lymphomas. Mortality from XLP is high with around 50% of patients developing fatal IM after primary infection with EBV and an additional 30% of patients developing malignant lymphomas. The defect responsible for XLP is mutation of an adaptor molecule, SAP, which mediates signaling in a wide range of immune cells and is involved, via its interaction with the SLAM family of receptors, in both innate and adaptive immune reactions.

Allograft recipients receiving immunosuppressive therapy and patients with acquired immune deficiency syndrome (AIDS) are also at increased risk for development of EBV-associated post-transplant lymphoproliferative disease (PTLD) and lymphomas. These lesions range from polyclonal EBV-driven lymphoproliferations, much like that observed *in vitro* in virus-transformed LCLs (latency III, **Figure 3**), to aggressive monoclonal non-Hodgkin's B-cell lymphomas (NHLs) in which additional cellular genetic changes (e.g., mutation of p53) are present. The incidence of both PTLD and B-cell lymphomas in allograft recipients varies with the type of organ transplanted and with the type of immunosuppressive regimen used. Allogeneic bone marrow or solid organ transplantation into EBV seronegative children is a particular risk factor for the development of these lesions. A proportion of these lesions in post-transplant patients resolve in response to a reduction in immunosuppression or to targeted therapies such as anti-CD20 monoclonal antibody or adoptive EBV-specific T cells.

The incidence of NHL in AIDS patients is increased approximately 60-fold compared to the normal population. Around 60% of these tumors are large B-cell lymphomas like those found in allograft recipients, 20% are primary brain lymphomas, and 20% are of the BL type. EBV

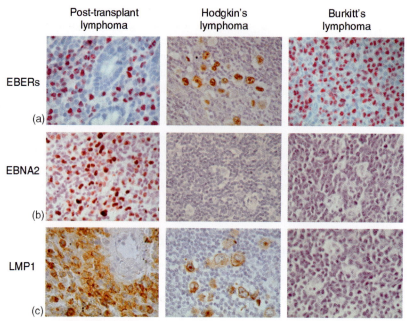

Figure 3 EBV-associated lymphomas display different patterns of virus latent gene expression. While the post-transplant lymphomas associated with EBV infection express a pattern of latent gene expression similar to that observed in virus-transformed B cells *in vitro* (latency III), Hodgkin's lymphoma (HL; Latency II) and Burkitt's lymphoma (BL; Latency I) express more restricted forms of EBV latency. (a) Upper panels show *in situ* hybridization for the abundant nonpolyadenylated EBER transcripts, which are expressed in all forms of EBV latent infection. (b) The middle panels depict immunohistochemical staining for the EBNA2 protein, which is only expressed in post-transplant lymphoma and not either HL or BL. (c) The bottom panels show immunohistochemical staining for the LMP1 protein, which is expressed in post-transplant lymphoma and HL but not in BL.

infection is present in approximately 50% of AIDS-related NHL, nearly all the primary brain and Hodgkin's lymphomas, and around 40% of the BL tumors.

Burkitt's Lymphoma

The endemic form of BL, which is found in areas of equatorial Africa and New Guinea, represents the most common childhood cancer (peak age 7–9 years) in these regions with an incidence of up to 10 cases per 100 000 people per year. This high incidence of BL is associated with holoendemic malaria, thus accounting for the climatic variation in tumor incidence first recognized by Denis Burkitt. More than 95% of these endemic BL tumors are EBV-positive compared with 20% of the low incidence, sporadic form of BL which occurs worldwide. In areas of intermediate BL incidence, such as Algeria and Malaysia, the increased number of cases correlates with an increased proportion of EBV-positive tumors. The pattern of EBV gene expression in BL is generally restricted to EBNA1 and the nonpolyadenylated EBER transcripts (latency I), although broader virus gene expression involving the EBNA3 proteins has been observed in a subset of tumors (**Figure 3**).

A consistent feature of BL tumors, irrespective of geographical location or EBV status, is chromosome translocations involving the long arm of chromosome 8 (8q24) in the region of the c-myc proto-oncogene and either chromosome 14 in the region of the Ig heavy-chain gene or, less frequently, chromosomes 2 or 22 in the region of the Ig light-chain genes. The cell surface phenotype of BL cells and the presence of ongoing Ig gene mutation supports the GC origin of this tumor.

Seroepidemiological studies have demonstrated elevated antibody titers to EBV VCA and EA in BL patients compared to children without the tumor. These elevated antibody titers have been found to precede the development of BL and can therefore be used to screen 'at risk' individuals. Most early cases of BL can be successfully treated using chemotherapeutic regimens that include cyclophosphamide. However, access to drugs and advanced tumor presentation have limited the efficacy of such therapies.

Hodgkin's Lymphoma

Epidemiological studies originally suggested a possible role for EBV in the etiology of HL. Thus, elevated antibody titers to EBV antigens have been detected in HL patients and these are present before the diagnosis of the disease. Furthermore, there is an increased risk of HL following IM. EBV has been demonstrated in around 40% of HL cases with both viral nucleic acid and virus-encoded latent proteins (EBNA1, LMP1, LMP2 – latency

II) localized to the malignant component of HL, the so-called Hodgkin's and Reed–Sternberg (HRS) cells (**Figure 3**). These malignant B cells tend to carry nonproductive Ig genes consistent with their GC origin and implicating EBV infection in the rescue of these crippled cells from their usual apoptotic fate. The association of HL with EBV is age-related; pediatric and older adult cases are usually EBV-associated whereas HL in young adults is less frequently virus-positive. The proportion of EBV-positive HL in developing countries is high, consistent with a greater incidence of HL in children and more frequent prevalence of the mixed cellularity histiotype; the histological subtype of HL in which EBV infection is most frequently detected. Although the incidence of HL is relatively low (1–3 per 100 000 per year), this tumor is not geographically restricted, making its association with EBV significant in world health terms.

Virus-Associated T-Cell and NK-Cell Lymphomas

EBV-positive monoclonal lymphomas of either CD4+ or CD8+ T-cell origin are more frequently found in Southeast Asian populations, arising as a consequence of either virus-associated haemophagocytic syndrome (VAHS) or in the setting of chronic active EBV infection. A more aggressive extranodal tumor predominantly expressing the CD56 natural killer (NK) cell marker is also EBV positive and manifests as an erosive lesion (lethal midline granuloma) of the nasal cavity.

Nasopharyngeal Carcinoma

The association of EBV with undifferentiated NPC (WHO type III) was first suggested by serological evidence and then confirmed by the demonstration of EBV DNA in NPC biopsy material. NPC is particularly common in areas of China and southeast Asia, reaching a peak incidence of 20–30 cases per 100 000 per year. Incidence rates are particularly high in Cantonese males, highlighting an important genetic predisposition as well as a role for environmental cofactors such as dietary components (e.g., salted fish). NPC tumors are characterized by a prominent lymphoid stroma, and the interaction between these activated lymphocytes and adjacent carcinoma cells appears to be crucial for the continued growth of the malignant component. EBV latent gene expression in NPC is restricted to EBNA1, the LMP2A/B proteins, the EBER transcripts, and the *Bam*HI-A rightward transcripts (BARTs), with around 20% of tumors also expressing the oncogenic LMP1 protein (**Figure 4** – latency II). The presence of monoclonal EBV episomes in NPC indicates that virus infection preceded the clonal expansion of the malignant cell population. More detailed analysis of the genetic changes in NPC suggests that some of these (particularly deletions in chromosomes 3p and 9p) occur early before EBV infection. Extensive serological screening for EBV-specific antibody titers in high-incidence areas, in particular IgA antibodies to VCA and EAs, have proved useful in diagnosis and in monitoring the effectiveness of therapy. More recent studies have demonstrated that the quantitative analysis of tumor-derived EBV DNA in the blood of patients with NPC using real-time polymerase chain reaction (PCR) is of both diagnostic and prognostic utility. EBV infection is also associated with the more differentiated forms of NPC (WHO types I and II) particularly in those geographical regions with a high incidence of the type III tumor. EBV-positive carcinomas, which morphologically resemble NPC (undifferentiated carcinomas of nasopharyngeal type), have been described at other anatomical sites (e.g., thymus, tonsil, lungs, skin) but the extent of the association of these tumors with EBV is geographically variable.

Gastric Carcinoma and Other Epithelial Tumors

EBV infection is also present in around 10% of typical gastric adenocarcinomas, accounting for up to 75 000 cases per year. These tumors resemble NPC in carrying monoclonal EBV genomes, having a restricted pattern of EBV gene expression (EBERs, EBNA1, LMP2A, BARTs, and BARF1) and in the appearance of virus infection as a relatively late event in the carcinogenic process. The geographical variation in the association of EBV with gastric adenocarcinomas probably reflects ethnic and genetic differences. EBV-positive tumors have distinct phenotypic and clinical characteristics compared with EBV-negative tumors.

A number of other more common carcinomas such as breast cancer and liver cancer have been reported to be infected with EBV. Difficulties in confirming these associations have raised concerns about the use of PCR analysis to detect EBV infection and about the specificity of certain monoclonal antibody reagents. It is possible, however, that a small proportion of tumor cells can be infected with EBV, perhaps sustaining a low-level replicative infection, and that this might contribute to the growth of the carcinoma.

Prevention and Control of EBV Infection

The significant burden of EBV-associated tumors worldwide has prompted the development of novel therapeutic strategies that either specifically target viral proteins or exploit the presence of the virus in malignant cells. Pharmacological approaches include the use of agents to induce the expression EBV lytic cycle antigens, including virus-encoded kinases (EBV thymidine kinase and

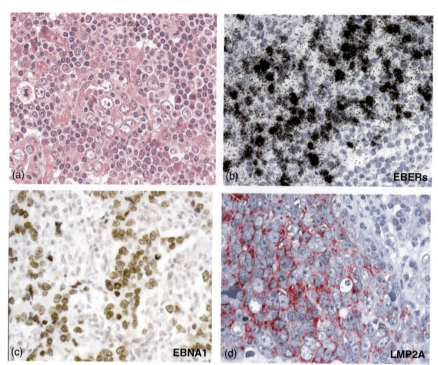

Figure 4 EBV gene expression in nasopharyngeal carcinoma (NPC). (a) shows a hematoxylin and eosin stain of a NPC tumor showing the intimate association between the undifferentiated carcinoma cells (large nuclei) and the activated lymphoid stroma (small intensely staining nuclei). (b–d) in situ hybridization for the EBER transcripts and immunohistochemical staining for EBNA1 and LMP2A.

BGLF4, a protein kinase) that will then phosphorylate the nucleoside analog gancyclovir to produce its active cytotoxic form. Demethylating agents such as 5′ azacytidine are able to de-repress lytic, as well as potentially immunogenic, latent genes and are currently in early-stage clinical trials in patients with NPC, HL, and AIDS-associated lymphoma. Hydroxyurea is a chemotherapeutic agent that can induce loss of EBV episomes and has shown some clinical efficacy in patients with EBV-positive AIDS-related central nervous system lymphoma. Other approaches are based on the more specific targeting of individual EBV proteins using either single-chain antibodies, siRNA, or dominant-negative molecules. The use of adoptive EBV-specific T-cell therapy for the treatment of existing PTLD and in the prophylactic setting has been extremely successful. This approach is also showing signs of clinical efficacy in the treatment of NPC and HL but in this setting the use of T cells enriched for reactivities to subdominant targets (e.g., LMP2A and EBNA1) is important. More direct vaccine approaches to treat patients with EBV-associated tumors or to prevent disease development are being examined including (1) whole EBV latent antigens delivered by a virus vector or in autologous dendritic cells; (2) peptide or polytope vaccination; and (3) prophylactic vaccination against MA (gp350/220) to generate a neutralizing antibody response.

Future Perspectives

EBV was discovered over 40 years ago and its DNA was fully sequenced in 1984. Much work has contributed to the unequivocal identification of EBV as oncogenic in humans. In more recent years, the nature of the interaction of EBV with the immune host has become clearer, illustrating the complex mechanisms that the virus exploits to persist in the memory B-cell pool. The fine detail of this interaction with normal B cells is still in its infancy, and a major question remains over the replicative lifecycle of EBV *in vivo*, particularly with regard to the relative role of B cells versus epithelial cells in this process. The development of more efficient *in vitro* systems for studying EBV infection and replication in different cell types will help to unravel the complex interplay between the virus and the cell. The use of EBV recombinants continues to shed light on the role of latent genes in the transformation process, on the requirements for the efficient production of progeny virus, and on the role of membrane glycoproteins in the infection process. Understanding the host cell–virus interaction will be dependent on the generation of appropriate *in vitro* models, particularly systems which allow a more detailed understanding of the tumor microenvironment and the role of the local cytokine milieu. There are many interesting aspects of EBV that remain to be understood, including the role of virus-encoded microRNAs and the possible contribution of lytic cycle antigens to virus persistence and oncogenesis. The challenge will be to exploit these new mechanistic insights both to gain a better understanding of EBV infection *in vivo* and to develop novel therapies for treating virus-associated disease.

See also: Epstein–Barr Virus: Molecular Biology; Herpesviruses: Discovery; Herpesviruses: General Features; Herpesviruses: Latency; Kaposi's Sarcoma-Associated Herpesvirus: General Features; Kaposi's Sarcoma-Associated Herpesvirus: Molecular Biology.

Further Reading

Kieff E and Rickinson AB (2001) Epstein–Barr virus and its replication. In: Knipe DM and Howley PM (eds.) *Fields Virology*, 4th edn., pp. 2511–2573. Philadelphia: Lippincott Williams and Wilkins.

Kuppers R (2005) Mechanisms of B cell lymphoma pathogenesis. *Nature Reviews Cancer* 5: 251–262.

Kutok JL and Wang F (2006) Spectrum of Epstein–Barr virus-associated diseases. *Annual Review of Pathology: Mechanisms of Disease* 1: 375–404.

Lo KW and Huang DP (2002) Genetic and epigenetic changes in nasopharyngeal carcinoma. *Seminars in Cancer Biology* 12: 451–462.

Rickinson AB and Kieff E (2001) Epstein–Barr virus. In: Knipe DM and Howley PM (eds.) *Fields Virology*, 4th edn., pp. 2575–2627. Philadelphia: Lippincott Williams and Wilkins.

Robertson ES (ed.) (2005) *Epstein–Barr Virus*. Norfolk, England: Caister Academic Press.

Tao Q, Young LS, Woodman CB, and Murray PG (2006) Epstein–Barr virus and its associated cancers – Genetics, epigenetics, pathobiology and novel therapeutics. *Frontiers in Bioscience* 11: 2672–2713.

Thorley-Lawson DA and Gross A (2004) Persistence of the Epstein–Barr virus and the origins of associated lymphomas. *New England Journal of Medicine* 350: 1328–1337.

Young LS and Rickinson AB (2004) Epstein–Barr virus: 40 years on. *Nature Reviews Cancer* 4: 757–768.

Epstein–Barr Virus: Molecular Biology

J T Sample and C E Sample, The Pennsylvania State University College of Medicine, Hershey, PA, USA

© 2008 Elsevier Ltd. All rights reserved.

Properties of the Virion

The morphology of the virion of Epstein–Barr virus (EBV; species *Human herpesvirus 4*, genus *Lymphocryptovirus*, family *Herpesviridae*) is typical of all herpesviruses, and comprises four major components. At the core is the viral genome, which is encased within an icosadeltahedral nucleocapsid that consists of 162 capsomeres and is approximately 100 nm in diameter. The tegument represents the amorphous proteinaceous region between the nucleocapsid and membrane envelope, the latter of which the virus acquires as the nucleocapsid buds through cellular membranes. Analysis of virion protein content by mass spectrometry has identified 6 viral capsid proteins, 12 glycoproteins, and as many as 17 tegument proteins. The tegument also contains several host proteins, for example, actin and Hsp70. The most abundant viral glycoprotein is gp350, which binds to the major cellular receptor for EBV attachment on B cells, CD21.

Properties of the Genome

Organization and Coding Potential

The EBV genome is a double-stranded (ds) DNA molecule of approximately 172 kbp. In its linear configuration, as within the virion, it is bounded on both ends by a variable number of 538-bp direct terminal repeats (TRs). The genome is separated into short (12 kbp) and long (134 kbp) predominantly unique sequence domains by 7–10 copies of a 3-kbp repeat, IR1. Moreover, the long unique domain is divided into four shorter domains (U2–U5) by repetitive elements containing short direct repeats (IR2–IR4) (**Figure 1**). Each of the noted internal repeats contributes coding information to a viral protein. Based on identification of potential translational open reading frames (ORFs), the existence of approximately 75 genes was initially predicted from the genomic nucleotide sequence of the B95–8 isolate of EBV. Further, a 12-kbp deletion from the B95–8 genome removes three ORFs present in wild-type isolates, and sequence analyses of EBV cDNAs have identified several additional genes that were not apparent from the genomic sequence due to the highly spliced nature of their mRNAs. Thus, currently there are an estimated 80–85 EBV proteins, though not all have been confirmed. In addition to these, EBV expresses two classes of noncoding RNAs: EBV-encoded small RNA (*EBER*) 1 and 2, which are highly structured RNAs of 167 and 172 nucleotides, respectively, and two clusters of micro-RNAs (miRNAs) derived from two different loci.

The majority of EBV genes contribute to the production of progeny virus, and their expression is limited to

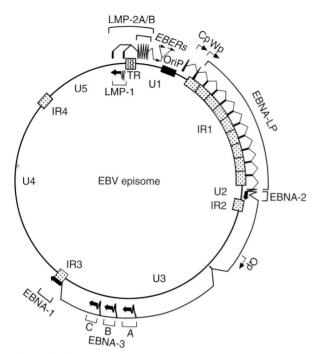

Figure 1 Structure and latency-associated transcription of the EBV genome. The episomal form of the genome that is maintained in latently-infected cells is shown; repeat elements (TR, IR1–IR4), unique sequence domains (U1–U5), and the latent-infection origin of DNA replication *oriP* are indicated, as are the latency-associated mRNAs and *EBERs*. The common EBNA-gene promoters Cp and Wp, and the EBNA-1 exclusive promoter Qp are depicted by bent arrows. Exons that contain ORFs are bracketed. The *BARTs* (not shown; see **Figure 2**) originate in U4 and terminate in U5.

the virus replicative or lytic cycle of infection. Up to an additional 12 genes, plus the *EBERs* and miRNAs, are expressed during the latent (nonproductive) programs of EBV infection. Few lytic-cycle mRNAs are spliced, and those that are in this class are relatively simple, generally containing only a single splice that removes a relatively short intron of no more than a few hundred base pairs. A discussion of the many lytic-cycle-specific genes is beyond the scope of this article, but their organization within the EBV genome and mRNA structures may be followed up via the 'Further reading' section. By contrast, all of the latency-associated gene mRNAs are spliced, and most splicing is extensive and occurs over 8–90 kbp. The nine known latency-associated proteins of EBV are grouped into two classes: the latent membrane proteins (LMP-1, LMP-2A, and LMP-2B) and the EBV-encoded nuclear antigens (EBNA-1, EBNA-2, EBNA-3A, EBNA-3B, EBNA-3C, and EBNA-LP). A transcriptional map of the EBV genome displaying the structures of the mRNAs encoding these proteins is presented in **Figure 1**. Notably, expression of the EBNA proteins is from a common transcription unit that spans 90 kbp of the viral genome, the primary transcripts from which are alternatively spliced to yield individual EBNA mRNAs. Moreover, the common highly spliced 5′ portion of the EBNA mRNAs contains an ORF composed primarily of a two-exon repeat, derived from each of the 7–10 IR1 elements, that encodes EBNA-LP (leader protein). A similar level of complexity is displayed by the LMP-2A and LMP-2B genes, each of which has a unique transcriptional promoter and 5′ exon, but share exons 2–9. A translational start codon is present in the unique exon of the LMP-2A mRNA, whereas the start codon for the LMP-2B ORF is within the common second exon, in frame with the LMP-2A ORF; consequently, LMP-2A differs from LMP-2B by the presence of 119 amino acid residues at its N-terminus. Further, because transcription of the LMP-2 genes traverses the fused TRs, they can only be expressed from the circular, latency-associated form of the EBV genome. The remaining latency-associated gene encodes for a family of alternatively processed and polyadenylated RNAs that co-terminate in the *Bam*HI-A restriction fragment of the EBV genome – hence their designation *Bam*HI-A rightward transcripts, or *BARTs* (**Figure 2**). These appear to arise from a common promoter, and contain several short ORFs. However, while there is ample circumstantial evidence to support protein expression from several of these ORFs, this has yet to be demonstrated with certainty in the context of EBV infection.

Genetic Diversity

There are two distinct genetic strains or subtypes of EBV, EBV-1 and EBV-2 (alternatively subtypes A and B), which were initially defined by specific sequence differences in their respective EBNA-2 ORFs. This was subsequently extended to the EBNA-3 ORFs (EBNA-3A, EBNA-3B, and EBNA-3C), which appear to have arisen through gene duplications from a common progenitor, and the *EBERs*. The respective degree of amino acid sequence identity between the EBNA-3 proteins of EBV-1 and EBV-2 ranges from 72% to 88%. The complete nucleotide sequence has recently been determined for the prototypic EBV-2 isolate, AG876, and no additional genetic loci were identified that would segregate with the type 2 genotype. Thus, the defining differences between EBV-1 and EBV-2 are limited to a subset of genes associated exclusively with the latent form of EBV infection. However, the biological significance of this divergence is currently unclear. *In vitro*, EBV-2 has an apparently lower transformation efficiency than EBV-1 isolates, such as B95-8. This is perhaps not surprising given that EBNA-2, EBNA-3A, and EBNA-3C are essential for B-cell immortalization. However, the molecular basis for this is unclear, as there is no strong evidence to indicate that these EBV-2 proteins differ significantly in function from their EBV-1 counterparts. The observation that EBV-2 isolates are more

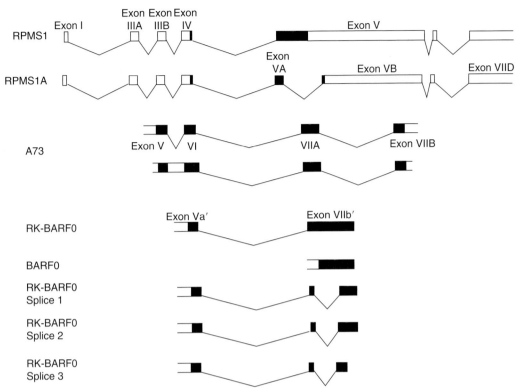

Figure 2 Structure of the latency-associated *BART* RNAs of EBV. Shown are portions of the 5′ and 3′ co-terminal *BARTs* that may encode protein. Potential ORFs within these highly spliced RNAs are shaded in black, and are designated RPMS1, A73, RK-BARF0 and BARF0; shorter versions are also indicated that arise through alternative splicing, but maintain the same translational reading frame. Transcription of the *BART* gene originates from a promoter upstream of exon I, and extends through the 3′-most exon (VII) that contains the ORF BARF0. Note that the three groups of *BARTs* are not shown as aligned relative to each other.

restricted geographically, notably to areas in which EBV-positive Burkitt lymphoma (BL) is endemic, initially raised the possibility of a link between this subtype and pathogenic potential. However, EBV-1 isolates also circulate in areas where EBV-2 is present, and the relative association of either subtype with BL more likely reflects its respective prevalence within the local population.

In addition to genetic differences that define EBV subtype, a number of sequence polymorphisms at various loci have been noted relative to the reference strain B95–8, and the LMP-1 ORF in numerous EBV isolates has a 30-bp deletion in the region encoding the C-terminal domain of this oncoprotein. Although EBV isolates with the LMP-1 deletion appear to be relatively widespread, the nondeleted form generally represents the more common strains circulating within any given population. Significant but non-type-specific differences have also been detected in the EBNA-1 ORF. Although clinical significance has at times been ascribed to some of these genetic variations, such reports are largely anecdotal. Another notable genetic anomaly within the EBV genome is a 6.8–7.4-kbp deletion that removes a portion of the EBNA-LP ORF and the entire EBNA-2 and BHLF1 ORFs (the latter encodes a protein of unknown function expressed during the lytic cycle of infection). Recent work has demonstrated that perhaps as many as 15% of BL tumors carry EBV with this deletion. Interestingly, these tumor cells often also carry the wild-type EBV genome, though it appears to be transcriptionally silent, such that EBV gene expression originates only from the deleted genome. This deletion likely reflects a selection against cells that express EBNA-2, the expression of which is normally repressed through epigenetic mechanisms within BL cells and during normal establishment of long-term latency within the B-cell reservoir of EBV (see below). Thus, deletion of the EBNA-2 locus may represent an alternative, albeit accidental, pathway in lymphomagenesis leading to BL.

Molecular Biology of Infection

Current knowledge of the EBV life cycle at the molecular level has come almost exclusively from *in vitro* studies, largely by necessity due to the lack of a representative and tractable animal model of EBV infection. These have primarily employed either lymphoblastoid cell lines (LCLs) generated by EBV immortalization of primary human B lymphocytes *in vitro*, or lines established from BL tumors. Regardless of their origin, these B-cell lines characteristically restrict EBV infection to its latent phase, through which *in vivo* the virus is able to persist within its host and potentially contribute to malignancy. Currently, there is a considerably better understanding of EBV latency than of the viral replicative cycle, at least from the perspective of viral protein functions. This is primarily the consequence of the more intense effort over the past four decades to define the oncogenic nature of EBV, which is derived from its latency-associated gene functions, and the lack of an experimental cell system that efficiently supports EBV replication *in vitro*. Nonetheless, virus replication is a fundamentally important aspect of the EBV life cycle, as it is essential for maintenance of the virus in the human population.

Virus Replication

Although some latently infected B-cell lines can be induced to replicate EBV in response to chemical agents such as phorbol esters and inhibitors of histone deacetylases, or by crosslinking of cell-surface immunoglobulin, the efficiency of induction is generally poor in that 20–30% of the treated cells, at best, will actually produce virus. Further, these do not necessarily represent physiological pathways of reactivation. While epithelial cells are thought to be significant sites of EBV replication *in vivo*, epithelial cell lines are difficult to infect. This may reflect alternative pathways of cell entry, such as cell-to-cell spread (e.g., through contact with productively infected B cells) or IgA-mediated entry into epithelial cells via the receptor for the dimeric form of IgA that is bound to EBV virions. Further, other factors, such as the differentiation state of the cell, influence whether infected epithelial cells will support virus replication or establishment of a latent infection.

Despite these drawbacks, the general process of EBV replication is reasonably well understood owing to parallels that can be drawn to the replication of other herpesviruses, notably the herpes simplex viruses that replicate well in cell culture and for which an extensive body of knowledge exists. This is aided by comparative sequence analyses of the mammalian herpesvirus genomes, which have revealed blocks of conserved genes that are expressed during the respective viral replication cycles. Thus, while EBV also possesses a subset of unique lytic-cycle genes that contribute to its distinct biology and ability to occupy a specific niche within the host, its replication is believed to occur through a basic process common to all herpesviruses. This occurs through a highly ordered cascade of viral gene expression that is initiated by the immediate-early class of genes (those that do not require *de novo* protein synthesis for transcription), followed by expression of the early and then the late genes. In general, immediate-early and early proteins regulate or contribute to viral gene expression and DNA replication, whereas late proteins are either structural components of the virion or are required for functions such as assembly and egress of the virus particle. A discussion of the unique aspects of EBV replication as they contribute to the EBV life cycle in the host is presented below.

Latency

Although EBV can exist as a latent infection in epithelial cells within the host, B lymphocytes are the major reservoir of latent EBV and the source of new virus (following reactivation) that is ultimately required to populate EBV-naïve hosts. Primary B cells can be readily infected by EBV *in vitro*, leading to direct establishment of a latent infection that promotes cell immortalization through the actions of the latency-associated EBV genes, leading to continuous growth and proliferation in culture. Shortly after infection, the TR elements of the EBV genome fuse to generate a closed circular DNA molecule or episome (**Figure 1**), a variable but relatively stable number of which (10–100) is ultimately established and maintained within the nuclei of latently infected B cells. Replication of the EBV episome in latently infected cells is mediated by the cellular DNA replication machinery, occurs once per cell division cycle, and is initiated within an origin of DNA replication, *oriP*, located in the short unique domain of the viral genome (**Figure 1**). Two functions of a single EBV protein, EBNA-1, are required for genome maintenance during latency. First, through binding to the dyad symmetry (DS) element of *oriP* that contains four EBNA-1-binding sites, EBNA-1 directs assembly of host replication proteins to initiate DNA synthesis. Second, through binding to the family of repeats (FR) element of *oriP* (20 EBNA-1-binding sites), EBNA-1 tethers the EBV genome to host chromosomes through its interaction with a cellular chromosome-associated protein, ensuring that EBV episomes are evenly partitioned upon cell division. By contrast, genome replication during the lytic cycle does not utilize *oriP*, but two origins of DNA replication (*oriLyt*) that function independent of EBNA-1 and lie 100 kbp apart within the long unique sequence domain. Further, replication of viral DNA during the lytic cycle is mediated predominantly by virally encoded DNA polymerase and accessory proteins.

Latency-Associated Gene Expression and Function

Immediately upon infection of a resting B cell, the EBNA promoter Wp is activated by cellular transcription factors, resulting initially in the expression of EBNA-LP and

EBNA-2, the ORFs for which are the first encountered within the multicistronic EBNA transcription unit. Shortly afterwards, there is an EBNA-2-mediated switch in transcription to the common EBNA promoter Cp, approximately 3 kbp upstream of Wp (**Figure 1**). In reporter gene assays, Cp can be regulated, either positively or negatively, by virtually all members of the EBNA family (see below). Thus, the switch to Cp-mediated EBNA expression is thought to afford the virus a means by which to finely regulate EBNA levels and ultimately to regulate LMP-1, LMP-2A, and LMP-2B, whose promoters are also controlled by members of the EBNA family. By approximately 24 h after infection, expression of the full complement of EBV latency-associated genes is evident, including the noncoding RNAs *EBER-1* and *EBER-2* (it is currently unclear at what point the *BARTs* and the newly discovered miRNAs are expressed). In response to expression of the latency-associated genes, the B cell is driven into the cell cycle, and thereafter is able to proliferate indefinitely as an LCL. *In vivo*, this expression of the full repertoire of latency-associated genes, known as the growth or Latency III program, serves to rapidly establish a pool of infected B cells. However, as discussed later, there is a reprogramming of EBV gene expression *in vivo* that substantially restricts viral gene expression to enable infected cells to persist in the face of the cellular immune response to EBV proteins.

EBNA-1

EBNA-1 is the only member of the EBNA family that binds to DNA in a sequence-specific manner, a property critical to its functional interaction with *oriP*. EBNA-1 also has a strong but nonspecific binding affinity for RNA, though the reason for this is unclear. As the genome maintenance protein, EBNA-1 is essential for the establishment and maintenance of latent infection, and thus the pathogenic potential of EBV. This is underscored by the fact that EBNA-1 is the only viral protein known to be expressed in all EBV-associated tumors. EBNA-1 also contributes to regulation of its own expression and that of other EBNA proteins. It does this by activating transcription from the common EBNA promoter Cp (via binding to the FR element of *oriP*) and negatively regulating its own expression from an EBNA-1 exclusive promoter, Qp (see below), which is approximately 43 kbp downstream of *oriP* and contains two EBNA-1-binding sites. The mechanisms by which EBNA-1 functions in these regards are undefined, and although it is possible that it may likewise regulate cellular gene expression, thus far there is a lack of conclusive evidence to support this. Whether EBNA-1 also contributes directly to tumorigenic potential has been examined in mice transgenic for EBNA-1. However, while the results of one group suggest that EBNA-1 can promote B-cell lymphomagenesis in this model, another group found, using independently derived transgenic lines, that EBNA-1 had no demonstrable oncogenic properties. Therefore, to date the principal functions of EBNA-1 appear limited to its roles in genome maintenance and the regulation of EBV gene expression.

EBNA-LP and EBNA-2

After EBNA-1, EBNA-2 has been one of the most extensively studied EBV latency proteins, largely due to the fact that it was the first viral protein shown to be essential for B-cell immortalization by EBV. EBNA-2's function has been studied extensively in EBV-negative B-cell lines, where it increases expression of numerous cellular genes, for example, the B-cell activation marker CD23, CD21 (the B-cell receptor for EBV attachment), the oncogenes *c-fgr* and *c-Myc*, and cyclin D2. This list has further grown with the advent of gene array technology. EBNA-2 is also a crucial transactivator of transcription from the EBV LMP-1, LMP-2A, and LMP-2B promoters, and the EBNA promoter Cp. EBNA-2 does not bind directly to DNA; instead, it is tethered to target promoters via interactions with cellular transcription factors such as CSL (alternatively known as RBP-Jκ or CBF-1), PU.1 and AUF1, allowing increased transcription through recruitment of basal transcription factors by EBNA-2's strong activation domain.

One protein targeting EBNA-2 to DNA is the ubiquitous cellular transcription factor CSL, which normally inhibits transcription by recruiting a co-repressor complex to the promoter. Through its interaction with CSL, EBNA-2 displaces this repressor complex, instead recruiting co-activators. Binding sites for CSL are found in all of the viral promoters responsive to EBNA-2 mentioned above and are key for their regulation by EBNA-2. CSL normally participates in the Notch signaling pathway. Upon ligand-mediated activation of Notch, its intracellular domain is released by proteolysis, enters the nucleus, and binds to CSL, displacing the co-repressor complex and recruiting transcriptional coactivators. Thus, EBNA-2 essentially acts as a constitutively activated Notch protein, though EBNA-2 and Notch likely possess unique functions as suggested by the observation that activated Notch can only partially rescue mutant EBV lacking an EBNA-2 gene in a B-cell immortalization assay. Although the full complement of genes targeted by CSL is unknown, Notch and EBNA-2 do appear to activate expression of many of the same EBV and cellular genes. An additional function of EBNA-2 and Notch is the ability to protect against apoptosis mediated by the nuclear hormone receptor family member Nur77 through a direct association. The fact that Notch is activated in several cancers is consistent with the belief that EBNA-2's interaction with CSL is an essential component of EBV-mediated immortalization.

A second cellular factor targeting EBNA-2 to DNA, specifically to the LMP-1/2B bidirectional promoter (**Figure 1**), is PU.1, and also the highly related protein Spi-B. Both are members of the Ets family of transcription factors, which recognize the same DNA sequence element. Whereas expression of PU.1 is restricted to and plays an important role in the development of B lymphocytes and myeloid cells, less is known about Spi-B, whose expression is limited to B and T lymphocytes. The Ets family of transcription factors, which are clearly important for B-cell development and differentiation, regulate a wide variety of cellular genes. The expression of at least some of these genes, therefore, is likely to be regulated by EBNA-2 and important in EBV-mediated control of B-cell growth and differentiation.

The activity of EBNA-2 is regulated by a number of the other EBNA proteins. Transcription mediated through CSL is repressed by the EBNA-3 proteins, which, by also binding to CSL, prevent its association with EBNA-2 and DNA, thus inhibiting activation of transcription. Relatively little is known of the function(s) of EBNA-LP other than its role as a coactivator of EBNA-2-mediated transcription. Although EBNA-LP is critical for the efficient immortalization of B lymphocytes by EBV, it is not absolutely essential. In primary B lymphocytes stimulated with the EBV gp350, expression of both EBNA-2 and EBNA-LP results in increased expression of cyclin D2. EBNA-LP also appears to specifically coactivate the viral LMP-1/2B bidirectional promoter and Cp, rather than all EBNA-2 target promoters. EBNA-LP augments transcription in a manner requiring only the activation domain of EBNA-2, suggesting that EBNA-LP may act as an adaptor between EBNA-2's activation domain and the basal transcription machinery, though only a weak direct interaction is observed between full-length EBNA-2 and EBNA-LP. Although EBNA-LP has been reported to associate with several cellular proteins, its association with Sp100, which displaces both Sp100 and heterochromatin protein 1 from PML nuclear bodies, is likely to play the most direct role in coactivation along with heat shock protein 72, also an EBNA-LP partner.

The EBNA-3 family

The EBNA-3 family (EBNA-3A, EBNA-3B, and EBNA-3C) apparently arose by gene duplication of a single progenitor, though little notable sequence homology now exists between the proteins. EBNA-3A and EBNA-3C are required for B-cell immortalization by EBV, whereas EBNA-3B is dispensable. One common function of the EBNA-3 proteins is their ability to bind CSL through a conserved domain in their respective amino termini, lending further support to the belief that these proteins are related. As mentioned in the previous section, interaction of CSL with the EBNA-3 proteins inhibits EBNA-2/CSL-mediated transcription in reporter gene assays, though this may not occur in EBV-infected cells, in which the EBNA-3 proteins are expressed at much lower levels. The fact that all three proteins bind to CSL suggests that this interaction is important, and studies with recombinant virus lacking EBNA-3 CSL-interacting domains have demonstrated that at least association of EBNA-3A with CSL plays an essential role in immortalization of B lymphocytes. Both EBNA-3A and EBNA-3C possess transactivation domains within their C termini and bind to a variety of cellular transcription factors, indicating that they likely function themselves as transcriptional activators. Indeed, EBNA-3C can maintain the levels of LMP-1 that otherwise fall in the G1 phase of the cell cycle, and reporter gene assays have demonstrated that this effect is mediated through the PU.1 site in the LMP-1 promoter. EBNA-3C can bind to PU.1 and Spi-B proteins *in vitro*, and results of chromatin immunoprecipitation assays have revealed that EBNA-3C is indeed present on the LMP-1/2B bidirectional promoter. The EBNA-3 proteins are therefore likely to influence gene expression in both CSL-dependent and -independent mechanisms.

Co-transformation assays employing rat embryo fibroblasts have demonstrated that both EBNA-3A and EBNA-3C can immortalize these cells in collaboration with an activated *ras* gene. This function is dependent on a site that mediates interaction with the cellular transcriptional repressor CtBP. The EBNA-3 proteins are implicated in preventing cell-cycle arrest that normally occurs in response to genotoxic agents, presumably by inactivating cell-cycle checkpoint controls. Thus, the EBNA-3 proteins are also likely to contribute to EBV-induced cell growth and proliferation during the Latency III program in a manner independent of their ability to regulate viral and cellular gene expression.

LMP-1

LMP-1 is unique among the EBV latency-associated proteins in that it alone induces the classic phenotypic changes associated with transformation when expressed in rodent fibroblasts. Expression of LMP-1 is essential to maintain proliferation of EBV-infected LCLs, and transgenic mice that express LMP-1 in their B cells develop lymphomas. However, LMP-1 alone is not sufficient to sustain B-cell proliferation, underscoring the importance of the contributions of the other EBV proteins expressed during the Latency III program.

LMP-1 is comprised of a small cytoplasm-orientated domain at its N-terminus, followed by six transmembrane domains and a larger cytoplasmic domain at the C-terminus. Two regions of the C-terminal domain, transformation effector sites (TES)-1 and -2, are essential for efficient B-cell immortalization by EBV. Through these domains, LMP-1 associates directly with intracellular signaling proteins of the tumor necrosis factor receptor

(TNFR) family. LMP-1 aggregates within lipid rafts of the plasma membrane, resulting in ligand-independent, constitutive activation of these signaling proteins, a critical outcome of which is activation of the transcription factor NFκB. JNK, p38/MAPK, and STAT proteins are also activated by LMP-1, though activation of STAT has been shown not to be important for LMP-1's effects in B lymphocytes. Through these pathways, LMP-1 affects the expression of cellular genes such as adhesion molecules associated with antigen activation of B lymphocytes, anti-apoptotic proteins such as Bcl-2, Bfl-1, and A20, as well as various cytokines. Additional genes activated in the presence of LMP-1 continue to be identified. The membrane proximal TES-1 site interacts with TNFR-associated factors or TRAFs, mediating a low-level activation of NFκB that drives the initial proliferation of B lymphocytes following EBV infection. TES-2, located at the very C-terminus of LMP-1, mediates the majority of NFκB activation through binding to the TNFR-associated death domain protein TRADD, and is required for the continued growth of EBV-infected B lymphocytes in culture. The TNFR family comprises members associated with the promotion of cell growth (CD40) as well as cell death (FAS receptor), and LMP-1 mimics many of the functions of CD40, partially rescuing the defects in CD40-null mice. Thus, LMP-1 plays a direct role in the immortalization associated with EBV infection.

Within epithelial cells, LMP-1 is able to activate expression of the epidermal growth factor receptor through TRAF signaling, resulting in inhibition of terminal differentiation. This pathway is likely to mediate the development of epithelial hyperplasia observed in LMP-1 transgenic mice, and suggests a significant role for LMP-1 in the more restricted Latency II program maintained in nasopharyngeal carcinoma (NPC).

LMP-2A and LMP-2B

LMP-2A and LMP-2B are integral plasma membrane proteins comprised almost entirely of 12 transmembrane domains, with both their N and C termini oriented toward the cell cytoplasm. The unique 5′ exon of LMP-2A encodes a 119-amino acid residue N-terminal domain not present in LMP-2B, which contains only the transmembrane domains and the common 27-residue C-terminal tail. It is through its N-terminal domain that LMP-2A, aggregated in lipid rafts within the plasma membrane, recruits the protein tyrosine kinase Lyn and subsequently Syk to activate a pro-survival signaling cascade that includes a number of signaling proteins in the B-cell receptor (BCR) pathway, for example, Akt/protein B kinase, phosphatidylinositol-3 kinase (PI3K), phospholipase-Cγ2 (PLCγ2), and mitogen-activated protein kinase (MAPK). Consequently, LMP-2A mimics a constitutively active BCR that provides a submitogenic survival signal important for long-term persistence of latently infected B cells. The level of signaling by LMP-2A appears to be tightly regulated through its association with Nedd4 ubiquitin ligases, which control levels of LMP-2A and its associated proteins through ubiquitin-mediated protein degradation. Also, LMP-2B, which is unable to interact with Lyn and Syk, is believed to regulate LMP-2A negatively by interfering with juxtapositioning of the N-terminal domains of LMP-2A (required for trans-phosphorylation of tyrosine motifs) within lipid rafts. LMP-2A also limits the reactivation of EBV within LCLs that can be induced by crosslinking of surface immunoglobulin, that is, the BCR. Inhibition of BCR signaling appears to occur as a result of LMP-2A-dependent reduction of Lyn available for BCR signaling. However, it is unclear whether this is a significant means of inhibiting EBV reactivation *in vivo*. LMP-2A is not essential for immortalization of B lymphocytes *in vitro*, probably because its pro-survival functions are superseded by those of LMP-1.

With the exception of the Latency I program in BL, LMP-2A is probably expressed in all latent EBV infections. Further, LMP-2A expression in NPC and EBV-positive cases of Hodgkin lymphoma implies a role for it in these human cancers. Indeed, despite being nonessential for B-cell immortalization, LMP-2A is tumorigenic when expressed in immortalized epithelial cell lines, and the pro-survival function of LMP-2A is believed to rescue the Reed–Sternberg cells of Hodgkin lymphoma – derived from germinal center (GC) B cells – from apoptosis. Transformation of epithelial cells by LMP-2A is thought to occur through stabilization of β-catenin via activation of the PI3K/Akt pathway. Inhibition of TGF-β-induced apoptosis by LMP-2A may further contribute to its tumorigenic potential within epithelial cells. LMP-2A is also able to promote migration of epithelial cells *in vitro* through upregulation of integrin-α-6 expression, suggesting that it plays an ancillary role in tumorigenesis as well by contributing to metastasis.

BARTs

The *BARTs* are expressed in all forms of EBV latency *in vitro*, and are detectable by PCR in the peripheral blood of healthy EBV carriers. While they are not essential to immortalization of B cells by EBV *in vitro*, their universal expression suggests that they play a critical role(s) in EBV biology. These are highly spliced transcripts that contain poly(A) tails on their common 3′ termini, and several putative ORFs have been identified within various members of the *BART* family, suggesting that they indeed function as mRNAs (**Figure 2**). Though proteins encoded by these ORFs can be expressed from eukaryotic expression vectors in cells following transfection, definitive identification of virally encoded proteins within infected cells has yet to be reported. There are four *BART* ORFs for which there is at least indirect evidence that they encode proteins: BARF0, RK-BARF0 (slightly longer than BARF0 due to an RNA splicing event), RPMS1, and A73 (**Figure 2**). While serum from NPC patients does not recognize endogenous protein by immunoblotting, it will detect bacterially produced BARF0 polypeptide. Three *BART*-encoded polypeptides have been shown in yeast two-hybrid screens to interact with cellular proteins. RPMS1 and RK-BARF0 interact with components of the Notch signaling pathway: RPMS1 with CSL and its associated co-repressor CIR, and RK-BARF0 with Notch4. A73 interacts with the cytoplasmic receptor for activated protein kinase 1 (RACK1), though the functional significance of this is unclear. The interaction of RPMS1 with CSL and CIR is believed to repress promoters with CSL-binding sites (e.g., the LMP-1 promoter) by stabilizing the CSL–CIR repressor complex on the DNA. Interestingly, though different from the mechanism of EBNA-3 repression of CSL-regulated transcription, the outcome is similar. Interaction of RK-BARF0 with Notch4 results in translocation of the unprocessed form of Notch4 directly to the nucleus, which is believed to then activate expression of LMP-1. Thus, one set of functions for RPMS1 and RK-BARF0 may be analogous to those of the EBNA-3 and EBNA-2 proteins, respectively, in the regulation of LMP-1, but during the Latency II program in which these EBNA proteins are absent.

EBERs

EBER-1 and *EBER-2* are small, highly structured noncoding RNAs whose genes lie adjacent to one another immediately upstream of *oriP* (**Figure 1**) and are transcribed by cellular RNA polymerase III (Pol III). They are expressed in all forms of EBV latency and are restricted to the cell nucleus, where they are present at approximately 10^7 copies per cell (levels of *EBER-1* are approximately tenfold higher than *EBER-2*). While *EBER* expression is not essential for immortalization of B lymphocytes by EBV *in vitro*, like EBNA-LP the *EBERs* enhance the efficiency of this process. The most interesting property of the *EBERs* is their ability to enhance the tumorigenic potential of some EBV-negative BL cell lines, suggesting that they may play a role in tumor maintenance. How the *EBERs* promote tumorigenic potential is unresolved. There are numerous reports that the *EBERs* can interact with and influence the actions of proteins within the type 1 (antiviral) interferon response pathway that bind to dsRNA, namely the dsRNA-activated protein kinase PKR and RIG-I. However, as the *EBERs* are present in ribonucleoprotein complexes and unlikely to be available for interaction with these predominately cytoplasmic proteins, these interpretations of *EBER* effects *in vitro* or under experimental conditions that take the *EBERs* out of their normal intracellular context must be viewed with caution.

The *EBERs* are present in RNA–protein complexes that include La/SS-B (*EBER-1* and *EBER-2*) and the ribosomal protein L22 (predominately *EBER-1*). La is critical for the synthesis and protection from exonucleolytic digestion of small cellular transcripts generated by Pol III (e.g., tRNAs) that have a UUU-OH moiety at their 3′ termini, which is bound by La. La also retains some Pol III transcripts within the nucleus via this interaction, as it probably does the *EBERs*. As Pol III transcripts themselves, La binding to the *EBERs* probably represents a normal physiological interaction. *EBER* interaction with L22 results in retention of approximately 50% of the cellular pool of L22 within the nucleoplasm. The functional consequence of this is unknown, but presumably it influences L22 function. In bacterial ribosomes, L22 lies within the exit tunnel and regulates the passage of nascent polypeptide by interaction with a motif within the secretion monitor protein, secM. L22 may perform an analogous function in mammalian cells that is affected by *EBER-1*, though it is possible that *EBER-1* interaction with L22 influences a nonribosomal function of L22 (as yet undescribed), or imparts upon it a gain of function.

miRNAs

Two clusters of EBV miRNAs have recently been identified that are derived from two loci within the viral genome: the BHRF1 locus encoding for a pro-survival Bcl-2 homolog, and a portion of the *BART* locus. Analysis of miRNA expression in various EBV-infected cells has revealed that the *BART* miRNAs are highly expressed in latently infected epithelial cells (Latency II program), but at much lower levels in B-cell lines. By contrast, the BHRF1 miRNAs are present in high levels within B cells maintaining the Latency III program, but are very low or undetectable during the restricted latency programs in B (Latency I) and epithelial (Latency II) cell lines. Three of the *BART* and one of the BHRF1 miRNAs are induced to higher levels upon induction of the lytic cycle. The identity of the genes post-transcriptionally silenced by these miRNAs (both viral and cellular) and the contribution of virus-mediated RNA interference to EBV biology and pathogenic potential are currently unclear, but are currently the focus of an intense research effort.

Infection in the Host

For two decades after the discovery of the immortalizing effect of EBV on B lymphocytes *in vitro*, latency within LCLs was widely accepted as the model of EBV persistence within the host (although the extent of EBV gene expression in LCLs was not appreciated at the time). Increasing evidence indicated, however, that such cells are highly susceptible to killing by $CD8^+$ cytotoxic T lymphocytes (CTLs), and subsequently that this cellular immune response is directed at epitopes derived from the latency-associated proteins of EBV, particularly members of the EBNA-3 family. Thus, it was unlikely that EBV could persist within an immunocompetent host solely in cells with the characteristics of an LCL. The subsequent observation that some BL cell lines are resistant to CTL killing, particularly at low passage in culture, was largely explained by the soon-to-follow discovery that expression of known latency-associated proteins in these cells was restricted to EBNA-1. EBNA-1, moreover, is not a good inducer of the CTL response. This is due in part to poor presentation of EBNA-1 peptides with HLA class I antigen on the infected cell surface as a consequence of self-inhibition of proteolytic processing of EBNA-1 by its substantial Gly-Ala repeat domain. Thus, B cells that limit EBV expression to EBNA-1 are not particularly vulnerable to the host's immune surveillance.

Restricted patterns of EBV gene expression were later described for other EBV-associated tumors, for example, NPC and Hodgkin lymphoma, though cells in these tumors express the LMP proteins in addition to EBNA-1. This led to the classification scheme of latency-associated gene expression patterns or programs: Latency I exemplified by BL cells (EBNA-1 only); Latency II, in which the LMP proteins are expressed in addition to EBNA-1; and Latency III or the growth program, in which expression of the full complement of latency-associated genes is supported. The *EBERs* and *BARTs* are also expressed in each of these programs, whereas the miRNAs are differentially expressed, as discussed above. Importantly, the realization that tumor cells can support alternative programs of latency-gene expression provided the basis for studies that have led to the current understanding of EBV persistence within the host. With the aid of reverse transcription-PCR and cell-isolation procedures, it has been possible to evaluate EBV gene expression in phenotypic subsets of B cells, and even approaching the single-cell level. Analyses of EBV gene expression in B cells from peripheral blood and tonsils have provided a reasonably good, though certainly not complete, picture of the pathway(s) taken by EBV to establish itself permanently within the B-cell pool of its host.

Establishment of Persistent Infection

The major cellular reservoir of EBV within the peripheral blood are resting memory B cells (IgD$^-$/CD20$^+$) that express the LMP-2 proteins, and probably the *EBERs* and *BARTs* (referred to as the Latency program). Periodic activation of cell division results in EBNA-1 expression via Qp (suppressed in resting cells by the retinoblastoma protein) to prevent loss of the viral genome, thereby maintaining EBV within this reservoir. The virus is believed to enter the memory cell pool not directly, but by either driving or participating in a GC reaction, the normally antigen-driven process that gives rise to the long-lived memory B cells and the plasma cells that produce high-avidity antibody. This is supported by the presence of EBV within BL cells, which are derived from GC B cells and are thought to acquire the virus early in the oncogenic process, quite possibly during the primary establishment of EBV latency. The pathway by which EBV is believed to gain entry into the memory B-cell pool is illustrated in **Figure 3**, support for which has largely come from PCR-based analyses of EBV gene expression in tonsil B cells following surgical resection associated with either acute infectious mononucleosis or tonsillitis of unknown etiology. The basic premise of this model is that EBV infects a naïve resting B cell (IgD$^+$), driving rapid proliferation and expansion of infected cells, much as it does *in vitro* through its Latency III program. Gradually, however, there is an epigenetic down-regulation of EBV gene expression (as in BL cells) that both enables infected B cells to evade the developing CTL response and is compatible with long-term persistence in a resting B cell.

Conclusive evidence that EBV enters the memory cell pool by participating in a GC reaction, however, has been elusive. EBV-positive B cells appear infrequently within the GCs themselves, and those that do are not undergoing somatic hypermutation of *Ig V* genes, arguing against EBV driving or participating in a classic GC reaction. These cells harbor somatically mutated *V* gene rearrangements, suggesting that EBV may gain entry into GC B cells by direct infection, but that this pool of infected cells expands without further hypermutation. However, it is unlikely that EBV directly inhibits somatic hypermutation, as BL cell lines can continue to undergo this process *in vitro* despite infection with EBV or, in EBV-negative BL cells, stable expression of the individual latency genes. An important caveat to these observations is that, as a consequence of the pathology that prompted tonsillectomy, EBV infection in these B cells may not represent entirely normal events in the establishment of long-term latent infection.

The Role of Virus Replication

It is unclear whether establishment of a persistent latent infection in B cells requires an initial round of productive infection in epithelial cells within the oropharynx, or results instead from direct infection of B cells, most likely within Waldeyer's ring of the oropharyngeal lymphoid tissue. Extended treatment of EBV-positive individuals with acyclovir, an inhibitor of herpesvirus replication, does not reduce the level of latently infected cells in the peripheral blood, suggesting that intermittent reactivation of EBV is not essential for persistence. By contrast, reactivation of EBV from latency is necessary to pass the virus onto naïve hosts to maintain the virus in the human population. This is believed to occur sporadically from latently infected memory B cells that begin to replicate virus in response to their terminal differentiation into plasma cells. Virus produced from B cells has a slightly higher tropism for epithelial cells, whereas virus produced in epithelial cells is considerably more infectious for B cells. This is due to the relative

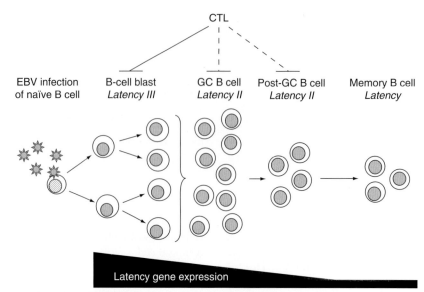

Figure 3 Establishment of persistent EBV latent infection. Shown is the currently favored model for the establishment of latency within memory B cells, the major long-term reservoir of EBV. As latently infected B cells progress from an initial stage of EBV-driven cell proliferation through a germinal center (GC) reaction, ultimately entering the memory B-cell pool, there is a restriction of latency-associated EBV genome expression. This occurs primarily through epigenetic silencing of the common EBNA promoters, Cp and Wp. The different latency-gene expression programs at the various stages are indicated in italics. The Latency program within memory B cells is similar to Latency I in BL cells, except for the expression of LMP-2A, and EBNA-1 expression is repressed; periodic cell division results in reactivation of EBNA-1 expression from the cell-cycle-regulated promoter Qp, as in Latency I. Solid to dashed lines indicate reduced potential for killing by cytotoxic T lymphocytes (CTLs) directed at epitopes derived from EBV proteins expressed in the Latency III and II programs.

amount of two EBV glycoprotein complexes on the virion envelope: gH/gL and gH/gL/gp42. B cells are preferentially infected by virions rich in gH/gL/gp42, whereas the presence of gp42 (prevalent on virus produced in epithelial cells) blocks interaction of gH/gL with its putative receptor on epithelial cells, which do not express the B-cell receptor for virus attachment, CD21, that is bound by gp350. B-cell-derived virus has lower levels of gp42 due to its interaction with HLA class II antigen (lacking in epithelial cells), which targets it for degradation. Recent evidence indicates that EBV present in saliva is programmed preferentially to infect B cells (high in gp42), suggesting that virus reactivated from B cells goes through an amplification step in tonsillar epithelium before it is secreted to be passed onto naïve hosts.

Activation of EBV replication is initiated by expression of the viral immediate early protein Zta – a protein related to the cellular AP-1 family of transcription factors – that binds to specific DNA motifs in a number of EBV early-gene promoters, initiating the cascade of lytic-cycle gene expression. This is done in collaboration with a second EBV transactivator, Rta, itself activated by Zta. Zta and Rta also participate in replication of EBV DNA during the lytic cycle through interaction with their response elements within *oriLyt* domains. An interesting property of Zta is its preference for binding DNA that has been methylated. Because the EBV genome is highly methylated within latently infected B cells *in vivo*, the propensity of Zta for methylated DNA gives further credence to the model in which new virus originates from reactivation within the memory B-cell reservoir of EBV.

To aid replication of virus in the host, EBV encodes a number of proteins that, while not directly involved in virion production, influence the intra- or extracellular milieu to enhance virus production. For example, the EBV BHRF1 ORF encodes a homolog of the cellular anti-apoptotic protein Bcl-2, and presumably enhances virus production by countering induction of programmed cell death in response to infection. Similarly, Zta and Rta, in addition to their direct contributions to EBV gene expression and DNA replication, interact with or regulate the expression of a number of cellular proteins, examples of which are p53, NFκB, pRB, p21, TGF-β and the retinoic acid receptor. These properties of Zta and Rta likely provide the optimal cellular environment for virus replication. EBV also expresses several proteins during the lytic cycle that are capable of modulating the host immune response. These include homologs of interleukin 10 (vIL-10), a negative regulator of the immune response, and a secreted form of the receptor for colony stimulating factor 1 (CSF-1) able to bind and neutralize CSF-1. Further, gp42, in addition to aiding in infection of

B cells, is able to inhibit HLA class II-mediated antigen presentation, and Zta is able to inhibit type 2 interferon signaling by downregulating expression of the γ interferon receptor.

See also: Epstein–Barr Virus: General Features; Herpesviruses: Discovery; Herpesviruses: General Features; Herpesviruses: Latency; Simian Gammaherpesviruses; Tumor Viruses: Human.

Further Reading

Cai X, Schäfer A, Lu S, *et al.* (2006) Epstein–Barr virus microRNAs are evolutionarily conserved and differentially expressed. *PLoS Pathogens* 2: 236–246.

Farrell PJ (2005) Epstein–Barr virus genome. In: Robertson ES (ed.) *Epstein–Barr Virus*, pp. 263–288. Norfolk, England: Caister Academic Press.

Kieff E and Rickinson AB (2006) Epstein–Barr virus and its replication. In: Fields BN, Knipe DM, Howley PM, *et al.* (eds.) *Fields Virology*, 5th edn., vol. 2, pp. 2603–2654. Philadelphia: Lippincott Williams and Wilkins.

Rickinson AB and Kieff E (2006) Epstein–Barr virus. In: Fields BN, Knipe DM, Howley PM, *et al.* (eds.) *Fields Virology*, 5th edn., vol. 2, pp. 2655–2700. Philadelphia: Lippincott Williams and Wilkins.

Robertson ES (ed.) (2005) *Epstein–Barr Virus*. Norfolk, England: Caister Academic Press.

Tselis A and Jenson HB (eds.) (2006) *Epstein–Barr Virus*. New York, NY: Taylor and Francis Group, LLC.

Equine Infectious Anemia Virus

J K Craigo and R C Montelaro, University of Pittsburgh School of Medicine, Pittsburgh, PA, USA

© 2008 Elsevier Ltd. All rights reserved.

History

Equine infectious anemia (EIA), colloquially known as swamp fever, has been documented in numerous diverse geographical areas and is currently considered a worldwide disease that occurs only in members of the family Equidae. EIA was first identified as an infectious disease of horses by veterinarians in France in 1843. In 1904, the infectious organism that caused EIA was identified as a 'filterable agent', making EIA one of the first animal diseases to be assigned a viral etiology.

Despite this early identification of the equine infectious anemia virus (EIAV), the characterization of this virus was extremely slow because of the difficulties experienced in the isolation and propagation of the virus in cell culture. Thus, the major focus on the control of EIA has been the development of regulatory policies that involve the identification and elimination of EIAV-infected horses. More recently, advances in animal vaccine strategies and the demand for animal models for AIDS vaccine development have provided renewed impetus to the development of an EIAV vaccine to prevent virus infection. EIAV also offers an important model for the role of antigenic variation in a persistent retrovirus infection.

Classification

EIAV is classified as a member of the genus *Lentivirus* based on criteria of virion morphology, serological properties, and genomic sequence homologies. There has been no formal further subdivision of EIAV isolates into subtypes.

Properties of the Virion

The EIAV particle has the general morphology of a lentivirus, including an oblong core enclosed in a viral envelope with surface projections (**Figure 1**). The oblong core observed in EIAV is characteristic of lentiviruses, in contrast to the icosahedral cores found in most oncoviruses. This distinctive structural feature was the initial indication that HIV-1 was related to EIAV and a member of the lentivirus rather than the oncovirus subfamily of retroviruses. The virus particles appear roughly spherical in the electron microscope (**Figure 2**), although there are various degrees of polymorphism depending on the sample preparation. The overall diameter of the virion is approximately 100 nm. The surface projections extend about 7 nm and appear to be distributed on the viral surface in a symmetrical pattern.

Properties of the Genome

The EIAV genome consists of a dimer of single-stranded, positive-sense RNA. The genomic organization of EIAV is characteristic of a complex retrovirus, but is the simplest and smallest of characterized human and animal

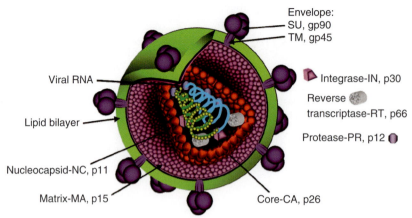

Figure 1 Virion morphology and protein organization of EIAV.

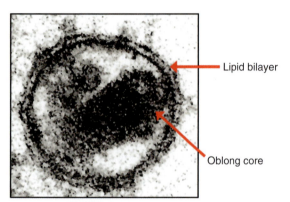

Figure 2 Transmission electron micrograph of an EIAV particle.

lentiviruses (**Figure 3**). The viral RNA contains about 8200 bp and contains three major genes (*gag, pol,* and *env*) encoding viral structural proteins and three minor genes (*tat, S2,* and *rev*) that encode nonstructural proteins that regulate various aspects of virus replication. The order of the EIAV genome is 5′-R-U5-*gag-pol-env*-U3-R-3′. The *tat* and *S2* genes are encoded as distinct alternate reading frames within the *pol–env* intergenic region, while the *rev* gene is encoded by alternate reading frames contained at the beginning of the *env* gene sequences and following the 3′ end of the *env* gene. The relatively small size of the EIAV genome compared to other lentiviruses is primarily due to the much smaller size of the *pol–env* intergenic region.

The EIAV *gag* gene encodes the four viral core proteins in the order of 5′-p15-p26-p11-p9–3′, while the *env* gene encodes the two envelope glycoproteins of the virus in the order of 5′-gp90-gp45–3′. The *pol* gene encodes a complex of enzymes with an organization of 5′-protease-reverse transcriptase-RNase H-dUTPase-integrase-3′. As with other lentiviruses, the EIAV *tat* and *rev* genes encode important regulatory proteins that either transactivate virus transcription (Tat) or control viral transcription patterns (Rev) after infection of host cells. The function of the 8 kDa protein encoded by the *S2* gene remains to be defined, although mutation studies of this gene indicate that it is not essential for *in vitro* viral replication and results in a single $\log_{10}$ reduction in viral replication in equines. The terminal LTR sequences of EIAV contain the usual complex of transcriptional regulatory domains distinctive of lentiviruses.

Properties of the Viral Proteins

The proteins encoded by the *gag* and *env* genes of EIAV constitute the major structural proteins of the virus (cf. **Figure 1**). The gp90 protein is a highly glycosylated, hydrophilic surface (SU) protein that forms the outermost knobs of the envelope projections, while the gp45 is a sparsely glycosylated, hydrophobic transmembrane (TM) protein that forms the membrane-spanning spike of the envelope projection. There are approximately 300 copies each of the envelope glycoproteins per virion, and the surface projections are composed of trimers of gp90 and gp45. The final component of the EIAV envelope structure is the fatty acylated p15 that forms a continuous matrix (MA) immediately beneath the lipid bilayer of the virus particle. The virion core shell or capsid (CA), composed predominantly of p26 molecules, encloses a helical ribonucleoprotein complex containing the basic nucleoprotein (NP), p11, and various polymerases (RT, IN, and DU) in close association with the viral RNA genome. The location of the final core protein, p9, is not certain, but it has been proposed as a linker protein between the core shell and envelope matrix. Recent experiments have indicated that the EIAV p9 protein mediates late stages of viral budding.

The *gag*-encoded proteins of the virus are present in molar amounts that are at least tenfold greater than the envelope glycoproteins, in the range of 3000–5000 copies

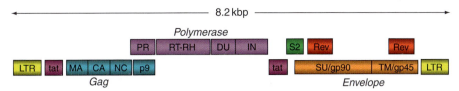

Figure 3 Organization of the EIAV genome indicating viral genes (italic print) and the respective proteins encoded by these genes (block print).

per virus particle. In contrast, the *pol*-encoded enzymes appear to be present in the virus at levels of about 10 molecules per virion.

Replication Strategy

The EIAV replication cycle is characteristic of retroviruses in general and lentiviruses in particular (summarized in **Figure 4**). Viral recognition, attachment, and penetration of target cells is believed to be mediated by specific interaction of the viral envelope glycoproteins and cellular receptor proteins contained in the plasma membrane. Until recently, there were no recognized receptors for EIAV. Recently, a cellular receptor for EIAV, designated ELR1, has been identified as a member of the tumor necrosis family of receptors (TNFRs). Once inside the target cell cytoplasm, the virus reverse transcriptase copies the single-stranded RNA genome into double-stranded DNA provirus that is then transported to the nucleus of the cell. There is no evidence for translation of the incoming EIAV genome at this stage of infection. Once inside of the cell nucleus, the EIAV integrase mediates apparently random but limited incorporation of the provirus DNA into cellular chromosomes, although some extrachromosomal viral DNA is always present in productively infected cells. There is typically a total of about 10 copies of proviral DNA per infected cell. Cytopathic infections by EIAV have been correlated with higher ratios of integrated to unintegrated proviral DNA.

Transcription of the proviral DNA by cellular polymerases produces a complex pattern of viral messenger RNA species whose relative proportions may differ depending on the virus strain and target cell. In all instances, the predominant EIAV transcripts are an 8.2 kbp transcript representing the full-length genomic RNA and translated to produce the *gag* and *pol* gene products, and a 3.5 kbp mRNA that is a singly spliced transcript translated to produce the viral envelope proteins. Lentivirus-infected cells usually contain in addition to these major viral transcripts, a heterogeneous population of small multiply spliced RNA that are used to produce the various regulatory proteins. In the case of EIAV, however, infected cells reveal only minute quantities of these small, multiply spliced RNA species. The relatively low abundance of small transcripts may in part reflect the relative genetic simplicity of EIAV and the

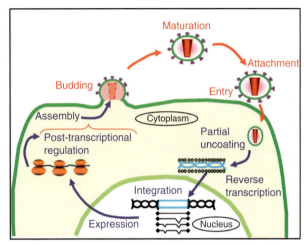

Figure 4 Schematic representation of the EIAV life cycle.

lack of extensive splicing to ensure production of all of the minor viral genes. On the other hand, it is intriguing that the 3.5 kbp transcript is in fact a tricistronic messenger that can produce by *in vitro* translation the viral Tat and Rev proteins in addition to the more abundant envelope glycoproteins. The use of a tricistronic messenger may represent a novel mechanism of maximizing genetic efficiency in EIAV replication.

The viral Gag and Env proteins are initially produced as polyproteins that are cleaved into the mature virion components by a combination of viral and cellular proteases. The only modification documented in the core proteins is a fatty acylation at the N terminus of the matrix protein, p15. The envelope proteins are modified by N linked glycosylation. Although the gp90 and gp45 polypeptides contain about 400 amino acid residues, the gp90 contains an average of 17 potential N linked glycosylation sites, while the gp45 contains only about five potential glycosylation sites. By comparison to HIV-1 gp120, it is assumed that all potential glycosylation sites are occupied by complex oligosaccharides.

The precise mechanisms of EIAV assembly have not been completely dissected, although it is assumed that it follows the general model for retrovirus assembly. It appears to be a highly concerted process that is mediated by interactions between the viral Gag polyproteins and host cofactors. Accordingly, the viral envelope glycoproteins are initially inserted into the plasma membrane to create distinct sites of virus assembly at which the Gag polyprotein is accumulated beneath the membrane lipid. Subsequently, Gag proteins recruit host cellular components that also regulate the endocytic pathway for assembly and budding. To date, early (AP-2) and late endocytic proteins (TSG101) as well as the actin cytoskeleton have been associated with EIAV virion production. The intermediate viral proteins are next cleaved into the mature virion proteins as the particle buds from the cell surface and is released to produce progeny virions.

Epidemiology, Geographic and Seasonal Distribution

EIAV has been diagnosed in many areas of the world and is considered a worldwide disease of horses. Although localized outbreaks of disease can occur, the incidence of EIAV-infected horses is the highest in tropical and subtropical climates, presumably due to the longer warm seasons and more abundant populations of insect vectors that may transmit EIAV among horses. During the past 20 years, the EIAV infection rate reported by the USDA has dropped from about 4% to less than 0.2%. However, these testing results do not reflect the general horse population as less than 10% of the horses in the United States are tested for EIA, usually because of requirements for transportation across state lines or for participation in organized shows or races. General surveys of unregulated herds in the Southeast United States demonstrate infection rates of up to 15%. EIAV infections are especially prevalent in Central and South America, where limited surveys of Latin American countries have demonstrated infection rates in unregulated herds frequently approach 50% indicating that EIAV infection is epidemic in these areas. Although the probability of EIAV infection by insect vectors is greatest during seasons that are warm, infections can occur throughout the year via mechanical transfers of blood by hypodermic needles and other veterinary instruments. Sexual transmission of EIAV has not been demonstrated to date.

Host-Range and Virus Propagation

EIAV appears to infect only members of the family Equidae. There is no evidence to support the concept of natural or experimental infections of humans or of other mammalian species. EIAV infection of horses results in high levels of virus replication, persistent infection, and clinical disease. EIAV infection of donkeys produces only limited virus replication, presumably a persistent infection, but no signs of clinical disease.

Field strains of EIAV can only be propagated *in vitro* in cultures of equine monocyte or macrophage cells, where virus infection typically produces a cytopathic effect within several days. Large-scale production of EIAV is limited to cell culture-adapted strains of virus that can be grown in primary cultures of equine dermal cells or fetal equine kidney cells and in a limited selection of nonequine continuous cell lines, including canine fetal thymus (Cf2th) cells and the Fea and FEF feline cell lines. The cell-adapted strains of EIAV are noncytopathic to these permissive cell lines. Field strains of EIAV retain their pathogenic properties when propagated in leukocyte cultures, but usually become avirulent when adapted to other types of cell cultures. Cell-adapted strains of EIAV that retain their virulence have been produced by back passage of avirulent cell-adapted strains in ponies or horses. There is no evidence for infection of cultured human cells by EIAV.

EIAV production in cell cultures is most easily detected by the presence of viral antigens or reverse transcriptase activity in culture media.

Genetics

Like other retroviruses, EIAV replication is mediated by a virion reverse transcriptase (RT) that copies the viral RNA genome into proviral DNA that is found in the nucleus of infected cells randomly integrated into the cellular chromosome and as extrachromosomal molecules. There are typically only about 10 copies of EIAV DNA per infected cell. EIAV replication in horses is characterized by relatively rapid and diverse genomic mutations that produce an apparently wide variety of variant virus strains. Analyses of sequential antigenic variants of EIAV from experimentally infected horses suggest that the rate of mutation in the envelope gene of the virus is greater than 10^{-2} base substitutions per site per year. The fidelity of DNA synthesis by purified EIAV RT has been measured *in vitro*, and an average error rate of 1/700 bp has been estimated. This value is similar to the *in vitro* error rate calculated for human immunodeficiency virus (1/700 bp), but is significantly higher than the rate observed for oncovirus RT such as avian myeloblastosis virus (1/3000 bp). The error-prone nature of EIAV RT produces significant biological diversity that is important in EIAV persistence and pathogenesis. As demonstrated for HIV and SIV, recombination between variant EIAV genomes in infected cells may also contribute to genetic diversification during persistent infection.

Evolution

Phylogenetic analyses based on the nucleotide sequences of various retroviruses indicate that EIAV is most closely related to the ungulate lentiviruses (visna-maedi virus,

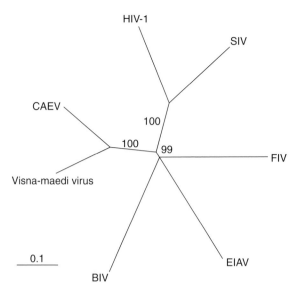

Figure 5 Phylogenetic tree of the complete genomes of identified lentiviruses.

caprine arthritis-encephalitis virus, and bovine immunodeficiency virus) and equally divergent from the human and simian immunodeficiency viruses (**Figure 5**).

Studies of EIAV evolution through analyses of sequence variation during persistent infection in experimentally infected equids have clearly identified dynamic changes in envelope sequences that alter viral antigenic properties, evidently as a result of immune selection. Variation of the envelope gene has therefore served as a distinct marker for analysis of viral population evolution. Detailed molecular characterization of envelope variation during sequential disease cycles in experimentally infected ponies revealed the presence of distinct EIAV envelope variants with each wave of viremia. Examination of inapparent stage viral populations from the plasma of ponies indicated that evolution of the viral quasispecies is continuous, even with relatively low levels of detectable virus replication in the periphery or tissues. These results suggest that even in the absence of detectable plasma virus, viral populations, most likely in tissue reservoirs, continue to replicate and evolve, seeding the plasma with new viral quasispecies.

Serologic Relationships and Variability

Field and laboratory isolates of EIAV display a remarkable variability in antigenic properties. The core proteins of the virus contain conserved antigenic determinants that are the basis of current serological diagnostic assays. The viral glycoproteins also contain a limited number of conserved antigenic determinants, but predominantly present an array of variable antigenic sites that can be distinguished by their reactivity with monoclonal antibodies or by neutralization with polyclonal immune serum from infected horses. The range of variation observed among EIAV isolates has precluded any classification of virus strains on the basis of serological properties.

Immune serum from EIAV-infected horses is reactive with the respective major core protein of most animal and human lentiviruses, but not with any of the major core proteins of oncoviruses. Immune serum taken from other species infected with a lentivirus generally do not reveal a cross-reactivity with EIAV or with other lentiviruses. This one-way serological reactivity suggests that horses infected with EIAV uniquely recognize a conserved lentivirus-specific, antigenic determinant.

Transmission and Tissue Tropism

Blood from persistently infected horses is the most important source of EIAV for transmission; *in utero* transmissions of EIAV from mare to foal are evidently very rare. This blood transfer can be affected by man or blood-feeding vectors. EIAV has been shown to remain infectious on hypodermic needles for up to 96 h, emphasizing the potential for transmission via routine animal husbandry or veterinary medical practices. However, the mechanical transmission of EIAV by arthropods, especially horseflies, is generally accepted as the major natural means of transmission in the field. Transmission of EIAV by a single horsefly carrying only approximately 10 nl of virus-infected blood has been documented under experimental conditions.

The target cell during persistent EIAV infections appears to be exclusively cells of the monocyte/macrophage lineage; there is no evidence for infection of lymphocytes as observed with some other lentiviruses. The virus burden in infected horses is predominantly in tissue macrophage found in liver, kidney, and spleen with much lower levels of virus found in lymph nodes, bone marrow, or in circulating monocytes. Thus, the relatively high levels of viremia (10^{4-6} $TCID_{50}$) observed during episodes of chronic EIA evidently result primarily from the production and release of virus from infected tissue macrophage, rather than an extensive infection of blood monocytes.

Pathogenicity

Field isolates and laboratory strains of EIAV differ markedly in their pathogenicity, ranging from avirulent to lethal strains of the virus. Little is known about the viral determinants or host factors that influence the course of virus replication and pathogenesis. In other lentiviruses, differences in viral pathogenesis have been mapped to specific changes in viral envelope genes or to changes in gene regulatory sequences in the viral genome. It is likely that variation in EIAV pathogenicity will follow a similar pattern.

Clinical Features of Infection

The clinical response of horses following artificial inoculation or natural exposure to EIAV is variable and depends in part on host resistance factors, viral virulence factors, and environmental factors (e.g., weather and work load). In general, EIAV infections can be apparent with distinctive clinical symptoms or inapparent without any clinical signs of EIA. The clinical disease is typically described as acute, chronic, or asymptomatic (**Figure 6**).

Acute EIA is most often associated with the first exposure to the virus, with fever and hemorrhages evident from 7 to 30 days after exposure. Acute disease is thought to be associated with massive virus replication in and destruction of infected macrophages. Horses in the initial phase of acute EIA will be seronegative, because the immune system has had insufficient time to respond to the viral antigens. During the peak of the febrile response in acute EIA, viremia of greater than 10^6 horse-infectious doses per ml of whole blood is often observed. The initial acute phase of EIA infection may not be seen by the veterinarian unless there is an epizootic of the infection in a group of horses. Even then, the horses must be under close supervision before the initial fever and anorexia are detected. Neither anemia nor edema is seen at this stage of disease.

The more classic clinical signs of EIA such as loss of weight, anemia, diarrhea, and edema are seen later during recurring cycles of the illness, which appear at irregular intervals ranging from a couple of weeks to several months. The frequency and severity of clinical episodes in horses with chronic EIA usually decline with time, about 90% occurring within 1 year of infection. Horses with chronic EIA are seropositive and have variable viremia levels which are the highest during the periodic febrile episodes. Although a small percentage of chronic EIA cases may result in death, the predominant clinical course of infection is a cessation of detectable clinical symptoms by the end of the first year post infection and the establishment of a lifelong inapparent carrier stage. However, animals can experience a recrudescence of viremia and disease due to stress or immune suppression from a contemporous bacterial or viral infection.

The highest percentage of EIAV-infected horses in the field are in fact inapparent carriers. These horses have no clinical illness associated with the viral infection, and viremia is usually undetectable. However, these inapparent carriers maintain high levels of EIAV-specific antibodies, suggesting a chronic low level of virus replication. Inapparent carriers can be shown to harbor infectious EIAV by transfusions of whole blood (200 ml) to recipient horses that become seropositive to EIAV within 3–6 weeks post inoculation. In addition, treatment of certain inapparent carriers with immunosuppressive drugs or exposure to extreme stress can cause the recrudescence of chronic EIA, even in some horses that have been free of clinical symptoms for years.

Pathology and Histopathology

Gross and histopathological lesions in EIAV-infected horses are variable and closely associated with levels of viral replication. In the acute stage of EIA, gross pathological lesions consist of swelling of the parenchymatous organs, and hemorrhages can be observed in most tissues.

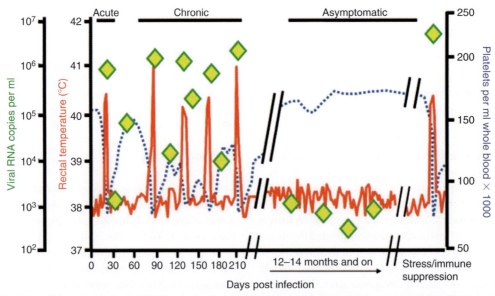

Figure 6 Clinical profile associated with EIAV infection of horses indicating the characteristic stages of EIA. Febrile episodes are defined as rectal temperatures above 39 °C (103 °F), and thrombocytopenia is defined as platelet levels below 105 000 μl^{-1} of blood.

The most pronounced histopathological lesions are hepatic and lymphoid necrosis in association with large numbers of activated macrophage and Kupffer cells. Hepatic necrosis is most severe near the central vein evidently resulting from degenerative changes in the parenchymal cells. Lesions in the spleen are characterized by degenerative erythrocytes, and small focal hemorrhages are found in the splenic capsule and adjacent tissue. Lymphocytic infiltrations can be observed in several organs including liver, spleen, lymph node, kidney, heart, and lung. The majority of these lesions are thought to be the combined effect of immune-mediated lysis of virus-infected cells and an immune complex-mediated inflammatory response.

The pathological changes in the chronic form of EIA include a developing immunological control of virus replication. Gross pathological lesions include splenomegaly, lymphadenopathy, and hepatomegaly. Microscopic changes are characterized by infiltration of lymphoid cells in almost all organs and tissues. Anemia has long been considered the hallmark of EIA. The two major causes of anemia, hemolysis and bone marrow depression, are closely associated with replicating virus. Hemolysis is immunologically mediated. Erythrocytes are coated with the viral surface glycoprotein which in the presence of specific antibodies and bound C3 induces erythrophagocytosis and complement-mediated hemolysis. Bone marrow suppression is less well characterized but appears to be associated with iron deficiency. Thrombocytopenia is frequently the earliest pathology observed during chronic EIA and can precede the detection of virus-specific antibodies. The mechanism for the marked reduction in blood platelets is unknown, but has recently been associated with cytokine dysregulation.

Immune Responses

Horses infected with EIAV typically become seropositive in standard serological assays within 21 days post infection. The humoral immune responses are predominantly against the viral envelope glycoproteins, gp90 and gp45, and the major core protein, p26. All currently approved diagnostic assays for EIAV infection are based on the detection of antibody to the major core antigen, although the antibody response to the envelope glycoproteins is at least tenfold greater than the antibody titer measured against the p26 protein. Significantly lower levels of antibody can be detected against virtually all of the other structural and nonstructural proteins encoded by EIAV. Interestingly, EIAV-specific antibody levels remain relatively constant throughout the course of chronic EIA and even in the unapparent stage of infection.

The immune responses generated during chronic EIA initially mediate significant pathogenesis in the presence of sufficient levels of EIAV antigenimia, but progressively evolve to establish a strict immunologic control over virus replication. Thus, the EIAV system is unique among lentiviruses in that the host immune responses routinely accomplish an effective control of aggressive virus replication and recurring clinical disease to maintain an indefinite inapparent stage of infection. The immune correlates of this protection remain to be defined. However, it has been shown that the neutralizing capacity of serum antibodies elicited to EIAV during the chronic stage of disease progressively increases, indicating an evolution of immune responses to the sequential generation of antigenic variants of virus. In addition, recent studies have demonstrated a lengthy and complex evolution of antibody and cellular immune responses to experimental EIAV infections of horses demonstrating a dynamic maturation process that apparently correlates with the development of protective immunity. The specific humoral or cellular immune correlates of protection have not yet been identified. The ability of the horse immune responses to overcome the array of persistence and escape mechanisms employed by EIAV suggests that a successful vaccine is feasible.

Prevention and Control

The transmission of EIAV infection has been controlled by improving animal husbandry techniques to prevent the spread of infected blood, by reducing the horsefly population in the vicinity of herds, and primarily by identifying and segregating or sacrificing horses that are seropositive for the virus. In the United States, the most common diagnostic assays are an agar gel immunodiffusion test, the Coggins assay, which was developed in the early 1970s and enzyme-linked immunoassay that was approved by the USDA as an EIA diagnostic assay in the 1980s and 1990s. Both diagnostic assay procedures are based on the detection of serum antibodies to the major viral core protein, p26. The diagnostic enzyme-linked immunosorbent assay (ELISA) assay has been engineered to produce a sensitivity that is equivalent to the previously established Coggins test. More sensitive diagnostic assays based on the detection of antibodies to the EIAV envelope glycoproteins have been developed, but have not yet been approved for commercial use. The most sensitive and specific assay for detecting EIAV infection is horse inoculation tests with 200 ml of whole blood from the horse being tested. The horse inoculation test is used only in rare cases where the standard serological assays may give ambiguous results.

There is currently no effective vaccine for the prevention of EIAV infection and disease. The primary challenge in developing an effective EIAV vaccine is overcoming the antigenic diversity intrinsic to this virus. An important practical requirement in the development of any EIAV vaccine is compatibility with established regulatory policies and diagnostic assays. The ability of EIAV-infected horses to routinely establish immunologic control over virus replication and disease suggests that an effective vaccine can indeed be developed, if the critical natural immune correlates of protection can be elicited by a candidate vaccine. An attenuated live EIAV vaccine with a reported protection efficacy of about 70% has been used in China since the early 1980s, but the effectiveness of this vaccine remains to be confirmed outside of that country. Evaluation of other candidate EIAV vaccines (live-attenuated, inactivated whole virus, subunit vaccines, synthetic peptides, etc.) under experimental conditions has revealed a spectrum of vaccine efficacy that ranges from 'sterile protection' (prevention of infection upon inoculation with EIAV) to severe elevation of EIAV replication, and exacerbation of disease. These results indicate that immune responses to EIAV are a double-edged sword that can either mediate protection or yield vaccine enhancement. Vaccine enhancement has previously been reported for other viral infections (dengue virus, respiratory syncitial virus, feline infectious peritonitis virus) and is of special concern with macrophage-tropic viruses. Similar examples of vaccine protection and enhancement have been reported in studies of experimental vaccines for other lentiviruses, including feline immunodeficiency virus, caprine arthritis-encephalitis virus, and visna-maedi virus. These observations in several diverse animal lentivirus systems suggest that the potential for immune enhancement may be a general property of lentiviruses, including HIV-1. Current efforts in the production of a commercial EIAV vaccine are focused on the development of a vaccine that can achieve sufficient maturation of immune responses to provide protection from virus infection, but allow the serological differentiation between vaccinated and infected horses. In this regard, DNA vaccine strategies appear to be well suited to accomplish these criteria for a commercial EIAV vaccine.

Future Research

EIAV provides a dynamic system for examining the interaction between virus populations and host immune responses that are evolving in response to each other. In addition, EIAV offers a remarkable model for studying the delicate balance between immune responses to a persistent virus infection that result in disease and those that have beneficial results. A characterization of the nature of protective and enhancing immune responses can provide important information about the mechanisms of lentivirus disease and the type of immune responses to be elicited or avoided by a vaccine. The results of these studies in the EIAV system should be applicable to other lentiviruses, including HIV-1.

See also: Human Immunodeficiency Viruses: Antiretroviral agents; Human Immunodeficiency Viruses: Molecular Biology; Human Immunodeficiency Viruses: Origin; Human Immunodeficiency Viruses: Pathogenesis; Immune Response to viruses: Antibody-Mediated Immunity; Vaccine Strategies; Viral Pathogenesis.

Further Reading

Cook RF, Issel CJ, and Montelaro RC (1996) Equine infectious anemia virus. In: Studdert R (ed.) *Viral Diseases of Equines*, pp. 295–323. Amsterdam: Elsevier.

Cordes TA and Issel CJ (1996) Equine infectious anemia: A status report on its control. USDA Animal and Plant Health Inspection Service Publication No. APHIS 91–55–032.

Montelaro RC, Ball JM, and Rushlow KE (1992) Equine retroviruses. In: Levy J (ed.) *The Retroviridae*, vol. 2, pp. 257–360. New York: Plenum.

Montelaro RC and Bolognesi DP (1995) Vaccines against retroviruses. In: Levy J (ed.) *The Retroviridae*, vol. 2. New York: Plenum.

Sellon DC, Fuller FJ, and McGuire TC (1994) The immunopathogenesis of equine infectious anemia virus. *Virus Research* 32: 111.

Evolution of Viruses

L P Villarreal, University of California, Irvine, Irvine, CA, USA

© 2008 Elsevier Ltd. All rights reserved.

Glossary

Dendrogram A schematic line drawing, often tree-like, that represents evolutionary relationships between species.

Error catastrophe A threshold at which a high error rate of genome replication can no longer maintain the integrity of essential genetic information.

ERV An endogenous retroviral-like genetic element, containing a long terminal repeat

that is found in the genomes of many organisms.
Fitness landscape A hypothetical representation of a three-dimensional surface that represents relative fitness of genetic variants.
Muller's ratchet A decrease in fitness resulting from a repetition of a genetically restricted founder population.
Quasispecies A population of viral genomes that are the product of error-prone replication usually envisioned as a swarm or cloud of related genomes.
Red Queen hypothesis An evolutionary concept in which an organism must evolve at high rates in order to maintain its competitive advantage.
Reticulated evolution A pattern of evolution in which elements are derived from different ancestors represented as a tree with cross-connections between distinct branches.
Sequence space A multidimensional representation of all possible sequences for a given genome.

Introduction

The initial study of virus evolution sought to explain how virus variation affects viral and host survival and to understand viral disease. However, we now realize that virus evolution is a basic issue, impacting all life in some way. In general, the principles of virus evolution are very much the same Darwinian principles of evolution for all life, involving genetic variation, natural selection, and survival of favorable types. However, virus evolution also entails features such as high error rates, quasispecies populations, and genetic exchange across vast reticulated gene pools that extend the traditional concepts of evolution. Evolution simply means a noncyclic change in the genetic characteristics of a virus; and viruses are the most rapidly evolving genetic agents for all biological entities. Principles of virus evolution provide an integrating framework for understanding the diversity of viruses and the relationships with host as well as providing explanations for the emergence of new viral disease. Most emerging viral diseases are due to species jumps from persistently infected hosts that have long-term virus–host evolutionary histories (called natural reservoirs). Since early human populations (i.e., small bands of hunter gatherers) could not have supported the great human viral plagues of civilization (e.g., smallpox virus/variola and measles), these viruses must also have originated from species jumps that adapted to humans. In recent history, the emergence of HIV demonstrates that virus evolution continues to impact human populations. Early observation established that viruses often show significant variation in virulence. Such variation was used as an early, yet risky, form of vaccination (i.e., variolation against smallpox). The variation that occurred with passage into alternate tissue and host was also used to make various vaccines (rabies in 1880s) or attenuate viral virulence, such as live yellow fever vaccine. But variation also allows some viruses, such as influenza A, to escape neutralization by vaccination. However, variation in viral disease or virulence does not provide a quantitative basis to study evolution.

The study of virus variation and evolution is an applied science that allows the observation of evolutionary change in real time. For example, human individuals (or populations) infected with either human immunodeficiency virus 1 (HIV-1) or hepatitis C virus (HCV) show progressive or geographical evolutionary adaptation associated with the emergence of specific viral clades that affect disease therapy and progression (such as resistance to antiviral drugs). **Figure 1** shows HIV variation in an individual human patient whereas **Figure 2** shows HCV variation in the human population. Virus evolution is also important for the commercial growth of various organisms, such as the dairy industry (lactose fermenting bacteria), the brewing industry, agriculture, aquaculture, and farming. In all these applied cases major losses can result from virus adaptation to the cultivated species, often from viruses of wild species. Some organisms appear much less prone to viral adaptations (e.g., nematodes, ferns, sharks). Virus evolution can also be applied to technological innovation, as in phage display. This is a process in which a terminal surface protein of some filamentous bacterial virus can be genetically engineered for novel surface protein expression. By generating diversity *in vitro* (with up to 10^{15} types), and applying the principles of evolution (random variation) to biochemical selection (such as binding to a chemical substrate), a reiterative amplification can find solutions to problems in biochemistry, such as surface interactions or catalytic activity.

Virus Evolution as a Basic Science

Virus variation is a global issue. In the last decade it has been established that viruses are the most numerous biological entities on the planet. The oceans and soil harbor vast numbers of viral-like particles (VLPs), mostly resembling the tailed DNA viruses of bacteria. In addition, some of these environmental viruses are unexpectedly large and complex, such as the phycodnaviruses of algae or mimivirus of amoeba, a 1.2 Mb DNA virus that can encode nearly 1000 genes. Thus viruses represent a vast and diverse source of novel genes. However, the evolutionary dynamics of this population and its effect on hosts is not well understood. It is likely that this virus gene pool also affects host evolution since prophage colonization is

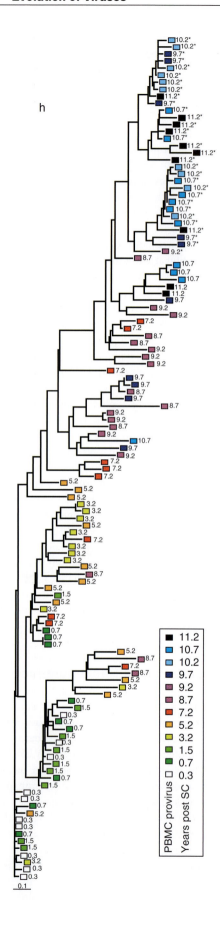

known for all prokaryotic genomes. Thus, this vast pool of viruses connects directly to prokaryotes and the 'tree of life'. The study of virus evolution has become an extension of all evolution.

Distinctions from Host Evolution

For the most part, virus evolution conforms to the same Darwinian principles as host evolution, involving variation and natural selection. However, viruses have multiple origins, and are thus polyphyletic. There are six major categories of viruses (+RNA, −RNA, dsRNA, retro, small DNA, large DNA) that have no common genes and hence have no common ancestor. However, these categories all have conserved hallmark genes (i.e., capsid proteins, Rd RNA pol, RT, primase, helicase, and RCR initiation proteins) which are all monophyletic and which may trace their origins to primordial host/viral gene pools. Many DNA viruses, for example, have replication strategies and polymerases that are clearly distinct from that of their host, which depends on such hallmark genes. Thus, virus evolution appears ancient but inextricably linked to its host. Also, in contrast to host evolution, viral quasispecies show a population-based adaptability that extends the selection of the fittest to include populations of otherwise unfit genomes (described below). Viruses can clearly cross the usual host-species barriers so that viral evolution can be reticulated in vast genes pools. For example, bacterial DNA viruses and +RNA viruses can show high rates of recombination across viruses that infect numerous host species. Accordingly, tailed DNA phage of bacteria appear to represent one single vast gene pool. Viruses can also violate concepts of death and extinction, reassembling genomes from parts and/or repairing lethal damage by multiplicity reactivation. In addition, damaged (defective) viruses can also affect virus and host evolution. Such defectives are found in many types of viruses and can also be found in many host genomes (as defective prophage or defective endogenous retroviruses). Such defective viruses can clearly affect host survival. In all these characteristics, viruses extend the Darwinian principles of host evolution.

Figure 1 HIV population analysis from an infected individual. Shown is a neighbor-joining phylogram derived from maximum likelihood distance between all sequences. Sequences are represented by a square for PBMC sequences or a triangle for plasma sequences. The arbitrary color gradient corresponds to the time of sampling. Adapted from Shankarappa R, Margolick JB, Gange SJ, et al. (1999) Consistent viral evolutionary changes associated with the progression of human immunodeficiency virus Type 1 infection. *Journal of Virology* 73: 10489–10502, with permission from the American Society for Microbiology.

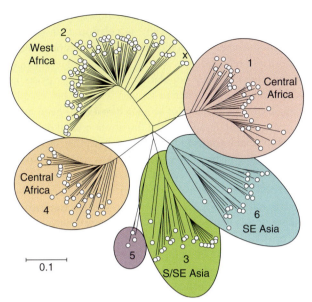

Figure 2 Unrooted phylogenetic analysis of HCV nucleotide sequences from globally distributed human isolates. Adapted from Simmonds P (2001) Reconstructing the origins of human hepatitis viruses. *Philosophical Transactions of the Royal Society of London* 356: 1013–1026, with permission from Royal Society Publishing.

A History of Virus Evolution

The coherent study of virus evolution awaited the development of sequence technology to measure mutations and genetic variation in viral populations. Concepts of natural selection, fitness, and propagation of favorable variation had long been established in the evolutionary biology literature prior to the growth of virology. Thus, mathematical models, such as Fisher population genetics, concerned with gene frequency in a (sexually exchanging) species population, had been well developed and seemed directly applicable to virus evolution, which resemble that of host genes. Here, viral fitness was typically expressed as relative replication rates (replicative fitness) but sometimes host virulence and disease were also used. However, as presented below, a comprehensive definition of viral fitness remains problematic. The first quantitative measurements of the rate of virus mutation was done in the 1940s with bacterial phage. The mutation rates were expressed as a set of ordinary differential equations that were subsequently used to develop the quasispecies equations as applied to error-prone RNA genomes (see below). However, the species definition of a virus poses a problem for evolutionary thinking and challenges how we define kinship in viruses. Unlike the sex-defined host, a virus species is currently defined as a polythetic class: a mosaic of related parts of which not all elements are shared (such as host range, genome relatedness, antigenic properties). No specific defining characteristic or gene exists for a virus species and sexual exchange need not be included. This is an inherently fuzzy definition, like defining a 'heap', which although clear, cannot be specified by its number of parts. The ensuing molecular characterization of many virus populations supports this species definition. The challenge then is to understand viral evolutionary patterns working with such fuzzy definitions. Yet, conserved patterns of virus evolution are still seen, some of which suggest viruses are indeed an ancient lineage, possibly extending into the primordial RNA world.

Error-Prone Replication and Quasispecies

In the 1970s, Manfred Eigen and also Peter Shuster developed a fundamental theoretical model of virus evolution. A set of ordinary differential equations was published that described what was called 'quasispecies'. Starting from measurements of phage mutation rates, they considered the consequences of high error rates as expected from RNA replication (an error-prone noncorrecting replication process). The resulting population shared many properties and was called quasispecies, a society (or community) of individuals that are the error products of replication. The name 'quasispecies' thus describes a chemically diverse set of molecules and was not intended to refer to a biological species (i.e., genetic exchanging). However, as discussed above, the fuzzy definition of virus species and quasispecies overlaps somewhat, which has been a source of confusion. Several premises were used to develop this theory: (1) the individual products ignore one another and interact only as individuals; (2) the system is not at equilibrium and resources are not limiting. Based on relative replication, the growth of favorable types is described which provides a mathematical definition of replicative fitness. The original equations represent an idealized generalized system of infinite population size and are not directly applicable to the real world, although they provide valuable insights into real world systems. The equations do not address variable mortality (longevity), interference, exclusion, competition, complementation, and persistence, or how such issues affect nonreplicative fitness definitions. The issue of mortality and fitness is interesting from the perspective of viruses. For example, an interfering defective virus can be considered dead, but can clearly interfere with and drive the extinction wild-type template replication in quasispecies. In some cases, the quasispecies equations appear to be mathematically equivalent to classical Wright and Fisher population genetics equations as applied by Kimura and Maroyama to asexual haploid populations at the mutation-selection balance. However, these two approaches begin with distinct perspectives, and it was the assumption of high error in the quasispecies equations that had a major impact on experimentation and our current understanding of virus evolution. This has also led to some counter intuitive conclusions, such as the concept of selection of 'the fittest' compared to the consensus character of the master template. Quasispecies from a virus with high error rates (such as HIV-1) might be composed of all mutant progeny RNAs such that the consensus template (the mean, the fittest, or the master template) may not actually exist. With classical population genetics, an asexual clonal population should fix the clonal sequence. With quasispecies, this is not observed. The first laboratory measurements of viral quasispecies were made using Qβ RNA polymerase *in vitro*. Error estimates ranged from 10^{-3} to 10^{-4} substitutions per site per year (an error rate applicable to most RNA viruses). With Qβ, the replication of many nonviable mutants generated a genetic spectra that had a characteristic makeup. For example, separate DNA clones of Qβ were initially distinct from each other but quickly generated the same RNA quasispecies as before cloning. Additional lab measurements have shown that quasispecies can have significant adaptive fitness (above the cloned master template) and display memory; that is, they can retain information of prior selections in a minority of the population. Complementation, interference, competition suppression, and extinction have all been measured in various quasispecies, thus indicating a collective form of evolution and violating an original premise of genome independence in these equations. In addition, the sequence diversity in a quasispecies is now seen as a source of adaptive potential, not simply error (see below). Despite these results, the concept of quasispecies has still been highly useful, and not simply as a theoretical development. For example, the live poliovirus vaccine is clearly a heterogeneous quasispecies. Within this population exists a minority of neurovirulent variants that are suppressed by the majority avirulent virus. A main point of the quasispecies concept is that it provides an understanding of high adaptability from a population that has many, even lethal, mutations. It is interesting that the proposed early RNA world would have also been a collective quasispecies world.

Error Catastrophe, Sequence Space

Quasispecies theory also predicts a situation known as error catastrophe, defined as an error rate threshold at which information is lost and the system decays. If error rates are too high, or the information content (genome length) too extensive, the system will be unable to maintain its information integrity. This predicts a basic limit on the size of RNA genomes, consistent with the observation that the largest RNA genomes are only about 27–32 kbp (coronaviruses). There is a possible therapeutic use of error catastrophe: drugs (possibly Ribavarin and 5-fluorouracil) that increase the error rates of RNA polymerase can potentially push a virus beyond its error threshold and induce a catastrophe. Quasispecies is an inherently fuzzy and dynamic population that has no sharp boundaries or specific members and has been metaphorically referred to as swarms and clouds. Here, cloud is a metaphor for the population landscape that exists in high-dimensional hyperspace and cannot be readily visualized. The concept of sequence space has been used to represent topography of the distribution of all mutants. Kinship relationships between mutants can be measured by Hamming distance; the minimal steps needed to specify the difference between two mutants. In spite of high error and adaptation rates (and sometimes high recombination rates), RNA viruses are not able to explore all potential sequence space. Selection significantly limits the quasispecies, since the potential sequence space is hyperastronomical even for a moderately sized virus. For example, an RNA virus with 10 000 nt would correspond to 10^{6000} possible sequences, well beyond what could be explored by even the potentially vast number of viruses over the lifetime of the world. In addition, there are clearly mechanistic constraints that prevent many possible sequences, such as necessary domains of ± strand RNA folds, physical association with ribonucleotide proteins, virion packaging and assembly – all in addition to usual selection for gene fitness (function) that all severely limit possible adaptations. This creates a multipeak 'fitness' landscape in hyperspace (see **Figure 3**). Assuming fitness itself can have a single definition (i.e., replicative fitness, not subjected to variable and stochastic competition), we can visualize this space as many steep valleys and ledges (in this case with 10 000 dimensions). Normally we think that adaptation by natural selection is the force to explore and move through fitness landscape. But as the deep valleys are often lethal, they cannot be explored via natural selection. Here we see the major adaptive power of the quasispecies

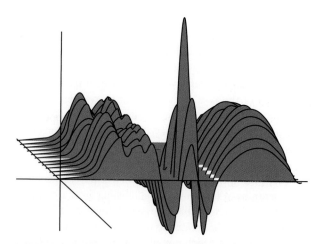

Figure 3 Hypothetical fitness landscape for an RNA virus. Assuming one definition for a nonrelativistic fitness (such as replicative fitness), the coordinates indicate relative fitness. Those below the *y*-axis are interfering or lethal variants.

collective. Since random, even lethal, errors and drift are inherent in a quasispecies, lethal valleys can be readily crossed by such variable genomes, allowing the master genome to adapt by natural selection to a new fitness peak. Thus, error-prone replication and the generation of mutant clouds allows for much better exploration of sequence space and eventual adaptability.

When viruses are transmitted to new hosts, they can experience a genetic bottleneck since a relatively small number of viral genomes could be involved (aka low multiplicity passage). If this process is serially repeated, a phenomenon known as 'Muller's Ratchet' can result in lost competitiveness as the essentially clonal RNA virus accumulates deleterious mutations (sometimes measured as pfu/plaque). However, in lab studies, virus extinction from serial passage does not occur, presumably due to plaque selection for a restored phenotype. Even a single plaque is in reality a small population (due to nonideal particle/pfu ratios and ID_{50}). However, lost competitiveness with other viruses is seen with clonal laboratory passage. However, if a quasispecies population is passed, this generally results in increased competitive fitness. Such passage can produce a seemingly never-ending better version of the virus that outcompetes all prior versions of the same virus (although virion yields and absolute replication are not necessarily improved). This has been likened to the Red Queen hypothesis in that the viruses are evolving at high rates, simply to maintain their competitive position, so as not to be displaced as the dominant viral type. Virus–virus competition is thus a crucial selection.

Never-Ending Adaptation

A real world example of the potentially never-ending virus adaptation is shown in **Figure 4**. The HA and NA genes of human influenza A virus have been monitored for several decades. As shown, the prevalent master template of the virus circulating in the human population has been continually changing, due to immune selection and stochastic viral immigration, necessitating yearly vaccine changes (also shown). Although such a population dynamic has been stably maintained in the human population, all prior versions have essentially become extinct, as they do not reappear.

Not all RNA virus populations show this dynamic of a continual change or even the diversity expected from quasispecies. Even in influenza virus A, avian isolates from natural host (waterfowl) can be genetically stable. Some RNA (and retro) viruses with high error rates can nevertheless maintain stable populations in specific hosts. For example, measles virus shows much less antigenic drift in human infections compared to influenza virus A. Hepatitis G virus (a human prevalent and distant relative of HCV) shows little variation in even isolated human populations. The filoviruses (Ebola virus and Marburg virus) have shown no genetic variation in Zaire isolates from 20 years apart. Hendra virus, isolated from Australian fruit bats, and Nipah virus from Malaysian fruit bats also show little genetic diversity. Arenaviruses and hantavirus are also genetically stable in their natural rodent host. The reasons for such population stability have not been well evaluated. In some cases (measles), purifying selection would seem likely. In other cases, persistence and low replication rates seem to apply. For example, simian foamy virus (SFV) and human T-lymphotropic virus II (HTLV-II) generate only about 10^{-8} substitutions per site per year, probably due to low replication rates.

Virus–Host Congruence and RNA Stability

There are now many examples of species-specific RNA virus/host coevolution, indicating very slow rates of virus evolution. Since error rates must be similar, this appears to be at odds with the quasispecies theory. For example, Hantavirus (genus *Bunyavirus*) coevolution with its rodent host, suggests a 20 million year association. Arenaviruses (ssRNA bisegmented ambisense) also coevolve in Old/New World rodents. These viruses are of special interest with regard to emergence as they represent the source of five hemorrhagic human fevers (such as Lassa virus). In all these examples, however, it appears that the virus causes a persistent unapparent infection in its natural host and that human disease is due to species jump.

Tools

Although viral genomes were the first to be sequenced, the initial focus was simply to identify similarity between viral genes, not to evaluate distant evolutionary relationships. The most popular tool for finding similarity is BLAST (Basic Local Alignment Search Tool) from the National Centers for Biotechnology Information (NCBI), which calculates similarity between query sequences and infers a probability based on a matrix database. Various versions of BLAST are the most used tool in bioinformatics to trace evolution. Although BLAST will identify similar genes, it is also necessary to compile and evaluate the similarities in sets of the related sequences. Multiple sequence alignment software, such as ClustalW, is used for this purpose. Phylogenetic relationships are then inferred from tree-building software. This software includes maximum parsimony, neighbor-joining, and maximum likelihood methods. The statistical significance of the tree (relative to all possible trees) can then be evaluated by algorithms such as bootstrap. More recently, Bayesian analyses, such as Bali-Phy, which implements a Markov Chain Monte Carlo (MCMC) method and calculates joint posterior

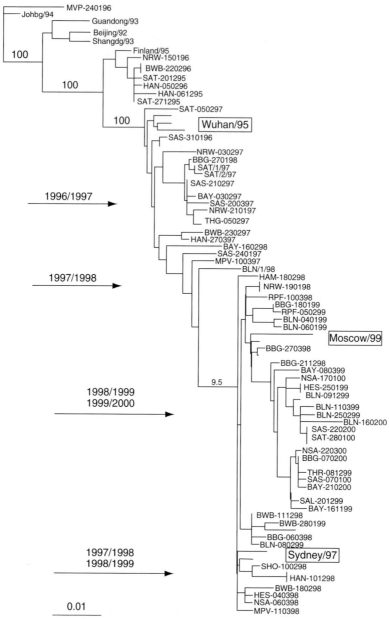

Figure 4 Phylogenetic tree of yearly influenza A/H3N2 viruses variants based on the hemaglutinin gene. Locations of specific vaccine strains are boxed. Reproduced from Schweiger B, Zadow I, and Heckler R (2002) Antigenic drift and variability of influenza viruses. *Medical Microbiology and Immunology* 191: 133–138, with permission from Springer-Verlag.

probabilities of phylogeny and alignment have become popular. This software has the added potential of using sliding windows and evaluating multiple trees, such as in virus–host coevolution. These methods, however, use the single master consensus sequence as its query and do not evaluate quasispecies, collective-based populations. Also, the clearly reticulated or hybrid character of some virus evolution is problematic. Also, distant evolutionary relationships are no longer preserved in the same sequences. Here, conservation of structural motifs, assembly patterns, gene order, and replication strategy are used to identify distant kinship.

Patterns of RNA Virus Evolution

The sometimes extreme variation of RNA virus sequence has led some to propose that most family lineages appear to be only about 10 000 years old, which is clearly at odds with much older estimates. The +RNA viruses in particular show a remarkable diversity of genomes and replicator mechanisms. These families also show much evidence of recombination and a tendency to cross host barriers. About 38 families of +RNA viruses with up to four segments are known. There are four distinct classes of replicase recognized in viruses that also share a common genetic plan. These RNA viruses have three helicase superfamilies, two protease superfamilies, and two jelly roll capsid domains. For the most part, capsid and RdRpol sequences are congruent except for members of the families *Luteoviridae* and *Tetraviridae*, which appear to have undergone recombination between these two gene lineages. The smallest +RNA virus is a member of the bacterial virus family *Leviviridae*, which has only four genes. This simple virus appears to represent the ancestral +RNA virus. Curiously, no RNA virus has yet been found to infect archaebacteria. The largest +RNA viruses belong to the genus *Coronavirus* (27–32 kbp). The most recently described +RNA viruses are members of the family *Marnaviridae* infecting bats and marine organisms sea life, which appear to be basal to evolution of picornaviruses. However, natural populations of some +RNA viruses can be stable. For example, dengue virus (*Flaviviridae*) shows low rates of amino acid substitution (e.g., nonsynonymous to synonymous ratios). Since it is an acute arbovirus infection with high error rates, strong selective constraints likely account for this stability involving multiple (systemic) tissues and vector transmission.

Negative-strand RNA viruses have distinct patterns of evolution, which is traced via their polymerase genes. Gene order tends to be highly conserved. The unsegmented viruses, such as rhabdoviruses, lyssaviruses, and paramyxoviruses, do not undergo significant recombination so their variation tends to be by point mutations and deletions. Although high error rates, variation, and quasispecies generation can be seen in laboratory settings, natural isolates, such as lyssaviruses and measles virus, tend to be relatively homogeneous. For example, lyssaviruses show a slow rate of evolution (5×10^{-5}/site/year). Lyssavirus persistent infections in natural host might contribute to this stability. However, measles virus is a strictly human-specific acute infection so its stability is likely due to purifying selection.

Patterns of DNA Virus Evolution: Tailed Phage

Large DNA viruses of bacteria, archaea, and eukaryotes appear to be evolutionarily linked. Although little sequence conservation can be identified between the T4 phage of bacteria, the halophage of archaea, the members of the family *Phycodnaviridae* infecting algae, and the herpes viruses of vertebrate eukaryotes, all show similarities in their gene programs, DNA polymerase types, capsid structures, and capsid assembly, consistent with a common ancestor. For example, both the Enterobacteria phage T4 (T4) and herpes simplex virus 1 (HSV-1) have $T = 1$ capsid symmetry with 60 copies of capsid protein. The bacterial DNA viruses would appear to represent the ancestor of all these viruses, but the origins of these phages now appear lost in the primordial gene pool. These DNA viruses can have large genomes that could not be sustained by error-prone replication. Thus, many DNA viruses do have error-correcting DNA replication, with error rates that approach or equal those of their host cells (10^{-8}). Giant bacterial phage genomes (*Bacillus megaterium* phage G, of about 600 genes), and algal phycodnaviruses have now been characterized. Even larger DNA viruses of amoeba (acanthamoeba polyphaga mimivirus) coding for more than 1000 genes are known to be abundant in some water habitats.

The tailed phage of bacteria have been called the dark matter of genetics, due to their numerical dominance ($\sim 10^{31}$ *en toto*). This corresponds to about 10^{24} productive infections per second on a global scale. Most host-restricted phage lineages clearly conserve sets of core proteins (especially capsid genes), but others (the broader T-even phages) do not conserve any hallmark genes. Hallmark genes, when present, are usually recognized by conserved domains within proteins, such as replication and structural proteins. Replicator strategy and gene order are also frequently conserved. Phage also tends to conserve genes that are active against other phage (i.e., DNA modification, lambda RexA, T4 rII). With the sequencing of numerous phage genomes, however, a large number of novel genes have been identified. Currently, 350 full genomes of tailed phage and 400 prophage from bacterial genomes have been sequenced. In general, large DNA viruses are tenfold overrepresented in small single-domain genes (~ 100 aa). Comparative genomics, especially of lactobacterial phages, suggest that most phage genomes evolve as mosaics, with sharp boundaries between genes as well as at protein domains within genes (see **Figure 5**). Recombination between lytic, temperate, and cryptic prophages appears to account for this gene and subgene domain variation. Some specific phages have mechanisms to generate specific gene diversity (such as bordetella phage using RT for surface receptor diversity), but most diversity is the product of recombination. Two broad patterns of phage variation have been observed corresponding to host-unassociated lytic and host-associated (congruent) temperate phage. In most bacterial genomes (ECOR *Escherichia coli* collection, cyanobacteria, *B. subtilis*), patterns of prophage colonization account for significant genetic distinctions between closely related host strains. The general picture for tailed phage of bacteria is that they are not the products of reduction of host genomes.

Large Eukaryotic DNA Viruses

As noted, evolutionary links between tailed phage and large DNA viruses of eukaryotes are apparent. The phycodnaviruses of unicellular green algae clearly have many phage-like characteristics, including the presence of

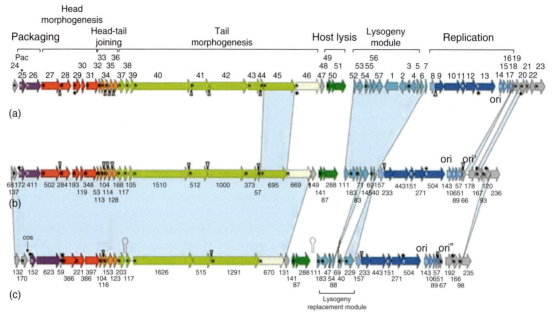

Figure 5 Genome comparison of temperate *S. thermophilus* phage 01205, virulent *S. thermophilus* Sfi11, and the virulent *S. thermophilus* Sfi19. Probable gene functions are indicated and genomes have been divided into functional units. Genes belonging to the same module are indicated with the same color. Areas of shading indicate regions of major difference. From Desiere F, Lucchini S, Canchaya C, Ventura M, and Brüssow H (2002) Comparative genomics of phages and prophages in lactic acid bacteria. *Antonie van Leeuwenhoek* 82: 73–91.

restriction/modification enzymes, homing endonucleases, the injection of viral DNA, and the external localization of the viral capsid. They also have many characteristics of eukaryotic DNA viruses, such as a clearly herpes virus-related DNA polymerase, PCNA proteins, nonintegrating DNA, and numerous signal-transduction proteins. Thus, phycodnaviruses show hybrid characteristics of prokaryotic and eukaryotic viruses.

The evolutionary pattern of the large DNA viruses of eukaryotes is generally best traced by comparing their respective DNA-dependent DNA polymerases (DdDp). These exist in distinct classes that are typically specific for each viral lineage and are usually the most highly conserved of the set of core genes within a viral lineage. However, some viruses, such as the white spot syndrome virus (WSSV) infecting shrimps, have almost no genes in common with other DNA viruses. Generally, the specific set of core genes is clade specific. The first fully sequenced viral genome tree was that of the baculoviruses (see **Figure 6**). The overall pattern of evolution shows the conservation of the core set in which most clades can be differentiated from one another mainly by acquisition of several novel viral genes (although some lineage-specific gene loss is also apparent). In another example, coccolithoviruses differ from related phycodnaviruses by the acquisition of 100 kbp gene set, including six subunits of DdDp core genes. Similar patterns of divergence can be seen with the herpesvirus family members. In addition, most herpesvirus clades also show coevolution with their host. However, the poxviruses (orthopoxviruses), show a different overall evolutionary pattern and are not congruent with host. The more ancestral orthopoxvirus members, such as cowpox virus and mousepox virus, have greater gene numbers that appear to have been lost in the human-specific and virulent smallpox virus. Avipoxviruses have even greater gene diversity but the entomopoxviruses are the most complex and diverse of all. The complexity and brick shape of the poxviruses originally inspired the view that these viruses might evolve from bacterial cells following the reduction of complexity. However, DNA sequencing makes it clear that viral core genes have no bacterial analogs. In some instances, viral lineages have clearly fused with other viral and host lineages. For example, the baculovirus *Autographa californica* MNP virus (AcMNPV) has acquired a gypsy-like retrovirus (e.g., TED), an endogenous retrovirus associated with moth development. The polydnaviruses (circular DNA viruses) are fused into their host genomes (as endogenous DNA viruses) of some parasitoid wasps, essential for survival of the wasp larvae.

Small DNA Viruses

The small, double-stranded, circular DNA viruses (*Papillomaviridae* and *Polyomaviridae*) show evolutionary patterns that are highly host linked. Virus and host evolution are mostly congruent, and virus evolution tends to be slow.

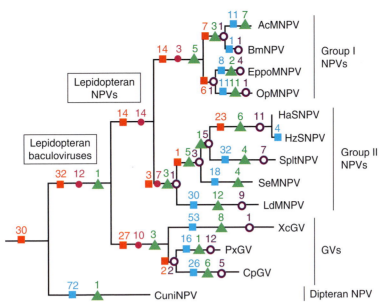

Figure 6 Gene content map of 13 complete sequences of baculoviruses, including the genus *Granulovirus*. The tree shows the most parsimonious hypothesis of changes in gene content during baculovirus evolution. Colors and shapes indicate gene conservation, acquisition, and loss. Reproduced from Herniou EA, Olszewski JA, Cory JS, O'Reilly DR (2003) The genome sequence and evolution of baculoviruses. *Annual Review of Entomology* 48: 211–234, with permission from Annual Reviews.

For example, approximately 100 human papillomaviruses show congruent evolution with human (and primate) host. This seems to be due to both a highly species- and tissue-specific virus replication, as well as a tendency to establish persistent infections. However, the rolling circular replicon (RCR) viruses, such as parvoviruses, can have distinct evolution patterns. Both mouse minute virus (MMV) and canine parvoviruses can have quasispecies-like populations, which can show evolutionary rates at 10^{-4} substitutions per site per year. Such rates are at the lower end of those seen with RNA viruses. In bacteria, RCR viruses and RCR plasmids appear to represent a common gene pool. Other poorly characterized small eukaryotic DNA viruses, such as human torque teno virus, are asymptomatic but show high variation during persistence for unknown reasons.

Endogenous and Autonomous Retroviruses

Retroviruses present a special problem in understanding patterns of eukaryotic virus evolution. Like prophage of bacteria, retroviruses both stably colonize their host as endogenous or genomic retroviruses (ERVs) that are often defective, but may also sometimes emerge from their host (especially rodents) to produce autonomous virus. In addition, retroviruses are polyphyletic and prone to generating quasispecies due to high error rates as well as high rates of recombination. The most common conserved retrovirus genome elements are domains within the long terminal repeats (LTRs), RT, integrase, protease, gag protein, and env protein. Of these, env are the most often altered or deleted in host genomes. In addition, tRNA primer sites (such a lys [K] tRNA) are also often conserved and used for classification (i.e., human endogenous retrovirus, HERV-K family). However, each of these retroviral elements can potentially have distinct patterns of evolution and conservation, generating distinct dendrograms. Vertebrates, especially mammals, seem to host many retroviral elements within their genomes. Their autonomous retroviruses have a tendency to infect cells of the immune system. Murine leukemia virus (MLV) is the best-studied simple autonomous retrovirus, but many endogenous MLV relatives also exist. Retroviruses are present in genomes of early eukaryotes but significantly expanded in vertebrates. Gypsy-like retroviruses (aka chromoviruses, defined via RT and gag similarity) are often found conserved as full-length elements including env genes in most lower eukaryote genomes (e.g., *Caenorhabditis elegans*), but were mostly lost from tetrapods. Some lower eukaryotes clearly prevent colonization by ERVs, such as *Neurospora* fungi (via the RIP exclusion system). Many endogenous retroviruses are congruent with host evolution, whereas other ERVs are recently acquired and highly host specific. In terms of gene diversity, the retroviral *env* are the most diverse. There are five RT-based families recognized such as *Retroviridae, Hepadnaviridae, Caulimoviridae, Pseudoviridae,* and *Metaviridae,* the latter three being especially prevalent as genomic elements in flowering plants (especially Gypsy). Yet not all retroviruses seem able to colonize host germ line. For example, lentiviruses (such as simian immunodeficiency virus (SIV) and HIV) show no examples of endogenization compared to the simpler MLV-related viruses that can be both autonomous (i.e., MLV, Gibbon ape leukemia virus) and endogenous (i.e., *Mus dunni* ERV, koala ERV). The converse can also be true, since no autonomous versions of HERV K, for example, are known.

Early views proposed that retroviruses evolved from nonviral retroposons (LTR RT elements, non-LTR LINE-like elements). These non-LTR elements, have distinct nonretroviral mechanistic features and core protein domains, but retain some virus-like domains of RT; thus, they appeared to predate retroviruses. However, we now know that gypsy-like retroviruses were present in the earliest eukaryotes. In addition, some LTR-containing elements, such as Gypsy, had initially been considered ancestral to retroviruses because all copies seemed to be defective. However, it is now established that complete gypsy retroviruses are conserved as ERVs in some yeast and *Drosophila* strains. Thus, although endogenous and exogenous retroviruses appear to evolve from each other, there is no evidence that exogenous retroviruses have emerged from non-LTR LINE-like elements.

The congruence between ERVs and host eukaryote evolution is sometimes striking. For example, all mammals have acquired their own peculiar versions of ERVs (and LINES). Recently, it has become clear that the placental mammals have conserved several families of ERV-derived *env* genes that provide an essential function for placental tissue (ERV W-syncytin 1, ERV FRD syncytin 2, enJSRV). Clearly, retrovirus evolution is highly intertwined with that of their hosts.

Never-Ending Emergence

A remaining concern of virus evolution is to understand the emergence of new viral pathogens. The unpredictable and stochastic nature of such virulent adaptations makes predictions difficult, as the link between virulence and evolution is vague. For example, the genetic changes that made the SARS virus (persisting in bats) into an acute human pathogen are still not predictable. Viral fitness and selection, and how they change from persistent states with acute species jumps, are not yet defined. However, some variables contribute to the likelihood of viral emergence, such as virus ecology. The population density and dynamics of the new host and the ecological interactions between new and stable viral host are often crucial. In addition, virus–virus interactions can be important, allowing for recombination and/or reassortments or lowering immunological selective barriers via immunosuppression. The emergence of HIV-1 from different, persistent SIVs of African monkeys through chimpanzees into a new human disease, for example, includes the same issues. Also, the potential emergence of pandemic human influenza from avian (Anatiformes) sources, such as H5N1, remains a great concern. Thus, virus evolution will continue to interest us as we seek to predict, control, or eradicate viral agents of disease.

See also: Antigenic Variation; Coronaviruses: Molecular Biology; Emerging and Reemerging Virus Diseases of Plants; Emerging and Reemerging Virus Diseases of Vertebrates; Emerging Geminiviruses; Origin of Viruses; Phylogeny of Viruses; Picornaviruses: Molecular Biology; Quasispecies; Retrotransposons of Vertebrates; Virus Databases; Virus Evolution: Bacterial Viruses; Virus Species.

Further Reading

Desiere F, Lucchini S, Canchaya C, Ventura M, and Brussow H (2002) Comparative genomics of phages and prophages in lactic acid bacteria. *Antonie Van Leeuwenhoek* 82(1–4): 73–91.

Domingo E (2006) *Current Topics in Microbiology and Immunology, Vol. 299: Quasispecies: Concept and Implications for Virology.* Berlin: Springer.

Herniou EA, Olszewski JA, Cory JS, and O'Reilly DR (2003) The genome sequence and evolution of baculoviruses. *Annual Review of Entomology* 48: 211–234.

Hurst CJ (2000) *Viral Ecology.* San Diego: Academic Press.

Schweiger B, Zadow I, and Heckler R (2002) Antigenic drift and variability of influenza viruses. *Medical Microbiology and Immunology* 191: 133–138.

Shankarappa R, Margolick JB, Gange SJ, et al. (1999) Consistent viral evolutionary changes associated with the progression of human immunodeficiency virus type 1 infection. *Journal of Virology* 73: 10489–10502.

Simmonds P (2001) Reconstructing the origins of human hepatitis viruses. *Philosophical Transactions of the Royal Society of London* 356: 1013–1026.

Roossinck MJ (2005) Symbiosis versus competition in plant virus evolution. *Nature Reviews Microbiology* 3(12): 917–924.

Villarreal LP (2005) *Viruses and the Evolution of Life.* Washington, DC: ASM Press.

Zanotto PM, Gibbs MJ, Gould EA, and Holmes EC (1996) A reevaluation of the higher taxonomy of viruses based on RNA polymerases. *Journal of Virology* 70(9): 6083–6096.

Feline Leukemia and Sarcoma Viruses

J C Neil, University of Glasgow, Glasgow, UK

© 2008 Elsevier Ltd. All rights reserved.

History

The description of murine leukemia viruses by Moloney and others stimulated an intensive search for similar viruses in other species. William Jarrett made the perceptive observation that lymphomas (lymphosarcomas) in cats often occurred at particularly high incidence in certain households, and in 1964 he showed that typical type C retroviruses could be demonstrated in the tumor cells by electron microscopy. He went on to show that these feline leukemia viruses (FeLVs) could be transmitted to cats where they induced lymphosarcomas and a range of degenerative diseases, including anemias and thymic atrophy.

Following these studies, Snyder and Theilen isolated a retrovirus from a feline fibrosarcoma that rapidly reproduced this tumor on inoculation into experimental cats. It is now recognized that feline sarcoma viruses (FeSVs) arise by recombination between FeLV and cellular protooncogenes and that, in contrast to FeLV, these viruses are not transmitted from cat to cat.

In clinical veterinary medicine, FeLV remains one of the most important viruses affecting the cat, despite advances in control of this infection. As a naturally occurring disease in an outbred host, FeLV has served as a paradigm for the natural history and molecular pathogenesis of the γ-retrovirus subfamily. It also played a foundational role in cancer genetics as a transducing agent which led to the discovery of novel cellular transforming genes.

Taxonomy and Classification

FeLVs are RNA viruses and belong to the family *Retroviridae*. They are further classified in the genus *Gammaretroviruses*.

Virion and Genome Structure

The virion particles are around 100 nm in diameter and consist of an outer membrane derived from the host cell surrounding a spherical core particle. The core encapsidates the viral genome which, as in other members of this viral family, is present as two linear single-stranded RNA molecules linked as a dimer. The virion RNA is positive-stranded and resembles cellular RNA having a 5′ cap and a 3′ poly(A) tract. As deduced from sequencing of proviral forms, the FeLV genome is around 8 kb long with a 67 base terminal redundancy and the gene order gag-pol-env.

The particles have surface spikes composed of multimers of the two env-coded proteins, the gp70 surface glycoprotein (SU), and the p15E transmembrane anchor protein (TM). Inside the envelope are the structural *gag* gene products which form a spherical core particle composed of p27, the major capsid protein (CA), with an outer layer formed by the p15 matrix protein (MA). Another gag product, the p10 nucleocapsid (NC), is associated with the virion RNA. Other minor virion proteins encoded by the *pol* gene comprise the protease (PR), reverse transcriptase (RT), and integrase (IN) enzymes.

Replication

Virus replication proceeds, following binding to specific host cell-surface receptors, internalization, and uncoating. The virion RNA is converted to a double-stranded DNA form by the virion RT which uses a proline tRNA primer and carries a ribonuclease H function that degrades the virion RNA. After nuclear translocation, viral integration is catalyzed by the IN protein and involves the creation of a staggered cut in cellular DNA with a consequent 4 bp duplication of host DNA at the insertion site.

As illustrated in **Figure 1**, the proviral form is flanked by long terminal repeats (LTRs) of 480–560 bp. These are generated during reverse transcription by duplication of unique sequences at the 3′ (U3) and 5′ (U5) ends of the RNA genome. The LTRs contain promoters and enhancers that drive transcription of viral RNA and also processing signals for cleavage and polyadenylation of the

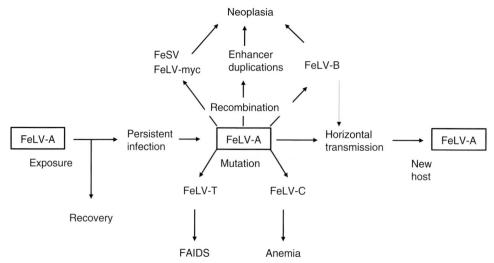

Figure 1 Life cycle of FeLV in its host, the domestic cat. FeLV-A is the most readily transmitted form and is found in 100% of isolates. In cats which become persistently viremic, this prototypic virus may evolve by recombination or point mutations to generate pathogenic variants that can lead to the rapid demise of the infected host animal. Multiple variants can arise in a single host. Few of these variants show any capacity for horizontal transmission to new hosts with the exception of FeLV-B. In this way, FeLV inflicts a substantial disease burden without significant reducing host numbers.

RNA transcripts. The 5′ LTR functions to initiate transcription while the 3′ LTR acts primarily as an RNA processing signal.

The virion RNA can function as a messenger RNA for Gag and Pol products, while a spliced subgenomic mRNA of around 3 kb encodes the Env products. Most full-length RNA translation products terminate at the 3′ end of *gag* to produce the Pr65gag precursor, while a small percentage read through into *pol* by misreading of an UAG termination codon, generating the Pr180Gag-pol precursor. During and after the budding process, the virion aspartyl PR catalyzes the cleavage of both precursor proteins to their mature forms. The envelope gene products are synthesized as a Pr80env precursor and processed by cellular PRs to the mature, disulfide-linked gp70 and p15E envelope proteins. The *gag* gene is also abundantly expressed in an alternative, glycosylated form via an upstream AUG codon. This product is expressed on the cell surface and shed after cleavage by cellular PRs. It is dispensable for *in vitro* virus replication but is highly conserved and may play a role *in vivo*.

Assembly of virus particles occurs at the cell surface by extrusion of cores which form at the budding site, concomitantly acquiring a host cell-derived outer membrane with virus-coded surface spikes. Virus replication and release is often noncytopathic.

Geographic Distribution

FeLV occurs worldwide in domestic cat populations although prevalence varies significantly and has declined in pet cats in areas where active control measures have been instigated. FeLV has also been isolated from the European wild cat (*Felis sylvestris*). Endogenous retroviral sequences closely matched to FeLV are found in the same species and in related small felids such as the sand cat (*F. margarita*) and the jungle cat (*F. chaus*). Although not a direct source of disease, these endogenous sequences can participate in recombination with exogenous FeLV to generate variant viruses with altered host range.

Epidemiology

The outcomes of FeLV infection fall into several categories, which vary in likelihood according to the age and immune status of the exposed host. The majority of cats undergo a transient infection lasting up to 3 months, during which they are viremic and shed virus. They then develop neutralizing antibody and concomitantly clear infectious virus and a little later, virus antigen from the blood stream. Some cats appear to clear virus infection successfully, while others may harbor latent virus in the bone marrow for some years. A recent study based on analysis of virus loads by polymerase chain reaction (PCR) suggests that animals that control viremia may be further subdivided into abortive, regressive, or latent infection. Even where latent infection persists, reactivation is not a frequent event and the vast majority of these cats do not develop an FeLV-related disease. However, at this stage virus can be reactivated by immunosuppression. The final group (persistent or progressive infection) remains actively infected, shedding virus from epithelial surfaces and displaying high titer plasma viremia. Such cats may remain apparently healthy for 2–3 years before succumbing to an FeLV-related disease.

The proportion of cases falling into these groups differs between multicat households and those households containing one or two free-ranging cats. In the former case, the introduction of an FeLV carrier results in repeated exposure of susceptible cats, often at a young age so that up to 30–40% become persistently viremic and at risk of disease. Approximately 50% of free-ranging urban and suburban cats have serological evidence of exposure to FeLV but only 1–5% of these cats are actively infected and the disease incidence is correspondingly lower.

Dual infection with FeLV and feline immunodeficiency virus (FIV) occurs and is associated with rapid disease onset, particularly if cats with preexisting FeLV infection encounter FIV. Rapid death of dually infected animals may reduce the apparent overlap of these agents in the field, but the populations at risk of infection also differ. FIV infection rates increase directly with age while persistent FeLV infection has a peak incidence in young cats.

FeLV Subgroups and Host Range

FeLV isolates were initially classified as subgroup A, B, C according to their viral interference properties in feline fibroblast cells *in vitro*. Viruses of a given subgroup prevent superinfection (interfere) with other viruses of the same subgroup (Table 1). This property is based on the use of three different host cell-surface receptors by FeLV-A, B, and C, and the blockade of these receptors in persistently infected cells by viral Env glycoproteins. The lymphotropic variant FeLV-T has more complex entry requirements, and does not replicate well in fibroblast cells. It appears that this isolate requires the FeLV-A receptor and an auxiliary mechanism in which a truncated *env* gene product encoded by endogenous FeLV sequences (FeLIX) is used as a co-receptor through binding to the FeLV-B receptor. Primary receptors for subgroups A, B, and C have been identified and shown to be transmembrane transporter molecules which the virus has subverted to gain entry to the host cell.

Natural isolates contain either subgroup A alone, or mixtures of subgroups A + B, A + C, or A + B + C. FeLV isolates of subgroup A are generally restricted to growth in feline cells, whereas subgroups B and C have a greatly expanded host range, infecting cat, human, mink, and canine cells. FeLV infection is generally noncytopathic and persistent and the virus is commonly propagated in long-term cultures of embryo-derived fibroblasts. Some strains such as FeLV-T or FeLV-C are cytopathic or induce apoptosis in lymphoid cells *in vitro*, reflecting their *in vivo* pathogenic properties.

Clinical Features and Pathology

Of those cats which become persistently viremic following FeLV exposure, over 80% die within 3.5 years. Most young cats infected with FeLV die from degenerative diseases rather than from tumors. Profound immunosuppression associated with thymic atrophy is a common finding in kittens. Other diseases seen in FeLV-infected cats include enteritis, immune complex glomerulonephritis, pancytopenia, and hemolytic anemia. Erythroid hypoplasia, an acute disease involving failure of red cell development past the burst-forming unit (BFU) stage, is specifically associated with FeLV subgroup C.

The most common neoplasm induced by FeLV is lymphosarcoma of T-cell origin, usually restricted to the thymus or sometimes occurring as a multicentric tumor in lymph nodes. The tumors often develop between 1 and 3 years after infection and the first signs may be chronic wasting and anemia. At presentation, the normal architecture of the lymphoid organ has usually been destroyed by a monomorphic infiltrate of lymphoblastic cells. The thymic tumors frequently display a rearrangement of the T-cell antigen receptor β-chain gene and may also express the co-receptor molecules CD4+ and/or CD8+. FeLV is also commonly associated with myeloid leukemias and a myelodysplastic syndrome-like disease, as well as with other forms of hematopoietic malignancy. Multicentric fibrosarcoma is a rare sequel to FeLV infection but these tumors are

Table 1 Properties of feline leukemia virus subgroups

Subgroup	Origin	Receptor	Function	Pathogenesis
A	Exogenous	feTHTR1?	Thiamine transport?	Minimally pathogenic to acute immunosuppression
B	Recombination FeLV-A × endogenous FeLV	Pit-1 (Pit-2)	Phosphate transport	More common in leukemic cats
				Some isolate-specific diseases, e.g., FeLV-GM1 myeloid leukemia
C	Mutation of FeLV-A (Env vrA)	FLVCR	Heme export	Erythroid hypoplasia
T	Mutation of FeLV-A (outside RBD)	? FeLV-A receptor +FeLIX, Pit-1		Acute immunosuppression

often associated with the *de novo* generation of a FeSV. FeLV is also found in association with 35% of alimentary tumors, primarily of B cell origin, but the virus is not always clonally integrated in these tumors and the role of the virus is, therefore, unclear. FIV can also increase the frequency of malignant diseases in cats, and some rare cases arise on a background of dual infection.

Envelope Gene Variation and Pathogenicity

The common infectious form of FeLV is FeLV-A which is a remarkably highly conserved virus as shown by sequence analysis of several strains and serotypic analysis of a much larger number. FeLV variants frequently arise from FeLV-A by mutation and recombination, and such variants are often implicated in the acute diseases which develop in persistently infected cats. The variant viruses generated from FeLV-A are generally dependent on the continued presence of the prototype for their propagation *in vivo*. However, as the variants are less efficiently transmitted and are in some cases rapidly fatal, they tend to die out with the host while the prototypic FeLV-A continues to colonize new hosts (**Figure 1**).

The most commonly isolated FeLV recombinants are subgroup B viruses (**Table 1**). These are derived by recombination between FeLV-A and endogenous FeLV-related proviruses which are found in the genome of the domestic cat and related small feline species. Although the endogenous FeLV-related proviruses all appear to be replication defective, their envelope genes can be rescued by the recombination process leading to the generation of FeLV-B viruses. FeLV-B can infect cells refractory to, or already containing, FeLV-A by virtue of their distinctive receptor specificity.

The anemia-inducing FeLV-C isolates are rarer and appear to be derived from FeLV-A by mutation of a single variable domain (VRA) of the *env* gene. The acute disease properties of these variants appear to be due to compromised viability of erythroid progenitor cells due to downregulation of the FeLV-C receptor, a vital heme exporter.

Minor *env* mutations also appear to give rise to the acutely immunosuppressive FeLV-FAIDS variants. The prevalence of acutely immunosuppressive viruses in nature is unknown, but immunosuppressive disease is a common manifestation of FeLV infection.

The relationship of subgroup variation to oncogenesis is complex. FeLV-B recombinants are more common in tumor bearing than in infected asymptomatic cats. This higher frequency might reflect merely longer-standing infection, but some FeLV-B-containing isolates have an altered spectrum of neoplastic disease. For example, FeLV-GM1, which contains a replication-defective FeLV-B component, induces mainly myeloid leukemia.

FeLV Oncogenesis: Virus Evolution and Mutagenesis of Cellular Oncogenes

Two modes of virus-induced host gene mutation have been described in FeLV-associated cancers. The first is 'transduction', where recombination leads to the generation of an acutely oncogenic variant in which viral gene sequences are replaced by a host-derived insert. Such viruses are replication defective and are found in nature in association with a replication-component FeLV helper virus.

Multicentric fibrosarcomas of young cats are relatively rare, but are generally FeLV positive and frequently involve a novel sarcoma virus. Similarly, transduction of c-*myc* has been observed in up to 20% of naturally occurring thymic lymphosarcomas in FeLV positive cats. In all, nine different host cell genes have been shown to be transduced by FeLV (**Table 2a**). The transducing viruses induce tumors with short latency in cats and in the case of FeSVs may transform cells in tissue culture.

Alternatively, host genes can be affected by proviral 'insertional mutagenesis' (*cis*-activation). Four known oncogenes and an uncharacterized novel integration locus have been identified as common tumor-specific insertion

Table 2a FeLV gene transduction in neoplasia

Gene	Normal function of host gene product	Associated tumor	Examples[a]
abl	Plasmamembrane protein kinase	Fibrosarcoma	FeSV-HZ2
fes	Plasmamembrane protein kinase	Fibrosarcoma	FeSV-GA,-HZ1,-ST
fgr	Plasmamembrane protein kinase	Fibrosarcoma	FeLV-GR,-TP1
fms	Receptor protein kinase (CSF-1 receptor)	Fibrosarcoma	FeSV-SM,-HZ5
kit	Receptor protein kinase (SCF receptor)	Fibrosarcoma	FeSV-HZ4
myc	Transcription factor	T-cell lymphoma	FeLV (T3, T17, FTT)
Notch2	Transmembrane receptor	T-cell lymphoma	(Inoculum FeLV-61E)[a]
sis	Growth factor (B chain PDGF)	Fibrosarcoma	FeSV-PI
tcr	T-cell antigen receptor (β-chain)	T-cell lymphoma	FeLV-T17

[a]Isolated from naturally occurring tumors apart from the indicated exception.

Table 2b Insertional mutagenesis and FeLV oncogenesis

Common integration site	Gene function	Tumor
fit-1	Transcription factor (c-myb)	T-cell lymphoma
flvi-1	Unknown	non-T, non-B lymphoma
flvi-2 (bmi-1)	Transcription factor	T-cell lymphoma
c-myc	Transcription factor	T-cell lymphoma
pim-1	Protein kinase	T-cell lymphoma

sites for FeLV in thymic lymphosarcomas (**Table 2b**). In this respect FeLV oncogenesis appears remarkably similar to that of the murine γ-retroviruses, and most studies of this process have been conducted recently in the laboratory mouse which has the advantages of complete genome sequence and the opportunity to manipulate the germline.

Changes within the LTR are also a feature of tumor-associated FeLV. In thymic lymphosarcomas, sequence duplications of the core enhancer domain are frequently found, and have been shown to arise *de novo* from infection with molecularly cloned FeLV isolate lacking such features. By analogy with the murine oncoretroviruses, the duplications are likely to increase the oncogenicity of the virus and reduce the latent period for tumor formation, possibly by increasing the potency of viral enhancer activity on nearby cellular promoters. There is evidence that these adaptive changes to the LTR operate tissue-specifically and proviruses carrying different duplications of LTR regions 5' and 3' to the core-enhancer region have been identified in myeloid leukemias and non-T, non-B splenic lymphomas, respectively. Chimeric murine retroviruses carrying the FeLV enhancer region have been generated, confirming that the tissue specificity is carried by this structure.

Immune Response

Unlike infection with the lentiviruses HIV and FIV, FeLV infection of cats may lead to recovery. There is evidence that both virus-specific cytotoxic T cells (CTLs) and neutralizing antibodies play a role in resistance as either can be used to prevent or limit infection by passive transfer. In natural infection, CTLs to virus structural protein epitopes can be detected as early as a week after infection and may be the primary means of control of infection as neutralizing antbodies are not detected until around 6 weeks post infection.

In the early literature on FeLV a distinction was made between antiviral immunity and antitumor immunity. Cats with antitumor (FOCMA) antibody were thought to be protected from tumor development. This antibody response is now believed to be directed to endogenous FeLV *env* proteins and its role in modulating tumor development is uncertain.

Transmission

Cats persistently infected with FeLV shed virus in their saliva, urine, and feces but, as the virus is fragile, close contact is required for transmission. The most frequent routes involve saliva and transplacental spread. Kittens infected *in utero* become persistently infected, but the consequences of infection in older cats depend on a number of factors. There is an age-related resistance to infection such that cats up to 12 weeks of age are highly susceptible, but above 16 weeks they are difficult to infect either naturally or experimentally.

FeLV subgroup A is always found in field isolates and about half also contain FeLV-B, whereas FeLV-C is present in only 1–2% of isolates. Although FeLV-B can arise *de novo* by recombination, it may also be transmitted between cats. This occurrence is dependent on pseudotype formation in which the genome of the B virus becomes enclosed in an envelope containing glycoproteins of the A subgroup.

Prevention and Control

Successful control measures can be adopted in multicat households by removing or isolating persistently infected animals. Productively infected cats are detected by virus isolation from plasma or more usually by enzyme-linked immunosorbent assay (ELISA) for virus antigen in the blood. A few cats remain persistently antigenemic but nonviremic. These cats do not usually transmit the virus unless they are shedding virus in the milk or saliva. Assays are conducted twice, 3 months apart to exclude cats that are transiently viremic.

Numerous vaccine strategies have been shown to offer protection against FeLV in laboratory conditions (**Table 3**) and FeLV was the first retrovirus for which commercial vaccines were developed. Vaccines in current use include whole inactivated virus preparations, subunit vaccines from recombinant viral Env protein expressed in bacterial cells, and a canarypox recombinant virus expressing Gag and Env. These vaccines offer a measure of protection against experimental challenge and are under evaluation for longer-term efficacy in the field. These vaccines do not generate sterilizing immunity but appear to prime the immune system to favor clearance of virus infection instead of persistent viremia.

Table 3 FeLV vaccines

Vaccine	Protection	Commercial use
Live attenuated FeLV	Yes	No (safety concerns)
Inactivated whole virus	Yes	Yes
Subunit vaccines		
SU from *E. coli*	Yes	Yes
ISCOM-Env (native)	Yes	No
Lymphoma cell extract	Yes (poor in some studies)	Withdrawn
Live vector vaccine		
Vaccinia-Env	No	No
Canarypox-Env-Gag	Yes	Yes
Feline herpesvirus-Env	Partial	No

Future Perspectives

There is continuing interest in control of FeLV infection due to its importance in veterinary medicine. In the future we can look forward to improvements in vaccine efficacy and further dissection of the host responses that confer protection. While the focus of attention of cancer genetics has moved on to more easily manipulated models, FeLV remains as a useful touchstone for our understanding of retroviral pathogenesis in an outbred, naturally infected host. Also, with the impetus of FIV as a model for human AIDS, the generation of reagents to probe the feline immune system offers new opportunities for comparative study of FeLV.

See also: Bovine and Feline Immunodeficiency Viruses.

Further Reading

Flynn JN, Dunham SP, Watson V, and Jarrett O (2002) Longitudinal analysis of feline leukemia virus-specific cytotoxic T lymphocytes: Correlation with recovery from infection. *Journal of Virology* 76: 2306–2315.

Hanlon L, Barr NI, Blyth K, *et al.* (2003) Long-range effects of retroviral activation on c-myb over-expression may be obscured by silencing during tumor growth *in vitro*. *Journal of Virology* 77: 1059–1068.

Mendoza R, Anderson MM, and Overbaugh J (2006) A putative thiamine transport protein is a receptor for feline leukemia virus subgroup A. *Journal of Virology* 80: 3378–3385.

Miyazawa T (2002) Infections of feline leukemia virus and feline immunodeficiency virus. *Frontiers in Bioscience* 7: D504–D518.

Roca AL, Nash WG, Menninger JC, Murphy WJ, and O'Brien SJ (2005) Insertional polymorphisms of endogenous feline leukemia viruses. *Journal of Virology* 79: 3979–3986.

Sparkes AH (2003) Feline leukaemia virus and vaccination. *Journal of Feline Medicine and Surgery* 5: 97–100.

Torres AN, Mathiason CK, and Hoover EA (2005) Re-examination of feline leukemia virus: Host relationships using real-time PCR. *Virology* 332: 272–283.

Tsatsanis C, Fulton R, Nishigaki K, *et al.* (1994) Genetic determinants of feline leukemia virus-induced lymphoid tumors: Patterns of proviral insertion and gene rearrangement. *Journal of Virology* 68: 8294–8303.

Filamentous ssDNA Bacterial Viruses

S A Overman and G J Thomas Jr., University of Missouri – Kansas City, Kansas City, MO, USA

© 2008 Elsevier Ltd. All rights reserved.

Glossary

Cloning vector The DNA molecule of a virus, plasmid, or cell into which a foreign DNA fragment can be integrated without loss of self-replicating activity. The vector introduces the foreign DNA fragment into an appropriate host cell for autonomous replication, usually in large quantity.

Phage library An ensemble of up to about 10^{10} phage clones, each harboring a different foreign coding sequence in-frame with either the N- or C-terminal region of a capsid protein gene. The clone thus allows display of a different 'guest' peptide on the virion surface.

Raman spectroscopy The branch of optical spectroscopy (named after its 1928 founder, C. V. Raman) concerned with measurements of the intensities of light scattered inelastically by molecules that have been excited by monochromatic radiation. The resulting spectrum, usually a plot of scattering intensity (in arbitrary units) versus energy (in wave number or cm^{-1} units), reflects the transfer of discrete energy quanta from the impingent photons to vibrational energy states of the molecules. The Raman spectrum, which is determined by both intramolecular bonding arrangements (covalency and conformation) and intermolecular interactions, provides a sensitive

signature of molecular structure and local environment.

Trans-envelope network A multiprotein complex that is located in the envelope of a bacterial cell and brings the inner and outer membranes in close proximity to one another.

Ultraviolet resonance Raman spectroscopy (UVRR) A type of Raman spectroscopy (see above) in which the wavelength of the exciting monochromatic radiation is in the ultraviolet region (i.e., wavelength $\lambda < 400$ nm, or wave number $\sigma > 25\,000$ cm^{-1}), so as to achieve resonance with electronic absorption processes of the molecules. In UVRR spectra, only the vibrational states of the chromophore are represented.

Introduction

The filamentous ssDNA viruses (genus *Inovirus*) are members of a genus of morphologically similar virions that infect different bacteria via molecular recognition of a host-specific pilin. The most well studied of these phages are the closely related M13, fd, and f1 virions, which infect *Escherichia coli* displaying a conjugative F-pilus. The genome sequences of these F-specific phages are sufficiently similar that they are collectively called Ff phage. Other filamentous phages that have been studied include IKe, which infects *E. coli* displaying a conjugative N-pilus, Pf1, which infects *Pseudomonas aeruginosa* strain PAK by binding to the bacterial type IV pilus, Pf3, which infects *P. aeruginosa* strain PAO by binding not to the inherent type IV pilus but to the conjugative RP4 pilus, and PH75, which infects *Thermus thermophilus*. These filamentous phages are nonlytic and nonlysogenic. On the other hand, the filamentous ssDNA bacteriophage CTXφ, which infects *Vibrio cholera* by recognition of a toxin-co-regulated pilus (TCP), is lysogenic.

Genetic, biochemical, and biophysical methods have been used to study the Ff bacteriophage life cycle, which is unique in its use of the bacterial cell envelope for virus assembly. Features of the nonlytic Ff assembly pathway are (1) the prolific production of phage particles, (2) the extrusion of progeny virions through the cell envelope, and (3) the requirement of stable transmembrane (TM) domains in virally encoded proteins that participate in the assembly process. Also noteworthy in Ff morphogenesis is that genomes of variable size can be packaged without deleterious consequences. For example, although the single-stranded DNA (ssDNA) genome of wild-type Ff (~6400 nt) is sheathed by 2750 copies of the major capsid protein, viable variants have been isolated containing as many as 12 000 nt and corresponding modifications in the filament length and number of subunits in the sheath.

Biophysical studies have shown that the Ff phage particles are highly thermostable and highly flexible. Polymorphism of the viral capsid is also revealed by electron cryomicroscopy.

The life cycle of the filamentous ssDNA bacteriophage enables its use as a model for membrane-associated nucleoprotein assembly and as a valuable tool for molecular cloning, phage display, and pharmacotherapy. The filamentous ssDNA bacteriophage also serves as a vehicle for orientation of small molecules in solution spectroscopic (nuclear magnetic resonance, NMR) applications and as a model for nanowire self-assembly. Details of the molecular structure and life cycle of the filamentous ssDNA bacteriophage are considered in this article.

Taxonomy and Classification

The filamentous ssDNA bacterial viruses, which are visualized in electron micrographs as thin cylindrical filaments about 7 nm in diameter and ranging from 700 to 2000 nm in length, belong to the genus *Inovirus* of the family *Inoviridae*. They are distinguished from rod-shaped members of the genus *Plectrovirus* of the same family (typically 85–280 × 10–15 nm dimensions) by their greater contour length, smaller diameter, and lower flexural rigidity. The many species of the genus *Inovirus* so far identified have been categorized into four broad groups on the basis of the types of bacteria infected. The species encompassed by these four groups are listed in **Table 1**, in accordance with the classification of the International Committee on Taxonomy of Viruses (ICTV). The best characterized of the filamentous viruses with respect to both biological and structural properties is the coliphage M13, which serves as the prototype of the species *Enterobacteria phage M13*. This species also includes the closely related phages f1, fd, AE2, dA, Ec9, HR, and ZJ/2.

Physical Properties

Table 2 lists selected physical properties of several well-studied filamentous ssDNA bacterial viruses. Although each of the viruses included in **Table 2** exhibits a characteristic filamentous shape, significant differences occur in their contour lengths and in the mass ratio of capsid protein to DNA. Accordingly, differences are also anticipated in the packing arrangement of capsid subunits with respect to the encapsidated genome. The overall length of the viral filament is dictated by both the genome size and the number of subunits required to electrostatically balance the negatively charged DNA phosphates, which is accomplished by the distribution of basic side chains (Lys and Arg) near the subunit C-terminus. On the basis of fiber X-ray diffraction results, two distinct symmetry classes (I and II) have been

Table 1 Species of the genus *Inovirus*[a]

Host type	Species
Enterobacteriaceae	*Enterobacteria phage M13,[b] C-2, If1, IKe, I2–2, PR64FS, SF, tf-1, X,* and *X-2*
Spirillaceae	*Vibrio phage 493, fs1, fs2, CTX, v6, Vf1, Vf33,* and *VSK*
Pseudomonadaceae	*Pseudomonas phage Pf1, Pf2,* and *Pf3*
Xanthomonadaceae	*Xanthomonas phage Cf16, Cf1c, Cf1t, Cf1tv, Lf, Xf, Xfo,* and *Xfv*

[a]The species are categorized by host type.
[b]Enterobacteria phages fd, f1, and several others have been classified as a type of *Enterobacteria phage M13*.

Table 2 Physical properties of filamentous ssDNA bacteriophages

Property	M13	Pf1	Pf3	Xf	PH75[a]
Virion length (nm)	880	1900	680	980	910
Percent mass protein	87	93	85	85	84
Capsid subunits per virion	2750	7370	2500	3500	2700
Residues per subunit	50	46	44	44	46
Nucleotides per genome	6407	7390	5830	7420	6760
Nucleotides per subunit	2.3	1.0	2.3	2.1	2.5
λ_{max} (nm)	269	270	264	263	267
Extinction coefficient ($cm^2\ mg^{-1}$)	3.84	2.07	4.53	3.52	3.77
Symmetry class[b]	I	II	II	II	II

[a]The thermophilic filamentous ssDNA bacteriophage, PH75, which infects *Thermus thermophilus*, has not yet been assigned to a species, genus, or family by the International Committee on Taxonomy of Viruses.
[b]Class I phages have capsid subunits arranged with C_5S_2 symmetry (fivefold rotational symmetry and approximately twofold screw axes) and class II phages have capsid subunits arranged with $C_1S_{5.4}$ symmetry (no rotational symmetry and a single-start superhelical array of approximately 5.4 subunits per turn).

defined to categorize subunit/DNA packing arrangements: subunits of the class I phages (e.g., M13, f1, and fd) are arranged with approximate C_5S_2 symmetry, while class II phages (e.g., Pf1, Pf3, and PH75) exhibit $C_1S_{5.4}$ symmetry. Further details of virion structure are discussed in the section titled 'Composition and structure'.

Genome Organization

The genomes of many filamentous viruses, including M13, f1, fd, IKe, Pf1, Pf3, PH75, and CTX, have been sequenced. Comparative analyses indicate that at least those of M13, f1, fd, IKe, and PH75 are similarly organized, as shown in **Figure 1** for fd.

Of the 11 virally encoded Ff genes, two (X and XI) overlap and are in-frame with larger genes (II and I, respectively). The Ff genome contains only one significant noncoding region (intergenic region, IG) consisting

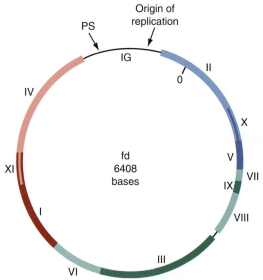

Figure 1 The genome organization of the Ff (fd) bacteriophage. All genes (I–XI) are labeled and color-coded by function of the gene product (blue, DNA synthesis; green, capsid; red, virus assembly). The single-stranded DNA genome of fd contains 6408 nt, which are numbered clockwise from the unique *Hind*II site located in gene II. The intergenic region (IG) is located between genes II and IV and includes the packaging signal (PS) and the origin of replication. Adapted from Petrenko VA and Smith GP (2005) Vectors and modes of display. In: Sidhu S (ed.) *Phage Display in Biotechnology and Drug Discovery*, pp. 63–110. Boca Raton, FL: CRC Press.

of about 500 nt. The IG contains the origin of replication, as well as a functionally and structurally distinct region called the packaging signal (PS), which is capable of forming a hairpin of 78 nt. During the phage life cycle, recognition of the genomic PS by virally encoded proteins initiates assembly of the viral particle. The viral genes are expressed from the origin of replication in a counterclockwise direction and are organized on the genome in functional groups, such that genes II, V, and X are grouped together and express proteins that facilitate DNA replication; genes III, VI, VII, VIII, and IX are grouped together and express structural proteins found in the mature infectious virion; and genes I, IV, and XI are grouped together and express proteins that direct virus assembly. The CTX genome lacks gene IV but contains genes that encode proteins specific to the CTX life cycle.

Composition and Structure

The typical filamentous virion consists of a covalently closed, ssDNA genome (5–10 kbp) sheathed by several thousand copies of the major capsid protein (pVIII, ~5 kDa) plus a few copies of minor proteins at the filament ends (**Figure 2**). The numbers of pVIII subunits and DNA nucleotides in several well-studied filamentous phages are given in **Table 2**. The pVIII subunits of the *Enterobacteria phage M13* species members (including phages M13, f1, and fd) have been studied extensively and can be discussed collectively. (Note that the amino acid sequence of the pVIII subunits in fd and f1 are identical and differ from that of M13 by a single amino acid change, namely Asp12 of fd (and f1) is replaced by Asn12 in M13; see **Table 3**.) It is evident from **Table 3** that the pVIII sequence in many filamentous viruses comprises three distinct components, an acidic N-terminal region, a hydrophobic central region, and a basic C-terminal region, suggesting that these fulfill distinct functions in phage morphogenesis.

Here, we consider in further detail the composition and structure of phage fd, which has served as the focus of many detailed and comprehensive biochemical and biophysical studies. The pVIII subunit is generally modeled as a gently curved α-helix that is tilted at a small average angle ($16 \pm 4°$) relative to the filament axis. The N-terminus of pVIII is exposed on the exterior of the capsid and the C-terminus lines the core, where it is presumed capable of contacting phosphates of the packaged ssDNA. Fiber X-ray diffraction studies indicate that pVIII subunits of the capsid lattice are arranged in a right-handed slew with near-fivefold rotational symmetry and twofold screw symmetry (**Figure 2**). Five copies of each of four minor proteins form the ends of the fd filament. The initially assembled end ('head') contains minor proteins pVII (3.6 kDa) and pIX (3.6 kDa), while the ultimately assembled end ('tail') contains minor proteins pIII (42.6 kDa) and pVI (12.3 kDa). The packaged genome is oriented with its IG region (PS hairpin) at the head, gene III region near the tail, and the intervening non-base-paired antiparallel strands of the DNA loop spanning the length of the filament core (**Figure 2**). Phage display and genetic studies suggest further that the minor proteins pVII and pVI at the head and tail, respectively,

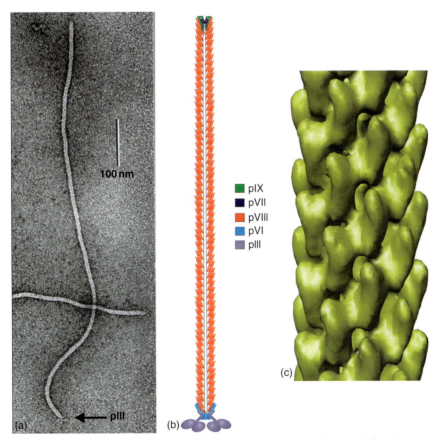

Figure 2 The fd filamentous ssDNA bacteriophage. (a) A negatively stained electron micrograph showing one complete virion (top to bottom) oriented with the pIII tail (arrow) at the bottom. The flexibility of the virion, which is apparent, is consistent with the experimentally determined equilibrium persistence length of 1.0 μm. (b) Cartoon of fd (not to scale) showing locations of the minor coat proteins. (c) Reconstruction of a 24 nm segment of fd at ~8 Å resolution from images obtained by electron cryomicroscopy. (a) Courtesy of Carla W. Gray, University of Texas, Dallas. (c) Courtesy of E. H. Egelman, University of Virginia.

Table 3 pVIII sequences for filamentous ssDNA bacteriophages

Phage	pVIII sequence[a]
M13	¹AEGDDPAKAAFNSLQASATEYIGYAWAMVVVIVGATIGIKLFKKFTSKAS⁵⁰
Pf1	¹GVIDTSAVESAITDGQGDMKAIGGYIVGALVILAVAGLIYSMLRKA⁴⁶
Pf3	¹MQSVITDVTGQLTAVQADITTIGGAIIVLAAVVLGIRWIKAQFF⁴⁴
Xf	¹SGVGDGVDVVSAIEGAAGPIAAIGGAVLTVMVGIKVYKWVRRAM⁴⁴
PH75	¹MDFNPSEVASQVTNYIQAIAAAGVGVLALAIGLSAAWKYAKRFLKG⁴⁶

[a]The hydrophobic segment of each pVIII sequence is underlined.

form contacts with the neighboring pVIII subunits, while pIX and pIII decorate the surfaces of the filament tips. Each pIII subunit is known to contain a globular N-terminal domain that is visible in electron micrographs (**Figure 2**).

In the mature fd virion, pVIII is predominantly α-helical and the C-terminal end of the α-helix (residues 40–50) consists of 'basic' and nonpolar side chains along opposite faces. Although direct experimental evidence is lacking, model-building exercises have revealed that it is possible to position the positively charged side chains of this amphipathic helix proximal to the surface of the capsid core, where interaction with packaged DNA would be facilitated. The net positive charge of the C-terminal region apparently functions to neutralize the negatively charged DNA phosphates. This is supported by the fact that mutating Lys 48 of pVIII to an uncharged side chain results in the assembly of an appropriately elongated (~35%) fd virion; that is, additional pVIII subunits are required to package the genomic complement. The opposite nonpolar (hydrophobic) face of the C-terminal helix provides a suitable surface for intersubunit contacts.

The N-terminal region of pVIII (residues 1–20) also forms an amphipathic helix, in this case with 'acidic' and nonpolar faces. The former is exposed on the capsid surface, while the latter is presumably engaged in inter-subunit contacts. In aqueous environments of low or moderate ionic strength, electron micrographs reveal extensive intervirion clustering, which may be mediated by the acidic (and/or polar) residues of the capsid surface. In some micrographs, liquid-crystalline phase formation is evident. Although high ionic strength conditions tend to disfavor such intervirion clustering, the bundles once formed are not readily dissociated.

The highly hydrophobic central region of the pVIII α-helix (residues 21–39) appears to serve as the pillar of capsid stability. This α-helical segment also appears to function as a distinct domain in the assembly process, as is discussed in the next section. Studies based upon site-directed and random mutations of residues throughout the central α-helix region have contributed significantly to an understanding of inter-subunit contacts in both the mature virion and its assembly precursors. As implied above, hydrophobic contacts between subunits involve not only the central region of the α-helix but also the nonpolar faces of the N- and C-terminal segments. The periodicity of small nonpolar side chains (Gly, Ala) along specific faces of the α-helix facilitates close packing of subunits. For example, small side chains appear to be required at pVIII positions 25, 34, 35, and 38 for fd capsid stability. A similar requirement applies to the capsid subunits of other filamentous bacteriophages.

Genetic and biophysical studies of proteins show that the pentapeptide motif Gly/Ala-X-X-X-Gly/Ala (where X is any residue) increases the thermostability of packed α-helices. Of the sequences listed in **Table 3**, all contain at least one Gly/Ala-X-X-X-Gly/Ala motif within the central hydrophobic region (residues 21–39). Interestingly, the corresponding segment of the pVIII subunit of the thermophilic *T. thermophilus* phage PH75 contains four Gly/Ala-X-X-X-Gly/Ala motifs among 11 small side chains (residues Ala and Gly). Conversely, the mesophile infecting fd phage contains two Gly/Ala-X-X-X-Gly/Ala motifs among six small side chains of the 19-peptide region. In summary, hydrophobic interactions lead to close packing of pVIII helices, which is likely the principal source of capsid stability.

The multifunctional minor capsid protein pIII, which directs host infection, progeny virus assembly termination, and virion tail stabilization, has been the subject of several structural studies. The 406-amino-acid protein comprises three domains, namely two closely associated and globular N-terminal domains (N1 and N2, consisting of residues 1–67 and 88–218, respectively) that protrude from the capsid, and a stalk-like C-terminal domain (residues 253–406) that is anchored to the capsid. The domains are linked by flexible, glycine-rich sequences. The crystal structure of the N1–N2 fragment indicates a largely β-stranded structure.

The conformation of the packaged genome is not known for fd or any other filamentous phage. Because DNA typically represents a very small percentage of the virion mass (6–16%, **Table 2**), most biophysical probes do not yield definitive structural information regarding the packaged genome. Methods that selectively probe the ssDNA bases, such as ultraviolet resonance Raman (UVRR) spectroscopy, reveal strong hypochromic effects in the fd genome, which implies close contact between the purine and pyrimidine bases. Conversely, UVRR studies of Pf1 suggest unstacked bases in its packaged genome.

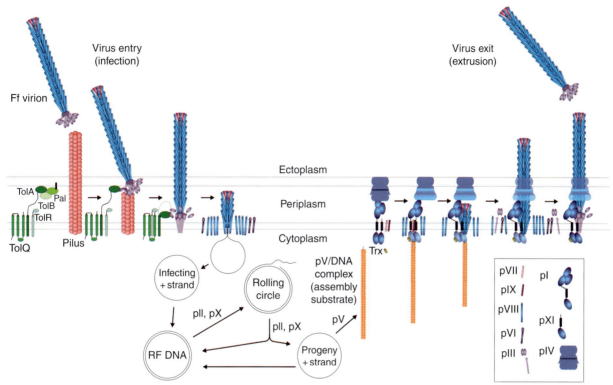

Figure 3 Schematic representation of the Ff life cycle. In the periplasmic space, the arrows show progressive stages of disassembly (left) and assembly (right) as discussed in the text. Viral and bacterial components are not drawn to scale (e.g., the actual dimensions of the Ff virion are 6.5 × 880 nm). Pilin biogenesis proteins are not shown.

Raman spectroscopy also indicates very different local deoxynucleoside conformations in the packaged ssDNA molecules of phages fd and Pf1.

Life Cycle

Infection

The mechanism of host cell infection by a filamentous bacteriophage has been studied in detail only for the bacteriophage Ff, which recognizes the conjugative F pilus of *E. coli* cells harboring the F plasmid. The process involves a trans-envelope network of host proteins – the Tol–Pal system – that is located in the *E. coli* cell envelope and serves to maintain the stability of the outer membrane. The Tol–Pal system includes the cytoplasmic membrane proteins TolA, TolR, and TolQ, which are anchored to the inner membrane, and the proteins TolB and Pal, which are associated with the outer membrane. TolA consists of an N-terminal TM domain, a periplasmic spanning α-helix, and a C-terminal globular domain also located in the periplasmic space. These domains are linked by glycine-rich segments. Like TolA, TolR has TM and periplasmic domains, whereas TolQ lacks a periplasmic domain and is located predominantly in the cytoplasm. TolA, TolR, and TolQ are associated with one another via their TM domains and all are necessary for viral infectivity. TolB is a periplasmic protein and Pal is a lipoprotein that is anchored to the outer membrane. Interaction of TolA with both TolB and Pal connects the outer and inner membranes of the *E. coli* cell. In the initial step of infection (**Figure 3**), the N2 domain of pIII binds the F-pilus, inducing a conformational change in N1–N2 that exposes a site of the N1 domain specific for TolA binding. The conformational change in N1–N2 allows a fast *cis–trans*-isomerization of the peptide bond between Gln212 and Pro213. When the peptide bond is in the more stable *trans*-form, the TolA binding site is exposed. Isomerization to the initial *cis*-form is a slow reaction, the rate of which is determined by the sequence of amino acids flanking Pro213. The rapid isomerization of the Gln212–Pro213 bond to the *trans*-form is a molecular switch controlled by the slower reverse reaction, which acts as a molecular timer. During the time that the N1 TolA binding site is exposed, the F-pilus is believed to retract through a secretin channel in the outer membrane of the host cell, by depolymerization of pilin, thus translocating pIII into the periplasm. The N1 domain of pIII binds the periplasmic C-terminal domain of TolA, the second phage receptor, forming a bridge between the adsorbed phage particle and the bacterial inner membrane. Subsequently, the C-terminal domain of pIII directs continuation of the infection process. Genetic studies suggest that the TolA–N1 interaction causes a conformational change in the pIII C-terminal domain, which 'unlocks' the capsid. Once unlocked, a TM region of the pIII C-terminal domain likely interacts with the inner membrane to initiate dissociation of the capsid proteins, pVIII, pVII, and pIX, into the inner membrane with concomitant release of the viral genome into the cytoplasm. The capsid proteins of the infecting virion are retained in the membrane for use in the assembly of progeny virions.

DNA and Protein Synthesis

Upon entry into the cytoplasm, the ssDNA viral genome is converted into a replicative double-stranded form (RF) by the activity of host-encoded polymerases (RNA Pol and DNA Pol III) and a host-encoded single-stranded DNA-binding protein (SSB). To initiate this event, Pol synthesizes a short RNA primer beginning at nucleotide 5753. Pol III extends this primer the length of the viral strand, and Pol I and ligase close the complementary strand to produce a circular, dsDNA molecule. Gyrase supercoils the circle, thus resulting in the RF DNA, which serves as the template for transcription of viral genes.

Synthesis of ssDNA viral molecules requires the virally encoded protein pII, which specifically nicks the viral strand ((+)-strand) of RF DNA. The resultant 3′ hydroxyl group is used by the host DNA synthesizing apparatus (Pol III, SSB, and Rep helicase) to generate a new (+)-strand by the rolling-circle mode of replication, thus displacing the parent (+)-strand, which is subsequently circularized by pII. The newly synthesized RF DNA is closed and supercoiled for additional rounds of replication. The displaced and circularized (+)-strand can serve as either a substrate for additional RF DNA formation (template for (−)-strand synthesis) or a virus genome (encapsidation). At early stages of infection, the former path is more likely; at later stages, as the concentrations of virally encoded proteins become large, encapsidation is favored.

The (−)-strand serves as the template for transcription of viral genes, beginning at gene II. The presence of two strong terminators, one of which is rho-independent between genes VIII and III and the other rho-dependent in the IG, divides the genome into regions that are relatively frequently (genes II–VIII) and infrequently (gene III–IG) transcribed. The former has three strong promoters that lead to three primary transcripts. Post-transcriptional modification of these primary transcripts results in a set of six smaller and more stable mRNA molecules, all of which encode pVIII and a subset (four) of which encode pV. Multiple transcripts facilitate synthesis of the large amounts of pV and pVIII required to coat the prolifically produced (+)-strands (up to 200–300 progenies per cell). The frequently transcribed region also includes genes for the minor capsid proteins pVII and pIX. However, the translational initiation sites for synthesis of these proteins are weak. Accordingly, pVII and pIX are not produced in abundance. Proteins pII and pX, which likewise are encoded in the frequently transcribed region but not required in abundance, are regulated at the translational level via pV binding to the mRNA transcripts. This inhibition is evident at high cellular concentrations of pV, when amounts of the protein exceed the threshold required to sequester (+)-strands targeted for viral assembly.

Transcription of the genome and translation of the mRNA transcripts produces 11 viral proteins, of which pII, pX, and pV are required for (+)-strand synthesis, pIII, pVI, pVII, pVIII, and pIX are incorporated into the capsid, and pI, pIV, and pXI facilitate membrane-associated assembly. pII (409 residues) is a strand-specific endonuclease required for (+)-strand and RF syntheses; pX (111 residues) is synthesized from an AUG site within and in-frame with the C-terminal coding region of pII; pV (87 residues) forms a stable homodimer, the crystal structure of which has been solved to 1.8 Å and gives insight into the mechanism of binding of antiparallel ssDNA strands.

Each of the five capsid proteins of Ff exhibits a sequence consistent with an α-helical TM domain. All reside in the cytoplasmic membrane prior to assembly into the virion. Nascent pVIII contains a 23-residue signal sequence, which is removed from the N-terminus by a periplasmic signal peptidase, yielding the 50-residue subunit of the mature capsid. The N-terminal segment (1–19) of the mature subunit resides on the periplasmic face of the inner membrane, the middle 20 or so residues span the membrane, and the C-terminal segment (40–50) is situated in the cytoplasm. Although Sec proteins of E. coli ordinarily facilitate membrane insertion and translocation, these processes for pVIII are Sec-independent and mediated instead by the host-encoded protein YidC. Maturation of the pIII precursor, which contains an 18-residue N-terminal signal, is also accomplished following membrane translocation. A C-terminal sequence of 23 hydrophobic residues anchors pIII to the inner membrane, while most of the protein is located in the periplasmic space. The three remaining minor capsid proteins, pVI (112 residues), pVII (33 residues), and pIX (32 residues), are membrane-inserted without the aid of a signal sequence.

The three morphogenetic proteins of Ff are also located in the bacterial cell envelope. pI (348 residues) contains an internal TM region and spans the inner membrane with residues 1–253 in the cytoplasm and 273–348 in the periplasm. pXI (108 residues), which is synthesized from within the pI coding region, shares the sequence of the pI C-terminal segment. Accordingly, pXI is anchored to the membrane by a short N-terminal TM region, while most of the protein resides in the periplasm. The only viral protein located in the outer membrane is pIV. The 426-residue precursor is translocated into the periplasm where a 21-residue signal sequence is removed. The resulting 405-residue protein is integrated into the outer membrane as an oligomer of 14 subunits, each with its N-terminus in the periplasm. A cryoelectron microscopy-based image reconstruction of detergent solubilized pIV at 22 Å resolution reveals a barrel-like cylindrical complex of three domains – an N-terminal ring (N-ring) of inner diameter 6.0 nm, a middle ring (M-ring) blocked by protein density, and a C-terminal ring (C-ring) of inner diameter 8.8 nm. In a recent model of the pIV multimer, it has been proposed that the C-ring is embedded in the outer membrane and the M- and N-rings are located in the periplasmic space, where a virus assembly-induced conformational change would allow opening of the gated M-ring and widening of the narrow N-ring.

Virus Assembly

Virus assembly occurs in the bacterial envelope at sites where the inner and outer membranes are associated by a trans-envelope network of proteins comprising the virally encoded pI, pIV, and pXI. This network is not transient and exists even in the absence of virus assembly. In addition to pI, pIV, and pXI, virus assembly is also dependent on host-encoded thioredoxin, ATP hydrolysis and the proton motive force across the cytoplasmic membrane. Thioredoxin must be in the reduced state; however, redox capability of the protein is unnecessary. pI likely catalyzes ATP hydrolysis, since its cytoplasmic domain contains an ATP-binding Walker motif, which is necessary for phage assembly.

The substrate for assembly is the pV-coated viral genome, in which the PS is exposed at one end. The PS likely forms an initiation complex with the cytoplasmic domains of pI, pVII, and pIX. pV is stripped from the genome concurrent with association of the positively charged C-terminal pVIII domain to the ssDNA molecule. The viral genome is translocated across the inner membrane as pVIII subunits assemble to it. A proposed hinge region between the TM and N-terminal amphipathic domains of pVIII may facilitate the transition of the capsid subunit from the inner membrane to the viral coat, although direct experimental evidence is lacking. Concurrent with assembly of the capsid, the progeny virion exits the bacterial cell through the pIV channel. The periplasmic N-terminal and middle domains of the pIV channel must undergo large conformational changes to allow passage of the 7 nm virus particle. The narrow N-terminal ring of the channel (6 nm i.d.) must expand and the gated middle ring must open. Since pI binds to both the initiation complex and the pIV channel, the initiation of virus assembly may be coupled to pIV conformational change by way of pI, which would avoid unintended channel opening. Additionally, if energy is necessary to widen the pIV channel or open its gate, pI may provide this energy via ATP hydrolysis.

Elongation of the progeny phage continues until the end of the viral genome is reached, at which time pVI and pIII associate with the tip of the viral capsid to terminate assembly and release the virion into the extracellular milieu. Without pVI and pIII, virus assembly continues by packaging additional genomes, one after the other, until the resultant polyphage is removed from the cell by mechanical shearing. Genetic studies have shown that separate regions of the C-terminal domain of pIII are important for termination of virus assembly, release of the viral filament from the host cell, and stabilization of the virion particle. An 83-residue C-terminal segment (residues 324–406) allows incorporation of pVI, but not release of the progeny phage. Ten additional residues (313–322) within the 313–406 segment confer the ability to release progeny phage from the bacterial cell envelope. A pIII C-terminal fragment of at least 121 residues is needed to release stable filamentous ssDNA bacteriophage progeny from an *E. coli* cell.

Biotechnology

Several aspects of the unique life cycle and morphology of the Ff bacteriophage have been exploited for biotechnological applications. (1) The size of the packaged genome is variable, such that it can easily tolerate an insertion of up to 6 kbp, which is compensated by the additional assembly of the proportionate number of pVIII subunits, as noted earlier. (2) The orientations of pVIII within the capsid lattice and pIII at the virion tail allow fusion of non-native peptides and proteins to exposed sites of the viral particle without significant disruption of phage viability or capsid stability. (3) The mature virion is very stable to changes in pH, temperature, and ionic strength. (4) Phage preparations yield high titers resulting in efficient and inexpensive large-scale phage production. (5) The phages are distinct from animal and plant viruses, and are generally not toxic to mammalian cells.

The filamentous ssDNA bacteriophage M13 has been used as a cloning vector for decades. The foreign gene to be cloned is usually inserted between functional sequences of the IG region of the genome. The virus assembly process allows up to 12 kbp of DNA to be packaged without adverse effects on phage viability. Inserts greater than 6 kbp, which result in a genome larger than 12 kbp, are possible with compensatory mutations elsewhere in the genome. One advantage of cloning with M13 is that both single-stranded and double-stranded products can be isolated from progeny phage and intracellular RF DNA, respectively.

Most biotechnological applications of M13 are byproducts of the revolutionary exploitation of the phage as a platform from which peptides (or proteins) can be displayed. To produce a phage that displays a foreign peptide, the coding sequence is inserted in-frame with that of a capsid protein, usually pVIII or pIII, so that the fused peptide is exposed on the surface of the virion. Minor capsid proteins pVI, pVII, and pIX have also been successfully used for phage display, although to a lesser extent than pIII and pVIII. The display of the peptide must not interfere with capsid protein function. In the case of pVIII tethering, the peptide must be compatible with translocation of the capsid subunit across the inner membrane and with capsid assembly. Accordingly, peptides are usually fused to the pVIII N-terminus. For pIII, the peptide must not occlude the pilin binding site. Interference of the displayed peptide with the life cycle and/or stability of the phage can often be overcome by using a helper phage to facilitate production of hybrid virions containing both wild-type and fused capsid proteins, thus minimizing the adverse effect of the display. When sequences from a DNA library are inserted into the genome for display, a phage library containing up to 10^{10} different displayed peptides is produced.

The applications of phage display for biotechnological progress range from identification of molecular recognition elements between interacting proteins and ligands to targeted drug delivery for treatment of diseased mammalian cells. Biomedical applications include (1) antibody selection using a phage-displayed epitope library to facilitate the design of vaccines with optimal binding affinity for target epitopes; (2) immunopharmacotherapy for cocaine addiction, in which cocaine-sequestering antibodies displayed on a filamentous ssDNA bacteriophage are delivered to the central nervous system intranasally; (3) *in vivo* visualization of β-amyloid plaques by delivery of phage-displayed anti-β-amyloid antibodies to the brain with subsequent staining with fluorescent-labeled anti-phage antibody; and (4) cell-targeted gene and drug delivery, in which the phage is essentially a therapeutic nanocourier. Filamentous ssDNA bacteriophage display systems have also been successfully applied for nonmedical biotechnological purposes, for example, in the synthesis and assembly of nanowires for lithium ion battery electrodes. In this case, the tetrapeptide Glu-Glu-Glu-Glu is fused to the N-terminus of pVIII and treated to bind cobalt oxide forming a Co_3O_4 nanowire. These Co_3O_4 viral nanowires electrostatically self-assemble to form a nanostructured monolayer, which is used as the negative electrode in the construction of a Li-ion battery.

See also: Icosahedral Tailed dsDNA Bacterial Viruses.

Further Reading

Bennett NJ and Rakonjac J (2006) Unlocking of the filamentous bacteriophage virion during infection is mediated by the C domain of pIII. *Journal of Molecular Biology* 356: 266–273.

Davis BM and Waldor MK (2005) Filamentous phages linked to virulence of Vibrio cholera. *Current Opinion in Microbiology* 6: 35–42.

Eckert B, Martin A, Balbach J, and Schmid FX (2005) Prolyl isomerization as a molecular timer in phage infection. *Nature Structural and Molecular Biology* 12: 619–623.

Frenkel D and Solomon B (2002) Filamentous phage as vector-mediated antibody delivery to the brain. *Proceedings of the National Academy of Sciences, USA* 99: 5675–5679.

Nam KT, Kim D-W, Yoo PJ, et al. (2006) Virus-enabled synthesis and assembly of nanowires for lithium ion battery electrodes. *Science* 312: 885–888.

Opalka N, Beckmann R, Boisset N, et al. (2003) Structure of the filamentous phage pIV multimer by cryo-electron microscopy. *Journal of Molecular Biology* 325: 461–470.

Petrenko VA and Smith GP (2005) Vectors and modes of display. In: Sidhu S (ed.) *Phage Display in Biotechnology and Drug Discovery*. pp. 63–110. Boca Raton, FL: CRC Press.

Samuelson JC, Chen M, Jiang F, et al. (2000) YidC mediates membrane protein insertion in bacteria. *Nature* 406: 637–641.

Tsuboi M and Thomas GJ, Jr. (2007) Polarized Raman and polarized infrared spectroscopy of proteins and protein assemblies. In: Uversky VN and Permyakov EA (eds.) *Protein Structures: Methods in Protein Structure and Stability Analysis*, ch. 3.4. Hauppage, NY: Nova Science Publishers.

Wang YA, Yu X, Overman SA, et al. (2006) The structure of the filamentous bacteriophage. *Journal of Molecular Biology* 361: 209–215.

Webster RE (1999) Filamentous phages. In: Granoff A and Webster RG (eds.) *Encyclopedia of Virology*, 2nd edn., pp. 547–552. London: Academic Press.

Filoviruses

G Olinger, USAMRIID, Fort Detrick, MD, USA
T W Geisbert, National Emerging Infectious Diseases Laboratories, Boston, MA, USA
L E Hensley, USAMRIID, Fort Detrick, MD, USA

© 2008 Published by Elsevier Ltd.

Glossary

D-dimers Fibrinolysis is mediated by plasmin, which degrades fibrin clots into D-dimers. Clinically, D-dimers can be used to diagnose DIC.

Disseminated intravascular coagulation (DIC) An acquired syndrome characterized by inappropriate accelerated systemic activation of coagulation with fibrin deposition in the microvasculature and consumption of procoagulants and platelets.

Fibrin degradation products (FDPs) The protein fragments produced after digestion of fibrinogen or fibrin by plasmin. Typically, a hallmark of disseminated intravascular coagulation. Also, referred to as fibrin-split products.

RNAi The silencing of the expression of a selected gene by a homologous portion of double-stranded RNA. Also, referred to as RNA interference.

Tissue factor (TF) A procoagulant protein responsible for activation of the extrinsic coagulation pathways.

Introduction

In 1967, the first filovirus outbreak occurred in Germany and Yugoslavia among laboratory workers handling African green monkeys and/or tissues from contaminated monkeys imported from Uganda. The causative agent was identified as a new virus called Marburg virus (MARV). MARV re-emerged in 1975 in Johannesburg, South Africa, when a man who had recently returned from Zimbabwe became ill. The infection was spread to the man's traveling companion and a nurse at the hospital. While the man died of the disease, the other two cases recovered from the illness after receiving supportive care. The following year in 1976, a mysterious outbreak swept through several remote villages in Africa. The disease was rapidly fatal (318 cases; mortality rate 88%). Examination of samples revealed a previously unknown virus similar to that of the MARV. This virus was named Ebola after a local river in central Africa and became the second member of the family *Filoviridae*. Since 1976, there have been a number of outbreaks of both types of filoviruses. Isolation and characterization of the causative agents of the Ebola outbreaks resulted in the identification of a number of related viruses. Today *Filoviridae* forms a separate family within the order *Mononegavirales* and can be divided into two genera: *Marburgvirus* and *Ebolavirus*. The latter is composed of four distinct species with varying degrees of lethality in man, *Zaire ebolavirus* (~90% lethal), *Sudan ebolavirus* (~50%), *Reston ebolavirus* (unknown lethality), and *Ivory Coast ebolavirus* (unknown lethality). In contrast, the MARV genus is composed of a single species, *Lake Victoria marburgvirus*. However, there have been a number of isolates of MARV identified with varying degrees of morbidity and mortality. The case–fatality (CF) rate from the most recent isolate, MARV-Angola, has been reported to be ~90%. Although there has been a tremendous amount of work performed on filoviruses, the source of these outbreaks remains undetermined and there remains no approved vaccine or therapeutic intervention.

Background

Ebola Virus History

Ebola virus (EBOV) was first discovered during near-simultaneous outbreaks in the former Zaire and Sudan. These outbreaks were due to serologically distinct viral species that would be designated ZEBOV and SEBOV, respectively. The CF was ~88% in the initial ZEBOV outbreak and ~53% in the SEBOV outbreak. The source of these outbreaks was never determined. These outbreaks were followed by several smaller outbreaks in the late 1970s. After an almost 20-year absence, a new EBOV outbreak was reported in 1989 in a colony of cynomolgus monkeys (*Macaca fascicularis*), imported from the Philippines to a holding facility in Reston (VA, USA). Although the Reston species appeared to be highly lethal in nonhuman primates (NHPs), no disease was reported in any of the documented human exposures. Although there have been a number of subsequent outbreaks of REBOV in NHP facilities both in and outside of the US, to date there have been no cases of human infections reported. Investigations traced the source of all REBOV outbreaks to a single export facility in the Philippines (Laguna Province); however, the original source and mechanism of entry into this facility remain unknown. In 1994, EBOV reemerged in Africa with the identification of a fourth species, ICEBOV, as well as the detection of ZEBOV for the first time in Gabon. The fourth species of EBOV, ICEBOV, was identified in Côte d'Ivoire following the exposure of a researcher during a necropsy to determine the cause of death of local chimpanzee populations. The source for the infection of the chimpanzees was undetermined. The outbreak in Gabon occurred in the Ogooue-Ivindo Province; a total of 51 cases and 31 deaths were documented (CF = 61%) in this outbreak.

The largest outbreak of EBOV occurred in 1995 in Kikwit, Democratic Republic of the Congo (DRC) (formerly Zaire). A total of 310 cases with 250 deaths (CF = 81%) were reported. Approximately one-fourth of all cases were reported among healthcare workers. This outbreak was quickly followed by two additional outbreaks that occurred in DRC and Gabon. Only one of the three primary cases identified could be linked to contact with a dead chimpanzee. A total of 31 cases with 21 deaths (CF = 68%) were documented. The second outbreak occurred as a series of unrelated cases in hunters during July and August 1996. Reminiscent of previous outbreaks (Gabon and Côte d'Ivoire), there were reports of several dead great apes (gorillas and chimpanzees) in the same area. A total of 60 cases with 45 deaths (CF = 75%) were reported. Interestingly, sequence analysis of the isolates from 1994 and 1996 differed by less than 0.1% in the glycoprotein and polymerase genes.

Ongoing, sporadic outbreaks of EBOV continue to be reported in central Africa. Most often, these outbreaks have been associated with infection of the great ape population. It is believed that these outbreaks are to blame for the widespread loss of these animals. It has been reported that the population of Western lowland gorillas has dropped by 50% and as much as 90% in some areas. Conservationists have stated that it will take decades for the populations of these animals to recover.

Marburg Virus History

MARV was first identified in 1967 during simultaneous outbreaks in Germany and Yugoslavia among laboratory workers handling African green monkeys and/or tissues from contaminated monkeys imported from Uganda. Secondary cases were reported among healthcare workers and family members. All secondary cases had direct contact with a primary case. A total of 32 cases and 7 deaths (CF = 21%) occurred during this outbreak. Over the next several decades MARV was associated with multiple sporadic and isolated cases among travelers and residents in southeast Africa.

From 1998 to 2000 a number of MARV cases were reported among young male workers at a gold mine in Durba, DRC. Cases were subsequently detected in the neighboring village. Although a few cases were detected among family members, secondary transmission appeared to be rare. Subsequent virological investigation revealed that these cases were due to multiple introductions of different strains from an as-of-yet unidentified source. In total there were 154 cases with 128 deaths (CF = 83%) reported.

The largest and most deadly outbreak of MARV occurred in 2004–05 in the Uige Province of Angola. A total of 252 cases and 227 deaths (CF = 90%) were reported. Unlike previous outbreaks, a large number of the victims were children under the age of five. Virological analyses suggest that this isolate is very close to the strain from the original cases in 1967 in Europe. The reason for the increased virulence of the Angola isolate remains unclear.

Transmission

Both EBOV and MARV are transmitted by direct contact with the blood, secretions, organs, or other bodily fluids from infected persons, or infected animals. Many of the initial cases in outbreaks can be traced to the handling of infected animals, particularly chimpanzees and gorillas. Amplification has most often been associated with hospitals, where the virus is spread through nosocomial routes, or through burial ceremonies, where it is not uncommon for mourners to have direct contact with the deceased. While the modes of transmission have been well documented, the true source of these outbreaks and the reservoir for these viruses remain unclear. Although NHPs have been a source of infection for humans, the highly virulent nature of filoviruses in these animals suggests that they are not the reservoir. Different hypotheses have been developed to try to explain the origin of EBOV outbreaks. The presence of bats or contact with bat guano is often a common theme during outbreaks of both MARV and EBOV. Laboratory observation has shown that fruit and insectivorous bats experimentally infected with ZEBOV can support viral replication without apparent signs of illness. In a recent survey of small animals in Africa near areas of human or great ape EBOV outbreaks, asymptomatic infection of three species of fruit bats, *Hypsignathus monstrosus*, *Epomops franqueti*, and *Myonycteris torquata*, was reported. The exact ecology of EBOV and MARV remains to be determined and may include multiple natural reservoirs.

Virion Structure and Composition

Filoviruses are enveloped, nonsegmented, negative-strand RNA viruses. The virus envelope is derived from the lipid membrane envelope from the host cellular plasma membrane and therefore the lipid and protein composition reflect that of the infected cell. Filovirus virions have a characteristic thread-like, filamentous morphology (**Figure 1**). Pleomorphic virus structures ranging from straight tube-like structures, branched structures, and curved filaments shaped like a 6 (shepherd's crook), a U, or a circle. The diameter of filovirus virions are typically 80 nm, while the length can vary between 130 and 14 000 nm. The average length of a virus particle ranges from about 800 to 1000 nm with MARV being closer to 800 nm and EBOV closer to 1000 nm. Structurally, the virus particle is composed of seven viral proteins. The glycoprotein (GP), a type I transmembrane protein complex composed of two proteins, a ∼140 kDa GP_1 and ∼26 kDa GP_2. Normally, the two GP subunits are disulfide linked; however, when GP_1 is not disulfide-linked with GP_2, the GP_1 can be released in a soluble form from infected cells. Glycosylation, both N- and O-linked, provides approximately 50% of the GP mass. Complex cellular processing including glycosylation and proteolytic cleavage of the GP_1–GP_2 precursor protein leads to the generation of the GP_1–GP_2 heterotrimeric spikes in the virus envelope. GP_1 allows for receptor binding while GP_2 contains the fusion domain necessary for viral entry into the host cell. Surface projections of GP are dispersed evenly over the entire surface at 10 nm intervals. Functionally, the virus GP allows for attachment, receptor-mediated endocytosis, fusion with endocytotic vesicles, and release of the virion core into the host cell cytoplasm. To date, no specific host cell receptor has been identified. A variety of putative receptors and cofactors have been identified including the asialoglycoprotein receptor, folate receptor-α, and C-type lectins. Given the wide cell tropism of the virus, or the receptors are likely either constitutively expressed proteins, that involve multiple proteins with conserved binding domains, or

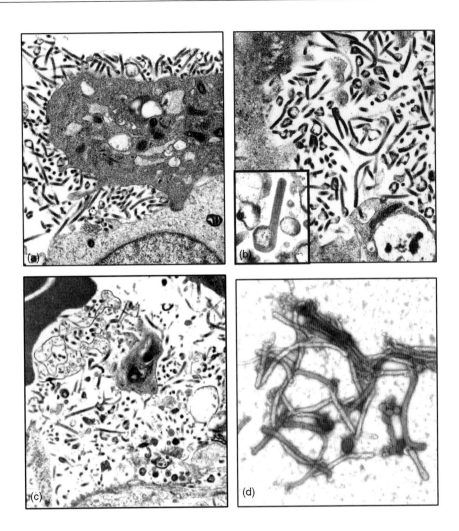

involve cofactors that facilitate virus entry. Viral particles may gain entry into the cell by co-opting common phagocytic pathways leading to a nonspecific entry mechanism.

The remaining six viral proteins are found within the virion, underlying the host-derived membrane. The viral protein 40 (VP40) functions as a matrix protein, and is essential for efficient viral assembly and budding. VP24 has been demonstrated to participate in the spontaneous formation of the nucleoprotein complex. Similar to VP40, studies indicate that VP24 behaves as a minor matrix protein. The nucleoprotein (NP) and VP30, VP35, polymerase (L) proteins, and the viral genomic RNA form the ribonucleoprotein complex. The nucleocapsids form a symmetrically helical filamentous structure underlying the viral envelope and the core structure has a striated appearance similar to that of rhabdoviruses. The L protein, the largest protein, functions as the virus-associated RNA-dependent RNA polymerase. VP30 functions as an EBOV-specific heterologous transcription activation factor that recognizes a secondary stem–loop structure during transcription. Similar to other transactivator factor proteins, VP30 exists in both phosphorylated and unphosphorylated forms within virions. In addition to their function as structural proteins, VP35 and VP24 have been shown to act as interferon antagonists.

Genome Organization and Expression

The linear genome is approximately 19 000 nt in length and consists of a single-stranded, negative-sense RNA (**Figure 2**). The genomic nucleic acid is not infectious. The genome contains highly conserved, transcription initiation and termination sites. Genomes contain intergenic sequence regions that vary in length, nucleotide sequences, and contain regions of gene overlap. While EBOV and MARV have a substantial degree of structural similarity in their respective genome organization, limited nucleotide sequence, homology is observed. Similar to the majority of other mononegaviruses, filovirus replication and transcription occurs in the cytoplasm of the infected

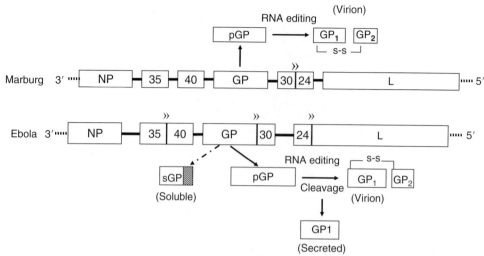

Figure 2 Filovirus genome organization and generation of glycoproteins. Linear, negative-sense, single-stranded, RNA monopartite genome, ~19 000 nt. Relative positions of the encoding regions for viral proteins depicted as boxes; NP, nucleoprotein; 35, VP35; 40, VP40; GP, glycoprotein; 30, VP30; 24, VP24; L, viral polymerase. Leader and trailer and intergenic regions (nontranscribed regions) shown as lines. Overlapping regions indicated by ≫ symbol. For Ebola, glycoprotein gene generates a soluble GP (sGP) and precursor GP (pGP). pGPs for Ebola and Marburg are cleaved by host-derived enzymes to generate a GP_1 and GP_2 subunit that are disulfide linked.

cell. The viral-encoded RNA-dependent RNA polymerase generates full-length positive-sense antigenomes that serve as templates for the synthesis of the progeny negative-sense genomes. Filoviral subgenomic mRNAs are monocistronic and polyadenylated. A single genome contains seven genes that are arranged linearly. Monocistronic subgenomic mRNAs complementary to the viral RNA lead to the generation of seven filovirus proteins found within the virion. For both MARV and EBOV, each gene encodes a single polypeptide, except for the GP. The structural glycoprotein GP_1 and GP_2 are generated as a single precursor polypeptide that is cleaved by host cell pro-protein convertases to generate the two subunits of the final protein. EBOV is different from MARV in that an additional viral protein is generated; a nonstructural, soluble form of GP, referred to as secreted GP (sGP). sGP is translated from an mRNA that has undergone RNA editing. Interestingly, the majority of GP mRNA encodes for the sGP which has been detected in cell culture and serum from EBOV-infected patients. Both soluble GP_1 and sGP have been implicated in the pathogenesis of EBOV by acting as decoy proteins that sequester neutralizing antibodies during infection; however, the exact role of the soluble forms of GP remains to be elucidated.

Pathogenesis

Clinical Features/Presentation

The clinical manifestations of EBOV and MARV are often compared to that of severe sepsis or septic shock. Following a 2–21 day incubation period, cases often present with a variety of nonspecific symptoms including high fever, chills, malaise, and myalgia. As the disease progresses, there is evidence of multisystemic involvement, and manifestations include prostration, anorexia, vomiting, nausea, abdominal pain, diarrhea, shortness of breath, hypotension, edema, confusion, maculopapular rash, and eventually coma. Patients normally progress rapidly with death occurring 6–9 days after the onset of symptoms. The development of petechiae, ecchymoses, mucosal hemorrhages, and uncontrolled bleeding at venipuncture sites are indicative of development of abnormalities in coagulation and fibrinolysis. Massive loss of blood is atypical and, when present, is largely restricted to the gastrointestinal tract. Fulminant infection typically evolves to shock, convulsions, and, in most cases, diffuse coagulopathy.

Animal Models

While only limited information is available for pathophysiology of human cases of EBOV or MARV, a lot has been learned from studies employing experimentally infected NHPs. Although the majority of studies in humans and NHPs to date have focused on ZEBOV, limited studies suggest that MARV may behave in a similar manner. EBOV and MARV are most often characterized by three common factors: development of lymphocyte apoptosis, aberrant production of pro-inflammatory cytokines, and the development of coagulation abnormalities. Sequential analysis of samples collected from ZEBOV-infected NHP shows early and prominent infection of monocyte/macrophages (MO/Φ) and dendritic cells (DCs). The importance of this selective targeting has been well documented.

Infection of the MO/Φ not only facilitates distribution of virus to the spleen and lymph nodes but also triggers a cascade of events including the upregulation and production of the procoagulant protein tissue factor (TF) as well as a number of pro-inflammatory cytokines/chemokines and oxygen free radicals. *In vitro* studies of human MO/Φ have confirmed these observations. Similar to septic shock caused by bacterial pathogens, disease triggered by EBOV or MARV infection is thought to involve inappropriate or maladaptive host responses. It is likely that these responses are more important to the development of the observed pathology than any structural damage directly induced by viral replication. Infection of DC likely facilitates disease progression by hindering the ability of the host to orchestrate an effective immune response to clear the virus.

Although lymphocytes do not appear to be productively infected by filoviruses, the importance of their role in the development of the disease pathology cannot be understated. The loss of lymphocytes through apoptosis has been well documented. Flow cytometry demonstrated a striking loss of circulating CD8+ T cells and natural killer (NK) cells prior to or corresponding with the onset of symptoms. The exact mechanism(s) responsible for triggering this widespread loss of lymphocytes remains unclear. RNA analysis of peripheral blood mononuclear cells (PBMCs) collected from NHPs has demonstrated the upregulation of a number of pro-apoptotic genes and anti-apoptotic genes.

Infection of MO/Φ also appears to be critical in the triggering and propagation of the coagulation abnormalities observed in filovirus infections. Recently, it was demonstrated that infection of MO/Φ *in vitro* or *in vivo* resulted in the upregulation of TF. Interestingly, this expression was observed only in infected cells, suggesting a direct correlation with the virus. Analysis of sera from infected animals revealed the presence of microparticles or small membrane vesicles with procoagulant and pro-inflammatory properties. Analysis of these microparticles showed that many of these expressed TF and may therefore have been derived from filovirus-infected MO/Φ.

The contribution of the endothelium as well as the infection of endothelium remains controversial. *In vitro* studies have demonstrated that endothelial cells are easily infected by either MARV or EBOV. However, analysis of tissues collected from human cases has produced conflicting results. Analysis of sequential tissues from NHP suggests that endothelial cells are not targeted until the later stages of disease. Electron microscopy examination of tissues showed activation of endothelial cells but no widespread damage to the vasculature. Analyses of clinical chemistries support this observation, with a drop in small molecular weight proteins but not total proteins. The cause for this vascular leakage is unclear and is likely multifactorial. Fibrin degradation products (FDPs), thrombin, reactive oxygen species, pro-inflammatory cytokines, as well as activation of the complement and kinin systems can all contribute to increased vascular permeability. Rapid increases in levels of D-dimers were observed in sera of EBOV-infected NHP with initial increases being detected before the onset of clinical symptoms. While the contributions of the complement and kinin systems in the development of EBOV and MARV pathogenesis are yet to be evaluated, the production of pro-inflammatory cytokines and oxygen free radicals has been well documented. In EBOV-infected NHP, IFN-α, interleukin-6 (IL-6), monocyte chemoattractant protein 1 (MCP-1), macrophage inflammatory protein (MIP)-1α, and MIP-1β were detected in sera at the early to mid-stages of disease. IFN-β, IFN-γ, IL-18, and TNF-α were also detected in sera but not until the later stages of disease.

In recent years, the importance of the interaction between coagulation and inflammation as a response to severe infection has become increasingly appreciated. In addition to driving upregulation of procoagulant proteins, inflammatory mediator, may inhibit fibrinolytic activity, and downregulate natural anticoagulant pathways, in particular, the protein C anticoagulant pathway. Uncontrolled activation of coagulation systems may lead to disseminated intravascular coagulation (DIC). DIC is a syndrome with both bleeding and thrombotic abnormalities characterized in part by the presence of histologically visible microthrombi in the microvasculature. These microthrombi may hamper tissue perfusion and thereby contribute to multiple organ dysfunction and high mortality rates. Before acute DIC can become apparent, there must be a sufficient stimulus to deplete or overwhelm the natural anticoagulant systems. During activation of the clotting system, the host regulates the process through the production and activation of a variety of inhibitors of the clotting system. In this process, however, the inhibitors are consumed, and if the rate of consumption exceeds the rate synthesized by liver parenchymal cells, plasma levels of inhibitors will decline. Analysis of tissue and sera from ZEBO-infected NHPs shows widespread fibrin deposition and marked decreases in plasma levels of protein C. The drops in protein C correlate with increase in D-dimers and disease progression.

Host Responses

The low sporadic disease incidence and remote locations of most outbreaks have led to limited careful study of the host responses to virus infection. Perturbations of the innate and acquired immune responses are consistent with fatal disease outcome. In a few studies patients that survived Ebola infections were reported to have generated a virus-specific immunoglobulin M (IgM) response and to have transient expression of pro-inflammatory cytokines. In sharp contrast, fatal EBOV cases have dysregulated cytokine expression levels, coagulation abnormalities, fibrin deposition, DIC, and evidence of cell death and the lack of an effective adaptive immune response. Through direct and indirect mechanisms, and pathways that are not clearly understood, the fatal cases are associated with the lack of B- or T-lymphocyte responses. Altered innate responses, both cellular and soluble, combined with early infection of MO/Φ and DC, likely impair the host's ability to initiate an appropriate adaptive immune response to clear filovirus-infected cells. Simultaneously, virus-induced or inappropriate host defenses lead to dysregulated innate immune responses. Filovirus proteins VP35 and VP24 interfere with and modulate these key immune pathways by inhibiting cellular antiviral defenses. Ultimately, these events result in immune dysregulation and uncontrolled explosive virus replication.

Animal models have offered some insight into protective immune responses. In EBOV infection in mice, early interferon responses are vital to controlling virus replication. In concert, EBOV-specific antibody and cellular responses help eliminate virus infection in this animal model. Protective filovirus vaccines have been shown to induce antibody and/or cellular responses that offer protection in rodents and NHPs. Therefore, an appropriate innate and adaptive immune response can offer immunity to both MARV and EBOV.

Treatment and Prevention

Currently, there are no approved therapies or vaccines for the treatment and/or prevention of EBOV and MARV. Control and treatment are limited to containment and basic supportive care. Suspected cases should be isolated from other patients and strict-barrier-nursing techniques implemented. Contact tracing and follow-up of people who may have been exposed through close contact with cases are essential. If not already implemented proper disinfection and handling of materials from infected patients must be initiated. Education of the community as to the presentation of the disease, the nature by which it is transmitted, and the sources by which patients are likely to come into contact with the virus are critical to preventing and containing future outbreaks.

Although there are no approved vaccines there has been considerable progress in the identification and successful testing of candidate vaccines in animal models. These platforms can be separated into several categories: replication-deficient platforms, replication-competent platforms, and virus-like particles (VLPs). The replication-deficient platforms are viral vectors such as adenovirus or Venezuelan equine encephalitis virus (VEEV) expressing filoviral proteins. The first successful strategy to protect NHP from EBOV employed a replication-deficient adenovirus expressing EBOV GP and NP. In this strategy, a DNA prime of GP and NP genes was followed by an adenovirus-expressing GP boost. Subsequent studies demonstrated protection using a single adenovirus GP vaccination

28 days before EBOV challenge. Questions have been raised about the durability of the replication-deficient strategies as well as the ability of these strategies to overcome any pre-existing immunity to the vaccine platform. Efficacy has also recently been demonstrated using a live-attenuated (replication-competent) vesicular stomatitis virus (VSV) expressing the GP of EBOV. Additional work demonstrated that by simply swapping out the EBOV GP for either the MARV or Lassa fever virus GP, this platform could provide full protection against lethal MARV or Lassa fever virus challenge, respectively. The VSV platform has the additional benefit in that it has been demonstrated to provide 100% protection against MARV when administered 30 min after MARV exposure and ~50% protection for EBOV when rapidly administered in an EBOV post-exposure setting. While it is likely that the replication-competent platforms may provide increased durability there is some concern about the safety of administration of any live vaccine. Therefore, additional studies are needed to fully evaluate the safety profile of this platform. Finally, filovirus-like particles generated by the co-expression of viral membrane proteins (GP and VP40) are currently being evaluated in NHPs. This strategy may be beneficial in that it will likely overcome the safety concern associated with the use of live-attenuated platforms and/or the issues of preexisting immunity. Consistent with other protein vaccines, there are concerns about the durability of the immunity, adjuvant choice, as well as the necessary dosing regimen to achieve long-lasting immunity.

In addition to the wealth of candidate vaccine platforms developed and tested over the last few years, there have also been a number of candidate therapeutics that been evaluated in animal models. These strategies have met with varying degrees of success and can be divided into two basic categories, those that directly target the virus and those that target the host and/or the disease manifestations in the host. Recently, investigators reported success using either antisense oligonucleotides or RNA interference (RNAi) corresponding to sequences in the viral genomic or mRNA *in vivo*. Additional studies will be needed to evaluate whether the antisense technology can provide any therapeutic value when used after exposure and whether the success of the RNAi strategies in guinea pig models will translate into the more stringent NHP models. Reduction in morbidity and mortality has also been observed using therapeutics targeting the clinical manifestations in the NHP models. Treatment of ZEBOV-infected NHP with a recombinant protein, recombinant nematode anticoagulant protein c2 (rNAPc2) that blocks the TF-mediated activation of the extrinsic pathway of blood coagulation, resulted in ~33% survival as well a substantial increase in the mean time to death. Interestingly, substantial reductions in plasma levels of IL-6, MCP-1, and viral load were noted in rNAPc2-treated animals. It is unknown if treatment with other anticoagulant strategies that block the intrinsic and/or common pathways may also offer comparable and/or improved therapeutic benefit.

Finally, it is important to mention therapeutic strategies that are yet to provide any survival benefit when tested in NHP models. Although type I IFNs have shown promise in the mouse model of EBOV infection, testing of IFN-α2 failed to protect NHPs against EBOV or MARV. In addition, S-adenosylhomocysteine, which had been hypothesized to protect mice against EBOV, at least in part through induction of a type I IFN response, also failed to protect NHP. Currently, alternative IFN therapies and formulations are under evaluation. It is unclear if these approaches will have any improved efficacy in NHPs. In addition, heparin, convalescent serum, and equine anti-EBOV immunoglobulin have been used to treat infections in humans and/or NHPs, but were inconsistent and appeared to be of little therapeutic value. While passive antibody treatment has been successful in mice, the benefit of antibody immunotherapy remains controversial. Although results in rodent models have been encouraging, initial studies in NHPs have not been as successful. Additional studies are needed to fully evaluate the utility of this therapeutic avenue.

Acknowledgment

Work on filoviruses at USAMRIID was supported by the Medical Chemical/Biological Defense Research Program, US Army Medical Research, and Material Command.

See also: Ebolavirus.

Further Reading

Bosio CM, Moore BD, Warfield KL, *et al.* (2004) Ebola and Marburg virus-like particles activate human myeloid dendritic cells. *Virology* 326: 280–287.

Feldmann H, Geisbert TW, Jahrling PB, *et al.* (2004) *Filoviridae*. In: Fauquet C, Mayo MA, Maniloff J, Desselberger U and Ball LA (eds.) *Virus Taxonomy: Eighth Report of the International Committee on Taxonomy of Viruses,*, pp 645–653. San Diego, CA: Elsevier Academic Press.

Feldmann H, Volchkov VE, Volchkova VA, and Kenk HD (1999) The glycoproteins of Marburg and Ebola virus and their potential roles in pathogenesis. *Archives of Virology Supplementum* 15: 159–169.

Geisbert TW, Young HA, Jahrling PB, Davis KJ, Kagan E, and Hensley LE (2003) Mechanisms underlying coagulation abnormalities in Ebola hemorrhagic fever: Overexpression of tissue factor in primate monocytes/macrophages is a key event. *Journal of Infectious Diseases* 1(188): 1618–1629.

Grolla A, Lucht A, Dick D, Strong JE, and Feldmann H (2005) Laboratory diagnosis of Ebola and Marburg hemorrhagic fever. *Bulletin de la Société de Pathologie Exotique* 98(3): 205–209.

Harlieb B and Weissenhorn W (2006) Filovirus assembly and budding. *Virology* 344: 64–70.

Hensley LE, Jones SM, Feldmann H, Jahrling PB, and Geisbert TW (2005) Ebola and Marburg viruses: Pathogenesis and development of countermeasures. *Current Molecular Medicine* 5: 761–772.

Ignatyev GM (1999) Immune response to filovirus infections. *Current Topics in Microbiology and Immunology* 235: 205–217.

Jones SM, Feldmann H, Stroher U, *et al.* (2005) Live attenuated recombinant vaccine protects nonhuman primates against Ebola and Marburg viruses. *Nature Medicine* 11: 786–790.

Mohamadzadeh M, Chen L, Olinger GG, Pratt WD, and Schmaljohn AL (2006) Filoviruses and the balance of innate, adaptive, and inflammatory responses. *Viral Immunology* 19: 602–612.

Paragas J and Geisbert TW (2006) Development of treatment strategies to combat Ebola and Marburg viruses. *Expert Review Anti-Infective Therapy* 4: 67–76.

Peterson AT, Bauer JT, and Mills JN (2004) Ecologic and geographic distribution of filovirus disease. *Emerging Infectious Diseases* 10: 40–47.

Reed DS, Hensley LE, Geisbert JB, Jahrling PB, and Geisbert TW (2004) Depletion of peripheral blood T lymphocytes and NK cells during the course of Ebola hemorrhagic fever in cynomolgus macaques. *Viral Immunology* 17: 390–400.

Schnittler HJ and Feldmann H (1999) Molecular pathogenesis of filovirus infections: Role of macrophages and endothelial cells. *Current Topics in Microbiology and Immunology* 235: 175–204.

Theriault S, Groseth A, Artsob H, and Feldmann H (2005) The role of reverse genetics systems in determining filovirus pathogenicity. *Archives of Virology Supplementum* 19: 157–177.

Fish and Amphibian Herpesviruses

A J Davison, MRC Virology Unit, Glasgow, UK

© 2008 Elsevier Ltd. All rights reserved.

This article is a revision of the previous edition articles by Andrew J Davison, volume 1, pp 553–557, and Allan Granoff, volume 1, pp 51–53, © 1999, Elsevier Ltd.

Glossary

Fingerling Young fish between the fry and adult stages.

Fry Newly hatched or very young fish.

Tegument The layer of proteins situated between the capsid and the envelope in herpesvirus particles; equivalent to the matrix in other viruses.

Classification and History

Most currently recognized fish herpesviruses infect species that are farmed or harvested from the wild for human consumption. The conditions used in aquaculture may enhance the disease potential of these pathogens and increase their likelihood of detection. It is probable, therefore, that many more fish herpesviruses await discovery. A similar prospect applies to amphibian herpesviruses, since the two known examples originated from a single host species. Herpesviruses of fish and amphibians, like those of higher vertebrates, are usually highly species specific in the natural setting. This indicates that they have evolved in close association with their hosts over long periods of time. Many of these viruses cause mortality only in young fish, and some are also agents of epidermal hyperplasia or neoplasia. Some have been studied to such a limited extent that causal links with the disease whose occurrence led to their identification have not yet been established adequately.

Several fish and amphibian herpesviruses have been classified as members of the family *Herpesviridae*. These include viruses that infect white sturgeon, *Acipenser transmontanus* (acipenserid herpesviruses 1 and 2), Japanese eel, *Anguilla japonica* (anguillid herpesvirus 1), carp, *Cyprinus carpio* (cyprinid herpesvirus 1), goldfish, *Carassius auratus* (cyprinid herpesvirus 2), northern pike, *Esox lucius* (esocid herpesvirus 1), rainbow trout, *Oncorhynchus mykiss* (salmonid herpesvirus 1), channel catfish, *Ictalurus punctatus* (ictalurid herpesvirus 1), Pacific salmon, *Oncorhynchus* species (salmonid herpesvirus 2), walleye, *Stizostedion vitreum* (percid herpesvirus 1), and turbot, *Psetta maxima* (pleuronectid herpesvirus 1). Only one of these viruses (ictalurid herpesvirus 1) has been classified further into a species (*Ictalurid herpesvirus 1*) and a genus (*Ictalurivirus*). Candidate herpesviruses have also been reported in black bullhead (*Ictalurus melas*), Japanese flounder (*Paralichthys olivaceus*), pilchard (*Sardinops sagax*), and redstriped rockfish (*Sebastes proriger*). The viruses representing the classified amphibian herpesviruses (ranid herpesviruses 1 and 2) originated from the leopard frog, *Rana pipiens*. The tenuous genetic relationships between herpesviruses of fish and amphibians and those of higher vertebrates have prompted the proposal that the former group be separated into a new family (*Alloherpesviridae*) and grouped with other herpesviruses into an order (*Herpesvirales*). Elucidation of the genetic relationships among these viruses, upon which detailed subclassification will depend, is at an early stage.

This article focuses on three principal viruses, one infecting an amphibian and two infecting fish. These are Lucké tumor herpesvirus (LTHV; ranid herpesvirus 1), channel catfish virus (CCV; ictalurid herpesvirus 1), and koi herpesvirus (KHV; proposed cyprinid herpesvirus 3). Comments are made on other viruses where appropriate.

In 1934, Lucké reported that intranuclear acidophilic inclusion bodies typical of certain virus infections were present in renal adenocarcinomas of the leopard frog. He concluded 4 years later that the tumor was caused by a virus. Almost two decades elapsed before herpesvirus particles were observed in frog kidney tumor cells by electron microscopy (**Figure 1**), and another decade followed before a causative relationship was established between LTHV and the tumor. The complete genome sequence of the McKinnell strain of LTHV was published in 2006 along with that of the other amphibian herpesvirus, the Rafferty strain of frog virus 4 (FV4; ranid herpesvirus 2).

In 1968, Fijan reported the first isolation of CCV from young channel catfish, in which it causes an acute disease of high mortality. The first detailed characterization of the virus was published 3 years later, and showed that capsids with the typical herpesvirus morphology are assembled in infected cell nuclei. Three-dimensional reconstructions

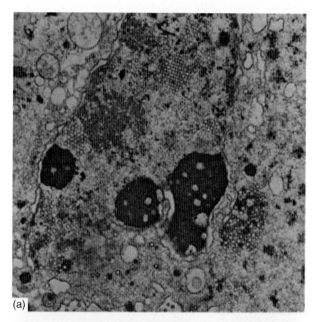

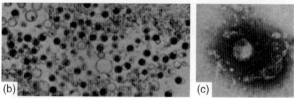

Figure 1 (a) Thin section of an inclusion-bearing Lucké tumor cell with typical herpesvirus particles in various stages of development in the nucleus. Magnification ×19 400. (b) Enveloped extracellular virions. Magnification ×24 000. (c) Negatively stained nonenveloped particles showing typical herpesvirus morphology. Magnification ×110 000. Reproduced from Granoff A (1972) Lucké tumor-associated viruses – a review. In: Biggs PM, de-The G, and Payne LN (eds.) *Oncogenesis and Herpesviruses*, pp. 171–182. Lyons: International Agency for Research on Cancer, with permission from WHO.

derived by cryoelectron microscopy have shown that, apart from minor dimensional differences, the CCV capsid is strikingly similar to that of herpes simplex virus type 1 (HSV-1) (**Figure 2**). The complete genome sequence of the Auburn strain of CCV was published in 1992.

In 1998, a herpesvirus-like agent was identified as the cause of high mortality among carp, including expensive ornamental fish known as koi. The first outbreaks of this new disease may have occurred as early as 1996. In initial studies, this agent (KHV) was also known as carp interstitial nephritis and gill necrosis virus. Consideration of developmental, morphological, and, most recently, genetic attributes have led to the virus being proposed formally as a member of the proposed family *Alloherpesviridae*. The complete genome sequences of three strains of KHV (from Japan, USA, and Israel) were published in 2007.

Geographic and Seasonal Distribution

The Lucké tumor occurs in north-central and northeastern parts of the USA and southern Canada. In the 1960s, the frequency of affected frogs reached 9% in some collections. However, leopard frog populations have since declined drastically, and the tumor has also become very rare. Production of LTHV within tumors is temperature dependent and therefore seasonal. Tumor cells contain acidophilic intranuclear inclusions and virus particles at temperatures below 11.5 °C but not at higher temperatures, and transition between these states can be induced by temperature shift. The mechanisms by which temperature regulates virus growth have not been elucidated, and their operation *in vivo* and *in vitro* (explanted tumors) indicates that the immune system is not a primary factor. Despite the absence of virus particles at higher temperatures, there is some evidence that tumors express virus-specific proteins.

The occurrence of CCV disease has paralleled the intensification of channel catfish farming in the southern USA and other warmer temperate regions. The lack of reported CCV isolations from wild channel catfish indicates that factors such as dense stocking and poor environmental conditions may predispose farmed fish stocks to outbreaks of disease. A key factor is water temperature, since epizootics occur in the summer months and the incubation period has been shown to be considerably shorter at higher temperatures up to 30 °C.

KHV has spread to many countries worldwide, presumably as a result of intensive aquaculture and unregulated transport of live fish in the absence of diagnostic assessment. Water temperature is an important seasonal factor influencing disease, with losses tending to occur during periods at 18–25 °C (spring and autumn). In one study, experimental infections of koi resulted in a cumulative mortality of 95% at 23 °C, with marginally fewer deaths occurring more rapidly at 28 °C and more slowly at 18 °C. Mortalities did not occur at 13 °C, but shifting virus-exposed fish from 13 to 23 °C resulted in rapid onset of mortality.

Pathology, Clinical Features, and Pathogenesis

Little information is available on the natural transmission of LTHV and the factors that influence the occurrence of tumors. It is possible that virus is transmitted by contact with urine or by infection of oocytes in tumor-bearing frogs. Although FV4 infects the embryos and larvae of *R. pipiens*, surviving animals do not develop tumors. Examinations of

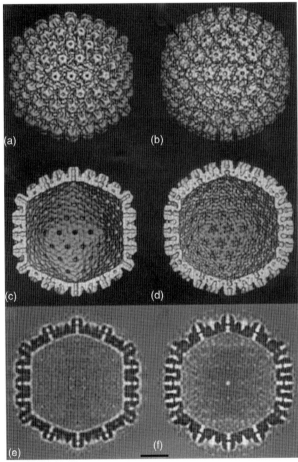

Figure 2 Comparison of the three-dimensional capsid structures of CCV (a, c, e) and HSV-1 (b, d, f). The capsids are viewed along a twofold symmetry axis. Outer surfaces (a and b); inner surfaces (c and d); central thin sections (e and f). The distribution of capsomers (a and b) is distinctive for the triangulation number, $T = 16$, with lines of three hexons connecting pentons along each edge of the icosahedral surface lattice. Scale = 25 nm. Reproduced from Booy FP, Trus BL, Davison AJ, and Steven AC (1996) The capsid architecture of channel catfish virus, an evolutionarily distant herpesvirus, is largely conserved in the absence of discernible sequence homology with herpes simplex virus. *Virology* 215: 134–141, with permission from Elsevier.

the gene complements of the two viruses have not provided a ready explanation of this biological difference.

CCV causes an acute hemorrhagic disease in juvenile channel catfish. It is readily transmitted from fish to fish, probably entering through the gills. In artificial settings, it can also be transmitted by injection or orally by ingestion of contaminated food. After experimental infection, CCV can be isolated from the kidneys and then from other organs, in some of which impressively high titers of virus may be attained. The primary route for virus shedding is probably via the urine.

CCV can be remarkably virulent in susceptible populations of channel catfish. Under optimal conditions the incubation period can be as short as 3 days, and mortality can rise rapidly to 100%. Signs of distress may be accompanied by convulsive swimming, including a 'head-up' posture, and lethargy and death follow. Externally, affected fish may display protruding eyes, a distended abdomen and hemorrhages, largely on the ventral surface. Internally, viscera may be enlarged and hemorrhagic, but certain organs, such as kidneys, liver, and spleen, may be pale. The digestive tract is empty of food, instead containing a mucoid secretion, and a fluid accumulation is present in the peritoneal cavity. Histopathological examination reveals widespread and profound changes. Initially, the kidneys show edema, hemorrhage, and necrosis, and then these features develop in the liver and digestive tract. Electron microscopic studies reveal virus particles in affected organs.

Given the virulence of CCV in young catfish, it is reasonable to suppose that the virus might persist in an inapparent or latent form in adult fish, as do higher vertebrate herpesviruses in their hosts. There is evidence in support of this. Viral mRNA and antigens have been detected in adult fish and, in more recent studies, virus was recovered by co-cultivation of tissues from wintering adult fish and virus DNA was detected by polymerase chain reaction (PCR) in various tissues of surviving fish many months after infection. There is evidence that the CCV genome may exist in circular or concatemeric form in surviving fish. However, the site of latency has not been identified. Reactivated CCV from adult fish could be transmitted horizontally, and there is circumstantial evidence that vertical transmission may also occur.

KHV is highly contagious, has an incubation period of 2–3 days, and exhibits mortality rates of 80–100%. Clinical signs include increased mucus production, pale patches on the skin and gills, labored breathing, and swelling and then necrosis of the gill filaments. Pathological lesions are evident primarily in the gills but may be present in the kidney, skin, liver, spleen, brain, and gastrointestinal tract. Gill disease is characterized by hyperplasia and then fusion of secondary lamellae, inflammation of the gill rakers, and subsequent necrosis. Kidney disease, if present, is characterized by a peritubular and interstitial nephritis. Virus particles are present in affected organs. The routes of infection and shedding are not known, though the gills and skin are obvious candidates, and cohabitation is an effective means of transmitting the disease. Depending on water temperature, virus may survive for periods of days to weeks. Immunohistochemistry and PCR analyses have demonstrated large amounts of the virus in the gills, skin, kidneys, and spleen with smaller amounts in liver, brain, and intestinal tract, mirroring the pathological involvement of these organs. Given its identity as a herpesvirus, it is likely that KHV has a latent aspect to its life cycle. There is some evidence for this, in that temperature-dependent reactivation of KHV has been shown to occur experimentally in surviving carp several months after initial exposure.

Two other carp herpesviruses have been characterized. From limited DNA sequence data, both appear to be closely related to, though distinct from, KHV. Carp pox herpesvirus (cyprinid herpesvirus 1) has been implicated as the cause of a localized epithelial hyperplasia, which manifests itself as benign smooth nodules on the skin and may be transferred by applying material from the lesions to the abraded skin of other fish or following bath exposures of young carp or koi to virus from cell culture. The agent is difficult to grow successfully in cell culture, and thus is isolated infrequently from clinical cases. Hematopoietic necrosis herpesvirus of goldfish (cyprinid herpesvirus 2) was first observed in Japan in 1992, and since then severe outbreaks of the associated disease have occurred in several countries. The virus affects goldfish and not carp, and the disease shares some pathological manifestations similar to that caused in carp by KHV, although the hematopoietic tissue in the kidney rather than the gills is the principal virus target. Similar to cyprinid herpesvirus 1, the goldfish virus is difficult to isolate, although it has been successfully cultured on occasion, with the experimental demonstration that upon bath exposure goldfish will succumb to typical signs of the disease and death.

Two salmonid herpesviruses have been shown to cause virulent disease either naturally or experimentally. Herpesvirus salmonis (salmonid herpesvirus 1; SalHV-1) was isolated on several occasions from a rainbow trout hatchery in the state of Washington. Oncorhynchus masou virus (salmonid herpesvirus 2; SalHV-2) was isolated from Japanese land-locked salmon, and has the property of causing epithelial tumors in survivors of experimental infection. Two additional herpesviruses, the NeVTA and Yamame tumor viruses, have been isolated from *Oncorhynchus* species. They appear to be related to each other, but it is not known whether they represent distinct salmonid herpesviruses or whether they are isolates of SalHV-2 or, as seems less likely, SalHV-1.

Diseases ostensibly due to several other fish herpesviruses have been described. Walleye herpesvirus is associated with epidermal hyperplasia, but causality has not been demonstrated. Similarly, the role of turbot herpesvirus has not been proven in episodes of substantial mortality among farmed fry.

Growth Properties

The Lucké tumor occurs naturally in *R. pipiens*. Tumors can be induced readily in *R. pipiens* embryos or larvae inoculated with LTHV-containing extracts of Lucké tumors or virus purified by density gradient centrifugation. The ascites frequently produced by tumor-bearing frogs contains virus that can induce tumors when inoculated into embryos. Tumors have also been induced experimentally in other species of the host genus. Despite attempts employing cell lines from a wide range of hosts, it has not proved possible to culture LTHV *in vitro* from homogenates of Lucké tumors or the urine of tumor-bearing frogs. Therefore, it has not been possible to investigate the replication cycle. In contrast, FV4 can be grown in cell culture, and produces cytopathic effects typical of herpesviruses within 10–21 days after infection of *R. pipiens* embryo or adult kidney cells at 25 °C. These include cell rounding and vacuolization, enlargement of nuclei and development of intranuclear inclusions, and formation of syncytia.

CCV causes acute infection only in young channel catfish up to about 6 months old, the degree of mortality depending on the strain of fish. Certain other closely related species, such as the blue catfish (*Ictalurus furcatus*), may be infected experimentally by injection, but other species are refractory even by this route. Host cell requirements are a little less stringent in cell culture, but virus growth still only occurs in ictalurid and clariid fish cell lines. Among commonly used cell lines BB (*Ictalurus nebulosus* or brown bullhead) and CCO (channel catfish ovary), the latter is more susceptible. Optimal virus yield is obtained at 25–30 °C, and about 50% of progeny virus is released into the culture medium. Virus infection causes the formation of syncytia, particularly at lower multiplicities of infection, followed eventually by disaggregation and lysis. Under optimal conditions, peak virus yield may be attained only 12 h after infection. Thus, CCV is one of the fastest growing herpesviruses *in vitro*.

Experimental determinations have shown that about 50 CCV proteins are expressed or induced in infected cell culture. As with other herpesviruses, expression appears to be coordinately regulated. The major immediate early protein, which is likely to regulate the expression of other virus genes, has an apparent molecular mass of 117 000 Da. The capsid contains an abundant constituent (the major capsid protein) with a molecular mass of 128 000 Da and at least two smaller proteins. The major capsid protein of higher vertebrate herpesviruses is invariably larger, having a molecular mass of about 150 000 Da. Different isolates of CCV may be differentiated by restriction endonuclease digestion of their DNAs, but are recognizably similar. CCV DNA replicates in the nuclei of infected cells via head-to-tail concatemers made up of a unit comprising the unique region linked to a single copy of the direct repeat.

KHV disease is restricted to koi and common carp, and other fish tested have been shown to be resistant even after extensive exposure to infected carp at permissive temperatures. A cell line derived from koi fin has been used to culture KHV, with optimal growth occurring at 15–25 °C and no growth at 10 or 30 °C. The highest virus yields are generated 7 days after infection at 20 °C, with most infectivity released into the culture medium. KHV strains are highly similar to each other as assessed by restriction endonuclease and sequence analyses. A total of 31 virion proteins have been detected. Studies of KHV gene expression are in their infancy.

SalHV-1 causes disease when injected into young rainbow trout maintained at 6–9 °C, but not in other salmonid species. A similar temperature optimum is characteristic of growth in the rainbow trout cell line RTG-2. SalHV-2 has a slightly wider host range, causing virulent disease in the young of several members of the genus *Oncorhynchus* and the rainbow trout. It has a higher temperature optimum for growth in RTG-2 cells than SalHV-1 (15 °C).

Genetics and Evolution

The complete genome sequences of two fish and two amphibian herpesviruses have been determined: those of CCV, KHV, LTHV, and FV4. Partial data are available for other viruses in this group. **Table 1** shows some of the basic characteristics of the sequenced genomes. All have a type A genome structure, consisting of a unique region (U) flanked by a direct repeat at the termini (TR). However, this structure is not universal among fish and amphibian herpesviruses, as the SalHV-1 genome has a type D structure. All of the sequenced genomes contain a number of families of genes that are related (usually distantly) to each other and have presumably arisen by gene duplication. KHV has the largest genome thus far reported for any herpesvirus.

Figure 3 shows the predicted gene layout in CCV, which has the smallest sequenced genome among the fish and amphibian herpesviruses. Analysis of the sequence indicates the presence of 62 genes in U (two spliced) and 14 genes in TR. Several gene functions have been predicted from computer-aided comparisons of predicted CCV proteins with proteins from other organisms. Ten genes belong to four gene families: two encoding related sets of protein kinases (PK1 with ORF14, ORF15, and ORF16; PK2 with ORF73 and ORF74), one encoding proteins similar to a

Table 1 Characteristics of sequenced fish and amphibian herpesvirus genomes

Virus	Genome size[a] (bp)	G+C composition (mol.%)	Genes[b]	Gene families
CCV	134 226 [18 556]	56.2	76	4
KHV	295 146 [22 469]	59.2	156	5
LTHV	220 859 [636]	54.6	132	15
FV4	231 801 [912]	52.8	147	15

[a]The total size of the DNA sequence is given, with the size of TR in square brackets.
[b]Genes that are duplicated by virtue of their presence in the terminal repeat are counted only once.

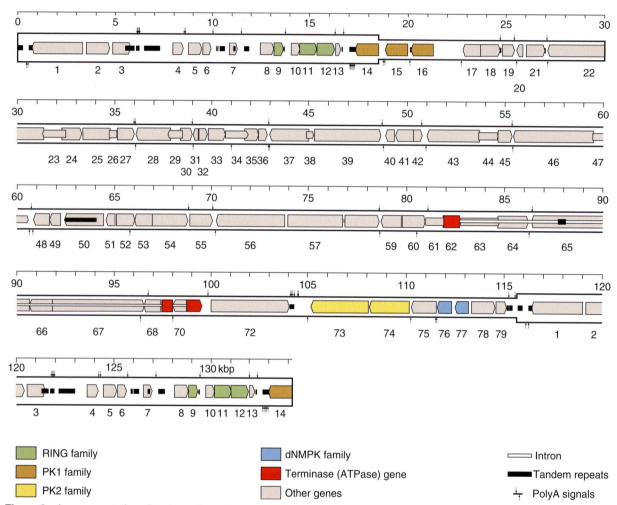

Figure 3 Arrangement of predicted protein-coding open reading frames (ORFs) in the CCV genome. ORFs are colored and numbered with the 'ORF' prefix omitted. The thinner portion of the genome denotes the unique region (U) and the thicker portions the terminal repeats (TR).

bacteriophage deoxynucleoside monophosphate kinase (dNMPK; ORF76 and ORF77), and one encoding proteins containing a potential zinc-binding domain (RING; ORF9, ORF11, and ORF12). Other genes encode a potential zinc-binding protein (ORF78), a protein containing a myosin-like domain (ORF22), a helicase (ORF25), a DNA polymerase (ORF57, composed of two exons), a deoxyuridine triphosphatase (ORF49), a thymidine kinase (ORF5), a subtilisin-like proprotein convertase (ORF47), the ATPase subunit of the putative terminase (ORF62, composed of three exons), a putative primase (ORF63), a protein containing an ovarian tumor (OTU)-like cysteine protease domain (ORF65), an envelope glycoprotein (ORF46), and seven other membrane-associated proteins (ORF6, ORF7, ORF8, ORF10, ORF19, ORF51, and ORF59). The principal constituent proteins of CCV virions have been identified by proteomic analysis. The capsid includes the major capsid protein (ORF39), a potential scaffold protein

(ORF28), and two other proteins that possibly constitute the intercapsomeric triplex (ORF27 and ORF53). Several virus proteins in the tegument have been identified, in addition to actin. The major envelope protein is specified by ORF59.

LTHV, FV4, and KHV have substantially more genes than CCV, and many of these are members of gene families or genes apparently captured from the cell or other viruses. For example, the frog herpesviruses, unlike the fish herpesviruses, encode a DNA (cytosine-5-)-methyltransferase and, presumably as a result, have extensively methylated genomes. Overall, the two frog herpesvirus genomes are the most closely related of the four, sharing 40 genes (when genes with marginal similarity are excluded). CCV shares 19 genes with the frog herpesviruses and only 15 with KHV. The modest number of conserved genes indicates that the evolutionary distance among lower vertebrate herpesviruses is substantially greater than that among higher vertebrate herpesviruses.

Where functions may be predicted for the conserved genes, they are central to herpesvirus replication, being involved in capsid morphogenesis, nucleotide metabolism, DNA replication, and DNA packaging. The arrangement of homologous genes in LTHV and FV4 is collinear, except for one block of genes that is in a different position and orientation in the two genomes. In contrast, homologous genes in the frog herpesviruses are situated in several rearranged blocks in the more distantly related fish herpesviruses. This parallels the situation among members of different subfamilies of higher vertebrate herpesviruses. As with higher vertebrate herpesviruses, homologous genes tend to be located centrally in the genomes.

The salmonid viruses SalHV-1 and SalHV-2 have been shown to be distinct viruses on the base of serological comparisons, DNA hybridization studies, and limited DNA sequencing. The SalHV-1 genome is approximately 174 kbp in size and has homologues of at least 18 CCV genes in several rearranged blocks. Sequence data indicate that these two viruses are related to each other and, more distantly, to CCV.

No proteins encoded by fish and amphibian herpesviruses are detectably related in their primary sequences to proteins that are specific to higher vertebrate herpesviruses (e.g., those that make up the capsid, of which the structure is conserved). This observation suggests that the two groups diverged from their common ancestor a long time ago. The best (and perhaps only) sequence-based evidence for a common evolutionary origin of the two groups depends on conservation of the putative ATPase subunit of the terminase, which is encoded by a spliced gene containing three exons (**Figure 3**). This protein is thought to be involved in packaging replicated DNA into the capsid and is also distantly conserved in T4-like bacteriophages, thus contributing to the notion that herpesviruses may share an ancient origin with bacteriophages.

Immune Response

Understanding of the immune responses to amphibian and fish herpesviruses is not well developed, and data obtained under relatively normal conditions of infection are limited largely to CCV. Adult channel catfish produce peak neutralization titers on average 8 or 9 weeks after primary immunization with CCV, and a further moderate increase in titer of short duration is apparent after boosting. The importance of virus growth in eliciting the immune response is indicated by the observation that heat-inactivated virus is poorly immunogenic. Serum neutralization indexes have been used to document a link between outbreaks of CCV disease and potential carriers of the virus. These studies, though not extensive, have given some indication that seroconverted adult fish are able to transmit CCV.

Control and Prevention

The devastating economic losses occasioned by certain of the fish herpesviruses prompt the need for intervention protocols. Avoidance and containment of outbreaks presently involve the use of ostensibly virus-free or partially resistant breeding stock, as clean environmental conditions as practicable, destruction of affected populations, and disinfection of ponds. The use of temperature variation regimes to aid the survival of infected fish has also been mooted. It is clear from the KHV situation, where individual fish may be very valuable, that effective diagnosis and monitoring of stock movements should be considered in order to avoid the spread of highly virulent diseases.

CCV is sensitive to nucleoside analogs that inhibit the growth of higher vertebrate herpesviruses. In general, these compounds are phosphorylated by the virus, but not the cellular, thymidine kinase. Further steps in phosphorylation probably involve cellular enzymes. The triphosphate form then inhibits DNA replication by direct interaction with the DNA polymerase or by incorporation into nascent DNA. These compounds include the widely used antiherpetic drug acyclovir. Phosphonoacetate, which inhibits the DNA polymerase by mimicking pyrophosphate, is also inhibitory. Against considerations of cost and administration, however, it is doubtful whether antiviral therapy will ever be generally practicable for the treatment of fish herpesvirus diseases.

The poor immunogenicity of inactivated CCV and KHV preparations suggests that the use of live, attenuated viruses is likely to be more effective in vaccination strategies. However, given the potential for latent vaccine virus to revert to wild type, thorough characterization of vaccine candidates and development of assays to differentiate vaccine from wild-type virus are essential (for a precedent in higher vertebrate herpesviruses, consider pseudorabies virus). The use of avirulent strains of virus in which virulence genes are replaced by detection markers may lead in the long term to safe, efficacious vaccines.

CCV attenuated by passage in a clariid fish cell line was able to provide substantial protection against the lethal effect of infection by virulent virus, particularly when the initial immunization was boosted. This virus exhibits several genomic differences from wild type, most notably a substantial deletion in ORF50, which potentially encodes a secreted glycoprotein. In this context, deletion of ORF5, which encodes thymidine kinase, has been shown to cause attenuation of virulence in channel catfish fingerlings. Attenuated versions of KHV produced by serial passage in koi fin cell culture and subjecting clones to further mutation *in vitro* did not induce lethal disease and protected immunized fish against challenge. This protection correlated with antibody production.

DNA vaccines also hold promise in combating fish herpesvirus diseases, particularly in their avoidance of utilizing intact virus. From seven CCV genes tested, ORF59 (encoding the major envelope glycoprotein) and ORF6 (encoding a predicted membrane protein) produced significant resistance to challenge, with combined vaccination even more effective.

Acknowledgment

The author is grateful to Ronald Hedrick for comments on the pathological features of KHV in light of his unpublished data.

See also: Herpesviruses: General Features; Pseudorabies Virus.

Further Reading

Aoki T, Hirono I, Kurokawa K, *et al.* (2007) Genome sequences of three koi herpesvirus isolates representing the expanding distribution of an emerging disease threatening koi and common carp worldwide. *Journal of Virology* 81: 5058–5065.

Booy FP, Trus BL, Davison AJ, and Steven AC (1996) The capsid architecture of channel catfish virus, an evolutionarily distant herpesvirus, is largely conserved in the absence of discernible sequence homology with herpes simplex virus. *Virology* 215: 134–141.

Davison AJ (1992) Channel catfish virus: A new type of herpesvirus. *Virology* 186: 9–14.

Davison AJ, Cunningham C, Sauerbier W, and McKinnell RG (2006) Genome sequences of two frog herpesviruses. *Journal of General Virology* 87: 3509–3514.

Gilad O, Yun S, Adkison MA, *et al.* (2003) Molecular comparison of isolates of an emerging fish pathogen, koi herpesvirus, and the effect of water temperature on mortality of experimentally infected koi. *Journal of General Virology* 84: 2661–2667.

Granoff A (1972) Lucké tumor-associated viruses – a review. In: Biggs PM, de-The G and Payne LN (eds.) *Oncogenesis and Herpesviruses*, pp. 171–182. Lyons: International Agency for Research on Cancer.

Granoff A (1983) Amphibian herpesviruses. In: Roizman B (ed.) *The Herpesviruses*, vol. 2, pp. 367–384. New York: Plenum.

Ilouze M, Dishon A, and Kotler M (2006) Characterization of a novel virus causing a lethal disease in carp and koi. *Microbiology and Molecular Biology Reviews* 70: 147–156.

McKinnell RG, Lust JM, Sauerbier W, *et al.* (1993) Genomic plasticity of the Lucké renal carcinoma: A review. *International Journal of Developmental Biology* 37: 213–219.

Naegele RF, Granoff A, and Darlington RW (1974) The presence of Lucké herpesvirus genome in induced tadpole tumors and its oncogenicity: Koch–Henle postulates fulfilled. *Proceedings of the National Academy of Sciences, USA* 71: 830–834.

Pikarsky E, Ronen A, Abramowitz J, *et al.* (2004) Pathogenesis of acute viral disease induced in fish by carp interstitial nephritis and gill necrosis virus. *Journal of Virology* 78: 9544–9551.

Plumb JA (1989) Channel catfish virus. In: Ahne W and Kurstak E (eds.) *Viruses of Lower Vertebrates*, pp. 198–216. Berlin: Springer.

Wolf K (1983) Biology and properties of fish and reptilian herpesviruses. In: Roizman B (ed.) *The Herpesviruses*, vol. 2, pp. 319–366. New York: Plenum.

Fish Retroviruses

T A Paul, R N Casey, P R Bowser, and J W Casey, Cornell University, Ithaca, NY, USA
J Rovnak and S L Quackenbush, Colorado State University, Fort Collins, CO, USA

© 2008 Elsevier Ltd. All rights reserved.

Glossary

Cytopathic effect Morphologic changes in cells resulting from virus replication.
Dysplasia An abnormality in the appearance of cells and indicative of a preneoplastic change.
Hyperplasia An increase in the number of cells of an organ or tissue.
Leiomyosarcoma Neoplasia of smooth muscle cells.
Malpighian cells Cells of the deepest layer of the epidermis.
Neoplasia An abnormal growth of cells.

> **Oncogene** Cellular or viral gene whose products are capable of inducing a neoplastic phenotype.
> **Proto-oncogene** A normal cellular gene which when inappropriately expressed or mutated becomes an oncogene.

Introduction

Retroviruses have been documented in a wide range of vertebrate species including lower vertebrates such as frogs, snakes, sharks, fish, birds, and turtles. To date, however, the majority of these reported sequences represent only partial fragments of the retrovirus. In recent years, the complete genomic sequence of exogenous fish retroviruses from Atlantic salmon (*Salmo salar*), walleye (*Sander vitreus*), and snakehead (*Ophicephalus striatus*), and an endogenous retrovirus of zebrafish (*Danio rerio*) have been determined. Analysis of these viral genomes indicates a high degree of diversity among the fish retroviruses as well as several unique features compared to mammalian and avian retroviruses.

The etiological relationship between retroviruses and cancer is well established in mammalian and avian systems. Simple retroviruses can promote cell proliferation by inducing the ectopic expression of captured cellular oncogenes (viral transduction for acutely transforming viruses) in which an oncogene recombines into the viral genome, or by integration near or within the coding sequence of cellular proto-oncogenes (nonacute transforming viruses). Transcription factors encoded by some complex retroviruses, like human T-cell leukemia virus-1 (HTLV-1), also have nonacute oncogenic potential by deregulating cellular pathways controlling cell proliferation.

Like their mammalian and avian counterparts, many fish neoplastic/proliferative diseases are suspected of having a retroviral etiology. Retroviruses are reported to be associated with 13 spontaneous proliferative diseases of fish based on the observation of retrovirus-like particles and, in some cases, reverse transcriptase activity in neoplastic lesions. Seven of these diseases with putative viral etiologies display a seasonal cycle, that is, they develop and regress annually. Retrovirus-associated tumors of the skin have received the most attention, since they are most visible and easily sampled. Skin tumors with suspected viral etiologies have been found in white sucker (*Castostomus commersoni*), walleye, yellow perch (*Perca flavescens*), and European smelt (*Osmerus eperlanus*). Retroviruses have also been linked to lymphoma in northern pike (*Esox lucius*), leukemia in chinook salmon (*Oncorhynchus tshawytscha*), and leiomyosarcoma of the swimbladder in Atlantic salmon. One noteworthy piscine retrovirus, the snakehead fish retrovirus (SnRV), has not been associated with tumor induction. In most of these systems, the etiologic relationship of retroviral infections and neoplasms rests on circumstantial evidence such as observations of viral particles by electron microscopy and the presence of reverse transcriptase activity. However, sequencing of fish retroviruses has facilitated use of molecular-based diagnostic reagents and, in some cases, shed light on potential modes of pathogenesis. In particular, the walleye retroviruses, walleye dermal sarcoma virus (WDSV) and walleye epidermal hyperplasia virus type 1 (WEHV-1) and type 2 (WEHV-2), express a subset of accessory genes exclusively during oncogenesis, specifically implicating the encoded proteins in the process of tumorigenesis.

Retroviruses and their Life Cycle

Retrovirus particles are composed of viral structural and enzymatic proteins and two copies of the positive-sense single-stranded RNA genome. The proteins are surrounded by a lipid bilayer, acquired from the host cell, in which viral envelope glycoproteins are embedded. Retroviral infection is initiated when the surface glycoprotein binds to its cognate receptor on the surface of a susceptible cell. Upon fusion of the viral and cellular membranes, the viral core enters the cytoplasm where the virally encoded enzyme, reverse transcriptase, copies the RNA genome into DNA. The viral integrase protein binds reverse transcribed DNA and mediates its integration into the genomic DNA of the host. In rare cases, retroviral infection of host germ-line cells and subsequent integration into the chromosomes can establish retroviral sequences as heritable genetic elements, known as endogenous retroviruses.

Based on the complexity of the viral genome, retroviruses can be divided into two broad categories: simple and complex. Simple retroviruses, including members of the *Gammaretroviruses* and *Alpharetroviruses*, contain three genes: *gag*, *pol*, and *env*. Expression of a full-length unspliced messenger RNA (mRNA) and a spliced mRNA serve as the templates for translation of the retroviral structural and enzymatic proteins. These proteins are first translated as polyproteins from *gag* and *pol*, respectively, and are processed by a virally encoded protease. The viral envelope glycoprotein is encoded from a singly spliced viral transcript containing the *env* gene.

Complex retroviruses are distinguished from simple retroviruses on the basis of their additional coding capacity and pattern of viral gene expression. This category of retroviruses includes *Lentiviruses* such as human immunodeficiency virus type 1 (HIV-1), *Spumaviruses*, and the *Deltaretroviruses*, human T-cell leukemia virus (HTLV-1), and bovine leukemia virus (BLV). Complex retroviruses produce multiply spliced transcripts that encode two classes of accessory/regulatory proteins. One class includes transcriptional regulatory proteins such as HTLV-1 Tax and HIV-1 Tat, which act *in trans* to directly regulate the activity of the viral promoter. A second class, including HTLV-1 Rex and HIV-1 Rev, act post transcriptionally, to facilitate transport of unspliced and singly spliced viral transcripts to the cytoplasm. The combined action of these proteins divides the replication cycle into two temporal phases: an early, regulatory phase, and a later, structural phase. During the regulatory phase, low levels of fully spliced transcripts encoding *trans*-activators increase levels of transcription leading to the accumulation of post-transcriptional regulatory proteins. These proteins allow production of unspliced and singly spliced mRNAs that encode the viral structural proteins.

Both simple and complex retroviruses have been identified in fish species. Retroviruses isolated from walleye and snakehead fish resemble complex retroviruses based upon their genome structure and transcriptional profile, while simple retroviruses have been identified in Atlantic salmon and zebrafish.

Skin Tumors in Walleye and Their Associated Retroviruses

Walleye dermal sarcoma (WDS) and walleye discrete epidermal hyperplasia (WEH) are neoplastic and hyperplastic skin lesions in walleye that have been etiologically associated with infection by three distinct retroviruses (**Figure 1**). WDS and WEH were first reported in 1969 on fish from Oneida Lake in New York. Since then, WDS and WEH have been reported on walleyes throughout North America. These diseases have a seasonal cycle; tumor incidence is highest in the late fall and early spring months at frequencies of 27% and 5% for WDS and WEH, respectively. Lower tumor prevalence in summer months is associated with regression on individual adult fish.

WDS appears as a cutaneous mesenchymal neoplasm arising multicentrically within the superficial dermis and overlaid with epidermis. Fall tumors appear highly vascularized due to a network of capillaries, while spring tumors are frequently white with ulcerated surfaces. Invasive tumors are rarely observed in feral fish. WEH lesions are benign mucoid-like plaques with distinct boundaries. They can range from 2 to 3 mm in diameter to large lesions with irregular borders that may be as large as 50 mm across. Histologically, this disease appears as an epidermal proliferation consisting primarily of Malpighian cells.

Experimental transmission of WDS and WEH has been achieved with cell-free filtrates. Interestingly, only cell-free filtrates prepared from the spring, regressing WDS, not fall, developing WDS, are able to transmit disease. Infection of walleye fingerlings less than 9 weeks of age results in the frequent development of invasive tumors.

Electron microscopic observation of retrovirus particles and cell-free transmission of both diseases suggest a retroviral etiology for WDS and WEH. WDSV was isolated from WDS tissue, and two independent retroviruses, WEHV1, WEHV2, were isolated from WEH lesions. Similar hyperplastic lesions on yellow perch in Oneida Lake are associated with two new retroviruses, perch epidermal hyperplasia virus types 1 and 2 (PEHV1, PEHV2).

Isolation and Sequencing of Retroviruses from WDS and WEH Lesions

WDSV, WEHV1, and WEHV2 have been molecularly cloned and sequenced (**Figure 2**). The genome structures of WDSV, WEHV1, and WEHV2 indicate large and complex viral genomes (12.7, 12.9, and 13.1 kbp, respectively).

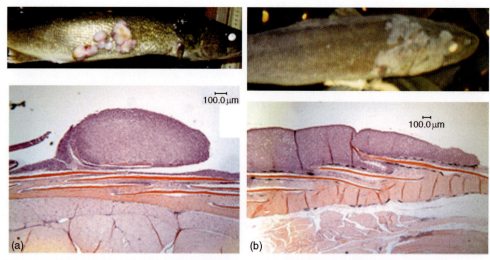

Figure 1 (a) Gross and histological images of walleye dermal sarcoma and (b) walleye epidermal hyperplasia.

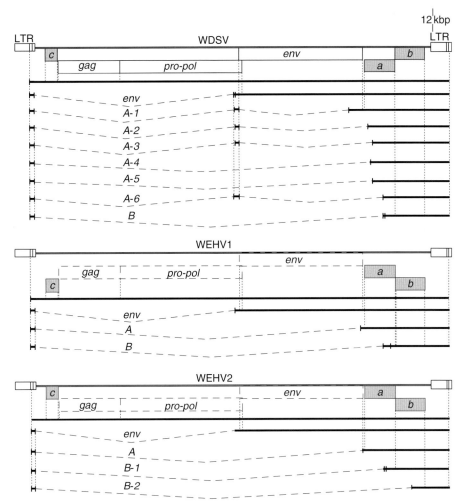

Figure 2 Genomic and transcriptional maps of WDSV, WEHV1, and WEHV2. Genomic organization of viral DNA with retroviral genes and accessory genes *orfa*, *orfb*, and *orfc* is depicted under each virus heading. The envelope-spliced transcript (*env*) and additional spliced transcripts (*A* and *B*), which are capable of expressing accessory genes, are illustrated. The vertical dashed lines indicate the boundaries of the exons, introns, and open reading frames.

All three viruses have intact open reading frames (ORFs) capable of encoding the structural and enzymatic genes *gag*, *pol*, and *env*. *gag* (viral capsid) and *pol* (reverse transcriptase) are in the same reading frame and synthesized as a polyprotein through a termination suppression mechanism. *pro* (viral protease), responsible for cleaving the polyprotein, is located in the same reading frame as *pol*. A unique feature of all three viruses is the presence of additional orfs that have tentatively been called *orfa*, *orfb*, and *orfc*. The *orfa* and *orfb* genes are located between *env* and the 3′ long terminal repeat (LTR), while *orfc* is located between the 5′ LTR and the start of *gag*. Database searches with OrfA protein amino acid sequences suggest limited homology with D-type cellular cyclins within the cyclin-box motif, and are referred to as rv-cyclins. WEHV1 rv-cyclin has a 20% amino acid identity and 35% similarity with human cyclin D-3, whereas WEHV2 and WDSV rv-cyclin are most similar to human cyclin D-1 (22%/35% and 19%/29% amino acid identity/similarity). Similarities with known walleye and other piscine cyclins are not significantly greater. The WEHV cyclin homologs are 37% identical within the cyclin box motif and 21–28% identical with the WDSV cyclin. Sequence homologies between rv-cyclin and OrfB in the walleye retroviruses suggest they arose by a gene duplication event.

Roles of the WDSV Accessory Proteins in Pathogenesis

Transcriptional mapping by RT-PCR and Northern blot analyses of developing and regressing WDS and WEH have demonstrated temporal gene expression profiles and complex splicing patterns analogous to those seen in the complex retroviruses of mammals. In developing WDS tumors, only low levels of multiply spliced subgenomic transcripts are detected. These transcripts predominantly contain the coding sequences for rv-cyclin and OrfB. Regressing spring tumors contain high levels of full-length and subgenomic transcripts as well as unintegrated viral DNA and transmissible virus. OrfC is likely encoded by the full-length transcript coincident with tumor regression.

A detailed transcriptional analysis of WDSV showed an alternative splicing pattern of the *orfa* transcript (**Figure 2**). In developing tumors, the *orfa* transcript contains the coding sequence for a full-length rv-cyclin protein that localizes to the nucleus of mammalian and walleye cells. Alternatively spliced forms of this transcript encode amino-terminal truncated forms of the rv-cyclin protein, which localize in the cytoplasm of cells. The different forms of rv-cyclin may play functionally different roles in developing and regressing tumors. A single, spliced transcript encodes the OrfB protein, which is predominantly localized in the cytoplasm in tumor explant cells, but is capable of shuttling into and out of the nucleus when expressed in piscine and mammalian cells.

The OrfC protein from WDSV has been shown to localize to mitochondria and to disrupt mitochondrial function, which results in apoptosis. Only the full-length viral transcript found in regressing tumors is capable of encoding OrfC. Apoptotic cells are present in regressing, but not developing, WDS tumor sections. Therefore, the OrfC protein may be responsible for a direct viral mechanism of tumor regression.

WDSV *orfa* and *orfb* are the only virus transcripts found in developing tumors, suggesting direct roles for rv-cyclin and OrfB in the process of tumorigenesis, and their oncogenic potential has been demonstrated by several experimental approaches. WDSV *orfa* supported growth of a yeast (*Saccharomyces cerevisiae*) strain conditionally deficient for the synthesis of the G1 to S cyclins that are necessary for cell cycle progression. WEHV *orfa* did not support yeast growth in this model. Transgenic expression of WDSV *orfa* in mice from a skin-specific promoter caused a moderate to severe squamous epithelial hyperplasia and dysplasia dependant on skin injury. This suggests that rv-cyclin is not oncogenic as a result of a 'single-hit' mechanism, but rather, that secondary genetic or epigenetic events (possibly wound repair) are necessary for tumor development and progression. A similar phenotype has been described in transgenic mice expressing *v-jun* oncogene from the H-2K^k major histocompatibility complex (MHC) class I antigen gene promoter. When expressed in mammalian or piscine cell culture, WDSV rv-cyclin localized in the nucleus and was found by co-immunoprecipitation to be associated with cyclin dependent kinase 8 and cyclin C, general transcription initiation factors, and RNA polymerase II transcription complexes. In piscine cells, rv-cyclin inhibited transcription from the WDSV promoter independent of *cis*-acting DNA sequences. However, rv-cyclin can activate other viral promoters in fish cells, and can activate the WDSV promoter in select mammalian cells; thus rv-cyclin activation and inhibition of transcription are dependent on both the promoter and the cell type. WDSV rv-cyclin contains a defined transcription activation domain in the carboxy end of the protein, and the isolated activation domain is capable of interacting with co-activators of transcription. WEHV rv-cyclins do not have a corresponding carboxy-region activation domain. WDSV rv-cyclin may exhibit oncogenic potential by differential regulation of host gene expression such as proto-oncogenes or tumor suppressors.

Less is known about the capabilities of the OrfB protein, but initial studies indicate its direct association with the regulation of signal transduction pathways. It is concentrated, in tumor explant cells, at focal adhesions, and along actin stress fibers. Established OrfB-expressing lines are resistant to the chemical induction of apoptosis, suggesting a role in oncogenesis.

The origins of the accessory genes of complex retroviruses are unclear. This includes the origins of the WDSV accessory genes encoding rv-cyclin, OrfB, and OrfC. In the case of rv-cyclin and OrfB, their extreme divergence from host cyclin sequences indicates that any transduction event was ancient and led, ultimately, to an exclusive, complex viral species. The OrfC protein, like the accessory proteins of other complex retroviruses, has no clear homology to host proteins.

Control of the Seasonal Cycle of Disease

An interesting aspect of WDS is the control of the seasonal switch in viral gene expression and associated tumor regression. Potential modulators of viral gene expression include host immunity and endocrine activity (accompanying spawning) and environmental factors such as water temperature, physical trauma, and sunlight. A number of *cis*-acting elements important for transcription activation have been identified in the WDSV promoter. There is differential binding of proteins from developing and regressing tumor nuclear extracts to a 15 bp repeat region in the WDSV promoter. This element may be critical to the induction of high levels of virus expression. The WDSV rv-cyclin protein negatively regulates the WDSV promoter in tissue culture cells. Presumably, it is advantageous for the virus to have lower levels of gene expression during tumor development to avoid immune surveillance as well as the cytopathic effects associated with virus production. While rv-cyclin may function in the repression of virus expression during tumor growth, the host, environmental, or viral signals that switch on full virus expression are yet to be determined.

Snakehead Retrovirus

The SnRV was isolated from a productively infected cell line derived from a Southeast Asian striped snakehead fish. Cell culture supernatant from these infected cells demonstrated high levels of RT activity and the presence of type C-like retrovirus particles. Additionally, supernatant from the infected cell line induced cytopathic effects in cultures of a bluegill cell line (BF-2).

Molecular approaches were utilized to identify and sequence the SnRV from infected cells. The large 11.2 kbp genome of SnRV contains intact coding regions for *gag*, *pol*, and *env* (**Figure 3**). An arginine tRNA primer binding site used for reverse transcription initiation distinguishes SnRV from other retroviruses. The structure and transcriptional profile of SnRV suggests a complex expression pattern capable of encoding an ORF located between *env* and the 3′ LTR and two very small ORFs termed ORF1 and ORF2. Within the leader sequence of SnRV resides a start codon located just upstream of the major splice donor site that could potentially encode a 14 amino acid peptide (LP). Expression of Env from a singly spliced transcript is predicted to utilize the LP initiation codon and fuses this 14 amino acid peptide in frame to downstream *env* sequences. A transcript with four exons and two initiator codons would encode ORF1 and ORF2 proteins. The 3′ ORF would be expressed from a transcript containing three exons and would encode a protein of 24 kDa. Finally, the fourth spliced transcript may encode 3′ ORF or possibly a LP-Env cytoplasmic domain (CD) fusion protein. The 3′ ORF contains an N terminal acidic domain, cysteine residues, and a basic region, motifs commonly found in transcriptional activators. These small ORFs have no significant homology to any known proteins in the databases, and their role in the viral life cycle is unknown. No endogenous copies of SnRV have been identified in the snakehead genome and an uninfected cell line has been established.

Although the presence of SnRV in wild populations has not been thoroughly examined, the virus has been independently isolated from two separate snakehead cell lines derived from whole fry tissue and from caudal peduncle tissue from juvenile fish. In all cases, fish from which these cell lines were derived appeared healthy.

Salmon Swimbladder Sarcoma-Associated Retrovirus

An outbreak of neoplastic disease of the swimbladder of Atlantic salmon was first reported at a Scottish commercial marine fish farm in 1975. Affected salmon were sluggish and in poor condition. A viral etiology was suspected, and electron microscopic evaluation of tumors revealed the presence of budding viral particles with retroviral morphology. A second outbreak occurred between 1995 and 1997 in brood stock salmon held at the North Attleboro National Fish Hatchery in Massachusetts. The fish exhibited skin discoloration and hemorrhages on the fins and body and showed a general debilitation and lack of vigor. In May of 1997, significant mortality was noted at the facility such that 35% of the population was affected by late spring of 1998. All of the affected fish displayed swollen abdomens due to multinodular masses on internal and external surfaces of the swimbladder. In several cases, these multinodular masses occupied the entire swimbladder (**Figure 4**). Histologic examination revealed that the tumors were composed of well-differentiated fibroblastic cells that were arranged in interlacing bundles, which were classified as leiomyosarcomas.

Isolation of a Retrovirus from Salmon Swimbladder Sarcoma

An exogenous retrovirus, termed Atlantic salmon swimbladder sarcoma virus (SSSV), was initially identified in tumors by degenerate RT-PCR of tumor RNA, and the entire viral sequence completed by DNA sequencing of

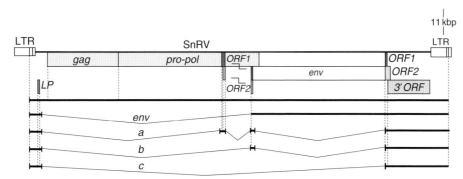

Figure 3 Genomic and transcriptional map of SnRV. Predicted coding regions are shown as open or shaded boxes. The envelope-spliced transcript (*env*) and additional spliced transcripts (*a*, *b*, *c*), which are capable of expressing accessory genes, are illustrated. LP indicates the location of a predicted leader peptide. The vertical dashed lines indicate the boundaries of the exons, introns, and open reading frames.

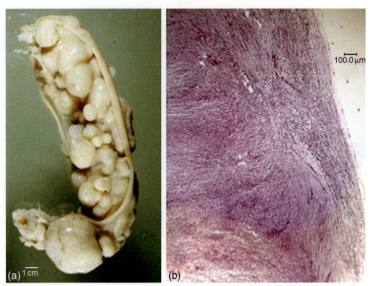

Figure 4 (a) Gross and (b) histological images of Atlantic salmon swimbladder sarcoma.

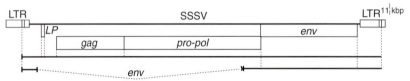

Figure 5 Genomic and transcriptional map of SSSV. Predicted coding regions are shown as open boxes. The envelope-spliced transcript (env) is illustrated. LP indicates the location of a predicted leader peptide. The vertical dashed lines indicate the boundaries of the exons, introns, and open reading frame.

proviral DNA. In contrast to the complex walleye retroviruses, SSSV is a simple retrovirus. The viral genome contains ORFs capable of expressing *gag*, *pol*, and *env* (**Figure 5**). Additionally, a short 25 amino acid leader peptide of unknown function is located upstream of the Gag–Pol polyprotein ORF. SSSV differs from other simple retroviruses by not having related endogenous sequences in the host genome, and SSSV is the only retrovirus to use a methionine-tRNA as a plus-strand primer. Additionally, sequences in *pol* display homology to central polypurine tract regions identified in complex retroviruses. Central polypurine tracts facilitate the formation of a single-stranded DNA region called a central DNA flap as a result of reverse transcription priming from this internal site. In HIV-1, this feature plays a major role in complex retroviral replication and allows efficient infection of nondividing cells. It is intriguing that SSSV displays a high proviral copy number (greater than 30 copies per cell) with a polyclonal integration pattern in swimbladder tumors. SSSV must be capable of initiating multiple rounds of infection within the same cell. The specific mechanisms leading to the high copy number and its implications in the pathogenesis of disease are of significant interest for future research.

Sequence Comparisons of SSSV with Other Retroviruses

A comparison of the sequence of SSSV viral proteins with other retroviruses indicates large regions of homology with mammalian type C and type D viruses in Pol and Env. Additionally, a 179 amino acid region in the C terminus of Gag and a 1064 amino acid region of Pro–Pol displays 23% and 33% identity to the related WDSV proteins, respectively. SSSV has striking homology to the sequence of the zebrafish endogenous retrovirus (ZFERV). BLAST analysis of the Gag, Pol, and Env ORFs of SSSV reveals a 25% identity with ZFERV over 533 amino acids within Gag, a 40% identity over 533 amino acids of Pol, and a 39% identity over 429 amino acids of Env.

Prevalence, Seasonality, and Transmission of SSSV

A PCR diagnostic assay was developed using sequences within the *pol* gene of SSSV. Prevalence at the North Attleboro fish facility has been found to be 52% and 5% of a natural stock of Pleasant River salmon were found to harbor SSSV.

Like other retroviral diseases of fish, observations suggest that salmon swimbladder sarcoma is seasonal. First, of the 34% fish mortality at the North Attleboro facility between 1997 and 1998, 57% occurred in June. Second, in the following year, the surviving salmon had an SSSV incidence (measured by PCR) that cycled with a peak in late summer to early winter and then diminished in late winter to early spring. The period of highest SSSV incidence correlated with salmon spawning runs in late fall. Interestingly, WDSV prevalence, which peaks during late spring, also correlates with the time of spawning. This suggests that endocrine changes during spawning runs may be a critical factor in the observed seasonal variation in disease incidence.

Pathogenesis of SSSV

The existence of SSSV sequences in association with an outbreak of the swimbladder sarcoma suggests a role for the retrovirus in pathogenesis of the disease. As a simple retrovirus with no transduced cellular oncogenes, one possible mechanism of tumorigenesis is the insertional activation of a cellular proto-oncogene. The high copy number of proviruses in tumors has made the analysis of insertional activation events difficult, but multiple insertions increase the likelihood of such a mechanism. It is also feasible that SSSV may express an oncogenic viral gene product. This mechanism of oncogenesis has been proposed for the Jaagsiekte sheep retrovirus (JSRV) in the induction of ovine pulmonary adenocarcinoma and in Friend spleen focus-forming virus (SFFV)-associated erythroid hyperplasia in mice. Additionally, the presence of common integration sites within JSRV-associated tumors suggests that insertional mutagenesis may act in concert with the envelope in tumor development. Thus, a multifactorial mechanism for the development of salmon swimbladder sarcoma must also be considered.

In addition to sarcoma, a number of other distinct pathologies are associated with SSSV infection. These diseases are frequently debilitating and are, therefore, of significance to the aquaculture industry. Atlantic salmon at the North Attleboro hatchery, during the SSSV outbreak, presented with multifocal hemorrhages, sloughing of the epidermis, lethargy, wasting, and failure to mature sexually in addition to swimbladder tumors.

Zebrafish Endogenous Retrovirus

Endogenous retroviruses have been identified in almost all vertebrate genomes, and most are defective due to mutations and deletions. An endogenous retrovirus, ZFERV with a genome of 11.2 kbp has been identified in the Tubingen stock of zebrafish, and contains intact coding regions for *gag*, *pol*, and *env*. The *gag* and *pol* genes are in the same reading frame. While the majority of endogenous retroviruses are transcriptionally silent because of mutation or methylation, ZFERV remains transcriptionally active. In addition to genomic transcripts that encode Gag and Pol proteins, an unusual multiply spliced *env* transcript is produced. Expression appears highest in the larval and adult zebrafish thymus, and no expression was detected in 2-day old embryos, suggesting that ZFERV expression may be tied to thymic development.

Phylogeny of Fish Retroviruses

Retroviruses have been classified into seven genera based largely on highly conserved amino acid sequences in the retroviral reverse transcriptase gene. While the majority of the viral sequences employed in this classification represent mammalian and avian retroviruses, a new genus termed *Epsilonretroviruses*, representing the fish retroviruses WDSV, WEHV-1, WEHV-2, has been added to the most recent classification. As more retroviral sequences from lower vertebrates have been identified, it has become apparent that this classification scheme may be inadequate to represent the apparent diversity.

Based on the unique characteristics of SnRV, including its genomic organization, tRNA primer, and complex transcriptional profile, the virus is yet to be definitively placed in the current classification. The large size of the genome, genetic organization, and presence of additional ORFs suggest that SnRV is closest to the spumaviruses and walleye retroviruses, but its limited sequence homology suggests SnRV is divergent from these groups.

Phylogenetic analysis indicates that, while the walleye retroviruses cluster in a group representing the *Epsilonretroviruses*, SSSV and ZFERV appear to represent a new branch of piscine retroviruses between the walleye retroviruses and the *Gammaretroviruses*, a genera that includes the murine leukemia virus (MLV)-related retroviruses (**Figure 6**). The SnRV appears quite divergent from the other fish retroviruses by its placement in a distinct branch near the *Spumaviruses*. This would suggest that SnRV is quite divergent from the genus *Epsilonretrovirus* and may represent yet another group of retroviruses. Interestingly, a more encompassing phylogenetic analysis using all known retroviral sequences from lower vertebrates, including partial endogenous retroviral *pol* fragments from Stickleback (*Gasterosteus aculeatus*), brook trout (*Salvelinus fontinalis*), Brown trout (*Salmo trutta*), freshwater whiting (*Corogonus lavaretus*), and puffer fish (*Fugu rubripes*), indicates that the majority of the fish viruses, excluding SnRV, cluster together with MLV-related viruses in a group

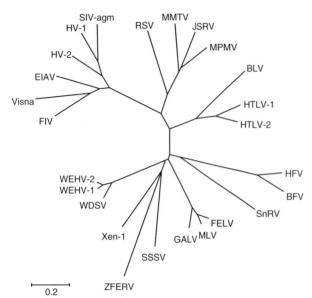

Figure 6 Unrooted phylogenetic tree of representative retroviruses based on an amino acid alignment of seven conserved domains in reverse transcriptase. Retroviruses are designated as follows: MPMV (Mason–Pfizer monkey virus), JSRV (Jaagsiekte sheep retrovirus), MMTV (mouse mammary tumor virus), RSV (Rous sarcoma virus), EIAV (equine infectious anemia virus), FIV (feline immunodeficiency virus), Visna (visna virus), HIV-2 (human immunodeficiency virus-2), HIV-1 (human immunodeficiency virus-1), SIV-agm (simian immunodeficiency virus-agm), HTLV-1 (human T-cell leukemia virus type 1), HTLV-2 (human T-cell leukemia virus 2), BLV (bovine leukemia virus), SSSV (salmon swimbladder sarcoma virus), SnRV (snakehead retrovirus), BFV (bovine foamy virus), HFV (human foamy virus), FeLV (feline leukemia virus), MLV (murine leukemia virus), GALV (gibbon ape leukemia virus), WDSV (walleye dermal sarcoma virus), WEHV-1 (walleye epidermal hyperplasia virus type 1), WEHV-2 (walleye epidermal hyperplasia virus type 2), ZFERV (zebrafish endogenous retrovirus), Xen-1 (Xenopus endogenous retrovirus-1).

separate from most non-MLV related mammalian retroviruses. This raises the possibility that some retroviral groups maybe restricted to particular vertebrate classes. However, it is evident from the diversity among the fish retroviruses, that there is a high degree of heterogeneity within this group.

Acknowledgment

This research was supported in part by USDA grants 99-35204-7485 and 02-35204-12777 to J.W.C., National Oceanic and Atmospheric Administration award no. NA86RG0056 to the Research Foundation of State University of New York for New York Sea Grant to P.R.B., American Cancer Society grant RPG-00313-01-MBC to S.L.Q., and National Institutes of Health grant CA095056 to S.L.Q. T.A.P. was supported by National Institutes of Health training grant 5T32CA09682.

See also: Equine Infectious Anemia Virus; Feline Leukemia and Sarcoma Viruses; Foamy Viruses; Human Immunodeficiency Viruses: Antiretroviral agents; Human Immunodeficiency Viruses: Molecular Biology; Human Immunodeficiency Viruses: Origin; Human Immunodeficiency Viruses: Pathogenesis; Human T-Cell Leukemia Viruses: General Features; Human T-Cell Leukemia Viruses: Human Disease; Mouse Mammary Tumor Virus; Reticuloendotheliosis Viruses; Retroviruses of Birds; Simian Retrovirus D; Visna-Maedi Viruses.

Further Reading

Bowser PR, Wolfe MJ, Forney JL, and Wooster GA (1988) Seasonal prevalence of skin tumors from walleye (*Stizostedion vitreum*) from Oneida Lake, New York. *Journal of Wildlife Diseases* 24: 292–298.

Coffin JM, Hughes SH, and Varmus HE (eds.) (1997) *Retroviruses*, Cold Spring Harbor, NY: Cold Spring Harbor Laboratory Press.

Hart D, Frerichs GN, Rambaut A, and Onions DE (1996) Complete nucleotide sequence and transcriptional analysis of the snakehead fish retrovirus. *Journal of Virology* 70(6): 3606–3616.

Hernious E, Martin J, Miller K, Cook J, Wilkinson M, and Tristem M (1998) Retroviral diversity and distribution in vertebrates. *Journal of Virology* 72(7): 5955–5966.

Holzschu DL, Fodor SK, Quackenbush SL, et al. (1995) Nucleotide sequence and protein analysis of a complex piscine retrovirus, walleye dermal sarcoma virus. *Journal of Virology* 69(9): 5320–5331.

Lairmore MD, Stanley JR, Weber SA, and Holzschu DL (2000) Squamous epithelial proliferation induced by walleye dermal sarcoma retrovirus cyclin in transgenic mice. *Proceedings of the National Academy of Sciences, USA* 97(11): 6114–6119.

LaPierre LA, Casey JW, and Holzschu DL (1998) Walleye retroviruses associated with skin tumors and hyperplasias encode cyclin D homologs. *Journal of Virology* 72: 8765–8771.

LaPierre LA, Holzschu DL, Bowser PR, and Casey JW (1999) Sequence and transcriptional analyses of the fish retroviruses walleye epidermal hyperplasia virus types 1 and 2: Evidence for a gene duplication. *Journal of Virology* 73(11): 9393–9403.

Paul TA, Quackenbush SL, Sutton C, Casey RN, Bowser PR, and Casey JW (2006) Identification and characterization of an exogenous retrovirus from Atlantic salmon swimbladder sarcomas. *Journal of Virology* 80(6): 2941–2948.

Poulet FM, Bowser PR, and Casey JW (1994) Retroviruses of fish, reptiles, and molluscs. In: Levy JA (ed.) *The Retroviridae*, vol. 3, pp. 1–38. New York: Plenum.

Quackenbush SL, Holzschu DL, Bowser PR, and Casey JW (1997) Transcriptional analysis of walleye dermal sarcoma virus (WDSV). *Virology* 237: 107–112.

Rovnak J, Hronek BW, Ryan SO, Cai S, and Quackenbush SL (2005) An activation domain within the walleye dermal sarcoma virus retroviral cyclin protein is essential for inhibition of the viral promoter. *Virology* 342(2): 240–251.

Rovnak J and Quackenbush SL (2002) Walleye dermal sarcoma virus cyclin interacts with components of the mediator complex and the RNA polymerase II holoenzyme. *Journal of Virology* 76: 8031–8039.

Shen CH and Steiner LA (2004) Genome structure and thymic expression of an endogenous retrovirus in zebrafish. *Journal of Virology* 78(2): 899–911.

Fish Rhabdoviruses

G Kurath and J Winton, Western Fisheries Research Center, Seattle, WA, USA

Published by Elsevier Ltd.

Glossary

Cytopathic effect (CPE) Damage caused in cultured cells due to insult such as infection with virus.

Genogroup Grouping of virus isolates within a species based on genetic data, such as nucleotide sequences and phylogenetic analyses.

Novirhabdovirus A member of the taxonomic genus *Novirhabdovirus*, within the family *Rhabdoviridae*. This genus contains many important fish rhabdoviruses.

Nucleocapsid For a rhabdovirus, the inner structure of the virion comprised of the viral genomic RNA wrapped in nucleocapsid (N) protein with associated phosphoprotein (P) and polymerase (L) proteins.

Spill-over and spill-back events Transmission of pathogens, and consequently disease, between wild and cultured populations of animals (in this case, fish).

Vesiculovirus A member of the taxonomic genus *Vesiculovirus*, within the family *Rhabdoviridae*. This genus currently contains only mammalian rhabdoviruses but several fish rhabdoviruses are very closely related to this genus.

Introduction

Some of the most significant viral pathogens of fish are members of the family *Rhabdoviridae*. The viruses in this large and important group cause losses in populations of wild fish as well as among fish reared in aquaculture. While many of the best known fish rhabdoviruses produce acute disease and high mortality, others have been isolated from chronic or asymptomatic infections. Fish rhabdoviruses often have a wide host and geographic range, and infect aquatic animals in both freshwater and seawater.

The diseases caused by fish rhabdoviruses are generally characterized as acute, hemorrhagic septicemias affecting multiple organs. Death is usually due to organ failure and subsequent loss of osmoregulation. In addition to petechial hemorrhages, gross signs include accumulation of ascites, darkening, and exophthalmia. Internally, necrosis of multiple organs is evident upon histological examination. Mortality is frequently highest in younger fish and can be explosive in aquaculture settings, resulting in losses approaching 100% in some cases. Fish that survive infection typically develop protective immunity. Transmission of fish rhabdoviruses occurs both horizontally between fish, and vertically through egg-associated transmission from an infected adult to its progeny. The role of long-term carriers and reservoirs is an area requiring further research.

Understanding of fish rhabdovirus infections improved significantly following the establishment of a variety of fish cell lines, beginning in the 1960s. Infection of cultured cell lines produces typical cytopathic effects (CPEs) consisting of cell rounding and necrosis following the release of mature virions by budding. The rate of the appearance of CPE is related to the incubation temperature, which is typically set to mirror the temperature optimum for the fish host (e.g., 10–25 °C). Highly reproducible laboratory challenge models have been developed for several important fish rhabdoviruses and their principal hosts. These tools have been used extensively to further our understanding of host–pathogen relationships and the role of environmental factors such as temperature on the virus replication cycle and the disease process.

The rhabdoviruses recovered from fish and described in the literature comprise a diverse collection of isolates (**Table 1**). Those studied in detail have been assigned to one of two quite different groups: isolates that are members, or likely members, of the established genus, *Novirhabdovirus*, and those fish rhabdovirus isolates that are most similar to members of the genus *Vesiculovirus* (**Figure 1**). Formal species names have been assigned to only four fish rhabdoviruses, all within the genus *Novirhabdovirus*, while several others are recognized as tentative members of a genus. There are also several partially characterized or unusual isolates that have been reported as rhabdoviruses based largely on their characteristic bullet-shaped morphology in electron micrographs. Upon further study, some of the fish rhabdoviruses listed in **Table 1** may be shown to be the same virus species.

Novirhabdoviruses of Fish

The genus *Novirhabdovirus* was established in 1998, and includes four official species as shown in **Table 1**. Infectious hematopoietic necrosis virus (IHNV) and viral hemorrhagic septicemia virus (VHSV) are economically significant viruses that cause epidemics in salmonid fish. IHNV is endemic in salmonids of northwestern North America and has become established in Europe and Asia through historical aquaculture-related activities. VHSV

Table 1 Fish rhabdoviruses from the published literature and their status in the *Eighth Report of the International Committee on Taxonomy of Viruses* (ICTV). Virus names shown as indented are isolates described in the literature that are either now considered to be indistinguishable from a recognized species (indicated in parentheses), or those that appear to be similar to the virus directly above but require further characterization

Named virus	Abbreviation	Status in 8th ICTV
I. Species and tentative species in the genus *Novirhabdovirus*		
Infectious hematopoietic necrosis virus	IHNV	Type species of the genus
Oregon sockeye virus (IHNV)	OSV	
Sacramento River Chinook virus (IHNV)	SRCV	
Viral hemorrhagic septicemia virus	VHSV	Formal species in genus
Egtved virus (VHSV)		
Brown trout rhabdovirus (VHSV)		
Cod ulcus syndrome rhabdovirus (VHSV)		
Carpione brown trout rhabdovirus (VHSV)	583	
Hirame rhabdovirus	HIRRV	Formal species in genus
Snakehead virus	SHRV	Formal species in genus
Eel virus B12	EEV-B12	Tentative species in genus
Eel virus C26	EEV-C26	Tentative species in genus
II. Vesiculovirus-like viruses from fishes		
Spring viremia of carp virus	SVCV	Tentative species in genus
Swim bladder inflammation virus (SVCV)	SBI	
Pike fry rhabdovirus	PFRV	Tentative species in genus
Grass carp rhabdovirus	GCV	None
Tench rhabdovirus		None
Eel virus American	EVA	Tentative species in genus
Eel virus European X	EVEX	None
Eel virus C30, B44 and D13	C30, B44, D13	None
Rhabdovirus anguilla	None	
Ulcerative disease rhabdovirus	UDRV	Tentative species in genus
Perch rhabdovirus		None
Pike perch rhabdovirus		None
Pike rhabdovirus	DK5533	None
Grayling rhabdovirus		None
European lake trout rhabdovirus	903/87	None
Swedish sea trout rhabdovirus	SSTV	None
Starry flounder virus	SFRV	None
III. Incompletely characterized fish rhabdoviruses		
Rio Grande cichlid virus		Unassigned animal rhabdovirus
Siniperca chuatsi rhabdovirus	SCRV	None
Scophthalmus maximus rhabdovirus	SMRV	None

was originally known as a major pathogen of trout in Western Europe, but in recent years it has also been revealed as a relatively ubiquitous pathogen of numerous species of marine fish in both the Pacific and Atlantic Oceans. Hirame rhabdovirus (HIRRV) and snakehead rhabdovirus (SHRV) are important pathogens of cultured fish in Asia. Hirame rhabdovirus has been isolated from several species of wild and cultured marine fish suffering from a hemorrhagic septicemia. In Japan and Korea, HIRRV causes economically significant disease, especially in cultured Japanese flounder (hirame). Snakehead rhabdovirus has been isolated from both wild and cultured snakehead in Southeast Asia where it causes a severe ulcerative disease. These four viruses have all been well studied, and each one has been developed as an important research model for studies of various aspects of fish rhabdovirus molecular biology. In addition to these formal species in the genus, there are two tentative members that have been isolated from eel, but these and several other fish rhabdoviruses reported in the literature are less well characterized.

The virions of novirhabdoviruses have the typical bullet-shaped morphology of rhabdoviruses and are $c.$ 150–190 nm in length and 65–75 nm in diameter (**Figure 2(a)**). Virions are comprised of five viral proteins. The single glycoprotein (G) on the surface of enveloped virus particles has been shown to be a major antigenic protein that is, by itself, capable of eliciting a protective immune response. The matrix protein (M) is an inner component of the virions that binds both the cytoplasmic domain of the G protein and the viral nucleocapsid and, for IHNV, has been shown to downregulate host transcription. The major structural component of the nucleocapsid is the viral nucleocapsid (N) protein that associates with

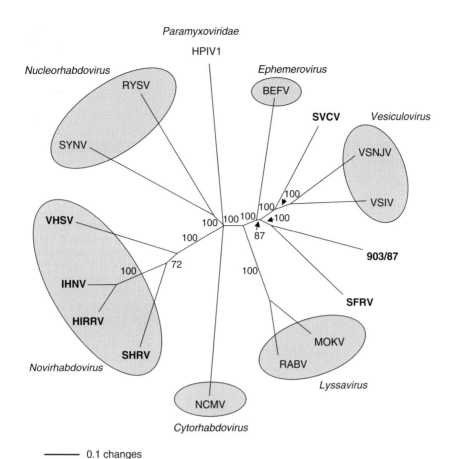

Figure 1 Phylogenetic tree showing the relationship of fish rhabdoviruses (in bold) with other representative members of the family *Rhabdoviridae*. Partial polymerase protein sequences comprising ~460 amino acids at the N-terminus of the L protein were used in a neighbor-joining distance program. Two groups of fish rhabdoviruses are shown either as members of the genus *Novirhabdovirus* or as vesiculovirus-like isolates. Abbreviations used are: BEFV, bovine ephemeral fever virus; SVCV, spring viremia of carp virus; VSNJV, vesicular stomatitis New Jersey virus; VSIV, vesicular stomatitis Indiana virus; 903/87, European lake trout rhabdovirus; SFRV, starry flounder rhabdovirus; MOKV, Mokola virus; RABV, rabies virus; NCMV, northern cereal mosaic virus; SHRV, snakehead virus; HIRRV, hirame rhabdovirus; IHNV, infectious hematopoietic necrosis virus; VHSV, viral hemorrhagic septicemia virus; SYNV, sonchus yellow net virus; RYSV, rice yellow stunt virus. The paramyxovirus human parainfluenza virus 1 (HPIV-1) was used as the outgroup. Phylogenetic analysis used 1000 bootstrapped data sets, and nodes with bootstrap values less than 70% were collapsed to polytomies.

full-length viral genomic RNA and facilitates its interaction with the viral polymerase. The large polymerase protein (L) is a complex protein that carries out most of the enzymatic functions of genome replication and transcription. The phosphoprotein (P) is an essential cofactor that is required for polymerase function. These five virion proteins are found in all known rhabdoviruses and, to date, it appears that their roles are conserved across genera. In addition to these familiar rhabdovirus proteins, the hallmark of the novirhabdoviruses is the presence of a sixth protein, the nonvirion (NV) protein, which has not yet been identified in any rhabdovirus of another genus. The NV protein is not found in virus particles, but it is synthesized, often in only trace amounts, within infected cells. The presence of an apparently functional NV gene is conserved in all novirhabdovirus isolates examined to date, but the role of this protein is not well defined.

Through the use of reverse genetics, it has been shown that the NV gene and its encoded protein product are not essential for replication in cultured cells or in fish, but in some cases it enhances replication and pathogenicity.

Fish rhabdovirus replication occurs in the cytoplasm of infected cells by processes common to all rhabdoviruses. During the replication cycle the viral RNA-dependent RNA polymerase (L) transcribes individual mRNAs for each viral gene, and also copies the RNA genome to produce progeny genomes. Viral mRNAs are translated into viral proteins that traffic to the outer cell membrane where assembly with viral genomic RNA occurs and budding through the cell membrane releases new enveloped virus particles. All novirhabdovirus genomes are single-stranded minus-sense RNA molecules *c.* 11 000 nt in length. The genome organization is conserved across the genus, and is shown for the type species *Infectious hematopoietic necrosis*

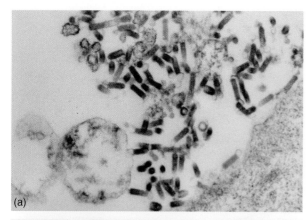

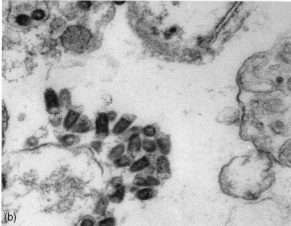

Figure 2 Electron micrograph showing a representative member of each of the two groups of fish rhabdoviruses. (a) *Viral hemorrhagic septicemia virus*, a species in the genus, *Novirhabdovirus*. (b) Starry flounder rhabdovirus, a vesiculovirus-like rhabdovirus of fish. Scale = 200 nm (a, b).

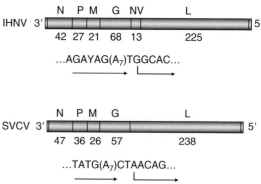

Figure 3 Genome organization of representatives of the two major subgroups of fish rhabdoviruses. IHNV represents the genus *Novirhabdovirus* and spring viremia of carp virus (SVCV) represents the vesiculovirus-like rhabdoviruses of fish. Both have negative-sense single-stranded RNA genomes of c. 11 000 nt containing nucleocapsid (N), phosphoprotein (P), matrix (M), glycoprotein (G), and polymerase (L) genes as shown. Species in the genus *Novirhabdovirus* have an additional gene encoding an NV protein. Approximate sizes (in kDa) of the encoded proteins are shown below each gene. Conserved gene junction sequences for each virus are shown in positive-sense (mRNA) orientation below the genome with arrows indicating putative transcription stop (ending with seven A's) and start signals.

virus in **Figure 3**. The six viral genes occur in the order 3′-N–P–M–G–NV–L-5′. This is identical to the genome organization of members of the genera *Vesiculovirus* and *Lyssavirus*, with the exception of the additional NV gene at the G–L junction. The genome termini have complementary leader and trailer regions of approximately 50–60 and 40–115 nt, respectively. The untranscribed, intergenic regions are single nucleotides, typically A or T, at all gene junctions except for the G–NV junction, which is slightly more variable. Putative transcription start and stop signals for each gene, shown in **Figure 3**, are highly conserved among individual genes. They are also conserved among novirhabdovirus species and differ from the conserved regulatory sequences of rhabdoviruses in other genera.

Genetic diversity within novirhabdovirus species has been well characterized for IHNV and VHSV. Genetic typing of hundreds of IHNV field isolates from North America, as well as from Europe, Russia, and Asia, has shown surprisingly low diversity, with a maximum of 9% nucleotide sequence diversity in a relatively variable region of the G gene. Despite this overall low diversity, genetic typing clearly resolves IHNV isolates into three major genogroups that correlate with their geographic origin in North America. Similar typing of extensive global collections of VHSV isolates has revealed much greater diversity, with up to 16% nucleotide diversity in full-length G gene sequences, and four phylogenetic subgroups that also correlate with geographic range. For both the IHNV and VHSV subphylogenies, additional studies with partial nucleocapsid (N) gene sequences have confirmed the genogroups defined by partial G gene analyses. Diversity between species of novirhabdoviruses can be illustrated with full G protein amino acid sequences. Within the genus, IHNV and HIRRV are most closely related with 74% identity (83% similarity) in their G proteins. Within the entire genus, the G proteins of the four member species are more divergent, with as little as 37% identity (54% similarity). Comparisons outside the genus reveal 13–23% identity (27–48% similarity) between the G proteins of novirhabdoviruses and the fish vesiculovirus-like isolates or members of other rhabdovirus genera.

Vesiculovirus-Like Viruses of Fish

At present, no fish rhabdoviruses are accepted as formal species within the genus *Vesiculovirus*, but many rhabdoviruses isolated from fish are clearly closely related to members of this genus, and distinct from the members of the genus *Novirhabdovirus*. Some vesiculovirus-like fish rhabdoviruses are considered tentative species in the genus *Vesiculovirus* (**Table 1**). The best studied vesiculo-like fish rhabdovirus is spring viremia of carp virus (SVCV), which has a long history of causing severe epidemics among cultured carp in Europe. SVCV is well characterized at both the biological and molecular levels, and it is considered the representative of the vesiculovirus-like fish rhabdoviruses. Full-length genome sequences are available for several strains of SVCV, and numerous other isolates are partially sequenced, facilitating phylogenetic analyses that confirm its close relationship to the mammalian vesiculoviruses. Many isolates of a similar virus, pike fry rhabdovirus, have also been described, and several other fish rhabdoviruses have been isolated from various cold- and warm-water hosts including various species of eel, snakehead, trout, and perch. By serological assays and phylogenetic analyses, at least some of these isolates appear to be distinct vesiculo-like viral species. Phylogenetic trees of partial G or L gene sequences indicate that these isolates form an emerging cluster of tentative species around the mammalian genus *Vesiculovirus*, as illustrated in **Figure 1**. Future work by the ICTV will likely establish some formal species among the vesiculo-like fish rhabdoviruses and clarify whether they should be accepted as members of the genus *Vesiculovirus*.

Virions of the fish vesiculo-like rhabdoviruses are typically shorter and wider than novirhabdoviruses, with bullet-shaped particles measuring *c.* 80–150 nm in length and 60–90 nm in diameter (**Figure 2(b)**). Particles are composed of five viral proteins that correspond to the five major virion proteins found in all rhabdoviruses. The virus replication cycle roles of the N, P, M, G, and L proteins are as described above for novirhabdoviruses. Among fish rhabdoviruses, the distinguishing feature of the fish vesiculo-like viruses is the absence of an NV gene at the G–L junction, indicating that, whatever the role of the NV protein in novirhabdoviruses, it is not essential for rhabdovirus replication in fish *per se*.

The negative-sense RNA genome of SVCV is just over 11 000 nt in length, and it has genes encoding the five major viral proteins in the same order found in all rhabdovirus genomes (**Figure 3**). The upstream untranslated region of each gene is a conserved 10 nt sequence and the downstream untranslated regions are more variable. The 3′ and 5′ termini of the genome have complementary leader and trailer regions of *c.* 20–60 and 12–20 nt, respectively. The untranscribed, intergenic regions between genes are dinucleotides, which, for SVCV, are all CT with the exception of the G–L junction that has the tetranucleotide CTAT. Putative transcriptional start and stop signals, shown in **Figure 3**, are highly conserved among individual genes and among SVCV isolates. These regulatory signals at the SVCV gene junctions are nearly identical to those of mammalian virus members of the genus *Vesiculovirus*, and they differ from the regulatory sequences conserved among members of other rhabdovirus genera, including novirhabdoviruses. As an example of the levels of sequence similarity, the SVCV G protein has 31–33% amino acid sequence identity (52–53% similarity) with the G proteins of members of the genus *Vesiculovirus*, but only 19–24% identity (40–47% similarity) with G proteins of 11 rhabdoviruses from all other genera including *Novirhabdovirus*. Among several viruses reported as isolates of SVCV, partial G gene nucleotide sequences have 83–100% identity, and phylogenetic analyses resolved four genogroups that correlated with geographic origin in Eurasia. Among SVCV and other tentative species of fish vesiculo-like rhabdoviruses, partial G or L gene nucleotide sequences typically have 70–80% identity.

Fish Rhabdovirus Detection and Control

Because the diseases caused by fish rhabdoviruses are important to aquaculture, diagnostic methods for the detection and identification of the most significant pathogens are well established. Many cultured cell lines derived from finfish support the replication of a broad range of fish rhabdoviruses and are used routinely in cell culture-based assays to detect the presence of infections or to quantify infectious virus. Serological methods using polyclonal or monoclonal antibodies are used for both diagnostics and research, and molecular techniques such as DNA probes, polymerase chain reaction (PCR) assays, and quantitative polymerase chain reaction (qPCR) assays have been developed for confirmatory diagnosis, identification, and quantification of the most significant rhabdoviruses affecting fish. In many regions of North America, Europe, and Asia, the majority of all cultured fish stocks are surveyed on a regular basis for viral pathogens by fish health professionals.

Understanding the potential impacts of fish rhabdovirus infections on the health of native species and fish reared in aquaculture has led to increased attention to biosecurity and in measures to prevent the spread of aquatic animal diseases. During the mid-1900s, the inadvertent spread of several fish rhabdoviruses by movement of infected eggs or juvenile fish was responsible for several intercontinental introductions of pathogens with serious economic impacts. National and international standards now require health inspections for certain fish species prior to international or interstate transport. Several fish rhabdoviruses are on the list of pathogens for which inspections are required. The global trade in aquaculture species is reasonably well regulated. However, the huge global trade in ornamental fish is essentially unregulated and is a documented source of transboundary virus movement.

Considerable attention has also been given to improving methods for the control or prevention of aquatic animal diseases. These include sound management practices such as improved sanitation, the use of pathogen-free well water sources for early life-stage rearing, the installation of water treatment technologies for facilities using large volumes of water from open sources, and more frequent fish health inspections that make use of improved technologies. In aquaculture settings, the ability to break the cycle of vertical transmission between generations by the disinfection of eggs with an iodine solution has been an important control approach for fish rhabdoviruses.

Research on vaccines to protect fish from rhabdovirus infections has been conducted for more than 30 years. Initially, this work focused on traditional approaches including inactivated viruses or live-modified vaccines. Later, recombinant DNA technology produced experimental vaccines using bacterial or baculovirus expression systems as well as peptide immunogens. While some of these traditional and molecular vaccines were reported to be highly efficacious in both laboratory and field trials, prohibitive development and licensing costs, inconsistent efficacy, or safety concerns have prevented their use on a commercial scale. More recently, DNA vaccines have been engineered for protecting fish against the novirhabdoviruses, IHNV, VHSV, and HIRRV. These vaccines have been studied extensively and show exceptionally high efficacy against severe viral challenges under a wide range of conditions. An IHNV DNA vaccine was licensed for use in Canada in 2005. With regulatory concerns and vaccine delivery methods currently being addressed, it is possible that these will be among the first DNA vaccines to be used commercially for preventative veterinary care.

Current Aspects

The number of reported fish rhabdoviruses continues to grow as a result of the growth of aquaculture, the increase in global trade, the development of improved diagnostic methods, and the expansion of surveillance activities. The increase in nucleotide sequence information available for many fish rhabdoviruses has not only improved our understanding of the genetic diversity among members of the group, but also complicated our ability to form neat taxonomic categories. Upon further analysis, the creation of several new virus species and, possibly, genera may be needed to accommodate the range of fish rhabdoviruses, particularly the unassigned or less well characterized members of the family.

Due to their economic importance, the rhabdoviruses of finfish are among the best studied of all aquatic animal viruses, providing models for their function, movement, and persistence in the environment. There is an increasing understanding that rhabdoviruses are important pathogens of wild or free-ranging fish in both freshwater and marine ecosystems. Extensive oceanographic surveys have revealed marine fish as important reservoirs for virus infections of fish in freshwater and marine aquaculture. In addition to causing natural mortality that may limit population size, anthropogenic stressors can act synergistically to amplify the effects of disease on the overall health and survival of aquatic species. Field studies of the relatively unique disease interface between cultured and wild aquatic host populations are providing insight into virus spill-over and spill-back events, and anthropogenic impacts on fish health in altered environments.

Fish rhabdoviruses are also serving as useful components of model systems to study vertebrate virus disease, epidemiology, and immunology. The availability of a variety of established fish cell lines, the creation of new-generation reagents and tools, and the well-established laboratory challenge models that can use statistically robust numbers of trout, catfish, or zebrafish make these model systems particularly attractive and powerful. Recent work includes the use of DNA vaccines against fish rhabdoviruses in mechanistic studies to determine how DNA vaccines elicit strong innate and adaptive immune responses in vertebrate hosts. Reverse genetics systems for both IHNV and SHRV have been developed and used to investigate the roles of individual viral genes and proteins. These studies have already shown that the G genes of different fish rhabdovirus species can be interchanged, and that the NV gene is not essential for virus replication. In the area of immunology, quantitative real-time PCR assays and microarray analyses have been used to profile the fish immune gene response following either viral infection or DNA vaccination. Future work will include the use of these models to address questions such as the genetics of host disease resistance, viral host specificity and virulence, viral fitness and competition, and host immune selection as a driver of virus evolution.

See also: DNA Vaccines; Fish Viruses; Vesicular Stomatitis Virus.

Further Reading

Ahne W, Bjorklund HV, Essbauer S, Fijan N, Kurath G, and Winton JR (2002) Spring viremia of carp (SVC). *Diseases of Aquatic Organisms* 52: 261–272.

Betts AM, Stone DM, Way K, et al. (2003) Emerging vesiculo-type virus infections of freshwater fish in Europe. *Diseases of Aquatic Organisms* 57: 201–212.

Bjorklund HV, Higman KH, and Kurath G (1996) The glycoprotein genes and gene junctions of the fish rhabdoviruses spring viremia of carp virus and hirame rhabdovirus: Analysis of relationships with other rhabdoviruses. *Virus Research* 42: 65–80.

Bootland LM and Leong JC (1999) Infectious hematopoietic necrosis virus. In: Woo PTK and Bruno DW (eds.) *Fish Diseases and Disorders, Vol. 3: Viral, Bacterial and Fungal Infections*, pp 57–121. New York: CABI.

Bremont M (2005) Reverse genetics on fish rhabdoviruses: Tools to study the pathogenesis of fish rhabdoviruses. *Current Topics in Microbiology and Immunology* 292: 119–141.

Einer-Jensen K, Ahrens P, Forsberg R, and Lorenzen N (2004) Evolution of the fish rhabdovirus viral haemorrhagic septicaemia virus. *Journal of General Virology* 85: 1167–1179.

Hoffmann B, Beer M, Schutze H, and Mettenleiter TC (2005) Fish rhabdoviruses: Molecular epidemiology and evolution. *Current Topics in Microbiology and Immunology* 292: 81–117.

Kim D-H, Oh H-K, Eou J-I, *et al.* (2005) Complete nucleotide sequence of the hirame rhabdovirus, a pathogen of marine fish. *Virus Research* 107: 1–9.

Kurath G (2005) Overview of recent DNA vaccine development for fish. In: Midtlyng PJ (ed.) *Fish Vaccinology, Developments in Biologicals,* vol. 121, pp. 201–213. Basel: Karger.

Kurath G, Garver KA, Troyer RM, Emmenegger EJ, Einer-Jensen K, and Anderson ED (2003) Phylogeography of infectious haematopoietic necrosis virus in North America. *Journal of General Virology* 84: 803–814.

Smail DA (1999) Viral haemorrhagic septicaemia. In: Woo PTK and Bruno DW (eds.) *Fish Diseases and Disorders, Vol. 3: Viral, Bacterial and Fungal Infections,* pp 123–147. New York: CABI.

Snow M, Bain N, Black J, *et al.* (2004) Genetic population structure of marine viral haemorrhagic septicaemia virus (VHSV). *Diseases of Aquatic Organisms* 61: 11–21.

Stone DM, Ahne W, Sheppard AM, *et al.* (2002) Nucleotide sequence analysis of the glycoprotein gene of putative spring viraemia of carp viruses and pike fry rhabdovirus isolates reveals four distinct piscine vesiculovirus genogroups. *Diseases of Aquatic Organisms* 53: 203–210.

Tordo N, Benmansour A, Calisher C, *et al.* (2004) Family *Rhabdoviridae*. In: Fauquet CM, Mayo MA, Maniloff J, Desselberger U and Ball LA (eds.) *Virus Taxonomy: Eighth Report of the International Committee on Taxonomy of Viruses,* pp. 623–644. San Diego, CA: Elsevier Academic Press.

Wolf K (1988) *Fish Viruses and Fish Viral Diseases.* Ithaca, NY: Cornell University Press.

Fish Viruses

J C Leong, University of Hawaii at Manoa, Honolulu, HI, USA

© 2008 Elsevier Ltd. All rights reserved.

Glossary

Anadromous Migrating from the sea to freshwater to spawn.

Cardiomyopathy A weakening of the heart muscle or a change in heart muscle structure.

Salmonid Belonging to, or characteristic of the family Salmonidae which includes the salmon trout and whitefish.

Swimbladder An air-filled sac near the spinal cord in many fishes that helps maintain buoyancy.

Introduction

Viruses that infect fish and cause disease are represented in 14 of the families listed for vertebrate viruses by the International Committee on the Taxonomy of Viruses (27 May 2005). The fish viruses containing DNA genomes are listed in the families *Iridoviridae, Adenoviridae,* and *Herpesvirdae* and those with RNA genomes are listed in the families *Picornaviridae, Birnaviridae, Reoviridae, Rhabdoviridae, Orthomyxoviridae, Paramyxoviridae, Caliciviridae, Togaviridae, Nodaviridae, Retroviridae,* and *Coronaviridae.* As more fish species are brought under culture, there will be additions to this list, and possibly new viruses will be assigned to families not previously characterized in vertebrates.

Viral diseases have had a tremendous economic impact on both wild and farm-reared fish. A herpesvirus outbreak in 1998 and early 1999 reduced the pilchard (*Sardinops sagax neophilchardus*) fishery in southern Australia by two-thirds. In 2001, infectious salmon anemia virus (ISAV) was discovered in cultured Atlantic salmon in Cobscook Bay, Maine. The discovery of this orthomyxovirus virus forced farmers to destroy about 2.6 million fish in an effort to contain the spread of the disease. The cost to the Maine salmon industry (valued at more than $100 million) was about $24 million (USDA Animal and Plant Health Inspections Service (APHIS) estimates). More recently, the rhabdovirus viral hemorrhagic septicemia virus (VHSV), a very serious pathogen of marine and freshwater fish in Europe, was detected in the Great Lakes of North America in 2005. On 24 October 2006, APHIS issued an emergency order that blocked the live export of 37 fish species from any of the eight Great Lakes states. The order caused strong protests from fish farmers who make their living with live bait shipments and fish-stocking programs that sustain the Great Lakes' $4.5 billion fishing industry. Since vaccines and/or therapeutics for fish viruses are not readily available, containing the spread of these viruses by restricting movement and destruction of the affected population has been the only effective control strategy. Thus, the World Animal Health Organization (OIE) lists nine reportable diseases of fish (**Table 1**), seven of which are viral diseases.

Known and characterized fish viruses numbered only 16 in 1981, with an additional 11 observed by electron microscopy. Now, there are over 125 described viruses of fish and countless reports of electron microscopic observations of

Table 1 Diseases of fish listed by the OIE, 2006

Epizootic hematopoietic necrosis	Iridovirus
Infectious hematopoietic necrosis	Novirhabdovirus
Spring viremia of carp	Vesiculovirus
Viral hemorrhagic septicemia	Novirhabdovirus
Infectious salmon anemia	Orthomyxovirus
Epizootic ulcerative syndrome	Fungal infection
Gyrodactyloses	Flat worm fluke
Red sea bream iridovirus diseases	Iridovirus
Koi herpesvirus disease	Herpesvirus

viruses in wild-caught and cultured fish. The dramatic increase in reports of new fish viruses correlates with growth of the aquaculture industry that has increased production more than fivefold since 1985 to now represent more than 30% of global fishery production.

RNA Viruses of Fish

Rhabdoviridae

Fish rhabdoviruses are considered among the most serious viral pathogens of aquacultured fish, affecting predominantly salmon and trout. The first fish rhabdovirus was described in 1938 by Schaperclaus in European rainbow trout and, 6 years later, the Sacramento River Chinook virus was reported by Rucker. Since then, these viruses have been isolated, grown in tissue culture cells, and the genomes have been cloned and sequenced. The salmon virus is now called *Infectious hematopoietic necrosis virus* (IHNV) which is the type species for a new genus *Novirhabdovirus* in the *Rhabdoviridae*. Other fish rhabdovirus species assigned to this genus include *Viral hemorrhagic septicemia virus* (VHSV), *Hirame rhabdovirus* (HIRRV), and *Snakehead rhabdovirus* (SHRV). All of these viruses have negative-sense ssRNA genomes with a physical map ordered from the 3'-end as follows: leader-N-P-M-G-NV-L, where N is the nucleoprotein gene, P is the phosphoprotein gene, M is the matrix protein gene, G is the glycoprotein gene, NV is the nonvirion protein gene, and L is the virion RNA polymerase gene. The presence of the NV gene distinguishes the rhabdoviruses in the genus *Novirhabdovirus*, 'novi' standing for nonvirion.

Reverse genetic analysis of the IHNV, SHRV, and VHSV NV genes has, to date, provided mixed results. For IHNV and VHSV, deletion of the NV gene ameliorated virus-induced cytopathic effect (CPE) in tissue culture cells and reduced pathogenicity in fish. However, deletion of the NV gene for SHRV did not affect virus production or virus-induced CPE in tissue culture cells, or reduce pathogenicity in live fish challenges.

IHNV is a virus of salmonid fish and outbreaks of this virus in rainbow trout (*Oncorhynchus mykiss*), sockeye salmon (*O. nerka*), Chinook salmon (*O. tshawytscha*), Atlantic salmon (*Salmo salar*), and masou salmon (*O. masou*) have been economically devastating to the fish farmers. This virus prefers colder temperatures with optimal growth at 8–15 °C. VHSV is also a serious pathogen of salmonid fish, but its host range is broader and it has been shown to kill Pacific herring (*Clupea pallasi*), Pacific cod (*Gadus macrocephalus*), whitefish (*Coregonus* sp.), European sea bass (*Dicentrarchus labrax*), and turbot (*Scophthalmus maximus*). This broad host range has caused a great deal of concern to authorities in the USA since VHSV was first reported there in 2006 and found to affect freshwater drum (*Aplodinotus grunniens*), round goby (*Neogobius melanostomus*), smallmouth bass (*Micropterus dolomieu*), bluegill (*Lepomis machrochirus mystacalis*), crappie (*Pomoxis nigromaculatus*), gizzard shard (*Dorsoma cepadianum*), and other species occurring in the Great Lakes. HIRRV affects hirame, the Japanese flounder (*Paralychthys olivaceus*), which is a highly prized food fish in Japan. Its host range includes ayu (*Pleuroglossus altivelis*) as well as salmonid fish. SHRV was isolated from snakehead fish (*Channa striatus*) suffering from epizootic ulcerative syndrome (EUS) in Thailand. It is not considered the etiologic agent of EUS which is caused by a fungal pathogen. However, zebrafish (*Brachydanio rerio*), exposed to tissue culture grown SHRV develop, will develop petecchial hemorrhages and die.

Spring viremia of carp virus (SVCV), unlike the fish rhabdoviruses described above, does not contain an intervening gene between its glycoprotein and L genes. Analysis of the SVCV genome sequence indicates that it clusters with the members of the genus *Vesiculovirus* that includes vesicular stomatitis virus. SVCV was first identified as the etiologic agent of an acute hemorrhagic disease in common carp (*Cyprinus carpio*) in Europe in 1972. The disease has since been found in Asia, the Middle East, and most recently in South and North America. Outbreaks in the USA have raised concerns that indigenous fish species in the minnow family (*Cyprinidae*), some of which are endangered species, may be susceptible to SVCV. Other virus isolates that are closely related in nucleotide sequence to SVCV are the pike fry rhabdoviruses (PFRV) from a variety of freshwater fish in Europe including pike (*Esox lucius*), common bream (*Abramis brama*), roach (*Rutilus rutilus*), eel virus Europe X (EVEX), eel virus American (EVA); ulcerative disease rhabdovirus (UDRV); the grass carp rhabdovirus isolated from grass carp (*Ctenopharyngdon idella*); trout rhabdovirus 903/87 (TRV 903/87); and sea trout rhabdovirus 28/97 (STRV-28/97).

Phylogenetic analyses comparing aligned data for the N and G genes of members of the family *Rhabdoviridae* have confirmed the classification of the six genera: *Lyssavirus*, *Vesiculovirus*, *Ephemerovirus*, *Cytorhabdovirus*, *Nucleorhabdovirus*, and *Novirhabdovirus*. To date, fish rhabdoviruses are restricted to the genera *Novirhabdovirus* and *Vesiculovirus*. Analyses using the viral P gene sequences indicate that the aquatic vesiculoviruses form a separate cluster from the arthropod-borne vertebrate vesiculoviruses.

Paramyxoviridae

The first description of a paramyxovirus-like virus in fish was reported in 1985. During a routine health assessment of Chinook salmon juveniles in Oregon, tissue culture cells inoculated with a cell-free homogenate of organ tissue exhibited syncytia formation. Electron micrographs of the infected cell line showed enveloped, pleomorphic virus particles with a diameter of approximately of 125–250 nm and a single helical nucleocapsid with a diameter of 18 nm and a length of 100 nm. No disease syndrome was observed in trout and salmon fingerlings in subsequent infectivity trials with the tissue cultured virus. A second fish paramyxovirus that caused epidermal necrosis in juvenile black sea bream (*Acanthopargrus schlegeli*) was identified in Japan by electron microscopy. This virus was never cultured *in vitro*. The most recent description of a fish paramyxovirus was from Atlantic salmon post-smolts suffering from inflammatory gill disease in Norway. The genome of this virus has been cloned and partial sequences from the viral L protein have been used to determine the phylogenetic placement of this virus among the *Paramyxoviridae*. The Atlantic salmon paramyxovirus (ASPV) clustered with the subfamily *Paramyxovirinae*, in the genus *Respirovirus* that includes human parainfluenza virus and Sendai virus.

Orthomyxoviridae

Infectious salmon anemia virus (ISAV) is the only fish orthomyxovirus that has been fully described to date. There are eight RNA segments in the ISAV genome. Segment 1 encodes PB2, a component of the virion RNA polymerase; segment 2 encodes PB1; segment 3, the nucleocapsid protein NP; segment 4, the RNA polymerase PA; segment 5, acetylcholinesterase P3 or fusion protein; segment 6, hemagglutinin; segment 7, protein P4 and P5; and segment 8, proteins P6 and P7. The proteins P4 and P5 may be the ISAV counterparts to the membrane proteins M1 and M2 of influenza A virus; proteins P6 and P7 may be related to the nonstructural proteins NS1 and NEP of influenza A virus. The ISAV hemagglutinin does agglutinate fish red blood cells or mammalian red blood cells.

A comparative sequence analysis of the PB1 gene of ISAV and other members of the *Orthomyxoviridae* led to its assignment as the type species of a new genus *Isavirus*. More recent comparative analyses of the fusion protein gene (segment 5) and the hemagglutinin gene (segment 7) indicate that ISAV isolates can be divided into two subtypes, a North American subtype and a European subtype.

ISAV causes a highly lethal disease with affected farmed Atlantic salmon displaying severe anemia, leucopenia, ascetic fluids, hemorrhagic liver necrosis, and petecchiae of the viscera. The virus also causes disease in sea trout (*Salmo trutta*), rainbow trout, and Atlantic herring.

Picornaviridae

The first reported observation of picorna-like viruses in fish was made in 1988 from rainbow smelt (*Osmerus mordax*) in New Brunswick, Canada. Since then, picornaviruses have been isolated from barramundi, turbot, sea bass, grass carp, blue gill, grouper (*Epinephelus tauvina*), Japanese parrotfish (*Oplegnathus fasciatus*), and salmonid fish. In most of these descriptions, the presumptive characterization of the etiologic agent as a picornavirus was based on growth in tissue culture cells and the observation of crystalline arrays in the cytoplasm of small virus particles with a size and morphology consistent with picornaviruses. Analysis of RNA extracted from purified blue gill virus has indicated that it is single-stranded RNA virus. Sequence characterization of the viral genomes has not been carried out and there is some suggestion that at least some of these viruses might actually be betanodaviruses. In many cases, diseased fish infected with these viruses contain picorna-like virus particles in the brain and medulla and the victims display corkscrew-like swimming and eventually die.

Nodaviridae

Members of the *Nodaviridae* that infect fish belong to the genus *Betanodavirus* for which the type species is *Striped jack nervous necrosis virus* (SJNNV). These viruses are nonenveloped with icosahedral symmetry and virion diameters of approximately 30 nm. The viral genome consists of two molecules of positive-sense ssRNA. RNA1, the largest RNA genome segment encodes the viral polymerase. RNA2 encodes the virion capsid protein. A third RNA, transcribed from the $3'$ terminal region of RNA1, encodes a 75 amino acid protein that bears little similarity with the B2 and B1 proteins encoded by a similar RNA3 in the alphanodaviruses. Despite this, the SJNNV B2 protein RNA has RNA silencing-suppression activity, as does the B2 protein of insect-infecting alphanodaviruses.

The betanodaviruses are the causative agents of viral nervous necrosis or viral encephalopathy and retinopathy in a variety of cultured marine fish. The disease affects young fish and produces a necrosis and vaculoation in the brain, spinal cord, and retina in most cases. It has been reported in striped jack (*Pseudocaranx dentex*), grouper (*Epinephelus* spp.), red drum (*Sciaenops ocellatus*), guppy (*Poicelia reticulate*), barfin flounder (*Verasper moseri*), red sea bream, tiger puffer (*Takifugu rubripes*), Japanese flounder, Atlantic halibut (*Hippoglossus hippoglossus*), amberjack (*Seriola* d*umerili*), sea bass, and barramundi. The recent detection of betanodaviruses in apparently healthy aquarium fish and invertebrates has raised concerns that the disease could be spread by trade in aquarium fish, particularly from Southeast Asia. Comparative sequence analyses of the coat protein genes for 25 isolates suggest that there are four genotypic variants: tiger puffer nervous necrosis virus (TPNNV), striped jack nervous necrosis virus (SJNNV), barfin flounder nervous necrosis virus (BFNNV), and red-spotted grouper nervous necrosis virus (RGNNV).

Nidovirales

The family *Coronaviridae* comprises two genera, *Coronavirus* and *Torovirus*, and is classified with the families *Arteriviridae* and *Roniviridae* in the order *Nidovirales*. Members of the *Coronaviridae* share the common feature of pleomorphic, enveloped virions with diameters of 126–160 nm and prominent surface projections. The nucleocapsid is helical and contains a single molecule of linear, positive-sense ssRNA. Coronavirus-like particles have been isolated from a common carp from Japan showing petecchial hemorrhages on the skin and abdomen. A similar virus has also been isolated from moribund colored carp (*Cyprinus carpio*) with ulcerative dermal lesions. The investigators were able to grow the virus in epithelioma papulosum cyprini (EPC) cells and produce the same disease in carp injected with the tissue culture grown virus. Further characterization of these virus isolates was never carried out and there was no confirmation that they are, indeed, coronaviruses.

Recently, a novel virus with morphological features resembling those found in rhabdo-, corona-, and baculoviruses has been detected during the routine diagnostic screening of white bream (*Blicca bjoerkna*) in Germany. Ultrastructural studies indicated that the cell-free virions contain of a rod-shaped nucleocapsid similar to that seen in baculoviruses. Virions are bacilliform-shaped structures somewhat reminiscent of plant rhabdoviruses with an envelope containing coronavirus-like spikes. Sequence analysis has indicated that the 26.6 kbp white bream virus (WBV) contains five open reading frames, ORF1a, -1b, -2, -3, and -4, which are produced from a 'nested' set of 3′-coterminal mRNAs. The largest mRNA is of genome length. ORF1a and ORF1b form the viral replicase gene. ORF1a encodes several membrane domains, a putative ADP-ribose 1′-phospatase, and a chymotrypsin-like serine protease. ORF1b encodes the putative polymerase, helicase, ribose methyltransferase, exoribonuclease, and endoribonuclease activities. These characteristics are consistent with classification of WBV in the order *Nidovirales*. Phylogenetic analyses of the helicase and polymerase core domains indicate that WBV is more closely related to toroviruses than to coronaviruses and it has been suggested that a new nidovirus genus *Bafinivirus* be established (from bacilliform fish nidoviruses).

Togaviridae

The family *Togaviridae* comprises the genera *Alphavirus* and *Rubivirus* among the vertebrate viruses. These viruses have spherical virions, 70 nm in diameter, with a lipid envelope containing glycoprotein peplomers and a ssRNA genome which is capped at the 5′-end and polyadenylated at the 3′-end. Salmonid alphaviruses (SAVs) cause mortality in salmon and trout in Europe (Norway, France, UK, and Ireland). At least three subtypes of SAV exist: Salmon pancreas disease virus (SPDV/SAV-1) in Atlantic salmon; sleeping disease virus (SDV/SAV-2) in rainbow trout; and Norwegian salmonid alphavirus (NSAV/SAV-3). An early study on the evolutionary relationships of the alphaviruses has indicated that SAVs represent a separate and distant group in the genus *Alphavirus*.

Pancreas disease, due to SPDV (SAV-1) infection, was first described in Scotland in Atlantic salmon. It occurs during the first year at sea following transfer of young fish from freshwater tanks. The fish become anorexic and exhibit sluggish swimming activity with mortality rates reaching 10–50%. Histological examination of the affected fish has shown pancreatic acinar necrosis, and cardiac and skeletal myopathy. In rainbow trout, SDV (SAV-2) infection is characterized by the unusual behavior that fish lie on their side at the bottom on the tank. The lesion responsible for this behavior is red and white muscle degeneration. The histological lesions are similar to those observed in SPDV infection with progressive pancreatic necrosis and atrophy, mulifocal cardiomyopathy and muscle degeneration.

Caliciviridae

San Miguel sea lion virus (SMSV) is classified in the species *Vesicular exanthema of swine virus* in the genus *Vesivirus* in the family *Caliciviridae*. Investigators have found that the serotype 7 strain of the virus (SMSV-7), isolated from the opaleye fish (*Girella nigrigans*), can produce vesicular exanthema in swine. Thus, it is a virus that can jump from fish to mammals. The same serotype has also been reported in elephant seals and a sea lion trematode. Tissue culture grown serotype 5 SMSV injected into opaleye replicated to high titer 15 °C, producing $10^{7.6}$ TCID$_{50}$ per gram of spleen. There is no apparent disease in opaleye caused by this virus.

SMSV virions are nonenveloped with icosahedral symmetry and are 27–40 nm in diameter. The genome consists of a 7.5–8.0 kbp linear, positive-sense ssRNA that contains a covalently linked protein (VPg) attached to its 5′-end. The 3′-end of the genome is polyadenylated. The nonstructural polypeptides are encoded as a polyprotein in the 5′-end of the genomic RNA, while the single structural protein is encoded in the 3′-end. The identity of nonstructural polypeptides 2C (RNA helicase), 3C (cysteine protease), and 3D (RNA-dependent RNA polymerase) has been suggested by similarity to highly conserved amino acid motifs in the nonstructural proteins of the picornavirus superfamily. Phylogenetic analysis of the capsid protein region of caliciviruses including the

Sapporo-like human caliciviruses indicate that the genus *Vesivirus* includes SMVL-1, SMSV-4, SMSV-13, SMSV-15, SMSV-17, three feline caliciviruses, and the primate calicivirus Pan-1. These viruses are distant from the human caliciviruses and the rabbit caliciviruses.

Retroviridae

The family *Retroviridae* consists of two subfamilies, the *Orthoretrovirinae*, containing six genera, and the *Spumaretrovirinae*, containing only one genus. The piscine retroviruses constitute the genus *Epsilonretrovirus*, a genus established within the *Orthoretrovirinae* to include the piscine retroviruses: walleye dermal sarcoma virus (WDSV), walleye epidermal hyperplasia virus type 1 (WEHV-1), walleye epidermal hyperplasia virus type 2 (WEHV-2), and snakehead retrovirus (SnRV). The genomes of all of these viruses have been sequenced. There are also numerous reports of C-type (retrovirus-like) particles of about 110–150 nm in epidermal papillomas of European smelt (*Osmerus eperlanus*) and in cells cultured from neurofibromas of damselfish (*Pomacentrus partitus*). A retrovirus has also been suggested as the etiological agent of plasmacytoid leukemia in Chinook salmon.

The first report of a retrovirus-like agent in fish was made in 1976 in lymphosarcoma of northern pike and muskellunge (*Esox masquinongy*). The lymphosarcoma lesions contained a reverse transcriptase-like DNA polymerase with a temperature optimum of 20 °C. The first molecular evidence for a piscine retrovirus was reported in 1992 for a type C retrovirus from dermal sarcomas that form on the surface of adult walleye. These tumors are formed on the surface of adult walleye (*Stizostedion vitreum*) in the fall and regress in the spring. The genome of the virus (13.2 kbp) was larger than all other known retroviruses at the time. Sequence analysis indicated that WDSV contained three additional open reading frames: ORF C at the 5′ terminal end; and ORF A and ORF B at the 3′ terminal end. ORF A encodes a D-cyclin homolog (retroviral cyclin) that locates in the nucleus of tumor cells in interchromatic granule clusters. ORF C encodes a cytoplasmic protein that targets the mitochondria and is associated with apoptosis. It is expressed in regressing tumors when full-length viral RNA is synthesized. The function of the protein encoded in ORF B, which is distantly related to ORF A, remains unknown. The WDSV protease cleavage sites have been identified to contain glutamine in the P2 position. The WDSV reverse transcriptase is rapidly inactivated at temperatures greater than 15 °C; a finding that is consistent with adaptation to growth in a coldwater fish species.

Two additional retroviruses have been cloned from epidermal hyperplasias on walleye. The genome sequences indicate that they are distinctly different from each other (77% identity) and from WDSV (64% identity). Walleye epidermal hyperplasia viruses 1 and 2 (WEHV-1 and -2) have genome organizations similar to WDSV. Each of the walleye retroviruses produces lesions when a cell-free filtrate from homogenized tumors is injected in naïve walleye juveniles.

Complete nucleotide sequence and transcriptional analyses of snakehead fish retrovirus have also been reported. The proviral genome is arranged in a typical 5′-LTR-gag-pol-env-LTR-3′ retrovirus organization. There are three additional ORFs: ORF1 encoding a 52 aa protein (5.7 kDa); ORF2 encoding a 94 aa protein (11 kDa); and ORF3 encoding a 205 aa protein (24 kDa). BLAST searches for possible homologs of these proteins have not produced any meaningful matches and their functions remain unknown. The SnRV genome differs from the retroviruses of walleye in that it has no ORF between the Unique region in the 5′ LTR (U5) and the gag region. The pathogenicity of SnRV has also not been determined.

In 2006, a novel piscine retrovirus was identified in association with an outbreak of leiomyosarcoma in the swimbladders of Atlantic salmon. The swimbladder sarcoma virus (SSSV) provirus is 10.9 kbp in length with a simple *gag, pro-pol, env* gene arrangement similar to that of murine leukemia viruses. Phylogenetic analysis of pol sequences suggests that SSV is most closely related to the sequenced zebrafish endogenous retrovirus (ZFERV) and that these viruses represent a new group of piscine retroviruses.

Reoviridae

Reoviruses that infect aquatic animals are grouped in the genus *Aquareovirus* in the family *Reoviridae* and are characterized by a nonenveloped double capsid shell, 11 segments of double-stranded RNA and seven structural proteins. John Plumb isolated the first finfish reovirus, golden shiner virus, GSRV, from golden shiner (*Notemigonus crysoleucas*) in 1979. Since then, several reovirus-like agents have been reported in piscine, molluscan, and crustacean hosts. Each has 11 segments of dsRNA and grow at temperatures that reflect their host range. The aquareoviruses have been divided by RNA–RNA hybridization kinetics into six groups (A–F) and several tentative species (**Table 2**). The type species of the genus *Aquareovirus* is *Aquareovirus A* which includes striped bass (*Morone saxatilis*) reovirus (SBRV). Like other reoviruses, the aquareoviruses are ether resistant and resistant to acid to pH3.

Most aquareovirus isolates are nonpathogenic or of low virulence in their host species. Grass carp virus (GCV; species *Aquareovirus C*) is the exception and appears to be the most pathogenic aquareovirus. GCV was isolated from grass carp in the People's Republic of China, causing severe hemorrhagic disease and affecting about 85% of infected fingerling and yearling populations.

Table 2 Species and tentative species of aquareoviruses

Aquareovirus A
    Angelfish reovirus AFRV
    Atlantic salmon reovirus HBR
    Atlantic salmon reovirus ASV
    Atlantic salmon reovirus TSV
    Chinook salmon reovirus DRC
    Chum salmon reovirus CSV
    Threadfin reovirus
    Herring reovirus HRV
    Masou salmon reovirus MSV
    Smelt reovirus
    Striped bass reovirus
Aquareovirus B
    Chinook salmon reovirus B
    Chinook salmon reovirus LBS
    Chinook salmon reovirus YRC
    Chinook salmon reovirus ICR
    Coho salmon reovirus CSR
    Coho salmon reovirus ELC
    Coho salmon reovirus SCS
Aquareovirus C
    Golden shiner reovirus[a]
    Grass carp reovirus
Aquareovirus D
    Channel catfish reovirus
Aquareovirus E
    Turbot reovirus
Aquareovirus F
    Chum salmon reovirus PSR
    Coho salmon reovirus SSR
Tentative species of Aquareoviruses
    Chub reovirus
    Landlocked salmon reovirus
    Tench reovirus

[a]Grass carp reovirus and Golden shiner reovirus are variants of the same virus. Table 2 taken from *ICTVdB-The Universal Virus Database*, version 4. http://www.ncbi.nlm.nih.gov/ICTVdb/ICTVdB.

Full-length and partial genome sequences for several members of the genus *Aquareovirus* have been reported. The complete sequence is available for several isolates of GCV, golden shiner reovirus (species *Aquareovirus C*), chum salmon (*O. keta*) reovirus (species *Aquareovirus A*), golden ide (*Leuciscus idus melanotus*) reovirus (tentative species), and striped bass reovirus (species *Aquareovirus A*). Segment 6 of the guppy reovirus has been determined and threadfin (*Eleutheronema tetradactylus*) reovirus (untyped) segments 10, 6, and 11 are available. Segment 1 encodes a putative guanylyl/methyl transferase; segment 2 encodes the RNA-dependent RNA polymerase; segment 3 encodes a dsRNA binding protein with NTPase and helicase activity; segment 4, a nonstructural protein; segment 5, a NTPase core protein; segment 6, the outer capsid protein; segment 7, a nonstructural protein; segment 8, a core protein; segment 9, a nonstructural protein; segment 10, the external capsid protein; and segment 11, a nonstructural protein. Phylogenetic comparisons of the available sequences support the current taxonomic classification of the aquareoviruses and orthoreoviruses in two different genera, a distinction that was made originally on their genome segment number and specific econiches.

Birnaviridae

Members of the *Birnaviridae* have single-shelled non-enveloped capsids and genomes comprising two segments of double-stranded RNA. There are three genera in this family: *Aquabirnavirus*, *Avibirnavirus*, and *Entomobirnavirus*. The names of each genus denote the host specificity. The larger genome segment A encodes a polyprotein containing the virion capsid protein VP2, an autocatalytic protease NS, and an internal capsid protein VP3 in the physical order 5′-VP2-NS-VP3-3′ in the positive sense. There is an additional 17 kDa protein encoded in a second reading frame at the 5′-end of RNA segment A and it has been shown to be a novel anti-apoptosis gene of the Bcl-2 family. Segment B encodes the virus RNA-dependent RNA polymerase. There is no evidence of 5′-capping of any of the viral mRNAs.

The type species of the genus *Aquabirnavirus* is *Infectious pancreatic necrosis virus* (IPNV). Infectious pancreatic necrosis is a highly contagious viral disease of salmonid fish. The disease most characteristically occurs in fry of rainbow trout, brook trout (*Salvelinus fontinalis*), brown trout (*S. trutta*), Atlantic salmon, and several species of Pacific salmon. In salmonid fish, the virus causes an acute gastroenteritis and destruction of the pancreas. The signs of the disease are typically darkening, a pronounced distended abdomen, and a spiral swimming motion. The virus has also been associated with disease in Japanese eels (*Anguilla japonica*) in which it causes nephritis, menhaden (*Brevoortia tryrranus*) in which it causes a 'spinning disease', and in yellowtail fingerlings (*Seriola quinqueradiata*). A birnavirus has been associated with hematopoietic necrosis, causing high mortalities in turbot with renal necrosis, and birnaviruses have been isolated from clams exhibiting darkened gills and gill necrosis. A nontypical apoptosis has been observed in cultured cells infected by IPNV.

Transmission of the virus can occur via the feces of piscivorous birds. Fish that survive an IPNV outbreak become IPNV carriers and continue to shed the virus for life. Most IPNV isolates are antigenically related and belong to one large serogroup A. There is only one virus in serogroup B, a clam Tellina virus. More recent studies using comparisons of the deduced VP2 amino acid sequence have identified six genogroups. Genogroup 1 (equivalent to serotype A1) comprises four subgroups: genotypes 1, 2, 3, and 4. With increased culture of marine species of fish, there have been increasing reports of mortalities in yellow tail and amberjack in Japan from marine aquabirnaviruses.

Several vaccines have been developed for IPNV, including bacterially produced capsid protein and a DNA vaccine. The protein vaccine has been moderately effective in reducing the lethal effects of IPNV infection in Atlantic salmon in Norway. However, control methods still reply on quarantine and certification of eggs/fry as disease free.

DNA Viruses of Fish

Iridoviridae

In the family *Iridoviridae*, the genera *Iridovirus*, *Lymphocystivirus*, *Ranavirus*, and *Megalocystivirus* contain all of the known iridoviruses that infect fish. Their common features are icosahedral virions, 120–350 nm in diameter, that may acquire an envelope, and a viral genome consisting of one molecule of linear dsDNA of 100–303 kbp. Lymphocystis disease was one of the first fish diseases to be described due, in large part, to the characteristic giant cells observed in the connective tissue and benign nodules in the skin of plaice (*Pleuronectes platessa*) and flounder (*Platichthys flesus*). The causative agent, lymphocystis disease virus (LCDV), has been detected in more than 140 species of freshwater, estuarine, and marine fishes.

Six iridoviruses genomes have been completely sequenced, including those of the *Lymphocystis disease virus 1* (LCDV-1, genus *Lymphocystivirus*), Chilo iridescent virus (CIV, species *Invertebrate iridescent virus 6*, genus *Iridovirus*), Tiger frog virus (TFV, species *Frog 3 virus*, genus *Ranavirus*), *Infectious spleen and kidney necrosis virus* (ISKNV, genus *Megalocystivirus*), *Abystoma tigrinum virus* (ATV, genus *Ranavirus*), and Singapore grouper iridovirus (SGIV, tentative species, genus *Ranavirus*). Comparisons of the different iridovirus genomes have revealed that many genes have been conserved during evolution and, among closely related species, the gene order is well preserved. The number of genes (ORFs) in viruses of this family range from 93 (ATV) to 468 (CIV). There are 195 ORFs in LCDV-1 and 120 ORFs in GIV.

Other iridovirus diseases of fish include epizootic hematopoietic necrosis which is caused by viruses in the species *Epizootic hematopoietic necrosis virus* (EHNV, genus *Ranavirus*) in perch and rainbow trout, European sheatfish virus (ESV, genus *Ranavirus*) in sheatfish (*Silurus glanis*), and European catfish virus (ECV, genus *Ranavirus*) in catfish. ESV and ECV are classified as the same species, *European catfish virus*. Santee–Cooper ranavirus (SCRV) is the species name given to three iridoviruses: largemouth bass iridovirus (LMBV), doctor fish virus (DFV-16), and guppy virus (GV-6). The white sturgeon iridovirus group (WSIV) is comprised of a group of unassigned viruses that infect sturgeon in North America and Russian sturgeon (*Acipenser guldenstadi*) in Europe. Red seabream iridoviruses (RSIV) (genus *Megalocystivirus*) causes mortality in cultured juvenile red sea bream in Japan. It has also been observed in grouper in Thailand. Two goldfish iridovirus-like viruses (goldfish virus 1 and 2, GFV-1 and -2) have been isolated from swimbladder tissue culture of healthy goldfish. Electron microscopic observations of iridoviruses in the cytoplasm of erythrocytes (viral erythrocytic necrosis, VEN) have been observed in many marine and anadromous bony fish.

Herpesviridae

Herpesviruses have been isolated from channel catfish (*Ictalurus punctatus*), common and koi carp (*C. carpio*), common goldfish (*Carassius auratus*), eel (*Anguilla* spp.), rainbow trout, masou salmon, lake trout (*S. namaycush*), sturgeon, walleye, and Japanese flounder. Channel catfish virus is the only fish herpesvirus assigned to the genus, *Ictalurivirus*, and this genus is not assigned to any of the three subfamilies (*Alphaherpesvirinae*, *Betaherpesvirinae*, and *Gammaherpesvirinae*) of the family *Herpesviridae*. The other fish herpesviruses, cyprinid herpesviruses 1 and 2 (CyHV-1 and CyHV-2), koi herpesvirus (CyHV-3), salmonid herpesvirus 1 and 2 (SalHV-1 and -2), eel herpesvirus (Anguilla herpesvirus, AngHV-1), and the acipenserid or white sturgeon herpesviruses remain as unassigned members of the family *Herpesviridae*. Electron micrographic evidence of herpesviruses has been found in sharks, eels, pike, flounder, perch, angelfish, grouper, and other fish.

The genomes of fish herpesviruses range in size from 134 to 295 kbp and the physical organization of the genome varies sufficiently to suggest that they have evolved separately from the herpesviruses of birds and mammals. The genome of *Ictalurid herpesvirus 1* (IcHV-1) or Channel catfish virus has a unique long (UL) region flanked by a substantial direct repeat that is similar to the betaherpesviruses of the genus *Roseolovirus*. The SalHV-1 genome is more similar in organization to the alphaherpesviruses of the genus *Varicellovirus* with a unique short (US) region flanked by a unique long (UL) region which is not flanked by a repeat. Phylogenetic comparisons of the individual genes including the DNA polymerase gene, the major capsid protein gene, the intercapsomeric triplex protein gene, and the DNA helicase gene indicate that the three cyprinid viruses are closely related and are distinct from IcHV-1.

These viruses are serious pathogens in their respective hosts. IcHV-1 outbreaks among juvenile catfish result in mortality and fish that survive the infection become carriers. The cyprinid herpes viruses produce a systemic disease with lesions in hematopoietic tissue in goldfish and papillomas on the caudal regions in koi carp. SalHV-2 was isolated from the ovarian fluid of masou salmon and it induces syncytia formation and lysis of infected cells. Epithelial papillomas are induced in young masou salmon injected with tissue culture-grown virus. The acipenserid herpesviruses cause serious losses in hatchery-reared young of white sturgeon (*Acipenser transmontanus*).

Adenoviridae

Adenovirus particles have been observed in lesions in a number of fish species and have been isolated from white sturgeon, dabs (*Limanda limanda*), cod, and Japanese red sea bream. The white sturgeon adenovirus has been isolated in tissue culture and its hexon protein and protease gene sequences are available. Based on an alignment of partial DNA polymerase gene sequences of the sturgeon adenovirus and 24 other adenovirus types, it is clear that the fish adenovirus is distantly related to the other adenoviruses, and might constitute a fifth genus of the *Adenoviridae*.

Adenovirus infection in cod produces an epidermal hyperplasia. In California, white sturgeon adenovirus affects young fish in hatcheries. Infection is characterized by epithelial hyperplasia and enlarged cell nuclei. A lympholeukemia has been observed in red sea bream and papillomas have been observed in dabs infected with adenoviruses.

See also: Fish and Amphibian Herpesviruses; Fish Rhabdoviruses; Fish Retroviruses; Aquareoviruses; Infectious Pancreatic Necrosis Virus; Infectious Salmon Anemia Virus.

Further Reading

Attoui H, Fang Q, Jaafar FM, *et al.* (2002) Common evolutionary origin of aquareoviruses and orthoreoviruses revealed by genome characterization of Golden shiner reovirus, Grass carp reovirus, Striped bass reovirus, and Golden ide reovirus (genus *Aquareovirus*, family *Reoviridae*). *Journal of General Virology* 83: 1941–1951.

Essbauer S and Ahne W (2001) Viruses of lower vertebrates. *Journal of Veterinary Medicine, Series B* 48: 403–475.

Hoffmann B, Beer M, Schutz H, and Mettenleiter TC (2005) Fish rhabdoviruses: Molecular epidemiology and evolution. *Current Topics in Microbiology and Immunology* 292: 81–117.

Lewis TD and Leong JC (2004) Viruses of fish. In: Leung KY (ed.) *Current Trends in the Study of Bacterial and Viral Fish and Shrimp Diseases. Molecular Aspects of Fish & Marine Biology*, vol. 3, 39–81.

Paul TA, Quackenbush SL, Sutton C, Casey R, Bowser P, and Casey JW (2006) Identification and characterization of an exogenous retrovirus from Atlantic salmon swimbladder sarcomas. *Journal of Virology* 80: 2941–2948.

Phelan PE, Pressley ME, Witten PE, Mellon MT, Blake S, and Kim CH (2005) Characterization of Snakehead rhabdovirus infection in zebrafish (*Danio rerio*). *Journal of Virology* 79: 1842–1852.

Schutze H, Ulferts R, Schelle B, *et al.* (2006) Characterization of White bream virus reveals a novel genetic cluster of Nidoviruses. *Journal of Virology* 80: 11598–11609.

Skall HF, Olesen NJ, and Mellergaard S (2005) Viral haemorrhagic septicaemia virus in marine fish and its implications for fish farming – A review. *Journal of Fish Diseases* 58: 509–529.

Troyer RM and Kurath G (2003) Molecular epidemiology of infectious hematopoietic necrosis virus reveals complex virus traffic and evolution within southern Idaho aquaculture. *Diseases of Aquatic Organisms* 55: 175–185.

Wolf K (1988) *Fish Viruses and Fish Viral Diseases*. Ithaca, NY: Cornell University Press.

Flaviviruses of Veterinary Importance

R Swanepoel, National Institute for Communicable Diseases, Sandringham, South Africa
F J Burt, University of the Free State, Bloemfontein, South Africa

© 2008 Elsevier Ltd. All rights reserved.

Glossary

Argasid Soft-skinned tick, member of the family *Argasidae*.

Arthropod Any member of the phylum Arthropoda, including insects and arachnids (spiders, scorpions, ticks, and mites).

Culicoid midge Blood-sucking midge of the genus *Culicoides*.

Instar Stage in the life cycle of arthropods, for example, egg, larva, nymph, and adult in mosquitoes; egg, larva, nymph, and adult in ticks.

Ixodid Tick with a hardened shell or scutum.

Microhabitat The environmental niche in which an organism is found.

Phlebotomine fly Blood-sucking sandfly of the genus *Phlebotomous*.

Phylogenetics The study of the genetic relatedness of organisms.

Transovarial transmission of virus Passage of virus infection through the eggs of arthropods to the succeeding generation, thus ensuring perpetuation of the virus.

Vectors Blood-sucking arthropods (mosquitoes, midges, sandflies, and ticks) which transmit viruses from one vertebrate to another.

Virion A complete virus particle with its protein coat and core of DNA or RNA nucleic acid.

Introduction

Approximately 94 viruses are currently assigned to three genera (*Flavivirus*, *Pestivirus*, and *Hepacivirus*) within the family *Flaviviridae* (L. *flavus* = yellow, from the type species *Yellow fever virus*, which causes jaundice); the number of viruses in the family changes with periodic discovery of new members or revision of the taxonomic status of existing viruses. Diseases of major medical and veterinary importance caused by hepaciviruses and pestiviruses are discussed elsewhere in this encyclopedia, while this article summarizes aspects pertaining only to the veterinary importance of members of the genus *Flavivirus*.

Viruses of the *Flaviviridae* have spherical virions 40–60 nm in diameter, with a tightly applied lipid envelope incorporating envelope (E) protein spikes which are glycosylated in pestiviruses and flaviviruses. The genomes consist of a single, linear molecule of positive-sense, single-stranded RNA (ssRNA), ranging in size from approximately 9.5 kbp in hepaciviruses, to 10.7 kbp in flaviviruses, and 12.5 kbp in pestiviruses. Replication takes place in the cytoplasm, with morphogenesis, transport, and maturation of virions occurring in cell membrane vesicles. A single open reading frame (ORF) in the genome is translated directly into a polyprotein which is cleaved into structural proteins, the RNA-associated internal capsid or core (C) protein, envelope proteins (E1, E2), and a membrane (M) protein, plus seven nonstructural proteins which exhibit viral polymerase and other enzymatic activities. The E proteins mediate binding of virus to host cells, are responsible for the hemagglutinating property exhibited by most flaviviruses, and constitute the major antigenic determinants of protective immunity. Viruses of the three genera are not closely related, with less than 30% nucleotide sequence homology and an absence of antigenic cross-reactivity, but they share genome organization and replication strategies. Viruses of the genera *Hepacivirus* and *Pestivirus* are most closely related structurally, and have in common a lack of arthropod vectors and the tendency to cause persistent infection in their vertebrate hosts.

The 82 recognized or tentative members of the genus *Flavivirus* include two inherent viruses of mosquitoes, cell fusion agent virus (CFAV) and Kamiti River virus (KRV), which have no known medical or veterinary significance and which may be assigned to a new genus, plus Tamana bat virus from Trinidad, which may also be assigned to a new genus. Most recently, there has been molecular identification of a tick-associated agent, Ngoye virus from Senegal, which has not as yet been successfully cultured in a laboratory host system, has no known medical or veterinary significance, and is only distantly related to existing flaviviruses. The remaining flaviviruses are antigenically related to each other and were originally assigned to serogroups on the basis of close antigenic affinities. Recent classification of the viruses based on phylogenetic analyses shows remarkable concordance with antigenic classification, and with the biological properties of the viruses, such as vertebrate host and arthropod vector associations.

About 21% of the members of the genus have been isolated from mammals only and have no known vectors (NKV viruses), a further 51% are usually transmitted by mosquitoes, and the remaining 28% appear to be transmitted principally by ticks, although definitive demonstration of transmission by arthropod vectors has been reported in a proportion of instances only. Depending on which genes are subjected to phylogenetic analysis, different conclusions may be reached with regard to the evolution of the flaviviruses, but from limited analyses of whole genome sequences it appears that NKV viruses diverged from a putative ancestor before mosquito-borne viruses, and that tick transmission is a recently acquired trait within the genus. Many flaviviruses were discovered in the course of surveys conducted on vertebrate and arthropod populations, rather than in the investigation of disease syndromes, and consequently many of the viruses have no known medical or veterinary significance. From the genetic distance which exists between the recently discovered Ngoye virus and other members of the genus, and between existing mosquito-borne viruses, it has been extrapolated that several thousand flaviviruses remain to be discovered.

Various mechanisms are postulated to ensure the persistence of arboviruses through prolonged periods of vector inactivity during winters and dry seasons, including the migration of infected vertebrate hosts, hibernation of infected adult vectors, and passage of virus through the eggs of vectors to infect the succeeding generation, so-called transovarial transmission of infection, which generally occurs with low frequency. Unlike mosquitoes, ticks feed on vertebrates in different stages (instars) of their life cycles, as larvae, nymphs, and adults, so that even in the absence of transovarial transmission of infection the perpetuation of virus can be ensured by the long intervals which intervene between the feeding of successive instars on different hosts.

Viruses transmitted by mosquitoes often cause sporadic infections which may pass unrecognized in endemic areas, but they are capable of causing explosive outbreaks when exceptionally heavy rainfall creates favorable breeding conditions for the vectors, or human manipulation of the environment, such as the building of dams and irrigation schemes, or the implementation of intensive livestock production systems, results in the juxtaposition of susceptible human or farm animal populations with vectors. Tick-borne virus infections tend to occur at endemic level within areas of vector distribution, and larger outbreaks are less immediately linked to climatic events which may favor population explosions of the wild hosts of immature ticks, but human intervention, such as changes in livestock farming systems, can also precipitate epidemics. A few arthropod-borne flaviviruses are named for the disease syndromes with which they are associated, but most are named for the geographic location from which the initial isolate was obtained.

Mosquito-Borne Flaviviruses Causing Livestock or Wildlife Diseases

Japanese Encephalitis Virus

Japanese encephalitis virus (JEV) was first isolated in 1935 in Japan as a cause of encephalitis in humans. Severe outbreaks of human disease occur at intervals in India, China, Korea, and Japan, and lesser outbreaks or sporadic cases are recorded in Nepal, Bangladesh, Sri Lanka, Laos, Cambodia, Vietnam, the Far East provinces of the Russian Federation, Thailand, Malaysia, Singapore, Philippines, Indonesia, Papua New Guinea, Taiwan, Guam, Saipan, and Myanmar. There are four genotypes of the virus, but they are all cross-protective. The virus is transmitted by *Culex tritaeniorhyncus* and other culicine mosquitoes and circulates between the mosquitoes, water birds, particularly egrets, herons and ducks, and domestic pigs. Overt outbreaks of disease are associated with rainfall and the irrigation of rice fields. The virus infects a wide range of vertebrates, including donkeys, cattle, and dogs, but most animals do not manifest disease. There is fetal wastage and abortion in sows. Humans and horses are incidental hosts which develop encephalitis, but play no significant role in the circulation of the virus. Tens of thousands of human infections occur each year, and while less than 1% of infected persons develop encephalitis the death rate is high in such patients. Case fatality rates may exceed 50% in horses with encephalitis. The virus is not known to cause disease in wild vertebrates, but bats can circulate virus following experimental infection, and low prevalences of antibody to JEV have been found in reptiles and amphibians.

St. Louis Encephalitis Virus

St. Louis encephalitis virus (SLEV) was isolated in 1933 from human brain as the causative agent of a large outbreak of encephalitis in St. Louis, Missouri, and was subsequently found to be transmitted by culicine mosquitoes, which are capable of breeding in drainage water in urban settings. The virus is widely distributed in North, Central, and South America, but the disease is seen mainly in the USA where it was considered to be the most important arbovirus pathogen prior to the introduction of West Nile virus (WNV) in 1999. There is some regional variation in the mosquitoes involved, but the main vectors are members of the *Cx. pipiens* complex, plus *Cx. tarsalis* and *Cx. nigripalpus*. Human infection with SLEV is usually asymptomatic but a low proportion of patients develop encephalitis with an approximately 10% case fatality rate. Outbreaks of varying magnitude occur at irregular intervals throughout the USA, but there are regional differences in occurrence of the infection depending on host and breeding-site preferences of the mosquito species concerned. Asymptomatic infection occurs in domestic animals such as horses, cattle, goats, sheep, pigs, cats, dogs and poultry, as well as in wild birds and small mammals including raccoons, opossums, rodents, and bats, but only birds are thought to play a significant role in the circulation of the virus. SLEV was isolated from the brain of a gray fox with encephalitis, and forest mammals including sloths are thought to be involved in transmission cycles in South America.

West Nile Virus

Following the intial isolation of WNV from a febrile patient in Uganda in 1937, sporadic cases and outbreaks of benign febrile disease were recorded in humans in Africa, the Near East and Asia, with the largest outbreaks occurring in Israel in 1950–54 and 1957, and in South Africa in 1974. It was found that WNV is widely endemic in southern Africa in localities where the vectors, ornithophilic (bird-feeding) culicine mosquitoes, and avian hosts of the virus are present, such as at dam sites with reed beds and heronries. Outbreaks of human infection, which occur when high rainfall or hot weather favors mosquito breeding, are associated with high seroconversion rates in wild birds, and 13 species of birds in South Africa were found to support replication of the virus without developing disease. It was deduced that the widespread dispersal of the virus was associated with migrating birds. Cattle and sheep appeared to undergo inapparent infection and could also theoretically serve as hosts for infection of mosquitoes. Subsequently, limited experimental evidence was obtained to indicate that WNV is capable of causing abortion in pregnant sheep.

The occurrence of meningoencephalitis was first observed in elderly WNV patients in Israel in 1957, and subsequently also in young children in India. The occurrence of horse encephalitis was recognized in Egypt in 1962, and in 1962–66 outbreaks of meningoencephalitis were observed in both humans and horses in France. There was a marked increase in the frequency and severity of outbreaks of human disease during the 1990s, often involving horses as well, and including epidemics in Algeria, Romania, Morocco, Tunisia, Italy, Russia, France, and Israel. Moreover, the outbreaks in Romania and Israel were characterized by concurrent mortality in birds. The vectors of WNV included *Cx. univitattus* and *Cx. antennatus* in Africa and Europe, members of the *Cx. pipiens* complex in Europe and Israel, and members of the *Cx. vishnui* complex plus *Cx. fatigans* in Asia.

The presence of WNV in the Western Hemisphere was first recognized in New York in 1999, and over the next few years it spread rapidly throughout the country and beyond into Canada and South America. Vectors in North America included members of the *Cx. pipiens* complex, plus *Cx. restuans*, *Cx. tarsalis*, and *Cx. nigripalpus*. The virus circulating in the USA was found to be genetically most closely related to a WNV isolate associated with goose mortality in Israel in 1998, suggesting that the virus was imported into America from the Near East, either in an infected bird, mosquito, human, or other animal, although the exact mechanism of the introduction will probably remain unknown.

In North America, WNV infection was confirmed in more than 300 species of birds, with death rates approaching 100% in corvids (crows, magpies, and jays), and with robins constituting important hosts for the infection of the mosquito vectors. It is notable that the virus spared African bird species in the New York zoo where deaths of local species were first observed. Infection was also recorded in 30 species of mammals, reptiles, and amphibians. Approximately 10% of infected horses manifested clinical illness, with a high proportion of these developing fatal encephalitis. Vaccine for horses became available in 2005. Although fatal infections were recorded in dogs and a wolf, infection rates were low in cats and dogs. Fatal disease was observed in farmed alligators in the USA and Mexico, with infection resulting from the feeding of infected horse meat, and also from transmission between alligators in crowded conditions. Serological evidence of infection was also found in small wild animals, such as raccoons, opossums, and squirrels, but the pathogenicity of the virus for these hosts is uncertain. Serological evidence of infection was obtained in farmed Nile crocodiles in Israel in the absence of disease. The Kunjin strain of WNV transmitted by *Cx. annulirostris* mosquitoes in Australia generally causes febrile illness in humans, seldom encephalitis, and was confirmed to be associated with encephalitis in horses on one occasion only.

The increase in the frequency of neurologic infections, and the occurrence of human, horse, and bird fatalities in the Northern Hemisphere, raised the question of whether there had been a change in the pathogenicity of WNV. Phylogenetic analyses revealed that WN isolates fell into two lineages. Lineage 1 included isolates from Africa north of the equator, Europe, Asia, and North America, with Kunjin virus from Australia constituting a subtype within this lineage. Lineage 2 consisted solely of viruses from Africa and Madagascar. The findings were interpreted to support the emergence of exalted virulence in lineage 1, with lineage 2 isolates considered to be associated with endemic infection of low virulence in Africa. However, WNV belonging to lineage 2 has recently been isolated from a goshawk with fatal encephalitis in Hungary, and three additional lineages of the virus have been described from the Czech Republic, Georgia, and India. Moreover, despite the apparently low level of virus activity observed in southern Africa in recent years, WNV was isolated from the brain of a suspected rabid dog in Botswana (initially described as having an infection with Wesselsbron virus), and in South Africa from humans and a foal with encephalitis, plus an ostrich chick in a commercial flock where extensive mortalities occurred. Experiments in mice confirmed that neuroinvasive WNV strains occur within both lineage 1 and 2. Thus, it is possible that the virulence of WNV in southern Africa may previously have been underestimated, while the perceived enhanced virulence of the virus in recent epidemics in the Northern Hemisphere may partly be due to the emergence or re-emergence of existing strains of WNV in geographic locations with immunologically naive populations, high medical alertness, and active surveillance programs.

Israel Turkey Meningoencephalomyelitis Virus

A disease caused by Israel turkey meningoencephalomyelitis virus (ITV) infection in turkeys was first observed in Israel in 1958 and the causative virus isolated in 1959. ITV was later found to be transmitted by culicine mosquitoes, but was also isolated from culicoid midges, and phlebotomine sandflies were shown to be capable of transmitting the virus. In 1978 the virus was isolated as a cause of disease of turkeys in South Africa. The virus has not been found in any other country, and disease has only been reported in turkeys. The disease generally occurs seasonally, only in birds older than 10 weeks, and it is characterized by progressive paresis and paralysis, with morbidity and mortality rates ranging from 15% to 80%. Turkey breeder hens exhibit a severe drop in egg production. Vaccines are available.

Sitiawan Virus

Sitiawan virus (SITV) was recently isolated as a cause of deaths of chick embryos and broiler chickens in Malaysia. It is closely related to the mosquito-borne Tembusu virus, which has no known medical or veterinary significance.

Wesselsbron virus

Wesselsbron virus (WESSV) was identified in 1955 in South Africa as a cause of deaths of newborn lambs in a flock where a modified live vaccine against the mosquito-borne Rift Valley fever virus (RVFV) (not a flavivirus) had been administered 2 weeks previously. There was concern that the RVFV vaccine may have caused the deaths, but it was concluded that the new virus, which also proved to be transmitted by *Aedes* species mosquitoes of the subgenera *Ochlerotatus* and *Neomelanoconion*, occurred in outbreaks together with RVFV and caused similar disease in sheep, cattle, and goats, namely, deaths of young animals and abortion in pregnant animals. Consequently, an attenuated WESSV vaccine was developed by serial intracerebral passage of virus in infant mice, and this was marketed for use together with RVFV vaccine on nonpregnant animals. During extended outbreaks of RVF in South Africa and Namibia in 1974–76 some 13.9 million doses of the live WESSV vaccine and 22 million doses of live RVFV were used indiscriminately on livestock, with resultant widespread occurrence of abortions and congenital malformations in sheep and cattle. It was confirmed experimentally that both of the vaccines were abortigenic and teratogenic. However, it subsequently became clear that WESSV is more widely endemic in southern Africa than RVFV, and although sporadic isolations of virus have been made from livestock in South Africa and Zimbabwe, mainly in association with disease of newborn lambs and calves, the virus has not been incriminated as the cause of large outbreaks of disease in farm animals, nor has a veterinary problem been described elsewhere. The vaccine, which confers lifelong immunity, continues to be marketed for use in nonpregnant sheep and goats.

Altogether, WESSV has been isolated from livestock and a gerbil in southern Africa, a camel in Nigeria, mosquitoes in South Africa, Zimbabwe, Uganda, Kenya, the Central African Republic (CAR), Cameroon, Ivory Coast, Senegal, Guinea and Thailand, a tick in CAR, naturally infected humans in South Africa, Senegal, and CAR, and from human laboratory infections in South Africa, Uganda, Nigeria, Senegal, CAR, and the USA. In addition, antibody has been found in humans and/or livestock in Malawi, Zambia, Botswana, Namibia, Angola, Mozambique, and Madagascar. The principal vectors of the virus appear to be aedine mosquitoes, and although the virus has been isolated from a tick, there is no evidence that ticks play a significant role in transmission of the virus. The few human infections recorded in South Africa were acquired from contact with infected animal tissues or mosquito bite, while human infections reported in Senegal and CAR seem to be consistently associated with mosquito bite. The virus generally causes benign febrile illness in humans, but a patient who acquired laboratory infection from the splashing of virus into an eye developed encephalitis from which he recovered without sequelae.

It cannot be excluded that WESSV or various other flaviviruses could emerge as important pathogens under particular circumstances, as occurred with WNV in North America in 1999. In this connection it should be noted that Bagaza virus has been isolated from aborted cattle fetuses on three occasions in South Africa, that WNV and Banzi virus were shown to cause abortion in experimentally infected pregnant sheep, and that Usutu virus, initially isolated from mosquitoes, a rodent and birds in the context of surveys in Africa, emerged as a cause of bird deaths in Austria in 2001.

Yellow Fever Virus

Descriptions of disease compatible with yellow fever date back to 1492. However, many medical conditions were confused in early records, and the name yellow fever was first applied to a fatal disease of humans in the Caribbean region in 1750. Shortly thereafter, the presence of the disease was recognized in South America and West Africa. Over the next 150 years the virus, introduced by infected people on ships, caused epidemics in many port cities in North America and Europe. From 1900 onwards the Walter Reed commission from the USA investigated yellow fever in Cuba and, using American military volunteers, demonstrated that the disease could be transmitted from person-to-person by *Ae. aegypti* mosquitoes, and that the infection was caused by a filterable agent present in the blood of patients (filtration was an early method used to distinguish viruses from bacteria). It was not until 1927 that yellow fever virus (YFV) was isolated in a laboratory host system, rhesus monkeys. Meanwhile, the identification of the *Ae. aegypti* vectors, which breed in small volumes of water, had facilitated the implementation of mosquito control methods which virtually eliminated circulation of YFV in urban settings. Initially, there was speculation as to whether YFV had originated in Africa or South America, but it was concluded that the virus and *Ae. aegypti* mosquitoes had been translocated from Africa in ships centuries ago. The occurrence of outbreaks of the disease on ships was well documented.

Resurgence of urban yellow fever in South America led to the discovery of jungle or sylvatic circulation of the virus in 1932. In South America, YFV circulates in the forest canopy between monkeys and *Haemagogus* species mosquitoes. Persons who live near or enter forests acquire infection and introduce it into urban environments where epidemics are generated by the circulation of YFV between humans and *Ae. aegypti* mosquitoes. In Africa, mosquitoes such as *Ae. africanus* transmit YFV between primates at forest canopy level, while mosquitoes which feed at intermediate levels, including *Ae. bromeliae* and members of the *Ae. furcifer-taylori* group, serve as link hosts to spread infection to humans which in turn serve to initiate urban transmission cycles involving *Ae. aegypti* mosquitoes. South American primates including tamarins, marmosets and howling, squirrel, owl, and spider monkeys undergo severe infection with variable mortality, while African monkeys and apes generally undergo viremic infection without overt illness. This was interpreted to indicate that there had been a longer period of co-evolution of YFV with African primates, but African YFV isolates were found to be of reduced virulence for neotropical monkeys. Asian monkeys are susceptible to severe and fatal YFV infection. Birds, reptiles, and amphibia are resistant to infection, and although several mammals including hedgehogs are capable of developing viremia, it appears that they have limited exposure to known vectors.

Effective vaccines were developed in the 1930s and used extensively in conjunction with mosquito control to reduce the occurrence of the human disease. However, the existence of sylvatic transmission cycles precludes eradication of the virus, and since 1980 there has been a marked resurgence of the disease in Africa with outbreaks being recorded in countries extending from Senegal in the north, to Angola in the south, and eastwards to Sudan, Ethiopia, and Kenya. No recent urban outbreaks have been recorded in South America, but fairly extensive outbreaks involving sylvatic vectors have occurred in Peru and Brazil.

Many of the remaining mosquito-borne flaviviruses are believed to circulate in birds and although a few, including Rocio, Spondweni, and Zika, have been associated with human disease on occasion, they have no known veterinary significance. It should be noted, however, that many of the flaviviruses which are pathogenic for humans produce symptomatic infection in a minority of people, and it can be surmised that infections which produce similar low morbidity in farm animals are likely to pass unrecognized.

Tick-Borne Flaviviruses

It has emerged from recent phylogenetic analyses that the tick-borne flaviviruses comprise three groups: a seabird group which includes three viruses with no known medical or veterinary significance, a Kadam virus group containing a single virus known only to have been isolated from ticks removed from cattle in Uganda and a camel in Saudi Arabia, but which represents a distinct evolutionary lineage, and the so-called mammalian tick-borne group of viruses. Within the mammalian group Gadgets Gully, Royal Farm, and Karshi viruses and are more distantly related to others and are known only to have been isolated from ticks, while the remaining viruses, which were formerly clustered together in a tick-borne encephalitis antigenic complex, include important pathogens of humans and livestock. However, genetic differences between some of these latter viruses are smaller than those which exist between other flaviviruses, and hence they have been tentatively designated as types and subtypes of a proposed tick-borne encephalitis virus (TBEV) species.

Powassan Virus and Deer Tick Virus

Powassan virus (POWV) was first isolated as a cause of fatal encephalitis in a child in Ontario, Canada, in 1958. The virus was subsequently found in western Canada, and northeast, north-central and western USA. The principal vector was found to be the tick *Ixodes cookei*, and the main host for the tick and virus is believed to be the woodchuck or groundhog (*Marmota monax*). Other *Ixodes* species and squirrels have also been implicated in the circulation of the virus, and there have been isolations from other ticks including *Dermacentor andersoni*. Twelve human cases of encephalitis with four fatalities have been reported in North America. Antibody has been found in a range of animals, and virus was isolated from two dead foxes, although it was not clear that the virus had caused their deaths. The virus was experimentally shown to be capable of causing encephalitis in horses. Reportedly, POWV was isolated in 1978 from the blood of a person who had been bitten by a tick in the Maritime Province (Primorsky) in the Far East of the former Soviet Union (currently the Russian Federation), and later the virus was isolated from *Ixodes persulcatus* and other ticks and mosquitoes, and antibodies were found in birds in Primorsky. In 1996, deer tick virus (DTV) was isolated from deer ticks, *Ixodes scapularis*, in the eastern USA and found to be closely related to POWV. The white-footed mouse (*Peromyscus leucopus*) is thought to be the main host for the tick and virus, although other mammals, particularly white-tailed deer, are important hosts for the tick. As yet, no disease associations have been made for DTV. Although POWV and DTV are closely related, the results of recent phylogenetic studies confirm that the viruses circulate independently. However, the studies did not include POWV isolates from the western USA or the Russian Federation.

Kyasanur Forest Disease Virus

Kyasanur forest disease virus (KFDV) was discovered in 1957 as a cause of fatal febrile disease of humans in Karnataka State, India. The vectors are ticks of the genus *Haemaphysalis*, particularly *H. spinigera*, and two species of *Ixodes* ticks, but the virus has also been isolated from the argasid (soft-skinned) tick *Ornithodoros crossi*. Rodents are suggested to be the main hosts of the virus. About 500 human cases of the disease are recorded each year in Karanataka State, with a mortality rate of 5–10%. Infection occurs mainly in people who enter forests to gather wood, and the disease is seasonal with tick activity. Fatal disease occurs in langur and macaque monkeys in the forests, but no disease manifestations have been recorded in domestic animals. The virus has been isolated from rodents, shrews, and bats.

Omsk Hemorrhagic Fever Virus

Omsk hemorrhagic fever virus (OHFV) was isolated in 1947 from the blood of a patient with hemorrhagic disease in western Siberia in the former Soviet Union. The disease was first observed in 1941, and large numbers of cases were recorded in 1945–49. Since that time only sporadic cases have been recognized in the Omsk, Novosibirsk,

Kurgan, and Tjumen regions of western Siberia. The disease affects mostly muskrat trappers and their families. Muskrats, or musquash (*Ondatra zibethicus*), are large rodents which are trapped for their fur. They are indigenous to North America and were introduced into Siberia in 1925. No equivalent virus or disease has been found in North America. The principal vector of the virus in Siberia was found to be the tick *Dermacentor reticulatus* (syn. *D. pictus*) but the virus was also isolated from *D. marginatus* and other ticks and mites. Muskrats and local voles, some of which undergo benign infection, serve as reservoir hosts for the infection of ticks. Muskrats live in swamps, and virus which is present in the excretions of infected muskrats, or leaches into water from dead muskrats, can infect other muskrats or humans. OHFV was also found in mosquitoes and antibody in water birds, but these do not appear to play a significant role in circulation of the virus. Humans acquire infection mainly from contact with infected muskrat tissues, but also from infected water, or from tick bite.

Tick-Borne Encephalitis Virus

A human encephalitis syndrome was recognized in the far eastern provinces of the former Soviet Union at least as far back as the late nineteenth century, and has been known under a variety of names including Russian spring–summer encephalitis and Far Eastern tick-borne encephalitis. The causative agent, now named tick-borne encephalitis virus-Far Eastern subtype (TBEV-FE), was isolated from human patients in 1937 and later from its *Ixodes persulcatus* tick vectors. The virus occurs in the Primorsky, Khabarovsk, Krasnoyarsk, Altai, Tomsk, Omsk, Kemerovo, Western Siberia, Ural and Priural regions of the Russian Federation, China, and eastern Europe. After World War II the existence of a similar disease was recognized in several countries in Central Europe and adjacent parts of the Soviet Union, and a second virus, tick-borne encephalitis virus-European subtype (TBEV-Eu), was isolated from humans and *Ix. ricinus* ticks. Slight antigenic differences were shown to occur between the two viruses, and TBEV-Eu had human case fatality rates of 0.5–2.0%, in contrast to 5–20% for TBEV-FE. It has since been established that TBEV-Eu occurs in Albania, Austria, Belarus, Bornholm Island of Denmark, Bosnia, Bulgaria, China, Croatia, Czech Republic, Estonia, Finland, France, Germany, Greece, Hungary, Italy, Kazakhstan, Latvia, Lithuania, Norway, Poland, Romania, Russia, Slovakia, Slovenia, Sweden, Switzerland, Turkey, and Ukraine, although the distribution tends to be focal, dependent on suitable tick habitat. The *Ix. persulcatus* tick vector of TBEV-FE is distributed across nontropical Asia from Japan to eastern Europe, where there is overlap with the distribution of the *Ix. ricinus* vector of TBEV-Eu which extends across western Europe to the British Isles and Ireland. Other tick species play a minor role in transmission of the TBEV viruses. Phylogenetic analyses in the 1990s revealed the existence of a third subtype of virus, tick-borne encephalitis virus-Siberian subtype (TBEV-Sib), which is transmitted by *Ix. persulcatus* in the Irkutsk region of Siberia and has similar virulence for humans as does TBEV-FE. It was deduced that the TBEVs evolved in a cline in a westerly direction across the Eurasian continent over the past few thousand years. However, TBEV-FE and TBEV-Sib infections have recently been detected in Europe.

All three subtypes of TBEVs are believed to circulate between ticks and small vertebrates such as field mice and voles which undergo benign viremic infection. Antibody is also found in larger vertebrates such as squirrels, deer, badgers, and farm animals which develop low-grade viremia and are believed not to constitute a significant source of virus for infection of ticks. About 5000–10 000 cases of human disease are recorded each year, mainly in the Russian Federation, with the lowest incidence occurring in Austria where there is widespread use of vaccine. Humans acquire TBEV infection from tick bite, or from the ingestion of unpasteurized infected milk of sheep, goats, and cows. Less than 5% of humans develop symptomatic infection, but high death rates occur in patients who develop nervous disease. There may be neurological sequelae in those who recover, and a few patients develop chronic infections. A fatal hemorrhagic syndrome of humans has been described in the Novosibirsk region of Siberia in association with TBEV-FE infection. Recently, fatal TBEV-Eu infection was reported in captive monkeys in Germany which were inadvertently exposed to ticks.

Louping Ill Virus

Descriptions of a sheep disease conforming to louping ill (Scottish dialect *louping* = leaping) date back more than two centuries in Scotland, but it was not until 1929 that the causative virus was isolated, and shortly thereafter shown to be transmitted by the tick *Ix. ricinus*. Other species of ticks can be infected but do not appear to play a significant role in the transmission of the virus. Louping ill virus (LIV) occurs in hill sheep farming areas of Britain and Ireland where there is suitable microhabitat for the ticks. Phylogenetic evidence suggests that the virus was probably originally introduced into Ireland from western Europe as a result of the livestock trade, and successively evolved as Louping ill virus-Ireland (LIV-Ir), and Louping ill virus-British (LIV-Brit) as it spreads through Wales, England, Scotland, and into Norway.

LIV causes fatal encephalomyelitis in a proportion of sheep depending on age and predisposing factors, including concurrent infection with agents such as the tick-borne parasite *Ehrlichia phagocytophila*. In areas with high challenge rates the disease is seen most frequently in lambs once they have lost their maternal immunity, but in areas with low tick burdens primary infection and disease may be seen in sheep of all ages. Infection is less frequently associated with disease in cattle, horses, pigs, farmed red deer, dogs, red grouse, and humans. Infection also occurs in roe deer, mountain hares, feral goats, voles, shrews, and field mice. The only recognized disease problem in wild vertebrates exists in red grouse (*Lagopus lagopus scotticus*), which have a commercial hunting value, and it has been demonstrated that the chicks can acquire infection from ingesting ticks. Sheep and red grouse are the only hosts which develop sufficiently intense viremia for the infection of ticks. However, it has been shown experimentally that the so-called phenomenon of nonviremic transmission of infection between ticks feeding on the same host, facilitated by factors present in tick saliva, occurs with LIV-infected and noninfected ticks co-feeding on mountain hares, which are important hosts of *Ix. ricinus* ticks. The same phenomenon has been demonstrated with TBEV-Eu, and this implies that there should be reassessment of the role of vertebrate hosts in the circulation of tick-borne viruses.

Spanish sheep encephalitis virus, Turkish sheep encephalitis virus, and Greek goat encephalitis virus are recently recognized subtypes of TBEV analogous to LIV that have been associated with encephalitis in sheep and goats in Spain, Greece, Bulgaria, and Turkey, but their distribution and epidemiology are as yet incompletely understood.

Negishi Virus

A virus given the name Negishi virus (NEGV) was isolated in 1948 from a cerebrospinal fluid sample from a child near Tokyo with suspected Japanese encephalitis, and was found to be antigenically related to the tick-borne encephalitis group of flaviviruses. There were no further isolations of NEGV in Japan and recently the virus was found to be genetically closely related to LIV. However, in 1993 a virus was isolated from a patient with encephalitis in Hokkaido, the northern island of Japan where *Ix. persulcatus* ticks occur, and subsequently the virus was also obtained from ticks and blood samples from dogs in Hokkaido. The new isolates proved to be phylogenetically most closely related to TBEV-FE.

See also: Flaviviruses: General Features; Tick-Borne Encephalitis Viruses; West Nile Virus; Yellow Fever Virus.

Further Reading

Billoir F, de Chesse R, Tolue H, de Micco P, Gould EA, and de Lamballerie X (2000) Phylogeny of the genus *Flavivirus* using complete coding sequences of arthropod-borne viruses and viruses with no known vector. *Journal of General Virology* 81: 781–790.

Cook S, Bennett SN, Holmes EC, de Chesse R, Moureau G, and de Lamballerie X (2006) Isolation of a new strain of the *Flavivirus* cell fusing agent virus in a natural mosquito population from Puerto Rico. *Journal of General Virology* 87: 735–748.

Cook S and Holmes EC (2006) A multigene analysis of the phylogenetic relationships among the *Flaviviruses* (family: *Flaviviridae*) and the evolution of vector transmission. *Archives of Virology* 151: 309–325.

Fauquet CM, Mayo MA, Maniloff J, Desselberger U and Ball LA (eds.) (2005) *Virus Taxonomy: Eighth Report of the International Committee on Taxonomy of Viruses.* San Diego, CA: Elsevier Academic Press.

Grard G, Lemasson JJ, Sylla M, et al. (2006) Ngoye virus: A novel evolutionary lineage within the genus *Flavivirus*. *Journal of General Virology* 87: 3273–3277.

Grard G, Moureau G, Charrel RN, et al. (2006) Genetic characterization of tick-borne *Flaviviruses*: New insights into evolution, pathogenetic determinants and taxonomy. *Virology* 361: 80–92.

Kono Y, Tsukamoto K, and Hamid MABD (2001) Encephalitis and retarded growth of chicks caused by Sitiawan virus, a new isolate belonging to the genus *Flavivirus*. *American Journal of Tropical Medicine and Hygiene* 63: 94–101.

Pybus OG, Rambaut A, Holmes EC, and Harvey PH (2002) New inferences from tree shape: Numbers of missing taxa and population growth rates. *Systematic Biology* 51: 881–888.

Flaviviruses: General Features

T J Chambers, Saint Louis University School of Medicine, St. Louis, MO, USA

© 2008 Published by Elsevier Ltd.

Glossary

Cirrhosis Disease of the liver, with fibrosis and distortion of the cellular architecture.

Enzootic A disease or maintenance transmission cycle which occurs in a continuous manner among nonhuman animals within a certain geographic zone.

Epizootic Above average activity or occurrence of disease or pathogenic agent among nonhuman animals.

Sylvatic Located in forested regions.

Thrombocytopenia A reduction in the normal quantity of circulating blood platelets.

Zoonosis A disease of animals communicable to man.

General Features

The family *Flaviviridae* includes three genera, *Flavivirus*, *Pestivirus*, and *Hepacivirus*, that together comprise approximately 80 known viruses (**Table 1**), many of which cause diseases in human or animal hosts. Viruses in these genera are antigenically distinct from viruses in other genera and overall sequence homology is approximately 20%. The flaviviruses constitute the largest genera within the family, including arthropod-borne viruses that cause encephalitis and hemorrhagic fever. Pestiviruses exhibit a host range that includes pigs and ruminants. Hepaciviruses cause chronic liver disease in humans. Flaviviruses were originally classified as the 'group A' members of the family *Togaviridae*, distinguished serologically from 'group A' arboviruses (e.g., equine encephalitis viruses) but later assigned as constituting a unique family of viruses. Although vector associations together with antigenic properties previously served to classify the *Flaviviridae*, modern taxonomy based on nucleotide sequencing has been used to establish their phylogenetic and evolutionary relationships.

Members of the *Flaviviridae* are enveloped viruses which share a common genome organization and replication strategy. Their virions are small, spherical particles containing the capsid–genome complex and membrane-associated envelope proteins, which confer antigenic properties. The genomes are small, single-stranded, plus-sense RNA molecules ranging in size from 9.6 to 12.3 kbp. Viral proteins are encoded within a single long open reading frame that contains the structural proteins, followed by nonstructural proteins (protease(s), RNA helicase, RNA-dependent RNA polymerase (RdRp), other enzymatic activities involved in RNA replication, and several small hydrophobic membrane proteins).

Flaviviruses and pestiviruses exhibit generally broad cellular host range *in vitro*, whereas growth of hepaciviruses is very restricted. Entry of viruses into host cells occurs by receptor-mediated endocytosis, requiring a low pH-dependent fusion activity of the major envelope proteins. Cellular receptors involved in the entry process have been difficult to definitively identify. Viral replication occurs in association with intracellular membranes via a negative-strand RNA intermediate. Characteristic ultrastructural changes are induced in the membranes. Mature viruses are generated by budding through host intracellular membranes followed by release at the cell surface. Further details of these processes and unique features of each genus are described below.

Flaviviruses

History and Classification

The genus includes >70 viruses, approximately 40 of which are mosquito-borne, 16 tick-borne, and 18 without a known vector. The prototypic flavivirus, yellow fever virus (from the Latin *flavus* for 'yellow'), is recognized as a human disease agent for at least the past 300 years. In the early twentieth century, the virus was isolated and characterized as an arthropod-transmitted agent, and a live-attenuated vaccine (yellow fever strain 17D) was developed. Other flaviviruses causing a range of human diseases were identified within the nineteenth and twentieth centuries. The most important of these are Japanese encephalitis (JEV), West Nile virus (WNV), the tick-borne encephalitis (TBEV), and the dengue (DENV) viruses, which cause outbreaks of human disease on a large scale. Many other flaviviruses not listed in **Table 1** are causes of acute fever, meningoencephalitis, or hemorrhagic fever in humans on a regional basis throughout the world.

Based on comparative genomic sequence analysis, it is hypothesized that flaviviruses evolved as separate

Table 1 Taxonomy of viruses of the family *Flaviviridae*

Genus *Flavivirus* (12 complexes[a]; 1 tentative species/subspecies)
  Tick-borne viruses
    Mammalian tick-borne virus complex (10 members)
    Seabird tick-borne virus complex (3 members)
  Mosquito-borne viruses
    Aroa virus complex (4 members)
    Dengue virus complex (5 members)
    Japanese encephalitis virus complex (10 members)
    Kokobera virus complex (2 members)
    Ntaya virus complex (6 members)
    Spondweni virus complex (2 members)
    Yellow fever virus complex (10 members)
  Viruses with no known arthropod vector
    Entebbe bat virus complex (3 members)
    Modoc virus complex (6 members)
    Rio Bravo complex (7 members)
  Tentative members of the genus
    Alkhurma virus (1 member)
Genus *Pestivirus* (4 species plus 1 tentative species)
  Border disease virus (2 serotypes)
  Bovine viral diarrhea virus-1 (4 serotypes)
  Bovine viral diarrhea virus-2 (2 serotypes)
  Classical swine fever virus (*Hog cholera virus*), (4 serotypes)
  Tentative species in the genus
    Pestivirus of giraffe
Genus *Hepacivirus* (1 species)
  Hepatitis C virus (6 genotypes)
Unassigned viruses in the family
  Tamana bat virus
  Cell-fusing agent
  GB virus A
  GBV A-like agents
  GB virus B
  GB virus C/Hepatitis G virus

[a]Complex refers to closely related viruses, representing genetic and antigenic clusters, and viral species in the complexes are enumerated in parentheses, or otherwise indicated. All others represent virus strains or serotypes.
Reproduced in revised form from Calisher CH and Gould EA (2003) Taxonomy of the virus family Flaviviridae. *Advances in Virus Research* 59: 1–19, with permission from Elsevier.

tick-borne, mosquito-borne, and no-known-arthropod vector lineages, which diverged at an early point after the origin of the genus *Flavivirus* from a common ancestor whose identity remains unknown. This ancestral virus is assumed to have originated in central Africa as long ago as 10 000 years. There has been a general agreement between the phylogenetic clustering of flaviviruses and classification on the basis of serological relationships. Viruses such as the cell-fusing agent (CFA) and Tamana bat virus display homology that is distant enough to suggest a possible link among the three genera, but exact relationships remain unclear. According to this scheme, tick-borne viruses diverge first, followed by the mosquito-borne viruses. Yellow fever virus is representative of so-called Old World mosquito-borne (species *Aedes*) viruses. Additional clades, such as the dengue viruses and those of the JEV antigenic complex, vectored primarily by *Culex* mosquitoes, are believed to have diverged subsequently.

Epidemiology and Clinical Disease

The epidemiologic situation for flaviviruses involves both the appearance of new viruses, with several described in recent years, as well the emergence and re-emergence of established viruses. A number of contributing factors, such as human activities (urbanization, transportation, and land-use practices) and also natural phenomena (bird migration, host–vector relationships, and climate change), drive this process. Flaviviruses exhibit both global and regional distribution based on these various factors. Dengue has become the most significant cause of viral hemorrhagic fever globally, with a dramatic increase in prevalence of dengue involving up to 2.5 billion persons at risk and an annual infection rate of 50 million cases. JEV is regarded as the most important cause of epidemic viral encephalitis worldwide, causing up to 50 000 cases annually, and has spread from Southeast Asia to the south and west in recent years. West Nile virus, originally recognized as a cause of febrile illness primarily in Africa, the Middle East, and India, caused outbreaks of severe encephalitis in Eastern Europe and the former Soviet Union in the 1990s prior to its due appearance in the US in 1999. Since 1999, it has become distributed across North America and parts of Central and South America. In North America, it has caused annual outbreaks of encephalitic disease involving thousands of cases. Yellow fever is currently restricted to tropical Africa and South America, where it regularly causes endemic and periodic epidemic disease involving thousands of cases on a regular basis. The true incidence is believed to be underreported and may be as high as 200 000 cases per year, mainly in Africa.

Most flaviviruses are zoonotic viruses maintained in enzootic transmission cycles which depend on birds, rodents, or nonhuman primates as reservoirs, with humans usually dead-end hosts. However, dengue viruses are exceptions, due to their adaptation to humans and maintenance in urban-based areas in transmission cycles that do not require animal reservoirs as they seem to exist in sylvatic cycles in Africa and Asia.

Structure and Replication

The structure and replication of the flaviviruses have served for many years as a model for the entire family, since the first report of the genomic sequence for yellow fever virus in 1985. The use of molecular clones, together with the ease of propagating most of these viruses in cell lines *in vitro*, and the advent of modern expression systems for production and purification of viral proteins, have facilitated progress for understanding the molecular biology of the family. A number of different molecules in different cell types have been characterized as candidate receptors, including glycosaminoglycans. The best characterized receptor interaction is between dengue virus and dendritic cell (DC)-specific intercellular adhesion molecule 3 (ICAM-3)-grabbing nonintegrin (DC-SIGN), present on human dendritic cells. Fc and complement receptors can also mediate entry of the dengue viruses and several other viruses through the process of antibody-dependent enhancement occurring in the presence of subneutralizing antibody levels.

The virions contain a lipid envelope, have a buoyant density of $1.19–1.3 \text{ g ml}^{-1}$, and are sensitive to inactivation by heat, acid pH, organic solvents, and detergents. Currently the virion structure is best understood for dengue viruses and TBEVs. Dengue viruses, as reconstructed from cryoelectron microscopy (cryo-EM) images at 24 Å resolution, exist as a smooth particle comprised of dimeric units of the envelope (E) protein arranged in a head-to-tail fashion on its surface with icosahedral symmetry, encasing a nucleocapsid composed of capsid protein and genomic RNA (**Figure 1**). These genomes are on the order of 11 kbp (42 S sedimentation) and differ from other flavivirus genera in having a 5′ cap. The 5′ untranslated region (UTR) encodes a unique secondary RNA structure. The 3′ UTR contains conserved sequences and secondary structures, including a unique terminal stem–loop structure required for viral RNA replication (**Figure 2**). Such structures are found among both mosquito and tick-borne viruses despite differences in nucleotide sequence, arguing for conserved mechanisms relating to viral replication.

The flavivirus polyprotein undergoes co- and post-translational processing to produce the mature viral proteins. Structural proteins (capsid, prM (precursor to M), and E (envelope)) are generated by cleavage by host signalase in association with membranes of the endoplasmic reticulum, whereas nonstructural proteins (NS1, -2A, -2B, -3, -4A, -4B, and -5) are generated primarily by cleavage reactions mediated by the viral NS3 serine protease together with its NS2B cofactor. NS1 is a soluble,

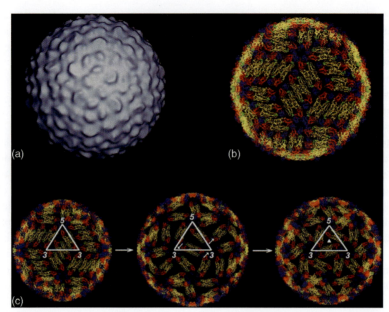

Figure 1 Structure of the dengue-2 virion as determined by cryo-EM. (a) An image reconstruction model of the virion in which the smooth surface is shown. (b) E protein dimers are displayed in the 'herringbone' arrangement, as fitted into the electron density map. (c) Acid-catalyzed rearrangement of the E protein dimers into the fusion-active state. Triangle represents the icosahedral asymmetric unit with three- and fivefold symmetry axes shown. The small arrows indicate the rotation of the E protein according to the proposed model. Reproduced from Lindenbach BD and Rice CM (2003) Molecular biology of the flaviviruses. *Advances in Virus Research* 59: 23–61, with permission from Elsevier, and adapted from Kuhn RJ, Zhang W, Rossmann MG, *et al.* (2002) Structure of dengue virus: Implications for flavivirus organization, maturation, and fusion. *Cell* 8: 717–725.

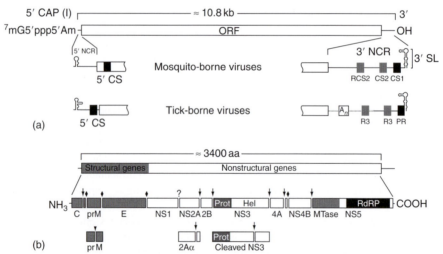

Figure 2 Organization and expression of the flavivirus genome. (a) RNA elements of the 5′ and 3′ noncoding regions (NCRs) flanking the open reading frame (ORF) for mosquito- and tick-borne viruses are indicated. The 5′ CAP structure and 5′ and 3′ stem–loop (SL) structure are shown. Conserved sequence elements for mosquito- and tick-borne viruses are denoted by CS, RCS, and CS, R, PR, respectively. A_n indicates a polyadenylate sequence contained within the variable region (dotted line) of some tick-borne isolates. (b) Expression and processing of the polyprotein. Cleavage events mediated by host signalase and furin-like enzyme are depicted by solid diamonds and triangle, respectively. Question mark indicates processing of NS1-2A by an unknown host enzyme. Downward arrows indicate cleaveage by the viral NS3 protease (PROT). HEL, helicase; MTase, methyltransferase. Reproduced from Lindenbach BD and Rice CM (2003) Molecular biology of the flaviviruses. *Advances in Virus Research* 59: 23–61, with permission from Elsevier.

cell surface, and intracellular protein which is involved in viral RNA synthesis, together with NS3 (protease, helicase, and NTPase activities), NS5 (methyltransferase and RdRP activities), and the hydrophobic proteins of the NS2 and NS4 regions. Viral replication occurs in association with intracellular membranes that are reorganized into packets of vesicles, convoluted membranes, and paracrystalline arrays. The capsid protein is responsible for formation of the viral nucleocapsid. During subsequent virus maturation, prM acts as a molecular chaperone for the E protein, and undergoes a furin-like cleavage to form the fusion-competent heterodimer prM–E complex. The E protein is the major virion surface protein that harbors receptor binding and fusion activity. Structural characterization of the TBEV E protein has revealed that this protein resembles the E1 protein of the alphaviruses, both of which are designated as 'class II' viral fusion proteins, distinguished from the class I proteins typified by the influenza hemagglutin by major differences in overall fold, and a flat elongated orientation on the virion surface rather than an extended spike.

Pathogenesis

Flaviviruses are readily propagated in a variety of cell lines in culture, including cells of mammalian origin, in which cytopathic effects and plaque formation are common, as well as in mosquito-derived cells. Many of these infected cell types exhibit cell fusion and syncytia formation and are capable of supporting persistent infection. *In vivo*, flaviviruses cause syndromes in humans ranging from febrile illness to acute meningoencephalitis and hemorrhagic fever. Flaviviruses are inherently neurotropic viruses, possibly reflecting the propensity of viruses of the early phylogenetic branches to propagate in the central nervous system of arthropod vectors, and in turn, in vertebrate hosts, in which disease manifestations occur. Arthropod vectors undergo chronic infections that serve to deliver virus into enzootic and zoonotic transmission cycles that lead to infections of humans and animals. Whereas many of the encephalitic flaviviruses which typically depend on avian species as reservoirs and for amplification cycles, dengue and yellow fever viruses have adapted to primate hosts, and cause hemorrhagic fever as a primary disease manifestation. Nonhuman primates have been used for characterization of yellow fever and dengue viral infections and for evaluation of experimental vaccines for dengue viruses. Most flaviviruses are pathogenic for laboratory rodents, particularly mice, which have been used as models for experimental studies of acute flavivirus encephalitis.

Flavivirus resistance is a phenomenon observed in certain inbred strains of mice, in which markedly reduced susceptibility to encephalitic strains, such as of WNV, have been observed. The genetic basis of this resistance has been mapped to a single autosomal dominant locus which encodes the Oas1B gene. It is not known yet if there is an analogous resistance mechanism that operates in humans to explain variability in susceptibility to these viruses. A recent study has also suggested a link between CCR5D32 (a common mutant allele of the chemokine and human immunodeficiency virus (HIV) receptor CCR5) and fatal WNV infection.

Flaviviruses are transmitted via feeding by chronically infected arthropod vectors. Following local replication at the site of inoculation and in regional lymph nodes, virus disseminates by viremia, with distribution to target organs in the periphery, including primarily visceral organs in the case of hemorrhagic fever viruses, and meninges, brain, and spinal cord in the case of encephalitic agents. Skin Langerhans (dendritic) cells are implicated as sites of initial virus replication as well as early spread of virus to lymphatic tissues. Monocytes and macrophages are also major targets of infection by dengue viruses. Flaviviruses also interact with many other cell types *in vivo* and elicit changes in cell surface molecules that are believed to influence pathogenesis of encephalitis and hemorrhagic fevers. Upregulation of surface molecules such as major histocompatibility complex (MHC) I and ICAM-1 has been observed.

Pathogenesis of both hemorrhagic fever due to yellow fever virus and acute encephalitis due to JEV complex and TBEV complex members are generally associated with cytopathic effects of the virus in target organs, although many viruses have been shown to induce apoptotic cell death as well. Flavivirus infections of the central nervous system, as typified by members of JEV complex or TBEV complex viruses, range from mild meningitis to severe destructive encephalitis, in which neurons of the brain and spinal cord are targeted. Mortality rates are as high as 40% and permanent neurologic effects may occur in survivors. Persistent and congenital infections have been described for some of the encephalitic viruses. Primary infection with dengue viruses either is generally characterized by flu-like illness or is asymptomatic. However, some secondary infections with heterologous serotypes result in a distinct form of hemorrhagic disease. This involves an immunopathogenic process of antibody-dependent enhancement of infection mediated by heterologous antibodies, in concert with activation of cross-reactive memory T-cells, cytokine release, and diffuse capillary leakage. This is referred to as dengue hemorrhagic fever and may be accompanied by a shock syndrome. Yellow fever causes a unique form of acute hemorrhagic fever with characteristic cytopathology in the liver and kidneys, and is fatal in up to 50% of severe cases.

Diagnosis and Prevention

Because flaviviruses may cause many different disease syndromes, disease manifestations are variable. Although diagnosis is often suspected on the basis of clinical findings, case history, and geographical distribution, a range of laboratory tests are available for determining infection with specific viruses, including virus isolation, serologic techniques (enzyme immunoassay, hemagglutination inhibition, immunofluorescence, complement fixation, dot blot immunoassays, and plaque neutralization), antigen detection, and reverse transcriptase/polymerase chain reaction (RT/PCR) amplification. Detection of immunoglobulin M (IgM) by immunoassay is a common diagnostic test for flaviviruses, but in some cases persistence of high IgM levels from previous infections with other flaviviruses can complicate the diagnosis of recent infections. Detection of IgM antibody in cerebrospinal fluid is of use in cases of flavivirus encephalitis. RT/PCR of serum is a sensitive and rapid technique that has increasing application for flaviviruses, but must be carried out during the viremic stage, which can differ from one flavivirus infection to another. For instance, RT/PCR for dengue viruses loses sensitivity after the first week of infection due to clearance of viremia.

Outside of mosquito control measures, vaccination is the only available means of preventing flavivirus infections in humans. The live-attenuated yellow fever 17D vaccine has been used since its original development in 1936 for prevention of yellow fever and is one of the safest and most effective vaccines ever developed for human use. Inactivated vaccines are available for prevention of diseases caused by JEV and TBEV. Development of vaccines for dengue viruses and WNVs have been pursued because of the increasing public health burdens associated with these agents, but are still in investigational stages. At this time, there are no antiviral agents or other therapies licensed for treatment of flavivirus infections.

Pestiviruses

History and Classification

The genus *Pestivirus* includes four accepted species that are important causes of veterinary disease in farm animals which are of great economic consequence worldwide: *Classical swine fever virus*, *Bovine viral diarrhea virus 1* and *2*, and *Border disease virus*. The host range of pestiviruses is restricted to 'artiodactyles' (cloven-hoofed ruminants and pigs). Classical swine fever virus (CSFV) infects members of the Suidae as the natural hosts. Bovine viral diarrhea virus (BVDV) is a pathogen primarily of cattle. Classical swine fever was characterized (e.g., 'hog cholera') in the 1800s, but the viral cause not established until *c.* 1904. The diseases caused by BVDV in cattle were described in the 1940s and 1950s. Border disease virus (BDV) was discovered in 1959 in the border region of England and Wales, as a congenital viral infection affecting sheep and goats. Pestiviruses exhibit great genetic diversity, and characterization of new isolates, including those from wildlife species, about which understanding is limited, will have future impact on the classification of viruses of this genus.

Prior to use of genomic sequencing, pestiviruses were differentiated by host species of origin, and antigenic cross-reactivity prevented assignments to different serogroups, even though differences among groups could be demonstrated with monoclonal antibodies. Comparison of genomic sequences later revealed the existence of two genotypes of bovine viral diarrhea virus (BVDV-1 and BVDV-2), which became significant after it was recognized during the 1980s that BVDV-2 strains could cause a severe acute hemorrhagic syndrome, with some of the first isolations of BVDV-2 arising during BVDV1 vaccine breaks. The two genotypes are distinguishable antigenically, but not to a level which allows classification into different serotypes. Subtypes (BVDV-1a, -1b, -2a, -2b) also exist. BVDVs also display pronounced heterogeneity in phenotype (e.g., biotype and virulence type). Biotypes (cytopathic vs. noncytopathic) are defined on the basis of cytopathic effects in cultured epithelial cells. However, virulence phenotype *in vivo* correlates with the noncytopathic rather than the cytopathic biotype.

The authentic form of BDV is predominantly an ovine pathogen, although pigs are also affected. Three major genotypes have recently been identified within the BDV species, referred to as BDV-1, BDV-2, and BDV-3.

Epidemiology and Clinical Disease

BVDVs cause infections throughout the world. In the US, both BVDV-1a and BVDV-1b isolates are common, with BVDV-1b more prevalent. The prevalence of BVDV-2 appears to be greater in North America than elsewhere, and is relatively rare in Europe and Asia. However, a number of isolates have been made in South America. BVDV-2a isolates also are more predominant than BVDV-2b, which are rare. BVDV is endemic in many cattle populations, with maximum rates of persistent infection sometimes reaching 1–2%, and seropositivity as high as 65%. BVDVs also cross species barriers among animal hosts, with or without evidence of disease.

Ruminant pestiviruses frequently cause persistent infections in cattle and sheep, which serve as reservoirs for maintenance of virus, resulting in transmission from asymptomatic viremic animals to naive animals upon introduction into new herds; transmission by acutely infected animals can also occur. Contact with waste products as well as with contaminated animal feed and farm equipment may also spread infection. Animals acquire pestiviruses through the nasal and oropharyngeal tracts. Direct contact is the most efficient means of transmission. Other transmission routes may apply, as in the case of closed, nonpasturing herds which sometimes develop BVDV infections. A number of factors including the density of cattle populations and regional practices of cattle trading affect the introduction of BVDV into naive herds.

Manifestations of illnesses caused by BVDV are diverse, ranging from mild transient infections to severe acute disease involving the enteric, respiratory, and hematologic systems. Postnatally acquired infections may cause fever, diarrhea, and cough associated with virus shedding, but seroconversion and clearance of infection leads to lifelong immunity. Persistent infections (PIs) occur after virus crosses the placenta in nonimmune animals to cause intrauterine infection early in fetal life. PI animals continuously excrete large amounts of virus and remain BVDV-antibody-negative due to immune tolerance. These animals may develop mucosal disease (MD), a fatal disease with profuse bloody diarrhea due to extensive ulceration of the gastrointestinal tract.

CSFV is the cause of a highly contagious and serious disease of pigs, and is of worldwide importance. Classical swine fever (CSF) is endemic in large portions of Asia, and has been reported in parts of Central and South America. CSF has been progressively eradicated from Western Europe, but periodic reintroduction has occurred. Eradication efforts have also generally led to control of CSF in North America and Australia. CSF can also be spread through pig products, with swill-feeding of pigs a particularly efficient means of transmission, in many regions now outlawed. Spread by semen of infected boars can also occur. CSF is an acute illness with significant mortality, but has become less common now than milder chronic forms of CSF and persistent infections following intrauterine transmission. The latter can result in stillbirth or yield piglets which develop late-onset disease.

BDV is an economically important cause of congenital infection in sheep and goats with a probable worldwide distribution; prevalence in sheep may be up to 50%. Vertical transmission results principally in abortion and stillbirth. Surviving animals are small and weak, and exhibit tremor and hairy fleece ('hairy shaker syndrome'). Infections are spread by these persistently infected animals, and can cause acute postnatal infections. Border disease can also be caused by bovine viral diarrhea disease types.

Structure and Replication

Pestivirus virions are 40–60 nm in size, contain three glycoproteins in the virion envelope complexed as disulfide-linked homo- and heterodimers, and have a buoyant density of $1.134\,g\,ml^{-1}$ in sucrose. They are sensitive to heat, organic solvents, and detergents, but in contrast to flaviviruses, BVDV and CSFV are highly stable in acidic environments, although virus entry is through a low-pH endosome compartment. CD46 has been identified as a receptor for BVDV; however, glycosaminoglycans, such as heparan sulfate, may act as a receptor for cell culture-adapted BVDV and CSFV. It is believed that pestiviruses invade host cells after binding by envelope glycoproteins E^{rns} and E2. Low-density-lipoprotein receptor has also been proposed as a BVDV receptor.

The pestivirus genome is c. 12.3 kbp, with variant genomes of larger and smaller size generated by recombination events (**Figure 3**). Pestivirus genomes lack a 5′ cap structure, and translation is mediated by a 5′ internal ribosome entry site (IRES) element that has similarities to the hepatitis C virus (HCV) IRES. The 3′ UTR contains a short polyC tract, as well as variable and conserved regions, the latter of which encodes hairpin structures. The open reading frame codes for 4000 amino acids, and is processed into at least 12 mature proteins: N^{pro}, C, E^{rns}, E1, E2, p7, NS2, NS3, NS4A, NS4B, NS5A, and NS5B. N^{pro} is a unique N-terminal leader autoprotease, which is followed by the four structural proteins C, E^{rns}, E1, and E2. E^{rns}, which is not found in other viruses of the family, is a heavily glycosylated protein that is secreted from cells and exhibits a ribonuclease activity and a lymphocytotoxic activity that is involved in pathogenesis. E1 and E2 are integral membrane proteins that form the heterodimeric complex which mediates virus entry. Antibodies

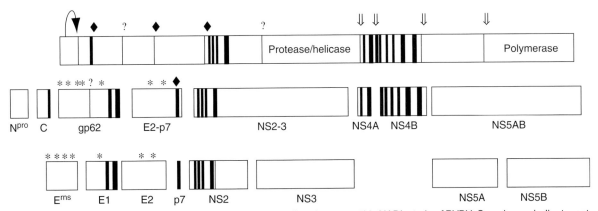

Figure 3 Pestivirus genome organization and processing, based on the cytopathic NADL strain of BVDV. Open boxes indicate mature proteins, with shaded regions representing stretches of hydrophobic amino acids. Diamonds and question marks indicate proteolytic processing by signalase and an unknown host enzyme, respectively. Downward arrows mark cleavages by the viral serine protease. The autoproteolytic cleavage by N^{pro} is shown as a curved arrow. Asterisks indicate proteins which undergo glycosylation. Reproduced from Lindenbach BD and Rice CM (2001) *Flaviviridae*: The viruses and their replication. In: Knipe DM and Howley PM (eds.) *Fields Virology*, 4th edn., vol. 1. Baltimore, MD: Lippincott Williams and Wilkins, with permission.

directed against E^{rns} and E2 confer protective immunity. The p7 protein serves as a membrane leader sequence for NS2. NS2, together with the N-terminus of NS3, is an autoprotease with specificity for the NS2–NS3 junction. This cleavage is necessary for RNA replication and pathogenesis of BVDV disease, but, paradoxically, the uncleaved NS2–NS3 precursor is required for virion formation. NS3 also has multiple functions, including protease (together with the NS4A cofactor), helicase, and nucleoside triphosphatase activities. By analogy with the hepaciviruses, NS4B is involved in cellular membrane reorganization and scaffolding for the replicase complex, which includes NS5B (RdRp). NS5A is a large, hydrophilic phosphorylated zinc metalloprotein analogous to the hepacivirus NS5A, and likely to function in the regulation of viral RNA synthesis.

Pathogenesis

Growth of pestiviruses in cell culture generates virus yields that are not robust, due to intracellular sequestration of the viral progeny and association with serum components that also cause difficulties with virus purification. Whereas susceptible cell culture systems are mainly derived from cells from the natural host species, lack of cytopathogenicity (e.g., 'noncp' biotypes) is common, in contrast to the cytopathic variants ('cp' biotypes), which arise spontaneously during infections in animal hosts.

In vivo, pestiviruses cause acute and persistent infections. BVDV causes a complex pathogenesis involving broad tissue tropism, and a propensity for high cytopathogenicity through mechanisms that remain poorly defined. There is evidence that cytopathic strains cause cell death by activating apoptosis through a double-stranded RNA (dsRNA)-dependent pathway, as well as by mechanisms related to oxidative stress. This process may underlie the development of MD associated with superinfection by a cytopathic strain.

Infection of cows by noncytopathic BVDV during the first 4 months of pregnancy can lead to birth of persistently infected calves immunologically tolerant to BVDV. Subsequently, the calves develop MD after superinfection by cp BVDV. Cytopathic strains associated with such cases are generated by mutations occurring in the genomes of noncytopathic strains, and require the production of NS3 which is cleaved from its NS2-3 precursor. This phenomenon is often a consequence of recombination events which introduce sequences allowing cleavage by host ubiquitin C-terminal hydrolase, or in other cases, by the N^{pro} autoprotease itself. Point mutations can also result in cytopathic variants. Production of NS3 is believed to be the basis for both BVDV cytopathogenicity in cell culture and pathogenesis of MD in immunologically tolerant PI calves.

Following inoculation of the mucosal surface, BVDV gains access to and replicates in lymphatic tissue, in many cases associated with leukopenia. Viremia has been experimentally demonstrated within 2–4 days. In severe cases, as in MD supervening in persistently infected calves, there is disseminated infection, mucosal ulceration, and hemorrhagic disease associated with platelet deficiency and dysfunction. Persistent viral infection is typically found in lymphoid tissue of the intestine, circulating mononuclear cells, and neurons in the central nervous system.

CSF is a disseminated infection that in its most severe form involves extensive thrombosis, endothelial damage, and severe thrombocytopenia resulting in hemorrhagic disease, necrosis, and multiorgan dysfunction. Viral infection of lymphoid tissue causes depletion of lymphocytes and germinal follicles and induces an immunosuppressive state with delayed appearance of CSF-specific antibodies. Superinfection with other pathogens can complicate the disease course.

BDV typically causes a less inflammatory disease, although a variant type of disease with necrosis of central nervous system has been described. Persistent infection of newborn lambs commonly involves widespread lymphoid infection and teratogenic lesions in multiple organs, with endocrine and neurologic abnormalities.

Diagnosis, Control, and Prevention

In the context of control programs for BVDV, diagnostic assays are employed for both surveillance at the herd and population levels, and also at the level of individual animals for identification of persistently infected animals that require elimination. Tests to determine immunity (antibody in serum or milk) as well as presence of virus are commonly done. Detection of antibodies by virus neutralization test or by antibody ELISAs are both used for diagnosis. Presence of virus can be demonstrated by virus isolation, antigen detection, or RT-PCR. Use of these assays depends on a variety of considerations, including age of animals to be tested, the epidemiological situation, and vaccination history.

Laboratory confirmation of CSFV infection has traditionally depended on virus isolation and detection of antigen in organ tissue, although more rapid techniques, such as antigen ELISA and RT-PCR, have been developed. Early diagnosis of infection based on antibody detection is limited by the slow appearance of virus-specific antibodies and lack of specificity of some antibody ELISA tests.

Vaccines have been developed against both CSFV and BVDV. However, control programs based on detection and eradication of persistently infected animals, as well as on quarantining and maintaining closed herds, have been employed to reduce the economic losses caused by these viruses. National eradication campaigns conducted in Scandinavia in the 1990s were particularly successful, facilitated by development of ELISA for detection of

BVDV antibodies. Due to the high effectiveness of these programs, the threat of reintroduction of virus into uninfected herds became apparent, and vaccination was recommended in some regions as a prophylactic measure against a resurgence of infections. Although some questions remain about the duration of protection and its efficacy in preventing fetal infection, newer vaccine regimens combined with comprehensive BVDV control programs are probably the best approach to lowering the incidence of PI in calves and reducing the problem of BVDV outbreaks. Commercial BVDV vaccines include either inactivated or modified live vaccines. Although there are some data to indicate that both antibody and cellular immune mechanisms are important for protection against BVDV infection, more studies are needed to define the protective mechanisms associated with these vaccines.

Live-attenuated vaccines for CSF have been available since the 1960s, but are not universally employed because of interference with traditional serologic methods to demonstrate absence of disease. Stamping-out policies are therefore employed as control efforts. Research into the use of marker vaccines, which do not exhibit interference in detection of infection, is ongoing, together with modeling studies to assess the impact of vaccine and eradication policies on control of disease.

Hepaciviruses

History and Classification

After serologic screening tests for the detection of hepatitis B and hepatitis A viruses were introduced in the 1970s to eliminate these viruses from human blood supplies, the role of HCVs in causing up to 90% of transfusion-associated hepatitis was recognized. The term hepatitis C virus (HCV) was coined in 1989 after identification of an RNA genome of this virus from cDNA libraries of human serum containing the infectious agent. In humans, infection with HCV commonly evolves into a chronic persistent hepatitis which can lead to progressive hepatic cirrhosis and hepatocellular carcinoma. Lack of a suitable cell culture system for propagating HCV had until only recently been a major impediment to progress in understanding the details of its replication. In addition, studies of HCV pathogenesis have been hampered by the lack of animal model systems, and progress has depended on use of clinical data from human cases, or from the use of a chimpanzee model.

Hepaciviruses display great genetic heterogeneity on a worldwide basis and are classified into six genotypes based on nucleotide sequence similarity, and further classified into multiple subtypes (identified by lowercase letters: e.g., 1a, 1b, 2a) as well as quasispecies for a given viral isolate. These divisions correspond to sequence divergence rates of $c.$ 30–35%, 20–25%, and 1–10%, respectively. This level of genetic diversity has implications for the epidemiology of HCV transmission, clinical manifestations of liver disease caused by HCV, response to treatment, and development of a prophylactic vaccine.

Epidemiology and Clinical Disease

It is estimated that there are at least 170 million chronically infected HCV subjects worldwide, with prevalence varying from 0.1% to as high as 18% (Egypt). In areas where HCV has been eliminated from blood supplies by donor screening, use of illicit injectable drugs remains the most important mode of transmission. Various other forms of exposure, including occupational and sexual exposures, are involved in some cases of transmission. HCV is a primary cause of cirrhosis, a significant cause of hepatocellular carcinoma, and in the United States is the leading reason for liver transplantation. There is great variability in the spectrum of disease. Acceleration of disease is observed in persons who are older at the time of infection, and in the setting of continuous alcohol exposure, and co-infection with HIV or hepatitis B virus (HBV). Individuals who sustain infection at a younger age tend to have slower progression.

Genotypes 1, 2, and 3 are found worldwide, but with regional differences with respect to distribution. Genotypes 1a and 1b are predominant in the United States and Europe, 2a and 2b are relatively common in North America and Europe, and type 3a common in the United States and Europe among intravenous drug abusers. Genotype 4 is notable in the Middle East/North Africa region, whereas 5 is found in South Africa, and 6 in Hong Kong. Additional genotypes (e.g., 7–11) have been identified but have been proposed as variants of genotype 6. Genotypes are useful in establishing genetic markers of HCV in epidemiologic investigations and for determining modes of transmission. Differences in terms of disease potential and treatment response among genotypes also have been described, such as the relative resistance of genotype 1 to interferon alpha (IFN-α) therapy. This genotype has been associated with more aggressive liver disease as well, but it is unclear if this results from a longer duration of infection among people infected with this genotype. Genotypic differences may affect usefulness of diagnostic immunoassays for HCV antibody which are based on various portions of the HCV genome.

Structure and Replication

In contrast to those of flaviviruses and pestiviruses, hepacivirus virions have been difficult to visualize, and exhibit larger size and lower buoyant density (1.03–1.1 g ml^{-1}) as a result of association with immunoglobulins and low-density lipoproteins. HCV genomes contain a 9.6 kb

RNA which differs in some structural features from flaviviruses and pestiviruses (**Figure 4**). The 5′ UTR is the most highly conserved region of the genome and contains an IRES which mediates cap-independent translation of viral RNA. The structure of the IRES has been resolved, including studies of complexes with the ribosomal 40S subunit and the human translation factor eIF3. These efforts may facilitate the design of small molecules capable of inhibiting this critical viral factor. Because of its high sequence conservation, the 5′ UTR has also been used for development of sensitive detection assays for HCV infection. The 3′ UTR exhibits a tripartite structure, containing several unique elements, including a short variable region, a poly(U/UC) tract, and a highly conserved RNA sequence (the X-tail). A unique RNA structure, the *cis*-acting replication element (CRE) within the NS5 region, has also been identified.

The HCV polyprotein (*c.* 3000 amino acids) is processed by cellular (signalase) and viral proteases into mature viral proteins, consisting of the core, E1 and E2, the hydrophobic p7 peptide, and six nonstructural proteins (NS2, NS3, NS4A, NS4B, NS5A, and NS5B). Unique properties of several of these proteins have been characterized. The RNA-binding core protein has been associated with a range of effects on host cells, including transactivation, but the relevance to HCV pathogenesis *in vivo* is unknown. Alternative forms of core, produced by frameshifting, have been described. The E1 and E2 proteins form the heterodimeric complex of the virion that serves receptor binding, membrane fusion, and entry. These proteins exhibit the highest mutation rate at both the nucleotide and amino acid level. E2 in particular contains a hypervariable region (HVR-1) which undergoes an extremely high mutation rate and, because antibody epitopes are found within this region, it is believed that the mutations provide a mechanism for HCV to escape antibody responses involved in controlling virus infection. The small membrane-associated p7 protein forms an amantadine-sensitive ion channel which may function in the secretory pathway during virus maturation.

HCV encodes two known proteases involved in polyprotein processing, the NS2-3 autoprotease, and the NS3 serine protease, that with its NS4A cofactor mediates cleavage of the remainder of the nonstructural region. The structure of the NS3-4A protease has been solved, promoting interest in a new class of antiviral agents that target its protease activity. NS3 can also cleave adaptor proteins (Toll-like receptor-3 adaptor TRIF (TICAM-1)

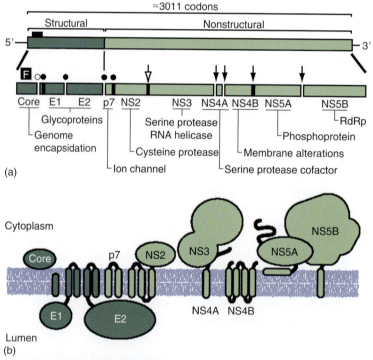

Figure 4 HCV genome organization and processing. (a) The HCV genome and polyprotein encoding structural and nonstructural genes and known functions. Closed and open circles indicate signal peptidase cleavage sites and signal peptide peptidase cleavage sites, respectively. F represents an alternative open reading frame protein produced by ribosomal frameshifting within the core region. Open and closed arrows indicate cleavage by the NS2 autoprotease and the NS3/4A serine protease, respectively. (b) Proposed topology of the HCV proteins with respect to the intracellular membrane. Reprinted by permission from Macmillan Publishers Ltd; *Nature* (Lindenbach BD and Rice CM (2005) Unravelling hepatitis C virus replication from genome to function. *Nature* 436(7053): 933–938), Copyright (2005).

and Cardiff (IPS-1, VISA, or MAVS)) required for activation of interferon regulatory factor 3, indicating a role for NS3 in downregulating the dsRNA signaling pathway that leads to induction of interferon beta and other genes involved in antiviral defense.

The NS4B protein is a mediator of the reorganization of intracellular membranes that together with other viral and host proteins generates the 'membranous web' scaffold for formation of RNA replication complexes. NS5A is a multidomain protein containing an RNA-binding activity. Differential phosphorylation of NS5A appears to have a role in regulation of HCV RNA synthesis, as a molecular switch for critical steps in the transition from RNA synthesis to virus packaging. NS5A also interacts with interferon-induced dsRNA-activated protein kinase PKR. Mutations in the domain of NS5A which affect this interaction (interferon sensitivity-determining region (ISDR) have been reported to influence the response of HCV to interferon alpha. The NS5B protein contains the RdRp activity that forms the core of the replication complex, and together with other NS3, NS5A, and cellular factors, synthesizes the genome-length plus- and minus-strand RNAs involved in the replication of the virus. NTPase and helicase activities participating in this process are encoded in the NS3 protein. Cellular proteins that have been implicated in function of the replication complex by interacting with NS5A and NS5B include VAP-A and -B (vesicle-associated membrane protein-associated proteins A and B), geranylgeranylated protein FBL-2, and cyclophilin B. These proteins appear to affect the assembly and function of the replication through such interactions such as membrane localization and RNA binding.

Recent progress in understanding HCV genome replication has been achieved through the use of HCV replicon systems. Initially established for genotype 1b, these were subgenomic constructs that encoded only the 5′ and 3′ structures and the nonstructural proteins required for viral RNA replication. Studies with additional iterations of replicons, including those for other genotypes, as well as full-length versions that exhibit efficient replicative functions *in vitro*, are excellent tools for studying the virus life cycle, in the absence of robust propagation of infectious HCV viruses in cultured cells. Crystal structures of several HCV proteins harboring key enzymatic functions, including NS3 protease, helicase, NS5A, and NS5B are known and are being applied to the development of HCV-specific antiviral agents.

The exact cell entry process for HCV has not been definitively characterized. It is believed that CD81, a tetraspanin family protein widely found on cell substrates, is involved in HCV entry, but requirements for other cell surface molecules on hepatocytes probably exist. Other proposed receptors include the LDL receptor, the class B type I scavenger receptor, mannose-binding lectins DC-SIGN and L-SIGN, heparan sulfate proteoglycans, and the asialoglycoprotein receptor. Further studies are needed to demonstrate the role of such molecules in HCV infections, and will be facilitated with new model systems such as retrovirus pseudoparticles that express HCV glycoproteins and exhibit low-pH-dependent infectivity for hepatic cell lines.

Pathogenesis and Clinical Disease

HCV causes acute and chronic forms of viral hepatitis. Acute hepatitis resembles that caused by other agents of viral hepatitis, but overt symptoms are not as frequent, despite biochemical evidence of liver disease in most subjects and appearance of HCV RNA in the serum. HCV causes persistent infection in most infected individuals, who then have chronic hepatitis throughout their lives in the absence of treatment and who remain at risk for cirrhosis and hepatocellular carcinoma. The infection is often diagnosed by screening of asymptomatic blood donors, or by the appearance of either chronic fatigue or features otherwise typical of viral hepatitis. Such individuals have persistent viremia and evidence of chronic inflammatory disease on liver biopsy.

Currently HCV is believed to cause persistent infection because primary and secondary immune responses are inadequate in breadth and magnitude to control the infection. Underlying mechanisms include the inherent difficulty in optimizing an immune response within the liver, and HCV-specific factors which suppress the host immune response, including effects of the core protein on host gene expression, inhibition of the interferon-induced protein kinase PKR by E2 and NS5A proteins, and NS3 inhibition of the dsRNA signaling pathway. The high genetic variability of HCV results in viral variants that may contribute to virus persistence through escape from antibody and from cytotoxic lymphocyte responses directed at specific epitopes.

The mechanism of liver injury associated with persistent HCV infection is believed to be immune-mediated and initiated by virus-specific T-lymphocytes which infiltrate the liver. Such cells are predominantly of a T_H-1 phenotype and produce interferon gamma and tumor necrosis factor alpha (TNF-α). Because these cytokines have been implicated in noncytolytic clearance of other viral agents in the liver, other mechanisms of hepatocyte injury and death associated with HCV have been proposed, including Fas-Fas ligand-mediated apoptosis, perforin-dependent killing by cytolytic T-cells, or TNF-α itself. HCV is not highly cytopathic for hepatocytes; however, fat deposits and eosinophilic inclusions are often seen in liver cells. In the absence of virus clearance, repeated episodes of hepatocellular injury and regeneration trigger development of hepatic fibrosis, cirrhosis, and eventual hepatocellular carcinoma. Additional factors that promote this process include coexistent infection with other agents of viral hepatitis, alcohol intake, and iron overload. Extrahepatic manifestations can also occur in association with HCV infection, typically manifesting as skin and renal disease caused by a systemic vasculitis involving immune complexes with viral proteins.

Diagnosis and Treatment

Diagnosis of chronic hepatitis C infection is established by screening enzyme immunoassay to detect circulating antibodies to HCV proteins, followed by confirmatory determination of HCV RNA in serum, using qualitative and quantitative tests based on target (RT-PCR and transcription-mediated amplification) and signal amplification (branched DNA (bDNA)) techniques. Quantitation of serum HCV RNA helps to identify patients likely to benefit from treatment and undergo virus clearance as defined by a sustained viral response (SVR). Determination of the HCV genotype guides treatment decisions, as genotypes 1 and 4–6 are less amenable to treatment than are genotypes 2 or 3. Therapeutic trials have shown that combinations of interferons and ribavirin are more effective than is interferon alone and pegylated (long-acting) interferons have yielded improved SVR rates. An SVR is less common in patients with genotype 1 infections, high pretreatment HCV RNA levels, or advanced stages of fibrosis. Liver biopsy is used to establish the pre-treatment stage of liver disease and to facilitate decisions concerning the need for therapy. Advances in the future diagnosis and management of HCV will require close monitoring of the transmission of this infection, as well as extending treatment to populations not usually evaluated in clinical trials, and the development of new therapies, including antiviral agents targeting the viral protease and RNA polymerase.

GB Viruses

GB viruses are unassigned members of the family that include GB-A and -B, which are marmoset viruses, and GB-C, which was detected in the serum of human chronic hepatitis cases. GB-C and hepatitis G were discovered coincidentally and currently given the taxonomic assignment GBV-C/HGV. Currently there are five genotypes of this virus.

GBV-C/HGV is related to HCV, sharing 29% amino acid homology, and similarities in genome organization and structure, including a $5'$ IRES, and a 3000 amino acid polyprotein but there are notable differences. GBV-C/HGV contains a capsid protein in the nucleocapsid; but this protein has not been identified. The E2 protein lacks both the extensive glycosylation and the hypervariable region found in the HCV E2 protein, which may in part explain the less frequent occurrence of persistent infection by GBV-C/HGV.

The $3'$ UTR also differs from that of HCV. GBV-C/HGV is not hepatotropic, but rather replicates in mononuclear cells and bone marrow. Infection of humans is common, but no disease association has yet been identified. Instead, a beneficial effect of GBV-C/HGV infection on the outcome of HIV infection has been observed, presumably as a result of altered cellular immune responses caused by persistent infection.

See also: Classical Swine Fever Virus; Flaviviruses of Veterinary Importance; Tick-Borne Encephalitis Viruses; Yellow Fever Virus.

Further Reading

Chambers TJ and Monath TP (eds.) (2003) *The Flaviviruses: Advances in Virus Research,* vols. 59–61. San Diego, CA: Academic Press.

Chang KM (2003) Immunopathogenesis of hepatitis C virus infection. *Clinics in Liver Disease* 7: 89–105.

Cocquerel L, Voisset C, and Dubuisson J (2006) Hepatitis C virus entry: Potential receptors and their biological functions. *Journal of General Virology* 87: 1075–1084.

Calisher CH and Gould EA (2003) Taxonomy of the virus family Flaviviridae. *Advances in Virus Research* 59: 1–19, with permission from Elsevier.

Gubler DJ and Kuno G (eds.) (1997) *Dengue and Dengue Hemorrhagic Fever.* London: CABI.

Heinz FX, Collett MS, Purcell RH, et al. (2000) Family *Flaviviridae*. In: Van Regenmortel MHV, Fauquet CM Bishop DHL, et al. (eds.) *Virus Taxonomy: Seventh International Committee for the Taxonomy of Viruses,* pp. 859–878. San Diego, CA: Academic Press.

Houe H (1999) Epidemiological features and economical importance of bovine virus diarrhoea virus (BVDV) infections. *Veterinary Microbiology* 64: 89–107.

Kovács F, Magyar T, Rinehart C, et al. (2003) The live attenuated bovine viral diarrhea virus components of a multi-valent vaccine confer protection against fetal infection. *Veterinary Microbiology* 96: 117–131.

Kuhn RJ, Zhang W, Rossmann MG, et al. (2002) Structure of dengue virus: Implications for flavivirus organization, maturation, and fusion. *Cell* 8: 717–725.

Lindenbach BD, Evans MJ, Syder AJ, et al. (2005) Complete replication of hepatitis C virus in cell culture. *Science* 309: 623–626.

Lindenbach BD and Rice CM (2001) *Flaviviridae*: The viruses and their replication. In: Knipe DM and Howley PM (eds.) *Fields Virology,* 4th edn., vol. 1. Baltimore, MD: Lippincott Williams and Wilkins.

Lindenbach BD and Rice CM (2003) Molecular biology of the flaviviruses. *Advances in Virus Research* 59: 23–61.

Lindenbach BD and Rice CM (2005) Unravelling hepatitis C virus replication from genome to function. *Nature* 436(7053): 933–938, Copyright (2005).

Radostits OM, Gay CC, Blood DC, and Hinchcliff KW (eds.) (2000) Bovine virus diarrhoea, mucosal disease, bovine pestivirus complex. In: *Veterinary Medicine, A Textbook of the Diseases of Cattle, Sheep, Pigs, Goats, Horses,* 9th edn., pp. 1085–1105. New York: WB Saunders.

Ridpath JF (2003) BVDV genotypes and biotypes: Practical implications for diagnosis and control. *Biologicals* 2: 127–133.

Tassaneetrithep B, Burgess TH, Granelli-Piperno A, et al. (2003) DC-SIGN (CD209) mediates dengue virus infection of human dendritic cells. *Journal of Experimental Medicine* 197: 823–829.

Tautz N, Meyers G, and Thiel HJ (1998) Pathogenesis of mucosal disease, a deadly disease of cattle caused by a pestivirus. *Clinical and Diagnostic Virology* 10: 121–127.

Flexiviruses

M J Adams, Rothamsted Research, Harpenden, UK

© 2008 Elsevier Ltd. All rights reserved.

Glossary

Triple gene block (TGB) protein A set of three proteins in overlapping reading frames that function together in facilitating cell-to-cell movement of plant viruses.

Introduction

The creation of the family *Flexiviridae* was approved by the International Committee on the Taxonomy of Viruses (ICTV) in 2004 to include a number of existing genera and some unassigned species of plant viruses. The family is so named because its members have flexuous, filamentous virions, and there are a number of characteristics that distinguish it from the families *Closteroviridae* and *Potyviridae*, the two other families of plant viruses that have virions of similar morphology (**Table 1**).

Main Characteristics

The chief characteristics shared by all members of the family are considered in this section.

Virion Morphology

Virions are flexuous filaments, 12–13 nm in diameter, and usually 500–1000 nm in length, depending on the genus. They are constructed from helically arranged coat protein subunits in a primary helix with a pitch of about 3.4 nm and nine to ten subunits per turn. Each particle contains a single molecule of the full-length genomic RNA. Smaller 3′ terminal subgenomic RNAs may also be encapsidated in some species, sometimes in shorter virions. In negative contrast electron microscopy the virions usually appear to be cross-banded. A typical electron micrograph is shown in **Figure 1** (for red clover vein mosaic virus, genus *Carlavirus*) but in the genera *Trichovirus* and *Vitivirus*, the virions appear to be rather more flexuous and hence resemble more those of the family *Closteroviridae*.

Genome

All members have a monopartite, positive-sense, single-stranded RNA (ssRNA) genome in the size range 5.5–9.5 kbp, depending on the species. This forms about 5–6% of the virion by weight. The RNA is capped at the 5′ terminus and there is a 3′ polyA tail.

Replication and Translation

In all members, there is a short 5′ nontranslated region followed by a single large alpha-like replication protein (145–250 kDa). This contains (in this order) conserved methyltransferase, helicase, and RNA-dependent RNA polymerase (RdRp) motifs and appears to be the only protein required for replication, which is believed to take place in the cytoplasm. Translation of the smaller open reading frames (ORFs) downstream occurs from one or more 3′ terminal subgenomic mRNAs. In some species, there is a 'leaky scanning' mechanism that ensures that two or three different proteins are translated from the same mRNA.

Genomic Organization

Genomic organization varies between genera but there are up to six ORFs ordered from 5′ to 3′:

1. a large alpha-like replication protein (see above);
2. one or more cell-to-cell movement proteins (MPs): in some genera there is a single MP of the '30K' superfamily

Table 1 Characteristics of the families of filamentous plant viruses[a]

Family	Genome size (kbp)	RNAs[b]	ORFs	Translation strategy[c]	5′ end VPg[d]	3′-polyA
Closteroviridae[e]	15–20	1, 2, or 3	10+	Subgenomic	No	No
Flexiviridae	5.5–9.5	1	2–6	Subgenomic	No	Yes
Potyviridae	9–12	1 or 2	1 or 2	Polyprotein	Yes	Yes

[a]All these have ssRNA genomes that encode in the virion (positive) sense.
[b]Numbers of genomic RNA components.
[c]Proteins are translated either from a set of nested subgenomic mRNAs or are cleaved by virus-encoded proteases from a polyprotein precursor.
[d]Presence or absence of a genome-linked protein covalently bound to the 5′ terminus of the RNA.
[e]Members of the family *Closteroviridae* have distinctive, tailed virions. The tail is constructed from a second, minor, coat protein.

(distantly related to the 30 kDa MP of tobacco mosaic virus (genus *Tobamovirus*)) while in the others there is a set of three overlapping proteins known as a 'triple gene block' (TGB);
3. a single capsid (coat) protein (CP) (21–45 kDa);
4. a final ORF (in some viruses only), which is thought to have RNA-binding properties and is probably a suppressor of gene silencing; it partially overlaps the 3′ end of the CP gene in some species.

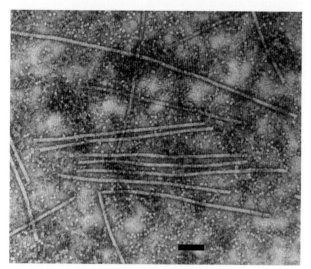

Figure 1 Electron micrograph of virions of red clover vein mosaic virus (genus *Carlavirus*), negatively stained. Scale = 100 nm.

Biological Properties

Biologically, the viruses are fairly diverse. They have been reported from a wide range of herbaceous and woody mono- and dicotyledonous plant species but the host range of individual members is usually limited. Natural infections by members of the genera *Capillovirus*, *Citrivirus*, *Foveavirus*, *Mandarivirus*, *Trichovirus*, and *Vitivirus* are mostly or exclusively of woody hosts. Many of the viruses have relatively mild effects upon their host. All species can be transmitted by mechanical inoculation, often readily. Many of the viruses have no known invertebrate or fungal vectors but allexiviruses and some trichoviruses are thought to be mite borne, most carlaviruses are transmitted naturally by aphids in a nonpersistent manner, and a range of vectors have been reported for different vitiviruses. Aggregates of virus particles accumulate in the cytoplasm but there are generally no distinctive cytopathic structures.

Component Genera and Species

Some properties of the nine genera included in the family and of six species that are not assigned to any genus are summarized in **Table 2**. These are mostly distinguished by the number of ORFs, the type of MP encoded, and the sizes of their virions and replication protein genes. In some genera, there are additional ORFs not present in other members of the family. These distinctions based on

Table 2 Genera and unassigned species included within the family *Flexiviridae*

Genus and unassigned species[a]	Virion length (nm)	ORFs	MP type[b]	Rep (kDa)	CP (kDa)
Potexvirus (28)	470–580	5	TGB	146–191	22–27
Mandarivirus (1)	650	6	TGB	187	36
Allexivirus (8)	~800	6	TGB	175–194	26–28
Carlavirus (35)	610–700	6	TGB	226–238	31–40
Foveavirus (3)	800+	5	TGB	244–247	28–45
Capillovirus (3)	640–700	2 or 3	30K	214–241	25–27
Vitivirus (4)	725–785	5	30K	195–196	18–23
Trichovirus (4)	640–760	3 or 4	30K	216–217	21–22
Citrivirus (1)[c]	960	3	30K	227	41
Banana mild mosaic virus (virus: BanMMV)	580	5	TGB	205	27
Banana virus X (virus: BanVX)	?	5	TGB	nd[d]	24
Cherry green ring mottle virus (virus: CGRMV)	1000+	5	TGB	230	30
Cherry necrotic rusty mottle virus (virus: CNRMV)	1000+	5	TGB	232	30
Potato virus T (virus: PVT)	640	3	30K	nd	24
Sugarcane striate mosaic-associated virus (virus: SCSMaV)	950	5	TGB	222	23
Possible species					
Botrytis virus X (BotV-X)	720	5	na[e]	158	44

[a]Numbers of species recognized within each genus are shown in parentheses (2006 data).
[b]Either a set of three overlapping proteins (TGB) or a single protein of the '30K' superfamily (phylogenetically related to the 30 kDa MP of *Tobacco mosaic virus*, genus *Tobamovirus*).
[c]Genus proposed in 2005 for the species *Citrus leaf blotch virus* (virus: CLBV).
[d]Not determined (only partial sequence data available).
[e]Not applicable (no MP encoded).

genome organization are also supported by phylogenetic analyses (see below). Genome organization is summarized in **Figure 2**.

A recently described mycovirus, botrytis virus X, shares many of the characteristics of the family and its CP and replication protein are phylogenetically related to members of the family *Flexiviridae*, in particular to members of the genera *Potexvirus* and *Allexivirus*. It is likely that a new genus will be proposed to accommodate this species in due course.

Phylogenetic Relationships

The relationships detected between the genera within the family and those between members of the family *Flexiviridae* and other viruses are dependent on the genome region used for analysis. This probably indicates that the viruses have arisen from reassortment of entire genes (or blocks of genes). These relationships are summarized below.

Replication Protein

A dendrogram based on the nucleotide sequence of the replication proteins (**Figure 3**) supports the taxonomic division of the family *Flexiviridae* into genera and falls into two major parts. One part contains the genera *Allexivirus*, *Mandarivirus*, and *Potexvirus* together with botrytis virus X. These proteins are the smaller ones in the family (145–195 kDa) and the plant-infecting members all have a TGB. The second part of the tree contains those viruses with larger proteins (195–250 kDa). It includes all those with a '30K'-like MP (*Capillovirus*, *Citrivirus*, *Trichovirus*,

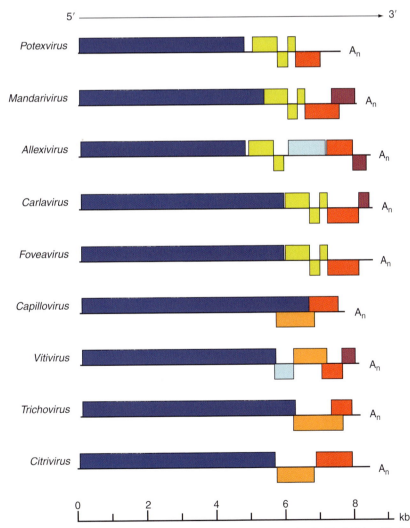

Figure 2 Diagram showing genome organization for the component genera of the family *Flexiviridae*. Blocks represent predicted ORFs. Replication proteins are shown in dark blue, '30K-like' MPs in orange, TGB proteins in yellow, CPs in red, and RNA-binding proteins in purple. ORFs of unknown function are shown in light blue.

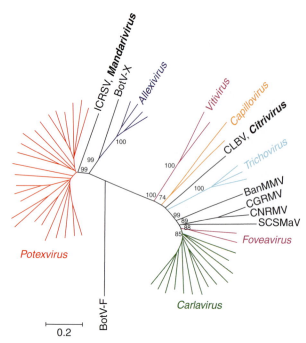

Figure 3 Phylogenetic tree of the replication proteins of all sequenced members of the family *Flexiviridae* with botrytis virus X and botrytis virus F. For abbreviations, see **Table 2**. The neighbor-joining tree is based on the codon-aligned nucleotide sequences. Values on the major branches show the percentage of trees in which this grouping occurred after bootstrapping the data (10 000 replicates). The scale bar shows the number of substitutions per base.

and *Vitivirus*), together with the TGB-containing species that make up the genera *Carlavirus* and *Foveavirus* and some unassigned viruses that have foveavirus-like genome organization.

Most studies of 'longer distance' relationships between viruses have been based upon the motifs within the replication protein as these are relatively conserved within ssRNA viruses. Such studies consistently place the family *Flexiviridae* within a larger group ('alpha-like' or 'Sindbis-like' with 'type III' RdRp) that includes the family *Tymoviridae* (plant-infecting viruses that have isometric particles) and, more distantly, the families *Togaviridae*, *Bromoviridae*, *Closteroviridae*, and the rod-shaped plant viruses including the genus *Tobamovirus*. Phylogenetic analysis of the entire replication protein shows that this protein in members of the *Tymoviridae* is related to that found in the genera *Allexivirus, Mandarivirus*, and *Potexvirus* but distinct from that found in the remaining members of the family *Flexiviridae* (**Figure 4**). This might suggest that the *Tymoviridae* and *Flexiviridae* could be placed within the same family or that the current family *Flexiviridae* could be divided into two separate families. However, the relationships between other genes (see below) and differences in virion morphology and cytopathology suggest that the current classification makes better biological sense. The

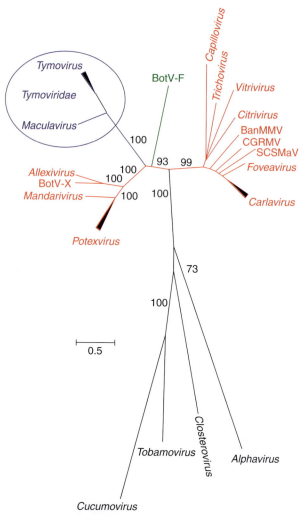

Figure 4 Phylogenetic tree of the replication proteins of selected members of the families *Flexiviridae* (in red) and *Tymoviridae* (in blue) with other related viruses. For abbreviations, see **Table 2**. The neighbor-joining tree is based on the aligned peptide sequences. Values on the major branches show the percentage of trees in which this grouping occurred after bootstrapping the data (500 replicates). The scale bar shows the number of substitutions per base.

replication protein of the mycovirus botrytis virus F also appears to belong to the *Flexiviridae–Tymoviridae* group but falls outside any of the major groups and is probably best left unclassified at present.

Movement Protein

'30K'-like MPs occur in the genera *Capillovirus, Citrivirus, Trichovirus*, and *Vitivirus* and also in the unassigned potato virus T. The '30K' superfamily proteins are largely recognized by structural similarities rather than by primary sequence comparisons. Nonetheless, phylogenetic analysis shows that those from the family *Flexiviridae* are related to one another (although *Vitivirus* is more distant) and that there are some similarities to the corresponding proteins in the rod-shaped plant virus genera *Furovirus*, *Tobravirus*, and *Tobamovirus* (**Figure 5**).

TGB Proteins

Within the family, the TGB proteins (for those genera that have them) are related in much the same way as their CPs. A TGB module is also found in members of the genera *Benyvirus*, *Hordeivirus*, *Pecluvirus*, and *Pomovirus*. However, the first and third proteins of the TGB of the rod-shaped viruses are considerably larger than the corresponding proteins in the family *Flexiviridae* and clearly form a separate grouping in phylogenetic analyses (**Figure 6**). In some species, notably all members of the genus *Allexivirus*, a sequence similar to that coding for the third TGB protein can be recognized but there is no standard AUG start codon and it is not known whether the protein is expressed.

Coat Protein

A phylogenetic tree of the CP sequences shows that most viruses with a '30K'-like MP in the genera *Capillovirus*, *Trichovirus*, and *Vitivirus* are only distantly related to those in the remainder of the family but that citrus leaf blotch virus, the single member of the genus *Citrivirus*, groups with a well-defined group containing all the TGB-containing members (**Figure 7**).

It is difficult to detect any significant relationship between the CPs of members of the *Flexiviridae* and those of other viruses beyond the modest similarity that occurs between the CPs of all viruses with filamentous particles.

RNA-Binding Protein

Putative RNA-binding proteins are found in the genera *Allexivirus*, *Carlavirus*, *Mandarivirus*, and *Vitivirus*. The proteins are clearly related and share a strictly conserved core amino acid sequence $GxSxxAxR/KRRAx_5Cx_{1-2}Cx_{5-13}C$. In phylogenetic analyses, the proteins from each genus cluster separately. The proteins probably act as silencing suppressors and are related to those of the closteroviruses (p23 of citrus tristeza virus and homologs in other species) (**Figure 8**).

General Comments

The limitations of phylogenetic analysis are demonstrated by the complex relationships within and outside the family

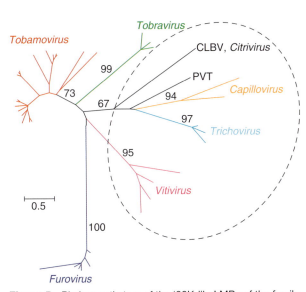

Figure 5 Phylogenetic tree of the '30K-like' MPs of the family *Flexiviridae* (in dotted oval) and the corresponding proteins of rod-shaped plant viruses. For abbreviations, see **Table 2**. The neighbor-joining tree is based on the aligned peptide sequences. Values on the major branches show the percentage of trees in which this grouping occurred after bootstrapping the data (500 replicates). The scale bar shows the number of substitutions per base.

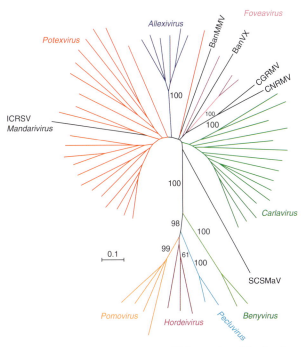

Figure 6 Phylogenetic tree of the TGB proteins of the family *Flexiviridae* and related rod-shaped plant viruses. For abbreviations, see **Table 2**. The neighbor-joining tree is based on the codon-aligned nucleotide sequences of all three proteins. Values on the major branches show the percentage of trees in which this grouping occurred after bootstrapping the data (10 000 replicates). The scale bar shows the number of substitutions per base.

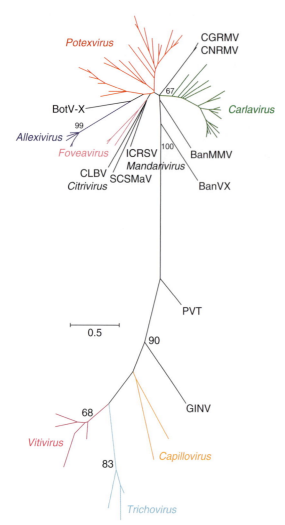

Figure 7 Phylogenetic tree of the CPs of all sequenced members of the family *Flexiviridae*. For abbreviations, see **Table 2**. The neighbor-joining tree is based on the aligned peptide sequences. Values on the major branches show the percentage of trees in which this grouping occurred after bootstrapping the data (500 replicates). The scale bar shows the number of substitutions per base.

that are revealed when different genes are used in the analysis (**Table 3**). In particular, there are big differences in the relative positions of the genera *Carlavirus*, *Citrivirus*, and *Foveavirus* to the rest of the family. The single virus in the genus *Mandarivirus* consistently groups with the genus *Potexvirus*, but it has a larger CP and an RNA-binding protein related to those of the genera *Allexivirus* and *Carlavirus*. Outside the family, the replication proteins are related to those found in the *Tymoviridae* and to botrytis virus F. By contrast, the movement protein(s), whether TGB or '30K'-like, are clearly related to those of the rod-shaped viruses, which are much more distantly related in their replication proteins.

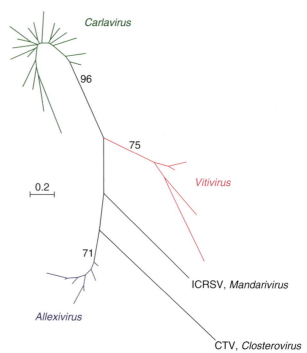

Figure 8 Phylogenetic tree of the nucleic acid-binding proteins of members of the family *Flexiviridae* and citrus tristeza virus (CTV, genus *Closterovirus*). The neighbor-joining tree is based on the aligned peptide sequences. Values on the major branches show the percentage of trees in which this grouping occurred after bootstrapping the data (500 replicates). The scale bar shows the number of substitutions per base.

Table 3 Summary of the main phylogenetic relationships within the family *Flexiviridae*

Genus	Rep[a]	MP[b]	CP[c]	RNABP[d]
Potexvirus	1	TGB	1	No
Mandarivirus	1	TGB	1	Yes
Allexivirus	1	TGB	1	Yes
(Botrytis virus X)	1	na[e]	1	No
Carlavirus	2	TGB	1	Yes
Foveavirus+[f]	2	TGB	1	No
Capillovirus	2	30K	2	No
Vitivirus	2	30K	2	Yes
Trichovirus	2	30K	2	No
Citrivirus	2	30K	1	No

[a]Type of replication protein.
[b]Either a set of three overlapping proteins (TGB) or a single protein of the '30K' superfamily.
[c]Type of coat protein.
[d]Presence of RNA-binding protein.
[e]Not applicable (no protein encoded).
[f]Including the unassigned members of the family: *Banana mild mosaic virus*, *Banana virus X*, *Cherry green ring mottle virus*, *Cherry necrotic rusty mottle virus*, and *Sugarcane striate mosaic-associated virus*.

See also: Allexivirus; Capillovirus, Foveavirus, Trichovirus, Vitivirus; Carlavirus; Carmovirus; Citrus Tristeza Virus; Phylogeny of Viruses; Potexvirus; Potyviruses; Tobamovirus; Tymoviruses.

Further Reading

Adams MJ, Antoniw JF, Bar-Joseph M, et al. (2004) The new plant virus family Flexiviridae and assessment of molecular criteria for species demarcation. Archives of Virology 149: 1045–1060.

Howitt RLJ, Beever RE, Pearson MN, and Forster RLS (2001) Genome characterization of Botrytis virus F, a flexuous rod-shaped mycovirus resembling plant 'potex-like' viruses. Journal of General Virology 82: 67–78.

Howitt RLJ, Beever RE, Pearson MN, and Forster RLS (2006) Genome characterization of a flexuous rod-shaped mycovirus, Botrytis virus X, reveals high amino acid identity to genes from plant 'potex-like' viruses. Archives of Virology 151: 563–579.

Koonin EV and Dolja VV (1993) Evolution and taxonomy of positive-strand RNA viruses: Implication of comparative analysis of amino acid sequences. Critical Reviews in Biochemistry and Molecular Biology 28: 375–430.

Martelli G, Adams MJ, Kreuze JF, and Dolja VV (2007) Family Flexiviridae: A case study in virion and genome plasticity. Annual Review of Phytopathology 45: 73–100.

Melcher U (2000) The '30K' superfamily of viral movement proteins. Journal of General Virology 81: 257–266.

Morozov SY and Solovyev AG (2003) Triple gene block: Modular design of a multifunctional machine for plant virus movement. Journal of General Virology 84: 1351–1366.

Wong S-M, Lee K-C, Yu H-H, and Leong W-F (1998) Phylogenetic analysis of triple gene block viruses based on the TGB 1 homolog gene indicate a convergent evolution. Virus Genes 16: 295–302.

Foamy Viruses

M L Linial, Fred Hutchinson Cancer Research Center, Seattle, WA, USA

© 2008 Elsevier Ltd. All rights reserved.

Glossary

Provirus A viral DNA that is incorporated into the genetic material of a host cell.
Syncytia Structures in a monolayer cell culture formed by fusion of multiple cells that form a large cell with multiple nuclei.
Zoonotic infection An infection of animals that is transmitted to humans.

Introduction

Foamy viruses (FVs, also known as spumaretroviruses or spumaviruses) have been known for over 50 years as the cause of cytopathic effects in cell cultures derived from monkey tissues. The original name, spumaviruses, is derived from the Latin word *spuma* for the foamy appearance of infected tissue culture cells. A foamy virus was isolated from a human-derived tissue culture in 1971 and called HFV for human foamy virus. The cell culture that yielded virus was obtained from a Kenyan patient with a nasopharyngeal carcinoma. At the time, this isolate was thought to be the first human retrovirus. Isolation of HFV generated a considerable amount of interest because of its possible link with human cancer. However, subsequent work over the past decade has definitively shown that the HFV isolate is a chimpanzee virus derived from the species of chimpanzee present in East Africa. This has led to the suggestion that the Kenyan patient had acquired FV through a zoonotic infection. This chimpanzee isolate is now designated PFV for prototype foamy virus. Unless otherwise indicated, the information in this article refers specifically to PFV.

Epidemiology

FVs have been isolated from a number of hosts (**Table 1**), including all nonhuman primates (NHPs) that have been examined, as well as cats, cows, and horses. It is interesting that all species naturally infected with lentiviruses are also infected with harbor foamy viruses, although the reverse is not true. FV infection of natural hosts has not been linked to any pathologies. The widespread presence of FV suggests that it is part of the natural host microbial flora. In a small number of cases examined, infection of other species such as mice or rabbits has also failed to yield any manifestation of disease.

A large number of publications have suggested links between FVs and a variety of human diseases. However, ultimately none of these have been verified. Currently, there is no evidence of widespread infection of humans with FV. However, human infections are well documented. All the known cases can be traced to interactions with NHP. Veterinarians, primate center workers, bush meat hunters, pet owners, and people associated with monkey parks and monkey temples in Asia have been found to be infected. The frequency of infection in these high risk groups is about 2–3% of those examined. Humans have been shown to be infected by the presence of specific antibodies, and in some cases by sequencing provirus from peripheral blood cells. In the few cases examined, human-to-human transmission has not been observed. Recent studies have shown that primate FVs can be transmitted to monkeys via blood transfusions from naturally infected animals. Whether or not this would also be true for infected human blood is not known. Limited studies have been made to determine whether humans can be infected by feline FVs from domestic cats. To date, there is no evidence for human infection by these agents.

Taxonomy and Classification

FVs are clearly retroviruses based on their genomic organization and general life cycle. However, they differ in fundamental ways from other retroviruses. These differences have led to their reclassification into a subfamily of *Retroviridae*. All other groups of retroviruses such as human immunodeficiency virus (HIV) are now called orthoretroviruses, or true retroviruses, while foamy viruses are spumaretroviruses (**Table 2**). Many aspects of their replication place FV closer to the hepadnaviruses such as human hepatitis B virus (HBV) than to orthoretroviruses.

Table 1 Foamy virus isolates

Host species	Virus isolate
Feline	FFV
Bovine	BFV
Equine	EFV
Rhesus macaque	SFVmac
African green monkey	SFVagm
Squirrel monkey	SFVsqu
Galago	SFVgal
Chimpanzee	SFVcpz
Spider monkey	SFVspm
Capuchin monkey	SFVcap
Baboon	SFVbab
Orangutan	SFVora
Gorilla	SFVgor
Marmoset	SFVmar
Human (not a natural host)	PFV (chimpanzee origin)

Viral Infections

Primate FV have been shown to have a very broad host range in tissue culture, with almost all vertebrate cell types being susceptible. In tissue culture, many cell types are lytically infected with virus. Such infected cells fuse to form large multinucleated cells, or syncytia. These syncytia rapidly die. Infected cells become filled with large numbers of vesicles, giving the cell cultures a foamy appearance (hence the name). An example of an FV-infected tissue culture is shown in **Figure 1**, using a hamster-derived indicator cell line in which infected cells stain blue. The arrows indicate multinucleate syncytia. However, some cell types in culture, including some human hematopoietic cell lines, are not killed after infection. Cell lines such as human Jurkat T cells are infected, evidenced by the presence of full-length viral DNA in the chromosomes. Few or no virions are produced from Jurkat cells, and there is no apparent effect on the health of the cells. These cell types are latently infected, and the infection persists indefinitely. The inability of virus to replicate in such persistent cultures may be caused by a transcriptional block. It appears that viral replication in tissue culture cells is invariably associated with cell death. However, it is not known whether or not this is true *in vivo*.

FV infections of organisms are lifelong in all species examined, including humans. There are no documented examples of viral clearance. Antibodies to viral proteins are present at all times after initial infection. Virus is believed to be transmitted between individuals through saliva. NHPs acquire the virus through biting or grooming. Young animals are not infected, and appear to be resistant to virus infection because of acquisition of maternal antibodies. Once such antibodies wane, serum conversion is rapid, and by 3 years of age, most animals are infected. In some populations of NHPs, infection reaches nearly 100% in adult animals. In naturally infected NHPs, viral DNA sequences can be found in most tissues, at very low levels. However, viral replication has only been documented in oral mucosal tissues, and virus is present in saliva. It is likely that FVs are latent in most tissues, and viral replication is confined to a small subset of cells in the oral mucosa. Such infected cells could secreted into saliva and be responsible for viral transmission.

Table 2 Classification of retroviruses (family *Retroviridae*)

Subfamily	Genus	Type species
Orthoretrovirinae (Orthoretrovirus)	Alpharetrovirus	Avian leukosis virus
	Betaretrovirus	Mouse mammary tumor virus
	Gammaretrovirus	Murine leukemia virus
	Deltaretrovirus	Bovine leukemia virus
	Epsilonretrovirus	Walleye dermal sarcoma virus
	Lentivirus	Human immunodeficiency virus
Spumaretrovirinae (Spumavirus)	Spumavirus	Simian foamy virus

Foamy Viruses

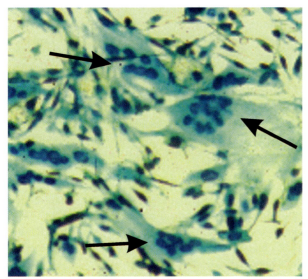

Figure 1 Baby hamster kidney indicator cells infected with PFV. These cells contain an indicator gene that turns the cell nuclei blue after staining when the viral transactivator Tas is present. The arrows point to multinucleate syncytia formed by PFV infection.

Virion Properties

FVs resemble other retroviruses in that they are fairly pleotropic in morphology, with large extracellular spikes. One distinct difference is that the FV core is electron lucent rather than condensed into a spherical core as in murine leukemia virus or into a cone-shaped core as in HIV (**Figure 2(a)**, red arrow). The morphology of mature, infectious FV resembles that of immature gammaretroviruses. Foamy virions have a density of about $1.16\,\text{gm}\,\text{ml}^{-1}$, as do orthoretroviruses. The infectious particles are surrounded by long spikes, composed of the viral glycoproteins encoded by the *env* gene (**Figure 2(a)**, purple dotted arrow). Like most retroviruses, FV bud from cellular membranes (**Figure 2(b)**, black arrow). Most often budding is from intracellular membranes, and the majority of virions remain cell associated. A cartoon of an FV particle is shown in **Figure 2(c)**. Virions are composed of two large structural proteins (Gag, blue ovals), surrounded by glycoprotein spikes (green). Particles contain the viral enzymes, including reverse transcriptase, encoded in the Pol protein (red). Like orthoretroviruses, FVs package RNA, but unlike

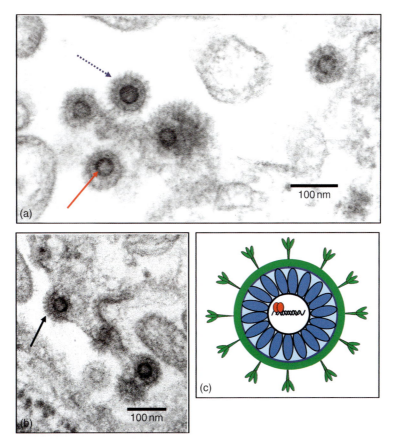

Figure 2 Morphology of foamy virus particles. (a, b) Negatively stained electron micrographs of tissue culture cells infected with PFV. (a) Intracellular particles; (b) budding particle; (c) depiction of a mature particle. The viral Gag proteins are shown in blue, the Env proteins in green, and the Pol proteins in red. The DNA genome is indicated by the black helix.

orthoretroviruses, the infectious genome of FVs is double-stranded DNA (dsDNA) (depicted by black lines), implying that reverse transcription occurs during assembly and/or budding (see below). Because of the mechanism of Pol synthesis, there are many fewer Pol molecules per foamy virion compared to the orthoretroviruses.

Genome and Protein Organization and Expression

Foamy viruses are complex retroviruses. Their genomes are greater than 12 kb (**Figure 3(a)**), placing them among the largest retroviruses. They encode the three genes common to all retroviruses (*gag*, *pol*, and *env*), as well as additional genes (*tas* and *tas/bel2* encoding Bet), whose gene products are not virion associated (**Figure 3(b)**). The Gag, Pol, and Env gene products are all polyproteins that are cleaved by proteases to smaller mature products. The Gag and Pol precursor proteins are cleaved by the viral protease (PR) encoded within the Pol protein (cleavage sites are indicated by the yellow arrows), whereas the Env precursor protein is cleaved by a cellular furin-like protease (indicated by the white arrows). The Gag protein is synthesized from an RNA that is indistinguishable from the viral genome. The Pol and Env proteins are synthesized from spliced mRNAs that utilize the same 5′ splice site (5′ss) but different 3′ splice sites (**Figure 3(c)**). Synthesis of Env from a spliced mRNA is common to all retroviruses. In contrast, synthesis of Pol from a spliced RNA is unique to FV. Orthoretroviruses synthesize Pol as a Gag–Pol read-through protein from genomic RNA. Since the orthoretrovirus precursor Pol protein has Gag determinants, it is co-assembled with Pol into virions. Lack of Gag determinants on FV Pol (**Figure 3(d)**) indicates that Pol encapsidation into virions occurs by a different mechanism. Tas and Bet are synthesized from a second promoter (IP) and Bet is made from a spliced mRNA joining part of *tas* to *bel2* (**Figure 3(c)**).

There are several noticeable differences in organization of the FV Gag, Pol, and Env proteins compared to orthoretroviruses (**Figure 3(b)**). The FV Gag is only cleaved once by PR at a site near the C-terminus and there are no mature cleavage products analogous to the matrix, capsid, or nucleocapsid proteins of orthoretroviruses.

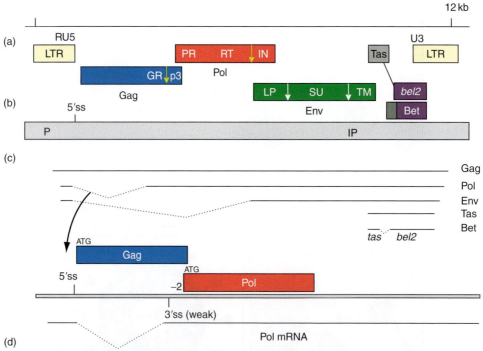

Figure 3 Viral genome. (a, b) The size of the PFV proviral genome and all of the gene products. The long terminal repeats (LTRs) contain the regulatory elements created by reverse transcription of the viral genomic RNA, duplicating RU5 and U3. The yellow arrows show sites of cleavage by the viral protease, the white arrows show sites of cleavage by a cellular protease. The Gag gene encodes the structural proteins. GR indicates three glycine-arginine-rich regions. One cleavage event occurs releasing a 3 kDa peptide (p3). The Pol gene contains the three enzymatic domains, protease (PR), reverse transcriptase (RT), and integrase (IN). The Env gene contains three domains, leader peptide (LP) unique to foamy viruses, surface (SU), and TM (transmembrane). Tas is the viral transactivator protein, and Bet is made from two exons, Tas and *bel2*. The function of Bet is not completely understood. (c) The location of the two viral promoters; The canonical retroviral promoter (P) in the LTR and the unique internal promoter (IP) encoded within the *env* gene. 5′ss indicates the major splice site. Indicated below are the major mRNAs. (d) The unique spliced mRNA that gives rise to the Pol protein. ATG indicates the start sites for translation. Adapted from Linial ML (1999) Foamy viruses are unconventional retroviruses. *Journal of Virology* 73: 1747–1755, with permission from American Society for Microbiology.

FV Gag lacks the hallmarks of other retroviral Gag proteins such as the cysteine–histidine boxes and the major homology region (MHR). The only distinct feature of FV Gag is the presence of three glycine-arginine-rich (GR) boxes near the C-terminus of the protein. These are involved in assembly and/or RNA binding. Most retroviral Gag proteins have a post-translational modification that adds a myristylation signal at the N-terminus. However, this does not occur in foamy viruses. The Pol protein encodes the four enzymatic activities associated with all retroviruses: PR, reverse transcriptase (RT) with its associated RNase H activity (RH), and integrase (IN). However, only one cleavage occurs in Pol, leading to a PR–RT (RH) fusion protein and an IN protein. The Env protein is composed of three subunits. In addition to the usual surface (SU) and transmembrane (TM) proteins, there is an N-terminal leader peptide (LP). The LP is equivalent to the signal peptide of other retroviruses that is cleaved and discarded. However, in FV, the LP is incorporated into particles with the same stoichiometry as SU and TM. In orthoretroviruses, the signal peptide is not incorporated into particles.

FVs also encode two proteins which are not found in virions, Tas and Bet. They are translated from mRNAs transcribed from the internal promoter located within the *env* coding region (IP; **Figure 3(c)**). Tas is a transcriptional transactivator, which binds to sequences in both the long terminal repeat (LTR) promoter and the IP, although the sites have different sequences. The IP has a higher affinity for Tas, and is activated first after infection. When additional Tas is synthesized from the IP, the level becomes high enough to allow transcription from the LTR and production of the virion proteins Gag, Pol, and Env. A block at the transcriptional level, preventing sufficient Tas to be synthesized to allow LTR transcription, is thought to occur in latent infections. Some latently infected cells can be activated by protein kinase C pathway activators, leading to viral production and ultimately to cell death. The second nonstructural protein, Bet, is an approximately 60 kDa protein with no obvious functional domains. Although all foamy virus Bet proteins have some similarities, there are no other proteins encoded by any organism with homology. Although the function of Bet is unknown, there are many tantalizing findings. Bet is both nuclear and cytoplasmic, and is also released from cells. It can be taken up by uninfected cells and there is some evidence that it can prevent infection. Although primate foamy virus Bet has not been found to be required for viral infectivity, it does appear to be required for replication of feline FV in some tissue culture cells. The role of Bet in infections of animals has not been assessed. Overexpression of Bet appears to have some effect on activation of the LTR promoter by Tas. Most recently, a number of groups have shown that Bet can inhibit the antiretroviral effects of the host protein APOBEC3, which is also an inhibitor of HIV and other retroviruses. However, the exact function(s) of Bet in viral replication *in vitro* and *in vivo* remains to be determined.

Viral Life Cycle

An overview of the FV life cycle is depicted in **Figure 4**. The steps in viral replication that differ in some aspects from those of orthoretroviruses are shown in red text. Some aspects of the viral life cycle are similar to that of the Hepadnaviruses such as hepatitis B virus (HBV), as indicated below.

1. *Viral entry.* The receptor(s) have not been identified for any foamy viruses. Identification is made difficult by the lack of cells that are resistant which could be used for selection. FV entry has been shown to use a pH-dependent endocytic pathway that is resistant to chloroquine, showing

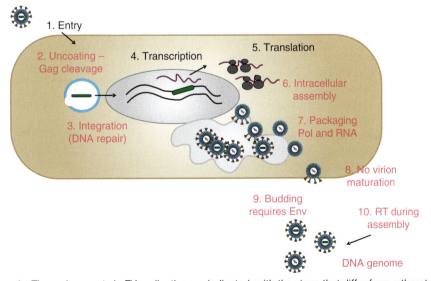

Figure 4 Viral life cycle. The major events in FV replication are indicated, with the steps that differ from orthoretroviruses indicated in red. For details, see text.

that the details of entry differ from that of the well-studied vesicular stomatitis virus.

2. *Uncoating.* FV Gag proteins are only cleaved once upon assembly/release by the viral protease PR. However, during viral entry, at least one additional PR cleavage occurs, which is required for infectivity. Presumably, this cleavage is important for uncoating of the viral capsid to release viral nucleic acid. Incoming particles traffic to the cellular microtubule organizing center (MTOC).

3. *Integration.* Although virions contain both RNA and DNA, the infectious genome is double-stranded linear DNA, reverse transcription having occurred late in the replication cycle (step 10, below). FVs share the property of packaging RNA but having DNA genomes with HBV. FV integration appears to occur in a manner similar to orthoretroviruses, although some repair of the dsDNA may occur at this step. The viral integrase (IN) is required for productive infection.

4 and 5. *Transcription and translation.* These occur as in other retroviruses using cellular machinery. As noted above, FV transcription from the two promoters requires the viral DNA binding protein Tas. Unlike other complex retroviruses in the genera *Deltaretrovirus* and *Lentivirus*, no viral proteins have been shown to be involved in RNA splicing and/or export of unspliced RNA from the nucleus. A region in the LTR which is present in both *gag* and subgenomic RNAs is necessary for efficient translation of Gag and Pol proteins.

6–8. *FV assembly and Pol and RNA packaging.* FV assembly occurs intracellularly, rather than at the plasma membrane, in a manner similar to that of Mason–Pfizer monkey virus (MPMV), a type D simian retrovirus. Like MPMV, FV Gag contains a cytoplasmic targeting and retention signal (CTRS) that is required for viral assembly. Newly synthesized Gag traffics to the MTOC which appears to be involved at some stage of viral assembly. RNA is incorporated into particles through a specific region of the RNA that interacts with one or more of the arginine-rich GR boxes. The RNA sequence is not exclusively located near the 5′ of the RNA where orthoretroviral packaging sequences (psi) are located. The FV psi sequence is more complex, and appears to include additional sequences encoded within the *pol* gene. The FV Pol protein is synthesized from a spliced mRNA and lacks Gag sequences. Thus, encapsidation is not through Gag assembly domains. Instead, viral RNA is required for Pol packaging. The Pol encapsidation sequences, like psi, are complex and involve both the U5 region of the LTR as well as sequences at the 3′ end of the *pol* gene. Protein–protein interactions of Gag and Pol may be required for Pol encapsidation. Because of the requirement for RNA sequences for binding, fewer Pol particles are packaged than in orthoretroviruses. As described above, cleavage of Gag into smaller subunits does not occur. This probably is responsible for the morphology of infectious virions that resemble immature retroviruses lacking a condensed core.

9. *Budding.* Orthoretroviral particles can bud from cells in the absence of the envelope glycoproteins, although the presence of Env can increase the efficiency of their release. In contrast, FVs absolutely require Env proteins to bud from cells, reminiscent of HBV. The component of FV Env that is required for particle egress has been mapped to the N-terminus of the LP protein. In the absence of LP intracellular particles accumulate. In general, FV budding is quite inefficient with a vast majority of infectivity remaining cell associated. It is likely that transmission from infected to naive animals occurs via cell-associated virus, rather than free virus.

10. *Reverse transcription.* Unlike orthoretroviruses but similar to HBVs, reverse transcription of the FV RNA genome occurs during assembly. The mechanism of reverse transcription appears to be identical to that of orthoretroviruses, using a $tRNA_{lys1,2}$ as a primer. Budded virions contain full-length dsDNA that it is required for infectivity. DNA extracted from virions has been shown to be infectious when fused into cells. The RT is unique in that it also contains the protease domain. FV RT is highly active and processive *in vitro*.

Viral Genetics and Evolution

FVs are the most ancient of retroviruses, and of all RNA viruses. It is estimated that they have been present in NHP populations for over 60 million years and in Old World primates for over 30 million years. They appear to have cospeciated with their hosts. The lack of significant homologies to orthoretroviruses in their Gag proteins, suggest that they may have evolved separately from orthoretroviruses. FV viral genomes are very stable and there is much less sequence diversity than seen with other retroviruses, either within or between species. The viral RT appears to be as error prone as that of other retroviruses, so the low sequence diversity may reflect the lower total viral load per animal, as replication is confined to the oral mucosa. It is assumed that FVs undergo recombination and reassortment as seen in the other genera, but this has not been studied.

Viral Vectors

Orthoretroviruses, including murine leukemia virus and HIV, have been developed into vectors for gene therapy applications. The large genome size, the broad host range, and especially the nonpathogenic nature of FV, have made them excellent candidates for gene therapy vector development. While simian and feline FVs have been developed as vectors, most work has been concentrated on PFV. Vector development has been the impetus for much of the present information about the replication and infection by FVs. FVs are rather stable. They cannot complete replication in nondividing tissue culture cells. If nondividing cells are infected, FVs can apparently persist in the quiescent cells until the cells become activated. At that time, viral infection proceeds and the FV genome integrates into the host chromosome. Viral integrations into normal human cells do not occur preferentially within genes although there is a significant preference for GC-rich regions of the genome.

Thus far, it has not been possible to create stable packaging cell lines which contain all of the viral gene products. Instead, as in the case of HIV, cells are transfected with three vectors encoding separately the *gag*, *pol*, or *env* genes, and a fourth vector encoding the gene of interest embedded in an RNA containing the RNA and Pol packaging sequences. FV cannot be pseudotyped with vesicular stomatis virus glycoproteins (VSV G), so the cognate FV Env protein is used. This has an equally broad host range as VSV G. The vectors do not use the viral promoters, so the Tas gene product is not required. Resulting replication-defective FV vectors have been successfully used in both mice and dogs to repopulate all hematopoeitic cell lineages with the gene encoded by the vectors. This indicates that the vectors can integrate into early hematopoietic stem cells. Work proceeds on using FV vectors to repair genetic defects, and also to interfere with the replication of pathogens, such as HIV. Thus far, no human trials with FV vectors have been reported.

Future Perspectives

There are currently two major foci of current foamy virus research. The first is the continued development and refinement of vectors for a variety of applications. The published results thus far have been encouraging and human trials are probably not far away. The second area is surveillance of human populations at risk for FV infection. The progenitors of HIV are apparently not pathogenic in their natural hosts, and the human pathogen arose by recombination between two monkey viruses with a chimpanzee intermediate host. There are obvious parallels with FVs, where humans are increasingly in contact with NHP harboring a variety of FV. Future studies need to determine whether or not humans shed virus in saliva, and if human-to-human transmission can be documented. But many other interesting questions remain to be determined. It is not known how FVs have adapted so perfectly to their hosts, while maintaining the ability to be highly cytopathic, at least *in vitro*. The intricate control mechanisms could reveal new strategies for dealing with other cytopathic retroviruses such as HIV. There are many unanswered questions about the unique FV replication pathway. How is Pol packaged into virions? Why is FV RT activated during assembly whereas HIV RT is not activated until after infection of new cells? What is the FV receptor? Where do the interactions between Gag and Env occur and how do they allow viruses to bud from cells? We have just begun to scratch the surface in unraveling the replication of these fascinating retroviruses.

See also: Assembly of Viruses: Enveloped Particles; Feline Leukemia and Sarcoma Viruses; Gene Therapy: Use of Viruses as Vectors; Hepatitis B Virus: General Features; Human Immunodeficiency Viruses: Molecular Biology; Simian Retrovirus D; Vesicular Stomatitis Virus; Viral Pathogenesis.

Further Reading

Lecellier C-H and Saïb A (2000) Foamy viruses: Between retroviruses and pararetroviruses. *Virology* 271: 1–8.

Linial ML (1999) Foamy viruses are unconventional retroviruses. *Journal of Virology* 73: 1747–1755.

Linial ML (2007) Foamy viruses. In: Knipe DM and Howley PM (eds.) *Fields Virology*, 5th edn., ch. 60, pp. 2245–2262. Philadelphia: Lippincott Williams and Wilkins.

Meiering CD and Linial ML (2001) Historical perspective of foamy virus epidemiology and infection. *Clinical Microbiology Reviews* 14: 165–176.

Murray SM and Linial ML (2006) Foamy virus infections in primates. Review. *Journal of Medical Primatology* 35: 225–235.

Trobridge G, Vassilopoulos G, Josephson N, and Russell DW (2002) Gene transfer with foamy virus vectors. *Methods in Enzymology* 346: 628–648.

Foot and Mouth Disease Viruses

D J Rowlands, University of Leeds, Leeds, UK

© 2008 Elsevier Ltd. All rights reserved.

Glossary

Serotype Animals recovered from infection with a virus belonging to a given serotype are resistant to subsequent infection with viruses belonging to that serotype but are still susceptible to infection with viruses belonging to any of the other serotypes.

Foot and Mouth Disease Viruses

History

The earliest account that clearly describes foot-and-mouth disease (FMD) was made by Fracastorius in 1546 and it is still today one of the most important diseases of domestic livestock.

It was the first animal disease demonstrated to be caused by a filterable agent in 1897 by Loeffler and Frosch, who also demonstrated the presence of neutralizing antibody in serum. It is also the first virus for which serotype differences were recognized. Further milestones were the demonstration by Waldemann and Pope in 1921 that guinea pigs could be infected and by Skinner in 1951 that it caused a lethal infection in suckling mice. Subsequently, cultivation of the virus in tissue culture cells has enabled studies of viral structure and replication and also the large-scale production of vaccines.

Taxonomy and Classification

Foot-and-mouth disease viruses (FMDVs) are a species within the genus *Aphthovirus* of the family *Picornaviridae*. The nature and organization of the genome, mode of replication, and structure of the virion are, in general, similar to other viruses in the family. The subdivision of the *Picornaviridae* into four genera, *Enterovirus*, *Rhinovirus*, *Cardiovirus*, and *Aphthovirus*, was originally based on physicochemical properties such as susceptibility to acid inactivation, buoyant density of CsCl solution, and the nucleotide composition. Analyses of evolutionary relationships by nucleotide sequence comparisons have largely endorsed the original classifications; however, the family has now expanded to include nine genera. Of special note is the recent inclusion of equine rhinitis A virus (ERAV) in the genus *Aphthovirus*. Properties which distinguish the FMDVs are: (1) extreme sensitivity to acid inactivation (<pH 6.8 in low ionic strength buffer); (2) high buoyant density in CsCl (1.43–1.50 g cm^{-3}); (3) possession of a poly(C) tract in the 5′ untranslated region (UTR) of the RNA (a property shared with some cardioviruses); (4) three separately encoded VPg proteins; (5) the use of two alternative in-frame protein translation initiation sites; and (6) a leader protease protein located in the N-terminal region of the polyprotein.

Properties of the Virion

As for other picornaviruses, FMDV particles are non-enveloped icosahedrons comprising 60 copies each of four structural proteins, VP1–VP4, encapsidating a single copy of the single-stranded positive-sense genomic RNA. The crystallographic structure of the virus (**Figure 1**) has shown that the three larger proteins (VP1–VP3) have the eight-stranded antiparallel β-barrel folding motif seen in

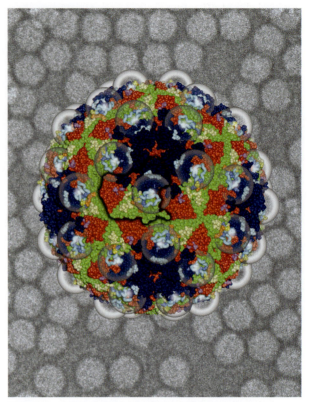

Figure 1 CPK rendition of reduced FMDV, serotype O superimposed on a cryoelectron micrograph of particles. Color codes: VP1, blue; VP2, green; VP3, red (VP4 is internal and not visible in these views). The VP1 G–H loop is shown as a worm in cyan with the Arg-Gly-Asp residues in orange CPK. Antigenic residues are color-coded according to their classification into sites: sites 1 and 5 (mid-blue), site 2 (pale yellow), site 3 (light blue), site 4 (magenta). The potential occupancy of the mobile VP1 G–H loop is modeled by a transparent sphere centered at the midpoint between the two ends of the loop. A protomeric subunit is outlined in black. Courtesy of E. Fry and D. Stuart.

other picornaviruses and some plant viruses. VP4 is disposed on the inner surface of the particle. In common with other picornaviruses, VP1 molecules are located around the axes of fivefold symmetry whereas VP2 and VP3 alternate around the two- and threefold symmetry axes. Heat or acid degradation of the particles results in dissociation into pentameric subunits, consisting of five copies each of VP1–VP3, with release of the RNA and VP4.

Several features of the structure are unique to FMDV. The protein shell is generally thinner and the external surface is smoother than in other picornaviruses. This is a result of the smaller sizes of VP1–VP3 (VP1 213, VP2 218, VP3 220 amino acids for serotype O1 virus) compared to other picornaviruses. The truncations are in the loop regions linking the core elements of the β barrels, which in other picornaviruses form prominent features at the outer surface. The deep grooves or pits encircling the fivefold axes of many picornaviruses are not present in FMDV and the position equivalent to these invaginations is occupied by the C-terminal portion of VP1. An important exception to the generally smooth contours of the surface of FMDV is provided by the G–H loop of VP1 (**Figure 1**). This large loop extends from about residue 130 to 160 of which residues around 135–158 are too disordered to be visible in electron density maps. In serotype O1 viruses, the disorder of the VP1 G–H loop is induced by a disulfide bond between the cystine residues VP2 130 and VP1 134. Under reducing conditions this bond is broken and the G–H loop collapses onto the surface of the virus in an ordered configuration. This feature includes an immunodominant antigenic site to which a high proportion of virus-neutralizing antibodies are directed. Also, synthetic peptides representing sequences from this region are immunogenic and can induce protective immunity. The sequence of this region is variable both in composition and length between different viruses with the exception of a highly conserved receptor binding motif which includes the triplet, Arg, Gly, Asp (see the section titled 'Cell attachment and entry').

Properties of the Genome

The genome consists of a single molecule of single-stranded positive-sense RNA, which is infectious. The order of the gene products on the genome is basically similar to other picornaviruses but there are some unique features (**Figure 2**). The genomic RNA terminates at the 5′ untranslated end with a small protein, VPg, linked by a phosphodiester bond through a Tyr. There is a variable length of poly(A) tract at the 3′ end. Uniquely, the FMDV genome encodes three distinct VPgs, which are used with equal efficiency. The reasons for this gene triplication are unknown. In fact, virus derived from an infectious cDNA clone from which one or two of the VPg copies had been deleted is still infectious, although RNA synthesis is reduced. The 5′ UTR is exceptionally long (c. 1300 nt), even by picornavirus standards (**Figure 3**). There is an uninterrupted poly(C) tract of 100–200 residues, depending on the virus isolate, located c. 400 nt from the 5′ end. The function of the poly(C) tract is unknown and it is not present in ERAV. There is evidence from FMDV and mengovirus (a cardiovirus) that the length of poly(C) is related to pathogenicity.

The sequence between the 5′ end and the poly(C) tract has a high degree of secondary structure and is predicted to fold into an almost complete hairpin. Its function is unknown but by analogy to the highly structured 5′ sequence of poliovirus RNA it is likely to be involved in the control of RNA replication. To the 3′ side of the poly(C) tract there are a variable number of repeat domains that are predicted to fold as pseudoknot structures. The 3′ 435 nucleotides of the 5′ UTR fold into a series of stem/loop structures similar to those present in the equivalent region of cardiovirus 5′ UTRs and function as an internal ribosome entry site (IRES) to initiate protein synthesis. A stem/loop structure (*cre* – *cis*-active replication element), located between the pseudoknots and the IRES, functions as a template for the uridylation of VPg, which is an essential step in the initiation of viral RNA synthesis.

Protein Products

The protein-coding region is a continuous open reading frame of 6999 or 6915 nt for FMDV serotype A[10],

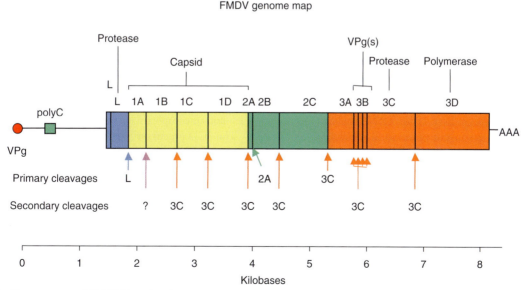

Figure 2 Genome map of FMDV. Boxed region is the polyprotein translation product with processing sites arrowed.

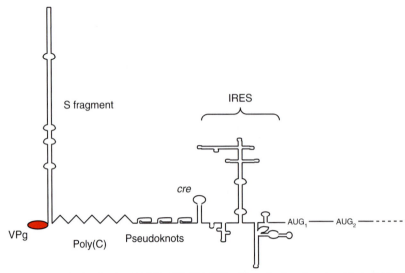

Figure 3 Secondary structure prediction for the 5' UTR of FMDV RNA with major functional motifs individually labeled.

depending on which of two functional in-frame initiation codons is used. The order of the gene products is shown in **Figure 2**.

Leader protein. The leader protein(s), Lab and Lb, which precede the structural proteins, have a proteolytic function which cleaves at the L–P1 junction and also affects the cell translation machinery (see the section titled 'Translation'). Lb can perform all of the known functions of L and the significance of Lab is unknown, although presence of two polyprotein initiation sites is completely conserved. The L proteins are not essential for virus viability since virus derived from an infectious clone with deleted L domain can replicate. However, it has an attenuated phenotype and has been proposed as a candidate live vaccine.

P1 region. The P1 region consists of the structural proteins 1A, 1B, 1C, and 1D, which are equivalent to VP4, VP2, VP3, and VP1, respectively.

P2 region. The 2A protein is vestigial in size compared to other picornaviruses, being only 18 amino acids long, but it enables the nascent separation of the polyprotein at the 2A–2B junction. 2C has a nucleoside triphosphate binding motif and it appears to have a role in RNA replication since amino acid changes in this protein can relieve the inhibition of RNA synthesis seen in the presence of guanidine. The precursor protein 2BC influences membrane trafficking in infected cells, a role performed by 3A in poliovirus.

P3 region. The role of 3A in FMDV is unclear but it does not seem to function in a similar manner to the 3A protein of poliovirus, which serves as a membrane-bound donor of VPg during viral RNA synthesis.

3B or VPg occurs as three tandem copies of 23, 24, and 24 amino acids. Although differing in sequence, each is rich in Pro, Arg, and Lys residues and has a single Tyr at position 3. Each of the individual VPg molecules is highly conserved between the A, O, and C serotype viruses. The VPg molecules are post-translationally modified to function as primers in RNA synthesis (see next section). All encapsidated RNA molecules terminate with a VPg molecule and each of the three forms is found in equal abundance. Actively translating viral RNA lacks VPg, and cell extracts contain an enzyme that cleaves the phosphodiester bond to produce RNA terminating with a 5'-monophosphate. The observation that all virion-associated RNA terminates in VPg suggests that it may have a role in the selection of molecules for encapsidation. Mutagenesis experiments with an infectious cDNA clone have provided some support for this conclusion.

$3C^{pro}$ is the protease responsible for the majority of the processing cleavages. In common with other picornaviruses, sequence analysis suggests that the catalytic site of $3C^{pro}$ is related to trypsin, a serine protease, but with the replacement of the nucleophilic serine residue with cysteine.

$3D^{pol}$ is the RNA-dependent RNA polymerase responsible for RNA replication and VPg uridylation (see next section). The crystal structure has been solved in the apo form and in complex with primer and template.

The 3CD precursor has a distinct function as a catalyst for the $3D^{pol}$-mediated uridylation of VPg.

The molecular structures of virus particles, L^{pro}, $3C^{pro}$, $3D^{pro}$, and precursor 3CD have all been determined by X-ray crystallography.

RNA Replication

RNA of infecting virus functions as a template for the synthesis of a negative-sense complementary strand, which serves, in turn, as template for the synthesis of positive-sense strands, identical to the original infecting molecule. Positive strands are synthesized in a complex structure (replicative intermediate (RI)) consisting of a single negative-strand template and several (around six) nascent positive strands. RNA synthesis is asymmetrical in favor of positive strands. A proportion of the negative-strand templates occur as full-length double-stranded hybrids (replicative form (RF)) and appear to take no further part in RNA synthesis. RF molecules accumulate in the cell during viral replication. Each RNA replication event is initiated by priming with a uridylated VPg molecule.

A single molecule of viral RNA is sufficient to initiate infection, which implies that it can function sequentially as a template for translation, to produce the polymerase enzyme(s), and as a template for RNA replication. Single-stranded viral RNA is infectious in the presence of inhibitors of host cell DNA-dependent RNA polymerase. Double-stranded viral RNA is also infectious but not in the presence of inhibitors of the cellular polymerases.

Translation

FMDV RNA is efficiently translated in a cell-free system (rabbit reticulocyte lysate) to produce protein products similar to those found in infected cells. The 435 nucleotides upstream of the first AUG initiation codon are folded into a complex structure similar to that of the equivalent region of cardiovirus RNAs. This sequence acts as an IRES, allowing initiation of translation in the absence of the host cell cap-binding complex. The rate of total protein synthesis in virus-infected cells does not change until its decline toward the end of the growth cycle, when cytopathic effects (CPEs) are apparent. There is, however, a marked change in the profile of proteins produced. When viral replication is maximal, virtually no host cell proteins are produced. This 'swap-over' of translation from host to viral products is similar to the situation in cardiovirus-infected cells and differs from the kinetics of translation following infection of cells with entero- or rhinoviruses. In the latter, infection results in a rapid shutdown of host cell protein translation, which is followed later by a resumption of protein synthesis due to the increasing production of viral proteins. The shutdown induced by entero- and rhinoviruses is largely, if not entirely, due to a virus-induced cleavage of a host protein, p220 or eIF4G, an important component of the cap-binding complex required for the initiation of translation of host mRNAs. In these viruses, eIF4G cleavage is indirectly induced by 2A^{pro} protease. Cardioviruses do not induce eIF4G cleavage and appear to simply outcompete host mRNAs for utilization of the translation machinery. Although the kinetics of protein translation in FMDV-infected cells resemble those of cardiovirus-infected cells, eIF4G is cleaved. In contrast to entero- and rhinoviruses, FMDV cleavage of eIF4G is not induced by 2A but by L. 3C^{pro} can also cleave eIF4G but at a different site from L^{pro} and later in the infection cycle. Inhibition of host cell protein synthesis is likely to be of advantage to the virus both by removing competition for access to the translation machinery and also by inhibiting the expression of innate immunity response genes in infected cells.

Post-Translation Processing

The polyprotein translation product of FMDV RNA is proteolytically processed by three of four virus-encoded enzyme activities (**Figure 2**). Three of the cleavages occur nascently on the growing polypeptide chain. The first separates L from P1, the structural protein precursor, and is carried out by leader protein. The cleavage is within a Lys-Gly dipeptide and probably occurs normally in *cis* but can also occur in *trans*. L–P1 cleavage is the only processing step which is inhibited by the tripeptide D-Val-Phe-Lys-CH$_2$Cl. Both Lb and Lab are proteolytically active.

The second primary cleavage occurs at the junction of 2A and 2B and is catalyzed by the 18-amino-acid 2A sequence. There is good evidence that the mechanism of 'cleavage' at the 2A/2B junction involves a novel process of interrupted translation, referred to as ribosomal skipping, rather than proteolytic cleavage.

The third primary cleavage is between 2C and 3A and is catalyzed by 3C^{pro} protease. This protease is responsible for all other processing cleavages, apart from that which generates VP2 (1B) and VP4 (1A) from the precursor VP0 (1AB). Many cleavages catalyzed by 3C^{pro} protease occur at Glu-Gly junctions but other dipeptides are recognized and 3C^{pro} of FMDV is the most promiscuous of the picornavirus proteases. The cleavage event to produce VP2 and VP4 from the precursor, VP0, occurs in the final stages of virus maturation. The mechanism of this cleavage is not known.

Virus Assembly and Release

A variety of assembly intermediates containing equimolar amounts of VP1, VP2, and VP0 are detected in infected cells. These correspond to monomer and pentamer subunits of the icosahedral capsid and 75S empty particles, which lack viral RNA but possess antigenic properties similar to mature viral particles. Following pulse-labeling experiments, empty particles can be 'chased' into viral particles, but it has not been shown that they are on the direct morphogenetic pathway. All encapsidated RNA terminates with VPg, suggesting that this plays a role in selection. Altered VPg molecules within an infectious cDNA clone are less efficiently encapsidated.

Paracrystalline arrays of virus are visible in infected cells and are released by lysis of the cell. There is evidence that some viral particles are secreted prior to cell disruption.

Cell Attachment and Entry

The large mobile G–H loop of VP1 located at the surface of FMDV particles mediates their binding to susceptible cells. The sequence of the loop contains a highly conserved motif, Arg, Gly, Asp, which is the hallmark of ligands for a number of heterodimeric cell surface molecules called integrins. Synthetic peptides including this sequence can compete with virus for cell attachment and treatment of the virus with proteolytic enzymes such as trypsin, which cleave within the G–H loop, also prevents virus binding. Although the virus can bind to a number of integrins there is good evidence that $\alpha_v\beta_6$ is the preferred receptor *in vivo*.

Following attachment the virus is internalized by endocytosis and release of the RNA is triggered by acidification in early endosomes. Reduction of the pH below ~6.8 seems to be all that is required to initiate the infection process and the mechanism by which the genome is delivered to the cytoplasm is unknown. Many tissue-culture-adapted strains of the virus have evolved *in vitro* to use heparin sulfate at cell surface as an alternative receptor.

Geographic Distribution

FMD occurs widely and is endemic in many countries, especially in tropical regions (**Figures 4** and **5**). North America, Australia, New Zealand, and Japan are free of the disease and maintain this status by rigorous application of import controls and quarantine. Mass vaccination campaigns have virtually eliminated the virus from some areas, for example, Europe, but have been less effective in others, largely due to logistical problems of vaccine distribution and the techniques of animal husbandry employed. The global distribution of the serotypes is shown in **Figures 4** and **5**.

Host Range and Viral Propagation

The virus typically infects cloven-hoofed species with domestic cattle being the most susceptible. Domestic pigs are also important hosts and are particularly effective in propagating the disease, since they secrete large quantities of virus in the form of aerosols. In sheep and goats, the clinical manifestations of infection are usually less

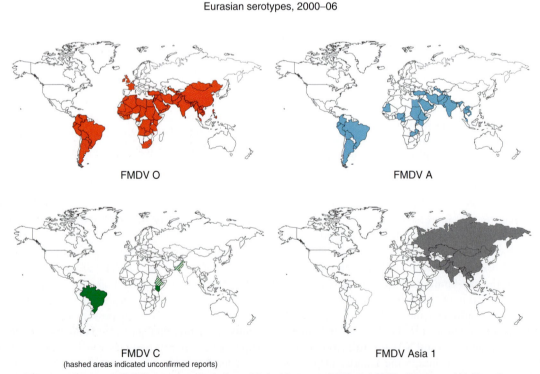

Figure 4 Global distribution of FMDV serotypes O, A, C, and Asia 1 between 2000 and 2006. Courtesy of N. Knowles.

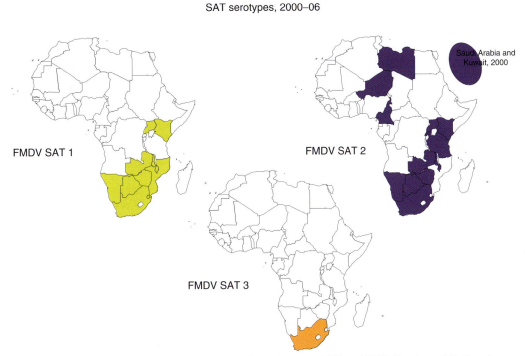

Figure 5 Distribution of FMDV serotypes SAT 1, SAT 2, and SAT 3 between 2000 and 2006. Courtesy of N. Knowles.

severe than those seen in cattle and pigs. Natural infection of Indian elephants and of camels has been reported. Many wild species of deer and antelope are susceptible to infection, and in African Cape buffalo infection is asymptomatic. Persistent infection with prolonged shedding of virus for months or years has been reported in wild and domestic species. A wide range of animals, including Australian marsupials and birds, have been infected under laboratory conditions and, very rarely, infection of human has been demonstrated.

The most important small animals for laboratory investigations are the guinea pig and the suckling mouse. In the former, injection of virus intradermally into plantar pads results in the formation of vesicular lesions both at the site of injection and in the mouth and the remaining feet, and so resembles the lesion distribution in naturally infected susceptible species. Intraperitoneal infection of suckling mice results in rapid death. The viruses can be propagated in primary cells and cell lines of bovine or porcine origin. Cells derived from the BHK-21 line are most widely used for research or vaccine production purposes. The virus can be titrated by plaque assay in cultured cell monolayers or by cytopathic end point dilution assay in microtiter plates.

Genetics and Evolution

In common with other RNA viruses, the mutation rate is extremely high and virus populations exist as quasispecies in which each individual genome is likely to differ from every other. Antigenic sites on the viral particle are tolerant of sequence variation, and antigenic diversity is a significant property of the virus.

In addition to evolution by the accumulation of point mutation, genomic recombination occurs at a high rate *in vitro*. The frequency and genomic location of recombinatorial events mirror the genetic relatedness of the parental viruses.

Serological Relationships and Variability

Seven serotypes of FMDV are recognized, the distinction of serotypes being that an animal convalescent from infection by virus of one serotype is fully susceptible to viruses of any of the remaining six. In addition to the major serotype differences, there is considerable antigenic variation between viruses within serotypes. The serological relationships between FMDV isolates are paralleled by genetic relationships as evidenced by RNA sequence analyses (**Figure 6**).

For epidemiological studies and vaccine strain selection, the serological relationships between field virus isolates or laboratory strains are expressed as r values, that is, the ratio of the neutralizing titers of immune sera against heterologous and homologous viruses. The serological relationships between virus isolates are frequently nonreciprocal, showing that closely related viruses may induce broadly cross-reactive or narrowly specific immune responses.

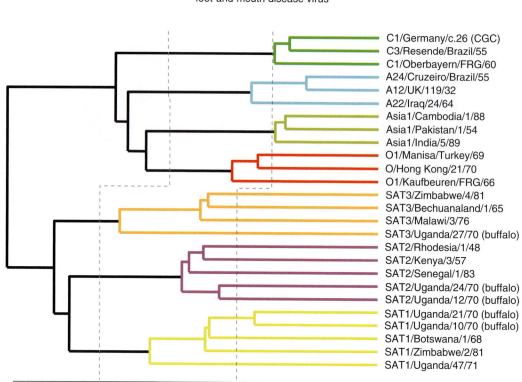

Figure 6 Genetic relatedness of FMDV strains and serotypes based on sequences within the VP1 gene. Courtesy of N. Knowles.

Complement fixation assay and enzyme-linked immunosorbent assay (ELISA) using polyclonal sera or monoclonal antibodies are also used for epidemiological studies.

Epidemiology

In regions endemic for FMD, the virus is most likely maintained in persistently infected animals. It has been shown experimentally that infected bovines can secrete virus for long periods after the initial episode of disease. In some areas the wild animal population may act as a reservoir for infection (e.g., Cape buffalo in Africa). In nonendemic areas, infection may be introduced from a variety of sources such as the importation of infected livestock, contaminated animal products such as carcasses containing bone (in contrast to meat, the postmortem acidification of bone marrow and other nonmuscular tissues is insufficient to inactivate the virus), or contaminated materials. More locally, transmission of infection is by direct transport of contaminated animals or materials or by wind-borne carriage of infectious aerosols. It is also suspected that the virus can be passively transmitted by migrating birds.

Transmission and Tissue Tropism

The principal route of infection appears to be via aerosol impinging on the pharynx and respiratory tract. Aerosols may be produced locally during feeding on contaminated foodstuff or may be transmitted over considerable distances under appropriate meteorological conditions. Pigs secrete particularly high levels of virus-contaminated aerosols. In addition to mucosal secretion, high levels of virus are found in milk. Pasture may be contaminated with virus from urine and feces.

Vesicular lesions appear in the mouth on the tongue, gums, and cheeks and later on interdigital mucosa and coronary bands of the feet. Virus can be isolated from many tissues in the body. The onset of clinical disease is usually very rapid and lesions can develop as early as 1–2 days after infection, depending on the virus strain and level of exposure.

Pathogenicity

The pathogenicity of FMDV varies according to the virus strain, host species, and age. The factors that govern the virulence of FMDVs are poorly understood and as with most viruses are probably multifactoral. Domestic cattle are usually the most susceptible species and morbidity is usually c. 100% in nonvaccinated animals. Wild bovines, such as African Cape buffalo, may produce no clinical manifestations. Although the disease is rarely fatal in adult domestic animals (<5%), significant mortality may occur in young animals (c. 50%).

Clinical Features of Infection

Infection typically produces a rapidly progressing febrile illness and the development of often massive vesicular lesions in the mouth and on the feet. The lesions rupture with considerable loss of epithelial tissue (**Figure 7**). The resulting discomfort discourages feeding until the lesions heal by the infiltration of fibrous tissue. The severity of the disease results in long-term loss of productivity in terms of meat and milk yield, and lameness may be a serious consequence for draught animals. Abortion and chronic subfertility are also common. Other organs infected include mammary glands, pancreas, and heart.

Immune Response

Infection elicits a vigorous humoral antibody response and, after recovery, immunity to reinfection with viruses of the same serotype is prolonged. The role of cytolytic T cells in recovery is unclear.

Four antigenic sites recognized by antibodies capable of neutralizing the virus have been described. One of these is an immunodominant linear sequence comprising the G–H loop of VP1, and synthetic peptides representing this tract are effective in inducing high levels of virus-neutralizing antibody and can protect animals from infection.

Figure 7 Severe symptoms of FMD showing extensive sloughing of linguinal epithelium. Crown copyright. Reproduced from "Foot and mouth disease: Ageing of lesions" (Defra, 2005) by permission of the Controller of HMSO.

Antibody responses to several of the nonstructural proteins are induced during infection and these are being developed as markers to distinguish vaccinated from infected animals.

Prevention and Control

In disease-free regions, natural protection afforded by geographical barriers is rigorously reinforced by strict controls on the importation of susceptible animals and potentially contaminated materials. Where outbreaks occur sporadically, due to occasional introduction from external sources, embargoes on animal movement and slaughter of infected herds have been successful in maintaining a disease-free national herd. In endemic areas, control is by mass vaccination. Ring or barrier vaccination is also used to limit the spread of infection.

Dramatic demonstrations of the vulnerability of naive livestock to large-scale epidemics of FMD were provided by the outbreaks of serotype O virus in the UK and of serotype A virus in South America in 2001. The outbreak in the UK resulted in the slaughter of over 10 million animals (most of which were uninfected) to control spread of the disease and the cost to the country was several billions of pounds. In South America, the outbreak was controlled by the resumption of blanket vaccination of all livestock.

The first vaccines against FMD were produced by formalin inactivation of lymph drawn from lesions on the tongues of infected cattle. This source of immunizing virus was replaced by the Frenkel method of culture in fragments of epithelium stripped from the tongues of slaughtered cattle. Most vaccine in use today is produced by growing the virus in suspensions of BHK-21 cells in fermentation vessels of up to 10 000 l capacity. Approximately 2×10^9 monovalent doses of vaccine are administered annually. Aziridines have largely replaced formalin as the inactivant, since the inactivation kinetics of the latter are nonlinear and residual live virus in vaccines has occasionally been the source of outbreaks of disease. The serotype and strain composition of vaccines have to be tailored for local requirements. Inactivated virus is usually adjuvanted by adsorption onto aluminum hydroxide gel, and saponin may also be added to enhance potency. In pigs, such vaccines elicit only immunoglobulin M responses, and for this species vaccines are formulated with oil adjuvants. The use of oil-adjuvanted vaccines are now being extended to cattle.

Solid protection requires high levels of neutralizing antibodies, and, to achieve this with inactivated vaccines, immunization is repeated two to three times a year. The development of live-attenuated vaccines has largely been abandoned mainly due to the complexity of antigenic diversity and the fear of reversion to a virulent phenotype.

Future Perspectives

There is scope for developments in FMD vaccines to improve stability, cross-protective efficacy, and duration of immunity. Peptide vaccines, recombinant viral capsids, and rationally designed nonreverting attenuated viruses are potential routes by which these goals may be achieved. The production of non-biodegradable peptides and the creation of an attenuated L protein deletion mutant may be important developments. Since the massive outbreaks in the UK and in South America in 2001, there is increased interest in the development of drugs to halt the rapid spread of infection. Improved rapid diagnostic methods are being explored as is the ability to accurately distinguish infected from vaccinated animals.

As to the molecular properties of the virus, both the determination of the crystallographic structure of the particle and several of the nonstructural proteins and the cloning and manipulation of full-length infectious cDNA molecules are important steps toward understanding the unique features of the virus.

See also: Antigenic Variation; Antigenicity and Immunogenicity of Viral Proteins; Cardioviruses; Evolution of Viruses; Immune Response to viruses: Antibody-Mediated Immunity; Picornaviruses: Molecular Biology; Replication of Viruses; Viral Receptors; Virus Particle Structure: Nonenveloped Viruses.

Further Reading

Barteling SJ and Vreeswijk J (1991) Developments in foot and mouth disease vaccines. *Vaccine* 9: 75–88.

Brooksby JB (1982) Portraits of viruses: Foot and mouth disease virus. *Intervirology* 18: 1–23.

Brown F (1989) The development of chemically synthesised vaccines. *Advances in Veterinary Science and Comparative Medicine* 33: 173.

Defra (2005) Foot and Mouth Disease – Ageing of Lesions. http://www.defra.gov.uk/footandmouth/pdf/ageing-lesions.pdf (accessed January 2008).

Fry E, Logan D, Fox G, Rowlands D, Brown F, and Stuart D (1990) Architecture and topography of an aphthovirus. *Seminars in Virology* 1: 439–451.

Mahy BWJ (ed.) (2005) *Foot-and-Mouth Disease Virus*. Heidelberg: Springer.

Rowlands DJ (ed.) (2003) Special issue: Foot-and-mouth disease virus. *Virus Research* 91: 1–161.

Sobrino F and Domingo E (eds.) (2004) *Foot-and-Mouth Disease – Current Perspectives*. Norfolk, UK: Horizon Bioscience.

Fowlpox Virus and Other Avipoxviruses

M A Skinner, Imperial College London, London, UK

© 2008 Elsevier Ltd. All rights reserved.

Taxonomy

Fowlpox virus, the prototype of the *Avipoxvirus* genus, is the best-studied species. There are currently nine other recognized species (*Canarypox virus, Juncopox virus, Mynahpox virus, Pigeonpox virus, Psittacinepox virus, Quailpox virus, Sparrowpox virus, Starlingpox virus,* and *Turkeypox virus*) and three tentative species (Crowpox virus, Peacockpox virus, and Penguinpox virus). Avipoxvirus infections have been observed in more than 230 of the known 9000 species of birds, spanning 23 orders, yet little is known about the genome diversity, host range, and host specificity of the causative agents.

History

With its characteristic lesions, ubiquitous distribution among domesticated poultry, and large virion size, fowlpox virus (FWPV) was one of the earliest recognized viruses. It also played important roles in the development of modern virological techniques during the middle decades of the twentieth century. Fowlpox was one of the first diseases of livestock and poultry for which effective vaccines were developed, as early as the late 1920s. These vaccines led to effective control of the disease and its virtual eradication from commercial poultry production in temperate regions. Fowlpox remains enzootic in most tropical and subtropical regions where poultry is produced. It represents, after Newcastle disease, the second largest virus infection of backyard poultry in Africa and is thus of considerable socioeconomic importance.

Virion Structure

Like other poxviruses, the brick-shaped avipoxviruses are large enough (at 330 nm × 280 nm × 200 nm) to be resolved by light microscopy. Indeed, the particles were originally observed following staining with basic dyes (such as basic fuchsin in Gimenez stain) and were termed 'elementary particles' or 'Borrel bodies'. Overall, the structure of the avipoxvirus virion is assumed to be similar to that of the much better studied vaccinia virus (VACV). Likewise, the replication cycle of the avipoxviruses is assumed to be essentially similar to that elucidated for the mammalian poxviruses. However, there are differences between avipoxviruses and the commonly studied mammalian poxviruses in the complement of structural genes and in morphogenesis of extracellular enveloped virions (described below), both indicative of significant structural differences at the molecular level. As a consequence, it is not clear which proteins are present in the envelope of avipoxvirus extracellular enveloped virus (EEV).

Genome Size and Organization

The avipoxviruses appear to have some of the largest known genomes of viruses of vertebrates, more than 300 kbp for canarypox virus (CNPV), encoding about 300 proteins. By way of comparison, this is 100 times the size of the smallest animal virus genomes and half the size of the genome of the smallest free-living bacterium. Among mammalian poxviruses, the central two-thirds of the genome generally encode conserved structural proteins and enzymes. In contrast, the terminal regions are more variable and divergent, encoding genus and species-specific, nonessential proteins, which are frequently involved in virus–host interactions affecting host–range, pathogenesis, and virulence. This general distribution holds in the avipoxviruses but, relative to the mammalian poxviruses, several large genome rearrangements (translocations, inversions, and transversions), which could have occurred in either avian or mammalian lineages (or both), have resulted in transfer of likely nonessential genes into more central locations.

Gene Complement

Although avipoxviruses appear to encode equivalents of most of the internal structural proteins of the VACV core, they lack some important proteins found on the surface of the intracellular mature virus (IMV) of VACV (such as D8 and A27), on the surface of the EEV (such as A33, A56, and B5), as well as on the surface of intracellular enveloped viruses, or IEV (such as A36).

Avipoxviruses also encode proteins not found in mammalian poxviruses, such as homologs of the host PC-1 nucleotide phosphodiesterase, DNaseII (DLAD), alpha-SNAP, and the lipid-pathway enzyme involved in Stargardt's macular dystrophy (ELOVL4). Such proteins appear to be nonessential for replication in tissue culture and are presumed to be nonstructural. Avipoxviruses also encode multiple members of several gene families, notably the 31 ankyrin-repeat proteins encoded by FWPV (51 by CNPV) and the six copies of the massive B22 ortholog (2000 amino acid residues) found only as a single copy in some mammalian poxviruses, such as variola virus (VARV), cowpox virus (CPXV), and molluscum contagiosum virus (MOCV). The latter gene family therefore accounts for about 35 kbp of the additional sequence found in the avipoxvirus genome.

Antigens

Three major immunodominant antigens of FWPV that are recognized by murine monoclonal antibodies have been identified. They correspond to immunodominant antigens recognized by hyperimmune serum from FWPV-infected chickens and to equivalent immunodominant antigens of mammalian poxviruses. They are the 30/35 kDa IMV surface protein fpv140 (corresponding to VACV H3), the 39 kDa virion core protein fpv168 (corresponding to VACV A4), and the 63 kDa 'virion occlusion' protein fpv191 (corresponding to the p4c protein retained in very few VACV strains). Neutralizing antisera or monoclonal antibodies have not been reported.

Survival Factors

A-Type Inclusion Protein

The avipoxviruses express a number of proteins that might be considered as factors to potentiate their survival in the face of various environmental stresses. These include the A-type inclusion (ATI) protein, which forms large cytoplasmic inclusions. These inclusions can contain large assemblies of embedded virions (analogous to spheroidin inclusions in the entomopoxviruses), which would probably be better protected from dessication in desquamated dermis. It is likely that entry of avipoxviruses into the ATI requires the presence of fpv191 (p4c, virion occlusion protein) on the surface of the IMV, assuming the mechanism is as elucidated for CPXV.

Photolyase

Avipoxviruses encode photolyases (e.g., fpv158), which are capable of repairing ultraviolet (UV)-induced pyrimidine dimers in DNA in a light-dependent manner. Such lesions might be induced in viral genomic DNA when the virion is exposed to the environment in desquamated epithelium, or while the virion is near the surface of a skin lesion.

Glutathione Peroxidase

Like MOCV, avipoxviruses encode a glutathione peroxidase. Such proteins might be able to protect the virus from environmental oxidative stress. However, the FWPV protein (fpv064) is probably not an ortholog of the MOCV protein (MC066), which is most closely related to cellular glutathione peroxidase. Instead it is probably a paralog, most closely related to cellular phospholipid hydroperoxide glutathione peroxidase. The avipoxvirus proteins might protect infected cells from the oxidative burst of immune cells or affect cell signaling pathways.

Immunomodulators

Unlike most mammalian poxviruses, but in common with MOCV, no obvious candidates for proteins interfering with the host type I interferon response have been identified. Thus the avipoxviruses lack orthologs of the VACV double-stranded RNA-binding protein (E3) and the eIF2 mimic (K3). No soluble homologs of type I interferon receptors (equivalent to VACV B18) have yet been identified in the avipoxviruses, although a soluble binding protein (e.g., fpv016) for type II interferon has been identified biochemically. Conversely, avipoxviruses also encode proteins not generally found in mammalian poxviruses, such as homologs of transforming growth factor β (e.g., fpv080), which has also been found in deerpox virus, and IL10 (e.g., CNPV018), also found in orf parapoxvirus.

No obvious soluble chemokine-binding proteins encoded by avipoxviruses have been identified, in contrast to the mammalian poxviruses. However, proteins resembling cellular serpentine chemokine receptors of the 7-transmembrane spanning G-protein-coupled receptor family are encoded by avipoxviruses (e.g., fpv021, fpv027, and fpv208), as are single copies by some mammalian poxviruses (capripoxviruses, suipoxviruses, yatapoxviruses, and deerpox virus). Like MOCV, avipoxviruses encode putative IL18-like proteins (e.g., fpv214) and chemokine-like proteins (e.g., fpv060, fpv061, fpv116, and fpv121).

Features of Avipoxviruses Shared with Molluscum Contagiosum Virus

The mammalian poxvirus to which the avipoxviruses are most closely related phylogenetically is MOCV. The avipoxviruses also share a number of other features with MOCV. For example, they lack orthologs of the VACV proteins E3 and K3 involved in evading the type I interferon response. Whereas avipoxviruses encode a family of proteins (e.g., fpv097, pfv098, fpv099, fpv107, fpv122, and fpv123) with homology to the large VARV protein B22, MOCV (like VARV and CPXV) encodes only a single protein (MC035). In CPXV, as in VARV, the gene encoding the B22 ortholog (CPXV219) is located in the variable region at the right terminus of the genome. The gene encoding the MOCV ortholog MC035 is, however, more central, located between orthologs of VACV E4L and E6R.

This is syntenic with the location of FWPV genes encoding B22 orthologs fpv097, fpv098, and fpv099, also found between orthologs of VACV E4L and E6R.

Like the avipoxviruses, MOCV lacks orthologs of VACV IMV surface proteins A27 and D8 as well as EEV proteins A56 and B5 (though, unlike the avipoxviruses, MOCV does encode orthologs of VACV EEV protein A33 and IEV protein A36).

Virus Replication

Host Range Restriction

Avipoxviruses are incapable of productive replication in mammalian cell lines, including those (such as BHK-21) permissive or semipermissive for the host-range-restricted MVA variant of VACV, and those defective in the type I interferon pathway, such as Vero cells. Otherwise, the replicative cycle is largely similar to that elucidated for VACV, although considerably slower (taking about 24 h rather than 12 h). Genome transcription and replication mechanisms are essentially the same, with relatively well conserved polymerases and other enzymes. VACV promoters function well in cells infected with avipoxviruses and are frequently used to drive foreign gene expression in recombinant avipoxviruses, and the inverse appears to be true. Virus-induced shut-off of host gene expression, clearly observed with VACV, is far less dramatic with FWPV.

Genetic Reactivation

The long-recognized phenomenon of 'genetic reactivation' illustrates that many of the essential replicative components of poxviruses are interchangeable. Thus the proteins of a UV-inactivated poxvirus can rescue the genome of a heat-inactivated poxvirus, even when the two viruses are from different genera. This is well illustrated by the modern practise of rescuing naked (and often genetically manipulated) genomic DNA of a mammalian poxvirus by infection of transfected cells with FWPV. The rescued mammalian poxvirus can be recovered and amplified by passage through mammalian cells, which eliminates the rescuing FWPV due to its host range restriction.

Virion Morphogenesis

The most obvious difference between the replication cycles of avipoxviruses and that of VACV is in the formation of EEV. VACV EEV formation predominantly involves wrapping of IMV with a double membrane derived from the *trans*-Golgi network (to form IEV) followed by subsequent exocytosis to form cell-associated enveloped virus (CEV) and EEV. Avipoxviruses, however, form EEV by a more conventional pathway in which IMV buds through the cell membrane to acquire an envelope. This difference, observed by electron microscopy (**Figure 1**), is consistent with the absence from the avipoxvirus genome of genes involved in wrapping and exocytosis in VACV.

Lipid Metabolism

There have been reports that FWPV has a profound effect on host cell lipid metabolism. It is tempting to speculate that this might be attributable to some of the unusual genes carried by FWPV, notably fpv048 encoding a protein with unusually high sequence similarity to the ELOVL4 host protein, a fatty acid elongation factor implicated in macular degeneration.

Virus Propagation

Avipoxviruses replicate only in avian cells; replication in mammalian cells is abortive and no infectious progeny are produced. Some vaccine strains of FWPV, such as the Cyanamid Webster FPV-M vaccine, even display a preference for chick embryo skin cells (CESs) over chick embryo fibroblast cells (CEFs). FWPV FP9 has been effectively adapted to CEFs but it displays a distinct preference for primary as opposed to secondary CEFs. Plaques on CEFs are not lytic, but the cytopathic effects manifest as changes in cell morphology, resulting in areas of altered refractive index with the plaques best viewed by dark field illumination. FWPV fails to plaque and replicates poorly in the recently derived chicken fibroblast cell line, DF-1. Replication is similarly poor in the chemically transformed cell line OU-2. It can be plaqued and replicated quite efficiently in quail cell lines, such as QT-35, but the presence in these cells of viable endogenous Marek's disease virus (a herpesvirus) means that their use for preparation of vaccines is not advisable. Of these possible cell substrates for avipoxvirus propagation, currently only CEFs and CESs are licensed for use in the production of human vaccines.

Publications reporting the replication of uncharacterized avipoxviruses of unknown origin in embryonic bovine tracheal cells or in BHK-21 are atypical and, until corroborated, should be viewed with caution.

Molecular Phylogenetics

The taxonomy and classification of avipoxviruses are important to the study of epidemiology and hence to

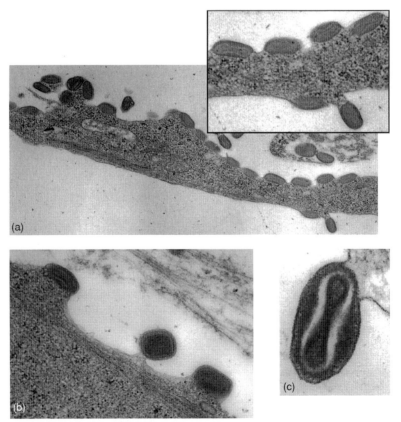

Figure 1 FWPV IMV particles budding through the membranes of infected chick embryo fibroblasts. Magnifications: (a) ×27 000 (inset ×51 000); (b) ×60 000; (c) ×170 000. Reproduced from Boulanger D, Smith T, and Skinner MA (2000) Morphogenesis and release of fowlpox virus. *Journal of General Virology* 81: 675–687, with permission from the Society for General Microbiology.

the future ability to control the diseases. Recent molecular studies confirm earlier impressions that avipoxviruses can jump species to cause disease. Thus isolation of a virus from a turkey, for instance, is not sufficient to demonstrate that turkeypox virus is the cause and, more importantly, that the outbreak can be controlled by a turkeypox virus vaccine.

Avipoxviruses can be highly diverged at the molecular level. For instance, the relatively highly conserved FWPV and CNPV P4b proteins (encoded by genes fpv167 and cnpv240, respectively) share only 64.2% amino acid sequence identity. This level of divergence is comparable to that seen between different genera of mammalian poxviruses (see **Figure 2**). In fact, recent molecular studies have demonstrated that avipoxviruses fall into three major diverged groups: (1) the FWPV-like viruses, (2) the CNPV-like viruses, and (3) the psittacinepox viruses (**Figure 2**). As their name suggests, the psittacinepox viruses have so far only been isolated from psittacines. In contrast, viruses of the FWPV-like and CNPV-like groups appear either inherently capable of infecting a wider range of birds or have evolved and adapted to be able to infect different birds. FWPV-like viruses have been isolated from turkeys, pigeons, ospreys, albatrosses, doves, and falcons. Viruses closely related to, and even barely distinguishable from, CNPV can cause disease in great tits, sparrows, stone curlews, and houbaras. CNPV-like viruses have been isolated from pigeons and starlings. A report in 2006 from Virginia in the USA indicated that one particular CNPV-like virus caused disease in a wide range of birds (robins, crows, herons, finches, doves, hawks, gnatcatchers, mockingbirds, and cardinals). It was not known whether the virus that caused the epornitic had a wider geographic distribution throughout the USA.

It is apparent that pigeonpox can be caused by two distinct types of virus, either FWPV-like or CNPV-like. The FWPV-like virus is very closely related to viruses from turkeys and osprey, and a little less closely related to those from an albatross, a falcon, and a dove. The CNPV-like virus is closely related to one from a starling. Infection of avian species outside the normal host range of the particular avipoxvirus can result in altered pathogenesis compared to that observed in the normal host. For instance, a virus which was isolated from an Andean condor (*Vultur gryphus*), in which it had caused an aggressive, diphtheritic form of the disease, produced only mild lesions in inoculated chickens. Conversely, viruses causing mild poxvirus lesions in wild birds, such as those found in 50% of short-toed larks (*Calandrella rufescens*) and 28% of Bertholet's pipits (*Anthus berthelotti*) in the Canary Islands, might cause more severe disease in naive populations of other avian species. It is, for instance, suspected that canaries might not be the natural reservoir for CNPV, which causes them a devastating disease.

Epizootiology

Evidence of poxvirus infection has been observed in many avian species but little is known about the natural host range of the causative agents, due to the lack of robust techniques for the identification and differentiation of isolates. Initially, the viruses were distinguished by their ability to cause disease in a range of test species, such

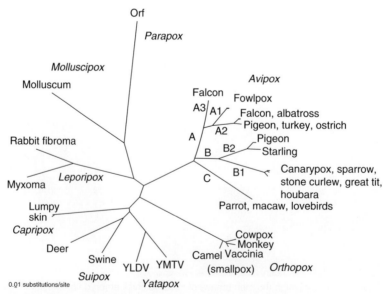

Figure 2 A phylogram showing phylogenetic relationships and evolutionary distances of avipoxvirus clades and mammalian poxvirus genera based on neighbor-joining analysis (supported by bootstrapping) of P4b orthologs.

as chickens, pigeons, turkeys, and quail. Such infectivity studies probably overestimate the degree of similarity between the viruses, as many isolates seem able to cause disease in the test species, possibly due to the inoculation of high doses. *In vivo* analysis of antigenic cross-reactivity allowed the viruses to be grouped or distinguished by their ability or inability to induce protective immunity against each other. Thus, psittacinepox virus offered no protection against FWPV or pigeonpox virus, or vice versa, nor did quailpox virus. Similarly, mynahpox virus showed no cross-protection with fowl, pigeon, psittacine, quail, or turkey poxviruses, and a condor poxvirus induced no protection against FWPV. *In vitro* serological techniques were applied to only a small proportion of the observed isolates because initial propagation in chick embryos or chick embryo fibroblast cells often failed. Possibly as a consequence of the limited characterization of isolates, there are only the ten recognized species of avipoxviruses.

Reintroduction programs appear to be particularly susceptible to problems caused by poxvirus infection. For instance, houbara bustard reintroduced into the Middle East have proved susceptible to uncharacterized avipoxvirus infections from unknown sources. Similar programs, such as that for the closely related great bustard on Salisbury Plain, Wiltshire, in the UK may face similar threats. There are also concerns that avipoxviruses might threaten marginal species such as some species of birds in the Hawaiian Islands and especially the kakapo, the flightless, nocturnal New Zealand owl parrot.

Transmission

Poxviruses produce two types of virion particles that are both infectious: IMV released into the environment only by desquamation of skin lesions and EEV released from the cell to spread to secondary sites of infection. In the wild, the primary route of infection is mechanical transmission of blood-borne EEV by biting insects (mosquitoes and midges). Lesions are therefore characteristically restricted to the featherless areas around the eyes and nares, on the comb, wattle, lower legs, and feet. Fowlpox remains a problem when control of biting insects is difficult but is generally not a problem for intensive production systems in the temperate areas of Northern Europe and the northern USA, even in the absence of vaccination. Problems with epizootics have emerged with free-range flocks, presumably because they are more exposed to biting insects, as well as to pecking by infected birds. In subtropical and tropical climates, fowlpox poses a significant problem requiring vaccination of commercial flocks.

FWPV can also transmit via inhalation or ingestion of dust (or dander, representing virus-infected cells shed from cutaneous lesions), or aerosols, leading to the 'diphtheritic form' of the disease, with lesions on the mucous membranes of the mouth, pharynx, larynx, and sometimes trachea. More likely to occur at high population density, this form of the disease is associated with commercial poultry flocks or collections of captive or domesticated birds (including quarantine facilities). However, such lesions may also form as a consequence of secondary viremia following primary cutaneous lesions, and viral genetic factors may play a role. Avipoxvirus infections of passerine birds (e.g., canaries and finches) can cause significantly higher rates of mortality than those of chickens or turkeys.

Diagnosis

Clinical Signs

FWPV and many other avipoxviruses primarily cause disease in mature birds. The cutaneous form of avipox, including fowlpox in chickens and turkeys, spread mechanically by biting insects or pecking, is normally relatively mild (**Figures 3** and **4**) but the diphtheritic form (**Figure 5**) causes higher mortality by occlusion of the oropharynx. A third form of poxvirus disease causes a pneumonia-like illness in canaries, with high mortality.

Histology

Smears of material scraped from cutaneous lesions or postmortem sections from diphtheritic lesions (or airway epithelium in the case of pneumonia) can be stained with hematoxylin/eosin, or Gimenez stain. The presence of eosinophilic inclusions (A-type inclusions, Bollinger bodies) in the cytoplasm is diagnostic of poxviruses. Basophilic (or B-type) inclusion bodies may also be seen in the cytoplasm (these represent the sites of virus replication, the so-called 'viral factories'), as may Borrel bodies (the virions themselves).

Virus Isolation and Detection

Virus may be passaged and amplified *in vivo* through specific pathogen-free chickens via wing web scarification or through embryonated eggs via chorio-allantoic membrane infection. The method of choice for virus propagation, however, is *in vitro* culture on CEFs for amplification through liquid culture or for plaque purification under semisolid overlay. Not all avipoxvirus isolates will form plaques, or even propagate, on CEFs. Primary cultures of other poultry species (duck, turkey, or quail) may also be used.

Serology

Serological methods such as enzyme-linked immunosorbent assays (ELISAs), available as commercial flock

Figure 3 Typical avipoxvirus lesions seen on the head and feet of a Laysan albatross. Reproduced from Hansen W (1999) Avian pox. In: Friend M and Franson JC (eds.) *Field Manual of Wildlife Diseases, General Field Procedures and Diseases of Birds*, pp. 163–169, with permission from USGS.

monitoring kits, can be used to provide evidence for ongoing or prior infection in the flock. Virus neutralization tests are more specific but are technically more demanding and the responsible antigenic epitopes have not been identified.

Monoclonal Antibodies

Monoclonal antibodies against the three major immunodominant structural antigens of FWPV have been isolated and characterized, as described above. None appears to be neutralizing but all can be used in ELISAs and western blotting.

Restriction Enzyme Digestion and Southern Blotting

Before extensive genome sequencing, Southern blotting was the only available molecular diagnostic method, involving examination of restriction enzyme digest profiles. Without the use of radioisotopes, the method was relatively insensitive.

PCR-Based Analysis

With the advent of DNA sequence data for the avipoxviruses, polymerase chain reaction (PCR) amplification became the molecular diagnostic method of choice to identify avipoxvirus infections. This has been based routinely on the sequence of the gene encoding the P4b protein (the A3 ortholog encoded by gene fpv167), with primers for a 578 bp fragment. However, all avipoxviruses produce fragments of the same length, so the avipoxvirus could only be identified either by sequencing the PCR product or by restriction enzyme fragment polymorphism profiling.

A second PCR locus (the H3 locus), which works with most (but not all) avipoxviruses, has the advantage that viruses of the two major groups (CNPV-like and FWPV-like) can be discriminated purely on the basis of the size of the PCR product (**Figure 6**).

The genome sequences of two different FWPV strains, one extensively culture-passaged and attenuated, and of a pathogenic CNPV have been determined. They confirmed

Figure 4 Extensive avipoxvirus lesions seen on the head of a bald eagle. Reproduced from Hansen W (1999) Avian pox. In: Friend M and Franson JC (eds.) *Field Manual of Wildlife Diseases, General Field Procedures and Diseases of Birds*, pp. 163–169, with permission from USGS.

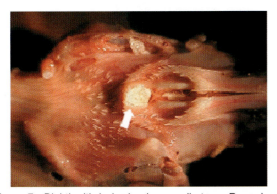

Figure 5 Diphtheritic lesion in a Laysan albatross. Reproduced from Hansen W (1999) Avian pox. In: Friend M and Franson JC (eds.) *Field Manual of Wildlife Diseases, General Field Procedures and Diseases of Birds*, pp. 163–169, with permission from USGS.

the divergence and differences between the avipoxviruses and the mammalian poxviruses typified by VACV. In fact, phylogenetic analysis showed that avipoxviruses are most closely related to MOCV of humans, a relationship confirmed by other specific aspects of their molecular biology (see **Figure 2**). Despite having sequence information for the prototypic members of the two major groups of avipoxviruses, it has still proved extremely difficult to identify conserved primer sequences that work in PCR reactions for all other members of the genus, further illustrating the extent of sequence divergence and diversity within this group of viruses.

Intraspecies and Interstrain Differentiation

Despite their large genome size, the high level of conservation within species of avipoxviruses makes it difficult to differentiate between different strains or isolates or between vaccines and pathogenic strains. Although numerous loci within FWPV have been surveyed, only one locus (the H9 locus), has been reported to allow clear differentiation. It appears to be an unstable region, with deletions of different lengths found in various FWPV strains, isolates, and vaccines. Three commercial fowlpox vaccines were found to have the same parental sequence found in two clinical isolates, while two other different commercial vaccines shared an identical deletion at the H9 locus (see **Figure 7**).

Control and Prevention

Vaccination

Since its introduction in the 1920s, vaccination against fowlpox using live FWPV or pigeonpox virus (now known to be closely related antigenically to FWPV) has become commonplace and extensive, such that the disease has been nearly eradicated from developed countries in temperate zones. Most currently available commercial vaccines originate from early isolates, though a minority were isolated in the mid-1960s.

Vaccination may be undertaken from 10 days or 4 weeks of age, depending on the residual pathogenicity of the vaccine, which is often related to whether the vaccine

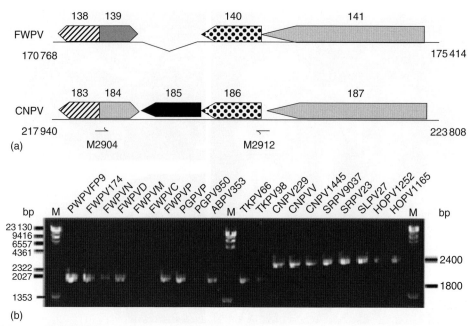

Figure 6 PCR-based differentiation of FWPV- and CNPV-like viruses. (a) Arrangement of FWPV genes fpv138 to fpv141 at the H3 locus aligned with their CNPV orthologs. The positions of primers used for PCR analysis are indicated. (b) The PCR analysis shows the length difference in the amplified DNA fragment obtained from FWPV-like viruses (1800 bp) and from CNPV-like viruses (2400 bp). Reproduced from Jarmin S, Manvell R, Gough RE, Laidlaw SM, and Skinner MA (2006) Avipoxvirus phylogenetics: Identification of a PCR length polymorphism that discriminates between the major clades. *Journal of General Virology* 87: 2191–2201, with permission from the Society for General Microbiology.

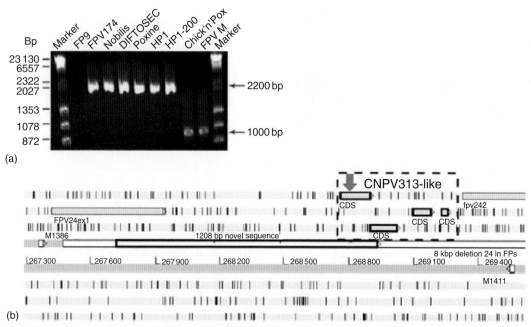

Figure 7 PCR analysis of the H9 locus of FWPV. (a) The 2200 bp product for field viruses (FPV174, HP1, and HP1-200) and three vaccines (Nobilis, DIFTOSEC, and Poxine) and identically deleted 1000 bp products for two other vaccines (Chick'n'Pox and FPV M). No product was seen for an extensively passaged and attenuated virus FP9. (b) Organization (to scale) of the H9 locus (FWPV genes fpv241 to fpv242), including the 1200 bp product, which is deleted from the vaccines, shown for FWPV HP1. The positions of primers used for the PCR are shown. Reproduced from Jarmin SA, Manvell R, Gough RE, Laidlaw SM, and Skinner MA (2006) Retention of 1.2 kbp of 'novel' genomic sequence in two European field isolates and some vaccine strains of fowlpox virus extends open reading frame fpv241, *Journal of General Virology* 87: 3545–3549 with permission from the Society for General Microbiology.

is propagated in cell culture or in embryonated eggs. Inoculation is into the wing web using a bifurcated needle or by scarification of the thigh. The formation of a small lesion at the site of vaccination 6–8 days later indicates a good take, with immunity developing in 8–15 days. Though desirable, alternative routes for vaccination (aerosol, intranasal, or via drinking water) have not generally proved successful.

Treatment

There are no demonstrated effective treatments for avipoxvirus infections other than palliative treatment and to prevent or treat secondary bacterial infections.

Variant FWPVs

Antigenic Variants

Outbreaks of fowlpox have been reported within flocks previously vaccinated with FWPV or pigeonpox virus vaccines. If the vaccination had been effective, which is often not the case, antigenic variation might be suspected, though serotype specificity would be novel among poxviruses. Little is known about protective epitopes for humoral or cellular immune responses in FWPV and there has only been preliminary characterization of the 'variant' isolates.

Tumorigenic Variants

FWPV has also been associated with cases of dermal squamous cell carcinoma in poultry in Brazil. It is not clear whether FWPV is a causative agent and, if so, whether a variant is involved.

The Role of Integrated Reticuloendotheliosis Virus Sequences

FPV-S, a commercial FWPV vaccine known to be contaminated with reticuloendotheliosis virus (REV), was found to carry a near full-length, infectious progenome of REV integrated into the poxvirus genome. The provirus has subsequently been found in most, if not all, pathogenic field isolates of FWPV, but the majority of vaccine strains of FWPV carry only noninfectious, long terminal repeat (LTR) sequences of REV. The provirus and the LTR sequences have only ever been found at the same single locus, although differences exist between the REV LTR sequences retained in different viruses. It appears likely that a single, ancestral event inserted the provirus into the FWPV genome between genes fpv201 and fpv203, in contrast to the multiple REV insertions that have been observed in Marek's disease virus. It is not known whether REV can reintegrate at the site of a retained LTR or whether this would regenerate a pathogenic virus. There are no REV sequences in the completely sequenced genome of CNPV, but they have been detected in a commercial CNPV vaccine as well as in some commercial pigeonpox virus vaccines. Full-length REV sequences have been detected in an isolate from a turkey.

Recombinant Avipoxvirus Vaccines

Recombinant FWPV as a Vaccine Vector in Poultry

Shortly after the development of methods for isolating recombinant VACV in the early 1980s, FWPV was developed as an equivalent recombinant vector for use in poultry. Several commercial vaccine and laboratory attenuated strains were used as vectors against a number of important poultry pathogens, especially avian influenza virus, Newcastle disease virus, infectious bronchitis virus, avian hemorrhagic enteritis virus, Marek's disease virus, turkey rhinotracheitis virus, REV, and infectious bursal disease virus, as well as *Mycoplasma gallisepticum*. Commercial recombinant FWPV vaccines against Newcastle disease virus and avian influenza virus have been licensed for commercial use in the USA. Those against avian influenza have also been licensed for use in Mexico; indeed, between 1997 and 2003, approximately 459 million doses of a recombinant fowlpox-H5 vaccine were used in Mexico as part of a program to control H5N2. The same recombinant and similar viruses developed in China are being used in Southeast Asia to counter the highly virulent avian influenza H5N1 strain.

Recombinant Avipoxviruses as Vaccine Vectors in Mammals

It was demonstrated in the late 1980s that recombinant avipoxviruses (initially FWPV and then CNPV) carrying antigens from mammalian viral pathogens could enter mammalian cells and express the foreign antigens, even though the recombinants could not replicate and spread to neighboring cells. Moreover, vaccination of mammals with the recombinants could elicit immune responses, which could be protective against viral challenge. This discovery, which was surprising because of the low level of antigen expressed from the single-round, abortive infections, offered the prospect of extremely safe recombinant vaccines. Several recombinant avipoxvirus vaccines, mainly based on CNPV, have been developed primarily for veterinary use. However, the more recent focus on T-cell-mediated immunity has seen more use of FWPV-based vectors. Many clinical trials have been conducted, are underway, or are planned, against viral diseases such as acquired immune deficiency syndrome (AIDS), parasitic diseases including malaria, and various cancers. In many of these trials, the recombinant avipoxviruses are used in combined (so-called 'prime-boost') regimes with the same antigens expressed by different vectors, such as DNA plasmids, the host-restricted MVA variant of VACV, or adenoviruses. These regimes allow immune responses to be boosted by multiple vaccinations without eliciting excessive responses against vector-specific proteins.

Recombinant avipoxvirus vaccines are yet to be licensed for agricultural use in Europe, though recombinant CNPV vaccines against equine influenza and feline leukemia have received European Medicines Agency (EMEA) approval. Both are licensed for use in the USA, besides similar recombinants against canine distemper and West Nile virus (for use in horses).

See also: Cowpox Virus; Entomopoxviruses; Herpesviruses of Birds; Leporipoviruses and Suipoxviruses; Molluscum Contagiosum Virus; Mousepox and Rabbitpox Viruses; Parapoxviruses.

Further Reading

Boulanger D, Smith T, and Skinner MA (2000) Morphogenesis and release of fowlpox virus. *Journal of General Virology* 81: 675–687.

Hansen W (1999) Avian pox. In: Friend M and Franson JC (eds.) *Field Manual of Wildlife Diseases, General Field Procedures and Diseases of Birds*, pp. 163–169. Madison, WI: US Geological Survey.

OIE (World Organisation for Animal Health) (2004) Page on Fowlpox diagnostic tests and vaccines from Manual of Diagnostic Tests and Vaccines for Terrestrial Animals (5th edn.) http://www.oie.int/eng/normes/mmanual/A_00113.htm (updated 23 July 2004).

Jarmin S, Manvell R, Gough RE, Laidlaw SM, and Skinner MA (2006) Avipoxvirus phylogenetics: Identification of a PCR length polymorphism that discriminates between the major clades. *Journal of General Virology* 87: 2191–2201.

Jarmin SA, Manvell R, Gough RE, Laidlaw SM, and Skinner MA (2006) Retention of 1.2 kbp of 'novel' genomic sequence in two European field isolates and some vaccine strains of fowlpox virus extends open reading frame fpv241. *Journal of General Virology* 87: 3545–3549.

Skinner MA, Laidlaw SM, Eldaghayes I, Kaiser P, and Cottingham MG (2005) Fowlpox virus as a recombinant vaccine vector for use in mammals and poultry. *Expert Review of Vaccines* 4: 63–76.

Skinner MA (2007) *Poxviridae*. In: Pattison M, McMullin P, Bradbury J, and Alexander D (eds.) *Poultry Diseases*, 6th edn. New York: Elsevier.

Tripathy DK and Reed WM (2003) Pox. In: Saif YM, Barnes HJ, Fadly A, Glisson JR, McDougald LR and Swayne DE (eds.) *Diseases of Poultry*, 11th edn., pp. 253–265. Ames, IA: Iowa State University Press.

Fungal Viruses

S A Ghabrial, University of Kentucky, Lexington, KY, USA
N Suzuki, Okayama University, Okayama, Japan

© 2008 Elsevier Ltd. All rights reserved.

Glossary

Hyphal anastomosis The union of a hypha with another resulting in cytoplasmic exchange.
Hypovirulence Attenuated fungal virulence mediated by virus infection, mitochondrial defects, or mutations in the fungal genome.
Mycoviruses Viruses that infect and multiply in fungi.
Vegetative incompatibility A genetically controlled self/nonself recognition system in fungi that determines the ability to undergo hyphal anastomosis.

Introduction

Relative to plant virology, animal virology, or bacterial virology, fungal virology is new. In 1948, an economically important disease of the cultivated mushroom, *Agaricus bisporus*, characterized by malformed fruiting bodies and serious yield loss, was first reported in a mushroom house owned by the La France Brothers of Pennsylvania. The disease was called 'La France disease', and similar diseases were reported shortly thereafter from Europe, Japan, and Australia. Different designations, such as 'X-disease', 'watery stripe', 'brown disease', and 'dieback' were given to basically the same disease as the La France. The significance of the 1948 report lies in the fact that it led to the discovery of fungal viruses. In 1962 Hollings noted the presence of at least three types of virus particles in diseased mushroom sporophores. This was the first report of virus particles in association with a fungus and is regarded as the dawn of mycovirology. The subsequent discovery that dsRNA of mycoviral origin was responsible for interferon-inducing activities of cultural filtrates of several species of *Penicillium* spp. greatly stimulated the search for fungal viruses.

Fungi, like other living organisms, can be infected by a number of viruses, and mycoviruses are found in all the major groups of fungi. Although mycoviruses are widely prevalent, only those infecting a limited number of fungal host species have been studied, for example, the yeast *Saccharomyces cerevisiae*, edible mushroom, and phytopathogenic fungi. Given the predicted vast number of fungal species (approximately 10 000 known species and many more unknown species), it is expected that a greater number of unrecognized mycoviruses occur in nature. Support for this idea comes from recent extensive searches of field fungal isolates that showed relatively high frequencies of virus infection, for example, approximately 65%, 20%, and 2–28% of *Helicobasidium mompa*, *Rosellinia necatrix*, and *Cryphonectria parasitica* isolates were found to be infected, respectively.

Biological Properties

Host Range

The natural host range of mycoviruses is likely to be restricted to the same or closely related vegetative compatibility groups that allow lateral transmission. Until recently, there were no known experimental host ranges for fungal viruses because of lack of suitable infectivity assays. Experimental host ranges for some mycoviruses, however, were recently demonstrated and shown to extend to different vegetative compatibility groups and even to different genera. For example, the prototype mycoreovirus (mycoreovirus 1) can replicate and induce phenotypic alterations in different vegetative compatibility groups of the chestnut blight fungus, *Cryphonectria parasitica*, similar to those exhibited by the original virus-containing strain. Furthermore, CHV1-EP713, the type member of the family *Hypoviridae*, can replicate in and confer hypovirulence onto members of several other fungal genera, for example, *Endothia gyrosa* and *Valsa ceratosperma*, in addition to its natural host, *C. parasitica*.

Symptom Expression

Although the majority of the viruses that infect phytopathogenic fungi have been reported to be avirulent, phenotypic consequences of infections with mycoviruses can vary from symptomless to severely debilitating, and from hypovirulence to hypervirulence. Mycoviruses that attenuate the virulence of plant pathogenic fungi provide excellent model systems for basic studies on development of novel biological control measures and for dissecting the mechanisms underlying fungal pathogenesis. In general, infections due to mycoviruses are both symptomless and persistent. Latency benefits the host for survival, and persistence helps the virus in the absence of extracellular modes of transmission. To ensure their retention, some mycoviruses have evolved to bestow selective advantage to their host (e.g., the killer phenotypes in yeasts and smuts). Because of their ability to secrete killer toxins, yeast killer strains have been utilized by the brewing industry to provide protection against contamination with adventitious sensitive strains. The genes encoding the smut killer toxins have been used for the development of novel transgenic approaches to control the corn smut.

Macroscopic symptoms caused by fungal viruses are a consequence of alterations in complex physiological processes that involve interactions between host and virus factors. Several virally encoded proteins are identified as symptom determinants including the papain-like protease, p29, of some hypoviruses. This protein acts to repress host pigmentation and conidiation regardless of whether it is expressed from host chromosomes or the homologous virus genome. Another interesting example is the proteins encoded by the totivirus *Helminthosporium victoriae* 190S virus. Co-expression of the capsid and RNA-dependent RNA polymerase (RdRp) proteins results in empty capsid production and phenotypic changes similar to those induced in virus-infected isolates, suggesting that viral replication is not required for symptom development. Host factors involved in symptom expression are not well studied except for host genes involved in the killer phenomenon in yeast infected with the totivirus L-A and associated satellite dsRNAs. Transcriptome analysis was performed for only limited virus/natural host combinations. Of the 2200 *C. parasitica* genes, 13.4% are either upregulated or downregulated upon infection with the prototype severe strain of the hypovirus CHV1, while only 7.5% are altered in their transcription levels by infection with a mild strain of CHV1. One-half of the genes responsive to the infection with the latter virus are commonly altered in transcription by the severe strain, which generally causes greater magnitude of transcriptional changes.

Transmission

Mycoviruses lack an extracellular phase to their life cycles. They are transmitted intracellularly during cell division, sporogenesis, and cell fusion. Lateral (horizontal) transmission usually occurs only between individuals within the same species, which belong to the same or closely related vegetative compatibility groups. Vegetative compatibility is governed genetically. Mycoviruses may be eliminated during sexual spore formation. Although the totiviruses and narnaviruses that infect the yeasts are effectively transmitted via ascospores, the mycoviruses infecting the ascomycetous filamentous fungi are eliminated during ascospore formation. Whereas the ssRNA and dsRNA mushroom viruses are transmitted efficiently via basidiospores, the virus-containing strains of the basidiomycete *H. mompa* are cured during the sexual sporulation processes. Therefore, whether a mycovirus is transmitted through sexual spores depends on the host/virus combination involved. Whereas mycovirus transmission through asexual spores occurs frequently, its rate varies greatly depending on the combination of viral and host strains. It was recently shown that the papain-like protease p29, encoded by CHV1, might play a role in virus transmission since it enhances the transmission of the homologous virus (CHV1) as well as heterologous viruses (mycoreoviruses) in *C. parasitica* conidia.

While rare in nature, interspecies transmission has been reported between members within the same genus including *Cryphonectria*, *Sclerotinia*, and *Ophiostoma* that share the same habitats. It remains unknown whether interspecies barrier is overcome by physical contacts or by vectors. Experimental evidence for vector transmission of fungal viruses, however, is lacking.

Mixed Infections

Mixed infections with two or more unrelated viruses and accumulation of defective dsRNA and/or satellite dsRNA are common features of mycovirus infections. Examples include the sphaeropsis sapinea totiviruses SsRV1 and SsRV2, saccharomyces cerevisiae totiviruses ScV-L-A and ScV-LB/C, and penicillium stoloniferum partitiviruses PsV-S and PsV-F.

There are a number of known examples of mixed infections with plant or animal viruses where one virus either interferes with or enhances the replication of the other one. As a consequence, reduction or increase in symptom severity may arise. Recent extensive searches for fungal viruses confirmed relatively high mixed infection rates in some phytopathogenic fungi like *C. parasitica*, *R. necatrix*, and *H. mompa*. In contrast to mixed virus infections in plants or animals, possible interactions between co-infecting mycoviruses in single hosts is little studied because of the limitation in experimental manipulation of viruses and hosts. Recent studies suggest synergistic interactions between a hypovirus (phylogenetically related to potyviruses and viruses belonging to the picorna-like superfamily) and a mycoreovirus through a hypovirally encoded protein. In this case, one-way synergism is observed in that only hypovirus infection transactivates the replication of a mycoreovirus, which has a replication strategy distinct from that of hypoviruses.

Fungal Virus Taxonomy

A list of the virus families and genera into which mycoviruses are classified is included in **Table 1**. Members of some fungal virus families, for example, the families *Narnaviridae*, *Chrysoviridae*, and *Hypoviridae*, infect only fungi, while members in other families, for example, the families *Metaviridae*, *Pseudoviridae*, *Reoviridae*, *Totiviridae*, and *Partitiviridae*, infect fungi, protozoa, plants, or animals. Except for the rhizidiomyces virus with dsDNA genome (the only member in the genus *Rhizidiovirus*), almost all other mycoviruses have RNA genomes and many have dsRNA genomes. Viruses with (−)-strand RNA or ssDNA genomes have yet to be found in fungi.

dsRNA Mycoviruses

Mycoviruses with dsRNA genomes represent the majority of the fungal viruses so far reported. With the exception of the mycoreviruses (genus *Mycoreovirus*, family *Reoviridae*), which have spherical double-shelled particles 80 nm in diameter, the dsRNA mycoviruses are typically isometric particles 25–50 nm in diameter. They are classified, based on the number of genome segments, into three families, *Totiviridae*, *Partitiviridae*, and *Chrysoviridae*. Viruses in the family *Totiviridae* have nonsegmented dsRNA genomes coding for a CP and an RdRp. At present, three genera have been placed in this family: *Totivirus*, *Giardiavirus*, and *Leishmaniavirus*. Viruses in the genus *Totivirus* infect fungi, whereas those belonging to the latter two genera infect parasitic protozoa. At least two distinct RdRp expression strategies have been reported for totiviruses: (1) those that express their RdRp as a fusion protein (CP-RdRp or Gag-Pol) by ribosomal frameshifting, such as the yeast L-A and the viruses that infect parasitic protozoa and (2) those that synthesize RdRp as a separate nonfused protein by an internal initiation mechanism (e.g., a coupled termination–reinitiation mechanism), as proposed for Hv190SV and others that infect filamentous fungi. Phylogenetic analysis of CP or RdRp sequences of totiviruses reflects these differences, and separate phylogenetic clusters can be generated. Hv190SV and other totiviruses that infect filamentous fungi are closer to each other than to viruses infecting yeast, smut fungi, and protozoa. The fact that independent alignments of CP and RdRp sequences give similar phylogenetic relationships supports the conclusion that totiviruses infecting filamentous fungi should reside in a genus of their own. The genus *Victorivirus* has been proposed to include the totiviruses that infect filamentous fungi with Hv190SV, as the type species (**Table 1**). The genomes of partitiviruses and chrysoviruses consist of two and four segments, respectively.

The unclassified dsRNA mycovirus agaricus bisporus virus 1 (AbV1), also designated La France isometric virus, causes a serious disease of cultivated mushroom (named La France disease). AbV1 is of special interest because of its historical and economic importance. The AbV1 virions, isolated from diseased fruit bodies and mycelia, are isometric 36 nm in diameter and co-purify with nine dsRNA segments (referred to as disease-associated dsRNAs). The size of dsRNA segments varies from 3.6 to 0.78 kbp, three of which are believed to be satellites. It is not clear at present whether the nine dsRNA segments are encapsidated individually, in various combinations, or all nine segments are packaged in single particles. Based on

Table 1 List of viral families and genera into which fungal viruses are classified

Family/genus/virus species[a]	Virus abbreviation	No. of segments	Accession number
Double-stranded DNA genome			
Unassigned			
Rhizidiovirus			
Rhizidiomyces virus	RhiV	1	
RNA reverse transcribing genome			
Pseudoviridae			
Pseudovirus			
Saccharomyces cerevisiae Ty1 virus	SceTy1V	1	M18706
Hemivirus			
Saccharomyces paradoxus Ty5 virus*	SceTy5V	1	U19263
Metaviridae			
Metavirus			
Saccharomyces cerevisiae Ty3 virus*	SceTy3V	1	M34549
Double-stranded RNA genome			
Reoviridae			
Mycoreovirus			
Cryphonectria parasitica mycoreovirus 1 (9B21)	MyRv1–9B21	11	AY277888; AY277889; AY277890; AB179636; AB179637; AB179638; AB179639; AB179640; AB179641; AB179642; AB179643
Totiviridae			
Totivirus			
Saccharomyces cerevisiae virus L-A (L1)	Sc V-L-A	1	J04692; X13426
Victorivirus[b]			
Helminthosporium victoriae 190SV	Hv190SV	1	U41345
Partitiviridae			
Partitivirus			
Atkinsonella hypoxylon virus	AhV	2	L39125; L39126; L39127
Chrysoviridae			
Chrysovirus			
Penicillium chrysogenum virus	PcV	4	AF296439; AF296440; AF296441; AF296442
Single-stranded (+) RNA genome			
Narnaviridae			
Narnavirus			
Saccharomyces 20S narnavirus	ScNV-20S	1	M63893
Mitovirus			
Cryphonectria mitovirus 1	CMV1	1	L31849
Barnaviridae			
Barnavirus			
Mushroom bacilliform virus	MBV	1	
Hypoviridae[c]			
Hypovirus			
Cryphonectria hypovirus 1-EP713	CHV-1/EP713	1	M57938
Endornavirus[c]			
Phytophthora endornavirus 1*	PEV1	1	AJ877914
Unassigned			
Botrytis virus F	BVF	1	AF238884
Sclerotinia sclerotiorum debilitation-associated RNA virus	SsDRV	1	AY147260
Diaporthe RNA virus	DRV	1	AF142094
Sclerophthora macrospora virus A	SmV A	2	AB083060; AB083061
Sclerophthola macrospora virus B	SmV B	1	AB012756

[a]The type species of the specified genus are listed. Otherwise, a fungal virus member of the genus is listed and marked with an asterisk.
[b]Proposed new genus in the family *Totiviridae*.
[c]Grouped with dsRNA viruses in the Eighth Report of ICTV.

the size of the particles, cesium sulfate gradient profile, and results of dsRNA and protein analyses of the gradient fractions, it is highly unlikely that all dsRNAs are packaged together in single particles. More realistically, AbV1 represents a multiparticle system in which the various particle classes have similar densities. Interestingly, phylogenetic analysis of the conserved motifs of AbV1 RdRp, encoded by dsRNA segment 1, and other dsRNA mycoviruses showed that AbV1 is closely related to the multipartite chrysoviruses.

ssRNA Viruses

There are a number of mycoviruses with apparent ssRNA genomes that do not code for capsid proteins and exist more or less predominantly as dsRNA 'replicative' forms in their hosts. Because of lack of true virions, these viruses were easier to isolate and study as their dsRNA forms, and some were grouped with dsRNA viruses (e.g., family *Hypoviridae*). However, there is ample evidence at present that many of these viruses replicate and express their genomes like (+)-strand RNA viruses and that the lineage of their RdRp and helicase genes are within the lineages of (+)-strand RNA viruses. The simplest types of these viruses include members of the genera *Narnavirus* and *Mitovirus* (family *Narnaviridae*), whose RNA genomes code only for RdRp and the viruses exist as RNA/RdRp nucleoprotein complexes. The corresponding dsRNAs can be isolated from infected tissues, usually in lesser molar amounts than the genomic ssRNA. Phylogenetic analysis of RdRps of members of the family *Narnaviridae* along with those of other fungal viruses and related taxa indicate a distant relationship between members of the family *Narnaviridae* and bacteriophages belonging to the family *Leviviridae*.

Lack of true virions is characteristic of two other groups of classified viruses, those belonging to the family *Hypoviridae* and the genus *Endornavirus*. Although these viruses were grouped with dsRNA viruses in the latest ICTV Report, their genome organization and expression strategies are indicative of (+)-strand RNA viruses. Viruses in the family *Hypoviridae* are phylogenetically related to the (+)-strand RNA viruses in the family *Potyviridae* (picorna-like virus supergroup). Comparisons of hypovirus-conserved motifs of RdRp, helicase, and protease with those of members of the family *Potyviridae* suggest that viruses in the genus *Bymovirus* are the closest relatives to hypoviruses. Phytophthora endornavirus (PEV1) is the only nonplant virus in the genus *Endornavirus* of plant viruses. Although *Phytophthora* species and other members of the class Oomycetes have many biological properties in common with fungi, they are currently classified, based on sequence similarities, in a protist group known as the Stramenopiles. Endornaviruses are believed to have evolved from an alpha-like virus that has lost its capsid gene. This is consistent with the recent finding that RdRps of PEV1 and other endornaviruses cluster with those of families and genera in the alpha-like virus superfamily of (+)-strand RNA viruses.

The mushroom bacilliform virus (MBV; genus *Barnavirus*, family *Barnaviridae*) is the only mycovirus known to have bacilliform virions. MBV has a (+)-strand RNA genome that contains seven open reading frames (ORFs), three of which encode a putative chymotrypsin-like serine protease, a putative RdRp, and a CP. The polypeptides encoded by the remaining four ORFs have no homology to known proteins. Amino-acid sequence comparisons of the putative protease and RdRp suggest that MBV is evolutionarily related to sobemoviruses and poleroviruses. Although double infections of cultivated mushrooms with MBV and ABV1 are of common occurrence, the role of MBV in the ensuing dieback disease of cultivated mushroom remains unknown.

Unassigned ssRNA Viruses

It is worth noting that a relatively large number of mycoviruses remain unassigned including some well-characterized ones like botrytis virus F (BVF), sclerotinia sclerotiorum debilitation-associated RNA virus (SsDRV), diaporthe RNA virus (DRV), and sclerophthora macrospora viruses A and B (SmV A and SmV B). BVF has flexuous rod-shaped particles comparable in size and morphology to ssRNA plant 'potex-like' viruses. Amino-acid sequence identities of the conserved helicase and RdRp regions and the coat protein genes are greatest to those of potex-like viruses. The main difference between BVF and these plant viruses is the lack of a movement protein. Although particle morphology along with amino-acid sequence similarities of both the replicase and coat protein genes support the classification of BVF in the family *Flexiviridae*, it is obvious that the mycovirus BVF is distinct enough to belong to a new genus in this family. It is proposed that a new genus tentatively designated '*Mycoflexivirus*' is created to include BVF (**Table 1**). The genome of SsDRV contains a single ORF encoding a protein with significant sequence similarity to the replicases of the 'alphavirus-like' supergroup of (+)-strand RNA viruses. The SsDRV-encoded putative replicase protein contains the conserved methyl transferase, helicase, and RdRp domains characteristic of the replicases of potex-like plant viruses (flexiviruses) and BVF. Although phylogenetic analysis of the conserved RdRp motifs verified that SsDRV is closely related to BVF and to the allexiviruses in the family *Flexiviridae*, SsDRV is distinct enough from these viruses, mainly based on the lack of coat protein and movement protein, to justify the creation of yet another genus in the family *Flexiviridae*.

DRV is another naked RNA mycovirus that is associated with hypovirulence of its fungal host. It has two large ORFs present in the same reading frame, which are most likely translated by readthrough of a UAG stop codon in the central part of the genome. The longest possible translation product has a predicted molecular mass of about 125 kDa, which shows significant homology to the nonstructural proteins of carmoviruses of the (+)-strand RNA virus family *Tombusviridae*. Interestingly, transcripts derived from full-length cDNA clones were infectious when inoculated to spheroplasts and the transfected isolates exhibited phenotypic traits similar to the naturally infected isolate.

Sclerophthora macrospora virus A (SmV A) found in *S. macrospora*, the pathogenic fungus responsible for downy mildew of gramineous plants, is a small icosahedral virus containing three segments of (+) ssRNAs (RNAs 1, 2, and 3). Whereas RNA 1 contains the RdRp motifs, RNA 2 codes for a capsid protein. RNA 3 is a satellite RNA. Whereas the deduced amino acid sequence of RdRp shows some similarity to RdRps of members of the family *Nodaviridae*, the amino acid sequence of the viral CP shows similarity to those of members in the family *Tombusviridae*. The capsid of SmV A is composed of two capsid proteins, CP 1 and CP 2, both encoded in ORF2. CP 2 is apparently derived from CP 1 via proteolytic cleavage at the N-terminus. The genome organization of SmV A is distinct from those of other known fungal RNA viruses, and suggests that SmV A should be classified into a new genus of mycoviruses.

Sclerophthora macrospora virus B (SmV B), which is also found in *S. macrospora*, has small icosahedral, monopartite virions containing a (+) ssRNA genome. The viral genome has two large ORFs: ORF1 encodes a putative polyprotein containing the motifs of chymotrypsin-related serine protease, and ORF2 encodes a capsid protein. The genome arrangement of SmV B is similar to those belonging to the genera *Sobemovirus*, *Barnavirus*, and *Polerovirus*. The putative domains for the serine protease, VPg, RdRp, and the CP are located in this order from the $5'$ terminus to the $3'$ terminus. SmV B, however, is distinctive since its genome has only two ORFs. The genome organization of the barnavirus MBV, on the other hand, resembles that of poleroviruses. These results suggest that SmV B, like SmV A, should also be classified into a new genus of mycoviruses.

The mycoviruses belonging to the families *Pseudoviridae*, *Metaviridae*, *Reoviridae* (genus *Mycoreovirus*), *Totiviridae*, *Partitiviridae*, *Chrysoviridae*, *Narnaviridae*, and *Barnaviridae* are discussed in more detail elsewhere in this encyclopaedia.

Replication and Gene Expression Strategy

Replication cycles of fungal viruses are not well studied except for a few cases including the *Saccharomyces cerevisiae* L-A virus. For dsRNA fungal viruses including members of the *Totiviridae*, *Partitiviridae*, *Reoviridae*, and *Chrysoviridae*, virus particles or subviral particles, containing RdRp, are believed to play pivotal roles in RNA transcription and replication. Replication of naked RNA mycoviruses represented by members of the family *Hypoviridae* may occur in infection-specific, lipid-membranous vesicles presumed to contain viral RdRp and RNA helicase. These vesicles are able to synthesize *in vitro* both plus and minus RNA at a ratio of 1:8. The narnavirus RNA, encoding only a single protein (RdRp), is associated with RdRp rather than being encapsidated. These RNA/RdRp complexes seem to play a key role in RNA replication.

Fungal viruses, like many RNA viruses of plants and animals, employ noncanonical translational strategies for expressing their genomes. These include −1 (*Totiviridae*) and +1 frameshifting (*Totiviridae*, *Pseudoviridae*, *Metaviridae*), termination-coupled initiation (*Totiviridae*, *Hypoviridae*), and IRES-mediated initiation (*Totiviridae*). Furthermore, a readthrough of a termination codon strategy is proposed for translation of the $3'$ proximal ORF of DRV. A noncanonical mechanism may also be required for efficient translation of mRNA of narnaviruses, which lack poly(A) tails. The (CAA)n repeats found at the $5'$ UTR of chrysoviruses are implicated in translation augmentation, as observed for the $5'$ UTR sequence of tobacco mosaic virus. Translation of the viral genes that are regulated by these mechanisms is considered critical for virus viability.

Recent Technical Advances in Fungal Virology

Fungal virology has been thwarted by many constraints on manipulation of fungal viruses. Many, if not all, plant and animal viruses can be inoculated into individuals/tissue cultures of plant and animal hosts. Some assays with those hosts allow quantitative detection of biologically active viruses. As for mycoviruses, it is rather rare to be able to inoculate into host fungi because of experimental limitations, which often makes the etiology of mycoviruses difficult to establish. Fungal cells have rigid cell walls and are usually difficult to digest for preparation of cell-wall-free protoplasts ready for transformation or transfection. Even if protoplasts are made, their maintenance like animal cell culturing is not possible. Furthermore, fungal hosts usually have self/nonself recognition systems operating at inter- and intraspecies levels. Intraspecies barriers are based on vegetative incompatibility/compatibility that is governed genetically. This is often regarded as one of the host defense barriers that inhibits virus transfer between individuals. To overcome these barriers, a few methods are available. The prototypic hypovirus cryphonectria parasitica hypovirus 1 (CHV1) is the first for which a reverse genetics is established. Infection with different CHV1 strains can be launched either from cDNA integrated into host chromosomes or *in vitro*-synthesized RNA viral cDNA. It is noteworthy that via bombardment of mycelia, not protoplasts, the infectious CHV1 cDNA clone can be integrated into chromosomes of fungi other than the natural host *C. parasitica*. cDNA-based transfection systems are now available for three other species of RNA viruses: *Diaporthe RNA virus*, *Saccharomyces cerevisiae 20S narnavirus*, and *Saccharomyces cerevisiae 23S narnavirus*.

Mycoviruses in general lack infectivity as purified virions. However, all members of the new genus *Mycoreovirus* including mycoreovirus 1 (MyRV1), MyRV2, and MyRV3 were found to be infectious as purified particles when applied to fungal protoplasts. It is of interest in this regard that treatment of purified virions with trypsin or chymotrypsin was not required for infectivity. Protoplast fusion provides an alternative approach to introduce mycoviruses into vegetatively incompatible fungal strains that are incapable of hyphal anastomosis. Intra- and interspecies virus transfer via protoplast fusion has been reported in *Aspergillus* spp. This method is particularly useful for viruses for which infectious particles or cDNA-derived RNA are unavailable. Another recent revelation is that monokaryotic strains are able to serve as an intermediate virus transmitter between different mycelial incompatibility groups within the same species of *R. necatrix*.

To complete Koch's postulates, virus curing is as important as virus inoculation, because the virus must be back inoculated into an isogenic, virus-free strain with the same genetic background as the original virus-infected strain. Virus-free isolates may be obtained from germlings of asexual spores if virus transmission through spores is less than 100%. Alternatively, virus-free strains may be isolated by hyphal tip culturing, as in the case for *H. mompa* and *R. necatrix*. This technique is applicable to fungi infected with a virus that is transmitted to 100% of the asexual spores or for fungi that produce little or no spores.

Future Perspectives

Yeast as a (Model) Host to Study Viral Replication

S. cerevisiae has provided an excellent system to investigate virus assembly and replication of the dsRNA totivirus L-A and the ssRNA narnaviruses that infect yeast. With the robust yeast genetics, a number of host factors involved in totivirus replication were identified, many of which are related to translation events. Furthermore, the yeast provides an 'artificial' viral host model system to explore host genes affecting viral replication on a genomewide basis. Genetic screens of a collection of 4500–4800 single-gene deletion yeast strains (Yeast Knockout strain collection) have been successfully conducted for identifying host factors involved in replication and recombination with two different plant ssRNA viruses: brome mosaic virus and tomato bushy stunt virus (TBSV). Each screen led to the identification of approximately 100 host genes (approximately 1.8% of the entire yeast genes) that affect virus replication. Interestingly, the replication of the two viruses in yeast is affected by a different set of genes. A similar approach was employed to identify genes affecting the recombination of TBSV. This type of use of the yeast system can be expanded to a (+)-strand RNA animal virus and a (−)-strand RNA vertebrate virus. It may be surprising that the yeast has yet to be used for viruses infecting filamentous fungi. The yeast should be able to serve as a model host for mycoviruses other than those that naturally infect yeast.

Host Defense against Fungal Viruses

Host defense responses against viruses have not been explored intensively. RNA silencing is regarded as one of the host defense strategies of eukaryotes to molecular parasites including viruses, and operates in a number of fungi including important model and phytopathogenic fungi. However, no direct evidence is shown for RNA silencing that target mycoviral RNA. Recently, the hypovirus p29 was shown to be a suppressor of RNA silencing targeting a transgene that functions in both plant and fungal cells. This suggests that RNA silencing may function as an antiviral mechanism in fungal cells. The observation that p29 enhances the replication of a heterologous virus supports this idea. There are genetic elements involved in RNA silencing, for example, that are conserved widely from fungi to vertebrates. Functional roles of these factors in RNA silencing as antiviral reactions will need to be elucidated. Unraveling the mechanism by which the hypovirus p29, or other mycovirus-encoded RNA silencing suppressors, may block the RNA silencing pathway will be an interesting challenge.

Role of Mycoviruses in Plant–Fungal Mutualistic Associations

The question of whether mycoviruses are involved in the mutualistic interactions between endophytic fungi and their host plants is of considerable interest because of the attractive beneficial features of these associations and because of the common occurrences of fungal viruses in all major groups of fungi. This question was recently addressed in an intriguing report that presented evidence for a dsRNA mycovirus being involved in the mutualstic interaction between a fungal endophyte (*Curvularia protuberata*) and a tropical panic grass. This association allows both organisms to grow at high soil temperatures. The virus in question, which was designated curvularia thermal tolerance virus (CThTV), has unusual genome organization with an unknown genome expression strategy. CThTV has apparently a bipartite genome (RNA 1 and RNA 2), but no evidence that these RNAs are packaged in the 27-nm isometric particles isolated from the fungal host. Although many questions pertinent to the fungal endophyte, CThTV, and the veracity of the evidence for viral etiology remain unanswered, this report will undoubtedly stimulate the search for mycoviruses in other mutualstic fungal endophytes. In this regard, it is noteworthy that the well-characterized mutualistic endophyte, *Epichloë festucae*, was found to harbor a totivirus, but no phenotypes were associated with virus infection.

Mycovirus as Biocontrol Agents and as Tools for Fundamental Studies

The hypovirulence phenotype in the chestnut blight fungus (*Cryphonectria parasitica*) is an excellent and well-documented example for a mycoviral-induced phenotype that is currently being exploited for biological control. The debilitating disease of *Helminthosporium victoriae*, the causal agent of Victoria blight of oats, and the disease phenotype of the Dutch elm disease fungus *Ophiostoma novo-ulmi* are examples of pathogenic effects of dsRNA fungal viruses. An understanding of the molecular basis of disease in these fungal-virus systems would provide excellent opportunities for development of novel biocontol strategies of plant pathogenic fungi. Mycoviruses also continue to serve as versatile tools to study the virulence of host fungi, as recognized in studies with the hypovirus/ *Cryphonectria parasitica* system, in which substantial advances in our understanding of the molecular basis of hypovirulence have been made.

See also: Barnaviruses; Hypovirulence; Hypoviruses; Metaviruses; Mycoreoviruses; Narnaviruses; Partitiviruses of Fungi; Partitiviruses: General Features; Pseudoviruses; Ustilago Maydis Viruses; Viral Killer Toxins; Yeast L-A Virus.

Further Reading

Buck KW (1986) Fungal virology-an overview. In: Buck KW (ed.) *Fungal Virology*, pp. 1–84. Boca Raton, FL: CRC Press.

Buck KW (1998) Molecular variability of viruses of fungi. In: Bridge PD, Couteaudier Y and Clarkson JM (eds.) *Molecular Variability of Fungal Pathogens*, pp. 53–72. Wallingford: CAB International.

Ghabrial SA (1994) New developments in fungal virology. *Advances in Virus Research* 43: 303–388.

Ghabrial SA (1998) Origin, adaptation and evolutionary pathways of fungal viruses. *Virus Genes* 16: 119–131.

Ghabrial SA (2001) Fungal viruses. In: Maloy O and Murray T (eds.) *Encyclopedia of Plant Pathology,* vol. 1, pp. 478–483. New York: Wiley.

Goodin MM, Schlagnhaufer B, and Romaine CP (1992) Encapsidation of the La France disease-specific double stranded RNAs in 36 nm isometric viruslike particles. *Phytopathology* 82: 285–290.

Hillman BI and Suzuki N (2004) Viruses in the chestnut blight fungus. *Advances in Virus Research* 63: 423–472.

Howitt RLJ, Beever RE, Pearson MN, and Forster RLS (2001) Genome characterization of Botrytis virus F, a flexuous rod-shaped mycovirus resembling plant 'potex-like' viruses. *Journal of General Virology* 82: 67–78.

Márquez LM, Redman RS, Rodriguez RJ, and Roossinck MJ (2007) A virus in a fungus in a plant: Three-way symbiosis required for thermal tolerance. *Science* 315: 513–515.

McCabe PM, Pfeiffer P, and Van Alfen NK (1999) The influence of dsRNA viruses on the biology of plant pathogenic fungi. *Trends in Microbiology* 7: 377–381.

Nuss DE (2005) Hypovirulence: Mycoviruses at the fungal–plant interface. *Nature Reviews Microbiology* 3: 632–642.

Sun L-Y, Nuss DL, and Suzuki N (2006) Synergism between a mycoreovirus and a hypovirus mediated by the papain-like protease p29 of the prototypic hypovirus CHV1-EP713. *Journal of General Virology* 87: 3703–3714.

Tavantzis S (ed.) (2001) *Fungal dsRNA Elements: Concepts and Application in Agriculture, Forestry and Medicine.* Boca Raton, FL: CRC Press.

Van der Lende TR, Duitman EH, Gunnewijk MGW, Yu L, and Wessels JGH (1996) Functional analysis of dsRNAs (L1, L3, L5, and M2) associated with isometric 34-nm virions of *Agaricus bisporus* (White Button Mushroom). *Virology* 217: 88–96.

Wickner RB (1996) Double-stranded RNA viruses of *Saccharomyces cerevisiae*. *Microbiological Reviews* 60: 250–265.

Furovirus

R Koenig, Institut für Pflanzenvirologie, Mikrobiologie und biologische Sicherheit, Brunswick, Germany

© 2008 Elsevier Ltd. All rights reserved.

Glossary

Furovirus Siglum derived from fungus-transmitted rod-shaped virus.

History

Soil-borne wheat mosaic virus (SBWMV) is the type species of the genus *Furovirus*. Originally it was classified as a possible member of the tobamovirus group, because its rod-shaped particles resemble those of the tobamoviruses. In 1991, the fungus-transmitted rod-shaped viruses which all have several genome segments were separated from the tobamoviruses with a monopartite genome to form a new group, named furovirus group, with SBWMV as the type member. Molecular studies performed in the following years revealed that the genome organization of many of these furoviruses greatly differed from that of SBWMV. Eventually four new genera were created, that is, the genus *Furovirus* with SBWMV as the type species, the genus *Benyvirus* with *Beet necrotic yellow vein virus*, the genus *Pomovirus* with *Potato mop-top virus*, and the genus *Pecluvirus* with *Peanut clump virus* as type species, respectively. The genus *Furovirus* presently comprises five species, that is, SBWMV, *Soil-borne cereal mosaic virus* (SBCMV), *Chinese wheat mosaic virus* (CWMV), *Oat golden stripe virus* (OGSV), and *Sorghum chlorotic spot virus* (SrCSV). SBCMV was simultaneously and independently described by two different working groups who had suggested the names European wheat mosaic virus (EWMV) and soil-borne rye mosaic virus (SBRMV), respectively. These names are no longer in use.

Host Ranges, Diseases, and Geographic Distribution

SBWMV has been known since the early 1920s. It causes mosaic, stunting, and severe losses of yield in winter wheat in the USA where it is widely distributed in the central parts. As with other soil-borne diseases the symptoms often occur in patches in the fields. SBWMV may also naturally infect barley. A deviating strain of SBWMV which differs from the type strain considerably in its nucleic acid sequences, but not in the proteins translated from them has been observed to be rapidly spreading in upper New York State since 1998. Diseases on wheat with symptoms similar to those described for SBWMV are caused by SBCMV in Europe, CWMV in China and by the distantly related Japanese strain (jap) of SBWMV in Japan. A virus closely related to SBWMV jap has recently been isolated from barley in France. SBWMV jap has also been obtained from barley in Japan. SBCMV is now widely distributed in Europe where it infects mainly wheat in Italy, France, and England, but mainly rye in Germany, Denmark, and Poland. SBWMV has been detected on a single field in Germany for the first time in 2002 with apparently no tendency to spread. OGSV has been detected in oats at various sites in Britain, France, and the USA (North Carolina). In oats it induces conspicuous chlorotic striping of leaves, but it fails to infect wheat. SBWMV, SBCMV, CWMV, and OGSV may be transmitted mechanically (sometimes only with difficulty) to some *Chenopodium* and *Nicotiana* species. SrCSV was isolated in 1986 from a single sorghum line from a breeder's plot in Kansas/USA. It is readily mechanically transmissible to maize where it produces a bright yellow mosaic and elongated ringspot symptoms several weeks after inoculation. Local infections are produced on mechanically inoculated *Chenopodium quinoa*, *C. amaranticolor*, and *N. clevelandii*. Attempts to transmit the virus back to *Sorghum bicolor* or to winter wheat either mechanically or by growing plants in soil in which infected sorghum was growing were unsuccessful.

Plants infected naturally by furoviruses are often also infected by bymoviruses, that is, wheat spindle streak mosaic virus in Europe and North America, wheat yellow mosaic virus in East Asia or oat mosaic virus in Europe and the USA. The symptoms caused by these bymoviruses are very similar to those caused by the furoviruses, and both the furoviruses and the bymoviruses are transmitted by *Polymyxa graminis* (see below).

Transmission in Nature and Long-Distance Movement in Infected Plants

Furoviruses, with the possible exception of SrCSV, are soil-borne and transmitted by the zoospores of *Polymyxa graminis*, a ubiquitous plamodiophorid protozoan formerly considered to belong to the fungi. *Polymyxa*-transmitted viruses (or their RNAs?) are taken up by the plasmodia of the vector in infected root cells, but there is no evidence that they multiply in the vector. The multinucleate plasmodia which are separated from the host cytoplasm by distinct cell walls may either develop into zoosporangia from which secondary viruliferous zoospores are released within a few days. Alternatively, the plasmodia may form cystosori which act as resting spores and may survive in the soil for many years even under extreme conditions. They may be distributed on agricultural equipment, by irrigation or even by wind blow. Upon germination they release primary zoospores transmitting the virus. Zoospores inject their contents into the cytoplasm of root cells where new plasmodia are formed. Zoospores treated with antisera to SBWMV or resting spores treated with 0.1 N NaOH or HCl retained their ability to transmit virus into plants indicating that the infectious viral material was carried inside. Immunolabeling and *in situ* RNA hybridization studies have revealed the presence of SBWMV movement protein and RNA but not of SBWMV coat protein in the resting spores of *P. graminis*. This might suggest that the vector does not transmit SBWMV in form of its particles but rather as a ribonucleoprotein complex possibly formed by the movement protein and the viral RNAs. The furovirus movement protein belongs to the '30 K' superfamily of movement proteins which are known to mediate cell-to-cell and vascular transport of viruses by binding viral nucleic acids and carrying them through plasmodesmata and through the vasculature. There is also strong evidence that two transmembrane regions found in the coat protein readthrough proteins of furoviruses are involved in the transmission process.

Immunogold-labeling studies have suggested that SBWMV uses the xylem in order to move from infected roots to the leaves. It may enter primary xylem elements before cell death occurs and then move upward in the plant after the xylem has matured into hollow vessels. There is also evidence for lateral movement between adjacent xylem vessels.

Control

Polymyxa-transmitted plant viruses may survive in soil for decades in the long-living resting spores of the vector. The diseases caused by these soil-borne viruses are, therefore, much more difficult to control than those caused, for instance, by insect-transmitted viruses. Chemical control, for example, by soil treatment with methyl bromide, is neither efficient nor acceptable for economic and ecological reasons. Growing resistant or tolerant varieties currently represent the only practical and environmentally friendly means to lower the impact of these diseases on yield. Immunity to the wheat-infecting furoviruses has not been found. The tolerant varieties which have been developed contain high virus levels in the root system. The partial resistance in some varieties may be overcome when the plants are grown at temperatures >23 °C. Cereal genotypes with resistance to *P. graminis* have so far not been identified.

Particle Properties

Furovirus particles are nonenveloped hollow rods which have a helical symmetry (**Figure 1**). The diameter of the particles is *c.* 20 nm and in leaves of freshly infected field grown plants in the spring the predominant lengths are *c.* 280–300 and 140–160 nm. Due to internal deletions in the coat protein readthrough protein genes shorter particles arise later in season in naturally infected plants and in laboratory isolates which may outcompete the 140–160 nm particles. The single coat protein species of furoviruses has a molecular mass of *c.* 20 kDa.

Antigenic Properties

Serological relationships exist between all five furoviruses. Monoclonal antibodies (MAb) prepared in several laboratories were either species or even isolate specific or allowed broad-range detection of several or all furoviruses except for SrCSV that was not included in such studies. MAbs to SBWMV have also been used to determine the accessibility of stretches of the coat protein amino acid chain on the virus particles. The C-terminus of the coat protein is apparently exposed along the length of the particles and is readily removed by treatment with trypsin.

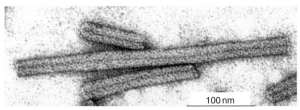

Figure 1 Particles of *Soil borne wheat mosaic virus* in a purified preparation negatively stained with uranyl acetate. Courtesy of Dr. D.-E. Lesemann and Dr. J. Engelmann, BBA Braunschweig.

Nucleic Acid Properties and Differentiation of Furoviruses

Furovirus genomes consist of two molecules of linear positive-sense ssRNAs. Their complete or almost complete nucleotide sequences have been determined for all five furoviruses described so far. They are 5′ capped and terminate in a functional tRNA-like structure with an anticodon for valine. This tRNA-like structure is preceded in the 3′ untranslated region by an upstream hairpin and an upstream pseudoknot domain (UPD) with two to seven possible pseudoknots. The genome organization is more or less identical for all furoviruses although there are considerable differences in nucleotide sequences (**Figure 2**). The percentages of sequence identities range from *c.* 60% to 75% for RNA1 and from *c.* 50% to 80% for RNA2 of different furovirus species. With strains of CWMV and SBCMV percentages of sequence identities are >94% for RNA1 and >85% for RNA2. Classification on the basis of sequence dissimilarities would suggest that SBWMV jap might be considered to be a separate species rather than a strain of SBWMV. However, the fact that in reassortment experiments a mixture of RNA1 of SBWMV jap and RNA2 of the type strain of SBWMV yielded an infectious progeny and the observation that the biological properties of all wheat-infecting furoviruses are very similar has been taken by some researchers as evidence that the type and the Japanese strains of SBWMV as well as CWMV and SBCMV should all be regarded as distantly related strains of the same species. As with viruses in other genera the decision whether related viruses should be considered as different strains of one virus or as separate species is not always easy, because these categories are men-made and many border-line cases exist in nature.

Organization of the Genome and Properties of the Encoded Proteins

Furoviral RNA1 codes for two N-terminally overlapping, presumably replication-associated proteins and for a *c.* 37 kDa movement protein (**Figure 3**). The shorter of the two replication-associated proteins contains methyltransferase and helicase motifs and the longer one, in addition, the RNA-dependent RNA polymerase motifs (**Figure 3**). The green fluorescent protein (GFP)-labeled 37 kDa protein of SBWMV, as opposed to its GFP-labeled coat protein, was shown to move from cell to cell in leaves of wheat but not in those of tobacco (a nonhost) and, similar to other viral movement proteins, to accumulate in the cell wall of both SBWMV-infected wheat leaves and transgenic wheat plants expressing the 37 kDa protein.

Furoviral RNA2 codes for the *c.* 20 kDa coat protein, a *c.* 84 kDa coat protein readthrough protein and a *c.* 19 kDa

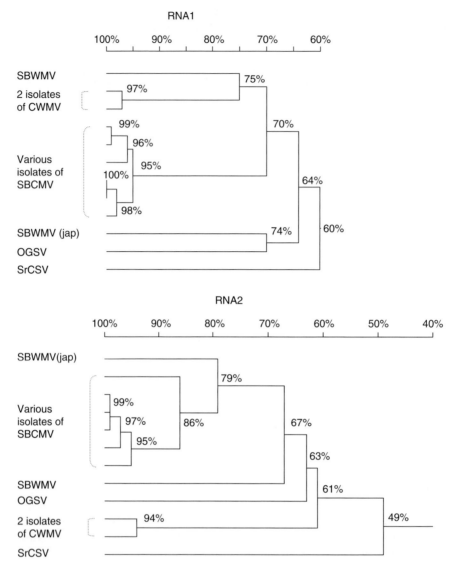

Figure 2 Percentages of nucleotide sequence identities among furoviral RNA1 and furoviral RNA2, respectively.

cysteine-rich protein. A 24 kDa protein found in *in vitro* translation experiments and also *in planta* has been shown to be coat protein with a 40-amino-acid N-terminal extension initiated at a conserved CUG codon 120 nt upstream of the AUG initiation codon for the coat protein gene. Neither the coat protein with this N-terminal extension nor the coat protein readthrough protein are required for SBWMV virion formation and systemic infection. SBWMV with full-sized RNA2 is found in naturally infected wheat in winter or early spring. Prolonged cultivation of field-infected plants, virus propagation by repeated mechanical inoculations or growth at elevated temperatures result in spontaneous deletions in the coat protein readthrough domain which may vary in size from 519 to 1030 nt. Computer analyses have revealed the presence of two complementary transmembrane domains (TM1 and TM2) in the coat protein readthrough proteins of furoviruses and also in those of the likewise *Polymyxa*-transmitted benyviruses and pomoviruses and in the P2 proteins of the *Polymyxa*-transmitted bymoviruses. The TM helices are apparently tightly packaged with ridge/groove arrangements between the two helices and strong electrostatic associations. It has been suggested that they facilitate the movement of the virus across the membrane surrounding the plasmodia in the plant cells. Nontransmissible deletion mutants lack the second transmembrane region partially or completely. The cysteine-rich 19 kDa protein of SBWMV is a suppressor of post-transcriptional gene silencing.

The products of the two 3′ proximal ORFs on RNA1 and RNA2, that is, the movement protein and the cysteine-rich protein, have not been found in *in vitro* translation experiments. It is assumed that they are expressed from subgenomic RNAs.

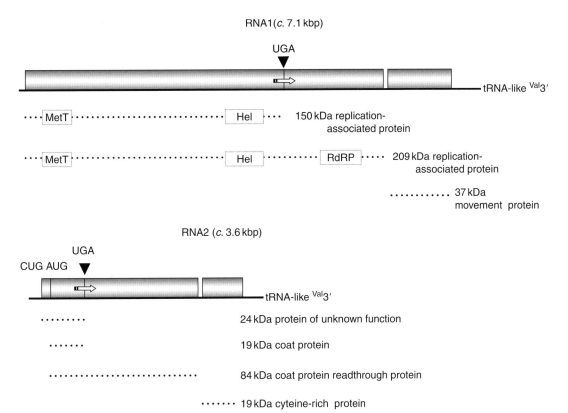

Figure 3 Organization of furoviral genomes (illustrated for SBWMV). The arrows in the open reading frames (boxed) indicate a readthrough translation due to a leaky stop codon.

Diagnosis

Symptoms caused by furoviruses may differ depending on environmental conditions, such as moisture and temperature. Nutrient deficiencies, winter injury or other viruses (especially bymoviruses), may produce symptoms that are easily confused with those caused by wheat-infecting furoviruses. Serological tests such as the enzyme-linked immunosorbent assay (ELISA), tissue print immunoassay, and immunoelectron microscopy as well as polymerase chain reaction (PCR) techniques allow a reliable detection. Real-time reverse transcriptase-PCR (RT-PCR) and PCR assays based on TaqMan chemistry have been developed for the detection and quantitation not only of SBCMV but also of its vector *P. graminis*. Real-time assays were found to be a 1000 times more sensitive than ELISA for the quantitation of SBCMV, and a 100 times more sensitive than conventional PCR for the quantitation of *P. graminis*.

Similarities and Dissimilarities with Other Taxa

The morphology of furoviruses resembles that of other rod-shaped viruses, that is, of benyviruses, pecluviruses, pomoviruses, hordeiviruses, tobraviruses, and tobamoviruses. The CPs of all these viruses have a number of conserved residues, for example, RF and FE in their central and C-terminal parts, respectively, which are presumably involved in the formation of salt bridges. Furoviruses, like pecluviruses and tobraviruses but unlike the other aforementioned viruses, have bipartite genomes. They differ from pecluviruses by having their movement function encoded on a single ORF rather than a triple gene block. They also differ from pecluviruses as well as from tobraviruses by having a coat protein readthrough protein gene. The gene for their cysteine-rich protein is located on RNA2, whereas with pecluviruses and tobraviruses it is located on RNA1. The cysteine-rich proteins of furoviruses, pecluviruses, and tobraviruses act as suppressors of post-transcriptional gene silencing and are phylogenetically interrelated, whereas those of beny- and pomoviruses are surprisingly unrelated to them. Furoviruses have two N-terminally overlapping replication-associated proteins (**Figure 2**) which in their amino acid sequences and structure are related to those of pomoviruses, pecluviruses, tobraviruses, and tobamoviruses. The 37 kDa movement protein of the furoviruses which belongs to the '30 K' superfamily of movement proteins relates them to the dianthoviruses and the tobamoviruses, whereas the valine-accepting tRNA-like structures on the 3' ends of their RNAs relate them to the tymoviruses.

In other properties, such as particle morphology and mode of transmission, dianthoviruses and tymoviruses are very different from furoviruses. Furoviruses like tobamoviruses have an upper pseudoknot domain in the 3′ untranslated regions of their genomic RNAs.

See also: Benyvirus; Cereal Viruses: Wheat and Barley; Hordeivirus; Pecluvirus; Pomovirus; Potyviruses; Vector Transmission of Plant Viruses.

Further Reading

Adams MJ, Antoniw JF, and Mullins JG (2001) Plant virus transmission by plasmodiophorid fungi is associated with distinctive transmembrane regions of virus-encoded proteins. Archives of Virology 146: 1139–1153.

An H, Melcher U, Doss P, et al. (2003) Evidence that the 37 kDa protein of soil-borne wheat mosaic virus is a virus movement protein. Journal of General Virology 84: 3153–3163.

Diao A, Chen J, Gitton F, et al. (1999) Sequences of European wheat mosaic virus and oat golden stripe virus and genome analysis of the genus furovirus. Virology 261: 331–339.

Driskel BA, Doss P, Littlefield LJ, Walker NR, and Verchot-Lubicz J (2004) Soilborne wheat mosaic virus movement protein and RNA and wheat spindle streak mosaic virus coat protein accumulate inside resting spores of their vector, Polymyxa graminis. Molecular Plant–Microbe Interactions 17: 39–48.

Goodwin JB and Dreher TW (1998) Transfer RNA mimicry in a new group of positive-strand RNA plant viruses, the furoviruses: Differential aminoacylation between the RNA components of one genome. Virology 246: 170–178.

Hariri D and Meyer M (2007) A new furovirus infecting barley in France closely related to the Japanese soil-borne wheat mosaic virus. European Journal of Plant Pathology 118: 1–10.

Kanyuka K, Ward E, and Adams MJ (2003) Polymyxa graminis and the cereal viruses it transmits: A research challenge. Molecular Plant Pathology 4: 393–406.

Koenig R, Bergstrom GC, Gray SM, and Loss S (2002) A New York isolate of Soil-borne wheat mosaic virus differs considerably from the Nebraska type strain in the nucleotide sequences of various coding regions but not in the deduced amino acid sequences. Archives of Virology 147: 617–625.

Koenig R, Pleij CW, and Huth W (1999) Molecular characterization of a new furovirus mainly infecting rye. Archives of Virology 144: 2125–2140.

Miyanishi M, Roh SH, Yamamiya A, Ohsato S, and Shirako Y (2002) Reassortment between genetically distinct Japanese and US strains of Soil-borne wheat mosaic virus: RNA 1 from a Japanese strain and RNA 2 from a US strain make a pseudorecombinant virus. Archives of Virology 147: 1141–1153.

Ratti C, Budge G, Ward L, et al. (2004) Detection and relative quantitation of soil-borne cereal mosaic virus (SBCMV) and Polymyxa graminis in winter wheat using real-time PCR (TaqMan). Journal of Virological Methods 122: 95–103.

Shirako Y, Suzuki N, and French RC (2000) Similarity and divergence among viruses in the genus Furovirus. Virology 270: 201–207.

Te J, Melcher U, Howard A, and Verchot-Lubicz J (2005) Soil-borne wheat mosaic virus (SBWMV) 19K protein belongs to a class of cysteine rich proteins that suppress RNA silencing. Virology Journal 2: 18.

Torrance L and Koenig R (2005) Furovirus. In: Fauquet CM, Mayo M, Maniloff J, Desselberger U and Ball LA (eds.) Virus Taxonomy: Eighth Report of the International Committee on Taxonomy of Viruses, pp 1027–1032. San Diego, CA: Elsevier Academic Press.

Verchot J, Driskel BA, Zhu Y, Hunger RM, and Littlefield LJ (2001) Evidence that soil-borne wheat mosaic virus moves long distance through the xylem in wheat. Protoplasma 218: 57–66.

Yamamiya A and Shirako Y (2000) Construction of full-length cDNA clones to Soil-borne wheat mosaic virus RNA 1 and RNA 2, from which infectious RNAs are transcribed in vitro: Virion formation and systemic infection without expression of the N-terminal and C-terminal extensions to the capsid protein. Virology 277: 66–75.

Fuselloviruses of Archaea

K M Stedman, Portland State University, Portland, OR, USA

© 2008 Elsevier Ltd. All rights reserved.

Glossary

Acidophile An organism whose optimal growth is at acidic pH, often 2 or below.

Archaea One of the three domains of life as defined by Carl Woese, completely microbial and separate from bacteria.

Archaeon Singular of Archaea or Archaebacteria.

Cryptic plasmid Plasmid with no known function.

Cryptic protein Protein with no known function.

Extreme thermophile An organism whose optimal growth temperature is above 80 °C.

Fusiform Having a spindle shape.

Thermoacidophile An organism that is thermophilic and acidophilic.

Thermophile An organism whose optimal growth temperature is above 50 °C.

Tyrosine recombinase One of a family of DNA recombining enzymes with an active site tyrosine.

Introduction

Fuselloviruses are unique spindle-shaped viruses with a short tail at one end that have so far only been observed in archaeal viruses or extreme environments dominated by

Archaea (**Figure 1**). The first fusellovirus to be found and characterized was SSV1, a virus-like particle of *Sulfolobus shibatae*, an extremely thermoacidophilic archaeon (optimal growth at 80 °C and pH 3). Fusellovirus genomes are relatively small double-stranded circular DNA molecules from 15 to 17 kbp. Only one gene, the viral integrase, shows clear similarity to genes outside the fuselloviruses. Their genomes persist in host cells both as episomes and integrated into the host genome. The genome is positively supercoiled when packaged in virus particles. Virus production can be induced by UV-irradiation. Virion production is usually constitutive at a low level and does not impede host growth or lyse the host cells. Viruses appear to be produced by budding at the cellular membrane.

Fuselloviruses have been found throughout the world and five complete genomes have been reported (**Table 1**). Fusellovirus-like DNA sequences have been found in many more locations by culture-independent techniques. Intriguing virus-plasmid hybrids have been discovered along with the SSV1-like viruses. SSV1, or portions thereof, has been used to make the first widely used vectors for genetic manipulation in thermophilic Archaea. Biochemical characterization has focused on the viral integrase with recent results showing that the integrase is similar to eukaryotic recombinases. Interestingly, the integrase gene and integration do not appear to be absolutely necessary for virus function. Nonetheless, the presence of the integrase gene appears to give such viruses a competitive advantage. Two structures of SSV1 encoded proteins have been solved to high resolution, but elucidation of the structure of the entire virion and many of the proteins remains elusive.

A number of archaeal viruses have been described as having a similar morphology to the SSV viruses. However, the spindle-shaped viruses of methanogens that were originally grouped with the fuselloviruses are very pleiomorphic and probably do not belong in this family. Additionally, the haloviruses His1 and His2 have very different genomes and replication and have been placed in the new genus *Salterprovirus*. Other spindle-shaped viruses of Archaea are much larger than the fuselloviruses.

History

The virus SSV1 was originally detected as a UV-inducible plasmid in a *Sulfolobus* isolate from Beppu Onsen in Kyushu, Japan by Wolfram Zillig and his co-workers in the early 1980s. Production of a 60 × 90 nm spindle-shaped virus-like particle was shown soon thereafter. The term 'virus-like particle' was used because infection of an otherwise uninfected strain could not initially be shown. It appears to be very difficult, if not impossible, to cure a *Sulfolobus* strain of a fusellovirus once infected. This may be due to the integration of the virus genome into a tRNA gene in the host genome by the virus integrase gene. SSV1 integrates into the CCG arginyl tRNA gene, whereas other fuselloviruses integrate into other tRNA genes. SSV1 became a model for the understanding of transcription in Archeaea. It was in the study of SSV1 genes and their promoters that Wolf-Dieter Reiter noticed that they resembled eukaryotic promoters with their canonical TATA-boxes. Together with previous work from the Zillig laboratory that had shown that the DNA-dependent RNA polymerases in Archaea were similar to eukaryotic DNA-dependent RNA polymerases, there was a strong evidence that the archaeal transcription machinery was eukaryote-like. The complete 15 495 bp genome sequence of SSV1 was determined in 1990. Genome analysis showed that other than the previously characterized virus coat protein genes, VP1 and VP3, an apparent DNA-binding protein, VP2, and the viral integrase (see below), none of the other 34 open reading frames (ORFs) showed any similarity to proteins in the known databases (**Figure 2**).

A major breakthrough in the study of SSV1 was made when Christa Schleper showed that SSV1 could infect *Sulfolobus solfataricus*, an uninfected *Sulfolobus* isolated from Pisciarelli, Italy, showing conclusively that SSV1

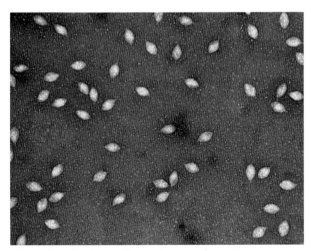

Figure 1 SSV1 image. Typical transmission electron micrograph of negatively stained SSV1 virions. Virus particles are 60 nm across and 90 nm long.

Table 1 Sequenced fuselloviruses

Virus	Genome size (bp)	Location of isolation
SSV1	15 495	Kyushu, Japan
SSV2	14 794	Reykjanes, Iceland
SSV3	15 230	Krisovik, Iceland
SSV-RH	16 473	Yellowstone, USA
SSV-K1	17 385	Kamchatka, Russia
pSSVx[a]	5 705	Reykjanes, Iceland

[a]Satellite.

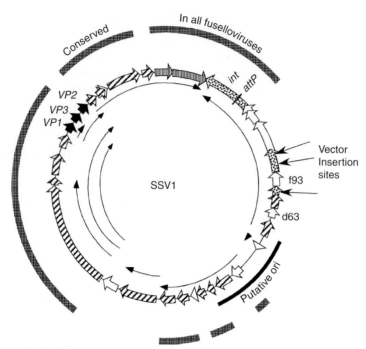

Figure 2 The SSV1 genome with its ORFs is shown, together with mapped transcripts. The attachment site in the viral integrase gene is shown as 'attP'. Known viral genes are labeled. VP1, VP3, and VP2 are virus structural genes. The location of the ORFs whose products have been crystallized, f93 and d63, are labeled. The putative origin of replication is labeled as 'Putative ori'. Insertion points for full-length shuttle vectors are shown with arrows outside the viral genome. Dotted ORFs have been shown to be not essential for virus function. Diagonally striped ORFs appear to be important for virus function. Vertically striped ORFs are conserved in pSSVx. ORFs conserved in all fusellovirus genomes are indicated by a stippled curve, outside of the ORF map.

was a virus. She was also able to transform *Sulfolobus* for the first time using this DNA. With this technique, plaque tests were established, the virus could be characterized, and large quantities could be purified. *Sulfolobus solfataricus* is one of the best studied of the thermophilic Archaea and a complete genome sequence and many other data are available. The development of the plaque test also allowed screening for fuselloviruses (and other viruses) in samples from throughout the world.

Fuselloviruses

Isolates

Using a plaque test and related spot-on-lawn halo techniques, first Wolfram Zillig's group and then others isolated fuselloviruses from Iceland, the USA, and Russia, in addition to SSV1 from Japan. The Zillig group found that approximately 8% of isolates from habitats with $T > 70\,°C$ and $pH^+ < 4$ contained SSVs that could be detected by their ability to infect *S. solfataricus*. First characterized was SSV2, from Iceland, followed by SSV-RH from Yellowstone National Park in the USA and SSV-K1 from the Kamchatka peninsula in Russia. All of these viruses had the typical spindle-shape and size of fuseloviruses and all contained c. 15 kbp genomes (**Table 1**). Those tested appeared to integrate into the host genome and were also inducible with UV-irradiation. However, the level of induction varied greatly between isolates as did the tRNA gene used as the site of virus integration into the host genome.

The complete genome sequences of all of these new fuselloviruses were determined and it was found that only about 50% of the genome is conserved (**Figure 2**). In some cases there is no sequence similarity whatsoever between parts of the virus genomes. The overall nucleotide identity is only about 55%, but the level of amino acid identity varies from undetectable to about 80%. The genomes have similar organization, with the exception of a large insertion and inversion in the SSV-K1 genome. Strangely, the putative DNA-binding protein, VP2, is missing in all genomes other than SSV1. The question then arose whether these genomic differences were due to geographical isolation of the viruses or to large amounts of local heterogeneity. To address this question the complete genome of a new SSV from Iceland, SSV3, was determined and compared to SSV2. It was found to have about 70% overall nucleotide identity to SSV2 and most of the ORFs were well conserved between the two viruses. However, there were a number of ORFs that were not conserved in SSV2, the other Icelandic virus, but were found in fuselloviruses from other parts of the world.

Culture-Independent Studies

The availability of multiple genome sequences for the fuselloviruses allowed the design of oligonucleotide probes that could be used to amplify conserved parts of fusellovirus genomes using the polymerase chain reaction. This technique was used on DNA samples collected directly from the environment, pioneered by Mark Young and his group. They showed that in a single spring that was known to harbor both *Sulfolobus* and fuselloviruses, the composition and relative abundance of fusellovirus sequences changed on at least a monthly basis. They also showed that a great deal of diversity was present in a single spring, much more diversity than seen in their hosts. Analysis of the sequence data over time indicated that there was a great deal of migration between springs and possibly a very large reservoir of fuselloviruses worldwide. Similar data from Lassen Volcanic National Park in the USA indicate this to be the case. These data contrast with the conservation seen between the whole SSV2 and SSV3 genomes from Iceland mentioned above. This conundrum remains to be resolved.

Plasmid Virus Hybrids

During the investigation of SSV2 from Iceland, a small virus-like particle was observed. This virus-like particle corresponded to a small plasmid, then named pSSVx, that was present in the same culture as SSV2 and appeared to be dependent on it or another complete fusellovirus for propagation. The genome sequence of pSSVx indicated that it was a fusion of a plasmid from the pRN family of cryptic *Sulfolobus* plasmids and two ORFs from a fusellovirus. It is not clear if activity of these ORFs is required for the plasmid to be packaged or if a *cis*-acting DNA sequence is required. At least one more of these virus-plasmid pairs has been reported in viruses from Iceland. This plasmid not only provides insight into virus and plasmid function but also is the basis for new genetic tools for *Sulfolobus*.

Fusellovirus-Based Vectors

One of the major reasons for sequencing the *S. solfataricus* genome and for isolation of viruses and plasmids of thermophilic Archaea was to be able to establish a genetic system for the analysis of these organisms. Many of the first and most successful steps have been taken using SSV1 or other fuselloviruses as the basis for plasmids that replicate in *Sulfolobus* and allow recombinant DNA to be transformed into *Sulfolobus*. A number of repeated sequences near the UV-inducible promoter, Tind, together with divergent promoters, led to the proposition that this region was the origin of replication (**Figure 2**). Pieces of the SSV1 genome containing this region have been incorporated into vectors that appear to replicate in *S. solfataricus* after transformation. However, these vectors have not been widely used. More successful vectors use the whole SSV1 genome with an insertion of an *Escherichia coli* plasmid in a region that was shown not to be critical for virus function to create infectious clones (**Figure 2**). Vectors based on this technique are now in their third generation. They have been successful in complementing mutants, overexpressing homologous and heterologous genes, and for preliminary gene-expression studies.

Fusellovirus Integrases

The only gene in fusellovirus genomes to show clear similarity to other genes is the virus integrase gene. It shows distant but clear similarity to the large family of site-specific tyrosine recombinases. Fusellovirus integrases are, however, unlike most other integrases in that the viral attachment site is within the integrase gene. Thus, integration disrupts the viral integrase gene. The attachment site in all fusellovirus integrases is in the N-terminal domain of the protein, and it is assumed, but not proved, that the remainder of the protein is not active, even though it contains all of the conserved catalytic amino acids responsible for recombination. The host attachment sites are in tRNA genes and proviral insertion preserves most of the tRNA gene so that it should remain functional. Intriguingly, all of the fuselloviruses sequenced to date integrate or are predicted to integrate into different tRNA genes. No additional host or virus genes appear to be required for either integration or excision. No excision of a provirus has been observed to date. Extensive *in vitro* studies with the SSV1 integrase indicate that the integrase itself is necessary and sufficient for both integrative or deletion reactions. Elegant molecular genetic studies have shown that the active site of the integrase is shared between two monomers and performs '*trans*-cleavage', similar to eukaryotic recombinases, but not bacterial ones.

Very recently, it was shown that the virus integrase and integration are not necessary for virus function. This was somewhat surprising due to the conservation of the integrase gene in all fusellovirus genomes sequenced to date. Additionally, a number of partial integrase genes, apparently made by past virus integration events, are present in not only the *S. solfataricus* genome but also in many other extremely thermophilic Archaea where they are thought to be highly involved in horizontal gene transfer.

Viruses lacking the integrase gene appeared not to integrate, but were stably maintained in laboratory cultures and under a number of stress conditions. However in head-to-head competition experiments between viruses containing and lacking the virus integrase gene respectively, the construct without the integrase was rapidly out-competed.

Structures

A structural genomics program to elucidate all of the structures of the products of all of the ORFs in the SSV1 genome has produced two high-resolution structures to date. The first, the product of ORF f93, is clearly a winged helix-domain containing protein, almost undoubtedly involved in DNA binding. However, its role in virus function is not clear. This ORF is also not conserved in most of the other SSV genomes. A high-resolution structure for the ORF product d63 has been solved. It is a very simple four-helix bundle which may be involved in protein–protein interactions but again its function is not clear. Recently developed genetic tools may help in the elucidation of these functions.

Other 'Fusiform' Viruses

The viruses of extremely halophilic Archaea, His1 and His2, have strikingly similar morphology to the *Sulfolobus* fuselloviruses, but their genomes are linear and their replication is protein primed. Therefore, they have been recently assigned to the new floating genus *Salterprovirus*. The virus-like particle reported from *Methanococcus voltae* strain A3 at first glance appears to be very similar to the *Sulfolobus* fuselloviruses, but it has a much larger genome and many different shapes in transmission electron microscopy. A virus-like particle with a similar shape to the *Sulfolobus* fuselloviruses, PAV1, has been isolated from a deep-sea *Pyrococcus abysii* strain. This virus has not yet been fully characterized and its infectivity has not yet been shown. Its genome has no sequence similarity to the *Sulfolobus* SSVs.

A number of other viruses with a spindle shape with or without projections at one or both ends have been isolated from a number of thermoacidophilic Archaea. Most of them, however, have either been assigned to other virus families or have not yet been classified. Generally, their genomes are very different both in size and sequence from the known fuselloviruses and often their virions have very different sizes.

See also: Crenarchaeal Viruses: Morphotypes and Genomes; History of Virology: Bacteriophages; Satellite Nucleic Acids and Viruses; Taxonomy, Classification and Nomenclature of Viruses; Viruses Infecting Euryarchaea.

Further Reading

Albers SV, Jonuscheit M, Dinkelaker S, et al. (2006) Production of recombinant and tagged proteins in the hyperthermophilic archaeon *Sulfolobus solfataricus*. Applied and Environmental Microbiology 72(1): 102–111.

Arnold HP, She Q, Phan H, et al. (1999) The genetic element pSSVx of the extremely thermophilic crenarchaeon *Sulfolobus* is a hybrid between a plasmid and a virus. Molecular Microbiology 34(2): 217–226.

Prangishvili D, Forterre P, and Garrett RA (2006) Viruses of the Archaea: A unifying view. Nature Reviews Microbiology 4(11): 837–848.

Schleper C, Kubo K, and Zillig W (1992) The particle SSV1 from the extremely thermophilic archaeon *Sulfolobus* is a virus: Demonstration of infectivity and of transfection with viral DNA. Proceedings of the National Academy of Sciences, USA 89(16): 7645–7649.

Stedman KM (2005) Fuselloviridae. In: Fauquet CM, Mayo MA, Maniloff J, Desselberger U, and Ball LA (eds.) Virus Taxonomy Eighth Report of the International Committee on Taxonomy of Viruses, pp 107–110. San Diego, CA: Elsevier Academic Press.

Stedman KM, Clore A, and Combet-Blanc Y (2006) Biogeographical diversity of archaeal viruses. In: Logan NA, Pappin-Scott HM, and Oynston PCF (eds.) SGM Symposium 66: Prokaryotic Diversity: Mechanisms and Significance, pp. 131–144. Cambridge: Cambridge University Press.

Stedman KM, Prangishvili D, and Zillig W (2005) Viruses of Archaea. In: Calendar R (ed.) The Bacteriophages, 2nd edn., pp. 499–516. New York: Oxford University Press.

Stedman KM, She Q, Phan H, et al. (2003) Relationships between fuselloviruses infecting the extremely thermophilic archaeon *Sulfolobus*: SSV1 and SSV2. Research in Microbiology 154(4): 295–302.

Wiedenheft B, Stedman KM, Roberto F, et al. (2004) Comparative genomic analysis of hyperthermophilic archaeal *Fuselloviridae* viruses. Journal of Virology 78(4): 1954–1961.

Gene Therapy: Use of Viruses as Vectors

K I Berns and T R Flotte, University of Florida College of Medicine, Gainesville, FL, USA

© 2008 Elsevier Ltd. All rights reserved.

Glossary

Biodistribution Tissues infected by a viral vector.
Episomal form Existence of the transgene as an extrachromosomal element.
Gene therapy Introduction of exogenous genes into cells to correct genetic or physiologic defects.
Nonenveloped virus A virus particle in which the viral coat is not surrounded by a lipid membrane.
Nonhomologous recombination Recombination between two unrelated DNA sequences.
Persistent infection An infection in which the viral genome is retained for extended periods, often for life, in host cells.
Replication competent A vector which can self-replicate and potentially spread from cell to cell.
Transgene A gene carried by a vector into a target cell.
Vector The means by which is a transgene is introduced into a target cell.

Introduction

Gene therapy represents the ultimate application of the evolution of molecular genetics from the elucidation of the structure of the double helix to patient treatment. There are only two requirements: the gene or DNA to be used to correct a genetic defect and a vehicle to introduce the DNA into the patient's cells. With the advent of cloning and recombinant DNA technology and the determination of the sequence of the human genome, a large number of genes are available for the therapy of genetic defects. Yet the clinical application of gene therapy remains in its infancy, albeit it has probably moved beyond the neonatal period. The major reason for the lack of more extensive progress has been the challenge of developing effective vectors. Successful vectors must be able to get the transgene (the new gene to be introduced) to the appropriate cell, transport the transgene into the cell, and then into the nucleus. Once present in the nucleus, a successful vector will express the transgene at the desired level and most often for an extended period of time. A major concern is toxicity, which may reflect either a host reaction to the vector itself, a commonly observed problem for early-generation vectors, or to the transgene, if it is perceived as nonself, which could occur in the case of a person with a null mutation. A further consideration is the ability of any DNA molecule introduced into the nucleus to potentially integrate into the host cell genome, most often by nonhomologous recombination. Nonhomologous recombination, which can occur at many sites in the genome, has the potential of disrupting normal gene expression by either altering the gene product or the regulation of the expression of the gene. Thus, important cellular functions may be lost which can impair the normal biology of the cell and in some cases lead to oncogenic transformation. Because of these considerations or hurdles, development of safe, effective vectors has proved to be much more challenging than initially appreciated.

Introduction of the transgene into patients can be done either *in vivo* or *ex vivo*. A variety of routes have been employed for administration *in vivo*; intramuscular, intravenous, via the airway, intraocular (either subretinal or into the vitreous humor), etc. The alternative approach is to remove the target cells from the body, introduce the transgene, and return the cells to the body. This method has been used primarily with the bone marrow. A second consideration is whether the vector will preferentially persist as an extrachromosomal element or will integrate into the host genome, either randomly or at a specific location. If the target consists of cells which divide rarely, if at all, then persistence as an extrachromosomal element may be an advantage. However, if the cell is a stem or progenitor cell which will undergo a sizable number of divisions, the vector is likely to be diluted out if it is not integrated into the genome.

A number of types of vectors have been tried, ranging from purified DNA administered either IV or via a gene gun, to liposomes, to vectors derived from viruses. The latter have been a favorite because viruses have evolved in nature to function by delivering and expressing genetic material within cells. All DNA viruses which replicate in the nucleus (including retroviruses) cause persistent infections which usually last for the life of the host. However, most of the viruses we know about also cause disease and thus the challenge is to engineer the virus so that it can function as a vector without causing disease. Interestingly, two of the viruses to be discussed below are not known to cause human disease. In any event, what has become clear is that to develop safe and effective vectors requires a detailed knowledge of the fundamental biology of the viruses to be used.

Despite the various challenges to be surmounted, gene therapy has developed to the point where there have been numerous clinical trials in which various levels of toxicity have been observed. In the most paradoxical example 12 children in France with severe combined immunodeficiency disease were cured by gene therapy; however, two of the children developed leukemia caused by the vector (happily, they have been successfully treated for this problem, as well). Clearly, much developmental work remains to be done, yet the promise is great. Several of the types of viral vectors which have been developed are described in the sections below.

There are several general considerations in the design of viral vectors. One is the host range or tissue targeting of the vector. A broad host range means that the vector can potentially be used to target many organs; however, if target specificity is desirable, a broad host range is not desirable. In other cases the desired target normally may not be infected by the vector. Thus, in many instances there have been attempts to modify the original host range. This has been achieved naturally by the use of different serotypes with differing host ranges, by modifying the genes encoding the capsid so that new epitopes are present on the surface of the virion, or, in some cases, by chemical linking of ligands to surface proteins. Another approach to altering host range is by creation of pseudotypes, enveloped viruses in which the envelope glycoprotein of a second virus is substituted for the normal constituent, for example, vesicular stomatitis glycoprotein has been substituted for a retrovirus envelope protein. Vector host range may also be affected by the alteration of the regulation of gene expression, so that it can occur or not in specific tissues.

Retrovirus Vectors

Retroviruses have long been considered to be good candidates to be developed as vectors for gene therapy. The viral capsid is surrounded by a lipoprotein envelope and contains two identical copies of a linear, single stranded RNA of plus polarity, in addition to at least two enzymes, reverse transcriptase and an integrase. As a consequence the viral life cycle passes through a double stranded DNA intermediate which integrates at a variety of sites in the cellular genome through the action of the virion integrase. Three different types of retroviruses have been used as vectors; the oncogenic retroviruses, lentiviruses, and spumaviruses. The oncogenic retroviruses have a wide host range (which can be extended by pseudotyping, in which the envelope proteins can be replaced by glycoproteins from other viruses), but can only integrate their genome in dividing cells after the nuclear membrane is removed before mitosis occurs. This property has limited their utility on the one hand; on the other, it has some desirable features because it limits the potential for horizontal spread since most cells in the body do not divide. Since the oncogenic retroviruses cause cancer, it is important that vector preparations not contain replication competent retrovirus (RCR). The viral genome contains three genes, *gag*, *pol*, and *env*. All of these are deleted from vector constructs; the only original sequences that are retained are the long terminal repeats (ltr), the packaging signal, and an ori. Vector production requires the normal gene products and the challenge is to make the probability of recombination to produce RCR very small so that quality checks can be passed. Since the viral genome integrates at a variety of sites, another potential concern is insertional mutagenesis in which either an essential gene is disrupted or a normally silent gene under tight regulation gains the ability to be expressed constitutively. The former seems relatively remote since most genes are present in at least two copies in a diploid genome. However, the latter could be potentially a one-hit phenomenon and lead to cancer. Indeed, this appears to have happened in a clinical trial, which otherwise was one of the great success stories of gene therapy. Twelve French infants with an X-linked form of severe combined immunodeficiency were treated with a vector derived from Moloney murine leukemia virus; nine of the children were 'cured' of this otherwise lethal disease. However, two children developed leukemia; in both cases the disease was clonal and the cells in each case contained the vector inserted next to the LMO gene. The conclusion was that the inserted vector had turned on gene expression, leading to the disease. Fortunately, it has been possible to successfully treat the children for this additional problem. Interestingly, the French government decided to continue the trial since the disease was otherwise lethal; that is, the benefit outweighed the proven risk.

Lentivirus vectors are derived from human immunodeficiency virus (HIV) and, thus, there is the real and psychological challenge of using a vector based upon a dangerous human pathogen. In addition to the three genes contained in the genomes of all retroviruses, the lentivirus genome contains six additional genes which are the products of splicing and function in a variety of regulatory roles. Again the approach has been to generate vectors from which all the viral genes have been deleted in producer cells where the helper genes required have a minimal amount of sequence homology so that there is very little chance of recombination to generate RCR. Lentivirus vectors are easier to produce in higher titers than the oncogenic retrovirus vectors and have the great potential advantage that they can successfully infect both dividing and nondividing cells and integrate the genome in both cases to establish persistence. Whether the psychological question can be successfully overcome remains to be seen. An alternative possibility is the development of spumavirus vectors. These have many of the same desirable features as lentivirus vectors without the problem of being derived from a serious pathogen. Spumavirus is a common human infection but has never been convincingly associated with human disease. They have been used successfully in a variety of animal models and indeed seem promising.

Adenovirus Vectors

Certain DNA viruses offer the potential to transduce both dividing and nondividing cells, with a much lower risk of integration-related carcinogenesis. Recombinant adenovirus (rAd) and recombinant adeno-associated virus (rAAV) have both been used in clinical gene therapy trials, as have recombinant herpes simplex virus (rHSV). Adenoviruses are nonenveloped, double stranded DNA viruses with a 36 kb genome that contains both early genes (encoding regulatory proteins expressed prior to DNA replication during a lytic life cycle) and late genes (encoding structural proteins expressed after DNA replication). Adenoviruses are present in over 60 serotypes in humans, and they commonly cause acute, self-limited infections of the respiratory and gastrointestinal tracts of humans. rAd vectors based on the group C adenoviruses, Ad2 and Ad5, were initially developed by deleting portions of the early genes E1a and E3 and inserting a therapeutic gene of interest. These first-generation rAd vectors are relatively easy to propagate and mediate robust short-term expression in a wide range of cell types, *in vitro* and *in vivo*. rAd vectors also trigger innate and adaptive immune responses in the host under most circumstances, which can limit both the duration of expression and the safety of *in vivo* gene therapy. An acute inflammatory response to a first-generation rAd vector resulted in the death of one patient with partial ornithine transcarbamylase (OTC) deficiency in a well-publicized incident. Nonetheless, clinical trials of rAd vectors continue in cancer patients and as recombinant vaccines.

Recognition of the immunogenicity of rAd vectors has led to the generation of later-generation versions of the vector, in which more viral coding sequences have been deleted. Second-generation vectors were produced by deletion of other early genes such as E2a and E4, in addition to the E1a and E3 deletions described above. The second-generation vectors mediate less expression of viral proteins, especially the late proteins, and trigger less adaptive immunity than first-generation vectors. The latest version of the rAd technology, known as the high-capacity or helper-dependent adenoviral (HD-Ad) vector, has a greater payload for therapeutic genes and expression elements, a lower risk of adaptive immune responses, and a longer duration of effect. HD-Ad vectors are currently beginning early-stage clinical trials.

Adeno-Associated Virus Vectors

Adeno-associated viruses (AAVs) are nonpathogenic, and include among others, human and primate parvoviruses, with a nonenveloped icosahedral capsid and a 4.7 kbp single stranded DNA genome. The genome contains genes required for replication (*rep*) and for the capsid components (*cap*). The use of internal promoters and alternate splicing allows for the production of a total of four Rep proteins (Rep78, Rep68, Rep52, and Rep40) and three capsid proteins (VP1, VP2, and VP3). These two genes are flanked by two palindromic inverted terminal repeats (ITRs), which contain all of the *cis*-acting elements required for replication and packing of the AAV genome.

AAVs are not adenoviruses, but were originally discovered as contaminants of adenovirus cultures. Wild-type AAV requires a helper virus (usually an adenovirus or herpesvirus) for efficient replication. In the absence of helper virus co-infection, AAV enters into the latent stage of its life cycle, either by maintaining itself as a stable episome or by integrating its DNA into the host genome, often into a specific site on human chromosome 19, the AAVS1 site. This site-specific integration is dependent upon the AAV Rep protein, which is deleted from rAAV vectors. Thus, the vectors have been found to persist primarily as episomal form, with a low frequency of non-site-specific integration. rAAV vectors are also capable of transducing nondividing cells and have been shown to persist long term *in vivo* in nondividing cells in animal models. One of the major limitations of the rAAV system is the small packaging capacity of the virion, with a payload of approximately 4.5 kb of exogenous DNA. This may be overcome in certain instances by taking advantage of the propensity of rAAV vector genomes to form hetero-multimers, thus allowing for *trans*-splicing between two different vector genomes within a single target cell.

rAAV vectors have been used in clinical trials in cystic fibrosis, hemophilia B, limb-girdle muscular dystrophy, Parkinson's disease, Batten's disease, and alpha-1 antitrypsin (AAT) deficiency. Results of these early phase trials confirm the fact that rAAV is safe over a wide dose range when administered to the airways, muscle, and central nervous system. Delivery to the liver in patients with hemophilia B resulted in transient elevations of liver enzymes that appeared to be associated with cell-mediated immune responses to rAAV capsid proteins retained in hepatocytes. While these immune responses may have limited the duration of therapeutic effect of the vector, no clinically important toxicity was observed.

All of the previous clinical trials experienced with rAAV have been using the type strain, AAV serotype 2. In recent years, over 100 new serotypes and genomic variants of AAV have been identified. Many other serotypes have been developed as rAAV vectors and some have substantially higher efficiency for transduction of specific tissues than rAAV2. One trial of an alternative serotype has just been initiated in AAT deficient patients with AAV serotype 1 capsids. The preclinical rAAV1 experience demonstrates that switching to a different AAV serotype capsid can still retain the basic property of long-term persistence of vector DNA, yet capsid–receptor interactions may lead to a distinct pattern of vector biodistribution to distant organs and to increased innate immunity as the distinct capsids interact with pattern recognition receptor proteins. Altered biodistribution could be advantageous if it leads to increased transduction of a target cell population. However, it could also lead to an increased risk of inadvertent germ line transmission of vector DNA, a phenomenon that is generally to be avoided. The clinical utility of alternative rAAV serotype vectors may become clearer in the near future as more clinical trials with these vectors are undertaken.

Both rAd and rAAV vector capsids have also been genetically engineered to create receptor-targeted versions of their respective vectors. Essentially, this approach involves insertion of specific peptides that may serve as ligands for specific cell-surface receptors. These 'designer' vectors have the potential for highly efficient and specific cell targeting. One key to the use of targeting technology is the identification of sites within the capsid which can be modified without a loss of the integrity of the capsid. Another important factor is the size of the peptide insert that may be tolerated within the insertion site. In the case of rAAV, one of the three capsid proteins, VP2, can either be completely deleted from the vector without loss of infectivity (and with a modest increase in packaging capacity), or can be modified at its N-terminus to include very large inserts, including single-chain antibodies. This form of rationally designed vector appears to hold much promise for future vector development, as does another variant of rAAV capsid alteration, an *in vitro* evolution, or DNA shuffling approach. In these types of approaches libraries of rAAV capsid variants are developed by mutagenesis and/or recombination of domains between distinct AAV serotypes. Serial passage on relatively nonpermissive cell types allows for positive selection for novel AAV variants capable of transducing those cell types, thus further expanding the repertoire of AAV capsids available as vehicles for therapeutic gene transfer.

Herpesvirus Vectors

Human herpesviruses are large double stranded DNA viruses with a complex structure. An envelope is separated from the nucleocapsid by the tegument which is composed of a large number of proteins which subserve various functions during viral infection, including turning off cell protein synthesis and initiating viral transcription. The herpes simplex virus (HSV) genome is composed of two parts, a long unique sequence and a short unique sequence, both bounded by terminal repeats. There are three classes of cellular genes: (1) immediate early, which control gene expression; (2) early, involved in DNA replication; and (3) late, which encode structural proteins. HSV preferentially infects cells of the central nervous system; after initial entry through epithelial cells of the skin or mucosa the virions travel retrograde through axons to cell bodies, particularly in sensory ganglia, where the virus most commonly establishes a latent infection. The viral genome is maintained as an extrachromosomal circle. Because the neurons which are persistently infected do not divide, the latent state is maintained unless a variety of environmental stimuli which can stress the cell activate viral replication. HSV vectors of two types have been developed and most often are intended to be lytic and cidal for tumors of the brain. The first type of vector has had several viral genes which are normally associated with virulence removed (immediate early genes and the gene for the tegument protein which shuts off cell protein synthesis) and replaced with genes specifically toxic to tumor cells. The second type is termed an amplicon, which consists of a cassette containing the transgene with its regulatory sequences, the HSV origin of replication and the packaging signal. The former type has a transgene capacity of almost 30 kbp while the latter can accommodate almost the full 160 kbp in the full genome. Since the latter expresses no viral genes, toxicity associated with wild-type virus and an inflammatory immune response can be avoided (unless caused by the transgene). HSV vectors of the first type have been used in several clinical trials and in phase I trials have been found to be nontoxic and to indicate the possibility of beneficial effects on the target tumors, although the number of patients in completed trials has been too small for definitive results.

Clinical Trials

The future of clinical gene therapy may lie with the appropriate matching of vector properties with the properties of the cell target and the desired timing of transgene expression. Situations in which transient, high-level expression is needed may be best addressed with rAd or liposomal vectors. First-generation rAd vectors may be particularly well suited for recombinant vaccine approaches since the innate responses to Ad may serve as an adjuvant for adaptive immune responses to the transgene product. Longer-term expression in nondividing cells might be better addressed with lentivirus or rAAV vectors. Finally, long-term expression in rapidly dividing cells might be best addressed with onco-retrovirus or lentivirus vectors since episomal rAAV vector genomes may be lost in such cells. The choice of vector and route of administration is often dictated by the disease process itself. While many applications of gene therapy have been for recessive genetic disorders, which usually require long-term gene expression within a particular organ, others have aimed to induce immune responses to malignant tumors, to stimulate a new vascular supply to ischemic limbs or segments of myocardium, to augment wound healing, or to relieve inflammatory diseases, such as rheumatoid arthritis. Each of these examples would require a different route of administration, as well as different timing and level of gene expression in order to achieve the desired therapeutic effect.

As more gene therapy vectors are developed into clinical products for specific diseases, a regulatory pathway to the clinic has emerged in USA and in other industrialized countries. The paradigms established for biological agents by the US Food and Drug Administration (FDA) provide the model for such regulatory approval. The pathway begins with proof-of-principle studies in which an appropriate vector is chosen to transfer the therapeutic gene into a cell culture or animal model. These studies generally address the efficiency and duration of gene expression, the biological effects of the transgene product, and functional correction of the defect. The gene transfer approach used in the proof-of-principle study generally mimics the ultimate clinical use of the vector, in terms of whether the vector is injected directly *in vivo* into a target organ, or is used for *ex vivo* transduction of a cell population that is amenable to reimplantation.

If the proof-of-principle studies demonstrate a potentially therapeutic effect, additional preclinical safety testing is usually undertaken in animals. These studies often are done in a larger number of animals, so that important potential toxicities can be understood and their frequency predicted. In some cases, both small animal (usually rodent) models and larger animal (occasionally nonhuman primate) models will be used. Formal safety studies should cover a fairly broad dose range and should include evaluations at both early and late time points. Safety studies in gene therapy have special considerations, including studies of the biodistribution of vector genomes to distant organs, particularly to germ cells within the gonads, studies of the immune responses to vector components and transgene products, and possibly a survey for vector-related tumorigenesis. The complete toxicology package should address potential toxicity of the vector over a specific dose range, and thus guide both the dosing and the safety studies within the early-phase clinical trials.

Traditionally, clinical trials have included phase I safety studies, phase II studies that focus on safety and biological effects, and phase III studies to prove efficacy and document lower frequency toxicities. In recent years, there has been a growing emphasis on phase IV studies, which occur after licensure of a drug or biological product for distribution and sale. In addition, there is a new concept of phase 0 studies in gene transfer, which are used as an experimental context to learn about properties of a given vector or vector class, without necessarily being part of the regulatory process for a specific gene therapy product. Once again, clinical gene therapy studies have certain special considerations, including sampling of semen (as an indicator of potential for inadvertent germ line transmission in males), studies of the immune responses to the vector and transgene product, and long-term follow-up of gene therapy study participants. The latter is designed to monitor the long-term effects of vector-mediated mutagenesis, that might result in secondary malignant changes in target cells at a later time. There are also special regulatory processes for gene therapy trials, particularly those that receive federal funding in USA. These include the involvement of the Office of Biotechnology Activities at the National Institutes of Health, and its Recombinant DNA Advisory Committee, as well as the use of Data and Safety Monitoring Boards by sponsoring NIH institutes for ongoing monitoring of clinical trials.

Given the fairly broad range of available vectors and the rapidly developing systems for clinical testing of vector candidates, it seems likely that a growing number of gene therapy agents will reach clinical application in the coming years. However, the long-term impact of gene therapy will likely take much longer to be fully realized. It remains to be seen just how broad that impact will be.

See also: Plant Virus Vectors (Gene Expression Systems); Viral Suppressors of Gene Silencing; Virus Induced Gene Silencing (VIGS).

Further Reading

Baum C, Schambach A, Bohne J, and Galla M (2006) Retrovirus vectors: Toward the plentivirus? *Molecular Therapy* 13: 1050–1063.

Epstein AL, Marconi P, Argnani R, and Manservigi R (2005) HSV-1-derived recombinant and amplicon vectors for gene transfer and gene therapy. *Current Gene Therapy* 5: 445–458.

Ghosh SS, Gopinath P, and Ramesh A (2006) Adenoviral vectors: A promising tool for gene therapy. *Applied Biochemistry and Biotechnology* 133: 9–29.

Hibbitt OC and Wade-Martins R (2006) Delivery of large genomic DNA inserts >100 kb using HSV-1 amplicons. *Current Gene Therapy* 6: 325–336.

Loewen N and Poeschla EM (2005) Lentiviral vectors. *Advances in Biochemical Engineering Biotechnology* 99: 169–191.

Mergia A and Heinkelein M (2003) Foamy virus vectors. *Current Topics in Microbiology and Immunology* 277: 131–159.

Morris KV and Rossi JJ (2006) Lentiviral-mediated delivery of siRNAs for antiviral therapy. *Gene Therapy* 13: 553–558.

Schambach A, Galla M, Modlich U, *et al.* (2006) Lentiviral vectors pseudotyped with murine ecotropic envelope: Increased biosafety and convenience in preclinical research. *Experimental Hematology* 34: 588–592.

Shen Y and Nemunaitis J (2006) Herpes simplex virus 1 (HSV-1) for cancer treatment. *Cancer Gene Therapy* 13: 975–992.

Wu Z, Asokan A, and Samulski RJ (2006) Adeno-associated virus serotypes: Vector toolkit for human gene therapy. *Molecular Therapy* 14: 316–327.

Genome Packaging in Bacterial Viruses

P Jardine, University of Minnesota, Minneapolis, MN, USA

© 2008 Elsevier Ltd. All rights reserved.

Glossary

Chromosome Encapsidated nucleic acid polymer that is the genomic component of the phage virion.

Concatamer A long nucleic acid polymer made up of tandemly linked genomes.

Packaging ATPase Enzyme complex that binds to, or is part of, the procapsid and translocates the chromosome from the outside to the inside of the capsid.

Procapsid The preformed, precursor capsid into which the genome is packaged.

Prohead See procapsid.

Virion The mature, infectious phage particle.

Introduction

The parasitic phase of the virus life cycle begins with the delivery of a viral genome into the host cell. Therefore, the bringing together of the virus chromosome with the viral capsid can be considered the culmination of the process of virus assembly that yields the infectious particles that will continue the next round of host-dependent replication. Bacteriophages have evolved complex mechanisms by which they ensure that this process is both effective and efficient.

As in other virus types, the encapsidation of the bacteriophage chromosome occurs by one of two distinct pathways. The first is the co-assembly of the components of the proteinaceous capsid with the chromosome at the terminus of viral replication. By this mechanism, capsid components assemble around the phage chromosome via the principles of self-directed assembly, with large classes of single-stranded DNA (ssDNA) and single-stranded RNA (ssRNA) phages being general examples. The ssDNA filamentous phages of the family *Inoviridae*, such as M13, achieve co-assembly of chromosome and capsid via an extrusion process by which ssDNA chromosome, coated with binding protein, is translocated through the cell membrane. The chromosome picks up external protein components of the capsid at membrane associated assembly sites. This extrusion/co-assembly is ATP dependent. In contrast, ssRNA phages of the family *Leviviridae*, such as MS2, present a co-assembly of an icosahedral capsid around a highly structured ssRNA chromosome.

In the second general pathway of genome encapsidation, the focus of this article, capsid assembly and genome replication remain separated until they converge in an event during which the phage chromosome is packaged into a preformed capsid using a process distinct from those that drive co-assembly. Unlike co-assembly, genome packaging is often mediated by mechanisms that energetically drive the chromosome from the outside to the inside of the capsid. In many phage systems, this event requires the assembly of specialized molecular machinery and the input of energy to push the nucleic acid chromosome into a confined space against a concentration gradient that result in the compaction of the nucleic acid polymer by several orders of magnitude.

There are three types of genome packaging that represent three large groups of phages and are distinguished by the type of nucleic acid substrate that is packaged. These consist of (1) the double-stranded DNA (dsDNA) phages of the order *Caudovirales*, which includes the families *Podo-*, *Sipho-*, and *Myoviridae*, and others groups including the family *Tectiviridae*, (2) the ssDNA phages of the family *Microviridae*, and (3) the double-stranded RNA (dsRNA) phages of the family *Cystoviridae*. All three of these groups of phages have evolved complex and highly efficient strategies to select and translocate viral chromosomes from the cytosol of the host cell into a preformed capsid.

General Considerations

One of the most crucial aspects of genome packaging in bacteriophage, as in other viruses, is the appropriate selection of virus chromosomal nucleic acid polymer from the milieu of nucleic acid inside the host cell. This selectivity is achieved by the binding and recognition of precursor viral chromosomes by components of the genome packaging apparatus and targeting them to the preassembled precursor capsid. This targeting event can be coupled to other events such as genome replication, as in the family *Microviridae*, be part of the chromosomal maturation process, as in the dsDNA phages, or precede the final replicative stages of the chromosome as found in the family *Cystoviridae*. Nomenclature can be diverse throughout the literature, with precursor capsids being referred to as procapsids, proheads, or preheads, depending on the system. This article refers to the precursor capsid targeted for packaging as the prohead or procapsid as is the convention for each system.

Additionally, distinction must be made between the bacteriophage genome and the packaged chromosome that becomes part of the infectious progeny virion. In some cases, these two terms are synonymous, as in the family *Microviridae* and the dsDNA phages lambda, ø29, and PRD1, since the chromosome is limited to a single copy of the genome. Exceptions arise for two reasons: (1) the genome is segmented into multiple chromosomes, as in the family *Cystoviridae*; and (2) the chromosome is terminally redundant as a result of a headful packaging mechanism using a concatamer as a substrate, as in caudoviruses such as the T-phages, P22, SPP1, etc. A singular exception is the Mu-like bacteriophage, in which the chromosome is comprised of a complete copy of the phage genome flanked by host-derived DNA sequences, a result of excision of the phage DNA from the host chromosome.

Finally, the inclusion of the packaging motor complex into the mature virion is variable, depending on the phage. In general, the enzyme complex responsible for chromosome translocation detaches from the packaged capsid in the dsDNA phages and is replaced by tail organelles. In the family *Microviridae*, the packaging complex is part of the DNA replication complex that is similarly not retained on the capsid surface. In contrast, the packaging complex of the family *Cystoviridae* remains attached to the capsid and is part of the mature virion. A summary of the key components and processes involved in packaging is presented in **Table 1**.

dsDNA Packaging

Since dsDNA phages are the most prevalent biological pathogens in the biosphere, the dominant type of genome packaging is the dsDNA packaging seen in the families *Podo-*, *Sipho-*, *Tecti-*, and *Myoviridae* (**Figure 1**). As with other packaging motifs, dsDNA phages pre-assemble an empty, immature capsid that docks with the phage chromosome for packaging. Universally, DNA packaging in dsDNA phages involves three viral components: (1) the preassembled prohead or procapsid, (2) the matured DNA chromosome, and (3) the ATP-dependent DNA-translocating enzymes.

The dsDNA phage chromosome is packaged into an immature, icosahedral prohead. The architecture of the capsid in these dsDNA phages is different from other icosahedral phages. At a unique fivefold vertex of the prohead, a portal structure is embedded, replacing one of the 12 pentamers of the icosahedron. This portal is a

Table 1 Packaging facts

Phage family	Genome type	Point of entry into procapsid	Enzymes involved	Nucleic acid polymer translocated	Completion events
Sipho-, Podo-, Myo-, Tectiviridae	dsDNA	Unique fivefold vertex via portal	Large and small packaging enzyme complex	Concatameric dsDNA (lambda, P22, SPP1, T-phages) Unit-length genome inserted into host chromosome (Mu) Unit-length chromosome with terminal proteins (phi29, PRD1)	Cleavage of concatamer, motor detachment, and tailing Cleavage of chromosome from host genome, motor detachment, and tailing Motor detachment and tailing
Microviridae	ssDNA	Twofold axis	Replication complex, J protein	ssDNA	
Cystoviridae	dsRNA	Fivefold vertex	Translocating NTPase complex	ssRNA	Replication of ssRNA to dsRNA

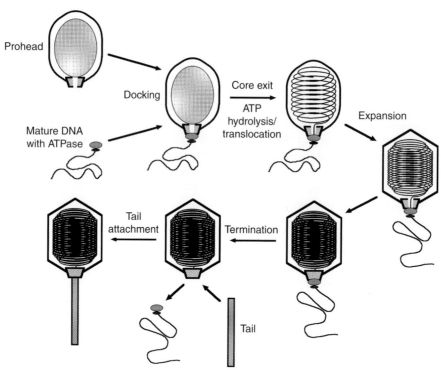

Figure 1 Schematic of generalized dsDNA phage assembly. A prohead interacts with the packaging ATPase holoenzyme–DNA complex via its head–tail connector. ATP hydrolysis powers translocation of the mature DNA, and at some point the scaffold core is ejected, either whole or following proteolysis. After an amount of DNA enters the head the shell capsomeres rearrange, making the head more angular and, in most phages, increasing the head volume. DNA translocation continues until a full complement of DNA enters the head, determined by either the unit length of the DNA, sequence recognition of the DNA length, or a headful mechanism. The ATPase–DNA complex detaches from the connector and is replaced by neck, tail, and/or tail fiber components, yielding a mature, infectious virion. Reproduced from Calendar R (ed.) (2006) *The Bacteriophages*. New York: Oxford University Press, with permission from Oxford University Press.

12-fold homo-oligomer of the phage-encoded portal protein forming a channel through the capsid wall. In phages with prolate heads, the portal vertex is always at one end of the long axis of the prohead. In most cases, it is believed that the portal is responsible for initiation of capsid polymerization in conjunction with an associated scaffold structure. The portal provides the binding site for the rest of the DNA packaging machinery and is the channel through which the DNA passes into, and ultimately out of, the capsid. After packaging, it is the axis around which the neck and tail components assemble, hence the designation connector for this structure in some systems.

The dsDNA phage prohead lattice is in an immature form prior to genome packaging. The immature prohead lattice is often physically less stable than in the mature virion. In some cases, accessory domains of the major capsid protein required for lattice assembly are proteolytically cleaved prior to packaging. In most phages, the conformational switch of the capsid lattice to the mature form occurs during packaging. This rearrangement of the capsid lattice is accompanied by an increase in the sixfold symmetry of the capsid hexons, and an overall increase, or expansion, of the head shell volume and consequent thinning of the shell thickness. Expansion ranges from the unperceivable, as in ø29, to the dramatic, as in HK97 where the capsid volume doubles.

Considerable diversity in the structural organization and maturation pathway of the DNA chromosome exists in these phages. DNA maturation converges at the point of chromosome translocation in that all dsDNA phages studied employ a common mechanism for genome translocation. The differences in DNA chromosome structure in the dsDNA phages arise as a result of differing strategies that have evolved to replicate phage dsDNA inside the cell. The central challenge, not faced by RNA phages, is the inability to initiate DNA-dependent DNA replication from a linear dsDNA without loss of information from the $5'$ ends of the replicated DNA polymer. This 'riddle-of-the-ends' is solved by dsDNA phages by using one of four strategies: (1) replication of a circular form of the genome (P1); (2) producing multiple, tandemly linked copies of the genome, termed concatamers, in a single replicating molecule such that complete chromosomes can be produced by cleavage (P22, T-phages, SPP1); (3) replication of the genome as part of integrated segment imbedded in the host cell DNA genome (Mu); and (4) initiate replication of linear DNA using a priming mechanism dependent on a covalently linked terminal protein that does not lead to degeneracy of the chromosome (ø29, PRD1). The first three replication strategies require a DNA cleavage reaction as part of the packaging initiation process, since only linear dsDNA of a certain size can be packaged. Bacteriophage lambda combines the first two replication strategies in that early replication is by theta replication of its circular genome that switches to the lambda rolling circle mechanism to produce long, multigenome concatamers which serve as the packaging substrate. The fourth replication mechanism above does not require a cleavage event since the DNA substrate is present in the cell in unit length form.

The DNA substrate is recruited from the cytosol by the enzymatic components of the packaging apparatus. In general, there are two proteins involved, one large and one small, that form a holoenzyme complex. Their stochiometry, relative to each other and to the procapsid, is still unclear, with reports from various systems ranging from four complexes per motor up to six. The 'small' subunit is a DNA binding and recognition protein, while the 'large' component is an ATPase. These proteins assemble on the DNA substrate, with the small subunit recognizing a specific sequence in the DNA termed the pac site. This process has been particularly well described in phage lambda (**Figure 2**). It is believed that initial DNA-binding step is mediated by other factors of cellular origin (integration host factor (IHF) in the case of lambda). (In phages with terminal proteins at the DNA ends, the terminal protein is analogous to the small subunit.) The specificity of this recognition ensures that only phage chromosomes will be packaged into the receptive phage prohead. In the case of phages with either circular- or concatameric-replicated DNA, the large subunit of these enzymes cut the DNA to produce a free end. This endonuclease activity is also responsible for production of a second endonuclease cut of the DNA at the end of packaging in phages that package segments of concatameric DNA, thus the designation 'terminase' for this enzyme ensemble in many phages (P22, lambda, T-even, T-odd, SPP1, Mu, etc). The designation terminase is not relevant to phages with unit length chromosomes such as ø29 and PRD1 since no cleavage is required; thus, these enzymes are more accurately termed packaging ATPases.

Once formed, the packaging ATPase–DNA complex docks with the prohead via the portal complex. The initial events that mediate the insertion of the DNA into the portal are unclear. Once initiated, DNA translocation occurs via an ATP hydrolysis-mediated process. The energy requirement for this process is clear when one considers the forces against which this molecular machine is working. First, entropy dictates that the concentration of DNA outside and inside the prohead should be equal. Thus, entropy works against dsDNA packaging throughout the process. Second, dsDNA is a relatively stiff polymer, with a persistence length (the minimum curvature of relaxed dsDNA) on the same scale as the capsid into which the DNA is being packaged in most phages. Therefore, energy is required to bend the DNA within the confines of the prohead. Finally, DNA has a net negative surface charge owing to the phosphate backbone of the polymer. As the DNA becomes compacted into the prohead, DNA strands are forced into close contact and electrostatic repulsion builds. Once again, energy must be invested into the DNA to achieve this end state. The end result is that the compacted DNA is pressurized inside the capsid to several atmospheres, requiring that the packaging motor complex must exert a translocating force on the order of 100 pN, making these molecular motors some of the strongest described in any biological systems. It is not clear how the DNA is organized to accommodate the degree of packing density, which approaches that of crystalline DNA. It is clear that the DNA must be

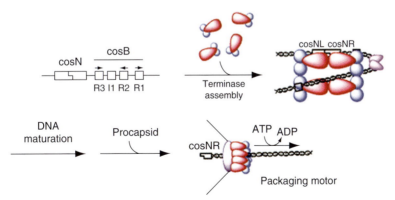

Figure 2 Model for terminase assembly at the cos site. The terminase protomer is a heterotrimer composed of one gpA subunit (red lobes) tightly associated with a gpNu1 dimer (blue spheres). Four protomers assemble at cos, resulting in bending of the duplex into a 'packasome' complex. Duplex bending by a gpNu1 dimer (purple lobes) bound to the R3 and R2 elements is central to the assembly of the packaging machinery. DNA maturation includes duplex nicking and separation of the DNA strands. Reproduced from Ortega M and Catalano CE (2006) Bacteriophage lamda gpNu1 and *E. coli* proteins cooperatively bind and bend viral DNA: Implications for the assembly of a genome packaging motor. *Biochemistry* 45: 5180–5189, with permission from ACS Publications.

pressed together in some form of hexagonal array that follows the contour of the capsid (**Figure 3**), which is the most energy-efficient conformation. Several models, including solenoids, liquid crystals, and folded toroids, have been suggested, but to date no structural study has been able to clearly resolve the structure of the packaged chromosome.

Packaging is terminated either by simply reaching the end of the unit-length chromosome (ø29, PRD1), or by a second DNA cleavage event necessitated by the translocation of a concatameric packaging substrate. In the latter, this cleavage event relies on the signaling from the packaged DNA density inside the capsid to the translocating complex on the outside of the capsid. In some cases, the near filling of the capsid permits the recognition and cleavage of the next approaching pac sequence, as with the lambda cos sequence, such that all progeny have identical genomes. In other instances, the second cleavage event is strictly 'headful', meaning the restriction of the DNA being packaged is sequence independent and relies on the detection of packaged DNA density inside the head. In this case, the capsid volume has evolved to exceed the capacity required to package a chromosome the length of the genome; thus the packaged DNAs are terminally redundant. This terminal redundancy is crucial to the progeny in that it will buffer the genome against loss during the next replicative infection. How this signal is transmitted from the portal to the packaging enzymes is unclear.

As described above, prohead assembly often involves the assembly of a precursor capsid that is structurally distinct from the mature capsid. In many cases, the prohead is yet to undergo ejection of the core-scaffold components that mediate head shell polymerization into the size and shape required by the virus. Similarly, final conformational rearrangements that stabilize the capsid shell allow the binding of accessory stabilizing proteins to the head surface, or release of the packaging motor and assembly of neck and tail components required to make

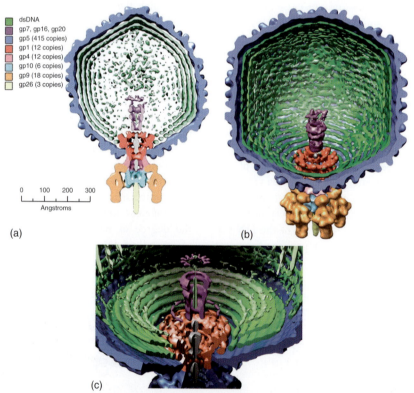

Figure 3 The interior features of the P22 virion. (a) The locations, deduced from many previous molecular biological studies, of the assembled gene products within a cutaway view of the reconstructed density of the P22 virion. Gene products 1, 4, 9, 10, and 26 make up the tail machine. Layers of dsDNA (green) are clearly visible as concentric shells within the capsid; they break into distinct rings of density near the portal vertex. Density (green) in the center of the channel formed by the ejection proteins (purple) could be the end of the P22 chromosome; however, density on this axis within the portal protein ring (red) does not appear to be consistent with DNA. (b) A cutaway view of the internal portion of the asymmetrically reconstructed particle contoured at 3, showing the 12-fold symmetry of the portal (red), the putative ejection proteins (purple), and individual strands of dsDNA (green). (c) Close-up view of the packaged interior upon 12-fold averaging along the tail tube axis. Although the E-proteins (purple) themselves in reality may or may not exhibit 12-fold symmetry, this view demonstrates the channel-like nature of the structure they form in the virion, as well as the dsDNA (green) that may be seated within their channel. Three concentric shells of spooled DNA are clearly visible. Reproduced from Lander GC, Tang L, Casjens SR, *et al*. (2006) The structure of an infections P22 vision shows the signal for headful DNA packaging. *Science* 312(5781): 1791–1795, with permission from American Association for the Advancement of Science.

the completed virion infectious. The end result of the separation of these final events in assembly permits the specific targeting of immature proheads by the DNA packaging machinery and completion of assembly (i.e., packaging motor detachment and neck/tail assembly) only after packaging has been completed. Upon completion of packaging, the ATPase complex detaches from the packaged head and neck and tail components assemble. This generally involves the assembly of several neck proteins, followed by the assembly of the tail in completed form (e.g., T4, lambda) or by the stepwise addition of tail components to the packaged head (P22, ø29).

The mechanism of DNA translocations in these motors remains unresolved, owing mostly to the complex nature of the portal motor complex. The role of the ATPases is clear in that they bind and hydrolyze ATP, transferring the energy of this reaction to the physical movement of the DNA substrate. It is widely held that the ATPases therefore have an active role in translocation, and function in a manner similar to the helicases, as is clear in the dsRNA phage packaging complex. What is unclear is whether the portal protein actively participates in this process. Regardless of a direct role in translocation, the portal protein ring does appear to play a role in signaling the end of translocation, as illustrated by mutants in the portal proteins of SPP1 and P22 phages which package altered chromosome lengths. This suggests the head-full sensor for these phages lies in the portal protein, and that the signal for the cleavage of the DNA upon completion of packaging is transmitted to the ATPase by the portal ring.

ssDNA Packaging

The archetype ssDNA phage assembly system is øX174 due to its extensive genetic and structural characterization over recent decades. Unlike the dsDNA phages, genome packaging in ssDNA phages is coupled to genome replication. The substrate genome for replication is a closed, circular dsDNA that is copied using a complex of host cell enzymes. This replicating complex switches to rolling circle replication upon interaction with viral proteins A and C, a host helicase (the *rep* protein), and the procapsid. As the positive polarity DNA is extruded from the rolling circle, it is transferred into the procapsid through a depression in the twofold axis of the preformed icosahedral shell. Unlike dsDNA phages, this point of entry is not structurally unique like a portal, but the replication/packaging complex binding confers exclusion to the other twofold axis on the capsid surface. Packaging in øX174 is dependent on an ssDNA binding protein J, which enters the capsid during packaging, and binds the DNA via charge interaction. J protein also mediates interaction of the packaged DNA with the inner surface of the capsid. Changes in capsid size and stability also occur during packaging, as with the dsDNA phages. In contrast to the dsDNA phages, phages of the family *Microviridae* do not require the high chromosome density inside the capsid that requires extensive input of energy for packaging. Thus, the energy of translocation harnessed from coupled DNA replication and DNA binding is sufficient in generating the required packaging force to efficiently produce infectious progeny.

ssRNA Packaging

A large class of dsRNA phages, the family *Cystoviridae*, translocate their genome into preformed capsids. These phages are analogous in their replication and structure to the eukaryotic reoviruses, and are best characterized in the ø6 group. Unlike the dsDNA phages, the cystoviruses translocate a single-stranded nucleic acid polymer into the procapsid. This circumstance is the product of the replication strategy of the virus, during which positive-strand RNA synthesized during infection serves as both template for translation and substrate for genome packaging. Conversion of the ssRNA to dsRNA occurs inside the capsid after translocation at the end of the assembly pathway. Also, unlike the dsDNA or ssDNA phages, the dsRNA genome is segmented into three chromosomes of different lengths (S, M, and L for small, medium, and large, respectively). This requires the translocation apparatus to not only select chromosomal nucleic acid polymer as in the dsDNA phages, but to select and process multiple ssRNAs in sequence (designated s, m, and l in the ssRNA form prior to conversion to S, M, and L) to allow for fidelity in the assembly of the mature virion.

The translocating NTPase of this class of viruses, the P4 enzyme, has been extensively characterized by biochemical and structural analysis. (Unlike the ATPases in the dsDNA phages, the packaging enzymes in the family *Cystoviridae* are not limited to ATP, but can use other NTPs depending on the phage.) A hexameric complex of this enzyme is assembled at each fivefold vertex of the precursor procapsid. The mechanism by which only one of the 12 vertices is selected to translocate a bound ssRNA substrate is unclear. The complex is believed to function in a manner analogous to helicases, translocating ssRNA through the hexamer and into the capsid. The P4 prohead complex recognizes stem loop structures at the 5′ end of the ssRNAs to be packaged, presumably through binding sites on the capsid surface adjacent to the ATPase. The order of binding, and thus packaging, is controlled by the alteration of the capsid binding site that occurs as RNA is packaged. Thus, a naive procapsid binds the s fragment specifically, packages it, and in doing so undergoes a conformational change that alters the procapsid RNA binding domain such that it now binds the m fragment,

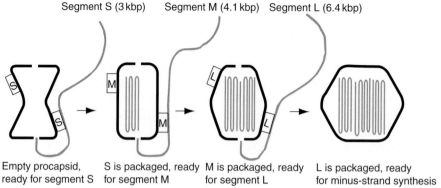

Figure 4 The packaging model of ø6. The empty procapsid shows only binding sites for S. After a full-size S is packaged, the S sites disappear and M sites appear. After a full-size M is packaged, the M sites disappear and L sites appear. After a full-size L is packaged, minus-strand synthesis commences. Reproduced from Calendar R (ed.) (2006) *The Bacteriophages*. New York: Oxford University Press, with permission from Oxford University Press.

which is then packaged, followed by the l fragment (**Figure 4**). The determinate of these conformational changes in the capsid to allow recognition of the variant 5′ ends of the chromosomes is the length of packaged RNA, rather than sequence.

Concluding Remarks

The phenomenon of genome packaging remains one of the most intensively studied processes of virion assembly, bringing together diverse but complementary fields of biochemistry, molecular genetics, structural biology, and biophysics. Since many of the phage studied have evolved mechanism analogous to eukaryotic viruses, efforts to resolve the mechanisms involved in genome packaging in phage promise to be reflected in their eukaryotic counterparts. Since genome translocation has no direct link to most cellular processes, inhibition of genome packaging has been long held to serve as a future target for antiviral therapy once the mechanisms of chromosome and capsid maturation and chromosome translocation are resolved. Regardless of this practical consideration, the efficiency, complexity, and power of this process will remain a target of investigation for its own sake.

See also: Filamentous ssDNA Bacterial Viruses; Icosahedral ssRNA Bacterial Viruses; Virus Particle Structure: Nonenveloped Viruses.

Further Reading

Calendar R (ed.) (2006) *The Bacteriophages*. New York: Oxford University Press.

Lander GC, Tang L, Casjens SR, *et al.* (2006) The structure of an infections P22 virion shows the signal for headful DNA packaging. *Science* 312(5781): 1791–1795.

Ortega M and Catalano CE (2006) Bacteriophage lamda gpNu1 and *E. coli* IHF proteins cooperatively bind and bend viral DNA: Implications for the assembly of a genome packaging motor. *Biochemistry* 45: 5180–5189.

Giardiaviruses

A L Wang and C C Wang, University of California, San Francisco, CA, USA

© 2008 Elsevier Ltd. All rights reserved.

History

In 1986, a 6.3 kbp linear double-stranded (ds) RNA molecule was observed in nucleic acid extracts of *Giardia lamblia* Portland I trophozoites. *Giardia lamblia* is an anaerobic parasitic flagellate that inhabits the upper gastrointestinal tracts of humans as well as many other mammals. When infecting humans, the parasitic protozoan causes giardiasis, an acute diarrhea that often progresses to chronic, carrier-stage for adults and severe malnutrition for children. Further examination of this 6.3 kbp dsRNA isolated from the extract of *G. lamblia* revealed it to be the genome of a small isometric virus, hence named giardiavirus (GLV), that specifically infects this protozoan.

It is interesting to note that although GLV was first detected in an isolate of *G. lamblia* Portland I (P1) obtained from Dr. D. G. Lindmark of Cleveland State University, the same P1 isolate from the American Type Culture Center is virus free. It therefore remains a mystery as to how the Cleveland P1 became exposed to the virus. All the information included here has been derived from studies of the virus originally isolated from this Cleveland P1 strain.

Taxonomy and Classification

GLV belongs to the family *Totiviridae*, genus *Giardiavirus*, of RNA viruses. This family is characterized by the non-segmented dsRNA genome and simple virion structure. Members of *Totiviridae* include the yeast dsRNA virus (ScV-L). Many of the viruses recently discovered from protozoa, such as the trichomonas vaginalis virus (TVV), leishmania RNA virus (LRV), and eimeria stiedae virus (ESV), also belong to this family.

Host Range and Geographic Distribution

Purified GLV readily infects many virus-free isolates of *G. lamblia* trophozoites, but not any other parasitic protozoa tested, including *Tritrichomonas foetus, Trichomonas vaginalis, Trypanosoma brucei brucei, Entamoeba histolytica,* and *Eimeria stiedae*. The virus has also been shown not to infect two transformed human intestinal cell lines. It is therefore believed that giardiavirus has a rather narrow host range. It probably infects only *G. lamblia* in nature. On the other hand, the cellular host for this virus, *G. lamblia*, parasitizes many mammals other than humans. Viruses that are identical in shape and size of genomic dsRNA and share dsRNA sequence homology with one another have been detected from many *G. lamblia* strains and isolates obtained from humans, guinea pigs, cats, beavers, llamas, and sheep. The human isolates were collected from Belgium, Poland, England, Israel, Ecuador, Puerto Rico, and various states in the USA. Since *G. lamblia* is found in almost all parts of the world, affecting developing as well as developed countries, it is expected that GLV follows its host and is distributed worldwide.

Physical and Biochemical Characteristics

Under the electron microscope, GLV appears as an icosahedron of 36 nm diameter, consisting of an electron-dense core encapsidated in a shell 5–6 nm thick (**Figure 1**). In addition to the 6.3 kbp linear dsRNA genome, the virion contains a major polypeptide of 100 kDa (p100) which is most likely the viral capsid. Purified virions are good antigens in mice. Polyclonal antibodies raised against whole virions react with p100 in Western blots and

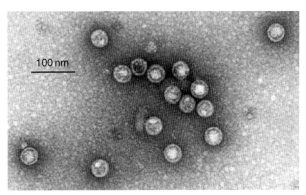

Figure 1 Giardiavirus.

can effectively block viral infection in the *in vitro* culture of *G. lamblia* trophozoites. The same antiserum also reacts positively with a minor component of the virion, a polypeptide of 190 kDa, which is believed to be the viral RNA-dependent RNA polymerase (RDRP). Indeed, GLV-encoded RDRP activity has been detected in the infected cell extract as well as the purified GLV fractions.

Antisera raised against synthetic dsRNA cross-react only with the viral dsRNA and are not protective against viral infection of *G. lamblia* trophozoites *in vitro*. Giardiavirus does not contain any lipids nor is the p100 polypeptide glycosylated.

The entire nucleotide sequence of the linear GLV dsRNA genome has been determined. It consists of 6277 bp and is flush at either end. As the dsRNA can be readily radiolabeled by [^{32}P]pCp and RNA ligase, this molecule must have a free 3′-OH group at one or both of its 3′ termini. The exact structure at the 5′ termini of the dsRNA has not yet been elucidated. It is known that the dsRNA molecule cannot be phosphorylated by T4 polynucleotide kinase, and that the denatured dsRNA molecule can be circularized with T4 RNA ligase. Therefore, the 5′ termini of the GLV genomic dsRNA are probably phosphorylated. Additionally, no covalently linked protein is found at these termini such as in the case of poliovirus.

Organization and Molecular Biology of the GLV dsRNA Genome

The 6277 nt GLV dsRNA genome contains only two large open reading frames (ORFs), both on the same strand of RNA (**Figure 2**). The first ORF (nts 368–3015) encodes the precursor polypeptide for the GLV capsid protein. Thirty-two amino acid residues from the N-terminus of this precursor protein are apparently removed by a cellular cysteine protease before the processed capsid protein is assembled into the virion.

The second ORF (nts 2806–5976) is −1 relative to ORF 1, and the two ORFs overlap by 220 nts. Amino acid

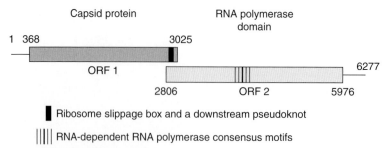

Figure 2 Organization of the giardiavirus dsRNA genome.

sequence motifs conserved for all RDRPs have all been found in ORF 2. It is now known that the 190 kDa GLV RNA polymerase is synthesized as a fusion protein of ORFs 1 and 2 at a level that is 2–5% of p100. Apparently, ribosomes carrying the growing nascent polypeptide chain are stalled when they encounter the pseudoknot structure residing within the 220 nt overlap of these two ORFs. At a 2–5% frequency, the ribosomes slip back one reading frame and proceed to translate ORF 2 as the C-terminus of a fusion protein. The ability of this GLV overlap fragment to induce −1 ribosomal frameshifting has been demonstrated in a reporter system in yeast.

Flanking the two ORFs, there are 367 and 301 nt untranslated regions respectively on the 5′ and 3′ sides. These two regions contain sequence elements that are critical for initiation of transcription and replication of GLV RNA. For example, deletion of a single nucleotide from the 5′ terminus of the (+)-strand GLV RNA totally abolishes transcription of GLV mRNA. Similarly, deletion or alteration of sequences in these two regions drastically reduces the level of progeny viral RNA.

In recent years, GLV has been successfully used as a vector for the introduction of foreign genes into *G. lamblia*. When a portion of the GLV genome is replaced with the firefly luciferase gene in a cDNA construct downstream from a T7 promoter, chimeric RNA can be synthesized *in vitro* using T7 polymerase. *Giardia* cells that are infected with wild-type GLV and electroporated with the chimeric RNA show luciferase activity of a millionfold above background. The chimeric RNA is replicated as dsRNA and packaged into recombinant virions that are shed into the culture supernatant. These recombinant viruses, together with wild-type GLV, can in turn infect naive *Giardia* trophozoites to produce luciferase activity. The list of foreign genes tested in this system include genes encoding neomycin phosphotransferase, hygromycin phosphotransferase, and green fluorescence protein. All can be delivered and expressed at high levels. Inclusion of a fragment, nts 368–631 in the coding region of p100, as the N-terminus of a fusion protein with the foreign gene product has been shown to enhance translation level by 5000-fold. This region therefore may contain elements that can promote interaction between GLV mRNA and ribosomes of *G. lamblia*.

Giardiavirus mRNA is uncapped and not adenylated. Translation is initiated at an internal ribosomal entry site (IRES) in the mRNA consisting of a 253 nt segment in the 5′ untranslated region (5′ UTR) and a 264 nt stretch in the ORF immediately downstream from the 5′ UTR. The initiation codon is localized at the center of an unstructured 31 nt segment flanked by complex secondary structures in the 5′ UTR and the ORF of the IRES. The precise position of the initiation codon is critical for translation initiation. It suggests a direct recruitment of the 40 S small ribosomal subunit to the initiation codon without ribosomal scanning, which is known to be absent from the translation machinery in *Giardia*.

Infection and Replication

GLV infects susceptible isolates of *G. lamblia* very efficiently. It has been estimated that demonstrable infection can be achieved at a multiplicity of infection as low as 10 GLV particles per cell. The exact mode of viral entry into the cell has not been elucidated. However, a number of virus-free *G. lamblia* strains have been found resistant to GLV infection. When trophozoites of these strains are electroporated with the single-stranded (ss) GLV RNA (see below), they become infected by GLV and can support and complete the cycle of replication, producing progeny virions that are fully infective to sensitive strains. It is therefore probable that certain cellular component(s) such as viral receptors are involved in the initial interaction between GLV and susceptible strains.

Using purified GLV to infect virus-free *G. lamblia* WB strain, it was shown by *in situ* hybridization that dsRNA replication was first detected in the cytoplasm. Toward the late stage of infection, viral dsRNA was found to spread into the twin nuclei of the infected flagellate. Transmission electron microscopic examination of thin sections of infected cells at stationary phase also reveals paracrystals of virus-like particles in the nuclei (**Figure 3**). It has been estimated that an infected cell may harbor as many as 10^5 GLV particles without lysis. Meanwhile, mature and infectious GLV particles begin to appear in the supernatant of *Giardia* culture medium 42 h after infection. It is not known whether a specific cellular process is involved in the release of giardiavirus into the culture supernatant, although cell lysis has not been observed as a consequence of viral infection.

The growth rate of newly infected cells decreases with increasing multiplicity of infection. As the ratio of the infecting virus to cell increases, the percentage of the nonadhering (nondividing) trophozoites also increases. However, in established infected cell lines or at moderate multiplicity of infection (<1000), the infected cell assumes normal appearance and maintains the same growth rate as its uninfected counterpart. Furthermore, GLV infection persists indefinitely throughout repeated subculturings.

In addition to the 6.3 kbp genomic dsRNA, an ssRNA (SS) of identical length that is homologous to only one of the two strands of GLV dsRNA is also found in the GLV-infected cellular extract. Studies of the infection time course showed that, in contrast to GLV dsRNA which increases steadily from 23 to 141 h after infection, SS becomes detectable at about the same time, peaks at 42–50 h after infection, then gradually declines. Electroporation of gel-purified SS into the uninfected WB cells resulted in the recovery of dsRNA from the cell extract and infectious GLV particles in culture supernatant, demonstrating that SS is the viral messenger RNA as well as the full-length replicative form of the viral genome. Recent nucleotide sequence analysis of GLV cDNA clones also verifies that it is the SS strand that encodes the large ORFs. There is no subgenomic viral RNA detected inside the infected cell.

The RNA products synthesized *in vitro* by the virion-associated RDRP are homologous to SS and complementary to the negative strand. Transcription of viral message therefore must proceed conservatively by utilizing the dsRNA as its template.

Evolution

The extent to which a person is affected by giardiasis varies widely. The outcome of the severity of parasitic infection depends on many variables derived not only from the parasite but also from the human host. Attempts at correlating the presence or absence of GLV with the severity of giardiasis have been inconclusive, and we still do not know the role GLV might play in the delicate balance of the host–parasite interaction, if any.

Some of the characteristics of GLV are listed with those of other dsRNA viruses of *Totiviridae* in **Table 1**. When compared with ScV-L, the best-known virus of this group, GLV and ScV are similar in many ways: (1) single molecules of genomic dsRNAs of similar sizes; (2) single major capsid polypeptides of similar sizes; (3) virion-associated RDRP with similar amino acid sequence motifs deduced from the genomic sequence, (4) ability to synthesize viral messages from intact virions *in vitro*, and (5) use of ribosomal frameshifting for the synthesis of viral polymerase. However, the two viruses do not cross-infect and the two dsRNAs do not cross-hybridize in Northern blots. GLV dsRNA has also been shown not to cross-hybridize with dsRNAs of TVV or LRV. Comparison of the nucleotide sequence from GLV

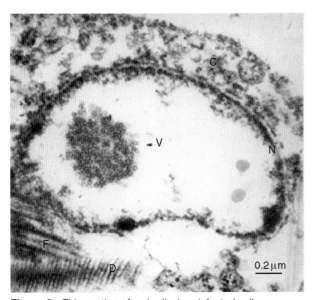

Figure 3 Thin section of a giardiavirus-infected cell. C, cytoplasm; N, nuclear envelope; F, flagella; D, ventral disk; V, aggregates of virus particles.

Table 1 Physical properties of some small dsRNA viruses

Virus	Shape	Diameter (nm)	Density (gm/ml^{-1})	dsRNA (μm)	dsRNA (kbp)	Capsid protein (kDa)
GLV	Isometric	33	1.368	1.50	6.277	100
TVV	Isometric	33	1.468	1.50	4.3–4.8	85
LRV	Isometric	32			5.3	
ESV	Isometric	35		1.63	6.5	
BBV	Isometric	38	1.358		5.5	
ScV-L	Isometric	33–41	1.368	1.31	4.3–4.8	88
UmV-P1	Isometric	41–43	1.418		6.3	73

GLV, giardia lamblia virus; TVV, trichomonas vaginalis virus; LRV, leishmania RNA virus; ESV, eimeria stiedae virus; BBV, babesia bovis virus; ScV-L, saccharomyces cerevisiae L; UmV-P1, ustilago maydis virus.

cDNAs with that of ScV-L, LRV or TVV indicates that, despite the similar organization of the viral genomes, GLV share very little overall sequence identity with any of these viruses whose genomic sequences have been completely determined. GLV is therefore not closely related to any of these viruses evolutionarily.

See also: Leishmaniaviruses; Totiviruses; Ustilago Maydis Viruses; Yeast L-A Virus.

Further Reading

Garlapati S and Wang CC (2005) Structural elements in the 5′-untranslated region of giardiavirus transcript essential for internal ribosome entry site-mediated translation initiation. *Eukaryotic Cell* 4 (no. 4): 742–754.

Wang AL and Wang CC (1991) Viruses of the protozoa. *Annual Review of Microbiology* 45: 251.

Wang AL and Wang CC (1995) The double-stranded RNA genome of giardiavirus In. In: Boothroyd JC (ed.) *Molecular Approaches to Parasitology*, 179pp. New York: Wiley.

Hantaviruses

A Vaheri, University of Helsinki, Helsinki, Finland

© 2008 Elsevier Ltd. All rights reserved.

Glossary

Coevolution Evolution of a virus together with its rodent/insectivore host.
Hantavirus A virus in the genus *Hantavirus*.
Reassortant A virus having genome segments of two different viruses.
Robovirus A virus transmitted from rodents to other vertebrates without arthropods.
Sympatric Occupying the same or overlapping geographic areas without interbreeding.

Historical Introduction

Hantavirus infections are not new to humankind. The first description of a hemorrhagic fever with renal syndrome (HFRS)-like disease can be found in a Chinese medical account written in about AD 960 and the earliest definite description of HFRS comes from Far East Russian clinical records dating back to 1913. During World Wars I and II, HFRS became an important military problem; for example, 'field nephritis' in Flanders during World War I may well have been caused by a hantavirus. In Manchuria in the mid-1930s, 12 000 Japanese soldiers caught the disease and military researchers were investigating the cause of the disease, sometimes using prisoners of war in infection experiments. Finnish and German soldiers encountered an HFRS-like epidemic in Finnish Lapland in 1943–44. During the Korean conflict in 1950–53, the disease again gained much attention when about 3000 United Nations troops contracted it – since then known as Korean hemorrhagic fever – with a 5–10% case–fatality rate. The milder form of HFRS, nephropathia epidemica (NE), common in Fennoscandia (Scandinavia and Finland), was first described by Swedish authors in 1934. However, the infecting agent of HFRS remained unknown until 1976, when the Korean researcher Ho Wang Lee discovered that cryostat-sectioned lungs of striped field mice (*Apodemus agrarius*), trapped near the Hantaan river, contained virus-specific antigen reactive with HFRS-patient sera. By a similar approach, the causative agent of NE was demonstrated in 1980 in bank voles, *Clethryonomys glareolus*, trapped in Puumala, Finland. Another agent, this one from urban rats in Seoul, Korea, was also found to cause HFRS. In the early 1990s, a further distinct European human-pathogenic hantavirus was isolated from the yellow-necked mouse, *Apodemus flavicollis*, near the village of Dobrava in Slovenia, and a few years later, the related, less pathogenic, Saaremaa virus from striped field mice on Saaremaa Island, Estonia. Thottapalayam virus from an insectivore (*Suncus murinus*, a shrew) in India, Prospect Hill virus from a meadow vole (*Microtus pennsylvanicus*) in USA, and Thailand virus from a bandicoot rat (*Bandicota indica*) already had been isolated in the early 1970s to the 1980s – they all apparently are nonpathogenic for humans.

Human disease had not been known to be caused by a hantavirus in the Americas until a cluster of acute respiratory distress syndrome cases with a high (60%) case–fatality rate in the Four Corners area of the American Southwest (where Arizona, Utah, Colorado, and New Mexico are contiguous) was recognized in May 1993. Subsequent studies led to the discovery of Sin Nombre virus and other hantaviruses that cause hantavirus pulmonary syndrome (HPS), all transmitted from sigmodontine rodents indigenous to the New World. An increasing number of pathogenic hantaviruses have been reported in South America, including Andes virus, which can be transmitted person to person, and which have high case–fatality rates. While only about 2000 HPS cases have been reported so far, approximately 150 000 HFRS cases are estimated to occur worldwide annually.

Virology

Hantaviruses are enveloped viruses with a tripartite negative-stranded RNA genome and belong to the genus

Hantavirus of the large virus family *Bunyaviridae*. The 6.4 kb L (large) genome segment encodes the ~250 kDa RNA polymerase, the ~3.6 kb M (medium) segment, the two glycoproteins 68–76 kDa Gn and 52–58 kDa Gc (formerly known as G1 and G2), and the ~1.7 kb S (small) segment the 50–54 kDa nucleocapsid protein (N). In addition, the S segment of hantaviruses carried by arvicolid and sigmodontine rodents has another overlapping (+1) open reading frame (ORF) named NSs, which has been shown to inhibit the interferon response. Viral messenger RNAs of the members of the *Bunyaviridae* are not polyadenylated and are truncated relative to the genomic RNAs at the 3' termini. Messenger RNAs have 5'-methylated caps and 10–18 nontemplated nucleotides derived from host cell mRNAs. The termini of all three segments are highly conserved and complementary to each other, a feature that has assisted in cloning and discovery of new hantaviruses.

Unlike most other bunyaviruses, hantaviruses are not arthropod borne (arboviruses), but are rodent borne (roboviruses). Each hantavirus is primarily carried by a distinct rodent/insectivore species, although several host switches seem to have occurred during the tens of millions of years of their co-evolution with their carrier hosts. We now know that the genetic diversity of hantaviruses is generated by (1) genetic drift (accumulation of point mutations and insertions/deletions) leading to mixtures of closely related genetic variants, quasispecies; (2) genetic shift (reassortment of genome fragments within a given virus genotype/species); and (3) according to recent findings, also by homologous recombination, a mechanism not previously observed for negative-strand RNA viruses.

Ecology and Epidemiology

Hantavirus infections are prime examples of emerging and reemerging infections. As are most of these infections, hantaviral diseases are zoonoses. With the exception of the South American Andes virus, hantavirus infections are thought to be transmitted to humans primarily from aerosols of rodent excreta (feces, urine, saliva). Only some hantaviruses cause disease in humans. In Asia, Hantaan and Amur viruses, carried by *Apodemus* spp. mice, cause severe HFRS, and Seoul virus, carried by rats, a milder disease. In Europe, there are three major hantaviral pathogens: Puumala virus carried by bank voles causes NE; Dobrava virus, carried by yellow-necked mice, causes severe HFRS; and the genetically and antigenically closely related Saaremaa virus carried by striped field mice causes mild NE-like disease. There also are reports that Seoul virus, carried by rats (*Rattus norvegicus* and *Rattus rattus*), causes HFRS of moderate severity both in Europe and North America. In addition, European common voles (*Microtus arvalis* and *Microtus rossiaemeridionalis*) carry Tula hantavirus, which can asymptomatically infect humans. Topografov hantavirus isolated from Siberian lemmings (*Lemmus sibiricus*) has not been detected in North European lemmings (*Lemmus lemmus*), although it can replicate in them. Infections of hantaviruses in rodents are asymptomatic and, often, persistent.

Hantavirus infections are quite common in Europe. Puumala virus occurs in Northern Europe, European Russia, and parts of central-western Europe. Dobrava virus is found mainly in the Balkans and the neighboring Central European areas. Saaremaa virus has been detected in Eastern and Central Europe, and in Slovenia the two viruses Dobrava and Saaremaa occur sympatrically. Apart from infections of laboratory rats, Seoul virus has been detected in wild rats in France and in several city harbors. It is apparent that many parts of Europe, such as Britain, Poland, and Byelorussia, and most of Africa, remain completely or relatively unstudied with regard to hantaviruses. This suggests either that HFRS is rare or nonexistent in these regions or is not generally recognized and is not diagnosed by local biomedical communities.

In Northern Europe, HFRS as well as the carrier rodents exhibit peaks in 3–4 year cycles, while in Central Europe the HFRS incidence follows the fluctuations of 'mast years', that is, the abundance of beech and oak seeds for the hantavirus-carrying rodents. In Central Europe, HFRS peaks in the summer whereas in Northern Europe most cases occur in late autumn and early winter, from November to January. Risk factors for acquiring hantavirus infections and HFRS include professions such as forestry, farming, and military, or activities such as camping, and the use of summer cottages. Cigarette smokers and males are more likely to be infected than are females. In the Americas, the increased precipitation during El Niño/Southern Oscillation in South and North America has been suggested as the main reason for the peaks in rodent population densities and for the consequent increased number of HPS cases.

Sigmodontine-borne hantaviruses circulating in North America (**Table 1**) form three phylogenetically distinct groups: those associated with *Peromyscus* spp. and *Reithrodontomys* spp. rodents, and the third carried by *Sigmodon* spp. and *Oryzomys* spp. rats. The first group carries both human pathogens (Sin Nombre virus, New York virus) and viruses not thus far associated with human disease (e.g., Blue River virus, Limestone Canyon virus). *Reithrodontomys*-borne viruses have not been shown to cause human disease. Hantaviruses discovered in South America are associated with several tribes of Sigmodontinae, mostly Oryzomyini-associated viruses, several of which are important human pathogens, including Andes, Lechiguanas, and Laguna Negra viruses. Results of phylogenetic analyses of South American hantaviruses suggest that several host-switching events have occurred during coevolution with their

Table 1 Hantavirus types

Virus	Host	Distribution (origin)	Disease
Murid-borne viruses	Mice and rats		
Hantaan (HTNV)[a]	Striped field mouse (*Apodemus agrarius*)	Asia (Korea)	HFRS
Da Bie Shan	Chinese white-bellied rat (*Niviventer confucianus*)	Asia (China)	NR
Seoul (SEOV)	Rat (*Rattus rattus, R. norvegicus*)	Worldwide (Korea)	HFRS
Dobrava (DOBV)	Yellow-necked mouse (*A. flavicollis*)	Europe (Slovenia)	HFRS
Saaremaa (SAAV)	Striped field mouse; western (*A. agrarius*)	Europe (Estonia)	HFRS
Thailand (THAIV)	Bandicoot rat (*Bandicota indica*)	Thailand (Thailand)	NR
Amur (AMRV)	Korean field mouse (*A. peninsulae*)	Asia (Far East Russia, China)	HFRS
Arvicolid-borne viruses	Voles and lemmings		
Puumala (PUUV)	Bank vole (*Clethrionomys glareolus*)	Europe (Finland)	HFRS
Hokkaido (HOKV)	Red bank vole (*C. rufocanus*)	Asia (Japan)	NR
Tula (TULV)	European common vole (*Microtus arvalis*)	Europe (Russia)	NR
Prospect Hill (PHV)	Meadow vole (*M. pennsylvanicus*)	North America (USA)	NR
Bloodland Lake (BLLV)	Prairie vole (*M. ochrogaster*)	North America (USA)	NR
Isla Vista (ISLAV)	Californian vole (*M. californicus*)	North America (USA)	NR
Khabarovsk (KHAV)	Reed vole (*M. fortis*)	Asia (Far East Russia)	NR
Topografov (TOPV)	Lemming (*Lemmus sibiricus*)	Siberia (Russia)	NR
Vladivostok (VLAV)	Reed vole (*M. fortis*)	Asia (Far East Russia)	NR
Sigmodontine borne	New World sigmodontine rodents		
Sin Nombre (SNV)	Deer mouse (*Peromyscus maniculatus*)	North America (USA)	HPS
Monongahela (MGLV)	Deer mouse (*P. maniculatus nubiterrae*)	North America (USA)	HPS
New York (NYV)	White-footed mouse (*P. leucopus*)	North America (USA)	HPS
Blue River (BRV)	White-footed mouse (*P. leucopus*)	North America (USA)	NR
Limestone Canyon (LCV)	Brush mouse (*P. boylii*)	North America (USA)	NR
Bayou (BAYV)	Rice rat (*Oryzomys palustris*)	North America (USA)	HPS
Black Creek Canal (BCCV)	Hispid cotton rat (*Sigmodon hispidus*)	North America (USA)	HPS
Muleshoe (MULV)	Hispid cotton rat (*S. hispidus*)	North America (USA)	NR
Andes (ANDV)	Long-tailed pygmy rice rat (*Oligoryzomys longicaudatus*)	South America (Argentina)	HPS
Lechiguanas (LECV)	Rice rat (*O. flavescence*)	South America (Argentina)	HPS
Oran (ORNV)	*O. longicaudatus*	South America (Argentina)	NR
Bermejo (BERV)	*O. chacoensis*	South America (Argentina)	NR
HU39694	Unknown	South America (Argentina)	HPS
Choclo (CHOV)	Pygmy rice rat (*O. fulvescens*)	Central America (Panama)	HPS
Calabazo	*Zygodontomys brevicauda*	Central America (Panama)	NR
Laguna Negra (LANV)	Vesper mouse (*Calomys laucha*)	South America (Paraguay)	HPS
Rio Mamore (RIOMV)	Small-eared pygmy rice rat (*O. microtis*)	South America (Bolivia)	NR
Caño Delgadito (CADV)	Cane mouse (*S. alstoni*)	South America (Venezuela)	NR
El Moro Canyon (ELMCV)	Western harvest mouse (*Reithrodontomys megalotis*)	North America (USA)	NR
Rio Segundo (RIOSV)	Mexican harvest mouse (*R. mexicanus*)	Central America (Costa Rica)	NR
Maciel (MACV)	Dark field mouse (*Necromys benefactus*)	South America (Argentina)	NR
Pergamino (PERV)	Grass field mouse (*Akodon azarae*)	South America (Argentina)	NR
Juquitiba	Unknown	South America (Brazil)	HPS
Araraquara	Unknown	South America (Brazil)	HPS
Castelo dos Sonhos	Unknown	South America (Brazil)	HPS
Insectivore-borne	Insectivores		
Thottapalayam (TPMV)	Asian house shrew (*Suncus murinus*)	Asia (India)	NR

Distinct hantavirus species listed in the 8th ICTV Report are shown in bold.
HFRS, hemorrhagic fever with renal syndrome; HPS, hantavirus pulmonary syndrome; NR, disease not recorded.
[a]Type species.

rodent hosts. The phylogenetic split between murine and sigmodontine rodents presumably dates to a divergence that occurred between subfamilies 30 million years ago, when the precursors of the sigmodontine rodents crossed the Bering Strait into the Americas. Notably, sigmodontine rodents are found only in the Americas and thus it is unlikely that HPS would occur in Eurasia, unless imported.

Clinical Picture and Pathogenesis

Amur, Dobrava, Hantaan, Seoul, Puumala, and Saaremaa viruses all cause HFRS but the infections differ considerably in severity. All are characterized by acute-onset, fever, headache, abdominal pains, backache, temporary renal insufficiency (first oliguria, proteinuria, and increase in serum creatinine, and then polyuria), and thrombocytopenia, but the extent of hemorrhages (hematuria, petechiae, internal hemorrhages), requirement for dialysis treatment, hypotension, and case–fatality rates are much higher in HFRS caused by Amur, Dobrava, or Hantaan viruses than in NE caused by Puumala or Saaremaa viruses. About a third of NE patients experience temporary visual disturbances (myopia), which is a very characteristic if not pathognomonic sign of the disease. Notably, the clinical consequences of all of the hantaviral pathogens in humans vary from none to fatal. Severe NE is associated with a certain haplotype, HLA-B8, DR3, DQ2 alleles, severe HPS with HLA-B35, and mild NE with HLA-B27. Yet, although Puumala virus infection is generally associated with mild HFRS, NE may have significant long-term consequences. A 5-year followup study demonstrated that 20% of NE patients had a somewhat increased systolic blood pressure and proteinuria. This is important, since the infection is so common in many areas of Europe. In addition, in some patients, Puumala virus infection may infect the pituitary gland and lead to mortality or at least to hypophyseal insufficiency requiring hormone-replacement therapy.

The pathogeneses of HFRS and HPS are poorly understood. However, it is known that β3 integrins can mediate the entry of pathogenic hantaviruses and that hantaviruses can regulate apoptosis. Also, there is evidence that increased capillary permeability is an essential component in the pathogenesis of both HFRS and HPS, although different target tissues, kidneys and lungs, respectively, are affected in the two diseases. HFRS and HPS patients show activation of tumor necrosis factor α (TNF-α) in the plasma and tissues and high levels of urinary secretion of the pro-inflammatory cytokine interleukin 6 (IL-6) are seen in NE. Studies with a monkey model mimicking human Puumala virus infection and HPS-like disease in Andes virus-infected Syrian hamsters may assist in elucidating the mechanism of pathogenesis.

HPS is characterized by pulmonary edema but death often results from cardiac failure; thus the term hantavirus cardiopulmonary syndrome (HCPS) has been proposed for the disease. HFRS and HPS, although primarily targeted at kidneys and lungs, respectively, share a number of clinical features, such as capillary leakage, TNF-α, and thrombocytopenia; notably, hemorrhages and alterations in renal function occur also in HPS and pulmonary involvement is not rare in HFRS.

Of the four structural proteins, both in humoral and cellular immunity, the nucleocapsid protein appears to be the principal immunogen. Cytotoxic T-lymphocyte responses are seen in both HFRS and HPS and may be important for both protective immunity and pathogenesis in hantavirus infections.

Diagnostics and Prevention

The diagnosis of acute hantavirus infection is primarily based on serology. Both immunofluorescence tests and enzyme immunoassays are widely used for detection of specific IgM or low-avidity IgG antibodies, characteristics of acute infection. In addition, immunochromatographic 5 min IgM-antibody tests have been developed. Hantaviruses show extensive serological cross-reactivity, especially within each of the three virus subgroups (murid, arvicolid, and sigmodontine borne; **Table 1**), but for accurate typing, neutralization tests are needed. For example, in Paraguay, a considerable seroprevalence of hantavirus antibodies, as high as 40%, is noted in people without a history of HPS. Average seroprevalences in Finland and Sweden suggest that only 10–25% of Puumala virus infections are diagnosed; thus, most infections either are subclinical or are mild or atypical and remain undiagnosed.

Viral RNA usually can be detected in the blood of HPS patients using polymerase chain reaction with samples collected during the first week of illness, which is useful because it also identifies the infecting virus genotype. The same is true for severe forms of HFRS but in the case of Puumala virus infections, viral RNA cannot be regularly detected in the blood or urine of NE patients.

Vaccines against hantaviral infections have been used for years in China and Korea, but not in Europe or the Americas. No specific therapy is used in Europe, but both ribavirin and interferon-α have been administered in trials in China. A major problem is that at the time the patients are hospitalized, the rate of virus replication is declining, so that reduction of virus replication no longer is necessary for the patient. Thus, prevention of hantaviral infections continues to rely on reduced contact with excreta from infected rodents.

See also: Bunyaviruses: General Features; Human Eye Infections; Zoonoses.

Further Reading

Brummer-Korvenkontio M, Vapalahti O, Henttonen H, Koskela P, Kuusisto P, and Vaheri A (1999) Epidemiological study of nephropathia epidemica in Finland 1989–96. *Scandinavian Journal of Infectious Diseases* 31: 427–435.

Calisher CH (ed.) (1990) *Hemorrhagic Fever with Renal Syndrome, Tick and Mosquito-Borne Viruses: Archives of Virology, Supplementum 1*, pp. 1–347. Vienna: Springer.

Fauquet CM, Mayo MA, Maniloff J, Desselberger U, and Ball LA (eds.) (2005) *Virus Taxonomy: Eighth Report of the International Committee on Taxonomy of Viruses, Bunyaviridae, Hantavirus*, pp. 704–707. San Diego, CA: Elseviver Academic Press.

Kallio-Kokko H, Uzcategui N, Vapalahti O, and Vaheri A (2005) Viral zoonoses in Europe. *FEMS Microbiology Reviews* 29: 1051–1077.

Kaukinen P, Vaheri A, and Plyusnin A (2005) Hantavirus nucleocapsid protein: A multifunctional molecule with both housekeeping and ambassadorial duties. *Archives of Virology* 150: 1693–1713.

Khaiboullina SF, Morzunov SP, and St. Jeor SC (2005) Hantaviruses: Molecular biology, evolution and pathogenesis. *Current Molecular Medicine* 5: 773–790.

Kukkonen SKJ, Vaheri A, and Plyusnin A (2005) L protein, the RNA-dependent RNA polymerase of hantaviruses. *Archives of Virology* 150: 533–555.

Maes P, Clement J, Gavrilovskaya I, and Van Ranst M (2004) Hantaviruses: Immunology, treatment, and prevention. *Viral Immunology* 17: 481–497.

Muranyi W, Bahr U, Zeier M, and van der Woude FJ (2005) Hantavirus infection. *Journal of the American Society of Nephrology* 16: 3669–3679.

Peters CJ and Khan AS (2002) Hantavirus pulmonary syndrome: The new American hemorrhagic fever. *Clinical Infectious Diseases* 34: 1224–1230.

Pini N (2004) Hantavirus pulmonary syndrome in Latin America. *Current Opinion in Infectious Diseases* 17: 427–431.

Plyusnin A (2002) Genetics of hantaviruses: Implications to taxonomy (review). *Archives of Virology* 147: 665–682.

Schmaljohn CS and Nichol ST (eds.) (2001) Hantaviruses. *Current Topics in Microbiology and Immunology* 256: 1–191.

Terajima M, Vapalahti O, Van Epps HL, Vaheri A, and Ennis FA (2004) Immune response to Puumala virus infection and the pathogenesis of nephropathia epidemica. *Microbes and Infection* 6: 238–245.

Vapalahti O, Mustonen J, Lundkvist Å, Henttonen H, Plyusnin A, and Vaheri A (2003) Hantavirus infections in Europe. *Lancet Infectious Diseases* 3: 653–661.

Henipaviruses

B T Eaton and L-F Wang, Australian Animal Health Laboratory, Geelong, VIC, Australia

© 2008 Elsevier Ltd. All rights reserved.

History

The first henipavirus was isolated in 1994 following an outbreak of acute respiratory disease in a stable in Brisbane, Australia. In less than 3 weeks, 21 of 30 horses became infected and 14 either died or were euthanized. Two people who worked at the stable also became infected and one died. The virus responsible was a previously undescribed paramyxovirus, subsequently called Hendra virus (HeV) after the suburb where the outbreak occurred. In the subsequent decade, HeV was identified as the cause of equine deaths at four separate locations in Queensland, spilling over on two occasions to cause single human infections, one of which was fatal. In the latter instance, the patient succumbed to encephalitis over a year after infection during the necropsy of HeV-infected horses. Initial serological evidence suggested that fruit bats (flying foxes) in the genus *Pteropus* were reservoir hosts and in 1996, HeV was isolated from several Australian flying fox species.

The second member of the genus, Nipah virus (NiV), emerged in 1998–99 in Perak State, Peninsular Malaysia, as the cause of an outbreak of respiratory and encephalitic disease of low morbidity and mortality in pigs. The virus spread to humans causing febrile encephalitis among pig farmers and those who had direct contact with pigs. The epidemic moved south to the intensive pig-farming areas of Negeri Sembilan in December 1998 and was only brought under control by the imposition of movement restrictions and the culling of over 1 million pigs. Two hundred and sixty-five human cases of encephalitis were documented with 105 deaths. A cluster of 11 cases with one death occurred among abattoir workers in Singapore. NiV was subsequently isolated from the urine of Malaysian flying foxes in 2002. NiV re-emerged in Bangladesh and an adjoining area of India (West Bengal) in 2001, and since then outbreaks of encephalitis caused by the virus have re-occurred in Bangladesh almost every year, with case–fatality rates approaching 75%.

Taxonomy and Classification

The full-length genome sequence of HeV revealed it to be a member of the family *Paramyxoviridae*, subfamily *Paramyxovirinae*, but one with unique genetic features that precluded its classification in the genera *Morbillivirus*, *Respirovirus*, or *Rubulavirus*, the three genera established in the subfamily at that time. Complete genome sequencing of NiV revealed a high degree of similarity to HeV and, in 2002, the genus *Henipavirus* was created to accommodate these novel paramyxoviruses. The genus *Henipavirus* is now one of five genera in the subfamily *Paramyxovirinae* (**Figure 1**).

Several molecular features distinguish henipaviruses from other paramyxoviruses. First, the genome length of NiV and HeV at 18 246 and 18 234 nt, respectively, is approximately 15% larger than most other paramyxoviruses. It is not the extra length *per se* that is unique to henipaviruses. The unclassified paramyxoviruses J virus and Beilong virus have genomes over 19 000 nt in length. The novel henipavirus genomic feature is the length of the noncoding regions at the 3′ end of five of the six genes, which in most cases are 3–13 times longer than their *Paramyxovirinae* counterparts. The function of these regions is unknown. A second unique feature of henipaviruses

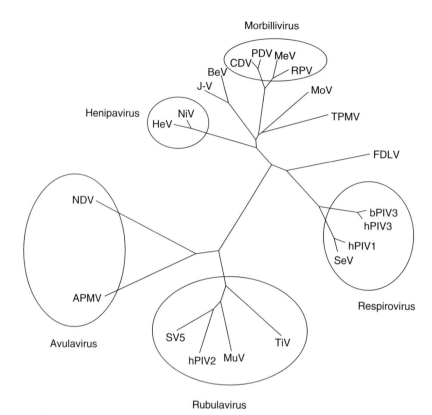

Figure 1 Classification of henipaviruses. The number of genera in the subfamily *Paramyxovirinae* was raised in 2002 from three (*Respirovirus*, *Morbillivirus*, and *Rubulavirus*) to five by the addition of two new genera, *Avulavirus* and *Henipavirus*. The phylogenetic tree shown here is based on an alignment of the deduced amino acid sequence of the L gene of selected family members using the neigbor-joining method. Viruses are grouped according to genus and abbreviated as follows. *Avulavirus* genus: APMV (avian paramyxovirus), NDV (Newcastle disease virus); *Respirovirus* genus: SeV (Sendai virus), hPIV1 (human parainfluenzavirus 1), hPIV3 (human parainfluenzavirus 3), bPIV3 (bovine parainfluenzavirus 3); *Morbillivirus* genus: MeV (measles virus), CDV (canine distemper virus), RPV (rinderpest virus), PDV (phocine distemper virus); *Henipavirus* genus: HeV (Hendra virus), NiV (Nipah virus); *Rubulavirus* genus: hPIV2 (human parainfluenzavirus 2), MuV (mumps virus), SV5 (simian parainfluenzavirus 5), TiV (Tioman virus); and unclassified viruses: MoV (Mossman virus), BeV (Beilong virus), FDLV (Fer-de-lance virus), TPMV (Tupaia paramyxovirus), and J-V (J virus).

commensurate with their status in a separate genus within the family *Paramyxoviridae* is the presence of 3′ and 5′ genomic terminal sequences 12 nt in length, that differ from all other paramyxoviruses and contain the genus-specific promoter elements for replication and transcription.

Molecular Biology

Although the genomes of henipaviruses, respiroviruses, and morbilliviruses are organized in an identical manner and, with the exception of the P protein, virus-encoded proteins in each genus are similar in size, analyses have revealed striking differences between some henipavirus proteins and their respirovirus and morbillivirus counterparts.

The attachment glycoproteins of paramyxoviruses are designated as hemagglutin-neuraminidase (HN), hemagglutinin (H), or glycoprotein (G) on the basis of their capacity to agglutinate red blood cells and remove sialic acid from carbohydrate moieties. The HN-positive status of respiroviruses, avulaviruses, and the vast majority of rubulaviruses reflects their use of sialic acids as cell surface receptors. In contrast, morbilliviruses and henipaviruses bind to cells using sialic acid-independent mechanisms. SLAM (signaling lymphocyte-activation molecule) is regarded as a universal morbillivirus receptor and ephrin B2 has been identified as a functional receptor for both HeV and NiV. Ephrin B2 is a member of a family of cell surface glycoprotein ligands that bind to surface-expressed tyrosine kinases known as Eph receptors. Binding of ephrins to Eph receptors leads to bidirectional signaling and cell-to-cell communication. The widespread distribution of ephrin B2 among mammals may help explain the extensive host range of henipaviruses (see below). Like the fusion protein of all other paramyxoviruses, the henipavirus F protein is activated by proteolytic cleavage of a precursor F_0 protein. Henipaviruses invariably generate systemic infections and most paramyxoviruses which do so synthesize F proteins which are cleaved at a multibasic cleavage site by furin, a cellular protease located in the *trans*-Golgi network. Surprisingly, the cleavage site of both HeV and NiV is a single basic amino acid, arginine and lysine, respectively. What is more, that basic residue and the amino acids in the vicinity of it are not required for proteolytic cleavage. A second unique feature of henipavirus F proteins is the presence of an endocytosis consensus motif YXXφ (where Y is tyrosine, X is any amino acid, and φ is an amino acid with a bulky hydrophobic group) in the cytoplasmic tail of the protein. Endocytosis is required for F protein cleavage activation, an observation consistent with the fact that cathepsin L, a lysosomal cysteine protease that lacks a distinctive cleavage recognition site, is responsible for cleavage of the HeV F protein.

In paramyxoviruses, mechanisms to inhibit the antiviral effects of interferon are encoded by the P gene, which generates multiple proteins (P, V, W, and C in henipaviruses) by means of internal translation initiation sites, overlapping reading frames, and a transcription process in which nontemplated G nucleotides are inserted at a conserved editing site, resulting in a reading frameshift during translation. Insertions of one and two G residues generate transcripts which encode the V and W proteins, respectively. Whereas the henipavirus V protein resembles that of other paramyxoviruses in having a conserved cysteine-rich C-terminal domain, the henipavirus W protein is unusual because the C-terminal domain is significantly longer than that of most *Paramyxovirinae*, where translation of the encoding P gene transcript terminates soon after the editing site.

Host Range and Viral Propagation

HeV and NiV replicate in a variety of mammalian cell lines. The rate of replication, the size of syncytia generated, and the location of nuclei in syncytia vary depending on cell type and virus species. Infected Vero cell monolayers yield titers in excess of 10^8 infectious virions per milliliter. The broad host range of henipaviruses has been corroborated by experiments in which cells infected with vaccinia virus recombinants and expressing henipavirus cell attachment (G) and fusion (F) proteins on the surface fuse with cells from a gamut of vertebrate sources. Susceptibility to virus infection and virus glycoprotein-mediated fusion *in vitro* are reflected in the wide range of mammalian hosts sensitive to virus infection *in vivo*. The host range of henipaviruses is considered uncommon for paramyxoviruses, which usually adopt a narrow host range. In addition to bats, NiV has been shown to infect animals in six other mammalian orders. Although the widespread distribution of ephrin B2 in vertebrates provides an explanation for the diverse host range of henipaviruses, there are mammalian species such as mice which express functional ephrin B2, shown to bind henipaviruses, but remain resistant to infection with HeV and NiV.

Serologic Relationships

NiV was initially described as Hendra-like on the basis of its reactivity with anti-HeV antibodies, a similarity corroborated by complete genome sequencing. No serological relationship between henipaviruses and other paramyxoviruses has been confirmed. The close serological relationship of HeV and NiV makes it difficult to identify antibodies to each virus and determine the virus species circulating in specific geographical areas. A four- to eightfold difference between homologous and heterologous neutralization titers is the basis upon which antibody specificity has been determined.

Four neutralizing epitopes have been mapped on the globular head of the HeV G protein. Two discontinuous epitopes are located on the base of the head and two on the top, in locations resembling those identified as neutralizing sites in other paramyxoviruses. The amino acid homology between HeV and NiV is relatively high at one of the discontinuous epitopes but decreases significantly at the other three sites.

Geographical Distribution

HeV has been isolated from horses, humans, and flying foxes in Australia and NiV from humans, pigs, and flying foxes in Malaysia, Bangladesh, and Cambodia. NiV has also been isolated from fruit partially eaten by flying foxes in Malaysia, and NiV RNA has been detected in human patients and bats in India and Thailand, respectively. Comparative neutralization studies using Australian bat sera confirm that HeV populates the southern domain of the henipavirus distribution. Conversely, Malaysian and Indonesian sera preferentially neutralize NiV, rather than HeV. Equivalent neutralization titers to both HeV and NiV suggest that a related virus may be circulating in the Cambodian bat population. The geographic distribution of *Pteropus* species which range from the east coast of Africa, through the Indian subcontinent and Southeast Asia, north to Okinawa and south to Australia, suggests that henipaviruses may also be found in flying fox populations in geographically more diverse locations.

Evolution and Genetics

Genetically diverse populations of NiV circulate in Southeast Asia. In Malaysia, three closely related lineages of NiV have been identified. A number of identical isolates from humans and pigs during the major outbreak in the pig-farming communities in central Malaysia in 1999 constitute the first lineage. A 1999 isolate from the initial focus of infection in northern Malaysia is the sole representative of a second lineage and NiV isolated from the urine of flying foxes on Tioman Island off the east coast of peninsular Malaysia constitutes the third lineage, one that is genetically closer to viruses isolated during the major outbreak. It has been suggested that there may have been two bat-to-pig spillover events in Malaysia, only one of which caused significant transmission to humans. However, the role played by virus evolution in pigs, particularly during the period following introduction into the pig population in 1998 and prior to the major outbreak in 1999, remains uncertain.

Unlike the Malaysian isolates which displayed an overall sequence divergence of less than 1%, regardless of source, NiV isolates from human patients in Bangladesh in 2004 differed from the Malaysian isolates by approximately 7% in the protein-coding regions and up to 25% in the non-coding regions. The heterogeneity of Bangladesh isolates suggested multiple spillovers of NiV from flying foxes into the human population. A further evolutionary lineage situated between the NiV-Malaysia isolates and NiV-Bangladesh isolates appears to exist in Cambodia.

The paucity of HeV isolates has restricted speculation on the evolution of this henipavirus in Australia. Partial sequencing suggests that viruses isolated from equine and human sources during the initial outbreak in Brisbane in 1994 and from flying foxes two years later were identical.

Epidemiology and Transmission

Human henipavirus infections in Australia and Malaysia occurred as a result of transmission of the viruses from flying foxes via horses and pigs, respectively. In contrast, NiV may have been directly transmitted to humans in Bangladesh, where it is believed that date palm juice contaminated with bat secretions constituted a potential transmission route. The mode of transmission from bats to spillover hosts in Australia and Malaysia is not known. It has been suggested that horses and pigs may have been infected by HeV and NiV through masticated fruit pulp spat out by flying foxes or by flying fox urine which contaminated pastures or pig sties. Alternatively, the source of virus may have been infected fetal tissues or fluids, a suggestion based on the fact that HeV outbreaks occurred during the birthing period of some species of flying fox, the isolation of HeV from a pregnant flying fox and its fetus, and the transplacental transmission of HeV in experimental infections.

HeV has been transmitted from horse to man on four occasions. The source of the virus may have been the saliva of infected horses, the nasal discharge commonly found at the terminal stages of the disease, or a wide range of infected tissues made accessible by necropsy. The paucity of virus in the bronchi or bronchioles of infected horses suggests transmission of HeV to either man or horse by aerosol is highly unlikely and experimental infection has confirmed the poor transmissibility of HeV in horses.

Risk factors for human infection by NiV in Malaysia were contact with pigs or fresh pig products and because the virus was present in a wide range of organs, the greatest likelihood arose for those in direct contact with sick or dying pigs on farms during farrowing or slaughtering or in abattoirs. NiV is readily observed in the respiratory epithelium of naturally and experimentally infected pigs, a feature suggesting that the virus probably spread to humans and within the pig population by aerosol or by direct contact with oropharyngeal or nasal secretions. Although NiV was present in the urine and respiratory secretions of patients, human-to-human transmission in Malaysia was extremely rare. In contrast, in Bangladesh, epidemiologic evidence indicates spread of the virus from person to person.

Tissue Tropism

Henipaviruses cause systemic infections displaying a predilection for vascular endothelial cells. The identification of ephrin B2 as a henipavirus receptor not only provides an explanation for the diverse host range of these viruses, but also helps to explain the observed systemic distribution of viral antigen particularly in arterial endothelial cells and smooth muscle. Ephrin B2 is found in arteries, arterioles, and capillaries in multiple tissues and organs but is absent from venous components of the vasculature. It is also found in arterial smooth muscle. Ephrins play critical roles in axonal guidance during vertebrate embryonic development and ephrin B2 is expressed on neurons. The availability of ephrin B2 or other receptors, the requirement for co-receptors, or the differing capacities of cells to support the production of infectious virus may determine whether henipaviruses act as respiratory or neurological pathogens.

Pathogenicity

The outcome of henipavirus infections can range from high mortality, as seen with HeV in horses and NiV in humans, through low mortality and morbidity, best exemplified by NiV in pigs, to asymptomatic as observed in flying foxes. The highly virulent nature of henipaviruses in humans and the lack of therapeutic modalities have led to the classification of HeV and NiV as Biosafety Level 4 (BSL4) pathogens. Experimental infections reveal that henipaviruses are infectious following either oronasal or parenteral administration. Figures for the minimum lethal dose of henipaviruses are known only for NiV in golden hamsters (*Mesocricetus auratus*) where the LD_{50} was 270 plaque-forming units (pfu) and 47 000 pfu following intraperitoneal and intranasal administration, respectively.

Clinical Features of Infection

Henipaviruses display either predominantly respiratory or neurological tropisms depending on the host. In natural infections of horses and young pigs with HeV and NiV, respectively, and in experimental infections of cats with either virus, respiratory symptoms predominate. Neurological symptoms were also observed in a proportion of HeV-infected horses. Experimental infection of horses and cats is usually fatal, with death or euthanasia occurring 5–10 days post infection. Following natural infection of horses, however, the observations during the initial outbreak indicated that some animals displayed respiratory symptoms but survived and others responded to infection asymptomatically. HeV-induced respiratory disease in horses may be accompanied by facial swelling and ataxia and *in extremis*, a copious frothy nasal discharge. Natural infection of pigs with NiV is usually asymptomatic, an outcome also observed after experimental administration of NiV by the ocular and oronasal route. When symptoms are present, they vary according to the age of the pig. Young animals present primarily with respiratory symptoms. Older animals display increased salivation and nasal discharge and on occasion develop neurological signs such as trembling, muscle spasms, and an uncoordinated gait. In Malaysia, pigs displaying symptoms of NiV infection, such as nasal discharge and rapid and labored respiration, had a harsh and nonproductive cough, which gave rise to the name 'barking-pig-disease'.

In contrast, NiV infection in humans is usually associated with severe acute encephalitis and although a proportion of cases presented with respiratory disease, particularly in Bangladesh, the majority displayed fever, headache, drowsiness, dizziness, myalgia, vomiting, and a reduced level of consciousness. Clinical signs such as the absence of reflexes and the irregular twitching of muscles or parts of muscles and an abnormal doll's eye reflex are indicative of brainstem and upper cervical spinal cord dysfunction. In the Malaysian outbreak, 105 of 256 patients died, a mortality rate of 41%. However, this figure reduces to *c.* 30% when individuals who experienced either a mild or asymptomatic infection are taken into account. In Bangladesh, 66 of 90 patients died in outbreaks in 2001, 2003, and 2004, giving a combined case–fatality rate of approximately 70%. Whereas HeV or NiV infection of cats presents a model of respiratory infection, the clinical and pathological features observed in NiV-infected hamsters resemble those found in human cases of encephalitis.

Both HeV and NiV can cause prolonged infection in humans before manifesting as causes of severe neurological disease. In Malaysia, NiV persisted in a proportion of patients (*c.* 10%) who either recovered from encephalitis or who had experienced an asymptomatic infection. Such cases of relapsed and late-onset encephalitis respectively presented from months to years after the initial infection. Only four cases of human HeV infection have been recorded. Two displayed influenza-like symptoms and one died. A third, fatal case of encephalitis occurred over a year after a self-limiting episode of meningitis, attributed to HeV infection.

Pathology and Histopathology

The respiratory disease caused by HeV in horses is characterized by pulmonary edema and congestion. Viral antigen is found in endothelial cells in a range of organs including lungs, lymph nodes, kidneys, spleen, bladder, and meninges, and virus can be recovered from a number of internal organs, including lung, and from saliva and urine. Young pigs infected with NiV present primarily with respiratory symptoms caused by tracheitis and bronchial and interstitial pneumonia. After experimental infection by the ocular and oronasal routes, virus replication occurs in the oropharynx and spreads to the respiratory tract and lymphoid tissues before appearing in the trigeminal ganglion and neural tissue. Viral antigen is found, particularly in clinically affected animals, in both lungs and meninges and virus is recovered from a range of tissues including tonsil, nasal, and throat swabs and lung. Detection of NiV in the urine of infected pigs is uncommon.

The primary site of replication and dynamics of henipavirus spread in humans are unknown but the distribution and time of appearance of lesions throughout the vasculature and in the brain in NiV encephalitis suggest that hematogenous spread delivers the virus from primary to secondary sites of replication in widely dispersed vascular endothelial cells. Inflammation of blood vessels, particularly small arteries, arterioles, and capillaries, occurs in most organs but is prominent in brain, lung, heart, and kidney. The interval between maximum vasculitis in the brain and parenchymal infection in NiV encephalitis suggests that infection of neurons occurs as a result of vascular damage and that neurological impairment may be due, not only to the effects of ischemia and infarction, but also viral infection of neurons. Although NiV antigen is found infrequently in epithelial cells of the bronchi and kidney, replication in such locations may play a role in virus dissemination. Recent outbreaks of NiV in Bangladesh strongly suggest human to human transmission.

No disease syndromes associated with henipaviruses have been documented in wild flying fox populations. Nor does HeV cause clinical disease in experimentally infected flying foxes and only a proportion of infected bats respond with sporadic vasculitis in the lung, spleen, meninges, kidney, and gastrointestinal tract. The mode of henipavirus transmission between flying foxes is unknown but the presence of virus in the placenta and transmission without apparent harm to the fetus in experimentally infected bats suggests that horizontal transmission is feasible.

Immune Response

Adaptive

In susceptible nonpteropid species, henipaviruses elicit strong humoral immune responses. No studies have been made of cellular responses to henipavirus infection. Antibodies to P, N, and M proteins are evident in Western blots and anti-F and anti-G antibodies are readily detected by enzyme-linked immunosorbent assay (ELISA), using both traditional and bead-based formats, and by immune precipitation. In contrast to the consistent antibody response in susceptible, nonvolant species, in flying foxes there appears to be little direct correlation between detectable virus replication and the appearance of antiviral antibody. Antibody production is irregular and of uncertain longevity.

Anti-NiV antibodies were present in a majority of patients with clinical NiV encephalitis and IgM antibodies occurred more frequently than IgG antibodies in both serum and cerebrospinal fluid (CSF). The appearance of specific IgM antibodies in serum preceded their appearance in the CSF, a sequence consistent with viremia preceding central nervous system infection.

Innate

Henipaviruses inhibit both interferon (IFN) induction and IFN signaling. In the induction phase, the detection of viral double-stranded (ds)RNA by both RNA helicase enzymes and Toll-like receptor 3 (TLR3) leads to the transcription of IFN. Henipaviruses inhibit this process in a number of ways. The V protein, like that of other paramyxoviruses, blocks the activity of the helicase sensor in the cytoplasm but does not inhibit the TLR3 pathway. In contrast, in a process unique to paramyxoviruses and by virtue of the nuclear localization signal in its C-terminal domain, the W protein inhibits IFN induction by blocking a late nuclear step in the activation process which is shared by both helicase and TLR3 pathways.

In the IFN signaling pathway, IFN binds to cell surface receptors in a paracrine manner and initiates a sequence of reactions that leads to activation of members of a family of proteins called signal transducers and activators of transcription (STAT). STAT proteins activate transcription of hundreds of genes some of whose products inhibit virus replication. In a strategy unique to paramyxoviruses, henipaviruses inhibit IFN signaling by sequestering STAT proteins in high molecular weight complexes. The V and P proteins bind STAT proteins in the cytoplasm, whereas the W protein co-localizes with STAT in the nucleus. The extra length of the henipavirus P gene compared with that of other *Paramyxovirinae* results in the encoded P, V, and W proteins having an N-terminal extension of c. 100 amino acids compared with their morbillivirus and respirovirus counterparts. This domain appears to provide henipaviruses with a multifaceted anti-IFN signaling activity.

Prevention and Control

Therapeutic options for treatment of infections caused by henipaviruses are limited and currently use of ribavirin appears to be the only recourse. The drug inhibits replication of HeV in cells in culture, and in an open-label study of 194 patients during the NiV outbreak in Malaysia, its use resulted in a 35% reduction in mortality. Duration of ventilation and total hospital stay were both significantly shorter in those receiving ribavirin. Human monoclonal antibodies to the henipavirus G protein also appear to have therapeutic potential. Antibodies which react with the soluble G protein of HeV have been generated from naive recombinant human antibody libraries and they neutralize both HeV and NiV *in vitro*.

A number of promising strategies to develop henipavirus vaccines are being explored. NiV F and G glycoproteins expressed from vaccinia virus elicit neutralizing antibodies in both mice and hamsters and in the latter both anti-F and anti-G antibodies protected from a lethal NiV challenge, although they did not prevent virus replication. Murine anti-F and anti-G antisera also inhibited membrane fusion mediated by F and G glycoproteins expressed in cell culture. More importantly, because of the attractiveness of a subunit compared with a recombinant vaccine, purified soluble HeV and NiV G proteins elicit potent neutralizing antibody responses in rabbits. Antibodies neutralized both HeV and NiV in cell culture and displayed a slightly higher titer against the homologous virus.

A further therapeutic option relies on the fact that henipavirus F_1 glycoproteins have α-helical heptad repeat (HR) domains proximal to the fusion peptide and transmembrane domain at the N- and C-terminal of the protein, respectively. HR domains form a trimer-of-hairpins structure during the fusion of virus and cell membranes. Peptides corresponding to the N- and C-terminal HR domains of HeV and NiV form trimer-of-hairpins structures *in vitro*. Addition of exogenous peptide from either HR domain blocks formation of the trimer-of-hairpins, inhibits cell fusion mediated by vaccinia virus-expressed F and G proteins, and prevents virus infection of cells in culture.

Future Perspectives

The presence of HeV or NiV in flying foxes throughout Australia, Southeast Asia, and part of the Indian subcontinent suggests that the distribution of henipaviruses may parallel that of pteropid species, which range from Madagascar to the South Pacific. Although the factors responsible for the emergence of henipaviruses have not been clearly elucidated, the destruction of native habitats, a process which is unlikely to abate, forces flying foxes into contact with man as they seek food in areas frequented by humans. Although documented outbreaks of disease caused by henipaviruses are relatively infrequent, lack of knowledge of the modes of transmission between fruit bats and from fruit bats to spillover hosts, the high infectivity of the viruses for certain species, their broad species tropism, and their zoonotic potential will ensure that further work on henipaviruses remains a priority.

See also: Measles Virus; Mumps Virus; Paramyxoviruses of Animals; Rinderpest and Distemper Viruses.

Further Reading

Eaton BT, Broder CC, Middleton D, and Wang LF (2006) Hendra and Nipah viruses: Different and dangerous. *Nature Reviews Microbiology* 4: 23–35.

Eaton BT, Mackenzie JS, and Wang LF (2006) Henipaviruses. In: Knipe DM and Howley PM (eds.) *Fields Virology*, 5th edn., pp. 1587–1600. Philadelphia: Lippincott Williams and Wilkins.

Field HF, Mackenzie JS, and Daszak P (2007) Henipaviruses: Emerging paramyxoviruses associated with fruit bats. *Current Topics in Microbiology and Immunology* 315: 133–160.

Hyatt AD, Zaki SR, Goldsmith CS, Wise TG, and Hengstberger SG (2001) Ultrastructure of Hendra virus and Nipah virus. *Microbes and Infection* 3: 297–306.

Lamb RA and Parks GD (2006) *Paramyxoviridae*: The viruses and their replication. In: Knipe DM and Howley PM (eds.) *Fields Virology*, 5th edn., pp. 1449–1496. Philadelphia: Lippincott Williams and Wilkins.

Wong KT, Grosjean I, Brisson C, et al. (2003) A golden hamster model for human acute Nipah virus infection. *American Journal of Pathology* 163: 2127–2137.

Wong KT, Shieh WJ, Zaki SR, and Tan CT (2002) Nipah virus infection, an emerging paramyxoviral zoonosis. *Springer Seminars in Immunopathology* 24: 215–228.

Hepadnaviruses of Birds

A R Jilbert, Institute of Medical and Veterinary Science, Adelaide, SA, Australia
W S Mason, Fox Chase Cancer Center, Philadelphia, PA, USA

© 2008 Elsevier Ltd. All rights reserved.

Introduction

The avian hepadnaviruses belong to the genus *Avihepadnavirus* in the family *Hepadnaviridae*. Within this genus, duck hepatitis B virus (DHBV) and heron hepatitis B virus (HHBV) are assigned to the species *Duck hepatitis B virus* and *Heron hepatitis B virus*, respectively. These avian viruses are related phylogenetically through similarities in genome sequence and organization of open reading frames (ORFs) to human hepatitis B virus (HBV) and other hepadnaviruses that infect mammals (genus *Orthohepadnavirus*). The major site of hepadnavirus infection and replication is the hepatocyte, the predominant cell type of the liver, constituting about 60–70% of liver cell mass. There are $\sim 5 \times 10^{10}$ hepatocytes in the liver of an adult duck.

Hepadnaviruses contain a relaxed circular double-stranded DNA (rcDNA) genome which is converted in the nucleus to covalently closed circular DNA (cccDNA). cccDNA acts as the virus transcriptional template and is used by host RNA polymerase II to produce a greater than genome length, 'pregenomic' RNA, referred to as the pregenome, and a number of other mRNA species. Hepadnaviruses have a unique method of replication that involves reverse transcription of pregenomic RNA into DNA. Virus replication and release are generally considered to be noncytopathic and disease activity is attributed to the host immune response to infected hepatocytes.

The avian hepadnaviruses are naturally transmitted *in ovo* via vertical transmission of virus from an infected female duck to the egg, with virus replication occurring in the yolk sac and liver of the developing embryo. Ducks infected *in ovo* develop widespread and persistent DHBV infection of the liver but have minimal or no liver disease as they are immune tolerant to the virus. Although experimental infection of newly hatched ducks also leads to widespread and persistent DHBV infection of the liver, DHBV-infected ducks, unlike HBV-infected humans, do not develop severe liver disease, cirrhosis, or liver cancer. However, they provide a reproducible experimental system for studying virus kinetics and immune clearance, and much of what we know of the hepadnavirus replication strategy has been discovered from studies of DHBV *in vitro* and *in vivo*. DHBV-infected ducks are also used to evaluate new antiviral therapies and vaccines with the ultimate aim of applying the same strategies to the treatment of HBV infections in humans.

Virion and Particle Structure

Like the mammalian hepadnaviruses, avian hepadnaviruses are enveloped viruses possessing an icosahedral nucleocapsid and ~3000 nt rcDNA genome (**Figure 1**). DHBV, the prototypic avian hepadnavirus, has a diameter of 40–45 nm, and different isolates contain genomes ranging in length from 3021 to 3027 nt. In infected ducks, mature enveloped DHBV virions are released from hepatocytes directly into the bloodstream. Virus titers in chronically infected ducks are up to 1×10^{10} virions per ml of serum.

The envelope of DHBV is composed of a lipid bilayer containing multiple copies of two transmembrane virus proteins, Pre-S/S and S (36 and 17 kDa, respectively). These two proteins are synthesized from the same virus ORF, with initiation of Pre-S/S occurring upstream of S. The two proteins share a common carboxy terminus. An additional 28 kDa protein antigenically related to Pre-S/S and S has been detected in DHBV-infected liver but it is unclear whether this is a degradation product of Pre-S/S or is translated from an AUG codon mapping between the start sites of Pre-S/S and S. As for HBV, the envelope proteins not only participate in virion formation but also self-assemble into noninfectious virus-like DHBV surface antigen particles (DHBsAg particles). This process occurs in the endoplasmic reticulum and results in the release from infected hepatocytes of a 500- to 1000-fold excess of DHBsAg particles over infectious virions. DHBsAg particles do not contain a virus nucleocapsid and lack virus nucleic acids. DHBsAg particles are pleomorphic and spherical, with a diameter of ~35–60 nm, and the larger particles are almost indistinguishable from DHBV virions by electron microscopy (EM). In contrast, during HBV infection, filamentous forms of surface antigen-containing particles (HBsAg) are produced as well as 22 nm spherical HBsAg particles, both of which are readily distinguished from those of HBV by EM.

The nucleocapsid of DHBV is icosahedral, as observed by cryoelectron microscopy, and is comprised of dimers of 30 kDa nonglycosylated DHBV core antigen (DHBcAg). In mature virions, the rcDNA genome and a virus-encoded DNA polymerase are contained in the virus nucleocapsid (**Figure 1**), which also contains cellular chaperones.

The DHBV polymerase is a 90 kDa molecule that functions as both an RNA-dependent DNA polymerase and a DNA-dependent DNA polymerase. The polymerase also has an RNase H activity that digests the RNA pregenome during virus replication. The protein sequence

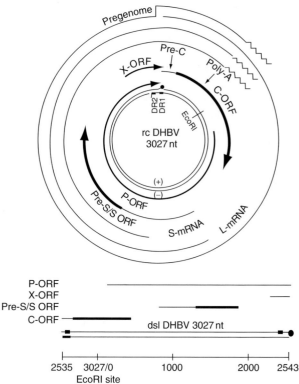

Figure 1 Organization of the DHBV genome. The rcDNA genome is shown at the top center, with the virus polymerase covalently attached to the 5′ end of the negative strand and an 18 nt RNA bound to the 5′ end of the positive strand. The DHBV genome contains two direct repeat (DR) sequences (DR1 and DR2). The locations of the virus ORFs, DR1 and DR2 and maps of the virus mRNAs are also shown. None of the major mRNAs is spliced. Thus, each has its own promoter. However, they share a common polyadenylation (poly-A) signal in the virus C-ORF. The pregenome RNA is terminally redundant and serves as the mRNA for both the C-ORF and, less frequently, the downstream P (polymerase) ORF. The pregenome lacks the AUG of the Pre-C/C-ORF, which instead is translated from an mRNA a few nt longer at the 5′ end than the pregenome (not shown). The L and S mRNAs encode the Pre-S/S and S envelope proteins, respectively. The more rarely synthesized double-stranded linear (dsl) virus genome is shown at the bottom. The common nucleotide numbering system is also displayed, proceeding in a clockwise direction from a conserved EcoRI restriction site.

of the DHBV polymerase has four separate domains listed from the N terminal end, that include a terminal protein (TP), 'spacer', reverse transcriptase (RT), and RNase H domain. Each of these domains has one or more key roles during virus replication, as described below.

Also released from hepatocytes and circulating in the blood of infected ducks is the so called DHBV e antigen (DHBeAg), a proteolytically processed form of the 35 kDa Pre-C/C protein. Different forms of DHBeAg range in size from a nonglycosylated 27 kDa to 30–33 kDa glycosylated forms. DHBeAg is thought to play an important role in either the initiation or maintenance of persistent infection by delaying or diverting the host immune response to DHBcAg. Both DHBsAg and DHBcAg are highly immunogenic. Anti-DHBc antibodies can be detected by ELISA in the serum of ducks during acute and persistent DHBV infections. Anti-DHBs antibodies are produced during the resolution phase of acute DHBV infection and are generally not detected in ducks with persistent infection except in complexes with circulating DHBsAg. Anti-DHBs antibodies bind virus particles and are able to block DHBV infection of cells. For this reason, ducks that have recovered from acute DHBV infection are immune to challenge with DHBV.

Genome Organization

As noted above, all hepadnaviruses have a similar genome organization, despite the low DNA sequence identity between the mammalian and avian hepadnaviruses (40%). The negative strand of the DHBV genome contains S-ORF, C-ORF, and P-ORF encoding the surface or envelope proteins, core and e antigen proteins, and polymerase protein, respectively (**Figure 1**). The DHBV genome was originally thought to contain only these three ORFs and to lack an ORF encoding a protein analogous to the X protein of the mammalian hepadnaviruses. It was subsequently discovered that avian hepadnaviruses, including HHBV and DHBV, have a fourth ORF, the X-ORF, which directs synthesis of a candidate X protein that is translated using an unconventional start codon. The significance of this observation is still unclear. Using woodchuck hepatitis virus (WHV), a close relative of HBV, the X gene was shown to be essential for successful WHV infections *in vivo*. In contrast, knockout of the X gene of DHBV did not alter the time course of DHBV infection. As with wild-type DHBV, the X gene knockout of DHBV spread rapidly through the liver of newly hatched ducks and caused persistent DHBV infection.

The positive strand of the DHBV genome does not contain any functional ORFs. All virus mRNA species are produced from the negative strand of the virus genome.

The difference in size between the genomes of the avian (e.g., DHBV with a genome of 3027 nt) and the mammalian hepadnaviruses (e.g., HBV with a genome of 3200–3300 nt), is due in part to deletion of a 150 nt stretch in S-ORF. This region in the HBV genome encodes the so-called 'a' determinant, a highly immunodominant region to which most of the neutralizing anti-HBV antibodies are directed. This region in HBV is present on the small form of HBsAg, the S protein. The absence of this region in DHBV results in a lack of immunodominance, and both DHBV Pre-S/S and S domains have been shown to induce high titer, neutralizing, anti-DHBs antibodies.

Replication Cycle

Hepatocytes, the major parenchymal cell of the liver, are the primary site of DHBV infection. The cell specificity of infection is determined by the presence of specific receptors required for virus binding and entry into hepatocytes. In early studies, DHBsAg particles, which appear to have the same envelope structure as DHBV virions, were shown to bind cell surface receptors on primary duck hepatocytes in a species-specific manner. Receptor binding occurs via the Pre-S/S region of DHBV and is followed by cell entry via receptor-mediated endocytosis. Recent studies have determined that the cellular protein carboxypeptidase D (180 kDa) binds DHBsAg particles and DHBV with high affinity and is found on both internal and external membranes of the cell. However, whereas DHBV has a narrow host range, carboxypeptidase D is expressed on many cell types that are not susceptible to DHBV infection. In addition, despite the observed binding of DHBsAg to carboxypeptidase D, transfection of cells with carboxypeptidase D cDNA does not confer susceptibility to DHBV, suggesting that other co-receptors or mechanisms may be operating. An additional DHBV-binding protein, glycine decarboxylase (120 kDa) has been identified. Its cellular expression is restricted to the liver, kidney, and pancreas. Interestingly, DHBV infection and replication has been detected in a few percent of cells in both the kidney and pancreas (see below and **Figure 5**). Thus, these proteins are potential components of a DHBV receptor complex and probably have a role in determining DHBV organ tropism. However, definite proof that these are the receptors that lead to DHBV binding, receptor-mediated endocytosis, and infection is still lacking.

Whatever the receptor, the immediate sequel to virus uptake is transport of the rcDNA genome to the nucleus to form cccDNA (**Figure 2**). The negative strand of the rcDNA DHBV genome is nicked, with a 9 nt terminal redundancy and with the virus polymerase molecule covalently attached to its 5′ end. Covalent attachment of the virus polymerase occurs during priming of reverse transcription through a tyrosine located in the TP domain of the polymerase. Similarly, the positive strand of the DHBV genome has an 18 nt RNA primer attached to its 5′ end (**Figure 1**). In addition, the positive strand is incomplete, with a minimum gap of 12 nt. Thus, following transport to the nucleus, the covalently linked polymerase protein, the RNA primer, and the terminal redundancy in the negative strand are removed, the positive strand is completed, and ligation of the ends of each strand takes place. As noted above and discussed in detail below, ∼10% of circulating DHBV virions contain double-stranded linear (dsl) genomes that result from a defect that occurs during synthesis of positive-strand DNA. Following infection these dsl genomes can also enter the nucleus to form cccDNA, which in this case is formed

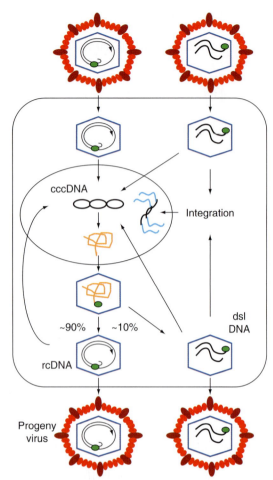

Figure 2 DHBV infection of hepatocytes. Virions, shown at the top, may have either an rcDNA or, more rarely, a dsl DNA genome. Upon infection virus nucleocapsids are uncoated and move to the nucleus. Virus DNA is released from nucleocapsids and probably enters the nucleus at nuclear pores. Both rcDNA and dsl DNA give rise to cccDNA, the virus transcriptional template. cccDNA is used by host-cell RNA polymerase II to produce pregenomic RNA that is reverse transcribed to rcDNA or dsl DNA. dsl DNA also has a propensity to integrate into host-cell DNA (see text). Early in infection, cccDNA copy number increases by intracellular amplification from rcDNA and dsl DNA made in the cytoplasm via reverse transcription. Later, this pathway is shut down and nucleocapsids containing rcDNA and dsl DNA bud into the endoplasmic reticulum and are released as progeny virus.

by illegitimate recombination between the ends of the dsl DNA. This recombination event typically involves some loss of sequences, and cccDNA formed from dsl genomes is generally defective.

DHBV cccDNA molecules exist within the nucleus of each infected hepatocyte as a population of virus minichromosomes that bind up to 20 nucleosomes per 3000 nt molecule. cccDNA does not undergo semiconservative replication. New copies are therefore formed from rcDNA and dsl DNA synthesized in the cytoplasm of infected hepatocytes. Once formed, cccDNA is transcribed by host RNA polymerase II to produce pregenomic and other virus mRNAs (**Figures 1** and **2**), followed by protein production, assembly of nucleocapsids from the viral core protein and packaging of pregenomic RNA. Each virus mRNA is produced from its own promoter and virus RNA molecules including the pregenome have a 5′ cap and are polyadenylated at a common site (nt 2778–2783) on the DHBV genome numbered according to the sequence of the Australian strain of DHBV (GenBank AJ006350) where the unique EcoRI site is nucleotide 1.

Packaging of pregenomic RNA is facilitated by the presence of an encapsidation signal, epsilon (ε), located at the 5′ end (nt 2566–2622) and at the 3′ end of the pregenome, within the terminal redundancy (**Figure 3(a)**). Pregenome packaging is dependent upon binding of the virus polymerase to the 5′ ε sequence. In fact, the pregenome is able to serve as the mRNA for both core and polymerase proteins (**Figure 1**). However, once polymerase is translated, it may bind to its own message and block its further translation. Once packaging into nucleocapsids has occurred, virus DNA synthesis takes place via reverse transcription of the pregenomic RNA, with production of replicative intermediate DNA (RI DNA) to produce rcDNA and dsl DNA genomes in a $\sim$10:1 ratio (**Figures 3(f)** and **3(h)**).

DNA synthesis begins with reverse transcription of 4 nt (5′ UUAC 3′) in the bulge in the stem–loop structure of ε, leading to synthesis of 4 nt of DNA (5′ GTAA 3′) (**Figure 3(b)**). As noted above, a tyrosine residue in the TP domain of the polymerase serves as the primer of reverse transcription. Following synthesis by reverse transcription of the first 4 nt, the complex is translocated to DR1, in particular the 3′ copy of DR1, DR1* (**Figure 3(c)**).

DR1 is a 12 nt sequence located 6 nt from the 5′ end of the pregenome at nt 2541–2552. It is therefore also present in the terminal redundancy of the pregenome. The same sequence known as DR2 is located about 50 nt upstream of the terminal redundancy at nt 2483–2494. Once transferred to DR1*, where the 4 nt can base pair due to sequence homology, reverse transcription re-initiates and continues to the 5′ end of the pregenome, to produce a full-length negative strand with a 9 nt terminal redundancy (**Figures 3(d)** and **3(e)**). Most of the pregenome is degraded during negative-strand elongation by the RNase H activity of the virus polymerase. However, the 5′ $\sim$18 nt of the pregenome, including the cap and all of the 5′ copy of DR1, escapes RNase H degradation and serves as the positive DNA strand primer. Positive-strand synthesis leading to rcDNA formation initiates near the 5′ end of the negative strand. The primer is first translocated from its original location at the 3′ end of the template to DR2 where it can hybridize due to the sequence identity of DR1 and DR2 (**Figure 3(e)**). Following positive-strand elongation to the 5′ end of the negative strand, circularization occurs to facilitate continuation

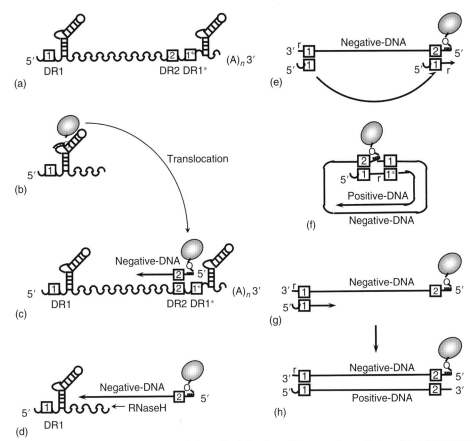

Figure 3 Mechanism of DHBV DNA synthesis. Virus DNA synthesis begins with reverse transcription of 4 nt (5′ UUAC3′) of the bulge in the stem–loop structure, ε, and leads to synthesis of the first 4 nt (5′ GTAA3′) of negative-strand DNA, as shown in (b). Polymerase with the nascent transcript then translocates to the right-hand copy of DR1, DR1*, and reverse transcription of the negative strand and degradation of the pregenomic RNA template by RNase H proceeds as shown in (c) and (d). Following completion of the negative strand, the 5′ 18 nt of the pregenome, including the cap, are typically translocated to DR2 and positive-strand DNA synthesis then initiates from this RNA primer. In this case, rcDNA formation is of necessity an early step in positive-strand synthesis ((e) and (f)). About 10% of the time, positive-strand synthesis initiates without translocation of the primer (g), leading to the formation of dsl DNA (h). The details of DNA synthesis are described in greater detail in the text.

of positive-strand synthesis to produce mature rcDNA (**Figure 3(f)**). Circularization is facilitated by the 9 nt terminal redundancy on the negative-strand template.

dsl DNA genomes reflect a failure of positive-strand primer translocation from DR1 to DR2, resulting in a phenomenon known as *in situ* priming to produce a DNA that is collinear with pregenomic RNA up to DR1* (**Figure 3(h)**). Interestingly, both the polymerase, the protein primer of negative-strand synthesis and the RNA primer of positive-strand synthesis remain associated with virus DNA throughout virion assembly. Biochemical evidence suggests that each nucleocapsid contains only a single copy of polymerase protein, implying that a single protein is simultaneously the primer of DNA synthesis and the polymerase for both RNA- and DNA-dependent DNA synthesis.

Newly made rcDNA and dsl DNA are enveloped by budding into the endoplasmic reticulum and exported as progeny virus via the Golgi apparatus, or are transported to the nucleus to make additional copies of cccDNA, typically found at 10–30 copies per hepatocyte. Nuclear transport is negatively regulated by the virus envelope proteins, which direct nucleocapsids into the pathway of virus assembly. This negative regulation is essential because excessive accumulation of cccDNA will kill the host cell. Strong negative regulation may also be important because of the stability of cccDNA, and new synthesis of cccDNA may only be necessary to restore cccDNA levels in the progeny following division of infected hepatocytes. Negative regulation of cccDNA synthesis also appears to occur during HBV infection, but it is still not known whether the viral envelope proteins have an essential role in this process.

dsl DNA also integrates randomly into chromosomal DNA and has been detected in ~0.01–0.1% of hepatocytes during acute infections by both DHBV and the mammalian hepadnaviruses. Higher levels of integrated virus DNA accumulate during chronic infections. This probably reflects ongoing import of rcDNA and dsl DNA into the nucleus to restore cccDNA copy number as infected hepatocytes divide to replace those killed by the host immune response, thereby providing new dsl DNA genomes for integration. Integrated forms of hepadnavirus DNA are usually unable to act as templates for transcription of pregenomic RNA since the promoter for this virus RNA is located at the 3' end of the integrant, not the 5' end, and integrated DNA is therefore not able to direct virus replication (see **Figure 1**). In addition, virus sequences may be lost during the integration process, particularly from the ends of the dsl molecule.

Since each integrated virus–cell DNA junction is unique, and will be present only once in the liver unless the infected hepatocyte divides, integration sites can be used to identify and track the fate of individual hepatocyte lineages. In particular, assays for integrated DNA were used to show that hepatocytes present in the liver following recovery from an acute WHV infection were derived from previously infected hepatocytes, indicating the existence of mechanisms for removal of cccDNA from infected hepatocytes.

Phylogenetic Information

Assigned species within the genus *Avihepadnavirus* include DHBV isolated from Pekin ducks (*Anas domesticus*) and HHBV from grey herons (*Ardea cinerea*). Many DHBV isolates have been found in domesticated ducks and, in the wild, in the mallard, the species from which most domesticated ducks are derived. Viruses less closely related to DHBV have been isolated from geese and other duck species and include the Ross's goose hepatitis B virus (RGHBV) from Ross's geese (*Anser rossii*), Mandarin duck hepatitis B virus (MDHBV) from Mandarin ducks (*Aix galericulata*), and the snow goose hepatitis B virus (SGHBV) from snow geese (*Anser caerulescens*). The stork hepatitis B virus (STHBV) has been isolated from white storks (*Ciconia ciconia*), with additional viruses isolated from demoiselle (*Anthropoides virgo*) and grey crowned cranes (*Balearica regulorum*) (**Figure 4**). By genome sequencing, HHBV and STHBV are the most distant from DHBV (**Figure 4**). HHBV was assigned as a species based both on genome divergence and a host range difference from DHBV. Current information does not provide clear evidence for designation of additional new species among the isolates shown in **Figure 4**.

Thus, the avihepadnaviruses are currently classified phylogenetically into 'Chinese' and 'Western Country' isolates as well as four highly distinct lineages that include SGHBV, RGHBV, plus MDHBV, STHBV, and HHBV. Sequence divergence within the 'Chinese' and 'Western Country' DHBV strains is 5.99% and 3.35%, respectively, while divergence between the strains is 9.8%. SGHBV diverges from DHBV by 11–13%, RGHBV and MDHBV by 17–19% and STHBV and HHBV by 22–24%. Additional closely related viruses have been isolated from the puna teal (*Anas puna*), ashy-headed sheldgoose (*Chloephaga poliocephala*), Orinoco sheldgoose (*Neochen jubata*), and Chiloe wigeon (*Anas sibilatrix*) (**Figure 4**).

Pathogenesis and Control of Diseases

In persistently DHBV-infected ducks, the natural route of virus transmission is from the bloodstream to the egg. Virus replication then occurs in the yolk sac and developing liver and pancreas of the embryo (**Figures 5(a)–5(c)**) and results in congenital DHBV infection. Congenitally DHBV-infected ducks remain persistently infected for life, with infection in >95% of hepatocytes (**Figure 5(e)**). Bile duct cells are also infected with DHBV and express high levels of DHBcAg and DHBsAg (**Figure 5(e)**). Virus replication also occurs in acinar and islet cells in the pancreas (**Figure 5(f)**), and in glomeruli (**Figure 5(d)**) and tubular epithelial cells in the kidney. Cells located in the germinal centers of the spleen contain nonreplicating DHBV DNA, which is thought to be associated with follicular dendritic cells. In ducks with congenital infection, levels of DHBV DNA and DHBsAg in the bloodstream gradually decrease over 800 days and anticore antibodies can be detected from ~90 days post hatch, consistent with a high degree of immune tolerance.

Persistent DHBV infection also results following experimental inoculation of newly hatched ducks, and inoculation of infected serum containing the equivalent of one virion is sufficient to initiate DHBV infection. Infection of newly hatched ducks with DHBV has allowed detailed studies of the kinetics and tissue specificity of DHBV infection. Inoculation of 2- to 4-week-old ducks with DHBV results in either acute or persistent infection depending on the age of the duck at the time of inoculation and the dose of DHBV administered. Ducks from this age group that develop persistent infection have histological changes in the liver with mild mononuclear cell infiltration of portal tracts but no evidence of lobular hepatitis or extensive liver damage. Experimental inoculation of 4-month-old ducks with high doses of DHBV leads to either persistent infections with mild to marked liver disease or to transient infections that are rapidly cleared.

Since hepadnavirus infection is noncytopathic, the liver damage seen during DHBV infection has been attributed to the immune response directed against infected hepatocytes. Although cirrhosis and primary hepatocellular carcinoma (HCC) are not reported to occur in DHBV infection, the clinical and serological events, hepatocyte specificity, and ability of the host to rapidly resolve an infection involving the entire hepatocyte population are similar for DHBV, WHV, and HBV.

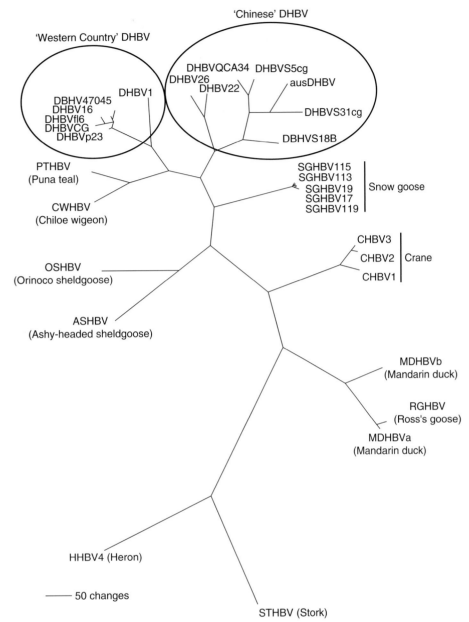

Figure 4 Phylogenetic relationships of avian hepadnaviruses. A dendrogram file was constructed using Clustal X and full-length virus sequences and displayed using Treeview. The following sequences were included: 'Chinese' DHBV isolates, accession numbers M32990, M32991, X60213, AJ006350, M21953, X58568, X58569; 'Western Country' DHBV isolates, K01834, M60677, X12798, X74623, AF047045, X58567; snow goose hepatitis B virus isolates: AF110996, AF110997, AF110998, AF110999, AF11000; Ross's goose hepatitis B virus: M95589; grey heron hepatitis B virus: M22056; stork hepatitis B virus: AJ251937; crane hepatitis B virus isolates: AJ441111, AJ441112, AJ441113; Mandarin duck hepatitis B virus isolates: AY494848, AY494849; Chiloe wigeon hepatitis B virus: AY494850; puna teal hepatitis B virus: AY494851; Orinoco sheldgoose hepatitis B virus: AY494852; and ashy-headed sheldgoose hepatitis B virus: AY49485.

Humans infected with HBV, especially when less than a year of age, often develop a healthy carrier state which can last for several decades without clinically evident liver disease. Similarly, persistent infection with avian hepadnaviruses, resulting either from *in ovo* infection or experimental inoculation after hatching, generally results in only mild hepatitis similar to the 'healthy' carrier state in humans. The failure to detect liver disease, cirrhosis and HCC in persistently DHBV-infected ducks may to be linked to the timing and mode of transmission of these viruses, since they are usually transmitted vertically by *in ovo* transmission resulting in congenital infection with immune tolerance and an absence of, or only mild, liver disease. The inability of DHBV-infected ducks to progress to HCC may also be affected by their limited life span in captivity, which is generally less than 5 years.

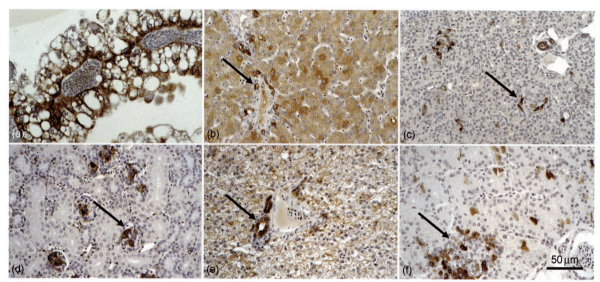

Figure 5 Hepatic and extrahepatic infection with DHBV of a 14-day-fertilized duck embryo (a–c) and a 2-week-old congenitally infected duck (d–f). Immunoperoxidase staining of DHBV core antigen in yolk sac (a), liver (b, e), pancreas (c, f), and kidney (d). DHBV is transmitted *in ovo* from the bloodstream of the persistently DHBV-infected female duck to the egg, resulting in DHBV replication in yolk sac cells (a), hepatocytes and bile duct cells (arrow) in the liver (b), and pancreatic acinar cells (c). DHBV infection persists in congenitally DHBV-infected ducks in glomerular cells in the kidney ((d); arrow), in >95% of hepatocytes and in bile duct cells of the liver ((e); arrow shows high levels of DHBV core antigen detection in bile duct cells), and in pancreatic acinar and islets cells ((f); arrow indicates a group of DHBV core antigen-positive pancreatic beta islet cells; alpha islets also positive (not shown)). Tissues were fixed in ethanol:acetic acid, immunostained with polyclonal anti-DHBV core antibodies and counterstained with hematoxylin.

HCC was detected in ducks in a Chinese province but in those areas aflatoxin exposure may have been common and the key contributory factor in the development of the liver cancers.

Conclusion

Although the avian hepadnaviruses share only 40% nucleotide sequence identity with HBV and the other mammalian hepadnaviruses, they are nearly identical in genome organization and replication strategy. Indeed, the molecular details of hepadnavirus replication were initially worked out with DHBV and later extended to HBV, and the similarities were found to be striking. The biology of infection of HBV and DHBV is also strikingly similar, with hepatocytes the primary target of infection and immune-mediated cell death the primary cause of liver disease in infected hosts.

The only major difference between the two viruses is how they are able to persist in their respective host populations. DHBV appears to be maintained primarily by vertical transmission to the developing embryo, whereas the primary route for maintenance of HBV in the human population is by horizontal transmission during the first year of life from mother to child, or among young children. Both routes lead to chronic infection, but it is generally believed that the degree of immune tolerance to the virus is much greater following vertical than horizontal transmission. This probably explains the greater disease activity in chronically infected humans than ducks. This difference is also seen in the woodchuck model. Chronic WHV infection results, as in HBV in humans, primarily from horizontal transmission, leading to a much more rapid disease progression than found in ducks.

HBV was discovered in 1967, WHV in 1978, and ground squirrel hepatitis B and DHBV viruses shortly thereafter. Among these, DHBV was the easiest to work with because of the ready availability of hosts from commercial sources. Thus, in the initial absense of a cell culture system, much of the early work on hepadnavirus replication focused on DHBV. As a result, DHBV provided much of the information on how the hepadnaviruses replicate as well as detailed early information on the biology of infection and virus spread, including possible identification of the cell surface receptor for the virus. Studies of DHBV have also provided important information on how cccDNA is made and on its high degree of stability in nondividing cells. In addition, the DHBV model has been used to demonstrate the reproducible and rapid kinetics of the spread of DHBV infection *in vivo* and to determine the high specific infectivity of DHBV, where one virus particle has been shown to be infectious in neonatal ducks.

Because of the reproducible nature of the kinetics of infection and the predictable outcomes of infection using defined doses of DHBV, the model is especially useful for the evaluation of new antiviral therapies and vaccine strategies for HBV infection in humans. Similarities in the replication strategy and polymerase enzymes of the avian and human hepadnaviruses allow evaluation of new antiviral drugs in *in vivo* models of HBV infection, including the development of antiviral drug-resistant mutants and competition between wild-type and mutant DHBV strains.

See also: Hepadnaviruses: General Features; Hepatitis B Virus: General Features; Hepatitis B Virus: Molecular Biology.

Further Reading

Foster WK, Miller DS, Scougall CA, Kotlarski I, Colonno RJ, and Jilbert AR (2005) The effect of antiviral treatment with Entecavir on age- and dose-related outcomes of duck hepatitis B virus infection. *Journal of Virology* 79: 5819–5832.

Funk A, Mhamdi M, Will H, and Sirma H (2007) Avian hepatitis B viruses: Molecular and cellular biology, phylogenesis, and host tropism. *World Journal of Gastroenterology* 13: 91–103.

Gong SS, Jensen AD, Chang CJ, and Rogler CE (1999) Double-stranded linear duck hepatitis B virus (DHBV) stably integrates at a higher frequency than wild-type DHBV in LMH chicken hepatoma cells. *Journal of Virology* 73: 1492–1502.

Guo H, Mason WS, Aldrich CE, *et al.* (2005) Identification and characterization of avihepadnaviruses isolated from exotic anseriformes maintained in captivity. *Journal of Virology* 79: 2729–2742.

Jilbert A and Locarnini S (2005) *Avihepadnaviridae*. In: Thomas H, Lemon S, and Zuckerman A (eds.) *Viral Hepatitis*, 3rd edn., pp. 193–209. Adelaide: Blackwell Publishing.

Jilbert AR, Freiman JS, Gowans EJ, Holmes M, Cossart YE, and Burrell CJ (1987) Duck hepatitis B virus DNA in liver, spleen, and pancreas: Analysis by *in situ* and Southern blot hybridization. *Virology* 158: 330–338.

Jilbert AR and Kotlarski I (2000) Immune responses to duck hepatitis B virus infection. *Developmental and Comparative Immunology* 24: 285–302.

Jilbert AR, Miller DS, Scougall CA, Turnbull H, and Burrell CJ (1996) Kinetics of duck hepatitis B virus infection following low dose virus inoculation: One virus DNA genome is infectious in neonatal ducks. *Virology* 226: 338–345.

Jilbert AR, Wu TT, England JM, *et al.* (1992) Rapid resolution of duck hepatitis B virus infections occurs after massive hepatocellular involvement. *Journal of Virology* 66: 1377–1388.

Meier P, Scougall CA, Will H, Burrell CJ, and Jilbert AR (2003) A duck hepatitis B virus strain with a knockout mutation in the putative X ORF shows similar infectivity and *in vivo* growth characteristics to wild-type virus. *Virology* 317: 291–298.

Miller DS, Bertram EM, Scougall CA, Kotlarski I, and Jilbert AR (2004) Studying host immune responses against duck hepatitis B virus infection. *Methods in Molecular Medicine* 96: 3–25.

Summers J, Smith PM, and Horwich AL (1990) Hepadnavirus envelope proteins regulate covalently closed circular DNA amplification. *Journal of Virology* 64: 2819–2824.

Yang W and Summers J (1995) Illegitimate replication of linear hepadnavirus DNA through nonhomologous recombination. *Journal of Virology* 69: 4029–4036.

Yang W and Summers J (1999) Integration of hepadnavirus DNA in infected liver: Evidence for a linear precursor. *Journal of Virology* 73: 9710–9717.

Zhang YY, Zhang BH, Theele D, Litwin S, Toll E, and Summers J (2003) Single-cell analysis of covalently closed circular DNA copy numbers in a hepadnavirus-infected liver. *Proceedings of the National Academy of Sciences, USA* 100: 12372–12377.

Hepadnaviruses: General Features

T J Harrison, University College London, London, UK

© 2008 Elsevier Ltd. All rights reserved.

Introduction

The infectious nature of viral hepatitis has been recognized throughout human history, including by documentation of epidemics of jaundice (campaign jaundice) associated with warfare. Parenteral transmission of hepatitis (serum hepatitis) was recognized from the end of the nineteenth century following transmission by routes such as tattooing, the reuse of syringes and needles, and contamination of vaccines by 'stabilization' with human 'lymph' or serum. The viral etiology of infectious and serum hepatitis was recognized in studies involving transmission to volunteers, during and after World War II. In addition, these studies established different incubation times for infectious (short incubation) and serum (long incubation) hepatitis. The terms hepatitis A and hepatitis B were coined and became accepted for infectious (or epidemic) and serum hepatitis, respectively.

Progress with hepatitis B began with the serendipitous discovery of the envelope protein of the virus, hepatitis B surface antigen (HBsAg), originally termed Australia antigen. This was discovered in 1965 by Blumberg, who precipitated an antigen–antibody complex by immunodiffusion of sera from a multiply transfused patient and an Australian aboriginal, during a study of blood and leukocyte antigens. The protein was later recognized to be associated with transmission of hepatitis and could be detected in the sera of a proportion of patients with viral hepatitis. Electron microscopic studies in the late 1960s led to the discovery of particles of around 20 nm diameter that are now known to be composed of membrane-embedded HBsAg, secreted from the hepatocytes as non-infectious, subviral particles. In 1970, Dane visualized larger, 42 nm particles with electron dense cores. Originally termed 'Dane particles', these are the hepatitis B virions and were shown later to contain a small circular

DNA genome that was partially single-stranded and associated with a DNA polymerase activity that could render the molecule fully double-stranded.

The development of tests for HBsAg, and for antibodies to hepatitis A virus, led to the recognition of transmission of hepatitis, termed non-A, non-B hepatitis (NANBH), potentially by other viruses. Post-transfusion (or parenterally transmitted), PT-NANBH is caused principally by hepatitis C virus and enterically transmitted, ET-NANBH, by hepatitis E virus. The genomes of these viruses were cloned in 1989. Earlier, in 1977, Rizzetto had described a novel antigen–antibody system, termed delta/anti-delta, in patients with hepatitis B. The delta antigen is now known to be the nucleocapsid protein of hepatitis D virus (HDV), a defective virus that requires HBsAg to form its envelope.

The Hepadnaviruses

Hepatitis B virus (HBV) is now recognized as the prototype of a family of viruses that infect mammals and birds. The *Hepadnaviridae* (named for *hepa*totropic *DNA* viruses) are divided into mammalian (*Orthohepadnavirus*) and avian (*Avihepadnavirus*) genera.

In the 1960s, it was noted by Snyder that eastern woodchucks (*Marmota monax*) in a colony in the Philadelphia Zoo were highly susceptible to the development of chronic hepatitis and hepatocellular carcinoma, and he proposed a viral etiology. The sera of animals with chronic hepatitis proved to contain a virus with similar biochemical properties to HBV and this was named woodchuck hepatitis virus (WHV). A search for similar viruses in relatives of the woodchuck turned up a closely related virus, ground squirrel hepatitis virus (GSHV) in the Beechey ground squirrel (*Spermophilus beecheyi*; *Spermophilus* and *Marmota* are genera of the family *Sciuridae*). Other related viruses are endemic in Richardson ground squirrels (*Spermophilus richardsonii*) in Canada (Richardson ground squirrel hepatitis virus, RGSHV) and arctic ground squirrels (*Spermophilus parryi kennicotti*) in Alaska (arctic squirrel hepatitis virus, ASHV).

Reports of hepatocellular carcinoma in farmed ducks in China led to the discovery of duck hepatitis B virus (DHBV; the 'B' is included to avoid confusion with the picornavirus, duck hepatitis virus, and is retained in the names of the other avian hepadnaviruses) in flocks of Pekin ducks (*Anas domesticus*) in China and the USA. In the USA, up to 10% of ducks in some commercial flocks are persistently infected. The discovery of DHBV was followed by the identification of related viruses in grey herons (*Ardea cinerea*) in Germany (heron hepatitis B virus, HHBV), in Ross's geese (*Anser rossii*; Ross's goose HBV, RGHBV) and snow geese (*Anser caerulescens*; snow goose HBV, SGHBV), in white storks (*Ciconia ciconia*; STHBV) and in grey crowned cranes (*Balearica regulorum*; CHBV). Recently, other hepadnaviruses have been isolated from a variety of exotic ducks and geese, including teal, widgeon, and sheldgeese. However, because these birds were kept in captivity, it is not clear that these novel viruses are native to the species from which they were first isolated. Indeed, RGHBV and SGHBV also were isolated from captive birds and are quite distinct viruses (**Figure 1**), despite the fact that the two avian species mix and may interbreed in the wild.

Biology of the Viruses

As noted above, hepadnavirus particles consist of a DNA-containing nucleocapsid enveloped with the surface proteins embedded in a lipid bilayer derived from the internal membranes of the host cell. Excess surface proteins are secreted as subviral particles, which may serve to subvert the immune response to the surface proteins. The surface open reading frame (ORF) in the mammalian viruses contains three, in-frame initiation codons leading to the synthesis of large (L, pre-S1+pre-S2+S), middle (M, preS2+S), and small (or major, S) surface proteins (**Figure 2(a)**). The pre-S1 region contains an endoplasmic reticulum-retention signal involved in virion assembly; its incorporation into subviral particles leads to the formation of tubular, rather than spherical, forms. The avian viruses have only large (pre-S+S) and small (S) surface proteins, the middle protein being absent (variants of HBV with small deletions that abrogate synthesis of the middle protein also seem to be viable). For both classes of virus, the ligand that interacts with the major cellular receptor seems to reside within the domain unique to the large surface protein and it is this interaction that seems to be responsible for the high degree of species specificity of this virus family. The primary receptor for DHBV is carboxypeptidase D but the receptors and co-receptors of the mammalian viruses remain to be identified.

Within the nucleocapsid, the genome (around 3.2 kbp for HBV, 3.3 kbp for the rodent viruses, and 3.0 kbp for the avian viruses) is composed of two linear strands of DNA held in a circular configuration by base-pairing of a short region where the 5′ ends overlap. As noted above, one strand (the plus strand) of the HBV genome is incomplete, usually being approximately 60–80% full length. On the other hand, the genome of DHBV tends to be almost completely double-stranded, while it has been reported that SGHBV produces a significant proportion of virions containing single-stranded DNA.

With its small size, the HBV genome was one of the first viral genomes to be cloned and sequenced completely.

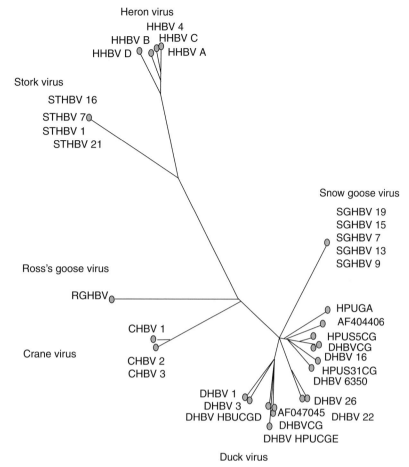

Figure 1 An unrooted phylogenetic tree showing the genetic relatedness between hepadnaviruses isolated from ducks, geese, herons, storks, and cranes. Reproduced from Prassolov A, Hohenberg H, Kalinina T, et al. (2003) New hepatitis B virus of cranes that has an unexpected broad host range. *Journal of Virology* 77: 1964–1976, with permission from American Society for Microbiology.

Four overlapping ORFs are conserved among the mammalian viruses (**Figure 2(a)**). These encode the surface and nucleocapsid proteins, the polymerase, and a small protein called x (HBxAg) for its unknown function. HBxAg is now known to be a transactivator of transcription and is presumed to act by upregulating the activity of the viral promoters and may be important in 'kick-starting' a new infection. The X ORF is absent from the genomes of the avian viruses (**Figure 2(b)**). However, the avian viruses do encode an ORF that resembles X but lacks an initiation codon; it is not know whether this is expressed during infection.

The hepadnaviruses replicate by reverse transcription of a pregenomic RNA and have been termed 'pararetroviruses'. The replication of HBV, many features of which were elucidated using as a model DHBV in its natural host, is discussed elsewhere in this encyclopedia. Briefly, the viruses are believed to enter the host cell via endosomes with delivery of the genomes to the nucleus, where they are converted to a covalently closed circular (ccc) form that is the template for transcription. The pregenomic RNA, which is also the mRNA for the nucleocapsid protein and polymerase, has a stem–loop structure near to the 5′ end. This is recognized by the polymerase, which also acts as a protein primer, and DNA synthesis begins, also signaling encapsidation. Reverse transcription and second-strand synthesis take place in immature cores in the cytoplasm; completion of the cores being followed by envelopment and exocytosis. Early in infection, some nascent cores cycle back to the nucleus to build up a pool of ccc DNA templates.

Genetic Variation and Epidemiology

HBV originally was typed serologically on the basis of the reactivity of HBsAg. All types share a common determinant known as *a*; this has been mapped to a hydrophilic region, roughly between amino acid residues (aa) 111 and 156 of the (226 aa) small surface protein, S. This highly

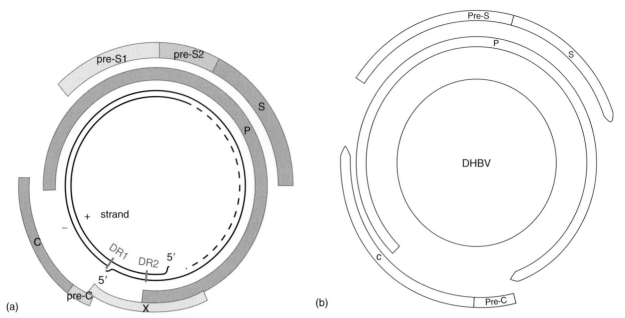

Figure 2 (a) Organization of the HBV genome. The inner circles depict the complete minus strand and incomplete plus strand and the positions of the direct repeats (DR) are indicated. The blocks surrounding the genome show the locations of the four overlapping ORFs; C and S contain two and three in-frame initiation codons, respectively. (b) Organization of the DHBV genome. A simplified view illustrating the single pre-S region and lack of an X ORF. (a) Reproduced from Kidd-Ljunggren K, Miyakawa Y, and Kidd AH, et al. (2002) Genetic variability in hepatitis B viruses. *Journal of General Virology* 83: 1267–1280, with permission from Society for General Microbiology.

conformational region, comprising a number of overlapping epitopes, is the main target of the humoral response, and antibodies synthesized during convalescence or in response to the hepatitis B vaccine are protective. Two pairs of mutually exclusive subdeterminants, *d* or *y* and *w* or *r*, correlate with variation (in both cases between lysine and arginine) at aa 122 and aa 160, respectively. Thus, four principal subtypes are recognized (*adw*, *adr*, *ayw*, and (rarely) *ayr*) and these show varying geographical distribution.

The error-prone nature of HBV DNA replication, via an RNA intermediate and without proofreading, is balanced by constraints on variation imposed by overlapping ORFs and the various *cis*-acting elements all being embedded in ORFs. Nonetheless, individual isolates of HBV vary by up to 10–14% of nucleotide positions and genotypes have been defined on the basis of >8% nucleotide sequence divergence. Currently, eight genotypes (A–H) are recognized and most of these have been divided further into subgenotypes. Genotype A is found in northern Europe and North America, B and C in east and Southeast Asia, D in a wide area though southern Europe and North Africa to India, E in western Africa, and F in South and Central America. Arguably, too few of the most recently described genotypes have been isolated to establish an epidemiological pattern, but G has been found in the USA and Europe, and H in the USA and Central America.

HBV has also been isolated from a variety of nonhuman primates. The first such complete sequence was derived from HBV from a captive chimpanzee and, at the time, it was not possible to determine whether this represented a divergent human isolate (i.e., the result of an unintentional human-to-chimpanzee transmission) or a genuine chimpanzee virus. Some years later, a related virus was isolated from a white-handed gibbon (*Hybolates lar*) and complete sequences now are available from a considerable number of chimpanzees, gorillas, orangutans, and gibbons. The most divergent of the primate hepadnaviruses sequenced at the time of writing is from the woolly monkey (*Lagothrix lagotricha*), a New World monkey. Woolly monkey HBV (WMHBV) shows around 20% nucleotide sequence divergence from HBV.

Figure 3 shows a dendrogram of the eight human HBV genotypes along with the primate viruses and rooted using WMHBV as an outgroup. Fewer genome sequences are available for the rodent hepadnaviruses. Isolates of WHV vary in fewer than 4% of nucleotide positions and differ from GSHV and ASHV in around 15% (no sequence is available for RGSHV). **Figure 4** shows a dendrogram of the primate and rodent hepadnavirus sequences, rooted using DHBV as an outgroup.

As noted above, the avian hepadnaviruses show somewhat less-restricted host specificity than those infecting mammals. **Figure 1** shows a phylogenetic tree of viruses isolated from ducks, geese, herons, storks, and cranes.

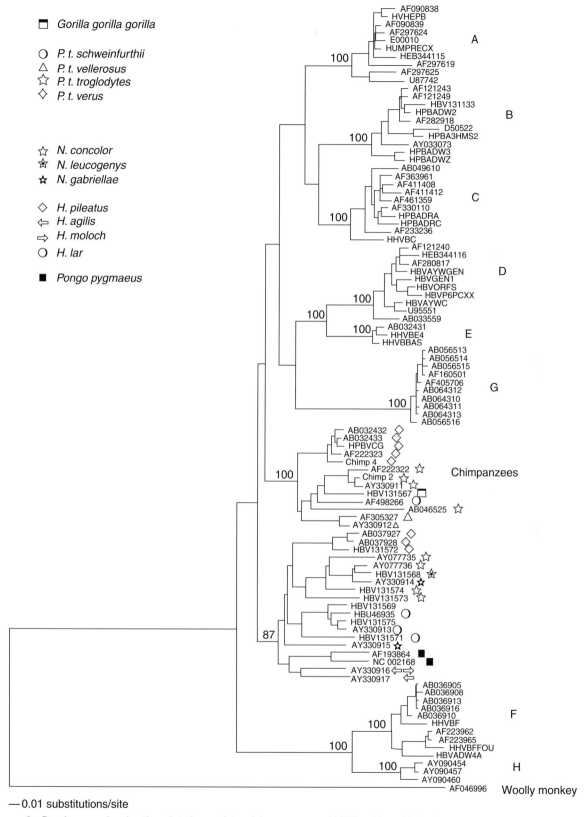

Figure 3 Dendrogram showing the relatedness of the eight genotypes of HBV and hepatitis B viruses isolated from nonhuman primates. The tree is rooted using the woolly monkey HBV sequence as an outgroup. Reproduced from Starkman SE, MacDonald DM, Lewis JCM, Holmes EC, and Simmonds P (2003) Geographic and species association of hepatitis B virus genotypes in non-human primates. *Virology* 314: 381–393, with permission from Elsevier.

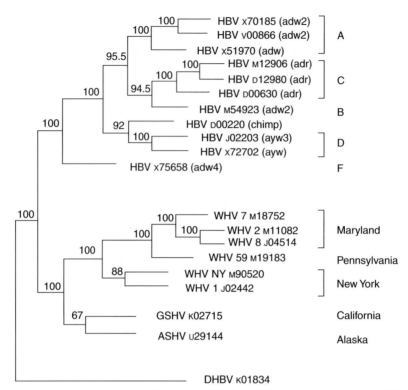

Figure 4 Dendrogram showing the relatedness of HBV and the rodent hepadnaviruses, rooted using DHBV as an outgroup. Reproduced from Testut P, Renard CA, Terradillos O, et al. (1996) A new hepadnavirus endemic in arctic ground squirrels in Alaska. *Journal of Virology* 70: 4210–4219, with permission from American Society for Microbiology.

Some of these viruses may originate from species of birds other than those from which they were first isolated, and it seems quite possible that many other varieties of avian hepadnaviruses remain to be described.

Clinical Features and Pathology

The clinical features of acute viral hepatitis in humans are nonspecific and are not dependent on the etiology of the infection; they include fatigue, anorexia, myalgia, and malaise. Jaundice may be evident in the more severe cases, but often the infections may be anicteric (without jaundice) or even asymptomatic. In hepatitis B, these clinical features are evidence of a robust immune response to the virus and a sign that the infection will be cleared by the immune system. In a minority of cases, less than 5% of immune competent adults, asymptomatic infections persist in individuals who do not mount a vigorous immune response. Such persistent infections, originally termed the chronic carrier state, are defined formally by the persistence of HBsAg in serum for more than 6 months.

When a persistently infected woman gives birth, she will almost invariably pass on the virus to her infant. HBV normally does not cross the placenta and transmission is believed to occur during or immediately after the birth process. Infants who are infected perinatally have a very high probability (>90%) of becoming persistently infected – they are extremely immune tolerant of the virus and the infection may persist even for life. A soluble protein, HBeAg, which is related to the nucleocapsid protein, is secreted from the infected hepatocytes and may cross the placenta and induce tolerance in the foetus. A protein equivalent to HBeAg is made by all of the hepadnaviruses, including the avian viruses.

HBV infections of children up to the age of 5 also are much more likely to progress to chronicity than those in adults. Thus, in regions of the world where HBV is highly endemic, and more than 8% of individuals are persistently infected, the virus is maintained in the population by mother-to-infant transmission and horizontal transmission from persistently infected young children to their peers. These regions include sub-Saharan Africa, China, and Southeast Asia (**Figure 5**). Areas of intermediate endemicity (2–7% of individuals persistently infected) include North Africa, Eastern Europe, and northern Asia. In areas of low endemicity, such as North America and Western Europe, important routes of transmission include sexual contact and parenteral exposure.

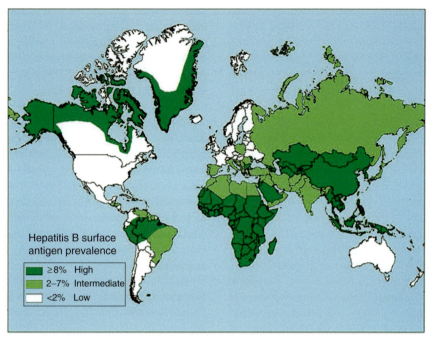

Figure 5 Global prevalence of persistent HBV infection (HBsAg in serum). Reproduced from http://www.cdc.gov/ncidod/diseases/hepatitis/slideset/hep_b/slide_9.htm, with permission from Central Food Technological Research Institute.

HBV seems not to be cytopathic and, in immune tolerant individuals, massive amounts of virus may be produced over long periods of time with little damage to the liver. However, such replication leads to the accumulation of hepatocytes with integrated (partial) copies of the viral genome and these seem to be at high risk of becoming neoplastic. Integration is not part of the HBV life cycle and seems to be a dead-end for the virus, but may be detected in almost all HBV-associated primary liver tumors.

In persistent infections, tolerance may break down over time, particularly with a cell-mediated immune response to the virus. Liver damage is attributable especially to the lysis of infected hepatocytes by cytotoxic T cells and peptides derived from the core (C) ORF seem to be a major target. In the long term, chronic active hepatitis may lead to the development of fibrosis, cirrhosis, and ultimately, end-stage liver failure. Cirrhosis also is a high risk for the development of primary liver cancer (hepatocellular carcinoma, HCC). However, almost all HCCs arising in HBV-infected livers contain chromosomally integrated HBV DNA and, in contrast to hepatitis C, in hepatitis B HCC may develop in the absence of cirrhosis and even on a background of almost normal liver histology.

The rodent hepadnaviruses can also become persistent in their natural hosts and chronic hepatitis, cirrhosis, and HCC may ensue. On the other hand, DHBV infection of ducks results in very little pathology, the infection may be passed vertically through the egg, and hatchlings are extremely immune tolerant of the virus. The avian and rodent systems have been studied experimentally. DHBV in its natural host has been especially useful. As noted above, the DHBV system had an important role in the elucidation of the mode of replication of the viral DNA. It has also been valuable for testing nucleoside and nucleotide analogs for potential therapeutic use. Despite their closer relatedness to humans, woodchucks are not best suited to experimentation; they are large, hibernate in the winter, and wild-caught animals have many parasites. Nevertheless, these animals have been used in various studies, including investigations of the actions of antiviral agents, the immune response in chronic infection, and the development of HCC.

Prevention and Treatment of Infection

HBV infection can be prevented by immunization, intramuscular administration of HBsAg leading to a protective anti-HBs response. The first vaccine (so-called plasma-derived vaccine) was produced by purifying HBsAg from the plasma of hepatitis B carriers. Second-generation vaccines contain HBsAg produced from yeast or (less commonly) mammalian cells by recombinant DNA technology. The World Health Organization recommends universal immunization of infants and this is now carried out in many countries. The key is to break the chain of transmission from infected mothers to their infants, and giving the vaccine within 24 h of birth protects 70–90% of such babies. Protective efficacy may be increased by giving passive protection with hepatitis B immune globulin (HBIG) at a contralateral site. In Taiwan, one of the first countries with a high prevalence of HBsAg to introduce universal immunization of infants, that prevalence has been reduced from more than 10% to less than 1% in the immunized cohort. Furthermore, although HCC is very rare in children, the incidence of childhood HCC has been reduced and all indications are that universal immunization against hepatitis B will, in time, lead to a marked decrease the incidence of HCC worldwide.

Prior to the introduction of nucleoside and nucleotide analogs, persistent HBV infection was treated most often by the administration of interferon. Up to 30% of individuals so treated cleared the virus. Interferon seemed to work by upregulating the cytotoxic T-cell response to the virus, rather than by a direct antiviral action. Responses were notably poor in patients from China and the Far East, presumably because these individuals had been infected early in life and were very immune tolerant.

As noted above, HBV replicates via reverse transcription of an RNA pregenome, and some of the nucleoside and nucleotide analogs introduced for the treatment of human immunodeficiency virus (HIV) infection are also active against HBV. The first to be licensed for such use was lamivudine (3TC); monotherapy reduces the viral load considerably but resistance develops in around 15% of treated patients per year. Resistance parallels that seen in HIV, with mutations affecting the 'YMDD' motif in the active site of the polymerase. Resistance to adefovir dipivoxil arises less frequently and the drug is active against lamivudine-resistant HBV. Other drugs also have been licensed but regimens of combined therapy are less well developed than those used for HIV. It should be noted also that, while such treatments reduce the viral load considerably, they rarely result in complete clearance of the infection.

Hepatitis D Virus

HDV merits a brief mention because of its particular association with HBV. As noted above, a novel antigen, termed delta, was discovered in patients with hepatitis B and this turned out to be the nucleocapsid protein of HDV. The HDV genome is a single-stranded circle of RNA that resembles the viroids and virusoids of plants; it is believed to be replicated by the host RNA polymerase II, with cleavage and rejoining of the circle mediated by a ribozyme activity. Unlike the viroids and virusoids, the RNA contains (in the antigenomic sense) an ORF, encoding the delta antigen. HDV requires HBsAg for envelopment and exocytosis, and for subsequent binding and entry to the target hepatocyte. HDV has been transmitted experimentally to WHV-infected woodchucks and is there enveloped by WHsAg.

HDV was first discovered in Italy and is found in the Mediterranean area, reportedly with a declining prevalence, and also in the Far East and South America. The virus may be acquired as a coinfection with HBV or by super-infection of someone already HBsAg-positive. In both cases, disease may be more severe than with HBV alone, and chronic delta hepatitis may progress to cirrhosis more frequently, and more rapidly, than chronic hepatitis B. The hepatitis B vaccine also protects against coinfection but there is no licensed vaccine to protect hepatitis B carriers against HDV infection.

See also: Hepatitis B Virus: General Features; Hepatitis B Virus: Molecular Biology; Hepadnaviruses of Birds.

Further Reading

Cougot D, Neuveut C, and Buendia MA (2005) HBV induced carcinogenesis. *Journal of Clinical Virology* 34(supplement 1): S75–S78.

Harrison TJ (2006) Hepatitis B virus: Molecular virology and common mutants. *Seminars in Liver Disease* 26: 87–96.

Kidd-Ljunggren K, Miyakawa Y, Kidd AH, et al. (2002) Genetic variability in hepatitis B viruses. *Journal of General Virology* 83: 1267–1280.

Lavanchy D (2005) Worldwide epidemiology of HBV infection, disease burden, and vaccine prevention. *Journal of Clinical Virology* 34(supplement 1): S1–S3.

McMahon BJ (2005) Epidemiology and natural history of hepatitis B. *Seminars in Liver Disease* 25(supplement 1): 3–8.

Norder H, Courouce A-M, Coursaget P, et al. (2004) Genetic diversity of hepatitis B virus strains derived worldwide: Genotypes, subgenotypes, and HBsAg subtypes. *Intervirology* 47: 289–309.

Prassolov A, Hohenberg H, Kalinina T, et al. (2003) New hepatitis B virus of cranes that has an unexpected broad host range. *Journal of Virology* 77: 1964–1976.

Schultz U, Grgacic E, and Nassal M (2004) Duck hepatitis B virus: An invaluable model system for HBV infection. *Advances in Virus Research* 63: 1–70.

Starkman SE, MacDonald DM, Lewis JCM, Holmes EC, and Simmonds P (2003) Geographic and species association of hepatitis B virus genotypes in non-human primates. *Virology* 314: 381–393.

Taylor JM (2006) Hepatitis delta virus. *Virology* 344: 71–76.

Tennant BC, Toshkov IA, Peek SF, et al. (2004) Hepatocellular carcinoma in the woodchuck mode of hepatitis B virus infection. *Gastroenterology* 127: S283–S293.

Testut P, Renard CA, Terradillos O, et al. (1996) A new hepadnavirus endemic in arctic ground squirrels in Alaska. *Journal of Virology* 70: 4210–4219.

Zoulim F (2006) Antiviral therapy of chronic hepatitis B. *Antiviral Research* 71: 206–215.

Hepatitis B virus, Division of Viral Hepatitis, Centers for Disease Control and Prevention. http://www.cdc.gov/ncidod/diseases/hepatitis/slideset/hep_b/slide_9.htm (accessed August 2007).

Hepatitis A Virus

A Dotzauer, University of Bremen, Bremen, Germany

© 2008 Elsevier Ltd. All rights reserved.

Glossary

Antibody prevalence The percentage of a population with antibodies against a certain disease at a given time.

Aplastic anemia Disease characterized by a decrease in blood cells resulting from underproduction due to bone marrow failure; also called hypoplastic anemia.

Incidence The rate of occurrence of new cases of a particular disease in a population in a certain period of time.

Polyprotein processing Cascade of proteolytic cleavage events resulting in release of mature proteins from the polyprotein.

Serine-like protease Proteolytic enzyme characterized by a catalytic triad similar to that in serine-type proteases with Ser, His, and Asp.

History

Jaundice has been known as an epidemic disease for centuries, but the earliest outbreaks of what was almost certainly hepatitis A were documented by Rayger, occurring in Preßburg, now Slovakia, in 1674 and 1697. Besides sporadic occurrence of the disease, characterized by a slow increase and slow decrease of the number of infected persons in the course of months (spread by person-to-person contact), with an overall small number of infections, larger vehement epidemics (spread by contaminated water and food) were reported in later times. The first pandemic wave occurred in the 1860s and the first considerable record of the disease was registered during the American Civil War.

In the second half of the nineteenth century, the disease became known as 'icterus catarrhalis' (catarrhal jaundice; inflammation of the biliary tract was supposed) or 'icterus epidemicus' (epidemic jaundice) as well as campaign jaundice, as epidemics are common in military medical history.

At the turn of the nineteenth to the twentieth century, the disease was recognized as infectious and transmissible by person-to-person contact. The terms 'hepatitis epidemica' (epidemic hepatitis) and 'hepatitis infectiosa' (infectious hepatitis) were introduced as synonyms for catarrhal jaundice.

Detailed epidemiologic recordings have been conducted since the beginning of the twentieth century.

Two large pandemic waves were observed during the twentieth century, the first one originating during World War I and reaching its summit between 1918 and 1922, and the second one originating in the early 1930s reaching its widest distribution with the beginning of World War II.

During World War II, hepatitis had been demonstrated to be caused by at least two separate filterable agents, and the resulting diseases were called hepatitis A (infectious hepatitis) and B (serum hepatitis), and the etiological agents hepatitis A virus (HAV) and hepatitis B virus (HBV), respectively.

After World War II, epidemiologic studies in human volunteers showed that hepatitis A is spread by the fecal–oral route, and provided information on the duration of viremia and shedding of virus in feces. In the late 1960s and early 1970s, it was shown that marmoset monkeys and chimpanzees could serve as animal models for human hepatitis A, and replication of HAV in cell cultures was established between 1979 and 1981.

The etiologic agent was identified through immune electron microscopy by Feinstone *et al.* in 1973.

The molecular cloning of the HAV genome in the 1980s revealed that the genomic organization of HAV is similar to that of picornaviruses, and the first infectious cDNA clone of HAV was reported by Cohen *et al.* in 1987.

The disease manifestations by hepatocellular destruction could be attributed to an immunopathogenic mechanism in the late 1980s by Vallbracht *et al.*

An inactivated vaccine has been available since 1992.

Taxonomy

HAV is the only member of the genus *Hepatovirus* within the family *Picornaviridae*.

All human HAV strains known belong to only one serological group, but phylogenetic analysis of the VP1–2A junction region and the VP1 coding region, respectively, revealed that several distinct HAV genotypes (seven in the case of VP1–2A junction analysis, five in the case of VP1 region analysis), which include several subgenotypes, can be distinguished by the degree in their genetic heterogeneity. These genotypes correlate with the geographic origin of the virus isolates. Genotype I is the most common type worldwide, particularly genotype IA. The cell culture-adapted viruses most commonly used are variants of the Australian strain HM175 (genotype IB) and the German strain GBM (genotype IA).

Morphology and Physicochemical Properties

The infectious spherical virion is a nonenveloped particle with a diameter of $c.$ 27 nm. The icosahedral capsid, which embodies the viral RNA genome, contains 60 copies of each of the three major proteins, VP1 (also known as 1D), VP2 (1B), and VP3 (1C) (see **Table 1**). It is not known whether the small protein VP4 (1A) is integrated into the capsid. The mature HAV virion seems to have a different structure than other picornaviruses, as a usually prominent feature of picornaviruses and one which represents the viral attachment site to cellular receptors, the canyon surrounding the fivefold symmetry axes, is obviously missing.

The HAV particle has a buoyant density of $c.$ 1.33 g cm^{-3} in CsCl and a sedimentation coefficient of $c.$ 160S. Empty capsids found in feces have a sedimentation constant of about 70S. HAV is extremely resistant to acid (pH 1.0 for 2 h at room temperature) and thermal inactivation (60 °C for 1 h).

Genome Organization and Expression, Replication, Morphogenesis

The linear, single-stranded, positive-sense RNA genome of HAV is ~7500 nt in length and is not capped but covalently linked by a tyrosine-O^4-phosphodiester bond to the 2.5 kDa viral protein 3B (also known as VPg; see **Table 1**). It encompasses a structurally complex 5′ nontranslated region (NTR) of ~740 bases, followed by a single open reading frame encoding a polyprotein ($c.$ 250 kDa; ~2230 amino acids) and a 3′ NTR of ~60 bases which terminates with a poly(A) tail of about 60 nt. In the polyprotein, the capsid proteins and those with functions during virion assembly represent the N-terminal third (VP4, VP2, VP3, VP1, also known as P1 region, and 2A) (see **Table 1** and **Figure 1**) with the remainder of the polyprotein comprising a series of proteins required for RNA replication (2B and 2C followed by 3A to 3D) (see **Table 1** and **Figure 1**).

The viral polyprotein is translated directly from the messenger-sense genomic RNA, which is released into the cellular cytoplasm after uncoating of the virion. An internal ribosomal entry site (IRES) located within the 5′ NTR (see **Table 1**) is involved in the cap-independent initiation of protein synthesis. The IRES of HAV differs from that of other picornaviruses and forms its own group (type III picornavirus IRES). Translation from the HAV IRES requires all of the initiation proteins, including eIF4E and intact eIF4G, and infection with HAV does not result in cleavage of the translation initiation factor eIF4G to block cap-dependent host protein synthesis, as featured by other picornaviruses. The HAV IRES directed initiation of translation, which depends on the entire 5′ NTR, is enhanced by sequences of the 5′ terminal coding region and by the cellular protein poly(A) binding protein (PABP), which mediates circularization of the template RNA and seems to be necessary for HAV protein translation. Involvement of the host cells poly(C) binding protein 2 (PCBP2), polypyrimidine tract binding protein (PTB), and glyceraldehyde 3-phosphate dehydrogenase (GAPDH) is described, but the functional significance of these interactions is uncertain.

Proteolytic processing of the primary polyprotein (see **Figure 1**) occurs simultaneously with translation and is largely carried out by the viral 3C protease, which is a serine-like protease in which cysteine replaces the nucleophilic serine in the catalytic triad of the active center and for which, in contrast to other picornaviruses, no additional cellular substrate has been described so far.

Synthesis of viral RNA by the viral 3D polymerase follows the accumulation of the nonstructural proteins spanning 2B to 3D, which induce the assembly of a macromolecular replication complex on membranes that are recruited from the endoplasmatic reticulum. The proteins 2B, 2C, and 3A may be involved in the structural rearrangements of the intracellular membranes. RNA transcription is most likely protein-primed, with the uridylylated VPg protein 3B representing the primer for the negative-sense RNA replication intermediate and the subsequent positve-sense genomic RNA synthesis. A participation of the 5′ terminal NTR structures in the switch from translation to replication on the same viral RNA is suggested.

HAV morphogenesis is poorly understood. The primary polyprotein cleavage event occurs at the 2A/2B junction mediated by the 3C protease resulting in the structural precursor P1–2A. The steps then resulting in particle formation are not entirely clear, but the following model is suggested (see **Figure 1**). The P1–2A structural precursors assemble to a pentameric structure and are further cleaved by the 3C protease to generate the precursors VP4–VP2 (also known as VP0) and VP1–2A, as well as VP3 resulting in the structural building block (VP0, VP3, VP1–2A)$_5$. The 2A C-terminal extension of VP1 is a critical structural intermediate in virion morphogenesis, maybe clamping the pentamer at the fivefold symmetry axis. After assembly of 12 such pentamers with the genomic RNA to provirions (12 × (VP0, VP3, VP1–2A)$_5$-RNA), two subsequent maturation cleavage events occur. First, the cleavage at the VP1/2A junction leading to the removal of 2A seems to result from the action of an unknown host protease and, second, VP0 is cleaved into VP4 and VP2 by an unknown mechanism. 2A has never been identified in infected cells, which may indicate an extracellular cleavage event for the VP1–2A processing, and the role of VP4 in virion morphogenesis is mysterious. While the HAV polyprotein appears to

Table 1 HAV regulatory genomic regions and HAV proteins; all numbering refers to HAV strain HM175 (accession no. NC_001489)

Regions/proteins	Nucleotide position	Length in amino acids	Proposed or known function	Remarks
5' nontranslated region (5' NTR)	1–734		5' stem-loop region (nucleotides 1–94): required for RNA replication Polypyrimidine tract pY1 (nucleotides 99–138): unknown function Internal ribosomal entry site (IRES) (nucleotides 152–734): directs cap-independent initiation of translation	Mechanisms regulating switch from translation to replication are not clear Picornaviral type III IRES
3' nontranslated region (3' NTR)	7416–7478		Involved in regulation of RNA replication Extended by 40–80 adenylates, which are structurally involved in translation (closed-loop model)	
1A	735–803	23	VP4	Not myristoylated So far not identified in mature virions
1B	804–1469	222	Capsid protein VP2	During morphogenesis, precursor VP4–VP2 (VP0), presumably cleaved autocatalytically
1C	1470–2207	246	Capsid protein VP3	Heterogeneous C-terminus, presumably depending on HAV strain and host cell
1D	2208–3023/ 3026/ 3029	272 273 274	Capsid protein VP1	During morphogenesis, precursor VP1–2A, presumably cleaved by cellular protease
2A	3024/3027/ 3030–3242	71/72/73	Morphogenesis	So far not identified in infected cells
2B	3243–3995	251	Structural rearrangements of intracellular membranes for replication	Peripheral association with membranes Influence on membrane permeability Mutations accompany adaptation to growth in cell culture
2C	3996–5000	335	Structural rearrangements of intracellular membranes for replication	Integral association with membranes Mutations accompany adaptation to growth in cell culture
3A	5001–5222	74	As 3AB VPg precursor (pre-VPg)	Interacts with membranes, may anchor pre-VPg to cellular membranes through a central region of 21 hydrophobic amino acids
3B	5223–5291	23	VPg (virus protein genomic) Supposed protein primer for RNA replication	Tyr^3 attached to 5' terminus of genomic RNA
3C	5292–5948	219	Sole protease	Serine-like protease, with replacement of Ser by Cys in catalytic triad: Cys^{172}, His^{44}, Asp^{84}
3D	5949–7415	489	RNA-dependent RNA polymerase	

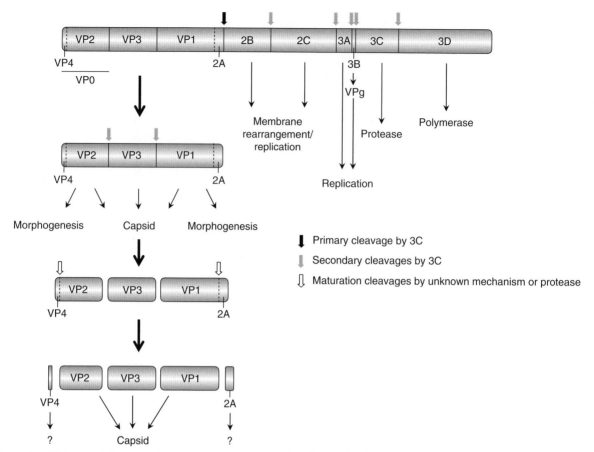

Figure 1 HAV polyprotein processing and known or supposed functions of the viral proteins during morphogenesis and replication. The polyprotein that results from IRES-directed translation is cleaved by the viral protease 3C to release certain precursor proteins with biologically relevant functions as well as the mature nonstructural proteins. Primary cleavage results in release of the structural precursor VP4–VP2–VP3–VP1–2A, which associate to assembly intermediates (pentamers), which are stabilized by 3C cleavage of the VP3 junctions (VP4–VP2 (VP0), VP3, VP1–2A)$_5$. After assembly of the intermediate protein building blocks and the viral genome to provirions, maturation cleavages occur through an unknown proteolytic activity to release VP4 from VP2 and by a so far unknown cellular protease to release 2A from VP1, resulting in the infective virion with 60 copies of each of the main structural proteins VP1, VP2, and VP3.

possess the short VP4 segment at its N-terminus, this putative VP4 moiety has never been identified in virions. Moreover, the HAV-VP4 sequence does not contain a myristoylation signal, which is important for the morphogenesis of other picornaviruses.

Host Range, Transmission, and Tissue Tropism

Under natural conditions, only humans and certain species of nonhuman primates seem to be susceptible to HAV. These primates, which are also used as animal models, include chimpanzees, marmosets, and owl monkeys.

HAV is transmitted via the fecal–oral route. As the virus is excreted in feces, it is typically acquired by ingestion of feces-contaminated food or water. Direct person-to-person spread occurs under poor hygienic conditions.

The site of replication is the liver. The events that occur during the passage of HAV across the intestinal epithelium into blood, in which the virus reaches the liver, are not clearly understood. Although HAV antigen and the T-cell immunoglobulin mucin 1 (TIM1; function so far unknown) protein, which has been identified as a cell surface protein binding HAV (HAV$_{cr1}$), could be detected in different organs, such as kidney, spleen, and gastrointestinal tract, no extrahepatic sites of HAV replication have been clearly identified. Furthermore, it was demonstrated that infection of polarized intestinal cells does not result in penetration of the epithelium. A functional cellular receptor, whose selective expression in the liver is assumed, could not be identified so far. Some studies suggest that the hepatotropism of HAV may be supported by immunoglobulin A (IgA)-virus complexes (HAV/IgA), as HAV/IgA uptake via the hepatocellular asialoglycoprotein receptor (ASGPR) results in infection of hepatocytes (IgA-carrier hypothesis).

The virus progeny produced in the liver is then released back into the intestinal tract via bile.

Cell Culture and Growth Characteristics

HAV can infect a variety of primate and nonprimate cell lines, including nonhepatic cells, *in vitro*. The virus exhibits a protracted replication cycle and normally establishes a noncytolytic, persistent infection with low virus yields, and there is no evidence that HAV notably interferes with the macromolecular synthesis of its host cell. After infection of cultivated cells with wild-type virus, a minimum of 8 weeks elapses before HAV can be isolated. Although a more rapid replication and higher virus titers are obtained after serial virus passages in cultivated cells resulting in cell culture-adapted viruses, even the replication of these virus variants is not detectable within the first few days after infection. Adaptation of HAV to growth in cultivated cells seems to be achieved by varying sets of multiple interacting mutations, with adaptive mutations within the IRES enhancing viral translation in a cell-type-specific fashion, and mutations clustering in the 2B and 2C proteins (see **Table 1**) increasing replication regardless of the cell line used.

The virus apparently downregulates its own replication and this may, for example by supporting the ability of HAV to inhibit innate cellular antiviral defense mechanisms, be important for the establishment of persistent infections. A large proportion of the virus progeny remain cell associated, but extensive release of HAV from the cells also occurs, caused by an unknown mechanism.

Several cytopathogenic variants of HAV have been isolated which induce apoptosis resulting in cell death. These variants are highly cell culture adapted and characterized by a rapid replication phenotype and high virus yields. In these variants, both the downregulation of viral replication and the ability to inhibit the innate defense mechanisms are less effective. The molecular mechanisms resulting in apoptosis are not known.

Clinical Features and Pathology

Infections with HAV may produce a wide spectrum of manifestations ranging from silent infections, over icteric courses to fatal fulminant hepatitis. The acute icteric course of infection varies between common, over prolonged to relapsing hepatitis A. Persistent infections or chronic disease have not been described.

The clinical presentation of the disease depends on the age of infection. The likelihood of having symptoms and the severity of the disease increases with the age of the patient. Inapparent infections (asymptomatic or at least anicteric) are normally observed in very young children, under the age of 2. However, clinically obvious disease can occur even in infancy (aged 2 weeks to 8 months) and may be characterized by prolonged courses. In children under 5 years of age *c.* 3%, in children 5–15 years old *c.* 30%, and in individuals over the age of 18 years as many as 70% may develop a clinically apparent disease.

Common Course of Infection

The incubation period ranges from 2 to 6 weeks with a mean duration of 4 weeks. The prodromal (preicteric) period of normally 4–6 days (which may vary from 1 day to more than 2 weeks) is characterized by nonspecific symptoms, like anorexia, nausea with vomiting, malaise, abdominal pain, loss of appetite, accelerated pulse, rash, headache, and fever (38–39 °C) as well as by gastrointestinal symptoms, normally in form of obstipation, but diarrhea is also observed.

The prodromal symptoms disappear with the onset of jaundice, which is seldom abrupt (in 15% of the cases no obvious prodrome is observed before appearance of jaundice). The icteric phase, which ranges from 2 to >22 days (mean duration 3 weeks), is marked by jaundice, which may start with scleral icterus, dark beer-colored urine (conjugated bilirubinuria), clay-colored stool, and clearly decelerated pulse.

The reconvalescence period ranges from 3 to 6 weeks, but fatigue, dullness, right upper quadrant tenderness, and fast exhaustion may remain for 2–4 months. In almost all cases the liver is enlarged.

The clinical symptoms are accompanied by several biochemical parameters (see **Figure 2**). Elevation of aminotransferase levels in serum (alanine aminotransferase (ALT) and to a lesser degree aspartate aminotransferase (AST)), which reflect hepatocellular damage with release of the liver enzymes into the circulation, roughly correlate with the severity of the disease. Elevation of serum alkaline phosphatase activity and in the serum bilirubin level relate to intrahepatic cholestasis.

At the onset of symptoms, seroconversion to anti-HAV occurs.

Large amounts of HAV, which are produced in the liver and released into the gastrointestinal tract via bile, already occur in the feces during the late incubation period when no clinical symptoms are observable and are shed for approximately three weeks until a few days after the onset of elevated levels of liver enzymes in the serum. Fecal shedding of HAV reaches its maximum just before the onset of hepatocellular injury.

Viremia occurs a few days before and during the early acute stage of the clinical and biochemical hepatitis, in which it roughly parallels the shedding of virus in feces, but at a lower magnitude.

More sensitive methods (especially nucleic acid amplification technologies) demonstrated that low levels of viral RNA may be present in feces and blood for many weeks.

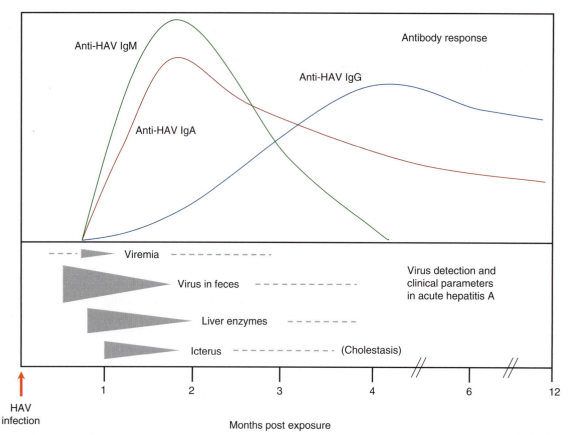

Figure 2 Course of clinically relevant events in acute hepatitis A. This figure schematically shows the mean duration and intensity of certain parameters. The dotted lines indicate that the duration and intensity of the events may vary.

Prolonged Hepatitis A

In 8.5–15% of the cases, jaundice lasts for up to 17 weeks. The biochemical abnormalities are resolved by 5 months. Occasionally, prolonged courses are accompanied by high serum bilirubin levels persisting for months (cholestatic hepatitis A). The cholestatic form is marked by extensive itching of the skin.

Relapsing Hepatitis A

After initial improvement in symptoms and liver test values (serum aminotransferase levels), one or more relapses of the disease (mostly biphasic) are described for up to 20% of the patients. These relapses occur between 30 and 90 days after the primary episode, when high titers of neutralizing antibodies are already present. The severity of symptoms, the biochemical abnormalities, and the immunoglobulin M (IgM) response are essentially the same as observed during the initial phase, with a tendency to greater cholestasis. The pathogenesis of relapsing hepatitis is not understood and basically two hypotheses are suggested: that the disease is a manifestation of a persistent viral infection, or, alternatively, that the relapse may represent a manifestation of an enterohepatic cycling of HAV, which may be supported by anti-HAV IgA (IgA-carrier hypothesis). But none of these hypotheses is currently supported by *in vivo* data.

Fulminant Hepatitis A

In rare cases, acute hepatitis A results in a fatal deterioration in liver function with massive destruction of liver cells. Surprisingly, no vigorous inflammatory response is observed. The fatality rate is below 1.5% of all hospitalized icteric cases. This course of the disease is accompanied with fever over 40 °C. This outcome is more frequent in adults, especially in patients over 50 years of age, than in children, and the risk is increased in patients with underlying chronic liver disease.

Extrahepatic Manifestations

Extrahepatic manifestations of the disease are rare and the etiology is uncertain. Besides frequently observed transient suppression of hematopoiesis, rare cases of aplastic anemia (pancytopenia) with a lethality rate of over 90% are described, and it was demonstrated that HAV infects monocytes and inhibits their further differentiation. In some patients, interstitial nephritis was observed. In connection with the finding that the HAV-binding receptor HAV_{cr1} is identical with the T-cell immunoglobulin mucin 1 (TIM1), a protein suggested to play a role in asthma susceptibility, and by statistical evaluation of medical records, it was suggested that HAV exposure may leave a protective effect on the development of asthma and allergic diseases.

Pathology

The pathology of the liver in acute hepatitis A has been studied in humans and several animal models. The pathological lesions, which are caused by an immunopathogenic mechanism, are characterized by hepatocellular necrosis, which is most prominent in periportal regions, accompanied with large inflammatory infiltrates of mononuclear cells.

Innate and Adaptive Immune Response

Innate Immune Response

HAV, which does not interfere with the replication of other viruses, prevents the synthesis of beta-interferon (IFN-β), but is not resistant to alpha- and beta-interferon (IFN-α/β) exogenously added to persistently infected cells. It could be demonstrated that HAV does inhibit dsRNA-induced transcription of IFN-β by blocking effectively interferon regulatory factor 3 (IRF-3) activation due to an interaction of HAV with the mitochondrial antiviral signaling protein MAVS (also known as IPS-1, VISA, or Cardif), which is a component of the retinoic acid-inducible gene I (RIG-I) and melanoma differentiation-associated gene 5 (MDA-5) signaling pathway. Signaling through the Toll-like receptor 3 (TLR-3) pathway may also be partially impaired.

HAV also has the ability to prevent apoptosis induced by accumulating dsRNA, but the underlying mechanism is not clear.

Gamma interferon (IFN-γ) produced by HAV-specific HLA-dependent cytotoxic T lymphocytes (CTLs) may contribute to the elimination of HAV infections by inducing an antiviral state in the later course of the infection.

Adaptive Immune Response

Neutralizing anti-HAV IgM antibodies are present in almost all patients at the onset of the symptoms (see **Figure 2**). These antibodies disappear in the course of 3 months, but in the case of prolonged courses IgM can be detected up to 1 year after onset of icterus. Anti-HAV IgA antibodies are also detectable at the onset of the symptoms (see **Figure 2**). This response reaches its peak titer 50 days post infection and may last for >5 years. The majority of the IgA remains as serum IgA in circulation and is not secreted into the intestinal tract as secretory IgA by the polymeric immunoglobulin receptor (pIgR) pathway. But a significant fraction of this serum IgA is released into the gastrointestinal lumen via bile by liver functions under participation of the hepatocellular IgA-specific asialoglycoprotein receptor (ASGPR). The role of IgA antibodies in the protection against HAV infections appears to be limited, and studies suggest that HAV-specific IgA may serve as a carrier molecule for a liver-directed transport of the virus, supporting the hepatotropic infection by uptake of HAV/IgA immunocomplexes via the ASGPR (IgA-carrier hypothesis). Neutralizing anti-HAV immunoglobulin G (IgG) antibodies are also detectable 3 weeks post infection for the first time, but this response develops slowly, reaching its peak titer 4 months post infection (see **Figure 2**). Anti-HAV IgG persists lifelong, although the titer may fall to undetectable levels after several decades. Neutralizing antibodies, which are effective in eliminating the virus from the blood, do recognize a conformational epitope clustered into a major, immunodominant antigenic site involving residues contributed by VP3 and VP1.

HAV-specific, HLA-restricted cytotoxic $CD8^+$ T lymphocytes (CTLs) have been detected within the liver during acute HAV infection and play prominent roles both in eliminating the virus and in causing liver injury (immunopathogenesis). Gamma interferon (IFN-γ), released by these CTLs, may stimulate HLA class I expression on hepatocytes and in the following promote upregulation of the normally low level display of antigen on liver cells. Specific T-cell epitopes have not been identified so far.

Diagnosis

Since the clinical presentation of hepatitis A cannot be distinguished from hepatitis due to the other hepatitis viruses, serologic tests or nucleic acid amplification techniques are necessary for a virus-specific diagnosis.

The routine diagnosis of acute hepatitis A is made by detection of anti-HAV IgM in the serum of patients (see **Figure 2**). A further option is the detection of virus in the feces.

In order to improve the safety of blood and blood products, blood screening with HAV-specific polymerase chain reaction (PCR) is performed, which reduces the window period of up to 3 weeks post infection during which HAV infection fails to be diagnosed by serologic assays.

Epidemiology

HAV occurs worldwide and accounts for over 1.5 million clinical cases reported annually. The seroprevalence pattern ranges from high endemicity, such as in Africa, South Asia, and Latin America, where infection normally occurs in childhood, over intermediate endemicity, such as in Eastern Europe and the northern parts of Asia, to low endemicity, such as in Western Europe and North

America, where the majority of the population remains susceptible to HAV infection. However, the epidemiology pattern is complex and continuously changing, with considerable heterogeneity among different countries. In general, the anti-HAV antibody prevalence inversely correlates with the quality of the hygienic standards, and the incidence declines in many populations through improvements in public sanitation and living conditions. These improvements result in an increase of the pool of susceptible adults, with a shift in the age of infection to older age groups, in which a more severe disease is observed, leading to an increased morbidity.

A minor seasonal distribution of HAV infections is observed, with a peak occurring during fall and winter, mentioned in almost all earlier and contemporary reports, nowadays possibly as a result of exposure during summer vacation spent in endemic countries.

At special risk for acquisition of hepatitis A are international travelers from areas of low endemicity to endemic areas, employees of child-care centers and sewage plants, gully workers, injecting drug users, homosexually active men, and persons with an increased risk of developing a fulminant disease, such as persons with chronic hepatitis C virus (HCV) infections.

The high physical stability of HAV provides a good opportunity for common-source transmission. Community-wide outbreaks are reported in association with infections of food handlers, and linked to contaminated food and drink, or uncooked clams from contaminated water. Hepatitis A is most commonly acquired by sharing the household with an infected person.

Prevention and Control

There is no specific treatment for hepatitis A. As almost all HAV infections are spread by the fecal–oral route, good personal hygiene, high-quality standards for public water supplies, and proper disposal of sanitary waste are important measures to reduce virus transmission.

Until the availability of an active prophylaxis, the disease could be prevented for up to 5 months with a certainty of 80–90% by passive immunization with pooled IgG of at least 100 IU anti-HAV. IgG is still used for postexposure prophylaxis. If administered within 2 weeks after exposure, either development of the disease is prevented or the severity of the disease as well as virus shedding is reduced.

Since 1992, inactivated vaccines for active immunization have been available. These vaccines contain purified, formalin-inactivated virions produced in cell culture, which are absorbed to an aluminum hydroxide adjuvant. They are highly immunogenic and protect against both infection and disease caused by all strains of HAV with 100% efficacy for at least 10 years, which is consistent with the finding that all human HAV strains belong to one single serotype.

Candidate live, attenuated HAV vaccines have been developed using virus adapted to growth in cell culture, but were poorly immunogenic. Nonetheless, such a vaccine has been widely used in China and appears to be capable of inducing protective levels of antibody.

See also: Hepatitis B Virus: General Features; Hepatitis C Virus; Hepatitis E Virus; Innate Immunity: Introduction.

Further Reading

Bell BP (2002) Global epidemiology of hepatitis A: Implications for control strategies. In: Margolis HS, Alter MJ, Liang JT, and Dienstag JL (eds.) *Viral Hepatitis and Liver Disease*, pp. 9–14. Atlanta: International Medical Press.

Cuthbert JA (2001) Hepatitis A: Old and new. *Clinical Microbiology Reviews* 14: 38–58.

Gerety RJ (ed.) (1984) *Hepatitis A*. London: Academic Press.

Gust ID and Feinstone SM (1988) *Hepatitis A*. Boca Raton, FL: CRC Press.

Hepatitis B Virus: General Features

P Karayiannis and H C Thomas, Imperial College London, London, UK

© 2008 Elsevier Ltd. All rights reserved.

Glossary

Icterus Jaundice, or yellowing of the skin and particularly the whites of the eyes, due to failure to excrete bilirubin, a bile pigment.

History

It was not until the mid-1960s that hepatitis A (HAV) and B (HBV) viruses were recognized as the causative agents for infectious and serum hepatitis, respectively. These studies, performed by Krugman and colleagues at the

Willowbrook State School for mentally handicapped children, were preceded a few years earlier by the description of the Australia antigen in the sera of patients with leukemia. The connection between Australia antigen, or hepatitis B surface antigen (HBsAg) as it is now known, and HBV became apparent in later studies performed by the teams of Prince and Blumberg. These studies set the groundwork for the subsequent serological tests for the diagnosis of HBV and allowed detailed investigations into the epidemiological and virological aspects of infection.

Electron microscopic studies in 1970 by Dane and Almeida and their colleagues led to the visualization of the infectious virion or Dane particle, and the nucleocapsid core, respectively. This was followed in the early 1970s by the characterization of the virus genome, the virion-associated proteins, and the detailed definition of the serological profiles in acute and chronic HBV infection, performed primarily by Robinson's group at Stanford. The connection between the virus and the development of hepatocellular carcinoma (HCC) followed soon after, but the absence of a cell culture system for propagation of the virus impeded the study of its molecular biology. This changed in the early 1980s with the development of genetic engineering techniques that allowed the cloning of the viral genome, the study of its protein funtions, and the unravelling of the fascinating mechanism of its replication strategy. The polymerase chain reaction (PCR) allowed the speedy amplification and sequencing of virus isolates that led to the identification of quasispecies, virus mutants, and genotypes by bioinformatic approaches. In 1987, Carman and colleagues described the molecular basis of HBe antigen negative viremia (precore stop mutation – see below) and described the first, and most common, vaccine escape variant (arginine 145) in vaccinated children born to HBV-infected mothers (see below).

The introduction of a plasma-derived vaccine, following its extensive evaluation by Smuzness and colleagues in chimpanzees, constitutes another historical landmark in HBV research. This was soon followed by the production of a recombinant vaccine, which has effectively reduced the prevalence of the infection in many countries of the world where the virus was endemic. Almost concurrently, interferons were used for the first time in the treatment of chronic hepatitis B. These remain, with the subsequently introduced nucleos(t)ide analogs, the main treatment options in order to prevent progression of chronic liver disease to cirrhosis and HCC.

Taxonomy and Classification

Hepatitis B is the prototype virus of the family *Hepadnaviridae*, a name that signifies the hepatotropism and DNA nature of the genome of its members. There are two genera within the family. The genus *Orthohepadnavirus* contains members that infect mammals, and, other than HBV, includes hepadnaviruses that infect rodents such as woodchucks (woodchuck hepatitis virus, WHV) and squirrels (70% nucleotide identity). In recent years, HBV-like isolates have also been obtained from primates such as chimpanzees, gibbons, gorillas, orangutans, and woolly monkeys. These are more closely related to HBV and may in fact represent progenitors of the human viruses (**Figure 1**). The *Avihepadnavirus* genus on the other hand contains members that infect birds such as ducks (duck hepatitis B virus, DHBV), herons, storks, and geese. Over the years, the woodchuck and duck animal models, as well as chimpanzees, which are susceptible to infection with human HBV isolates, have proved invaluable in the study of the replication of these viruses, the natural history of

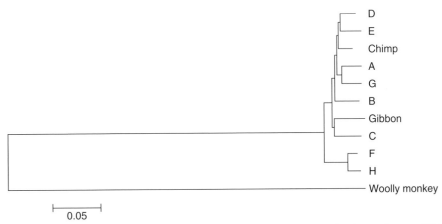

Figure 1 Phylogenetic tree based on the nucleotide sequences from the HBsAg region of all known human HBV genotypes and isolates from a chimp, gibbon, and woolly monkey. The tree was constructed using the Mega 2 software and rooted to the woolly monkey sequence.

infection, and the testing for efficacy of vaccines and antiviral drugs.

Distribution and Epidemiology

Conservative estimates place the number of persons chronically infected with HBV at over 350 million worldwide. The prevalence of HBV infection varies by geographical region, so that in northwestern Europe, North America, and Australia it is 0.1–2%. In the Mediterranean region, Eastern Europe, Middle East, Indian subcontinent, and Central and South America, it is 3–7%, while in Africa and the Far East it is 10–20%. This geographical distribution of HBV infection is mirrored by the incidence of HCC in the same regions. In areas of high endemicity, the virus is transmitted perinatally from carrier mothers to their infants, or horizontally from infected siblings and other children in early childhood. In areas of intermediate endemicity apart from perinatal transmission, household and sexual contact, as well as percutaneous exposure, are likely routes of infection. Finally, in Western countries, transmission is nowadays through sexual contact or intravenous drug use.

Virion Structure and Genome Organization

The infectious virion or Dane particle measures about 42 nm in diameter and consists of an outer envelope containing hepatitis B surface proteins (HBsAg) in a lipid bilayer (**Figure 2**). This in turn encloses the nucleocapsid core of the virus, within which lies the viral genome and a copy of its polymerase. Apart from virions, liver hepatocytes release into the circulation subviral particles devoid of nucleic acid and consisting entirely of HBsAg. These are the 22 nm spheres and the filamentous forms of similar diameter, which outnumber the virions by 100–10 000-fold.

The viral genome is a relaxed circular, partially double-stranded DNA molecule of 3.2 kbp in length, and contains four wholly or partially overlapping open reading frames (ORFs) (**Figure 3**). In addition, all regulatory elements such as enhancers, promoters, and encapsidation and replication signals lie within these ORFs. The Pre-S/S ORF encodes the three envelope glycoproteins, which are known as the large (L), middle (M), and small (S) HBsAgs, produced by differential initiation of translation at each of three in-frame initiation codons. The proteins are therefore co-terminal and the sequence of the more abundant small protein is shared by the other two. The M protein has an additional 55 amino acid residues at its N-terminus encoded by the Pre-S2 region, while the L protein includes in addition another 125 residues from the Pre-S1 region. The Pre-S1 protein is thought to contain the region responsible for the virus interaction with the hepatocyte receptor. All three proteins are glycosylated, while the L and S proteins may also be present in an unglycosylated form in particles.

The precore/core ORF contains two in-frame initiation codons and therefore yields two translation products. Initiation of translation from the first results in synthesis of the precore polypeptide, which forms the precursor of the soluble hepatitis B e antigen (HBeAg). This protein contains a signal peptide at its N-terminus that anchors the protein in the endoplasmic reticulum membrane. Cleavage by signal peptidase in the lumen is followed by further processing of the C-terminus. The resulting protein is the HBeAg, a nonstructural protein and marker of active virus replication. Moreover, the protein is thought to have a tolerogenic effect on the immune response to the virus. The nucleocapsid or core protein is synthesized following initiation of translation at the second initiation codon. The HBeAg and core proteins are translated from two separate transcripts known as the precore mRNA and the pregenomic RNA (pgRNA), respectively. The

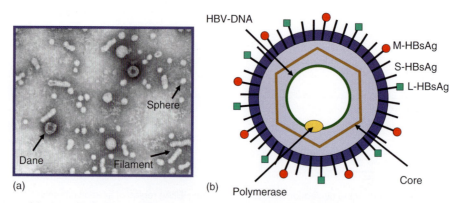

Figure 2 Structure of the hepatitis B virion. (a) Electron migrograph of HBV purified from plasma showing the infectious Dane particle and the spherical and filamentous subviral particles. (b) Cartoon of the virion and its components.

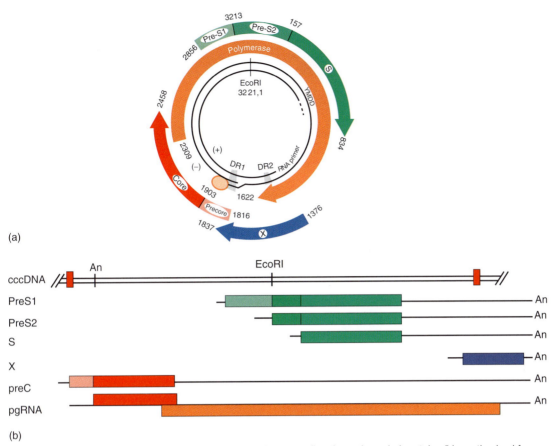

Figure 3 Genome organization of the virus (a) and the transcripts encoding the various viral proteins (b), synthesized from cccDNA. All of them terminate at the common polyadenylation site. Reproduced from Hunt CM, Mc Gill JM, Allen MI, *et al.* (2000) Clinical relevance of hepatitis B viral mutations. *Hepatology* 31: 1037–1044. American Association for Liver Diseases. Reprinted with permission of Wiley-Liss, Inc., a subsidiary of John Wiley & Sons, Inc.

latter also encodes the polymerase ORF of the virus, which covers almost all of the genome. The polymerase has multifunctional enzymatic actions as described below. Finally, the fourth ORF encodes for the X protein, which modulates host cell signal transduction and acts as a gene transactivator under experimental conditions.

Replication Strategy

The life cycle of the virus begins with its attachment to the appropriate hepatocyte receptor, which still remains unknown. In contrast, the region between residues 21 and 47 of Pre-S1 has long been known to be involved in virus binding to the hepatocyte membrane (**Figure 4**). The virion is internalized and uncoated in the cytosol, whence the genome translocates to the nucleus, where it is converted into a double-stranded covalently closed circular DNA (cccDNA) molecule, following completion of the shorter positive (+)-strand and repair of the nick in the negative (−)-DNA strand. The cccDNA constitutes the template for viral transcript synthesis by the host RNA polymerase II. All the transcripts terminate at a common polyadenylation signal situated within the proximal end of the core ORF, and their synthesis is controlled by individual promoters and the two enhancer elements, Enh 1 and 2.

The pgRNA is longer than genome length (3.5 kbp) and, apart from encoding the core and polymerase proteins, also forms the template for (−)-DNA strand synthesis (**Figure 5**). The polymerase has three functional domains, each one in turn involved in DNA priming (terminal protein), reverse transcription (rt), and pgRNA degradation (RNase H). There is also a spacer region of unknown function between the terminal protein and the rt domain. Once synthesized, the polymerase engages epsilon (ε), a secondary RNA structure at the 5′ end of the pgRNA, triggering encapsidation of the complex by the core protein. The subsequent steps in virus nucleic acid synthesis then take place within the nucleocapsid. Host cell factors including chaperones from the heat shock protein family are thought to be instrumental in aiding encapsidation, stabilization, and activation of the polymerase.

As pgRNA is longer than genome length, its terminal sequence duplicates the elements contained in its 5′ end

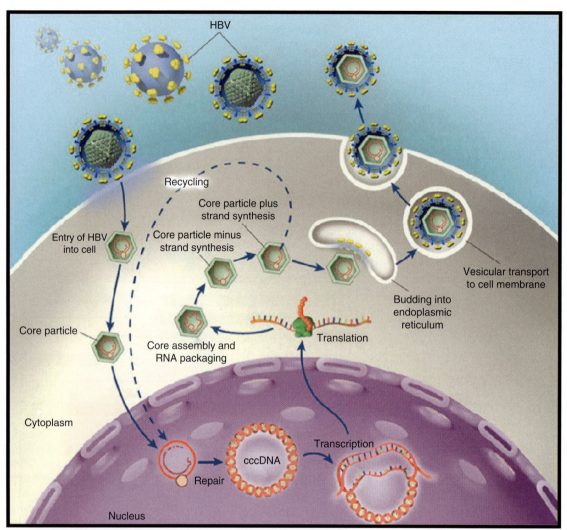

Figure 4 Diagrammatic representation of the life cycle of the virus. Reproduced from Ganem D and Prince AM (2004) Hepatitis B virus infection – Natural history and clinical consequences. *New England Journal of Medicine* 350: 1118–1129, with permission from Massachusetts Medical Society.

and includes the direct repeat 1 (DR1) and ε (**Figure 5**). The bulge of the ε structure serves as a template for the synthesis of a 3–4-nt-long DNA primer, which is covalently attached to the polymerase through a phosphodiester linkage between dGTP and the hydroxyl group of a tyrosine residue in the terminal protein (96). This event involves the ε structure at the 5′ end of the pgRNA, and is then followed by the translocation of the polymerase–primer complex to the 3′ end, where it hybridizes with the DR1 region with which it shares similarity. How this translocation occurs remains unknown. As the complex proceeds toward the 5′ end of the pgRNA, the (−)-DNA strand is synthesized by reverse transcription and the RNA template is concurrently degraded by the RNase H activity of the polymerase, except for the final 18 or so ribonucleotides. A second translocation event then occurs during which the ribonucleotide primer hybridizes with the DR1 region at the 5′ end of the newly synthesized (−)-DNA strand. A template exchange occurs that allows the (+)-DNA strand synthesis to proceed along the 5′ end of the complete (−)-DNA strand, effectively circularizing the genome. (−)- and (+)-DNA strand synthesis occurs within the nucleocapsid as already mentioned, and this is facilitated through pores allowing entry of nucleotides. Once the maturing nucleocapsid is enveloped by budding through the endoplasmic reticulum membrane, the nucleotide pool within the capsid cannot be replenished, hence the incomplete nature of the (+)-DNA strand.

Subtypes

The S protein sequence shared by all three envelope proteins contains the major immunogenic epitope of the virus referred to as the '*a*' determinant. This epitope, shared by all isolates of the virus, is recognized by

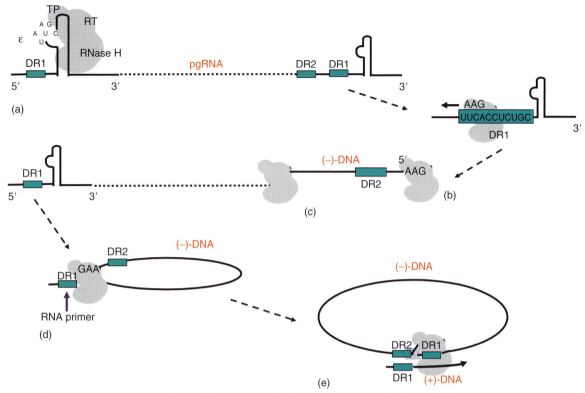

Figure 5 Replication strategy of the virus. (a) Primer synthesis; (b) translocation and binding to DR1; (c) synthesis of the (−)-DNA strand; (d) RNA primer fragment preserved from the degradation of the pgRNA; (e) (+)-DNA strand synthesis. Reproduced from Karayiannis P (2003) Hepatitis B virus: Old, new and future approaches to antiviral treatment. *Jouranl of Antimicrobial Chemotherapy*. 51: 761–785, with permission from Oxford University Press.

neutralizing antibodies and is conformational in nature, probably encompassing residues 110–160. In addition, there are subtypic specificities originally detected by antibodies. The presence of lysine (K) or arginine (R) at position 122 confers *d* or *y* specificity, respectively. Similarly, specificities *w* and *r* are conferred by the presence of K or R at position 160. Moreover, the *w* subdeterminant can be further divided into *w1–w4* specificities. There are nine subtypes of the virus (**Table 1**) depending on the presence of other subtype-determining residues elsewhere in the '*a*' determinant region.

Genotypes

Nucleotide sequencing studies soon established that the virus could be divided into genotypes based on sequence divergence of >8% (**Table 1**). There are currently eight genotypes, designated A–H, with characteristic geographical distribution. Genotypes A and D occur frequently in Africa, Europe, and India, while genotypes B and C are prevalent in Asia. Genotype E is restricted to West Africa, and genotype F is found in Central and South America. The distribution of genotypes G and H is less clear, but these have been described from isolates in central America and Southern Europe. These genotypes can be further subdivided into a total of 21 subgenotypes; A1–3, B1–5, C1–5, D1–4, and F1–4. Subgenotypes of B and C differ in their geographical distribution, with B1 dominating in Japan and B2 in China and Vietnam, while B3 and B4 are confined to Indonesia and Vietnam, respectively. Subgenotype C1 is common in Japan, Korea, and China, C2 in China, Southeast Asia, and Bangladesh, C3 in Oceania, and C4 in Australian aborigines. Recombinants between genotypes A and D, as well as between B and C, have also been described.

Likely differences between genotypes in relation to pathogenesis and response to antiviral treatment are beginning to emerge. Genotype C is more frequently associated than B with abnormal liver function tests, lower rates of seroconversion to anti-HBe, higher levels of serum HBV-DNA, cirrhosis, and HCC. Moreover, there is a better sustained response to interferon treatment in patients with genotype B than those with C, and in patients with genotype A than those with D. Genotype A infection appears to be associated with biochemical remission and clearance of HBV-DNA more frequently than genotype D, and has a higher rate of HBsAg clearance compared with genotype D.

Table 1 Genotype determining amino acid variation over the 'a' determinant region of HBsAg, and relationship between genotypes and subtypes

Genotype	Subtype	Amino acid sequence of 'a' determinant (positions 122–160)
		K T C T T P A Q G N S M F P S C C C T K P T D G N C T C I P S S W A F A K
A	adw2/ayw1	a^a -
B	adw1/ayw1	a - - - - - - - - - - T -
C	ayr/adrq+/adrq-/adr	a - - - I - - - - - T - R
D	ayw2/ayw3/ayw4	- - - - I - - - - - T - V R
E	ayw4	R - - M - - - - - - T - - - - - - - - - - - - S - - - - - - - - - - - - G -
F	adw4q-/adw2/ayw4	R - - - L - - - - - T - - - - - - - - - S - S - - - - - - - - - - - - - G -
G	adw2	R - - - L - - - - - T - - - - - - - - - - - S - - - - - - - - - - - - L G -
H	adw4	a - - - - - - - - - - Y - - - - - - - - - S - - - - - - - - - - - - - - G -
		- - - T V - - - - - Y - - - - - - - - - S - - - - - - - - - - - - - - - G -

^{a}R or K.

In contrast to the differences observed in response to interferon therapy, treatment with nucleos(t)ide analogs does not show differential responses between genotypes.

Variants

The HBV genome is not as invariant as originally thought. As HBV replicates through an RNA intermediate that is reverse-transcribed, this step in the replication cycle of the virus is prone to errors made by the viral reverse transcriptase. It is estimated that the HBV genome evolves at a rate of 1.4–3.2×10^{-5} nucleotide substitutions/site/year. The virus therefore circulates in serum as a population of very closely related genetic variants, referred to as a quasispecies. Although a lot of these variants would have mutations that would be deleterious to the virus, as a result of constraints imposed by the overlapping ORFs, some would be advantageous, either offering a replication advantage or facilitating immune escape. These are discussed below in the clinical settings in which they have been described.

Markers of Infection

Serological diagnosis of HBV infection relies on the detection of HBsAg in serum, and its persistence for longer than 6 months indicates progression to chronic infection. Appearance of anti-HBs (antibody to HBsAg) indicates recovery from infection, or acquired immunity after preventive vaccination. Detection of HBeAg denotes active viral replication, as does the detection of serum HBV-DNA by qualitative or quantitative PCR tests. Seroconversion to anti-HBe occurs after recovery from acute infection, and less often during the chronic phase, either spontaneously or after therapeutic intervention. In the latter case, this leads to a quiescent phase of disease that can be long term, but does not necessarily mean that virus replication ceases completely (see below). Detection of antibody to core antigen (anti-HBc) of IgM class at high level is a marker of acute infection, whereas total anti-HBc (primarily IgG) is detectable during both acute and chronic infection.

Natural History of the Disease

Exposure to HBV may result in asymptomatic, acute icteric, or, in some instances, fulminant hepatitis (0.1–0.5%). Approximately 5% of adults and 95% of perinatally infected young children become persistently infected. The outcome depends on the age of the patient and genetic factors determining the efficiency of the host immune response. Genetic factors influencing outcome (in more than one study) include polymorphisms of the MHC class II glycoproteins, which influence presentation of viral peptides during induction of the cellular immune response, and mannin-binding lectins, which bind to mannose-terminated carbohydrate residues such as those present on the C-terminus of the Pre-S2 region of the middle envelope protein facilitating phagocytosis. The risk of chronicity in children decreases with increasing age. A small proportion of carriers each year may become HBsAg negative (0.05–2%, depending on age of infection), thus leading to resolution of the hepatitis.

Acute HBV Infection

The incubation period following exposure is 3–6 months. In the week before icterus appears, some patients develop a serum sickness-like syndrome including arthralgia, fever, and urticaria. The clinical picture varies from asymptomatic anicteric infection to protracted icterus and, in some patients (<1%), liver failure (fulminant hepatitis). The acute infection is self-limiting and most patients recover within 1–2 months after the onset of icterus.

Chronic HBV Infection

This is defined as persistent viremia of more than 6 months duration and accompanied by hepatic inflammation. The latter is based on histological examination of liver biopsy material that is followed by assigning of scores for necroinflammatory activity (out of 18) and stage of fibrosis (out of 6), which are used to decide whether a patient needs therapy.

Course of Chronic Infection

Chronic HBV infection is quite variable and is typically characterized by four phases. These phases are the immune tolerant, the immune clearance, the nonreplicative (immune-controlled low-level infection), and the reactivation phase that may be seen in some patients, particularly in Southern Europe and the Far East (**Figure 6**). During the immune-tolerant phase, the patient is HBsAg- and HBeAg-positive with high levels of HBV-DNA, but with near-normal or minimally elevated alanine aminotransferase (ALT) levels. Children infected at birth or soon after are more likely to go through this phase, which may last for 2–3 decades. During the immune clearance phase more commonly seen in those infected in adult life, HBeAg and HBV-DNA levels decrease, ALT levels increase, and necroinflammatory changes are seen in liver biopsies. Loss of HBeAg may be accompanied by an ALT flare, culminating in seroconversion to anti-HBe and entry to the nonreplicative phase when the infection is under

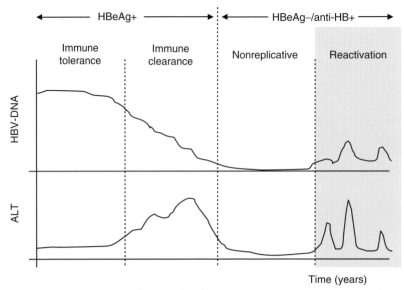

Figure 6 Diagram of the natural history of chronic HBV infection showing the immune tolerant, immune clearance, nonreplicative, and reactivation phases. Reproduced from Karayiannis P, Carman WF, and Thomas HC (2005) Molecular variations in the core promoter, precore and core regions of hepatitis B virus, and their clinical significance. In: Thomas HC, Lemon S, and Zuckerman AJ (eds.) *Viral Hepatitis* 3rd edn., pp. 242–262. London: Blackwell, with permission from Blackwell Publishing.

immune control. Viral DNA integration into the host genome may take place during chronic infection and persist during the nonreplicative (low replicative) phase. During this phase, plasma HBV-DNA may or may not be detectable, while ALT levels return to normal. This serological profile characterizes the 'inactive carrier state', which is maintained thereafter. Some patients, however, for reasons that still remain unknown, show disease reactivation accompanied by ALT rises and return of viremia. Such patients may exhibit fluctuations in ALT levels with occasional severe exacerbations. Continued necroinflammatory activity may lead to fibrosis, and faster development of cirrhosis and HCC.

Mutant Viruses and Chronic Infection

Anti-HBe-positive patients in the reactivated phase of the disease are also referred to as the HBeAg-negative viremic group. Genomic analyses has revealed that such patients carry natural mutants of the virus that have either reduced levels (core promoter variants) or complete abrogation of HBeAg (precore variants) production. These variants are selected at the time of, or soon after, seroconversion, and become dominant during the reactivation phase. The most common precore mutation is the G1896A substitution, which creates a premature stop codon in the precursor protein from which HBeAg is elaborated. This mutation affects the stem of the ε encapsidation signal, but leads to stronger base pairing with the A1896 change in genotypes with a T at position 1858 of the precore region, such as B, C, D, and E. The double mutation affecting the core promoter region (A1762T, G1764A) is thought to result in decreased transcription of the precore mRNA, with a knockon effect on HBeAg production, while pgRNA production remains the same or is even upregulated. It is now apparent that additional mutations in this region may contribute to this phenotype.

Vaccination

Prophylactic vaccination offers the only means of interrupting the transmission of the virus. Vaccines currently used consist of recombinantly expressed HBsAg in yeast such as *Saccharomyces cerevisiae*. In adults, the vaccine is administered intramuscularly into the deltoid at 0, 1, and 6 months. In countries of high and medium seroprevalence, universal vaccination programmes have been instituted, and HBV vaccination is recommended for infants born to carrier mothers within 12 h of birth, given together with hepatitis B immune globulin (HBIg). The response to the vaccine is determined by measuring anti-HB levels 1–4 months after the last dose of the vaccine, and the minimum protection level is set at 10 mIU ml^{-1}.

Development of anti-HBs following vaccination has been recorded in 90–95% of healthy individuals, with lower response rates in hemodialysis and hemophiliac patients (70%). Recent studies on the duration of antibodies have shown maintenance of levels above the 10 mIU ml^{-1} cutoff for 12 years, in up to 80% of individuals immunized at a young age. Booster immunizations therefore may not be required for at least 10 years after vaccination, and some countries are reconsidering the necessity for this.

The beneficial effects of HBV vaccination are becoming increasingly apparent, particularly in reducing new infections. There has been a dramatic drop in HBV prevalence in populations where the disease was endemic. In Taiwan, 15 years after the start of the vaccination programme, the prevalence of HBsAg in children under 15 years of age has decreased from 9.8% in 1984 to 0.9% in 1999. Similarly, the incidence of HCC has been on the decline from 0.7 per 100 000 children between 1981 and 1986, to 0.57 and 0.36 in 1986–90 and 1990–94, respectively.

Vaccine Escape Mutants

In spite of vaccination and the presence of a satisfactory antibody level, it has been observed that in some instances breakthrough infections occur, the commonest of which involves a mutant with a G145R substitution in the 'a' determinant region of HBsAg. This and additional mutations in this region have been shown to result in altered antigenicity, accompanied by failure of HBsAg recognition by neutralizing antibody. Such mutant viruses have also been described in the liver transplantation setting, where use of HBIg or monoclonal anti-HBs is recommended in an attempt to prevent infection of the new liver graft.

Treatment

The agents currently available for the treatment of chronic HBV infection are divided into two main groups: the immunomodulators, which include interferon-alpha (IFN-α), and nucleos(t)ide analogs such as lamivudine (3TC), adefovir dipivoxil (Hepsera), entecavir (Baraclude), and telbivudine (Tyzeka or Sebivo), which are currently approved for this purpose. The immunomodulators act by promoting cytotoxic T-cell activity for lysis of infected hepatocytes and by stimulating cytokine production for control of viral replication. Nucleos(t)ide analogs on the other hand are chain terminators acting at the stage of DNA synthesis.

HBeAg-Positive Patients

Treatment with pegylated IFN-α-2, the current standard treatment administered once a week for up to 1 year, achieves seroconversion to anti-HBe in 32% of patients, compared to 18%, 12%, 21%, and 23% with lamivudine, adefovir, entecavir, and telbivudine, respectively. The latter drugs are taken orally daily, in contrast to the weekly intramuscular injections of pegylated interferon. These responses are sustainable in over 95% of patients.

HBeAg-Negative Patients

Treatment of these patients with pegylated interferon for a year results in virologic remission (HBV-DNA <20 000 IU ml^{-1}; equivalent to 105 copies ml^{-1}) in about 44% of them, followed by ALT normalization. This response appears durable in around 20%. Similarly, treatment with nucleos(t)ide analogs for a year leads to HBV-DNA becoming undetectable in between 65% and 90% of treated patients. However, on stopping therapy, only a small minority have a sustained response at the end of 24 weeks of follow up.

Protracted Treatment

To manage such relapses following initial interferon treatment, nucleos(t)ide therapy should be started and continued long term in both HBeAg-positive and -negative patients not achieving a sustained response after a trial of pegylated interferon for 6–12 months. In HBeAg-positive patients, prolonged treatment with lamivudine, for example, leads to increased seroconversion rates from 17% in year 1 to 27% and 40% for years 2 and 3, respectively. Besides, prolonged treatment leads to normalization of ALT levels and an obvious improvement in the histological findings of liver biopsy material. Unfortunately, in many cases, there are breakthrough infections which are attributed to the development of resistance as described below. In such cases, virological breakthrough is soon accompanied by biochemical (ALT rise) and histological relapse. The latter can be avoided by switching to a different nucleos(t)ide analogue that has no cross-resistance with the previous one. Monitoring at three monthly intervals for viral resistance, using molecular assays, is essential.

Resistance

Lamivudine resistance develops in about 24% of patients at year 1 rising to >70% by year 5. Adefovir resistance on the other hand is delayed, being 0% at year 1, 3% at year 2, and rising to 28% by year 4. Entecavir resistance has only been seen so far in patients with lamivudine resistant strains. Longer term follow-up with this nucleoside analogue is ongoing. Nevertheless, it appears that this analog has a high genetic resistance barrier while lamivudine has a low one. Resistance to telbivudine is already a problem after a year of treatment but it appears initially to be more potent than lamivudine.

Molecular Basis of Drug Resistance

The rt domain of the HBV polymerase contains six subdomains (A–F) that are spatially separated, but closely associated with, the normal function of the protein. The characteristic YMDD (tyrosine-methionine-aspartate-aspartate) motif of the catalytic site is located within subdomain C. Subdomains A, C, and D are most likely involved with dNTP binding and catalysis, whereas subdomains B and E interact with the pgRNA template and primer. Amino acid substitutions that confer resistance to lamivudine predominantly affect the YMDD motif, so that the methionine (M) at position 204 is changed either to valine (YVDD, rtM204V) or isoleucine (YIDD, rtM204I). The former mutation is almost always associated with a second one in subdomain B, involving a substitution of leucine with methionine at position 180 (rtL180M). Adefovir resistance is conferred by mutations rtN236T in subdomain D and rtA181V in subdomain B. In the small number of entecavir-resistant cases detected so far, in addition to the lamivudine-resistant substitutions additional ones that include rtI169T, rtT184G, rtS202I, and rtM250V have been identified.

See also: Hepatitis A Virus; Hepadnaviruses of Birds; Hepadnaviruses: General Features; Hepatitis B Virus: Molecular Biology.

Further Reading

Carman WF, Jazayeri M, Basune A, Thomas HC, and Karayiannis P (2005) Hepatitis B surface antigen (HBsAg) variants. In: Thomas HC, Lemon S, and Zuckerman AJ (eds.) *Viral Hepatitis,* 3rd edn., pp. 225–241. London: Blackwell.

Ganem D and Prince AM (2004) Hepatitis B virus infection – Natural history and clinical consequences. *New England Journal of Medicine* 350: 1118–1129.

Hadziyannis SJ and Papatheodoridis GV (2006) Hepatitis B e antigen-negative chronic hepatitis B: Natural history and treatment. *Seminars in Liver Disease* 26: 130–141.

Hunt CM, Mc Gill JM, Allen MI, et al. (2000) Clinical relevance of hepatitis B viral mutations. *Hepatology* 31: 1037–1044.

Karayiannis P (2003) Hepatitis B virus: Old, new and future approaches to antiviral treatment. *Journal of Antimicrobial Chemotherapy* 51: 761–785.

Karayiannis P, Carman WF, and Thomas HC (2005) Molecular variations in the core promoter, precore and core regions of hepatitis B virus, and their clinical significance. In: Thomas HC, Lemon S, and Zuckerman AJ (eds.) *Viral Hepatitis,* 3rd edn., pp. 242–262. London: Blackwell.

Thomas HC (2006) Hepatitis B and D. *Medicine* 35: 39–42.

Zoulim F (2006) Antiviral therapy of chronic hepatitis B. *Antiviral Research* 71: 206–215.

Hepatitis B Virus: Molecular Biology

T J Harrison, University College London, London, UK

© 2008 Elsevier Ltd. All rights reserved.

Glossary

Pseudogene Nonfunctional DNA sequence that is very similar to that of a known gene.

Tolerogen A substance that produces immunological tolerance.

Introduction

Hepatitis B virus (HBV) is the prototype of the family *Hepadnaviridae*, a group of viruses that infect mammals (primates and rodents) and birds. These viruses are distantly related to the retroviruses and, although their genomes are DNA, they are replicated by the reverse transcription of an RNA pregenome.

Structure and Replication of HBV

The 42 nm hepatitis B virion is composed of the DNA genome packaged together with a copy of the virus-encoded polymerase in an icosahedral nucleocapsid made up of dimers of the hepatitis B core antigen, HBcAg. In turn, the nucleocapsid is covered by an envelope composed of a lipid bilayer derived from internal cellular membranes and embedded with the hepatitis B surface protein, HBsAg. The open reading frame (ORF) encoding HBsAg has three in-frame initiation codons (**Figure 1**) which are used for the translation of the large (L, pre-S1 + pre-S2 + S), middle (M, pre-S2 + S), and small (S or major) surface proteins. All three proteins, which share the same C-terminus, are found in the virions.

The early events in infection of the hepatocyte are not well understood, but begin with the virion binding to an unidentified receptor (and, most likely, co-receptors) on the plasma membrane. The first contact seems to involve a domain located near to the N-terminus of L (and not present in M or S), and other interactions involving S also seem to be important. Virus entry likely is via the endosomal route and results in the delivery of the genome, with or without the nucleocapsid, to the nucleus.

The 3.2 kbp genome is composed of two linear strands of DNA held in a circular configuration by base pairing of a short region ('cohesive end region') where the 5′ ends overlap (**Figure 1**). The plus strand is incomplete so that the circle is partially single-stranded. After delivery to the

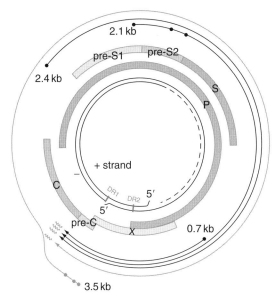

Figure 1 Organization of the HBV genome. The inner circles depict the complete minus strand and incomplete plus strand, and the positions of the direct repeats (DRs) are indicated. The blocks surrounding the genome show the locations of the four overlapping ORFs; C and S contain two and three in-frame initiation codons, respectively. The outer arrows show the viral transcripts, solid circles indicate the positions of the 5′ ends, and the arrowheads indicate the common polyadenylation site. Reproduced from Kidd-Ljunggren K, Miyakawa Y, and Kidd AH (2002) Genetic variability in hepatitis B viruses. *Journal of General Virology* 83: 1267–1280, with permission from Society for General Microbiology.

nucleus, the plus strand is completed (probably by a host enzyme), the primers that remain attached to the 5′ ends of both strands are removed, and the ends are ligated with the introduction of superhelical turns to yield covalently closed circular (ccc)-DNA. This cccDNA associates with histones and other host proteins and resides in the nucleus as a 'minichromosome', the template for transcription of the viral RNAs.

The genome encodes four overlapping ORFs, specifying the core and surface proteins, the polymerase, and a small protein known as HBx (**Figure 1**). All of the *cis*-acting elements, including two enhancers, four promoters, and the single polyadenylation signal, are embedded within the ORFs. Two families of RNAs (2.1 and 3.5 kb in size) are sufficiently abundant to be detectable by northern blotting of RNA from HBV-infected human liver. The 2.1 kb family is translated from a promoter (surface promoter 2, SP2) in the pre-S1 region and encodes M and S. The 3.5 kb RNAs, transcribed from the core promoter, include the precore RNA, which is translated to yield hepatitis B e-antigen (HBeAg), described in detail below, and the pregenomic RNA, which also encodes HBcAg and the polymerase. Surface promoter 1 (SP1), upstream of the surface ORF, transcribes a less-abundant 2.4 kb RNA that encodes L, and the X promoter transcribes an RNA of ~0.7 kb, HBx. All of the RNAs are 3′ co-terminal, being polyadenylated in response to a polyA signal in the core (or C) ORF (**Figure 1**).

The pregenomic RNA is the 3.5 kb transcript that is reverse-transcribed to the HBV genome and also acts as the mRNA for HBcAg and the polymerase. HBcAg forms dimers and has arginine-rich motifs at the C-terminal end that are believed to interact with the viral nucleic acid in the nucleocapsid. Less frequently, the downstream ORF is translated to yield the polymerase, probably following a ribosomal translocation that bypasses the core initiation codon. There is a stem–loop structure, termed epsilon (ϵ), near to the 5′ end of the pregenomic RNA (**Figure 2(a)**), and the polymerase binds to a bulge on the stem, priming first strand synthesis and signaling packaging of the pregenome and polymerase into 'precores', as the dimers of HBcAg self-assemble around the RNA–protein complex. The primer is the N-terminal domain of the polymerase, in which a tyrosine residue is covalently linked to guanosine, and this is extended by a further three residues on the bulge. The extended primer then translocates to a complementary sequence near to the 3′ end of the RNA (**Figure 2(b)**).

The direct repeats (DR1 and DR2) are motifs of 11 nt that occur twice in the HBV genome (**Figure 1**). Located at the positions of the 5′ ends of the two strands, they play an important role in template switches during viral DNA synthesis. The translocated primer binds within DR1, and minus-strand synthesis commences with concomitant degradation of the template by an RNaseH activity in the C-terminal domain of the polymerase (**Figure 2(c)**). Completion of minus-strand synthesis leaves a capped oligoribonucleotide (which was the 5′ end of the pregenome) containing the DR1 sequence (**Figure 2(d)**). This translocates to the copy of DR2 near to the 5′ end of the minus strand and primes plus-strand synthesis, which then proceeds to the 5′ end of the minus strand (**Figures 2(e)** and **2(f)**). The short (~8 nt) terminal redundancy of the minus strand allows another template switch and circularization of the genome. Completion of the nucleocapsid during plus-strand synthesis starves the polymerase of the substrate, leaving the characteristic partially single-stranded structure that typifies the HBV genome (**Figures 2(g)** and **2(h)**).

Mature cores then bud through cellular membranes containing HBsAg (S, M, and L) and are exocytosed. A vast excess of subviral particles, which lack nucleocapsids and are noninfectious, are also secreted from the hepatocyte and are presumed to overwhelm the immune response of the host. These comprise 22 nm spheres, composed of S and M, and tubular forms of the same diameter, composed of S, M, and L (**Figure 3**). An endoplasmic-retention signal within the pre-S1 domain of L is involved in the formation of the virions and tubular subviral particles.

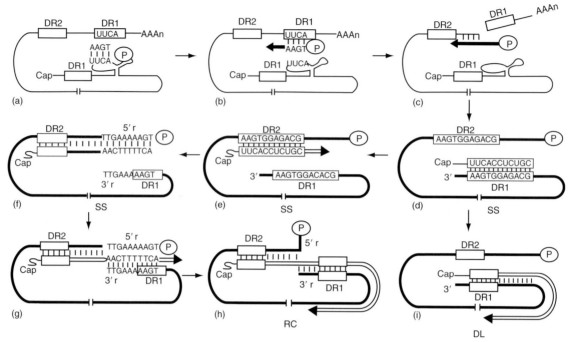

Figure 2 Replication of the HBV genome. See text for details. Reproduced from Liu N, Ji L, Maguire ML, and Loeb DD (2004) cis-Acting sequences that contribute to the synthesis of relaxed-circular DNA of human hepatitis B virus. *Journal of Virology* 78: 642–649, with permission from American Society for Microbiology.

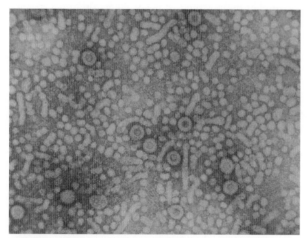

Figure 3 Electron micrograph of particles from the plasma of an infectious carrier. The 42 nm virions (Dane particles) are outnumbered by 22 nm spherical and tubular noninfectious, subviral particles.

The Surface Protein, HBsAg

The 226-amino-acid (aa) residue S protein is quite hydrophobic and structural models suggest that it contains four membrane-spanning regions (**Figure 4(a)**). The major hydrophilic loop is exposed on the surface of viral particles and is the main target of the immune response to HBsAg. This domain is rich in cysteine residues, which are believed to be linked by disulfide bridges, forming a complex structure that includes the *a* determinant and *d/y* and *w/r* subdeterminants. Antibody (anti-HBs) responses to this domain give immunity to infection in convalescence or following immunization with hepatitis B vaccine. Approximately half of the S residues are N-glycosylated at aa 146 (in the *a* determinant) so that the protein exists in two forms, p24 and gp27.

The N-terminus of S is exposed on the surfaces of the virions and subviral particles and, in M, is extended by 55 aa (the pre-S2 region). The pre-S2 region is glycosylated in M to give two proteins, gp33 and gp36, depending upon the glycosylation of asparagine 146 in the S domain. M is much less abundant than S in the virion and subviral particles. The SP2 promoter lacks a 'TATA box' and the 5′ ends of the 2.1 kb RNAs are heterogeneous, most being located downstream of the pre-S2 initiation codon (**Figure 1**). Furthermore, that codon is in a poor 'Kozak' context for recognition by the ribosome; consequently, much more S than M is produced.

The large surface protein, L, has a further extension of the N-terminus by the pre-S1 region. Again, there are two forms, p39 and gp42, depending on glycosylation at aa 146 of S (the pre-S2 region is not glycosylated in L). The glycine at aa 2 of the pre-S1 region is myristylated and embedded in the membrane. L has a dual topology: upon expression, the pre-S loop may translocate into the endoplasmic reticulum (ER) lumen, so that it is exposed on the exterior of the particle and presents the

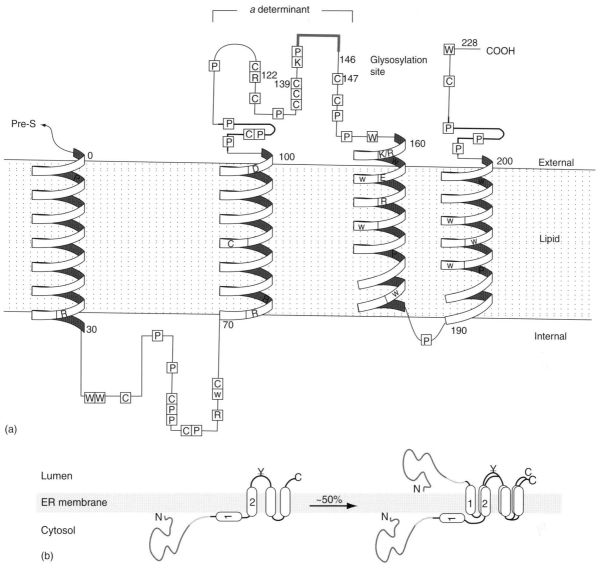

Figure 4 (a) Predicted structure of the major surface protein, S. Note the abundance of cysteine residues in the major hydrophilic region. (b) Dual topology of the large surface protein, L. (a) Reproduced from Stirk HJ, Thornton JM, and Howard CR (1992) A topological model for hepatitis B surface antigen. *Intervirology* 33: 148–158, with permission from Karger Publishers. (b) Reproduced from Lambert C, Mann S, and Prange R (2004) Assessment of determinants affecting the dual topology of hepadnaviral large envelope proteins. *Journal of General Virology* 85: 1221–1225, with permission from Society for General Microbiology.

receptor-binding domain. However, in around 50% of the L molecules, the loop does not translocate but remains in the cytosol and is believed to interact with the nucleocapsid during budding and maintain this contact in the virion (**Figure 4(b)**).

The Hepatitis B Vaccines

The hepatitis B vaccines contain HBsAg and are given via an intramuscular injection to stimulate a protective anti-HBs response. The first-generation (so-called 'plasma derived') vaccine was produced by purifying HBsAg from donated blood. Despite some worries that blood-borne viruses (particularly human immunodeficiency virus (HIV)) potentially might contaminate particular batches, this method of vaccine production is safe provided that manufacturing protocols, including steps that inactivate potential infectivity, are adhered to strictly. Even today, plasma-derived vaccines constitute the majority of doses given worldwide. Currently, most doses of hepatitis B vaccine given in the West are produced by expression of HBsAg in yeast (*Saccharomyces cerevisiae*), and several vaccines based on expression in mammalian cells also have been licensed.

Most individuals produce a protective immune response with the standard, three-dose regimen of immunization, but there is a problem of nonresponsiveness in up to 5% of recipients, particularly the immune-suppressed and older individuals, and especially males. The vaccines contain only S and there has been interest in whether vaccines containing pre-S epitopes may circumvent this problem of nonresponsiveness. Up to 50% of nonresponders may sero-convert following immunization with a vaccine containing pre-S epitopes and the nonresponsiveness of many of the remainder seems to be genetically based.

Most countries worldwide have now introduced universal immunization of infants. Although the vaccine may be incorporated into the Expanded Programme of Immunization, it is essential that it is given as soon as possible after birth to the infants of infectious mothers, in order to break the chain of mother-to-infant transmission (see below). Immunization within 12 h of birth protects around 70% of the infants of carrier mothers and up to 90% success may be achieved when passive immunization with hepatitis B immune globulin (HBIG) also is given, at a contralateral site, to neutralize any maternal virus that reaches the newborn infant. Rarely, however, the antibody may select variants with mutations that alter the amino acid sequence of the *a* determinant. Such 'antibody escape' variants may also be selected in liver-transplant recipients given HBIG in an attempt to prevent HBV infection of the graft and, indeed, in the natural course of infection, with seroconversion to anti-HBs. The most common mutations affect the highly conserved codons 144 and 145 (for aspartic acid and glycine, respectively) of S. Fortunately, there is no evidence that these 'antibody escape' variants can be transmitted to individuals with vaccine-induced immunity.

Hepatitis B e-Antigen, HBeAg

This antigen circulates as a soluble protein in the plasma of some persistently infected individuals and is recognized as a marker of infectivity. The synthesis of HBeAg is nonessential for virus replication and its function is thought to be as a tolerogen; it may be especially important in perinatal transmission. HBV does not cross the placenta, but HBeAg is believed to do so and to circulate in the fetus. The infants of viremic mothers almost invariably become infected at or around the time of birth. At least 90% of such infants become persistently infected and often are extremely immune-tolerant of the virus, with very high viral loads that may persist for decades or even for life. When they are females, they infect their children in turn; thus, HBeAg seems to be important in maintaining HBV in the human population through the generations.

As noted above, the pregenomic RNA is translated from the second initiation codon in the core ORF to produce the nucleocapsid protein. The precore RNA has its cap site further upstream, so that it contains the precore initiation codon and the entire core ORF is translated (**Figure 5**). There are an additional 29 aa, encoded by the precore region, at the N-terminus of the precursor to HBeAg (p25) and these include a signal sequence that targets the protein to the ER. Here, the cellular signal peptidase cleaves this sequence and the protein is secreted with further proteolysis, removing the arginine-rich domain at the C-terminus.

Seroconversion from HBeAg positivity to anti-HBe, with clearance of HBV replication, is often a feature of recovery from persistent infection. However, a significant proportion of patients with circulating anti-HBe remain viremic, infected with HBV variants that do not synthesize, or synthesize reduced quantities of, HBeAg. The most common mutation affects codon 28 in the precore region, changing it from a tryptophan to a termination codon and preventing the synthesis of p25. Because this mutation affects base-pairing in ε, it is constrained in some genotypes and subgenotypes of HBV. Other, less common 'precore mutations' introduce termination codons elsewhere in the precore region, destroy the precore initiation codon, or modify the signal peptidase cleavage site.

Another class of mutations found in HBV from anti-HBe-positive individuals affects the core promoter. These mutations decrease the transcription of the precore RNA (and may increase the transcription of the pregenomic RNA), resulting in reduced synthesis of HBeAg. Core promoter mutations may be associated with a particularly high risk of developing hepatocellular carcinoma (HCC), especially in genotype C.

Treatment of Chronic Hepatitis B

Prior to the introduction of specific antiviral therapeutics, persistent HBV infection was treated with interferon, with mixed success. In some studies, up to one-third of treated individuals cleared the virus, but the treatment was rather less successful in certain populations, particularly individuals from the Far East. Interferon seems to work by modulating the immune response of the host, rather than a direct antiviral effect, and was found to be most successful in individuals who had evidence of hepatitis or, in other words, already were mounting a cellular immune response to the virus.

Because the HBV genome is replicated via an RNA intermediate, treatment with nucleoside and nucleotide analogs is an option. The predicted amino acid sequence of the HBV polymerase is similar to those of retroviruses and a motif, Tyr-Met-Asp-Asp (YMDD), in the active site is highly conserved. The HBV polymerase has not been crystallized but its structure is inferred from those of other RNA-dependent DNA polymerases, particularly that of HIV type 1 (HIV-1). Several conserved regions,

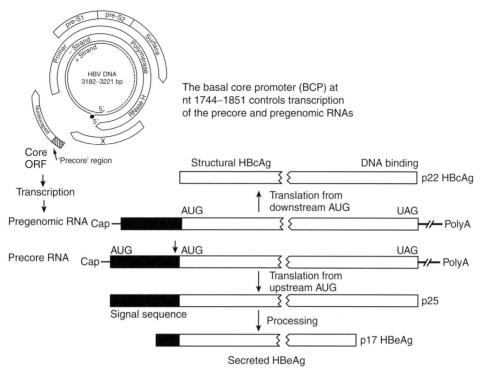

Figure 5 Expression of the HBV core ORF. The pregenomic RNA is translated from the second initiation codon in the ORF to yield HBcAg. The precore RNA is translated from the upstream initiation codon to p25, which is processed to HBeAg. The arrow indicates the position of the most common precore mutation, changing a tryptophan to a termination codon. Reproduced from Harrison TJ (2006) Hepatitis B virus: Molecular virology and common mutants. *Seminars in Liver Disease* 26: 87–96, with permission from Thieme International (Stuttgart).

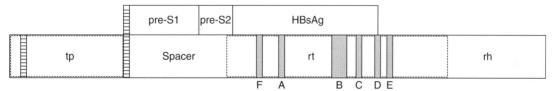

Figure 6 Conserved domains in the HBV reverse transcriptase. The polymerase and overlapping surface ORFs are shown. tp, terminal protein (primase); rt, reverse transcriptase; rh, RNaseH. Reproduced from Stuyver L, Locarnini S, Lok A, *et al*. (2001) Nomenclature for antiviral-resistant human hepatitis B virus mutations in the polymerase region. *Hepatology* 33: 751–757, with permission from the American Association for the Study of Liver Diseases.

designated A–F, are recognized within the reverse transcriptase domain and the YMDD motif is located within conserved region C (**Figure 6**).

Despite the similarity between the HIV and HBV polymerases, many of the nucleoside analogs first used against HIV, such as zidovudine, were found not to be effective against HBV. Lamivudine, a deoxycytidine analog which acts as a chain terminator during reverse transcription of the pregenome and can result in a 4–5 $\log_{10}$ suppression of viral load, was the first nucleoside analog to be used successfully and licensed for therapy of hepatitis B. However, prolonged monotherapy with lamivudine results in the emergence of resistant virus in approximately 40% of patients after 2 years of therapy (and 65% after 5 years). Resistance to lamivudine is associated with point mutations which result in substitution of the methionine residue in the YMDD motif by either valine or isoleucine, changes which parallel lamivudine resistance in HIV-1. The valine (but usually not the isoleucine) substitution is associated with a second, leucine-to-methionine substitution in the upstream, conserved region B and this seems to compensate partially for the adverse effect of the former on replication efficiency.

Another nucleoside analog, adefovir dipivoxil, inhibits the replication not only of wild-type HBV but also of lamivudine-resistant mutants. Resistance to adefovir emerges at a slower rate than to lamivudine, reaching around 22% after 2 years for patients initially treated with lamivudine, and perhaps less frequently for nucleoside-naive individuals. Adefovir resistance is associated particularly with substitutions in conserved regions D or B and these mutants are susceptible to lamivudine. The absence of substitutions in conserved region C is consistent with the concept that adefovir inhibits the priming of reverse transcription, rather than chain elongation, and suggests that combination therapy with lamivudine and adefovir may inhibit the emergence of resistant mutants.

In vitro, the deoxyguanosine analog entecavir inhibits priming of reverse transcription as well as chain elongation of both the minus and plus strands. The drug is active against lamivudine- and adefovir-resistant HBV. Resistance to entecavir has been described in patients who had been treated previously with lamivudine and in whom lamivudine-resistant mutants had already been selected, but resistance seems to arise in otherwise treatment-naive patients at a rate of less than 1% per year. Other analogs, including telbivudine, which is similar in structure to lamivudine but seems to have better efficacy, have also been licensed recently for use in hepatitis B but the design of regimens for combination therapy lags way behind that for HIV.

HBV and HCC

When tests for HBsAg were first developed, it became clear that regions of the world with a high prevalence of chronic hepatitis B and a high annual incidence of HCC were coincident and that, in those regions, most individuals with HCC were HBsAg-positive. In Taiwan, a large prospective study confirmed the increased risk of HCC for carriers of HBsAg, and there is now evidence that the hepatitis B vaccine will reduce considerably the risk of developing the tumor. Although the importance of hepatitis C virus as a cause of HCC has increased considerably over recent decades, HBV remains a critical and preventable cause of one of the most common cancers of humans.

Cloning of the HBV genome enabled the use of virus-specific DNA sequences as probes in Southern hybridization assays, leading to evidence of chromosomal integration of viral sequences in at least 80% of HBV-associated HCCs. The chromosomal sites of integration of viral DNA seem to be random. Nevertheless, the tumors are clonal with respect to the integrated HBV DNA (often with multiple copies or partial copies of the genome), suggesting they arise from a single cell and that integration is the first step in the oncogenic process. In fact, integrated HBV DNA often can be detected by Southern hybridization of DNA from liver biopsies of infected patients who do not have HCC, suggesting that clonal expansion of hepatocytes containing integrated HBV DNA may be common in persistent infection.

Although the chromosomal sites of integration are random, there are hot spots in the viral genome, particularly around the direct repeats, for recombination with human DNA. A double-stranded linear form of HBV DNA, an aberrant product (and not a replicative intermediate) generated when the plus-strand primer fails to translocate and primes *in situ* (**Figure 2(i)**), may be that which recombines with the host DNA. Integration seems to be of no value to the virus, it is not a part of the replication cycle, and the integrants often are rearranged and never have sufficient redundancy to act as templates for the transcription of the 3.5 kb pregenome. On the other hand, the surface ORF often is expressed from integrants and many HBV-associated HCCs secrete HBsAg.

Despite considerable study, there is no consensus regarding the mode of HBV oncogenesis. Interest has focused on the product of the X ORF; this acts as a transactivator of transcription and seems to be important in 'kick-starting' the infection by stimulating transcription from the viral promoters. The protein seems to have pleiotropic effects in the cell, including stimulating various signal pathways and perhaps also preventing apoptosis. Transgenic mice with liver-specific expression of HBx develop HCC. Truncated middle surface proteins, produced following transcription of incomplete surface ORFs in integrated HBV DNA, also may act as transactivators, perhaps by locating in the plasma membrane and interacting with the receptors that initiate various signaling cascades. Some mice with the entire HBV genome as transgenes show accumulation of the large surface protein (L) in the ER of the hepatocytes and the resultant ER stress also seems to lead to the development of HCC.

Researchers have also investigated the hypothesis that HBV acts as an insertional mutagen, perhaps by inactivating a tumor suppressor gene or causing the inappropriate expression of a normally quiescent gene. Early studies, based on the analysis of genomic libraries of tumor DNA, identified insertions in or next to human genes, such as for cyclin A and the retinoic acid receptor β gene. However, these were isolated instances and no common targets were identified. More recently, amplification of virus–host junctions using polymerase chain reaction (PCR)-based techniques has identified further examples, particularly the human telomerase reverse transcriptase (hTERT), which seems to be a target of integration in several tumors. It is worth noting that, while the initial sites of HBV DNA integration are random, the integrants in tumors are a selected subset that may be implicated in the oncogenic process.

Woodchuck hepatitis virus (WHV) also causes HCC in its natural host. In contrast to HBV, in woodchuck tumors WHV DNA often is integrated within or adjacent to myc family oncogenes (c-myc or N-myc). In fact, woodchucks possess an additional copy of N-myc, a retrotransposed pseudogene that lacks introns. Insertion of the

WHV enhancer 5′ or 3′ of the myc-coding region activates transcription from the myc promoter.

See also: Hepadnaviruses: General Features; Hepadnaviruses of Birds; Hepatitis B Virus: General Features.

Further Reading

Bertoletti A and Gehring AJ (2006) The immune response during hepatitis B virus infection. *Journal of General Virology* 87: 1439–1449.

Bouchard MJ and Schneider RJ (2004) The enigmatic X gene of hepatitis B virus. *Journal of Virology* 78: 12725–12734.

Harrison TJ (2006) Hepatitis B virus: Molecular virology and common mutants. *Seminars in Liver Disease* 26: 87–96.

Kidd-Ljunggren K, Miyakawa Y, and Kidd AH (2004) Genetic variability in hepatitis B viruses. *Journal of General Virology* 83: 1267–1280.

Kremsdorf D, Soussan P, Paterlini-Brechot P, and Brechot C (2006) Hepatitis B virus-related hepatocellular carcinoma: Paradigms for viral-related human carcinogenesis. *Oncogene* 25: 3823–3833.

Lambert C, Mann S, and Prange R (2004) Assessment of determinants affecting the dual topology of hepadnaviral large envelope proteins. *Journal of General Virology* 85: 1221–1225.

Liu N, Ji L, Maguire ML, and Loeb DD (2004) *cis*-Acting sequences that contribute to the synthesis of relaxed-circular DNA of human hepatitis B virus. *Journal of Virology* 78: 642–649.

Norder H, Courouce A-M, Coursaget P, *et al.* (2004) Genetic diversity of hepatitis B virus strains derived worldwide: Genotypes, subgenotypes, and HBsAg subtypes. *Intervirology* 47: 289–309.

Stirk HJ, Thornton JM, and Howard CR (1992) A topological model for hepatitis B surface antigen. *Intervirology* 33: 148–158.

Stuyver L, Locarnini S, Lok A, *et al.* (2001) Nomenclature for antiviral-resistant human hepatitis B virus mutations in the polymerase region. *Hepatology* 33: 751–757.

Zoulim F (2006) Antiviral therapy of chronic hepatitis B. *Antiviral Research* 71: 206–215.

Hepatitis C Virus

R Bartenschlager and S Bühler, University of Heidelberg, Heidelberg, Germany

© 2008 Elsevier Ltd. All rights reserved.

Glossary

Pseudotypes Retroviral vector particles bearing heterologous glycoproteins on their surfaces.

Replicon DNA or RNA molecule capable of self-replication in a cell or in an adequate *in vitro* system (e.g., cell lysate).

Sustained virological response (SVR) Continuous absence of HCV RNA from the serum starting 6 months after cessation of antiviral therapy.

Introduction

In the 1970s, when blood tests for the detection of hepatitis A virus (HAV) and hepatitis B virus (HBV) became available, many blood samples responsible for post-transfusion hepatitis were negative when tested for these two viruses. Therefore, this third form of transfusion-associated hepatitis was named non-A, non-B hepatitis. However, in 1989, Choo and co-workers discovered hepatitis C virus (HCV) as the causative agent of parenterally transmitted non-A, non-B hepatitis. Initially, they isolated a viral complementary DNA clone from the serum of a chimpanzee experimentally infected with non-A, non-B hepatitis and used it to establish a screening test for antibodies in patients infected with this agent. By using this initial cDNA clone as a hybridization probe, Choo and colleagues then isolated a near full-length viral genome that was readily classified as a close relative of the animal pathogenic pestiviruses based on similarities of genome organization and virion properties.

Taxonomy and Geographical Distribution

HCV has been classified as the only member of the genus *Hepacivirus* and grouped together with the genera *Pestivirus*, *Flavivirus*, and tentatively the GB-viruses, in the family *Flaviviridae*. According to phylogenetic analyses, HCV is more closely related to the pestiviruses than to the flaviviruses.

Based on genomic heterogeneity, six major genotypes, having more than 30% nucleotide sequence divergence, and more than 70 subtypes differing from each other by 10–30% at the nucleotide sequence level, have been defined. Subtypes are designated by lowercase letters following the number of the genotype (e.g., genotype 1 subtype b = 1b). While genotype 1 and 2 viruses are prevalent almost worldwide, HCV genotypes 3–6 are to a large extent restricted to distinct geographical regions, including the Indian subcontinent and Southeast Asia (genotype 3), Africa and Middle East (genotype 4), South Africa (genotype 5), and Southeast Asia (genotype 6). Individual genotypes have not been ascribed to particular disease manifestations, except for a higher prevalence of steatosis with patients infected with genotype 3 viruses, but genotypes are important predictors of therapy outcome (see below).

Transmission

HCV is mainly transmitted by parenteral exposure to blood and blood products. The development of effective screening tests for blood and blood products and the implementation of viral disinfection procedures have almost excluded this route of transmission in countries where these measures are in place. Thus, the major remaining risk factor for acquiring HCV infection in developed countries is the use of contaminated needles in injection drug use. In some countries, HCV infection has been spread primarily by the use of inadequately sterilized medical instruments. In contrast to HBV infection, sexual transmission and maternal–infant spread of HCV are much less frequent.

Clinical Manifestation

Acute HCV infections are usually asymptomatic or, in about 30% of cases, associated with nonspecific symptoms such as abdominal pain, fatigue, weakness, poor appetite, and nausea. During an incubation period of 15–75 days, HCV RNA becomes detectable in serum by reverse transcription-polymerase chain reaction (RT-PCR), and virus titers usually peak at 10^5–10^7 genomes/ml between weeks 6 and 10, irrespective of disease outcome (**Figure 1(a)**). Two to four weeks after onset of viremia serum alanine aminotransferase (ALT) levels begin to rise, indicative of hepatocellular injury. Due to these nonspecific signs and symptoms, acute infection often remains unrecognized. The duration of viremia in acute hepatitis C is unpredictable and can vary from 2 to more than 4 months. Some patients even become HCV RNA negative during early convalescence but later on viremia rebounds. Overall 50–80% of HCV infections lead to a chronic carrier state (**Figure 1(b)**). About 30% of these chronically infected persons progress to liver cirrhosis 10–30 years after primary infection and hepatocellular carcinoma occurs in up to 2.5% of these patients. In contrast to chronic hepatitis B, in case of persistent HCV infection a hepatocellular carcinoma only develops on the basis of prior cirrhosis.

Pathology and Histopathology

Hepatocytes are the primary target of HCV. Therefore, the histological alterations of chronic hepatitis C are hepatocellular injury, portal and parenchymal inflammation, and necrosis. The injury of the hepatocyte is thought to be induced primarily by the immune reaction rather than by viral cytopathogenicity. Liver damage is typically spotty and focal with accompanying chronic inflammatory cells, macrophages, and, eventually, variable degrees of fibrosis. The progression rate of hepatic fibrosis is the major determinant for the outcome of chronic hepatitis C in terms of developing cirrhosis and hepatocellular carcinoma. Unfortunately, there are only a few histological markers that are more often associated with hepatitis C than with other causes of hepatitis, such as steatosis. Although the liver is the primary target, persistent HCV infection is often associated with extrahepatic symptoms, such as renal complications, lymphoma, and diabetes. A high proportion of patients with chronic hepatitis C develop cryoglobulinemia, which may account for some of these extrahepatic manifestations.

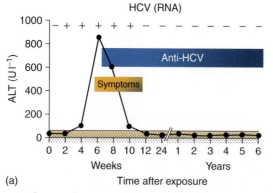
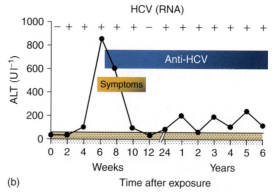

Figure 1 Course of acute and chronic HCV infection. (a) In acute HCV infection, viral RNA is typically detectable between weeks 2 and 10 after virus exposure. ALT levels rise between weeks 4 and 10 peaking around week 6. This is also the time when the first symptoms manifest. HCV-specific antibodies arise late in infection. In self-limiting infection HCV is cleared as measured by RNA levels and ALT returns to normal levels. (b) In chronic hepatitis C the early phase is similar to acute infection but later on the virus persists. In the chronic phase, ALT levels fluctuate as does viremia.

Genome Organization

The genome of HCV is a 9.6 kb single-stranded RNA molecule of positive polarity (**Figure 2(a)**). It is flanked by two highly structured nontranslated regions (NTRs). The 5′ NTR contains an internal ribosome entry site (IRES) (**Figure 2(b)**), mediating cap-independent translation of a polyprotein of c. 3000 amino acids in length. The 3′ NTR is composed of a 40-nt-long variable region, a polypyrimidine tract (heterogeneous length), and the 98-nt-long highly conserved 3′ terminal X-tail (**Figure 2(b)**). Both the 5′ and the 3′ NTRs contain cis-acting RNA elements (CREs) that are required for viral replication. Within the 3′ terminal part of the NS5B gene, three additional CREs are localized (5BSL3.1–5BSL3.3), wherein 5BSL3.2 is absolutely required for viral RNA replication by forming a long-distance RNA–RNA interaction with the middle loop in the X-tail.

The organization of the HCV polyprotein and the functions of the individual gene products are depicted in **Figure 2(a)**. The structural proteins, core and the envelope proteins 1 (E1) and 2 (E2), are located within the N-terminal part of the polyprotein preceding the p7 protein which appears to be an ion channel and therefore was grouped into the viroporin protein family. The nonstructural (NS) proteins NS2, NS3, NS4A, NS4B, NS5A, and NS5B are encoded in the remainder. The NS2 protein together with the N-terminal protease domain of NS3 is responsible for the autocatalytic cleavage of the NS2–NS3 junction (**Figure 2(a)**). It is a dimeric cysteine protease with a composite active site. The same NS3 protein domain carries a serine-type protease that after association with its cofactor NS4A cleaves the residual junctions between the NS proteins. Moreover, the same protease also cleaves two cellular signaling molecules involved in the induction of the innate immune response (see below). Two additional enzymatic

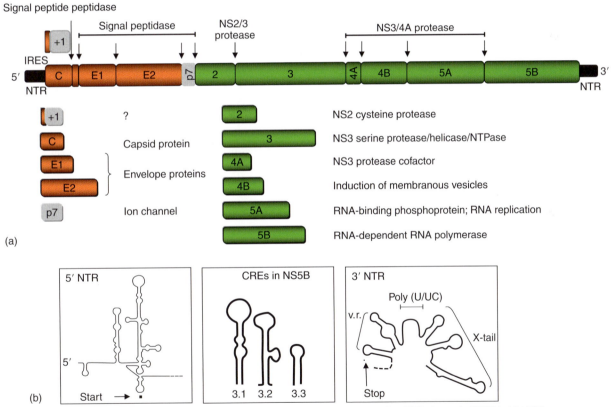

Figure 2 Illustration of the HCV genome organization, gene products, and cis-acting RNA elements (CREs). (a) HCV genome organization, gene products, and their functions in the replication cycle. The HCV coding region is shown as bar with arrows above indicating the positions of polyprotein cleavage and involved proteases. The structural region in the N-terminal third of the polyprotein is drawn in orange, and the region encoding the nonstructural proteins is drawn in green. The core+1 protein(s) and the p7 protein are drawn in gray. The HCV coding region is flanked by two nontranslated regions (5′ and 3′ NTR), indicated in dark gray. The 5′ NTR contains an internal ribosome entry site (IRES). Functions of the individual gene products are given in the lower panel. (b) Schematic representation of CREs. The 5′ NTR (left panel) contains the IRES as well as structures important for viral RNA replication. Three CREs are located in the 3′ terminal part of the NS5B coding region (middle panel). The loop region of stem–loop 5BSL3.2 forms a long-distance RNA–RNA interaction essential for RNA replication with the loop region of the middle stem–loop in the X-tail. The right panel displays the organization of the 3′ NTR with variable region (v.r.), poly(U/UC) tract, and X-tail. Positions of start and stop codons are indicated with dots.

activities (RNA helicase and nucleoside triphosphatase) reside in the C-terminal two-thirds of NS3. Alterations of intracellular membranes, in particular membranous vesicles, are mainly induced by the 27 kDa integral membrane protein NS4B. These vesicles accumulate in the perinuclear region, are called the membranous web, and are the site of viral RNA replication. So far, the role of the phosphorylated zinc metalloprotein NS5A in the viral replication cycle is unclear. NS5A is composed of an N-terminal amphipathic α-helix serving as a membrane anchor and three largely cytosolic domains. The X-ray crystal structure of RNA-binding domain I was resolved and shown to form homodimers resulting in a basic groove. NS5A phosphorylation is mediated by cellular kinases, in particular the α isoform of the protein kinase CKI. The NS5A phosphorylation state appears to affect replication efficiency indicating that NS5A is an important replication factor. Furthermore, NS5A may also contribute to interferon-alpha (IFN-α) resistance that is often observed with genotype 1 and 4 viruses. The 68 kDa protein NS5B is the RNA-dependent RNA polymerase (RdRp). It is a membrane-associated enzyme with a structural organization similar to that of other polymerases with palm, finger, and thumb subdomains. However, it differs from most other RdRps by having a fully encircled active site, which is due to tight interactions between the finger and thumb subdomains.

In addition to the polyprotein, a heterogeneous group of HCV proteins is expressed either by ribosomal frameshifting into the +1 ORF or by internal translation initiation. The resulting proteins, collectively designated core +1, are not essential for replication and virus production in cell culture, and their role *in vivo*, if any, remains to be elucidated.

HCV Replication Cycle

HCV infection starts by binding the envelope glycoprotein E1/E2 complex on the surface of the virus particle to its cognate receptor(s) presumably leading to clathrin-mediated endocytosis and a subsequent fusion step from within an acidic endosomal compartment (**Figure 3**). Cellular factors implicated in virus binding and entry are glycosaminoglycans, scavenger receptor class B type 1 (SR-B1), CD81, and low-density lipoprotein (LDL) receptor. However, for most of these factors, the precise role is not well understood and one or several additional factors may be required for productive entry. Upon release of the RNA genome, the polyprotein is expressed by IRES-dependent translation occurring at the rough endoplasmic reticulum (rER) where host cell signal peptidases, signal peptide peptidases, and viral proteases catalyze polyprotein cleavage.

During or after cleavage, the membrane-associated replication complex, which catalyzes the RNA amplification via negative-strand RNA intermediates, is formed (**Figure 3**). These membrane-associated complexes are composed of viral RNA, viral proteins, and most likely host cell factors. Newly synthesized positive-strand RNAs either are used for translation or serve as templates for further RNA synthesis or interact with the core protein to form the viral nucleocapsid. The E-proteins are retained at rER membranes indicating that viral envelopes are generated by budding into the lumen of this organelle. Progeny particles are thought to be exported by the secretory pathway and after fusion of the transport vesicle with the plasma membrane, virions are released. However, most of these steps are poorly understood and several assumptions were made that are based on studies with heterologous expression systems and analogies to closely related viruses.

Virion Properties

HCV particles are enveloped and spherical and have a diameter of 55–60 nm as determined by filtration and electron microscopy (**Figure 4**). By analogy to other flaviviruses, HCV particles are composed of at least the genomic RNA, the core protein, and the two envelope proteins E1 and E2 which are embedded into the lipid envelope. The core protein forms the internal viral capsid (presumably 30–35 nm in diameter) that shelters the single-stranded RNA genome (**Figure 4**). The S_{20w} is approximately 200S and infectious HCV virions isolated from the plasma sample of infected patients and chimpanzees have low buoyant densities in the range of $1.05-1.10\,\mathrm{g\,ml^{-1}}$.

Tissue Tropism and Host Range

Hepatocytes are considered to be the natural target cells for HCV. Viral RNA was also detected in peripheral blood mononuclear cells (PBMCs) and bone marrow cells but it is unclear whether productive infection occurs in these cells. In cell culture, HCV RNA replication was demonstrated in non-liver cells like human T- and B-cell lines or embryonic kidney cells (293). Furthermore, certain mouse cell lines can support replication of HCV replicons demonstrating that the viral replication machinery is also functional in a murine host cell environment.

Experimental Systems

Animal Models

Possibilities to propagate HCV *in vivo* are rare. For many years HCV could be propagated only in chimpanzees after experimental inoculation with virus containing samples from patients or synthetic *in vitro* transcripts derived from cloned infectious genomes. More recently, transgenic mice

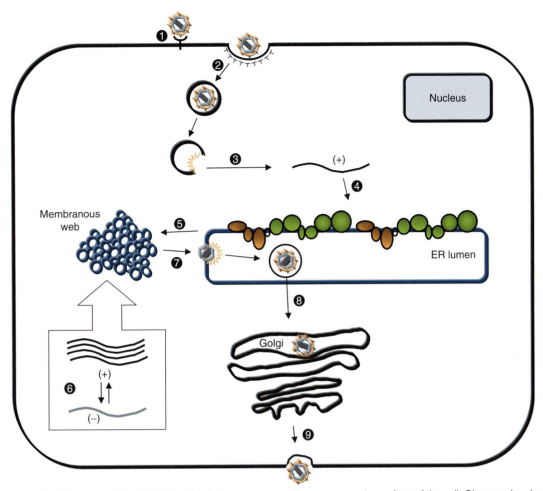

Figure 3 HCV replication cycle. (1) HCV virion binds to one or several receptors on the surface of the cell. Glycosaminoglycans, SR-B1, CD 81, and eventually LDL receptor are required for or contribute to virus binding and entry. (2) The virus particle supposedly enters the cell via clathrin-mediated endocytosis. (3) After a low-pH-mediated fusion step from within an acidic endosome and uncoating, the HCV genome is liberated into the cytoplasm of the host cell. (4) The viral RNA genome is translated at the rough endoplasmic reticulum (rER). (5) The membranous web presumably originating from ER membranes is formed. (6) It is the site of viral RNA amplification which occurs via negative-strand RNA intermediates. Newly synthesized positive-strand RNA is either used for translation or replication or the RNA is packaged into nascent capsids (7). This may occur at the ER where the E-proteins are retained. It is assumed that virions are generated by budding into the ER lumen. (8) Progeny virions are thought to be exported by the secretory pathway and, after fusion of the transport vesicle with the plasma membrane, virions are released (9).

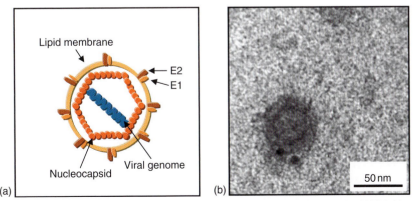

Figure 4 Composition of HCV particles. (a) Schematic of the HCV virion. (b) Electron micrograph of HCV virion produced in cell culture. The spherical particle has an outer diameter of about 60 nm and an inner core of about 30 nm. E2 was detected by immunoelectron microscopy using an E2-specific antibody and a secondary antibody conjugated to 10 nm gold particles.

xenografted with primary human hepatocytes were found to be susceptible to HCV infection. Virus replication occurs in the transplanted human tissue but, given the technical challenge, this mouse model is not widely available. Alternative models to study HCV *in vivo* are the closely related GB-viruses, especially GBV-B that replicates in tamarins (*Sanguis* sp.), causing an acute self-limiting infection of the liver or, under certain experimental conditions, a persistent infection.

Cell Culture Systems

Development of efficient cell culture systems for HCV propagation was difficult and was not successful until more than 10 years after the discovery of the virus. The first breakthrough was the development of subgenomic replicons composed of the NTRs, a selectable marker (*neo*), and a heterologous IRES directing translation of the HCV replicase (NS3–NS5B) (**Figure 5(a)**). When transfected into a human hepatocarcinoma cell line (Huh-7) and subjected to selection (e.g., with G418 in case of replicons containing the *neo* gene), stable cell lines were established that carry autonomously replicating HCV replicons. These viral RNAs replicate to very high levels and are maintained persistently when the cells are passaged under conditions of continuous selective pressure (e.g., G418). Owing to its high efficiency, this replicon system was of enormous value for studying HCV replication and HCV–host interaction, and for the development of antivirals targeting any of the viral replicase components (e.g., NS5B RdRp or the NS3 protease).

Since the first description of this system in 1999 numerous improvements have been made. These include the identification of replication-enhancing mutations (so-called adaptive mutations), different replicon formats allowing short-term replication analyses, and high-throughput screening assays. Thus far, replicons from different HCV isolates and two genotypes (1 and 2) are available, as are various cell lines, including two of murine origin.

Studies of the early steps of the HCV replication cycle are often performed using HCV pseudoparticles (HCVpp) (**Figure 5(b)**). These are retroviral capsids harboring a retroviral vector RNA into which a reporter gene has been inserted and which are surrounded by a lipid envelope carrying mature HCV E1/E2 glycoprotein complexes. This HCVpp system is an important tool to analyze the infection process and to measure HCV neutralization.

Robust production of infectious HCV particles in cell culture finally became possible in 2005 (**Figure 5(c)**). Key to this achievement is a particular HCV isolate designated JFH-1 (abbreviation for Japanese fulminant hepatitis) that was cloned from the serum of a patient with fulminant hepatitis C. When the JFH-1 genome is introduced into Huh-7 cells, virus particles are released that are infectious for naive Huh-7 cells, chimpanzees, and mice with human liver xenografts. Efficiency of this virus system has been increased by the construction of chimeric JFH-1 genomes, cell-culture-adapted virus variants, and by using highly permissive Huh-7 cell clones for HCV replication and entry. Very recently, a highly cell-culture-adapted variant of a genotype 1a HCV isolate has been

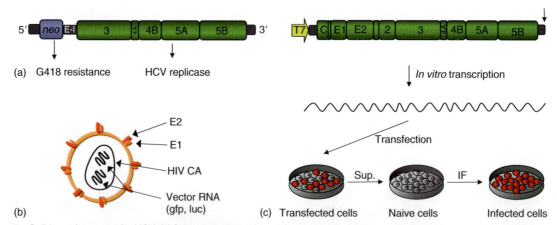

Figure 5 Cell-based systems for HCV. (a) Schematic illustration of the structure of a subgenomic HCV replicon carrying the selectable marker *neo* that confers G418 resistance. Translation of *neo* is directed by the 5′ NTR of HCV whereas translation of the replicase (NS3 to NS5B) is directed by the encephalomyocarditis virus (EMCV)-IRES. (b) Schematic of an HCV pseudoparticle (HCVpp). The retroviral vector RNA carries a reporter gene encoding, for example, for the green fluorescent protein (gfp) or the luciferase (luc). Two copies of vector RNA are packaged into the viral capsids surrounded by a lipid envelope carrying functional E1/E2 complexes. (c) Principle of the HCV infection system. Viral genomes are synthesized by *in vitro* transcription using a plasmid encoding a DNA copy of the viral genome. The T7 promoter at the 5′ end and a ribozyme or a restriction site (arrow) at the 3′ end are used to obtain run off transcripts with authentic termini. *In vitro* transcripts are transfected into Huh-7 cells and virus containing supernatant is transferred onto naive Huh-7 cells. Infection of these cells is detected, for example, by immunofluorescence (IF).

constructed that also supports virus production in Huh-7 cells, but virus titers are extremely low.

HCV–Host Interaction

Replication Factors

Several cellular proteins appear to contribute to HCV replication. For example, the ubiquitously expressed human vesicle-associated membrane protein-associated protein A (VAP-A) and its isoform VAP-B were identified as interaction partners of NS5A and NS5B. NS5A hyperphosphorylation seems to disrupt the VAP-A association and thereby negatively regulates HCV RNA replication. It is thought that VAP-A directs HCV nonstructural proteins to cholesterol-rich, detergent-resistant membranes, which are the presumed sites of HCV RNA replication.

Another host cell factor interacting with NS5A is FBL-2 belonging to the family of proteins that contain an F box and multiple leucine-rich repeats. FBL-2 is modified by geranylgeranylation which is important for NS5A interaction. FBL-2 appears to be required for HCV RNA replication. Likewise, Cyclophilin B (CyPB), a cellular protein interacting with the NS5B RdRp, seems to contribute to replication of genotype 1 isolates by promoting RNA-binding capacity of NS5B. CyPs are a family of peptidyl-prolyl *cis–trans*-isomerases which catalyze the *cis–trans*-interconversion of peptide bonds N-terminal of proline residues, facilitating changes in protein conformation. Cyclosporine or derivatives thereof potently block HCV RNA replication and it is assumed that this is due in part to sequestration of CyPB, which binds to cyclosporine with high affinity.

Also the host cell lipid metabolism plays a fundamental role in the HCV replication cycle. Treatment of replicon cells with lovastatin, an inhibitor of 3-hydroxy-3-methylglutaryl CoA reductase, or with an inhibitor of protein geranylgeranyl transferase I induced the disintegration of the HCV replication complex. Fatty acids can either stimulate or inhibit HCV replication, depending on their degree of saturation. Saturated and monounsaturated fatty acids stimulate viral replication, whereas polyunsaturated fatty acids impair replication.

Innate Immunity

Both in cell culture and in the majority of patients treated with IFN-α, a rapid and efficient block of HCV replication occurs. This result is somewhat surprising given the high rate of persistence of HCV infections (50–80%) and the finding that, in infected liver, type 1 IFN-induced genes are activated. In several studies it was concluded that HCV proteins, such as core, interfere with the various steps of the IFN-α/β-induced signaling and that some HCV proteins appear to block individual IFN-α/β-induced effectors. One prominent example is NS5A, assumed to block activity of the double-strand RNA-activated protein kinase PKR by binding to PKR via a particular NS5A region. This region overlaps with the so-called interferon-sensitivity-determining region in NS5A, assumed to correlate with outcome of antiviral therapy. However, this original assumption is still contradictory and it is still unclear whether HCV indeed interferes with one or several type 1 IFN-induced effector molecule(s). It also remains to be clarified by which mechanism interferons block HCV RNA replication.

Much less controversial are the mechanisms by which HCV interferes with the induction of innate antiviral defense. Several studies have shown that the NS3 protease proteolytically cleaves two signal-transducing molecules: TRIF, linking the activation of Toll-like receptor 3 to kinase complexes responsible for the phosphorylation of interferon response factor-3 (IRF-3) and CARDIF (also called MAVS, ips-1, VISA) relaying the activation of retinoic-acid-inducible gene 1 (RIG-1) also to IRF-3 phosphorylation. As a result, IRF-3-dependent genes are not expressed including IFN-β and IRF-7 and cells remain sensitive to virus infection. Although linking the block of IFN-β expression to persistence is attractive, it is unclear if and to what extent that is the case. On one hand blocking IRF-3 activation would not affect the antiviral program induced by type 1 IFN (e.g., produced by activated dendritic cells or administered during therapy), whereas on the other hand this block may affect the secretion of cytokines required for the development of a vigorous adaptive immune response and its attenuation may facilitate persistence.

Adaptive Immunity

The role of HCV-specific antibodies in controlling viral infection is not clear. They appear to be dispensable for viral clearance and do not protect from reinfection, neither in experimentally infected chimpanzees nor in humans after multiple exposure to HCV. However, there is evidence that the presence of HCV-specific antibodies at least partially attenuates infection. For instance, antibodies neutralizing HCV virions of different genotypes have been detected in sera of chronic hepatitis C patients but the frequency of these antibodies appears to be low. Control of acute HCV infection is primarily achieved by a rigorous and multispecific T-cell response. Thus, successful antiviral response generally encompasses multiple major histocompatibility complex (MHC) class-I and class-II restricted T-cell epitopes and a profound expansion of $CD8^+$ and $CD4^+$ T cells. In contrast, persistent infections are characterized by oligoclonal T-cell responses and a low frequency of HCV-specific T cells. The underlying reasons for the weak response in the majority of patients are not clear but several possibilities have been suggested: (1) an impaired antigen presentation that might be due to interference of HCV with dendritic cell function; (2) $CD4^+$ T-cell failure due to deletion or anergy; (3) mutational escape in important T- (and B-)cell epitopes; and (4) functional impairment of HCV-specific $CD8^+$ T cells. How T-cell impairment is brought about is unclear but one attractive possibility is that the defect induced in innate immunity results in a defect in $CD4^+$ T-cell help. In fact, HCV-induced loss of T-cell help appears to be the key event of immune evasion.

Diagnosis

Routine screening tests for detecting HCV infections are based on serological assays measuring HCV-specific antibodies (most often by enzyme-linked immunosorbent assay (ELISA)) and nucleic acid-based tests to determine viral RNA. Current ELISA assays have a specificity of >99% and they are positive in 99% or more of immunocompetent patients in whom viral RNA is detectable. Given the higher sensitivity, the diagnostic window can be reduced by using nucleic acid-based tests. Qualitative RNA detection assays have been implemented in many blood banks in European Union countries and in the US. The risk to acquire transfusion-associated hepatitis C in such countries has been reduced to <1/million blood donations. Determination of HCV genotypes, which is an important parameter for current antiviral therapy, is based on analyzing viral RNA either by hybridization or direct sequence analysis or by using genotype-specific primers for PCR.

Treatment

Current antiviral therapy is based on the combination of a polyethylene glycol conjugated form of IFN-α with ribavirin resulting in an overall sustained virological response (SVR) of about 60%. However, the success rate depends very much on the genotype of the infecting virus. While up to 85% of genotype 2- and 3-infected patients develop SVR, only about 45% of patients infected with genotype 1 viruses do so. In addition, this therapy has numerous side effects, such as flu-like symptoms, including increased body temperature, headache and muscle pain; neuropsychiatric alterations and hemolytic anemia are also severe side effects of this combination therapy. Thus, therapy often has to be discontinued and many patients are not eligible for this treatment. Numerous efforts are therefore undertaken to develop selective drugs targeting viral functions without causing side effects. The first promising candidates are currently in clinical trials and have shown potent antiviral efficacy. Most advanced are inhibitors of the NS3 protease and the NS5B RdRp. However, monotherapy with these compounds leads to rapid selection for therapy-resistant HCV variants. Future therapy of HCV infection most likely will be based on a combination therapy, which may include a selective drug and IFN.

See also: Hepatitis A Virus; Hepatitis B Virus: General Features.

Further Reading

Appel N, Schaller T, Penin F, et al. (2006) From structure to function: New insights into hepatitis C virus RNA replication. *Journal of Biological Chemistry* 281: 9833–9836.

Bartenschlager R (2006) Hepatitis C virus molecular clones: From cDNA to infectious virus particles in cell culture. *Current Opinion in Microbiology* 9: 416–422.

Bartenschlager R, Frese M, and Pietschmann T (2004) Novel insights into hepatitis C virus replication and persistence. *Advances in Virus Research* 63: 71–180.

Bartosch B and Cosset FL (2006) Cell entry of hepatitis C virus. *Virology* 348: 1–12.

Blight KJ, Kolykhalov AA, and Rice CM (2000) Efficient initiation of HCV RNA replication in cell culture. *Science* 290: 1972–1974.

Bowen DG and Walker CM (2005) Adaptive immune responses in acute and chronic hepatitis C virus infection. *Nature* 436: 946–952.

Choo QL, Kuo G, Weiner AJ, et al. (1989) Isolation of a cDNA clone derived from a blood-borne non-A, non-B viral hepatitis genome. *Science* 244: 359–362.

Foy E, Li C, Sumpter R, et al. (2003) Regulation of interferon regulatory factor-3 by the hepatitis C virus serine protease. *Science* 300: 1145–1148.

Gale M Jr. and Foy M (2005) Evasion of intracellular host defence by hepatitis C virus. *Nature* 436: 939–945.

Lindenbach BD and Rice CM (2005) Unravelling hepatitis C virus replication from genome to function. *Nature* 436: 933–938.

Lindenbach BD, Evans MJ, Syder AJ, et al. (2005) Complete replication of hepatitis C virus in cell culture. *Science* 309: 623–626.

Lohmann V, Körner F, Koch J, et al. (1999) Replication of subgenomic hepatitis C virus RNAs in a hepatoma cell line. *Science* 285: 110–113.

Manns MP, Wedemeyer M, and Cornberg M (2006) Treating viral hepatitis C: Efficacy, side effects, and complications. *Gut* 55: 1350–1359.

Wakita T, Pietschmann T, Kato T, et al. (2005) Production of infectious hepatitis C virus in tissue culture from a cloned viral genome. *Nature Medicine* 11: 791–796.

Zhong J, Gastaminza P, Cheng G, et al. (2005) Robust hepatitis C virus infection *in vitro*. *Proceedings of the National Academy of Sciences, USA* 102: 9294–9299.

Hepatitis Delta Virus

J M Taylor, Fox Chase Cancer Center, Philadelphia, PA, USA

© 2008 Elsevier Ltd. All rights reserved.

Glossary

Antigenome For hepatitis delta virus (HDV), the antigenome refers to an exact complement of the genome. It also is a single-stranded circular RNA.

Editing In recent years it has become clear that many RNAs undergo nucleotide sequence changes relative to the nucleic acid templates from which they are derived. There are many different forms of this process, collectively known as editing. It can occur during or after the process of RNA transcription. HDV RNA undergoes a specific form of post-transcriptional editing in which certain adenosines are deaminated to inosine.

Genome For a virus the genome is that nucleic acid species present within virus particles. For HDV, the genome is a single-stranded circular RNA.

Ribozyme When it was realized that certain RNA molecules could have enzymatic activities similar to certain proteins, such RNAs were defined as ribozymes. Both the genomic and antigenomic RNAs of HDV contain regions that undergo specific self-cleavage, and are thus defined as ribozymes.

Rolling-circle replication For agents that have a circular genome, whether it be of DNA or RNA, there is the possibility that replication can initiate at one or more locations, leading to the synthesis of species longer than the original circle. Such multimers may then undergo processing (e.g., by cleavage and ligation) to form new unit-length products. Such a mechanism, which is referred to as rolling-circle replication, applies to certain agents and has been implicated for the replication of HDV.

Viroid Among the infectious agents of plants there are small circular single-stranded RNAs that seem to replicate without the aid of a helper virus or the synthesis of any encoded protein. These agents do not fulfill the definition of a virus and so have been named viroids.

Classification

Hepatitis delta virus or hepatitis D virus (HDV) was discovered in patients with a more severe form of human hepatitis B virus (HBV) infection. HDV is a subviral satellite of HBV. HDV is often called a virus but strictly speaking it does not satisfy the definition of a virus and should be called subviral; HDV infection and assembly of new virus particles depends upon the envelope protein provided by the natural helper virus, HBV. HDV is also called a satellite of HBV because there is no nucleotide homology between the genomes of HBV and HDV.

No other infectious agents of animals resemble HDV. There are agents in plants that share several important characteristics of RNA genome structure and replication. These plant agents include the viroids and certain satellite RNAs and satellite viruses. Nevertheless, there are also enough major differences between HDV and these plant agents that the International Committee on Taxonomy of Viruses has agreed to assign HDV as a separate genus, *Deltavirus*, with only itself as a member.

Structure

HBV, the natural helper virus of HDV, encodes three envelope proteins. These are also used for the assembly of HDV, but in a different way.

Infectious HBV particles are roughly spherical with a diameter of about 42 nm. Inside these is an icosahedral nucleocapsid of about 27 nm diameter. HBV assembly is inefficient in that infectious particles are found in serum in the presence of a 1000–1 000 000-fold excess of empty particles. These empty particles consist of 25 nm spheres and filaments with a 22 nm diameter and a heterogenous length.

Infectious HDV particles contain a ribonucleoprotein composed of the HDV RNA genome in a complex with more than 70 copies of the delta antigen, the only protein encoded by HDV. It is considered from electron microscopy and filtration studies that infectious HDV particles are spherical and at least 38 nm in diameter. However, such studies are not as clear-cut as for HBV. First, they do not show which particles actually contain RNA. Second, they do distinguish which of the RNA-containing particles are infectious. What seems clear is that the same HBV envelope protein domains are needed for HDV and HBV infectivity. That is, it is likely that for infection HDV uses the same as yet unidentified receptor as HBV.

The internal ribonucleoprotein complex of the HDV RNA genome and delta proteins has been demonstrated but the actual structure has not been clarified yet.

Replication

The RNA genome (**Figure 1**) is a single-stranded RNA of about 1700 nt in length. It is unique for many reasons

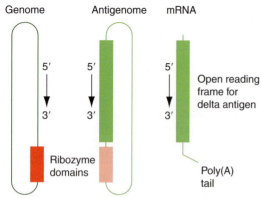

Figure 1 A representation of the three RNAs detected during the replication of HDV. The first is the 1679 nt single-stranded circular RNA found within virions, and thus defined as the genome. The second is an exact complement of the genome, defined as the antigenome, and it along with the genome is found in infected cells. A third RNA, the least abundant, is cytoplasmic, 5′-capped and 3′-polyadenylated, and is the mRNA for the translation of the delta protein. The genome and antigenomic circles are drawn as elongated in order to represent their ability to form extensive intramolecular base pairing and fold into an unbranched rod-like structure. Note that both the genome and the antigenome each contain a domain that will act as a self-cleaving ribozyme.

including its very small size and its circular conformation. The replication takes place by a mechanism that is fundamentally different from the reverse transcription pathway used by the helper virus HBV. HDV replicates with RNA rather than DNA intermediates. Inside infected cells we find not only the circular RNA genome but also an exact complement, the antigenome, which is also circular (**Figure 1**). A third RNA is also complementary to the genome. It is about 800 nt in length, 5′-capped and 3′-polyadenylated, and acts as the mRNA for the translation of the delta protein (**Figure 1**).

HDV genome replication seems to take place in the nucleus and involves redirection of the host RNA polymerase II. Because the genomic RNA is circular, it is possible for transcription to make multimeric RNAs from this template, in what is called a rolling-circle mechanism. Both the genomic and antigenomic RNAs contain a ribozyme that cleaves the RNA at a unique location. A detailed structure for one of these ribozymes has been reported. Ribozyme cleavage allows the processing of the multimeric RNAs into unit-length species, that are then somehow ligated to makes circles. The RNases of animal cells are mainly exonucleases and so circular RNAs are much more stable than the corresponding linear species. In infected liver tissue, the circular genomic RNAs accumulate to around 300 000 copies per average cell.

The mRNA of an infectious HDV needs to be translated into what is known as the small delta protein. This 195 amino acid protein is highly basic. It can dimerize using a region near the N-terminus, through what is known as an antiparallel coiled-coil domain. It can also bind with specificity to HDV RNAs, and localize to the nucleus. For a combination of these and maybe additional reasons, the small delta protein is essential for genome replication. However, during this replication, some of the antigenomic RNAs are post-transcriptionally modified at a specific location, by a form of RNA-editing. The editing involves a double-stranded RNA-specific adenosine deaminase, now known as ADAR. The editing occurs in the middle of the amber termination codon for the open reading frame of the small delta protein. This leads to the translation of a protein that is 19 amino acids longer. This 214-amino-acid long delta protein is a dominant negative inhibitor of genome replication. However, it also has a positive role. Most of it becomes farnesylated at a cysteine located four amino acids from the new C-terminus. This modified protein is essential for the late phase of HDV replication, the assembly of new virus particles, using the envelope proteins of the helper virus, HBV.

Epidemiology

HDV infections are transmitted via infected blood or blood products. There are two main classifications of HDV infection. A co-infection is one in which the individual receives both HBV and HDV at the same time. A superinfection is one in which the individual has a prior infection with HBV, usually a chronic infection, and is then infected with HDV. Such superinfections have a much greater chance of being more extensive and also a greater chance of going on to chronicity.

It was not realized until 1980 that HDV was an infectious agent. Convenient tests for HDV were soon developed and epidemiological information gathered. The initial test was for antibody directed against the delta antigen and then tests were developed for the antigen itself. Now we have much more sensitive tests for the RNA, but such tests are not routinely used.

HDV was first studied among populations in southern Italy. Of course it is only found in situations where HBV is also present, but the converse is not always true. There are geographic populations in which HBV infections are more likely to be associated with HDV. Data indicate that the fraction of HBV infections that are also HDV-associated can range from <1% to >10%. The high levels have been reported for areas such as the Amazon basin in South America and southern regions of the former Soviet Union. Also, within a geographic population there may be individuals, such as intravenous drug users who share needles, where the fraction is very high, >70%.

Levels of HDV infection have been greatly reduced in recent years. This is in part due to changes in the behavior of susceptible individuals, especially the intravenous drug users, and also due to screening of blood supplies.

In addition, the introduction of widespread immunization against HBV, which simultaneously protects against HDV, will ultimately reduce both the number of chronic HBV carriers and HDV infections.

Clinical Features

The replication of HBV is not directly cytopathic. The liver damage arises because of the involvement of the host immune response to these infections. The same is probably also true for HDV. In addition, there are experimental situations where HDV infections can be directly cytopathic.

In certain geographic areas, such as the Amazon basin, HDV infections can have a very high risk of producing a fulminant hepatitis, sometimes with death in less than 1 week.

Control

Immunization against the natural helper virus, HBV, confers protection against HDV. However, for nonimmunized individuals and especially for those already chronically infected with HBV, the risks remain high.

For an individual infected with HDV, there is no good therapy that directly targets the HDV infection. In one study, treatment with high doses of interferon was associated with some cures. Of course the interferon therapy also attacks the HBV infection.

Recently, antiviral therapies based upon inhibitors of the HBV reverse transcriptase are being directed against HBV infections. One might expect that these therapies would indirectly inhibit cycles of HDV replication, but such has not been the case.

See also: Hepatitis B Virus: General Features; Hepatitis B Virus: Molecular Biology.

Further Reading

Casey JL (ed.) (2006) *Current Topics in Microbiology and Immunology, Vol. 307: Hepatitis Delta Virus.* Berlin: Springer.

Farci P, Roskams T, Chessa L, et al. (2004) Long-term benefit of interferon alpha therapy on chronic hepatitis D: Regression of advanced hepatic fibrosis. *Gastroenterology* 126: 1740–1749.

Handa H and Yamaguchi Y (eds.) (2006) *Hepatitis Delta Virus.* Georgetown, TX: Landes Bioscience.

Taylor JM (2006) Hepatitis delta virus. *Virology* 344: 71–76.

Hepatitis E Virus

X J Meng, Virginia Polytechnic Institute and State University, Blacksburg, VA, USA

© 2008 Elsevier Ltd. All rights reserved.

Glossary

Animal reservoir A large animal population serving as a source of virus supply for the transmission of virus(es) to other animals including humans.

Enzootic An epidemiology term describing multiple continuous disease presence in an animal population in a defined geographic region and time period.

Zoonosis Infectious diseases transmissible under natural conditions from vertebrate animals to humans.

Introduction

The initial evidence for the existence of a new form of enterically transmitted viral hepatitis came from serological studies of waterborne epidemics of hepatitis in India in 1980, as cases thought to be hepatitis A were tested negative for antibodies to the hepatitis A virus (HAV). To conform to the standard nomenclature of viral hepatitis, this new form of viral hepatitis was named hepatitis E. In 1983, viral hepatitis E was successfully transmitted to a human volunteer via fecal–oral route with a stool sample collected from a non-A, non-B hepatitis patient, and virus-like particles were visualized by electron microscope in the stool of the infected volunteer. Subsequently, the agent was also successfully transmitted to nonhuman primates. However, the identity of the virus was not known until 1990 when the genomic sequence of the new virus, named hepatitis E virus (HEV), was determined. The recent identification and characterization of animal strains of HEV, swine hepatitis E virus (swine HEV) from pigs in 1997 and avian hepatitis E virus (avian HEV) from chickens in 2001, have broadened the host ranges and diversity of the virus, and also provided two unique homologous animal model systems to study HEV replication and pathogenesis.

Convincing evidence indicates that hepatitis E is a zoonotic disease, and pigs and perhaps other animal species are reservoirs for HEV.

Properties and Structure of Virions

HEV is a symmetrical, icosahedral, spherical virus particle of approximately 32–34 nm in diameter without an envelope (**Figure 1**). The capsid protein encoded by the open reading frame (ORF) 2 gene of HEV is the only known structural protein on the virion. The N-terminal truncated capsid protein, which contains amino acid residues 112–660, can self-assemble into empty virus-like particles when expressed in baculovirus. The buoyant density of HEV virions is reportedly 1.35–1.40 g cm^{-3} in CsCl, and 1.29 g cm^{-3} in potassium tartrate and glycerol. The virion sedimentation coefficient is 183 S. HEV virion is sensitive to low-temperature storage and iodinated disinfectants but is reportedly stable when exposed to trifluorotrichloroethane. HEV virion is more heat labile than is HAV, another enterically transmitted hepatitis virus. HAV was only 50% inactivated at 60 °C for 1 h but was almost totally inactivated at 66 °C. In contrast, HEV was about 50% inactivated at 56 °C and almost totally inactivated (96%) at 60 °C. The fecal–oral route of transmission indicates that HEV is resistant to inactivation by acidic and mild alkaline conditions in the intestinal tract.

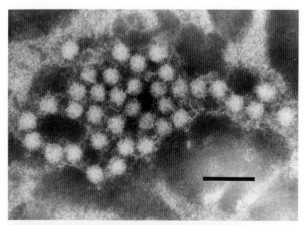

Figure 1 Electron micrograph of 30–35 nm diameter particles of an avian strain of the HEV. The virus particles were detected from a bile sample of a chicken with hepatitis–splenomegaly syndrome. Scale = 100 nm. Reproduced from Haqshenas G, Shivaprasad HL, Woolcock PR, Read DH, and Meng XJ (2001) Genetic identification and characterization of a novel virus related to human hepatitis E virus from chickens with hepatitis–splenomegaly syndrome in the United States. *Journal of General Virology* 82: 2449–2462, with permission from Society for General Microbiology.

Taxonomy and Classification

HEV was initially classified in the family *Caliciviridae* on the basis of its superficial similarity in morphology and genomic organization to caliciviruses. However, subsequent studies revealed that the HEV genome does not share significant sequence homology with caliciviruses, and that the codon usage and genomic organization of HEV are also different from that of caliciviruses. Therefore, in the *Eighth Report of the International Committee on Taxonomy of Viruses* (ICTV), HEV was declassified from the family *Caliciviridae*, and was placed in the sole genus *Hepevirus*. The proposed family name *Hepeviridae* has yet to be officially approved by ICTV.

The species in the genus *Hepevirus* includes the four recognized major genotypes of HEV (**Figure 2**): genotype 1 (primarily Burmese-like Asian strains), genotype 2 (a single Mexican strain), genotype 3 (strains from rare endemic cases in the United States, Japan, and Europe, and swine strains from pigs in industrialized countries), and genotype 4 (variant strains from sporadic cases in Asia, and swine strains from pigs in both developing and industrialized countries). A tentative new species in the genus *Hepevirus* is the recently discovered avian HEV from chickens (**Figure 1**). The nucleotide sequence identity differs by 20–30% between genotypes, and by c. 40–50% between avian HEV and mammalian HEVs. Despite the extensive nucleotide sequence variations, however, it appears that there exists a single serotype of HEV.

Genome Organization and Gene Expression

The genome of HEV, which possesses a cap structure at its 5′ end, is a single-stranded, positive-sense RNA molecule of approximately 7.2 kb in length. The viral genome consists of a short 5′ noncoding region (NCR) of approximately 26 bp, three open reading frames (ORFs 1–3), and a 3′ NCR. ORF3 overlaps ORF2, but neither ORF2 nor ORF3 overlaps with ORF1 (**Figure 3**). The ORF1 at the 5′ end encodes viral nonstructural proteins that are involved in viral replication and protein processing.

Functional motifs, characteristic of methyltransferases, papain-like cystein proteases, helicases, and RNA-dependent RNA polymerases (**Figure 3**), were identified in ORF1 based upon sequence analyses and analogy with other single-strand positive-sense RNA viruses. ORF2, located at the 3′ end, encodes the major capsid protein that contains a typical signal peptide sequence, and three potential glycosylation sites. The capsid protein contains the most immunogenic epitope, induces neutralizing antibodies against HEV, and is the target for HEV vaccine development. ORF3 encodes a small

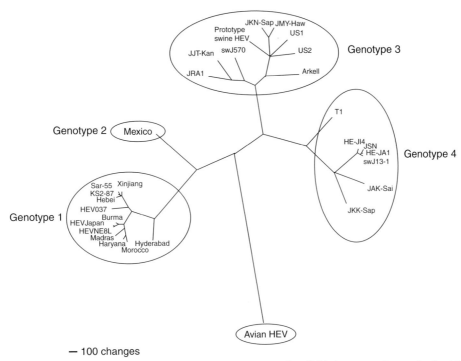

Figure 2 A phylogenetic tree based on the complete genomic sequences of available human, swine, and avian HEV strains. The tree was constructed with the aid of the PAUP program by using heuristic search with 1000 replicates. A scale bar, indicating the number of character state changes, is proportional to the genetic distance. The four recognized major genotypes of HEV are indicated. The avian HEV in a branch distinct from mammalian HEV is tentatively classified as a separate species. Reproduced from Huang FF, Sun ZF, Emerson SU, et al. (2004) Determination and analysis of the complete genomic sequence of avian hepatitis E virus (avian HEV) and attempts to infect rhesus monkeys with avian HEV. *Journal of General Virology* 85: 1609–1618, with permission from Society for General Microbiology.

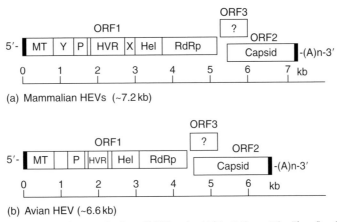

Figure 3 Schematic diagram of the genomic organization of HEV: a short 5′ (solid box at the 5′ end), a 3′ NCR (solid box at the 3′ end), and three ORFs. ORF2 and ORF3 overlap each other but neither overlaps ORF1. ORF1 encodes nonstructural proteins including putative functional domains, ORF2 encodes putative capsid protein, and ORF3 encodes a small protein with unknown function. MT, methyltransferase; X, 'X' domain; Y, 'Y' domain; P, a proline-rich domain that may provide flexibility; HVR, a hypervariable region; Hel, helicase; RdRp, RNA-dependent RNA polymerase.

cytoskeleton-associated phosphoprotein. The N-terminus of ORF3 has a cysteine-rich region, binds to HEV RNA, and enters into a complex with the capsid protein. The C-terminal region of the ORF3 protein is a multifunctional domain, and may be involved in HEV virion morphogenesis and viral pathogenesis. Antibodies to ORF3 protein have been detected in some animals (rhesus monkeys and chickens) experimentally infected with HEV; however, it remains unknown if ORF3 is a structural protein in the virion.

The transcription and translation mechanisms of HEV genes are still poorly understood, largely due to the lack of an efficient cell culture system for the propagation of HEV. By using an *in vitro* HEV replicon system, a bicistronic subgenomic mRNA was identified and found to encode both ORF2 and ORF3 proteins. Subgenomic mRNAs were also detected in infected liver tissues. The ORF3 gene contains a *cis*-reactive element and encodes a protein that is essential for virus infectivity *in vivo*. However, recent studies showed that the expression of ORF3 protein is not required for virus replication, virion assembly, or infection of hepatoma cells *in vitro*. The translation and post-translational processes of the ORF1 polyprotein remain largely unknown. When expressed in baculovirus, the nonstructural ORF1 polyprotein was found to process into smaller proteins that correlate with predicted functional domains in ORF1. However, no processing of the ORF1 polyprotein was detected when it was expressed in bacterial or mammalian expression systems.

Epidemiology

HEV is transmitted primarily through the fecal–oral route, and contaminated water or water supplies are the main sources of infection. Hepatitis E, generally affecting young adults, is an important public health disease in many developing countries of the world where sanitation conditions are poor. Rare epidemics are associated with fecal contamination of water, and are generally restricted in developing countries of Asia and Africa, and in Mexico. Most cases of acute hepatitis E occurred as endemic or sporadic forms worldwide in both industrialized and developing countries. The genotypes 1 and 2 HEV strains are associated with epidemics, whereas genotypes 3 and 4 strains cause sporadic cases of hepatitis E. A unique feature of HEV infections is the high mortality rate, up to 25%, in infected pregnant women, although the mechanism of fulminant hepatitis during pregnancy is still not known.

Seroprevalence of HEV antibodies is age dependent, with the highest rate occurring in young adults of 15–40 years of age. In endemic countries, the seroprevalence rate, generally ranging from 3% to 27%, is much lower than expected, although higher seroprevalence rates have been reported in some endemic countries such as Egypt. Surprisingly, a significant proportion of normal blood donors in the United States (4–36%) and other industrialized countries were tested seropositive for immunoglobulin G (IgG) anti-HEV antibodies, suggesting prior infections with HEV (or an antigenically related agent). The source of seropositivity in individuals from industrialized countries is not clear but a zoonotic origin via direct contact with infected animals or consumption of raw or undercooked animal meats has been suggested.

Swine HEV

The identification and characterization of swine HEV from pigs in the United States lend credence to the zoonosis concept for HEV. Since the initial report in 1997, swine HEV has now been identified from pigs in more than a dozen countries. Thus far, the viruses identified from pigs worldwide belong to either genotype 3 or 4, which both cause sporadic cases of acute hepatitis E. Recently, a genotype-1-like swine HEV was reportedly identified from a pig in Cambodia but independent confirmation of this finding is still lacking. Genetic and phylogenetic analyses of the complete genomic sequences of swine HEV revealed that swine HEV is very closely related, or identical in some cases, to the genotype 3 and 4 strains of human HEV (**Figure 2**). Seroepidemiological studies demonstrated that swine HEV is ubiquitous in pigs in the Midwestern United States, and that about 80–100% of the pigs in commercial farms in the United States were infected. Similar findings were also reported in more than a dozen other developing and industrialized countries, indicating that swine HEV infection in pigs is common worldwide. The ubiquitous nature of swine HEV infection in pigs provides a source of virus supply for zoonotic infection. Swine HEV infection generally occurs in pigs of 2–4 months of age. The infected pigs remain clinically normal, although microscopic evidence of hepatitis has been observed.

Avian HEV

Avian HEV was first isolated and characterized in 2001 from bile samples of chickens with hepatitis–splenomegaly (HS) syndrome in the United States, although a big liver and spleen disease virus (BLSV) with 62% nucleotide sequence identity to HEV (based on a 523 bp genomic region) was reported in 1999 from chickens in Australia. HS syndrome is an emerging disease of layer and broiler breeder chickens in North America characterized by increased mortality and decreased egg production. Dead birds have red fluid or clotted blood in their abdomens, and enlarged livers and spleens. Avian HEV shared *c.* 80% nucleotide sequence identity with the Australian BLSV, suggesting that big liver and spleen disease (BLS) in Australia and HS syndrome in North America are caused by variant strains of the same virus.

Avian HEV is morphologically, genetically, and antigenically related to human HEV. The complete genomic sequence of avian HEV is about 6.6 kb in length (**Figure 3**), which is about 600 bp shorter than that of human and swine HEVs. Although avian HEV shares only c. 50–60% nucleotide sequence identity to mammalian HEVs, the genomic organization, functional motifs, and common antigenic epitopes are conserved between avian and mammalian HEVs (**Figure 3**). Phylogenetic analyses indicated that avian HEV is distinct from mammalian HEV (**Figure 2**) and may represent either a fifth genotype of HEV or a separate species within the genus *Hepevirus*. Like swine HEV, avian HEV infection in chickens is widespread in the United States: c. 71% of chicken flocks and 30% of chickens are positive for IgG antibodies to avian HEV. Like human and swine HEVs, avian HEV antibody prevalence is also age dependent: c. 17% of chickens younger than 18 weeks are seropositive, as compared to c. 36% of adult chickens. Avian HEV isolates recovered from chickens with HS syndrome in the United States displayed 78–100% nucleotide sequence identities to each other, and 56–61% identities to known strains of human and swine HEVs.

Host Range and Cross-Species Infection

Swine HEV has been shown to cross species barriers and infect both rhesus monkey and chimpanzee. Infection of nonhuman primates, the surrogates of man, with swine HEV demonstrated the possibility of human infection with swine HEV. Conversely, genotype 3 human HEV has been shown to infect pigs under experimental conditions. However, attempts to experimentally infect pigs with genotypes 1 and 2 strains of human HEV were unsuccessful. Cross-species infection of HEV has also been reported in other animal species. Lambs were reportedly infected with human HEV isolates. Similarly, Wistar rats were reportedly infected with a human stool suspension containing infectious HEV, although others failed to confirm the rat transmission results. Cross-species infection by avian HEV has also been demonstrated, as avian HEV recovered from a chicken with HS syndrome was able to successfully infect turkeys. However, an attempt to experimentally infect two rhesus monkeys with avian HEV was unsuccessful. Thus, it appears that, unlike swine HEV, avian HEV may not readily infect humans.

There is potentially a wide host range of HEV infection. In addition to swine and chickens, antibodies to HEV have been reportedly detected in several other animal species including rodents, dogs, cats, sheep, goats, cattle, and nonhuman primates, suggesting that these animals have been exposed to HEV (or a related agent) and thus might serve as reservoirs. However, the source of seropositivity in these animal species, with the exception of pigs and chickens, could not be definitively identified since virus was either not recovered from these species or the recovered virus could not be sequenced to confirm its identity. A significant proportion of wild rats caught in different geographic regions of the United States (ranging from 44% to 90%) and other countries tested positive for IgG anti-HEV antibodies. Rodents are frequently found in both urban and rural environments and, thus, could potentially play an important role in HEV transmission. The existence of a wide host range of animal species positive for HEV antibodies further supports the zoonosis concept for HEV.

Pathogenesis and Tissue Tropism

The pathogenesis of HEV is largely unknown. It is believed that HEV enters the host through the fecal–oral route. However, the primary site of HEV replication is not known. In primates and pigs experimentally infected with human and swine HEVs, virus replication in the liver has been demonstrated. It is believed that, after replication in liver, HEV is released to the gallbladder from hepatocytes and then is excreted in feces. In pigs experimentally infected with human and swine HEVs, in addition to the liver, HEV replication was also identified in extrahepatic tissues including small intestine, colon, hepatic, and mesenteric lymph nodes. Although the clinical and pathological significance of these extrahepatic sites of virus replication is not known, it is believed that HEV may first replicate in the gastrointestinal tract after oral ingestion of the virus, and subsequently spreads to the target organ liver via viremia. It has been well documented that pregnancy increases the severity and mortality of the disease. The overall mortality rate caused by HEV in infected pregnant women can reach up to 25%, although, under experimental conditions, fulminant hepatitis E could not be reproduced in infected pregnant rhesus monkeys or infected pregnant sows. Therefore, the mechanism of fulminant hepatitis E in infected pregnant women remains unknown.

Zoonotic Risk

Hepatitis E is now considered a zoonotic disease, and pigs (and possibly other animal species) act as reservoirs. It has been demonstrated that pig handlers such as pig farmers and swine veterinarians in both developing and industrialized countries are at increased risk of HEV infection. For example, swine veterinarians in the United States are 1.51 times (swine HEV antigen, $p = 0.03$) more likely to be positive for HEV antibodies than age- and geography-matched normal blood donors. Also, it has been reported that HEV antibody prevalence in field workers from the Iowa Department of Natural Resources is significantly higher than that in blood donors, suggesting that human populations with occupational exposure to wild animals also have a higher risk of HEV infection. Transmissions of hepatitis E from a pet cat and a pet pig to human owners were also reported.

As a fecal–orally transmitted disease, waterborne epidemics are the characteristic of hepatitis E outbreaks in humans. Large amounts of viruses are excreted in feces of infected pigs and other animals. Therefore, swine and other animal manure and feces could be a source of contamination of irrigation water or coastal waters with concomitant contamination of produce or shellfish. Strains of HEV of both human and swine origin have been detected in sewage water. Consumption of contaminated shellfish has been implicated in sporadic cases of acute hepatitis E. Sporadic and cluster cases of acute hepatitis E have also been epidemiologically and genetically linked to the consumption of raw or undercooked pig livers. Approximately 2% of the pig livers sold in local grocery stores in Japan and 11% in the United States were tested positive for swine HEV RNA. The contaminating virus in commercial pig livers sold in the grocery stores of the United States remains fully infectious. Most importantly, the virus sequences recovered from pig livers in grocery stores are closely related, or identical in a few cases, to the viruses recovered from human hepatitis E patients in Japan. Recently, a cluster of four cases of acute hepatitis E were definitively linked to the consumption of raw deer meat in two families in Japan. The viral sequence recovered from the leftover frozen deer meat was 99.7–100% identical to the nucleotide sequence of viruses recovered from the four human patients. Family members who ate none or little deer meats were not infected. Taken together, these data provide convincing evidence of zoonotic HEV transmission either via direct contact with animals or animal wastes or via consumption of infected animal meats. Recently, numerous novel strains of HEV genetically closely related to swine HEV have been identified from patients with acute hepatitis E in many countries, suggesting that these novel strains of human HEV may be of swine origin.

Transmission and Natural History

The primary transmission route for HEV is fecal–oral. Feces from infected humans and other animals contain large amounts of viruses and are likely the main source of virus for transmission. However, under experimental conditions, reproduction of HEV infection in nonhuman primates and pigs via the oral route of inoculation has proven to be difficult, even though the animals can be readily infected by HEV via the intravenous route of inoculation. Chickens were successfully infected via the oral route of inoculation with avian HEV. Other route(s) of transmission cannot be ruled out.

The natural history of HEV is not known. The identification of HEV strains from swine and chickens and their demonstrated ability of cross-species infection and the existence of a wide host range of seropositive animals has further complicated the understanding of HEV natural history. Since human-to-human transmission is uncommon for HEV, therefore a speculative scenario is that HEV is constantly circulating among different animal species, and that each species is infected by a unique strain that is enzootic in respective animal species. Some species, but not all, carry HEV strains that are genetically very similar to human HEV and are transmissible to humans. Therefore, HEV can infect humans through direct contact with infected domestic or farm animals or with feces-contaminated water or via consumption of infected raw or undercooked animal meats. If this happens in a developing country, it may result in an endemic or epidemic due to the poor sanitation conditions. However, if this happens in an industrialized country with good sanitation measures, it will occur only as a sporadic case. It will now be important to determine whether there are basic differences in host range or pathogenicity between epidemic or endemic strains in humans and enzootic strains that are mainly circulating in animal species.

Immunity

The immune response to HEV infection, characterized by a transient appearance of immunoglobulin M (IgM) HEV antibodies followed by long-lasting IgG antibodies, appears late during the period of viremia and virus shedding in stool. The HEV capsid protein is immunogenic and induces protective immunity. The capsid proteins between mammalian and avian HEV strains share common antigenic epitopes, and all HEV strains identified thus far appear to belong to a single serotype. Cross-challenge experiments in primates have demonstrated cross-protection following infection with different genotypes of human HEV strains. The cell-mediated immune response against HEV infection is largely unknown.

Prevention and Control

A vaccine against HEV is not yet available. The experimental recombinant HEV vaccines appear to be very promising; however, their efficacies against the emerging strains of HEV including animal strains with zoonotic potential need to be thoroughly evaluated. In the absence of a vaccine, important preventive measures include the practice of good hygiene, and avoiding drinking water of unknown purity or consuming raw or undercooked pig livers. The demonstrated ability of cross-species infection by swine HEV raises a public health concern, especially for high-risk groups such as pig handlers. An effective measure to prevent potential zoonotic transmission for pig and other animal handlers is to wash hands thoroughly after handling infected animals.

See also: Hepatitis A Virus.

Further Reading

Emerson SU, Anderson D, Arankalle A, et al. (2004) Hepevirus. In: Fauquet CM, Mayo MA, Maniloff J, Desselberger U, and Ball LA (eds.) *Virus Taxonomy, Eighth Report of the International Committee on Taxonomy of Viruses*, pp. 851–855. San Diego, CA: Elsevier Academic Press.

Emerson SU and Purcell RH (2003) Hepatitis E virus. *Reviews in Medical Virology* 13: 145–154.

Haqshenas G, Shivaprasad HL, Woolcock PR, Read DH, and Meng XJ (2001) Genetic identification and characterization of a novel virus related to human hepatitis E virus from chickens with hepatitis–splenomegaly syndrome in the United States. *Journal of General Virology* 82: 2449–2462.

Huang FF, Sun ZF, Emerson SU, et al. (2004) Determination and analysis of the complete genomic sequence of avian hepatitis E virus (avian HEV) and attempts to infect rhesus monkeys with avian HEV. *Journal of General Virology* 85: 1609–1618.

Meng XJ (2000) Novel strains of hepatitis E virus identified from humans and other animal species: Is hepatitis E a zoonosis? *Journal of Hepatology* 33: 842–845 (editorial).

Meng XJ (2003) Swine hepatitis E virus: Cross-species infection and risk in xenotransplantation. *Current Topics in Microbiology and Immunology* 278: 185–216.

Meng XJ (2005) Hepatitis E as a zoonosis. In: Thomas H, Zuckermann A, and Lemon S (eds.) *Viral Hepatitis*, 3rd edn., pp. 611–623. Oxford: Blackwell.

Meng XJ and Halbur PG (2005) Swine hepatitis E virus. In: Straw BE, Zimmerman J, D' Allaire S, and Taylor DJ (eds.) *Diseases of Swine*, 9th edn., pp. 537–545. Ames, IA: Blackwell/Iowa State University Press.

Meng XJ, Purcell RH, Halbur PG, et al. (1997) A novel virus in swine is closely related to the human hepatitis E virus. *Proceedings of the National Academy of Science, USA* 94: 9860–9865.

Meng XJ, Shivaprasad HL, and Payne C (in press) Hepatitis E virus infections. In: Saif M (ed.) *Diseases of Poultry*, 12th edn., ch. 14. Ames, IA: Blackwell/Iowa State University Press.

Purcell RH and Emerson SU (2001) Hepatitis E virus. In: Knipe D, Howley P, Griffin D, et al. (eds.) *Fields Virology*, 4th edn., pp. 3051–3061. Philadelphia: Lippincott Williams and Wilkins.

Herpes Simplex Viruses: General Features

L Aurelian, University of Maryland School of Medicine, Baltimore, MD, USA

© 2008 Elsevier Ltd. All rights reserved.

History

The word 'herpes' as applied to spreading, ulcerative skin manifestations is traced to Hippocrates. In 1736, Jean Astruc classified the condition as sexually transmitted. Its infectious nature was demonstrated in 1921, when lesion material was serially transmitted in rabbits. In 1929 Goodpasture concluded that a latent state is established in ganglionic neurons, and in the 1930s it became evident that the manifestation is gingivostomatitis in seronegative subjects and recurrent 'fever blisters' in those who are seropositive. We now know that two distinct viruses preferentially cause facial or genital lesions. A critical advance has been the development of antiviral chemotherapy.

Classification

Herpes simplex virus serotypes 1 and 2 (HSV-1 and HSV-2) are members of the family *Herpesviridae*. They belong to the subfamily *Alphaherpesvirinae* and the genus *Simplexvirus*. Features of the *Alphaherpesvirinae* include a variable host range, relatively short reproductive cycle, rapid spread in culture, efficient destruction of infected cells, and latency establishment in sensory ganglia.

Structure

Virus particles are 150–200 nm in diameter and consist of four components: core, capsid, tegument, and envelope (**Figure 1(b)**). The core contains the 152 kbp linear, double-stranded DNA genome. It has two covalently linked components consisting of unique sequences (U_L and U_S) bracketed by inverted repeats (**Figure 1(a)**). Open reading frames (ORFs) are located within the unique and repeat sequences, the latter in two copies per genome. Some genes (e.g., R1 (large subunit of ribonucleotide reductase) and $\gamma_1 34.5$) contain regions with cellular homology, and were presumably captured during evolution.

The capsid is a 100–110 nm protein shell with $T = 16$ icosahedral symmetry (**Figure 1(c)**). The faces and edges of the icosahedron are formed by 150 hexons. The vertices are made by pentons. The triplexes are present in the threefold positions between hexons and between hexons and pentons. One of the 12 vertices is believed to be the portal complex through which the genome enters and leaves the capsid, replacing one of the pentons. Over 30 proteins are in the mature particle, eight of which form the capsids. VP5 forms the bulk of the pentons and hexons. VP26 occupies the outer surfaces of the hexons. VP19C and VP23 form the triplexes, which stabilize the

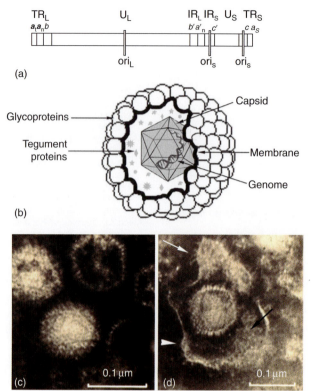

Figure 1 Structure. (a) The HSV genome has two covalently linked components consisting of unique sequences (U_L and U_S) bracketed by inverted repeats (IR_L/TR_L and IR_S/TR_S) and contains three origins of replication (one Ori_L and two Ori_S). (b) The virus particle consists of four components: core, capsid, tegument, and envelope. (c) Negatively stained virus capsids. Capsomers are $9.5 \times 12.5 \, nm^2$ (longitudinal section) and evidence a 4 nm in diameter channel that runs from the surface along the long axis. (d) Enveloped virus particle. The capsid is surrounded by an asymmetrical fibrous-like tegument (black arrow). The envelope (arrowhead) is decorated with spikes projecting from its surface that consist of glycoproteins (white arrow).

capsid shell through interactions with adjacent VP5 molecules. The tegument surrounds the capsid. It is asymmetrically distributed and may appear fibrous on negative staining. Tegument proteins are involved in initiating virus replication (e.g., the transactivator VP16) and inhibiting host cell translation (e.g., the virion host shutoff (vhs) protein). The envelope is the outer covering, which is acquired from cellular membranes. It is decorated with spikes that contain viral glycoproteins (**Figure 1(d)**).

Host Range

Humans are the natural host, but HSV also infects experimental animals and embryonated chicken eggs. The animal species, virus type, route of inoculation, and state of immune competence affect the outcome of infection. In the mouse, intracerebral, footpad, intranasal, and intraperitoneal inoculations are used as models of human fatal neurological or visceral disease. Eye infection is used as a model of encephalitic and latent disease. Certain mouse strains (viz., C57BL/6) exhibit natural resistance, while newborn mice are particularly sensitive. Skin infection was described in mice, rabbits, and guinea pigs, the latter providing the best approximation of human recurrent disease. A wide variety of cell lines support HSV growth. Human primary and secondary cultures are generally more sensitive than established nonhuman cell lines and are used for virus isolation.

Virus Replication

The replicative cycle is represented schematically in **Figure 2**.

1. *Entry.* Virus glycoproteins gC and/or gB bind heparan sulfate chains on the cell surface, followed by fusion of the viral envelope with the cell membrane. Fusion requires viral glycoproteins gB, gD, gH, and gL and one cellular entry receptor that binds gD. Receptors include the herpesvirus entry mediator (HVEM), a member of the tumor necrosis factor receptor superfamily, and nectins 1 and 2, which are members of the immunoglobulin superfamily. The same receptor may also initiate endocytosis, in a cell-type specific context. A cellular entry receptor that binds gB was postulated.

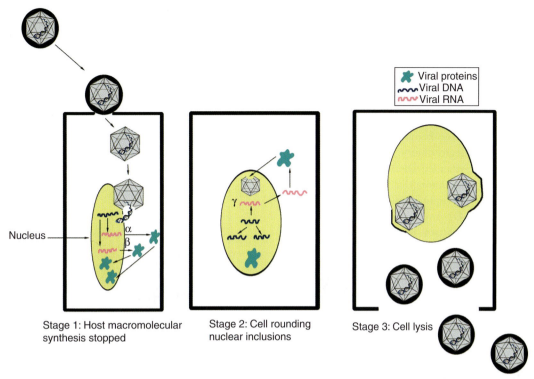

Figure 2 Replicative cycle. Schematic representation includes temporally regulated transcription of IE (α), E (β), and L (γ) genes and the stages of cellular alterations.

2. *Transcription.* Entering capsids dock at the nuclear pore complex and release their genomes. The viral DNA traverses the nuclear pore channel as a condensed, densely packed rod-like structure. In the nucleus, the HSV-1 genomes associate with promyelocytic leukemia protein nuclear bodies (or nuclear domain 10 (ND10)), giving rise to virus replication compartments. The genome encodes up to 84 proteins, and 38 HSV-1 ORFs are essential for virus replication in cell culture. HSV transcripts can have common initiation signals but different termination sites, different initiation sites but co-terminal ends, or differ in their initiation and termination sites but have partially collinear sequences. At least three gene pairs are known to occupy both DNA strands at complementary positions, generating antisense transcripts, the expression of which is mutually exclusive. Only a few HSV mRNAs are spliced (e.g., ICP0).

Three classes of genes are coordinately transcribed and temporally regulated during lytic infection: immediate early (IE), early (E), and late (L) (also known as α, β, and γ, respectively). IE (α) genes (e.g., ICP4, ICP27, ICP0, and ICP22) are transcribed first and play crucial roles in the regulation of productive infection. ICP4 is required for transcription of the E (β) and L (γ) genes. Early after infection, ICP27 co-localizes with ICP4 and has transcriptional and post-transcriptional regulatory activities. Late in infection, it functions in the processing and efficient export of viral mRNAs. ICP0 is not essential, but it is a major transactivator of HSV gene expression and stimulates virus growth, particularly in cells infected at a low multiplicity of infection (moi). ICP22 and its in-frame N-terminal truncation US1.5 are required for virus replication in most cell types. E gene products include enzymes (e.g., thymidine kinase (TK) and DNA polymerase (Pol)), which are required for virus DNA replication. Their expression requires functional IE genes. An exception is R1, which is regulated with IE kinetics. HSV-2 R1 responds to cellular AP-1 transcription factors. L gene expression requires viral DNA synthesis.

The promoters of the IE and R1 genes contain reiterated core enhancer elements (TAATGARAT) that are targeted by VP16 in conjunction with the cellular protein Oct-1 and the transcriptional co-activator HCF-1 in order to mediate transcription. The presence of VP16 is associated with the recruitment of transcription factors (e.g., TFIIB and TFIIH) and the TATA-binding protein (TBP), and assembly of an RNA polymerase II preinitiation complex. Oct-1 is critical for virus replication at low moi.

3. *DNA replication.* Upon reaching the nucleus, the viral genome circularizes and serves as template for DNA replication. DNA replication initiates in an origin-independent manner and continues through an origin-dependent process, perhaps by a rolling-circle and/or homologous recombination-driven mechanism. Homologous recombination may help maintain the integrity of the viral genome through repair of mutated/damaged DNA, and is likely to

contribute to the intratypic molecular diversity of the viral genome. The majority of the replicative intermediates are long concatemers that appear to be generated through sequence replacement or insertion. They contain genomic units with U_L and U_S segments in different orientations that give rise to four possible isomers in equimolar ratios. The biologic relevance of isomerization is unclear. Viral proteins required for DNA replication in cell culture are UL30/UL42, a highly processive heterodimeric DNA polymerase; UL5/UL8/UL52, a heterotrimeric helicase–primase complex; UL29, a single-stranded DNA-binding protein; and UL9, an origin-binding protein that is only required early post infection (pi). Host recombination–repair proteins are also recruited to viral DNA replication sites.

The HSV genome contains three origins of replication with high-affinity binding sites for UL9: a single copy of Ori_L located in the middle of the U_L segment and two copies of Ori_S located in the repeats flanking the U_S segment (**Figure 1(a)**). Mutants lacking Ori_L or both copies of Ori_S are replication competent, indicating that Ori_L and Ori_S can compensate for each other. However, differences in their respective functional efficiencies have been reported. Ori_L may be involved in HSV-1 morbidity and mortality and latency reactivation in mice.

4. *Assembly and egress.* Assembly occurs in the nucleus (**Figure 3(a)**). The viral genome is inserted into a spherical intermediate capsid (procapsid) by a molecular motor made up of viral proteins UL6 together with UL15 and UL28, which are transiently associated with the capsids, recognize packaging signals on DNA, and cleave replicated concatameric DNA into monomeric units. During packaging, the protein scaffold within the procapsid is cleaved and removed, and the procapsid undergoes morphological changes that result in a more robust and angularized structure. Proteins UL17 and UL25 stabilize capsid-DNA structures. DNA-containing capsids exit the nucleus by budding through the inner and outer nuclear membranes (**Figure 3(b)**). This involves the viral protein complex UL31/UL34, which recruits protein kinase (PK) C to the nuclear membrane, thereby causing lamin B phosphorylation and lamina permeabilization.

The mechanism of egress of the alphaherpesviruses has been the subject of a long-standing controversy. According to one model, primary envelopment occurs by budding through the inner nuclear membrane and is followed by transport of primary virions within vesicles from the outer nuclear membrane through the endoplasmic reticulum and secretory Golgi pathway to the cell surface. To account for the lack of identity between primary and mature virions, the second model proposes that enveloped capsids undergo de-envelopment at the outer nuclear membrane and secondary re-envelopment at the *trans*-Golgi network, where they acquire their mature envelope by budding into a cytoplasmic organelle that transports them to the plasma membrane (**Figure 3(b)**). Release is by

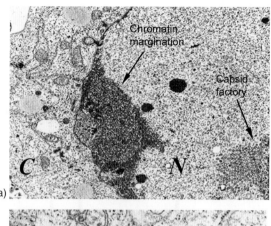

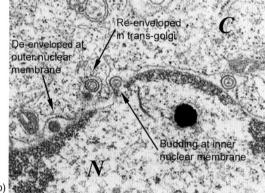

Figure 3 Virus replication. Thin sections of HSV-2-infected cells (12 h after infection) stained with uranyl acetate and lead citrate show nuclear (N) and cytoplasmic (C) compartments. (a) Intranuclear capsids are in different stages of assembly and the nucleus shows chromatin margination. (b) Virion assembly documents budding through the inner nuclear membrane, de-envelopment at the outer nuclear membrane, and re-envelopment in the *trans*-Golgi network. Magnification ×55 700.

exocytosis. During latency reactivation, HSV-1 envelopment likely occurs in neuronal varicosities and growth cones, where the primary site is the *trans*-Golgi network, and traffic/egress is along microtubules. In epithelial cells and neurons, newly assembled virions are directed to specialized cell surface domains (viz., epithelial cell junctions and neuronal synapses) for rapid entry into adjacent connected cells (cell-to-cell spread). Glycoproteins gE/gI bind ligands or receptors at these sites, promoting spread.

5. *Host cell effects.* The IE proteins inhibit cellular DNA synthesis and cause cell cycle arrest. Onset of E gene expression coincides with irreversible shutoff of host cell macromolecular synthesis, which is primarily mediated by the tegument phosphoprotein vhs released into the cytoplasm of the infected cells. Vhs degrades cellular mRNA through an endoribonuclease activity regulated, at least in part, through complexing with VP16. Virus replication is associated with cell rounding, chromatin margination, and the formation of nuclear inclusions (**Figures 2** and **3**).

Serologic Relationships

The HSV-1 and HSV-2 genomes have been sequenced. The genomes share a similar ORF organization, and overall sequence identity in coding region is 83%. Regions of low identity encode glycoprotein gG (used in HSV type-specific serology) and the R1 N-terminal domain, which has PK activity in HSV-2 but not HSV-1. Other biologic differences between HSV-1 and HSV-2 include virus yields in rabbit kidney cells, pock size on the chorioallantoic membrane of embryonated eggs, thermal stability, sensitivity to 5-iododeoxyuridine and interferon (IFN), and neoplastic potential.

Genetics

Epidemiologically unrelated isolates of the same HSV serotype are not identical (intratypic polymorphism). Differences result from base substitutions that may add or eliminate a restriction endonuclease cleavage site or substitute an amino acid, occasional deletions, or variable numbers of repeated sequences at various genome sites. The restriction endonuclease cleavage patterns of a given strain are relatively stable even after many years of serial passage *in vitro*, and may be related to neurovirulence.

Epidemiology and Associated Risks

HSV infects 80% of the adult population globally. Primary HSV-1 infection usually occurs by 5 years of age and is generally asymptomatic. It is estimated that 50 million adults have oral herpes in the USA. A recent cross-sectional survey of the general population in eight European countries revealed intercountry and intracountry differences in age-standardized HSV-1 seroprevalence (52–84%), but a significant proportion of recently screened adolescents in the USA and Europe were HSV-1 seronegative.

The current age-adjusted prevalence of HSV-2 in the USA is 20–25%. Individuals with genital HSV in the USA range between 40 and 60 million (one in five adolescents and adults) and new cases occur at a rate of *c.* 0.5–1 million per year. These estimates may be low, because the disease is not reported and many individuals do not seek medical attention, are misdiagnosed, or shed virus while asymptomatic. Factors that influence acquisition of HSV-2 include gender (women greater than men), race (blacks more than whites), marital status (divorced versus single or married), place of residence (city greater than suburb), and number of sexual partners (higher number has increased risk). Female prostitutes and homosexual men have the highest seroprevalence (75–98%). Women are about 45% more likely than men to be infected with HSV-2. The probability of infection is less than 10% for US women having one partner. It increases to 40%, 62%, and >80% as the number of partners increases to 2–10, 11–50, and >50, respectively. For heterosexual men, these risks are 0, 20%, 35%, and 70% for each of the risk groups, respectively. For homosexual men, the risks are greater than 60% and 90% for those with 11–50 and >50 partners, respectively. HSV-2 infection was also associated with a threefold increased risk of human immunodeficiency virus (HIV) acquisition by meta-analysis calculation of age- and sexual behavior-adjusted relative risks. Recently, HSV-1 became the most common cause of symptomatic first episode anogenital HSV among young people. Among US college students, the proportion of newly diagnosed genital HSV-1 infections increased from 31% in 1993 to 78% in 2001 ($P < 0.001$). HSV-1 was more common in females than males and in persons aged 16–21 than >22.

The major risk to the fetus is maternal primary or initial genital infection, the rate of which is 0.5–10% per year. The incidence of HSV-2 shedding at delivery is 0.01–0.6%, irrespective of past history and time of gestation. Most women whose children are infected (60–80%) are asymptomatic at the time of delivery, and they have neither a past history of genital HSV nor a sexual partner reporting genital infection. The incidence of cervical shedding in asymptomatic pregnant women averages 3%. Clinical diagnosis of genital HSV infection during pregnancy is a significant independent predictor of perinatal HIV transmission. HSV suppression during pregnancy may be a strategy to reduce perinatal HIV transmission.

HSV-induced encephalitis (HSE) is the most common viral encephalitis with an estimated incidence of 2.3 cases per million population per year. The age distribution is biphasic with one peak at 5–30 years of age and a second peak at >50 years of age. HSV-1 accounts for 95% of adult cases and the majority are due to reactivation from latency. HSV-2 does not generally cause encephalitis in adults, but the virus can cause aseptic meningitis, predominantly with primary infection. HSV-2 was isolated from the cerebrospinal fluid in 0.5–3% of patients with aseptic meningitis.

Transmission and Tissue Tropism

HSV is transmitted during close personal contact. The risk of transmission is directly related to virus load. Virus shedding in the saliva was reported in 20% of children 7–24 months old, 18% of those 3–14 years old, 2.7% in those >15 years old, and 2–9% of asymptomatic adults. HSV-1 was isolated from the mouth for 7–10 days after primary infection. Oral secretions from 36–45% of patients examined at prodrome (24–48 h before clinical recurrence) were HSV-1 positive by polymerase chain reaction (PCR). Genital HSV-1 results from self-inoculation or from oral sexual practices.

Sexual contact is the primary route of HSV-2 transmission. Asymptomatic shedding was reported for the majority of patients. Half of the time it occurs more than 7 days before or after a symptomatic recurrence, and it is believed to be the major cause of transmission. However, cervical lesions are generally unnoticed, and they may be confused with asymptomatic shedding. Most studies of asymptomatic patients used PCR, the sensitivity of which in these subjects is at least four- to fivefold higher than that of virus isolation. When virus titers were measured, they were 100–1000 times higher in clinical lesions than in secretions from asymptomatic subjects, suggesting that transmission due to contact with active lesions is likely to be more efficacious. In discordant couples, 50% of seronegative partners remained seronegative for 16 months of contact, but use of condoms or daily suppressive therapy may be an option to reduce the risk of transmission. Intrapartum transmission accounts for 75–80% of neonatal HSV cases. Approximately 30 babies were identified in the world literature with symptomatic congenital HSV disease. This frequently resulted from ascending infections in women who have had prolonged rupture of membranes before delivery.

In children and young adults, HSE usually results from primary infection by virus entering the central nervous system (CNS) through neurotropic spread by way of the olfactory bulb. The hematogenous route of infection is common in immunosuppressed patients and neonates. Murine studies suggest that gender influences hematogenous HSV-1 infection in the CNS, and apolipoprotein E (ApoE) is involved in HSV-1 colonization of the brain from the blood. HSV infects the skin, eyes, liver, lungs, adrenal glands, pancreas, small and large intestine, CNS, and bone marrow. The trigeminal, vagal, and sacral dorsal root ganglia are the major sites of latent HSV infection.

Pathogenicity

The HSV-1 portal of entry is generally the oropharyngeal mucosa. Following replication at the site of infection, the trigeminal ganglia are invariably colonized (latency). Initial HSV-2 replication is at genital sites with colonization of the sacral ganglia. Animal studies suggest that sex hormones modulate susceptibility to HSV-2, with estradiol, but not progesterone, delaying vaginal pathology. Virus growth in the CNS (after intracerebral inoculation of mice), and its invasion after peripheral inoculation (neurovirulence) were ascribed to several genes, notably TK and $\gamma_1 34.5$. The two serotypes have distinct pathogenicity in neurons. HSV-1 induces apoptosis through ICP0-mediated activation of the JNK/c-Jun pathway, and apoptosis is a component of HSE. HSV-2 does not trigger apoptosis in CNS neurons or cause encephalitis in human adults. This is likely due to the PK activity of the HSV-2 R1 protein (also known as ICP10), which is poorly conserved in HSV-1 and overrides apoptotic cascades through activation of survival pathways (viz., Ras/Raf-1/MEK/ERK and PI3-K/Akt) (**Figure 4**).

An unusual aspect of HSV pathogenicity is the infection of stem cells, both epithelial (**Figure 5(a)**) and CD34+ bone marrow derived. HSV-2 infection of human cervical cells causes atypia (**Figure 5(b)**) and the virus also infects spermatozoa (**Figure 5(c)**). Potential risks associated with these infection patterns require better elucidation, particularly as related to stem cells, an area of current research interest.

Clinical Features of Infection

HSV-1 and HSV-2 exhibit similar clinical manifestations. They infect neonates, children, and adults, and produce a wide spectrum of diseases including mucous membrane and skin lesions, ocular, visceral, and CNS disease. The incubation period is 1–26 days (median 6–8 days). Age, gender, host genetic factors, immune competence, associated illnesses, and virulence of the infecting virus strain influence severity.

Most (70–90%) childhood HSV-1 infections are asymptomatic. In children 1–3 years of age, the major manifestion is gingivostomatitis, a serious infection of the gums, tongue, mouth, lip, facial area, and pharynx, often accompanied by high fever, malaise, myalgias, swollen gums, irritability, inability to eat, and cervical lymphadenopathy (**Figure 6(d)**). Later in life, the major HSV-1 clinical manifestation is an upper respiratory tract infection, generally pharyngitis, and a mononucleosis-like syndrome. Reactivated HSV-1 is associated with mucosal ulcerations or lesions at the mucocutaneous junction of the lip presenting as small vesicles that last 4–7 days (known as herpes labialis, cold sores, or fever blisters). Other HSV-1 skin diseases include primary herpes dermatitis (a generalized vesicular eruption), eczema herpeticum (usually a manifestation of a primary infection in which the skin is the portal of entry) (**Figure 6(e)**), and traumatic herpes (resulting from traumatic skin breaks due to burns or abrasions). Herpetic whitlow is an occupational hazard (dentists, hospital personnel, wrestlers) resulting from infection of broken skin (often on fingers). Herpes folliculitis is a rare HSV-1 manifestation that often presents with lymphocytic folliculitis devoid of epithelial changes characteristic of virus infection. Acute HSV-1 rhinitis is a primary infection of the nose

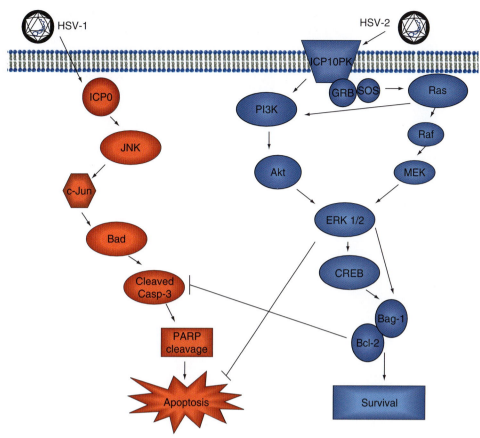

Figure 4 Distinct HSV-1 and HSV-2 pathogenicity. Schematic representation of signaling pathways activated by neuronal cell infection with HSV-1 or HSV-2. HSV-1 activates the pro-apoptotic JNK/c-Jun pathway and triggers apoptosis, likely mediated by the ICP0 protein. HSV-2 activates the MEK/ERK and PI3K/Akt survival pathways and blocks apoptosis, mediated by the PK function of the R1 protein (also known as ICP10).

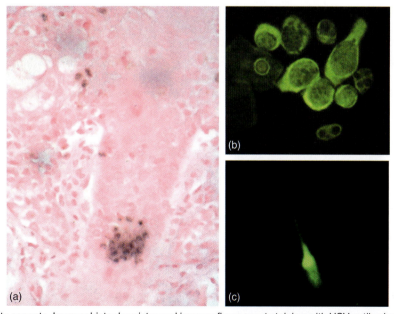

Figure 5 Pathogenicity aspects. Immunohistochemistry and immunofluorescent staining with HSV antibody reveals HSV infection of human epidermal (follicular) stem cells (a), cervical cells (resulting in mild to moderate atypia) (b), and a sperm cell (c).

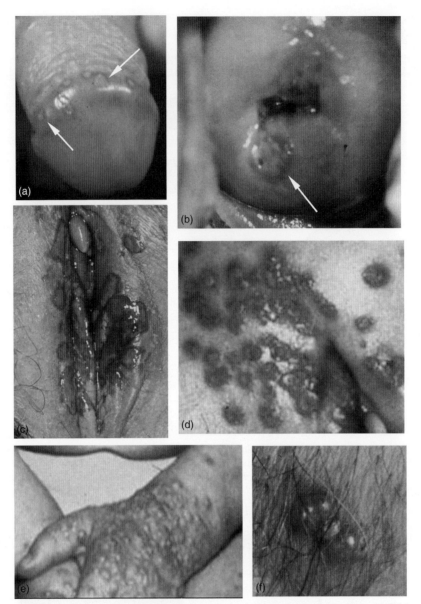

Figure 6 Clinical manifestations. HSV-2 induced lesions of penis (a), cervix (b), and vulva (c, f). Clinical HSV-1 manifestations include gingivostomatitis (d) and eczema herpeticum (e). Arrows in (a) and (b) identify vesicular lesions.

recognized by the appearance of tiny vesicles in the nostrils usually associated with fever and enlarged cervical lymph nodes. HSV-1 infections of the eye can lead to stromal keratitis, one of the leading causes of blindness in the Western world. This disease has a virus-specific T-cell (CD4+) component. Chorioretinitis is a manifestation of disseminated HSV infection that may occur in neonates or in patients with AIDS. HSV-1 accounts for nearly all cases of HSE in adults (and 10–20% of all viral encephalitis cases). Both serotypes can cause encephalitis in infants. With treatment, HSE mortality rates are 15% in newborns and 20% in others. Survivors have neurological sequelae involving impairments in memory, cognition, and personality. Untreated, mortality rates are 60–80%. Genital HSV-1 infections may be less severe and prone to recur than HSV-2 infections.

In females, HSV-2 infection is manifested by vesicles on the mucous membranes of the labia and the vagina (**Figure 6(f)**). Severe forms result in ulcers that cover the entire area surrounding the vulva (**Figure 6(c)**). Symptoms of primary infection include itching, pain, and lymphadenopathy. Cervical involvement is common, although it generally passes unnoticed (**Figure 6(b)**) and the patient is often mistaken for having asymptomatic shedding. Common sites of primary HSV-2 infection in males are the shaft of the penis, the prepuce, and the glans penis (**Figure 6(a)**). Urethritis is the main local expression accompanied by a watery discharge often resulting in dysuria. Symptomatic urethritis is rare in recurrent disease, but virus can often be cultured from the urethra. The infection is more severe in females than in males and systemic symptoms are more common. They often accompany the appearance of primary lesions and include fever, headache, photophobia, malaise, and generalized myalgias. Dysuria, urinary retention, urgency and frequency, pain, and discharge are also seen. Systemic symptoms are generally not seen in recurrent disease. A prodrome often signals a recurrence. It is characterized by a tingling sensation that may precede the lesions by a few hours to 2 days, could be accompanied by radiating radicular pain, and is related to loss of sodium conductance in neurons that support replication of reactivated virus. Anal and rectal infections are seen primarily in homosexuals, as a result of anal intercourse.

HSV-1- and HSV-2-associated erythema multiforme (HAEM) is a recurrent skin disease that follows reactivation of latent virus and occurs at distal sites. HAEM lesional skin contains HSV DNA fragments (most often comprising the viral Pol gene) that are delivered by migrating infected CD34+ stem cells. Lesion development is initiated by virus gene expression and contains virus-specific and autoimmune components. The virus-specific component consists of a restricted population of Th1 cells (primarily, CD4+ Vβ2) that produce IFN-γ. The autoimmune component is in response to antigens released by lysed/apoptotic virus-infected keratinocytes. A genetic predisposition is implicated.

Immunocompromised adults can develop a severe generalized disease that is occasionally responsible for herpetic hepatitis. HSV pneumonitis accounts for 6–8% of cases of interstitial pneumonia in recipients of bone marrow transplants. Mortality due to HSV pneumonia in immunosuppressed patients is >80%. Generalized HSV with involvement of adrenal glands, pancreas, small and large intestine, and bone marrow was reported in the immunocompromised patient. HSV has also been isolated from 40% of patients with acute respiratory distress syndrome. HSV-1 DNA was recently found by PCR in gastrointestinal sensory neurons and in the geniculate ganglion and adjacent areas, respectively, implicating HSV in recurrent gastrointestinal disorders and Bell's palsy. HSV-1 seropositivity was also associated with myocardial infarction and coronary heart disease in older adults, and with Alzheimer's disease (AD). AD is believed to result from frequent episodes of mild encephalitis and neuronal apoptosis resulting from virus reactivation in individuals homozygous or heterozygous for ApoE4. Virus DNA was detected in the brain from elderly people in some, but not other, studies. Uncommon complications of HSV infection include monoarticular arthritis, adrenal necrosis, idiopathic thrombocytopenia, and glomerulonephritis.

Latency

Following replication in the skin, sensory or autonomic nerve endings are infected by cell-to-cell spread (retrograde transport). Virions, or more likely capsids, are thought to be transported intraxonally to the nerve cell bodies in the ganglia (primarily trigeminal and sacral for HSV-1 and HSV-2, respectively), where viral DNA is maintained as an episome that is largely transcriptionally silent. Human trigeminal ganglia contain approximately 20 viral genome copies per latently infected neuron. Virus isolates obtained at various times from the same patients have different restriction endonuclease patterns, indicating that ganglia can be colonized by multiple HSV strains. Certain stimuli cause virus reactivation with concomitant reverse axonal transport of virus progeny to a peripheral site at or near the portal of entry (anterograde transport) (**Figure 7(a)**). Reactivating stimuli include fever, axonal injury, exposure to ultraviolet irradiation (sunlight), emotional stress, and possibly hormonal irregularities. Development of clinically apparent recurrent lesions is probably influenced by the amount of virus, virus–cell interactions, and the rapidity/efficacy of immune-mediated clearance. In the guinea pig model, latent virus reactivates spontaneously. In other animal models, virus is reactivated by neurectomy, ganglionic trauma, electrical stimulation, epinephrine iontophoresis, cadmium treatment, or ganglia explantation and organ culture.

The mechanism of latency establishment and reactivation is still unclear. One possibility is that they are controlled by cellular factors required for virus replication, such as HCF, which translocates to the nucleus at reactivation. However, the general consensus is that viral genes control latency, although their identity is still controversial. The only HSV transcript abundantly expressed during latency is the 2.0 kbp intron derived from the 8.3 kbp primary latency-associated transcript (LAT). Because LAT overlaps the 3′ end of ICP0, it was assumed to contribute to latency establishment by acting as antisense to ICP0. An alternative hypothesis was that ICP0 splicing and expression are suppressed by the ORF P protein (apparently encoded by a minor LAT). However, both spliced and intron-containing ICP0 transcripts were found in latently infected ganglia, and LAT or ORF P did not increase their levels. Others concluded that LAT promotes efficient latency reactivation. However, mutants lacking LAT were shown to reactivate as well as the wild-type virus. One group has reported the expression of a LAT protein encoded by an ORF within the 2.0 kbp intron, which acts somewhat like ICP0 in that it enhances lytic genes and promotes interaction with cellular transcription factors. Most studies attribute the role of LAT in latency reactivation to its anti-apoptotic activity, which ensures a large pool of latently infected neurons that contribute to efficient virus reactivation. It is thought that LAT

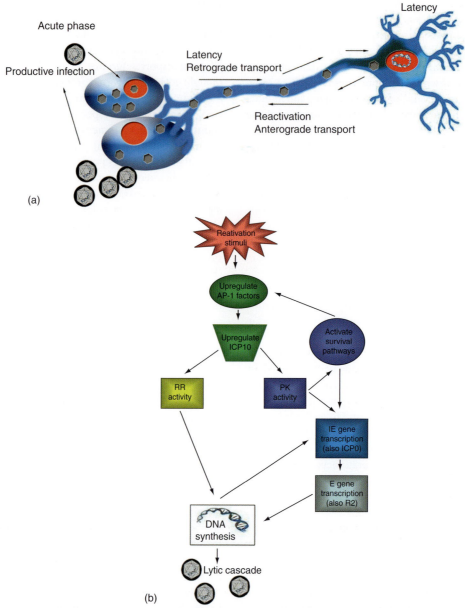

Figure 7 Schematic representation of latency establishment and reactivation. (a) Following replication in the skin, capsids are axonally transported to nerve cell bodies where viral DNA is maintained as a circularized episome. Reactivating stimuli cause reverse axonal transport of virus progeny. (b) Model for involvement of R1 (ICP10) in HSV-2 latency reactivation. Reactivating stimuli upregulate AP-1 factors, thereby causing ICP10 overload. ICP10 supplies the PK activity, which is required for IE gene transcription and a feedback amplification loop through activation of the Ras survival pathway. It also supplies the RR activity that is required for DNA synthesis. The outcome is initiation of the lytic cascade.

counteracts apoptosis induced by corticosteroids, which are upregulated by reactivation-inducing stress stimuli. However, controversy remains. While LAT enhances reactivation in the rabbit eye model, it does not seem to be required in small animal models. LAT anti-apoptotic activity was not studied in neurons, although apoptosis is cell-type specific. LAT anti-apoptotic activity may be unique to certain HSV-1 strains. Apoptosis was also associated with LAT downregulation by stress-induced cAMP early repressors (ICER). The presence of STAT1 response elements in the LAT promoter suggests that cytokines may initiate or contribute to LAT-mediated virus reactivation.

The HSV-2 LAT fails to substitute for its HSV-1 counterpart in promoting latency reactivation, and it does not modulate HSV-2 latency reactivation or its establishment. An HSV-2 mutant deleted in the PK domain of R1 (ICP10) was shown to be significantly impaired in latency establishment/reactivation, and virus reactivation (by ganglia co-cultivation) was inhibited by an ICP10 antisense oligonucleotide, indicating that ICP10 might be involved in HSV-2 latency. In this model, latency-reactivating stimuli upregulate AP-1 transcription factors, thereby inducing expression of ICP10, which is the only virus protein the basal expression of which is regulated by AP-1. The PK function of ICP10 induces IE gene expression and, thereby, the entire viral transcriptional cascade. It also activates the Ras/MEK/ERK pathway that blocks apoptosis, ensuring a large pool of surviving neurons that contribute to efficient virus reactivation, and provides a feedback amplification loop for ICP10 upregulation The ribonucleotide reductase activity of ICP10 contributes to viral DNA synthesis. The outcome is virus replication (**Figure 7(b)**).

HSV Infection during Pregnancy and Neonatal Disease

Localized genital infection is the most common form during pregnancy, but rare visceral involvement with approximately 50% mortality was reported. Fetal deaths occurred in 50% of cases, although mortality did not correlate with the death of the mother. Maternal primary infection prior to 20 weeks of gestation was associated with spontaneous abortion in 25% of infected women. Fetal infection is generally due to virus shed at the time of delivery. Skin lesions are the most commonly recognized features, but at least 70% of untreated cases have disseminated disease. Neonatal infection is decreased by surgical abdominal delivery when membranes are ruptured for less than 4 h. For women with a past history of genital infection, HSV culture can establish whether virus is shed at the time of delivery.

Pathology and Diagnosis

Histopathologic changes are a combination of virus-induced alterations (cell ballooning, condensed nuclear chromatin, and nuclear degeneration) and associated inflammatory responses. Virus-induced changes are generally within the parabasal and intermediate cells of the epithelium. Multinucleated giant cells are also formed and a clear (vesicular) fluid containing large quantities of virus, cell debris, and inflammatory cells appears between the epidermis and the dermal layer. An intense inflammatory response is seen in dermal structures, particularly in primary infection, and is accompanied by an influx of mononuclear cells, as host defenses are mounted. With healing, the vesicular fluid becomes pustular and, then, scabs. Scarring is uncommon. Similar histopathologic findings are seen in mucous membranes. Perivascular cuffing and areas of hemorrhagic necrosis are also seen in the area of infection, particularly in organs other than skin. In the brain oligodendrocytic involvement, gliosis, and astrocytosis are common. Virus culture is the gold standard for diagnosis, but results require approximately 2 weeks. Accordingly, in severe infections, treatment initiation should not be delayed. The increased sensitivity and rapidity of PCR are desirable for diagnosis of severe infections.

Host Immune Responses

An early nonspecific containment phase (innate immunity) and a later HSV-specific effector phase (adaptive immunity) contribute to protection. Natural killer (NK) cells and rapid production of IFN type I provide a threshold of resistance to HSV-1 infection and were associated with the natural resistance of certain mouse strains. In mice, IFNα/β inhibit the onset of IE gene expression, limit virus spread into the nervous system, and activate NK cells. Dendritic cells (DCs) are required for activation of NK cells and CD4+ and CD8+ T cells in response to HSV-1. Conventional CD11c+CD8α+ DCs are the principal antigen-presenting cells during acute HSV-1 infection. The plasmocytoid CD11c+B220+ DCs secrete large amounts of type I IFN *in vitro* after exposure to HSV-1 and help lymph node DCs to induce HSV-specific cytotoxic T cells (CTLs) optimally during a primary immune response. Various reports have incriminated or refuted the association of HLA with HSV infections.

HSV antibody is first seen on days 5–10 pi and persists indefinitely. Antibody titers are higher in subjects with a history of recurrent disease, reflecting virus reactivation. Antibodies to structural proteins are followed by those directed against gD, gB, ICP4, gE, gG, and gC. IgM and IgG antibodies can be demonstrated at different times pi. In children, antibodies to ICP4 may be predictive of long-term neurologic outcome. Infected newborns produce IgM antibodies (particularly against gD) within the first 3 weeks pi. Transplacentally acquired antibodies do not protect the newborn.

T cells play the major role in recovery from HSV infection. Their development is delayed in newborns (2–4 weeks), presumably explaining the increased sensitivity to disease progression in this age group. Effector mechanisms include: (1) delayed-type hypersensitivity (DTH), an early (days 5–10 pi) HSV-type common response that remains inducible for at least 2 years after infection; (2) effector lymphokines, such as IFN-γ, interleukin 2 (IL-2), and other soluble factors; and (3) CTL. CTL clones are specific for cell surface glycoproteins, primarily gD or gE, and intracellular IE proteins.

Naturally acquired HSV immunity reduces the severity of the disease, but it does not prevent latency establishment, recurrent disease, or reinfection. HSV has developed various mechanisms of immune evasion, notably inhibition of antigen presentation by the MHC class I complex through binding of the cellular protein that transports antigenic peptides into the endoplasmic reticulum (TAP) by the viral IE protein ICP47 (**Figure 8**). Other immune-evasion mechanisms include resistance to complement attack (mediated by gC), prevention of antibody binding (mediated by gE/gI), and blocking of CD8+ CTL activity through US3-mediated inhibition of caspase activation. HSV-1 also blocks the IFN response at multiple sites, including decreased Jak1 kinase and Stat2, which are associated with IFN signaling. However, the exact role of immune evasion in recurrent disease is unclear. It is believed that reactivated virus circumvents immune inhibition by replicating before immunity is alerted, or by shifting the immunological balance toward tolerance. For example, induction of Th2 cytokines (e.g., IL-10) could downregulate the Th1 antiviral response by inhibiting DC maturation/function. Generation of downregulatory factors, such as prostaglandin E2 (PGE2) and TGFβ, DC infection, and/or induction of T regulatory cells (Tregs) could also be involved. Our longitudinal studies of HSV patients, done approximately 20 years ago, implicated Tregs in the immune downregulation that enables recurrent disease development. Tregs have made a comeback recently and were again implicated in recurrent disease, now in the mouse model.

Neoplastic Transformation

Seroepidemiologic studies have associated HSV-2 infection with an increased risk of human squamous cervical cancer. Current opinion is that HSV-2 is only a co-carcinogen. Notwithstanding, a wealth of experimental evidence indicates that HSV-2 is a tumor virus. It causes neoplastic transformation of immortalized cells (including those of human origin) and causes tumors in animals, both under conditions that interfere with virus replication. Transformation is mediated by R1 (ICP10) PK through activation of the Ras/MEK/ERK pathway, a function that is conserved because it is required for latency reactivation. The HSV-1 R1 gene does not have oncogenic potential, but HSV-1 causes mutagenesis and gene amplification. In humans, HSV-2 can cause severe hyperproliferative lesions that may have an increased risk for neoplastic conversion. We reported the case of a patient with common variable immunodeficiency who presented with hyperproliferative lesions (pseudo-epitheliomatous hyperplasia) caused by an acyclovir-resistant HSV-2 mutant through overexpression of ICP10PK (**Figure 9**).

Therapy and Prevention

A recent critical advance has been the development of antiviral chemotherapy. Agents currently used are acyclovir, valaciclovir, and famciclovir, which have superior oral bioavailability (**Table 1**). Episodic oral therapy speeds healing and resolution of primary and recurrent HSV lesions, but subsequent recurrence rates are not affected. Long-term daily suppressive therapy may reduce the frequency of reactivation in patients with frequent recurrent episodes, but it does not eliminate ganglionic latency and reactivation occurs after therapy is discontinued. Its effect on asymptomatic virus shedding is still unclear. Daily oral administration for 6–12 months appears to be safe, but a potential concern of long-term treatment is that nonreplicating HSV-2 is oncogenic. Intravenous acyclovir is the agent of choice for treatment of neonatal HSV and HSE. These drugs are phosphorylated by the HSV TK at a 10^6-fold faster rate than the cellular enzyme. After additional phosphorylation by cellular enzymes, the triphosphate (active form) blocks viral DNA synthesis by acting as an inhibitor and substrate of viral DNA polymerase, becoming incorporated into the growing DNA chain and causing termination of chain growth. HSV DNA polymerase has a much higher affinity for the

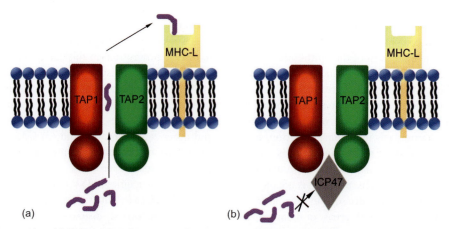

Figure 8 Immune evasion. (a) TAP-mediated transport of antigenic peptides for presentation by MHC class I. (b) ICP47 binds TAP, thereby inhibiting antigen presentation.

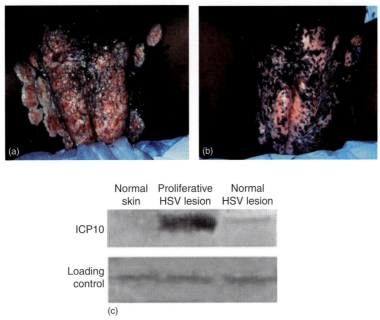

Figure 9 HSV-2 causes human hyperproliferative lesions. Hyperproliferative lesion caused by acyclovir-resistant HSV-2 (a) was treated with intravenous foscarnet (b). The lesion was associated with increased expression of the proliferation-inducing ICP10 protein (c). (a) Reprinted from *Journal of the American Academy of Dermatology*, 37, Beasley KL, Cooley GE, Kao GF, Lowitt MH, Burnett JW, and Aurelian L, Herpes simplex vegetans: Atypical genital herpes infection in a patient with common variable immunodeficiency, 860–863, Copyright (1997), with permission from The American Academy of Dermatology, Inc.

Table 1 CDC treatment recommendations for genital HSV

		Recurrent disease	
	First clinical episode (mg for 7–10 days)	*Episodic (mg × days)*	*Suppressive (mg, daily)*
Acyclovir	400 tid 200 5×/day	400 tid (5 d) 800 bid (5 d) 200 5×/day (5 d)	400 bid
Valaciclovir	1000 bid	500 bid (3–5 d) 1000 qd (5 d)	500 qd 1000 qd
Famciclovir	250 tid	125 bid (5 d)	250 bid up to 1 year

drug than the cellular enzyme. As DNA synthesis is required for the drugs to function, they cannot destroy the virus during latency. Acyclovir resistance emerges with moderate frequency and it crosses to valaciclovir and famciclovir.

Foscarnet and cidofovir are used to treat infections caused by acyclovir-resistant virus. Cidofovir undergoes two stages of phosphorylation, via monophosphate kinase and pyruvate kinase, to form an active metabolite, cidofovir diphosphate (CDVpp), that bears structural similarity to nucleotides and acts as a competitive inhibitor and an alternative substrate for DNA polymerase. Nephrotoxicity is a potential complication with parenteral use of cidofovir, and it should be administered with oral probenecid after prehydration to minimize adverse effects. A topical cidofovir formulation is currently under investigation. Acyclovir-resistant HSV can become sensitive after treatment with cidofovir, suggesting that alternating acyclovir and cidofovir therapies may provide a strategy for managing the emergence of alternating acyclovir-resistant and -sensitive infections.

It is becoming increasingly evident that prevention of HSV infection through vaccination may be an unrealistic goal. A more modest goal has been to reduce the clinical symptoms of primary infection. In two Phase III trials, an HSV-2 gD fragment formulated with a mixture of alum and 3-deacylated monophosphoryl lipid A (3-dMPL) was shown to reduce clinical symptoms of primary HSV-2 infection (approximately 70% efficacy) in women that were HSV-1 and HSV-2 seronegative. Males mounted similar immune responses, but they were not protected and the activity of the vaccine was attributed to the adjuvant used in its composition. Continued research is focusing on HSV-1 recombinants engineered to express immunostimulatory molecules, such as IL-12, which bridges innate and adaptive immunity.

Another vaccination goal has been to prevent/reduce the frequency of recurrent disease episodes in infected patients (therapeutic vaccine). One such vaccine is an HSV-2 mutant rendered replication incompetent and Th1 polarizing through deletion of ICP10PK. This vaccine caused a 75–90% reduction in the number of guinea pigs with spontaneous recurrent disease and reduced the frequency and duration of recurrent episodes experienced by the rare animals that were not absolutely protected. In Phase I/II clinical trials, the vaccine was tolerated well and recurrences were completely eliminated in 44% of the vaccinated patients at 1 year of follow up. This compares to 13% without recurrences in the placebo group ($p = 0.024$). There was also a significant reduction in the number of episodes relative to the previous year ($p < 0.001$) and in the number of recurrent episodes in the vaccinated relative to the placebo group ($p = 0.04$). Efficacy was gender independent, and was seen in both HSV-1 and HSV-2 seropositive subjects. Virus-specific CD8+ CTL and CD4+ Th1 cells mediated protection.

HSV Vectors and Oncolytic Viruses for Gene Therapy

One vector type, known as an amplicon, consists of a plasmid engineered to contain an HSV-1 origin of replication and packaging site along with a bacterial origin of replication in which a selected therapeutic gene is introduced. Amplicons are packaged by co-transfection with a defective HSV-1 helper virus, giving rise to a mixed population of virus particles that consist of the packaged amplicon and the helper virus. Helper-free amplicons were also developed. Another vector type consists of HSV-1 mutants rendered avirulent through deletion of relevant genes. Therapeutic genes delivered with these vectors limit neuronal death (e.g., Bcl-2), restore energy metabolism (e.g., glucose transporter), or promote neuronal survival (e.g., growth factors). HSV vectors were also developed to deliver short hairpin RNA to inhibit various transcripts (e.g., Aβ precursor). Drawbacks include residual toxicity, loss of immune-evasion mechanisms required for repeat treatment, and the loss of function by surviving neurons. HSV-2, which does not cause encephalitis or apoptosis in the CNS, may be a better candidate for the development of vectors for gene therapy of neurologic disorders. ΔRR, a recently developed growth-compromised and avirulent HSV-2 mutant (even after intracerebral inoculation), delivers the viral anti-apoptotic protein ICP10PK and has been shown to have protective activity in various paradigms of neuronal cell death *in vitro* and in animal models. The surviving neurons retain functional (synaptic) activity.

Oncolytic viruses are promising tools for cancer gene therapy. HSV-1 mutants G207 and MGH1 have double deletions in both copies of the neurovirulence gene $\gamma_1 34.5$ and an insertional mutation in the R1 gene. They are avirulent upon intracerebral inoculation, and were shown to kill CNS tumor cells (e.g., glioma), decrease tumor growth, and prolong survival in tumor-bearing nude mice. Tumor specificity appears to be related to increased MEK activity in tumor as compared to normal cells. Function requires intact T-cell activity. Additional HSV-1-based oncolytic viruses were constructed to express IL-4, GM-CSF, or the CD40 ligand. Others use the glioma-specific nestin promoter to drive $\gamma_1 34.5$ (to increase tumor specificity), or lack ICP47 (to reduce immune evasion). An HSV-2-based oncolytic virus, which is deleted in ICP10PK, was recently shown to eradicate tumors in >80% of breast cancer mouse xenografts. However, mutations may arise in untargeted regions of the viral genome, and their clinical significance is still unclear. Another cancer gene therapy approach (known as suicide) is based on the ability of HSV TK to preferentially activate ganciclovir (GCV), thereby killing tumor cells that contain the TK gene (delivered with an HSV-1, retrovirus, or adenovirus vector) as well as surrounding tumor cells (bystander effect).

Future

Studies of the molecular pathogenesis of HSV neuronal and stem cell infections and the role and modulation of HSV immunity in disease pathogenicity (notably recurrent disease) and protection remain principal goals for future research efforts. Prophylactic and therapeutic vaccines are badly needed. Further research must also focus on the mechanism of latency establishment and virus reactivation, and overcome problems of drug resistance. Risks associated with reinfection, particularly by drug-resistant virus strains, need elucidation. Significant progress may be achieved by developing novel oncolytic viruses, and further research is needed to elucidate the potential of HSV as a vector for gene therapy of neurodegenerative diseases.

See also: Herpes Simplex Viruses: Molecular Biology; Herpesviruses: General Features; Herpesviruses: Latency.

Further Reading

Aurelian L (2004) Herpes simplex virus type 2 (HSV-2) vaccines: New ground for optimism? *Clinical and Vaccine Immunology* 11: 437–445.

Aurelian L (2005) HSV-induced apoptosis in herpes encephalitis. *Current Topics in Microbiology and Immunology* 289: 79–112.

Aurelian L, Ono F, and Burnett J (2003) Herpes simplex virus (HSV)-associated erythema multiforme (HAEM): A viral disease with an autoimmune component. *Dermatology Online Journal* 9: 1.

Beasley KL, Cooley GE, Kao GF, Lowitt MH, Burnett JW, and Aurelian L (1997) Herpes simplex vegetans: A typical genital herpes infection in a patient with common variable immunodeficiency. *Journal of the American Academy of Dermatology* 37: 860–863.

Burton EA, Fink DJ, and Glorioso JC (2005) Replication-defective genomic HSV gene therapy vectors: Design, production and CNS applications. *Current Opinion in Molecular Therapy* 7: 326–336.

Carr DJ and Tomanek L (2006) Herpes simplex virus and the chemokines that mediate the inflammation. *Current Topics in Microbiology and Immunology* 303: 47–65.

Kimberlin DW and Whitley RJ (2005) Neonatal herpes: What have we learned. *Seminars in Pediatric Infectious Diseases* 16: 7–16.

Leone P (2005) Reducing the risk of transmitting genital herpes: Advances in understanding and therapy. *Current Medical Research and Opinion* 21: 1577–1582.

Novak N and Peng WM (2005) Dancing with the enemy: The interplay of herpes simplex virus with dendritic cells. *Clinical and Experimental Immunology* 142: 405–410.

Ono F, Sharma BK, Smith CC, Burnett JW, and Aurelian L (2005) CD34+ cells in the peripheral blood transport herpes simplex virus (HSV) DNA fragments to the skin of patients with erythema multiforme (HAEM). *Journal of Investigative Dermatology* 124: 1215–1224.

Shen Y and Nemunaitis J (2006) Herpes simplex virus 1 (HSV-1) for cancer treatment. *Cancer Gene Therapy* 13: 975–992.

Smith CC (2005) The herpes simplex virus type 2 protein ICP10PK: A master of versatility. *Frontiers in Bioscience* 10: 22820–22831.

Herpes Simplex Viruses: Molecular Biology

E K Wagner[†] **and R M Sandri-Goldin,** University of California, Irvine, Irvine, CA, USA

© 2008 Elsevier Ltd. All rights reserved.

The Virion

The Capsid

Like all herpesviruses, herpes simplex virus (HSV) type 1 (HSV-1) and the closely related HSV type 2 (HSV-2) have enveloped, spherical virions. The genome of HSV-1 is densely packaged in a liquid-crystalline, phage-like manner within a 100 nm icosahedral capsid. The capsid comprises 162 capsomeres of 150 hexons and 12 pentons. Hexons contain six molecules of the 155 kDa major capsid protein (VP5 or UL19) with six copies of the vertex protein VP26 (UL35) at the tips. Pentons contain five copies of VP5. Hexons are coordinated in threefold symmetry with a triplex structure made up of two other proteins: one copy of VP19C (UL38) and two copies of VP23 (UL18) per triplex. Small amounts of other viral proteins are also associated with the capsid; these include VP24 (derived from UL26), a maturational protease, and the UL6 gene product. The capsid portal formed from the UL6 gene product is now thought to occupy one of the penton positions, which leaves the capsid with 11 rather than 12 VP5-containing pentons.

The Tegument

The capsid is surrounded by a layer consisting of approximately 20 tegument or matrix proteins. These include the α-*trans* inducing factor (αTIF, VP16, or UL48), and a virion-associated host shutoff function (vhs or UL41).

The Envelope

The trilaminar viral lipid envelope forms the outer surface of the virion. This envelope has a diameter of 170–200 nm, although exact dimensions vary depending upon the method of visualization. The envelope is derived from the host cell nuclear membrane and contains at least 10 virally encoded glycoproteins. One glycoprotein (gG–US4) has sufficient differences in amino acid sequence to serve as a type-specific immune reagent to differentiate HSV-1 and HSV-2. At least four (gL–UL1, gH–UL22, gB–UL26, and gD–US6) are involved in virus penetration into the host cell. The sequence of gB is highly conserved among a broad range of herpesviruses. Another glycoprotein (gC–UL44) facilitates the initial attachment of the virion to glycosaminoglycans (GAGs) on the cell surface. Two others (US7–gI and US8–gE) function together as a heterodimer that binds the Fc region of immunoglobulin G (IgG) and influences cell-to-cell spread of virus.

The Genome

The complete genomic sequences of prototype strains of HSV-1 and HSV-2 are available in standard databases. Homology between HSV-1 strains is >99% for most regions of the genome while overall homology between HSV-1 and HSV-2 approximates 85% within translational reading frames and is significantly less outside them. The HSV-1 genome is shown schematically in **Figure 1**. It is a linear, double-stranded (ds) DNA duplex, 152 000 bp in length, with a base composition of 68% G+C. The genome circularizes upon infection. Because

---

[†] Deceased

the genome circularizes, the genetic/transcription map is conveniently shown as a circle. The genome contains at least 77 open reading frames (ORFs) translated into funtional proteins. Therefore, the map is complex.

The organization of the genome can be represented as $a_n b U_L b' a'_m c' U_S ca$. The genome is made up of five important regions. These are:

1. The ends of the linear molecules, with the left end comprised of multiple repeats of the 'a' sequence (a_n), which are important in circularization of the viral DNA upon infection and in packaging the DNA in the virion. Variable numbers of the 'a' sequence in inverted form (a'_m) are present internally in the genome.
2. The 9000 bp long inverted repeat (R_L – 'b' region). This region encodes an important immediate early regulatory protein (ICP0), and most of the 'gene' for the latency-associated transcript (LAT), including the promoter. It also encodes the ORF for the ICP34.5 gene, which is important in neurovirulence, and two partially overlapping ORFs antisense to that of ICP34.5, namely ORF-O and ORF-P. Two other short translational reading frames 5′ of the LAT cap site, termed ORF-X and ORF-Y, are also present, but have unknown functions.
3. The long unique region (U_L), which is 108 000 bp long, and encodes at least 58 distinct proteins. It contains an origin of DNA replication (ori_L) and genes encoding DNA replication enzymes, including the DNA polymerase. A number of capsid proteins and glycoproteins are encoded in U_L as well as many other proteins.
4. The 6600 bp short inverted repeats (R_S – 'c' region) encode the very important immediate early transcriptional activator, ICP4. R_S also encodes another origin of DNA replication (ori_S) and the promoters for two other immediate early genes.
5. The 13 000 bp short unique region (U_S) contains 13 ORFs, a number of which encode glycoproteins important in viral host range and response to host defense.

The arrangement of the inverted repeat elements allows the unique regions of the viral genome to invert relative to each other into four 'isomers': prototype (P), inversion of U_L relative to U_S (I_L), inversion of U_S relative to U_L (I_S), and inversion of both U_L and U_S (I_{LS}). Inversion is facilitated by the 'a' sequence elements noted above.

Viral Replication

Entry

Initial association of HSV with a host cell is mediated by the association with the envelope glycoproteins gB (UL27) and gC (UL44) and glycosamminoglycans (GAGs), such as heparin sulfate, on the cell surface. Virus internalization involves the mediation of four essential glycoproteins, gB (UL27), gH (UL22), gD (US6), and gL (UL1). In many cells, this is mediated by interaction of gD with a cellular surface protein named the herpesvirus entry mediator (HVEM). HVEM is a member of the TFN/NGF family of proteins. Other cellular surface proteins may also function in a similar capacity since some cells lacking this receptor still allow efficient viral entry. Following entry, HSV exists as a de-enveloped virion in the cytoplasm. This particle must then be transported to the nucleus by a process that requires the cell microtubule machinery, and then the viral DNA released into the nucleus (uncoating). The details of these processes are poorly understood.

Transcription and DNA Replication

Once in the nucleus, the viral DNA becomes available for transcription by host RNA polymerase II into mRNAs. The viral genome is accompanied into the nucleus by the VP16 or αTIF protein (UL48), which functions in enhancing immediate early viral transcription via cellular transcription factors. Gene expression occurs in three phases – immediate early or α (the first phase), early or β (which requires the immediate early protein ICP4), and late or γ (which requires the immediate early proteins ICP4 and ICP27 and viral DNA replication for maximal expression). DNA replication has a significant influence on viral gene expression. Early expression is reduced following the start of DNA replication, while late genes begin to be expressed at high levels. Immunofluorescence studies show that DNA replication occurs at discrete sites or replication compartments in the nucleus. Many of the early proteins are involved in viral DNA replication. Some of these early proteins, such as the two-subunit DNA polymerase and the three-subunit helicase–primase, participate directly in DNA synthesis, while others, such as thymidine kinase and the two-subunit ribonucleotide reductase, are involved indirectly by increasing the pools of deoxynucleoside triphosphates.

RNA Export, Translation, and Post-Translational Processing

HSV mRNAs are capped at the 5′ end and processed to form the 3′ end by polyadenylation, but the majority of viral transcripts do not undergo splicing. HSV transcripts are exported to the cytoplasm using the cellular TAP/NXF1 mRNA export pathway, and this process is facilitated by the HSV immediate early protein ICP27, which binds to viral RNAs as well as to the TAP/NXF1 receptor. HSV mRNA is translated on polyribosome structures using the host translational machinery.

Extensive proteolytic processing of HSV proteins does not occur, but the maturation of the capsid absolutely requires the activity of the maturational protease encoded by the UL26 gene product. Many viral proteins are extensively phosphorylated, including the immediate early regulatory proteins ICP4, ICP0, and ICP27, as well as a number of structural proteins. Other modifications such as arginine methylation of ICP27 have also been described. The precise role of such post-translational modifications upon the function of these proteins has not been well characterized.

Assembly and Release

Following DNA replication in the nucleus, the completed viral genomes are packaged into preformed capsids as the scaffolding proteins (the products of UL26.5 and the 3′ portion of UL26) are displaced. This assembly process entails several proteins that comprise the DNA encapsidation/cleavage machinery. The packaging process also involves the 'a' sequence at the other end of the genome-sized DNA fragment, leading to cleavage of the growing DNA chain and resulting in encapsidation of a full-length genome. Assembly is accompanied by a major change in capsid structure (maturation), which entails the action of the UL26 viral protease. This is followed by a very complicated process of egress. While there is controversy on how this process occurs, in the currently favored model, the nucleocapsid first buds through the inner nuclear membrane acquiring an envelope. The virion then enters the cytoplasm by fusing with the outer nuclear membrane, thereby losing its initial envelope. Subsequently, it becomes re-enveloped by budding into intracellular membranous compartments and proceeds through the cytoplasm. Whatever the exact mechanism, infectious virus can be recovered from cells many hours before cellular disintegration and release of virions into the extracellular medium. Virus spread by cell to cell contact and/or cell fusion is probably an important feature of the pathogenesis of infection in the host and can be readily observed with infections in cultured cells.

Viral Transcripts and Proteins

Each viral protein is expressed from its own independently regulated transcript. The viral transcription map is included in **Figure 1**. Independent starts of overlapping transcripts, (rare) splicing, and temporally differentiated polyadenylation site utilization result in the number of independent transcripts expressed exceeding 100.

General Properties of Transcripts and Promoters

Generally, the ORFs of HSV-1 are expressed as unique, intronless mRNA molecules that are controlled by temporal class-specific promoters. Nested, partially overlapping transcripts, each encoding a unique ORF but utilizing a shared polyadenylation site, are common (e.g., transcripts encoding UL24, UL25, and UL26 and those encoding US5, US6, and US7). Some complementary reading frames also exist, such as the mRNAs encoding ORF-P and ORF-O with that encoding ICP34.5 and transcripts encoding reading frames complementary to UL43.

Most HSV promoters are recognizable as eukaryotic RNA polymerase II promoters with obvious 'TATA' box homologies 20–30 nucleotides (nt) 5′ of the mRNA cap sites. The *cis*-acting elements mediating transcription of immediate early (α) transcripts include upstream enhancer elements at sites distal to and extending to several hundred nucleotides upstream of the cap sites, as well as several transcription factor binding sites located within the region from −120 to −50. The enhancers contain multiple copies of a 'TAATGARAT' sequence, which interacts with cellular transcription factors of the POU family (notably, Oct-1). Enhancement of transcription occurs through the action of virion-associated tegument protein VP16 or αTIF (UL48), which is a powerful transcriptional activator. The *cis*-acting elements mediating transcription of early (β) promoters include several transcription factor binding sites and the TATA box at −30, but the 'TAATGARAT' enhancer elements are not present. Leaky-late ($\beta\gamma$) and late transcript (γ) promoters contain critical sequence elements mediating full levels of transcription at or near the cap site, and at least some contain transcription factor binding sites in the proximal part of the leader sequence downstream of the start site of transcription. HSV-1 promoters representing the temporal classes of productive infection are shown in **Figure 2**.

Nomenclature of Proteins

The proteins encoded by HSV-1, their functions (if known), and the locations of their coding regions on the genome are given in **Table 1**.

Currently accepted protein nomenclature is a mix of historic and systematic systems. Proteins encoded in the repeats and a number of other proteins of well-established function are referred to by their historic or functional names. Proteins encoded by ORFs in the U_L and U_S regions are often referred to by the location of the ORF. Both nomenclatures are used in **Table 1**.

Functional Classification of Proteins

Regulatory proteins

The cascade of viral gene expression begins with the expression of immediate early or α proteins. There are five immediate early genes and their expression is highly activated by the action of VP16 or αTIF (UL48), which interacts with cellular transcription factors, Oct1 and

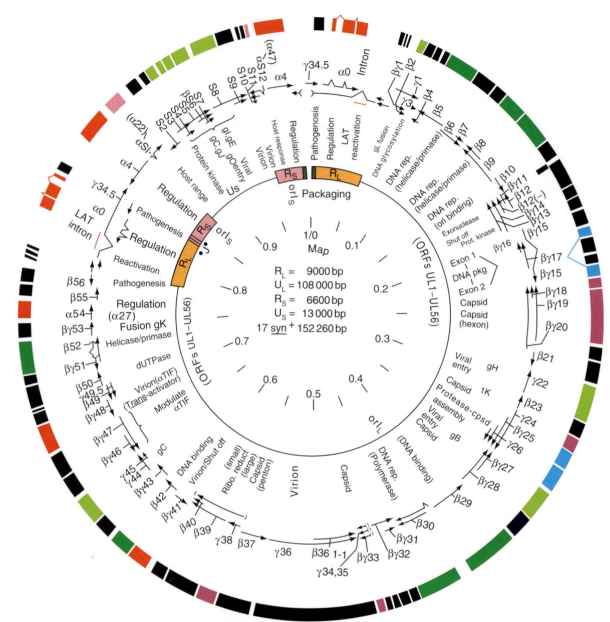

Figure 1 The genetic and transcriptional map of HSV-1 (strain 17syn+). The map is shown as a circle to indicate the fact that the linear genome circularizes upon infection. ORFs predicted from sequence data are named numerically according to their position on the viral genome, and transcripts expressing all or part of them are indicated with the arrowheads showing the transcription termination/polyadenylation sites. Known kinetic classes of transcripts are indicated; the LATs in the R_L regions are the only transcripts expressed during latent infection. Also shown are the location of the origins of replication and the 'a' sequences at the genome ends which are involved in encapsidation. In addition, and where known, the functions of individual genes are shown. In the literature, gene names sometimes contain a subscript (e.g., U_L19 rather than UL19).

HCF. The resulting complex binds to the 'TAATGARAT' sequences upstream of immediate early gene promoters. Three of the five immediate early genes, ICP4, ICP0, and ICP27, have roles in regulating viral gene expression at the level of transcription. ICP4 is a transcriptional activator that is required for the transcription of early and late viral mRNAs. Multiple phosphorylated forms of the protein have been identified in the infected cell. ICP4 apparently stabilizes the formation of the transcription initiation complexes at the TATA box of viral promoters.

The essential ICP27 protein is multifunctional, acting at both the transcriptional and post-transcriptional levels to regulate viral and cellular gene expression. Different functional domains of ICP27 are required for its activities, which include recruiting cellular RNA polymerase II to sites of viral DNA, inhibiting cellular splicing, promoting

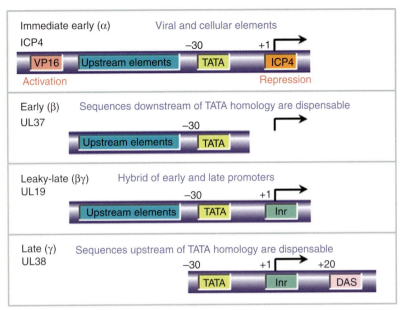

Figure 2 The architecture of HSV promoters. Features of the promoters controlling representative members of the kinetic classes of viral transcripts expressed during productive infection are shown. Inr, initiator element; DAS, downstream activator sequence.

export of viral intronless mRNAs, and enhancing translation of viral transcripts. ICP27 interacts with myriad cellular and viral proteins, and it binds to viral RNAs. ICP27 is phosphorylated and methylated on some arginine residues, and these modifications presumably help regulate its various activities.

ICP0 is also a multifunctional protein that is, first and foremost, a transcriptional activator. ICP0 appears to play a central role in regulating whether HSV infection goes into the lytic phase or the virus becomes latent. ICP0 is a member of the family of E3 ubiquitin ligase enzymes that have a so-called RING finger zinc-binding domain. This domain confers on ICP0 the ability to induce the proteasome-dependent degradation of a number of cellular proteins. This process results in multiple consequences, including the disruption of cellular nuclear substructures known as ND10 or PML nuclear bodies. ICP0 interacts with or forms complexes with a large number of proteins. It also undergoes extensive post-translational modification, including phosphorylation by viral and cellular kinases, and ubiquitination. The phosphorylation status of ICP0 alters as infection progresses, suggesting that the change in ICP0 localization from nuclear to predominantly cytoplasmic at later times of infection may in part be regulated by phosphorylation.

DNA replication proteins

Seven early proteins are required for the formation of the viral DNA replication complex: The DNA polymerase (UL30) functions in a complex with UL42, a processivity factor. The ori binding protein (UL9) binds to the critical core sequence 'GTTCGCAC' in ori_S and may function as a dimer. Equimolar amounts of proteins UL5, UL8, and UL52 make up the helicase–primase complex; and ICP8 (UL29), the major single-stranded DNA-binding protein functions in a manner analogous to phage T_4 gene 32 protein in keeping the replication fork open.

A number of other early proteins are involved in genome replication by altering the pool of deoxyribonucleoside precursors and for repair and proofreading functions. These include thymidine kinase (UL23), the large and small subunits of ribonucleotide reductase (UL39 and UL40), and deoxyuridine triphosphatase (UL50). An enzyme with a potential repair function is uracil-DNA glycosylase (UL2).

Structural proteins

More than 30 HSV late proteins are associated with mature virions. These include five capsid proteins, of which VP5 is the major capsid protein. A large number of tegument proteins and 10 or more glycoproteins are also found in mature virions.

Capsid assembly can be readily accomplished both *in vitro* and in insect cells with proteins expressed from a panel of recombinant baculoviruses expressing the four principal capsid proteins (UL19–VP5, UL38–VP19C, UL18–VP23, and UL35–VP26), the scaffolding protein (UL26.5), and the maturational protease (UL26). Other proteins have been implicated genetically in the capsid assembly process, but are dispensable in minimal *in vitro* systems.

Packaging of the viral DNA into the mature capsid requires a large number of viral proteins, including UL6, UL15, UL17, UL25, UL28, UL32, and UL33. The protein encoded by UL15, in conjunction with that encoded by UL28, may function as a terminase, cleaving

Table 1 Genetic functions encoded by HSV-1

Location	Name	Function
R_L	'a'	cis genome cleavage, packaging signal
	ICP34.5	RL1, neurovirulence, inhibits cellular apoptosis
	ORF-O	Modulates ICP0, ICP22 – inhibits splicing
	ORF-P	Modulates ICP4
	ICP0	RL2 (α0, IE1) immediate early trans-activator, E3 ubiquitin ligase, disrupts nuclear ND10 structures
	LAT-intron	Stable accumulation in nuclei of latently infected neurons
	LAT	c. 600 nt in 5' region facilitates reactivation
	ORF-X	Unknown
	ORF-Y	Unknown
U_L	UL1	gL, essential glycoprotein that functions as a heterodimer with gH – involved in viral entry
	UL2	Uracil DNA glycosylase, DNA repair
	UL3	Nuclear protein
	UL4	Nuclear protein
	UL5	Part of helicase–primase complex, essential for DNA replication
	UL6	Capsid portal protein, capsid maturation, DNA packaging in capsid
	UL7	Unknown, dispensable in cell culture
	UL8	Part of helicase–primase complex, essential for DNA replication
	UL9	Ori-binding protein, essential for DNA replication
	UL10	gM, glycoprotein of unknown function
	UL11	Tegument protein, capsid egress, and envelopment
	UL12	Alkaline exonuclease, DNA packaging (?), capsid egress
	UL12.5	C-terminal 2/3 of UL12, expressed by separate mRNA – specific function unknown
	UL13	VP18.8, tegument-associated protein kinase
	UL14	Unknown
	UL15	DNA packaging, spliced mRNA, exons flank UL16 and UL17
	UL16	Tegument protein
	UL17	Cleavage and packaging of DNA
	UL18	VP23, capsid protein, triplex
	UL19	VP5, major capsid protein, hexon
	UL20	Membrane-associated, virion egress
	UL21	Tegument protein, auxiliary virion maturation function (?)
	UL22	gH, glycoprotein involved in viral entry as heterodimer with gL
	UL23	TK, thymidine kinase
	UL24	Nuclear protein
	UL25	Tegument protein, capsid maturation, DNA packaging
	UL26	VP24 (N-terminal half), protease, minor scaffolding protein (C-terminal half)
	UL26.5	Scaffolding protein
	UL27	gB, glycoprotein required for virus entry, mediates binding through interaction with GAGs in plasma membrane
	UL28	ICP18.5, capsid maturation, DNA packaging
	UL29	ICP8, single-stranded DNA-binding protein, essential for DNA replication
	ori_L	Origin of DNA replication
	UL30	DNA polymerase
	UL31	Nuclear lamina protein involved in nuclear egress
	UL32	Capsid maturation, DNA packaging
	UL33	Capsid maturation, DNA packaging
	UL34	Nuclear membrane protein involved in nuclear egress
	UL35	VP26, capsid protein
	UL36	VP1/2, tegument protein
	UL37	Tegument phosphoprotein
	UL38	VP19C, capsid protein, triplex
	UL39	Large subunit of ribonucleotide reductase
	UL40	Small subunit of ribonucleotide reductase
	UL41	vhs, virion host shutoff protein, degrades mRNA
	UL42	Polymerase accessory protein, essential for DNA replication
	UL43	Multiple membrane spanning protein
	UL43.5	Antisense to UL43
	UL44	gC, glycoprotein involved in initial stages of virion-cell association with GAGs, complement binding protein
	UL45	Membrane associated
	UL46	VP11/12, tegument associated

Continued

Table 1 Continued

Location	Name	Function
	UL47	VP13/14, tegument associated
	UL48	αTIF, VP16, virion-associated transcriptional activator, enhances immediate early transcription
	UL49	VP22, tegument protein
	UL49.5	Membrane protein associated with gM, glycosylated in other herpesviruses (gN) but not in HSV-1
	UL50	dUTPase, nucleotide pool metabolism
	UL51	Unknown
	UL52	Part of helicase–primase complex, essential for DNA replication
	UL53	gK, glycoprotein involved in virion egress
	UL54	ICP27, immediate early regulatory protein, multifunctional regulator, inhibits splicing, mediates viral RNA export, enhances translation initiation
	UL55	Unknown
	UL56	Tegument protein, affects pathogenesis
R_L		See R_L above
	R_L/R_S junction	Joint region, site of inversions of long and short segments, contains 'a' sequences
R_S	LAT-poly(A) site	Polyadenylation site for the primary latency associated transcript and the transcripts encoding ORF-O and ORF-P
	ICP4	RS1, essential immediate early transactivator, required for early and late gene expression
	ori$_S$	Origin of DNA replication
U_S	ICP22	US1, nonessential immediate early protein, affects host range
	US2	Unknown
	US3	Tegument-associated protein kinase, phosphorylates UL34 and US9
	US4	gG, glycoprotein of unknown function
	US5	gJ, glycoprotein of unknown function
	US6	gD, glycoprotein involved in virus infectivity and entry, binds HVEM
	US7	gI, glycoprotein, which as heterodimer with gE binds IgG-Fc and influences cell-to-cell spread of virus
	US8	gE, glycoprotein, which as heterodimer with gI binds IgG-Fc and influences cell-to-cell spread of virus
	US8.5	Nucleolar protein
	US9	Tegument-associated phosphoprotein (type II membrane protein)
	US10	Tegument-associated protein
	US11	Tegument-associated phosphoprotein, RNA binding, post-transcriptional regulation
	US12	ICP47, immediate early protein, inhibits major histocompatibility complex type 1 antigen presentation in human and private cells

genome-sized fragments from the growing concatemeric replication complex. The UL15–UL28 complex also appears to mediate the initial entry of viral DNA into the capsid, that is, it serves a 'docking' function. The alkaline exonuclease gene (UL12) also plays a role in DNA packaging.

Proteins involved in pathogenesis and cytopathology

After the onset of viral DNA synthesis, a large fraction of the viral genome is represented in abundant amounts of complementary mRNA that can form stable dsRNAs, which serve to activate protein kinase R (PKR). Activated PKR phosphorylates the α subunit of the translation initiation factor 2 (eIF-2α), which shuts off protein synthesis. To combat this response, HSV encodes a protein named $\gamma_1 34.5$ (ICP34.5), which binds phosphatase 1α and redirects it to dephosphorylate eIF-2α, thereby preventing the shutoff of protein synthesis by activated PKR. One of the immediate early proteins, ICP47, appears to have a role in modulating host response to infection by specifically interfering with the presentation of viral antigens on the surface of infected cells by major histocompatibility complex class I (MHC-I), precluding immune surveillance by $CD8^+$ T lymphocytes at the very onset of infection.

A number of the envelope glycoproteins found in the intact virion appear to have a significant role in controlling host response. The complex of glycoproteins gE (US9) and gI (US8) functions as an IgG Fc receptor blocking the presentation of this region of the antibody molecule and precluding activation of the complement cascade. In addition, gC (UL44) binds several components of the complement complex, and viruses with mutations in gD (US6) display altered pathogenesis in mice.

Other viral gene products also have a role in cytopathology and pathogenesis. While dispensable for productive infection in some cells, the immediate early protein ICP22 is required for HSV replication in others – perhaps by mediating the expression of a set of late transcripts. Viruses lacking thymidine kinase (UL23) have significantly altered pathogenic patterns in experimental animals while certain mutations in the DNA polymerase gene (UL30) have altered patterns of neurovirulence.

Viral mutants defective in the virion host-shutoff protein (vhs; UL41) also have altered virulence in mice.

Latent Infection

HSV-1 latency is characterized by the persistence of viral genomes as episomes in the nuclei of sensory neurons. During latency, lytic functions are repressed and only one region of the genome is transcribed abundantly: the region encoding the LATs. The LAT domain is transcriptionally complex, and while the predominant species that accumulates during latency is a 2.0 kb stable intron, other RNA species are transcribed from this region of the genome, including a number of lytic or acute-phase transcripts. The LAT promoter (**Figure 3**) has not been characterized in quite as much detail as those promoters controlling productive cycle transcripts. It appears to have a number of regulatory elements important in neuronal expression over extended periods of time.

Latent infection with HSV can be viewed as having three separable phases: establishment, maintenance, and reactivation. In the establishment phase, the virus must enter a sensory neuron, and, following entry, there must be a profound restriction of viral gene expression so that the cytopathic results of productive infection do not occur. Thus, productive cycle genes are quiescent transcriptionally and functionally and only LAT is expressed. The maintenance of the HSV genome in latently infected neurons requires no viral gene expression apart from the LATs. Latent HSV genomes are harbored within the nucleus of a nondividing sensory neuron and do not need to replicate. HSV DNA is maintained as a nucleosomal, circular episome in latent infections and low levels of genome replication might occur or be necessary for the establishment or maintenance of a latent infection from which virus can be efficiently reactivated.

Reactivation of HSV results in the appearance of infectious virus at the site of entry in the host. Expression of only a 350 bp region of LAT near its 5′ end is both necessary and sufficient to facilitate reactivation in several animal models. Whatever the specific mechanism for this facilitation, it does not involve expression of a protein; rather, the active region appears to function by a *cis*-acting mechanism, the details of which are currently obscure. The process of reactivation is triggered by stress as well as other signals that may transiently lead to increased transcriptional activity in the harboring neuron, which may achieve a threshold that leads to lytic infection. The latency and reactivation cycle is shown in **Figure 4**.

Latent Phase Transcripts

Latently infected neurons represent 1–10% of the total neurons in the ganglia. In some neurons, an 8.5 kb polyadenylated primary LAT is expressed, which extends through the joint region (across the 'a' sequences) to a polyadenylation signal next to that of the ICP4 gene. The primary LAT is extensively processed, forming a stable 2 kb poly(A-) intron in lariat form, which accumulates in the latently infected neuronal nucleus. The mechanism by which LATs contribute to the establishment of and reactivation from latency has not been elucidated; in part, this is because LAT mutants behave somewhat differently in the different animal models used to study HSV latency. Because LAT does not become translated into protein, the RNA is believed to be the regulatory factor. It has been proposed that LAT may function as an antisense RNA or a small interfering RNA (siRNA). Future studies will reveal whether these possibilities are involved.

Future Trends

HSV possesses a number of features that will make it a continuing research model in the future. Among the most obvious are: (1) the neurotropism of the virus, along with the ability to generate recombinants containing foreign

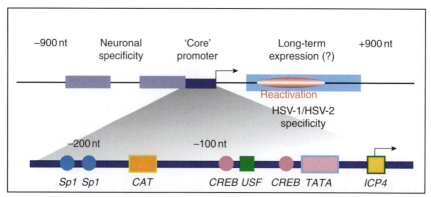

Figure 3 The architecture of the HSV LAT promoter. The transcriptional initiation site (arrow) and binding sites for transcriptional factors are shown in the expansion of the 'core' promoter.

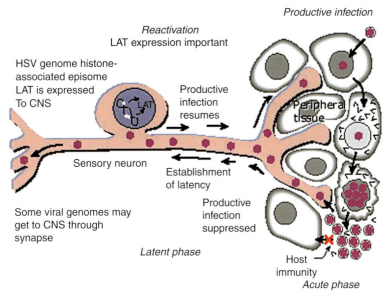

Figure 4 Schematic representation of HSV latent infection and reactivation. CNS, central nervous system.

genes, recommends HSV as a vector for introducing specific genes into neural tissue; (2) the fact that replication of HSV is restricted in terminally differentiated neurons suggests its potential as a therapeutic agent against neural tumors, and several studies have suggested real promise; (3) the abundance of viral genes dedicated to the efficient replication of viral DNA in the productive cycle provides a number of potential targets for antiviral chemotherapy; and (4) the number of multifunctional viral proteins that interact with many different cellular proteins makes HSV a useful model for detailed analysis of specific aspects of eukaryotic gene expression and cellular functions.

See also: Herpes Simplex Viruses: General Features; Herpesviruses: General Features; Herpesviruses: Latency.

Further Reading

Advani SJ, Weichselbaum RR, Whitley RJ, and Roizman B (2002) Friendly fire: Redirecting herpes simplex virus-1 for therapeutic applications. *Clinical Microbiology and Infection* 89: 551–563.

Bloom DC (2004) HSV LAT and neuronal survival. *International Reviews of Immunology* 23: 187–198.

Campadelli-Fiume G, Cocchi F, Menotti L, and Lopez M (2000) The novel receptors that mediate the entry of herpes simplex viruses and animal alphaherpesviruses into cells. *Reviews in Medical Virology* 10: 305–319.

Enquist LW, Husak PJ, Banfield BW, and Smith GA (1998) Infection and spread of alphaherpesviruses in the nervous system. *Advances in Virus Research* 51: 237–347.

Roizman B and Knipe DM (2001) Herpes simplex viruses and their replication. In: Knipe DM and Howley PM (eds.) *Fields Virology* 4th edn., pp. 2399–2459. Philadelphia: Lippincott Williams and Wilkins.

Taylor TJ, Brockman MA, McNamee EE, and Knipe DM (2002) Herpes simplex virus. *Frontiers in Bioscience* 7: 752–764.

Wagner EK and Bloom DC (1997) The experimental investigation of herpes simplex virus latency. *Clinical Microbiology Reviews* 10: 419–443.

Herpesviruses of Birds

S Trapp and N Osterrieder, Cornell University, Ithaca, NY, USA

© 2008 Elsevier Ltd. All rights reserved.

Introduction

A large number of herpesviruses are found in avian host species of several different orders. Similar to their mammalian counterparts, they generally have a relatively narrow host range *in vitro* and *in vivo*. At present, only the gallid herpesviruses (GaHV-1, GaHV-2, and GaHV-3), meleagrid herpesvirus 1 (MeHV-1; also referred to as herpesvirus of turkeys, HVT), and psittacid herpesvirus 1 (PsHV-1) have been classified taxonomically by the International Committee on Taxonomy of Viruses (ICTV) as members of the *Alphaherpesvirinae* subfamily. Genomic sequences have been reported for all of these alphaherpesviruses and deposited in the GenBank sequence database. Avian beta- or gammaherpesviruses have not been identified.

Taxonomically unclassified herpesviruses have been discovered in a wide variety of avian hosts; including the bald eagle (acciptrid herpesvirus 1, AcHV-1), duck (anatid herpesvirus 1, AnHV-1), black stork (ciconiid herpesvirus 1, CiHV-1), pigeon (columbid herpesvirus 1, CoHV-1), falcon (falconid herpesvirus 1, FaHV-1), crane (gruid herpesvirus 1, GrHV-1), bobwhite quail (perdicid herpesvirus 1, PdHV-1), cormorant (phalacrocoracid herpesvirus 1, PhHV-1), black-footed penguin (sphenicid herpesvirus 1, SpHV-1), and owl (strigid herpesvirus 1, StHV-1). Of these viruses, only AnHV-1 causes a known disease, duck virus enteritis (DVE), which accounts for significant economic losses. DVE, also referred to as duck plague, is a highly contagious acute disease of ducks, geese, and swans, and is characterized by hemorrhagic lesions in the blood vessels, gastrointestinal mucosa, and lymphoid tissues. The clinical symptoms of DVE are unspecific, and often mortality is the first observation in infected waterfowl flocks. Morbidity and mortality in domestic ducks may range from 5% to 100%.

Gallid herpesvirus 1 (GaHV-1), frequently referred to as infectious laryngotracheitis virus (ILTV), belongs to the type species of the genus *Iltovirus* and is the causative agent of a highly contagious, acute respiratory tract disease of domestic fowl termed infectious laryngotracheitis (ILT). ILT causes considerable economic losses to the poultry industry worldwide owing to mortality and decreased egg production associated with the disease. Epizootic forms of ILT are characterized clinically by severe respiratory symptoms such as dyspnea and hemorrhagic expectorations, and are generally associated with high mortality (5–70%). Enzootic infections, predominantly those occurring in poultry flocks of developed countries, are associated with mild symptoms (nasal discharge, conjunctivitis, and decreased egg production) or subclinical symptoms. Modified-live vaccines against ILTV infections are available, but their use is commonly restricted to areas where ILT is enzootic. However, most of the current modified-live virus vaccines have residual virulence, and several ILT outbreaks have been attributed to reversions of the ILTV vaccine strains to a virulent phenotype.

ILTV is the only virus allocated to the genus *Iltovirus*. However, recent genomic information indicates clearly an affiliation of PsHV-1 with this genus. PsHV-1 causes Pacheco's disease, a generally fatal disease of psittacine birds characterized by acute onset and massive necrotizing lesions of the crop, intestines, pancreas, and liver. Parrot species from multiple geographic regions are highly susceptible to Pacheco's disease. However, captive parrots such as macaws, Amazon parrots, and cockatoos account for the majority of clinical cases of Pacheco's disease; hence, the disease is of great concern to pet traders and exotic bird breeders worldwide. An association of subclinical PsHV-1 infections with mucosal papillomatosis in neotropical (Central and South American) parrots has been reported, but causality has not yet been demonstrated. Recently, genomic DNA of a novel psittacine alphaherpesvirus has been amplified from mucosal and cutaneous papillomas of African grey parrots. This virus is related to, but phylogenetically distinct from, PsHV-1, and has been designated psittacid herpesvirus 2 (PsHV-2).

Gallid herpesvirus 2 (GaHV-2) or Marek's disease virus (MDV) is the etiologic agent of Marek's disease (MD), an economically important, neoplastic, and neuropathic disease of chickens. MDV is highly infectious and routinely causes >90% morbidity and mortality in susceptible, unvaccinated animal populations. MD manifests as various clinical syndromes of varying severity. Of these, T-cell lymphoproliferative syndromes, including fowl paralysis (classic MD) and fatal MD lymphoma (acute MD), are most frequently associated with the disease. However, infections of chickens with the most recent clade of MDV strains are characterized by massive inflammatory brain lesions that are clinically manifested as transient paralysis and/or peracute death. Owing to its biological features, namely its lymphotropism and oncogenic potential, MDV was long thought to be related to Epstein–Barr virus, a lymphomagenic gammaherpesvirus. However, based on its genomic organization, MDV was reclassified as an alphaherpesvirus; thus, MDV is genetically more closely related to herpes simplex virus 1 (HSV-1) and varicella-zoster virus (VZV) (**Figure 1**). MDV belongs to the type species of the genus *Mardivirus*, into which two other closely related but distinct species have been grouped, represented by gallid herpesvirus 3 (GaHV-3) and MeHV-1 (HVT). Although the three viruses are closely related, only MDV causes MD while GaHV-3 and HVT are nonpathogenic. Therefore, the old nomenclature referring to GaHV-3 and HVT as MDV (serotypes) 2 and 3 is misleading and should be avoided.

Control of MDV-caused syndromes by vaccination was started in the late 1960s using an attenuated MDV strain (HPRS-16att) or the nonpathogenic HVT. Because MDV field strains are evidently evolving toward greater virulence in the face of vaccination, combination vaccines consisting of HVT and GaHV-3 or an attenuated MDV strain (CVI988-Rispens) are currently used to immunize chickens against the latest clade of MDV strains. Based on their ability to cause breaks of vaccinal immunity, MDV strains are subdivided into four different pathotypes: mildly virulent (m), virulent (v), very virulent (vv), and very virulent plus (vv+).

History

ILT was first described in 1925 as tracheolaryngitis, and shortly thereafter its viral etiology was demonstrated. The term infectious laryngotracheitis was adopted in 1931

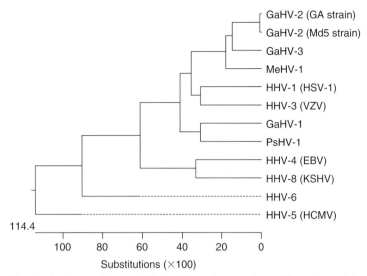

Figure 1 Genetic relationship of avian herpesviruses. A ClustalV-based comparison of the amino acid sequences of the DNA polymerase proteins of the viruses is shown. Human herpesviruses 5 (human cytomegalovirus) and 6, representatives of the *Betaherpesvirinae*, were chosen as outgroups. It is evident that the two representatives of the *Iltovirus* genus, GaHV-1 (ILTV) and PsHV-1 are closely related, as are the three members of the genus *Mardivirus*, GaHV-2 (MDV), GaHV-3, and MeHV-1 (HVT). These results also indicate independent evolution of *Alphaherpesvirinae* (GaHV-1 and GaHV-2) within one bird species (*Gallus gallus*). EBV, Epstein–Barr virus; HCMV, human cytomegalovirus; HHV, human herpesvirus; HSV-1, herpes simplex virus type 1; KSHV, Kaposi's sarcoma-associated herpesvirus; VZV, varicella-zoster virus.

by the Special Committee on Poultry Diseases of the American Veterinary Medical Association. ILT was the first major viral avian disease against which an effective vaccine was developed, being achieved in the 1930s by Hudson and Beaudette at Cornell University. Pacheco's disease was first recognized by Pacheco and Bier in the late 1920s in Brazil, but was not seen again until the 1970s, when outbreaks of the disease occurred in wild-caught neotropical parrots and a psittacine herpesvirus was identified as the etiologic agent.

The first account of MD dates from 1907, when József Marek (after whom the disease was named in 1960), a preeminent clinician and pathologist at the Royal Hungarian Veterinary School in Budapest, reported a generalized neuritis interstitialis or polyneuritis in four adult cockerels, which were affected by paralysis of the limbs and wings. Marek's detailed histological examination revealed that the sciatic nerve and areas of the spinal cord of the affected birds were infiltrated by mononuclear cells, an observation that is still made today after infection with most MDV strains. In the late 1920s, Pappenheimer and colleagues at Rockefeller University recognized the lymphoproliferative character of MD, and proposed that polyneuritis and visceral lymphoma were manifestations of the same disease. The identification of a highly cell-associated herpesvirus as the etiological agent of MD in the late 1960s by Churchill and colleagues led to the development of vaccines that achieved unparalleled success in preventing the disease, and provided the first example of immune prophylaxis against a cancer induced by an oncogenic virus.

Host Range and Virus Propagation

Like most avian herpesviruses, ILTV exhibits a very narrow host range. The chicken is the primary natural host, but pheasants and peafowl have also been reported to be susceptible. Turkeys have been shown to be susceptible to infection under experimental conditions, but other avian species are probably refractory. Embryonated chicken eggs are also susceptible, and can be used to propagate the virus. Chorioallantoic membrane (CAM) plaques resulting from necrosis and proliferative tissue reactions can be observed as early as 2 days post inoculation (pi). The virus can also be propagated in a variety of primary avian cell cultures, including chicken embryo liver cells (CELCs), chicken embryo kidney cells (CEKCs), and chicken kidney cells (CKCs). In addition, a permanent chicken cell line (LMH) derived from a chemically induced liver tumor was shown to permit replication of cell culture-adapted ILTV. Viral cytopathology *in vitro*, characterized by the formation of syncytia and nuclear inclusion bodies, can be observed as early as 4–6 h pi.

PsHV-1 infects a broad spectrum of psittacine species. In addition, it has been isolated from a superb starling, and some evidence suggests that other passerine species, such as the common cardinal, zebra finch, and canary, are also susceptible to PsHV-1 infections. Embryonated chicken eggs and primary chicken embryo cells (CECs) derived from 10- to 12-day-old chicken embryos are used for virus propagation *in vitro*.

MDV infections and MD-like symptoms have been observed in quail, turkeys, and pheasants, but the chicken is considered the natural host of MDV. Primary chicken cells (CKCs and CECs) or duck embryo fibroblasts (DEFs) are used routinely to propagate the virus *in vitro*. However, for primary isolation and propagation of low-passage MDV isolates, the use of CKC or DEF is recommended to retain wild-type properties of the virus. Permanent avian cell lines such as OU2, DF-1, and QM7 allow only low level or abortive MDV replication. On permissive cells, MDV-induced plaques (≤1 mm in diameter) develop very slowly (in 5–14 days pi) on primary isolation and 3–7 days pi after cell culture adaptation. In addition, 1-day-old chicks can be used for primary isolation and propagation of MDV *in vivo*.

Morphology

All herpesvirus particles share a complex and characteristic multilayered structure in which the nucleocapsid containing the viral DNA is separated from the envelope by a proteinaceous layer called the tegument. While both MDV and ILTV capsids have a diameter of approximately 125 nm, the incorporation of highly variable amounts of tegument into ILTV virions results in diameters of enveloped ILTV particles that vary between 195 and 350 nm. In contrast, enveloped MDV particles have a diameter of 150–160 nm. The morphologies of GaHV-3 and HVT closely resemble that of MDV, but, in thin sections, HVT capsids commonly show a characteristic cross-shaped appearance. The morphology of PsHV-1 has not been studied in detail, but typical herpesvirus particles have been visualized.

Genetics

The genomes of herpesviruses are linear, double-stranded DNA molecules that range in size from approximately 120 to 300 kbp. A total of six different genome organizations, referred to as classes A through F, are distinguished in the *Herpesviridae*, and only class D and E genomes are found in the *Alphaherpesvirinae*. ILTV and PsHV-1 have class D genomes and the three members of the *Mardivirus* genus (MDV, GaHV-3, and HVT) have class E genomes. Class D genomes consist of two unique sequence regions (unique-long, U_L; unique-short, U_S) with inverted repeats (internal repeat, IR; terminal repeat, TR) flanking U_S. Class E genomes consist of U_L and U_S, each bracketed by inverted internal (IR_L, IR_S) and terminal repeats (TR_L, TR_S).

The ILTV genome sequence, as assembled from 14 overlapping genome fragments from different virus strains, is approximately 147 kbp in length and has a G+C content of 48.2%. The PsHV-1 genome is 163 kbp in length and has a nucleotide composition of 61% G+C. ILTV and PsHV-1 contain 76 and 73 protein-coding open reading frames (ORFs), respectively, and the majority of these ORFs exhibit significant homologies to genes of the prototypic alphaherpesvirus, HSV-1. However, ILTV and PsHV-1 exhibit several unusual structural features, for example, a large internal inversion of the region within U_L containing genes UL22 to UL44 (a suffix is sometimes used in gene name, e.g., U_L 22). Also, U_L in both viruses contains a unique cluster of five ORFs (designated A to E) that do not share significant sequence similarity with known viral or cellular genes. The function of these ORFs is unknown, but they have been shown to be dispensable for ILTV replication *in vitro*. Another striking feature is the localization of the UL47 gene, which encodes a tegument protein and is usually localized in the U_L region of alphaherpesvirus genomes. UL47 is absent from the corresponding position in the ILTV and PsHV-1 genomes, but both viruses are still predicted to encode a homolog of UL47 (ILTV: UL47; PsHV-1: sORF1) in U_S.

One major structural difference between the closely related viruses ILTV and PsHV-1 can be found in the inverted repeat regions of the genomes. The ILTV repeats are considerably (18.5%) shorter than those of PsHV-1, and each harbors three genes: ICP4, US10, and sORF4/3. In contrast, ICP4 is the only gene located in the PsHV-1 repeat, and the US10 and sORF4/3 genes are located in U_S. Moreover, PsHV-1 U_L contains a homolog of UL16, an ORF that is conserved throughout the herpesvirus subfamilies but is absent from the ILTV genome, having presumably been lost during evolution. Conversely, ILTV U_L contains the UL48 gene and ILTV-specific UL0 gene, neither of which is found in the PsHV-1 genome.

The complete genomic sequences of representatives of all three species in the genus *Mardivirus* have been determined. The gene contents and linear arrangements of the genomes of MDV, GaHV-3, and HVT are similar in general, but vary considerably with regard to G+C content and size. Whereas the ~180 kbp MDV genome has a G+C content of 44.1%, the ~165 kbp GaHV-3 genome has a nucleotide composition of 53.6% G+C; the ~160 kbp HVT genome has a G+C content in between these two values (47.2%). A total of 103 (MDV), 102 (GaHV-3), and 99 (HVT) genes have been clearly identified in the mardivirus genomes. The majority of these genes are homologous to genes encoded by other alphaherpesviruses, but genus- and species-specific genes are also present.

Most of the differences between the three mardiviruses concern the TR_L and IR_L regions. In pathogenic MDV, they harbor a unique cluster of three species-specific genes, which are designated vTR, vIL-8, and meq. vTR, which encodes a viral homolog of telomerase RNA (vTR), exhibits 88% sequence identity to chicken telomerase RNA, and was presumably pirated from the chicken genome. The vIL-8 ORF encodes a CXC chemokine (viral interleukin-8, vIL-8) of 18–20 kDa in size, which functions as a chemoattractant for chicken mononuclear cells. The meq gene encodes the 339-residue oncoprotein Meq (MDV EcoRI Q), a transcriptional regulator that is characterized by an N-terminal bZIP domain and a proline-rich C-terminal transactivation domain.

In addition, a genus-specific ORF (pp38), which is located in the junctions between the U_L and the adjacent repeat regions of all three mardivirus genomes, encodes a pp38 phosphoprotein of variable size dependent on the virus and the phosphorylation status of the protein. Another mardivirus-specific ORF, vLIP, is located in U_L in close proximity to pp38, and encodes a 120 kDa N-glycosylated protein, a portion of which shows significant similarity to the α/β hydrolase fold of pancreatic lipases and was thus termed viral lipase (vLIP). The HVT-specific vN-13 ORF encodes a 19 kDa protein (vN-13) that exhibits 63.7% amino acid sequence identity to cellular N-13, an anti-apoptotic member of the Bcl-2 family. While Bcl-2-like sequences have been identified in several gammaherpesviruses, HVT is the only alphaherpesvirus that encodes a Bcl-2 homolog.

One striking feature of all three mardiviruses is the presence of heterogenic telomeric repeat sequences at the genome termini. Interestingly, similar telomeric repeat sequences have been identified in the genomes of two betaherpesviruses, human herpesviruses 6 and 7 (HHV-6 and HHV-7), and one gammaherpesvirus, equid herpesvirus 2 (EHV-2). The structural or functional significance of these telomeric repeats is unknown, but they might be responsible for genome integration and maintenance of virus genomes in latently infected (tumor) cells.

Pathobiology

Pathogenesis, Clinical Features, and Pathology of Iltovirus Infections

ILTV is readily transmitted from infected to susceptible chickens, and virus shedding and spread mainly occur via the respiratory and ocular routes. Early cytolytic replication of ILTV in the epithelia of the upper respiratory tract results in syncytia formation and subsequent desquamation. Following the acute phase of infection, which lasts for approximately 6–8 days, ILTV establishes latency in the central nervous system (CNS), in particular in trigeminal ganglia. No clear evidence exists for a viremic phase in the course of lytic infection, latency, or reactivation. Sporadic reactivations from the latent state are usually asymptomatic, but generally lead to productive replication in the upper respiratory tract and virus shedding, which can result in infection of susceptible contact animals. The severity of clinical symptoms of ILT depends on the virulence of a particular ILTV strain or isolate, and mortality rates range from 0% to 70%. Severe epizootic forms of ILT are characterized clinically by marked dyspnea and hemorrhagic expectorations. Clinical signs of the milder, enzootic forms include nasal discharge, conjunctivitis, sinusitis, gasping, and decreased egg production. The incubation period of ILT ranges from 6 to 12 days. Pathomorphological alterations may be found in the conjunctiva and throughout the respiratory tract, but are most consistently detected in the larynx and trachea. Typical gross lesions include mucoid to hemorrhagic tracheitis, conjunctivitis, infraorbital sinusitis, and necrotizing bronchitis. However, in the case of the mild enzootic forms of ILT, conjunctivitis and sinusitis are often the only detectable gross lesions. Microscopic lesions include epithelial syncytia, desquamation, and submucosal edema. Eosinophilic intranuclear inclusion bodies in epithelial cells are detectable from day 1 to 5 pi, but disappear as desquamation of infected epithelial cells progresses.

Similar to ILTV, PsHV-1 is shed from the upper respiratory tract but also in the feces of infected parrots. Transmission of the virus occurs by direct contact between infected and susceptible animals or indirectly by contact with contaminated fomites or environmental contamination. The target sites for primary lytic viral replication and the establishment of latency are unknown. The systemic character of Pacheco's disease strongly argues for a viremic phase of infection during which the virus disseminates into multiple organ sites. Anorexia, depression, diarrhea, nasal discharge, and ataxia are among the most common clinical signs of the disease, but a variety of symptoms may be observed, largely depending on which organ system is affected. However, the majority of diseased animals die peracutely, that is, before the onset of clinical symptoms. Latently infected parrots that survive the disease play an important epidemiological role as asymptomatic virus carriers and shedders. The incubation time of Pacheco's disease ranges from 3 to 14 days. Gross pathological lesions are unspecific and may be found in a multitude of organ systems, including the respiratory and gastrointestinal tract, liver, and CNS. Microscopically, Pacheco's disease is characterized by a necrotizing hepatitis with minimal associated inflammation and the presence of intranuclear inclusion bodies (Cowdry type A).

Pathogenesis, Clinical Features, and Pathology of MD

MDV shed from the skin and associated with feathers and dander is highly infectious and can persist for extended periods in the environment. Transmission of the virus occurs by inhalation, either through direct contact or through virus present in dust and dander. In the current model of MDV pathogenesis, phagocytic cells in the lower respiratory tract become infected either directly or after an initial round of replication in epithelial cells. Within 24 h of uptake, the virus is detectable in the primary and secondary lymphoid organs such as thymus, bursa of Fabricius, and spleen. Following primary productive replication, which takes place in B-lymphocytes, the virus infects activated CD4$^+$ lymphyocytes, and, rarely, CD4$^-$CD8$^-$ T-cells or CD8$^+$ T-cells. Infected CD4$^+$ T-cells serve as the target for the establishment of MDV latency and are also the means of virus dissemination within an infected animal. Besides epithelial layers in visceral organs, MDV enters the feather follicle epithelium, where cell-free infectious virus is assembled and released into the environment.

In latently infected T-cells, viral DNA is commonly integrated into the cellular genome. MDV is one of the few herpesviruses that achieve genome maintenance during the quiescent stage of infection through this mechanism, the details of which are entirely enigmatic. It is well known that latency is a prerequisite for oncogenic transformation by MDV, but only small subsets of latently infected T-cells become ultimately transformed and proliferate to generate tumors. Depending on the virulence of the virus strain and the susceptibility of the chicken population, mortality rates range from 0% to >90%. Clinical and pathological signs of MD vary according to the specific syndrome. The leading symptom of the classical form of the disease (fowl paralysis) is flaccid or spastic paralysis of the limbs and/or wings caused by lymphoproliferative peripheral nerve lesions. Chickens affected by MD lymphoma (acute MD) generally exhibit unspecific symptoms such as weight loss, anorexia, depression, and ruffled plumage, but (transient) paralysis may also be seen. Unlike the classic form of paralysis, transient paralysis is caused by inflammatory lesions of the CNS and peripheral nervous system. Most v and vv MDV strains induce MD lymphoma and/or (transient) paralysis in susceptible chicken lines. Recent outbreaks of MD caused by vv+ MDV strains are characterized clinically by transient paralysis, a massive rash affecting mainly the extremities, and/or peracute death.

Enlarged peripheral nerves and lymphomatous lesions are the most frequently observed gross pathological findings in MD. Lymphomatous lesions can develop as early as 14 days after infection and generally manifest as diffuse infiltrations and/or solid lymphomas, which affect a variety of organs, including the viscera (heart, liver, spleen, kidney, gonads, adrenal gland, etc.), skeletal muscle, and skin. Two main peripheral nerve pathologies are described: neoplastic proliferation that sometimes involves secondary demyelination (type A) and primary inflammatory cell-mediated demyelination (type B). MD lymphomas are cytologically complex and essentially comprised of lymphocytes and macrophages. MD tumors mainly consist of T-cells, but only a minority of these are transformed, the majority representing immune T-cells that try to contain the neoplasm.

Tumors induced by avian retroviruses represent a differential diagnosis for MDV-induced tumors, but onset of disease is delayed and nerve lesions are generally absent. Histologically, intranuclear inclusions are always absent from retrovirus-infected cells. A hallmark of MDV-transformed cells is expression of the viral oncoprotein Meq, and upregulation of the Hodgkin's antigen, CD30, has also been reported.

Prevention and Control

Control of Iltovirus Infections

ILTV-infected chickens produce peak virus-neutralizing antibody titers around 21 days post infection or vaccination. ILTV-neutralizing antibodies decline over the following months, but remain detectable for years. Although the detection of ILTV-specific antibodies by virus neutralization test is an important means of serological diagnosis, neutralizing antibody titers do not reflect the immune status of infected chickens or vaccinees, as immunity against ILT largely rests on local and cell-mediated immune response in the upper respiratory tract. Maternal antibodies to ILTV are transmitted to the chicken embryo via the egg yolk, but do not confer passive immunity against ILT or interfere with vaccination. For many years, modified-live virus (MLV) vaccines have been used to immunize chickens against ILT. They are administered via the intraocular route and for rapid mass vaccination via the drinking water or aerosolization. Immunization of chickens with attenuated ILTV vaccine strains results in latently infected carrier animals and spread of vaccine virus to nonvaccinates. However, spread to nonvaccinates results in consecutive bird-to-bird passages during which the vaccine virus may revert to a virulent phenotype. Therefore, the use of ILTV MLV vaccines has commonly been restricted to areas where infectious laryngotracheitis is enzootic. Experimental inactivated whole-virus vaccines and subunit preparations containing affinity-purified immunogenic ILTV glycoproteins were tested successfully as an immune prophylactic alternative to MLV vaccines. However, due to high costs of production and individual administration, the practical use of these vaccines in the field is debatable. Thus, the generation of MLV vaccines based on genetically engineered ILTV strains appears to be a more promising approach to the development of safe and efficacious ILTV vaccination protocols for the future.

Immune prophylaxis against Pacheco's disease is available in the form of an inactivated vaccine that is administered subcutaneously or intramuscularly. However, the vaccine is only protective against certain PsHV-1 serotypes, and its administration has been associated with adverse side effects, including granuloma formation and paralysis. Treatment with antiherpetic nucleoside analogues such as acyclovir can be used for pro- and metaphylactic measures in aviaries affected by Pacheco's disease. Strict quarantine protocols together with diagnostic screenings and isolation of birds that have been exposed to PsHV-1 are the most effective control measures for parrot breeding facilities and pet stores.

Control of MD

MD is controlled worldwide by vaccination of 18-day-old embryos *in ovo* or 1-day-old chickens by subcutaneous/intramuscular injection into the neck. Vaccine practices, however, vary between countries and, for example, broilers usually remain unvaccinated in Europe whereas all chickens are vaccinated in the USA. These different vaccine regimens exist largely because of differences in the production practices of broilers, which live longer in the Americas to produce carcasses with higher body mass as preferred by the consumer. Prior to the introduction of vaccines against MD in the late 1960s and early 1970s, the disease had an economically devastating impact on the poultry industry, and mortality rates reached as high as 30–60%. All MD vaccines used today are MLV preparations, and some of them have been used for almost 40 years. The first vaccines were based on HVT isolate FC126 (used mainly in the USA) and on HPRS16att (Europe), a formerly virulent strain that was rendered nonpathogenic by serial passage in cultured cells. Shortly after the introduction of vaccination in Europe, a naturally avirulent MDV strain, CVI988, was isolated, which has formed the basis of the MDV-based vaccines that are currently in use worldwide.

It became evident early on that MD vaccination is able to reduce and delay tumor development, but does not induce sterile immunity, which leads to a situation where constant evolutionary pressure is on field viruses to adapt to, and ultimately escape, vaccination. Consistent with this hypothesis, new MDV strains have been isolated that are able to break vaccine protection provided by HVT. The appearance of new, more virulent strains led to the introduction of so-called bivalent vaccines consisting of HVT and GaHV-3 strain SB-1. The latter was isolated from chickens, is completely avirulent, and is closely related genetically and antigenically to HVT and MDV. The bivalent vaccine was able to protect against more virulent (vv) viruses, but in the early 1990s even more virulent (vv+) MDV strains began to emerge. Most recently, repeated vaccinations using various combinations of vaccines (HVT + GaHV-3, HVT + MDV, MDV alone) are employed mainly in the USA to keep MD in check.

With the threat of new, even more virulent strains breaking through the current vaccination protocols, the development of novel vaccines is a huge challenge to the scientific community. Next to the efficacy of the vaccine preparations, the production process of vaccines in primary chicken embryo cultures is a major problem. The cost of producing these vaccines is high for several reasons, most notably the maintenance of pathogen-free chicken flocks, and approaches to find alternatives have been undertaken. The availability of infectious clones of various MDV strains and of HVT holds promise for a novel generation of rationally designed and efficacious vaccines.

See also: Herpesviruses of Horses; Herpesviruses: General Features; Retroviruses of Birds; Human Herpesviruses 6 and 7.

Further Reading

Davison F and Nair V (eds.) (2004) *Marek's Disease: An Evolving Problem.* London: Academic Press.

Guy JS and Bagust TJ (2003) Laryngotracheitis. In: Saif YM, Barnes HJ, Fadly A, Glisson JR, McDougald LR, and Swayne DE (eds.) *Diseases of Poultry,* 11th edn., ch. 5, pp. 121–134. Ames, IA: Iowa State.

Kaleta EF (1990) Herpesviruses of free-living and pet birds. In: Purchase HG, Arp LH, Domermuth CH, and Pearson JE (eds.) *A Laboratory Manual for the Isolation and Identification of Avian Pathogens,* pp. 97–102. Athens, GA: American Association of Avian Pathologists.

Osterrieder N, Kamil JP, Schumacher D, Tischer BK, and Trapp S (2006) Marek's disease virus: From miasma to model. *Nature Reviews Microbiology* 4: 283–294.

Panigraphy B and Grumbles LC (1984) Pacheco's disease in psittacine birds. *Avian Diseases* 28: 808–812.

Thureen DR and Keeler CL, Jr. (2006) Psittacid herpesvirus 1 and infectious laryngotracheitis virus: Comparative genome sequence analysis of two avian alphaherpesviruses. *Journal of Virology* 80: 7863–7872.

Herpesviruses of Horses

D J O'Callaghan, Louisiana State University Health Sciences Center, Shreveport, LA, USA
N Osterrieder, Cornell University, Ithaca, NY, USA

© 2008 Elsevier Ltd. All rights reserved.

History

The first disease attributed to an equine herpesvirus (EHV), the agent that is now referred to as equid herpesvirus 1 (EHV-1, equine abortion virus; equine rhinopneumonitis virus), was documented at the University of Kentucky Agricultural Experiment Station in Lexington, KY. EHV-1 was first shown to be associated with spontaneous abortions in pregnant mares in 1932. In 1941, equine abortions were found to be associated with mild respiratory disease with clinical signs similar to those associated with equine influenza virus infections. EHV-1 was also shown to be the etiological agent of epizootic respiratory disease in young horses. Based on these investigations, the disease was termed viral rhinopneumonitis, and the agent was called equine rhinopneumonitis virus. In the 1980s, however, the two disease manifestations, viral abortion and viral rhinopneumonitis, were shown to be caused by two closely related but clearly distinct viruses now classified as EHV-1 (equine abortion virus) and EHV-4 (equine rhinopneumonitis virus).

Equid herpesvirus 2 (EHV-2) was first isolated from horses in 1963. The cytopathology caused by this virus closely resembled that of cytomegalovirus infections, which were first described in 1921 in humans. EHV-2 is a ubiquitous, slow-growing virus that infects horses at a very young age (<2 years) and establishes a lifelong chronic infection such that the horse becomes a continuous shedder of the virus. To date, no major disease has been attributed to EHV-2. However, an association of EHV-2 with chronic throat infections (the 'lumpy bumpies') or recurrent eye disease has been established by some investigators, although the causal relationships have not been proved unequivocally. Also, EHV-2 may be a cofactor in EHV-1 and/or EHV-4 infections in that it is able to modulate EHV-1 and EHV-4 replication by immunosuppression, causing general malaise, or by modulation of EHV-1 gene expression through EHV-2-specific transcriptional transactivators.

Equid herpesvirus 3 (EHV-3; equine coital exanthema virus, ECE virus) was first isolated independently in 1968 in Canada, Australia, and the USA. EHV-3 is the etiological agent of equine coital exanthema, a generally mild genital infection of mares and stallions that is transmitted venereally and has been largely eradicated.

Equid herpesvirus 4 (EHV-4) is associated mainly with respiratory disease and has also been associated occasionally with equine abortions. On the other hand, EHV-1 is associated primarily with equine abortions, but frequently causes respiratory disease in young animals, highly fatal neurological disease, fulminating neonatal pneumonitis, and, very rarely, an exanthematous condition involving the external genitalia of the mare.

Equid herpesvirus 5 (EHV-5) is closely related to EHV-2 and no information on possible disease(s) caused by this virus is currently available.

Equid herpesvirus 9 (EHV-9), previously referred to as gazelle herpesvirus 1, was described very recently. It has been shown to be closely related to EHV-1 (and asinine herpesvirus 1) and exhibits a broad host range *in vivo*. While clinical signs in horses seem to be relatively mild, EHV-9 can cause lethal encephalitides in other animals, such as gazelles and goats.

Taxonomy and Classification

Eight of the nine EHVs have been classified as species in the family *Herpesviridae*. The species names are *Equid herpesvirus 1* through *Equid herpesvirus 9*, with the exception of *Equid herpesvirus 6*, which is not yet in existence as EHV-6 has not yet been fully assigned to a species. EHV-6, EHV-7, and EHV-8 are donkey herpesviruses that are also referred to as asinine herpesviruses 1, 2, and 3, respectively, and will not be discussed in detail in this article. The morphology of all six members is typical of the herpesviruses in that they are enveloped, contain an icosadeltahedral capsid, and have a proteinaceous coat, the so-called tegument, which surrounds the nucleocapsid. EHVs are composed of six distinct species: (1) EHV-1 is the major equine pathogen causing fetal abortions, respiratory illness, and neurological disease; (2) EHV-2 (and perhaps EHV-5) establish mainly asymptomatic, long-term persistent infections; (3) EHV-3 is the causative agent of mild progenital exanthema; (4) EHV-4 is a major respiratory pathogen that differs significantly from EHV-1 at the DNA level and is associated occasionally with equine abortions; and (5) EHV-9 can cause mostly subclinical encephalitides in horses, while infections of other hosts are often lethal. In the latter respect, EHV-9 possesses biological properties very akin to an alphaherpesvirus of pigs, pseudorabies virus (PRV). EHV-1, EHV-3, EHV-4, and EHV-9 are members of subfamily *Alphaherpesvirinae*, genus *Varicellovirus*. EHV-2 and EHV-5 are members of subfamily *Gammaherpesvirinae*, genus *Rhadinovirus*.

Geographic and Seasonal Distribution

EHV-1 and EHV-4 are distributed worldwide, and infections can occur year-round. Over the past 30 years in the USA, EHV-1 'abortion storms' have occurred in many areas. Major outbreaks have also been reported in Australia and England. Due to the nature of the disease manifestation, viral abortions caused by EHV-1 exhibit a seasonal cumulation in the spring on broodmare farms, whereas EHV-1 and EHV-4 infections of the upper respiratory tract (rhinopneumonitis), as well as the neurological form of EHV-1 disease, are observed mainly on race tracks and after crowding of large numbers of animals.

EHV-2 has also been isolated in many countries, including England, Switzerland, Germany, the USA, and South Africa. The existence of the closely related EHV-5 was described for England and Australia. More thorough studies, however, have led to the assumption that both EHV-2 and EHV-5 are distributed worldwide.

To date, EHV-3 has been isolated in five countries: Germany, USA, Australia, Canada, and England. EHV-9 so far has only been described in captive gazelles in Japan, although zoo animals in other parts of the world have been shown to harbor antibodies against the agent.

Host Range and Virus Propagation

Although the horse is the natural host of the EHVs, a variety of animals and tissue culture systems can be used to propagate the viruses. Regarding the major equine pathogen, EHV-1, experimental animals include Syrian hamsters and baby hamsters, chick embryos, baby mice and adult mice, and kittens. Primary tissue culture systems used to propagate EHV-1 include cells from a variety of equine tissues such as fetal lung, dermis, spleen, and kidney, as well as cells from domestic cats, dogs, hamsters, rabbits, mice, sheep, and swine. In the laboratory, permanent tissue culture systems commonly used to cultivate EHV-1 include primate HeLa, Vero, and CV-1 cells, rabbit kidney (RK) cells, mouse L–M cells, and equine NBL-6 and Edmin337 cells.

The host range for EHV-2, EHV-3, EHV-4, and EHV-5 is more restrictive than that for EHV-1. Except for RK cells and primary cat cells, EHV-2 and EHV-5 growth appears to be restricted to cells of equine origin. The host range for EHV-3 and EHV-4 is limited to cells of equine origin, although Vero cell culture-adapted EHV-4 strains have been described. EHV-9 is a notable exception of the rule of narrow host range of EHVs *in vivo*. As mentioned above, this virus was first isolated from an outbreak of fatal encephalitis in zoo gazelles and was later characterized in detail and shown to be most closely related to EHV-1. EHV-9 has a wide host range *in vivo* and *in vitro*.

Genetics

All six EHVs contain a linear, double-stranded DNA genome ranging between 140 and 184 kbp. The reported sizes of the genome are: EHV-1, 150 kbp; EHV-2, 184 kbp; EHV-3, 144 kbp; EHV-4, 145 kbp; and EHV-5, 179 kbp. The size of the EHV-9 genome has not yet been determined, but is likely to be close to that of EHV-1. The genomes of EHV-1, EHV-3, EHV-4, and probably EHV-9 exist in two isomeric forms, since the short region (S) can invert relative to the fixed orientation of the unique long region (U_L). The S region is composed of a central segment of unique sequences (U_S) bracketed by a pair of inverted sequences (IRs). In the case of the Kentucky A (KyA) tissue culture-adapted strain of EHV-1, each IR is 12.8 kbp. In contrast, the genomes of EHV-2 and EHV-5 exist as one isomer and are comprised of a large (149 kbp) central segment of unique sequences that is bracketed by a pair of direct repeat sequences. Each of the terminal direct repeat segments is 18 kbp, and the total genome size is 179–184 kbp. Other characteristics of the genomes are shown in **Table 1**.

There are varying degrees of homology at the DNA level among the six EHVs. The sequences shared by EHV-1, EHV-3, EHV-4, and EHV-9 appear to be arranged collinearly and are dispersed throughout the genome. EHV-1 and EHV-4 exhibit 55–84% identity at the DNA level and are antigenically very closely related, and antibodies can cross-neutralize. Levels of identity between EHV-1 and EHV-9 are even higher than those between EHV-1 and EHV-4. EHV-1 and EHV-2 show negligible identity, as do EHV-2 and EHV-3. EHV-1 and EHV-3 exhibit approximately 10% identity at the DNA level. Lastly, EHV-2 and EHV-5 show approximately 60% identity at both the DNA and protein levels.

The cloning of the genomes of a number of EHV-1 strains, among them the attenuated KyA and pathogenic RacL11 strains, as bacterial artificial chromosomes (BACs) has been achieved. These clones are infectious and provide a basis for rapid and efficient mutagenesis of the EHV-1 genome in prokaryotic cells. A variety of recombinant viruses that lack nonessential genes or portions of an essential gene have been generated and used in experiments to elucidate the functions of these genes in virus replication and/or pathogenesis.

Genome Structure

The entire genomes of EHV-1 strains, Ab4 and V592, EHV-4 strain NS80567, and EHV-2 strain 86/67 have been sequenced. The EHV-1 strain Ab4 genome is 150 224 bp in size, while that of EHV-4 is 145 597 bp. Both genomes contain 76 open reading frames (ORFs) potentially encoding proteins, with 4 duplicated in

Table 1 Properties of equine herpesvirus genomes

Member	Subfamily	Isomers	S value	G+C content	Clinical manifestations
EHV-1	Alpha	2	49–55	56%	Abortion, respiratory infection, paralysis
EHV-2	Gamma	1	61.8	57%	Chronic throat infection; perhaps recurrent eye infection
EHV-3	Alpha	2	55.4	66%	Equine coital exanthema
EHV-4	Alpha	2	ND[a]	50%	Respiratory infections
EHV-5	Gamma	1	ND	ND	Not known; possibly pneumonia
EHV-9	Alpha	2	ND	ND	Mild infections in horses; encephalitis in other animals (e.g., gazelles and goats)

[a]ND, not determined.

repeated regions in EHV-1 and -3 in EHV-4, giving a total of 80 ORFs in EHV-1 and -79 in EHV-4. The 63 ORFs in the U_L region are arranged colinearly between EHV-1 and EHV-4 and with those of herpes simplex virus (HSV) and varicella-zoster virus (VZV). Several genes mapping within IR and U_S of the S region differ in arrangement from those of other alphaherpesviruses. In addition, EHV-1 and EHV-4 contain a limited number of unique genes that are present in neither HSV-1 nor VZV and which might represent the viruses' gene repertoire involved in determining host specificity.

Each identical IR of the S region of the KyA tissue culture-adapted strain of EHV-1 is composed of 12 777 nucleotides. Six genes and the origin of DNA replication (ORI) have been mapped to IR: (1) the IR1 gene (gene 64) is an immediate-early (IE) gene encoding a spliced 6.0 kbp mRNA and a major phosphoprotein (1487 amino acid residues) with an apparent molecular mass of 203 kDa; (2) the IR2 gene is an early gene that is embedded within the IR1 gene, and its 4.4 kbp mRNA encodes a 130 kDa polypeptide (1165 amino acid residues) – the protein product of the IR2 gene represents a truncated form of the IE polypeptide; (3) the IR3 gene is a delayed-early gene encoding a 0.9 kbp mRNA that overlaps the IE promoter region on the opposite DNA strand; (4) the EHV-1 ORI maps downstream of IR3 and exhibits 60% identity to the corresponding ORI of HSV-1 and HSV-2; (5) the IR4 gene (gene 65) is a homolog of the ICP22 gene of HSV-1 and is differentially regulated to encode a 1.4 kbp early mRNA and a 1.7 kbp late mRNA; (6) the IR5 gene (gene 66) is a homolog of the US10 gene of HSV-1; and (7) the EHV-1 unique IR6 gene (gene 67) is a 'very early' gene encoding a 1.2 kbp mRNA and a 31/33 kDa phosphoprotein that has been shown to be a capsid constituent and a major determinant of virulence in some EHV-1 strains.

Additional EHV-1 genes in the S region map in the U_S segment and include homologs of the HSV-1 US2 gene, the protein kinase gene, the US9 gene, and genes that encode the glycoproteins (g) gG, gD, gI, and gE. In addition, a unique gene (EUS4 or gene 71) was mapped in U_S. This gene encodes a highly O-glycosylated protein referred to as gp2. Some size variations of gp2 were documented for EHV-1 strains KyA, Ab4, RacL11, and an EUS4 gp2 null mutant was apathogenic in a murine model of EHV-1 infection. Replacement of the truncated EUS4 gene of the apathogenic KyA strain with the EUS4 gene of the pathogenic RacL11 strain, which encodes the full size gp2 of 791 amino acid residues, resulted in a 'transfer' of pathogenic properties, indicating that gp2 is a major determinant of virulence. While gD is essential for virus replication in cultured cells, gI and gE are not, but do play a role in virulence. Restoration of the gI and gE sequences to a gI/gE deletion mutant or to vaccine strain KyA, which has an attenuated phenotype, restores virulence in the equine and murine models of infection.

Functional analyses of genes encoded in U_L have also been performed. Among the genes investigated are those encoding thymidine kinase, gB, gC, and gM, which show strong homology to their HSV-1 and VZV counterparts. Functional homology between HSV-1 and EHV-1 genes has been demonstrated for gM and gB.

The genomic sequences of EHV-2 and EHV-4 have been determined, but intensive research on gene functions has not yet been performed. Nothing is known about EHV-3 or EHV-9 gene functions.

Replication

The receptor for EHV-1 entry has not been identified, but recent studies showed that the virus utilizes a unique entry receptor, as it efficiently entered and replicated in cells that lack the entry receptors HveA, HveB, and HveC used for entry by other alphaherpesviruses, such as HSV-1, HSV-2, and PRV. As with other alphaherpesviruses, EHV-1 entry is mediated by gD.

The genes of the tissue culture-adapted KyA strain of EHV-1 are regulated at the transcriptional and translational levels in a temporal fashion, and three kinetic classes of genes designated IE, early, and late have been described (see **Table 2**). The sole IE gene (IR1) maps in both IR segments and gives rise to a spliced 6.0 kbp

Table 2 Regulatory genes of EHV-1

Gene	Temporal class	Gene product	Function in replication
IE (gene 64)	Immediate early	1487 aa[a]	Transactivates early genes and activates some late genes
IR4 (gene 65)	Early	293 aa	Enhances IE protein DNA-binding and binds TBP
UL5 (gene 5)	Early	470 aa	Binds the IE protein, TFIIB, and TBP
EICP0 (gene 63)	Early	419 aa	Promiscuous transactivator, which antagonizes IE protein function; binds IE protein, TFIIB, TBP
IR2	Early	1165 aa	Dominant negative regulator, which blocks IE protein binding to promoters
IR3	Early	0.9 kb RNA	Antisense to IE mRNA; possible precursor of microRNA
ETIF (gene 12)	Late	479 aa	Transactivates the IE promoter; essential for virus egress

[a]aa, amino acid residues in the primary translation product.

mRNA. Multiple IE polypeptide species have been observed, and the major IE protein (IE1, 203 kDa) is a nuclear-localized phosphoprotein that is capable of transactivating other viral genes and autoregulating its own transcription. The transactivation domain (residues 3–89), the DNA-binding domain (residues 422–597), and the nuclear localization domain (PPAPKRRV; residues 963–970) of the IE protein have been mapped. Recent studies have revealed that the IE protein harbors domains for binding general transcription factors, such as TFIIB and TATA-binding protein (TBP), and thus serves to promote the formation of pre-initiation complexes that mediate viral transcription.

Following IE polypeptide synthesis, approximately 45 early transcripts can be detected. Four of the early proteins serve as regulatory proteins and are designated IR4 (EICP22), UL5 (EICP27), EICP0 (UL63), and IR2. The IR4 protein interacts physically with the IE protein and serves to enhance the DNA binding of the IE protein to its target sequence (ATCGT) present within the promoters of EHV-1 genes characterized to date. The IR4 protein also binds to TBP and is present at viral early promoters in association with the IE protein and TBP. These interactions explain the synergistic effect on the transactivation of viral genes mediated by the IE and IR4 proteins. The EHV-1 UL5 protein exhibits limited identity to HSV-1 ICP27, and is essential for virus replication in cell culture. It acts synergistically with either the IE protein or the EICP0 early regulatory protein to activate expression of both early and late viral gene expression. This early regulatory protein also interacts physically with both the IE protein and TBP, and serves to enhance formation of transcriptional complexes on viral promoters. The third early regulatory protein is EICP0, which is a powerful and promiscuous transactivator that can independently activate expression of viral genes of all three temporal classes. Ironically, EICP0 cannot activate its own promoter, possibly due to a 28 bp negative regulatory element that maps at nucleotides (nt) −204 to −177 within the EICP0 promoter. The EICP0 protein binds to the IE protein and to cellular TFIIB and TBP, but is not a DNA-binding protein. The EICP0 and IE proteins have an antagonistic relationship that may result from their physical interaction in the nucleus and/or from competition for binding to TFIIB and TBP. Both viral proteins bind to the same domain in TFIIB. Deletion of the EICP0 gene greatly impairs virus replication and severely retards late gene expression, suggesting that this regulatory protein is important in the switch from early to late transcription. The fourth early regulatory protein is the IR2 protein, which is a truncated form of the IE protein lacking its essential transactivation and serine-rich domains. The IR2 protein serves a negative regulatory role as it downregulates viral gene expression by acting as a dominant negative protein that blocks IE protein binding to viral promoters and/or by squelching the limited supply of TFIIB and TBP. In addition to these four early auxiliary regulatory proteins, the EHV-1 unique IR3 gene contributes a regulatory role as it encodes a small transcript that is antisense to a portion of the IE transcript. In transient transfection assays, the IR3 transcript downregulates IE gene expression and is only minimally expressed as a protein. Initial studies suggest that the IR3 transcript is processed to a microRNA, and this indicates that the IR3 gene may use novel mechanisms to downregulate IE gene expression at late times of infection.

Early gene expression is followed by viral DNA replication and the production of approximately 29 late transcripts has been detected. Although these transcripts have been positioned on the viral genome, only a small number of protein products have been identified and characterized (see above).

Viral DNA replication initiates at approximately 4 h post infection and requires the virus-encoded DNA polymerase. DNA replication is thought to occur by the rolling circle mechanism whereby long concatemers of the viral genome are generated, cleaved, and then packaged into the maturing virions. The UL15 homolog of HSV-1, one of the two spliced EHV-1 genes known to date, appears to be essentially involved in the generation of unit length genomes and their packaging into mature capsids. Sequences at the L terminus of the EHV-1 genome are composed of direct repeats (DR1 = 18 bp and DR4 = 16 bp) as well as unique sequences (Uc = 60 bp), while sequences at the terminus of the S region contain a 54 bp region designated Ub. Thus, the sequence arrangement at the concatameric junction following replication of the EHV-1 genome is Ub-DR1-Uc-DR4, which represents a functional cleavage/packaging signal, similar, but not identical, to that of HSV-1.

The start of viral DNA synthesis initiates late gene expression and the synthesis of the late regulatory protein EHV transinducing factor (ETIF), a counterpart to the alpha-transinducing factor of HSV-1. This 60 kDa protein is multifunctional and plays at least three roles in EHV-1 replication. First, ETIF is present in several molecular sizes in the tegument and contributes to overall virion structure. Second, after virus entry and uncoating, ETIF serves to transactivate the IE promoter by binding to cellular factors that mediate its association with the TAATGARATT sequence at nt−630 to −620 in the IE promoter. This transactivation function to initiate viral gene programming is important but not essential, as EHV-1 DNA is infectious and virus progeny are produced in cells transfected with plasmids carrying the EHV-1 genome. The activation of a viral promoter by ETIF is specific for the IE promoter as ETIF has not been shown to transactivate any early or late promoter tested to date. Recent experiments with an ETIF-deleted virus and an ETIF-complementing cell line revealed a third function for ETIF, and one that is essential for virus replication. Ultrastructural studies of cells infected with ETIF-deleted virus showed a marked defect in secondary envelopment of viral nucleocapsids at cytoplasmic membranes, such that few enveloped virions are produced. Thus, this transactivator protein also plays a key role in secondary envelopment.

Three different EHV-1 capsid species have been identified and probably correspond to the forms found in HSV-1, which are designated type A, B, and C capsids. The EHV-1 capsid species were designated: (1) L capsids, which appear to be empty; (2) I capsids which possess an electron-lucent, immature core structure in the shape of a cross; and (3) H capsids, which contain an electron-dense, mature core. All three capsids appear at approximately 6 h post infection. I capsids are believed to be a major precursor in the formation of mature capsids. The major capsid protein has an apparent size of 148 kDa, and other structural proteins have also been identified. As reported for other herpesviruses, EHV-1 maturation occurs by interaction of mature nucleocapsids with the inner portion of the nuclear membrane resulting in the formation of enveloped particles. As noted above, recent studies with ETIF-deleted EHV-1 support the model of the sequential envelopment/de-envelopment/re-envelopment pathway for egress of EHV-1 from the cell.

Infectious EHV-1 particles contain an envelope whose protein component is composed mainly of glycoproteins. To date, 12 EHV-1 glycoproteins have been identified and an association of gB, gC, gD, gG, gH, gL, gM, gp2, and the tegument protein VP13/14 with purified virions has been demonstrated. The glycoproteins have been shown in other alphaherpesviruses to be involved in virus binding, virus penetration, and cell-to-cell spread of infection. In the case of EHV-1, these functions were confirmed for gB, gD, gM, and gp2, and the latter two proteins were shown to be nonessential for virus growth. In contrast, gB- or gD-deleted EHV-1 mutants are unable to grow in cell culture. The defect in replication of gB- and gD-negative viruses is caused by an entry defect and the inability to spread from infected to uninfected cells. Detailed functional analyses for the other glycoproteins have not been performed, but are now facilitated by the use of engineered virus mutants produced by targeted gene deletion or disruption using infectious DNA clones of the virus. Even less is known for proteins that make up the third component of the mature virion, the tegument. Only one protein that is related to an HSV-1 tegument protein has been analyzed in detail. As discussed above, recent studies have shown that ETIF is essential for virus maturation and egress.

Defective Interfering Particles and Persistent Infection

EHV-1, EHV-2, and EHV-5 have been shown to mediate persistent infection. In the case of EHV-1, defective interfering particles (DIPs) have been shown to initiate and maintain this outcome. EHV-1 DIPs have been generated *in vivo* in the Syrian hamster model, and therefore may be relevant during EHV-1 infection of the natural host. DIPs are replication defective and require standard EHV-1 as a helper. The overwhelming majority of EHV-1 DNA sequences are absent from DIPs. The packaged DIP DNA molecule is a concatamer of EHV-1 sequences ranging in size from 5.9 to 7.3 kbp, repeated head to tail until it is approximately the size of the standard viral genome. DNA sequencing has revealed that sequences from three regions of the EHV-1 genome are conserved in DIPs: (1) the L terminus, including genes UL3, UL4 and the 3′ portion of UL5; (2) the junction between U_L and the internal IR; and (3) the central portion of IR, including ORI and the 5′ portion of gene IR4. The UL3 and UL4 genes in DIP genomes are 100% identical to those of infectious virus, but their functions in virus replication remain to be elucidated. The DIP genome also contains a perfectly conserved cleavage/packaging signal. The sequences at the L terminus and IR are joined by a homologous recombination event mediated by a conserved 8 bp sequence present at both the L terminus and within the IR4 gene to generate a unique ORF present only in DIPs. This ORF is expressed as a 31 kDa 'hybrid protein' comprising the N-terminal 196 amino acid residues of the IR4 protein (the homolog of HSV-1 ICP22) linked in frame to the C-terminal 68 amino acids of the UL5 protein (the homolog of HSV-1 ICP27). Unique to EHV-1 persistently infected cells (not detected in EHV-1 cytolytic infection) is a 2.2 kbp transcript that maps to the U_L/IR junction and is antisense to the IE mRNA. Interestingly, this transcript exhibits significant homology to the latency associated transcripts of HSV-1, which appear to be associated with HSV-1 reactivation rather than establishment of latency.

Lastly, in EHV-1 persistently infected cells, transcription of certain viral genes appears delayed compared with cytolytically infected cells. Recent findings reveal that expression of the 31 kDa IR4/UL5 hybrid protein downregulates expression of specific EHV-1 promoters. Moreover, altered forms of the EHV-1 IE polypeptides have been observed only in persistently infected cells. Taken together, these studies indicate that altered or aberrant viral regulatory mechanisms may be involved in establishing or maintaining persistent infection. Ongoing studies with recombinant forms of the DIP genome indicate that the hybrid gene is not essential for DIP replication, but is important in the ability of EHV-1 DIPs to establish persistent infection.

Evolution

The six EHVs are biologically distinct. Initial DNA sequence analyses have revealed that genes identified to date are collinearly arranged in the genomes of EHV-1, EHV-3, and EHV-4, all of which possess a two-isomer genomic structure. Evolutionary relationships have become more apparent now that EHV-1, EHV-2, and EHV-4 have been sequenced and data on EHV-5 sequences and genomic organization are available. It is clear that EHV-1, EHV-4, and EHV-9 are closely related and may have arisen from the same ancestor. The same is true for EHV-2 and EHV-5. However, it is not possible to determine precise details of the EHV ancestor since additional sequence data (especially on EHV-3) are not available.

Serological Relationship and Variability

The six EHVs share certain antigens, but are antigenically distinct. EHV-1, EHV-4, and EHV-9 are closely related antigenically, such that cross-neutralizing antibodies are generated. Also, EHV-2 and EHV-5 are closely related. Almost no data are available on the relationship of EHV-3 to other members of the EHVs. However, all of the EHVs are believed to share complement-fixing antigens.

Epidemiology

Rhinopneumonitis caused by EHV-1 and EHV-4 is spread by direct or indirect contact (ingestion and inhalation). The viruses are most commonly shed in nasal droplets for up to 3 weeks after initial infection and are present in large amounts in aborted fetuses and the placenta.

EHV-1 infection can also result in spontaneous abortions in pregnant mares. Horses are most susceptible to EHV-1 infection between the eighth and eleventh months of pregnancy. The peak incidence is in the ninth and tenth months, at which time approximately 70% of abortions occur.

EHV-2 and EHV-5 have been isolated from the respiratory tract, kidneys, spleen, testicles, genital tract, and rectum. Once infected, the horse is a lifelong carrier and excreter of the virus. The exact modes of spread of EHV-2 and EHV-5 are unknown.

EHV-3 causes a mild coital exanthema that is spread by genital contact and – rarely – the respiratory route. An EHV-3 infection is usually cleared after 14 days and is not associated with equine abortions.

Transmission and Tissue Tropism

EHV-1, EHV-4, and EHV-9 are spread mainly by nasal discharge. EHV-2 and EHV-5 establish a chronic infection and may be spread by the respiratory route. EHV-3 is spread by genital contact. EHV-1 has a wide host range *in vitro* as described above, while EHV-4 is more limited. EHV-2, EHV-3, and EHV-5 are restricted mainly to cells of equine origin.

Pathogenicity

EHV-1 and EHV-4 cause rhinopneumonitis, which is often transient and mild but can be complicated by secondary bacterial infections and become more severe and long lasting. EHV-1 is also associated with spontaneous abortions as well as neurological disease. Although EHV-1 and EHV-4 respiratory infections are clinically indistinguishable, their pathogenesis is quite different. EHV-1 infection results in a systemic viremia that can lead to abortion and/or neurological disease. Alternatively, EHV-4 infection usually remains restricted to tissues of the upper respiratory tract. EHV-2 and also EHV-5 infections are acquired horizontally early in life usually by inhalation. EHV-2 and EHV-5 establish a chronic lifelong infection in peripheral blood mononuclear cells. EHV-3 causes an acute coital exanthema in both the mare and the stallion, and infection remains localized to the genitalia.

Clinical Features of Infection

EHV-1 and EHV-4 cause outbreaks of upper respiratory disease ('common cold') in young horses (mainly EHV-4 in adult horses) with no previous exposure to the viruses. Infection is characterized by a short incubation time (<1 day) followed by fever (39–41 °C), which can last between 1 and 4 days, sometimes with a second spike approximately 1 week after the primary pyrexia, and animals suffer from serous nasal discharge and congestion of the nasal mucosa and conjuctiva. Less frequently, one can detect a transitory period of anorexia, enlargement of the submandibular lymph nodes, and edematous swelling of the lower parts of the body and the extremities. An initial leukopenia is followed by leukocytosis before the temperature falls. Recovery is usually uneventful and occurs within 1 week. Older horses show few or no clinical signs, although increased sensitivity is seen in stressed horses. Death is not uncommon from natural acquired infection resulting in neurological disease.

Equine abortions induced by EHV-1 usually occur late in gestation (8–11 months). The foals are usually born dead, but, if alive, often succumb to pneumonia within the first few days. Clinical signs of neurological disease caused by EHV-1 are highly variable and include head pressing, ataxia, and paralysis with complete recumbence.

Both EHV-1 and EHV-4 have been isolated from the central nervous system (CNS) of infected horses, but EHV-1 is by far the leading cause of EHV-induced neurological disease. The virus strains do not invade the extravascular nervous tissue and are not neurotropic *per se*. Rather, the virus spreads from the respiratory tract to the CNS via infected leukocytes and infects endothelial cells of the blood vessels supplying the spinal cord. While the neurological form of the disease is usually lethal, some horses have fully recovered from it with no permanent neurological sequelae.

The roles of EHV-2 and EHV-5 in clinical disease are virtually unknown. However, the viruses have been associated occasionally with chronic throat infections and have been isolated from the respiratory tract of horses with respiratory disease.

An incubation period as long as 10 days can be observed following natural infection with EHV-3. The initial lesions are small (1–2 mm), raised, reddened papules. The lesions then progress rapidly to the pustular form, and there is a general reddening of the vaginal mucosa in the mare. The number of lesions increases in the first few days, and by day six, many of the lesions form ulcers up to 20 mm in diameter and 5 mm deep. Lesions can also be seen on the vulva, perineal skin, as well as on the penis and prepuce of the stallion. The disease is usually mild, with temperature, pulse, appetite, and respiration remaining close to normal. The severity of the disease can be increased by secondary bacterial infections; however, uneventful cases are usually cleared within 2 weeks. Lastly, EHV-3 is not abortigenic and does not lead to infertility.

Infections with EHV-9 lead to encephalitis in horses, but, more severely, in other species as well. Among the susceptible hosts are gazelles and goats, and also carnivores such as dogs and cats.

Pathology and Histopathology

Respiratory disease caused by EHV-1 and EHV-4 results in inflammation, congestion, and sometimes necrosis of the tissues of the upper respiratory tract. Extensive swelling of the nasal mucosa may occur and, in later stages, the lungs may become involved. One can find typical herpesvirus inclusion bodies in the nuclei of the respiratory epithelium. The respiratory infection can become more serious if followed by a secondary bacterial infection, which may lead to bacterial pneumonia.

Fetuses that are aborted as a consequence of EHV-1 infection present with widespread hemorrhages and edema, as well as a yellowish discoloration of the fetal conjunctiva and splenomegaly – if virus is transmitted from mother to animal. However, abortions without infection of the fetus also occur, and the pathogenesis in the uterine vasculature largely resembles that of the vasculitis observed in the neurological form of the disease. The classic histopathology includes the presence of eosinophilic intranuclear inclusion bodies in various organs of the aborted fetuses. Gross pathological and histopathological alterations are sometimes not as obvious, and only sensitive methods (virus isolation or PCR) are able to confirm the EHV-1 abortion. Classically, features associated with EHV-1-induced abortions differ in those fetuses aborted during the first 6 months of gestation as compared with those aborted after 6 months. Those before 6 months present with widespread cell necrosis and inclusion bodies in the liver and lung. Those after 6 months exhibit jaundice, subcutaneous edema, excessive pleural fluid, pulmonary edema, splenomegaly, and necrosis of the liver.

In experimental EHV-1 infections, a severe hepatitis is observed following intraperitoneal infection of the Syrian hamster, and a model of respiratory infection is also available in the hamster. Intranasal infection of mice with EHV-1 results in respiratory disease and subsequent cell-associated viremia, thus serving as a model that somewhat mimics the disease in equines. The murine model of EHV-1 infection is widely used in virulence and immunogenicity studies. Recent studies with a library of EHV-1 mutants reveal the importance of full-size gp2 in respiratory disease. DNA-array analyses of lung tissues confirmed earlier findings from RNase protection assays and showed that the massive influx of inflammatory cells into the lung is preceded by 2–13-fold increases in >30 inflammatory genes. Expression of these genes increased as early as 8–12 h post infection, and they encode major cytokines and chemokines, including interleukin-1β (IL-1β), IL-6, tumor necrosis factor alpha, macrophage inflammatory protein 1α (MIP-1α), MIP-1β, and MIP-2. Future investigations should give insight into specific domains of gp2 that activate this constellation of inflammatory genes and the signal transduction pathways that are involved.

EHV-2 (and also EHV-5) infection becomes widespread throughout the body, and the viruses have been isolated from a variety of tissues. The infected animal becomes a lifelong carrier, and the viruses remain highly cell associated.

Tissues affected following an infection with EHV-3 include the vaginal and vestibular mucosa, penis, prepuce, and the skin of the perineal region. One of the characteristics of an infection with EHV-3 is the sloughing of the surface epithelial cells. On occasion, the skin of the lips and mucus membranes of the respiratory tract may become involved, but the exanthema is usually mild.

Immune Response

Neutralizing antibodies are detected in the serum soon after an EHV-1 or EHV-4 infection. The antibodies can first be detected from a week following infection and are most abundant after several weeks. However, the immunity is short lived, in that horses can be re-infected and exhibit respiratory symptoms just 3 months after the initial infection. Multiple exposures to either EHV-1 or EHV-4 will result in the development of neutralizing, cross-reactive antibodies. However, cell-mediated immune responses are likely to be primarily responsible for induction of a sustainable immunity.

Immunity to EHV-2, EHV-3, and EHV-5 is poorly understood. However, virtually all horses have antibodies to EHV-2 and EHV-5, confirming the general apathogenicity of these viruses.

Prevention and Control

A number of vaccine approaches are followed to combat EHV-1 infections; among them are modified live vaccines (e.g., Rhinomune and Prevaccinol), inactivated vaccines (e.g., Pneumabort K and Prestige), inactivated combination vaccines, which – among others – also contain EHV-4 (e.g., Innovator, Resequin, and Duvaxyn1,4), and subunit vaccines also covering both viruses (e.g., Cavalon IR). Unfortunately, many EHV vaccines cause undesirable side effects in the form of massive local reactions while not affording acceptable levels of protection, especially when inactivated combination vaccines are considered. All vaccines are given repeatedly to pregnant mares usually in the third, fifth, seventh, and ninth months of pregnancy, since, to ensure protection, a good level of population immunity is imperative. To protect against viral rhinopneumonitis outbreaks, the vaccine is usually given to all horses every 3–6 months. There is considerable ongoing discussion as to proper vaccination against the neurological disease, which has become more prevalent during past years. Recent studies seem to suggest that modified live virus vaccines are superior, especially to multivalent inactivated vaccines, with respect to duration of fever, virus excretion from the nasal mucosa, and development of neurological symptoms. These studies, however, need to be corroborated by further comparative studies and/or larger numbers of animals in experimental groups.

Clinical management often involves the use of antibiotics to prevent severe bacterial complications following the viral rhinopneumonitis. Control of EHV-1 and EHV-4 infections involves isolation and quarantine of infected horses (for at least 3 weeks) and sound hygiene for prevention of viral infection, since the viruses are highly contagious. Quarantine procedures are required more often with EHV-1 infections, since EHV-1 can lead to more serious diseases of the CNS. Since many of the EHV vaccines provide unacceptable levels of protection, the first step in the prevention and control of EHV-1 and EHV-4 infections involves specific management practices and adequate day-to-day care of the animals. The viruses can be spread easily in contaminated feed and water. In addition, minimizing stress and close contact of large groups of horses can prevent the spread of disease.

Future Perspectives

Considerable progress in unraveling the nucleotide sequences of EHV-1, EHV-2, EHV-4, and EHV-5 has been made during recent years, and EHV-1, EHV-2, and EHV-4 have been entirely sequenced. With this information in hand, it will be possible to pursue studies on gene expression and on those proteins that are involved in virulence of EHV. These studies will in turn open the possibility for a rational design of anti-EHV vaccines, especially against the most important pathogens, EHV-1 and EHV-4. These goals may be achieved by the use of viral deletion mutants that carry targeted gene deletions, which is greatly facilitated by the advent of a number of infectious DNA clones during the past years. Using novel molecular approaches, a better understanding of EHV biology will be possible and in the future will open new perspectives in the understanding of diseases caused by EHVs.

See also: Herpesviruses: General Features; Herpes Simplex Viruses: General Features; Herpes Simplex Viruses: Molecular Biology; Pseudorabies Virus; Varicella-Zoster Virus: General Features; Varicella-Zoster Virus: Molecular Biology.

Further Reading

Albrecht RA, Kim SK, and O'Callaghan DJ (2005) The EICP27 protein of equine herpesvirus 1 is recruited to viral promoters by its interaction with the immediate-early protein. *Virology* 333: 74–87.

Buczynski KA, Kim SK, and O'Callaghan DJ (2005) Initial characterization of 17 viruses harboring mutant forms of the immediate early gene of equine herpesvirus 1. *Virus Genes* 31: 229–239.

Goodman LB, Wagner B, Flaminio MJBF, et al. (2006) Comparison of the efficacy of inactivated combination and modified-live virus vaccines against challenge infection with neuropathogenic equine herpesvirus type 1 (EHV-1). *Vaccine* 24: 3636–3645.

Kim SK, Albrecht RA, and O'Callaghan DJ (2004) A negative regulatory element (bp −204 to −177) of the EICP0 promoter of equine herpesvirus 1 abrogates the EICP0 protein's *trans*-activation of its own promoter. *Journal of Virology* 78: 11696–11706.

Pagamjav O, Sakata T, Matsumura T, Yamaguchi T, and Fukushi H (2005) Natural recombinant between equine herpesvirus 1 and 4 in the ICP4 gene. *Microbiology and Immunology* 49: 167–179.

Paillot R, Ellis SS, Daly JM, et al. (2006) Characterisation of CTL and IFN-gamma synthesis in ponies following vaccination with a NYVAC-based construct coding for EHV-1 immediate early gene, followed by challenge infection. *Vaccine* 24: 1490–1500.

Rudolph J, O'Callaghan DJ, and Osterrieder N (2002) Cloning of the genomes of equine herpesvirus type 1 (EHV-1) strains KyA and RacL11 as bacterial artificial chromosomes (BAC). *Journal of*

Veterinary Medicine. B, Infectious Diseases and Veterinary Public Health 49: 31–36.

Smith PM, Kahan SM, Rorex CB, von Einem J, Osterrieder N, and O'Callaghan DJ (2005) Expression of the full length form of gp2 of equine herpesvirus 1 (EHV-1) completely restores respiratory virulence to the attenuated EHV-1 strain KyA in CBA mice. *Journal of Virology* 79: 5105–5115.

Soboll G, Whalley JM, Koen MT, *et al.* (2003) Identification of equine herpesvirus-1 antigens recognized by cytotoxic T lymphocytes. *Journal of General Virology* 84: 2625–2634.

Telford EAR, Studdert MJ, Agius CT, Watson MS, Aird HC, and Davison AJ (1993) Equine herpesviruses 2 and 5 are γ-herpesviruses. *Virology* 195: 492–499.

Telford EAR, Watson MS, Aird HC, Perry J, and Davison AJ (1995) The DNA sequence of equine herpesvirus 2. *Journal of Molecular Biology* 249: 520–528.

Telford EAR, Watson MS, McBride K, and Davison AJ (1992) The DNA sequence of equine herpesvirus-1. *Virology* 189: 304–316.

Herpesviruses: Discovery

B Ehlers, Robert Koch-Institut, Berlin, Germany

© 2008 Elsevier Ltd. All rights reserved.

History of Herpesvirus Discovery

Diseases caused by herpesviruses have been known since ancient times, but it was not until the beginning of the twentieth century that the viruses causing these diseases were detected. Human alphaherpesviruses (members of subfamily *Alphaherpesvirinae*) such as herpes simplex virus (HSV) and varicella-zoster virus (VZV) were the first to be propagated in cell culture and visualized by electron microscopy. Later, betaherpesviruses (members of subfamily *Betaherpesvirinae*) such as human cytomegalovirus (HCMV) and gammaherpesviruses (members of subfamily *Gammaherpesvirinae*) such as Epstein–Barr virus (EBV) were discovered and studied in detail. EBV was the first herpesvirus whose genome was cloned in bacteria and completely sequenced. In the middle of the last decade of the twentieth century, the complete genome sequences of roughly 40 herpesvirus species had accumulated in public databases. These human and animal herpesviruses had previously all been cultured, and their physical, morphological, and antigenic characteristics had been studied. This was a cumbersome process and usually carried out only with viruses causing important diseases.

For the primary characterization of uncultured viruses, alternative, more-straightforward approaches were needed. Immunological cross-reaction and nucleic acid hybridization were the first tools used for the detection of unknown viruses closely related to known species. Later, nonspecific amplification methods were developed that detected nucleic acids of unknown DNA and RNA viruses directly in clinical samples. Expression libraries of genomic DNA or cDNA were screened immunologically with sera from patients infected with the putative etiologic agent. Then subtractive approaches were developed to isolate sequences in one DNA sample that were absent from another, and one of these, representational difference analysis, led to the discovery of human herpesvirus 8 (HHV-8; also known as Kaposi's sarcoma-associated herpesvirus). Although very powerful, these nonspecific amplification methods are most appropriate for the detection of agents in defined samples from diseased individuals where a microbiologic etiology is strongly indicated. They are too laborious for broad screening of large sample numbers.

In 1988, Mack and Snisky were the first to describe a polymerase chain reaction (PCR)-based method for the identification of uncharacterized viruses related to known virus groups, using degenerate primers (a pool of all the possible combinations of sequences that could code for a given amino acid sequence). Later, and following the same principles, a PCR method for the universal detection of herpesvirus genomes was published by VanDevanter and co-workers. Despite its restriction to herpesviruses, this method was a hallmark because it had the potential to detect virtually all vertebrate herpesviruses by a straightforward approach. Since then, nearly all novel herpesviruses have been discovered with the help of this method (in its original version or with modifications), and many were found in individuals lacking a recognized disease history.

Panherpes PCR with Degenerate Primers

The method is based on the fact that herpesvirus genomes contain highly conserved genes, which are present in all because they code for proteins essential for viral growth. One of the most conserved is the herpesvirus DNA polymerase (DPOL) gene. Investigators aligned the DPOL genes of all herpesviruses for which sequence data were available, in order to identify short blocks of greatest amino acid identity that encode domains essential for DPOL function. This procedure allowed the design of five degenerate primers that were used in a nested PCR – three primers in the first round and two primers in the second round (**Figure 1**). The region amplified in between the primers displayed a lower degree of conservation. This allowed investigators to assess (1) whether the detected herpesvirus was known or novel, and (2) to which herpesvirus subfamily the novel species could be assigned. In some cases, even a tentative assignment to a virus genus was possible.

The primers are degenerate in their 3′ part and nondegenerate in their 5′ part. As an example, the sense primer (TGV) of the second round is shown in **Figure 2**. The degenerate 3′ part enables it to bind universally to the DPOL gene of higher vertebrate herpesviruses in the first PCR cycle. The nondegeneracy of the 5′ part ensures amplification efficiency in subsequent cycles. The product amplified in the second round has a length of ∼210–235 bp, depending on the virus. Only HCMV and chimpanzee cytomegalovirus, as well as some herpesviruses of insectivoral shrews, give rise to significantly larger amplification products of >300 bp (**Figure 3**).

Several variations of the original method were subsequently published, using (1) deoxyinosine-substituted primers, (2) a nondegenerate consensus sequence as 5′-part of the primer (also called consensus-degenerate hybrid oligonucleotide primers (CODEHOPs)), or (3) other conserved primer binding sites in the DPOL gene. Also, targeting of other highly conserved herpesvirus genes such as the glycoprotein B (gB) gene or the terminase gene was attempted. Degenerate consensus primers were also used to amplify a larger number of conserved genes ($n = 8$) of a novel porcine herpesvirus (porcine lymphotropic herpesvirus 3, PLHV-3). These sequence 'islands' were then connected with the help of long-distance PCR to yield a contiguous sequence of ∼100 kbp (accession AY170316). This is the longest part of a herpesvirus genome of unknown sequence ever determined directly from primary virus-positive organ material solely through PCR (i.e., without virus cultivation).

How Many Herpesviruses Have Been Discovered?

Since the publication of these methods, the number of novel herpesviruses discovered has risen very fast. Prior to writing this article, more than 200 potential herpesvirus species had been detected, belonging to more than 20 mammalian orders (**Table 1**). These are many more than currently accepted by the International Committee on Taxonomy of Viruses (ICTV), which lists some 120 herpesvirus species (VIIIth Report). Sequences of more than 160 herpesviruses are available under the taxonomy browser of the NCBI, most detected by DPOL consensus primers. In addition, sequences indicating the existence of more than 100 additional herpesvirus species, not yet available in public databases, await publication. By combining all the data currently available, a total of approximately 300 detected

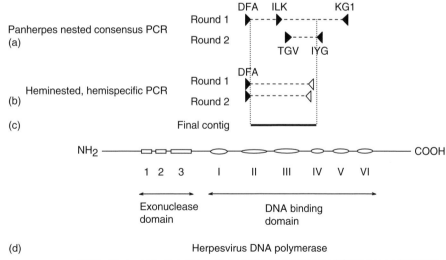

Figure 1 Panherpes consensus PCR with degenerate primers. Schematic diagram of the PCR strategy. (a) Initially, panherpes-nested PCR with degenerate and deoxyinosine-substituted primers (black triangles) is performed for amplification of novel herpesvirus DPOL sequences. In the first round, the sense primers DFA and ILK are combined with the antisense primer KG1. In the second round, the sense primer TGV and the antisense primer IYG are used. (b) Subsequent heminested PCR with the degenerate sense primer DFA and two virus-specific antisense primers (open triangles). The exact position of the virus-specific antisense primers is slightly different for each individual DPOL sequence but is always located between primers TGV and IYG. (c) The final contig is represented by the solid line (450–480 bp, without primer-binding sites). (d) The DPOL gene is shown schematically. Highly conserved, functionally important regions are represented by boxes and ovals, respectively, and numbered.

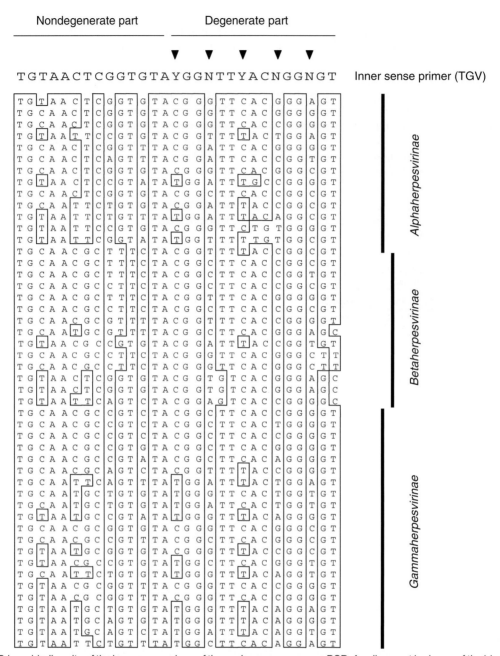

Figure 2 Primer-binding site of the inner sense primer of the panherpes consensus PCR. An alignment is shown of the binding regions of the inner sense primer (TGV) in alpha-, beta-, and gammaherpesviruses. Above, the TGV primer is shown, originally designed by VanDevanter and co-workers. In its degenerate part, the primer does not match perfectly the sequences of some alpha- and betaherpesviruses. Nonetheless, the DPOL sequences of these viruses are amplifiable by panherpes consensus PCR.

herpesvirus species can be presumed, the vast majority infecting a broad spectrum of mammalian hosts such as primates, elephants, ungulates, rodents, dolphins, carnivores, insectivores, manatees, and bats. Herpesviruses have also been found by consensus primer PCR in nonmammalian hosts such as reptiles, birds, and fishes (**Table 1**). Even a mollusk (Pacific oyster) has been shown to host a herpesvirus. The fact that herpesviruses are not yet known for many animal hosts is most probably due to lack of investigation.

Most of the novel viruses from which newly detected sequences originate have never been isolated or propagated in tissue culture, and a large number probably never will. On the other hand, some of the herpesviruses that were cultured in the premolecular era and classified formally have never been characterized genetically and are

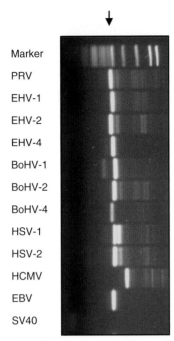

Figure 3 Amplified fragments obtained by panherpes consensus PCR. Partial DPOL sequences were amplified from various herpesviruses: porcine (PRV), equine (EHV-1, EHV-2, and EHV-4), bovine (BoHV-1, BoHV-2, and BoHV-4), and human (HSV-1, HSV-2, HCMV, and EBV). The arrow marks the amplified fragments, most of which have a length of ~230 bp. Note the considerably larger band (310 bp) obtained by amplification of HCMV. DNA of the polyomavirus SV40 was used as a negative control.

no longer available. Therefore, some of the new PCR sequences may represent rediscovery of a herpesvirus whose existence was reported many years earlier.

The mammalian herpesviruses may all be assigned to one of the subfamilies *Alpha-*, *Beta-*, and *Gammaherpesvirinae*, and many mammals host one or more herpesviruses from each subfamily. Humans, for example, are infected by three alpha-, three beta-, and two gammaherpesviruses (HSV-1, HSV-2, VZV, HCMV, human herpesviruses 6 and 7, EBV, and HHV-8, respectively). Other host species apparently lack herpesviruses of a certain subfamily (**Table 1**). For example, in dogs, birds, and reptiles, only alphaherpesviruses have been found, but no beta- or gammaherpesviruses. In Equidae and Bovidae, alpha- and gammaherpesviruses have been found, but no betaherpesviruses. These 'blind spots' may be due to evolutionary developments; that is, some herpesvirus species may have existed but died out later, or have never developed in certain host lineages. On the other hand, certain herpesvirus species may be low in prevalence, replicate to low titers and reside in narrow ecological niches with few or no pathological consequences, which would make their detection difficult. In addition, panherpes consensus PCR may be inefficient in detecting certain sequences because of insufficient primer binding. The latter could be due to the fact that the DPOL consensus primers were originally designed on the basis of a limited set of complete DPOL genes. This selection of genes may have been too small to ensure that the primers do not encounter mismatches in their binding sites within certain herpesvirus genomes. For example, betaherpesviruses are more heterogeneous in their DPOL sequences than alpha- or gammaherpesviruses, and some currently available sequences reveal mismatches with the consensus primers. Therefore, the sensitivity of the PCR may be suboptimal for some betaherpesviruses. Inclusion of novel DPOL sequences as a basis for redesign of current panherpes PCR systems may improve their universality and enable the discovery of additional herpesviruses that otherwise might remain undetected.

It is a common phenomenon that hosts are infected with more than one herpesvirus. The detection of all herpesviruses present in such samples is a technical challenge. Panherpes consensus PCR usually favors the amplification of one viral sequence and concurrently neglects others. This is a result of different binding efficiencies of the degenerate primers to each viral genome and the genome copy number of each virus in the sample. One solution for this problem is to design degenerate primers with a more limited spectrum of detection. These can be derived from less-conserved sites in the DPOL gene or from other, less well-conserved genes. For example, the gB gene allows the derivation of subfamily-specific or genus-specific degenerate primers only. Thus, samples which are positive for gammaherpesviruses in the panherpes DPOL PCR can be additionally screened with betaherpesvirus-specific gB primers. Such detection strategies often reveal the presence of several herpesviruses in a given sample, and this has resulted in the discovery of previously unknown herpesvirus sequences in primates and perissodactyls (odd-toed ungulates such as horse, zebra, and donkey).

Tentative Classification of Novel Herpesvirus Sequences

According to ICTV, "related herpesviruses are classified as distinct species if (a) their nucleotide sequences differ in a readily assayable and distinctive manner across the entire genome and (b) they occupy different ecological niches by virtue of their distinct epidemiology and pathogenesis or their distinct natural hosts." These requirements are only very rudimentarily fulfilled after determination of a short partial DPOL sequence from a primary sample. However, it is possible to make an initial attempt tentatively to classify the new virus entity.

It is a general rule (with some important exceptions such as herpes B virus and pseudorabies virus (PRV)) that

Table 1 Herpesviruses amplified through consensus PCR and their hosts

Host class	Host order	Host species (examples)	Number of viruses in the host taxonomic order	Alpha-herpesvirinae	Beta-herpesvirinae	Gamma-herpesvirinae
Mammalia						
	Primates	Humans and other Old World primates; New World primates	~100	x	x	x
	Scandentia	Tree shrew	3		x	x
	Proboscidea	Asian and African elephant	4		x	x
	Carnivora	Lion, hyena, cat, dog, badger, sea lion, seal	7	x		x
	Cetacea	Dolphins, whales	>10	x		x
	Rodentia	Mice, rats, guinea pig	>40		x	x
	Perissodactyla	Horse, donkey, zebra, tapir, pygmy hippopotamus	>10	x		x
	Artiodactyla	Cattle, rhinoceros, sheep, goat, and other ruminants; pig, babirusa, warthog	>35	x	x	x
	Insectivora	Shrews	3		x	
	Sirenia	Florida manatee	1			x
	Chiroptera	Bats	>5		x	x
Reptilia						
	Crocodilia	Alligator	1			
	Squamata	Monitor, chuckwalla, plated lizard	>5	x		
	Testudines	Tortoises and turtles	>15	x		
Aves						
	Galliformes	Chicken, turkey	4	x		
	Columbiformes	Pigeon	1	x		
	Psittaciformes	Parrot	2	x		
	Falconiformes	Vulture, falcon	2	x		
	Passeriformes	Finch	1	x		
Actinopterygii						
	Cypriniformes	Koi	3			
	Siluriformes	Channel catfish	1			
	Acipenseriformes	Sturgeon	3			
	Anguilliformes	Japanese eel	1			
	Perciformes	Tilapia larvae	1			

herpesviruses are specific for their hosts. Thus, related, albeit distinct, herpesvirus sequences from different hosts can be assigned as originating from distinct herpesvirus species. In line with this, analysis of the novel sequence is primarily performed by pairwise and multiple alignment with sequences of known and classified viruses.

When a more detailed view is desired, phylogenetic tree construction is performed. The most reliable phylogenetic trees are constructed with the proteins encoded by sets of conserved herpesvirus genes. This requires the determination of the complete genome sequence of the novel species in question (or at least a considerable part of it). For noncultured herpesviruses and with the amplification techniques presently at hand, this is either impossible or at least very time consuming. As a compromise, the short DPOL sequence, amplified with panherpes consensus PCR, is extended by many investigators in an upstream direction. The final contiguous sequence extends from the outer degenerate sense primer to the inner degenerate antisense primer (450 to >700 bp; the length depends on the virus) (**Figure 1**).

This DPOL region is sufficient for a first attempt to construct a phylogenetic tree in order to (1) identify the closest relatives among the herpesviruses already sequenced, and (2) classify tentatively the virus from which the novel sequence originated. As an example, a phylogenetic tree of all HHV-8-related gammaherpesviruses, for which sequences >450 bp are available, is shown in **Figure 4**. The tree contains 63 distinct sequences and reveals that a large group of Old World primate rhadinoviruses (RHVs) exists, and that this can be divided into two subgroups. One subgroup contains a human RHV (HHV-8), but a human member of the other subgroup has not been identified. Furthermore,

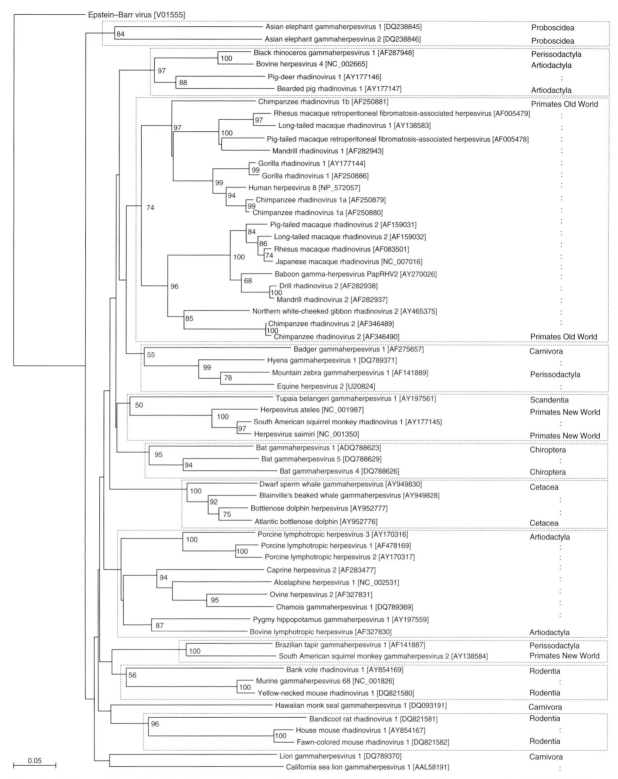

Figure 4 Phylogenetic analysis of well-known and recently discovered RHVs. A phylogenetic tree was constructed with DPOL sequences of 150–160 amino acid residues. They were aligned using the ClustalW module of MacVector™, and gaps were removed. The resulting alignment was analyzed using the neighbor-joining tree-construction module. A rooted phylogram is shown, with EBV as the outgroup. Bootstrap values of >50% are indicated at the nodes of the tree. All sequences are available in GenBank, and the accession numbers are given following the virus names. Host families are indicated.

two distinct RHV groups from artiodactyls (even-toed ungulates, such as cattle, sheep, goat, and pig) and rodents are present. Two mixed groups of carnivore/perissodactyl RHVs are seen, the latter also containing a New World primate RHV. Finally, distinct groups of cetacean (dolphin and whale) and chiropteran (bat) RHVs exist, and a small group of New World primate RHVs clusters loosely with a tree shrew RHV (host: *Tupaia belangeri*) (**Figure 4**).

In phylogenetic trees constructed using short conserved sequences, the deeper nodes close to the center of the tree are often not supported by statistical significance (e.g., see **Figure 4**). In addition, when a great number of herpesviruses from closely related hosts are very similar on the DPOL sequence level, tree resolution is insufficient. A good example is the huge number of EBV-like lymphocryptoviruses (LCVs) in Old World (African and Asian) primates. A satisfactory solution is the consensus amplification of a part of the gB gene, which is located immediately upstream of the DPOL gene in beta- and gammaherpesviruses. Subsequent long-distance PCR and sequencing of all amplification products results in a contiguous sequence of 3–4 kbp, covering large parts of the gB and DPOL genes – a 'bigenic approach' (**Figure 5**). This region codes for approximately 1100 amino acid residues and generally gives rise to considerably improved resolution in phylogenetic trees. This is exemplified by a tree in which 20 EBV-like LCVs of Old World and New World primates are included (**Figure 6**). In this tree, the New World LCVs are clearly separated from two groups of Old World LCVs. In the first group of Old World LCVs, EBV clusters with

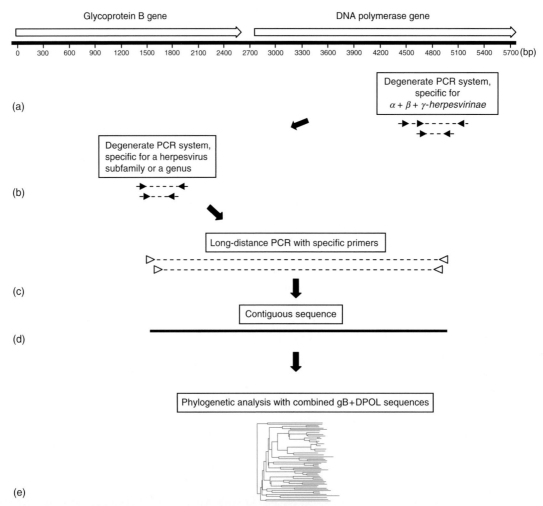

Figure 5 Bigenic amplification of beta- and gammaherpesviruses. Schematic diagram of the analytical strategy. (a) Initially, panherpes-nested PCR with degenerate and deoxyinosine-substituted primers (black triangles) is performed for amplification of the herpesvirus DPOL sequence. (b) According to the sequence of the amplification product and a tentative assignment of the novel DPOL sequence to a herpesvirus subfamily or genus, a degenerate PCR strategy for amplification of the gB gene is chosen, suited to the novel virus. (c) After both gB and DPOL sequences are determined, long-distance PCR is performed with specific primers binding to gB (sense) and DPOL (antisense). (d) A final contiguous sequence of >3 kbp is obtained (solid line) and (e) used for phylogenetic analysis.

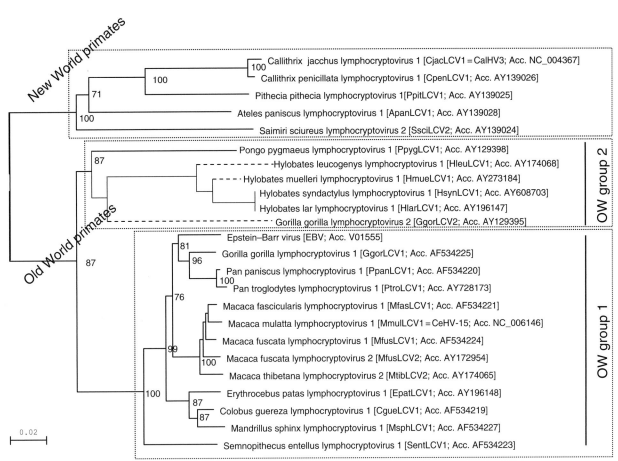

Figure 6 Phylogenetic analysis of well-known and recently discovered LCVs. 400 amino acid residues of gB and 700 residues of DPOL were concatenated for each LCV and analyzed as described in **Figure 2**. For most of the gibbon (genus *Hylobates*) LCVs and the second gorilla LCV (GgorLCV2), the analyzed gB to DPOL region was not available. Therefore, the branches of these viruses were taken from a tree that was constructed on the basis of short DPOL sequences (<160 residues), and are shown by broken lines. OW, Old World.

high probability with LCVs of great apes (chimpanzee, bonobo, and gorilla), as expected from co-evolution of the herpesviruses with their hosts. This subgroup is well separated from a subgroup of macaque LCVs and a subgroup consisting of other cercopithecine LCVs and a colobine LCV. The second group of great ape LCVs contains several distinct gibbon (genus *Hylobates*) LCVs, a second gorilla LCV, and an orangutan (*Pongo pygmaeus*) LCV (**Figure 6**). It can be speculated that this second group should also contain a second human LCV, but this virus, if it exists, remains to be discovered.

The 'bigenic approach' yields more sequence data and improves considerably the reliability of phylogenetic analyses. It has so far been applied to more than 30 novel herpesviruses from nine different orders of vertebrate hosts. It has the potential to become the gold standard for initial genetic characterization of novel beta- and gammaherpesviruses. However, it cannot be applied directly to alphaherpesviruses, in which the gB and DPOL genes are ordered differently, being separated by several kilobase pairs of sequence spanning two genes and, in some cases, the origin of DNA replication.

Pathogenic Impact of Novel Herpesviruses

The herpesviruses that were first discovered and studied in detail were those infecting either humans or animals of economical importance (pigs, PRV; cattle, bovine rhinopneumonitis virus (BoHV-1); horses, equine abortion virus (EHV-1); chickens, Marek's disease virus). Most of these viruses are alphaherpesviruses with overt primary pathogenicity. However, beta- and gammaherpesviruses such as HCMV or EBV often cause primary infections with mild or subclinical pathogenicity. Therefore, it is not surprising that for many of the more recently discovered herpesviruses a disease association is unknown. This is especially the case for all sequences discovered through screening of healthy individuals, and even the detection of herpesviruses in autopsy specimens of animals suffering from lethal disease can be incidental. Therefore, further investigations are generally needed to confirm etiological involvement. Despite these difficulties, for some recently discovered animal herpesviruses, an association with a specific disease is probable.

One of the more exotic examples is the endotheliotropic elephant herpesvirus, which is a serious threat to the endangered Asian elephant. This virus was first described in 1999, and was found in benign skin lesions of African elephants. In several Asian elephants of American and European zoological gardens, this virus was found to cause a severe endotheliotropic infection with multiorgan failure and death within days or even hours after onset of symptoms. It is thought that the virus is well adapted to the African elephant but can cross the species barrier where both elephant species are housed together. Sequence analysis of the complete gB and DPOL genes (GenBank accession AF322977) placed the virus in the subfamily *Betaherpesvirinae*, and further classification into a new genus (*Proboscivirus*) has been proposed.

Crossing of a species barrier is also an attribute of several ruminant gammaherpesviruses. Ovine herpesvirus 2 (OvHV-2) and alcelaphine herpesvirus 1 (AlHV-1), two gammaherpesviruses infecting sheep and wildebeest, respectively, without clinical symptoms, are known also to infect cattle (OvHV-2 and AlHV-1) and pigs (OvHV-2). In these hosts, a lymphoproliferative disease with high mortality is caused, called malignant catarrhal fever (MCF). More recently, a number of novel gammaherpesviruses were discovered in several wildlife ruminants with an apparently similar pathogenic potential (e.g., GenBank accessions AY237360–AY237366 and AY212111–AY212114).

A lymphoproliferative disease of high mortality has been observed in pigs after pig-to-pig blood stem cell transplantation combined with immunosuppressive drug treatment. The disease was apparently associated with a novel porcine lymphotropic gammaherpesvirus (PLHV-1; GenBank accession AF478169), and resembled the post-transplantation lymphoproliferative disease (PTLD) which is associated with EBV after human-to-human allotransplantation. In pig-to-human xenotransplantation, PLHV-1 is discussed as a risk factor. Xenotransplantation is under investigation because it is seen as a solution for the shortage of human organs for allotransplantation.

Gammaherpesviruses are also closely associated with abnormal proliferation and cancer in primates. Several primate LCVs are known that are very similar to EBV in genome organization and gene content. Some have already been studied in detail and found to be associated with lymphoproliferative diseases that resemble those caused by EBV. Therefore, these LCV infections may be viewed as potential models for EBV infection. For the LCVs more recently detected by panherpes PCR methods (examples are in **Figure 6** and GenBank accessions AY608702–AY608714, AY28171–AY28177, and AF534219–AF534229), such associations remain to be elucidated.

A large number of primate RHVs have been discovered. The first to be described and one of the best studied is herpesvirus saimiri (HVS), which naturally infects a New World primate, the squirrel monkey (*Saimiri sciureus*), without producing symptoms. Upon experimental infection of other monkey species, it causes fatal acute T-cell lymphoma. In Old World primates, several RHVs have since been discovered that are genetically very similar to HHV-8. One, rhesus monkey rhadinovirus (RRV), has been completely sequenced by two research groups and found to induce hyperplastic lymphoproliferative disease in rhesus macaques upon experimental co-infection with simian immunodeficiency virus. This disease resembles a variant of multicentric Castleman's disease, which is a rare HHV-8-associated disease of the human lymphatic system that involves hyperproliferation of certain cytokine-producing B-cells. RRV infection of rhesus monkeys was therefore proposed as a model for certain aspects of HHV-8 pathogenicity in humans. Retroperitoneal fibromatosis in rhesus and pig-tailed macaques is associated with another group of macaque RHVs. Panherpes consensus PCR revealed that they are even more closely related to HHV-8 at the sequence level (**Figure 4**). In recent years, both groups of primate RHVs have been populated with many additional viruses, in particular those infecting great apes (chimpanzee, gorilla, and gibbon), and also other Old World monkeys (e.g., mandrill and baboon) and a New World monkey (squirrel monkey) (**Figure 4**). For the majority of these viruses, a disease association remains to be described.

The first alphaherpesvirus from great apes was isolated very recently from oral and pharyngeal ulcers of chimpanzees. It is a close relative of HSV-2. Serological evidence suggests that gorillas may host a similar alphaherpesvirus.

Fibropapillomatosis of marine turtles is an emerging neoplastic disease associated with infection by a novel virus, fibropapilloma-associated turtle herpesvirus (FPTHV). The DPOL gene of this virus has been amplified through panherpes consensus PCR and originates from an alphaherpesvirus. Later, several turtle species were reported to come down with the disease. These findings underline the fact that tumorigenicity is not an exclusive property of gammaherpesviruses.

In the world of reptiles, novel herpesviruses have also been reported to be associated with proliferative stomatitis in green tree monitors (*Varanus prasinus*) and plated lizards (*Gerrhosaurus major* and *Gerrhosaurus nigrolineatus*).

Several marine mammals of the family Cetaceae (dolphins and whales) have been described as suffering from lesions on oral and genital mucosa as well as from systemic, fatal disease. These were attributed to herpesviruses, and their assignment as alpha- and gammaherpesviruses, respectively, was reported recently on the basis of consensus PCR and sequence analysis.

Among fish, mass mortality in common carp (*Cyprinus carpio*) and colored carp or koi (*Cyprinus carpio koi*) has been observed during the last decade and was reported to be caused by a herpesvirus that differs substantially in sequence and gene content from mammalian herpesviruses but is related to a number of other fish herpesviruses from the families Cyprinidae (carps and minnows), Acipenseridae (ray-finned fish, including sturgeons), and Ictaluridae (catfish).

Finally, sporadic mortality of the Pacific oyster (*Crassostrea gigas*) is associated with a novel herpesvirus, ostreid herpesvirus 1. This is the only known herpesvirus that has an invertebrate host, and is the sole representative of a third major class of herpesvirus in addition to the two classes infecting vertebrates (mammals, birds, and reptiles, and bony fish and amphibians).

Outlook

For the vast majority of potential animal hosts, herpesviruses are completely unknown. This holds true in particular for mammalian families that are populated with a huge number of species, such as those encompassing rodents or bats. Also, many nonmammalian hosts remain to be investigated. The number of putative novel herpesvirus species is growing exponentially, most being characterized exclusively from limited sequence information. The novel sequences obtained can be used to refine the available panherpes consensus PCR methods in order to improve their general detection potential. Recent developments in diagnosing known and novel pathogens by microarray technology may evolve into an additional powerful tool for herpesvirus discovery. Thus, the goal of obtaining a much more extensive list of herpesviruses can be approached, and exciting questions tackled, such as the possible existence of as yet unidentified human herpesviruses.

See also: Epstein–Barr Virus: General Features; Fish and Amphibian Herpesviruses; Herpes Simplex Viruses: General Features; Herpesviruses: General Features; Human Cytomegalovirus: General Features; Kaposi's Sarcoma-Associated Herpesvirus: General Features; Varicella-Zoster Virus: General Features.

Further Reading

Chmielewicz B, Goltz M, Franz T, et al. (2003) A novel porcine gammaherpesvirus. *Virology* 308: 317–329.

Damania B and Desrosiers RC (2001) Simian homologues of human herpesvirus 8. *Philosophical Transactions of the Royal Society of London B: Biological Sciences* 356: 535–543.

Davison AJ (2002) Evolution of the herpesviruses. *Veterinary Microbiology* 86: 69–88.

Davison AJ, Eberle R, Hayward GS, et al. (2005) Family *Herpesviridae*. In: Fauquet CM, Mayo MA, Maniloff J, Desselberger U, and Ball LA (eds.) *Virus Taxonomy: Eighth Report of the International Committee on Taxonomy of Viruses*, pp. 193–212. San Diego, CA: Elsevier Academic Press.

Ehlers B, Borchers K, Grund C, Frölich K, Ludwig H, and Buhk H-J (1999) Detection of new DNA polymerase genes of known and potentially novel herpesviruses by PCR with degenerate and deoxyinosine-substituted primers. *Virus Genes* 18: 211–220.

Ehlers B, Ochs A, Leendertz F, Goltz M, Boesch C, and Mätz-Rensing K (2003) Novel simian homologues of Epstein–Barr virus. *Journal of Virology* 77: 10695–10699.

Huang CA, Fuchimoto Y, Gleit ZL, et al. (2001) Posttransplantation lymphoproliferative disease in miniature swine after allogenic hematopoietic cell transplantation: Similarity to human PTLD and association with a porcine gammaherpesvirus. *Blood* 97: 1467–1473.

Kellam P (1998) Molecular identification of novel viruses. *Trends in Microbiology* 6: 160–165.

Lacoste V, Mauclere P, Dubreuil G, Lewis J, Georges-Courbot MC, and Gessain A (2000) KSHV-like herpesviruses in chimps and gorillas. *Nature* 407: 151–152.

Li H, Gailbreath K, Flach EJ, et al. (2005) A novel subgroup of rhadinoviruses in ruminants. *Journal of General Virology* 86: 3021–3026.

Luebcke E, Dubovi E, Black D, Ohsawa K, and Eberle R (2006) Isolation and characterization of a chimpanzee alphaherpesvirus. *Journal of General Virology* 87: 11–19.

Mack DH and Sninsky JJ (1988) A sensitive method for the identification of uncharacterized viruses related to known virus groups: Hepadnavirus model system. *Proceedings of the National Academy of Sciences, USA* 85: 6977–6981.

McGeoch DJ, Dolan A, and Ralph AC (2000) Towards a comprehensive phylogeny for mammalian and avian herpesviruses. *Journal of Virology* 74: 10401–10406.

Rose TM (2005) CODEHOP-mediated PCR – A powerful technique for the identification and characterization of viral genomes. *Virology Journal* 2: 20.

VanDevanter DR, Warrener P, Bennett L, et al. (1996) Detection and analysis of diverse herpesviral species by consensus primer PCR. *Journal of Clinical Microbiology* 34: 1666–1671.

Wang D, Coscoy L, Zylberberg M, et al. (2002) Microarray-based detection and genotyping of viral pathogens. *Proceedings of the National Academy of Sciences, USA* 99: 15687–15692.

Relevant Websites

http://blocks.fhcrc.org – CODEHOP to Design PCR Primers from Blocks, Fred Hutchinson Cancer Research Center.

http://www.ncbi.nlm.nih.gov – National Center for Biotechnology Information (NCBI) Home Page.

Herpesviruses: General Features

A J Davison, MRC Virology Unit, Glasgow, UK

© 2008 Elsevier Ltd. All rights reserved.

Introduction

The family *Herpesviridae* consists of a substantial number of animal viruses that have large, double-stranded DNA genomes and share a defining virion structure. Herpesviruses have been discovered in vertebrates from fish to humans, and one has been found in invertebrates (bivalves). They exhibit a wide range of pathogenic properties, ranging from inapparent in many instances to life-threatening (including cancer) in some, and have the ability to establish lifelong latent infections that can reactivate periodically. This chapter introduces the fundamental molecular properties of the herpesviruses: structure, replication, classification, genome, genes, and evolution.

Virion Structure

Herpesvirus particles (virions) are spherical and have an approximate diameter of 200 nm. They have a unique morphology consisting of four basic components (**Figure 1**).

Core

The core is occupied by the virus genome, which is a large, linear, double-stranded DNA molecule packaged at high density.

Capsid

The capsid is an icosahedron of diameter 125–130 nm. It is fashioned from 161 protein capsomeres, which are contributed by 150 hexons and 11 pentons, plus the portal in the 12th pentonal position. The hexons and pentons contain six and five copies each, respectively, of the major capsid protein, each copy of which in the hexons is decorated by a small, external protein. The portal consists of 12 copies of the portal protein and forms the vertex through which DNA enters and leaves the capsid. The capsomeres in the capsid shell are joined together via complexes known as triplexes, which contain two copies of one protein and one copy of another.

The immature capsid shell is constructed in the cell nucleus around a scaffold that consists of approximately 1200 copies of the scaffold protein plus approximately 150 copies of an N-terminally extended form of this protein that contains a protease domain in the additional region. The scaffold is replaced by the virus genome during DNA packaging in a process that involves proteolytic cleavage and loss of all the scaffold protein fragments except for the protease domain.

Tegument

The capsid is embedded in the tegument, and is often not located centrally therein. The tegument contains many virus protein species and is poorly defined structurally except for the region close to the capsid, where a degree of icosahedral symmetry exists.

Envelope

The tegument is wrapped in the lipid envelope, which contains several virus membrane glycoproteins as well as some cellular proteins.

Replication Cycle

The lytic replication cycle has been studied in detail for relatively few herpesviruses, most information having come from herpes simplex virus type 1 (HSV-1). As a consequence, there is a substantial degree of generalization – and even controversy – about applying conclusions from one virus across the whole family. This is especially pertinent to aspects of virus growth involving functions that are not genetically conserved. The mechanisms involved in establishment of, and reactivation from, latency are in this class, with different groups of herpesviruses having evolved different ways of ensuring that the virus stays with the host for life. Herpesviruses possess elaborate means of modulating the host responses to infection, and the genetic functions involved are largely not conserved across the family.

In broad outline, the phases of the lytic replication cycle are as follows. Details on the lytic and latent replication cycles (and pathogenic properties) of specific viruses are available in the relevant articles.

Entry

The virion envelope fuses to the cell membrane via the interactions of cellular receptors with envelope glycoproteins, and the capsid and tegument enter the cytoplasm. Some – perhaps many – of the incoming tegument proteins have specific roles in modulating the replication cycle, for example, in inhibiting host macromolecular synthesis and initiating virus transcription. Some tegument proteins also reach the nucleus, perhaps independently of the capsid.

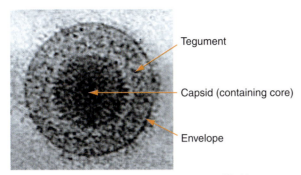

Figure 1 Structure of the herpesvirion exemplified by a cryoelectron microscope image of HSV-1. Adapted from Rixon FJ (1993) Structure and assembly of herpesviruses. *Seminars in Virology* 4: 135–144, with permission from Elsevier.

Expression

Capsids are transferred via microtubules to nuclear pores, through which the virus genome enters the nucleus. The genome is transcribed into mRNA by host RNA polymerase II in three major, regulated phases of transcription. Transcription of immediate early genes, unlike that of later phases, is not dependent on protein synthesis. Some of the immediate early proteins regulate the expression of early and late genes. Functions involved in DNA replication or nucleotide metabolism, and a subset of structural proteins, are in the former class, and many virion proteins are in the latter class.7

Some polyadenylated RNAs, or introns derived therefrom, appear to function via pathways not requiring translation into proteins. In some herpesviruses, small, noncoding RNAs are transcribed by RNA polymerase III. The functions of these RNAs are largely unknown.

Replication

In the nucleus, DNA replication initiates at one or more specific sequences (origins of replication) to generate head-to-tail concatemeric genomes, probably by a rolling circle mechanism. Unit-length genomes are cleaved from the concatemers and packaged into preformed capsids.

Maturation

Capsids acquire some tegument proteins in the nucleus and gain a temporary envelope by budding through the inner nuclear membrane. This envelope is lost upon transit across the outer nuclear membrane into the cytoplasm. Further tegument proteins are added in the cytoplasm, and the tegumented capsids gain their final envelope by budding into a post-Golgi compartment.

Exit

Mature virions are released from the cell by reverse endocytosis.

Classification

The task of classifying viruses, including herpesviruses, commenced in 1966 and is undertaken by the International Committee on Taxonomy of Viruses (ICTV). Various criteria have been utilized, as enabled by developments in knowledge and technology. The current taxonomy of herpesviruses, as laid out in the *Eighth Report of the ICTV* (2005), is summarized in **Table 1**. The family *Herpesviridae* is divided into three subfamilies populated by herpesviruses of higher vertebrates (mammals and birds): *Alphaherpesvirinae* (containing four genera: *Simplexvirus*, *Varicellovirus*, *Mardivirus*, and *Iltovirus*), *Betaherpesvirinae* (containing three genera: *Cytomegalovirus*, *Muromegalovirus*, and *Roseolovirus*), and *Gammaherpesvirinae* (containing two genera: *Lymphocryptovirus* and *Rhadinovirus*). The family also contains a genus (*Ictalurivirus*) that is not linked to any subfamily and has a single fish herpesvirus as a member.

In proposals under consideration by the ICTV (**Table 1**), an order (*Herpesvirales*) is introduced as the highest taxon. The family *Herpesviridae* is redefined as restricted to viruses of higher vertebrates (which in this context will eventually include reptiles as well as mammals and birds). Within this redefined family, an additional genus (*Proboscivirus*) is added to the *Betaherpesvirinae*, and two new genera (*Macavirus* and *Percavirus*) are split from the *Rhadinovirus* genus in the *Gammaherpesvirinae*. Two new families are also created: *Alloherpesviridae* to accommodate viruses of lower vertebrates (amphibians and fish), and *Malacoherpesviridae* for the single herpesvirus of invertebrates (bivalves).

The formal names of herpesvirus species follow a taxonomic designation derived from the natural host followed by the word 'herpesvirus' and a number (e.g., *Human herpesvirus 4* – the fourth of the eight species whose members infect humans). However, many herpesviruses are better known by common names (e.g., Epstein–Barr virus rather than human herpesvirus 4), and in this respect are characterized by a dual nomenclature. In addition, vernacular terms are often used to imply members of the family (e.g., herpesvirus), subfamily (e.g., gammaherpesvirus), or genus (e.g., rhadinovirus).

Herpesviruses are defined as separate species if their nucleotide sequences differ in a readily assayable and distinctive manner across the entire genome and if they occupy different ecological niches by virtue of their distinct epidemiology and pathogenesis or their distinct natural hosts. This definition implies that several criteria are employed in identifying a new species, and their application is to some extent arbitrary. The criteria are outlined below.

Morphological Criteria

Assignment of a virus to the family *Herpesviridae* depends primarily upon virion morphology, as described above.

Table 1 Present and proposed herpesvirus classification schemes. Species other than the type species of each genus are not included

Taxon level	Current taxon	Proposed taxon	Common name of virus
Order		Herpesvirales	
Family	Herpesviridae	Herpesviridae	
Subfamily	Alphaherpesvirinae	Alphaherpesvirinae	
Genus	Simplexvirus	Simplexvirus	
Type species	Human herpesvirus 1	Human herpesvirus 1	Herpes simplex virus type 1
Genus	Varicellovirus	Varicellovirus	
Type species	Human herpesvirus 3	Human herpesvirus 3	Varicella-zoster virus
Genus	Mardivirus	Mardivirus	
Type species	Gallid herpesvirus 2	Gallid herpesvirus 2	Marek's disease virus type 1
Genus	Iltovirus	Iltovirus	
Type species	Gallid herpesvirus 1	Gallid herpesvirus 1	Infectious laryngotracheitis virus
Subfamily	Betaherpesvirinae	Betaherpesvirinae	
Genus	Cytomegalovirus	Cytomegalovirus	
Type species	Human herpesvirus 5	Human herpesvirus 5	Human cytomegalovirus
Genus	Muromegalovirus	Muromegalovirus	
Type species	Murid herpesvirus 1	Murid herpesvirus 1	Murine cytomegalovirus
Genus	Roseolovirus	Roseolovirus	
Type species	Human herpesvirus 6	Human herpesvirus 6	Human herpesvirus 6
Genus		Proboscivirus	
Type species		Elephantid herpesvirus 1	Endotheliotropic elephant herpesvirus
Subfamily	Gammaherpesvirinae	Gammaherpesvirinae	
Genus	Lymphocryptovirus	Lymphocryptovirus	
Type species	Human herpesvirus 4	Human herpesvirus 4	Epstein–Barr virus
Genus	Rhadinovirus	Rhadinovirus	
Type species	Saimiriine herpesvirus 2	Saimiriine herpesvirus 2	Herpesvirus saimiri
Genus		Macavirus	
Type species		Alcelaphine herpesvirus 1	Wildebeest-associated malignant catarrhal fever virus
Genus		Percavirus	
Type species		Equid herpesvirus 2	Equine herpesvirus 2
Family		Alloherpesviridae	
Genus	Ictalurivirus	Ictalurivirus	
Type species	Ictalurid herpesvirus 1	Ictalurid herpesvirus 1	Channel catfish virus
Family		Malacoherpesviridae	
Genus		Ostreavirus	
Type species		Ostreid herpesvirus 1	Oyster herpesvirus

Biological Criteria

The three subfamilies *Alphaherpesvirinae*, *Betaherpesvirinae*, and *Gammaherpesvirinae* were defined initially on the basis of biological criteria, in the application of which a degree of generality is implicit: host range in cell culture, length of replication cycle, cytopathology, and characteristics of latency. Alphaherpesviruses exhibit a variable host range, have a short replication cycle, spread quickly with efficient destruction of infected cells, and establish latent infections primarily in sensory ganglia. Betaherpesviruses have a narrow host range and a long replication cycle with slow spread and cell enlargement. Viruses may become latent in secretory glands and lymphoreticular cells. Gammaherpesviruses are associated with lymphoproliferative diseases, and latent virus is present in lymphoid tissue. They infect B- or T-lymphocyte cells *in vitro*, in which infection is frequently arrested so that infectious progeny are not produced. However, some gammaherpesviruses are able to cause lytic infections in fibroblastoid or epithelial cell lines.

Serological Criteria

Closely related viruses may be detected using antisera or specific antibodies.

Genomic Criteria

The general layout of a genome in terms of unique and repeat regions (see below) is limited as a criterion since, although particular structures are found more commonly in certain subfamilies or genera, similar structures have evidently evolved more than once. The DNA sequences of herpesviruses are sufficiently diverged to limit extensive nucleic acid similarity (detectable by hybridization) to closely related viruses.

Sequence Criteria

Phylogenetic analysis based on sequence comparisons between conserved genes has become the predominant criterion in classification of herpesviruses at all taxonomic levels. **Figure 2** shows one example of a phylogenetic tree supporting the current classification scheme. In recent years, many herpesviruses have been detected in animal tissues solely by polymerase chain reaction (PCR) of short sequences. Given the multifactorial nature of the herpesvirus species definition (see above), the absence of other information renders formal classification problematic in these cases.

Genome Organization

The herpesvirus genomes studied to date range in size from approximately 124 kbp (simian varicella virus) to 295 kbp (koi herpesvirus) and in nucleotide composition from 32% to 74% G+C. Alphaherpesviruses are not deficient in the 5′ CG dinucleotide, betaherpesviruses are deficient in very limited regions, and gammaherpesviruses are deficient throughout their genomes. 5′ CG deficiency is thought to be a consequence of spontaneous deamination of methylated cytosine residues in DNA over evolutionary time to produce 5′ TG. There is evidence that methylation of certain regions of the Epstein–Barr virus genome resident in peripheral blood lymphocyte cells may be involved in perpetuating infection. Extensive methylation of the 5′ CG dinucleotide has been found in the virion DNA of two frog herpesviruses, both of which encode the enzyme DNA (cytosine-5-)-methyltransferase.

Herpesvirus genomes differ in the arrangements of direct and inverted repeat regions with respect to unique

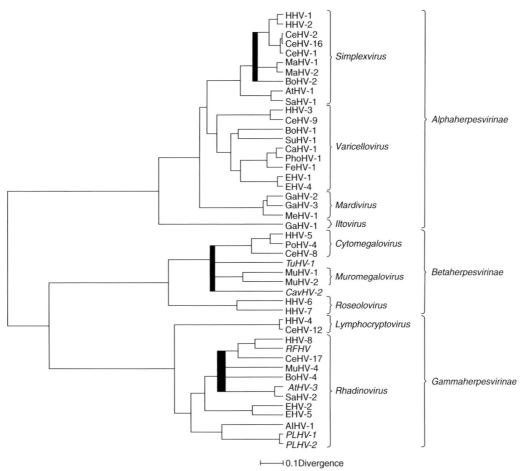

Figure 2 Phylogenetic tree for herpesviruses. The scope is limited to the three subfamilies that comprise the revised family *Herpesviridae*; see **Table 1**. The tree is a composite derived from sequence alignments of up to eight sets of core proteins, analyzed using the maximum-likelihood method with a molecular clock imposed. Thick lines denote regions of uncertain branching. Prefixes denote host species: H, human; Ce, cercopithecine; Ma, macropodid; Bo, bovine; At, ateline; Sa, saimiriine; Su, suid; Ca, canid; Pho, phocid; Fe, felid; E, equid; Ga, gallid; Me, meleagrid; Po, pongine; Tu, tupaiid; Mu, murid; Cav, caviid; and Al, alcelaphine. Other virus name abbreviations are: RFHV, retroperitoneal fibromatosis herpesvirus of macaques; and PLHV-1 and PLHV-2, porcine lymphotropic herpesviruses 1 and 2. Viruses that are not yet incorporated into genera are in italics. Genera and subfamilies are on the right. Adapted from McGeoch DJ, Dolan A, and Ralph AC (2000) Toward a comprehensive phylogeny for mammalian and avian herpesviruses. *Journal of Virology* 74: 10401–10406, with permission from American Society for Microbiology.

(nonrepetitive) regions. **Figure 3** illustrates the five types of structure that have been confirmed adequately. The type A structure (e.g., human herpesvirus 6) consists of a unique region (U) flanked by a direct terminal repeat (TR) at the genome ends. Type B genomes (e.g., human herpesvirus 8, also known as Kaposi's sarcoma-associated herpesvirus) contain variable numbers of a TR at each end of the genome. The type C structure (e.g., Epstein–Barr virus) has in addition a variable number of copies of an internal direct repeat (IR), which is unrelated to TR, and the presence of IR splits U into two unique regions (U_L and U_S). The type D structure (e.g., pseudorabies virus) also has two unique regions (U_L and U_S), with U_S flanked by an inverted repeat (TR_S/IR_S). In some type D genomes (e.g., varicella-zoster virus), U_L is flanked by a very small, unrelated inverted repeat (TR_L/IR_L). By virtue of recombination between inverted repeats, the two orientations of U_S are present in equimolar amounts in virion DNA populations, and U_L is present completely or predominantly in a single orientation. In the type E structure (e.g., HSV-1), TR_L/IR_L is larger and the two orientations of U_L and U_S are each present in equimolar amounts in virion DNA populations, giving rise to four genome isomers. A direct repeat (the *a* sequence) is also present at the termini, and an inverted copy (*a'*) is present internally.

In addition to large-scale repeats, regions consisting of tandem repeats of short sequences are found at various locations in many herpesvirus genomes. The elements that make these up may be identical to each other or merely similar, giving rise to simple or complex repeats, and their presence in variable numbers causes genome size heterogeneity.

Gene Content

To date, the genome sequences of herpesviruses belonging to 42 species have been published, with more than one strain determined for several of these species. Details are available via the genomic biology Web pages at the National Center for Biotechnology Information. In addition, a large amount of data is available for portions of other herpesvirus genomes. Analyses of these sequences provide detailed views of genetic content and phylogeny.

Herpesvirus genomes consist mostly of protein-coding sequences. Gene complements range from about 70 to 170 and are arranged about equally between the two DNA strands. They are not generally located in functionally related groups. Overlap between protein-coding regions in different reading frames on the same or opposing strands is infrequent and, where it does occur, is not extensive. In instances where extensive or complete overlap of two protein-coding regions has been mooted, evidence that both encode functional proteins is invariably unconvincing.

Most herpesvirus genes are transcribed from their own promoters and thus specify mRNAs with unique 5' ends. Groups of genes are frequently arranged tandemly on the same DNA strand and share a common polyadenylation site, with mRNAs sharing the same 3' end. Most genes are not spliced, particularly among the alphaherpesviruses. RNA polymerase II transcripts that do not encode proteins but may have other functions, and micro-RNAs, have been identified in representatives of all three subfamilies, and RNA polymerase III transcripts in certain gammaherpesviruses. The functions of such transcripts are largely unknown. Herpesvirus genomes also contain

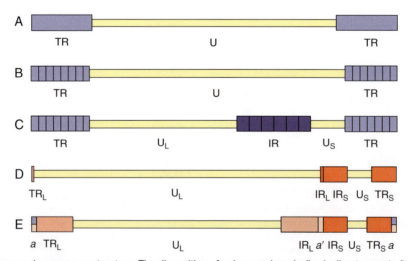

Figure 3 Types of herpesvirus genome structure. The disposition of unique regions (yellow), direct repeats (blue), and inverted repeats (red) are shown (not to scale). Within a structure, unique regions are unrelated to each other, and only the repeats depicted in the same color and shade are identical. Among the structures, repeats represented in the same color and shade are not necessarily related to each other. In the type E structure, an inverted copy of a direct repeat (*a*) at the genome termini is present internally (*a'*).

cis-acting signals in addition to those involved in transcriptional control, such as origins of DNA replication.

Sequence comparisons among members of the revised family *Herpesviridae* indicate that a set of 43 core genes has been inherited from the ancestor of the three subfamilies. This number of core genes is an approximation, since its derivation assumes that the *Alphaherpesvirinae* was the first subfamily to diverge (as indicated by molecular phylogeny; see **Figure 2**). Moreover, it is based not simply on primary sequence conservation but, in some cases, on positional and functional similarities. Three of the core genes derived from the ancestor have been lost subsequently from certain lineages. The core genes are generally arranged collinearly among members of the same subfamily, but in members of different subfamilies blocks of genes have been rearranged and sometimes inverted.

The most strongly represented functional categories among the core genes are involved in central aspects of lytic replication. Particularly prominent are genes encoding capsid proteins (six genes) or proteins involved in replication (seven genes) or processing and packaging of DNA (eight genes). Also featured are genes with roles in nucleotide metabolism or DNA repair (five genes), and, to a lesser extent, genes encoding tegument (seven genes) or envelope proteins (five genes). Genes with roles in niche-specific aspects of the life cycle, such as modulation of the host response and regulation of latency, do not generally belong to the core set. This feature illustrates the high degree of evolutionary flexibility that herpesviruses have exhibited in colonizing particular ecological niches. The fact that over half (in some cases well over half) of the genes in a herpesvirus genome may be removed individually without eliminating viral growth in cell culture, when all presumably have a role *in vivo*, testifies to the extensive interactions that the virus has with its host during natural growth and transmission.

The comments above refer to the revised family *Herpesviridae*, where most information is available. Similar points may be made about the proposed family *Alloherpesviridae*, though the number of core genes is considerably fewer (probably not more than 20), presumably due to the greater antiquity of the common ancestor.

Herpesvirus Evolution

The processes that have occurred during evolution of the three subfamilies of the *Herpesviridae* from their common ancestor are the same as in other organisms, and include the gradual effects of nucleotide substitution, deletion, and insertion to modify existing genes or produce genes *de novo*, and the recombinational effects of gene capture, duplication, and rearrangement. These processes are also evident in the proposed family *Alloherpesviridae*.

Herpesviruses are usually restricted to a single species in the natural setting, and severe symptoms of infection are often limited in the natural host to young or immunosuppressed individuals. This indicates that a substantial degree of co-adaptation and co-evolution have taken place over the long term between herpesviruses and their hosts. Although some exceptions suggesting ancient interspecies transfer have been registered, phylogenetic studies provide strong support for co-evolution as a general historic characteristic, in some instances supporting the idea that the viruses have speciated along with their hosts. This correspondence has allowed a timescale for herpesvirus evolution to be proposed on the basis of accepted dates for host evolution. For example, the most recent analyses suggest divergence dates of 400 million years (before present) for the three subfamilies, 120 million years for the genera *Simplexvirus* and *Varicellovirus* of the *Alphaherpesvirinae*, and 8 million years for HSV-1 and HSV-2. These dates are approximate, especially the further back they go, and vulnerable to the vicissitudes of estimated host dates and developments in analytical approaches.

The three lineages populated by the revised family *Herpesviridae* and the two proposed families *Alloherpesviridae* and *Malacoherpesviridae* share similarities in capsid structure and replication that indicate evolution from a common ancestor, the forerunner of all herpesviruses. However, in comparison with the situation within the families, detectable amino acid sequence similarity among proposed families is very limited, with evidence for a common ancestor focused on a single gene (encoding a subunit of the terminase complex responsible for DNA packaging). These features imply that the divergence events that gave rise to the three lineages are very ancient. Moreover, structural considerations focused on the capsid continue to stimulate the notion that herpesviruses might share an even earlier ancestor with T4-like bacteriophages.

Note: The proposals for new taxa mentioned in this article have been accepted by the ICTV.

See also: Herpesviruses: Discovery.

Further Reading

Davison AJ (2002) Evolution of the herpesviruses. *Veterinary Microbiology* 86: 69–88.

Davison AJ, Eberle R, Hayward GS, et al. (2005) Family *Herpesviridae*. In: Fauquet CM, Mayo MA, Maniloff J, Desselberger U, and Ball LA (eds.) *Virus Taxonomy: Eighth Report of the International Committee on Taxonomy of Viruses*, pp. 193–212. San Diego, CA: Elsevier Academic Press.

Grunewald K, Desai P, Winkler DC, et al. (2003) Three-dimensional structure of herpes simplex virus from cryo-electron tomography. *Science* 302: 1396–1398.

Kieff E and Rickinson AB (2001) Epstein–Barr virus and its replication. In: Knipe DM and Howley PM (eds.) *Fields Virology,* 4th edn., pp. 2511–2573. Philadelphia, PA: Lippincott Williams and Wilkins.

McGeoch DJ, Dolan A, and Ralph AC (2000) Toward a comprehensive phylogeny for mammalian and avian herpesviruses. *Journal of Virology* 74: 10401–10406.

McGeoch DJ and Gatherer D (2005) Integrating reptilian herpesviruses into the family *Herpesviridae*. *Journal of Virology* 79: 725–731.

McGeoch DJ, Rixon FJ, and Davison AJ (2006) Topics in herpesvirus genomics and evolution. *Virus Research* 117: 90–104.

Mocarski ES and Tan Courcelle C (2001) Cytomegaloviruses and their replication. In: Knipe DM and Howley PM (eds.) *Fields Virology*, 4th edn., pp. 2629–2673. Philadelphia, PA: Lippincott Williams and Wilkins.

Moore PS and Chang Y (2001) Kaposi's sarcoma-associated herpesvirus. In: Knipe DM and Howley PM (eds.) *Fields Virology*, 4th edn., pp. 2803–2833. Philadelphia: Lippincott Williams and Wilkins.

Rixon FJ (1993) Structure and assembly of herpesviruses. *Seminars in Virology* 4: 135–144.

Roizman B and Pellett PE (2001) Herpes simplex viruses and their replication. In: Knipe DM and Howley PM (eds.) *Fields Virology*, 4th edn., pp. 2399–2459. Philadelphia, PA: Lippincott Williams and Wilkins.

Zhou ZH, Dougherty M, Jakana J, *et al.* (2000) Seeing the herpesvirus capsid at 8.5 Å. *Science* 288: 877–880.

Herpesviruses: Latency

C M Preston, Medical Research Council Virology Unit, Glasgow, UK

© 2008 Elsevier Ltd. All rights reserved.

Glossary

Ganglion A small organ containing the cell bodies, including nuclei, of neurons.

Iontophoresis Introduction through the skin or cornea by applying an electrical charge.

Introduction

Members of the family *Herpesviridae* exhibit the ability to remain latent in tissues of the host following primary infection. Latent virus is retained for the lifetime of the host, and can be reactivated to cause recurrent disease. Latency is a crucial property for the survival of herpesviruses, overcoming the requirement for rapid reinfection of new individuals in order to spread within a population. As described below, the characteristics of latency show variations when the different herpesvirus subfamilies are considered. Operationally, latency is defined as the presence of the viral genome without detectable virus production, coupled with the potential for resumption of virus replication in response to reactivation signals. For descriptive purposes, latency is characterized by three phases: establishment, maintenance, and reactivation.

Early thoughts on latency focused on two basic ideas, named the static and dynamic models. In the static model, the viral genome is considered to be nonreplicating due to a failure to undergo the normal program of gene expression, and reactivation is viewed as a change in intracellular conditions such that replication resumes. Dynamic models of latency envisage a slow persistent production of virus that is normally controlled by the host immune system and does not cause overt disease until a reactivation stimulus diminishes host defenses. There is currently no consensus that either of these models is correct; instead, research into latency continually reveals greater complexity in the interaction of virus with host. It is now thought that interference with virus replication occurs at many levels, ranging from repression of gene expression to control by immunological defenses. The requirement for long-term retention of the viral genome, coupled with an ability to spread within a population, demands that a complex and intimate relationship exists between virus and host. This relationship can be considered at two levels, namely the individual cell and the organism. At the cellular level, there must be mechanisms for sequestering the viral genome in a host cell that is potentially permissive for replication, and for retaining the ability to resume replication in response to appropriate stimuli. In terms of the organism, the virus must evade detection by the immune system during latency and must overcome host defenses when reactivating.

Latency will be considered in terms of the three subfamilies of the family *Herpesviridae*, focusing on human viruses since these are understood in the greatest detail.

Alphaherpesvirinae

The prototype of the subfamily *Alphaherpesvirinae*, herpes simplex virus (HSV) type 1 (HSV-1), is the most intensively studied in terms of latency. In humans, its natural host, primary exposure usually occurs during infancy and is characterized by a mild infection of the oropharynx that is frequently unnoticed. After the initial infection has resolved, a proportion of individuals experience periodic reactivation in response to stressful stimuli, such as excessive exposure to sunlight, resulting in the appearance of 'cold sore' lesions around the lip and less frequently on other areas of the face. HSV type 2 (HSV-2) is more frequently associated with genital herpes, although nowadays genital HSV-1 is detected with increasing frequency. Genital herpes and facial cold sores are treated by application of acyclovir, but this agent is only effective in preventing lesions once reactivation has occurred. There is currently no antiviral that eradicates latent virus or prevents reactivation.

HSV establishes latency in sensory neurons that innervate the site of initial infection. The viral genome is retained in neurons of the relevant ganglia for the lifetime of the host, and upon reactivation virus is released into tissues served by neurons extending from the ganglion. During initial infection, replicating HSV enters nerve termini and virus particles are transported along axons in a retrograde manner to the ganglia. Upon reactivation, virus moves in an anterograde direction from the ganglion to the surface, where replication causes disease.

Animal models have been developed for the study of HSV latency, and although each has limitations, they have been important in linking molecular investigations with *in vivo* studies. Latency is established efficiently after infection of mice with HSV-1, but reactivation is difficult to achieve *in vivo*. The most common method of reactivating latent virus from mice is to explant ganglia, a process in which ganglia are removed by dissection and cultured in the laboratory. This invariably results in the appearance of infectious virus within a few days but its physiological relevance is debatable. Subjecting mice to transient hyperthermia reliably reactivates virus replication in ganglia *in vivo*, albeit at low efficiency. The best animal for the study of reactivation is the rabbit, since virus is spontaneously released in tear films after ocular infection and release can be stimulated by iontophoresis of epinephrine. Genital HSV-2 infection can be reproduced, to some extent, by intravaginal inoculation of guinea pigs.

In considering establishment of latency, a crucial question concerns the way in which the normally inexorable progression to lytic replication and cell death is interrupted. Transcripts and proteins characteristic of lytic infection cannot be detected easily during latency, indicating a global repression of gene expression. Furthermore, studies with HSV mutants have failed to identify a gene product that is dispensable for lytic replication but required for latency, thus the prevailing hypothesis is that establishment of latency results by default when viral gene expression is somehow arrested in neurons. It is currently thought that the block occurs early in the virus life cycle, and the most likely point for intervention by the cell is at the immediate early (IE) stage. A favored hypothesis contends that IE transcription is compromised in sensory neurons, possibly due to inefficient transport of the virion transactivator protein VP16 (virion protein number 16), which activates IE transcription, to the neuronal nucleus, or to lack of the cellular transcription factors Oct-1 (a cellular protein that binds sequences with the consensus ATGCAAAT) or host cell factor (HCF, a large cell protein), with which VP16 is able to form a complex. Alternatively, sensory neurons may contain proteins that act as competitive inhibitors of Oct-1 and/or HCF, blocking their interactions with VP16. The possibility of arrest at the IE stage is supported by experiments with tissue culture cells. HSV-1 mutants that are severely impaired for IE gene expression can be retained in cells for extended periods in a nontranscribed 'quiescent' state that apparently mimics the transcriptional silence characteristic of latency.

Despite the overall absence of viral lytic gene products, latently infected neurons can be readily identified by the presence of a 2 kbp viral RNA known as the latency-associated transcript (LAT). Thousands of molecules of this RNA are found in the nucleus, enabling its detection by *in situ* hybridization or RNA blots. The observed species is an unusually stable intron spliced from a larger precursor that is present in neurons at much lower levels. Structurally, LAT is a lariat, that is, a splicing intermediate of a type typically cleaved and rapidly degraded. The LAT lariat has an unusual branch point sequence that is not recognized by cellular debranching enzymes. A smaller 1.5 kbp RNA, derived by further splicing, is found only in neurons. Intriguingly, LAT is complementary to sequences encoding the C-terminal portion of the IE protein ICP0 (infected cell protein number 0), an important activator of gene expression, leading to the suggestion that LAT blocks ICP0 production by an antisense mechanism. At present, there is no evidence for protein products encoded by HSV-1 LAT or the longer precursor in latently infected neurons, although there is evidence that the bovine herpesvirus 1 latency-related transcript does encode one or more proteins.

The significance of LAT for latency is controversial. Viral mutants that are unable to produce LAT nonetheless establish latency and can be reactivated; thus, LAT does not have an essential role in any of the animal models currently available. In detail, however, LAT mutants are deficient in reactivation in a number of contexts, suggesting a modulatory role for the transcripts. In some animal models, the absence of LAT appears to reduce the efficiency of reactivation directly, whereas in others LAT mutants exhibit a defect in establishment, thereby indirectly reducing reactivation due to the smaller pool of latent genomes. It is relevant that LAT mutants cause greater destruction of neurons during the initial stages of infection. This may be because one of LAT's normal roles is to exert antisense inhibition of ICP0 expression, thereby restricting productive infection, but currently the favored interpretation is that LAT has anti-apoptotic properties that reduce neuronal death after infection. Therefore, LAT could be important for improving establishment of latency by preventing loss of infected neurons, but equally a direct effect on reactivation is possible by keeping reactivating neurons healthy for long periods.

During maintenance of latency, which can last for many decades in humans, it is thought that cellular factors contribute to the stable repression of transcription. The genome is believed to be organized into a chromatin-like structure, and indeed histones carrying post-translational modifications characteristic of inactive cellular chromatin are associated with latent HSV-1 genomes. Interestingly, the LAT region, which escapes repression, is associated with histones normally found on actively transcribed genes. Therefore, cellular mechanisms for global control of transcription may operate on latent HSV DNA.

The latent viral genome exists as a circular episome, structurally distinct from the linear molecule found in the virion. This suggests that the gene expression program is arrested at an early stage, since circularization occurs shortly after virus entry into the cell as a prelude to replication. Alternatively, circular molecules may be remnants from replication during productive infection. It is possible that circularization is important for the assembly of the latent genome into a chromatin structure. There is no evidence for integration of HSV DNA into the host genome during latency.

During latency in humans or experimental animals, only a fraction of neurons in the ganglion (typically 0.1–5% but in some cases up to 30%) harbor latent HSV-1, and neurons themselves constitute only about 5% of the cells in the ganglion. However, viral genomes can be detected by Southern hybridization of total ganglion DNA, demonstrating that neurons harbor many copies of HSV DNA. Analysis of single cells by polymerase chain reaction (PCR) confirms this conclusion and further reveals a wide disparity in the viral copy number per neuron. Most neurons harbor fewer than 100 viral genomes, but a minority can contain thousands.

Although it is clear that most latent genomes do not express lytic gene products at detectable levels, sensitive analysis of mouse ganglia by reverse transcriptase-PCR demonstrates the presence of viral transcripts from loci outside the LAT region. The transcripts are present in low amounts; they may signify infrequent viral lytic gene expression, but they may simply represent a small amount of background transcription of the latent genome. In an extensive analysis of mouse ganglia, a very small number of neurons (fewer than one per mouse) were undergoing an apparently productive infection, again showing that silencing of the genome is not absolute. Therefore, a low level of viral lytic gene expression may occur during latency.

Immunological studies support the view that virus lytic gene expression may occasionally occur during latency. Leukocytes, predominantly CD8+ T cells, can be found in ganglia from humans or latently infected experimental animals, often in intimate association with neurons. In mice, the infiltrating T cells are HSV-1 specific, with the majority recognizing a single immunodominant epitope on glycoprotein B. Furthermore, latently infected ganglia contain cytokines such as RANTES (an attractant for T cells), γ-interferon, and tumor necrosis factor (both produced by T cells). The continued presence of CD8+ T cells and derived cytokines is indicative of persistent stimulation of the immune system during latency. The favored interpretation of the data is that T cells respond to and eliminate a low level of virus that is produced from infected neurons.

The exact nature of the stimuli that provoke reactivation is unclear, and at present the general term 'stress' is used. The stress can be delivered at the periphery, directly to the neuron, or systemically. In experimental animals, explantation of ganglia efficiently reactivates latent HSV, but virus is only released from a small proportion of infected neurons. Procedures that provoke reactivation *in vivo*, such as hyperthermic treatment of mice, are even less efficient. The low efficiency imposes constraints on studying the molecular basis of reactivation, since it is not possible to identify the small number of reactivation-competent genomes among the total latent population. The normal program of HSV gene expression probably does not operate during reactivation because VP16 is absent during latency. It is suspected that stress causes alterations to the transcription factor profile of neurons, resulting in changes at the ICP0 promoter and synthesis of ICP0 protein which then de-represses the entire viral genome. An alternative view of reactivation is that stresses impair immunological mechanisms that normally prevent a low level of reactivating virus from causing disease. The impairment may occur in the ganglion, at the periphery, or both, with the outcome that virus is able to replicate. The application of PCR has revealed that low-level asymptomatic shedding of HSV in humans is more common than previously thought, supporting the view that local immunity prevents reactivated virus from causing disease.

A summary of HSV latency is shown in **Figure 1**.

Varicella-zoster virus (VZV) causes varicella (chickenpox) as primary infection and zoster (shingles) upon reactivation. VZV is latent in neurons within ganglia throughout the body because the primary infection is widespread rather than localized as in the case of HSV. Studies on VZV latency have proceeded slowly, relying on analysis of human tissue in the absence of a satisfactory animal model to reproduce latency. Although rats can be infected with VZV and harbor genomes in ganglia, virus cannot be reactivated. In humans, latent VZV genomes appear to be retained as circular episomes. The pattern of gene expression during latency, however, differs from that found with HSV. Transcripts representing four lytic genes (ORF21, ORF29, ORF62, and ORF63) have been detected in human ganglia, and there is evidence that protein products, particularly that of ORF63, are also present in infected human and rat neurons. Thus, despite many similarities, there may be significant differences in the details of VZV and HSV latency.

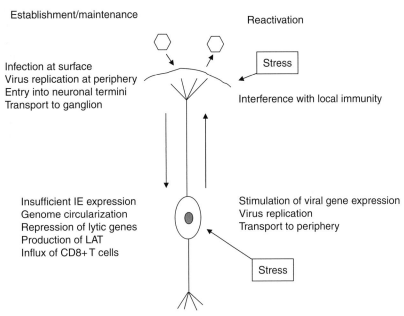

Figure 1 Schematic representation of HSV latency and reactivation.

Betaherpesvirinae

Most studies on latency of the subfamily *Betaherpesvirinae* are concerned with the cytomegaloviruses, particularly human cytomegalovirus (HCMV). In adults, initial infection with HCMV is normally asymptomatic unless it occurs *in utero*, but in immunocompromised individuals it may be serious. Reactivation of latent HCMV can cause rejection of organ transplants, due to replication of virus originating from either the recipient or the donated organ. In addition, AIDS patients can suffer a widespread HCMV infection that has serious consequences. HCMV can be transferred by blood transfusion, suggesting that circulating lymphocytes harbor latent virus, and sensitive PCR analysis, together with cell sorting, has detected the HCMV genome in a small proportion (1 in 10^4 cells) of peripheral blood monocytes. In addition, the HCMV genome can be detected in precursors of monocytes, the CD34+ myeloid progenitors in bone marrow, suggesting that these cells may be the primary reservoir of latent virus.

In view of the low proportion of myeloid cells containing latent HCMV, and the absence of animal models for this virus, molecular studies on HCMV latency have relied heavily on the use of cell culture models. The myelomonocytic line THP-1 is nonpermissive for HCMV replication, but becomes permissive upon differentiation by chemical treatment *in vitro*. Similarly, the teratocarcinoma line NT-2 is nonpermissive until induced to differentiate. The block in undifferentiated cells is due to failure of the major IE transcription unit to be expressed, resulting in arrest of the entire productive infection program. Plasmid-based transfection assays have indicated that specific sequence elements within the major IE promoter (MIEP) are targets for repression. Whereas the entire MIEP is inactive in undifferentiated cells but active in differentiated cells, a truncated MIEP, containing only the 300 base pairs proximal to the mRNA start site, is active in both situations. Mutational analysis has identified critical DNA elements in the MIEP at positions upstream of −300 that contain binding sites for two cellular proteins known to act as repressors. These proteins, known as Yin-Yang 1 (YY1) and Ets-2 repressor factor (ERF), mediate repression of the MIEP in transfection assays, and, furthermore, their levels decline after differentiation by chemical treatment. The implication is that differentiation changes the levels of repressors and hence permits IE gene expression in THP-1 and NT-2 cells.

Chromatin structure at the MIEP is a further important factor in determining permissiveness in cell culture models of HCMV latency. In undifferentiated cells, the MIEP is associated with methylated histones and heterochromatin protein 1 (HP1), both markers of inactive chromatin, whereas in differentiated cells the relative amount of HP1 is reduced and acetylation of histones, a correlate of active chromatin, is increased. It is suspected that YY1 and ERF exert their repressive effects by recruiting histone deacetylases (HDACs) to the MIEP, thereby promoting the formation of inactive chromatin, and this idea is supported by the fact that undifferentiated cells become permissive for IE transcription when treated with chemical inhibitors of HDACs.

The findings from the analysis of cell culture models have received support from studies on monocytes isolated from seropositive individuals and on CD34+ myeloid precursors from human bone marrow. Monocytes do not support HCMV replication, but virus is produced upon differentiation to macrophages in the laboratory. HCMV can be reactivated *ex vivo* by inducing CD34+ cells from seropositive individuals to differentiate to a mature dendritic cell phenotype. Furthermore, differentiation of CD34+ precursors to dendritic cells reproduces the switch from inactive to active chromatin structure at the MIEP. Thus it is thought that HCMV remains latent in CD34+ myeloid precursors and that differentiation to macrophages or dendritic cells leads to intracellular changes that activate the MIEP and ultimately result in virus replication. Although the precise signals that provoke reactivation *in vivo* are not known, one reported means of inducing virus replication is by allogeneic stimulation of blood monocytes, achieved by mixing cells from histo-incompatible individuals. This may reproduce events that occur during transplantation.

The latent HCMV genome is thought to exist as an episome, but it is not known how it is retained in the host. It may be replicated during cell division, by cellular factors or with the participation of viral proteins, or alternatively a low level of persistent virus production may continually seed new latently infected cells. There have been reports of HCMV-specific transcripts produced during latency, but there is little consensus between different laboratories and thus the possibility that viral gene products control latency remains an open question.

At the level of the whole organism, reactivation of HCMV generally causes disease only when the immune system is compromised. A high proportion (>1%) of circulating CD8+ T cells is specific for HCMV, suggesting frequent stimulation of the immune system, presumably through virus reactivation. Indeed, differentiation of monocytes occurs constantly and if such an event frequently reactivates HCMV, immune clearance must be crucial to prevent disease. The potential significance of the immune response in limiting HCMV infection is best demonstrated in studies with murine cytomegalovirus (MCMV). In its natural host, the mouse, latent MCMV is found in numerous tissues, in contrast to the situation with HCMV. During MCMV latency, CB8+ T cells specific for a major IE protein are present, suggesting that immune control eliminates cells containing reactivating virus at an early stage of the gene expression program. Damage to the immune system can result in widespread infection, and although the simplest interpretation of this finding is that immune surveillance is critical for maintaining latency, it must be remembered that immunosuppression also represents a significant stress that may itself activate the latent genome.

HCMV latency is summarized in **Figure 2**.

Gammaherpesvirinae

Most latency studies with the subfamily *Gammaherpesvirinae* have focused on Epstein–Barr virus (EBV), a member of the genus *Lymphocryptovirus*. This virus is usually transmitted by exchange of saliva, with cells of lymphoepithelial structures such as the tonsils initially infected. Infection is generally asymptomatic but infectious mononucleosis, a self-limiting lymphoproliferative disease, can result if EBV is first acquired during adolescence or later. A defining feature of EBV is its ability to activate resting cultured B cells into unchecked proliferation with retention of the viral genome but without virus production, and indeed the distinction between latency and transformation is often blurred in discussion of this virus. It is clear, however, that EBV is latent in B cells, and that the virus utilizes many aspects of normal B cell development to establish and control latency. EBV, in contrast to HSV and HCMV, possesses a number of genes that are specific to the nonreplicating state and indeed are crucial for coordinating the establishment of latency. These encode the EBV nuclear antigens (EBNAs) 1, 2, 3A, 3B, 3C, and LP, and the latent membrane proteins (LMPs) 1, 2A, and 2B. Other loci specifying small RNAs, known as EBERs, and Bam A rightward transcripts are active during B cell transformation *in vitro*, although their significance for latency is unclear at present. During establishment of

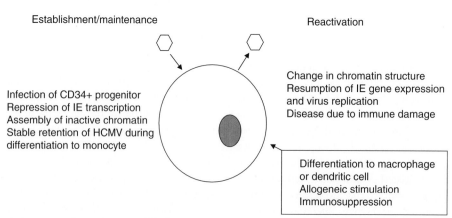

Figure 2 Stages in latency and reactivation of HCMV.

latency, the EBNAs and LMPs participate in a complex interaction with the host cell that results in utilization of the normal B cell development, to the advantage of the virus.

After penetration of the mucosal layers of the tonsils, virus replicates and naive B cells are infected and activated in a manner similar to that observed in culture. At this stage, no lytic gene transcription occurs but all of the EBNAs and LMPs are expressed in an interaction known as the growth program, or latency III. The combined action of the viral proteins gives the resting cell the characteristics of an activated B-lymphoblast, mimicking the first stage that occurs naturally after antigen recognition. Activated B cells are transported to the lymph node follicles, in which they replicate to form germinal centers. Again, following the normal pattern of B cell development after antigen stimulation, the infected cells undergo differentiation into memory B cells. At some stage during this process, the gene expression pattern changes to the default program, or latency II, in which only EBNA1, LMP1, and LMP2A are produced. Protein EBNA1 is crucial for maintaining the EBV genome in dividing cells, through stimulating its replication and by tethering it to cellular chromosomes. LMP1 and LMP2A provide survival signals that prevent the activated B cell from undergoing apoptosis. In the final step, the infected memory B cells are released from the germinal centers to become circulating memory B cells, and during this process gene expression changes to the latency program, or latency I, in which no viral gene products can be detected. During latency, the EBV genome is maintained as a circular episome. To retain the virus within the host, EBNA1 is produced during cell division, enabling the viral genome to replicate and persist in the memory compartment. Interestingly, each of the identified interactions with B cells is reflected in EBV-associated human tumors: viral gene expression in Burkitt's lymphoma (BL) cells resembles the latency program (but with EBNA1 continually expressed), Hodgkin's lymphoma has the default program, and nasopharyngeal carcinoma exhibits the growth program, albeit in epithelial-derived cells.

Reactivation of latent EBV occurs when a memory B cell differentiates into an antibody-secreting plasma cell, an event triggered by signaling from the B cell antigen receptor (BCR). The plasma cells migrate to the mucosal epithelium and virus is released into the saliva, possibly after replication in epithelial cells. The molecular mechanism of lytic cycle activation, and hence virus replication, has largely been studied by the use of BL cell lines, which express only EBNA1 and thus resemble the latent state in B cells. Cross-linking the BCR with immunoglobulin, possibly mimicking the natural differentiation signals provided by T cells, can activate EBV in certain BL lines. This process switches on expression of an IE protein, BZLF1, a transcription factor that triggers further viral gene expression and ultimately productive replication. The signal transduction pathways that respond to cross-linking the BCR result in activation of two transcription factors, MEF2D and ATF, which in turn induces histone acetylation and relief of repression by remodeling the chromatin structure. The BZLF1 protein activates its own promoter, providing a positive feedback loop that ensures rapid commitment to productive replication. In some BL cell lines, BZLF1 expression, and hence EBV replication, is activated in response to alternative agents, such as phorbol esters, but the relevance of these treatments to natural stimuli is unclear.

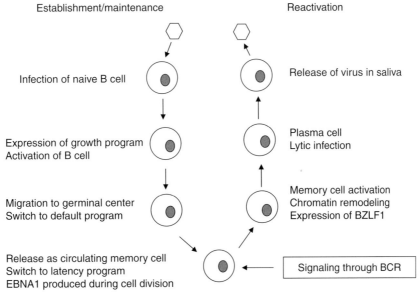

Figure 3 Establishment, maintenance, and reactivation of EBV latency in B cells.

The interaction between EBV and host B cells is summarized in **Figure 3**.

Members of the genus *Rhadinovirus* of the *Gammaherpesvirinae* have been studied in less detail than EBV. Murine herpesvirus 68 establishes latency in a number of cell types, including B cells, in mice. Kaposi's sarcoma-associated herpesvirus (KSHV) is also found in a variety of cell types, again including B cells. The natural pathways of latency and reactivation for these viruses are currently under investigation.

Common Themes

Although the examples described here deal with different viruses and host cells, some common themes emerge from the latent interactions. In establishment, specific features of the natural host cell are critical in preventing productive virus replication. Blocking IE protein production, an effective mechanism for inhibiting virus replication while minimizing cytopathology and immune recognition, appears to be common to establishment of latency. During maintenance, genomes are retained as circular episomes and viral gene expression is minimized to hide the virus from the immune system. Histones associated with the latent genome exhibit modifications characteristic of inactive chromatin. If repression of replication is not complete, low-level virus production may be neutralized by the immune system. Reactivation probably involves the virus utilizing normal cellular responses to a stimulus, but the stimulus may also damage the immune system and thereby cause disease.

See also: Bovine Herpesviruses; Human Cytomegalovirus: Molecular Biology; Epstein–Barr Virus: Molecular Biology; Herpes Simplex Viruses: Molecular Biology; Varicella-Zoster Virus: Molecular Biology; Herpesviruses: General Features; Kaposi's Sarcoma-Associated Herpesvirus: Molecular Biology.

Further Reading

Amon W and Farrell PJ (2004) Reactivation of Epstein–Barr virus from latency. *Reviews in Medical Virology* 15: 149–156.

Efstathiou S and Preston CM (2005) Towards an understanding of the molecular basis of herpes simplex virus latency. *Virus Research* 111: 108–119.

Jones C (2003) Herpes simplex virus type 1 and bovine herpesvirus 1 latency. *Critical Microbiology Reviews* 16: 79–95.

Kent JR, Kang W, Miller CG, and Fraser NW (2003) Herpes simplex virus latency-associated gene function. *Journal of Neurovirology* 9: 285–290.

Khanna KM, Lepisto AJ, Decman V, and Hendricks RL (2004) Immune control of herpes simplex virus during latency. *Current Opinion in Immunology* 16: 463–469.

Sinclair J and Sissons P (2006) Latency and reactivation of human cytomegalovirus. *Journal of General Virology* 87: 1763–1779.

Streblow DN and Nelson JA (2003) Models of HCMV latency and reactivation. *Trends in Microbiology* 11: 293–295.

Thorley-Lawson DA (2001) Epstein–Barr virus: Exploiting the immune system. *Nature Reviews* 1: 75–82.

Thorley-Lawson DA (2005) EBV the prototypical human tumor virus – just how bad is it? *Molecular Mechanisms in Allergy and Clinical Immunology* 116: 251–261.

Wagner EK and Bloom DC (1997) Experimental investigation of herpes simplex virus latency. *Clinical Microbiology Reviews* 10: 419–443.

History of Virology: Bacteriophages

H-W Ackermann, Laval University, Quebec, QC, Canada

© 2008 Elsevier Ltd. All rights reserved.

Glossary

Bacteriophage 'Eater of bacteria' (Greek, *phagein*).
Eclipse Invisible phase during phage replication.
Lysogenic (1) Bacterium harboring a prophage. (2) Bacteriophage produced by such a bacterium.
Lysogeny Carriage of a prophage.
Metagenomics Sequencing of total genomic material from an environmental sample.
One-step growth Single cycle of phage multiplication.
Phage typing Testing bacteria with a set of phages of different host ranges.
Prophage Latent bacteriophage genome in a lysogenic bacterium.
T-even Phage of the T-series with an even number (T2, T4, T6).

Introduction

Phages or bacteriophages, also referred to as bacterial viruses, include eubacterial and archaeal viruses and are now more appropriately called 'viruses of prokaryotes'. Phages are ubiquitous in nature and may be the most numerous viruses on this planet. They have a long evolutionary history, are considerably diverse, and include some highly evolved and complex viruses. Approximately 5500 phages have been examined in the electron microscope and they constitute the largest of all virus groups. The present system of phage taxonomy comprises 13 officially recognized families.

Following the discovery of phages, their study entered a symbiosis with the nascent science of molecular biology and then expanded to cover a multitude of subjects. The knowledge on phages grew by accretion and represents the efforts of thousands of scientists. For example, it took over 50 years and hundreds of publications to establish the genome map and sequence of coliphage T4. Phage literature is voluminous and comprises over 11 400 publications for the years 1915–65 and, to the author's reckoning, about 45 000 for 1965 to today. The volume of phage research is evident in the fact that there are two books each for single viruses, λ and T4, and that, by mid-2006, the journal *Virology* has published over 2000 phage articles since its start in 1955. The history of phages may be divided into three, partly overlapping periods. Because of the volume of phage research and the many scientists involved, the development of phage research is largely presented in tabular form. Any mention of individual scientists and facts is selective.

The Early Period: 1915–40

Discovery

Several early microbiologists had observed bacterial lysis. The British bacteriologist Ernest H. Hankin reported in 1896 that the waters of the Jumna and Ganges rivers in India killed *Vibrio cholerae* bacteria. The responsible agent passed bacteriological filters and was destroyed by boiling. In 1898, the Russian microbiologist Nikolai Gamaleya reported the lysis of *Bacillus anthracis* bacteria by a transmissible 'ferment'. It is probable that these scientists would have discovered bacterial viruses, if suitable techniques had been available.

Frederick William Twort (1877–1950), a British pathologist and superintendent of the Brown Institution in London, tried to propagate vaccinia virus on bacteriological media. He observed that 'micrococcus' colonies, in fact staphylococci, often became glassy and died. The agent of this transformation was infectious, passed porcelain filters, and was destroyed by heating. Twort proposed several explanations: an amoeba, an ultramicroscopic virus, a living protoplasm, or an enzyme with the power of growth. Subsequently, Twort joined the army and participated in the British Salonika campaign. Back at the Brown Institution, he published five more articles on phages and spent the rest of his career trying to propagate animal viruses on inert media.

Félix Hubert d'Herelle (1873–1949), born in Montreal, went to France as a boy and returned to Canada, styled himself a 'chemist', studied medicine without obtaining a degree, and finally went to Guatemala as a bacteriologist with minimal knowledge of this science. Later in Yucatan, he isolated the agent of a locust disease, the *Coccobacillus acridiorum* (now *Enterobacter aerogenes*), was sent to the Pasteur Institute of Paris, returned to Mexico and France, and went to fight locusts in Argentine, Turkey, and Tunisia. He noticed the appearance of holes ('plaques') in coccobacillus cultures. In 1915, again in Paris, he investigated an epidemic of bacillary dysentery in military recruits. He noted that his *Shigella* cultures were destroyed by a filterable, plaque-forming agent, whose appearance coincided with convalescence of the recruits. His discovery was published in 1917. D'Herelle understood immediately that he had found a new category of viruses and he used phages for the treatment of bacterial infections. D'Herelle then went to Vietnam, Holland, Egypt, India (Assam), the USA, the Soviet Union (Tbilisi), and finally to Paris. He coined the term 'bacteriophage', devised the plaque test, and stated that there was only one kind of phage, with many races, the *Bacteriophagum intestinale*. He promoted and practiced phage therapy in many countries and founded several phage institutes. The Tbilisi Institute, which he co-founded with G. Eliava in 1934, still exists.

Twort and D'Herelle could not have been more different personalities. Twort was a medical doctor and a self-effacing person; he traveled little, and showed minimal interest in his discovery. He published only five more articles on phages. D'Herelle was self-educated, peripatetic, worldly, and combative. He devoted the rest of his career to phages and wrote five books and nearly 110 articles on them. The almost simultaneous discovery of bacteriophages by two different scientists is a remarkable coincidence.

Phage Research: General

The years 1920–40 were characterized by research on the nature of phages and lysogeny, phage therapy, and the detection of phages in a wide variety of bacteria and a wide range of habitats (**Table 1**). D'Herelle and his followers asserted that bacteriophages were viruses. A major argument was that they formed plaques. Others, observing that some bacteria produced lytic agents that could be transmitted from one culture to another, held that 'bacteriophages' were generated spontaneously and were endogenous in nature. The famous Belgian immunologist, J. Bordet, proposed that phages were enzymes produced by 'lysogenic' bacteria. He stated bluntly: "The invisible virus of d'Herelle does not exist." When Bordet discovered Twort's paper from 1915, a nasty controversy on the nature of phages and the priority of phage discovery ensued.

Table 1 Phage discovery: phages of selected host organisms

Year	Investigators	Host species or group
1915	Twort	*Staphylococcus* sp.
1917	D'Herelle	*Shigella dysenteriae*
1918	D'Herelle	*Salmonella typhi*
1921	D'Herelle	*Bacillus subtilis, Corynebacterium diphtheriae, Escherichia coli, Pasteurella multocida, Vibrio cholerae, Yersinia pestis*
1934	Cowles	Clostridia
1937	Whitehead and Hunter	Lactic acid bacteria
1947	Reilly and coll.	Streptomycetes
1963	Safferman and Morris	Cyanobacteria
1970	Gourlay	Mycoplasmas
1974	Torsvik and Dundas	Halophilic archaea
1983	Janekovic and coll.	Hyperthermophilic archaea

Phages Become Viruses

The viral nature of phages was demonstrated by physicochemical studies. W. J. Elford and C. H. Andrewes (1932) showed that phages could differ in size. M. F. Burnet and colleagues (1933) demonstrated that phages differed in antigenicity and resistance to inactivation by heat, urea, and citrate. M. Schlesinger (1934) found that phages were made of DNA and protein. The publication of phage electron micrographs (1940) settled the matter of phage nature and made the enzyme theory untenable. It was now clear that phages were particulate and morphologically diverse. In retrospect, however, both d'Herelle and his opponents were right: phages are viruses with a dual nature.

Phage Therapy

Phage therapy was practiced by D'Herelle himself (1919) soon after phage discovery. He treated human dysenteria, chicken cholera, bovine hemorrhagic fever, bubonic plague, cholera, and a variety of staphylococcal and streptococcal infections. His phage laboratory in Paris was devoted to phage therapy. Many people embarked on phage therapy. Spectacular successes and dismal failures ensued. The failures had many reasons: use of crude, uncontrolled, inactivated, or endotoxin-containing lysates, narrow host specificity of some phages; absence of bacteriological controls; and plain charlatanism. The literature on phage therapy peaked in 1930 and totaled 560 publications for 1920–40. At this time, antibiotics were introduced and interest in phage therapy quickly waned.

Phage Typing and Ecology

Host range tests had shown that bacteria, even of the same species, usually differed in phage sensitivity. Some strains were lysed and some were not. By testing bacteria with a battery of phages, they could be subdivided into 'phage types', which greatly facilitated the tracing of bacterial infections to their sources and gave epidemiology an enormous boost. The first phage typing scheme, introduced in 1938, was for *Salmonella typhi* bacteria, which were subdivided by means of a series of adapted derivatives of phage ViII. For the first time, viruses were found to have a practical use. Phage typing developed rapidly in the following decades.

New phages were sought and found everywhere. It appeared that most bacterial pathogens had phages (**Table 1**) and that these viruses were omnipresent in water and soil, on plants, in food, on the skin of humans and animals, and in their body cavities and excreta. It also was discovered that bacteriophages caused faulty fermentations in the dairy industry. A few bacteriophages isolated during this early period are still available today: a streptococcal and a T4-like phage, two microviruses, and, possibly, the original phage isolated by Twort himself (**Table 2, Figure 1**).

The early period ended with the rise of dictatorships and World War II, first in the Soviet Union and later in Europe. Scientists were silenced, persecuted, forced into exile, or outright killed (G. Eliava was shot in 1937 by Stalin's secret police and Eugène and Elisabeth Wollman disappeared in 1943 in a Nazi concentration camp). Phages now had their martyrs. D'Herelle himself, a Canadian citizen, was interned by the Vichy Government and prevented from pursuing phage therapy. Phage research was obliterated in much of Europe.

The Intermediate Period: 1939–62

Phages as Tools and Products of Molecular Biology

Phage research survived and attained new heights in the USA. It helped to start molecular biology and provided crucial insights into vertebrate and plant virology. In turn, phage research benefited enormously from both disciplines. The next two decades became a legendary period of phage research, as a small number of researchers produced results of fundamental scientific importance (**Tables 3** and **4**).

In 1939, E. L. Ellis devised the 'one-step growth' experiment to measure the kinetics of phage infection and to establish the viral nature of bacteriophages. He was joined by Max Delbrück (1906–81), a German immigrant and physicist. Delbrück proposed to concentrate phage research on the seven coliphages T1–T7, collected and studied by F. Demerec and U. Fano in 1945 and all easily propagated. Fortunately, these phages were both diverse and partially related, thus allowing useful comparisons. Electron microscopy showed that the T phages (T for type) belonged to four morphotypes: T1, T3–T7,

T2–T4–T6, and T5 (**Figure 2**). Morphotypes correlated with growth parameters and antigenic properties. The T phages became the raw material for fundamental studies which, although limited to tailed phages of the T series, provided a general understanding of the phage life cycle. It became clear that:

1. Phages have different morphologies.
2. Phages contain DNA and transmit it from parent to progeny. The DNA is the carrier of heredity and genetic information. Phage coats consist of proteins.
3. Phages adsorb to bacteria by their tails.
4. Phage DNA enters the bacteria while the protein coat remains outside.
5. Phage synthesis occurs in stages, one of which is the 'eclipse'. Neither phages nor particulate precursors can be detected during the eclipse period. Novel phages are liberated ready-made in a single event ('burst').

Lysogeny

The problem of lysogeny, a leftover from the previous period, was solved by the French bacteriologist André Lwoff (1902–94) from the Pasteur Institute of Paris. The Wollmans had already postulated the dual nature of some phages: they were infectious outside the bacterium and noninfectious inside. Starting in 1949, Lwoff isolated single cells of a phage-carrying strain of *Bacillus megaterium* with a micromanipulator and transferred them into microdrops of fresh medium. After 19 successive generations, there was still no lysis. When lysis occurred, it was sudden and phages were plentiful in the microdrops. Lysis and phage production were inducible, especially by ultraviolet (UV) light. It was concluded that lysogenic bacteria perpetuate latent phages as 'prophages' in a repressed state. They do not secrete phages. Spontaneously or after induction, the prophage enters a vegetative phase, new phages are produced, and the bacterium dies. A major review article published in 1953 clarified the nature of lysogeny. The prophage became the origin of the 'provirus' concept in vertebrate virology.

In 1951, Esther Lederberg isolated the lysogenic coliphage λ. Small and easily manipulated, it was to have an extraordinary future and became perhaps the best studied of all viruses. As the prototypic lysogenic phage, it was the subject of countless publications on the nature of

Table 2 Phage discovery: individual phages and phage groups

Year	Investigators	Phage	Family[a]	Host genus[b]
1915	Twort	Twort?	M	*Staphylococcus*
1926	Clark and Clark	C1	P	*Streptococcus*
1933	Burnet and McKie	C16	M	Enterics
		S13	*Microviridae*	Enterics
1935	Sertic and Boulgakov	φX174	*Microviridae*	Enterics
1945	Demerec and Fano	T1 to T7	M, P, S	Enterics
1951	E. Lederberg	λ	S	Enterics
	G. Bertani	P1 and P2	M	Enterics
	Zinder and J. Lederberg	P22	P	Enterics
1960	Loeb	f1	*Inoviridae*	Enterics
1961	Loeb and Zinder	f2	*Leviviridae*	Enterics
	Takahashi	PBS1	M	Bacillus
1963	Taylor	Mu	M	Enterics
1964	Okubo	SPO1	M	Bacillus
1965	Reilly	φ29	P	Bacillus
1967	Schito	N4	P	Enterics
1968	Espejo and Canelo	PM2	*Corticoviridae*	*Pseudoalteromonas*
	Riva and coll.	SPP1	S	*Bacillus*
1970	Gourlay	L1 (MVL1)	*Plectrovirus*	*Acholeplasma*
1971	Gourlay	L2 (MVL2)	*Plasmaviridae*	*Acholeplasma*
1973	Six and Klug	P4	M	Enterics
	Vidaver and coll.	φ6	*Cystoviridae*	*Pseudomonas*
1974	Olsen and coll.	PRD1	*Tectiviridae*	Enterics
1982	Martin and coll.	SSV1	*Fuselloviridae*	*Sulfolobus*
1983	Janekovic and coll.	TTV1	*Lipothrixviridae*	*Thermoproteus*
1994	Zillig and coll.	SIRV1	*Rudiviridae*	*Sulfolobus*
2002	Rachel and coll.	ATV	*Bicaudaviridae*[c]	*Acidianus*
2004	Häring and coll.	PSV	*Globuloviridae*[c]	*Pyrobaculum*
2005	Häring and coll.	ABV	*Ampullaviridae*[c]	*Acidianus*

[a]Tailed phages: M, *Myoviridae*; P, *Podoviridae*; S, *Siphoviridae*.
[b]Enterobacteria are considered as a single host 'genus' because of frequent cross-reactions.
[c]Archaeal viruses awaiting formal classification.

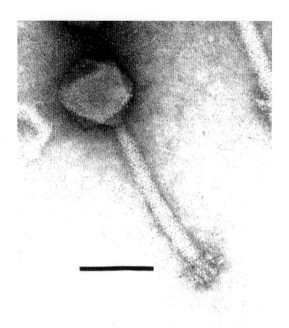

Figure 1 Phage Twort, uranyl acetate. Scale = 100 nm. The phage was reportedly deposited in 1948 by Twort himself in the collection of the Pasteur Institute of Paris.

lysogeny, genetic regulation, and transduction. It was shown that defective prophages could carry specific bacterial genes and insert them into bacteria (specific transduction). Certain phages such as P22 could transfer any bacterial genes and thus mediate 'general transduction'. This led to an understanding of phage conversion. It appeared that the converting phages were lysogenic and that their prophages converted bacteria by coding for novel properties, such as the production of antigens. As early as 1951, V. J. Freeman reported that the production of diphtheria toxin depended on the presence of prophages. The study of lysogeny culminated in 1962 when A. Campbell proposed his model of prophage insertion by a crossover process.

Varied and Interesting

In 1959, Brenner and Horne revolutionized electron microscopy by introducing negative staining with phosphotungstic acid. This produced pictures of unprecedented quality. They showed that T4 tails were contractile and that tail contraction was part of the T4 infective process. The same year, single-stranded (ss) DNA was discovered in the minute phage φX174 *(Microviridae)*. Filamentous phages *(Inoviridae)* and phages with ssRNA *(Leviviridae)* were isolated in 1960 and 1961, respectively. True marine phages and phages of cyanobacteria, then called 'blue-green algae', were also found. This offered some hope to control 'algal' blooms.

While phage therapy disappeared, phage typing became a major epidemiological tool (**Table 5**). In addition to the *S. typhi* typing scheme, international typing sets for *S. paratyphi* B and *Staphylococcus aureus* were devised and are still in use. Typing schemes for many medically important bacteria (e.g., corynebacteria, various *E. coli* and *Salmonella* serotypes, shigellae, *Pseudomonas aeruginosa*, and *Vibrio cholerae*) appeared. A literature survey by H Raettig lists no less than 2000 articles on phage typing, mostly in the years 1940–65. Phage research was considerably promoted by M H Adams' book *Bacteriophages*, which summarized phage research until 1957 and introduced the agar-double-layer method of phage propagation

Diversification and the Ongoing History of Bacteriophages: 1965–Today

Phage research now expanded explosively and divided into many, partly overlapping fields. No particular time periods are discernible; all fields developed gradually and are still expanding. As judged from literature listings, the present number of phage publications is 800–900 articles or monographs per year. The following is a selection of research activities.

Basic Phage Research

1. The present period of phage research started in 1965 (**Table 3**) after D. E. Bradley published a seminal, still cited paper on phage morphology (republished in 1967). Bradley classified phages into six basic types defined by morphology and nature of nucleic acid. Phages were divided into tailed phages (with contractile, long and noncontractile, or short tails), isometric phages with ssDNA or ssRNA, and filamentous phages. This became the basis of present-day phage classification.
2. Novel phages were isolated at a rate of about 100 per year. A first phage survey, published in 1967, listed 111 phages examined by electron microscopy, 102 of which were tailed and 9 isometric or filamentous. Most infected enterobacteria. Presently, the number of phages with known morphology exceeds 5500 and the number of host genera has grown to 155. It includes mycoplasmas, cyanobacteria, and archaea.
3. Replication and assembly of most phage families were studied. A highlight was the description of the assembly pathway of phage T4 (1967), obtained by a combination of electron microscopy and genetics.
4. Many phage genomes were sequenced. The first was that of ssRNA phage MS2 *(Leviviridae)*, the first viral genome ever to be sequenced. This was followed by the sequences of various small viruses of the families *Microviridae, Inoviridae,* and *Leviviridae*. A landmark was the sequencing of the complete λ genome (1982), followed by that of coliphage T7. The complete

Table 3 Basic phage research

Year	Investigators	Phages	Event or observation
(a) The rise of molecular biology			
1939	Ellis and Delbrück	Coliphage	One-step growth experiment
1940	Pfankuch and Kausche; Ruska	Coliphage	Phages seen in the electron microscope
1945	Demerec and Fano	T1–T7	Differentiation by host range
1942	Luria and Anderson	T1, T2	Two different morphologies
1942–53	Anderson and coll.	T1–T7	Morphological studies
1946	Cohen and Anderson	T2	Contains 37% of DNA
1947	Luria	T1–T7	Multiplicity reactivation
1948	Putnam and Cohen	T6	Transfer of DNA phosphorus from parent to progeny phage
1949	Hershey and Rothman	T2	Genetic recombination between phage mutants
1950	Anderson	T4	DNA is located in the head and separated from protein by osmotic shock
	Lwoff and Gutmann		Clarification of the nature of lysogeny
1951	Cohen	T-even	Phage synthesis in stages: protein first, DNA later
	Herriott	T2	Empty phages consist of protein
1951	Doermann	T3, T-even	Eclipse period in phage multiplication
1952	Anderson	T2, T4, T5	Phages adsorb tail first
	Cohen	T-even	DNA contains 5-hydroxymethylcytosine
	Hershey and Chase	T2	DNA carries genetic information
	Zinder and J. Lederberg	P22	General transduction
1953	Lwoff		Review paper on lysogeny
1955	Lennox	P1	General transduction
1959	Adams		Summary of phage research
	Brenner and Horne		Introduction of negative staining
1960	Doty and coll.		DNA–DNA hybridization
1962	Campbell	λ	Model of prophage insertion
1963	Taylor	Mu	Acts as a transposon
(b) After 1965			
1965, 1967	Bradley	General	Phage classification
1967	Eisenstark	General	First phage count
	Wood and Edgar	T4	Assembly pathway completed
	Ikeda and Tomizawa	P1	Prophage is a plasmid
	Ptashne	λ	Isolation of repressor
1968	Signer	λ	Prophage may be a plasmid
1969	Six and Klug	P2	Depends on helper phage P2
1973	Arber	P1	Restriction endonucleases identified
1992	Young	General	Clarification of phage lysis
2000	Wikoff and coll.	HK97	Capsid structure
2005	Fokine and coll., Morais and coll.	T4, HK97, φ29	Common ancestry of tailed phages
(c) Phage-coded bacterial toxins			
1951	Freeman	β	Diphtheria toxin
1964	Zabriskie	A25	Erythrogenic toxin of streprococci
1970	Inoue and Iida	CEβ	Botulinus toxin C
1971	Uchida	β	Diphtheria toxin gene is part of phage
1984	O'Brien and coll.	H-19J, 933W	Shiga-like toxins of E. coli
1996	Waldor and Mekalanos	CTXφ	Cholera toxin

sequence of the T4 genome was established in 1994. This was the beginning of large-scale phage genome sequencing (**Table 3(b)**) and of the novel science of 'genomics', devoted to computer-assisted comparisons of phage and other genomes. By 2006, the genomes of representatives of all phage families and of many individual tailed phages had been sequenced.

5. Bacteriophages appeared as major factors of bacterial virulence, coding for bacterial exotoxins (e.g., botulinus and Shiga toxins), antigens, and other virulence factors. A momentous discovery made in 1996 was that filamentous phages in their double-stranded replication form were able to convert nontoxigenic cholera strains to toxigenic forms. Virulence genes were found to be present in converting phages and integrated prophages. Sequencing of bacterial genomes showed that prophages with virulence factors were generally organized into 'pathogenicity islands'.

Table 4 Basic research: Milestones in genome sequencing

Year	Investigators	Phages	Family[a]	Host
1976	Fiers and coll.	MS2	*Leviviridae*	Enterics
1978	Sanger and coll.	φX174	*Microviridae*	Enterics
1980	Van Wezenbeek and coll.	M13	*Inoviridae*	Enterics
1981	Mekler	Qβ	*Leviviridae*	Enterics
1982	Sanger and coll.	λ	S	Enterics
1983	Dunn and Studier	T7	P	Enterics
1986	Vlcek and Paces	φ29[b]	P	*Bacillus*
1988	Mindich	φ6[b]	Cystoviridae	*Pseudomonas*
1994	Kutter and coll.	T4[b]	M	Enterics
1996	Grimaud	Mu	M	Enterics
1998	Christie and coll.	P2[b]	M	Enterics
1999	Mannisto and coll.	PM2	Corticoviridae	*Pseudoalteromonas*
2000	Vanderbyl and Kropinski	P22[b]	P	Enterics
2002	Mesyanzhinov and coll.	φKZ	M	*Pseudomonas*
2004	Lobocka and coll.	P1[b]	M	Enterics

[a]Tailed phages: M, *Myoviridae*; P, *Podoviridae*; S, *Siphoviridae*.
[b]Sequence completed.

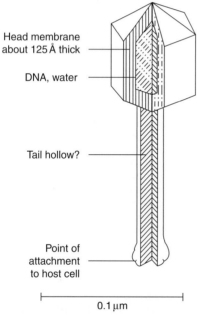

Figure 2 Drawing of phage T2 by T. F. Anderson (1952). Reproduced from Anderson TF, Rappaport C, and Muscatine NA (1953) On the structure and osmotic properties of phage particles. *Annales de l'Institut Pasteur* 84: 5–15, with permission from Elsevier.

long-sought evidence that tailed phages are a monophyletic evolutionary group.

Applied Research

See **Table 5** for a listing of research activities, arranged chronologically.

1. In the dairy industry, phages of lactococci and *Streptococcus thermophilus* were shown to cause major economic losses by destroying starter cultures and causing faulty fermentations. The phages involved were studied in detail by electron microscopy and genome sequencing.

2. Phages were investigated for use as indicators of fecal pollution. Phages with ssRNA (*Leviviridae*), somatic coliphages, and tailed phages of *Bacteroides fragilis* were proposed as indicators, but their usefulness was not conclusively proven. On the other hand, phages of many types proved useful as tracers of groundwater movements.

3. Phages were harnessed for the detection of bacteria, particularly by the construction of 'reporter' phages carrying luminescent (luciferase) genes, which are expressed when these phages infect a bacterium.

4. In biotechnology, phages are used as cloning and sequencing vectors, notably the filamentous phage M13 (*Inoviridae*), phages λ and P1, and hybrids of phages and plasmids (named 'cosmids' or 'phagemids'). Phage products such as T4 ligase and T7 promoters are used in cloning reactions and phage lysozymes are used to free useful intracellular bacterial enzymes.

5. The technique of 'phage display', introduced in 1985, mainly uses filamentous phages (M13, fd). Foreign peptide genes are fused to phage coat proteins. This

6. Cryoelectron microcopy and three-dimensional image reconstruction revealed the capsid structure of various isometric phages (*Microviridae, Leviviridae, Tectiviridae*) and of the tail and base plate structure of phage T4. It was also shown that the capsids of adenoviruses and tectiviruses and of representatives of all three tailed phage families (T4, HK97, φ29) were structurally related. This indicated phylogenetic relationships between adeno- and tectiviruses and provided the

Table 5 Applied and environmental phage research

Year	Investigators	Phages	Event or observation
(a) Phage therapy			
1919	D'Herelle		Treats human dysentery and chicken cholera
1921	Bruynoghe and Maisin		Treat human furunculosis
1925	D'Herelle		Treats plague in Egypt
1927–28	D'Herelle		Treats cholera in Assam
	Moore		Phages for control of plant disease
1938	Fisk		Treats septicemia in mice
1981–87	Slopek and coll.		Phage therapy in Poland
1982–87	Smith and Huggins		Treat diarrhea in calves, lambs, piglets
2001	Nelson and coll.		Phage lytic enzymes for elimination of streptococci in mice
(b) Diagnostics, industry, and biotechnology			
1936	Whitehead and Cox		Faulty fermentations in the dairy industry
1938	Craigie and Yen	ViII	Phage typing of *Salmonella typhi*
1942	Fisk		Phage typing of *Staphylococcus aureus*
1954	Cherry and coll.	O1	Identification of salmonellae
1955	Brown and Cherry	γ	Identification of *Bacillus anthracis*
1973	Lobban and Kaiser	T4	Ligase for cloning
	Studier	T7	Promoter for cloning
1977	Blattner and coll.	λ	Charon vectors for cloning
1977–83	Messing and coll., others	M13, fd	Cloning and sequencing vectors
1985	Smith	M13	Phage display
1987	Ulitzur and Kuhn	λ	Luciferase reporter phages
1990	Sternberg	P1	Cloning of large DNA fragments
(c) Phages and the environment			
1920	Dumas		Phages in soil
1923	Gerretsen and coll.		Phages in rhizobia
	Van der Hoeden		Phages in meat
1924	Zdansky		Phages in sewage
1955	Spencer		Indigenous marine phage
1966	Coetzee		Phages as pollution indicators
1974	Martin and Thomas		Phages as groundwater tracers
1987	Goyal and coll.		Book on phage ecology
1989	Bergh and coll.		Large numbers of phages in seawater
2002	Brüssow and Hendrix		Global phage population estimated at 10^{31}

powerful technique led to the development of antibody technology and has many medical and pharmaceutical applications.

6. Phage therapy reappeared. In France, it had always been practiced on a small scale. In the mid-1980s, a team in Wroclaw, Poland, conducted phage treatments of various human diseases (mainly pyogenic infections). After the fall of communism, the West learned with surprise that phage therapy had been practiced in the Soviet Union since the 1940s and that the institute founded in Tbilisi par d'Herelle and Eliava produced phages for the Soviet Union with up to 1200 employees engaged in production. Phage therapy is still practiced in Georgia.

Phage Ecology

1. Marine phages were investigated worldwide, including Arctic and Antarctic waters, largely by electron microscopy and the novel epifluorescence technique. Marine phages were found in vast numbers and appeared as part of a food web of phages, bacteria, and flagellates. Many marine phages were cyanophages and, all those investigated so far have tailed particles. Sequencing of uncultured virus DNA ('metagenomics') detected known entities such as T7-like phages and showed that only 25% of sequences matched known genes. Seawater appears as an immense phage reservoir (**Table 5**).

2. Phages were sought and found in extreme habitats such as volcanic hot springs, alkaline lakes, or hypersaline lagoons. This led to the discovery of several families of archaeal viruses (**Table 2**).

See also: Assembly of Viruses: Nonenveloped Particles; History of Virology: Vertebrate Viruses; Icosahedral dsDNA Bacterial Viruses with an Internal Membrane; Icosahedral Enveloped dsRNA Bacterial Viruses; Icosahedral ssDNA Bacterial Viruses; Icosahedral ssRNA Bacterial Viruses; Icosahedral Tailed dsDNA Bacterial Viruses; Inoviruses; Transcriptional Regulation in Bacteriophage.

Further Reading

Ackermann H-W (2006) 5500 bacteriophages examined in the electron microscope. *Archives of Virology* 152: 227–243.
Ackermann H-W and DuBow MS (1987) *Viruses of Prokaryotes, Vol. I. General Properties of Bacteriophages.* Boca Raton, FL: CRC Press.
Adams MH (1959) *Bacteriophages.* New York: Interscience Publishers.
Anderson TF, Rappaport C, and Muscatine NA (1953) On the structure and Ormotic properties of phage particles. *Annales de l'Institut Pasteur* 84: 5–15
Cairns J, Stent GS, and Watson JD (eds.) (1992) *Phage and the Origins of Molecular Biology.* (expanded edn.) Cold Spring Harbor, NY: Cold Spring Harbor Laboratory Press.
Calendar R (2006) *The Bacteriophages,* 2nd edn., New York: Oxford University Press.
Goyal SM, Gerba CP, and Bitton G (eds.) (1987) *Phage Ecology.* New York: Wiley.
Kutter E and Sulakvelidze A (eds.) (2004) *Bacteriophages: Biology and Applications.* Boca Raton, FL: CRC Press.
Marks T and Sharp R (2000) Bacteriophages and biotechnology: A review. *Journal of Chemical Technology and Biotechnology* 75: 6–17.
Raettig H (1958) *Bakteriophagie 1917 Bis 1956. Vol. I. Einführung, Sachregister, Stichwortverzeichnis. Vol. II. Autorenregister.* Stuttgart: Gustav Fischer.
Raettig H (1967) *Bakteriophagie 1957–1965. Vol. I. Einführung, Sachregister, Stichwort-Index. Introduction, Index of Subjects, Reference Word Index. Vol. II. Bibliography.* Stuttgart: Gustav Fischer.
Summers WC (1999) *Felix d'Herelle and the Origins of Molecular Biology.* New Haven, CT: Yale University Press.
Tidona CA, Darai G, and Büchen-Osmond C (2001) *The Springer Index of Viruses.* Berlin: Springer.
Twort A (1993) *In Focus, Out of Step: A Biography of Frederick William Twort F.R.S. 1877–1950.* Stroud, UK: A. Sutton.

History of Virology: Plant Viruses

R Hull, John Innes Centre, Norwich, UK

© 2008 Elsevier Ltd. All rights reserved.

Introduction

History shows that the study of virus infections of plants has led the overall subject of virology in the development of several major concepts including that of the entity of viruses themselves. To obtain a historical perspective of plant virology, five major (overlapping) ages can be recognized.

Prehistory

The earliest known written record describing what was almost certainly a plant virus disease is a poem in Japanese written by the Empress Koken in AD 752 and translated by T. Inouye as follows:

> In this village
> It looks as if frosting continuously
> For, the plant I saw
> In the field of summer
> The color of the leaves were yellowing

The plant, identified as *Eupatorium lindleyanum*, has been found to be susceptible to tomato yellow leafcurl virus, which causes a yellowing disease.

In Western Europe in the period from about 1600 to 1660, many paintings and drawings were made of tulips that demonstrate flower symptoms of virus disease. These are recorded in the Herbals of the time and in the still-life paintings of artists such as Johannes Bosschaert in 1610. During this period, blooms featuring such striped patterns were prized as special varieties leading to the phenomenon of 'tulipomania'. The trade in infected tulip bulbs resulted in hyperinflation with bulbs exchanging hands for large amounts of money or goods (**Table 1**).

In describing an experiment to demonstrate that sap flows in plants, Lawrence reported in 1714 the unintentional transmission of a virus disease of jasmine by grafting. The following quotation from Blair in 1719 describes the procedure and demonstrates that even in this protoscientific stage, experimenters were already indulging in arguments about priorities of discovery.

> The inoculating of a strip'd Bud into a plain stock and the consequence that the Stripe or Variegation shall be seen in a few years after, all over the shrub above and below the graft, is a full demonstration of this Circulation of the Sap. This was first observed by Mr. Wats at Kensington, about 18 years ago: Mr. Fairchild performed it 9 years ago; Mr. Bradly says he observ'd it several years since; though Mr. Lawrence would insinuate as if he had first discovered it.

Recognition of Viral Entity

In the latter part of the nineteenth century, the idea that infectious disease was caused by microbes was well established, and filters were available that would not allow the passage of known bacterial pathogens. Mayer in 1886 showed that a disease of tobacco (*Mosaikkrankheit*; now

Table 1 Tulipomania: the goods exchanged for one bulb of Viceroy tulip

4 t of wheat	4 barrels of beer
8 t of rye	2 barrels of butter
4 fat oxen	1000 lb cheese
8 fat pigs	1 bed with accessories
12 fat sheep	1 full dress suit
2 hogsheads of wine	1 silver goblet

known to be caused by tobacco mosaic virus; TMV) could be transmitted to healthy plants by inoculation with extracts from diseased plants; Iwanowski demonstrated in 1892 that sap from such tobacco plants was still infective after it had been passed through a bacteria-proof filter candle. This work did not attract much attention until it was repeated by Beijerinck who in 1898 described the infectious agent as *contagium vivum fluidum* (Latin for contagious living fluid) to distinguish it from contagious corpuscular agents. Beijerinck's discovery is considered to be the birth of virology. In 1904 Baur showed that an infectious variegation of *Abutilon* could be transmitted by grafting, but not by mechanical inoculation. Beijerinck and Baur used the term 'virus' in describing the causative agents of these diseases, to contrast them with bacteria; this term had been used as more or less synonymous with bacteria by earlier workers. As more diseases of this sort were discovered, the unknown causative agents came to be called 'filterable viruses'.

Between 1900 and 1935, many plant diseases thought to be caused by filterable viruses were described, but considerable confusion arose because adequate methods for distinguishing one virus from another had not yet been developed.

The original criterion of a virus was an infectious entity that could pass through a filter with a pore size small enough to hold back all known cellular agents of disease. However, diseases were soon found that had virus-like symptoms not associated with any pathogen visible in the light microscope, but that could not be transmitted by mechanical inoculation. With such diseases, the criterion of filterability could not be applied. Their infectious nature was established by graft transmission and sometimes by insect vectors. Thus, certain diseases of the yellows and witches'-broom type, such as aster yellows, came to be attributed to viruses on quite inadequate grounds. Many such diseases are now known to be caused by phytoplasma or spiroplasma, and a few by bacteria or rickettsia.

The Biological Age

During most of the period between 1900 and 1935, attention was focused on the description of diseases using the macroscopic symptoms and cytological abnormalities as revealed by light microscopy, host ranges, and methods of transmission which were the only techniques available. The influence of various physical and chemical agents on virus infectivity was investigated, but methods for the assay of infective material were primitive. Holmes showed that the local lesions produced in some hosts following mechanical inoculation could be used for the rapid quantitative assay of infective virus. This technique enabled properties of viruses to be studied much more readily and paved the way for the isolation and purification of viruses a few years later. Until about 1930, there was serious confusion by most workers regarding the diseases produced by viruses and the viruses themselves. This was not surprising, since virtually nothing was known about the viruses except that they were very small. In 1931 Smith made an important contribution that helped to clarify this situation. Working with virus diseases in potato, he realized the necessity of using plant indicators, plant species other than potato, which would react differently to different viruses present in potatoes. Using several different and novel biological methods to separate the viruses, he was able to show that many potato virus diseases were caused by a combination of two viruses with different properties, which he named virus X (potato virus X, PVX) and virus Y (potato virus Y, PVY). As PVX was not transmitted by the aphid *Myzus persicae*, whereas PVY was, PVY could be separated from PVX. He obtained PVX free of PVY by needle inoculation of the mixture to *Datura stramonium* which does not support PVY. Furthermore, Smith observed that PVX from different sources fluctuated markedly in the severity of symptoms it produced in various hosts leading to the concept of strains.

An important practical step forward was the recognition that some viruses could be transmitted from plant to plant by insects. Fukushi recorded the fact that in 1883 a Japanese rice grower transmitted what is now known to be rice dwarf virus, RDV) by the leafhopper *Recelia dorsalis*. However, this work was not published in any available form and so had little influence. In 1922 Kunkel first reported the transmission of a virus by a planthopper; within a decade, many insects were reported to be virus vectors leading to the recognition of specific virus–vector interactions.

Since Fukushi first showed in 1940 that RDV could be passed through the egg of a leafhopper vector for many generations, there has been great interest in the possibility that some viruses may be able to replicate in both plants and insects. It is now well established that plant viruses in the families *Rhabdoviridae* and *Reoviridae* and the genera *Tenuivirus*, *Tospovirus*, and *Marafivirus* multiply in insects as well as in plants.

The Biochemical/Biophysical Age

Beale's recognition in 1928 that plants infected with TMV contained a specific antigen opened the age in which the biochemical nature of viruses was elucidated. In the 1930s Gratia showed that plants infected with different viruses contained different specific antigens and Chester demonstrated that different strains of TMV and PVX could be distinguished serologically.

The high concentration at which certain viruses occur in infected plants and their relative stability turned out to be of crucial importance in the first isolation and chemical characterization of viruses, because methods for extracting and purifying proteins were not highly developed. In the early 1930s, various attempts were made to isolate and purify plant viruses using methods similar to those that had just been developed for purifying enzymes. Following detailed chemical studies suggesting that the infectious agent of TMV might be a protein, Stanley announced in 1935 the isolation of this virus in an apparently crystalline state. At first Stanley considered that the virus was a globulin containing no phosphorus but in 1936 Bawden et al. described the isolation from TMV-infected plants of a liquid crystalline nucleoprotein containing nucleic acid of the pentose type. They showed that the particles were rod-shaped, thus confirming the earlier suggestion of Takahashi and Rawlins based on the observation that solutions containing TMV showed anisotropy of flow. Electron microscopy and X-ray crystallography were the major techniques used in early work to explore virus structure, and the importance of these methods has continued to the present day. Bernal and Fankuchen applying X-ray analysis to purified preparations of TMV obtained accurate estimates of the width of the rods. The isolation of other rod-shaped viruses, and spherical viruses that formed crystals, soon followed. All were shown to consist of protein and pentose nucleic acid.

Early electron micrographs confirmed that TMV was rod-shaped and provided approximate dimensions, but they were not particularly revealing because of the lack of contrast between the virus particles and the supporting membrane. The application of shadow-casting with heavy metals greatly increased the usefulness of the method for determining the overall size and shape of virus particles but not structural detail. With the development of high-resolution microscopes and of negative staining in the 1950s electron microscopy became an important tool for studying virus substructure. From a comparative study of the physicochemical properties of the virus nucleoprotein and the empty viral protein shell found in TYMV preparations, Markham concluded in 1951 that the RNA of the virus must be held inside a shell of protein, a view that has since been amply confirmed for this and other viruses by X-ray crystallography. Crick and Watson suggested that the protein coats of small viruses are made up of numerous identical subunits arrayed either as helical rods or as a spherical shell with cubic symmetry. Subsequent X-ray crystallographic and chemical work has confirmed this view. Caspar and Klug formulated a general theory that delimited the possible numbers and arrangements of the protein subunits forming the shells of the smaller isodiametric viruses.

Until about 1948, most attention was focused on the protein part of the viruses. Quantitatively, the protein made up the larger part of virus preparations. Enzymes that carried out important functions in cells were known to be proteins, and knowledge of pentose nucleic acids was rudimentary. No function was known for them in cells, and they generally were thought to be small molecules primarily because it was not recognized that RNA is very susceptible to hydrolysis by acid, by alkali, and by enzymes that commonly contaminate virus preparations. In 1949 Markham and Smith isolated turnip yellow mosaic virus (TYMV) and showed that purified preparations contained two classes of particles, one an infectious nucleoprotein with about 35% of RNA, and the other an apparently identical protein particle that contained no RNA and that was not infectious. This result clearly indicated that the RNA of the virus was important for biological activity. Analytical studies showed that the RNAs of different viruses have characteristically different base compositions while those of related viruses are similar. About this time, it came to be realized that viral RNAs might be considerably larger than had been thought. A synthetic analog of the normal base guanine, 8-azaguanine, when supplied to infected plants was incorporated into the RNA of TMV and TYMV, replacing some of the guanine. The fact that virus preparations containing the analog were less infectious than normal virus gave further experimental support to the idea that viral RNAs were important for infectivity. However, it was the classic experiments in the mid-1950s of Gierer and Schramm, and Fraenkel-Conrat and Williams that demonstrated the infectivity of naked TMV RNA and the protective role of the protein coat. These discoveries ushered in the era of modern plant virology.

In the early 1950s Brakke developed density gradient centrifugation as a method for purifying viruses. Together with a better understanding of the chemical factors affecting the stability of viruses in extracts, this procedure has allowed the isolation and characterization of many viruses. The use of sucrose density gradient fractionation enabled Lister to discover the bipartite nature of the tobacco rattle virus genome. Since that time, density gradient and polyacrylamide gel fractionation techniques have allowed many viruses with multipartite genomes to be characterized. Their discovery, in turn, opened up the possibility of carrying out genetic reassortment experiments with plant viruses leading to the allocation of functions to many of the viral genes. Density gradient fractionation of purified preparations of some other viruses revealed noninfectious nucleoprotein particles containing subgenomic RNAs. Other viruses have been found to have associated with them satellite viruses or satellite RNAs that depend on the 'helper' virus for some function required during replication.

Further developments in the 1970s included improved techniques related to X-ray crystallographic analysis and a growing knowledge of the amino acid sequences of the coat proteins allowed the three-dimensional structure of the protein shells of several plant viruses to be determined in molecular detail.

For some decades, the study of plant virus replication had lagged far behind that of bacterial and vertebrate viruses mainly because there was no plant system in which all the cells could be infected simultaneously to provide the basis for synchronous 'one-step growth' experiments. However, following the initial experiments of Cocking in 1966, Takebe and colleagues developed protoplast systems for the study of plant virus replication. Although these systems had significant limitations, they greatly increased our understanding of the processes involved in plant virus replication.

Another important technical development has been the use of *in vitro* protein-synthesizing systems such as that from wheat germ, in which many plant viral RNAs act as efficient messengers. Their use allowed the mapping of plant viral genomes by biochemical means to begin.

The Molecular Biology Age

The molecular age opened in 1960 with the determination of the full sequence of 158 amino acids in the coat protein of TMV. The sequence of many naturally occurring strains and artificially induced mutants was also determined at about the same time. This work made an important contribution to establishing the universal nature of the genetic code and to our comprehension of the chemical basis of mutation.

Our understanding of the genome organization and functioning of viruses has come from the development of procedures whereby the complete nucleotide sequence of viruses with RNA genomes can be determined. In 1982 the genomes of both the first plant RNA virus (TMV) and DNA virus cauliflower mosaic virus (CaMV) were sequenced. Since then, the genomes of representatives of all the plant virus genera have been sequenced and there are many sequences of virus species.

The late 1980s and 1990s was a period when molecular biological techniques were applied to a wide range of aspects of plant virology. These included the ability to prepare *in vitro* infectious transcripts of RNA viruses derived from cloned viral cDNA allowing techniques such as site-directed mutagenesis to be applied to the study of genome function and reverse genetics being used to elucidate the functions of viral genes and control sequences. These approaches, together with others such as yeast systems for identifying interacting molecules, the expression of viral genes in transgenic plants, and labeling viral genomes in such a manner that their sites of function within the cell are known, were revealing the complexities of the interactions between viruses and their hosts. Nucleotide sequence information has had, and continues to have, a profound effect on our understanding of many aspects of plant virology, including (1) the location, number, and size of the genes in a viral genome; (2) the amino acid sequence of the known or putative gene products; (3) the molecular mechanisms whereby the gene products are transcribed; (4) the putative functions of a gene product, which can frequently be inferred from amino-acid-sequence similarities to products of known function encoded by other viruses; (5) the control and recognition sequences in the genome that modulate expression of viral genes and genome replication; (6) the understanding of the structure and replication of viroids and of the satellite RNAs found associated with some viruses; (7) the molecular basis for variability and evolution in viruses, including the recognition that recombination is a widespread phenomenon among RNA viruses and that viruses can acquire host nucleotide sequences as well as genes from other viruses; and (8) the beginning of a taxonomy for viruses that is based on evolutionary relationships. On the host side, advances in plant genome sequencing are identifying plant genes that confer resistance to viruses.

During the 1980s, major advances were made on improved methods of diagnosis for virus diseases, centering on serological procedures and on methods based on nucleic acid hybridization. Since the work of Clark and Adams reported in 1977, the enzyme-linked immunosorbent assay (ELISA) technique has been developed with many variants for the sensitive assay and detection of plant viruses. Monoclonal antibodies against TMV lead to a very rapid growth in their use for many kinds of plant virus research and for diagnostic purposes. The late 1970s and the 1980s also saw the start of application of the powerful portfolio of molecular biological techniques to developing other approaches to virus diagnosis, to a great increase in our understanding of the organization and strategy of viral genomes, and to the development of techniques that promise novel methods for the control of some viral diseases. The use of nucleic acid hybridization procedures for sensitive assays of large numbers of samples and the polymerase chain reaction, also dependent on detailed knowledge of genome sequences, are being increasingly used in virus diagnosis. Most recently DNA chips are being developed for both virus diagnostics and studying virus infection.

In the early 1980s, it seemed possible that some plant viruses, when suitably modified by the techniques of gene manipulation, might make useful vectors for the introduction of foreign genes into plants. Some plant viruses have been found to contain regulatory sequences that can be very useful in other gene vector systems, notably the widely used CaMV 35S promoter. Another practical application of molecular techniques to plant viruses has been the modification of viral genomes so that products of interest to industry and medicine can be produced in plants. This is being done either by the introduction of genes into the viral genome or by modification of the coat protein sequence to enable epitopes to be presented on the virus.

Early attempts (early to mid-1900s) to control virus diseases in the field were often ineffective. They were mainly limited to attempts at general crop hygiene, roguing of obviously infected plants, and searching for genetically resistant lines. Developments since this period have improved the possibilities for control of some virus diseases. Heat treatments and meristem tip culture methods have been applied to an increasing range of vegetatively propagated plants to provide a nucleus of virus-free material that then can be multiplied under conditions that minimize reinfection. Such developments frequently have involved the introduction of certification schemes. Systemic insecticides, sometimes applied in pelleted form at the time of planting, provide significant protection against some viruses transmitted in a persistent manner by aphid vectors. It has become increasingly apparent that effective control of virus disease in a particular crop in a given area usually requires an integrated and continuing program involving more than one kind of control measure.

However, such integrated programs are not yet in widespread use. Cross-protection (or mild-strain protection) is a phenomenon in which infection of a plant with a mild strain of a virus prevents or delays infection with a severe strain. The phenomenon has been used with varying success for the control of certain virus diseases, but the method has various difficulties and dangers. In 1986 Powell-Abel and co-workers considered that some of these problems might be overcome by the application of the concept of pathogen-derived resistance of Sandford and Johnston. Using recombinant DNA technology, they showed that transgenic tobacco plants expressing the TMV coat-protein gene either escaped infection following inoculation or showed a substantial delay in the development of systemic disease. These transgenic plants expressed TMV coat-protein mRNA as a nuclear event. Seedlings from self-fertilized transformed plants that expressed the coat protein showed delayed symptom development when inoculated with TMV. Thus, a new approach to the control of virus diseases emerged. However, this approach revealed some unexpected results which led to the recognition that plants have a defense system against 'foreign' RNA. This defense system, initially termed post-translational gene silencing and now called RNA silencing or RNA interference (RNAi) was first recognized in plants and had, and is still having, a great impact on molecular approaches as diverse as disease control and understanding gene functions. Among the tools that have arisen from understanding this new phenomenon is virus-induced gene silencing which is being used to determine the functions of genes in plants and animals by turning them off.

However, the RNAi defense system in plants which targets double-stranded RNA, an intermediate RNA virus replication, raised the question of how RNA viruses replicated in plants. Studies on gene functions revealed that many plant viruses contain so-called 'virulence' genes. Many of these have been shown to suppress host RNA silencing, thus overcoming the defense system. Suppression of gene silencing is widespread among plant viruses and examples are being found in animal viruses. Thus, as noted at the beginning of this article, plant virology is still providing insights onto phenomena applicable to virology in general.

See also: Diagnostic Techniques: Plant Viruses; Nature of Viruses; Plant Virus Diseases: Economic Aspects; Tobacco Mosaic Virus; Vector Transmission of Plant Viruses; Vaccine Production in Plants; Viral Suppressors of Gene Silencing; Virus Induced Gene Silencing (VIGS).

Further Reading

Baulcombe DC (1999) Viruses and gene silencing in plants. *Archives of Virology* 15(supplement): 189–201.

Caspar DLD and Klug A (1962) Physical principles in the construction of regular viruses. *Cold Spring Harbor Symposium on Quantitative Biology* 27: 1–24.

Clark MF and Adams AN (1977) Characteristics of the microplate method of enzyme-linked immunosorbent assay for the detection of plant viruses. *Journal of General Virology* 34: 475–483.

Fischer R and Emans N (2000) Molecular farming of pharmaceutical proteins. *Transgenic Research* 9: 279–299.

Hull R (2001) *Matthews' Plant Virology.* San Diego: Academic Press.

Li F and Ding SW (2006) Virus counterdefense: Diverse strategies for evading the RNA silencing mechanism. *Annual Review of Microbiology* 60: 507–531.

Lindbo JA and Dougherty WG (2005) Plant pathology and RNAi: A brief history. *Annual Review of Phytopathology* 43: 191–204.

Pavord A (1999) *The Tulip.* London: Bloomsbury.

Van der Want JPH and Dijkstra J (2006) A history of plant virology. *Archives of Virology* 151: 1467–1498.

History of Virology: Vertebrate Viruses

F Fenner, Australian National University, Canberra, ACT, Australia

© 2008 Elsevier Ltd. All rights reserved.

Introduction

As in other branches of experimental science, the development of our knowledge of viruses has depended on the techniques available. Initially, viruses were studied by pathologists interested in the causes of the infectious diseases of man and his domesticated animals and plants and these concerns remain the main force advancing the subject. The idea that viruses might be used to probe fundamental problems of biology arose in the early 1940s, with the development of the knowledge of bacterial viruses. These studies helped establish the new field of molecular virology in the period 1950–70, and this has revolutionized the study of viruses and led to an explosion of knowledge about them.

The Word Virus

Since antiquity the term virus had been synonymous with poison, but during the late nineteenth century it became a synonym for microbe (Pasteur's word for an infectious agent). It did not acquire its present connotation until the 1890s, after the bacterial or fungal causes of many infectious diseases had been discovered, using the agar plate, effective staining methods, and efficient microscopes. It then became apparent that there were a number of infectious diseases of animals and plants from which no bacterium or fungus could be isolated or visualized with the microscope. After the introduction in 1884 of the Chamberland filter, which held back bacteria, Loeffler and Frosch demonstrated that the cause of foot-and-mouth disease was a filterable (or ultramicroscopic) virus. The first compendium of all then known viruses was edited by T. M. Rivers of the Rockefeller Institute and published in 1928. Entitled *Filterable Viruses*, this emphasized that viruses required living cells for their multiplication. In the 1930s chemical studies of the particles of tobacco mosaic virus and of bacteriophages showed that they differed from all cells in that at their simplest they consisted of protein and nucleic acid, which was either DNA or RNA. Gradually, the adjectives filterable and ultramicroscopic were dropped and the word viruses developed its present connotation.

Early Investigations

Foot-and-Mouth Disease Virus

In 1898 F. J. Loeffler and P. Frosch described the filterability of an animal virus for the first time, noting that "the filtered material contained a dissolved poison of extraordinary power or that the as yet undiscovered agents of an infectious disease were so small that they were able to pass through the pores of a filter definitely capable of retaining the smallest known bacteria." Although the causative agent of foot-and-mouth disease passed through a Chamberland-type filter, it did not go through a Kitasato filter which had a finer grain. This led to the conclusion that the causative virus, which was multiplying in the host, was a corpuscular particle. Loeffler and Frosch gave filtration a new emphasis by focussing attention on what passed-through the filter rather than what was retained and established an experimental methodology which was widely adopted in the early twentieth century in research on viral diseases.

Yellow Fever Virus

Following the acceptance of the notion of filterable infectious agents, pathologists investigated diseases from which no bacteria could be isolated and several were soon shown to be caused by viruses. One of the most fruitful investigations, in terms of new concepts, was the work of the United States Army Yellow Fever Commission headed by Walter Reed in 1900–01. Using human volunteers, they demonstrated that yellow fever was caused by a filterable virus which was transmitted by mosquitoes and that the principal vector was *Aedes aegypti*. They also showed that infected persons were infectious for mosquitoes only during the first few days of the disease and that mosquitoes were not infectious until 7–10 days after imbibing infectious blood, thus defining the extrinsic incubation period and establishing essentially all of the basic principles of the epidemiology of what came to be called arboviruses (arthropod-borne viruses).

Physical Studies of Viruses

Further advances in understanding the nature of viruses depended on physical and chemical studies. As early as 1907 H. Bechhold in Germany developed filters of graded permeabilities and a method of determining their pore size. Subsequently, W. J. Elford, in London, used such membranes for determining the size of animal virus particles with remarkable accuracy.

In Germany, on the eve of World War II, H. Ruska and his colleagues had produced electron microscopic photographs of the particles of tobacco mosaic virus, bacteriophages, and poxviruses. Technical improvements in instrumentation after the war, and the introduction of negative staining for studying the structure of viruses by Cambridge scientists H. E. Huxley in 1957 and S. Brenner and R. W. Horne in 1959 resulted in photographs of the particles of virtually every known kind of virus. These demonstrated the variety of their size, shape, and structure, and the presence of common features such as the icosahedral symmetry of many viruses of animals.

Following a perceptive paper on the structure of small virus particles by F. H. C. Crick and J. D. Watson in 1956, in 1962 D. L. D Caspar of Boston and A. Klug of Cambridge, England, produced a general theory of the structure of regular virus particles of helical and icosahedral symmetry. Structure of the virus particle became one of the three major criteria in the system of classification of viruses that was introduced in 1966.

The Chemical Composition of Viruses

If the ultramicroscopic particles found in virus-infected hosts were the pathogens, what did these particles consist of? Following observations on plant and bacterial viruses in 1935, in 1940 C. L. Hoagland and his colleagues at the Rockefeller Institute found that vaccinia virus particles contained DNA but no RNA. Thus evidence was accumulating that viruses differed from bacteria not only in size and their inability to grow in lifeless media, but in that they contained only one kind of nucleic acid, which could be either DNA or RNA.

The development of restriction-endonuclease digestion of DNA, based on studies of phage restriction by W. Arber (Nobel Prize, 1978) of Basel in the 1960s, and then elaborated by biochemist H. O. Smith of Baltimore in 1970, has simplified the mapping of the genomes of DNA viruses, a study initiated by D. Nathans (Nobel Prize, 1978), also of Baltimore, using Simian virus 40. The development of the polymerase chain reaction in 1985 by K. B. Mullis (Nobel Prize, 1993) revolutionized sequencing methods for both DNA and RNA, leading to the availability of the complete genomic sequences of most viruses and the ability to make diagnoses using minute amounts of material.

Investigations of bacterial viruses were motivated by scientific curiosity, but animal virology was developed by pathologists studying the large number of diseases of humans and livestock caused by viruses, and it has retained this practical bias. Over the first two decades of the twentieth century, testing of filtrates of material from a number of infected humans and animals confirmed that they were caused by viruses. Among the most important was the demonstration in 1911 by P. Rous (Nobel Prize, 1966), at the Rockefeller Institute, that a sarcoma of fowls could be transmitted by a bacteria-free filtrate.

The Cultivation of Animal Viruses

The first systematic use of small animals for virus research was Pasteur's use of rabbits, which he inoculated intracerebrally with rabies virus in 1881. It was not until 1930 that mice were used for virus research, with intracerebral inoculations of rabies and yellow fever viruses. By using graded dilutions and large numbers of mice, quantitative animal virology had begun. The next important step, initiated by E. W. Goodpasture in 1931–32, was the use of chick embryos for growing poxviruses. This was followed a few years later by the demonstration by F. M. Burnet that many viruses could be titrated by counting the pocks that they produced on the chorioallantoic membrane, whereas others grew well in the allantoic and/or amniotic cavities of developing chick embryos.

Tissue cultures had first been used for cultivating vaccinia virus in 1928, but it was the discovery by J. F. Enders, T. H. Weller, and F. C. Robbins of Harvard University (Nobel Prize, 1954) in 1949 that poliovirus would grow in non-neural cells that gave a tremendous stimulus to the use of cultured cells in virology. Over the next few years their use led to the cultivation of medically important viruses such as those causing measles (by J. F. Enders and T. C. Peebles in 1954) and rubella (by T. H. Weller and F. C. Neva in 1962). Even more dramatic was the isolation of a wide variety of new viruses, belonging to many different families. Also the different cytopathic effects produced by different viruses in monolayer cell cultures were found to be diagnostic.

The next great advance, which greatly increased the accuracy of quantitative animal virology, occurred in 1952, when the plaque assay method for counting phages was adapted to animal virology by R. Dulbecco (Nobel Prize, 1975), using a monolayer of chick embryo cells in a petridish. In 1958 H. Temin and H. Rubin applied Dulbecco's method to Rous sarcoma virus, initiating quantitative studies of tumor viruses. Biochemical studies of animal virus replication were simplified by using continuous cell lines and by growing the cells in suspension.

Biochemistry

During the 1950s virus particles were thought to be 'inert' packages of nucleic acid and proteins, although in 1942 G. K. Hirst of the Rockefeller Institute had shown that particles of influenza virus contained an enzyme, later identified by A. Gottschalk of Melbourne as a neuraminidase.

Further advances in viral biochemistry depended on methods of purification of virus particles, especially the technique of density gradient centrifugation. In 1967 J. Kates and B. R. McAuslan demonstrated the presence of a DNA-dependent RNA polymerase in purified vaccinia virus virions, a discovery that was followed the next year by the demonstration of a double-stranded (ds) RNA-dependent RNA polymerase in reovirus virions and then of single-stranded (ss) RNA-dependent RNA polymerases in virions of paramyxoviruses and rhabdoviruses. In 1970 came the revolutionary discovery by D. Baltimore of Boston (Nobel Prize, 1975) and H. Temin of Wisconsin (Nobel Prize, 1975), independently, of the RNA-dependent DNA polymerase, or reverse transcriptase, of Rous sarcoma virus. Many other kinds of enzymes were later identified in the larger viruses; for example, no less than 16 enzymes have now been identified in vaccinia virus virions.

Just as investigations with bacterial viruses were of critical importance in the development of molecular biology, studies with animal viruses have led to the discovery of several processes that have proved to be important in the molecular biology of eukaryotic cells, although many of these discoveries are too recent for serious historical appraisal. Thus, in 1968 M. Jacobson and D. Baltimore of Boston showed that the genomic RNA of poliovirus was translated as a very large protein, which was then cleaved by a protease that was subsequently found in the viral replicase gene. RNA splicing was discovered in 1977, independently by P. A. Sharp (Nobel Prize, 1993) and L. T. Chow and their respective colleagues during studies of adenovirus replication. Work with the same virus by Rekosh and colleagues in London in 1977 resulted in the definition of viral and cell factors for initiation of new DNA strands using a novel protein priming mechanism. Capping of mRNA by m^7G and its role in translation was discovered in 1974, during work with reoviruses by A. J. Shatkin. Other processes first observed with animal viruses and now known to be important in eukaryotic cells were the existence of 3′ poly A tracts on mRNAs, the pathway for synthesis of cell surface proteins, and the role of enhancer elements in transcription.

Analysis of viral nucleic acids progressed in parallel with that of the viral proteins. Animal viruses were found with genomes of ssRNA, either as a single molecule, two identical molecules (diploid retroviruses), or segmented; dsRNA; ssDNA, dsDNA, and partially dsDNA. The so-called unconventional viruses or prions of scrapie, kuru, and Creutzfeld–Jacob disease, which have been extensively investigated since 1957 by D. C. Gajdusek of the US National Institutes of Health (Nobel Prize, 1976) and S. B. Prusiner of the University of California at San Francisco (Nobel Prize, 1997), appear to be infectious proteins. Soon after the discovery that the isolated nucleic acid of tobacco mosaic virus was infectious (see below), it was shown that the genomic RNAs of viruses belonging to several families of animal viruses were infectious; namely those with single, positive-sense RNA molecules. Then in 1973 came the discovery of recombinant DNA by P. Berg of Stanford University and his colleagues, using the animal virus SV40 and bacteriophage l.

Structure of the Virion

Using more sophisticated methods of electron microscopy, many animal virus virions have been shown to be isometric icosahedral structures, or roughly spherical protein complexes surrounded by a lipid-containing shell, the envelope, which contains a number of virus-coded glycoprotein spikes. X-ray crystallography of crystals of purified isometric viruses has revealed the molecular structure of their icosahedra; similar revealing detail has been obtained for the neuraminidase and hemagglutinin spikes of influenza virus.

Tumor Virology

After the discovery of Rous sarcoma virus in 1911, there was a long interval before the second virus to cause tumors, rabbit papilloma virus, was discovered by R. E. Shope in 1933. However, after that viruses were found to cause various neoplasms in mice, and in 1962 J. J. Trentin of Yale University showed that a human adenovirus would produce malignant tumors in hamsters. Since then, direct proof has been obtained that several DNA viruses (but not adenoviruses) and certain retroviruses can cause tumors in humans. Molecular biological studies have shown that oncogenicity is largely caused by proteins they produce that are encoded by viral oncogenes, a concept introduced by R. J. Huebner and G. Todaro of the US National Institutes of Health in 1969 and corrected, refined, and greatly expanded since the mid-1970s by J. M. Bishop, H. Varmus (Nobel Prize, 1989), and R. A. Weinberg of Boston.

Only one group of RNA viruses, the retroviruses, which replicate through an integrated DNA provirus, cause neoplasms, whereas viruses of five groups of DNA viruses are tumorigenic. Study of these oncogenic viruses has shed a great deal of light on the mechanisms of carcinogenesis. Oncogenic DNA viruses contain oncogenes as an essential part of their genome, which when integrated into the host cell DNA may promote cell transformation, whereas the oncogenes of retroviruses are derived from proto-oncogenes of the cell.

Impact on Immunology

Immunology arose as a branch of microbiology and several discoveries with animal viruses were important in the development of important concepts in immunology. It was the discovery in the 1930s by E. Traub in Tübingen of persistent infection of mice with lymphocytic choriomeningitis virus that led to the development by F. M. Burnet (Nobel Prize, 1960) of the concept of immunological tolerance in 1949. It was work with the same virus that led to the discovery of MHC restriction by P. C. Doherty and R. Zinkernagel in Canberra (Nobel Prize, 1996) in 1974.

Vaccines and Disease Control

From the time of Pasteur's use of rabies vaccine in 1885, a major concern of medical and veterinary virologists has been the development of vaccines. Highlights in this process have been the development of the 17D strain of yellow fever virus by M. Theiler (Nobel Prize, 1951) of the Rockefeller Foundation laboratories in 1937, the introduction of influenza vaccine in 1942, based on Burnet's 1941 discovery that the virus would grow in the allantois, the licensing in 1954 of inactivated poliovaccine developed by J. Salk and in 1961 of the live poliovaccine introduced by A. B. Sabin, both based on the cultivation of the virus in 1949 by Enders, and development of the first genetically engineered human vaccine with the licensing of yeast-grown hepatitis B vaccine in 1986. Finally, 1977 saw the last case of natural smallpox in the world and thus the first example of the global eradication of a major human infectious disease, the result of a 10-year campaign conducted by the World Health Organization with a vaccine directly derived from Jenner's vaccine, first used in 1796.

Recognition of HIV-AIDS

In 1981 a new disease was described in the male homosexual populations of New York, San Francisco, and Los Angeles. It destroyed the immune system, and the disease was called the acquired immune deficiency syndrome (AIDS) and the causal virus, human immunodeficiency virus (HIV), a retrovirus, was first isolated by L. Montagnier, of the Pasteur Institute in Paris, in 1983. A sexually transmitted disease, it has now spread all over the world, with about 40 million cases worldwide, and without very expensive chemotherapy is almost always fatal. It is now clear that it arose from an inapparent but persistent infection of chimpanzees in Africa. In spite of enormous efforts over the past 20 years, it has so far been impossible to devise an effective vaccine.

Arthropod-Borne Viruses of Vertebrates

Arthropods, mainly insects and ticks, were early shown to be important as vectors of virus diseases of vertebrates (the arboviruses) and of plants. Sometimes, as in myxomatosis, carriage was found to be mechanical, but in most cases the virus was found to multiply in the vector as well as the vertebrate concerned. The development of methods of growing insect cells in culture by T. D. C. Grace in Canberra in 1962 opened the way for the molecular biological investigation of the replication of insect viruses and arboviruses in invertebrate cells.

Taxonomy and Nomenclature

After earlier tentative efforts with particular groups of viruses, viral nomenclature took off in 1948, with the production of a system of latinized nomenclature for all viruses by the plant virologist F. O. Holmes. This stimulated others, in particular C. H. Andrewes of London and A. Lwoff of Paris, to actions which resulted in the setting up of an International Committee on Nomenclature (later Taxonomy) of Viruses in 1966. Adopting the kind of viral nucleic acid, the strategy of replication, and the morphology of the virion as its primary criteria, this committee has now achieved acceptance of its decisions by the great majority of virologists. One interesting feature of its activities is that its rules avoid the controversies about priorities that plague taxonomists working with fungi, plants, and animals.

The Future

The future will see an explosive expansion of understanding and knowedge of viruses of vertebrates, with the application of techniques of genetic engineering, nucleic acid sequencing, the polymerase chain reaction, and the use of monoclonal antibodies. All these discoveries are addressed in the appropriate sections of this encyclopedia. Molecular biology, which was initially conceived during studies of bacterial viruses, is now being used to study all the viruses of vertebrates and the pathogenesis and epidemiology of animal viral diseases. The expansion in knowledge of these subjects in the next decade can be expected to exceed that of the previous century.

See also: History of Virology: Plant Viruses; History of Virology: Bacteriophages; Nature of Viruses.

Further Reading

Bos L (1995) The embryonic beginning of virology: Unbiased thinking and dogmatic stagnation. *Archives of Virology* 140: 613.
Cairns J, Stent GS, and Watson JD (eds.) (1962) *Phage and the Origins of Molecular Biology.* Cold Spring Harbor, NY: Cold Spring Harbor Laboratory of Quantitative Biology.
Fenner F and Gibbs AJ (eds.) (1988) *Portraits of Viruses. A History of Virology.* Basel: Karger.
Horzinek MC (1995) The beginnings of animal virology in Germany. *Archives of Virology* 140: 1157.
Hughes SS (1977) *The Virus. A History of the Concept.* London: Heinemann Educational Books.

Joklik WK (1996) Famous institutions in virology. The Department of Microbiology, Australian National University and the Laboratory of Cell Biology, National Institute of Allergy and Infectious Diseases. *Archives of Virology* 141: 969.

Matthews REF (ed.) (1983) *A Critical Appraisal of Viral Taxonomy*. Boca Raton, FL: CRC Press.

Porterfield JS (1995) Famous institutions in virology. The National Institute of Medical Research, Mill Hill. *Archives of Virology* 141: 969.

van Helvoort T (1994) History of virus research in the twentieth century: The problem of conceptual continuity. *History of Science* 32: 185.

Waterson AP and Wilkinson L (1978) *An Introduction to the History of Virology*. Cambridge: Cambridge University Press.

Hordeivirus

J N Bragg, H-S Lim, and A O Jackson, University of California, Berkeley, Berkeley, CA, USA

© 2008 Elsevier Ltd. All rights reserved.

Introduction

Hordeiviruses represent a unique genus of serologically related viruses with rigid rods that are composed of 96% protein and 4% RNA. Although members of the hordeiviruses have similar particle structure, they collectively infect both monocots and dicots and exhibit considerable biological diversity. The type member of the group, barley stripe mosaic virus (BSMV), has been known to cause serious worldwide disease problems in cultivated barley for more than 75 years, but during this time has been isolated only infrequently from wheat and wild oats. Because of the yield losses caused in barley, a number of strains of the virus have been isolated and their biological properties evaluated. The survival of BSMV in nature depends solely on direct plant-to-plant contact and seed transmission; consequently, virus spread can be prevented by planting virus-free seed. The advent of sensitive virus detection methods has permitted seed to be screened prior to planting, thus virus-free seed production has resulted in eradication of the virus in most areas of North America and Europe.

In addition to *Barley stripe mosaic virus*, three additional species, *Poa semilatent virus* (PSLV), *Lychnis ringspot virus* (LRSV), and *Anthoxanthum latent blanching virus* (ALBV) have been included in the genus *Hordeivirus* in the *Eighth Report of the International Committee on Taxonomy of Viruses*. This assignment is based on serological relatedness of the coat proteins, genome organization, and sequence relatedness of BSMV, LRSV, and PSLV. The coat proteins of BSMV, PSLV, and LRSV are the same size, but have different mobilities on polyacrylamide gels and are distantly related serologically. Serology and analyses of the coat protein sequences indicate that BSMV and PSLV are more closely related to each other than to LRSV. Serological studies also show that ALBV is closely related to BSMV, and may be a strain of BSMV. However, ALBV has not been investigated by hybridization or sequence analysis to determine whether it is a strain of BSMV or is a distinct virus.

PSLV has been recovered on two occasions from native grasses in Canada from widely separated locations, so the virus may be relatively abundant among members of the Poaceae in Western Canada. LRSV was first isolated from *Lychnis divaricata* seeds imported into the USA, and an additional LRSV strain, Mentha, has been isolated in Hungary. ABLV has been described only in Britain and little is known about its biological or biochemical properties. As is the case with BSMV, LRSV is highly seed transmitted, but seed transmission has not been reported for PSLV or ABLV. There are no known biological vectors for the four viruses, but all reach high concentrations in infected plants and are easily transmitted from plant to plant by mechanical contact.

In addition to their limited range of natural hosts, the experimental hosts of hordeiviruses have been expanded substantially by mechanical inoculation. BSMV has been shown to infect several monocots and a few dicots. Local lesions on *Chenopodium* species have been used for several genetic studies of BSMV disease phenotype, and systemic infections in *Nicotiana benthamiana* have provided a basis for analyses of movement functions and RNA silencing. PSLV also infects numerous species of the Poaceae and several dicots, while LRSV has been shown to infect several experimental dicot hosts, including members of the families. Caryophyllaceae and Labiatae. Reverse genetic systems for BSMV, PSLV, and LRSV have resulted in considerable information about biological properties of the viruses and the molecular biology of infection processes.

Genome Structure and Expression

Hordeivirus genomes are composed of three RNAs that have been designated α, β, and γ based on hybridization patterns of the RNAs. The sizes of the BSMV α, β, and γ genomic (g) RNAs vary between strains, with the ND18 strain sizes being 3.8, 3.2, and 2.8 kb, respectively. RNAs α and β are similar in size among strains of BSMV, but the sizes and complexity of the γ RNAs vary considerably.

The type strain gRNAγ has a 366 nucleotide (nt) sequence duplication at the 5′ end of the genome and comigrates with RNAβ in agarose gels. The Argentina mild strain contains mixtures of three gRNAγ species of 3.2, 2.8, and 2.6 kb. The 3.2 kb RNA has a duplicated sequence similar to that of the type strain, and the 2.6 kb RNA is defective as the result of a deletion in the 3′ terminus of the γa gene. The PSLV and LRSV α, β, and γ gRNAs are similar in size (3.9, 3.6, and 3.2 kb, and 3.7, 3.1, and 2.6 kb, respectively) and sequence to those of BSMV. Strain-specific comparisons are not available for PSLV and LRSV.

Each of the hordeivirus RNAs has a 7-methylguanosine cap at the 5′ terminus, and among the gRNAs within a species, the 5′ untranslated regions (UTRs) that are necessary for replication have very little sequence similarity. Obvious sequence similarity is not evident among the 5′ UTRs of the α and γ RNAs of BSMV, PSLV, and LRSV, but the RNAβ components of the three viruses have considerable sequence similarity. However, the 5′ UTR of LRSV RNAβ is 65 nt longer than that of BSMV or PSLV due to a 5′ terminal extension that is identical to 53 nt at the 5′ end of LRSV RNAα. Interestingly, a 70 nt 5′ UTR recombination between the α and γ RNAs of the BSMV CV17 strain has also been described. These results and the sequence variability exhibited by the Argentina mild strain of BSMV indicate that recombination has played an important role in evolution of the hordeiviruses.

The hordeiviruses gRNAs also contain a conserved 3′ terminal nontranslated sequence that is required for replication. The 3′ termini are variable in length between the different viruses, ranging from 148 nt in LRSV to 330 nt in PSLV. The sequences form tRNA-like structures that in BSMV and PSLV are capable of binding tyrosine *in vitro*. Sequence comparisons indicate that the structures are very similar except that the 3′ UTRs upstream of the conserved tRNA regions of the PSLV and BSMV RNAs have a large stem loop element and an array of possible pseudoknot stem loop regions that make the 3′ UTR longer than that of LRSV. An internal poly(A) sequence of varying length is located directly after the stop codon of the 3′ proximal gene of each RNA, and resides immediately upstream of the conserved 3′ tRNA termini in the BSMV and LRSV genomes. The poly(A) sequence is absent in the PSLV genome.

Hordeivirus gRNAs (**Figure 1**, BSMV) encode seven proteins. The replicase protein subunits are encoded by

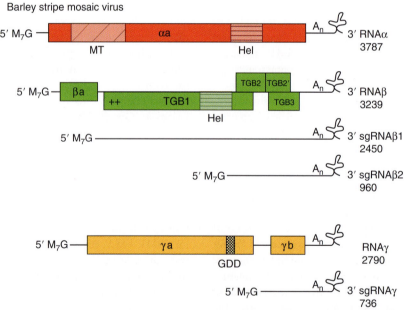

Figure 1 The tripartite genome of BSMV and sgRNAs used for expression of the genes encoded by the β and γ RNAs. Each RNA contains a 5′ cap structure (M_7G), and each 3′ proximal open reading frame (ORF) terminates with a UAA codon initiating an internal polyadenylate sequence (A_n) that precedes the 238 nt tRNA-like terminus. The three genomic RNAs (gRNAs), with ORFs represented by rectangular blocks, are designated α, β, and γ. The αa protein, which is the methyltransferase subunit of the RNA-dependent RNA polymerase (RdRp), is translated directly from RNAα. The αa protein is required for replication and contains methyltransferase (MT) and helicase (Hel) motifs that identify it as the 'helicase subunit' of the viral replicase complex. RNAβ encodes four major proteins. The coat protein, βa, is translated directly from the gRNAβ. The overlapping triple gene block (TGB) proteins, TGB1, TGB2, and TGB3, are each required for virus movement and are expressed from two subgenomic RNAs, sgRNAβ1 and sgRNAβ2, that are illustrated below gRNAβ. TGB1 contains two positively charged regions toward the N-terminus of the protein and an Hel domain, while TGB2 and TGB3 are small hydrophobic transmembrane proteins. RNAγ is bicistronic, and the 5′ ORF encodes the γa protein, which is the GDD-containing polymerase subunit of the RdRp. The cysteine-rich γb protein is involved in pathogenesis.

the α and γ gRNAs. The α RNAs encode a single protein (αa) that contains methyltransferase and helicase domains that are highly conserved between Sindbis-like viruses. The bicistronic γRNAs encode the polymerase (γa) subunit, which contains the conserved GDD motif found in RNA-dependent RNA polymerase (RdRp) proteins of RNA viruses, plus a 3′ encoded cysteine-rich protein (γb).

RNAβ encodes the coat protein in the first open reading frame (ORF), and is followed by a series of overlapping ORFs termed the 'triple gene block' (TGB) that are also present in allexi-, beny-, carla-, fovea-, peclu-, pomo-, and potexviruses. The first ORF of the TGB encodes the TGB1 protein (formerly designated βb). This protein contains a conserved helicase domain similar to the helicase domain present in the αa protein. The remaining two TGB ORFs encode the small hydrophobic proteins, TGB2 and TGB3 (formerly βd and βc, respectively).

The αa, βa (coat protein), and γa proteins encoded by RNAα, RNAβ, and RNAγ, respectively, are translated directly from the gRNAs (**Figure 1**). Expression of the TGB proteins is mediated by two subgenomic (sg) RNAs, designated sgRNAβ1 and sgRNAβ2. TGB1 is translated from sgRNAβ1, and the other overlapping proteins, TGB2 and 3, are expressed from sgRNAβ2. The γb protein is translated from sgRNAγ. Each of the encoded proteins is expressed at different levels and times during the replication cycle, and the three sgRNAs have similar variances in their abundance during replication. High-level constitutive expression of sgRNAγ occurs throughout the replication cycle, but expression of sgRNAβ1 and sgRNAβ2 is temporal. In addition, sgRNAβ1 is considerably more abundant than sgRNAβ2. The three sgRNA promoters have little obvious sequence relatedness; however, the individual BSMV, LRSV, and PSLV sgRNA promoters share a number of blocks of conserved sequence. Pairwise combinations indicate that the BSMV and PSLV promoters have the highest conservation, whereas the LRSV promoters have undergone more divergence. These results and the protein comparisons described below support the existing evidence for a common origin of the hordeiviruses and buttress biological evidence suggesting that BSMV and PSLV are more closely related to each other than to LRSV. Additional comparisons of the TGB promoter sequences of several other viruses possessing TGB movement proteins have revealed no strong correlations in sequence or secondary structure between these putative promoter sequences and those of BSMV.

Replication of BSMV gRNAs requires the 5′ and 3′ terminal regions, but additional internal *cis*-elements are required for replication of each of the RNAs. RNAα replication is *cis*-preferential, with replication appearing to be coupled to translation of a functional αa protein. Each of the proteins encoded by RNAβ is dispensable for replication; however, the 117 nt intergenic region separating the βa (coat protein) ORF and the TGB1 ORF has a *cis*-acting function required for replication. In contrast, RNAγ replication is not dependent upon the 42 nt intergenic region separating γa and γb, but essential *cis*-acting elements are present in the first 500 nt of the γa gene.

Function and Relatedness of the Hordeivirus Proteins

Replicase Proteins (αa and γa)

BSMV gRNAs expressing αa and γa are able to replicate in protoplasts. The αa and γa proteins are essential subunits of the RdRp and are members of the tobamo-lineage of supergroup III RdRps. A histidine-tagged γa protein recovered from infected barley fractionates with the αa protein. The recovered complex exhibits BSMV RNA-specific polymerase activity and does not contain detectable levels of other BSMV-encoded proteins. Therefore, the αa and γa proteins constitute the helicase and polymerase subunits of the RdRp complex.

The αa protein contains amino-terminal methyltransferase and carboxy-terminal NTPase/helicase domains. The methyltransferase domain likely functions in capping of the viral RNAs and the DEAD box helicase domain consists of at least six amino acid motifs that are conserved among the RdRps of different groups of RNA viruses.

The γa RdRp subunit contains a GDD polymerase motif toward its C-terminus that is characteristic of the supergroup III polymerases, and this domain falls into the 'tobamo-lineage' among the plant viruses. Phylogenetic comparisons of hordeivirus polymerase subunits reveal more than 75% conservation within the genus and lower levels of conservation extend into members of the genera *Pecluvirus*, *Furovirus*, and *Tobravirus* (>65%), and *Tobamovirus* (>30%). Sequence comparisons of the NTPase, helicase, and GDD motifs have revealed significant conservation among the polymerase subunits of Sindbis-like viruses (supergroup III). Among the plant viruses within this group, the conserved domains of BSMV and PSLV are more closely related than those of LRSV, and these sequences also share close relatedness with viruses in the genera *Furovirus* and *Pomovirus*. Lower levels of sequence relatedness are observed when comparisons are extended to the genera *Tobravirus* and *Tobamovirus* and members of the family *Bromoviridae*.

Coat Protein (βa)

The coat protein (CP) is translated directly from gRNAβ and is the most abundant of the viral proteins in infected plants. The CPs of all hordeiviruses are approximately 22 kDa (~200 amino acids) in size, but exhibit different electrophoretic mobilities. The CP interacts with the genomic RNAs to form rod-shaped particles that appear to contain 3 nt per protein subunit, with approximately

24 protein subunits per turn of the helix. In agreement with their serological relatedness, the BSMV and PSLV coat protein sequences are more similar to each other than to the CP of LRSV. Although the CP shares several conserved motifs with other rod-shaped viruses, the hordeivirus CPs are closely related to that of members of the fungal-transmitted virus species *Peanut clump virus* and *Indian peanut clump virus* but show more limited similarity to SBWMV.

Studies of BSMV and PSLV show that the CP is not essential for infectivity of two systemic hosts (barley and *N. benthamiania*), or hosts that form local lesions (*Chenopodium* species). Moreover, infections elicited by CP-deficient BSMV mutants appear to be more aggressive, and the phenotype is more severe and protracted in barley than derivatives expressing the CP. Plants infected with CP deletion mutants remain stunted and appear to lack the recovery phase of infection that is usually observed in barley inoculated with BSMV. These results suggest that the CP is a multifunctional protein that may affect replication and gene expression through a feedback mechanism rather than acting to suppress host RNA silencing.

TGB Movement Proteins

The hordeiviruses encode Class I TGB proteins that differ in several features from the Class II TGB proteins of PVX and other viruses of the family *Flexiviridae*. The coordinated actions of each of the three TGB proteins are required for cell-to-cell and systemic movement. The proteins are expressed transiently and simultaneously during the early stages of infection and decline in abundance at later stages of infection.

TGB1

The TGB1 protein, encoded by the 5′ proximal gene of the TGB, is expressed from sgRNAβ1 and accumulates to high levels early in infection. Among the hordeivirus species, the TGB1 proteins range from 50 to 63 kDa in size. These proteins can be distinguished from the ∼25 kDa Class II TGB proteins of the flexiviruses by having substantially larger N-terminal domains preceding the helicase domain. The N-terminal half of the hordeivirus TGB1 proteins contains two positively charged regions rich in lysine and arginine residues, while the C-terminal half of the protein contains an NTPase/helicase domain with seven conserved motifs (I, IA, II, III, IV, V, and VI) that are characteristic of superfamily I helicases of alpha-like viruses. The TGB1 protein has RNA helicase activity *in vitro* that is dependent on Mg^{2+} and ATP and can unwind RNA duplexes in both 5′- to -3′ and 3′- to -5′ directions. The TGB1 protein also binds ATP and dATP, and exhibits ATPase activity that maps to the helicase-containing, C-terminal half of the protein.

The TGB1 protein has RNA-binding activity at high ionic strength and exhibits a higher binding affinity for double-stranded (ds) RNAs than single-stranded (ss) RNAs. RNA sequence specificity has not been detected in binding studies, and TGB1 has little detectable affinity for DNA. The TGB1 proteins of both BSMV and PSLV contain multiple RNA-binding regions, and in PSLV, the N-terminal half of TGB1 containing the positively charged regions is able to bind RNA under high-salt conditions, whereas the C-terminal half containing the helicase motif binds RNA only under low-salt conditions.

A BSMV TGB1/RNA ribonucleoprotein (RNP) complex has been purified from infected barley, demonstrating that TGB1 also binds to RNA *in vivo*. The complex contains positive-sense α, β, and γ RNAs, but additional BSMV-encoded proteins are not present in detectable amounts. However, the TGB1 protein participates in both homologous interactions and in associations with TGB3. The TGB1 protein is membrane associated in infected cells, and visualization of a GFP:TGB1 fusion protein reveals paired foci on opposite sides of the cell wall, suggesting localization at plasmodesmata (PD). When TGB1 is transiently expressed in the absence of TGB2 or TGB3, the TGB1 protein is localized in cytoplasmic membranes. Due to its RNA-binding activities and presence in the RNP complex recovered from infected plants, the TGB1 protein appears to function in formation of viral RNA complexes that are transported to adjacent cells during interactions with TGB2 and 3.

TGB2 protein

Among the hordeivirus TGB proteins, the TGB2 protein contains the highest amount of sequence similarity, and the degree of sequence similarity extends to the class I TGB2 proteins encoded by PVX and other members of the family *Flexiviridae*. The TGB2 sequence contains two hydrophobic stretches and a conserved hydrophilic region that separates the hydrophobic residues. The hydrophobic regions are predicted to integrate into membranes with a U-shaped topology that directs the termini of the proteins to the cytoplasmic side of the membrane, and the strong conserved central portion of the protein is exposed to the ER lumen.

The TGB2 protein is associated with cellular membranes, and associates with the cortical ER and motile vesicles resembling Golgi stacks. In the presence of RNAα and RNAγ, a small portion of BSMV TGB2 fluorescent fusion proteins also associates with chlorophyll-containing vesicles that appear to be chloroplasts or derived from chloroplasts. From a functional perspective, a GFP:TGB2 fusion protein moves through several cells during transient expressions lacking TGB3, suggesting that TGB2 can modify the size exclusion limits of PD.

TGB3 protein

The TGB3 protein is encoded by the 3′ proximal ORF of sgRNAβ2 and is translated by leaky scanning of ribosomes past the TGB2 AUG, which is in a poor context for translation initiation. This mechanism normally produces about a 10:1 ratio of TGB2 to TGB3 during *in vitro* translation. The TGB3 proteins are the most poorly conserved of the TGB proteins, and the class 1 and 2 proteins are thought to have a polyphyletic origin, in contrast to the TGB2 proteins that are thought to have originated from a common ancestor.

The N-terminus of the approximately 17 kDa TGB3 protein from each of the hordeivirus species contains a stretch of conserved amino acids that includes one histidine and three cysteine residues. The TGB3 protein has two hydrophobic domains separated by a hydrophilic region that contains a conserved tetrapeptide (QDLN) sequence. The transmembrane regions are predicted to direct the termini of TGB3 toward the ER lumen. This topology is quite distinct from that of the class II TGB3 proteins, which have only a single membrane-spanning domain that results in location of the C-terminus on the cytoplasmic side of the ER.

When expressed alone, TGB3 localizes to paired peripheral bodies at the cell wall (CW) that appear to lie on opposite sides of the PD channels connecting adjacent cells. The targeting of PSLV TGB3 to the CW requires the central hydrophilic and the C-terminal transmembrane regions, whereas the BSMV TGB3 protein requires the five C-terminal residues, LSSKR, for PD targeting. Furthermore, TGB3 redirects the targeting of TGB1 and 2 to paired peripheral bodies at the CW, and protein interaction studies demonstrate that TGB3 binds with both TGB1 and TGB2.

Pathogenesis Protein (γb)

The 3′ proximal ORF of the hordeivirus γRNAs encodes a cysteine-rich γb protein that ranges in size from 16 to 20 kDa. The hordeivirus γb proteins share some structural similarities, but have little direct amino acid sequence similarity with the cysteine-rich proteins of tobra- and carlaviruses. Studies of the BSMV ND18 strain demonstrate that γb is not strictly required for infectivity in plants, but that it affects pathogenicity, viral RNA accumulation, and expression of genes encoded by RNAβ. Extensive mutational analyses of γb have revealed a number of distinct phenotypes in barley and *Chenopodium* that can ameliorate or exacerbate disease symptoms.

The majority of the cysteine residues in γb are concentrated in the N-terminal half of the protein. These residues are clustered in two zinc-finger-like motifs, designated C1 and C2. A basic motif (BM) rich in lysine and arginine residues is located between the cysteine-rich regions. The BSMV γb protein has been expressed and purified from *Escherichia coli* and shown to bind ssRNAs and zinc *in vitro*. The basic motif mediates RNA-binding activity, and the C1, BM, and C2 regions each show independent zinc-binding activity. A coiled-coil structure near the C-terminus of the γb protein mediates self-interactions that affect pathogenicity.

The PSLV γb protein is targeted to peroxisomes via a C-terminal SKL signal sequence, but deletion of the tripeptide does not affect infection phenotype. This sequence is also found in LRSV and in the BSMV type and China strain γb proteins, but is absent in the ND18 and CV17 strains. The BSMV ND18 γb:GFP fusion protein fluorescence is dispersed throughout the cytoplasm, and subcellular fractionation of infected barley tissue confirms that the majority of the protein is located in the soluble fraction. However, in the presence of the α and γ RNAs, a small portion of the γb protein is associated with vesicles surrounding the nucleus that contain chlorophyll. These associations with chloroplast membranes suggest that the γb protein may have a previously unrecognized role in virus replication or that it may protect viral replication factories against host RNA defense mechanisms.

Several lines of evidence indicate that a major function of γb is to serve as an inhibitor of host defense mechanisms that target RNA replication. A recombinant BSMV derivative that lacks γb expression compromises systemic virus infection and the γb function can be complemented by expression of other known silencing suppressor proteins. Similarly, expression of the γb protein is able to suppress silencing in heterologous virus-induced silencing systems. An agrobacterium-mediated transient silencing assay also provides strong support for a role of γb in suppression of RNA silencing. In this assay, γb interferes with silencing of GFP fluorescence that was induced by expression of a dsRNA. The BSMV γb coiled-coil region is required for silencing suppression in this system, suggesting that homologous interactions of γb are required for suppression activity. These results thus provide a persuasive argument that a major function of γb is to counter host innate RNA interference defenses.

Cytopathology and Replication

Microscopy studies of BSMV-infected barley indicate that most of the visible changes in leaf appearance can be attributed to the disruption of organelles, most notably chloroplasts, in the rapidly expanding cells of the growing leaf. The morphology and shape of chloroplasts are altered drastically during infection as grana become disorganized, and numerous small vesicles appear between the inner and outer membranes of the chloroplast envelope. Reduced amounts of chloroplast ribosomal RNA and decreased rates of synthesis of chloroplast proteins are also noted as BSMV RdRp activity increases. Some evidence also indicates that chloroplasts or chloroplast membrane vesicles may become reoriented around the nuclei of infected leaf cells. Additional membranes often proliferate and form large globular inclusions of unknown function. Late in infection, virions accumulate to high levels in the cytoplasm in disordered arrays, but a small proportion of rod-shaped particles presumed to be of viral origin accumulate in the nuclei. Biochemical studies undertaken to extend the ultrastructural studies show that BSMV infection inhibits chlorophyll accumulation, and suggest that alterations in photosystem II may result in decreased light absorbance. Reductions in lipids and galactolipids are also correlated with infection and proplastid degradation.

Natural plant-to-plant transmission of hordeiviruses involves mechanical entry of virions into cells. A model for replication posits that virions are disassembled by ribosomes and that RdRp subunits are translated from the α and γ gRNAs (**Figure 2**). Indirect evidence indicates that the polymerase subunits assemble with host components and establish sites of replication on invaginated membrane vesicles originating from chloroplasts or proplastids. Based on dsRNA serological assays, replicating RNAs also appear to be associated with the vesicles. The requirement for translation of a functional αa protein for replication of RNAα raises the hypothesis that the protein may function to recruit RNAγ to sites of replication and initiate vesicle formation. During the replication cycle, the replicase initiates at the 3′ termini of each of the gRNAs during synthesis of dsRNAs, and also initiates sgRNA transcription at internal promoters on the minus strands of RNAβ and RNAγb. Recent results indicate that in the presence of RNAs α and γ, small amounts of TGB2 and the γb protein are targeted to abnormal chloroplast-like vesicles in barley and *N. benthamiana*. These results suggest that the γb protein and the TGB proteins may be recruited to the chloroplast vesicles to interfere with host defenses and to interact with viral RNAs destined for cell-to-cell movement. Serological studies also suggest that small amounts of CP are present in chloroplast-associated vesicles, so it is possible that the CP has regulatory functions related to replication. These findings thus fit a model whereby replication is initiated in vesicles associated with proplastids that mature into chloroplasts.

During the initial stages of replication, the TGB1 movement protein probably functions to unwind and encapsidate positive-sense gRNAs to form nucleoprotein complexes in the vesicles. Biochemical experiments have shown that the resulting TGB1 nucleoprotein complexes can be recovered from infected plants and that TGB1 associates with TGB2 and TGB3. These interactions are thought to mediate transit of the nucleoproteins through the cortical ER to the lumen of the PD. An attractive model postulates that during transit to the PD, the nucleocapsid is associated with vesicles containing RdRp and that these vesicles are targeted by actin to the periphery of the cell. Although cytochalasin D fails to noticeably affect the membrane localization of TGB1, recent work shows that the diameter of actin cables increases significantly in infected cells and in cells expressing TGB2 and TGB3. Transient expression studies with PSLV and BSMV also show that TGB3 forms opposing pairs of punctate foci across PD, so it is plausible that TGB3 interacts with TGB2 to target TGB1 nucleoprotein movement complexes, and possibly replication complexes, through the PD.

In contrast to viruses such as PVX, which contain class II TGB proteins, the coat protein of hordeiviruses is dispensable for cell-to-cell and vascular movement and, therefore, nucleocapsid complexes must have a primary role in movement. During the early phases of cell-to-cell spread, GFP:TGB1 forms brightly fluorescent concentric rings at the advancing infection front and greatly reduced fluorescence behind the front. These results suggest that the TGB proteins are rapidly downregulated after movement, as is the case with TMV where the 30 kDa protein appears to be targeted to the host 26S proteosome for degradation.

Pathogenesis

Natural mutants in both the noncoding and coding regions of BSMV RNAs have been shown to have strain-specific effects on pathogenesis. For example, the ND18 strain is unable to infect oats, whereas the CV42 strain is pathogenic in this host. ND18 virulence in oats can be engineered by altering a single amino acid in the αa protein, so this phenotype may result from relatively simple associations of αa with a host protein. Natural variation among BSMV strains has also facilitated identification of factors affecting seed transmission. Primary determinants influencing the efficiency of seed transmission reside in RNAγ sequences in the 5′ UTR and in the γb gene. Therefore, it appears that RNAγ sequences that are crucial to seed transmission influence replication and have effects that counter host RNA interference mechanisms. In a fourth case of strain-specific virulence, ND18 is able to systemically infect *N. benthamiana*, while the type strain is able to replicate and move in the inoculated leaf but is unable to invade systemically. This aberrant phenotype appears to be caused by the presence of a short ORF in the 5′ UTR that reduces γa translation. Infections with the ND18 or type strains also result in different local lesion phenotypes on *C. amaranticolor*. These differences are complex, but are primarily due to the presence of an amino-terminal 372-nucleotide duplication in RNAγ of the type strain that downregulates expression of γa, and to differences in amino acids of the γb proteins.

Advances in understanding pathogenesis have also been obtained by induction of mutations by reverse

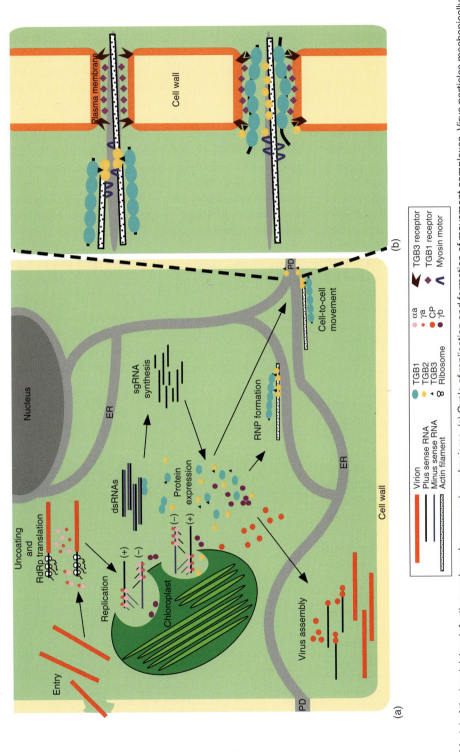

Figure 2 Model of the hordeivirus infection cycle and movement mechanisms. (a) Cycle of replication and formation of movement complexes. Virus particles mechanically enter the cell through breaks in the plant cell wall. Entry is followed quickly by uncoating of the gRNAs in the cytoplasm and translation of the αa and γa proteins to form the RdRp. The RdRp associates with host factors and recruits positive-sense gRNAs to chloroplast membranes where genome replication occurs on invaginated membrane vesicles. Production of positive- and negative-sense gRNAs results in the formation of dsRNA replicative intermediate species. The RdRp also initiates transcription at internal promoters located on the negative-strand RNAs to generate sgRNAs that function as messengers for the translation of proteins encoded by the 3′ ORFs of the gRNAs. Small amounts of the TGB2 and γb proteins are also targeted to abnormal chloroplast-like vesicles where they may function in movement and protection against host defenses. The TGB1 protein binds to positive-sense gRNA to create hordeivirus ribonucleoprotein (RNP) movement complexes that associate with the TGB2 and 3 proteins for transport to the plasmodesmata (PD) to facilitate virus cell-to-cell spread. The CP, which is translated directly from gRNAβ, binds positive-sense gRNAs to form progeny virions. (b) Expanded view of events occurring during entry into PD. RNP movement complexes containing TGB1 and positive-sense gRNA associate with the TGB2 and TGB3 integral membrane proteins for transport along actin filaments that are associated with the ER. The TGB3 protein targets the complex to PD and is presumed to interact with host receptor proteins in the PD desmotubule. TGB2 mediates RNP entry into PD, possibly through remodeling of actin filaments within the desmotubule. Host proteins within the PD are predicted to interact with TGB1 to convey RNP complexes to the adjacent cell. The TGB2 and 3 proteins may move through PD along with the RNP, or be recycled by the endocytic pathway to facilitate the movement of additional RNP complexes.

genetics. The early results showed that γb is not strictly required for infectivity in plants, but that it has important pathogenicity effects. For example, null mutations in γb inhibit infections of the BSMV type strain in barley, but the ND18 strain is able to establish systemic infections albeit with an altered symptom phenotype. However, the failure of the type strain to establish systemic infections in the absence of γb can be reversed by mutations that increase the translation of the γa replicase or that remove an amino-terminal extension of the γa protein that may reduce the function of the protein. Individual site-specific mutations within the cysteine-rich region of γb can also result in a variety of phenotypes ranging from striking white regions to necrotic streaks on systemically infected leaves of barley. In addition, plants infected with site-specific γb mutants have secondary effects that culminate in greatly reduced abundance of RNAβ, the CP, and the TGB1 protein. Although the CP is not required for infectivity in barley, *N. benthamiania*, or *Chenopodium* species, infections elicited by CP-deficient BSMV mutants appear to be more aggressive, because recovery is protracted and the symptoms are more severe than in the presence of the CP. These results suggest that in addition to being essential for formation of virions, the coat protein may have a critical role in downregulating kinetics of replication late in infection. Overall, these genetic analyses suggest that complex interactions may affect virulence and disease phenotype, and that the levels of replication may have important effects on movement, host range, and seed transmission.

Natural genes that condition resistance to the ND18, type, and other BSMV strains have been identified in several barley cultivars. The recessive nature of these genes indicates that resistance is due to the absence of a function required for virus replication rather than an active response to infection. The fact that related strains of BSMV can overcome resistance suggests that resistance may result from mutation of a factor required for replication rather than the complete absence of a host factor. This hypothesis is supported by evidence that virus strains targeted by resistance genes are unable to replicate in protoplasts from resistant varieties, whereas the protoplasts fully support the replication of strains able to overcome the resistance genes. Accordingly, single amino acids or short stretches of amino acids in the αa and γa proteins appear to have been targeted by host resistance responses. As more resistance genes are identified, it will be interesting to see whether dominant genes target other functions such as the CP and movement proteins to elicit dominant protective mechanisms.

Hordeivirus Applications to Cereal Genomic Analyses

Virus-induced gene silencing (VIGS) has provided useful approaches to improve our understanding of the functional genomics of dicots, and several virus vectors have been developed for this purpose. Recently, BSMV has been developed into an effective VIGS vector by modifying sequences downstream of the γb gene to enable expression of untranslatable foreign inserts. This strategy was first applied to barley by targeting phytoene desaturase (PDS) for VIGS. Subsequently, applications of VIGS to downregulate genes known to be involved in disease resistance have also enabled identification of genes required for powdery mildew resistance in barley. Extension of the VIGS system to hexaploid wheat has also been used to effectively downregulate PDS, subunit H of the magnesium–protoporphyrin chelatase complex, and the β7 subunit of the 20S proteosome complex. An additional study has identified genes required for rust resistance in wheat. These studies indicate that BSMV provides a powerful and robust system for gene silencing that may have numerous applications for genetic analyses of a large number of cereal crops.

Future Perspectives

The broad host range of hordeiviruses and the information accumulated about virus genes involved in replication, movement, and responses to host resistance afford a rich resource for obtaining insight into the mechanics of infection. Although considerable effort has recently focused on the requirements and interactions of hordeivirus-encoded proteins, very little is known about how virus and host proteins interact to facilitate replication and movement. New genomics findings arising from host sequencing projects, development of advanced tools for analysis of host genes and cell biology, and the ability to modulate expression of individual genes are accelerating the pace of research findings and enabling dissection of individual steps in many of these processes. Of particular importance is characterization of host genes regulated during infection and the effects of these genes on pathogenesis. An understanding of events leading to the establishment of replication factories will provide important information on hordeivirus replication and the interplay between chloroplasts and cytoplasmic elements during the early stages of infection. Additional cell biological analyses of the TGB-encoded proteins and their interactions with host proteins are required to hone our understanding of the cell-to-cell and vascular movement mechanisms that permit systemic virus infections. In addition to enhancing the understanding of how viruses invade vegetative tissue, seed, and pollen, these studies may also provide insights into global aspects of subcellular transport, chloroplast, and PD functions in plants.

See also: Cereal Viruses: Wheat and Barley; Virus Induced Gene Silencing (VIGS).

Further Reading

Boevink P and Oparka K (2005) Virus–host interactions during movement processes. *Plant Physiology* 138: 1815–1821.

Bragg JN, Solovyev AG, Morozov SYu, Atabekov JG, and Jackson AO (2005) Genus *Hordeivirus*. In: Fauquet CM, Mayo MA, Maniloff J, Desselberger U, and Ball LA (eds.) *Virus Taxonomy: Eighth Report of the International Committee on Taxonomy of Viruses*, pp. 1015–1026. San Diego,CA: Elsevier Academic Press.

Carroll TW (1986) Hordeiviruses: Biology and pathology. In: Van Regenmortel MHV and Fraenkel-Conrat H (eds.) *The Plant Viruses, The Rod-Shaped Plant Viruses*, vol. 2, pp. 373–395. New York: Plenum.

Edwards MC, Bragg JN, and Jackson AO (2005) Natural resistance mechanisms to viruses in barley. In: Loebenstein G and Carr JP (eds.) *Natural Resistance Mechanisms of Plants to Viruses*, pp 465–501. Dordrecht, The Netherlands: Springer.

Holzberg S, Brosio P, Gross C, and Pogue GP (2002) Barley stripe mosaic virus-induced gene silencing in a monocot plant. *Plant Journal* 30: 315–327.

Jackson AO and Lane LC (1981) Hordeiviruses. In: Kurstak E (ed.) *Handbook of Plant Virus Infections and Comparative Diagnosis*, pp. 565–625. Amsterdam: Academic Press.

Jackson AO, Petty ITD, Jones RW, Edwards MC, and French R (1991) Analysis of barley stripe mosaic virus pathogenicity. *Seminars in Virology* 2: 107–119.

Lawrence DM and Jackson AO (1998) Hordeiviruses. In: Foster G and Taylor S (eds.) *Plant Virology Protocols: From Virus Isolation to Transgenic Resistance*, vol. 81, pp. 99–106. Totowa, NJ: Humana Press.

Morozov SYu and Solovyev AG (2003) Triple gene block: Modular design of a multifunctional machine for plant virus movement. *Journal of General Virology* 84: 1351–1366.

Host Resistance to Retroviruses

T Hatziioannou and P D Bieniasz, Aaron Diamond AIDS Research Center, The Rockefeller University, New York, NY, USA

© 2008 Elsevier Ltd. All rights reserved.

Glossary

APOBECs A family of cytidine deaminases, the prototypic member is a component of apolipoprotein B editing complex.

Restriction factor A cellular protein whose major role is to inhibit retrovirus replication.

Vif Virion infectivity factor encoded by many lentiviruses which induces degradation of some APOBEC proteins.

Vpu Viral protein U, a small transmembrane protein encoded by HIV-1 and its close relatives.

Xenotropic Able to infect cells only from other species.

Introduction

The evolution of organisms and the viruses that colonize them is, obviously, closely linked. Often, pathogenic viruses impose a negative selection pressure on a host that results in the survival of a subpopulation in which infection is resisted, attenuated, or tolerated. In the case of retroviruses, the evolutionary association of virus and host can be exceptionally close, because of the unusual degree to which retroviral infection persists. Retroviral infection is normally lifelong, in part due to the propensity of retroviral genomes to irreversibly integrate into host DNA. Sometimes the targets of retroviral infection include germline cells, and infection of these cells, which generate so-called endogenous proviruses, enables persistence, not only for the host's own lifetime, but also that of its progeny. This persistence and inextricable association of host and virus may allow retroviruses to impose a uniquely sustained selection pressure and may explain why hosts have evolved unique and specific ways for their cells to resist colonization by retroviruses.

In principle, hosts could evolve resistance to the negative consequences of retroviral infection through changes in components of their innate and adaptive immune systems, but it has also become clear that host factors which directly inhibit the ability of retroviruses to replicate in host cells evolve under pressure of retroviral infection. Host resistance to retroviral infection can be acquired in many forms and be manifested at various stages of the retroviral life cycle (**Figure 1**). Of particular note are a collection of autonomously and constitutively active intracellular proteins, termed restriction factors, whose major function appears to be to provide intrinsic immunity to retroviral infection. As such, they are uniquely capable of actively preventing retrovirus replication at the very earliest stages of host colonization. Naturally, some retroviruses have adapted to cellular host resistance in its various forms. Since exogenous retroviruses evolve much more rapidly than their hosts, it seems obvious that they would do so. Nevertheless, some of the obstacles that cells have placed in the path of retroviral replication are not easily escaped by simple mutation, and in some cases retroviruses have learned

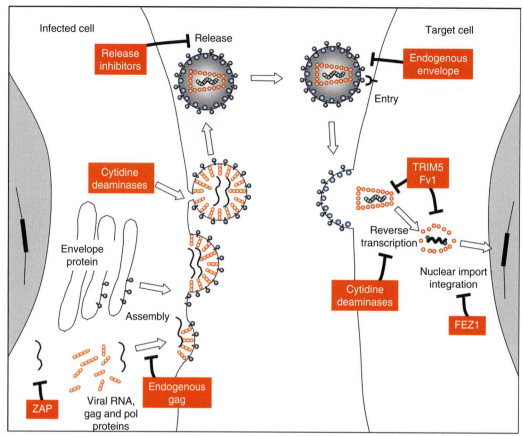

Figure 1 Overview of the retrovirus life cycle and activities encoded by cells that are known to inhibit retrovirus replication.

how to avoid or ablate host restriction factors in interesting and unique ways.

Host Cell Resistance Defined Early in the Retrovirus Life Cycle

Resistance to Retrovirus Entry

Imagine a cell attempting to thwart an infection event by a retrovirus. Optimal strategies would involve aborting the viral life cycle before integration into the genomic DNA and avoiding permanent residence of the retrovirus in the cell and all of its progeny. This could be achieved in numerous ways, an obvious approach being removal or alteration of the receptors that are exploited by retroviruses to enable entry. Several occurrences of this phenomenon are known. A striking example is the occurrence of endogenous murine leukemia viruses (MLVs) that are termed xenotropic because they are unable to replicate in their own hosts. Xenotropism generally occurs as a consequence of mutations that are acquired in the gene encoding the relevant receptor, presumably after colonization of the germline by the xenotropic MLV. Another important example of a receptor mutation that confers resistance to infection is present in humans. In this case a defective allele of the CCR5 co-receptor (*CCR5Δ32*) for HIV-1 exists at high frequency in certain human populations, and individuals that are homozygous for this mutation are resistant to infection by many strains of HIV-1. Although it is quite unlikely that HIV-1 itself is the source of the original evolutionary pressure that led to the dissemination of the *CCR5Δ32* allele, widespread HIV-1 infection in humans would clearly sustain or even increase the prevalence of the *CCR5Δ32* allele in populations where it is present. Other polymorphisms in the human *CCR5* gene also appear to have more modest effects on the acquisition and consequences of HIV-1 infection.

While receptor inactivation could clearly confer resistance to infection by certain retroviruses, some species have acquired or amplified specific genes that actively inhibit retroviral infection at the level of receptor–envelope interaction. Endogenous retroviruses themselves have proved to be a good source of such inhibitors and a few endogenous retroviral genes can efficiently control infection and inhibit the progression of disease caused by exogenous retroviruses. Certain of these preserved endogenous retroviral gene products, for example, Fv4 and Rmcf in mice, are viral envelope glycoproteins that bind and block the receptors shared by exogenous viruses, and constitute especially effective and specific entry inhibitors. The natural cellular ligands of viral receptors can also provide resistance to infection. For example, the chemokine ligands of CCR5 act as competitive inhibitors of HIV-1 envelope binding and one of the genes encoding a CCR5 ligand (*CCL3-L1*) exists in variable copy numbers in humans. Individuals with higher copy numbers tend to be less likely to acquire HIV-1 infection, and once infected seem less susceptible to its effects. While it is not yet certain that these effects are not immunologically mediated, it seems most likely that the *CCL3-L1* gene product acts as a direct inhibitor of HIV-1 infection and replication in humans.

Post-Entry Resistance Factors Targeting Incoming Retroviruses

In addition to providing a source of receptor-blocking viral envelope proteins, endogenous retroviruses have provided additional factors that restrict retroviral replication. A particularly intriguing restriction factor of endogenous retroviral origin is Friend virus susceptibility factor 1 (Fv1), a protein with about 60% homology to human and mouse endogenous retroviral Gag proteins. *Fv1* was first characterized as a dominant, heritable trait in laboratory mice that conferred resistance to infection by particular MLV strains. The two principal alleles of Fv1 are functionally distinguished by the spectrum of MLV strains to which they confer resistance: The $Fv1^n$ allele renders cells permissive to infection by N-tropic MLV but restricts B-tropic MLV infection, whereas $Fv1^b$ renders cells permissive to B-tropic MLV but resistant to N-tropic MLV. Heterozygous animals that carry both *Fv1* alleles are resistant to both N- and B-tropic MLV strains, while certain MLV strains are not susceptible to restriction by either *Fv1* allele and are known as NB-tropic. Additional *Fv1* alleles that induce partly overlapping patterns of sensitivity/resistance exist, as do inactive variants. While Fv1 restriction is not absolute, it can be very substantial at low multiplicities of infection and Fv1 restriction can have extremely dramatic effects on MLV pathogenesis.

An important feature of Fv1-mediated resistance is that it is saturable and infection by any given virus particle is greatly facilitated by the presence of large numbers of additional restricted virus particles, even if they are inactivated. Most interestingly, the Fv1-imposed block in MLV infection occurs after reverse transcription has been completed but before integration of the provirus in the host genome. The viral determinant governing Fv1 sensitivity is the capsid protein and a single-amino-acid change from arginine to glutamate at position 110 is sufficient to convert an N-tropic MLV strain to B-tropism. Other residues in proximal positions can also affect Fv1 sensitivity and together these data suggest that Fv1 is an inhibitor that inactivates incoming MLV capsids in the cytoplasm of target cells.

Several non-murine mammalian cells, including those from humans, African green monkeys (AGMs), and cows, restrict infection by N-tropic, but permit infection by B-tropic, MLV. Remarkably, this occurs despite the absence of an Fv1-like gene in species other than mice. The characteristics of restriction of N-MLV in these cells are similar to those mediated by Fv1 in mouse cells, that is, restriction can be saturated and amino acid 110 of capsid determines restriction sensitivity. However, in most cases of nonmurine mammalian cells, capsid-specific blocks to infection appear to occur before reverse transcription is completed.

In addition, lentiviruses are also subject to a similar type of post-entry restriction in primate cells. Again, dominant saturable factors are responsible and the viral determinant governing restriction is the capsid protein. Cells from different primate species often exhibit distinct retrovirus restriction specificities. For example, human cells restrict infection by equine infectious anaemia virus (EIAV) but not primate lentiviruses. Conversely, AGM cells restrict multiple primate lentiviruses and EIAV but not simian immunodeficiency viruses (SIVs) naturally found in AGMs. Remarkably, in many cases, restriction of one retrovirus can be abolished by saturation with another even if the two viruses are not closely related, provided that both viruses are restricted in the target cell. For example, in AGM cells, which restrict N-tropic MLV as well as HIV-1, HIV-2, SIVmac, and EIAV, saturation with lentivirus particles can completely abrogate restriction of N-tropic MLV infection. These findings demonstrate the divergence in specificity of mammalian restriction factors but also predict the presence of a single restriction factor capable of recognizing retroviruses whose capsids share little sequence homology.

Indeed, recent work shows that the gene that is largely responsible for the retrovirus restriction properties of primate cells is *TRIM5* and that variation in its sequence can account for most of the variation in species-specific retrovirus restriction properties. Unlike Fv1, TRIM5 has no homology with any known retroviral sequences. Rather, *TRIM5* is one member of a family of dozens of cellular genes that share a similar architecture. Indeed, *TRIM5* itself exists as one of a small cluster of closely related *TRIM* genes on human chromosome 11. Each *TRIM* gene encodes a protein with an amino-terminal tripartite motif, comprised of a RING domain, one or two additional zinc-binding or 'B-box' domains, and a coiled-coil domain (**Figure 2**). The TRIM domain can be linked to one of a number of C-terminal domains, and a single *TRIM* gene can encode several variant proteins with different C-terminal domains as a result of alternative splicing. The α-spliced variant of *TRIM5* gene is responsible for retrovirus restriction and encodes a C-terminal SPRY domain, which is related to domains found in other proteins of diverse organization and function, as well as in

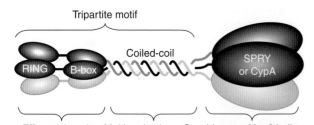

Figure 2 Domain and functional organization of the TRIM5 class of retrovirus restriction factors that block viral infection by targeting incoming capsids. Three major domains that are required for restriction are the N-terminal effector domain, of unknown function, a cenrtal coiled-coil that mediates trimerization, and a C-terminal domain that is responsible for capsid recognition.

several other members of the TRIM protein family. Functional dissection of the TRIM5α protein has revealed that the amino-terminal RING and the B-box sequences comprise a so-called effector domain that is required for restriction but not for recognition of incoming capsids. Conversely, the central coiled-coil is necessary for TRIM5 trimerization while the C-terminal SPRY domain is the principal specificity determinant that governs which retroviruses are inhibited by a given TRIM5α variant. The fact that TRIM5α exists as a trimer suggests a mode of capsid recognition that involves threefold symmetry, since the viral target of TRIM5 consists primarily of a hexameric array of capsid molecules. Moreover, residues on the outer surface of the hexameric capsid lattice of both HIV-1 and MLV have been shown to be important determinants of cell tropism and, in some cases, sensitivity to TRIM5α and Fv1.

The sequences of *TRIM5* genes in various primate species reveal rapid evolution and an excess of nonsynonymous mutations, specifically in the SPRY domain that determines restriction specifically. This finding is consistent with the notion that TRIM5 has been placed under evolutionary pressure by a variable selective force, most likely viral infections. Remarkably, in one particular New World monkey species, namely owl monkeys, a retrotransposition event has resulted in the almost precise replacement of the C-terminal SPRY domain with a cyclophilinA (CypA) domain. The resulting protein, termed TRIMCyp, inhibits HIV-1 and certain other retroviruses whose capsids bind to CypA.

The ability of certain retroviral capsid proteins, particularly that of HIV-1, to bind CypA can influence their sensitivity to TRIM5α as well as TRIMCyp, and there is a complex relationship between CypA:capsid interactions and sensitivity to restriction in various cell types. CypA binds to a flexible loop that is exposed on the surface of the viral capsid and overlaps with viral determinants that govern sensitivity to TRIM5α. The CypA:capsid interaction can be inhibited by drugs such as cyclosporine A (CsA), and treatment of certain human cells with CsA inhibits replication of HIV-1 while lentiviruses whose capsids do not bind CypA are unaffected. While CsA does not appear to render HIV-1 highly sensitive to human TRIM5α, mutations in the CypA-binding site do confer partial sensitivity to it and allow HIV-1 capsids to saturate human TRIM5α. These findings suggest that HIV-1 capsid may have specifically adapted to bind CypA in order to avoid inhibition by restriction factors, like TRIM5α, in humans.

However, the situation is more complex than these findings would suggest. The requirement for CypA for replication in human cells is HIV-1 strain and host cell-type dependent. Indeed, there are several examples of viral strains whose capsid binds CypA but in which disruption of the interaction has no detectable effect on virus replication. This phenotype is conferred by both naturally occurring and *in vitro* selected mutations that occur within the CypA binding loop but do not affect CypA binding. Remarkably, some of these mutations can confer CsA dependence in other human cell types. These differences among human cell types are not due to variation in TRIM5α sequence and suggest the existence of additional capsid-based restriction factors in human cells, whose activity is actually facilitated by the capsid:CypA interaction. Moreover, it appears that HIV-1 restriction by TRIM5α in Old World monkeys is facilitated by CypA:CA interactions.

Generally, the events immediately following entry of a retrovirus into the target cell are not well understood, complicating efforts to determine molecular details of the mechanism by which Fv1 and TRIM5α prevent retroviral infection. Although the hexameric capsid lattice forms the outer shell of the subviral complex that is delivered into the cell, biochemical studies suggest that the capsid shell may be discarded, in a process known as uncoating, shortly after entry. Therefore, there may be a relatively short time frame in which restriction factors such as TRIM5α and Fv1 can initiate events that inhibit infection. Experimental evidence is partly consistent with this notion, and irreversible TRIM5-mediated restriction events can indeed be initiated very rapidly, within minutes of virus entry into the target cell cytoplasm. TRIM5 can interact specifically with capsid proteins of restricted viruses *in vitro*, and direct binding of TRIM5 to capsids is almost certainly the initiating event in restriction. Ensuing events may involve accelerated capsid uncoating and/or degradation. While restriction by TRIM5 is often manifested as apparent inhibition of reverse transcription, this is not a necessary part of restriction since some TRIM5 variants do not affect viral DNA accumulation, nor does Fv1. The known involvement of RING domains in ubiquitin/sumoylation reactions is suggestive of potential effector mechanisms but neither of these pathways appears to be required for restriction activity, although proteasome inhibition can restore DNA synthesis.

Other Cellular Inhibitors of Incoming Retroviruses

In addition to the intensively studied TRIM5 and Fv1 proteins, cell-type-dependent effects on the early post-entry event in retroviral replication suggest the existence of additional factors that may inhibit incoming retroviruses. For example, certain HIV-1 and HIV-2 strains appear restricted in particular human cells in a manner that depends on both the envelope and capsid. This finding suggests that the route of retroviral entry may affect the sensitivity to unknown capsid-specific restriction factors, although TRIM5 and Fv1 restriction is clearly independent of entry route. Moreover, studies of chemically mutagenized cell lines have revealed that overexpression of fasciculation and elongation protein zeta-1 can be detrimental to retroviral infection. Whether these additional apparent inhibitors of retroviral infection are physiologically relevant and the mechanisms by which they work has not yet been established. However, it is entirely possible that there remain more, and perhaps many, undiscovered inhibitors of incoming retroviruses.

Host Resistance to Infectious Virus Generation

Cells have also evolved alternative antiretroviral activities that act after retroviral DNA has integrated into the host genome. Several inhibitors that act at the level of viral genome accumulation, or affect particle assembly, release, or infectivity are known. While such activities cannot preserve the uninfected state of the cell that expresses them, these intrinsic immune functions can attenuate or prevent the spread of infection to subsequent generations of target cells.

Cellular Cytidine Deaminases as Retrovirus Resistance Factors

One such defense strategy relies on a group of genes, exemplified by *APOBEC3G*, that act primarily by deaminating cytidines in nascent retroviral DNA. *APOBEC3G* is one of a group of genes that exhibit cytidine deamination activity, and other members of this family (e.g., *APOBEC1* and *AID*) are involved in editing of specific cellular mRNAs and genomic loci in order to change their coding potential. APOBEC3G is incorporated into retroviral particles and deaminates nascent viral DNA during reverse transcription in subsequently infected target cells by virtue of its single-stranded DNA-specific deaminase activity (**Figure 3**). Cytidine deamination results in the generation of uracil-containing viral DNA, which likely becomes targeted by DNA repair/degradation enzymes. However, if the viral DNA escapes degradation then uracil is replicated to adenosine during second strand synthesis, resulting in viral DNA with characteristic G-to-A hypermutation. Generally, the burden of G-to-A mutations is high, rendering the proviral DNA incapable of encoding replication-competent progeny.

APOBEC3G is promiscuous with respect to its antiretroviral effect, because it is packaged into diverse retrovirus particles. The mechanism by which this is achieved involves a substantial nonspecific component, whereby APOBEC3G binds in an apparently sequence-independent manner to the RNA that is packaged into virions. Thus, APOBEC3G exploits an essential property of retroviruses to infiltrate particles and therefore makes it difficult for retroviruses to evolve mutations that avoid packaging APOBEC3G into virions. Perhaps because of this difficulty, primate lentiviruses have arrived at an unusual and highly effective solution to the problem imposed on them by APOBEC3G. Indeed, all primate lentiviruses encode an accessory gene, *vif*, whose only function appears to be to counteract APOBEC3G. Vif proteins bind simultaneously to APOBEC3G and to a ubiquitin ligase complex containing Cul5, elongin B, and elongin C to induce ubiquitination of APOBEC3G. Predictably, this leads to proteasome-dependent APOBEC3G degradation, effectively depletes APOBEC3G from the cell, and abolishes incorporation into virion particles. Interactions between APOBEC3G and Vif proteins form the basis of additional species tropism restrictions in primate lentiviruses. Although HIV-1 Vif can efficiently bind to and induce the degradation of human APOBEC3G, it is inactive against its rhesus macaque or AGM counterparts. In contrast, rhesus macaque and AGM APOBEC3G can be recognized by Vif proteins from SIVs found in macaques and AGMs. A single-amino-acid difference between human and monkey APOBEC3G proteins (K128D) is responsible for differential sensitivity to Vif proteins and imparts a second barrier to cross-species transmission for at least some primate lentiviruses. Foamy viruses that are also common in primates have also acquired a gene, Bet, whose function is to exclude APOBEC3 proteins from virion particles. Moreover, other retroviruses appear to avoid the APOBEC3 proteins without encoding accessory genes to exclude it from particles. However, in these cases the molecular mechanism of particle exclusion is not clearly defined.

Like *TRIM5*, the locus containing *APOBEC3G* has been under evolutionary pressure during mammalian speciation, most likely as a consequence of viral infections. Indeed, while mice have only one *APOBEC3* gene, humans have no less than seven that have arisen as a result of *APOBEC3* gene duplication and recombination events.

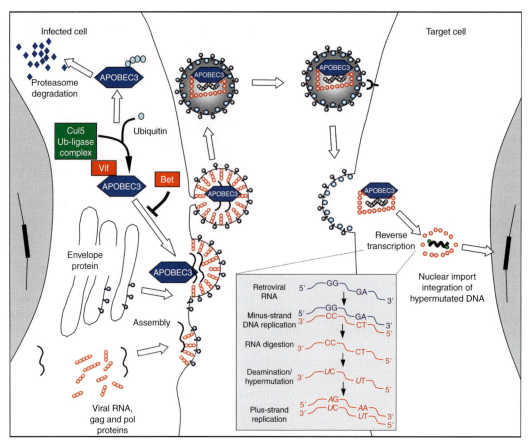

Figure 3 Major mode of action and alternative fates of APOBEC3 proteins that become incorporated into retrovirus particles and catalyzed cytidine deamination during infection of target cells. Some APOBEC3 proteins can be recruited by lentiviral Vif proteins to a ubiquitin ligase complex and degraded in a proteasome-dependent manner. Alternatively, foamy virus Bet proteins prevent incorporation of APOBEC3 into virus particles by unknown mechanisms.

Most of the *APOBEC3* genes, like TRIM5, exhibit excess nonsynonymous polymorphisms in primates, again suggesting evolutionary pressure. Several members of the *APOBEC3* gene cluster exhibit varying levels of activity against numerous retroviruses and retroelements, and some are targeted for degradation by primate lentiviral Vif proteins. In some cases, the APOBEC3 proteins are distinguishable by their substrate preference. For example, APOBEC3G-mutated retroviral DNA carries G-to-A changes primarily in the context of GG dinucleotides, while its close relative, APOBEC3F, preferentially targets GA dinucleotides. In the context of HIV-1-infected humans, APOBEC3G and its close relative APOBEC3F appear to be the major contributors to cytidine deamination-based host defense. Indeed, viral DNA carrying G-to-A hypermutation can be found quite easily in some patients, and the context in which the G-to-A changes occur suggests that both APOBEC3G and APOBEC3F are responsible.

While APOBEC3G and its relatives can clearly inhibit retrovirus replication by inducing catastrophic hypermutation, it may be that lentiviruses have turned its cytidine deamination activity to their advantage and are evolved to only partially resist APOBEC3 proteins. Less dramatic changes in viral DNA sequence may result from partial destruction of APOBEC3G by Vif, and infiltration of viral particles by a small number of APOBEC3 molecules. Indeed, Vif proteins with a range of potencies can be found in HIV-1-infected patients. Primate lentiviruses are capable of extremely rapid sequence divergence, and this is probably required to maintain their characteristic high levels of chronic replication in the presence of strong adaptive immune responses. Certainly, the A-rich nature of their genomes suggests that cytidine deaminases have played a role in influencing the rate and direction of sequence divergence in primate lentiviruses.

While cytidine deamination is thought to be the major mechanism by which APOBEC3 proteins inhibit retrovirus replication, some studies suggest that this class of proteins may also inhibit retrovirus and retroelement replication through mechanisms that do not involve hypermutation. In addition, it may be that inhibition does not require APOBEC packing into retroviral particles and that its presence in the cytoplasm of particular target cells is sufficient to inhibit the early post-entry steps of the retrovirus life cycle. However, these additional apparent activities of APOBEC proteins are poorly understood at present.

Other Host Resistance Factors That Act during Virus Generation

The genomes of retroviruses appear to be the target of a further host defense activity that likely recognizes newly synthesized RNA. Zinc-finger antiviral protein (ZAP) inhibits the cytoplasmic accumulation of the genomes of some retroviruses and, more impressively, alphaviruses. Its mechanism of action may be related to that of the tristetraprolin-like proteins that it resembles, which induce the degradation of cytokine mRNAs by binding to AU-rich sequences in their 3′ untranslated regions. By doing so, ZAP may recruit viral RNAs to a collection of ribonucleolytic activities referred to as the exosome. ZAP obviously must distinguish between viral and cellular mRNAs but the molecular basis for this distinction is currently unknown.

It is also possible for cells to interfere with the assembly of retrovirus particles. There exists at least one example where defective Gag proteins encoded by an endogenous retrovirus, namely Jaagsiekte sheep retrovirus, can form heteromultimers with and interfere with the assembly of the Gag protein of its exogenous cousin. Yet another intrinsic immune mechanism targets retrovirus particle release. In some human cells, retrovirus particles are released very inefficiently, unless a protein that is expressed by a subset of primate lentiviruses, termed Vpu, is also present. Vpu is not required for the efficient release of retroviruses from some cell lines, but fusion of cells that do or do not require the presence of Vpu for efficient release results in heterokaryons that exhibit the Vpu-dependent virus release phenotype. While many retroviruses do not encode a Vpu protein, some envelope proteins appear to exhibit a similar activity. These findings suggest the existence of a cellular inhibitor of retrovirus release that is counteracted by Vpu or envelope proteins. Because Vpu can facilitate the release of a number of divergent retroviruses, the presumed release inhibitor must target particle release in a rather nonspecific way. A current model that explains many of the effects of Vpu invokes the existence of a release inhibitor that endows cells and/or nascent virions with an adhesive property. Virion–cell adhesion would result in the retention of viral particles on the cell surface, from where they would be subsequently endocytosed. Consistent with this model, retroviral particles generated by cells in which release is inefficient in the absence of Vpu accumulate at the cell surface and later within intracellular endosomes.

Perspectives

Historically, the ability or inability of retroviruses to colonize a particular species or cell type was thought to be defined almost entirely by its ability to parasitize host functions present therein that are necessary to complete its life cycle. However, recent findings have shown that evolution has equipped cells with an often robust and multicomponent defense against retroviruses that seek to colonize them. These activities constitute an equally important determinant of retroviral cell tropism. Studies of restriction factors are in their infancy, but these new insights have given a deeper understanding of the evolutionary relationships between retroviruses and their hosts and what governs the ability of retroviruses to engage in zoonosis. Moreover, the discovery of antiretroviral activities within cells that target pathogenic retroviruses may provide new opportunities for therapeutic intervention.

See also: Equine Infectious Anemia Virus; Foamy Viruses; Human Immunodeficiency Viruses: Pathogenesis; Jaagsiekte Sheep Retrovirus; Simian Immunodeficiency Virus: General Features.

Further Reading

Best S, Le Tissier P, Towers G, and Stoye JP (1996) Positional cloning of the mouse retrovirus restriction gene Fv1. *Nature* 382: 826–829.

Bieniasz PD (2004) Intrinsic immunity: A front-line defense against viral attack. *Nature Immunology* 5: 1109–1115.

Goff SP (2004) Genetic control of retrovirus susceptibility in mammalian cells. *Annual Review of Genetics* 38: 61–85.

Harris RS, Bishop KN, Sheehy AM, et al. (2003) DNA deamination mediates innate immunity to retroviral infection. *Cell* 113: 803–809.

Hatziioannou T, Perez-Caballero D, Yang A, Cowan S, and Bieniasz PD (2004) Retrovirus resistance factors Ref1 and Lv1 are species-specific variants of TRIM5{alpha}. *Proceedings of the National Academy of Sciences, USA* 101: 10774–10779.

Sayah DM, Sokolskaja E, Berthoux L, and Luban J (2004) Cyclophilin A retrotransposition into TRIM5 explains owl monkey resistance to HIV-1. *Nature* 430: 569–573.

Sheehy AM, Gaddis NC, Choi JD, and Malim MH (2002) Isolation of a human gene that inhibits HIV-1 infection and is suppressed by the viral Vif protein. *Nature* 418: 646–650.

Stremlau M, Owens CM, Perron MJ, Kiessling M, Autissier P, and Sodroski J (2004) The cytoplasmic body component TRIM5alpha restricts HIV-1 infection in Old World monkeys. *Nature* 427: 848–853.

Towers G, Bock M, Martin S, Takeuchi Y, Stoye JP, and Danos O (2000) A conserved mechanism of retrovirus restriction in mammals. *Proceedings of the National Academy of Sciences, USA* 97: 12295–12299.

Human Cytomegalovirus: General Features

E S Mocarski Jr., Emory University School of Medicine, Emory, GA, USA
R F Pass, University of Alabama School of Medicine, Birmingham, AL, USA

© 2008 Elsevier Ltd. All rights reserved.

Introduction

Cytomegaloviruses, also known as salivary gland viruses, are widely distributed, species-specific herpesviruses, with an evolutionarily related representative identified in most mammalian hosts when cells of the same species have been used for isolation. Human cytomegalovirus (HCMV), formally called human herpesvirus 5 (HHV-5), infects a majority of the world population by adulthood, but causes acute disease in only a small proportion of immunocompetent individuals. Developing areas of the world typically exhibit widespread transmission early in life, whereas more developed areas show a broader range of patterns. Individuals may escape infection early in life and remain susceptible during the childbearing years. Once rubella was controlled by vaccination, transplacental transmission of HCMV emerged as the major infectious cause of congenital hearing loss preventable by vaccination. HCMV is an opportunistic pathogen associated with disease in immunocompromised hosts, predominating in the settings of genetic or acquired immunodeficiency, allograft tissue and organ transplantation, and pregnancy. Disease pathogenesis requires active viral replication and focuses on different target tissues and organs in different clinical settings, particularly in circumstances where the ability to mount a cellular immune response has been compromised. Transmission of this virus in the general population depends on direct contact with infected bodily secretions.

History and Classification

HCMV is designated formally in the species *Human herpesvirus 5* and belongs to genus *Cytomegalovirus* in subfamily *Betaherpesvirinae* of the family *Herpesviridae*. Human herpesvirus 6 (HHV-6; often considered as two variants, HHV-6A and HHV-6B) and human herpesvirus 7 (HHV-7) belong to the species *Human herpesvirus 6* and *Human herpesvirus 7*, respectively, and represent genus *Roseolovirus* in the same subfamily. All betaherpesviruses replicate slowly in cell culture, remain cell-associated, and exhibit species specificity.

Starting in the early 1930s, the most severe form of HCMV-associated congenital disease, cytomegalic inclusion disease (CID), was recognized by an 'owl's eye' cytopathology in salivary gland, liver, lung, kidney, pancreas, and thyroid autopsy materials from infants. By the early 1950s, CID diagnosis was based on the presence of inclusion-bearing cells in urine, and viral etiology was established by isolation on cultured human fibroblasts. By the early 1970s, the host species specificity of viral replication, widespread distribution of HCMV-like agents in mammals, and tissue and cell type distribution in the diseased host were well recognized. Most importantly, the relationship between transplacental transmission and neurological damage in newborns was established and the social cost of the major sequela, progressive sensorineural hearing loss, placed a high priority on control of this infection through vaccination, which has not been realized to this day. Although primary infection results in the most severe disease with congenital infection, less frequent transmission following recurrent maternal infection nevertheless remains important in the overall epidemiology of disease. Placental infection appears to be considerably more frequent than transplacental transmission, which overall is on the order of 0.5–1% of live births in the USA or Europe. Transplacental transmission is less frequent during recurrent infection (<1% of newborns) than during primary infection (~ 30–40% of newborns). Young children are a major reservoir of this virus, often responsible for transmitting primary infection in pregnant women.

The second prominent setting of HCMV disease is in the immunocompromised host, and this has driven initiatives for therapeutic intervention. HCMV is associated with various diseases in which T-lymphocyte immune surveillance is compromised. Infection of immunocompetent individuals, ranging from immediately postpartum through adulthood, does not generally lead to significant illness other than occasional (heterophile negative) mononucleosis. A remarkably strong and broad T-cell response, particularly the cytotoxic T-cell response, contributes to lifelong suppression of active virus replication. Immunosuppressive therapy necessary to prevent T-cell-mediated allograft rejection reduces HCMV immunosurveillance, allowing the virus to replicate to levels that lead to various diseases. The incidence of HCMV pneumonia in allogeneic hematopoietic cell transplantation (HCT) and the widespread incidence of HCMV retinitis in acquired immune deficiency syndrome (AIDS) contributed to the development of the antiviral drugs ganciclovir, foscarnet, and cidofovir, and to the later development of the orally administered drug, valganciclovir, as well as ongoing investigation of maribavir. HCMV plays a suspected role in chronic vascular diseases, and this is best illustrated by cardiac allograft vascular disease incidence in heart transplantation and the contribution of this virus to incidence of bacterial and fungal infections in allogeneic HCT settings.

Geographic and Seasonal Distribution

As a ubiquitous virus, HCMV does not show a particular seasonal distribution, although initial acquisition varies with living circumstances. In general, prevalence is more widespread in younger individuals in developing countries than in developed countries, and, within developed countries, it is more widespread in urban than in suburban groups, and low-socioeconomic groups than in high-socioeconomic groups. In a US-based survey, overall HCMV prevalence was approximately 60%, with a higher prevalence in females (64%) than males (54%) and in non-Hispanic blacks (76%) and Mexican Americans (82%) than in non-Hispanic whites (51%). There is a doubling in the prevalence of HCMV infection between childhood and old age, which rises to include 75–85% of the population regardless of demographic considerations. In addition to age and socioeconomic level, early acquisition of HCMV is observed in immigrants and is associated with fewer years of education, large family size, and residence in the southern states.

Host Range and Virus Propagation

Cytomegaloviruses are most readily propagated in fibroblasts cultured from the origin host species, although efficient replication of HCMV in human fibroblasts is facilitated by mutations in the viral genome that reduce the ability to replicate in other host-cell types. Studies on autopsy materials from immunocompromised individuals with acute HCMV disease have shown evidence of virus replication in epithelial, endothelial, macrophage, and dendritic cells. HCMV antigens are also readily detected in nonpermissive peripheral blood (PB) neutrophils of diseased individuals, probably as a result of phagocytosis, and this has become the basis of an important diagnostic assay (antigenemia) specific for a major viral structural antigen (pp65). Fresh clinical isolates replicate efficiently on primary or secondary cultures of epithelial, endothelial, macrophage, and dendritic cells, but this tropism is lost when virus is propagated in fibroblasts due to mutations that accumulate in the viral genome. Extensively passaged laboratory strains acquire a variety of large and small deletions that remove or disrupt tropism genes as well as other genes that are dispensable for replication in cultured fibroblasts. Variants of common laboratory strains, all used under a single name (e.g., AD169, Towne), have arisen via independent propagation since their isolation in different laboratories around the world. Although cellular receptors for HCMV are important determinants of entry into host cells, species specificity and human cell tropism is not determined solely by the distribution of receptors for this virus; nonpermissive fibroblasts from other species generally allow viral attachment and penetration, but are blocked at an early postentry step. An exception appears to be HCMV isolated and adapted to fibroblasts, during which variants arise that lack the ability to enter primary human endothelial cells, epithelial cells, and myeloid cells due to acquisition of mutations.

Genetics and Replication

Betaherpesviruses have linear DNA genomes with direct terminal repeats that contain the *cis*-acting signals for cleavage and packaging of progeny viral DNA during replication. Based on a subset of protein-coding genes conserved across mammalian members of the *Herpesviridae* (the core genes), all herpesviruses likely follow common pathways of DNA replication, DNA encapsidation, and virion maturation. Otherwise, betaherpesvirus genomes exhibit some divergent characteristics. Primate cytomegaloviruses undergo genome rearrangement similar to some alphaherpesviruses. HCMV packages any one of four isomers and has a genome organization that includes a unique long (U_L) component surrounded by a set of inverted repeats (*ab–b'd'*, or TRL–IRL) and a unique short component (U_S) also surrounded by a second set of inverted repeats (*d'c'–ca*, or TRS–IRS) (**Figure 1**). HHV-6 and HHV-7 do not rearrange, and have genomes with a large terminal repeat. Murine or rat CMV genomes also do not rearrange and have terminal repeats that are the smallest (<50 bp) of any characterized herpesvirus.

Genome sequence analysis has revealed evolutionary relationships and codified the classification of herpesviruses. Betaherpesvirus genomes are collinear and exhibit more extensive genetic similarity with each other than with other herpesviruses, where the 40 core genes are arranged in different clusters. Betaherpesviruses encode approximately 170 protein-coding gene homologs, 40 of which are common to alphaherpesviruses and 46 of which are common to gammaherpesviruses. **Figure 1** depicts these gene sets along with an estimate of the overall genomic coding capacity.

The HCMV virion has typical herpesvirus structure, although somewhat larger than alphaherpesviruses or gammaherpesviruses, at 200–300 nm diameter. The 125 nm icosahedral nucleocapsid is composed of five herpesvirus core proteins: the major capsid protein (MCP) comprising the hexons and pentons, the minor capsid protein (mCP or TRI2) together with the minor capsid protein-binding protein (mCP-BP or TRI1) comprising the intercapsomeric triplexes, the small capsid protein (SCP) that decorates the hexon tips, and the portal protein (PORT) that comprises one penton position and is used for encapsidation of viral DNA. The nucleocapsid encloses a single linear molecule of DNA as well as origin of lytic DNA replication (*ori*Lyt)-associated RNA. The nucleocapsid is embedded in a tegument (or matrix) containing at least 27

Human Cytomegalovirus: General Features

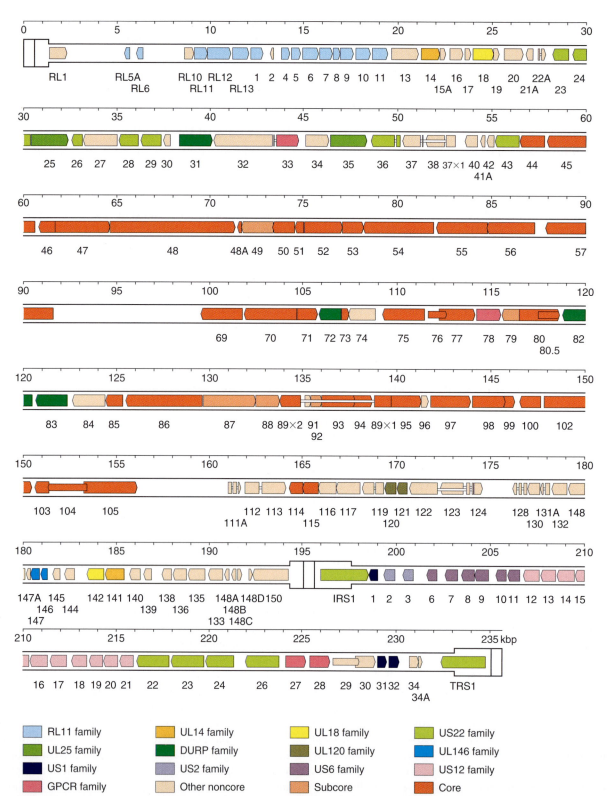

Figure 1 Genetic organization and content of wild-type human HCMV, based on low passage strain Merlin. The inverted repeats TRL/IRL (ab–b'a') and TRS/IRS (a'c'–ca) are shown in a thicker format than U_L and U_S. Protein-coding regions are indicated by arrows and gene names are listed below. Introns are shown as narrow bars. Genes corresponding to those in TR_L/IR_L and TR_S/IR_S of strain AD169 are given their full nomenclature, but the UL and US prefixes have been omitted from UL1–UL150 and US1–US34A. Herpesvirus core genes (inherited from the ancestor of the alpha-, beta-, and gammaherpesviruses), subcore genes (inherited from the ancestor of beta- and gammaherpesviruses), other unique genes (homologs present only in betaherpesviruses), and members of families of related genes are color-coded as depicted in the key. Adapted from Dolan A, Cunningham C, Hector RD, et al. (2004) Genetic content of wild type human cytomegalovirus. *Journal of General Virology* 85: 1301–1312, with permission from Society for General Microbiology.

relatively abundant virus-encoded phosphoproteins as well as many more proteins and RNAs that are present in small amounts. About a dozen herpesvirus core as well as a dozen betaherpesvirus-specific structural proteins are considered tegument components. These appear to be involved in the earliest steps of infection and uncoating and release of the viral genome into the nucleus, as well as at late stages of replication for encapsidation of progeny viral DNA, capsid egress from nuclei, and, importantly, final envelopment and egress from the cytoplasm. Tegument proteins also modulate events in the host cell to block intrinsic resistance mechanisms that impede viral replication. The tegument is surrounded by a lipid bilayer envelope derived from cellular endoplasmic reticulum (ER)–Golgi intermediate compartment (ERGIC) membranes and into which some 20 or more virus-encoded glycoproteins are inserted. Unlike other herpesviruses that encode subgroup-specific envelope glycoproteins central to entry, the most prominent HCMV envelope glycoproteins are herpesvirus core gene products. These are known by common nomenclature as glycoprotein (g)B, gH:gL, and gM:gN, and all play essential replication functions. None of the numerous additional glycoproteins encoded by HCMV are essential for replication in cultured fibroblasts. Overall, HCMV is the most structurally complex herpesvirus, reflecting the large number of gene products encoded by this virus. During infection in cell culture, HCMV produces an abundance of noninfectious particles, including dense bodies that lack a nucleocapsid as well as other defective, capsid-containing particles. These particles constitute at least 99% of virion preparations, producing particle-to-PFU ratios of 100 or greater.

Complete or largely complete genome sequences have been determined for a number of laboratory and low-passage HCMV strains. All HCMV strains exhibit an average of >95% DNA sequence identity, although large deletion mutations are present in many highly propagated strains. Wild-type HCMV is estimated to carry a minimum of 166 protein-coding genes, the great majority of which are conserved in the closest relative of HCMV, chimpanzee CMV. A liberal upper estimate of 252 ORFs has been suggested. In addition to protein-coding genes and potential miRNA-like elements, the HCMV genome has three types of resident signals that act to control gene expression, DNA replication, and DNA packaging during replication. The genes on the viral genome are controlled by promoter-regulatory signals that are recognized by host-cell RNA polymerase II and influenced by viral regulatory proteins, and are expressed in a temporal cascade that initiates following penetration of the host cell.

Entry into cells depends upon envelope glycoproteins and release of virion DNA into the nucleus depends upon tegument proteins. Entry initiates with a series of distinct steps involving (1) binding to specific cell surface receptors, (2) fusion of the virion envelope with the cellular membrane to release nucleocapsids into the cytoplasm, (3) nucleocapsid association with cytoskeletal elements and translocation toward the nucleus, (4) nucleocapsid interaction with nuclear pores, and (5) release of the viral genome into the nucleus (**Figure 2**). Expression of the immediate early (IE) class of viral genes follows entry and is influenced by virion tegument proteins (virion transactivators). IE genes include viral functions that regulate gene expression by activating transcription and suppressing host repression systems, cell death, and major histocompatibility complex class I gene expression. As in all herpesviruses, different temporal classes of genes are interspersed across the HCMV genome. The IE genes are not clustered and map to diverse locations in the U_L, U_S, and IR_S/TR_S regions of the viral genome. Expression of the next set of genes, the delayed early (DE) genes, depends on the expression of regulatory IE gene products. DE genes control the initiation of viral DNA replication and a wide range of host-cell modulatory functions. Viral DNA synthesis initiates on the viral genome at oriLyt and relies on six core virus-encoded DNA synthesis proteins well as additional functions that facilitate this process. Expression of late (L) genes occurs last and includes a majority of virion structural proteins, a great many of which are herpesvirus core functions. DE and L gene classes also include as many as 100 immunomodulatory functions that impact the host cell and the immune response of the host, only a fraction of which have been studied (**Table 1**).

The replication cycle of HCMV is slow, requiring 48–72 h to begin the release of progeny, with a switch from E to L phase being delayed until 24–36 h post infection (hpi) and maximum levels of virus release starting at 72–96 hpi. The basic features of HCMV capsid formation and maturation are likely to be common to all herpesviruses. Encapsidation of viral DNA is under the control of a set of viral proteins that package and cleave viral DNA molecules into preformed capsids. Seven herpesvirus core encapsidation proteins likely act as a complex to process a concatemeric DNA template via cleavage/packaging (*pac*) sites located within the terminal *a* sequences, and following the insertion of genome-length viral DNA, cleave to produce a filled nucleocapsid. The nucleocapsid follows a complex two-stage envelopment and egress process that starts in the nucleus and leads to virion release by exocytosis at the plasma membrane. This process depends upon many core herpesvirus functions and starts with primary envelopment at the inner nuclear membrane followed by a de-envelopment event at the outer nuclear membrane, a process that releases the nucleocapsid into the cytoplasm (**Figure 2**). Secondary envelopment occurs in the cytoplasm at ERGIC membranes with the resulting vesicles transporting mature virions to the cell surface using the cellular exocytic pathway. Dense bodies form in the cytoplasm,

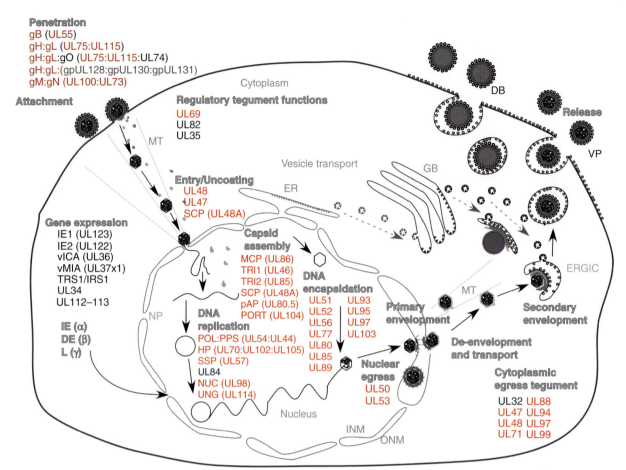

Figure 2 Summary of the HCMV replication pathway with focus on herpesvirus core (red) and betaherpesvirus-conserved (black) gene products playing known or predicted roles in replication. Major steps in productive replication are indicated in large gray outlined font, and black arrows indicate the procession of steps with individual functions identified by abbreviated names. The entry pathway shown employs direct fusion at the plasma membrane (attachment and penetration) although an endocytic pathway may also be important in certain cell types. Entry requires gB, gH:gL, possibly with gO, or gpUL128-gpUL130-gpUL131, and gM:gN. Virion tegument functions (UL69, UL82, UL35) are involved in regulation. UL47, UL48, and SCP are predicted to mediate transport on microtubules, docking at nuclear pores, and release of virion DNA from the capsid into the nucleus. Transcriptional regulation of viral and host-cell gene expression is mediated by IE genes (IE1, IE2) or DE genes (UL34, UL35, UL112-UL113). Cell death suppression is mediated by IE gene products vICA and vMIA. DNA replication depends on several core proteins (POL:PPS, HP, SSB, NUC, and UNG) as well as one betaherpesvirus-specific function (UL84). Capsid assembly uses core functions MCP, TRI1:TRI2, SCP, PORT, and pAP. Preformed capsids likely translocate to sites of DNA replication where several core proteins (UL51, UL56, UL77, UL80.5, UL85, UL89, UL93, UL95, UL97, UL103 gene products) are likely to be involved in encapsidation of viral DNA. Nuclear egress is likely to be controlled by UL50 and UL53 gene products, which are core proteins. Cytoplasmic egress (black arrows) and secondary (final) envelopment are controlled by core functions (UL47, UL48, UL71, UL88, UL94, UL97, UL99 gene products) as well as one betaherpesvirus-specific protein (UL32 gene product). Nucleocapsids are likely transported on microtubules and virion envelope glycoproteins follow vesicle transport to sites of final envelopment in the cytoplasm. Golgi body (GB), microtubules (MTs), and ER and ERGIC, as well as both virus particle (VP) and dense body (DB) final release steps are identified in the cytoplasm. Nuclear pores (NPs), inner nuclear membrane (INM), and outer nuclear membrane (ONM) are identified in the nucleus. The cellular vesicle transport pathway from the ER to GB is also designated (dashed gray arrow).

following only the cytoplasmic steps (**Figure 2**). Once maturation starts, infected fibroblasts continue to produce virus at peak levels for several days, depending on the viral strain and constellation of cell-death suppressors encoded by the strain. Based on systematic mutagenesis, about 80 viral genes play important roles in replication, with roughly 45 being essential for replication in fibroblasts.

Evolution

HCMV has a 236 kbp genome and the capacity to encode in excess of 166 gene products, including several gene families (**Figure 1**). This is considerably more than HHV-6 and HHV-7, whose smaller genomes (145–162 kbp) encode an estimated 84 or 85 gene products. Cytomegaloviruses from a wide variety of host species conserve the same 70

Table 1 HCMV gene characteristics and functions.

HCMV[a,b]	HCMV gene family/gene name[c]	Function/comments/abbreviation[d]
RL1		
RL5A	RL11 fam	
RL6	RL11 fam	
RL10	Virion env-gp	
RL11	RL11 fam; IgG Fc-binding glycoprotein	Modulation of antibody activity
RL12	RL11 fam; put mem-gp	
RL13	RL11 fam; put mem-gp	
UL1	RL11 fam	
UL2	Put mem	
UL4	RL11 fam; env-gp	
UL5	RL11 fam; virion envelope protein?	
UL6	RL11 fam; put mem-gp	
UL7	RL11 fam; put mem-gp	
UL8	RL11 fam; put mem-gp	
UL9	RL11 fam; put mem-gp	Temperance
UL10	RL11 fam; put mem-gp	Temperance
UL11	RL11 fam; mem-gp	
UL13	Put sec	
UL14	UL14 fam; put mem-gp	
UL15A	Put mem	
UL16	Mem-gp	Inhibits NK cell cytotoxicity via NKG2D ligands MICA and ULBPs; temperance
UL17		
UL18	UL18 fam; put mem-gp; MHC class I homolog	LIR-1 ligand
UL19		
UL20	T-cell receptor γ chain homolog	
UL21A[¥]		
UL22A	Virion env, sec-gp	CC chemokine-binding protein (also called UL21.5); possibly temperance
UL23	US22 fam; teg	Temperance
UL24	US22 fam; teg	
UL25	UL25 fam: teg	Temperance
UL26[¥]	US22 fam; teg	Activator of major IE promoter; regulates teg phosphorylation
UL27		Maribavir resistance
UL28[¥]	US22 fam	
UL29[¶]	US22 fam	Temperance
UL30[¥]		
UL31[¶]	dUTPase-related protein (DURP) fam	
UL32[†]	Major teg-pp (pp150); highly immunogenic	Binds to capsids, cytoplasmic envelopment of virions
UL33x1/ UL33x2	GPCR fam, virion env	Constitutive signaling
UL34[†]		Represses US3 transcription
UL35[¶]	UL25 fam; teg-pp	Interacts with UL82 protein; regulates virion transactivation and assembly
UL36x1/ UL36x2	US22 fam; IE, teg; inhibitor of caspase-8-induced apoptosis	Cell death suppression (vICA)
UL37x1[¶]	IE; mitochondrial inhibitor of apoptosis	Cell death suppression (vMIA)
UL37	IE glycoprotein	Gene regulation, vMIA domain at amino terminus
UL38[¶]	Mem-gp	Cell death suppression
UL40	Mem-gp	Signal peptide binds HLA-E to inhibit NK cell cytotoxicity via CD94:NKG2A
UL41A	Virion env-gp	Also called UL41.5
UL42	Put mem	
UL43	US22 fam; teg	
UL44[†]	(core) DNA polymerase processivity subunit	Increases DNA pol product length (PPS)
UL45	(core) Teg	Large subunit ribonucleotide reductase homolog (enzymatically inactive); virion protein (RR1)
UL46[†]	(core) Capsid	Component of capsid triplexes (minor capsid binding protein; TRI1)
UL47[¶]	(core) Teg	Intracellular capsid transport; binds to UL48 protein? (LTPbp)

Continued

Table 1 Continued

HCMV[a,b]	HCMV gene family/gene name[c]	Function/comments/abbreviation[d]
UL48[†]	(core) Largest teg	Intracellular capsid transport? (LTP)
UL48A[†]	(core) Smallest capsid	Located on tips of hexons in capsids; capsid transport? (SCP) (also called UL48.5)
UL49[†]		
UL50[†]	(core) Inner nuclear mem	Nuclear egress of capsids; virion protein? (NEMP)
UL51[†]	(core) Terminase component	DNA packaging (TER3)
UL52[†]	(core)	Capsid transport in nucleus
UL53[†]	(core) Teg	Nuclear matrix protein; capsid egress; nuclear egress lamina protein
UL54[†]	(core) DNA polymerase	DNA polymerase catalytic subunit (POL)
UL55[†]	(core) Virion env-gp B (gB)	Homomultimers; heparan-binding; role in entry and signaling
UL56[†]	(core) Terminase component	Binds to DNA packaging motif, exhibits nuclease activity (TER2)
UL57[†]	(core)	Single-stranded DNA-binding protein (SSB)
ori Lyt[†]		DNA replication origin for productive infection (cis-acting); positional conservation, sequence divergence in betaherpesviruses.
UL69[¥]	(core) Multiple regulatory protein; teg	Tegument protein; contributes to cell cycle block; exhibits nucleocytoplasmic shuttling; promotes nuclear export of unspliced mRNA (MRP)
UL70[†]	(core) DNA helicase primase subunit	Unwinding DNA, primase homology (HP2)
UL71[†]	(core) Teg	Cytoplasmic egress
UL72	(core) DURP fam	Deoxyuridine triphosphatase homolog (enzymatically inactive) (dUTPase)
UL73[†]	(core) Virion env-gp N (gN)	Complexes with gM; entry
UL74[¶]	Virion env-gp O (gO)	Complexes with gH:gL
UL75[†]	(core) Virion env-gp H (gH)	Associates with gL; complexes with gO or UL128-UL130-UL131; entry role
UL76[†]	(core) Virion-associated regulatory	
UL77[†]	(core) Portal capping	DNA packaging (PCP)
UL78	GPCR fam; put chemokine receptor, env-gp	
UL79[†]		
UL80[†]	(core)	Precursor of maturational protease (PR; N terminus) and capsid assembly (scaffold) protein (AP; C terminus)
UL80.5[†]	(core)	Precursor of capsid assembly (scaffold) protein (pAP)
UL82[¥]	DURP fam; teg-pp (pp71; upper matrix)	Virion transactivator; ND10 localized; degrades Daxx
UL83	DURP fam; major teg-pp (pp65; lower matrix)	Suppresses interferon response
UL84[†]	DURP fam	Role in organizing DNA replication; exhibits nucleocytoplasmic shuttling; binds IE2
UL85[†]	(core) Capsid	Component of capsid triplexes (minor capsid protein; TRI2)
UL86[†]	(core) Major capsid	Component of hexons and pentons (MCP)
UL87[†]		
UL88[¶]	(core) Teg	Cytoplasmic egress
UL89x1[†]/UL89x2	(core) Terminase component	ATPase subunit (TER1)
UL91[†]		
UL92[†]		
UL93[†]	(core) Teg	Capsid transport
UL94[†]	(core) Teg	Binds single-stranded DNA; cytoplasmic egress
UL95[†]	(core)	Encapsidation chaperone protein
UL96[†]	Teg	
UL97[¶]	(core) Viral serine-threonine protein kinase; teg	Phosphorylates ganciclovir; inhibited by maribavir; roles in DNA synthesis, DNA packaging and nuclear egress; mimics cdc2/CDK1 (VPK)
UL98[†]	(core) Deoxyribonuclease	(NUC)
UL99[†]	(core) Myristylated teg-pp pp28	Cytoplasmic egress tegument protein
UL100[†]	(core) Virion env-gp M (gM)	Complexes with gN
UL102[†]	(core) DNA helicase primase subunit	Unwinding DNA (HP3)
UL103[¶]	(core) Teg	Nuclear egress
UL104[†]	(core)	Portal protein; DNA encapsidation (PORT)
UL105[†]	(core) DNA helicase primase subunit	Unwinding DNA; helicase homology (HP1)
UL111A	CMV interleukin 10	(cmvIL-10) and latency-associated (LA) vIL-10

Continued

Table 1 Continued

HCMV[a,b]	HCMV gene family/gene name[c]	Function/comments/abbreviation[d]
UL112[¥]/ UL113[¥]		Transcriptional activation, orchestration of DNA replication
UL114[¶]	(core) Uracil-DNA glycosylase	Roles in excision of uracil from DNA and temporal regulation of DNA replication (UNG)
UL115[†]	(core) Virion env-gp L (gL)	Associates with gH; complexes with gO or UL128-UL130-UL131; entry
UL116	Put mem-gp	
UL117[¥]		
UL119	IgG Fc-binding glycoprotein; virion env-gp	Mem related to OX-2; modulation of antibody activity
UL120	UL120 fam; put mem-gp	
UL121	UL120 fam; put mem-gp	
UL122[†]	IE2 transactivator	Interacts with transcriptional machinery; repression via specific DNA-binding activity; cell cycle modulation (IE2)
UL123[¥]	Major immediate early 1 co-transactivator	Enhances activation by IE2; indirect effect on transcription machinery; binds HDACs; disrupts ND10 (IE1)
UL124	Mem-gp, latent protein	
UL128	Put sec	Endothelial and epithelial cell tropism, env-gp complexes with gH:gL
UL130	Put sec	Endothelial and epithelial cell tropism, envelope gp complexes with gH:gL
UL131A	Put sec	Endothelial and epithelial cell tropism; envelope gp complexes with gH:gL
UL132[¶]	Virion env-gp	Temperance
UL148	Put mem-gp	
UL147A	Put mem	
UL147	UL146 fam; put sec-gp;	Put CXC chemokine
UL146	UL146 fam; sec-gp	hCXCR2-specific CXC chemokine
UL145		
UL144	Mem-gp; TNF receptor homolog	Regulates lymphocyte activation via BTLA
UL142	UL18 fam; put mem-gp; MHC class I homolog	Inhibits NK cytotoxicity
UL141	UL14 fam; mem-gp	Inhibits NK cell cytotoxicity by downregulating CD155 (CD226 ligand)
UL140	Put mem	Inhibits NK cytotoxicity
UL139	Put mem-gp	
UL138	Put mem	
UL136	Put mem	
UL135	Put sec	
UL133	Put mem	
UL148A	Put mem	
UL148B	Put mem	
UL148C	Put mem	
UL148D	Put mem	
UL150	Put sec	
IRS1	US22 fam; IE protein; teg	Transcriptional activator, blocks PKR-mediated shutoff of translation
US1	US1 fam	
US2	US2 fam; mem-gp	Degradation of MHC class I and possibly MHC class II
US3	US2 fam; IE gene; mem-gp	Inhibits processing and transport of MHC class I and possibly MHC class II
US6	US6 fam; put mem-gp	Inhibits TAP-mediated ER peptide transport
US7	US6 fam; mem-gp	
US8	US6 fam; mem-gp	Binds to MHC class I
US9	US6 fam; mem-gp	Cell-to-cell spread
US10	US6 fam; mem-gp	Delays trafficking of MHC class I
US11	US6 fam; mem-gp	Selective degradation of MHC class I
US12	US12 fam; put 7TM, GPCR?	
US13[¶]	US12 fam; put 7TM, GPCR?	
US14	US12 fam; put 7TM, GPCR?	
US15	US12 fam; put 7TM, GPCR?	
US16	US12 fam; put 7TM, GPCR?	Temperance
US17	US12 fam; put 7TM, GPCR?	Nuclear, fragmented
US18	US12 fam; put 7TM, GPCR?	
US19	US12 fam; put 7TM, GPCR?	Temperance
US20	US12 fam; put 7TM, GPCR?	
US21	US12 fam; put 7TM, GPCR?	

Continued

Table 1 Continued

HCMV[a,b]	HCMV gene family/gene name[c]	Function/comments/abbreviation[d]
US22	US22 fam; teg	Released from cells
US23[¶]	US22 fam; teg	
US24	US22 fam; teg	
US26[¥]	US22 fam	
US27	GPCR fam; virion env-gp	
US28	GPCR fam; env-gp	CC and CX3C chemokine receptor
US29	Put mem-gp, necessary in RPE cells	
US30	Put mem-gp	Temperance
US31	US1 fam	
US32	US1 fam	
US34	Put sec	
US34A	Put mem-gp	
TRS1[¶]	US22 fam; IE protein; teg pp	Transcriptional activator; blocks PKR-mediated shutoff of translation

[a]Commonly annotated ORFs of HCMV (see **Figure 1**). Based on the replication properties of HCMV mutant viruses, most genes are dispensable in cell culture (normal type). Viruses with mutations that influence replication efficiency exhibit one of three broad growth deficient phenotypes: failure to replicate ([†]), very poor replication ([¥]), or slight replication defect ([¶]) when assayed in human fibroblasts. Mutations of putative HCMV ORFs UL60 and UL61 are interpreted as disrupting *ori*Lyt rather than any protein-coding ORF. Certain mutant virus phenotypes vary depending on viral strain and strain variant tested.
[b]Betaherpesvirus-common genes are in italics, and include 40 herpesvirus core functions, labeled as 'core'. Gene family, characteristics, and functional information are provided for HCMV and a blank indicates the lack of information.
[c]Abbreviations: fam, family; pp, phosphoprotein; teg, tegument; gp, glycoprotein; mem-gp, membrane gp; env-gp, envelope gp, TM-gp, transmembrane gp; sec, secreted; put, putative; GPCR, G-protein-coupled receptor; 7TM, seven transmembrane-spanning; IE, immediate early.
[d]Functional information is speculative in some cases.

betaherpesvirus-common genes, as do HHV-6 and HHV-7. Betaherpesviruses exhibit the greatest divergence observed in the herpesviruses, with even the most closely related primate cytomegaloviruses exhibiting differences in gene content and different CMV strains containing a subset of strikingly variable genes. Human roseoloviruses and HCMVs lack detectable DNA sequence homology. The deduced protein-coding capacity of cytomegaloviruses is also highly diverged, although a set of betaherpesvirus-conserved proteins emerges from bioinformatic analysis of potential protein-coding regions and greater levels of conservation are observed in cytomegaloviruses of closely related host species. The diversity of genes observed in biologically related viruses may be a product of immune clearance pressure through co-evolution with particular hosts. Each betaherpesvirus appears to have co-evolved with its host species and exhibits little genetic evidence of cross-species transmission.

Pathogenesis and Latency

HCMV transmission is via contact with virus-positive body fluids and requires direct contact with infectious material, as is the case with most human herpesviruses. HCMV is shed for prolonged periods of time in urine, saliva, tears, semen, and cervical secretions, as well as in breast milk. HCMV transmission rates are highest where contact with body fluids occurs, such that HCMV is an important sexually transmitted disease as well as an infection that is transmitted among children in day-care centers and from young children to parents and other care providers. Thus, two distinct types of exposure are associated with horizontal transmission of HCMV: sexual activity among adults and direct contact with urine or saliva from young children. Children in day-care centers, who frequently transmit virus to each other through casual contact, continue to shed virus in saliva and urine for years without symptoms. Day-care workers have markedly increased rates of HCMV infection as do parents of young children who attend day-care centers, where transmission rates of 10% or greater per year have been reported. Hygiene is thought to play a role in transmission, and regular hand washing has been recommended to reduce transmission. Although shedding of HCMV is common among hospitalized individuals, medical-care professionals who are known to use good hygiene practices do not exhibit any increased risk of HCMV infection. Horizontal transmission may also occur through blood transfusion, blood and marrow (hematopoietic cell) allograft transplantation, and in the clinically important setting of solid organ transplantation from HCMV seropositive donors to naive recipients.

HCMV is transmitted vertically during pregnancy, as well as during the birthing process or from mother's milk. Transplacental transmission is a characteristic of HCMV and possibly roseoloviruses, and is not observed with alphaherpesviruses or gammaherpesviruses. Primary infection has been reported to occur in 1–4% of pregnant women, and is associated with a significant risk for transplacental transmission. Between 20% and 40% of newborns delivered by women with primary infection are HCMV infected, compared to between 1% and 2% of newborns delivered by women with past infection. Primary infection is associated with more severe forms of congenital disease, although disease occurrence has been well documented in both settings. Transmission at birth as well as through mother's milk helps maintain HCMV infection in the population, with 25–50% of infants nursed by seropositive mothers acquiring HCMV by 1 year of age. Importantly, there are no known disease sequelae associated with intrapartum or postpartum acquisition of HCMV in healthy term newborns.

Although serotypes of HCMV have not been useful for epidemiology, various molecular genetic approaches that assess viral genome sequence variation have been used to differentiate viral strains in circulation. Comparison of polymerase chain reaction (PCR)-amplified segments of specific, highly variable genomic regions and proteincoding genes has been most useful, notably the noncoding genomic L–S junction (*a* sequence) as well as genes UL144, UL146, UL73, and UL74. To date, no particular viral strain has been unambiguously linked to any clinical disease.

HCMV appears to reside latent in progenitor cells of the myeloid lineage and to reactivate when these progenitors are stimulated by pro-inflammatory cytokines. HCMV is persistently detected in saliva approximately 3–5% of the time. Based on the distribution and levels of viral DNA, natural latency occurs in bone marrow-derived myelomonocytic cells that give rise to macrophages and dendritic cells, with approximately 0.01–0.001% of cells infected and each cell carrying two to ten genome copies of HCMV DNA. Little is known about sporadic reactivation in immunocompetent individuals, which leads to shedding in saliva and other bodily fluids. However, reactivation in allogeneic transplant settings appears to follow the elaboration of pro-inflammatory cytokines that are a component of tissue rejection. Patients with compromised immunity to HCMV amplify virus to levels that become a clinical threat. HCMV reactivation also occurs in settings without allogeneic stimulation. Pregnancy appears to allow reactivation at levels that are not a threat to women but may be a threat to a developing fetus. Severely immunocompromised patients, such as terminal AIDS patients, appear to support sufficiently high levels of HCMV replication for disease to occur.

Clinical Features and Pathology

HCMV infection in the immunocompetent host is typically clinically silent. A small percentage of mononucleosis is due to HCMV, which is typically less severe than that caused by Epstein–Barr virus. Severe or systemic HCMV disease affecting one or more organs only occurs in immunocompromised hosts and has been known to unveil underlying immunodeficiency in individuals with a previously unrecognized immune deficit.

Congenital disease is the hallmark of HCMV pathogenesis and follows transplacental transmission to the fetus any time during pregnancy, and is a particular risk for woman with primary infection. Congenital disease remains the most important medical and public health concern for HCMV, and is difficult to identify because primary infection is typically asymptomatic. A significantly lower rate of congenital disease follows recurrent compared to primary maternal infection during pregnancy, providing the key evidence that universal vaccination would benefit the population. In addition, primary infection at an earlier gestational age is most likely to manifest as disease in the newborn. HCMV disease recognized at birth often reflects neuronal damage (hearing loss, retinitis, optic neuritis, microcephaly, or encephalopathy) as well as damage to the reticuloendothelial system (hepatomegaly, splenomegaly, petechiae, or jaundice). Prematurity and poor intrauterine growth are common among newborns with symptomatic congenital HCMV infection. Although the majority of infected newborns are asymptomatic at birth, some still develop sequelae, notably sensorineural hearing loss, over the first few years of life. Most newborns with congenital HCMV infection, even those severely affected at birth, survive; congenital HCMV infection takes its toll by affecting hearing, eyesight, and mental capacity.

Very low birth weight premature infants who acquire HCMV infection during birth or postpartum (from human milk or transfusion) may develop systemic disease that is distinct from conventional congenital disease. Risk factors include very low birth weight, exposure to multiple units of blood and birth to a seronegative mother (lack of passive immunity). Transfusion-associated HCMV disease in newborns has been controlled by the use of seronegative (or leukocyte-depleted) blood products. Whether or not transmission of HCMV from a mother's milk to a very low birth weight premature newborn should be prevented remains an area of controversy.

HCMV has become one of the most common opportunistic pathogens as allograft transplantation has grown and as the AIDS epidemic has unfolded. Solid organ and hematopoietic cell allograft recipients all receive strong post-transplant immunosuppressive therapy and AIDS patients progress to a T-lymphocyte deficit that provides an opportunity for HCMV to replicate to high levels and cause disease. Active infection arises from primary infection of naive individuals, reinfection with additional strains of virus, and reactivation of latent virus. This aspect of disease pathogenesis is complicated because HCMV shedding and viremia are common in the face of impaired cellular immunity even when HCMV disease is absent. Clinically silent HCMV infection may predominate even in the face of antiviral prophylaxis, and may only progress to disease as T-cell-mediated immune surveillance becomes severely compromised, such as occurs in AIDS patients with very low CD4 T-cell counts.

Clinical manifestations of disease in solid organ transplant patients include systemic, febrile illness (CMV syndrome, with malaise, arthralgia and rash; neutropenia, thrombocytopenia, and elevated liver enzymes), as well as a direct or indirect impact on specific organs. This disease risk has been reduced but not eliminated by preemptive therapy or prophylaxis with antiviral drugs. Clinical detection of infection using assays for viral antigens or nucleic acids without the need for virus culture successfully reduced the incidence of acute HCMV disease in the first few weeks following transplantation. Preemptive therapy has been beneficial in other high-risk patients such as allogeneic blood and marrow (hematopoietic cell) transplantation where late-onset disease and indirect effects have become more prominent. The detection of HCMV antigens (usually pp65 antigenemia) or nucleic acids (usually PCR detection of viral DNA in plasma or leukocytes) in blood is used to implicate this virus in disease. Solid organ transplant recipients may acquire HCMV pneumonitis, or gastrointestinal lesions, hepatitis, retinitis, pancreatitis, myocarditis, and, in rare circumstances, encephalitis or peripheral neuropathy, whereas hematopoietic cell transplant recipients suffer HCMV pneumonitis or gastrointestinal disease. Establishing an etiologic role for HCMV requires detection of virus in affected tissue or bronchoalveolar lavage or quantitation of HCMV in blood or plasma. The principal indirect effects of HCMV, particularly during solid organ transplantation, include allograft rejection, vascular disease complications, and increased risk of opportunistic fungal and bacterial infections. Primary infection is most likely to lead to disease when a naive recipient receives an organ from an HCMV seropositive donor, although the particular organ and immunosuppression regimen also influence outcome.

Although antiviral prophylaxis has proved to be very effective in preventing HCMV disease in solid organ transplant recipients, late-onset HCMV disease may follow prophylaxis in as many as 5% of patients. Prophylactic or preemptive antiviral treatment in hematopoietic cell transplant recipients reduces the incidence of HCMV disease during the first 3–4 months to around 5%; however, late-onset disease, usually more than 100 days after transplant, remains a problem. In all clinical transplant settings, HCMV has been associated with chronic disease consequences that prophylaxis and preemptive therapy have not completely alleviated. In the nontransplanted individual, HCMV has been a suspected cofactor in atherosclerotic vascular disease and in autoimmune vasculitis.

Highly active antiretroviral therapy (HAART) reduces incidence of HCMV retinitis and gastrointestinal disease in AIDS patients due to reconstitution or preservation of cellular immune function, although HCMV remains a significant risk to human immunodeficiency virus (HIV)-positive individuals when CD4 T-cell counts drop to low levels.

Four antiviral agents are currently approved in the USA for treatment of HCMV disease, ganciclovir, valganciclovir, cidofovir, and foscarnet, with intravenous ganciclovir and orally administered prodrug valganciclovir being first choices in most settings. These two are also common choices for prophylaxis and preemptive therapy. All of the anti-HCMV drugs have potentially significant toxicity and are limited to use in patients with disabling or life-threatening disease risk. Thus, additional drugs are under development. Prolonged use of any investigational or licensed drug may select for resistant virus, but drug-resistant mutants have not been observed to circulate in the population.

Immune Response, Prevention, and Control

Prevention of HCMV infections during pregnancy to reduce congenital disease risk is an important public health goal, and universal vaccination during childhood is believed to be the appropriate route to achieve this control. Attempts to develop live, attenuated, killed, and subunit vaccines over the past 30 years have not yet resulted in success. The Centers for Disease Control and Prevention (CDC) currently recommends that parents and caregivers of young children be informed of how HCMV is transmitted and of hygienic measures that reduce transmission, particularly to women who care for young children and may become pregnant.

Natural HCMV infection is highly immunogenic, resulting in broad humoral antibody and cellular T-lymphocyte responses that persist for life. The parameters of CD4 and CD8 T-lymphocyte responses vary with age and other characteristics, but the continued high levels of response in a majority of infected individuals suggests that virus antigen provides a continuous immunogenic stimulus throughout life. In addition to the adaptive immune response, innate cellular clearance by natural killer (NK) cells appears to be important in the initial stages of infection.

HCMV encodes a very large array of gene products that deflect intrinsic host-cell antiviral responses mediated at the transcriptional level through activation of the interferon pathways or cell death. In addition, this virus encodes a wide range of functions aimed at disarming the innate NK cell response as well as the effectiveness of adaptive humoral and T-cell responses. These functions provide an impression that this virus survives in tight balance with host immune clearance mechanisms, and are summarized in **Table 1**. In addition, accumulating evidence from animal models of HCMV pathogenesis implicates pro-inflammatory viral gene products in creating the appropriate environment to allow efficient growth and dissemination to important tissue sites to assure transmission to new hosts.

Future Perspectives

Although there have been tremendous gains in knowledge of HCMV molecular biology and pathogenesis, this virus continues to be a very important cause of disease especially for immunocompromised hosts and for the fetus. To date, antiviral chemotherapy has been only partially successful in controlling HCMV infection in transplant and other immunocompromised patients and there remains a need for vaccination to reduce the incidence of congenital HCMV infection. An important goal for future research will be to translate the large and growing body of basic knowledge of HCMV biology into improved treatments and effective vaccines.

See also: Cytomegaloviruses: Murine and Other Non-primate Cytomegaloviruses; Cytomegaloviruses: Simian Cytomegaloviruses; Herpesviruses: General Features; Human Cytomegalovirus: Molecular Biology; Human Herpesviruses 6 and 7.

Further Reading

Davison AJ and Bhella D (2007) Comparative betaherpesvirus genome and virion structure. In: Arvin AM, Mocarski ES, Moore P, *et al.* (eds.) *Human Herpesviruses: Biology, Therapy and Immunoprophylaxis*, pp 177–203. Cambridge: Cambridge University Press.

Dolan A, Cunningham C, Hector RD, *et al.* (2004) Genetic content of wild type human cytomegalovirus. *Journal of General Virology* 85: 1301–1312.

Mocarski ES, Jr. (2007) Betaherpesvirus-common genes and their functions. In: Arvin AM, Mocarski ES, Moore P, *et al.* (eds.) *Human Herpesviruses: Biology, Therapy and Immunoprophylaxis*, pp 202–228. Cambridge: Cambridge University Press.

Mocarski ES Jr., Shenk T, and Pass RF (2006) Cytomegalovirus. In: Knipe DM, Howley PW, Griffin DE, *et al.* (eds.) *Fields Virology*, 5th edn., pp. 2701–2772. Philadelphia, PA: Lippincott Williams and Wilkins.

Human Cytomegalovirus: Molecular Biology

W Gibson, Johns Hopkins University School of Medicine, Baltimore, MD, USA

© 2008 Elsevier Ltd. All rights reserved.

Introduction

This chapter provides a general description of the molecular biology of human cytomegalovirus (HCMV) and highlights a number of specific features concerning the virus and its replication cycle. It is necessary to point out that most studies on the molecular biology of HCMV have utilized highly passaged, genetically deficient laboratory strains grown in human fibroblast cells. Although the findings are anticipated to apply in large part to wild-type strains grown in fibroblast and other cell types, and therefore to be relevant to the clinical situation, there are probably important differences to be discovered in the future.

Properties of the Virion

Typical of the herpesvirus group, the virion of HCMV (species *Human herpesvirus 5*, subfamily *Betaherpesvirinae*, family *Herpesviridae*) is approximately 230 nm in diameter and is composed of a DNA-containing capsid, surrounded by a less structured tegument layer, and bounded by a trilaminate membrane envelope. The capsid is isosahedral, has a diameter of approximately 110 nm, and is made up of four principal protein species that are organized into 162 capsomeres (150 hexamers plus 12 pentamers) and 320 triplexes located between the capsomeres. By analogy with herpes simplex virus type 1 (HSV-1), one of the pentamer positions is occupied by a portal complex (a dodecamer of the portal protein) through which DNA enters and leaves the capsid. The capsomeres are approximately 20 nm in length and 15 nm in diameter, have a channel about 3 nm in diameter that is open at the exterior end, possess favored cleavage planes along the longitudinal axis, and have short, spicule-like protrusions, probably representing the triplexes, extending out symmetrically and resulting in a pinwheel appearance of the capsomere viewed end on. The tegument region is approximately 50 nm thick and includes seven relatively abundant virus-encoded protein species, five of which are phosphorylated. The virion envelope is estimated to be 10 nm thick and contains at least 10 abundant protein species. Both the tegument and envelope contain a substantial number of less abundant virus-encoded and host-cell proteins. In addition, the virion has been reported to contain phospholipids (phosphatidylcholine, phosphatidylethanolamine, and phosphatidylinositol) and polyamines (spermidine and spermine). The properties of the virion DNA and proteins are described below.

Properties of the Genome

The genome of HCMV is composed of a linear, double-stranded DNA molecule with a size of 236 kbp in wild type virus. Thus, it is over 50% larger than that of HSV-1, and the largest among the human herpesviruses. It is a class E genome, like that of HSV-1, consisting of a long unique (U_L) and a short unique (U_S) sequence, both of which are flanked by much shorter inverted repeat sequences that enable U_L and U_S to invert relative to each other and give rise to the four structural isomers of the genome found in virions. The origin of DNA replication (*ori*-Lyt) has been localized to a 3–4 kbp region near the center of U_L. Like *ori*-L, one of three duplicated origin sequences in the HSV-1 genome, the HCMV *ori*-Lyt region is located adjacent to the promoter for the early, single-stranded DNA-binding protein (encoded by gene UL57). HCMV *ori*-Lyt includes a 2.4 kbp sequence that has no homology with other described virus DNA replication origins and contains direct and inverted repeat sequences. This region contains consensus cyclic AMP (cAMP) response elements and other transcription factor-binding sites, and 23 copies of the sequence AAAACACCGT that are conserved near the homologous *ori*-Lyt of simian cytomegalovirus (SCMV).

The complete nucleotide sequence of the high passage strain AD169 was the first to be determined for HCMV, with a size of 229 345 bp. It has a G + C content of 57.2 mol%, and was assessed as containing 189 protein-coding open reading frames (ORFs), some of which are spliced and some of which are duplicated in the inverted repeats. The low passage strain Toledo was found to contain an additional ∼15 kbp at the right end of U_L containing 19 additional ORFs. Analysis of the complete 235 645 bp sequence of the low passage strain Merlin and a reevaluation of the genetic content of strain AD169 have indicated that wild-type HCMV contains ∼165 protein-coding genes. The wild-type genome contains 13 families of related genes, many with members clustered into blocks and encoding predicted or recognized glycoproteins (e.g., the RL11, US6, and US27 families). Taken together, these gene families represent ∼40% of the total gene number and in part account for the comparatively large size of the HCMV genome. It is also worth noting that the HCMV genome contains homologs of several cellular genes, including surface receptors such as class I major histocompatibility (MHC) antigens, G protein-coupled receptors (GPCRs), and a tumor necrosis factor receptor.

Properties of the Proteins

HCMV proteins share the general pattern of expression characteristic of the herpesvirus group. Immediate early (IE or α), early (E or β), and late (L or γ) proteins are synthesized sequentially from corresponding mRNAs whose transcription is regulated in a temporal cascade. IE proteins are required to regulate transcription from their own promoters and those of subsequently expressed genes. E proteins include many of the enzymes and regulatory factors needed to carry out the synthesis of progeny DNA and proteins. L proteins include most of the virion structural proteins. Members of all three classes have been described for HCMV, though only a small number of the proteins encoded by HCMV have been identified. Those that have been reported include both nonvirion and virion species and their properties are briefly described below.

Nonvirion Proteins

Many of the recognized nonvirion proteins are made at early times after infection and localize to the nucleus. The first of these to appear is the major IE protein (IE1, encoded by UL123), a 72 kDa phosphoprotein that, together with products of a second IE gene, IE2 (UL122), can transactivate HCMV E promoters. Both IE1 and IE2 are expressed from spliced transcripts generated using the major IE promoter. In addition, the IE2 gene gives rise to a family of proteins ranging in size from 23 to 86 kDa that are generated by differential splicing and translational start sites. IE2 proteins can repress the major IE promoter by acting on a sequence immediately upstream of the transcription initiation site. Other IE proteins are encoded by genes IRS1, TRS1, US3, and UL36–38, and can act synergistically with IE1 and the US3 protein to regulate both cellular and virus gene transcription.

The E proteins include the 140 kDa virus DNA polymerase (encoded by UL54), a 140 kDa nuclear single-stranded DNA-binding protein (DB140; UL57), and a set of closely related nuclear phosphoproteins (UL112). Like its homologs in other herpesviruses, the DNA polymerase has an associated 3′-specific nuclease activity. Like its homolog (UL29) in HSV-1, UL57 is located adjacent to *ori*-Lyt, and its product DB140 localizes to discrete foci within the nuclei of infected cells and accumulates in the cytoplasm under conditions of inhibited virus DNA synthesis. UL112 encodes a set of four E, nuclear,

DNA-binding phosphoproteins (34, 43, 50, and 84 kDa) that appear to be produced by alternate splicing, are related to each other by their common N-terminal sequence, and are differentially phosphorylated. Expression of these proteins appears also to be regulated at both the transcriptional and post-transcriptional levels. The functions of these proteins are not known, and they may have a role in DNA replication or gene regulation.

Another nonvirion nuclear DNA-binding phosphoprotein is distinguished by its high abundance and kinetics of synthesis. The ~52 kDa species, DB52, is the product of gene UL44. On the basis of its reduced but not eliminated synthesis under conditions of inhibited virus DNA replication, DB52 is classified as a delayed-early (DE or γ_1) protein. Expression of this gene is under different promoter control at early and late times of infection, and there is evidence that its expression is also post-transcriptionally regulated. DB52 and its SCMV homolog bind to DNA *in vitro* and *in vivo* and associate with the virus DNA polymerase to enhance its activity, analogous to the HSV-1 UL42 protein. DB52, like the HSV-1 UL42 protein, has the structure of a homotrimeric sliding clamp similar to that of the cellular processivity factor, proliferating cell nuclear antigen. Three other proteins that function in virus DNA replication or processing are a phosphotransferase (encoded by UL97), and the ATPase (encoded by UL89) and its partner subunit (encoded by UL56) that constitute an apparent homolog of the bacteriophage DNA cleavage/packaging complex, terminase.

Several other nonvirion HCMV proteins have recently been shown to interfere with the MHC class I antigen presentation system (i.e. the US3, US6, and US11 proteins). These proteins are thought to help the virus-infected cell escape recognition and destruction by immune surveillance, and provide a mechanism for virus persistence in the immune-competent host. It is viewed as likely that wild-type HCMV encodes many proteins involved in modulating the host response to infection.

Virion Proteins

Estimates of the number of protein species in HCMV virions depend on the techniques used. Classical approaches utilizing sodium dodecyl sulfate polyacrylamide gel electrophoresis (SDS-PAGE) have detected a total of 30–35 protein species. The characteristics of the most abundant of these proteins, including size, charge, carbohydrate and phosphate content, and deduced locations in the particle, are summarized in **Table 1** and described below. Virions have also been reported to contain several enzymatic activities (including protein kinase, DNA polymerase, DNase, and topoisomerase II), and several host proteins (including PP1 and PP2A phosphatases and a 45 kDa cellular actin-related protein), although it is not known whether these are specific associations. A larger number of proteins have been detected by a mass spectrometric analysis of virions, amounting to species encoded by 59 recognized HCMV genes and over 70 host proteins. The sensitivity of this technique is sufficiently high to raise questions about whether all these proteins are specifically incorporated into virions, and whether they are present in all virions.

The capsid shell is composed of four abundant protein species referred to as the major capsid protein (MCP), the minor capsid protein (mCP), the mCP-binding protein (mCP-BP), and the smallest capsid protein (SCP). Formation of the capsid shell is coordinated by precursors of an internally located maturational protease (assemblin) and assembly protein (pAP), which serve a scaffolding role and are present in nascent intranuclear capsids (e.g., B capsids) and in noninfectious enveloped particles (see below), but not in virions. These internal scaffolding proteins are the only abundant capsid proteins known to be post-translationally modified (**Table 1**).

The bulk of the protein mass of the virion tegument is contributed by its seven principal protein species. The largest of these, called the high molecular weight protein (HMWP), is the homolog of HSV-1 VP1/2 (encoded by HSV-1 UL36), and is an active ubiquitin-specific cysteine protease. HMWP forms heterooligomers with the HMWP-binding protein (HMWP-BP) encoded by the adjacent gene. Neither HMWP nor HMWP-BP is detectably phosphorylated or glycosylated. The basic phosphoprotein (BPP or pp150) and the upper matrix (UM or pp71) and lower matrix (LM or pp65) proteins account for the greatest protein mass of the tegument constituents and are highly phosphorylated, both *in vivo* and *in vitro*, by the virion-associated protein kinase. BPP is distinguished among the virion proteins by having the highest density of O-linked *N*-acetylglucosamine residues, which are attached at a single site or tight cluster of sites. UM is a transacting inducer of IE gene transcription that may be functionally analogous to the α-*trans*-inducing factor of HSV-1. LM appears to be an important component in cellular immunity and there are reports that it has an associated protein kinase activity. The '24K protein' (pp28) is also phosphorylated and, like its HSV-1 homolog UL11, forms oligomers (with the protein encoded by UL94). HMWP and BPP are thought to make direct contact with the capsid.

The virion envelope contains at least 10 protein species. The high degree of size and charge heterogeneity among this group has complicated their analysis. The most abundant species are distributed among three glycoprotein complexes (gCI, gCII, and gCIII), each of which is disulfide linked. The complex gCI consists of three forms

Table 1 Principle proteins of the HCMV virion

Descriptive name	Predicted M_r[a]	Observed M_r[b]	Modification[c]	Charge[d]	Gene[e]
Capsid					
Major capsid protein (MCP)	153.9	153		N	UL86
Portal protein	78.5			B	UL104
Precursor assembly protein (pAP)[f]	38.2	50	P, C	B	UL80.5
Minor capsid protein (mCP)	34.6	34		B	UL85
mCP-binding protein (mCP-BP)	33.0	35		N	UL46
Protease assemblin (A, A_N, A_C)[g]	28.0	28		A	UL80a
Smallest capsid protein (SCP)	8.5	8		B	UL48A
Tegument					
High mol. wt. protein (HMWP)	253.2	212		N	UL48
Basic phosphoprotein (BPP; pp150)	112.7	149	G, P	B	UL32
HMWP-binding protein (HMWP-BP)	110.1	115		N	UL47
80 K		80	P		
Upper matrix protein (UM; pp71)	61.9	74	P	N	UL82
Lower matrix protein (LM; pp65)	62.9	69	P	B	UL83
24 K (pp28)	28.0	24	P	B	UL99
Envelope					
Glycoprotein B (gB)	102.0	102	G, P, C	N	UL55
N-terminal portion (gB_N)	52.6	130	G	A	UL55
C-terminal portion (gB_C)	49.5	62	G, P	A	UL55
Glycoprotein H (gH)	84.4	84	G	N	UL75
Glycoprotein M (gM)	42.9	43			UL100
Glycoprotein N (gN)	14.5	57	G	A	UL73
Glycoprotein O (gO)	54.2	125	G		UL74
GPCR27	42.0	42	G	B	US27
GPCR33	46.3	44	G	B	UL33

[a]Molecular masses (kDa).
[b]Molecular masses (kDa) given as determined by SDS-PAGE.
[c]G, glycosylation; P, phosphorylation; C, proteolytic cleavage.
[d]Relative net charges were determined by 2-D PAGE (charge/size separation) with respect to the major capsid protein, which is approximately neutral in such separations. N, neutral; A, acidic; B, basic.
[e]Gene locations are shown in figure 1 of the article on Human Cytomegalovirus: General Features.
[f]Present in immature capsids and noninfectious enveloped particles but absent from virions.
[g]Assemblin (A) is derived from a 74 kDa precursor (encoded by UL80a), and is autoproteolytically converted to amino (A_N) and carboxyl (A_C) portions.

of glycoprotein B (gB). HCMV gB is synthesized as a 130–160 kDa, cotranslationally N-glycosylated precursor that undergoes at least four further modifications: (1) its 17–19 predicted N-linked oligosaccharides are converted from nontrimmed to trimmed high-mannose forms, and then a portion is further modified to complex structures; (2) O-linked carbohydrates are added; (3) the C-terminal region of the molecule is phosphorylated; and (4) the protein is cleaved between Arg460 and Ser461 to yield an intramolecularly disulfide-linked heterodimer composed of a highly glycosylated 115–130 kDa N-terminal fragment (gB_N), and a less extensively modified 52–62 kDa C fragment (gB_C). This gB_N–gB_C heterodimer can form intramolecular disulfide crosslinks, yielding a complex with the composition (gB_N–gB_C)$_2$.

Early studies of the complex gCII identified proteins with M_rs of 50–52 and >200 kDa representing forms of the integral membrane protein glycoprotein M (gM). These proteins have similar peptide maps indicating that they are closely related, and the 52 kDa component is comparatively highly O-glycosylated and sialylated. Subsequently, the 50–60 kDa glycoprotein N (gN, UL73) was recognized as a second component of gCII.

The complex gcIII was initially characterized as composed of glycoprotein H (gH) and glycoprotein L (gL; not listed in **Table 1**). HCMV gH is an 86 kDa species that is less extensively N-glycosylated than gB and its oligosaccharides are not processed beyond the high-mannose structure. It has been proposed to mediate virus–host membrane fusion during the initial steps of virus infection, and its intracellular trafficking associates with gL. A third component, the highly sialylated, O-glycosylated virion glycoprotein O (gO), was later shown to be specified by UL74. Recently, using strains of HCMV that more closely reflect wild-type virus, an alternative complex consisting of gH, gL, and two small proteins encoded by UL128 and UL130 has been implicated in tropism for nonfibroblast cell types.

Physical Properties

Infectivity of HCMV is eliminated by exposure to 20% ether for 2 h, by heating at 65 °C for 30 min, by exposure to ultraviolet light (4000 erg s^{-1} cm^{-2}; 400 µJ s^{-1} cm^{-2}) for 4 min, by acidic pH (<5), and by treatment with low concentrations of either ionic or nonionic detergents. Infectivity is also reduced following sonication, pelleting by ultracentrifugation, and repeated cycles of freezing and thawing.

Replication

General features of the HCMV replication cycle are typical of the herpesvirus group. Following virus adsorption and entry into the cell, the DNA is transported to the nucleus where it is replicated and packaged into preformed capsids. Maturation of the nucleocapsid involves acquisition of the tegument constituents followed by envelopment, which has been observed at both nuclear and cytoplasmic membranes.

DNA Replication

HCMV DNA is replicated in the nucleus. Parental DNA appears to circularize soon after entering the cell, but maximal rates of progeny DNA synthesis do not occur until approximately 3 days after infection. The 'endless' concatemeric nature of progeny DNA is consistent with a rolling circle mechanism of DNA replication. However, the finding that virus DNA synthesis proceeds bidirectionally from *ori*-Lyt under conditions inhibition of replication by ganciclovir ([9-[2-hydroxy-1-(hydroxymethyl)-ethoxy]methyl]guanine) is not easily reconciled with this model, and perhaps indicates an intermediate stage of replication based on a theta mechanism. Unit-length genomic DNA is released from the concatemers by cleavage at a specific site located in the *a* repeat elements, 30–35 bp from a herpesvirus group-common cleavage/packaging region referred to as the herpes pac homology sequence.

Characterization of Transcription

The pattern and control of HCMV gene expression shares many features with that of HSV-1. As during translation, transcription of the HCMV genome is temporally regulated into three major kinetic classes. The first class of mRNAs made are transcribed from input DNA and are referred to as IE (or α) RNAs. Their expression requires no preceding virus protein or DNA synthesis, and they are operationally defined as those RNAs that can be made in the presence of inhibitors of protein synthesis. The most abundant IE RNAs are transcribed from the major IE promoter. These transcripts are related to each other by alternate splicing patterns and encode IE1 and multiple forms of IE2. The promoter region for this transcription unit is one of the strongest known, presumably due to its high content of consensus binding sites for host-cell transcription factors such as nuclear factor 1, CAAT-binding protein, and SP1, and the large number of repeat sequences that contain cyclic AMP response elements and that may interact with additional DNA-binding proteins. Transcription from this region is enhanced by UM, whose function appears to be analogous to that of the HSV α-*trans*-induction factor. Lower abundance IE RNAs arise from genes IRS1, TRS1, US3, and UL36–38.

Expression of E (or β) RNAs requires the preceding synthesis of IE proteins, but does not require virus DNA synthesis. Thus, E RNAs are made in the presence of inhibitors of virus DNA synthesis, but not in the presence of inhibitors of protein synthesis. Both IE and E genes appear to be transcribed by host-cell RNA polymerase II. Expression of L (or γ) RNAs requires the preceding synthesis of both IE and E virus proteins, as well as synthesis of virus DNA. The possibility that post-transcriptional events may be involved in controlling HCMV RNA processing and translation is suggested by the presence of some E RNAs that are not translated until late times of infection.

Characterization of Translation

HCMV RNAs are translated by the host-cell protein-synthesizing system. Virus-specific proteins that may augment the host system (e.g., initiation or elongation factors) have not yet been identified. Each virus protein species appears to be encoded by a unique mRNA. Cells infected with HCMV do not show the generalized early shut-off of cell protein synthesis that occurs in HSV-1-infected cells. There are, nevertheless, very early changes in the metabolism of HCMV-infected cells that include stimulated transcription of some host genes (e.g., heat shock protein 70, ornithine decarboxylase, thymidine kinase, creatine kinase, and cyclooxygenase 2), decreased transcription of other host genes (e.g., fibronectin), and changes similar to those induced by the GPCR signaling pathway (e.g., decreased intracellular Ca^{2+} stores and increased levels of intracellular cAMP).

Post-Translational Modifications

Modifications of HCMV proteins include phosphorylation, glycosylation, and proteolytic cleavage. Phosphorylation has been detected on IE (e.g., IE1 and IE2), E (e.g., UL112 protein) and L (e.g., DE nonvirion UL44 protein; late virion tegument (UL32, UL82, and UL83) proteins), and capsid (AP) proteins, all of which localize to the infected cell nucleus. At least some of these phosphorylations are likely to be catalyzed by reported virus-coded protein kinases. Glycosylation typifies many of the virion envelope glycoproteins and includes both N-linked (high-mannose and complex structures) and O-linked oligosaccharides. The only nonenvelope virion protein that has been demonstrated to be glycosylated is BPP, which has O-linked N-acetylglucosamine. Glycosylation of HCMV proteins is most likely carried out by host cell systems. The third modification that has been described for HCMV proteins is proteolytic cleavage. There are two maturational endoproteolytic cleavages, in addition to glycoprotein signal sequence cleavages. The first of these cuts between Arg460 and Ser461 in envelope gB and is catalyzed by a calcium-dependent cellular, furin-like proteinase. The other is a cleavage between residues Ala308 and Ser309 of the capsid assembly protein precursor (pAP). This latter cleavage is catalyzed by the genetically related, virus-encoded proteinase, assemblin (A), and is essential for the production of infectious virus.

Cytopathology

The mechanisms of cell recognition, binding, and penetration by HCMV are not yet understood, but there are at least five envelope proteins that may be involved: (1) a 'disintegrin-like' domain in virion envelope gB interacts with cellular integrins; (2) β_2-microglobulin is present on virions and may interact with cell-surface MHC proteins; (3) virion envelope gH interacts with a 90 kDa cell protein that may also be involved in binding or penetration; and (4) a highly O-glycosylated envelope glycoprotein complex (gcII) is able to bind heparin, and this interaction could promote early virus–cell interaction as proposed for cellular heparin and HSV-1. Different entry mechanisms may operate for different cell types.

Assembly of the capsid and DNA replication and packaging take place in the nucleus. As with HSV-1, some tegument constituents are presumed to be added to the maturing nucleocapsid in the nucleus, with envelopment and de-envelopment occurring as the tegument-coated capsid transits the two leaflets of the nuclear membrane into the cytoplasm. Further addition of tegument proteins is then thought to precede secondary envelopment at vesicle membranes in the infected cell cytoplasm.

The cytopathic effects of HCMV are distinctive. Early after infection, a transient cell rounding occurs, followed by overall enlargement (cytomegalia) and by the appearance of basophilic intranuclear inclusions, which gradually enlarge to fill and distort the nucleus into an elongated or kidney-shaped form late in infection. These intranuclear inclusions contain virus DNA, proteins, and capsids, and are thought to be nucleocapsid assembly sites. Late in infection, the cytoplasm contains numerous enveloped and nonenveloped capsids, large (>500 nm 'black holes') and smaller (250–500 nm 'dense bodies') electron-dense aggregates, and a large eosinophilic spherical region adjacent to the nucleus that exhibits strong F_C binding.

Release of HCMV from the cell results in three kinds of enveloped particles being freed into the growth medium of cultured fibroblasts: (1) virions; and two aberrant particles referred to as (2) noninfectious enveloped particles, which closely resemble virions in structure and composition but have no DNA and contain the capsid assembly protein that is absent from virions; and (3) dense bodies, which are solid, 250–600 nm spherical aggregates of LM that contain no DNA and are surrounded by an envelope layer.

Acknowledgment

I am grateful to Andrew Davison for his advice and kind assistance in updating the previous edition of this article.

See also: Herpesviruses: General Features; Human Cytomegalovirus: General Features.

Further Reading

Brignole EJ and Gibson W (2007) Enzymatic activities of human cytomegalovirus maturational protease assemblin and its precursor (pUL80a) are comparable: Maximal activity of pPR requires self-interaction through its scaffolding domain. Journal of Virology 81: 4091–4103.

Chee MS, Bankier AT, Beck S, et al. (1990) Analysis of the protein coding content of the sequence of human cytomegalovirus strain AD169. Current Topics in Microbiology and Immunology 154: 125–169.

Dolan A, Cunningham C, Hector RD, et al. (2004) Genetic content of wild type human cytomegalovirus. Journal of General Virology 85: 1301–1312.

Mocarski ES and Tan Courcelle C (2001) Cytomegaloviruses and their replication. In: Knipe DM and Howley PM (eds.) Fields Virology, 4th edn., pp. 2629–2673. Philadelphia: Lippincott Williams and Wilkins.

Varnum SM, Streblow DN, Monroe ME, et al. (2004) Identification of proteins in human cytomegalovirus (HCMV) particles: The HCMV proteome. Journal of Virology 78: 10960–10966.

Human Eye Infections

J Chodosh, A V Chintakuntlawar, and C M Robinson, University of Oklahoma Health Sciences Center, Oklahoma City, OK, USA

© 2008 Elsevier Ltd. All rights reserved.

Glossary

Blepharoconjunctivitis Inflammation of the eyelid and conjunctiva.
Dacryoadenitis Inflammation of the lacrimal gland.
Iridocyclitis Inflammation of the iris and ciliary body.
Meibomian gland Modified sebaceous glands lining the eyelid margin that provide the lipid layer of the preocular tear film.
Retinochoroiditis Inflammation of the retina and choroid (uveal) layers of the eye.

Introduction

The eye contains diverse tissues intricately linked to subserve visual function (**Figure 1**). The ocular adnexae – periorbita, eyelids and lashes, lacrimal and meibomian glands – produce, spread, and drain the preocular tear film, physically protect the sensitive ocular mucosa, and cushion the globe. The redundant conjunctiva with its low-viscosity tear film allows rapid multidirectional eye movements. Lymphoid tissues within the conjunctiva and lacrimal glands furnish acquired immune defense. The cornea and its tear film fashion the major refractive surface of the eye. The sclera forms the wall of the globe and scaffolds the intraocular tissues. The eye's lens provides additional refractive power and filters ultraviolet light. The iris diaphragm dynamically regulates the amount of light incident upon the retina, and together with the choroid and optic nerve head provides immune effector cells to the interior of the eye. The retina transduces light energy into neural signals; retinal function is requisite for vision. The vascular choroid nourishes the outer layers of the retina. The anterior (aqueous) and posterior (vitreous) humors provide internal pressure sufficient for maintenance of normal anatomic relationships, nourish the interior ocular tissues, provide immunosuppressive factors necessary to the maintenance of immune deviation, and during infection act as conduits for the distribution of inflammatory cells derived from the iris, ciliary body, and optic nerve head.

Eye infection by viruses most often follows direct contact with virus externally, either from infected secretions in the birth canal (herpes simplex virus, human papillomavirus), on fomites (adenovirus), or airborne particles (rhinovirus), or is acquired during viremia (human cytomegalovirus, measles virus). Other mechanisms of ocular viral infection include extension from contiguous adnexal disease (herpes simplex virus), spread from the upper respiratory tract via the nasolacrimal duct (rhinovirus), and transplacental passage of infectious virus (rubella virus). Rarely, ocular infection may disseminate elsewhere (enterovirus 70).

Acute viral infection produces stereotypic changes in ocular target tissues. Infection of the eyelid skin induces the formation of vesicles and ulcers. Viral infection of the conjunctiva results in vasodilatation, serous discharge, hyperplasia of conjunctival lymphoid follicles, and enlargement of the corresponding draining lymph nodes. Severe conjunctival infection can cause permanent scarring of the globe to the eyelids and turning in of the eyelashes against the eye. Viral infection of the corneal epithelium induces punctate epithelial cytopathic effect evident biomicroscopically as isolated swollen epithelial cells (punctate epithelial keratitis) and loss of individual epithelial cells (punctate epithelial erosions). When extensive, the punctate erosions may coalesce to form confluent epithelial ulcers with dendritic, dendritiform, or geographic morphology. With herpetic infection, corneal anesthesia can ensue, and in the absence of epitheliotropic neural growth factors, corneal epithelial integrity is impaired. Reduced corneal clarity and progressive sterile ulceration may result. Corneal stromal infection induces white blood cell recruitment; subsequent corneal scarring, vascularization, and lipid deposition may permanently reduce vision. Intraocular infection manifests in inflammatory cell deposits on the posterior surface of the cornea and on the vitreous scaffold, and in free-floating leukocytes and biomicroscopically visible protein spillage into the normally cell-free and protein-poor aqueous humor. Iridocorneal and iridolenticular adhesions may develop and lead to glaucoma and cataract. Retinal infection concludes with necrosis and lost function. Viral encephalitis and meningitis can result in cranial nerve inflammation and secondary dysfunction of vision and extraocular motility.

Classical viral pathogenic mechanisms of latency, reactivation, and carcinogenesis all can be demonstrated in the eye. Herpes simplex virus causes recurrent lytic epithelial keratitis when viral reactivation within sensory ganglia of the first division of the fifth cranial nerve gives rise to virus in the preocular tear film. Necrotizing herpes stromal keratitis follows viral reactivation within the cornea stroma. Intraepithelial neoplasia and invasive squamous cell carcinoma of the conjunctiva and cornea

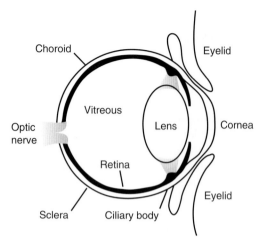

Figure 1 Cross section of the human eye.

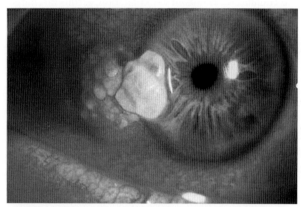

Figure 2 Squamous cell carcinoma of the corneal limbus is associated with infection by human papilloma virus types 16 and 18.

(**Figure 2**) have been associated with human papilloma virus types 16 and 18. When infected with oncogenic human papillomaviruses, corneal limbal stem cells can provide a persistent source of dysplastic ocular surface epithelium. Molecular mimicry has also been demonstrated as an immunopathogenic mechanism in ocular disease. Systemic infection with hepatitis C virus is associated with autoimmunity against a corneal stromal antigen and peripheral ulcerative keratitis. In a murine model of herpes simplex infection, non-necrotizing stromal keratitis accompanies T-cell reactivity against a corneal protein antigenically similar to a herpes simplex coat protein.

Ocular Immunology of Relevance to Viral Infection

Tissue diversity within the eye and adnexa compel varied means of innate immune defense. The eye's external surfaces (conjunctiva and cornea) encounter viruses by both airborne and contact routes. The eyelids, an intermittent barrier, periodically wipe the eye's surface free of debris and spread and drain the preocular tear film. The ability of the tear film to nonspecifically impede primary infection by viruses is unknown, although such mechanisms are well established for bacterial pathogens. An inhibitory effect of goblet cell-derived and intrinsic mucins and meibomian gland-derived lipids on viral adsorption to the ocular surface is speculative. Early in infection, aqueous tears from the main and accessory lacrimal glands furnish proinflammatory cytokines, and the conjunctival blood vessels provide both soluble and cellular components of innate immunity. After viral infection is established, aqueous tears carry lacrimal gland-derived monospecific secretory immunoglobulin A.

The constitutive defense armaments of the cornea and conjunctiva differ. The normal cornea is considered an immune-privileged site due to the high success rate of corneal transplantation; it lacks blood vessels, lymphatics, resident lymphoid cells, and Langerhans cells, expresses Fas ligand on its surface epithelium, and demonstrates reduced delayed hypersensitivity responses. Because corneal inflammation and subsequent scar reduce vision, corneal function is best served by its reduced immunologic responsiveness, also known as immune deviation. Necrotizing inflammation presupposes infection beneath the surface epithelium, and follows chemokine synthesis by infected corneal stromal fibroblasts. In contrast to the cornea, the conjunctiva is well endowed with blood and lymphatic channels, lymphoid cells, and Langerhans cells, and demonstrates classical delayed hypersensitivity responses. The immunology of the interior eye is less well established, but immune deviation appears to extend beyond the cornea to the aqueous and vitreous humors and to the central retina.

Ocular Disease Caused by RNA Viruses

Conjunctivitis is probably the most common viral ocular syndrome, and typically accompanies upper respiratory infections due to RNA viruses (**Table 1**). Rhinovirus, influenzavirus, respiratory syncytial virus, and parainfluenzavirus conjunctivitis typically are mild and self-limited, and most patients do not seek medical attention. More serious are the keratitis, uveitis, and retinitis caused by some RNA viruses. For example, influenzavirus infection of the respiratory tract, usually associated with a mild and short-lived conjunctivitis, less commonly causes inflammation in the lacrimal gland, cornea, iris, retina, optic and other cranial nerves.

Like influenzavirus, other RNA viruses can infect virtually every ocular tissue. For instance, rubella virus when acquired *in utero* may have devastating consequences for the eye. Characteristic features include

Table 1 Ocular targets of human RNA viruses

Virus	Family	Subfamily/genus	Nuc. acid	Env.	Ocular target
Rift Valley fever virus	Bunyaviridae	Phlebovirus	ss (−)	+	Conjunctiva
					Retina
Human coronavirus	Coronaviridae	Coronavirus	ss (+)	+	Conjunctiva
Dengue virus	Flaviviridae	Flavivirus	ss (+)	+	Conjunctiva
Hepatitis C virus	Flaviviridae	Hepatitis C virus	ss (+)	+	Cornea
					Lacrimal Glands
					Retina
West Nile virus	Flaviviridae	Flavivirus	ss (+)	+	Retina
					Uvea
					Optic nerve
					Cranial nerves
Yellow fever virus	Flaviviridae	Flavivirus	ss (+)	+	Conjunctiva
Influenzavirus	Orthomyxoviridae	Influenzavirus (A, B, C)	ss (−)	+	Lacrimal gland
					Conjunctiva
					Episclera
					Cornea
					Uvea
					Retina
					Optic nerve
					Cranial nerves
Measles (rubeola) virus	Paramyxoviridae	Morbillivirus	ss (−)	+	Conjunctiva
					Cornea
					Uvea
					Retina
					Optic nerve
					Cranial nerves
Mumps virus	Paramyxoviridae	Paramyxovirus	ss (−)	+	Lacrimal gland
					Conjunctiva
					Sclera
					Cornea
					Trabecular meshwork
					Uvea
					Optic nerve
					Cranial nerves
Newcastle disease virus	Paramyxoviridae	Paramyxovirus	ss (−)	+	Conjunctiva
					Cornea
Parainfluenza virus(es)	Paramyxoviridae	Paramyxovirus	ss (−)	+	Conjunctiva
Respiratory syncitial virus	Paramyxoviridae	Pneumovirus	ss (−)	+	Conjunctiva
Enterovirus(es): (includes poliovirus, coxsackievirus, echovirus, enterovirus)	Picornaviridae	Enterovirus	ss (+)	−	Conjunctiva
					Cornea
					Cranial nerves
Rhinovirus	Picornaviridae	Rhinovirus	ss (+)	−	Conjunctiva
Colorado tick fever virus	Reoviridae	Coltivirus	ds (+/−)	−	(?: reported to cause photophobia, retro-ocular pain)
Human T-cell lymphotropic virus-1	Retroviridae	Deltaretrovirus	ss (+)	+	Cornea
					Uvea
Human immunodeficiency virus	Retroviridae	Lentivirus	ss (+)	+	Lacrimal gland
					Retina
Rabies virus	Rhabdoviridae	Lyssavirus	ss (−)	+	(Transmission via corneal button)
Rubella virus	Togaviridae	Rubivirus	ss (+)	+	Cornea
					Uvea
					Lens
					Trabecular meshwork
					Retina
					Globe

+, Enveloped; −, nonenveloped; ss, single stranded; ds, double stranded; (+), positive-sense RNA genome; (−), negative-sense RNA genome.

Table 2 Ocular targets of human DNA viruses

Virus	Family	Subfamily/genus	Nuc. acid	Env.	Ocular target
Adenovirus	Adenoviridae	Mastadenovirus	ds	−	Conjunctiva Cornea
Herpes simplex virus, type 1 (HHV1)	Herpesviridae	Alphaherpesvirinae/Simplexvirus	ds	+	Eyelid Conjunctiva Cornea Trabecular meshwork Uvea Retina
Herpes simplex virus, type 2 (HHV2)	Herpesviridae	Alphaherpesvirinae/Simplexvirus	ds	+	Eyelid Conjunctiva Cornea Trabecular meshwork Uvea Retina
Varicella zoster virus (HHV3)	Herpesviridae	Alphaherpesvirinae/Varicellovirus	ds	+	Eyelid Conjunctiva Cornea Trabecular meshwork Uvea Retina Optic nerve
Epstein–Barr virus (HHV4)	Herpesviridae	Gammaherpesvirinae/Lymphocryptovirus	ds	+	Lacrimal gland Conjunctiva Cornea Uvea Retina Optic nerve
Human cytomegalovirus (HHV5)	Herpesviridae	Betaherpesvirinae/Cytomegalovirus	ds	+	Retina Optic nerve
Human herpes virus 6 (HHV6)	Herpesviridae	Betaherpesvirinae/Roseolovirus	ds	+	Retina
Human herpes virus 8 (HHV8)	Herpesviridae	Gammaherpesvirinae	ds	+	Conjunctiva (Kaposi sarcoma)
Human papillomavirus	Papovaviridae	Papillomavirus	ds	−	Eyelid Conjunctiva Cornea
Molluscum contagiosum virus	Poxviridae	Molluscipoxvirus	ds	+	Eyelid Conjunctiva Cornea
Orf virus	Poxviridae	Parapoxvirus	ds	+	Eyelid
Smallpox (variola) virus	Poxviridae	Orthopoxvirus	ds	+	Eyelid Conjunctiva Cornea Uvea Optic nerve
Vaccinia virus	Poxviridae	Orthopoxvirus	ds	+	Eyelid Conjunctiva Cornea

ds, Double stranded; +, enveloped; −, nonenveloped; HHV, human herpes virus.

microphthalmos, corneal haze, cataracts, iris hypoplasia, iridocyclitis, glaucoma, and 'salt-and-pepper' pigmentary retinopathy. Rubella virus can be cultured from the lens of infected neonates at the time of cataract extraction. Congenital ocular abnormalities due to rubella, like those in other organ systems, are much worse when maternal infection ensues earliest in pregnancy.

In contrast to rubella virus, measles (rubeola) virus infection *in utero* rarely causes significant ocular disease. The classic triad of postnatally acquired measles – cough, coryza, and follicular conjunctivitis – can be accompanied by Koplik spots on the conjunctiva and a mild epithelial keratitis. Less common are optic neuritis, retinal vascular occlusion, and pigmentary retinopathy. Measles keratopathy, a major source of blindness in the nonindustrialized world, typically presents as corneal ulceration in a malnourished child. A rare and fatal complication of measles virus infection, subacute sclerosing panencephalitis (SSPE), occurs in about 1 per 100 000 cases, and often years after clinically apparent measles. Along with devastating central nervous system damage, ocular abnormalities occur commonly in SSPE, including central retinal (macular) hyperpigmentation and inflammation, optic nerve atrophy, peripheral retinitis, and ocular motility disorders. Cortical blindness can occur in the absence of ocular involvement.

The most common ocular complication of mumps virus infection is dacryoadenitis, and this may occur concurrently with parotid gland involvement. Aseptic meningitis, associated oculomotor palsy, and optic neuritis also occurs. Follicular conjunctivitis, epithelial and stromal keratitis, iritis, trabeculitis, and scleritis have all been reported within the first 2 weeks after onset of parotitis.

Acute hemorrhagic conjunctivitis (AHC), caused predominantly by enterovirus type 70 and coxsackievirus A24 variant, but also by adenovirus type 11, is one of the most dramatic ocular viral syndromes. Sudden onset of follicular conjunctivitis associated with multiple petechial conjunctival hemorrhages characterizes AHC. The hemorrhages may become confluent and appear post-traumatic. In approximately 1 out of every 10 000 cases due to enterovirus type 70, a polio-like paralysis can ensue. Neurologic deficits are permanent in up to one-third of the affected individuals.

Human immunodeficiency virus (HIV) is the etiologic agent of the acquired immune deficiency syndrome (AIDS). Although HIV can be cultured from the retinas of individuals with AIDS, and has been shown to be present in the donated corneas of deceased AIDS patients, a direct relationship between local viral infection and ocular disease remains to be established. One example is the dry eye so common in AIDS patients. It is not known whether primary HIV infection of the lacrimal gland, immune deficit-induced potentiation of another virus such as Epstein–Barr virus within the lacrimal gland, or a putative HIV-induced neuro-immune-endocrine defect can account for AIDS-related dry eye. However, the severe immunosuppression of AIDS results in a host of other ocular diseases (discussed below).

Ocular Disease Caused by DNA Viruses

DNA viruses (**Table 2**) are responsible for most significant ocular viral infections in the industrialized world. Even the protean ocular manifestations of the HIV, an RNA virus, result largely from reduced immunity to DNA viruses.

Adenovirus is probably the most common DNA virus to cause eye disease. Three common ocular syndromes have been identified. Simple follicular conjunctivitis occurs with infection by many adenovirus types and may be subclinical. Pharyngoconjunctival fever typically follows infection with adenovirus types 3, 4, and 7. As the name implies, patients have pharyngitis, conjunctivitis, and fever, and may be misdiagnosed as having influenza. Epidemic keratoconjunctivitis, most often caused by adenovirus types 8, 19, and 37, is a highly contagious syndrome with significant morbidity. The conjunctivitis can be severe (**Figure 3**); associated inflammatory conjunctival membranes can permanently scar the eyelids to the globe. Corneal involvement begins as a punctate epithelial keratitis and may proceed to a large central epithelial ulcer. Stromal keratitis presents about 2 weeks after the conjunctivitis as multifocal subepithelial corneal infiltrates, and causes both foreign body sensation and reduced vision. The stromal infiltrates may resolve spontaneously, but can become chronic, require long-term treatment with corticosteroids, and cause persistent visual morbidity. A fourth ocular syndrome occasionally associated with adenovirus infection, AHC (discussed above), may be caused by adenovirus type 11. Interestingly, adenovirus type 11 also causes acute hemorrhagic cystitis. Follicular conjunctivitis (clinically indistinguishable from adenovirus conjunctivitis) can also be caused by Newcastle disease virus, an RNA virus that gives rise to fatal epidemics in poultry and infects the birds' human handlers.

The human herpes viruses are preeminent among DNA viruses in eye disease with at least seven of the eight known human herpes viruses associated with ocular disorders. Herpes simplex virus type 1 (HSV-1) is the most common herpes virus to cause eye disease, and herpes simplex keratitis is the most common cause of infectious blindness in the industrialized world. HSV-1 causes self-limited and relatively benign infections of the

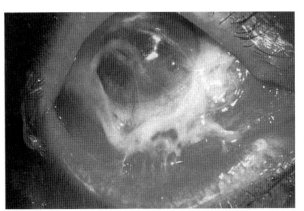

Figure 3 Epidemic keratoconjunctivitis. Infection with adenovirus serotype 19 has resulted in severe ocular surface inflammation.

eyelids, the conjunctiva, and the corneal epithelium, but infections of the corneal stroma, uvea, and retina may result in chronic or recurrent blinding stromal keratitis, uveitis, and retinal necrosis, respectively. Elevation of intraocular pressure due to involvement of the trabecular meshwork is not uncommon and may help to differentiate herpetic uveitis from noninfectious causes. Postnatally acquired HSV-2 ocular infection, less common than HSV-1, causes disease similar in most respects to HSV-1. Neonatal herpes simplex infection, acquired during transit through the birth canal and usually due to HSV-2, commonly causes vesicular blepharitis and conjunctivitis, but can also cause permanent visual loss due to keratitis, chorioretinitis, optic neuritis, and encephalitis of the visual cortex.

Varicella zoster virus, the etiologic agent of chickenpox and shingles, rarely causes keratouveitis with primary infection (chickenpox). However, vision-threatening keratitis, uveitis, and, less commonly, retinal necrosis are complications of varicella zoster virus reactivation in the distribution of the fifth cranial nerve (zoster ophthalmicus). Lid ulceration with frank tissue loss or lid malposition leads to corneal exposure and ulceration. Optic neuritis and cranial nerve paresis can accompany onset of the zoster rash. Sectoral iris atrophy is pathognomonic for zoster ophthalmicus. Postinfectious corneal anesthesia and secondary sterile corneal ulceration may follow herpes simplex types 1 and 2, but are most severe in zoster ophthalmicus. Chronic scleritis, keratitis, uveitis, and glaucoma may ultimately limit the visual acuity.

Acute systemic infection with Epstein–Barr virus may cause conjunctivitis and epithelial keratitis. Stromal keratitis occurs but is difficult to differentiate clinically from herpes simplex keratitis, and the true incidence of Epstein–Barr viral keratitis is unknown. Reports of uveitis and retinochoroiditis are unconfirmed. Delayed-onset optic neuritis following infectious mononucleosis is not uncommon.

Human cytomegalovirus (CMV) typically causes infectious retinitis (**Figure 4**) in immunocompromised patients with $CD4^+$ T-cell counts of less than $50\,cells\,ml^{-1}$. Although not the most common ocular complication of AIDS, CMV retinitis is the most common cause of blindness in AIDS patients. CMV retinitis in AIDS patients can be controlled but not cured. In contrast, congenital CMV infection in an otherwise normal fetus results in various degrees of retinochoroiditis, but is not progressive postnatally.

Human papillomavirus (HPV) causes a range of conjunctival tumors ranging from venereally acquired benign papillomas (HPV types 6 and 11) to invasive squamous cell carcinoma (**Figure 2**) (HPV types 16 and 18). Venereal papillomas are clinically similar to those of the larynx and anogenital tract. Ocular surface squamous neoplasia (conjunctival intraepithelial neoplasia and invasive squamous cell carcinoma) are most similar to dysplastic intraepithelial and invasive squamous lesions of the uterine cervix. Papillomatous eyelid neoplasms due to HPV also occur, and can be benign or malignant.

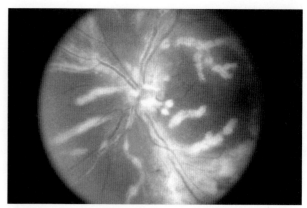

Figure 4 Cytomegalovirus retinitis. Discrete areas of perivascular necrosis and hemorrhage are typical.

Molluscum contagiosum virus is a poxvirus that may infect the eyelid skin or less commonly the conjunctiva. Skin lesions typically appear as elevated nodules with umbilicated centers, and may be multiple and quite large in HIV-infected patients. Molluscum lesions of the eyelid are fairly common in children, and can be associated with a follicular conjunctivitis that resolves with incisional or excisional biopsy of the lid lesion.

Prior to eradication, smallpox virus infection was associated with pustular blepharoconjunctivitis, secondary lid scarring, and stromal keratitis. In nonindustrialized nations, secondary bacterial infection of smallpox keratitis was a major source of blindness. Vaccination against smallpox virus with vaccinia virus was occasionally complicated by inadvertent autoinoculation of vaccinia into the eye, with potential for a severe blepharoconjunctivitis, keratitis, and globe perforation.

Ocular Complications of AIDS

Tay-Kearney and Jabs (1996) classified the ocular complications of HIV infection into five broad categories: (1) HIV retinopathy, (2) opportunistic ocular infections, (3) ocular adnexal neoplasms, (4) neuro-ophthalmic lesions, and (5) drug-induced manifestations.

HIV retinopathy is seen in over half of AIDS patients; cotton wool patches, or multifocal infarcts of the retinal nerve fiber layer, are the most common ocular sign of AIDS. Intraretinal hemorrhages occur less often. HIV can be cultured from the retina of AIDS patients, but a direct relationship between retinal infection and AIDS retinopathy has not been established.

Some ocular infections, including CMV retinitis (**Figure 4**), *Pneumocystis carinii*, fungal, and mycobacterial choroiditis, and microsporidial keratoconjunctivitis are seen almost exclusively in AIDS. CMV retinitis is a major cause of morbidity in AIDS patients. Other infections, such as toxoplasmosis retinochoroiditis, ocular syphilis, herpes zoster ophthalmicus, and molluscum contagiosum of the eyelids are seen in immunocompetent as well as immunosuppressed individuals, but may be more severe and leave more profound deficits in HIV-infected patients. Herpes zoster ophthalmicus in young patients may be the first clinical clue to HIV infection. Acute retinal necrosis due to herpes simplex virus types 1 or 2, or varicella zoster virus, occurs more commonly in HIV-infected than in otherwise normal patients and can result in unilateral or bilateral blindness despite antiviral therapy.

Kaposi sarcoma of the eyelids or conjunctiva, associated with human herpes virus 8 infection, is exceedingly uncommon in immunocompetent individuals, but is probably the most common adnexal tumor in AIDS patients. Non-Hodgkin's lymphomas of the orbit, although rare overall, occur more frequently in AIDS patients than in the general population. Recently, squamous cell carcinoma of the ocular surface (conjunctiva and cornea) has been suggested as a marker for AIDS, but whether HIV infection potentiates HPV-induced carcinogenesis in the eye remains speculative.

Neuro-ophthalmic lesions in AIDS may occur directly due to HIV infection of the central nervous system, but most commonly are caused by cryptococcal meningitis or other opportunistic infections. Retinitis and uveitis due to anti-HIV medications can be confused with opportunistic intraocular infections.

Conclusion

Diverse ocular tissues act in concert to create vision. All of the tissues and structures within the eye are susceptible to viral infection, with consequences ranging from mild discomfort to severe pain and blindness, and almost all known human viruses cause ocular disease. Often, the same virus can infect widely disparate tissues within an eye. Classical viral pathogenic mechanisms are readily demonstrated in the eye, but the fine functions of ocular tissues within the visual axis (cornea, anterior chamber, lens, vitreous, and macula) compel altered immune responsiveness. The eye is uniquely affected by viral infection and provides an exceptional model for studies of viral pathogenesis and immunity.

Acknowledgments

This work is supported by US Public Health Service grants, R01 EY13124 and P30 EY12190, and a Physician-Scientist Merit Award (to J. Chodosh) from Research to Prevent Blindness, New York, NY.

See also: Equine Infectious Anemia Virus; Feline Leukemia and Sarcoma Viruses.

Further Reading

Biswas PS and Rouse BT (2005) Early events in HSV keratitis – setting the stage for a blinding disease. *Microbes and Infection* 7: 799–810.

Bonfioli AA and Eller AW (2005) Acute retinal necrosis. *Seminars in Ophthalmology* 20: 155–160.

Brandt CR (2005) The role of viral and host genes in corneal infection with herpes simplex virus type 1. *Experimental Eye Research* 80: 607–621.

Chodosh J and Stroop WG (1998) Introduction to viruses in ocular disease. In: Tasman W and Jaeger EA (eds.) *Duane's Foundations of Clinical Ophthalmology*, pp. 1–10. Philadelphia: Lippincott-Raven.

Cunningham ET, Jr., and Margolis TP (1998) Ocular manifestations of HIV infection. *New England Journal of Medicine* 339: 236–244.

Darrell RW (ed.) (1985) *Viral Diseases of the Eye.* Philadelphia: Lea & Febiger.

Garg S and Jampol LM (2005) Systemic and intraocular manifestations of West Nile virus infection. *Survey of Ophthalmology* 50: 3–13.

Goldberg DE, Smithen LM, Angelilli A,, and Freeman WR (2005) HIV-associated retinopathy in the HAART era. *Retina* 25: 633–649.

Green LK and Pavan-Langston D (2006) Herpes simplex ocular inflammatory disease. *International Ophthalmology Clinics* 46: 27–37.

Liesegang TJ (2004) Herpes zoster virus infection. *Current Opinion in Ophthalmology* 15: 531–536.

Natarajan K, Shepard LA,, and Chodosh J (2002) The use of DNA array technology in studies of ocular viral pathogenesis. *DNA and Cell Biology* 21: 483–490.

Pepose JS, Holland GN and Wilhelmus KR (eds.) (1995) *Ocular Infection & Immunity.* New York: Mosby.

Streilein JW (1996) Ocular immune privilege and the Faustian dilemma. *Investigative Ophthalmology & Visual Science* 37: 1940–1950.

Tay-Kearney ML and Jabs DA (1996) Ophthalmic complications of HIV infection. *Medical Clinics of North America* 80: 1471–1492.

Human Herpesviruses 6 and 7

U A Gompels, University of London, London, UK

© 2008 Elsevier Ltd. All rights reserved.

Introduction

The roseoloviruses human herpesviruses 6 and 7 (HHV-6 and HHV-7) infect almost all babies to give an 'infant fever', sometimes with rash (then termed 'exanthema subitum' or 'roseola infantum'), and persist throughout the host's lifetime. Infections with these viruses are generally regarded as benign and self-limiting, although severe complications have been recorded, with occasional associated fatalities, during some primary as well as secondary, reactivated infections in immunosuppressed (human immunodeficiency virus/acquired immune deficiency syndrome (HIV/AIDS) or transplantation) patients. These viruses infect and can remain latent in leukocyte stem cells and neuronal cell types, with lytic replication in CD4+ T-lymphocytes. Immunomodulatory gene products can aid persistence and contribute to pathogenesis. There are two sides to this. On the one hand, studies on such a widespread, well-adapted virus allow a greater understanding of human immunity to infection and the possible development of new immunotherapies. On the other hand, recent associations of HHV-6 (in particular) with neurological disease, including 'status epilepticus', encephalitis, and multiple sclerosis (MS), have led to reevaluations of the pathogenic potential of the roseoloviruses. To date, there are no licensed vaccines or antiviral treatments available.

History and Classification

The first isolates of HHV-6 were characterized in 1986–88 in the USA (strains GS and Z29) and the UK (strain U1102), followed by Japan (strain HST). Using reagents derived from these laboratory strains, it was found that there are two strain groups, termed variants, HHV-6A (strains GS and U1102) and HHV-6B (strains Z29 and HST). Subsequently, HHV-6B was shown to be more prevalent in these countries, and most of the available data concern this variant. In comparison, HHV-6A appears to be an emergent infection. Complete genome sequences were derived from plasmid clones for HHV-6A (strain U1102; 159 kbp; accession X83413) and HHV-6B (strain Z29; 162 kbp; accession AF157706) in 1995 and 1999, respectively. These laboratory strains have subsequently been denoted as the prototype reference strains. They were isolated from viruses reactivated from adult immunosuppressed HIV/AIDS patients from African countries: HHV-6A strain U1102 from Uganda and HHV-6B strain Z29 from the Democratic Republic of Congo. Other laboratory strains include HHV-6A strain GS, the first report of HHV-6 infection, which was isolated from an adult HIV/AIDS patient in the USA; HHV-6A strain AJ, from an adult HIV/AIDS patient in the UK; and HHV-6B strain HST, from a pediatric patient with 'exanthema subitum' in Japan. HHV-6B strain HST is the only isolate from primary childhood infection to have been sequenced completely (accession AB021506), and partial genome sequences are available for HHV-6A strains GS and AJ.

HHV-7 was isolated in 1990 from an immunocompetent blood donor. Infections with this virus appear less severe, although the virus is as widespread as HHV-6 and persists at higher levels in the blood (possibly lacking 'true latency'), with frequent secretions in saliva as with HHV-6. HHV-7 can reactivate latent HHV-6 infections. Both viruses appear to give rise to 'exanthema subitum', although HHV-6 infection is earlier and predominates. The genomes of two strains of HHV-7 have also been sequenced: JI (145 kbp; U43400) and RK (153 kbp; AF037218).

The species *Human herpesvirus 6* and *Human herpesvirus 7* are classified in genus *Roseolovirus*, subfamily *Betaherpesvirinae*, family *Herpesviridae*, and are related to genus *Cytomegalovirus* in the same subfamily. The human virus that is most closely related to HHV-6 and HHV-7 is therefore human cytomegalovirus (HCMV). The DNA sequences of the HHV-6A and HHV-6B variants are very closely related overall (>95%), with most variation located in repetitive sequences at the ends of the genomes. Hypervariable loci are also present in other regions of the genome mostly near the ends, and include genes encoding the viral chemokine vCCL (U83), immediate early transcriptional regulators (U86 and U90), glycoprotein gQ (U100), and, at the center of the genome, glycoprotein gO (U47). These regions of substantial variation presumably contribute to the different biological properties of the variants. Comparisons of the genomes of HHV-6B strains Z29 and HST show less variation (1–2%).

Geographical Distribution

Serological studies show that HHV-6 and HHV-7 are widespread throughout the world, with over 95% of adults infected. However, the geographical distribution of HHV-6A and HHV-6B is difficult to establish. There is currently no serological assay that can distinguish between the variants, since the genes encoding immunodominant proteins belong largely to the well-conserved group. The variable genes, which are diagnostically attractive, do not encode consistently immunogenic proteins. Thus, current estimates of the distribution of HHV-6A and HHV-6B have been derived using conventional polymerase chain reaction (PCR) of virus DNA or reverse transcription-polymerase chain reaction (RT-PCR) of virus RNA from tissue, blood, saliva, or cerebrospinal fluid (CSF) samples, with the use of additional techniques such as quantitative PCR, restriction endonuclease digestion, and sequencing of variable genes. In analyses of specific patient groups in North America (USA), Europe (UK), and East Asia (Japan), HHV-6B causes most (>97%) of the primary pediatric infections that give rise to HHV-6-associated infant fever with or without the 'exanthema subitum' rash. HHV-6A is detected, but as a minor component in mother–child pairs. In studies of adult bone marrow transplantation (BMT) and solid organ transplantation patients in the USA and the UK, the distribution of the variants seems similar to that in childhood infections, with primarily HHV-6B identified. In contrast, HHV-6A and HHV-6B were shown to be equally prevalent in hospitalized febrile infants in southern Africa (Zambia). Thus, there appear to be geographical differences in the distribution of HHV-6A and HHV-6B.

Biological differences may also be driven by certain variable genes, thus compounding any differences in geographic distribution. For example, in comparison to HHV-6B, HHV-6A infections are more neurotropic and more prevalent in CSF samples and congenital infections, as well as in early infections of infants less than 3 months old. In analyses of adult lung samples in the USA, both HHV-6A and HHV-6B were detected with equal prevalence. Some analyses have detected HHV-6A DNA in sera but not in peripheral blood mononuclear cells (PBMCs). Thus, the biological compartmentalization of the variants *in vivo* may differ, contributing an additional dimension to the geographical distribution of the variants.

Cellular Tropism, Laboratory Culture, and Latency

HHV-6 and HHV-7 are T-lymphotropic and neurotropic viruses. They have both been adapted to grow in CD4+ T-leukemic cell lines: for example, J-JHAN (Jurkat), HSB2, and Molt-3 cells for HHV-6; and SupT-1 cells for HHV-6 and HHV-7. However, higher titers of both HHV-6 and HHV-7 are obtained in activated cord blood lymphocytes or mononuclear cells (CBLs or CBMCs) or in peripheral blood lymphocytes (PBLs) or PBMCs. Screening of these cells is required prior to use, as infection with laboratory strains can result in reactivation of resident latent virus from adult blood and occasionally from cord blood.

CD4+ T-lymphocytes are permissive for replication and virus production, as shown *in vitro* and *in vivo* during viremia from acute infection. Infection results in cell death by necrotic lysis, with associated apoptosis in bystander cells. Lytic replication results in a characteristic cytopathic effect of ballooning cells (cytomegalia) and fused, multinucleated cells (syncytia). Studies *in vitro* show that CD4+, CD8+, and γδ T-lymphocytes can be infected, particularly by HHV-6A, which may have a wider tropism. HHV-7 binds to CD4 on T-lymphocytes, contributing to cellular tropism. HHV-6 binds to CD46, a ubiquitous receptor, and there may also be a specific co-receptor.

Similar to HCMV, and a feature of the betaherpesviruses in general, latency occurs within myeloid subsets or bone marrow progenitor cells. There is also evidence, particularly for HHV-6A, for replication in glial cells, including oligodendrocytes, astrocytes and a microglial cell type with similarities to monocytic/macrophage cells, which may also be a site of persistence in the central nervous system (CNS). HHV-7 circulates at higher levels than HHV-6, and can reactivate HHV-6. Both HHV-6 and HHV-7 can be shed asymptomatically in saliva, and salivary glands may be a site of persistence.

Recent studies indicate that 0.2% or 0.8% of blood donors in Japan or the UK, respectively, exhibit germline integration of HHV-6 genomes (either HHV-6A or HHV-6B). How this relates to latency or pathogenesis is being evaluated, but it can lead to high levels of virus DNA being detected in the blood (in cells and in sera due to cell breakage) in the absence of virus production, although a limited number of genes may be expressed. Germline integration can confound diagnoses by quantitative PCR, since in these individuals high levels of DNA do not correlate with viremia or symptoms. However, gene expression from the integrated genomes may contribute to chronic disease or immunity to HHV-6 infection.

Genome Organization

The roseolovirus genomes consist of a large unique region (U) flanked by a terminal direct repeat (DR).

The variation in size among the genomes is primarily due to differences in the size of DR. DR is itself bounded by smaller repeats (t) that are related to human telomeric repeats, and there is some suggestion that this arrangement may mediate a latent chromosome-like state or facilitate the rare integrations into the human genome that are observed (see above). At the ends of DR are 'pac' sequences for cleavage and packaging of unit-length genomes from replicative head-to-tail concatamers. Other repetitive sequences with known functions include the origin of lytic DNA replication (ORI) in the center of the genome. This combines features of ORIs in betaherpesviruses (e.g., HCMV) and alphaherpesviruses (e.g., herpes simplex virus (HSV)). HHV-6 ORI is relatively large, as in HCMV, and contains binding sites for an origin-binding protein (OBP), as in HSV. HHV-6 and HSV each encode an OBP, whereas HCMV lacks a homolog to this gene.

The protein-coding regions predicted from the genome sequences (see **Figure 1**) are available in the sequence database accessions. The prototypic HHV-6 genome layout from HHV-6A strain U1102 is shown in **Figure 1**, with main differences between HHV-6 and HHV-7 as marked and cited below. Approximately 30% of the genome is

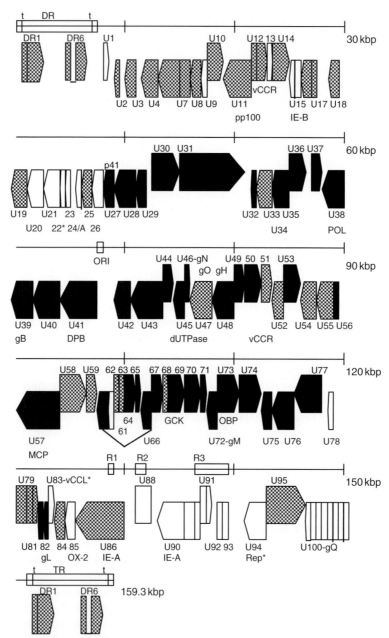

Figure 1 HHV-6 genome organization and locations of conserved genes. Protein-coding regions whose amino acid sequences are conserved among herpesviruses are indicated in black, those conserved only among betaherpesviruses are indicated as patterned, and those conserved in the roseoloviruses are indicated in white. HHV-6-specific genes are U22, U83, and U94 as asterisked, with additional possibly specific genes being U1, U9, U61, and U78. The U prefix is omitted from some gene names to enhance clarity. Repetitive sequences are marked: DR, terminal direct repeat; t, telomeric repeat; ORI, origin of lytic DNA replication; and R1, R2, and R3, locally repetitive sequences. Ancillary information is provided for certain genes mentioned in the text: p101, the major immunodominant tegument phosphoprotein; U94 Rep, the parvovirus Rep homolog and gene/replication regulatory latency gene; IE-A and IE-B, immediate early regulatory genes; POL, DNA polymerase; gB, gH, gL, gO, gM, gN, and gQ, envelope glycoproteins; vCCR, viral chemokine receptors; GCK, ganciclovir kinase; and chemokine vCCL.

taken up with genes encoding proteins that are conserved among mammalian and avian herpesviruses. These proteins dictate common features of replication, and include glycoproteins mediating virus entry by cell fusion, enzymes and accessory factors in the DNA replicase complex, and proteins for making the characteristic herpesvirus icosahedral nucleocapsid and packaging the genome into it. Proteins that are specific to betaherpesviruses are generally transcription factors controlling the gene expression cascade, functions for modulating the specificity of glycoproteins mediating virus entry by cell fusion, or proteins for modifying DNA virus replication. Functions specific to roseoloviruses that affect gene control and, in particular, immunomodulation presumably mediates latency and persistence.

Replication Cycle

Infection is initiated by virus binding to the cell, followed by penetration via virus-mediated cell fusion. As with other herpesviruses, HHV-6 and HHV-7 are first loaded onto the cell via proteoglycan interactions involving heparin or heparan sulfate. The herpesvirus-conserved gB and roseolovirus-specific gQ are involved at this stage. This is followed by membrane fusion mediated by the herpesvirus-conserved gH/gL complex. In HHV-6 and HHV-7, the gH/gL dimer also forms multimers, one with betaherpesvirus-specific gO, and the other with gQ (in two forms, gQ1 and gQ2). So far, no receptor has been identified for the gH/gL/gO complex, but the gH/gL/gQ1/gQ2 complex binds to the ubiquitous cellular receptor CD46 (which is also the measles virus co-receptor). The HHV-6A gH/gL/gQ1/gQ2 complex binds with much higher affinity than that of HHV-6B. HHV-7 additionally binds to the CD4 molecule (like HIV), but the virus attachment protein has not been identified.

Following transport of the nucleocapsid to the nucleus and release of the genome, expression of immediate early genes initiates the transcription cascade. The two spliced IE-A genes (U90 and U86) are positional counterparts of the well-characterized HCMV transcription factors, IE1 and IE2. U90 is particularly variable in sequence between HHV-6A, HHV-6B, and HHV-7. Like the corresponding region in HCMV, this major immediate early gene region is subject to CpG suppression, which is indicative of gene control by DNA methylation. This appears to be a common strategy for betaherpesviruses, which are latent in myeloid cells.

DNA replication is initiated at ORI utilizing a conserved replicase complex modified by the U55 protein, which corresponds to HCMV replication protein UL84, and also by the HHV-6-specific replication modulator, the U94 Rep protein. U94 Rep appears to be a latency-specific factor, and its absence from HHV-7 possibly contributes to the 'leaky latency' observed for this virus. The pac sites at the ends of the genome mediate cleavage of concatamers into unit-length genomes, which are then packaged into preformed nucleocapsids composed of herpesvirus-conserved capsid proteins. The virus buds through the nuclear membrane utilizing herpesvirus-conserved functions, and probably exits via the de-envelopment–re-envelopment pathway involving the *trans*-Golgi network (TGN), as has been shown for alphaherpesviruses.

Immunomodulation

HHV-6 and HHV-7 can have both direct and indirect modulatory effects on the immune system to favor persistence in immune cells. Direct effects include lysis of infected CD4+ lymphocytes during virus replication. In addition, apoptosis of bystander cells has been shown for both HHV-6- and HHV-7-infected cells. Indirect effects include modulation of the receptors on immune cells, thus affecting their function or specificity in immune activation or signaling. HHV-6 downregulates CD3 expression. Whereas HHV-7 downregulates CD4 expression, HHV-6 upregulates it. A protein encoded by both HHV-6 and HHV-7 (U21) downregulates major histocompatibility complex class I expression, which can lead to immune evasion of the antiviral TH1 cytotoxic T-cell response. A number of cytokines are dysregulated by HHV-6 infection, with IL-2 downregulated and IL-10, IL-12, TNFα, and IL-1 upregulated.

Like other primarily blood-borne herpesviruses, roseoloviruses engage the chemokine inflammatory system either to effect immune evasion or enhance virus dissemination. Homologs of chemokine receptors (vCCRs) are encoded by genes U12 and U51, and mimic properties of the human receptors in mediating immune cell traffic either constitutively, in homing to the lymph node, or inducibly, in the inflammatory response to infection. The HHV-6 vCCRs are betachemokine receptors that bind novel combinations of betachemokine ligands, including those with specificity for monocytic/macrophage cell types. Signaling by the U51 vCCR is affected by binding of different ligands and by G-protein levels in the cell. Early in infection when the U51 vCCR is produced, signaling leads to downregulation of the human CCR1-, CCR3-, and CCR5-specific chemokine Rantes/CCL5. In contrast, the HHV-7 vCCRs appear to mimic the constitutive homing chemokine receptors. HHV-6 infection can also mimic the constitutive receptor CCR7, which could allow homing of infected cells to sites of ligand secretion in the lymph node.

The HHV-6 U83 gene encodes a chemokine, vCCL. This gene is hypervariable between HHV-6A and HHV-6B but absent from HHV-7, and thus is a candidate for contributing to differences in virulence. HHV-6A vCCL has a high potency for human CCR1, CCR4,

CCR5, CCR6, and CCR8, which are present on mature monocytic/macrophage and T-cell subsets (TH2, skin homing) and immature dendritic cells. HHV-6B vCCL has a low potency for human CCR2 present on monocytes. These activities could aid immune evasion and virus dissemination, thus contributing to differences in tropism between the variants. They may also aid persistence via the chemoattraction of cell types for efficient antigen presentation and immune control.

In vivo reactivation of HHV-6 in BMT patients can lead to inhibition of outgrowth of cellular lineages. This is supported by *in vitro* data. HHV-6 infection is associated with monocyte dysfunction and suppression of macrophage maturation from bone marrow cultures. Similarly, HHV-6 infection can result in suppression of differentiation and colony formation from hematopoietic progenitor cells. For HHV-7, infection results in inhibition of megakaryocyte survival or differentiation.

Pathogenesis and Disease Associations

Infant Fever with or without Rash

HHV-6 and HHV-7 infect up to 95% of infants. Both HHV-6A and HHV-6B cause primary infant fever, and most infections occur following a drop in maternal immunity after 6 months of age. In the USA, HHV-6B accounts for 97% and HHV-6A for 3% of these infant infections. However, further studies in the USA and the UK show that in less common, earlier infections acquired either congenitally, during delivery, or heritably from germline integration, HHV-6A infections are more frequent, with a prevalence similar to that of HHV-6B. In southern Africa, infant febrile infections detected by PCR analysis of whole blood also indicate equal prevalence. Studies on acquisition of HHV-6B show that infection is symptomatic, with a fever of longer duration than other infections being the most common symptom (3–4 days, 39.4–39.7 °C in various studies). This is a frequent cause of referrals to physicians or hospital. HHV-7 is acquired later, in early childhood, with similar symptoms. As HHV-7 can reactivate HHV-6, the infections can be difficult to distinguish.

The main route of transmission appears to be spread by secretion of reactivated or persistent virus in the saliva from parents or siblings. Congenital infections with HHV-6, but not HHV-7, occur in 1% of births, a prevalence similar to that of HCMV, although HHV-7 has been detected in the placenta. Infections appear as asymptomatic compared to HCMV, although there is one report of neonatal HHV-6B infection associated with neurological disease, seizures, and mental retardation.

A minority of infants have other symptoms or complications, which may be mild or severe. Up to 10% have a skin rash similar to that of measles or rubella, which can confound diagnoses of those infections. This macular or papular rash spreads on face or trunk or both, and has previously been called 'exanthema subitum', 'roseola infantum', or sixth disease. The mean age for acquiring HHV-6 with rash is 7 months. Although this is primarily a benign, self-limiting disease, severe and occasionally life-threatening complications can develop, including lymphadenopathy, diarrhea, myocarditis, myelosuppression, and neurological disease. Indeed, HHV-6 has been reported to account for a third of childhood febrile seizures and direct or indirect mechanisms (via fever) are being evaluated. In the USA, where childhood infections by HHV-6B predominate, links with 'status epilepticus' have been suggested, HHV-6-associated childhood encephalitis cases can occasionally be fatal. The complications previously associated solely with HHV-6 rash are also found in HHV-6 infant fever without rash, and thus up to 90% of symptomatic HHV-6 infections may be undiagnosed or unrecognized.

Immunocompromised People

Solid organ transplantation

Owing to efficient and widespread early childhood transmission of HHV-6 and HHV-7, infections in the adult are primarily reactivations of latent virus, particularly in immunocompromised people such as transplantation patients. Where monitored, co-infections with HCMV present with more severe pathology ('CMV disease'), with a subgroup of delayed graft rejections in liver or kidney transplantation patients. HHV-6-linked encephalitis has occasionally been recorded in these patients, and also in cardiac and heart/lung transplantation patients. Immune suppression using monoclonal antibody to CD3 enhances HHV-6 reactivation and CMV disease; this treatment can also enhance HHV-6 replication *in vitro*. Where symptomatic correlates have been made, HCMV has a greater influence than HHV-6, and HHV-6 has a greater effect than HHV-7. Anecdotal correlates of HHV-6 infection in this context include fever, rash, pneumonitis, encephalitis, hepatitis, and bone marrow suppression. However, it is rare that HCMV reactivates in an HHV-6- or HHV-7-negative patient, as HHV-6 usually reactivates first, then HHV-7 concurrent with HCMV. Thus CMV disease is usually contingent on roseolovirus reactivation. An immunomodulatory role has been presented for HHV-6 reactivations that correlate with increased opportunistic infections, including those associated with HCMV and fungi. Occasional primary infections in the adult have been observed in rare HHV-6- or HHV-7-naive patients who become infected via the donor organ and experience more severe symptoms.

Bone marrow and hematopoietic stem cell transplantation

HHV-6 reactivation is associated with disease in BMT and hematopoietic or cord blood stem cell transplantation (SCT). HHV-6B predominates in Europe and the USA, where typing has been done, reflecting the prevalence of this variant in primary infections of children there. Half of the patients have HHV-6 reactivations 2–4 weeks post-transplantation, with both direct and indirect immunomodulatory effects. Direct effects of reactivation can lead to HHV-6-associated encephalitis and reduced outgrowth of progenitor lineages, called 'bone marrow suppression' or stem cell inhibition. Indirect immunomodulatory effects leading to HCMV reactivation have also been recorded. The presence of virus DNA in the blood has been demonstrated by quantitative PCR, prior to encephalitis or CNS disease developing by 2–3 weeks post-transplantation. The highest levels were shown for cord blood SCT. Retrospective studies have shown that both HHV-6 reactivation and the use of monoclonal antibody to CD3 are independent risk factors for the development of encephalitis. Amnesia, delirium, or 'confusion' following HHV-6 reactivation have been reported, with estimates of 40% mortality for HHV-6-linked encephalitis in SCT patients. Further investigations are required, and should include possible complications in future stem cell therapy; prospective studies are underway to assess neurological involvement in BMT and SCT.

HIV/AIDS

Like HIV, both HHV-6 and HHV-7 can infect and kill or remain latent in CD4+ T-lymphocytes as well as monocytic/macrophage cells. In addition to direct effects via reactivation of HHV-6 and HHV-7, there may also be indirect interactions between these viruses. As the CD4 count in blood decreases with AIDS progression, so does detection of HHV-6 and HHV-7. However, in studies of disseminated reactivated infections, increased levels of HHV-6 DNA in various organs correlate with higher levels of HIV, suggesting interactions during AIDS progression. HHV-6 reactivations have been identified with HIV in lymph nodes, and have been associated with the early phases of lymphadenopathy syndrome. Individual case studies have recorded HHV-6A-associated fatalities in HIV/AIDS patients from pneumonitis and encephalitis. Furthermore, HHV-6-associated retinitis (on its own or in association with HCMV) has been identified in HIV/AIDS patients. However, the use of antiretroviral therapy in some countries has restored immune control of opportunistic infections such as HCMV, and probably HHV-6 and HHV-7.

Studies *in vitro* and in animal models point to possible mechanisms of interaction between HIV and HHV-6. HHV-6A has been shown to reactivate HIV from latency in monocytes *in vitro*. Ongoing studies of HHV-6A co-infection with simian immunodeficiency virus (SIV) in monkey models show enhanced disease. HHV-7 binds to CD4 and can compete with HIV *in vitro* for interaction with this receptor, and so inhibit infection. HHV-6A can induce CD4, thus enhancing infection with HIV in different cell types *in vitro*, whereas HHV-6A chemokine U83A can bind to CCR5, the HIV co-receptor, and inhibit infection. However, the HHV-6A U51 vCCR can bind, sequester, and downregulate the CCR5 ligand and the HIV-1 inhibitor, Rantes/CCL5, and thus enhance HIV-1 infection. Expression of the other HIV-1 co-receptor, CXCR4, is downregulated by HHV-6 infection and may inhibit progression. Thus, complex interactions between HIV-1 and HHV-6 may either inhibit or enhance HIV-1, depending on virus strain, specific gene expression, cell type, and stage of the lytic/latent life cycle.

Neuroinflammatory Disease and Persistence in the Brain

HHV-6 and HHV-7 have been identified as commensals in the brain, with latent or persistent infection in the CNS, and it is a major challenge to distinguish these silent infections from any locally activated ones associated with neurological disease. This problem is further complicated by the rare occurrence of germline integration of HHV-6A or HHV-6B as noted above. In studies of postmortem brain samples, HHV-6 has been identified more frequently than HHV-7 (*c.* 40% compared to 5%). Where genotyped, 75% of the HHV-6 was HHV-6B, reflecting the greater prevalence of this variant in childhood infections in the populations studied (China and the UK).

In the case of encephalitis, most data concern HHV-6, where active infections have been detected by a combination of identifying virus DNA in the CSF and sera and, where biopsies are available, by *in situ* hybridization in order to detect expression of lytic genes in various temporal classes. RT-PCR of virus RNA in blood may also identify specific viremia, but may often miss relevant localized reactivations in the CNS. Encephalitis associated with HHV-6A and HHV-6B has been identified as a rare complication during primary infection in childhood, and in adults. As described above, SCT and BMT patients are at most risk, since frequent HHV-6 infection/reactivation is linked with life-threatening HHV-6-associated encephalitis; further prospective studies are required.

Links with other neurological conditions have been described, including that of HHV-6B with 'status epilepticus' noted above, and prospective studies are underway. Evidence for active lesions associated with mesial temporal lobe specific regions of the brain, particularly the hippocampus, has been presented and evaluated for links with mesial temporal lobe epilepsy. Interestingly, this region of the brain is also associated with memory, and virus reactivation here may correlate with observations of amnestic periods linked with HHV-6 reactivations in some SCT and BMT patients. Neurological symptoms associated particularly with HHV-6A reactivations have also been linked with subsets of chronic fatigue syndrome patients (and in some cases termed HHV-6A encephalopathy), but the data are conflicting and prompt further studies.

Multiple sclerosis

Roseoloviruses have also been linked with a subset of MS patients, in particular the active HHV-6A infections in a subset of the relapsing remitting form of the disease, and in some studies they have been correlated with exacerbations. These HHV-6A infections occur in countries where the most prevalent childhood infections would be HHV-6B, although there are also some data for multiple reactivations along with HHV-6B or HHV-7. Thus, HHV-6A infections linked with MS could represent rare primary infections occurring during adulthood with this more neurotropic variant. Alternatively, an underlying condition may permit infection with multiple strains.

There is some evidence supporting mechanisms that link HHV-6 with MS. In one study, T-cells from MS patients had lower frequencies of activity against the immunodominant HHV-6 p101 (or pp100) protein (U11), suggesting defects in virus clearance. Furthermore, *in situ* hybridization and immunostaining of biopsy material from MS lesions showed the presence of HHV-6 DNA in oligodendrocytes, lymphocytes, and microglia, plus antigen expression in astrocytes and microglia. Studies on an oligodendrocyte cell line showed some productive infection with HHV-6A but abortive infection with HHV-6B; similar results were found using progenitor-derived astrocytes. *In vitro* studies on glial precursor cell infection by HHV-6 suggest that early virus proteins act in cell cycle arrest, thus inhibiting differentiation to the myelin-producing oligodendrocytes essential for repair of demyelination. In a study conducted in the USA, β-interferon treatment of MS patients resulted in lower amounts of detectable HHV-6 in sera. β-interferon also inhibited growth of HHV-6 cell culture. It has been suggested that tests with antiviral agents such as ganciclovir (GCV) will be required to examine the role of HHV-6 in MS. This will require careful selection of patients and use of appropriate controls, particularly if HHV-6 contributes only to a subsection of this disease.

Immune Response, Diagnosis, and Control

Protective Immunity

In immunocompetent people, HHV-6 and HHV-7 infections give rise to lifelong protective immunity. However, subsequent infections with multiple strains, either HHV-6A or HHV-6B, have been recorded even during early primary infection, with milder or no symptoms. Prior infection with HHV-6 does not prevent infection with HHV-7, but it may provide some cross-protective immunity.

Neutralizing antibodies can be generated after primary infection. Targets include gB as well as the gH/gL/gQ complex. As with other infections, not all antibodies are neutralizing, and thus high serological titers do not always correspond to protection against infection.

Cellular immunity also develops in response to these infections. It is usually HHV-6- or HHV-7-specific, but sometimes cross-reactive. HHV-6- or HHV-7-specific CD4+ T- and NK-cell clones have been isolated, as well HHV-6 antigen-specific clones to p101. Thus, efficient cellular immunity develops after infection, and immunomodulation of lymphoid and myeloid cells may contribute to a possible strategy for effective persistence.

Control

Compounds with activity specifically against HHV-6 are not available and require development. However, HHV-6 and HHV-7 share with HCMV genes encoding proteins involved in replication, and these serve as targets for certain antiviral compounds. Thus, many drugs that are active against HCMV and other herpesviruses also show some activity against HHV-6 and HHV-7. These include GCV, foscarnet, and cidofovir. Acyclovir, which is used to treat alphaherpesvirus, and sometimes gammaherpesvirus, infections, is not effective *in vitro* against HHV-6 or HHV-7, as these viruses do not have the thymidine kinase gene required for activity. The oral prodrug form of GCV (valganciclovir, vGCV) has some anecdotal efficacy during virus reactivations in BMT patients, but is not as efficacious as it is with HCMV, and some HHV-6 or HHV-7 antigenemia can still be observed after vGCV prophylaxis. However, as for HCMV, HHV-6 GCV-escape mutants have been observed *in vivo* in transplantation patients, indicating that there is some efficacy against virus replication. Mutations occur in gene U69, which encodes the homolog of the HCMV GCV kinase (GCK), or in gene U38, which encodes the DNA polymerase (POL). *In vitro*, the HHV-6 GCK is tenfold less active than that of HCMV, which perhaps explains the differences in response between these two betaherpesviruses. Prospective clinical trials are required to demonstrate clinical efficacy *in vivo*, and are underway with vGCV for herpesvirus-associated CNS disease.

Acknowledgments

The author is grateful for support from the Biotechnology and Biological Sciences Research Council (BBSRC, UK) and the Bill and Melinda Gates Foundation.

See also: Central Nervous System Viral Diseases; Cytomegaloviruses: Murine and Other Nonprimate Cytomegaloviruses; Cytokines and Chemokines; Defective-Interfering Viruses; Emerging and Reemerging Virus Diseases of Plants; Herpes Simplex Viruses: General Features; Herpesviruses: General Features; Human

Cytomegalovirus: General Features; Human Cytomegalovirus: Molecular Biology; Human Eye Infections; Organ Transplantation, Risks; Viral Receptors.

Further Reading

DeBolle L, Michel D, Mertens T, et al. (2002) Role of the human herpesvirus 6 U69-encoded kinase in the phosphorylation of ganciclovir. *Molecular Pharmacology* 62: 714–721.

Dewin DR, Catusse J, and Gompels UA (2006) Identification and characterization of U83A viral chemokine, a broad and potent beta-chemokine agonist for human CCRs with unique selectivity and inhibition by spliced isoform. *Journal of Immunology* 176: 544–556.

Dietrich J, Blumberg BM, Roshal M, et al. (2004) Infection with an endemic human herpesvirus disrupts critical glial precursor cell properties. *Journal of Neuroscience* 24: 4875–4883.

Dominguez G, Dambaugh TR, Stamey FR, Dewhurst S, Inoue N, and Pellett PE (1999) Human herpesvirus 6B genome sequence: Coding content and comparison with human herpesvirus 6A. *Journal of Virology* 73: 8040–8052.

Fotheringham J, Akhyani N, Vortmeyer A, et al. (2007) Detection of active human herpesvirus-6 infection in the brain: Correlation with polymerase chain reaction detection in cerebrospinal fluid. *Journal of Infectious Diseases* 195: 450–454.

Gompels UA (2004) Roseoloviruses: Human herpesviruses 6 and 7. In: Zuckerman AJ, Banatvala JE, Pattison JR, Griffiths PD and Schoub BD (eds.) *Principles and Practice of Clinical Virology*, 5th edn., pp. 147–168. Chichester, UK: Wiley.

Gompels UA and Kasolo FC (2006) HHV-6 genome: Similar and different. In: Krueger G and Ablashi D (eds.) *Perspectives in Medical Virology, Vol. 12: Human Herpesvirus-6*, 2nd edn., pp. 23–46. Elsevier: New York.

Gompels UA, Nicholas J, Lawrence G, et al. (1995) The DNA sequence of human herpesvirus-6: Structure, coding content, and genome evolution. *Virology* 209: 29–51.

Goodman AD, Mock DJ, Powers JM, Baker JV, and Blumberg BM (2003) Human herpesvirus 6 genome and antigen in acute multiple sclerosis lesions. *Journal of Infectious Diseases* 187: 1365–1376.

Hall CB, Caserta MT, Schnabel KC, et al. (2006) Characteristics and acquisition of human herpesvirus (HHV)-7 infections in relation to infection with HHV-6. *Journal of Infectious Diseases* 193: 1063–1069.

Hall CB, Long CE, Schnabel KC, et al. (1994) Human herpesvirus-6 infection in children: A prospective study of complications and reactivation. *New England Journal of Medicine* 331: 432–438.

Isegawa Y, Mukai T, Nakano K, et al. (1999) Comparison of the complete DNA sequences of human herpesvirus 6 variants A and B. *Journal of Virology* 73: 8053–8063.

Ward KN, Leong HN, Thiruchelvam AD, Atkinson CE, and Clark DA (2007) Human herpesvirus 6 DNA levels in cerebrospinal fluid due to primary infection differ from those due to chromosomal viral integration and have implications for diagnosis of encephalitis. *Journal of Clinical Microbiology* 45: 1298–1304.

Zerr DM, Corey L, Kim HW, Huang ML, Nguy L, and Boeckh M (2005) Clinical outcomes of human herpesvirus 6 reactivation after hematopoietic stem cell transplantation. *Clinical Infectious Diseases* 40: 932–940.

Zerr DM, Meier AS, Selke SS, et al. (2005) A population-based study of primary human herpesvirus 6 infection. *New England Journal of Medicine* 352: 768–776.

Human Immunodeficiency Viruses: Antiretroviral Agents

A W Neuman and D C Liotta, Emory University, Atlanta, GA, USA

© 2008 Elsevier Ltd. All rights reserved.

Glossary

Host cell A cell that has been infected by a virus.
Provirus A virus that has integrated itself into the DNA of a host cell.
Viral load A measure of the severity of a viral infection which is reported in nucleic acid copies per milliliter of blood.
Viral tropism The specificity of a virus for a particular host tissue.
Virion A single virus particle.

Introduction

In 2006, there were an estimated 39.5 million people worldwide living with human immunodeficiency virus (HIV). An estimated 4.3 million new infections arose in 2006, with 65% of those in sub-Saharan Africa. Acquired immune deficiency virus (AIDS)-related infections accounted for about 2.9 million deaths in 2006 and over 25 million deaths since the syndrome was first recognized in 1981. AIDS is now recognized as one of the most destructive pandemics in recorded history, and with no effective vaccine or cure, treatment of the disease relies on antiretroviral therapy.

The life cycle of HIV begins with the attachment of the virion to the target cell via CD4 binding followed by binding to co-receptor, CCR5 or CXCR4. The viral lipid membrane then fuses to the host cell membrane, allowing the viral core, encapsidating HIV RNA and enzymes, to enter the host cell cytoplasm. HIV RNA subsequently undergoes reverse transcription into DNA, a process mediated by reverse transcriptase. The viral DNA enters the host cell nucleus and is integrated into the cellular DNA by the enzyme integrase. At this point, the provirus may remain latent for up to several years. When the host cell receives a signal to become active, the proviral DNA is transcribed into mRNA. The mRNA is transported out of the nucleus and is translated into polyproteins. The polyproteins assemble to form an immature virus particle, which then buds out from the host cell. The enzyme protease cleaves these polyproteins into functional proteins, creating a mature virus.

Currently, 25 antiretroviral drugs are available, targeting three steps of the HIV life cycle: a fusion inhibitor, reverse transcriptase (RT) inhibitors (including two classes, nucleoside reverse transcriptase inhibitors (NRTIs) and non-nucleoside reverse transcriptase inhibitors (NNRTIs)), and protease inhibitors (PIs). Additional agents are being developed to target other steps in the viral life cycle. See **Table 1** for a list of antiretroviral drugs in development.

Most HIV patients are treated with highly active antiretroviral therapy (HAART), usually a combination of three antiretroviral agents. This treatment paradigm can often reduce a patient's viral load to below detectable levels for prolonged periods. The use of HAART over the past decade has led to significant declines in HIV-associated morbidity and mortality.

Entry Inhibitors

HIV entry into target cells is mediated by the envelope protein (Env), which is comprised of two glycoprotein subunits, the surface protein gp120 and transmembrane subunit gp41. Three distinct steps are required for the entry of HIV into host cells. First, the virus attaches to the cell surface via CD4 binding, changing the conformation of Env and exposing a co-receptor binding site on gp120, allowing the viral glycoprotein to bind to one of the co-receptors, CCR5 or CXCR4. The three gp41 glycoproteins in the trimeric envelope complex subsequently form a 6-helix bundle from interaction of the heptad repeat (HR-) 1 and HR2 domains present in the ectodomain of each subunit. The 6-helix bundle brings the viral and cellular membranes into close apposition and mediates their fusion. These three steps – attachment, co-receptor binding, and fusion – can be targeted by attachment inhibitors (including CD4 binding inhibitors), co-receptor binding inhibitors, and fusion inhibitors, respectively.

HIV entry inhibitors have recently exploded as potential therapeutic agents. The main reason for the great interest in this field is that entry inhibitors offer the opportunity to prevent infection of new target cells.

The CD4 receptor was identified in 1984 as the primary cellular receptor for HIV. Because gp120 is responsible for binding to the CD4 receptor on the surface of the host cell, the CD4–gp120 interaction is an attractive target for HIV therapy.

Several years after the discovery of CD4, it was determined that CD4 alone is insufficient to permit HIV infection. In 1996, CCR5 and CXCR4 were identified as the major co-receptors. It was determined that R5 tropic viruses, those which bind to CCR5, are involved in HIV transmission based on the discovery that CCR5Δ32 homozygotes are highly resistant to HIV infection. These individuals, mostly of northern European origin, have a 32 bp deletion in both copies of their CCR5 gene, which confers a recessive phenotype that is associated with resistance to HIV-1 infection and antibody production.

The CXCR4 and CCR5 co-receptors are appealing targets; however, their implementation will require viral phenotyping for co-receptor tropism. Patients infected with an X4 virus, for example, would need to be treated with a CXCR4-specific inhibitor.

Maraviroc is a small molecule which became the first-in-class CCR5 antagonist and the first oral entry inhibitor with its FDA approval in August 2007 (**Table 2**). Maraviroc monotherapy including twice-daily doses of 300 mg led to a viral load reduction of 1.6 $\log_{10}$ copies ml^{-1}. In phase III studies, twice as many patients taking maraviroc along with optimized combination therapy as those taking a placebo with combination therapy had an undetectable viral load (less than 40 copies ml^{-1}) after 48 weeks.

Several other attachment and co-receptor binding inhibitors are currently in development, and biological data have been reported for the drugs in more advanced stages. TNX-355 is a monoclonal antibody which inhibits the interaction between gp120 and CD4. Intravenous infusion of 10 mg kg^{-1} led to a viral load reduction of 1.33 $\log_{10}$ copies ml^{-1}. Vicriviroc and INCB9471 are small-molecule antagonists of the CCR5 receptor. Twice-daily dosing of 50 mg of vicriviroc resulted in a viral load reduction of 1.62 $\log_{10}$ copies ml^{-1}. INCB9471 has recently been reported to show a 1.9 $\log_{10}$ copies ml^{-1} viral load reduction with 14 days of monotherapy. Additionally, the viral load continued to drop for several days after discontinuation of therapy.

The field of fusion inhibitors is led by enfuvirtide, a biomimetic 36-amino acid peptide that was approved for use in March 2003 (**Table 2**). Enfuvirtide is based on the sequence of the HR2 region of gp41. When the drug binds to HR1, HR2 is unable to bind and fusion cannot occur.

Enfuvirtide was shown to block virus-mediated cell-to-cell fusion with an IC_{90} of 1.5 ng ml^{-1}. A 15-day monotherapy study, with patients receiving 100 mg twice daily, resulted in an average plasma viral load reduction of 1.9 $\log_{10}$ copies ml^{-1}. The main disadvantage of enfuvirtide is its dosing regimen consisting of twice-daily subcutaneous injections. Furthermore, due to the complexity of its 106-step chemical synthesis, enfuvirtide is the most expensive antiretroviral drug on the market. Because of these obstacles, enfuvirtide is generally prescribed to patients for whom other antiretroviral agents have failed.

Table 1 HIV antiretroviral agents in clinical trials

Class	Name	Structure	Phase of development
Entry inhibitor (gp120-CD4)	BMS-378806		Phase I
Entry inhibitor (gp120-CD4)	TNX-355	Monoclonal antibody	Phase II
Entry inhibitor (CCR5)	Vicriviroc (SCH-D)		Phase II
Entry inhibitor (CCR5)	INCB9471	Structure not reported	Phase II
Entry inhibitor (CCR5)	PRO140	Monoclonal antibody	Phase I
Entry inhibitor (CCR5)	HGS004	Monoclonal antibody	Phase I
Entry inhibitor (CCR5)	AK602		Phase I
Entry inhibitor (CCR5)	TAK-652		Phase I
Entry inhibitor (CCR5)	TAK-220		Phase I
Fusion inhibitor	Suc-HSA	Succinylated human serum albumin	Phase I
Nucleoside reverse transcriptase inhibitor (NRTI)	Apricitabine (AVX754)		Phase II completed
NRTI	Racivir ±-FTC		Phase II completed

Continued

Table 1 Continued

Class	Name	Structure	Phase of development
NRTI	Elvucitabine		Phase II
NRTI	Alovudine		Phase II
NRTI	KP1461		Phase I
Non-nucleoside reverse transcriptase inhibitor (NNRTI)	BILR 355 BS		Phase II
NNRTI	Rilpivirine (TMC278)		Phase II
Integrase inhibitor	GS-9137		Phase II
Integrase inhibitor	GSK-364735	Structure not reported	Phase I
Protease inhibitor (PI)	PPL-100		Phase I

Continued

Table 1 Continued

Class	Name	Structure	Phase of development
Maturation inhibitor	Bevirimat (PA-457)		Phase II

Table 2 FDA-approved HIV entry inhibitors

Name and structure	Dosing schedule[a]	Major toxicity[a]	Key mutations conferring resistance[b]
Acetyl-YTSLIHSLIEESQNQQEKN EQELLELDKWASLWNWF-amide Enfuvirtide (T-20, Fuzeon)	One subcutaneous injection of 90 mg b.i.d. into the upper arm, thigh, or abdomen	Injection site reactions (itching swelling, redness, pain or tenderness, hardened skin or bumps)	gp41 single point mutations between positions 36 and 45; gp41 double and triple point mutations between positions 36 and 45; gp41 mutations outside of positions 36–45
Maraviroc (Selzentry)	One 300 mg tablet b.i.d.	Liver toxicity, cough, pyrexia, upper respiratory tract infections, rash, musculoskeletal symptoms, abdominal pain, dizziness	co-receptor tropism Changes in the V3 loop, including A316T, and A319A/S in the CC1/85 strain and a deletion in RU570[c]

[a]The Body: The Complete HIV/AIDS Resource. http://www.thebody.com (accessed October 2007).
[b]Stanford University HIV Drug resistance Database. http://hivdb.stanford.edu (accessed October 2007).
[c]Westby M, Smith-Burchnell C, Mori J, et al. (2007) Reduced maximal inhibition in phenotypic susceptibility assays indicates that viral strains resistant to the CCR5 antagonist maraviroc utilize inhibitor-bound receptor for entry. *Journal of Virology* 81: 2359–2371.

Reverse Transcriptase Inhibitors

Since the identification in 1983 of HIV as the etiological agent of AIDS, HIV RT has been one of the major targets for the development of antiretroviral drugs. HIV RT is a heterodimeric, RNA-dependent DNA polymerase (RdDp) that contains two subunits: the p66 catalytic unit and the p51 structural unit. Two classes of antiretroviral agents inhibit RT. These include NRTIs, which compete with normal 2′-deoxynucleoside triphosphates, and NNRTIs, which are allosteric, noncompetitive inhibitors of RT.

NRTIs are synthetic analogs of natural nucleosides that are incorporated into a nascent viral DNA chain. Because the NRTIs lack a 3′-hydroxyl group, incoming nucleotides are unable to form new phosphodiester linkages, therefore halting DNA replication (chain termination). NRTIs are administered as prodrugs, which must be phosphorylated to the active triphosphate before being incorporated into the DNA by HIV RT. Since these compounds enter the cell via passive diffusion, the prodrugs must be sufficiently lipophilic to penetrate the cell membrane.

Because NRTIs inhibit DNA polymerase, they can also affect normal cells. The drugs show much higher affinity for HIV reverse transcriptase compared to most human DNA polymerases; however, mitochondrial DNA polymerase is significantly inhibited by NRTIs. Various side effects of NRTIs can be attributed to mitochondrial dysfunction, including polyneuropathy, myopathy, cardiomyopathy, pancreatitis, bone-marrow suppression, and lactic acidosis.

Zidovudine, commonly known as AZT, was the first HIV drug approved by the FDA. AZT is a thymidine analog with an azido group replacing the 3′-hydroxyl group.

The azido group not only prevents binding of subsequent nucleosides to the DNA chain but also increases the lipophilicity of AZT, allowing it to easily cross the cell membrane.

Since the approval of AZT in 1987, six other NRTIs have come into the market: didanosine (ddI), zalcitabine (ddC), stavudine (d4T), lamivudine (3TC), abacavir (ABC), and emtricitabine (FTC). Zalcitabine is no longer manufactured as of February 2006. Although each of the NRTIs works by competitive inhibition of RT, they have significantly different dosing, toxicity, and resistance profiles (**Table 3**).

Several more NRTIs are in clinical trials. Apricitabine demonstrated a 1.65 $\log_{10}$ copies ml^{-1} viral load reduction after 10 days of 800 mg b.i.d. monotherapy. Other compounds in early clinical trials have not released efficacy data.

Nucleotide analogs, another type of RT inhibitors, are also available. While several nucleotide analogs are approved to treat various viruses, only tenofovir is approved for the treatment of HIV. Tenofovir is an acyclic nucleoside phosphonate analog that displays increased efficacy owing to its ability to bypass the nucleoside kinase step of activation. Tenofovir and other nucleoside phosphonate analogs also show activity against a broad range of DNA viruses and retroviruses (**Table 3**).

NNRTIs, on the other hand, are noncompetitive inhibitors that bind to a hydrophobic pocket near the polymerase active site. NNRTIs inhibit HIV-1 RT by locking the active catalytic site in an inactive conformation. Despite their structural diversity, most NNRTIs assume a similar butterfly-like structure consisting of two hydrophobic wings connected to a central polar body. The wing portions of the molecules contain significant π-electron systems and can act as π-electron donors to aromatic side chains of RT residues around the NNRTI binding pocket. Although NNRTIs bind at a common site on RT, they differ with regard to the exact amino acids of the binding site with which they interact. Therefore, NNRTIs show differences in their pharmacology and pharmacokinetic profiles, interactions with other drugs, and safety and toxicity profiles.

Four NNRTIs are currently licensed for clinical use: nevirapine, delavirdine, efavirenz, and etravirine. Nevirapine, a dipyridodiazepinone derivative, has been shown to reduce perinatal HIV transmission by 47% compared to treatment with AZT. However, the compound can cause life-threatening liver toxicity. Therefore, it is generally restricted to patients at lower risk for liver failure. Delavirdine is a bisheteroarylpiperazine derivative; it is not recommended as part of a first-line antiretroviral therapy program due to its inconvenient dosing regimen and lower efficacy compared to other NNRTIs. Efavirenz, a dihydrobenzoxazinone derivative, is the most commonly prescribed NNRTI due to its high efficacy, once-daily dosing, and relatively lower toxicity. Etravirine is second-generation NNRTI, a highly flexible diarylpyrimidine derivative with a higher genetic barrier to resistance than its precursors. It is the first NNRTI to demonstrate significant antiviral potency in patients with resistance to other NNRTIs.

Several other NNRTIs are currently in clinical trials and promise higher potency while eliciting fewer drug-resistant HIV strains. BILR 355 BS is a second-generation dipyridodiazepinone derivative that shows antiviral activity against a wide range of recombinant NNRTI-resistant viruses. Rilpivirine, like etravirine, is a diarylpyrimidine derivative. It shows a median viral load reduction of 1.20 $\log_{10}$ copies ml^{-1} regardless of dose.

Integrase Inhibitors

Following reverse transcription, HIV integrase mediates the integration of proviral cDNA by catalyzing two reactions. First, integrase mediates the endonucleolytic hydrolysis of the 3′ ends of the viral DNA. This 3′-processing step generates reactive nucleophilic hydroxyl groups. Integrase remains bound to the viral cDNA as a multimeric complex that bridges both ends of the viral DNA with intracellular particles called preintegration complexes (PICs). In the strand-transfer reaction, integrase mediates the nucleophilic attack of the viral 3′-hydroxyl cDNA across the major groove of the host DNA, resulting in a 5 bp stagger between the integrated viral and host DNA strands. The integration process is completed by DNA gap repair including a series of DNA polymerization, dinucleotide excision, and ligation reactions.

While integrase is essential for retroviral replication, there is no host-cell analog. Because integrase inhibitors do not interfere with normal cellular processes, they are an attractive target for antiretroviral agents. Integrase inhibitors were discovered as early as 1992, but not until 2007 was the first drug approved.

Raltegravir is a first-in-class integrase inhibitor that gained FDA approval in October 2007 (**Table 4**). Because it is the first integrase inhibitor on the market, raltegravir is efficacious even in patients with multidrug-resistant HIV. In phase III trials, raltegravir was found to reduce the HIV viral load to undetectable levels in nearly two-thirds of highly treatment-experienced patients after 16–24 weeks of combination therapy. While the initial approval is only for treatment-experienced patients, raltegravir may also prove to be an option for treatment-naive patients.

Two drugs in clinical trials, MK-0518 and GS-9137, and GSK364735 are shown in **Table 1**. GS-9137 is a dihydroquinoline carboxylic acid derivative which also acts by inhibition of strand transfer. GS-9137 displays a serum-free IC_{50} of 0.2 nM and has activity against viral strains resistant to NRTIs, NNRTIs, and protease inhibitors.

Table 3 FDA-approved reverse transcriptase inhibitors

Name and structure	Dosing schedule[b]	Major toxicity[b,c]	Key mutations conferring resistance[d]	Impact of monotherapy[e]
Zidovudine (AZT, Retrovir)	300 mg tablet b.i.d.	Pancreatitis; lactic acidosis; hepatomegaly; anemia; myopathy; neutropenia	Thymidine analog mutations (TAMs): M41L, D67N/G, K70R, L210W, T215F/Y, K219E/Q/N T215 revertants: T215C/D/E/S/I/V T69 insertion mutations Q151M complex usually in combination with V75I, F77L, F116Y	$1.0 \log_{10}$ copies ml^{-1} decrease in viral load
Didanosine (ddI, Videx)	200 mg tablet b.i.d. or 250 mg powder b.i.d.	Pancreatitis; lactic acidosis; hepatomegaly; peripheral neuropathy; optic neuritis	L74V/I K65R TAMs: M41L, D67N/G, L210W, T215F/Y, K219E/Q/N M184V/I	$0.8 \log_{10}$ copies ml^{-1} decrease in viral load
Zalcitabine (ddC, Hivid[a])	0.75 mg tablet t.i.d.	Pancreatitis; lactic acidosis; hepatomegaly; peripheral neuropathy; mouth and throat ulcers; neutropenia; stomatitis	T69A[f], L74V[g], K65R[h]	Less effective than either ddI or AZT
Stavudine (d4T, Zerit)	40 mg capsule b.i.d.	Pancreatitis; lactic acidosis; hepatomegaly; peripheral neuropathy; lipodystrophy; hyperlipidemia; rapidly progressive ascending neuromuscular weakness	TAMs: M41L, D67N/G, K70R, L210W, T215F/Y, K219E/Q/N T215 revertants: T215C/D/E/S/I/V T69 insertion mutations Q151M complex usually in combination with V75I, F77L, F116Y	$0.8 \log_{10}$ copies ml^{-1} decrease in viral load
Lamivudine (3TC, Epivir)	300 mg tablet daily or 150 mg tablet b.i.d.	Lactic acidosis; hepatomegaly	M184V/I K65R Q151M complex usually in combination with V75I, F77L, F116Y TAMs: M41L, D67N/G, L210W, T215F/Y, K219E/Q/N	Limited monotherapy data available
Abacavir (ABC, Ziagen)	300 mg capsule b.i.d.	Hypersensitivity reaction which can be fatal	K65R L74V/I TAMs: M41L, D67N/G, L210W, T215F/Y, K219E/Q/N T69 insertion mutations	$1.8 \log_{10}$ copies ml^{-1} decrease in viral load
Emtricitabine (FTC, Emtriva)	200 mg tablet q.d.	Lactic acidosis; hepatomegaly; hyperpigmentation	M184V/I K65R Q151M complex usually in combination with V75I, F77L, F116Y TAMs: M41L, D67N/G, L210W, T215F/Y	$2.0 \log_{10}$ copies ml^{-1} decrease in viral load

Continued

Table 3 Continued

Name and structure	Dosing schedule[b]	Major toxicity[b,c]	Key mutations conferring resistance[d]	Impact of monotherapy[e]
Tenofovir (Viread)	300 mg capsule q.d.	Lactic acidosis; hepatomegaly; liver and kidney damage; reduction of bone mineral density	K65R TAMs: M41L, L210W, T215Y T215 revertants: T215C/D/E/S/I/V T69 insertion mutations	1.2 $\log_{10}$ copies ml^{-1} decrease in viral load
Nevirapine (Viramune)	200 mg capsule q.d. for first two weeks, then 200 mg b.i.d.	Skin rash; liver damage	K103N/S Y181C/I G190A/S/E Y188L/H/C	1.5 $\log_{10}$ copies ml^{-1} decrease in viral load[i]
Delavirdine (Rescriptor)	400 mg (two 200 mg tablets) t.i.d.	Skin rash; proteinuria; lipodystrophy	K103N/S Y181C/I P236L G190A/S/E	1.0 $\log_{10}$ copies ml^{-1} decrease in viral load[j]
Efavirenz (Sustiva, Stocrin)	600 mg capsule q.d.	Skin rash; depression and other psychiatric symptoms; high triglyceride levels	K103N/S (± L100I, K101P, P225H, K238T/N) G190S/A/E Y188L/H/C Y191C/I	Limited monotherapy data available
Etravirine (Intelence)	200 mg (two 100 mg tablets) b.i.d.	Moderate to severe skin reactions	Three or more of the 13 specific NNRTI mutations: V90I, A98G, L100I, K101E/P, V106I, V179D/F, Y181C/I/V, G190A/S[k]	2.4 $\log_{10}$ copies ml^{-1} decrease in viral load[k]

[a]Zalcitabine is no longer manufactured as of February 2006.
[b]The Body: The Complete HIV/AIDS Resource. http://www.thebody.com (accessed October 2007).
[c]Painter GR, Almond MR, Mao S, and Liotta DC (2004) Biochemical and mechanistic basis for the activity of nucleoside analogue inhibitors of HIV reverse transcriptase. *Current Topics in Medicinal Chemistry* 4: 1035–1044.
[d]Stanford University HIV Drug Resistance Database. http://hivdb.stanford.edu (accessed October 2007).
[e]Sharma PL, Nurpeisov V, Hernandez-Santiago B, Beltran T, and Schinazi RF (2004) Nucleoside inhibitors of human immunodeficiency virus type 1 reverse transcriptase. *Current Topics in Medicinal Chemistry* 4: 895–919.
[f]Fitzgibbon JE, Howell RM, Haberzettl CA, Sperber SJ, Gocke DJ, and Dubin DT (1992) Human immunodeficiency virus type 1 *pol* gene mutations which cause decreased susceptibility to 2′,3′-dideoxycytidine. *Antimicrobial Agents and Chemotherapy* 36: 153–157.
[g]St. Clair MH, Martin JL, Tudor-Williams G, et al. (1991) Resistance to ddI and sensitivity to AZT induced by a mutation in HIV-1 reverse transcriptase. *Science* 253: 1557–1559.
[h]Gu Z, Gao Q, Fang H, et al. (1994) Identification of a mutation at codon 65 in the IKKK motif of reverse transcriptase that encodes human immunodeficiency virus resistance to 2′,3′-dideoxycytidine and 2′,3′-dideoxy-3′-thiacytidine. *Antimicrobial Agents and Chemotherapy* 38: 275–281.
[i]Lange JMA (2003) Efficacy and durability of nevirapine in antiretroviral drug naïve patients. *Journal of Acquired Immune Deficiency Syndromes* 34: S40–S52.
[j]Para MF, Meehan P, Holden-Wiltse JH, et al. (1999) ACTG 260: a randomized, phase I–II, dose-ranging trial of the anti-human immunodeficiency virus activity of delavirdine monotherapy. *Antimicrobial Agents and Chemotherapy* 43: 1373–1378.
[k]Madruga JV, Cahn P, Grinsztejn B, et al. (2007) Efficacy and safety of TMC125 (etravirine) in treatment-experienced HIV-1-infected patients in DUET-1: 24-Week results from a randomized, double-blind, placebo-controlled trial. *Lancet* 370: 29–38.

Table 4	FDA-approved integrase inhibitors		
Name and structure	Dosing schedule[a]	Major toxicity[a]	Key mutations conferring resistance[b]
Raltegravir (Isentress)	One 400 mg tablet b.i.d.	Elevated levels of creatine phosphokinase (CPK) in muscles; diarrhea; nausea; headache	N155H, Q148K/R/H

[a]The Body: The Complete HIV/AIDS Resource. http://www.thebody.com (accessed January 2008).
[b]Markowitz M, Nguyen B-Y, Gotuzzo E, et al. (2007) Rapid and durable antiretroviral effect of the HIV-1 integrase inhibitor raltegravir as part of combination therapy in treatment-naïve patients with HIV-1 infection. *Journal of Acquired Immune Deficiency Syndromes* 46: 125–133.

A 10-day study in treatment-naive and treatment-experienced patients resulted in mean HIV RNA decreases of 1.91 $\log_{10}$ copies ml^{-1} in patients receiving 400 or 800 mg monotherapy, or 50 mg boosted with 100 mg of ritonavir.

Protease Inhibitors

HIV protease is an aspartyl proteinase responsible for cleaving the Gag and Gag-Pol polyproteins in a late stage of the viral life cycle. Because there is no corresponding aspartyl protease that cleaves the Gag polyprotein in mammalian cells, HIV protease has been a popular target for antiretroviral drug development.

HIV PIs are peptidomimetic products that prevent cleavage of Gag and Gag-Pol protein precursors, preventing virions from maturing and becoming infectious. PIs bind to the protease site with high affinity due to their structural similarity to the tetrahedral intermediate formed during hydrolytic cleavage of a peptide bond in the natural substrate. First-generation PIs have a hydroxyethylene core which acts as a nonhydrolyzable transition state isostere. These compounds bind in the protease active site in a manner that mimics the transition state formed during peptide cleavage.

Although many peptidomimetic inhibitors of HIV protease have been reported since the first one in 1990, researchers have faced challenges regarding the physicochemical and pharmacokinetic properties of PIs. Many PIs exhibit poor bioavailability, a short plasma half-life, poor aqueous solubility, and high protein binding, and often require frequent dosing or high pill burden in order to achieve the necessary drug concentration. The newer PIs are starting to resolve these problems. The currently available PIs are shown in **Table 5**.

Saquinavir was the first HIV PI studied clinically; it was approved by the FDA in 1995. It was discovered as part of a strategy to find transition-state mimetics of the Phe-Pro peptide bond. The original formulation exhibited potent *in vitro* viral activity but very poor bioavailability in humans (3–5%). In 1997, the FDA approved a soft gelatin capsule formation that provided higher plasma levels and bioavailability.

Ritonavir was discovered as a result of efforts to design inhibitors based on the C_2-symmetric structure of HIV protease. It shows better bioavailability than saquinavir and also inhibits cytochrome P450-3A4, a liver enzyme that normally metabolizes PIs. Therefore, ritonavir is administered in combination with other PIs in order to enhance their pharmacokinetic profiles.

Indinavir resulted from a research program of mechanism-based drug design. The lead compounds preceding discovery of indinavir were modified to create smaller, less peptide-like structures with high water solubility and bioavailability. Nelfinavir was also discovered by rational design, which sought to maximize potency while improving on the pharmacokinetics of its predecessors.

An analysis of the molecular weight distribution of marketed drugs revealed that most drugs with acceptable pharmacokinetic profiles have molecular weights under 600 Da. Amprenavir therefore emerged from a program seeking to maintain high potency while reducing inhibitor size. This compound has a sulfonylated secondary amino hydroxyethyl core, with one of the sulfonyl oxygens playing a key structural role in binding to the enzyme. While amprenavir was the first FDA-approved PI for twice-daily dosing, the 1200 mg dose required by its poor aqueous solubility was very inconvenient, and amprenavir was later superseded by its prodrug fosamprenavir.

Lopinavir is a second-generation PI inhibitor based on the structure of ritonavir. It is potent against ritonavir-resistant virus and tenfold more potent than ritonavir in the presence of human serum. Because lopinavir is rapidly metabolized, it must be co-dosed with ritonavir in order to achieve good oral bioavailability.

Atazanavir, an azapeptide PI inhibitor, is the first PI approved for once-daily dosing. It also appears to have a reduced effect on cholesterol and triglyceride levels as compared to other PIs.

Table 5 FDA-approved protease inhibitors

Name and structure	Dosing schedule[a]	Major toxicity[a]	Key mutations conferring resistance[b]	Impact of monotherapy[c]
Saquinavir (Invirase)	Two 500 mg tablets b.i.d., boosted with ritonavir	Lipodistrophy; diabetes; increased bleeding in hemophiliacs; hyperlipidemia	G48V L90M ± G73S/C/T I84V I54V/T/L/ M/A	0.7 $\log_{10}$ copies ml^{-1} decrease in viral load
Ritonavir (Norvir)	Generally used as a boosting agent for other PIs, 100–400 mg q.d. or b.i.d. Maximum approved dose is 600 mg b.i.d.	Lipodystrophy; diabetes; increased bleeding in hemophiliacs; pancreatitis; paresthesias; hyperlipidemia; hepatitis	V82A/T/F/S I84V V32I I54V/T/ L/M[d]	Limited monotherapy data available
Indinavir (Crixivan)	400 mg capsule or 800 mg capsule b.i.d., boosted with ritonavir	Lipodystrophy; diabetes; increased bleeding in hemophiliacs; liver toxicity; kidney stones; jaundice; hyperlipidemia	V82A/T/F/S M46I/LI54V/ T/M/L/A L90M	1.2 $\log_{10}$ copies ml^{-1} decrease in viral load[e]
Nelfinavir (Viracept)	Two 600 mg tablets b.i.d.	Lipodystrophy; diabetes; increased bleeding in hemophiliacs; hyperlipidemia	D30N±N88D/S L90M±M46I/ L I84V V82A/ T/F/S	1.5 $\log_{10}$ copies ml^{-1} decrease in viral load
Amprenavir (Agenerase)	24 50 mg capsules (1200 mg total) b.i.d.	Lipodystrophy; diabetes; increased bleeding in hemophiliacs; skin rash; oral paresthesias; hyperlipidemia		2.0 $\log_{10}$ copies ml^{-1} decrease in viral load[f]
Lopinavir (co-formulation with ritonavir is Kaletra)	Administered with ritonavir in a single tablet. Two 200/50 mg (lopinavir/ ritonavir) tablets b.i.d. or four 200/ 50 mg q.d.	Lipodystrophy; diabetes; increased bleeding in hemophiliacs; liver damage; hyperlipidemia	V82A/T/F/S I54V/L/M/ A/T/SM46I/L I50V	1.9 $\log_{10}$ copies ml^{-1} decrease in viral load[g]

Continued

Table 5 Continued

Name and structure	Dosing schedule[a]	Major toxicity[a]	Key mutations conferring resistance[b]	Impact of monotherapy[c]
Atazanavir (Reyataz)	300 mg capsule or two 200 mg capsules q.d., boosted with ritonavir	Lipodystrophy; diabetes; increased bleeding in hemophiliacs; jaundice	I50L N88S/D V82A/T/F/S I84V	1.6 $\log_{10}$ copies ml^{-1} decrease in viral load[h]
Fosamprenavir (Lexiva)	Two 700 mg tablets q.d. or one 700 mg tablet b.i.d., boosted with ritonavir	Increased levels of cholesterol and triglycerides; lipodystrophy; diabetes; increased bleeding in hemophiliacs; skin rash	I50V I84V I54M/L/V/T/A M46I/L	2.4 $\log_{10}$ copies ml^{-1} decrease in viral load[i]
Tipranavir (Aptivus)	Two 250 mg capsules b.i.d., boosted with ritonavir	Lipodystrophy; diabetes; increased bleeding in hemophiliacs; liver damage; cerebral hemorrhage; hyperlipidemia; skin rash	V82A/T/F/S/M/L I84V/A/C L90M M46I/L/V	1.6 $\log_{10}$ copies ml^{-1} decrease in viral load[j]
Darunavir (Prezista)	Two 300 mg tablets b.i.d., boosted with ritonavir	Lipodystrophy; diabetes; increased bleeding in hemophiliacs; skin rash; hyperlipidemia; hyperglycemia	I50V V82A/T/F/S/M I8V/A/C I47V/A	1.4 $\log_{10}$ copies ml^{-1} decrease in viral load[k]

[a]The Body: The Complete HIV/AIDS Resource. http://www.thebody.com (accessed October 2007).
[b]Stanford University HIV Drug Resistance Database. http://hivdb.stanford.edu (accessed October 2007).
[c]Eron JJ, Jr. (2000) HIV-1 protease inhibitors. *Clinical Infectious Diseases* 30(supplement 2): S160–S170.
[d]De Mendoza C and Soriano V (2004) Resistance to HIV protease inhibitors: Mechanisms and clinical consequences. *Current Drug Metabolism* 5: 321–328.
[e]Gulick RM, Mellors JW, Havlir D, et al. (1997) Treatment with indinavir, zidovudine, and lamivudine in adults with human immunodeficiency virus infection and prior antiretroviral therapy. *New England Journal of Medicine* 337: 734–739.
[f]Adkins JC and Faulds D (1998) Amprenavir. *Drugs* 55: 837–842.
[g]Oldfield V and Plosker GL (2006) Lopinavir/ritonavir: A review of its use in the management of HIV infection. *Drugs* 66: 1275–1299.
[h]Barreiro P, Rendón A, Rodríguez-Nóvoa S, and Soriano V (2005) Atazanavir: The advent of a new generation of more convenient protease inhibitors. *HIV clinical trials* 6: 50–61.
[i]Chapman TM, Plosker GL, and Perry CM (2004) Fosamprenavir: A review of its use in the management of antiretroviral therapy-naïve patients with HIV infection. *Drugs* 64: 2101–2124.
[j]Dong BJ and Cocohoba JM (2006) Tipranavir: A protease inhibitor for HIV salvage therapy. *Annals of Pharmacotherapy* 40: 1311–1321.
[k]Arastéh K, Clumeck N, Pozniak A, et al. (2005) TMC114/ritonavir substitution for protease inhibitor(s) in a non-suppressive antiretroviral regimen: A 14-day proof-of-principle trial. *AIDS* 19: 943–947.

Because of the poor aqueous solubility of amprenavir, a more suitable analog was sought. Fosamprenavir, a phosphate ester prodrug of amprenavir, has excellent aqueous solubility due to the presence of the phosphate salt. Fosamprenavir is rapidly and extensively converted to amprenavir after oral administration and therefore maintains its positive pharmacokinetic profile.

Tipranavir, a nonpeptidic PI, shows excellent antiviral activity. However, its undesirable toxicity profile, including reports of hepatitis and hepatic failure, generally relegate this drug to salvage therapy for patients with resistance to other PIs.

Darunavir, the most recently approved PI, resulted from a drug discovery program aimed at replacing peptide segments of PIs with nonpeptidic isosteres – in particular, a bis-tetrahydrofuran system. The oxygen atoms in this portion of the molecule appear to form hydrogen bonds with aspartate residues in HIV protease. Because of these interactions, darunavir shows activity against a broad range of drug-resistant HIV strains.

Maturation Inhibitors

Recently, a new class of antiretroviral agents has been identified which blocks HIV-1 replication at a late step in the virus life cycle. PIs prevent the essential proteolytic processing of the Gag and Gag–Pol polyproteins, which leads to the structural maturation of the virus particle and activation of viral enzymes. Maturation inhibitors, on the other hand, act directly on the Gag protein by disrupting the conversion of the HIV capsid precursor p25 to the mature capsid protein p24. This results in the production of immature viral particles that have lost infectivity.

Bevirimat, also known as PA-457, is currently the only maturation inhibitor in clinical trials (**Table 1**). Bevirimat is a betulinic acid derivative, which exhibits a mean IC_{50} of 10.3 nM in assays using patient-derived WT virus isolates. The compound also retains a similar level of activity against a panel of virus isolates resistant to the three classes of drugs targeting the viral RT and PR enzymes. With an average 50% cytotoxicity value of 25 μM, the therapeutic index for bevirimat is *c.* 2500.

Combination Therapy

Antiretroviral agents are rarely used alone in treatment. Not only does monotherapy demonstrate inferior antiviral activity to combination therapy, but it also results in a rapid development of resistance. The most common antiretroviral regimens in treatment-naive patients generally contain two NRTIs along with one NNRTI or a single or ritonavir-boosted PI. Preferred regimens are shown below; patients who do not tolerate these combinations may try one of the alternative regimens. While these combinations are recommended by the Office of AIDS Research Advisory Council, each patient is encouraged to seek the optimal combination for his or her situation.

- Preferred
  - Atripla: efavirenz, tenofovir, emtricitabine
  - Atazanavir + ritonavir, tenofovir, emtricitabine
  - Fosamprenavir + ritonavir, tenofovir, emtricitabine
  - Lopinavir/ritonavir, tenofovir, emtricitabine
  - Efavirenz, zidovudine, lamivudine
  - Atazanavir + ritonavir, zidovudine, lamivudine
  - Fosamprenavir + ritonavir, zidovudine, lamivudine
  - Lopinavir/ritonavir, zidovudine, lamivudine

- Alternative
  - Nevirapine, abacavir, lamivudine
  - Atazanavir (unboosted), abacavir, lamivudine
  - Fosamprenavir (unboosted), abacavir, lamivudine
  - Fosamprenavir + ritonavir, abacavir, lamivudine
  - Lopinavir/ritonavir, abacavir, lamivudine
  - Nevirapine, didanosine, lamivudine
  - Atazanavir (unboosted), didanosine, lamivudine
  - Fosamprenavir (unboosted), didanosine, lamivudine
  - Fosamprenavir + ritonavir, didanosine, lamivudine
  - Lopinavir/ritonavir, didanosine, lamivudine

Conclusion

Significant progress in antiretroviral therapy has been made since AZT first hit the market in 1987. With 22 antiretroviral agents in four classes, patients have more choices than ever. Many of these new drugs also improve quality of life with more convenient dosing schedules and greatly improved tolerability.

Despite the important success of HAART, the evolution of many drug-resistant HIV strains has diminished the efficacy of antiretroviral therapy, and an ever-increasing number of patients have progressed to salvage therapy. An estimated 13% of adults receiving care in the USA exhibit resistance to all three drug classes, and 76% of patients show resistance to one or more drugs. Because so many individuals currently living with HIV infection are highly treatment experienced, there is a strong need for newer and more effective antiretroviral therapies.

See also: Human Immunodeficiency Viruses: Molecular Biology; Human Immunodeficiency Viruses: Origin; Human Immunodeficiency Viruses: Pathogenesis.

Further Reading

Adkins JC and Faulds D (1998) Amprenavir. *Drugs* 55: 837–842.
Arastéh K, Clumeck N, Pozniak A, *et al.* (2005) TMC114/ritonavir substitution for protease inhibitor(s) in a non-suppressive antiretroviral regimen: A 14-day proof-of-principle trial. *AIDS* 19: 943–947.

Barreiro P, Rendón A, Rodríguez-Nóvoa S, and Soriano V (2005) Atazanavir: The advent of a new generation of more convenient protease inhibitors. *HIV Clinical Trials* 6: 50–61.

Castagna A, Biswas P, Beretta A, and Lazzarin A (2005) The appealing story of HIV entry inhibitors: From discovery of biological mechanisms to drug development. *Drugs* 65: 879–904.

Chapman TM, Plosker GL, and Perry CM (2004) Fosamprenavir: A review of its use in the management of antiretroviral therapy-naïve patients with HIV infection. *Drugs* 64: 2101–2124.

Chrusciel RA and Strohbach JW (2004) Non-peptidic HIV protease inhibitors. *Current Topics in Medicinal Chemistry* 4: 1097–1114.

De Clercq E (2004) Antiviral drugs in current clinical use. *Journal of Clinical Virology* 30: 115–133.

De Clercq E (2004) Non-nucleoside reverse transcriptase inhibitors (NNRTIs): Past, present, and future. *Chemistry and Biodiversity* 1: 44–64.

De Mendoza C and Soriano V (2004) Resistance to HIV protease inhibitors: Mechanisms and clinical consequences. *Current Drug Metabolism* 5: 321–328.

Dong BJ and Cocohoba JM (2006) Tipranavir: A protease inhibitor for HIV salvage therapy. *Annals of Pharmacotherapy* 40: 1311–1321.

Eron JJ, Jr. (2000) HIV-1 protease inhibitors. *Clinical Infectious Diseases* 30(supplement 2): S160–S170.

Fitzgibbon JE, Howell RM, Haberzettl CA, Sperber SJ, Gocke DJ, and Dubin DT (1992) Human immunodeficiency virus type 1 *pol* gene mutations which cause decreased susceptibility to 2′,3′-dideoxycytidine. *Antimicrobial Agents and Chemotherapy* 36: 153–157.

Gu Z, Gao Q, Fang H, et al. (1994) Identification of a mutation at codon 65 in the IKKK motif of reverse transcriptase that encodes human immunodeficiency virus resistance to 2′,3′-dideoxycytidine and 2′,3′-dideoxy-3′-thiacytidine. *Antimicrobial Agents and Chemotherapy* 38: 275–281.

Gulick RM, Mellors JW, Havlir D, et al. (1997) Treatment with indinavir, zidovudine, and lamivudine in adults with human immunodeficiency virus infection and prior antiretroviral therapy. *New England Journal of Medicine* 337: 734–739.

Lange JMA (2003) Efficacy and durability of nevirapine in antiretroviral drug naïve patients. *Journal of Acquired Immune Deficiency Syndromes* 34: S40–S52.

Madruga JV, Cahn P, Grinsztejn B, et al. (2007) Efficacy and safety of TMC125 (etravirine) in treatment-experienced HIV-1-infected patients in DUET-1: 24-Week results from a randomized, double-blind, placebo-controlled trial. *Lancet* 370: 29–38.

Markowitz M, Nguyen B-Y, Gotuzzo E, et al. (2007) Rapid and durable antiretroviral effect of the HIV-1 integrase inhibitor raltegravir as part of combination therapy in treatment-naïve patients with HIV-1 infection. *Journal of Acquired Immune Deficiency Syndromes* 46: 125–133.

Meadows DC and Gervay-Hague J (2006) Current developments in HIV chemotherapy. *ChemMedChem* 1: 16–29.

Office of AIDS Research Advisory Council (2006) Guidelines for the use of antiretroviral agents in HIV-1-infected adults and adolescents.

Oldfield V and Plosker GL (2006) Lopinavir/ritonavir: A review of its use in the management of HIV infection. *Drugs* 66: 1275–1299.

Painter GR, Almond MR, Mao S, and Liotta DC (2004) Biochemical and mechanistic basis for the activity of nucleoside analogue inhibitors of HIV reverse transcriptase. *Current Topics in Medicinal Chemistry* 4: 1035–1044.

Para MF, Meehan P, Holden-Wiltse JH, et al. (1999) ACTG 260: A randomized, phase I–II, dose-ranging trial of the anti-human immunodeficiency virus activity of delavirdine monotherapy. *Antimicrobial Agents and Chemotherapy* 43: 1373–1378.

Pereira CF and Paridaen JTML (2004) Anti-HIV drug development – an overview. *Current Pharmaceutical Design* 10: 4005–4037.

Piacenti FJ (2006) An update and review of antiretroviral therapy. *Pharmacotherapy* 26: 1111–1133.

Pommier Y, Johnson AA, and Marchand C (2005) Integrase inhibitors to treat HIV/AIDS. *Nature Reviews Drug Discovery* 4: 236–248.

Randolph JT and DeGoey DA (2004) Peptidomimetic inhibitors of HIV protease. *Current Topics in Medicinal Chemistry* 4: 1079–1095.

Sharma PL, Nurpeisov V, Hernandez-Santiago B, Beltran T, and Schinzai RF (2004) Nucleoside inhibitors of human immunodeficiency virus type 1 reverse transcriptase. *Current Topics in Medicinal Chemistry* 4: 895–919.

St. Clair MH, Martin JL, Tudor-Williams G, et al. (1991) Resistance to ddI and sensitivity to AZT induced by a mutation in HIV-1 reverse transcriptase. *Science* 253: 1557–1559.

Temesgen Z, Warnke D, and Kasten MJ (2006) Current status of antiretroviral therapy. *Expert Opinion in Pharmacotherapy* 7: 1541–1554.

Westby M, Smith-Burchnell C, Mori J, et al. (2007) Reduced maximal inhibition in phenotypic susceptibility assays indicates that viral strains resistant to the CCR5 antagonist maraviroc utilize inhibitor-bound receptor for entry. *Journal of Virology* 81: 2359–2371.

Relevant Websites

http://hivdb.stanford.edu – Stanford University HIV Drug Resistance Database.

http://www.thebody.com – The Body: The Complete HIV/AIDS Resource.

Human Immunodeficiency Viruses: Molecular Biology

J Votteler and U Schubert, Klinikum der Universität Erlangen-Nürnberg, Erlangen, Germany

© 2008 Elsevier Ltd. All rights reserved.

Glossary

Pandemic An outbreak of an infectious disease that spreads across countries, continents, or even worldwide.

Zoonosis Transmission of a pathogen from one species to another (e.g., humans).

Introduction

Since the beginning of the 1980s, approximately 40 million people have been infected with the human immunodeficiency virus (HIV), the causative agent of the acquired immune deficiency syndrome (AIDS). To date, this multisystemic, deadly, and so far incurable disease has caused more than 20 million deaths. AIDS was first described in 1981 in a group of homosexual men suffering from severe opportunistic infections. Two years later, a retrovirus was isolated from lymphocytes of AIDS patients and was later termed HIV. The two types, HIV-1 and -2, together with the simian immunodeficiency viruses (SIVs) found in nonhuman primates, delineate the genus of primate lentiviruses. Viruses of the genus *Lentivirus* preferentially replicate in lymphocytes and differentiated macrophages and often cause long-lasting and mostly incurable chronic diseases. In contrast to other retroviruses that require the breakdown of the nuclear membrane during mitosis for integration of the viral genome into host cell chromosomes, lentiviruses are uniquely capable of infecting nondividing cells, preferentially terminally differentiated macrophages and resting T cells.

Despite anti-retroviral therapy becoming more effective, infections with HIV remain one of the most devastating pandemics that humanity has ever faced. Unfortunately and despite enormous efforts, there is only limited hope that an effective vaccine will be developed in the near future and there are no other mechanisms available to stimulate the natural immunity against HIV. Another major drawback is the fact that the virus continues to persist even during prolonged therapy in latently infected host cells, classified as the viral reservoirs *in vivo*. Current anti-retroviral treatment is based on drugs that target either the viral enzymes, protease (PR) and reverse transcriptase (RT), or the envelope (Env) protein-mediated entry of the virus into the target cell.

Although introduction of highly active anti-retroviral therapy (HAART) in the mid-1990s, which involves the combination of different classes of both, PR and RT inhibitors, has led to a significant reduction in morbidity and mortality, an eradication of the virus from HIV-1-infected individuals has never been achieved. In addition, these antiviral drugs can induce severe adverse effects, particularly when administered in combination and over prolonged medication periods. A drawback to these treatments is that with the high mutation rate of HIV and replication dynamic, drug-resistant mutants are evolving. Cellular genes have much lower mutation rates, and a potential solution to this problem is to target cellular factors, enzymes, or complex mechanisms that are essential for replication of HIV in the host cells.

Genomic Organization and Replication of HIV

The *Retroviridae* belong to the huge group of eukaryotic retro-transposable elements. Retroviruses are distinguished from other viruses by their ability to reverse-transcribe their RNA genomes into DNA intermediates by using the enzyme RT. The provirus DNA genome is then integrated into the host cell chromosomes by action of the viral enzyme integrase (IN). All retroviruses contain at least three major genes encoding the main virion structural components that are each synthesized as three polyproteins that produce either the inner virion interior (Gag, group specific antigen), the viral enzymes (Pol, polymerase), or the glycoproteins of the virion Env. The genomic organization of HIV-1 is outlined in **Figure 1**. In addition to virus structural proteins and enzymes, some retroviruses code for small proteins with regulatory and auxiliary functions. In the case of HIV-1, these proteins comprise the two essential regulatory elements: Tat, which activates transcription, and Rev, which modulates viral RNA transport. In contrast, Nef, Vpr, Vif, and Vpu are nonessential for replication in certain tissue culture cells and are generally referred as accessory proteins (**Figure 1**).

According to the nomenclature defined by the International Committee on Taxonomy of Viruses (ICTV) convention, members of the *Orthoretrovirinae* contain RNA inside the virus particle and are released from the plasma membrane. However, the assembly of orthoretroviruses follows two different strategies: alpha-, gamma-, delta-, and lentiviruses (ICTV nomenclature) assemble on the inner leaflet of the plasma membrane and are released as immature virions. In contrast, betaretroviruses and spumaviruses first assemble in the cytoplasm to form immature particles that are then transported to the plasma (betaretroviruses) or internal (spumaviruses) membrane where virus budding occurs.

As with other retroviruses, the HIV replication cycle begins with virus attachment and penetration through the plasma membrane. The schematic replication cycle of HIV-1 is depicted in **Figure 2**. HIV binds to different cell receptors, among them the differentiation antigen, CD4, acting as primary receptor, as well as different specific chemokine receptors that act as co-receptors after binding to CD4. The first retroviral receptor ever identified was the CD4 T-cell receptor that was established in 1984 as the primary receptor for HIV-1 and -2 and SIV. However, the CD4 molecule alone is not sufficient to allow Env-mediated membrane fusion and virus entry, as certain primary isolates of HIV-1 preferentially replicate in T-cell lines (T-cell-line tropic, 'TCL'-tropic), while others establish productive replication only in macrophage cultures ('M'-tropic). Thus, a second class of HIV receptors was predicted and finally identified as a member of the G protein-coupled seven transmembrane domain receptor superfamily acting as so-called 'co-receptors' during HIV entry. Co-receptors, in general, function as cellular receptors for α- and β-chemokines. Two types of co-receptors arbitrate the disparity in cell-type tropism: the α-chemokine receptor CxCR4 that is typically present in T-cell lines, and the β-chemokine receptor CCR5 that is present on macrophages. It was shown that the genetic heterogeneity in co-receptor alleles determines the vulnerability to HIV infection and disease

Human Immunodeficiency Viruses: Molecular Biology

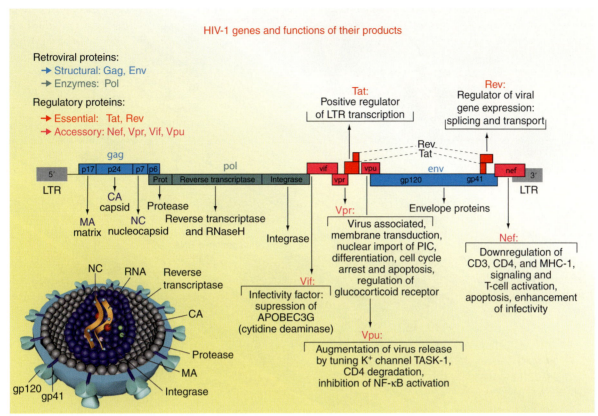

Figure 1 Genomic organization of HIV. This figure illustrates the complex organization of the HIV-1 genome comprising the standard retroviral elements for virus structure, *gag* and *env* (blue), the enzymes *pol* (green), and the six additional genes. These additional genes are either essential as *tat* and *rev*, or accessory as *vpr*, *vpu*, *vif*, and *nef*. The schematic model of the structure of the mature HIV-1 virus particle is given underneath.

progression. A well-characterized example is the identification of a truncated form of CCR5 (termed the 'CCR5/d32' mutation) that in its homozygous form almost completely protects individuals from HIV infection.

Virus entry starts with the so-called 'uncoating' of the virus which is the fusion between the virus and host cell membrane followed by the release of the viral core into the cytoplasm. Following this process, the viral RNA genome is transcribed by the reverse transcription complex (RTC) into double-stranded DNA, which, as part of the preintegration complex (PIC), is transported to the nucleus where the proviral genome is integrated into chromosomes. While there is evidence that the capsid (CA) molecules of the incoming virus are degraded via the ubiquitin proteasome system (UPS), the three virus proteins IN, matrix (MA), and accessory Vpr molecules together with the provirus genome form the PIC. Although the PIC components comprise various nuclear localization signals (NLSs), it is assumed that additional cellular and viral factors, including a central DNA flap that is formed during reverse transcription, facilitate the import of the PIC into the nucleus. It is worth noting that virus entry involves a number of consecutive and highly organized multistep events that might explain why approximately only 1 out of a 1000 mature HIV-1 particles are capable of establishing a productive infection.

After reverse transcription of the viral genome and nuclear import of the PIC, the proviral DNA is integrated into chromosomal DNA. Activation of HIV-1 long terminal repeat (LTR) promoter-driven retroviral gene expression, followed by export, transport, and splicing of the viral mRNA, the so-called late steps of the HIV-1 replication cycle are initiated at this point. These steps involve membrane targeting of Gag and Gag–Pol polyproteins, assembly, as well as budding and maturation of progeny virions. Upon activation of integrated provirus, viral mRNAs are processed and transported into the cytosol for translation of newly synthesized structural proteins that assemble at the plasma membrane into budding particles. The Gag polyprotein in HIV-1 and HIV-2 consists of different functional domains that mediate the recognition and binding of viral RNA, the membrane targeting of Gag and Gag–Pol polyproteins, virion assembly, and efficient particle release from the plasma membrane as the final step of virus budding. In general, the Gag polyprotein constitutes viral components that are both sufficient

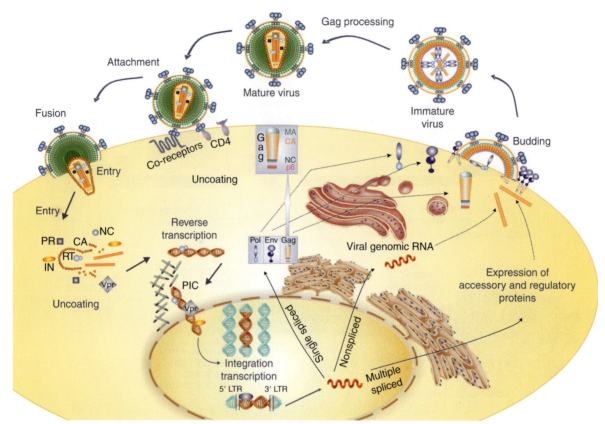

Figure 2 Replication cycle of HIV. Schematic overview of the different stages of the HIV-1 replication cycle.

and strictly required for virus particle assembly and budding, although further viral components such as the genomic RNA, the envelope, and the viral enzymes are required for production of infectious virions. The processing of the HIV-1 Gag polyprotein Pr55 by the viral PR generates the MA, CA, nucleocapsid (NC), as well as p6 proteins and the two spacer peptides, p2 and p1. The HIV-1 *pol*-encoded enzymes, PR, RT, and IN, are proteolytically released from a large polyprotein precursor, Pr160, the expression of which results from a rare frameshift event occurring during translation of Pr55. Like the Gag proteins, the Env glycoproteins are also synthesized from a polyprotein precursor protein. The resulting surface (SU) gp120 and transmembrane (TM) gp41 glycoproteins are produced by the cellular *protease* during trafficking of gp160 to the cell membrane. While gp120 contains the domains that mediate virus binding to CD4 and co-receptors, the gp41 anchors the trimetric complexes of TM/SU in the virus membrane and includes the determinants that regulate fusion between cellular and virus membranes during virus entry. The N-terminus of the extracellular ectodomain of gp41 harbors the so-called 'fusion peptide' that consists of a hydrophobic domain in concert with two helical motifs and regulates membrane fusion by formation of a six-helix bundle. The highly polymorphic SU protein gp120 is organized into five conserved domains (C1 to C5), and five highly variable domains that in most of the known SU sequences are concentrated near loop structures that are stabilized by disulfide bond formation.

In general, Gag proteins of different retroviruses exhibit a certain structural and functional similarity: MA mediates the plasma membrane targeting of the Gag polyprotein and lines the inner leaflet of the lipid bilayer of the mature virus particle, CA regulates assembly of Gag and forms the core shell of the infectious virus, and NC regulates packaging and condensation of the viral genome. In addition to these canonical mature retrovirus proteins, other Gag domains have been described, such as the HIV-1 p6 region that directs the incorporation of the regulatory protein Vpr into budding virions and governs efficient virus budding.

HIV particles bud from the plasma membrane as immature noninfectious viruses consisting predominantly of uncleaved Gag polyproteins. After virus release and in concert with PR activation which is autocatalytically released from the Gag–Pol polyprotein, processing of Gag and Gag–Pol polyproteins into its mature proteins and condensation of the inner core structure occurs that ultimately results in the formation of a mature infectious virus. Besides PR and Env, at least two other viral factors are known to promote efficient virus release: the HIV-1-specific accessory protein Vpu and the p6 domain. While Vpu supports virus release by an ion channel activity, p6 contains at least two distinct late assembly (L) domains that are required for efficient separation of assembled virions from the cell surface by a yet-undefined mechanism that somehow involves the cellular multivesicular body (MVB) as well as the UPS.

Role of Cellular Factors in HIV Replication

From entry to release and maturation into infectious progeny virions, each individual step in the HIV replication cycle exploits cellular pathways. As for most other intracellular parasites, the replication of HIV-1 depends on the interaction with specific host cell factors, and some of these proteins are specifically incorporated into progeny virions. Conversely, replication of HIV-1 is blocked in cells of the infected host by the action of restriction factors that function as barriers to retroviral replication. These factors are part of the so-called 'innate immune system' for which several mechanisms are known to interfere with replication at different stages of the viral life cycle. HIV-1, however, has evolved strategies to undermine these antiviral responses and, as a consequence, successfully propagates in the specific host environment.

Well-characterized examples for host–virus protein interaction are the virus receptors (CD4 and chemokine receptors) which enable virus entry into specific host cells, the role of the chromatin-remodeling system, and the HMG I family proteins for proviral DNA integration, as well as the requirement of different factors of the endosomal protein trafficking and ubiquitination systems for virus release.

An important case of host–virus interaction is the role of the UPS in virus budding. Recent work has provided intriguing insight into the mechanism of how virus budding exploits the cellular machinery that is normally involved in vacuolar lysosomal protein sorting and MVB-biogenesis. In the case of HIV-1, the recruitment of these cellular factors to the virus assembly site is facilitated by the interaction between the primary L-domain of p6 with at least one important host factor, the tumor susceptibility gene product 101 (Tsg101), an E2-type ubiquitin ligase-like protein. The second L-domain at the C-terminus of p6 mediates the binding of Gag to AIP1/ALIX, a class E vacuolar protein sorting factor that also interacts with Tsg101. AIP1/ALIX also binds to late-acting components of the endosomal sorting complexes required for transport (ESCRTs) and is necessary for the formation of MVB at endosomal membranes. Further, in more recent studies, VPS37B was identified as a new component of the ESCRT that binds to Tsg101. There has been accumulating evidence that HIV-1 recruits the components of the MVB system to the budding machinery which follows two separate and cell-type-specific pathways: in T cells HIV-1 buds primarily from the cell surface, while in monocytes/macrophages the virus buds into vacuoles of the MVB system.

An example for the ability of HIV-1 to escape host cell restriction is the recently discovered relationship between the HIV-1 accessory protein Vif (virus infectivity factor), and APOBEC3G (apolipoprotein B mRNA editing enzyme catalytic polypeptide-like 3G). APOBEC3G is a member of the cellular cytidine deaminase DNA/RNA-editing enzyme family that has the unique capability of hypermutating retroviruses, including HIV-1, with terminal consequences. However, the antiviral effect of APOBEC3G does not correlate with the frequency of mutation induced by APOBEC3G, suggesting that cytidine deaminase activity is not the only underlying mechanism of the antiviral activity of APOBEC. Nevertheless, the resulting innate block in virus replication is counteracted by Vif, which in the virus producer cell binds to APOBEC3G, induces its polyubiquitination by an SCF-like E3 ubiquitin ligase (Cullin5-ElonginB+C), and finally initiates proteasomal degradation of APOBEC3G. It is assumed that by this mechanism Vif is precluding the presence of APOBEC3G in progeny virions.

A second family of proteins with anti-retroviral activity are proteins of the tripartite interaction motif (TRIM) family, exemplified by TRIM5α. TRIM proteins contain a series of three motifs comprising a RING-finger (really interesting new gene), a B-box, and a coiled-coil (CC) domain. Only TRIM5α contains an additional SPRY domain which mediates interaction with the viral CA and is mainly responsible for the species-specific restriction activity. It has been shown that TRIM5α blocks incoming retroviruses at an early step of virus replication, occurring sometime after virus entry and before reverse transcription is initiated, and this antiviral activity of TRIM5α is clearly CA dependent. The RING-finger acts as an E3-type ubiquitin ligase, suggesting that ubiquitination of incoming CA molecules leading to proteasomal degradation might be responsible for the antiviral activity. However, the molecular mechanisms employed by TRIM5α to restrict retroviral infection are still poorly understood and a matter of intensive debate.

HIV Regulator Proteins

While retroviruses share the same fundamental replication cycle and have the same basic genomic organization (e.g., the canonical *gag*, *pol*, and *env* genes), they vary in the content of additional small regulatory genes. These proteins, except for Rev and Tat, dispensable for HIV-1 replication in certain cell lines *in vitro*, contribute enormously to pathogenesis and spread of HIV-1 *in vivo*.

The *trans*-Activator of Transcription (Tat) of HIV-1

It is now generally accepted that the HIV *trans*-activator (Tat) plays an important role in the pathogenesis of AIDS. Although originally described as an activator of the HIV-1 LTR promoter, Tat was later shown to regulate reverse transcription, to affect the expression of various cellular and viral genes, and to be released from infected cells. This so-called extracellular Tat, which acts as a cell membrane transducing peptide in the sense of a so-called 'trojan molecule', can affect neighboring cells, that are both uninfected and infected target cells. Indeed, there is accumulating evidence that Tat in its extracellular form plays a major role in AIDS-associated diseases like Karposi's sarcoma and HIV-associated dementia.

Tat can be expressed in two forms, as the 72-amino-acid one-exon Tat and as the 86–101-amino-acid (depending on the HIV-1 isolate) two-exon Tat expressed primarily early during infection. The 14–15 kDa Tat binds to an RNA stem–loop structure forming the Tat-responsive element (TAR) at the 5' LTR region. It activates transcriptional elongation by stimulating the protein kinase TAK (Tat-associated kinase) resulting in hyperphosphorylation of the RNA polymerase II. In general, Tat stimulates the production of full-length HIV transcripts and is, therefore, essential for HIV replication.

The Regulator of Expression of Virion Proteins (Rev)

The compact organization of the HIV-1 genome and expression of all structural and regulatory proteins from a single promoter in the 5'-LTR requires a complex splicing regime of the primary transcript. The fully spliced mRNAs encoding Tat, Rev, and Nef are readily transported from the nucleus to the cytoplasm and, in consequence, these proteins are synthesized early in infection. However, the nuclear export of unspliced or single spliced mRNAs encoding structural proteins Gag and Env, the viral enzymes, as well as the accessory proteins Vif, Vpr, and Vpu, all require the activity of Rev. Hence, these proteins are expressed later during viral infection as the Rev concentration in the infected cell increases.

The Rev protein binds to the viral mRNAs via an arginine-rich RNA-binding motif that additionally serves as NLS required for the transport of Rev from the cytosol to the nucleus. Rev recognizes a stem–loop structure in the viral transcripts known as the Rev response element (RRE) that is located in *env*. After binding to RRE, Rev forms multimeric complexes of up to 12 monomers that are exported from the nucleus by interaction with nuclear export factors. These factors are recruited by a leucine-rich nuclear export signal (NES) located in the C-terminal region of Rev and promote the shuttling of the Rev–RNA complexes to the cytosol. Hence, Rev is required for the synthesis of viral proteins and is therefore essential for HIV replication.

The Lentivirus Protein R (Vpr)

Vpr is a virion-associated, nucleocytoplasmatic shuttling regulatory protein that is encoded by (and conserved among) primate lentiviruses, HIV-1, HIV-2, and the SIVs. Although dispensable for growth of HIV-1 in activated and dividing T cells, Vpr appears to play an important role in virus replication *in vivo*, since deletion of *vpr* and the related *vpx* genes in SIV severely compromises the pathogenic properties in experimentally infected rhesus macaques. Furthermore, HIV-2 and SIV also encode an additional Vpr-related protein, Vpx, that is believed to function synergistically with Vpr. The Vpr of HIV-1 is reported to exhibit numerous biological activities, including nuclear localization (based on the presence of at least two NLS), ion channel formation, transcriptional activation of HIV-1 and heterologous promoters, co-activation of the glucocorticoid receptor, regulation of cell differentiation, induction of apoptosis, cell cycle arrest, and transduction through cell membranes. Although significant amounts of Vpr (approximately 0.15-fold molar ratio to viral core proteins) are packaged into budding HIV-1 particles in a process dependent on Vpr's interaction with the C-terminal p6 domain of the Gag, the biological role(s) of virion-associated Vpr still remains to be fully elucidated.

The highly conserved 96-amino-acid Vpr has received considerable attention, and a number of biological functions have been attributed to its presence in various cellular and extracellular compartments. The most intensively investigated biological functions of Vpr are those affecting the translocation of the PIC of the incoming virus from the cytoplasm to the nucleus, and the arrest in the G_2 phase of the cell cycle. The nuclear targeting function of Vpr has been associated with productive infection of terminally differentiated macrophages by mediating integration of the proviral DNA into the host genome. It is assumed that Vpr causes dynamic disruptions in the nuclear envelope architecture that enables transport of the PIC across the nuclear membrane. Regarding the second function of Vpr, which leads to G_2 cell cycle arrest in HIV-1-infected T cells, it was suggested that this activity provides an intracellular milieu favorable for viral gene expression. Numerous cellular binding partners of

Vpr have been identified, and for some of these a specific role in G_2 arrest has been proposed. Nevertheless, the precise molecular mechanism underlying the Vpr-induced G_2 arrest remains unclear. A potential explanation for the obvious paradigm that Vpr prevents proliferation of infected T cells by arresting them in the G_2 phase was provided by the observation that viral gene expression is optimal in the G_2 phase and that Vpr can increase virus production by delaying cells at this stage of the cell cycle. Interestingly, there are sufficient amounts of Vpr in incoming virus particles to induce G_2 cell cycle arrest even prior to the initiation of *de novo* synthesis of viral proteins. The secondary structures in Vpr emerging from several analyses indicate the presence of an α-helix–turn–α-helix motif at the N-terminus and an extended amphipathic helical region at the C-terminus which might play a key role in self-association and the interaction of Vpr with heterologous proteins, such as p6, NC, Tat of the virus and the adenine nucleotide translocator of the mitochondrial pore.

Other studies suggest that the prolonged G_2 arrest induced by Vpr ultimately leads to apoptosis of the infected cell. Conversely, early anti-apoptotic effects of Vpr have been described which are superseded by its pro-apoptotic effects. These pro-apoptotic effects of Vpr may result from either effects on the integrity of the nuclear envelope or direct mitochondrial membrane permeabilization, perhaps involving Vpr-mediated formation of ion channels in biological membranes.

The HIV-1-Specific Virus Protein U (Vpu)

Vpu is exclusively encoded by HIV-1, with one exception – the HIV-1 related isolate SIV_{cpz} that encodes a Vpu-like protein similar in length and predicted structure to the Vpu from HIV-1 isolates. Vpu of HIV-1 represents an integral membrane phosphoprotein with various biological functions: first, in the endoplasmic reticulum (ER), Vpu induces degradation of CD4 in a process involving the ubiquitin proteasome pathway and casein kinase 2 (Ck-2) phosphorylation of its cytoplasmic tail (Vpu_{CYTO}). Second, Vpu augments virus release from a post-ER compartment by a cation-selective ion channel activity mediated by its transmembrane anchor (Vpu_{TM}). It was shown previously that Vpu can regulate cationic current when inserted into planar lipid bilayers and in *Xenopus* oocytes. Recent results indicate that the virus release function of Vpu_{TM} involves the mutually destructive interaction between Vpu and the K^+ channel TASK-1. Thus, the virus release function of Vpu might be mediated by tuning the activity of host cell ion channels at the cell membrane.

Structurally, Vpu is a class I oligomeric membrane-bound phosphoprotein composed of an amphipathic sequence of 81 amino acids comprising a hydrophobic N-terminal membrane anchor proximal to a polar C-terminal cytoplasmic domain. The latter contains a highly conserved region with the CK-2 phosphorylation sites that regulate Vpu's function in the ER. The current model of the structure and the orientation in the membrane of the N-terminal hydrophobic Vpu_{TM} provides evidence for water-filled five-helix bundles. However, more work is required to understand Vpu in its functional forms *in vivo* where it exists as the phosphoprotein in mulitprotein complexes involving CD4 and β-TrCP, a component of the SkpI, Cullin, F-box protein (SCF^{TrCP}) E3 ubiquitin ligase complex that regulates the ubiquitination and degradation of cellular proteins by the proteasome.

The Viral Infectivity Factor Vif

The 23 kDa phosphoprotein Vif represents a viral infectivity factor that is highly conserved among primate lentiviruses. Vif regulates virus infectivity in a cell-type-dependent fashion. Permissive cells like Hela, Cos, 293T, SupT1, CEM-SS, and Jurkat cells support HIV replication independent of Vif. In contrast, primary lymphocytes, macrophages, and certain cell lines like H9 cells require Vif for production of fully infectious viruses.

Until recently, the mechanism of how Vif supports formation of infectious virions remained mostly enigmatic. Several findings now support the intriguing hypothesis that at least one main function of Vif is to neutralize cellular cytidine deaminase APOBEC3G, which, in the absence of Vif, is encapsidated into progeny virions. It has been a long-known fact that Vif can act *in trans* when it is expressed in the virus-producer cell, but not in the target cell, indicating that Vif must be situated at the virus-assembly site. This model of a post-entry function for Vif is further supported by the observation that Vifs from diverse lentiviruses function in a species-specific mode. Several hypotheses have been put forward to explain the mechanism of Vif function: the activity of Vif might regulate reverse transcription and proviral DNA synthesis, Vif can support the formation of stable virus cores, or Vif might bind to viral RNA in order to support post-entry steps of virus replication. The discovery of the APOBEC3G–Vif interaction provides at least some explanation for these previous observations.

The Multifunctional Nef Protein

Originally, the 27 kDa N-terminal myristoylated membrane-associated phosphoprotein Nef was described as a 'negative factor' that suppresses the HIV LTR promoter activity. As one of the most intensively investigated HIV-1 accessory proteins, Nef has been shown to fulfill multiple functions during the viral replication cycle. In general, it is now widely accepted that inconsistent with its originally 'negative' nomenclature, Nef plays an important stimulatory role in the replication and pathogenesis of HIV-1. The broad spectrum of Nef-associated activities so far described can be summarized in distinct classes: (1) modulation of cell activation and apoptosis; (2) change in intracellular trafficking of cellular proteins, particularly of cell receptor proteins like CD4 and major histocompatibility complex I (MHC-I); and (3) increase of virus infectivity.

Well characterized and one of the earliest observations in the course of studying the biology of Nef is the Nef-induced downregulation of cell surface CD4. This mechanism occurs on a post-translational level by augmentation of receptor internalization, where Nef binds to the adapter protein complex in clathrin-coated pits, followed by lysosomal degradation. Another influence on immune receptors is the Nef-mediated block in the transport of MHC-I antigen complexes to the cell surface, leading to the disturbance in the recognition of HIV-1 infected cells by cytotoxic T_{CD8}^+ lymphocytes (CTLs). More recent studies also described downregulation of the co-stimulatory protein CD28 induced by Nef. It appears that the mechanisms by which Nef blocks cell surface expression of CD4, MHC-I, and CD28 are independent and can be genetically separated.

Nef is expressed relatively early in the HIV replication cycle, and there are reports that Nef is even incorporated into budding virions where it can be cleaved by the viral protease. Nef is present during almost all steps of the viral life cycle, which might explain the multiple functions deployed by this accessory protein. Major attention has been focused toward the Nef-mediated modulation of the T-cell signal transduction and activation pathways. Several lines of evidence supported the hypothesis that, by affecting T-cell receptors, Nef may support several aspects of the virus replication cycle, including regulation of apoptosis and immune evasion. Earlier studies in transgenic mice indicated Nef-dependent activation in T-cell signaling. Later studies indicate that Nef imitates T-cell receptor signaling motifs. The Nef-induced T-cell activation requires the tyrosine kinase Zap70 and the ζ-chain of the T-cell receptor. Further, Nef was shown to activate T cells by binding to the Nef-associated kinase, also known as the p21-activated kinase 2 (PAK2). In addition to the activation of T cells, Nef also induces cell death, since it was shown that expression of the death receptor Fas ligand (FasL) is activated by Nef, and that this upregulation might increase killing of lymphocytes attacking HIV infected cells.

A remarkable turn to our understanding of Nef function was offered recently when it was demonstrated that Nef proteins from the great majority of primate lentiviruses, including HIV-2, efficiently downregulate the T-cell receptor CD3 (TCD3), thereby suppressing activation and apoptosis in T cells. In contrast, *nef* alleles from HIV-1 and its closest related SIV isolates do not encode this activity. Hyperactivation of the immune system as caused by T-cell activation is now generally accepted as an important clinical marker of AIDS progression. Hence, downregulation of TCD3 and the resulting avoidance preceding T-cell activation yields viral persistence without damaging the host immune system and disease progression to AIDS. Thus, in contrast to HIV-1, SIV-derived Nef proteins are better described as persistence factors than as pathogenesis factors.

HIV Subtypes

Analysis of viral sequences allows HIV-1 to be classified into three distinctive groups, representing the HIV-1 lineages main (M), outlier (O), and non-M-non-O (N). It is assumed that each group resulted from an independent zoonotic transfer from chimpanzees, which were infected with SIV_{CPZ} into humans. In addition, eight HIV-2 lineages arose from separate zoonotic transfers from sooty mangabeys infected with SIV_{SM}. Most intriguingly, of these 11 zoonotic events that transferred primate lentiviruses from nonhuman primates to humans, only those resulting in the HIV-1 M-group finally led to the current global AIDS pandemic. Most of group O and group N strains originate from West Africa, especially Cameroon. However, the propagation of these strains was relatively restricted. HIV-2, predominantly A and B groups, is also for the most part endemic in West Africa, particularly in Côte d'Ivoire. In comparison to HIV-1, HIV-2 led to a relatively small number of infections in humans. It is conceivable that more transmissions from nonhuman primates to humans occurred, which, however, were not able to spread efficiently in the human population and therefore have never been detected. At least one reason for the low frequency of successful zoonotic events is that transmissions of primate lentiviruses across the species barrier can be controlled by the existence of species-specific restriction factors (e.g., TRIM5α, as detailed above) that act as barriers against retroviral transfers.

See also: Human Immunodeficiency Viruses: Origin; Human Immunodeficiency Viruses: Pathogenesis; Retroviruses: General Features; Simian Immunodeficiency Virus: Animal Models of Disease; Simian Immunodeficiency Virus: General Features; Simian Immunodeficiency Virus: Natural Infection.

Further Reading

Anderson JL and Hope TJ (2004) HIV accessory proteins and surviving the host cell. *Current HIV/AIDS Reports* 1(1): 47–53.

Coffin JM, Hughes SH, and Varmus HE (1997) *Retroviruses*. New York: CSHL Press.

Flint SJ, Enquist LW, Krung RM, Racaniello VR, and Skalka AM (2000) *Principles of Virology*. Washington, DC: ASM Press.

Freed EO (2002) Viral late domains. *Journal of Virology* 76(10): 4679–4687.

Freed EO and Mouland AJ (2006) The cell biology of HIV-1 and other retroviruses. *Retrovirology* 3: 3–77.

Goff SP (2004) Genetic control of retrovirus susceptibility in mammalian cells. *Annual Reviews of Genetics* 38: 61–85.

Klinger PP and Schubert U (2005) The ubiquitin–proteasome system in HIV replication: Potential targets for antiretroviral therapy. *Expert Review of Anti-Infective Therapy* 3(1): 61–79.

Li L, Li HS, Pauza CD, Bukrinsky M, and Zhao RY (2005) Roles of HIV-1 auxiliary proteins in viral pathogenesis and host–pathogen interactions. *Cell Research* 15(11–12): 923–934.

Morita E and Sundquist WI (2004) Retrovirus budding. *Annual Reviews of Cell and Developmental Biology* 20: 395–425.

Schubert U and McClure M (2005) Human immunodeficiency virus. In: Mahy BWJ and Ter Meulen V (eds.) *Virology*, 10th edn., pp. 1322–1346. London: Topley and Wilson.

Human Immunodeficiency Viruses: Origin

F van Heuverswyn and M Peeters, University of Montpellier 1, Montpellier, France

© 2008 Elsevier Ltd. All rights reserved.

Glossary

Circulating recombinant forms These forms represent recombinant HIV-1 genomes that have infected three of more persons who are not epidemiologically related.

Endemic A classification of an infectious disease that is maintained in the population without the need for external inputs.

Epidemic A classification of a disease that appears as new cases in a given human population, during a given period, at a rate that substantially exceeds what is expected, based on recent experience.

Neighbor-joining method This clustering method constructs trees by sequentially finding of pairs of operational taxonomic units (OTUs) or neighbors that minimize the total branch length at each stage of clustering OTUs starting with a starlike tree.

Pandemic An epidemic that spreads through human populations across a large region (for e.g., a continent), or even worldwide.

Phylogenetic tree A phylogenetic tree, also called evolutionary tree, is a graphical diagram, showing the evolutionary relationships among various biological species of other entities that are believed to have a common ancestor, comparable to a pedigree showing which genes or organisms are most closely related. In a phylogenetic tree, each node with descendants represents the most common ancestor of the descendants, and the edge lengths in most trees correspond to time estimates. External nodes are often called operational taxonomic units (OTUs), a generic term that can represent many types of comparable taxa. Internal nodes may be called hypothetical taxonomic units (HTUs) to emphasize that they are the hypohetical progenitors of OTUs.

Prevalence The prevalence of a disease in a statistical population is defined as the total number of cases of the disease in the population at a given time, or the number of cases in the population, divided by the number of individuals in the population.

Introduction

Infectious diseases have been an ever-present threat to mankind. A number of important pandemics and epidemics arose with the domestication of animals, such as influenza and tuberculosis. Whereas the cause of some of the historic pandemics, such as the bubonic plague (the Black Death) that killed at least 75 million people, have been successfully eradicated, many others still cause high mortality especially in developing countries. Emerging infectious diseases continue to represent a major threat to global health. As such, HIV/AIDS is one of the most important diseases to have emerged in the past century. When on 5 June 1981, a report was published, describing five young gay men infected with *Pneumocystis carinii* pneumonia (PCP), no one could have imagined that 25 years later, more than 40 million people all over the world would be infected with the human immunodeficiency virus (HIV), the cause of the acquired immunodeficiency syndrome (AIDS). With more than 25 million deaths, HIV/AIDS continues to be one of the most serious public health threats facing humankind in the twenty-first century.

It is important therefore to identify where HIV came from, whether a natural host reservoir exists and how it was introduced into the human population. Today it is well established that human immunodeficiency viruses HIV-1 and HIV-2 are the result of several cross-species transmissions from nonhuman primates to humans.

West-Central African chimpanzees (*Pan troglodytes troglodytes*) are now recognized as the natural reservoir of the simian immunodeficiency viruses (SIVcpz*Ptt*), that are the ancestors of HIV-1. Similarly, HIV-2, which has remained largely restricted to west Africa, is the result of cross-species transmissions of SIVsmm from sooty mangabeys (*Cercocebus atys*).

Although it is clear now that HIV has a zoonotic origin, it remained for a long time less certain where, when, and how often these viruses entered the human population. In this article, we will describe in more detail the latest findings on the origin of HIV, more specifically of the three groups of HIV-1 (M, N, and O) and HIV-2.

Taxonomy, Classification, and Genomic Structure

Taxonomy

Human and *Simian immunodeficiency viruses* (HIV and SIV) belong to the genus *Lentivirus* of the family *Retroviridae*, characterized by their structure and replication mode. These viruses have two RNA genomes and rely on the reverse transcriptase (RT) enzyme to transcribe their genome from RNA into a DNA copy, which can then be integrated as a DNA provirus into the genomic DNA of the host cell. This replication cycle is common for all members of the family *Retroviridae*. As the name suggests, the genus *Lentivirus* consists of slow viruses, with a long incubation period. Five serogroups are recognized, each reflecting the vertebrate hosts with which they are associated (primates, sheep and goats, horses, cats, and cattle). A feature of the primate lentiviruses, HIV and SIV, is the use of a CD4 protein receptor and the absence of a dUTPase enzyme.

Classification

Classification of simian immunodeficiency viruses (SIVs)

SIVs isolated from different primate species are designated by a three-letter code, indicating their species of origin (e.g., SIVrcm from red-capped mangabey). When different subspecies of the same species are infected, the name of the subspecies is added to the virus designation, for example, SIVcpz*Ptt* and SIVcpz*Pts* to differentiate between the two subspecies of chimpanzees *P. t. troglodytes* and *P. t. schweinfurthii*, respectively. For chimpanzee viruses, the known or suspected country of origin is often included; for example, SIVcpzCAM and SIVcpzGAB are isolates from Cameroon and Gabon, respectively.

Currently, serological evidence of SIV infection has been shown for 39 different primate species and SIV infection has been confirmed by sequence analysis in 32 (see **Table 1**). Overall, complete SIV genome sequences are available for 19 species. Importantly, 30 species of the 69 recognized Old World monkey and ape species in sub-Saharan Africa have not been tested yet or only very few have been tested. Knowing that the vast majority (90%) of the primate species tested are SIV infected, many of the remaining species would be expected to harbor additional SIV infections. Only Old World primates are infected with SIVs, and only those from the African continent; no SIVs have been identified in Asian primate species. It is important also to note that none of the African primates naturally infected with SIV develop disease.

Classification of human immunodeficiency viruses (HIVs)

AIDS can be caused by two related lentiviruses; human immunodeficiency virus types 1 and 2 (HIV-1 and HIV-2). On the basis of phylogenetic analyses of numerous isolates obtained from diverse geographic origins, HIV-1 is classified into three groups, M, N, and O. Group M (for Major) represents the vast majority of HIV-1 strains found worldwide and is responsible for the pandemic; group O (for Outlier) and N (non-M–non-O) remain restricted to West-Central Africa. Group M can be further subdivided into nine subtypes (A–D, F–H, J, K), circulating recombinant forms (CRFs, CRF01–CRF32), and unique recombinants. The geographic distribution of the different HIV-1 M variants is very heterogeneous and differs even from country to country. Compared to HIV-1, only a limited number of HIV-2 strains have been genetically characterized and eight groups (A–H) have been reported.

Genomic Structure

All primate lentiviruses have a common genomic structure, consisting of the long terminal repeats (LTRs), flanking both ends of the genome, three structural genes, *gag*, *pol*, and *env* and five accessory genes, *vif*, *vpr*, *tat*, *rev*, and *nef*. Some primate lentiviruses carry an additional accessory gene, *vpx* or *vpu*, in the region between *pol* and *env*. Based on this genomic organization, we can distinguish between three patterns (**Figure 1**): (1) SIVagm, SIVsyk, SIVmnd1, SIVlho, SIVsun, SIVcol, SIVtal, and SIVdeb display the basic structure with three major and five accessory genes; (2) SIVcpz, SIVgsn, SIVmus, SIVden, SIVmon, and also HIV-1 harbor an additional accessory gene, *vpu*; HIV-1 and SIVcpz differ from the other members of this group by the fact that *env* and *nef* genes are not overlapping; (3) SIVsmm, SIVrcm, SIVmnd2, SIVdrl, and SIVmac, and HIV-2 harbor a supplemental accessory gene, *vpx*. For the remaining SIVs, SIVolc, SIVwrc, SIVasc, SIVbkm, SIVery, SIVblu, SIVpre, SIVagi, and SIVgor, full-length sequences are not yet available.

Table 1 Serological and/or molecular evidence for SIV infection in the African nonhuman primates

Genus	Species	Common name	SIV	Geographic distribution
Pan	troglodytes	Common chimpanzee	SIVcpz	West to East: Senegal to Tanzania
Gorilla	gorilla	Western gorilla	SIVgor[a]	Central: Cameroon, Gabon, Congo, Central Africa Republic
Colobus	guereza	Mantled guereza	SIVcol	Central: Nigeria to Ethiopia/Tanzania
Piliocolobus	badius	Western red colobus	SIVwrc[a]	West: Senegal to Ghana
Procolobus	verus	Olive colobus	SIVolc[a]	West: Sierra-Leone to Ghana
Lophocebus	albigena	Gray-cheeked managabey	?	Central: Nigeria to Uganda/Burundi
	aterrimus	Black crested mangabey	SIVbkm[a]	Central: Democratic Republic of Congo (DRC)
Papio	anubis	Olive baboon	?	West to East: Mali to Ethiopia
	cynocephalus	Yellow baboon	SIVagm-Ver[a]	Central: Angola to Tanzania
	ursinus	Chacma baboon	SIVagm-Ver[a]	South: southern Angola to Zambia
Cercocebus	atys	Sooty mangabey	SIVsmm	West: Senegal to Ghana
	torquatus	Red-capped mangabey	SIVrcm	West Central: Nigeria, Cameroon, Gabon
	agilis	Agile mangabey	SIVagi[a]	Central: northeast Gabon to northeast Congo
Mandrillus	sphinx	Mandrill	SIVmnd-1, SIVmnd-2	West Central: Cameroon (south of Sanaga) to Gabon, Congo
	leucophaeus	Drill	SIVdrl	West Central: southeast Nigeria to Cameroon (north of Sanaga)
Allenopithecus	nigroviridis	Allen's swamp monkey	?	Central: Congo
Miopithecus	talapoin	Angolan talapoin	SIVtal[a]	West Central: East coast of Angola into DRC
	ogouensis	Gabon talapoin	SIVtal	West Central: Cameroon (south of Sanaga)-Gabon
Erythrocebus	patas	Patas monkey	SIVagm-sab[a]	West to East: Senegal to Ethiopia, Tanzania
Chlorocebus	sabaeus	Green monkey	SIVagm-Sab	West: Senegal to Volta river in Burkina Faso
	aethiops	Grivet	SIVagm-Gri	East: Sudan, Erithrea, Ethiopia
	tantalus	Tantalus monkey	SIVagm-Tan	Central: Ghana to Uganda
	pygerythrus	Vervet monkey	SIVagm-Ver	South: South Africa to Somalia and Angola
Cercopithecus	diana	Diana monkey	?	West: Sierra-Leone to Ivory Coast
	nictitans	Greater spot-nosed monkey	SIVgsn	Central: forest blocks from West Africa to DRC
	mitis	Blue monkey	SIVblu[a]	East Central: East Congo to Rift-valley
	albogularis	Sykes's monkey	SIVsyk	East: Somalia to Eastern Cape
	mona	Mona monkey	SIVmon	West: Niger delta to Cameroon (north of Sanaga)
	campbelli	Campbell's mona	?	West: Gambia to Liberia
	pogonias	Crested mona	?	West Central: Cross River in Nigeria to Congo (east)
	denti	Dent's mona	SIVden	Central: south of Congo River
	cephus	Mustached guenon	SIVmus	West Central: Cameroon (south of Sanaga) to east of Congo River
	erythrotis	Red-eared monkey	SIVery[a]	West Central: Cross River in Nigeria to Sanaga in Cameroon, Bioko
	ascanius	Red-tailed monkey	SIVasc[a]	Central: South-East Congo to West Tanzania
	lhoest	l'Hoest monkey	SIVlho	Central: eastern Congo–Zaire to western Uganda
	solatus	Sun-tailed monkey	SIVsun	West Central: tropical forest of Gabon
	preussi	Preuss's monkey	SIVpre[a]	West Central: Cross river in Nigeria to Sanaga in Cameroon, Bioko
	hamlyni	Owl-faced monkey	?	Central: eastern DRC to Ruanda
	neglectus	de Brazza's monkey	SIVdeb	Central: Angola, Cameroon, Gabon to Uganda, western Kenya

[a]only partial sequences are available, ? only serological evidence for SIV infection.

Evolutionary History

HIV-2 and SIVsmm from Sooty Mangabeys

Shortly after the identification of HIV-1 as the cause of AIDS in 1983, the first SIV, SIVmac, was isolated from rhesus macaques (*Macaca mulatta*) at the New England Regional Primate Research Center (NERPRC). Retrospective research revealed that the newly identified SIV was introduced to the NERPRC by rhesus monkeys, previously housed at the California National Primate Research Center (CNPRC), where they survived an earlier (late 1960s) disease outbreak, characterized by immune suppression and opportunistic infections. A decade after the first outbreak, the story has been repeated in

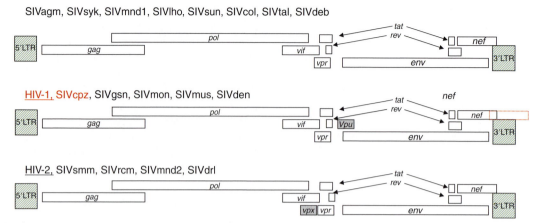

Figure 1 Genomic organization of the primate lentiviruses.

stump-nailed macaques (*Macaca arctoides*) in the same settings and 15 years later a lentivirus, called SIVstm, was isolated from frozen tissue from one of these monkeys. In both cases, the infected rhesus macaques had been in contact with healthy, but retrospectively shown SIVsmm seropositive sooty mangabeys at the CNPRC. The close phylogenetic relationship between SIVmac, SIVstm, and SIVsmm identified mangabeys as the plausible source of SIV in macaques. Since SIVmac induced a disease in rhesus macaques with remarkable similarity to human AIDS, a simian origin of HIV was soon suspected. The discovery in 1986 of HIV-2, the agent of AIDS in West Africa, and the remarkable high relatedness of HIV-2 with SIVsmm, naturally infecting sooty mangabeys in West Africa, reinforced this hypothesis.

In addition, the similarities in viral genome organization (the presence of *vpx*), the geographic coincidence of the natural range of sooty mangabeys and the epicenter of the HIV-2 epidemic in West Africa, as well as the fact that sooty mangabeys are frequently hunted for food or kept as pets, allowed the identification of SIVsmm from sooty mangabeys as the simian source of HIV-2.

HIV-1 and SIVcpz from Chimpanzees

The first SIVcpz strains, SIVcpzGAB1 and SIVcpzGAB2, were isolated from chimpanzees (*P. troglodytes*) in Gabon more than 15 years ago; of 50 wild-caught chimpanzees initially tested for SIVcpz infection, two (GAB1 and GAB2) harbored HIV-1 cross-reactive antibodies. Analysis of the SIVcpzGAB1 genome revealed the same genomic organization as HIV-1, including an accessory gene, named *vpu*, so far only identified in HIV-1. Furthermore, phylogenetic analysis indicated that SIVcpzGAB1 was more closely related to HIV-1 than to any other SIV. A few years later, a third positive chimpanzee, confiscated upon illegal importation into Belgium from the Democratic Republic of Congo (DRC, ex-Zaire), was identified among 43 other wild-caught chimpanzees. A virus, SIVcpzANT, was isolated and characterized, but this virus showed an unexpected high degree of divergence from the others. A fourth SIVcpz strain (SIVcpzUS) was obtained from a chimpanzee (Marilyn) housed in an American primate center and it was shown that SIVcpzUS was closely related to the SIVcpz strains from Gabon. Subspecies identification of the chimpanzee hosts revealed that the SIVcpzANT strain was isolated from a member of the *P.t. schweinfurtii* subspecies, whereas the other chimpanzees belonged to the *P.t. troglodytes* subspecies. These findings suggested two distinct SIVcpz lineages according to the host species: SIVcpz*Ptt*, from the West-Central African chimpanzees, and SIVcpz*Pts*, from eastern chimpanzees. HIV-1 strains are classified into three highly divergent clades, groups M, N, and O, each of which was more closely related to SIVcpz from *P. t. troglodytes* than to SIVcpz from *P. t. schweinfurthii*. These data pointed to the West-Central African subspecies, *P. t. troglodytes* as the natural reservoir of the ancestors of HIV-1.

Until recently, only a handful of complete or partial SIVcpz genomes have been derived. In addition to the four previously mentioned viruses, three other strains, SIVcpzCAM3, SIVcpzCAM5, and SIVcpzCAM13, have been identified in Cameroon and all clustered with the previously identified SIVcpz*Ptt* strains consistent with their species of origin. The discovery of HIV-1 group N in a Cameroonian patient in 1998 showed that HIV-1 N is closely related to SIVcpz*Ptt* in the envelope region of the viral genome, suggesting an ancient recombination event. This finding demonstrated the cocirculation of SIVcpz and HIV-1 N viruses in the same geographic area and provided additional evidence that the western part of Central Africa is the likely site of origin of HIV-1.

However, the number of identified SIV strains in chimpanzees was low, compared to that of other naturally occurring SIV infections. The major problem in studying SIVcpz infection in chimpanzees is their endangered status and the fact that they live in isolated forest regions. All previously studied chimpanzees were among wild-caught but young and captive animals. They were initially captured as infants, mainly as a by-product of the bushmeat trade. Since the age of maturation of chimpanzees is around 9 for males and 10 for females, these infections were most probably the result of vertical transmission and do not reflect true prevalences among wild-living adult animals. Apparently, the frequency of vertical transmission is low among naturally infected primates, which can explain the low prevalence rates initially observed.

The recent development of noninvasive methods to detect and characterize SIVcpz in fecal and urine samples from wild ape populations boosted the search for new SIVcpz strains in the vast tropical forests of Central Africa. The first report of a full-length SIVcpz sequence obtained from a fecal sample, SIVcpzTAN1, was from a wild chimpanzee from the *P. t. schweinfurtii* subspecies in Gombe National Park, Tanzania. Subsequently, additional cases of SIVcpz*Pts* infections were documented in Tanzania (SIVcpzTAN2 to SIVcpzTAN5) and around Kisangani, northeastern DRC (SIVcpzDRC1). All the new SIVcpz*Pts* viruses formed a separate lineage with the initially described SIVcpzANT strain and suggest that the SIVcpz*Pts* strains are not at the origin of HIV-1 (**Figure 2**). A recent study in wild chimpanzee communities in southern Cameroon documented a prevalence of SIVcpz infection ranging from 4% to 35%, and identified 16 new SIVcpz*Ptt* strains. All of these newly identified viruses were found to fall within the radiation of SIVcpz*Ptt* strains from captive *P. t. troglodytes* apes, which also includes HIV-1 groups M and N, but not group O or SIVcpz*Pts* (**Figure 2**). The new SIVcpz*Ptt* viruses are characterized by high genetic diversity and SIVcpz*Ptt* strains were identified which are much more closely related to HIV-1 groups M and N than were any previously identified SIVcpz strains. Interestingly, the new SIVcpz*Ptt* viruses exhibited a significant geographic clustering and made it possible to trace the origins of present-day human AIDS viruses to geographically isolated chimpanzee communities in southern Cameroon.

HIV-1 and SIVgor from Gorillas

SIV infection has recently been described in western gorillas (*Gorilla gorilla*) in Cameroon. Surprisingly, the newly characterized gorilla viruses, termed SIVgor, formed a monophyletic group within the HIV-1/SIVcpz*Ptt* radiation, but in contrast to SIVcpz*Ptt*, they were most closely related to the HIV-1 group O lineage (**Figure 2**). However, the phylogenetic relationships between SIVcpz, SIVgor, and HIV-1 indicate that chimpanzees represent the original reservoir of SIVs now found in chimpanzees, gorillas, and humans. Given their herbivorous diet and peaceful coexistence with other primate species, especially chimpanzees, it remains a mystery by what route gorillas acquired SIVgor.

The data on SIV in wild chimpanzee and gorilla populations showed that distinct chimpanzee communities in southern Cameroon transmitted divergent SIVcpz to humans giving rise to HIV-1 groups M and N; and that chimpanzees transmitted HIV-1 group O-like viruses either to gorillas and humans independently, or first to gorillas which then transmitted the virus to humans. Additional studies are needed to determine if the other African great ape species, the eastern gorillas (*Gorilla berengei*) and the bonobo (*Pan paniscus*) harbor any SIV.

Cross-Species Transmission

Where?

HIV-1

Based on mitochondrial DNA sequences, common chimpanzees (*P. troglodytes*) are classified into four subspecies: *P. t. verus* in West Africa; *P. t. vellerosus*, restricted to a geographical area between the Cross River in Nigeria and the Sanaga River in Cameroon; *P. t. troglodytes* in southern Cameroon, Gabon, Equatorial Guinea, and the Republic of Congo; and *P. t. schweinfurtii* in the Democratic Republic of Congo, Uganda, Rwanda, and Tanzania. No evidence of SIV infection is found in *P. t. verus*, despite testing of more than 2000 chimpanzees, mostly captive animals exported to zoos or research centers in the US. Wild *P. t. vellerosus* apes have not been found to harbor SIVcpz, but only about 100 samples have been screened. The single reported SIV infection of *P. t. vellerosus*, SIVCAM4, was most probably the result of a cage transmission from a *P. t. troglodytes* ape, infected with SIVcpzCAM3. Since the three groups of HIV-1 (M, N, and O) all fall within the HIV-1/SIVcpz*Ptt*/SIVgor lineage, the cross-species transmissions giving rise to HIV-1 most likely occurred in western equatorial Africa. Furthermore, no human counterpart is found for SIVcpz*Pts* from *P. t. schweinfurtii*, which undermines the idea of a human-induced origin of HIV-1 by oral polio vaccine (OPV) programs in East-Central Africa in the late 1950s. It has been suggested that tissues derived from SIVcpz-infected chimpanzees, captured in the northeastern part of DRC were used for the polio vaccine production. However, this geographical region is situated in the middle of the *P. t. schweinfurtii* range and the characterization of a partial SIV genome (SIVcpzDRC1) from a wild chimpanzee in this region proved once more the inconsistency of the OPV theory (**Figure 2**).

The recent studies in wild chimpanzee communities in Cameroon, not only confirm the West-Central African origin of HIV-1, but also indicate that HIV-1 group M and N arose from geographically distinct chimpanzee populations. Phylogenetic analysis showed that all SIVcpz strains collected in southeast Cameroon formed a cluster with HIV-1 group M, whereas SIVcpz isolates from chimpanzee communities of a well-defined region in

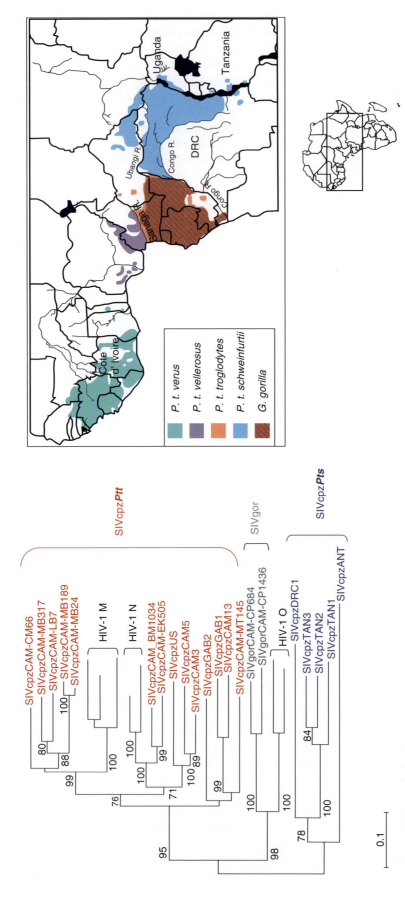

Figure 2 Natural range of chimpanzee subspecies and phylogeny of the HIV-1/SIVcpz/SIVgor lineage. The ranges of the four recognized chimpanzee subspecies are color-coded. The SIVcpz*Ptt* and SIVcpz*Pts* sequences in the phylogenetic tree are colored in red and blue, respectively, in accordance to the chimpanzee subspecies, illustrating that HIV-1 is more closely related to SIVcpz from West-Central African chimpanzees. The natural range for western gorillas (*Gorilla gorilla*) overlaps with the *P. t. troglodytes* range in West-Central Africa; SIVgor sequences are indicated in gray. The country of origin where SIVcpz strains were identified is indicated with a three-letter code: CAM for Cameroon, GAB for Gabon, TAN for Tanzania, DRC for the Democratic Republic of Congo, US for an animal in a primate center in the USA, and ANT for a captive animal in Antwerp, Belgium. This tree was derived by neighbor-joining analysis of partial *env/nef* nucleotide sequences. Horizontal branch lengths are drawn to scale.

south-central Cameroon clustered with the nonpandemic HIV-1 group N. It is also interesting to note that there is an uneven dissemination of SIV infection among chimpanzee populations, with the absence of SIV infection in some of them and with major geographical elements, like rivers, that can serve as important barriers.

As discussed above, HIV-1 group N resulted from an ancestral recombination event between divergent lineages. The discovery of such a recombinant virus in a geographically isolated chimpanzee community in southern Cameroon shows that HIV-1 N was already a recombinant in its natural hosts prior to its transmission to humans.

The origin of the third group of HIV-1, group O, remained uncertain until the recent identification of HIV-1 group O-like viruses in two different wild-living gorilla populations, in southern Cameroon (SIVgor). So all three HIV-1 groups seem to have their seeds in West-Central Africa. While HIV-1 group O infections remained restricted to West-Central Africa (Cameroon, Nigeria, Gabon, Equatorial Guinea) and HIV-1 N to Cameroon only, HIV-1 group M strains have spread across Africa and all the other continents.

HIV-2

A close phylogenetic relatedness is also observed between SIVsmm from sooty mangabeys (*Cercocebus atys*) and HIV-2 in West Africa. Sooty mangabeys are indigenous to West Africa, from Senegal to Ivory Coast, coinciding with the endemic center of HIV-2 (**Figure 3**). Eight groups (A–H) of HIV-2 have been described so far, but only subtypes A and B are largely represented in the HIV-2 epidemic, with subtype A in the western part of West Africa (Senegal, Guinea-Bissau) and subtype B being predominant in Ivory Coast. The other subtypes have been documented in one or few individuals only. Except for groups G and H, groups C, D, E, and F were isolated in rural areas in Sierra Leone and Liberia and these viruses are more closely related to the SIVsmm strains obtained from sooty mangabeys found in the same area than to any other HIV-2 strains. This suggests that the

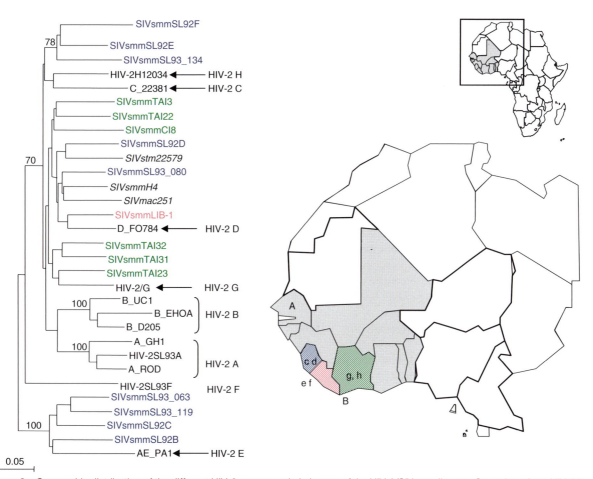

Figure 3 Geographic distribution of the different HIV-2 groups and phylogeny of the HIV-2/SIVsmm lineage. Countries where HIV-2 is endemic are colored in grey and overlap with the range of sooty mangabeys (*Cercocebus atys*) in West Africa. SIVsmm strains obtained from mangabeys in different regions are colored: green for Ivory Coast, blue for Sierra Leone, and pink for Liberia. SIVsmm strains isolated from mangabeys or macaques in US zoo's or primate centers are indicated in italic. This tree was derived by neighbor-joining analysis of partial *gag* nucleotide sequences from HIV-2 and SIVsmm sequences from West Africa. Horizontal branch lengths are drawn to scale.

different groups of HIV-2 must be the result of multiple independent cross-species transmissions of SIVsmm into the human population. Importantly, HIV-2 prevalence remains low and is even decreasing, since HIV-1 M is now predominating also in West Africa.

When?

It is clear now that each group of HIV-1 (M, N, and O) and HIV-2 (A–H) resulted from an independent cross-species transmission, followed by different viral evolution rates and possibly by one or more recombination events. HIV-1 group M can be further subdivided into subtypes, circulating recombinant forms and many unique recombinants, which had a heterogeneous geographical spread. The highest genetic diversity, in number of co-circulating subtypes and intrasubtype diversity has been observed in Central Africa, more precisely the western part of DRC, suggesting this region being the epicenter of HIV-1 M. Moreover, the earliest known HIV-1 virus, ZR59, isolated from an individual in Léopoldville (now Kinshasa) in 1959. The virus has been characterized as a member of HIV-1 M, subtype D. So, the common ancestor of this group should be dated prior to 1959. Molecular clock analyses estimated the date of the most recent common ancestor of HIV-1 group M around 1930 with a confidence interval of 1915–1941. A similar time frame is estimated for the origin of the HIV-1 group O radiation: 1920 with a range from 1931 to 1940. The oldest HIV-1 group O sample has been documented in a Norwegian sailor in 1964 infected in Cameroon (Douala), both groups show an exponential increase of the number of HIV infections during the twentieth century. However, the growth rate of HIV-1 group O ($r = 0.08$ [0.05–0.12]) is slower than the rate estimated for HIV-1 group M in Central Africa ($r = 0.17$). This is not surprising, considering the much lower prevalence of group O in Cameroon. Since the first identification of HIV-1 group N in 1998, about 10 group N infections have been described and all were from Cameroonian patients. The intra-group genetic diversity is significant lower for group N than for group M or O, which suggests a more recent introduction of the HIV-1 N lineage into the human population.

Similar analyses traced the origin of the pandemic HIV-2 groups A and B to be around 1940 with a confidence interval of ±16 years, and around 1945 with a confidence interval of 1931–59, respectively.

How?

Although the precise conditions and circumstances of the SIVcpz and SIVsmm transmissions remain unknown, human exposure to blood or other secretions of infected primates, through hunting and butchering of primate bushmeat, represents the most plausible source for human infection. Also bites and other injuries caused by primates kept as pets can increase the probability of viral transmission. Direct evidence of human infection with other SIVs is not yet reported, but that retroviruses can jump from primates to humans has been documented for simian foamy viruses (SFV) and simian T-cell leukemia viruses (STLVs). For example, SFV infection has been detected in rural Central Africa among individuals who hunt monkeys and apes. The infection with SFV in an Asian man, who regularly visited an ancient temple, was most probably the result of a bite from a macaque. An analysis of the man's blood indicated the presence of an SFV strain similar to a strain from macaques living around the temple. So far, SFV has not been shown to cause disease in humans, but the long-term effects of SFV in humans are not yet known. These effects are well documented for SIV and its human counterpart HIV; the HIV-1 group M epidemic clearly illustrates the devastating results of a single cross-species transmission.

Other SIV Cross-Species Transmissions, Risk for Novel HIV?

A recent study on primate bushmeat in Cameroon revealed that about 10% is SIV infected and illustrates an ongoing exposure to a plethora of different SIVs. Today, 69 different nonhuman primate species are identified in Africa and for 32 species (47%) SIV infection could already be proved by genomic amplification of the virus. This means that, in addition to sooty mangabeys, chimpanzees, and possibly gorillas, at least 29 other primate species harbor SIV's, which poses a potential risk for transmission to humans, especially for those in direct contact with infected blood and tissues. Therefore, efforts to reduce exposure to SIV-infected primates should be a primary concern. However, the opposite is occuring; the bushmeat trade has increased significantly during the last few decades, especially through the development of the logging industry. As a consequence, roads are now penetrating the formerly isolated forest areas and create a free passage for transport of wood, bushmeat, people, and subsequently different pathogens. The surrounding villages change from modest, little communities to real trade centers, with up to several thousand inhabitants. In addition, human migration and social and economic networks support this industry. It is very likely that the estimated prevalence of SIV infection of wild monkey species is underestimated for three reasons: first, only half of the recognized species have been tested; second, for some species only a few monkeys were tested; and third, and the most important reason, the sensitivity of the available diagnostic tools, initially developed for HIV detection, may be inaccurate for the detection of divergent SIVs. The increasing magnitude of human exposure to SIVs, combined with socioeconomic changes, which favor the dissemination of a plausible SIV transmission into the human population, could be the basis of novel zoonoses that lead in turn to novel HIV epidemics.

It is important to note that there is more needed than transmission of a virus to initiate an epidemic. After the virus has crossed the species barrier, adaptation to the new host is necessary to be able to spread efficiently within a population. The pathogenic potential of the virus and host genetic differences between individuals, as well as between species, determine susceptibility or resistance to further disease progression. In addition, environmental, social, and demographic factors play a major role in the further spread of new viruses.

Evolution of SIVs in their Natural Primate Hosts: The Origin of SIVcpz

With the increasing number of full-length SIV sequences from different primate species it becomes clear that cross-species transmissions and subsequent recombinations have occurred frequently among primates lentiviruses. Cross-species transmissions among co-habiting species in the wild has been observed between African green monkeys and patas monkeys in West Africa; between African green monkeys and baboons in southern Africa; and between different *Cercopithecus* species in Central Africa, that is, greater spot-nosed guenons (*C. nictitans*) and moustached (*C. cephus*) monkeys in Cameroon.

One of the most striking examples of cross-species transmission, followed by recombination, is SIVcpz in chimpanzees. Chimpanzees are known to hunt other primates for food, such as red-capped mangabeys (*Cercocebus torquatus*), greater spot-nosed guenons (*Cercopithecus nictitans*), and colobus monkeys (*Colobinae*). The isolation and characterization of the SIV genomes from the former monkey species revealed an unexpected high level of similarity between some parts of the SIVcpz genome and SIVrcm and SIVgsn. The 5′ region of SIVcpz (*gag, pol, vif,* and *vpr*) is most similar to SIVrcm, except for the accessory gene *vpx*, which is characteristic for the SIVrcm lineage, but absent in the HIV-1 and SIVcpz strains. The 3′ region of SIVcpz (*vpu, env,* and *nef*) is closely related to SIVgsn. Furthermore, SIVgsn is the first reported monkey virus to encode a *vpu* gene, the accessory gene characteristic of the HIV-1/SIVcpz lineage. Most probably, the recombination of these monkey viruses occurred within chimpanzees and gave rise to the common ancestor of today's SIVcpz lineages, which in turn were subsequently transmitted to humans. The cross-species transmission of this recombinant virus, or its progenitors, happened some time after the split of *P. t. verus* and *P. t. vellerosus* from the other subspecies, but possibly before the divergence between *P. t. troglodytes* and *P. t. schweinfurtii*.

The fact that gorillas are infected with an SIV from the SIVcpz lineage, represents a mystery, since only peaceful encounters have been documented among these sympatric apes. The evolutionary history of primate lentiviruses is complex and likely involved a series of consecutive interspecies transmissions, the timelines and directions of which remain to be deciphered.

Conclusion

We now have a clear picture of the origin of HIV and the seeds of the AIDS pandemic. SIVcpz, the progenitor of HIV-1, resulted from a recombination among ancestors of SIV lineages presently infecting red-capped mangabeys and *Cercopithecus* monkeys in West-Central Africa. HIV-1 groups M and O resulted from independent cross-species transmissions early in the twentieth century. The SIVcpz*Ptt* strains that gave rise to HIV-1 group M belonged to a viral lineage that persists today in *P. t. troglodytes* apes in south Cameroon. Most likely this virus was transmitted locally, but made its way to Kinshasa where the group M pandemic was spawned. HIV-1 group N, which has been identified in only a small number of AIDS patients from Cameroon, derived from a second SIVcpz*Ptt* lineage in south-central Cameroon and remained geographically more restricted. HIV-1 group O-related viruses are present in a second African great ape species, the western gorilla (*Gorilla gorilla*), but chimpanzees were the original reservoir of SIVgor.

Similarly, only HIV-2 group A and B play a major role in the HIV-2 epidemic, and most other groups (C–H) represent unique sequences found in a single patient. A possible explanation could be that some viruses were not able to adapt to the new host or that the environment was not suitable for epidemic spread.

Viral adaptation to the new host is one of the requirements for the generation of an epidemic, but also the interaction of sociocultural factors, such as deforestation, urbanization, and human migration, have been crucial in the emergence of the HIV-1 pandemic. While the origin of the HIV-1 and HIV-2 viruses has become clearer, important questions concerning pathogenicity and epidemic spread of certain SIV variants needs to be further elucidated.

See also: AIDS: Disease Manifestation; AIDS: Global Epidemiology; AIDS: Vaccine Development; Human Immunodeficiency Viruses: Antiretroviral agents; Human Immunodeficiency Viruses: Molecular Biology; Human Immunodeficiency Viruses: Pathogenesis; Simian Immunodeficiency Virus: Animal Models of Disease; Simian Immunodeficiency Virus: General Features; Simian Immunodeficiency Virus: Natural Infection.

Further Reading

Aghokeng AF and Peeters M (2005) Simian immunodeficiency viruses (SIVs) in Africa. *Journal of Neurovirology* 11(supplement 1): 27–32.

Hahn BH, Shaw GM, De Cock KM, and Sharp PM (2000) AIDS as a zoonosis: Scientific and public health implications. *Science* 287: 607–614.

Keele BF, Van Heuverswyn F, Li Y, et al. (2006) Chimpanzee: Reservoirs of pandemic and nonpandemic HIV-1. *Science* 313: 523–526.

Korber B, Muldoon M, Theiler J, et al. (2000) Timing the ancestor of the HIV-1 pandemic strains. *Science* 288: 1789–1796.

Lemey P, Pybus OG, Rambaut A, et al. (2004) The molecular population genetics of HIV-1 group O. *Genetics* 167: 1059–1068.

Santiago ML, Range F, Keele BF, et al. (2005) Simian immunodeficiency virus infection in free-ranging sooty mangabeys (*Cercocebus atys atys*) from the Tai Forest, Cote d'Ivoire: Implications for the origin of epidemic human immunodeficiency virus type 2. *Journal of Virology* 79: 12515–12527.

Sharp PM, Shaw GM, and Hahn BH (2005) Simian immunodeficiency virus infection of chimpanzees. *Journal of Virology* 79: 3891–3902.

VandeWoude S and Apetrei C (2006) Going wild: Lessons from naturally occurring T-lymphotropic lentiviruses. *Clinical Microbiological Reviews* 19: 728–762.

Van Heuverswyn F, Li Y, Neel C, et al. (2006) Human immunodeficiency viruses: SIV infection in wild gorillas. *Nature* 44: 164.

Worobey M, Santiago ML, Keele BF, et al. (2004) Origin of AIDS: Contaminated polio vaccine theory refuted. *Nature* 428: 820.

Human Immunodeficiency Viruses: Pathogenesis

N R Klatt, A Chahroudi, and G Silvestri, University of Pennsylvania School of Medicine, Philadelphia, PA, USA

© 2008 Elsevier Ltd. All rights reserved.

Glossary

Apoptosis Programmed cell death.

Bystander cell death The death of HIV-uninfected cells.

CCR5 A G-protein-coupled, seven transmembrane spanning receptor for the chemokines RANTES (CCL5), MIP-1α (CCL3), and MIP-1β (CCL4) that is primarily expressed on T lymphocytes, dendritic cells, macrophages, and microglial cells.

CD4+ T lymphocytes (also called T-helper cells) Cells bearing CD4, a co-receptor of the T-cell receptor (TCR) complex, that recognize peptide antigens bound to MHC class II molecules and are important for both humoral and cellular immunity.

Cytotoxic T lymphocytes (CTLs) Cells that are capable of killing other cells. Most CTLs express CD8, a co-receptor of the TCR complex, and recognize antigenic peptides from cytosolic pathogens, particularly viruses, that are bound to MHC class I molecules.

Generalized immune activation The activation of all lymphocyte subsets in HIV infection, leading to increased cell death.

Mucosal-associated lymphoid tissue (MALT) The system of lymphoid cells found in the epithelia and lamina propria of the body's mucosal sites, including the gastrointestinal tract, lungs, eyes, nose, and the female reproductive tract.

Neutralizing antibodies Antibodies that can limit the infectivity of a pathogen or the toxic effects of a toxin by binding to the receptor-binding site on the pathogen/toxin and thus block entry into the target cell.

Simian immunodeficiency virus (SIV) Like HIV it is a single-stranded, positive-sense, enveloped RNA virus that is classified as a member of the genus *Lentivirus* of the family *Retroviridae*. The virus infecting chimpanzees (SIV_{cpz}) is the origin of HIV-1 and the virus infecting sooty mangabeys (SIV_{smm}) is the origin of HIV-2.

Introduction

The AIDS pandemic, with an estimated 40 million individuals infected with the causative agent, human immunodeficiency virus (HIV), is without question one of the key medical challenges of modern times. Twenty-four years after the first identification of HIV, the situation in the fields of AIDS research, prevention, and treatment reflects several major advances as well as a number of areas where the progress has been slow or nonexistent (**Table 1**). While the #1 challenge (low rate of treatment in developing countries, particularly sub-Saharan Africa) is, in essence, a reflection of the sociopolitical climate arising from a lack of stability, development, and healthcare infrastructure in these countries, challenges #2 and #3 (absence of a preventative vaccine or a cure for HIV infection and AIDS) are a direct consequence of our incomplete understanding of the pathogenesis of HIV infection. Here we summarize the key advances that have improved our understanding of the mechanisms underlying the immunodeficiency that follows HIV infection.

While it is clear and universally accepted that HIV is the etiologic agent of AIDS and that the main

Table 1 Major achievements and remaining challenges in AIDS research

Major achievements in AIDS research:
1. Discovery and characterization of the etiologic agent.
2. Development of lab tests to monitor and prevent the infection via blood and blood products.
3. Definition of the origin of the epidemics.
4. Development of a large array of potent anti-HIV drugs, with consequent major reduction in mortality and MTCT in Western countries.

Major remaining challenges:
1. Abysmally low rate of treatment in developing countries.
2. Absence of a vaccine or a long-lasting microbicide.
3. Absence of a treatment that can eradicate infection.
4. Incomplete understanding of the pathogenesis of infection.

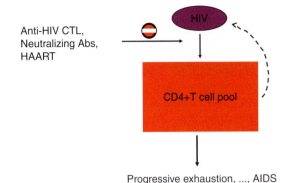

Figure 1 The classical or virus-centric model of AIDS pathogenesis.

pathophysiologic assault during HIV infection is the development of a state of chronic, progressive, and ultimately fatal immunodeficiency, several questions remain regarding the exact sequence and cause of the pathogenic events that define the progression from HIV infection to the development of AIDS. In particular, the precise mechanisms that underlie the progressive and generalized CD4+ T-lymphocyte depletion that represents the most striking and consistent laboratory finding in HIV-infected individuals as well as the best predictor of progression to AIDS are still unclear. While it is well established that HIV infects and kills CD4+ T lymphocytes, it is not known to what extent this direct effect of the virus contributes to the overall *in vivo* loss of these cells when measured against a series of indirect mechanisms of CD4+ T-lymphocyte depletion that are related to the state of generalized immune activation associated with HIV infection. Understanding the mechanisms responsible for the HIV-associated CD4+ T-lymphocyte depletion is a matter of utmost importance, as it is likely that greater knowledge of the immunopathogenesis of AIDS will pave the way for the implementation of novel therapeutic strategies.

Direct Cytopathic Effect of HIV on Infected CD4+ T Lymphocytes

A series of crucial discoveries made between 1983 (when HIV was first identified and proposed as the etiologic agent of AIDS) and 1996 (when highly active antiretroviral therapy, HAART, became available) resulted in a model for the pathogenesis of HIV infection that explains a large number of experimental and clinical findings (**Figure 1**). This model is based primarily on the following observations: (1) after transmission of HIV through either the sexual or intravenous route, the virus infects (using the CD4 molecule as well as one of the chemokine receptors CCR5 or CXCR4 as entry receptors) and then kills a subset of memory/activated CD4+ T lymphocytes; (2) the level of virus replication and the severity of CD4+ T-lymphocyte depletion are the key markers of disease progression from the initial asymptomatic phase of infection (lasting 2–15 years) to the phase characterized by constitutional symptoms and, eventually, the onset of opportunistic infections and cancer; (3) if HIV replication is controlled following the administration of antiviral drugs or spontaneously (in a small subset of HIV-infected individuals referred to as long-term nonprogressors, LTNP), the immunological damage induced by HIV is limited and survival improves dramatically. Collectively, these findings suggested that the pathogenesis of AIDS is mainly related to the direct, virus-mediated killing of CD4+ T lymphocytes that results in a slow but continuous erosion of the pool of these cells (**Figure 1**). This virus-centric model of AIDS pathogenesis was strengthened by the observations that HIV infection is associated with a high rate of virus turnover and a short lifespan of virus-infected cells. These studies were interpreted to mean that, in infected individuals, HIV kills large numbers of CD4+ T lymphocytes at a very fast pace, to the point that the compensatory drive to reconstitute the pool fails to maintain sufficient CD4+ T-lymphocyte numbers. Ultimately, this progressive depletion of CD4+ T lymphocytes results in the permanent loss of a key function of the human immune system, which in turn manifests itself as an increased susceptibility to opportunistic infections and neoplasms. It is important to note that, according to this model, the increased level of CD4+ T-lymphocyte proliferation that is consistently observed in HIV-infected patients is interpreted to be a homeostatic mechanism aimed at compensating for the loss of CD4+ T lymphocytes induced directly by HIV.

Further refinement of this model comes from studies defining the co-receptors used by HIV, in addition to CD4, to infect human cells. The main HIV co-receptor is the chemokine receptor CCR5 (CD195) that is used by the non-syncytium-inducing (NSI), macrophage-tropic viruses that are preferentially transmitted from person-to-person and represent the majority of strains found in infected patients. CCR5 is expressed in a relatively small fraction (approximately 10–15%) of circulating and lymph node CD4+ T lymphocytes that mainly display a memory phenotype (i.e., CD45RO+), but is present on a much larger fraction of mucosal-associated memory CD4+ T lymphocytes. This latter finding is particularly important as it may explain why CD4+ T lymphocytes may be more severely depleted in mucosal tissues than in peripheral blood and lymph nodes (see below). The second important HIV co-receptor is CXCR4 (CD184), that is used by syncytium-inducing (SI), T-lymphocyte tropic viruses that are found in a minority of HIV-infected individuals, usually during advanced stages of disease. Interestingly, CXCR4 is expressed primarily by naïve CD4+ T lymphocytes, thus potentially accounting for the relative preservation of naïve CD4+ T lymphocytes during the early stages of HIV infection. These data are important not only for their elucidation of potential targets for therapeutic interventions aimed at blocking HIV entry (e.g., CCR5 inhibitors), but also because they help to explain the finding that not all subsets of CD4+ T lymphocytes are equally sensitive to HIV infection, such that depletion of memory/activated CD4+ T lymphocytes is associated with infection with CCR5-tropic viruses, and depletion of naïve CD4+ T lymphocytes is associated with CXCR4-tropic viruses.

Indirect Mechanisms of Immunopathogenesis during HIV Infection

The virus-centric model of HIV/AIDS pathogenesis outlined in the previous section has the unquestionable advantage of being fairly straightforward while still encompassing several important observations. More recently, however, this model, at least in its simplest formulation, has been challenged by a series of new insights into the experimental data. First, the long period of time between virus transmission and the development of AIDS in which the level of CD4+ T lymphocytes appears to be in a quasi-steady state is not compatible with the kinetics of a typical viral infection that should rapidly and exponentially consume all available target cells. The original assumption was that the long duration of the disease process is due to the regenerative capacity of the CD4+ T-lymphocyte compartment that is able to at least partially (or temporarily) compensate for the virus-induced killing of these cells. However, it is now believed that it would be too striking of a coincidence that in the vast majority of patients the rate of killing by HIV happens to be just slightly faster than the rate of production of new CD4+ T lymphocytes (and never slower, in which patients would remain healthy, nor significantly faster, in which patients would die within days to weeks). It was then shown that the fraction of infected CD4+ T lymphocytes was remarkably low (i.e., 0.01–1%) at any given time during the course of infection, thus raising the question of why the immune system cannot simply ramp up the production of new CD4+ T lymphocytes to compensate for the relatively limited losses induced by direct viral infection and killing. The further observations that, in fact, the majority of CD4+ T lymphocytes that die *in vivo* in HIV-infected patients appear to be uninfected (termed bystander cell death) and that, when rates of cell turnover are measured *in vivo* in HIV-infected patients, both CD4+ and CD8+ T lymphocytes show similarly elevated rates of death, together support a pathogenic model, whereby nonviral (or indirect) factors are primarily responsible for HIV disease progression. It is important to note that the increased rate of death of uninfected CD4+ T lymphocytes clearly indicates that direct killing by HIV cannot be the only mechanism involved in the depletion of CD4+ T lymphocytes seen in HIV-infected patients.

Taken together, the above-mentioned observations suggest that the pathogenesis of the HIV-associated CD4+ T lymphocyte depletion is a complex phenomenon and that the host response to the virus must play a key role. In particular, it has been proposed that the presence of a chronic generalized immune activation that involves all lymphocyte subsets is an important driving force in the loss of CD4+ T lymphocytes that is associated with progression to AIDS. This concept of HIV infection having a general effect on the host immune system is confirmed by the fact that a large number of functional abnormalities involving virtually all other immune cell types are associated with HIV disease progression. Importantly, several clinical studies have shown that, in HIV-infected patients, disease progression correlates better with markers of immune activation than with viral load. In this perspective, the observation of the short *in vivo* lifespan of HIV-infected CD4+ T lymphocytes has been re-interpreted as the result of their heightened state of activation that dooms them to die (via activation-induced apoptosis) rather than as a consequence of direct infection. As such, the impact of HIV infection on the lifespan of infected CD4+ T lymphocytes may be less dramatic than was originally proposed and it has now been hypothesized that HIV, in essence, jumps from one activated cell (that is destined to die of apoptosis) to another without changing the fate of these infected cells. It could thus be conceived that the number of available activated CD4+ T lymphocytes dictates, independently, both the level of virus replication and the overall rate of CD4+ T lymphocyte death, and that these two factors are correlated with one another only because they reflect the prevailing level of immune activation.

Further key support for the immune activation hypothesis derives from studies of the nonpathogenic simian immunodeficiency virus (SIV) infections of natural hosts, such as sooty mangabeys (*Cercocebus atys*, SMs), African green monkeys (AGMs), mandrills, and numerous other nonhuman primate species. These natural SIV hosts are the origin of the human HIV epidemics through multiple episodes of cross-species transmission, with HIV-1 originating from the chimpanzee (*Pan troglodytes*) SIV_{cpz}, and HIV-2 originating from the SM SIV_{smm}. Naturally SIV-infected SMs are of interest also because SIV_{smm} is the source of the various rhesus macaque (*Macaca mulatta*)-adapted SIV_{mac} viruses (e.g., SIV_{mac239}, SIV_{mac251}) that are commonly used in laboratory studies of AIDS pathogenesis and vaccines. In natural SIV hosts, chronic high levels of virus replication and *in vivo* killing of CD4+ T lymphocytes are not associated with any signs of immunodeficiency in the setting of typically low levels of both immune activation and cellular immune responses to the virus. This phenotype suggests that natural SIV hosts have evolved to be disease-resistant without the acquisition of any special ability to control virus replication (via either immune or other mechanisms), but rather by an adaptation that attenuates the immune response to this retroviral infection (**Table 2**).

Interestingly, some recent findings may help to harmonize the classical (i.e., virus-centric) and immune activation models in that they support a more comprehensive model whereby the pathogenesis of HIV/AIDS involves both direct and indirect mechanisms (**Figure 2**). There are now thought to be two sequentially distinct phases in infection that involve very different pathogenic mechanisms. The first phase (acute infection) is characterized by a rapid, massive, directly HIV-mediated loss of memory CD4+ T lymphocytes that takes place mainly in mucosal tissues and may leave a profound scar on the overall function of the immune system. It is still unclear what determines the cessation of this brief early phase, although likely explanations include the consumption of available target cells and the generation of HIV-specific immune responses. The second phase (chronic infection) lasts typically for several years, and is characterized, in essence, by the immune system's struggle to both recover from the early injury (via activation of mechanisms of immune homeostasis) and control virus replication (via activation of humoral and cellular immune responses). Unfortunately, this latter endeavor proves almost always to be insurmountable as HIV is a pathogen of extraordinary genetic and structural plasticity allowing it to develop a series of highly effective strategies for immune evasion. The struggle of the second phase is characterized by the continuous generation of strong, though ultimately ineffective, immune responses to HIV, and by the attempt to maintain T-lymphocyte homeostasis in the face of massive T-lymphocyte loss due to both the direct effect of the HIV and bystander cell death. Sadly, this epic battle seems to end up inducing nothing but the very state of chronic immune activation that, as mentioned above, ultimately causes more damage than repair to the host immune system.

Determinants of HIV Transmission and Early Replication

A clearer understanding of the host and viral factors determining the likelihood of virus transmission upon exposure and/or the course of infection after transmission will likely provide useful information for the design of new strategies to prevent HIV infection and AIDS. However, while some host factors regulating the risk of infection and disease progression have been identified, predicting the natural course of the infection remains very difficult at the level of the individual patient. Among the host genetic factors that have been shown to induce resistance to HIV infection, a predominant role has been ascribed to the 32 bp deletion in the CCR5 gene seen in 1% of Caucasian individuals. This deletion results in a frameshift mutation that causes the CCR5 protein to be completely absent from the surface of all cells. Thus, individuals homozygous for the Δ32 allele are resistant to transmission of CCR5-tropic viral strains (the overwhelming majority of viruses involved in horizontal sexual transmission). Other genes whose polymorphisms have been associated with some effect on HIV disease progression include SDF-1, CX3CR-1, CCL3, and many others. The clinical consequence of these discoveries is that small molecule antagonists or a monoclonal antibody directed against CCR5 are increasingly recognized as promising strategies to prevent HIV transmission.

The search for host factors associated with more effective immune responses to HIV (and, in fact, the identification of any type of correlate of immune protection) has proved much more frustrating. As of early 2007, it is still unclear whether and to what extent the immune response to HIV is capable of inducing absolute (i.e., by preventing infection) or relative (i.e., by lowering virus replication) resistance to virus transmission and early dissemination. While stronger and/or broader T-lymphocyte responses to the virus have been associated with better course of disease, it is still unclear whether this apparently more

Table 2 Immunologic and virologic features of natural, nonpathogenic SIV infections vs. pathogenic HIV infection

	Natural SIVs	HIV
Chronic virus replication	Yes	Yes
Lack of immune control	Yes	Yes
Short lifespan of infected CD4+ T cells	Yes	Yes
Depletion of MALT CD4+ T cells	Yes	Yes
Systemic loss of CD4+ T cells	Rare	Yes
Generalized immune activation	No	Yes
Expression of CCR5 on CD4+ T cells	Low	High

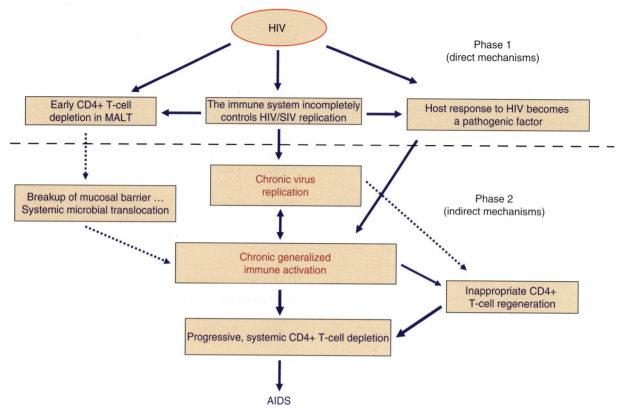

Figure 2 The biphasic model of AIDS pathogenesis. This model summarizes the key immunopathogenic events in acute and chronic infection that lead to progression to AIDS in HIV-infected individuals, taking into account both direct (virus-mediated) and indirect (immune activation-mediated) mechanisms.

effective immune response is cause or consequence of a more benign course of infection. Furthermore, the observation that natural SIV hosts, who are resistant to AIDS, show a low-level T-lymphocyte response to the virus and, more generally, an attenuated immune system activation clearly indicates that a strong host antiviral immune response is not necessary to lessen disease. While more studies are needed to fully elucidate the mechanisms underlying the lack of disease progression in natural SIV hosts, one clear message thus far is that co-evolution of SIV and its natural hosts have resulted in a happy equilibrium whereby the host immune function is preserved without any significant immune-mediated control of virus replication (**Table 2**).

Some recent advances in understanding what factors influence HIV-1 transmission come from recent studies of heterosexual transmission in Africa. One such study showed that the viruses that establish new infections encode an envelope surface glycoprotein (gp120) with a compact structure and reduced glycosylation that appears to be significantly more sensitive to antibody neutralization. To this end, it should be noted that the vast majority of gp120s isolated from patients with chronic HIV infection are resistant to neutralization by autologous sera. The implication of these findings for HIV vaccine development is significant, as they reveal a relative vulnerability of the HIV-Env glycoprotein in the very early phase of HIV infection. Also interesting is the finding that, in a cohort of discordant heterosexual partners (i.e., one HIV-infected and one HIV-uninfected) in Zambia, the sharing of HLA-B alleles is associated with a significant enhancement in the rate of transmission. This observation suggests that the foreign HLA molecules on virus-infected cells or the virus itself (acquired from the cell membrane) trigger protection against infection with a mechanism analogous to allogeneic transplant rejection; alternatively, transmission of viruses that have already escaped the cytotoxic T-lymphocyte (CTL) response may make infection more efficient in individuals with shared HLA alleles. While these reports shed some light on what type of adaptive immune response may control the spread of HIV infection, it is obvious that the virus appears to rapidly win the battle with the host immune system in the overwhelming majority of infected individuals.

When HIV first enters the body (most commonly via mucosal transmission but in certain instances through the bloodstream), the type and magnitude of the subsequent antiviral immune response is largely determined by immature dendritic cells (DCs), a cell type that is a component of the innate defense against mucosal penetration and widespread dissemination of many pathogens. Importantly, the role of DCs in the early events of HIV infection is complex and not always beneficial (**Figure 3**). On the one hand, DCs take up HIV-1 via C-type lectin binding receptors and migrate to the draining lymph node where they prime HIV-specific immune responses by presenting HIV antigens to CD4+ T lymphocytes. On the other hand, DCs may also contribute to early HIV dissemination throughout the body, due to their migratory potential and the fact that the DC–CD4+ T lymphocyte contact often results in productive infection of CD4+ T lymphocytes. While manipulation of the DC–T lymphocyte interaction may conceivably improve the host immune response to HIV, our limited understanding of the molecular mechanisms responsible for this interaction hampers the design of simple interventions aimed at blocking the early dissemination of HIV infection.

The Acute/Early Phase of HIV Infection

The acute phase of HIV infection is characterized by a peak of virus replication that occurs coincident with a transient decline of circulating CD4+ T lymphocytes and the presence of a flu-like clinical syndrome. This acute infection lasts for approximately 2–3 weeks and ends at approximately the time of antibody seroconversion. Some recent elegant studies have highlighted the role of the mucosal-associated lymphoid tissue (MALT) in acute HIV infection. The MALT houses the majority of CD4+ T lymphocytes in the body, a large fraction of which express CCR5 and have an activated memory phenotype. A key feature of the early phases of HIV and SIV infections is the massive infection and depletion of MALT-associated CD4+CCR5+ memory T lymphocytes (**Figure 2**). HIV replication in activated CD4+CCR5+ memory T lymphocytes proceeds relatively unchecked for 2–3 weeks after transmission, resulting in large-scale depletion of these cells. During early pathogenic SIV infection of Asian macaques (i.e., non-natural hosts), up to 60% of memory CD4+ T lymphocytes in the intestinal lamina propria contain SIV-RNA at the peak of infection (day 10) and the majority of these cells are eliminated by day 14. Importantly, though, the best predictor of rapid disease progression may not be the extent of the early CD4+ T lymphocyte depletion from mucosal sites, but rather the ability to reconstitute the pool of memory CD4+ T lymphocytes in the MALT. MALT-associated memory CD4+ T lymphocytes are an integral component of the mucosal immunological barrier against invading pathogens,

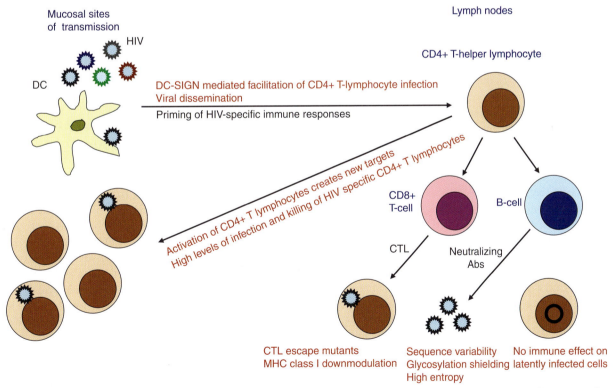

Figure 3 HIV transmission and generation of HIV-specific immune responses. HIV entry at mucosal sites is followed by capture by DCs that subsequently migrate to lymph nodes to present HIV antigens to CD4+ T-helper lymphocytes. In this process, DCs may also facilitate the infection of activated CD4+ T lymphocytes. CD4+ T lymphocytes help CD8+ T lymphocytes and B cells in mounting cellular and humoral anti-HIV immune responses. Legend in black indicates events of the HIV-specific immune response; legend in red indicates events leading to HIV immune evasion and dissemination.

and it is likely that their massive (and potentially irreversible) depletion may significantly affect the proper function of the mucosal barrier. This setting then allows for the emergence of numerous subclinical infections, thus inducing further microenvironment destruction and contributing to a state of chronic immune system activation (**Figure 2**). It should be noted, however, that to date the specific consequences of the early depletion of MALT-associated memory CD4+ T lymphocytes remain unknown, and it is still unclear why the opportunistic infections that are typical of full-blown AIDS do not appear until peripheral blood CD4+ T lymphocyte counts decline. Particularly puzzling is the observation that a rapid, severe, and persistent depletion of mucosal CD4+ T lymphocytes also occurs during nonpathogenic SIV infection of natural hosts such as SMs and AGMs (**Table 2**). However, in the context of the typically low levels of systemic and mucosal immune activation found in these species, no signs of mucosal immune dysfunction are observed, further emphasizing the potential protective role of an attenuated immune response during retroviral infection. Alternatively, it is possible that the early and rapid depletion of mucosal CD4+ T lymphocytes observed in pathogenic HIV/SIV infections simply reflects the systemic virus dissemination that is necessary to establish a chronically productive infection.

Generation of HIV-Specific Immune Responses and the Acute-to-Chronic Phase Transition

As mentioned above, the determinants of the relatively abrupt cessation of the early phase of HIV infection and the onset of the long chronic phase that occur at 2–3 weeks post infection are still poorly understood. However, several investigators have put forward the hypothesis that HIV-specific CTL responses as well as humoral responses directed against the virus envelope protein (i.e., neutralizing antibodies) play a central role in reducing the level of virus replication from its peak to set-point level. The evidence in favor of a significant role of CD8+ CTL-mediated responses reducing HIV replication include: (1) the temporal association between decline of viremia and emergence of cellular immune responses during primary HIV/SIV infection; (2) the increase in virus replication after experimental depletion of CD8+ T lymphocytes during pathogenic SIV infection of rhesus macaques; and (3) the observation that CTL escape mutations present in viruses transmitted between individuals with different HLA haplotypes are rapidly replaced by wild-type sequences. While CD8+ CTL-mediated responses play a significant role early on in protecting from unchecked HIV-1 replication and disease progression, their impact is ultimately limited by the extreme genetic variability of HIV that favors the emergence, during both acute and chronic infection, of viral escape mutants (**Figure 3**). The fact that HIV preferentially infects and kills HIV-specific memory CD4+ T lymphocytes (that are more likely to be activated, **Figure 3**) over memory T lymphocytes specific for other pathogens further complicates the situation, and results in a functional deficit in the anti-HIV response as compared to that directed against other pathogens. What must now be elucidated is whether preexisting, AIDS vaccine-induced HIV-specific CTLs will be more effective in controlling virus replication than those generated during natural infection. HIV-specific autologous neutralizing antibodies also exert potent and continuous selective pressure on the viral quasispecies. Unfortunately, HIV is able to escape antibody-mediated neutralization as efficiently as it does CTL control. A series of recent studies have clarified some of the mechanisms of HIV evasion from the humoral immune response (**Figure 3**). First, the envelope glycoprotein (Env) is highly variable, particularly in specific regions that form the surface of the virus. Second, the outer surface of gp120 is covered with a highly dynamic carbohydrate shield, in which glycans shift through point mutations to mask adjacent and distant epitopes. Third, conserved structural domains of Env are either sterically hindered or only transiently exposed. Fourth, receptor binding of Env is characterized by a high level of intrinsic entropy that makes it a moving target for antibody neutralization. The failure of both HIV-specific CTLs and neutralizing antibodies to completely suppress HIV replication during the transition between acute-to-chronic infection likely sets the stage for the following pathogenic events that are characterized by continuous, generalized immune activation in the face of chronic virus replication (**Figure 2**).

The Chronic Phase of Infection

Chronic HIV infection is a process lasting for several years that is characterized by a very slow decline of CD4+ T lymphocytes in peripheral blood, a small fraction of infected CD4+ T lymphocytes, and increased death rates of both CD4+ and CD8+ T lymphocytes. This last finding is related to the state of chronic, generalized immune activation that is now recognized as a major driving force behind CD4+ T lymphocyte depletion and a strong predictor of disease progression. The causes of the HIV-associated generalized immune activation are complex, and involve activation of T lymphocytes by specific antigens (e.g., HIV and a plethora of opportunistic pathogens), as well as bystander activation of T lymphocytes due to the presence of high levels of lymphotropic cytokines and other as yet poorly defined factors. But how does this increased immune activation induce CD4+ T lymphocyte depletion? First, CD4+ T lymphocyte activation provides new targets for HIV replication (that occurs preferentially in activated CD4+ T lymphocytes, **Figure 3**), thus creating an environment conducive to further virus-mediated damage to the immune system. Second, the increased T-lymphocyte turnover driven by this chronic immune activation (and, to some extent, by the increased compensatory homeostatic proliferation of T lymphocytes) results in chronic consumption of the pools of naïve and resting-memory cells, with a resultant relative expansion of short-lived effector CD4+ T lymphocytes. Third, the chronic activation and proliferation of T lymphocyte results in perturbations of cell-cycle control and an increased propensity to undergo activation-induced apoptosis. It is also possible that the HIV-associated immune activation is, at least in part, responsible for the changes in the lymph node architecture and thymic function that likely limit the proper reconstitution of the CD4+ T lymphocyte pool. It is hypothesized that depletion is unique to CD4+ T lymphocytes (and involves CD8+ T lymphocytes only in the very late stages of HIV disease) not only because these cells are targeted by HIV, but also because they are more sensitive to the deleterious effects of immune activation and/or are less efficiently regenerated.

Evidently, the assertion that chronic T-lymphocyte activation is a key pathogenic mechanism during HIV infection is in an apparent conflict with data suggesting that cellular immune responses to HIV/SIV infections may limit primary infection and/or disease progression. While T-lymphocyte responses to HIV likely play a protective role early on, T-lymphocyte immunity virtually always fails to control HIV infection due to the emergence of viral escape mutants. Once HIV-specific cellular immune responses are unable to control the virus, other mechanisms may determine the net effect that this ongoing T-lymphocyte activation has on the course of disease.

Is it possible that there is a direct connection between the early damage that HIV causes to the MALT-based pool of memory CD4+ T lymphocytes and the generation of the HIV-associated generalized immune activation? It is known that MALT-associated memory CD4+ T lymphocytes are an integral component of the mucosal immunological barrier that protects the host from both invading pathogens as well as commensal enteric bacteria that may also be recognized as pathogens if they cross the intestinal epithelium. It is likely that a massive depletion of these cells may affect the proper function of this mucosal barrier and thus may open the door to a series of subclinical infections (i.e., micro-invasion of the intestinal mucosa) that will, in turn, induce further micro-environment destruction and contribute to the generation of a state of chronic immune system activation (**Figure 2**). While this model delineates an interesting mechanistic link between direct and indirect (i.e., virus vs. immune activation-mediated) mechanisms of HIV-associated CD4+ T-lymphocyte depletion, the specific consequences of the early depletion of MALT-associated memory CD4+ T lymphocytes are in fact unknown, and it remains unclear why the opportunistic infections that are typical of full-blown AIDS do not manifest until peripheral blood CD4+ T-lymphocyte counts decline.

Implications for HIV Therapy and Vaccine Development

As described above, AIDS pathogenesis likely recognizes two distinct phases, a brief acute phase dominated by the direct killing of large numbers of memory CD4+ T lymphocytes followed by a long chronic phase characterized by a slow attrition of the CD4+ T lymphocyte pool that is related mainly to the prevailing level of immune activation (**Figure 2**). However, many questions remain to be answered. First, a precise quantification of the relative contribution of the direct and indirect mechanisms of AIDS pathogenesis is still lacking. Second, we do not have a clear explanation as to why HIV appears to be so uniquely powerful in inducing a chronic state of immune activation (as opposed to other chronic viral infections), and why the HIV-induced immune activation is so disruptive of the immune system homeostasis. The fact that generalized immune activation follows HIV infection of humans, which is a very recent disease, but does not follow infections that date back many thousands or even millions of years (such as HCV infection of humans or SIV infection of natural hosts), would suggest that attenuation of the immune response to a chronically replicating virus is the result of an advantageous evolutionary adaptation of the host immune system. Unfortunately, the mechanisms underlying the ability of natural hosts for SIV infection to avoid the chronic immune activation seen in HIV-infected humans are still poorly understood.

This biphasic model of AIDS pathogenesis provides a few important implications for HIV therapy and vaccine design. In terms of therapy, an important observation is that a complete recovery of the mucosal CD4+ T-lymphocyte system is unlikely with the currently available antiretroviral drugs even in the presence of prolonged viral suppression. As such, additional immune-based interventions, in particular using CD4+ T-lymphocyte tropic cytokines (e.g., IL-2, IL-7) as well as factors promoting the CD4+ T-lymphocyte repopulation of the MALT, may be useful to improve the immune system recovery from the severe and persistent damages inflicted during the early phase of the infection. In addition, given the well-recognized pathogenic role of the generalized immune activation during the chronic infection, it is conceivable that, in many HIV-infected patients, immune-based interventions aimed at reducing immune activation, preserving proper cell-cycle control, and preventing excessive levels of bystander cell death may restore the overall CD4+ T-lymphocyte homeostasis.

Among these potentially interesting new interventions are those that interfere directly with the signaling pathways involved in establishing the generalized immune activation or cell-cycle dysregulation (i.e., toll-like receptors, cytokine receptors, nuclear factor kappa B (NF-κB), cell cycle-dependent kinases) and those that improve the regeneration and/or the differentiation of T lymphocytes (i.e., cytokines and growth factors). Finally, a clearer understanding of the reason(s) why SIV infection is nonpathogenic in natural hosts (and, in particular, how virus replication can proceed in the absence of immune system damage and chronic immune activation) will likely also reveal further immune-based therapies to delay progression of disease.

Advances in the understanding of the early events of HIV transmission, with special emphasis on the immunogenic vulnerabilities of Env and the correlates of immune protection at the mucosal level, may lead to the development of vaccine and microbicide strategies that can limit the acute HIV-induced damage. In terms of vaccine development, the main implication of the accumulated pathogenesis research is that there may be only a short window of opportunity to control this rapidly disseminating virus that produces such extensive damage to the host immune system. This premise is especially disheartening for CTL-based vaccines that, in order to be effective, will have to induce a persistently high level of mucosa-associated, long-lived, HIV-specific memory CD8+ T lymphocytes that can rapidly proliferate and differentiate into effectors in the event of viral infection. Recent findings are more encouraging for a vaccine that aims to induce a high titer of broadly neutralizing antibodies, particularly if one considers that structural bottlenecks seem to be required for the virus to be efficiently transmitted to a new host in certain settings. In addition, it should be noted that even a vaccine that is only partially effective in containing the early viral dissemination (and resultant immune system damage) may end up having a significant impact on shaping the course of disease.

In summary, there have been many advances in the field of HIV immunopathogenesis over the past 24 years, but there is much that remains to be elucidated, particularly at the mechanistic level. It is critical that, with the help of new and highly sophisticated tools of analyses, we endeavor to more fully understand the dynamic interaction between HIV and the host immune system in order to eradicate this disease.

See also: AIDS: Disease Manifestation; AIDS: Global Epidemiology; AIDS: Vaccine Development; Human Immunodeficiency Viruses: Antiretroviral agents; Human Immunodeficiency Viruses: Molecular Biology; Human Immunodeficiency Viruses: Origin; Simian Immunodeficiency Virus: Animal Models of Disease; Simian Immunodeficiency Virus: General Features; Simian Immunodeficiency Virus: Natural Infection.

Further Reading

Gordon S, Pandrea I, Dunham R, Apetrei C, and Silvestri G (2005) The call of the wild: What can be learned from studies of SIV infection of natural hosts? In: Leitner T, Foley B, Hahn B, et al. (eds.) *HIV Sequence Compendium 2005*, pp. 2–29. Los Alamos, NC: Theoretical Biology and Biophysics Group.

Grossman Z, Meier-Schellersheim M, Paul WE, and Picker LJ (2006) Pathogenesis of HIV infection: What the virus spares is as important as what it destroys. *Nature Medicine* 12(3): 289–295.

Guatelli JC, Siliciano RF, Kuritzkes DR, and Richman DD (2002) Human immunodeficiency virus. In: Richman DD, Whitley RJ, and Hayden FG (eds.) *Clinical Virology,* 2nd edn., pp. 685–730. Washington: ASM Press.

Lederman MM, Penn-Nicholson A, Cho M, and Mosier D (2006) Biology of CCR5 and its role in HIV infection and treatment. *JAMA* 296(7): 815–826.

Human Respiratory Syncytial Virus

P L Collins, National Institute of Allergy and Infectious Diseases, Bethesda, MD, USA

Published by Elsevier Ltd.

Glossary

Fractalkine The sole known CX3C cytokine. It is present as a membrane-bound form that is upregulated by inflammation and promotes adhesion of monocytes and lymphocytes and as a secreted form that promotes chemotaxis of monocytes and lymphyocytes.

Furin A subtilisin-like cellular endoprotease present in the exocytic pathway with the consensus cleavage sequence Arg-X-Arg/Lys-Arg↓.

Glycosaminoglycan Long, unbranched, highly negatively charged polysaccharides that consist of repeating disaccharide subunits and are located on the cell surface or in the extracellular matrix.

Humanized antibody Substitution, by recombinant DNA techniques, of the antigen binding sequences of a nonhuman antibody molecule into the backbone of a human antibody molecule.

Mucin A family of large, heavily glycosylated, secreted or transmembrane proteins that form a protective barrier on the respiratory, gastrointestinal, and reproductive tracts.

Polylactosaminoglycan Carbohydrate of unknown function formed in the Golgi apparatus by the addition of a variable number of repeating units of lactosamine (galactose-β-1,4-N-acetylglucosamine-β-1,3) to an N-linked core oligosaccharide.

Rhinorrhea Runny nose.

Ribavirin A nucleoside analog (1-β-D-ribofuranosyl-1,2,4-triazole-3-carboxamide) that has antiviral activity against a number of RNA viruses.

Syncytium A multinucleated cell, formed in the case of human respiratory syncytial virus by fusion of the plasma membrane of an infected cell with those of its neighbors.

Tachykinin A family of biologically active peptides (prototype, substance P) that can be potent vasodilators and can induce smooth muscle contraction directly or indirectly.

History

Human respiratory syncytial virus (HRSV) was first isolated in 1956 from a laboratory chimpanzee with upper respiratory tract disease. Shortly thereafter, an apparently identical virus was recovered from two children ill with pneumonia or croup and was identified as a human virus. HRSV quickly became recognized as the leading viral agent of pediatric respiratory tract disease worldwide. It has also gained recognition as a significant cause of morbidity and mortality in the elderly and severely immunocompromised individuals. Its inefficient growth in cell culture and propensity to easily lose infectivity due to physical instability impede research. HRSV lacks an approved vaccine or clinically effective antiviral therapy, although, as described below, passive immunoprophylaxis is available for infants at high risk for serious HRSV disease.

Classification

HRSV is a member of family *Paramyxoviridae* of order *Mononegavirales*, the nonsegmented negative-strand RNA viruses. *Paramyxoviridae* has two subfamilies: (1) subfamily *Paramyxovirinae* contains five genera whose members include Sendai virus, the human parainfluenza viruses (HPIVs), and mumps and measles viruses among others; and (2) subfamily *Pneumovirinae* contains genus *Metapneumovirus*, which includes avian and human metapneumoviruses, and genus *Pneumovirus*, which includes HRSV, bovine respiratory syncytial virus (BRSV), ovine respiratory syncytial virus (RSV), and pneumonia virus of mice (PVM). **Figure 1**(a) compares gene maps of the two genera of *Pneumovirinae* with those of a representative genus (*Respirovirus*) of *Paramyxovirinae*. **Figure 1**(b) illustrates the relatedness among members of the two genera of *Pneumovirinae* based on the nucleotide sequence of the N gene. HRSV is most closely related to BRSV. One might speculate that HRSV originally emerged in the past from infection of humans with BRSV-like virus.

Virion Structure and Viral Proteins

HRSV virions consist of spherical particles of 80–350 nm in diameter and filamentous particles that are 60–200 nm in diameter and up to 10 μm in length (**Figure 2**). Filaments appear to be the predominant form produced in cell culture. The virion contains a helical nucleocapsid (diameter 12–15 nm compared to 18 nm for *Paramyxovirinae*) packaged in a lipoprotein envelope acquired from the host cell plasma membrane during budding. The outer surface of the envelope contains a fringe of surface projections or spikes of 11–12 nm. HRSV lacks a hemagglutinin or neuraminidase.

HRSV encodes 11 proteins. Four are associated with the nucleocapsid: the major nucleocapsid protein N (56 kDa), the nucleocapsid phosphoprotein P (33 kDa), the transcription processivity factor M2-1 (22 kDa), and the large L protein (250 kDa). The N protein binds tightly along the entire length of genomic RNA as well as its positive-sense replicative intermediate that is called the antigenome. L is the major polymerase subunit containing the catalytic domains. P is thought to associate with free N and L to maintain them in soluble form for association with the nucleocapsid and probably also participates as a cofactor in RNA synthesis. N, P, and L are the viral proteins that are necessary and sufficient to direct RNA replication. They also direct transcription, but this is poorly processive unless the M2-1 protein is also present.

There are three transmembrane viral glycoproteins that are expressed on the surface of infected cells and are packaged in the virion: the large G glycoprotein (90 kDa), the fusion F glycoprotein (70 kDa), and the small hydrophobic SH protein (7.5–60 kDa, depending on the amount of added carbohydrate). These assemble separately into homo-oligomers that constitute the surface spikes: F assembles into trimers, G assembles into trimers or tetramers, and SH has been detected as a pentamer.

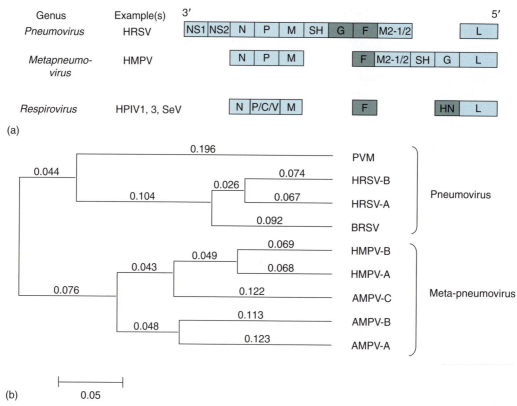

Figure 1 Comparison of human respiratory syncytial virus (HRSV) with other selected members of family *Paramyxoviridae*. (a) Alignment of the gene maps of HRSV, human metapneumovirus (HMPV), human parainfluenza virus (HPIV) types 1 and 3, and Sendai virus (SeV). (b) Relationships among the members of the two genera of subfamily *Pneumovirinae* based on the nucleotide sequence of the N gene. The scale bar represents approximately 5% nucleotide difference between pairs.

The G glycoprotein mediates viral attachment. It is a type II glycoprotein: there is a signal/anchor sequence near the N-terminus, with the C-terminal two-thirds of the molecule oriented extracellularly. G is heavily glycosylated, with several N-linked carbohydrate side chains and an estimated 24–25 O-linked side chains. The unglycosylated protein has an M_r of 32 500 compared to 90 000 for the mature protein. Most of the ectodomain is thought to have an extended, heavily glycosylated, mucin-like structure. In the middle of the ectodomain there is a predicted disulfide-linked tight turn (cystine noose) in the secondary structure. The cystine noose is overlapped on its N-terminal end by a 13-amino-acid conserved sequence of unknown significance and on its C-terminal end by a C–X3–C motif that it is embedded in a region of limited sequence relatedness with the CX3C chemokine fractalkine. A synthetic peptide containing this sequence has fractalkine-like chemotactic activity *in vitro*. The role of this apparent chemokine mimicry in the biology of HRSV is unclear. The tight turn, the conserved domain, and the fractalkine domain can be deleted from recombinant HRSV without significantly reducing its replication efficiency in cell culture or in the respiratory tract of mice. Thus, the region of G that is necessary for attachment has not been identified.

The F glycoprotein mediates cell-surface membrane fusion, which is responsible for delivering the viral nucleocapsid to the cytoplasm. In addition, F expressed at the cell surface mediates fusion of infected cells with their neighbors, resulting in the formation of syncytia. F is a type I glycoprotein, with an N-terminal cleaved signal anchor and a C-proximal membrane anchor. It has four or five N-linked carbohydrate side chains. F is synthesized as a precursor, F_0, which is cleaved intracellularly at two sites (amino acids 109/110 and 136/137) by a furin-like cellular protease. This yields the following fragments, in N- to C-terminal order: F_2 (109 amino acids), p27 (27 amino acids), and F_1 (438 amino acids). F_2 and F_1 remain linked by a disulfide bond and represent the active form of F. In BRSV, the p27 fragment has tachykinin activity and may play a role in promoting inflammation, but p27 of HRSV does not resemble a tachykinin.

The function of the SH glycoprotein is not known. Molecular modeling suggests that it might be an ion-channel-forming protein, although the significance of this to HRSV biology is unclear. SH is anchored in the

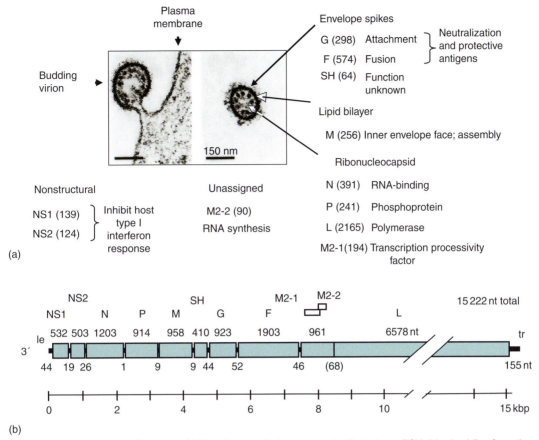

Figure 2 HRSV virion, proteins, and genome. (a) The electron photomicrographs illustrate an RSV virion budding from the plasma membrane of an infected cell (left) and a free virion (right). The viral proteins are indicated, with the amino acid length of the unmodified protein in parentheses. (b) The RNA genome is illustrated 3' to 5'. Each large rectangle indicates a gene that is transcribed into a separate mRNA, with the gene nucleotide length shown above. The overlapping open reading frames of the M2 gene are indicated. The nucleotide lengths of the extragenic leader (le), intergenic, and trailer (tr) regions are shown underneath. The overlap between the M2 and L genes is indicated by its nucleotide length in parentheses. Nucleotide and amino acid lengths are for strain A2.

membrane by a centrally located signal/anchor sequence, with the N-terminus oriented intracellularly and the C-terminus extracellularly. A portion of SH present intracellularly and in virions is unglycosylated (M_r 7500), a portion has a single N-linked side chain (M_r 13 000–15 000), and a portion has the further modification of polylactosaminoglycan added to the N-linked sugar (M_r 21 000–60 000 or more). This multiplicity of forms is conserved in *Pneumovirinae* but its significance is unknown.

Remarkably, the G and SH genes can be deleted individually or in combination from viable HRSV. Deletion of SH does not reduce the efficiency of replication in cell culture and has only a modest attenuating effect on replication in mice, chimpanzees, or humans (the last observation is based on trials of experimental live vaccines). Deletion of G was attenuating in HEp-2 cells but not Vero cells, and was highly attenuating in mice and humans. Since SH and G are dispensable, at least in cell culture, it is thought that F can function as an alternative attachment protein in that setting.

The nonglycosylated M protein is thought to be located on the inner surface of the envelope and to have a central role in organizing the envelope and directing packaging of the nucleocapsid.

The M2-2 protein is expressed at a low level and its status as structural or nonstructural is unknown. Recombinant HRSV from which the M2-2 coding sequence was deleted replicates more slowly than wild-type HRSV in cell culture and is attenuated *in vivo*. This deletion virus exhibits an increase in transcription and a decrease in RNA replication, suggesting that M2-2 normally is involved in downregulating transcription and upregulating RNA replication.

NS1 and NS2 are small proteins and do not appear to be packaged significantly in the virion. Both proteins inhibit the host interferon response. NS1, and to a lesser extent NS2, inhibit the induction of interferon α/β by blocking activation of interferon regulatory factor 3. In addition, NS2, and to a lesser extent NS1, inhibit interferon α/β-induced signaling through the Janus kinases/ signal transducers and activators of transcription (JAK/STAT) pathway, thus inhibiting the amplification of the interferon response, the upregulation of interferon-stimulated genes, and the development of the antiviral state. This appears to involve enhanced degradation of STAT2. Deletion of NS1 and NS2 individually or in combination yields viable viruses that replicate almost as efficiently as wild-type HRSV in cells that do not make interferon, but which are attenuated in interferon-competent cultured cells and *in vivo*.

Genome Organization, Transcription, and Replication

The HRSV genome is a single negative-sense strand of RNA that ranges in length from 15 191 to 15 226 nt for the strains that have been sequenced to date (**Figure 2(b)**). Its organization, expression, and replication follow the general pattern of *Mononegavirales*. The genome is neither capped nor polyadenylated. It is tightly encapsidated with N protein both in the virion and intracellularly. Based on the prototype A2 strain, the genomic RNA has a 44 nt 3′-extragenic leader at the 3′ end. This is followed in order by the NS1, NS2, N, P, M, SH, G, F, M2, and L genes. The L gene is followed by a 155 nt extragenic trailer region that comprises the 5′ end. The first nine genes are separated by intergenic regions of 1–52 nt. The last gene, L, initiates within the downstream nontranslated region of the M2 gene, and thus these two genes overlap by 68 nt.

In transcription, the viral polymerase initiates at a single promoter at the 3′ end of the genome and copies the genes by a sequential stop–start mechanism that yields subgenomic mRNAs. The upstream and downstream boundaries of each gene contain transcription signals: the upstream end consists of a conserved 10 nt gene-start motif that directs transcriptional initiation, and the downstream end consists of a moderately conserved 12–13 nt gene-end motif that directs termination and polyadenylation. The intergenic regions appear to lack any conserved motifs and seem to be nonspecific spacers. Between genes, the majority of polymerase remains template-bound and scans for the next gene. However, some polymerase dissociates from the template during transcription, yielding a polar gradient such that upstream genes are expressed more efficiently than downstream ones.

The HRSV mRNAs contain a virally synthesized 5′ cap [$m^7G(5′)ppp(5′)Gp$] and a 3′ polyadenylate tail. The latter is produced by reinterative copying by the viral polymerase on a tract of 4–7 U residues at the downstream end of each gene-end signal. Each mRNA encodes a single major viral protein except for the M2 mRNA, which contains two open reading frames (ORFs) that overlap by approximately 32 nt and encode two distinct proteins, M2-1 and M2-2 (**Figure 2(b)**). Expression of the downstream M2-2 ORF appears to be coupled to translation of the upstream M2-1 ORF and presumably involves a ribosomal stop–restart mechanism.

In RNA replication, the viral polymerase initiates in the same promoter region and ignores the gene-start and gene-end signals to make a full-length positive-sense copy called the antigenome. The antigenome is neither capped nor polyadenylated. It is tightly encapsidated with N protein and serves as the template for the synthesis of progeny genomes.

The first 11 nt at the 3′ end of the viral genome are essential for both transcription and RNA replication. The two processes involve overlapping sets of residues. For replication, nucleotides 16–32 are required in addition for encapsidation and the synthesis of complete antigenome. For transcription, nucleotides 36–43 are required in addition for full activity. The first 24 nt at the 3′ end of the antigenome have 81% sequence identity with the 3′ end of the genome, indicative of a conserved promoter element.

Viral Infection in Cell Culture and Experimental Animals

HRSV can infect and produce virus, with varying degrees of efficiency, in a surprisingly wide array of cell lines from various tissues and hosts. Replication is most efficient in epithelial cells, in particular the human HEp-2 line. HRSV that has been propagated *in vitro* generally retains its virulence.

Efficient infection by HRSV of established cell lines *in vitro* involves binding to cellular glycosaminoglycans, in particular heparan sulfate and chondroitin sulfate. Both G and F can mediate this binding. It is not known whether this represents attachment in total or whether it is an initial interaction that is followed by a second, higher-affinity step that remains to be identified. It is also not known how closely this mirrors the attachment process *in vivo*.

All events in HRSV infection occur in the cytoplasm. Viral mRNAs and proteins can be detected intracellularly by 4–6 h post infection. The accumulation of mRNAs has been reported to plateau by 14–18 h. This apparent shut-off of transcription might be due to accumulation of the M2-2 regulatory factor, since it does not appear to occur when the M2-2 coding sequence has been deleted. Apart from this, there is no apparent temporal regulation of gene expression. The release of progeny virus begins by 10–12 h, reaches a peak after 24 h, and continues until the cells deteriorate by 30–48 h. Although most of the infected cells in the culture are killed, persistent infections *in vitro* can readily be established.

In an *in vitro* model of human airway epithelium, HRSV infection was strictly limited to ciliated cells of the apical surface. Remarkably, HRSV infection caused little gross cytopathic effect despite ongoing infection over several weeks. In particular, infection did not result in the formation of syncytia. The expression of the F protein was polarized to the apical surface, which might have prevented contact with neighboring cells and might explain the lack of syncytia. Thus, HRSV is not inherently a highly cytopathic virus in the absence of syncytia formation, and much of the damage to infected cells that occurs *in vivo* might be a consequence of immune attack rather than direct viral effects. HRSV infection causes a modest reduction in cellular DNA, RNA, and protein synthesis. Apoptosis does not occur until late in infection. The virus buds at the plasma membrane. In polarized cells, this occurs at the apical surface.

HRSV does not replicate to high titer *in vitro*. Most of the virus remains cell associated and is released by freeze–thawing or sonication. HRSV can rapidly lose infectivity during unfrozen storage or freeze–thawing. Stability can be improved by adjusting the harvested culture supernatants to pH 7.5 and 0.1 M magnesium sulfate, or by including 15% w/v sucrose. The yield of virus is relatively low: 10 PFU per cell is typical. Most of the infectious particles produced *in vitro* were trapped by a 0.45 μm filter, suggesting that the infectious unit is in the form of large filaments or aggregates. The instability and size heterogeneity of the virus produced *in vitro* makes purification and concentration difficult.

The most permissive hosts for HRSV are the human and the chimpanzee, and these are the only hosts in which respiratory tract disease can reliably be observed. HRSV can also replicate in the respiratory tract of several species of monkey as well as in hamsters, guinea pigs, infant ferrets, cotton rats, and mice. However, infection in these animals is only semipermissive, particularly in the case of rodents. In the BALB/c mouse, the most commonly used experimental model, intranasal infection with 10^6 PFU yields only 10^6 PFU in lungs harvested at the peak of virus replication 4 days post infection.

Genetics

Like RNA viruses in general, HRSV has a high misincorporation rate (approximately one substitution in 10^4 nt), providing the potential for diversity. RNA recombination involving exchange between viruses appears to be exceedingly rare. Defective interfering particles presumably occur but have not been described molecularly.

Negative-sense RNA is not directly infectious alone. However, as for *Mononegavirales* in general, complete infectious recombinant virus can be produced entirely from cloned cDNAs (reverse genetics). This involves transfecting cells with plasmids that express a complete antigenomic RNA and the proteins of the nucleocapsid, which in the case of HRSV are the N, P, M2-1, and L proteins. These components assemble (inefficiently) into a nucleocapsid that initiates a productive infection. This method can be used to introduce desired changes into infectious virus via the cDNA intermediate.

Another system that has been very useful for basic studies involves helper-dependent mini-replicons. These are short, internally truncated versions of genomic or antigenomic RNA in which the viral genes have been replaced by one or more reporter genes under the control of HRSV transcription signals. When complemented by the appropriate mix of plasmid-encoded HRSV proteins, a mini-replicon can be encapsidated, transcribed, replicated, and packaged into virus-like particles. The small size and simplicity of a mini-replicon makes it ideal for detailed structure–function studies.

Antigens and Antigenic Subgroups

The F and G proteins are the only known HRSV neutralization antigens. They have markedly different antigenic properties. Many of the available F-specific monoclonal antibodies (MAb's) efficiently neutralize HRSV *in vitro*, whereas most of those for G neutralize weakly or not at all. However, polyclonal antibodies specific to G neutralize infectivity efficiently. The high content of O-linked carbohydrate in the G protein may be a factor in its unusual antigenic properties, and indeed the sugars have been shown to be important, directly or indirectly, in the binding of many, but not all, MAb's. The sugar side chains might mask the viral polypeptide. In addition, microheterogeneity might exist in the placement and composition of the side chains, which might alter or mask epitopes. In seropositive humans, most of the HRSV proteins have been shown to stimulate HRSV-specific memory CD8+ T lymphocytes.

HRSV has a single serotype. However, isolates can be segregated into two antigenic subgroups A and B. These exhibit a fourfold difference in cross-neutralization *in vitro* by postinfection sera. Epitopes in F tend to be conserved whereas those for G are not, such that antigenic relatedness between the two subgroups is greater than 50% for the F protein compared to only 5% or less for G. At the amino acid level, F is 89% identical between subgroups while G is the most divergent of the proteins and is only 53% identical, with most of the diversity occurring in the ectodomain. Other proteins range from 76% (SH) to 96% identical (N). Thus, the two antigenic subgroups have substantial differences throughout the genome and represent two divergent lines of evolution, as opposed to differing at only a few antigenic sites.

In a 2 year window following infection, there was a 64% reduction in the incidence of infection by the same subgroup versus a 16% reduction against the heterologous subgroup. In epidemics, typically, there is an alternating pattern with a 1–2 year interval with regard to the predominant subgroup. These observations indicate that antigenic dimorphism contributes to the ability of HRSV to reinfect, although its effect is modest rather than absolute.

Viral isolates obtained over successive years or decades exhibit some evidence of progressive amino acid changes and some changes in reactivity detectable with MAb's. However, all isolates retained a high level of antigenic relatedness and thus HRSV did not exhibit significant antigenic drift during this time frame.

Epidemiology and Clinical Factors

HRSV is the most important cause of serious viral respiratory tract disease in the pediatric population worldwide. In many areas it outranks all other microbial pathogens as a cause of bronchiolitis and pneumonia in infants less than 1 year of age. HRSV is conservatively estimated to cause 64 million infections and 160 000 deaths annually in the pediatric population. In the United States, HRSV is estimated to account for 73 400–126 300 hospitalizations annually for bronchiolitis and pneumonia in children of under 1 year of age, and is responsible for 90–1900 pediatric deaths annually. Hospitalization rates in developed countries are approximately one in 100–200 infections.

In a 13 year study of infants and children in the United States, HRSV was detected in 43%, 25%, 11%, and 10% of those hospitalized for bronchiolitis, pneumonia, bronchitis, and croup, respectively, and in aggregate accounted for 23% of pediatric hospitalizations for respiratory tract disease as a conservative estimate. Middle ear infection is a common complication. As another complication, approximately 40–50% of infants hospitalized with HRSV bronchiolitis have subsequent episodes of wheezing during childhood. HRSV infection also can exacerbate asthma.

Infection and reinfection with HRSV are frequent during the first years of life. Sixty to seventy percent of infants and children are infected by HRSV during the first year of life. The peak of serious disease is approximately 1–2 months of age (**Figure 3**). By age 2, 90% of children have been infected once with HRSV and 50% have been infected twice. The greatest incidence of serious disease occurs between 6 weeks and 6 months of age. Maternal antibodies account for the relative sparing of newborns. The major risk factors for serious HRSV disease in the pediatric population include chronic lung disease, congenital heart disease, prematurity, and young age (<6 months). While underlying disease is an important risk factor, the majority of infants hospitalized for HRSV disease were previously healthy.

Reinfection of older children and adults is common. For example, hospital staff members on pediatric wards

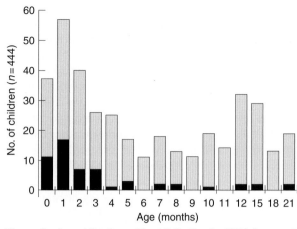

Figure 3 Age at the time of hospitalization for RSV disease at the Johns Hopkins Hospital during 1993–96. The filled bars indicate more severe disease. Adapted from Karron RA, Singleton RJ, Bulkow L, et al. (1999). Severe respiratory syncytial virus disease in Alaska native children. *Journal of Infectious Diseases* 180: 41–49.

have an infection rate of 25–50% during HRSV epidemics. Healthy individuals usually suffer only a common cold syndrome. However, HRSV is also an important cause of morbidity and mortality in the elderly, with an impact that is estimated to approach that of nonpandemic influenza virus. HRSV has been estimated to cause 17 000 deaths annually in the elderly in the United States. HRSV infection is also associated with a high mortality rate in children and adults with immunodeficiency or immunosuppression, particularly hematopoietic cell transplant recipients.

HRSV is highly contagious and is readily spread in day-care settings, to family members, in nursing homes, and in the hospital. Spread involves close contact and inoculation of conjunctival or mucosal surfaces by large droplets or by contaminated hands or objects. In experimental infections, HRSV had an incubation period of 4–5 days. Virus typically is shed for 7–12 days coincident with clinical disease, although sometimes shedding is longer and can continue after recovery.

Most primary infections are symptomatic, with upper respiratory tract disease and sometimes a fever. Twenty-five to forty percent of primary infections progress to the lower respiratory tract with the primary manifestations of serious disease being bronchiolitis or pneumonia. There is profuse rhinorrhea, coughing, intermittent fever, expiratory wheezes, and, frequently, middle-ear disease. Seriously ill infants have increased coughing and wheezing, rapid respiration, and hypoxemia, requiring the administration of humidified oxygen.

HRSV causes yearly epidemics that are centered in the winter months in temperate climates (**Figure 4**) or in the rainy season in the tropics, although there can also be local variations to this pattern.

The most definitive diagnosis is by virus isolation in cell culture. Rapid diagnosis can be made by (1) detection of viral antigen using an antigen-capture enzyme-linked immunosorbent assay (ELISA) or by an immunofluorescence analysis of exfoliated cells with HRSV-specific antibody or (2) detection of nucleic acid by reverse transcriptase polymerase chain reaction (RT-PCR). The sensitivity of detection can be over 90%.

Pathogenesis and Immunity

HRSV replicates primarily in the epithelial cells that line the lumen of the respiratory tract. It can also infect macrophages and dendritic cells and modify their functioning. Immunohistochemistry of airway tissues from infected individuals suggests a patchy distribution of infection, with only superficial cells expressing viral antigen. There is destruction of the epithelium and an influx of mononuclear cells, lymphocytes, plasma cells, and macrophages. Tissues become edematous and the secretion of mucus is excessive. Mucus, inflammatory cells, and debris from dead infected cells accumulate in the airways and can cause obstruction, which is a particular problem in infants due to the small diameter of their airways. With pneumonia, the alveolar spaces may fill with fluid and the interalveolar walls may thicken due to mononuclear cell infiltration.

HRSV is an acute infection that typically is completely cleared by host immunity, although sometimes shedding can persist for weeks. Studies in experimental animals and in the clinic indicate that virus-neutralizing secretory IgA antibodies, virus-neutralizing serum antibodies, and cytotoxic CD8+ T lymphocytes participate in resolving infection. These same effectors also confer protection against reinfection. Protection conferred by cytotoxic T lymphocytes appears to diminish within several weeks or months and thus is more important in the short term. Virus-neutralizing secretory IgA antibodies are particularly effective in restricting infection; this response is short-lived following primary infection but can increase in duration following reinfection. Virus-neutralizing serum antibodies provide long-lasting protection. However, they gain access to the respiratory lumen by the inefficient process of transudation, and thus protection is not efficient, particularly in the upper respiratory tract.

Perhaps the most important determinant of HRSV pathogenesis is its ability to infect young infants early in life despite the presence of maternal serum antibodies (**Figure 3**). Why HRSV seems more able than most other respiratory pathogens to evade maternal antibodies and infect very early in life is not understood. The fact that the virus can infect in early infancy when disease is less easily supported and airways are small and more easily obstructed probably accounts for much of its impact.

The immune response during the first months of life is reduced due to immunologic immaturity as well as the immunosuppressive effects of maternal antibodies on humoral responses. This probably contributes to difficulty in controlling the virus and facilitates reinfection early in life. The Th2-biased nature of immune responses in early infancy has also been speculated to contribute to HRSV disease and reduced immune responses. In addition, the tropism of HRSV to the superficial epithelium might reduce its exposure to immune effectors and might also provide reduced immune stimulation.

There is some evidence that HRSV might have suppressive effects on cell-mediated immunity, although the significance and magnitude of these effects are unclear. The cystine noose of the G protein has been shown to inhibit innate immune responses in human monocytes. The inefficiency of virus-neutralizing serum antibodies in controlling HRSV presumably is also a factor in its ability to reinfect throughout life.

Treatment and Immunoprophylaxis

Treatment of severe disease is largely supportive, including mechanical ventilation in the most severe cases. Ribavirin has potent antiviral activity in cell culture and experimental animals, but it has not been clearly beneficial clinically and is not routinely used in most hospitals. Anti-inflammatory therapies have not been beneficial to date

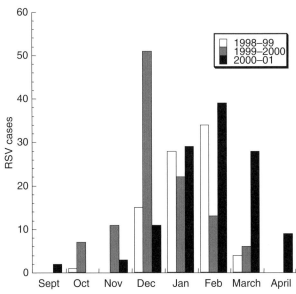

Figure 4 Variation in the timing of the RSV epidemic during three successive years at the Johns Hopkins Hospital. Unpublished data kindly provided by Karron RA, Johns Hopkins School of Hygiene and Public Health and School of Medicine.

but are still being investigated. Animal studies suggest a benefit of combining antiviral and anti-inflammatory therapies. Bronchodilators are commonly used but have, at most, modest benefits.

Passive immunoprophylaxis is available for infants who are at high risk for serious HRSV disease. This involves monthly intramuscular injections during the HRSV season of an HRSV-neutralizing MAb called palivizumab (Synagis), developed by MedImmune, Inc. This is a humanized version of a mouse MAb specific to the F protein.

A vaccine is not available. A major obstacle to developing a pediatric vaccine is that immunization ideally should start during the first weeks of life. As already noted, immune responses during the first few months of life are reduced due to immunologic immaturity and the immunosuppressive effects of maternal antibodies.

A vaccine consisting of formalin-inactivated, concentrated HRSV (FI-RSV) was evaluated in the 1960s. FI-RSV was found to be poorly protective and, unexpectedly, primed for immune-mediated enhanced HRSV disease that occurred on subsequent natural infection. In subsequent studies, FI-RSV was shown to induce antibodies that bound efficiently to viral antigen but did not efficiently neutralize infectivity. Antigen–antibody complexes might contribute to enhanced disease. In addition, compared to natural infection, FI-RSV caused a disproportionate stimulation of the Th2 subset of CD4+ T lymphocytes, whose role in enhanced disease was confirmed by depletion studies in experimental animals. Disease enhancement appears to be specific to killed or protein subunit HRSV vaccines and might occur because these nonreplicating vaccines do not efficiently stimulate natural killer cells and CD8+ T lymphocytes, which in turn play an important role in regulating the T-helper-cell response via secreted interferon γ. Disease enhancement appears to be an issue only for immunization early in life, and thus protein subunit vaccines appear to be safe in HRSV-experienced individuals.

Because of the phenomenon of vaccine-related disease enhancement, pediatric vaccine development is focused on the development of live-attenuated HRSV strains for intranasal administration. A number of candidate live vaccines developed by reverse genetics are presently in clinical trials. There is also a need for an HRSV vaccine for the elderly. Subunit vaccines are presently being developed for that age group.

See also: Human Respiratory Viruses; Paramyxoviruses.

Further Reading

Collins PL and Crowe JE, Jr. (2006) Respiratory syncytial virus and metapneumovirus. In: Knipe DM and Howley PM (eds.) *Fields Virology*, 5th edn., pp. 1601–1646. Philadelphia, PA: Lippincott Williams and Wilkins.

Collins PL and Murphy BR (2005) New generation live vaccines against human respiratory syncytial virus designed by reverse genetics. *Proceedings of the American Thoracic Society* 2: 166–173.

Crowe JE, Jr. and Williams JV (2003) Immunology of viral respiratory tract infection in infancy. *Paediatric Respiratory Reviews* 4: 112–119.

Hall CB (2001) Respiratory syncytial virus and parainfluenza virus. *New England Journal of Medicine* 344: 1917–1928.

Hussel T, Baldwin CJ, O'Garra A, and Oppenshaw PJ (1997) T cells control Th2-driven pathology during pulmonary respiratory syncytial virus infection. *European Journal of Immunology* 27: 3341–3349.

Karron RA, Singleton RJ, Bulkow L, et al. (1999). Severe respiratory syncytial virus disease in Alaska native children. *Journal of Infectious Diseases* 180: 41–49.

McGivern DR, Collins PL, and Fearns R (2005) Identification of internal sequences in the 3 leader region of human respiratory syncytial virus that enhance transcription and confer replication processivity. *Journal of Virology* 79: 2449–2460.

Melero JA, Garcia-Barreno B, Martinez I, Pringle CR, and Cane PA (1997) Antigenic structure, evolution, and immunobiology of human respiratory syncytial virus attachment (G) protein. *Journal of General Virology* 78: 2411–2418.

Murphy BR, Sotnikov AV, Lawrence LA, Banks SM, and Prince GA (1990) Enhanced pulmonary histopathology is observed in cotton rats immunized with formalin-inactivated respiratory syncytial virus (RSV) or purified F glycoprotein and challenged with RSV 3–6 months after immunization. *Vaccine* 8: 497–502.

Polack FP, Irusta PM, Hoffman SJ, et al. (2005) The cysteine-rich region of the respiratory syncytial virus attachment protein inhibits the innate immune response elicited by the virus and endotoxin. *Proceedings of the National Academy of Sciences, USA* 102: 8996–9001.

Singleton RJ, Bulkow L, et al. (1999). Severe respiratory syncytial virus disease in Alaska native children. *Journal of Infectious Diseases* 180: 41–49.

Spann KM, Collins PL, and Teng MN (2003) Genetic recombination during co-infection of two mutants of human respiratory syncytial virus. *Journal of Virology* 77: 11201–11211.

Teng MN and Collins PL (2002) The central conserved cystine noose of the attachment G protein of human respiratory syncytial virus is not required for efficient viral infection *in vitro* or *in vivo*. *Journal of Virology* 76: 6164–6171.

Tripp RA, Jones LP, Haynes LM, et al. (2001) Cytokine mimicry by respiratory syncytial virus G glycoprotein. *Nature Immunology* 2: 732–738.

Wright PF, Karron RA, Madi SA, et al. (2006) The interferon antagonist NS2 protein of respiratory syncytial virus is an important virulence determinant in humans. *Journal of Infectious Diseases* 193: 573–581.

Young J (2002) Development of a potent respiratory syncytial virus-specific monoclonal antibody for the prevention of serious lower respiratory tract disease in infants. *Respiratory Medicine* 96 (supplement B): S31–S35.

Zhang L, Peeples ME, Boucher RC, et al. (2002) Respiratory syncytial virus infection of human airway epithelial cells is polarized, specific to ciliated cells, and without obvious cytopathology. *Journal of Virology* 76: 5654–5666.

Human Respiratory Viruses

J E Crowe Jr., Vanderbilt University Medical Center, Nashville, TN, USA

© 2008 Elsevier Ltd. All rights reserved.

Glossary

Bronchiolitis A disease condition characterized by trapping of air in the lungs with difficulty expiring (i.e., wheezing), caused by inflammation or infection of the bronchioles, the smallest and highest-resistance airways.

Croup A disease condition characterized by a difficulty in inspiration, associated with a barky cough, caused by inflammation or infection of the larynx, trachea, and bronchi.

Lower respiratory tract The anatomical region below the vocal cords, including the trachea, bronchi, bronchioles, and lung.

Pneumonia Infection of the alveolar space of the lungs.

Introduction

Respiratory virus infections of humans are the most common and frequent infections of man. Hundreds of different viruses can be considered respiratory viruses. Viruses that enter through the respiratory tract include viruses that replicate and cause disease that is restricted to the respiratory epithelium, and other viruses that enter through the mucosa but also spread by viremia causing systemic disease. An example of the latter is measles virus. SARS coronavirus is another. In general, viruses that do not cause viremia are capable of reinfecting the same host multiple times throughout life. In contrast, infections with systemic viruses induce lifelong immunity. Probably, the high rate of reinfection of mucosally restricted viruses reflects the difficulty and metabolic cost of maintaining a high level of immunity at the vast surface area of the mucosa. Virus-specific IgA levels are maintained at high levels generally only for several weeks or months after infection.

The Human Respiratory Tract

The anatomy and the cell types of the respiratory tract dictate to a large degree the type of disease observed during respiratory virus infection. The demarcation between the upper and lower respiratory tracts is the vocal cords. The structures of the upper respiratory tract, which are all interconnected, include the nasopharynx, the larynx, the Eustachian tube and middle ear space, and the sinuses. Significant collections of lymphoid tissue reside in the upper respiratory tract, the tonsils and the adenoids. The lower respiratory tract structures include the trachea, bronchi, bronchioles, alveoli, and lung tissue. The cell types that line the respiratory tract are complex, and exhibit different susceptibilities to virus infection. The predominant cell types are ciliated and nonciliated epithelial cells, goblet cells, and Clara cells. Smooth muscle cells are prominent features of the airways down to the level of the bronchioles, and the lung possesses type I and II pneumocytes.

Disease Syndromes

The disease syndromes that are associated with respiratory viruses generally follow the anatomy of the respiratory tract. Different viruses appear to have tropisms for different cells or regions of the respiratory tract; therefore, there are particular associations of viruses with clinical syndromes. The clinical diagnoses for infections with disease manifestations in the respiratory tract are rhinitis and the common cold, sinusitis, otitis media, conjunctivitis, pharyngitis, laryngitis, tracheitis, acute bronchitis, bronchiolitis, pneumonia, and exacerbations of reactive airway disease or asthma. Clinical syndromes with more systemic illness due to respiratory viruses include the influenza syndrome, measles, severe acute respiratory syndrome (SARS), and hantavirus pulmonary syndrome (HPS).

Viruses That Cause Respiratory Illness in Immunocompetent Humans

The principal causes of acute viral respiratory infections in children became apparent through large epidemiologic studies conducted soon after cell culture techniques became available. The landmark studies of association of viruses with clinical syndromes were performed in the 1960s and 1970s. Recent studies have increased our understanding of the causes of viral respiratory infection in infants, especially because of the advent of molecular tests such as the polymerase chain reaction (PCR), which is more sensitive than cell culture. Respiratory syncytial virus (RSV), parainfluenza viruses (PIVs), adenoviruses, and influenza viruses were identified initially as the most common causes of serious lower respiratory tract disease in infants and children. More recently, human metapneumovirus (hMPV) was identified as a major cause of serious illness. In the last 10 years, a number of additional viruses have been associated with respiratory illness, as discussed below. However, still, infectious agents are not identified in 30–50% of clinical illnesses in large surveillance studies, even using sensitive diagnostic techniques such as viral culture on multiple cell lines, antigen detection assays, and RT-PCR based methods. It is not known if these illnesses are due to identified pathogens that are simply not detected due to low titers of virus in patient samples or if there are novel agents that are yet to be identified.

Immunocompromised Hosts

Reactivation of latent viruses, such as the herpesviruses HSV and CMV, and adenoviruses occurs in immunocompromised humans, particularly subjects with late-stage HIV infection, those with organ transplantation, and patients with leukopenia and neutropenia caused by chemotherapy. CMV is the most frequently recovered virus from diagnostic procedures such as bronchoalveolar lavage in immunosuppressed patients with pneumonia. These patients also suffer more frequent and more severe disease including mortality with common respiratory viruses, including RSV, hMPV, PIV, influenza viruses, rhinoviruses, and adenoviruses. Nosocomial transmission including large unit outbreaks is not uncommon, and can result in high frequency of transmission.

Specific Viral Causes of Respiratory Disease

Picornaviridae

A wide variety of picornaviruses cause respiratory disease, including rhinoviruses, the enteroviruses A to D including coxsackieviruses A/B, echoviruses, non-polio enteroviruses, and parechoviruses 1–3. Enterovirus infections occur most commonly in the summer months in temperate areas, which differs from the season of many of the other most common respiratory viruses such as paramyxoviruses and influenza virus. Rhinovirus infections occur year-round.

Rhinoviruses

Rhinovirus is a genus of the family *Picornaviridae* of viruses. Rhinoviruses are the most common viral infective agents in humans, and a causative agent of the common cold. There are over 105 serologic virus types that cause cold symptoms, and rhinoviruses are responsible for approximately half of all cases of the common cold. Rhinoviruses have single-stranded positive-sense RNA genomes. The viral particles are icosahedral in structure, and they are nonenveloped. Rhinovirus-induced common colds may be complicated in children by otitis media and in adults by sinusitis. Most adults, in fact, have radiographic evidence of sinusitis during the common cold, which resolves without therapy. Therefore the primary disease is probably best termed rhinosinusitis. Rhinovirus infection is associated with exacerbations of reactive airway disease in children and asthma in adults. It is not clear whether rhinovirus is restricted to the upper respiratory tract and induces inflammatory responses that affect the lower respiratory tract indirectly, or whether the viruses spread to the lower respiratory tract. In the past, it was thought that these viruses did not often replicate or cause disease in the lower respiratory tract. However, recent studies discern strong epidemiological associations of RVs with wheezing and asthma exacerbations, including episodes severe enough to require hospitalization. Likely, rhinoviruses can infect the lower airways to some degree, inducing a local inflammatory response. Another possibility is that significant local infection of the upper respiratory tract might induce regional elaboration of mediators that causes lower airways disease. Association of rhinovirus infection with lower respiratory tract illness is difficult to study because cell diagnosis by cell culture is not sensitive. RT-PCR diagnostic tests are difficult to interpret because they are often positive for prolonged periods of time and even asymptomatic individuals may have a positive test. Comprehensive serologies to confirm infection are difficult because of the large number of serotypes. Nevertheless, most experts believe rhinoviruses are a common cause of lower respiratory tract illness.

Coxsackieviruses

These viruses cause oral lesions and often are associated in children with a disease syndrome termed 'hand-foot-and-mouth disease'. The pharyngitis associated with this infection often is marked by the very characteristic findings of herpangina, a clinical syndrome of ulcers or small vesicles on the palate and often involving the tonsillar fossa associated with the symptoms of fever, difficulty swallowing, and throat pain. Outbreaks commonly occur in young children, in the summer.

Enteroviruses

Non-polio enteroviruses are common and distributed worldwide. Although infection often is asymptomatic, these viruses cause outbreaks of clinical respiratory disease, sometimes with fatal consequences. Studies have associated particular types with clinical syndromes, as enterovirus 68 with wheezing and enterovirus 71 with pneumonia.

Echoviruses

The term 'echo' in the name of the virus is an acronym for enteric cytopathic human orphan, although this may be an archaic notion since most echoviruses are associated with human diseases, most commonly in children. There are at least 33 echovirus serotypes. Echoviruses can be isolated from many children with upper respiratory tract infections during the summer months. Echovirus 11 has been associated with laryngotracheitis or croup. Epidemiology studies also have associated echoviruses with epidemic pleurodynia, an acute illness characterized by sharp chest pain and fever.

Parechoviruses

These viruses have been assigned a new genus of the family *Picornaviridae* because of distinctive laboratory-based molecular properties. The most common member of the genus *Parechovirus*, human parechovirus 1 (formerly echovirus 22) is a frequent human pathogen. The genus also includes the closely related virus, human parechovirus 2 (formerly echovirus 23). Human parechoviruses usually cause mild respiratory or gastrointestinal illness. Most infections occur in young children. There is a high seroprevalence for parechoviruses 1 and 2 in adults, and a few clear descriptions of neonatal cases of severe disease.

Paramyxoviridae

Respiratory syncytial virus

RSV is a single-stranded negative-sense nonsegmented RNA genome virus of the family *Paramyxoviridae*, genus *Pneumovirus*. It is one of the most infectious viruses of humans and infects infants at a very young age, often in the first weeks or months of life. It is the most common viral cause of severe lower respiratory tract illness in children and one of the most important causes of hospitalization in infants and children throughout the world. There is one serotype, but circulating viruses exhibit an antigenic dimorphism such that there are two antigenic subgroups designated A and B. Reciprocal cross-neutralization studies using human sera showed that the antigenic groups are about 25% related. Reinfection is common and can be caused by viruses of the same subgroup. Yearly, epidemics of disease often peak between January and March in temperate regions. RSV infection causes mild upper respiratory tract infection in most infants and young children, but results in hospitalization in 0.5–1% of infants. Most children have been infected by the age of 2 years. There is an association of RSV infection early in life and subsequent asthma, although a causal relationship is controversial. Most hospitalized infants are otherwise healthy, but some groups are considered high risk for severe disease such as premature infants especially those with chronic lung disease and infants born with congenital heart disease. Immunocompromised patients of any age are at risk of severe disease.

Human parainfluenza viruses

These viruses constitute a group of four distinct serotypes (types 1–4) of single-stranded RNA viruses belonging to the family *Paramyxoviridae*. When considered as a group, they are the second most common cause of lower respiratory tract infection in young children. PIV3 is the most common cause of severe disease. Repeated infection throughout life is common. First infections are more commonly associated with lower respiratory tract disease, especially croup, while subsequent infections typically are limited to the upper respiratory tract. PIVs are detected using cell culture with hemadsorption or immunofluorescent microscopy, and RT-PCR.

Human metapneumovirus

In 2001, investigators in the Netherlands described a new human respiratory virus, hMPV. Evidence of near universal seroconversion was found in the general population by 5 years of age, suggesting ubiquitous infection in early childhood. This virus, a member of the genus *Pneumovirus* with RSV, differs from RSV in that it lacks the NS1 and NS2 nonstructural genes that counteract host interferons and it possesses a slightly different gene order. Studies of the role of hMPV in pediatric lower respiratory tracts infection (LRI) in otherwise healthy children in the United States, using a prospectively collected 25-year database and sample archive representing about 2000 children, revealed that nearly 12% of LRI in children was associated with a positive hMPV test. This and similar studies suggested that the virus is one of the major respiratory pathogens of early childhood. The clinical features of hMPV LRI were similar to those of other paramyxoviruses, most often resulting in cough, coryza, and a syndrome of bronchiolitis or croup. Interestingly, hMPV seemed to be clinically intermediate between RSV and PIV in that it tended to cause bronchiolitis with similar frequency to RSV but more frequently than PIV, while causing croup less often than the latter. Studies in subjects with conditions predisposing to increased risk of respiratory illness suggest that hMPV plays a significant role in exacerbations of asthma in children and adults, LRI in immunocompromised subjects, and in the frail and elderly.

Measles virus

Measles virus, a paramyxovirus of the genus *Morbillivirus* causes infection with systemic disease, also known as rubeola. The virus is spread both by direct contact/fomite transmission and by aerosol transmission, and therefore is one of the most highly contagious infections of man. The classical symptoms of measles include 3 or more days of fever that is often quite high and a clinical constellation of symptoms termed 'the three Cs': cough, coryza, and conjunctivitis. A characteristic disseminated maculopapular rash appears soon after onset of fever. Transient mucosal lesions in the mouth of a characteristic appearance (Koplik's spots) are considered diagnostic when identified by an experienced clinician. The virus causes a number of systemic effects and can be complicated by severe pneumonia, especially when primary infection occurs in an unvaccinated adult or immunocompromised person of any age. Mortality in developing countries is high, especially when infection occurs in the setting of malnutrition.

Hendra and Nipah viruses

These emerging pathogens that are grouped in their own new genus *Henipaviruses* may not be respiratory pathogens in a conventional sense, but they are paramyxoviruses that probably infect humans by the respiratory route. Nipah virus is a newly recognized zoonotic virus, named after the location in Malaysia where it was first identified in 1999. It has caused disease in humans with contact with infectious animals. Hendra virus (formerly called equine morbillivirus) is another closely related zoonotic paramyxovirus that was first isolated in Australia in 1994. The viruses have caused only a few localized outbreaks, but their wide host range and ability to cause high mortality raise concerns for the future. The natural host of these viruses is thought to be a certain species of fruit bats present in Australia and the Pacific. Pigs may be an intermediate host for transmission to humans in Nipah infection, and horses in the case of Hendra. Although the mode of transmission from animals to humans is not defined, it is likely that inoculation of infected materials onto the respiratory tract plays a role. The clinical presentation usually appears to be an influenza-like syndrome, progressing to encephalitis, may include respiratory illness, and causes death in about half of identified cases.

Orthomyxoviridae

Influenza viruses

Influenza is a single-stranded segmented negative-sense RNA genome virus of the family *Orthomyxoviridae*. There are three types of influenza viruses: influenza virus A, influenza virus B, and influenza virus C. Influenza A and C infect multiple species, while influenza B infects humans almost exclusively. The type A viruses are the most virulent human pathogens among the three influenza types, and cause the most severe disease. The influenza A virus can be subdivided into different subtypes based on the antibody response to these viruses. The subtypes that have been confirmed in humans in seasonal influenza, ordered by the number of known human pandemic deaths, are: H1N1 which caused the 1918 pandemic, and H2N2 which caused the 1957 pandemic of avian influenza that originated in China, H3N2 which caused the pandemic of 1968. Currently, H3N2, H1N1, and B viruses cause annual seasonal epidemics. In addition, H5N1 virus infection of humans occurred during an epizootic of H5N1 influenza in Hong Kong's poultry population in 1997. The disease affected animals of many species and exhibited a high rate of mortality in humans. The virus is spreading throughout Asia, carried by wild birds. Human-to-human transmission does not occur efficiently at this time; however, there is widespread current concern about the potential for an H5N1 pandemic if the virus acquired transmissibility among humans. The H7N7 avian virus also has unusual zoonotic potential. In 2003 this virus caused an outbreak in humans in the Netherlands associated with an outbreak in commercial poultry on several farms. One death occurred and 89 people were confirmed to have H7N7 influenza virus infection. H1N2 virus appears to endemic in pigs and humans. H9N2, H7N2, H7N3, and H10N7 human infections have been reported. Influenza B virus is almost exclusively a human pathogen, and is less common than influenza A. It mutates less rapidly than influenza A, and there is only one influenza B subtype. In humans, common symptoms of influenza infection and syndrome are fever, sore throat, myalgias, headache, cough, and fatigue. In more serious cases, influenza causes pneumonia, which can be fatal, particularly in young children and the elderly. Influenza pneumonia has an unusually high rate of complication by bacterial superinfection with staphylococcal and streptococcal bacterial pneumonia occurring in as many as 10% of cases in some clinical series.

Adenoviridae

Viruses of the family *Adenoviridae* infect both humans and animals. Adenoviruses were first isolated in human lymphoid tissues from surgically removed adenoids, hence the name of the virus. In fact, some serotypes establish persistent asymptomatic infections in tonsil and adenoid tissues, and virus shedding can occur for months or years. These double-stranded DNA viruses are less than 100 nm in size, and have nonenveloped icosahedral morphology. The large dsDNA genome is linear and nonsegmented. There are six major human adenovirus species (designated A through F) that can be placed into 51 immunologically distinct serotypes. Human respiratory tract infections are mainly caused by the B and C species. Adenovirus infections can occur throughout the year. Sporadic outbreaks occur with many of the serotypes, while others appear to be endemic in particular locations. Respiratory illnesses include mild disease such as the common cold and lower respiratory tract illness, including croup, bronchiolitis, and pneumonia. Conjunctivitis is associated with infection by species B and D. There is a particular constellation of symptoms called 'pharyngoconjunctival fever' which is very frequently associated with acute adenovirus infection. In contrast, gastroenteritis has been associated most frequently with the serotype 40 and 41 virus of species F. Immunocompromised subjects are highly susceptible to severe disease during infection with respiratory adenoviruses. The syndrome of acute respiratory disease (ARD), especially common during stressful or crowded living conditions, was first recognized among military recruits during World War II and continues to be a problem for the military following suspension of vaccination. ARD is most often associated with adenovirus types 4 and 7.

Coronaviridae

Members of the genus *Coronavirus* also contribute to respiratory illness including severe disease. There are dozens of coronaviruses that affect animals. Until recently, only two representative strains of human coronaviruses were known to cause disease, human coronavirus 229E (HCoV-229E) and HCoV-OC43. A recent outbreak of SARS-associated coronavirus (SARS-CoV) showed that animal coronaviruses have the potential to cross species to humans with devastating effects. There has been one major epidemic to date, between November 2002 and July 2003, with over 8000 cases of the disease, and mortality rates approaching 10%. SARS-CoV causes a systemic illness with a respiratory route of entry. SARS is a unique form of viral pneumonia. In contrast to most other viral pneumonias, upper respiratory symptoms are usually absent in SARS, although cough and dyspnea occur in most patients. Typically, patients present with a nonspecific illness manifesting fever, myalgia, malaise, and chills or rigors; watery diarrhea may occur as well. Recently, investigators reported the identification of a fourth human coronavirus, HCoV-NL63, a new group 1 coronavirus. Evidence is emerging that HCoV-NL63 is a common respiratory pathogen of humans, causing both upper and lower respiratory tract illness. Human coronavirus (HCoV) HKU1 was first described in January 2005 following detection in a patient with pneumonia. Several cases of respiratory illness have been associated with the virus, but the infrequent identification suggests to date that this putative group 2 coronavirus causes a low incidence of illness.

Herpesviridae

Several herpes viruses cause upper respiratory infections, especially infection of the oral cavity. Herpes simplex pharyngitis is associated with characteristic clinical findings, such as acute ulcerative stomatitis and ulcerative pharyngitis. Herpes simplex virus 1 (HSV-1) and herpes simplex virus 2 (HSV-2), also called human herpesvirus 1 (HHV-1) and human herpesvirus 2 (HHV-2), respectively, cause oral lesions, although over 90% of oral infections are caused by HSV-1. Primary oral disease can be severe, especially in young children, who sometimes are admitted for rehydration therapy due to poor oral intake. A significant proportion of individuals suffer recurrences of symptomatic disease consisting of vesicles on the lips. Epstein–Barr virus (EBV) mononucleosis syndrome is often marked by acute or subacute exudative pharyngitis; in some cases, the swelling of the tonsils in EBV pharyngitis is so severe that airway occlusion appears imminent. Most of the viruses of the family *Herpesviridae* can cause severe disease in immunocompromised patients (especially hematopoietic stem cell transplant patients), including cytomegalovirus (CMV), EBV, varicella-zoster virus, herpesvirus 6, herpesvirus 7, and herpesvirus 8.

Parvoviridae

Human bocavirus

A new virus was identified recently in respiratory samples from children with lower respiratory tract disease in Sweden. Sequence analysis of the viral genome revealed that the virus is highly related to canine minute virus and bovine parvovirus and is a member of the genus *Bocavirus*, subfamily *Parvovirinae*, family *Parvoviridae*. This virus was tentatively named human bocavirus (HBoV). HBoV has been identified as the sole agent in a limited number of respiratory samples from children hospitalized with respiratory tract disease. It remains to be seen whether the virus is causative of or merely associated with disease in these preliminary studies.

Bunyaviridae

Hantavirus

Over 400 cases of HPS have been reported in the United States. The disease was first recognized during an outbreak in 1993. About a third of recognized cases end in death. The Four Corners area outbreak is well known; however, cases now have been reported in 30 states. Patients with HPS usually present with a febrile illness beginning with symptoms of a flu-like illness. Physical examination is not specific, often only with findings of fever, and increased heart and respiratory rates. In addition to the respiratory symptoms, abdominal pain and fever are common. Diagnosis is often delayed until a severe illness occurs requiring mechanical ventilation.

Reoviridae

Rotavirus

Rotaviruses are dsRNA enteric viruses that are the most common cause of severe viral infectious gastroenteritis in children. Clinical series suggest that some children with gastroenteritis suffer upper respiratory symptoms during the prodrome of disease manifestation, and virus can be recovered from respiratory secretions. Some reports suggest that rotavirus infection is associated with lower respiratory tract illness, although this association is unclear.

Reovirus

These dsRNA viruses (named using an acronym for respiratory enteric orphan virus) are not clearly associated with respiratory disease, but seroconversion rate is high in the first few years of life, and they probably cause minor or subclinical illness.

Retroviridae

Human immunodeficiency virus

Pharyngitis occurs with primary HIV infection and may be associated with mucosal erosions and lymphadenopathy.

Papovaviridae

Polyomaviruses

Polyomaviruses are small dsDNA genome nonenveloped icosahedral viruses that may be oncogenic. There are two polyomaviruses known to infect humans, JC and BK viruses. Eighty percent or more of adult US subjects are seropositive for these viruses. JC virus can infect the respiratory system, kidneys, or brain. BK virus infection causes a mild respiratory infection or pneumonia and can involve the kidneys of immunosuppressed transplant patients.

Co-Infections

Given the overlap in the winter season of these viruses in temperate areas, it is not surprising that co-infections with two or more viruses occur. In general, when careful studies using cell culture techniques were used for virus isolation, more than one virus was isolated from respiratory secretions of otherwise healthy subjects with acute respiratory illness in about 5–10% of cases in adults and 10–15% in children. There is little evidence that more severe disease occurs during co-infections, although there is insufficient evidence on this point to be definitive. The incidence of two molecular diagnostic tests being positive (generally RT-PCR, for these RNA viruses) is expected to be higher than that of culture, because molecular tests can remain positive for an extended period of time after virus shedding has ended.

Transmission

Respiratory viruses generally have two main modes of transmission, large particle aerosols of respiratory droplets transmitted directly from person-to-person by coughing or sneezing, or by fomites. Fomite transmission occurs indirectly when infected respiratory droplets are deposited on hands or on inanimate objects and surfaces with subsequent transfer of secretions to a susceptible subject's nose or conjunctiva. Most respiratory viruses, unlike measles virus or varicella zoster virus, do not spread by small particle aerosols across rooms or down halls. Therefore, contact and droplet precautions are sufficient to prevent transmission in most settings; handwashing is critical in healthcare settings during the winter season.

Antiviral Drugs for Respiratory Viruses

Ribavirin is a nucleoside antimetabolite pro-drug that is activated by kinases in the cell, resulting in a 5′ triphosphate nucleotide form that inhibits RNA replication. The drug was licensed in an aerosol form in the US in 1986 for treatment of children with severe RSV lower respiratory tract infection. The efficacy of aerosolized ribavirin therapy remains uncertain despite a number of clinical trials. Most centers use it infrequently, if ever, in otherwise healthy infants with severe RSV disease. Intravenous ribavirin has been used for adenovirus, hantavirus, measles virus, PIV, and influenza virus infections, although a good risk/benefit profile has not been established clearly for any of these uses.

A humanized mouse monoclonal antibody directed to the F protein of RSV, 'palivizumab', is licensed for prevention of RSV hospitalization in high-risk infants. It is efficacious in half or more of high-risk subjects. A more potent second-generation antibody is being studied in clinical trials. Experimental treatment of both immunocompetent and immunocompromised RSV-infected subjects has been reported but the efficacy of this approach is not established.

There are four licensed drugs in the US for treatment or prophylaxis of influenza. 'Amantadine' and 'rimantadine' are two of the drugs that interfere with the ion channel activity caused by the viral M2 protein of influenza A viruses, which is needed for viral particle uncoating following endocytosis. The other two drugs, 'oseltamivir' and 'zanamivir', are neuraminidase inhibitors that act on both influenza A and B viruses by serving as transition state analogs of the viral neuraminidase that is needed to release newly budded virion progeny from the surface of infected cells. The cell surface normally is coated heavily with the viral receptor sialic acid. Resistance to the ion channel inhibitors arises rapidly during prophylaxis or treatment, and in 2006 resistance levels became so common in circulating viruses that the CDC no longer recommends use of these drugs.

'Interferon-α' has been shown to protect against rhinovirus infections when used intranasally. This biological drug causes some side effects, such as nasal bleeding, and resistance to the drug developed during experimental use, so the molecule is no longer being developed for this purpose. 'Pleconaril' has been tested for treatment of rhinovirus infection, as it is an oral drug with good bioavailability for treating infections caused by picornaviruses. This drug acts by binding to a hydrophobic pocket in the VP1 protein and stabilizing the protein capsid, preventing release of viral RNA into the cell. The drug reduced mucus secretions and other symptoms and is being further examined.

'Acyclovir' and related compounds are guanine analog antiviral drugs used in treatment of herpes virus infections. HSV stomatitis in immunocompromised patients is treated with 'famciclovir', or 'valacyclovir', and immunocompetent subjects with severe oral disease compromising oral intake are sometimes treated. These compounds have also been used prophylactically to prevent recurrences of outbreaks, with mixed results. Intravenous acyclovir is effective in HSV or varicella zoster virus pneumonia in immunocompromised subjects. 'Ganciclovir' with human immunoglobulin may reduce the mortality associated with CMV pneumonia in hematopoietic stem cell transplant recipients and has been used as monotherapy in other patient groups.

'Cidofivir' is a nucleotide analog with activity against a large number of viruses, including adenoviruses. Intravenous cidofovir has been effective in the management of severe adenoviral infection in immunocompromised patients but may cause serious nephrotoxicity.

Vaccines

There are licensed vaccines for influenza viruses. In the US, both a trivalent (H3N2, H1N1, and B) inactivated intramuscular vaccine and a live attenuated trivalent vaccine for intranasal administration is available. The efficacy of these vaccines is good when the vaccine strains chosen are highly related antigenically to the epidemic strain. Antigenic drift caused by point mutations in the HA and NA molecules leads to antigenic divergence, requiring new vaccines to be made each year. The influenza genome is segmented, which allows reassortment of two viruses to occur during co-infection, which sometimes leads to a major antigenic shift resulting in a pandemic. Pandemics occur every 20–30 years on average. There is current concern about the potential for an H5N1 pandemic, and experimental vaccines are being tested for this virus. To date, H5N1 vaccines have been poorly immunogenic compared to comparable seasonal influenza vaccines. Vaccines were developed for adenovirus serotypes 4 and 7, and these were approved for preventing epidemic respiratory illness among military recruits. Essentially, these were unmodified viruses given by the enteric route in capsules, instead of the respiratory route, which is the natural route of infection leading to disease. Inoculation by the altered route resulted in an immunizing asymptomatic infection. All US military recruits were vaccinated against adenovirus from 1971 to 1999 with near complete prevention of the disease in this population, but the sole manufacturer of the vaccine halted production in 1996 and supplies ran out 3 years later. Since 1999, adenovirus infection has reemerged as a significant problem in the military with approximately 10% of all recruits suffering illness due to adenovirus infection during basic training; some deaths have occurred. Live attenuated vaccine candidates are under development and being tested in phase I and II clinical trials for RSV and the PIVs. Mutant strains with reduced pathogenicity were isolated in the laboratory, tested, and sequenced. Now, vaccine candidates are being optimized by combining mutations from separate biologically derived viruses into single strains using recombinant techniques for generating RNA viruses from cDNA copies, a process called reverse genetics. Subunit vaccines have been developed for RSV, but there are safety concerns about their use in young infants because formalin inactivated vaccine induced a more severe disease response to infection in the 1960s. There are no vaccines against rhinoviruses as there is little or no cross-protection between serotypes, and it is not feasible to develop a vaccine for over 100 serotypes. Efforts to develop coronavirus vaccines are in the preclinical stage.

Summary

Viruses are the leading causes of acute lower respiratory tract infection in infancy. RSV is the most common pathogen, with hMPV, PIV-3, influenza viruses, and rhinoviruses accounting for the majority of the remainder of acute viral respiratory infections. Humans generally do not develop lifelong immunity to reinfection with these viruses; rather, specific immunity protects against severe and lower respiratory tract disease.

See also: Human Respiratory Syncytial Virus.

Further Reading

Booth CM, Matukas LM, Tomlinson GA, *et al.* (2003) Clinical features and short-term outcomes of 144 patients with SARS in the greater Toronto area. *JAMA* 289: 2801–2809.

Booth CM, Matukas LM, Tomlinson GA, *et al.* (2003) Clinical features and short-term outcomes of 144 patients with SARS in the greater Toronto area – Erratum. *JAMA* 290: 334.

Collins PL and Crowe JE Jr. (2006) Respiratory syncytial virus and metapneumovirus. In: Knipe DM and Howley PM (eds.) *Fields Virology*, 5th edn., pp. 1601–1646. Philadelphia: Lippincott, Williams and Wilkins.

Fisher RG, Gruber WC, Edwards KM, *et al.* (1997) Twenty years of outpatient respiratory syncytial virus infection: A framework for vaccine efficacy trials. *Pediatrics* 99: E7.

Glezen WP, Frank AL, Taber LH, and Kasel JA (1984) Parainfluenza virus type 3: Seasonality and risk of infection and reinfection in young children. *Journal Infectious Diseases* 150: 851–857.

Glezen WP, Paredes A, Allison JE, Taber LH, and Frank AL (1981) Risk of respiratory syncytial virus infection for infants from low-income families in relationship to age, sex, ethnic group, and maternal antibody level. *Journal of Pediatrics* 98: 708–715.

Heymann PW, Carper HT, Murphy DD, et al. (2004) Viral infections in relation to age, atrophy, and season of admission among children hospitalized for wheezing. *Journal of Allergy and Clinical Immunology* 114: 239–247.

Karron RA and Collins PL (2006) Parainfluenza viruses. In: Knipe DM and Howley PM (eds.) *Fields Virology*, 5th edn., pp. 1497–1526. Philadelphia, PA: Lippincott Williams and Wilkins.

Martinez FD (2002) What have we learned from the Tucson Children's Respiratory Study? *Paediatric Respiratory Reviews* 3: 193–197.

Parrott RH, Kim HW, Arrobio JO, et al. (1973) Epidemiology of respiratory syncytial virus infection in Washington, DC. Part II. Infection and disease with respect to age, immunologic status, race, and sex. *American Journal of Epidemiology* 98: 289–300.

Subbarao K, Klimov A, Katz J, et al. (1998) Characterization of an avian influenza A (H5N1) virus isolated from a child with a fatal respiratory illness. *Science* 279: 393–396.

Williams JV, Harris PA, Tollefson SJ, et al. (2004) Human metapneumovirus and lower respiratory tract disease in otherwise healthy infants and children. *New England Journal of Medicine* 350: 443–450.

Winther B, Hayden FG, and Hendley JO (2006) Picornavirus infections in children diagnosed by RT-PCR during longitudinal surveillance with weekly sampling: Association with symptomatic illness and effect of season. *Journal of Medical Virology* 78: 644–650.

Wright PF, Neumann G, and Kawaoka Y (2006) Orthomyxoviruses. In: Knipe DM and Howley PM (eds.) *Fields Virology*, 5th edn., pp. 1691–1740. Philadelphia, PA: Lippincott Williams and Wilkins.

Human T-Cell Leukemia Viruses: General Features

M Yoshida, University of Tokyo, Chiba, Japan

© 2008 Elsevier Ltd. All rights reserved.

Glossary

pX region HTLV sequence between the env and 3′ LTR, encoding Tax, Rex and other small regulatory proteins.

Rex Trans-modulator of viral RNA splicing and transport.

Tax Pleiotropic regulator activating viral and cellular replication interacting with cellular transcription factors, tumor suppressor proteins, and cell cycle checkpoints.

Introduction

Human T-cell leukemia virus 1 (HTLV-1) is the first established tumorigenic retrovirus of humans; exogenous to humans this virus is classified as the species *Human T-cell leukemia virus*, in *Deltaretroviridae*, within the family *Retroviridae*. HTLV-1 infection is associated with leukemia and neural disease, adult T-cell leukemia (ATL) and HTLV-1-associated myelopathy/tropical spastic paraparesis (HAM/TSP), respectively. The genomic structure of the virus with genes for nonstructural proteins established a distinct viral genus that includes *Bovine leukemia virus*. HTLV-1 has no oncogene, but nevertheless transforms T cells rather efficiently and is identified as the etiologic agent of ATL. HTLV-1 has unique regulatory proteins, Tax and Rex, and Tax has been identified as a critical molecule not only in regulation of viral replication but also in induction of ATL.

History and Classification

After long and enormous efforts to identify a retrovirus in human tumors, HTLV was described in T-cell lines as a convincing human retrovirus. The first report of the virus (HTLV) was from a patient with Mycosis (MF) in the US, and another (adult T-cell leukemia virus (ATLV)) was from a patient with ATL in Japan. Subsequently, the MF case was characterized as ATL and the two isolates were established to be the same following a comparison of their genomes.

A prototypical retroviral genome contains the *gag*, *pol*, and *env* genes encoding the virion proteins including core proteins, reverse transcriptase, and surface glycoprotein, respectively. Acute leukemia viruses generally have an oncogene acquired from cellular genes that substitutes a part of the *gag*, *pol*, and *env* sequences. In contrast to these genomes, HTLV has additional genes in an extra pX region between *env* and the 3′ LTR (LTR – long terminal repeat). This unique genomic structure classified HTLV as a member of a distinct genus of the *Retroviridae*, which includes HTLV-1, and -2, bovine leukemia virus (BLV), and simian T-cell leukemia viruses (STLV-1, -2, and -3). HTLV-2 was isolated from a patient with hairy T-cell leukemia and its genome similarity to the type 1 is about 60%.

STLVs have been isolated from various species of Old World nonhuman primates, including the Japanese macaque, African green monkey, pig-tailed macaque, gorilla, and chimpanzee. Their genomes share 90–95% homology.

BLV infects and replicates in B cells of cows and sheep and induces B-cell lymphoma.

Geographic Distribution

Geographic Clustering

HTLV-1 carriers are defined by virus-specific antibodies. Nationwide surveys in Japan revealed the following: almost all ATL and HAM/TSP patients are infected with HTLV-1, 5–15% of adults in southwestern Japan are infected with HTLV-1 but this percentage is less in other areas of Japan. Worldwide, virus carriers are localized in the Caribbean islands and South America, Central Africa, and Papua New Guinea and the Solomon Islands. The prevalence of asymptomatic, seropositive adults varies significantly from district to district and even from village to village within these endemic areas. The unique clustering of virus infection in such remote areas is considered to reflect its mode of transmission and familial close contact such as sexual relations and breastfeeding. The presence of ATL and HAM/TSP is overlapping and clustered with HTLV-1 infection. Although some patients with ATL and healthy HTLV-1-infected carriers are found in nonendemic areas, they have frequently moved from an endemic area.

HTLV-2 is endemic in South America and is also detected in populations infected with HIV. STLVs are distributed in Old World monkeys in various areas, but not in New World monkeys. Although viruses were isolated from most species of Old World monkeys, some colonies are found to be virus-free even in the same geographic area.

Age-Dependent Infection

The frequency of HTLV-1 carriers increases with age after 20, reaching a maximum between 40 and 60 years of age. The prevalence is slightly (1.6 times) higher in females than in males, which is attributed to sexual transmission.

Familial aggregation

Familial aggregation of the HTLV-1 infection is apparent. If a mother is HTLV-1 positive, her children are frequently positive. This aggregation is due to viral transmission from husband to wife and mother to her children.

Host Range and Transmission

Receptor, Infection, and Transformation

The receptor for HTLV-1 infection is a glucose transporter, Glut-1, which is ubiquitously expressed on most cell types. Glut-1 is required for viral binding and infection, but it alone is not sufficient to understand preferential infection of CD4+ T cells *in vivo*. Cell-free viral particles of HTLV-1 released from established cell lines show extremely low infectivity *in vitro*, and co-cultivation with virus-producing cells is generally used to establish infection. A variety of cell types are infected including human T and B lymphocytes, fibroblasts, and epithelial cells, as well as cells from monkeys, rats, rabbits, and hamsters, but curiously not cells from mice. These cells infected by co-cultivation contain multiple integrated proviral copies of complete and defective genomes. Only T cells of the CD4+ phenotype are immortalized upon infection *in vitro*.

Despite the broad cell type host range *in vitro*, the cells infected *in vivo* are T cells. HTLV-1 infects both T cells with CD4+ and CD8+ phenotypes, although preferentially those of the CD4+ phenotype, but ATL cells are exclusively CD4+ T cells. Inoculation of infected cells can establish infection in rats, rabbits, and mice providing animal models for studies of the virus.

Natural Transmission

HTLV-1 is transmitted through (1) nursing of infants by infected mothers, (2) sexual contact, and (3) blood transfusion.

Mother to child. About 30% of the children born to seropositive mothers were seropositive. Infected T cells in breast milk are a source of transmissible virus. This mode of transmission was initially suggested by epidemiological studies and then direct evidence was obtained through cessation of breastfeeding by seropositive mothers that drastically reduced the seropositive rates in their children.

Sexual transmission. Wives with seropositive husbands are usually seropositive. Conversely, the husbands of seropositive wives show the same frequency of seropositivity as those in the region under study, indicating transmission occurs from husband to wife but not vice versa. Infected T cells in semen are thought to mediate transmission. This pathway may explain the higher (1.6 times) prevalence in females than in males, and is in sharp contrast to HIV infection. Infection in adulthood is not thought to be linked to subsequent ATL.

Blood transfusion. Transfusion of fresh and total blood transmits HTLV-1 into 60–70% of the recipients. The transfer of infected cells is critical for transmission. This mode of transmission does not lead to ATL, but does to HAM/TSP.

Zoonotic infections with retroviruses endemic in most Old World primates are described in people living in central African forests who have had contact with blood and body fluids of wild nonhuman primates. Such interspecies infection may explain the unique similarities between HTLVs and STLVs.

Genetics and Replication

Genome

The virion is spherical, enveloped with a membrane similar to that of the host cell and contains RNA genomes, core proteins, and reverse transcriptase, and envelope glycoproteins on the surface (**Figure 1**). The RNA genome is copied into a 'proviral' DNA upon infection and analysis of the retroviral genome is generally performed with the integrated proviral genome.

The integrated HTLV-1 proviral genome is 9032 bp long and its organization is LTR-*gag-pol-env*-pX-LTR. The pX sequence contains multiple overlapping genes encoding for the regulatory factors for the viral replication, but none of them has any homology to cellular genes. The major regulatory factors are $p40^{tax}$ and $p27^{rex}$. Alternative splicing of the pX sequence also expresses various proteins such as p12(I), p10(I), p11(V).

Antisense viral transcripts from the pX region code for HBZ proteins (HTLV-1 bZIP with a leucine zipper motif), which represent additional regulatory proteins, and also function at the RNA level.

Replication with Feedback Regulation

Reverse transcription and integration

Upon infection, the viral RNA genome is reverse-transcribed into complementary DNA (cDNA) by viral reverse transcriptase in the virion and further copied into double-stranded DNA, defined as 'proviral DNA'. The proviral DNA is integrated into the host cell DNA establishing the viral infection.

Transcription and splicing

RNA transcription from the 5′ LTR to 3′ LTR generates the viral genomic RNA, and this step represents a potential regulatory stage of the viral replication cycle. HTLV genomic RNA serves as an mRNA for Gag and Pol proteins, but has to be spliced to express other viral proteins: into a 4.2 kbp (singly spliced) env mRNA for Env expression, and into a 2.1 kbp (doubly spliced) mRNA for the regulatory proteins, Tax and Rex (**Figure 1**). Various alternative splicing events also take place to express alternative proteins.

Genomic transcription depends on cellular factors that respond to 21 bp enhancers in the HTLV LTR, but initial transcriptional activity is weak. The low levels of viral transcripts are fully spliced into Tax/Rex mRNA. The Tax thus produced *trans*-activates transcription further enhancing viral transcription to produce more Tax/Rex mRNA. This *trans*-activation is mediated by Tax binding to a transcriptional factor, cAMP-responsive element binding protein (CREB) that responds to the 21 bp enhancers in the

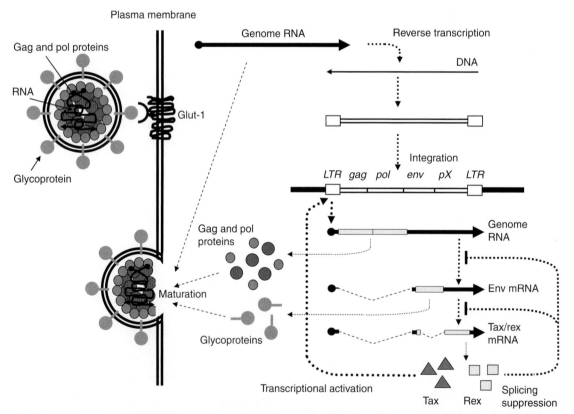

Figure 1 Replication cycle of HTLV. HTLV-1 binds to glucose transporter (Glut-1) and the genomic RNA is transcribed into complementary DNA by reverse transcriptase. The circularized proviral DNA is finally integrated into host chromosomal DNA establishing infection. The integrated provirus is transcribed into genomic RNA and expression of the viral proteins is regulated by Tax and Rex at transcriptional and splicing level making the expression transient. Genomic RNA and virion proteins mature at specific sites under plasma membrane and bud as the matured particles.

LTR. To express the virion proteins, Gag, Pol, and Env, the accumulated Rex protein specifically suppresses splicing of the viral RNA, and thus upregulates expression of unspliced genomic RNA and singly spliced *env* mRNA. In return, this Rex regulation reduces the level of spliced Tax/Rex mRNA, resulting in a lower level of Tax and ultimately reducing viral transcription. The combination of Tax and Rex functions exerts a feedback control on viral expression, making viral expression transient and resulting in the escape of infected cells from host immune surveillance. This feedback mechanism is unique to HTLV among oncogenic retroviruses and explains why HTLV is so repressed in expression and replication.

HTLV replication has also been reported to be regulated by small proteins such as p12(I), p10(I), p11(V), that are expressed by alternative splicing of the pX sequence. The regulatory mechanisms are not well understood but seem to be important for *in vivo* viral replication.

Maturation and budding

Genomic and *env* RNAs are efficiently accumulated by Rex and thus virion proteins, Gag, Pol, and Env, are transiently expressed. These virion proteins and genomic RNA come together and assemble at specific sites under the plasma membrane, finally budding by being enveloped with cellular membrane. The molecular mechanisms of these processes are not well characterized.

Variability and Evolution

The retroviral genome is rather labile in general since the reverse transcription process has no proofreading mechanism. In sharp contrast to HIV for example, however, the HTLV genome is highly stable and conserved among isolates from Japan, the Caribbean, and Africa. The viral isolates from Papua New Guinea may vary somewhat more but are still 90–95% homologous. This stable property is a reflection of an HTLV survival strategy that replicates the viral genome through infected cell replication rather than virus replication.

It has been suggested that HTLV and STLV are transmitted across species. Each of the isolates from various monkeys (STLV-1, -2, and -3) and humans (HTLV-1 and -2) are highly homologous and their variability is sometimes higher within the species than among species. Such unusual genomic conservation under different conditions is also well explained by a viral survival strategy through proliferation of infected cells.

Pathogenicity

Adult T-Cell Leukemia

Clinical features

Symptoms of ATL are variable and frequently complicated by skin lesions, enlargement of lymph node, liver and/or spleen, and infiltration of leukemic cells into various organs. Patients usually have antibodies to HTLV-1 proteins, show an increased level of serum lactate dehydrogenase (LDH), and most suffer from hypercalcemia.

The onset of ATL is observed in individuals between 20 and 70 years of age, the highest frequency being observed in people in their 40s and 50s. The male/female ratio of ATL incidence is 1.4/1.

Smoldering, chronic, acute forms of ATL and a lymphoma type have been recognized. Patients with smoldering ATL have a few or several percent of morphologically abnormal T cells in their peripheral blood, but do not show signs of severe illness for a long period. Patients with chronic ATL have rather high levels of HTLV-1-infected leukemic cells, but can maintain stable phenotypes for a certain period of time. Acute ATL is aggressive and resistant to any treatment; consequently, most patients die within one-half year of its onset.

Leukemic cells are exclusively CD4+ T cells with usually a highly lobulated nucleus. Leukemic cells generally have one complete integrated copy of an HTLV-1 provirus in their genome. Less frequently two copies are found, sometimes with defective forms. The site of integration is clonal in a given ATL patient, but different among patients. Leukemic cells carry aberrant chromosomes, frequently with multiple abnormalities, and express a high level of IL-2Ralpha, PTHrP, IL-1alpha, but no common abnormality has been described.

Etiology of ATL

An etiologic linkage of HTLV-1 with ATL is apparent from the observations: (1) ATL and HTLV-1 show identical geographic distribution, (2) almost all ATL patients are infected with HTLV-1, (3) leukemic cells are all infected with HTLV-1 but a vast majority of normal T cells are not, (4) leukemic cells are clonally integrated with an HTLV-1 provirus indicating their origin from a single infected cell, and (5) infection by HTLV-1 can immortalize T cells *in vitro* and the phenotypes of immortalized T cells are similar to those of leukemic cells.

There are approximately 1 million carriers of HTLV-1 in Japan. About 2–5% of all carriers of HTLV-1 are thought to develop ATL during their lifetime. While the vast majority of cases of ATL are associated with HTLV-1 infection, a form of ATL 'unrelated to HTLV-1 infection' has been described. The etiologic factor of these cases has not been identified.

Molecular mechanism of leukemogenesis

ATL cells have clonally integrated HTLV-1 proviruses; however, no common site for integration was observed among ATL patients. In this respect, HTLV-1 differs from animal chronic leukemia viruses, wherein integration was commonly adjacent to a proto-oncogene for its activation. Thus, the Tax protein has been focused on as a critical transforming protein.

Tax protein

Tax is able to transform fibroblasts and immortalize CD4+ T cells *in vitro*, and can induce tumors in transgenic mice. Tax exerts pleiotropic effects including (1) transcriptional activation of specific genes, (2) transcriptional repression of some other specific genes, (3) functional inactivation of tumor suppressor proteins, and (4) attenuation of cell-cycle checkpoints (**Table 1**). Cooperation of these pleiotropic functions is thought to contribute to ATL induction.

Transcriptional activation. Tax activates HTLV-1 genome transcription, which in turn is regulated by an enhancer binding protein, cyclic AMP-responsive element binding (CREB) protein. CREB has to be phosphorylated for active transcription, while Tax binds to CREB and activates it without a phosphorylation signal leading to a constitutively active protein. Similarly, Tax binds to other enhancer binding proteins such as nuclear factor kappa B (NF-κB) and serum responsive factor (SRF) and activates many specific cellular genes including IL-2Ralpha, IL-6, c-fos, and Bcl-x. NF-κB is alternatively activated by Tax through activation of IκB kinase (IKK) which disrupts inactive IκB–NF-κB complexes. The genes finally activated are linked to enhancement of cell proliferation or suppression of apoptosis.

Transcriptional repression. Tax also binds to transcription factors, CBP and P300, which interact with various enhancer binding proteins. The binding of Tax to CBP or p300 interferes with their interaction with the corresponding enhancer binding protein unless Tax is able to bind to enhancer binding protein, and consequently inhibits formation of a transcriptional initiation complex. The targets include p53-dependent transcription, DNA polymerase beta, p18ink4, Bax, and many others. The genes eventually repressed are linked to downregulation of p53-dependent stress responses, DNA repair and apoptosis.

Functional inactivation of tumor suppressor proteins. Tax directly binds to or modifies some tumor suppressor proteins and inactivates their negative regulation of the cell cycle. RB, APC, and p53 are targets. (a) Tax binds to and inactivates p16ink4 and p15ink4, which normally inhibit CDK4 and maintain an active RB pathway; consequently, Tax activates CDK4 which results in inactivation of RB and promotes cells to move from G1 arrest into S phase. Tax is also reported to bind directly to CDK4 to activate kinase activity. (b) Tax also binds to another tumor suppressor protein, hDlg, through its PDZ domain and inactivates its growth-retarding signal through APC and β-catenin. An abnormality in this pathway has been shown to play a critical role in colorectal tumors. (c) Tax inactivates p53 through phosphorylation and complex formation of p53–p300. Inactivation of the p53 pathway implies suppression of DNA repair resulting in the frequent fixation of mutations.

Attenuation of cell-cycle checkpoints. Check and review mechanisms for genomic processes before entering the next phase of the cell cycle is important to avoid genetic fixation of mutations. Tax interacts with hMad-1 and Chek-1 and attenuates the S- and G2-checkpoint functions. Tax also binds to Ran and Ran-binding protein and induces centrosome fragmentation and aneuploidy. These properties would explain a higher mutation rate in Tax-positive cells and may be the basis for why ATL cells have highly frequent choromosomal abnormalities.

Collectively, it has been proposed that these pleiotropic effects of Tax cooperate for abnormal cell proliferation. Significance of the Tax roles are twofold: (1) enhancing proviral replication and (2) promoting cells into leukemogenesis. Infected T cells are abnormally stimulated to proliferate through transcriptional activation/repression and inactivation of tumor suppressor proteins.

Table 1 Typical examples of pleiotropic Tax effects on cellular activities

Category/primary target	Targeted gene or process	Targeted cellular activity
Transcriptional activation		
CREB–CBP	HTLV-1 provirus	Viral replication/activation
	Unidentified	Transformation/activation
NF-κB-p300 and IKK-IκB	IL-2Ralpha, IL-6	Proliferation/activation
	Bcl-X	Apoptosis/suppression
SRF-CBP	c-Fos	Transformation/activation
Transcriptional repression		
p300–p53	p53-dependent transcription	Stress response/Suppression
	p18ink4	Proliferation/activation
CBP	DNA polymerase beta	DNA repair/suppression
	Bax	Apoptosis/suppression
Tumor suppressor protein		
p16ink4-CDK4	RB signaling	Cell cycle/promotion
p15ink4-CDK4		
hDLG-APC	APC signaling	Proliferation/activation
Kinase-p53	p53 signaling	Stress response/suppression
Checkpoint		
hMad-1	G2 checkpoint	Fidelity/suppression
Check-1	S, G2 checkpoint	Fidelity/suppression

Concomitantly, attenuation of cell-cycle checkpoints, reduction of DNA repair, and apoptosis would result in more mutations and their fixation, eventually leading to the leukemogenesis. In this respect, the pleiotropic effects of Tax may be equivalent to a multistep process for tumor formation.

Paradox in the leukemogenic mechanism

Tax plays a central role in the induction of ATL. Even its functional similarity to transforming proteins of DNA tumor viruses has been pointed out. However, it remains to be answered: (1) Why HTLV-1 transformation is selective for CD4+ T cells? and (2) Why ATL cells *in vivo* maintain tumor phenotypes in the absence of Tax? Tax/Rex mRNA is expressed only in a small percentage of tumor cells even using highly sensitive RT-PCR assays, yet tumor phenotypes are maintained in all ATL cells. The simplest answer for the latter question may be that Tax was critical for induction of tumors, but after establishment of tumors Tax is no longer required. If this is the case, what is the mechanism to maintain the tumor phenotypes?

An antisense gene, HTLV basic leucine zipper (HBZ) protein, may account for the paradox. This gene is transcribed from the pX region and codes for a DNA-binding protein with a bZIP domain. HBZ is able to moderately enhance T-cell proliferation *in vitro* and counteracts Tax *trans*-activation through dimer formation with other bZIP proteins. The antisense transcripts also operate at the RNA level. An exciting aspect of HBZ is that the antisense transcript is expressed in almost all ATL cases tested so far. It is therefore proposed to play a role after Tax *in vivo*. Tax is a potent antigenic protein; thus, its downregulation may make sense for tumor cell to escape from immune responses.

HAM/TSP and other diseases

Clinical features

HTLV-1 also induces a slowly progressive myelopathy known in tropical zones as tropical spastic paraparesis (TSP) and, in endemic areas of Japan, as HTLV-1-associated myelopathy (HAM). The unique phenotypes of HAM/TSP are chronic, symmetrical, bilateral involvement of the pyramidal tracts, at mainly the thoracic level of the spinal cord, and include progressive spastic paresis with spastic bladder and minimal sensory deficits. HTLV-1-infected T cells infiltrate into the spinal fluid and cord.

Most patients with HAM/TSP have much higher titers of HTLV-1 antibodies than those of asymptomatic carriers or ATL patients. This might be associated with particular types of human leukocyte antigens (HLAs). Despite their strong immunological responses to HTLV-1 infection, most HAM/TSP patients have larger populations of infected cells than do HTLV-1 carriers.

Etiology and other features

In endemic Japanese, all HAM/TSP patients are infected with HTLV-1. After screening for seropositive blood, the incidence of HAM/TSP has greatly decreased, clearly indicating that HTLV-1 is an etiologic agent of HAM/TSP. In contrast, TSP patients in the tropical areas are not always infected with HTLV-1, but the etiology in these cases is unknown. Indirect immunological reaction has been proposed as a pathogenic mechanism, but further studies are required.

The two HTLV-1-associated diseases, HAM/TSP and ATL, are mutually exclusive. The reason for this phenomenon is not well understood, but the route of primary viral infection may affect the pathogenic course: mucosal exposure to HTLV-1 for ATL, while primary infection of peripheral blood for HAM/TSP.

HTLV-1 infection is also proposed to be associated with some other diseases including uveitis, chronic lung disease, monoclonal gammopathy, and rheumatoid arthritis, but further systematic studies are required.

Prevention and Control of Infection

Transfusion of seropositive blood transmits HTLV-1 to two-thirds of the recipients. With the introduction of HTLV-1 screening systems in blood banks, viral transmission through transfusion has been greatly reduced. Application of these systems to populations in all endemic areas is critical to prevent HTLV-1 infection.

The major, natural route of viral transmission is from mother to child via infected T cells in breast milk. In Nagasaki City, Japan, pregnant women are tested for HTLV-1 antibodies by consent and those who are seropositive are recommended to avoid breastfeeding. A trial of this approach resulted in a drastic reduction in the incidence of seropositive children, from about 30% to just a few percent. The early success of this trial provides direct evidence for viral transmission through milk and suggests the possibility of eliminating ATL in the next few generations. Unfortunately, not all of children of seropositive mothers who did not breastfeed remained seronegative.

Future

Studies on HTLV-1 infection linked to ATL and HAM/TSP have progressed, but several very basic questions are still not answered: Why does only a small fraction of the infected population develop ATL or HAM/TSP? Why are ATL and HAM/TSP mutually exclusive? Why are only CD4+ T cells transformed into leukemic cells? What is required for ATL induction in addition to Tax? On the other hand, HTLV-1 transmission is now preventable in certain countries. However, it may not be feasible everywhere in the world. For example, cessation of breastfeeding might result in more serious problems in children in certain environments. So, original questions should be asked: Is a vaccine possible for complete eradication of the HTLV-1? Is vaccination possible for preventing disease development after infection?

Modern technologies developed in conjunction with genome research are now becoming available and make it possible to revisit these questions utilizing new appraoches.

See also: Human Immunodeficiency Viruses: Origin; Human Immunodeficiency Viruses: Pathogenesis; Human T-Cell Leukemia Viruses: Human Disease.

Further Reading

Gallo RC (2002) Human retroviruses after 20 years: A perspective from the past and prospects for their future control. *Immunological Reviews* 185: 236–265.

Matsuoka M and Jeang KT (2005) Human T-cell leukemia virus type I at age 25: A progress report. *Cancer Research* 65: 4467–4470.

Yoshida M (2001) Multiple viral strategies of HTLV-1 for dysregulation of cell growth control. *Annual Review of Immunology* 19: 475–496.

Yoshida M (2005) Discovery of HTLV-1, the first human retrovirus, its unique regulatory mechanisms and insights into pathogenesis. *Oncogene* 24: 5931–5937.

Human T-Cell Leukemia Viruses: Human Disease

R Mahieux and A Gessain, Pasteur Institute, CNRS URA 3015, Paris, France

© 2008 Elsevier Ltd. All rights reserved.

Introduction

In 1980, Dr. Gallo's laboratory (National Institutes of Health, USA) reported the isolation of HTLV-1, the first oncoretrovirus to be discovered in humans. HTLV-1 was present in the peripheral blood cells obtained from an Afro-American patient suffering from a lymphoproliferative disease, originally considered as a cutaneous T-cell lymphoma, with a leukemic phase. The virus was thus named human T-cell leukemia/lymphoma virus (HTLV). Later, it was recognized that this cutaneous lymphoma was in fact an adult T-cell leukemia/lymphoma (ATLL), a severe T-cell lymphoproliferation, originally described in Japan in 1977 by Takatsuki. The epidemiological characteristics of ATLL in Japan suggested a strong environmental factor, which prompted researchers to characterize the tumor cells and to search for an oncogenic virus. In 1981, a virus was isolated in Japan and termed adult T-cell leukemia/lymphoma virus (ATLV). Japanese and American scientists rapidly demonstrated that both isolates represented the same virus, and agreed to name it HTLV-1. In parallel, the causal association between ATLL and HTLV-1 was established. In 1983, the authors initiated a series of studies in the French West Indies in order to investigate the epidemiological and clinical impact of HTLV-1 in this area. This led them to demonstrate the etiological association between this virus and a chronic neuromyelopathy originally named tropical spastic paraparesis (TSP) that is endemic in the Caribbean. A similar neurological entity was then uncovered in Japan and labeled as HTLV-1-associated myelopathy. These two diseases were further shown to be identical and this myelopathy is now referred to as HAM/TSP. HTLV-1 infection has also been associated with other clinical conditions including uveitis, infective dermatitis, and myositis (**Table 1**).

HTLV-1, which is not a ubiquitous virus, is present throughout the world, with clusters of high endemicity located often nearby areas where the virus is nearly absent. These highly endemic areas are the southwestern part of the Japanese archipelago; the Caribbean area and its surrounding regions; foci in South America including Colombia, French Guyana, parts of Brazil; some areas of intertropical Africa (such as South Gabon); and of the middle East (such as the Mashad region in Iran); and isolated clusters in Melanesia. The origin of this puzzling geographical or rather ethnic repartition is not well understood but is probably linked to a founder effect in certain ethnic groups, followed by the persistence of a high viral transmission rate due to favorable local environmental and cultural situations. Interestingly, in all the highly endemic areas, and despite different socioeconomic and cultural environments, HTLV-1 seroprevalence increases gradually with age, especially among women. This might be either due to an accumulation of sexual exposures with age, or due to a cohort effect.

The worldwide infected population is estimated around 15–20 million. Two to ten percent of infected persons will develop an HTLV-1-associated disease (ATLL, HAM/TSP, uveitis, infective dermatitis, etc.) during their life (**Table 1**). Three modes of transmission have been demonstrated for HTLV-1: (1) Mother to child transmission, which is mainly linked to prolonged breastfeeding after 6 months of age. Ten to twenty-five percent of the

Table 1 Diseases associated with HTLV-I infection

Adult disease	Association
Adult T-cell leukemia/lymphoma	++++
Tropical spastic paraparesis/HTLV-I-associated myelopathy	++++
Intermediate uveitis	+++
Infective dermatitis	+++
Myositis (polymyositis and SIBM)	+++
HTLV-I-associated arthritis	++
Pulmonary infiltrative pneumonitis	++
Invasive cervical cancer	+
Small cell carcinoma of lung	+
Sjögren disease	+
Childhood	*Association*
Infective dermatitis	++++
Tropical spastic paraparesis/HTLV-I-associated myelopathy (very rare)	++++
Adult T-cell leukemia/lymphoma (very rare)	++++
Persistent lymphadenopathy	++

The strength of association is based on epidemiological studies as well as molecular data, animal models, and intervention trials. ++++, proven association; +++, probable association; ++, likely association; +, possible association.
SIBM: sporadic inclusion body myositis.

breast-fed children born from HTLV-1 infected mothers will become persistently infected. (2) Sexual transmission, which mainly occurs from male to female, and is thought to be responsible for the increased seroprevalence with age in women. (3) Transmission with contaminated blood products (containing infected lymphocytes), which is responsible for an acquired HTLV-1 infection among 15–60% of the blood recipients.

From a molecular point of view, HTLV-1 possesses remarkable genetic stability, an unusual feature for a retrovirus. Viral amplification via clonal expansion of infected cells, rather than by reverse transcription, could explain this striking genetic stability, which can be used as a molecular tool to follow the migrations of infected populations in the recent or distant past and thus to gain new insights into the origin, evolution, and modes of dissemination of such retroviruses and their hosts. The few nucleotide substitutions observed among virus strains are indeed specific to the geographic origin of the patients rather than the pathology. Four major geographic subtypes (genotypes) have been reported. The origin of most of these geographic HTLV-1 subtypes appears to be linked to episodes of interspecies transmission between STLV-1-infected monkeys and humans, followed by variable period of evolution in the human host.

Adult T-Cell Leukemia/Lymphoma

Epidemiological Aspects

After the initial discovery and characterization of ATLL in Japan, the disease was reported in the USA and in Caribbean immigrants living in the United Kingdom. ATLL cases have now been reported in all HTLV-1 endemic areas including intertropical Africa, South and Central America, Iran, and Melanesia. Sporadic cases were also described in areas of low HTLV-1 endemicity, often in immigrant patients originating from regions where HTLV-1 is endemic. In Japan, the ATLL incidence rate shows a steep increase with age, the mean age of disease onset being 57 years old and the sex ratio (male/female) being 1.4. In Japan, the annual incidence of ATLL cases is approximately 700, while the number of HTLV-1 carriers reaches 1.2 million. This gives an estimated yearly incidence of ATLL of 0.6–1.5 for 1000 HTLV-1 carriers older than 40. Lastly, the additive life-time risk of developing ATLL was estimated to be 1–5 % among Japanese HTLV-1 carriers. Studies performed in Brazil, Gabon, and French Guyana concluded that ATLL prevalence is usually underestimated until a specific disease research is performed. This is mostly due to the severity of the disease, its rapid evolution, and to confusion between ATLL and similar diseases, such as Sézary's syndrome, Mycosis fungoides, or other T cell non-Hodgkin lymphomas. In addition, laboratory tests such as Western blot and molecular investigations are not easily available in most tropical countries.

Diagnosis Criteria and Classification

Because of the extensive diversity in the clinical presentation and evolution of the disease, Japanese clinicians and researchers have defined stringent diagnosis criteria for ATLL as well as a classification into four major subtypes (**Table 2**). These diagnostic criteria are the following: (1) Histologically and/or cytologically proven lymphoid malignancy with expression of specific T-cell surface antigens (mostly CD2+, CD3+, and CD4+). (2) Abnormal T lymphocytes (flower cells and small/mature T lymphocytes with incised or lobulated nuclei) present in the peripheral blood, except in the ATLL lymphoma type. (3) Antibodies against HTLV-1 antigens present in the patient's serum at diagnosis.

The four major subtypes, as defined by the Japanese lymphoma study group, are the smoldering, the chronic, the lymphoma, and the leukemic/acute form. Both chronic and smoldering types can progress toward the acute or lymphoma forms. This classification was proved to be useful to discriminate ATLL from other type of leukemias/lymphomas. Patients who develop the smoldering form and about 30% of those suffering of the chronic ATLL have a fairly 'good prognosis' as compared to those with a leukemic or lymphoma type whose survival median does not exceed 6–9 months (**Table 2**). Finally, criteria for determining the diagnosis of ATLL during epidemiological studies have also been proposed (**Table 3**).

Table 2 Diagnostic criteria for clinical subtype of HTLV-I associated ATLL

	Smouldering	Chronic	Lymphoma	Acute
Anti-HTLV-I antibody	+	+	+	+
Lymphocyte ($\times 10^3/\mu l$)	< 4	$\geq 4^a$	< 4	*
Abnormal T lymphocytes	$\geq 5\%^c$	$+^b$	<1%	$+^b$
Flower cells of T cell marker	#	#	No	+
LDH	$\leq$1.5N	$\leq$2N	*	*
Corrected Ca (mEq/liter)	< 5.5	<5.5	*	*
Histology-proven lymphadenopathy	No	*	+	*
Tumor lesion				
Skin and/or lung	*c	*	*	*
Lymph node	No	*	Yes	*
Liver	No	*	*	*
Spleen	No	*	*	*
Central nervous system	No	No	*	*
Bone	No	No	*	*
Ascites	No	No	*	*
Pleural effusion	No	No	*	*
Gastrointestinal tract	No	No	*	*

^{a}Accompanied by T lymphocytosis (3.5x10^3/μl or more).
^{b}If abnormal T lymphocytes are less than 5% in peripheral blood, histology-proven tumor lesion is required.
^{c}Histology-proven skin and/or pulmonary lesion(s) is required if abnormal T lymphocytes are less than 5% in peripheral blood.
*no essential qualification except terms required for other subtype(s).
N: normal upper limit.
#: typical flower cells seen occasionally.
Adapted from Shimoyama M and The Lymphoma Study Group (1991) Diagnostic critreria and classification of clinical subtypes of adult T-cell leukemia-lymphoma. *British Journal of Haematology* 79: 428–437.

Clinical, Cytological, and Immuno-Virological Features

The predominant clinical findings of ATLL at onset are lymphadenopathy, hepatomegaly, splenomegaly, specific skin lesions, and hypercalcemia (**Table 4**). The symptoms may include abdominal pain, diarrhea, pleural effusion, ascites, and cough. White blood count ranges from normal up to a very high number of abnormal peripheral blood lymphocytes, especially in acute and chronic ATLL patients, while anemia, neutropenia, or thrombocytopenia are occasionally observed. ATLL cells differ in size and characteristics, depending on the subtype. As an example, at the typical acute/leukemic stage, most abnormal cells exhibit multilobulated nuclei and are named 'flower cells' (**Figure 1**). At the terminal stage, cells often significantly vary in size. They display a cytoplasmic basophilia and a marked nuclear lobulation. Chronic ATLL cells are generally small and of uniform shape with minor nuclear abnormalities such as indentation or convolutions. Smoldering ATLL cells are often relatively large with a bi- or trifoliate nucleus.

Most ATLL cells are mature T cells of helper/inducer phenotype (CD2+, CD3+, CD4+, CD7–, CD8–) and express activation markers (CD25+, HLA DP+, DQ+, DR+). CD4– and CD8– ATLL cells are uncommon, but a decrease of the CD3/T cell receptor expression is frequent. Changes in cell activation marker expression (CD4+CD8– to CD4+CD8+) have been observed in some patients during the course of the disease. Lymph-node histology analysis frequently demonstrates infiltration by medium and large T cells with irregular nuclei effacing the nodal architecture. This is consistent with the diagnosis of pleiomorphic large T-cell lymphoma. However, there is no specific histological pattern for ATLL.

Cytogenetic abnormalities are not specific for ATLL, but are frequently reported in acute and lymphoma patients. They include chromosome 14 translocations (14q 32, 14q 11) and 6q deletions. Trisomy 3, 7, and 21 as well as X chromosome monosomy or the loss of the Y chromosome were also reported. Mutations in the p53 encoding tumor-suppressor gene are detected in 20–30% of all patients, mostly those in advanced stages, suggesting the involvement of p53 mutation in a late phase of leukemogenesis, or as a consequence of the cellular transformation.

Detecting (by Southern blot) the clonal integration of HTLV-1 provirus(es) in the tumor cells represents the gold-standard for establishing that a patient suffers from ATLL (**Figure 2**). In most cases, only one copy of the provirus is integrated in the tumor cells. Nevertheless, two or more proviruses can infrequently be detected (5–20% of all ATLL cases). It is also worth noting that 5–20% of ATLL patients carry defective HTLV-1 proviruses. The 3'pX region is commonly conserved, while deletions occur frequently in *gag*, *pol*, and/or *env* open reading frames. Inverse-PCR, a technique that is more sensitive

Table 3 Registry criteria for definition of ATLL

Definition of ATLL	
Clinical/routine laboratory criteria	
Hypercalcemia	1 point
Skin lesions[a]	1 point
Leukemic phase[b]	1 point
Research laboratory criteria	
T-cell lymphoma or leukemia	2 points
HTLV-I antibody	2 points
TAC-positive tumor cells	1 point
HTLV-I-positive tumors[c]	2 points
ATLL classification	
Classical	≥7 points
Probable	5 or 6 points
Possible	3 or 4 points
Inconsistent with ATLL	<3 points
Exclusion criteria	
B-cell positivity, nodular or follicular lymphoma, lymphoblastic lymphoma, small lymphocytic lymphoma	

[a]Lymphomatous cells documented morphologically.
[b]More than 2% abnormal lymphocytes.
[c]Determined by PCR or Southern blot analysis of the DNA of tumoral cells and indicating a monoclonal integration of HTLV-1 provirus(es).
Adapted from Levine PH, Cleghorn F, Manns A, et al. (1994) Adult T-cell leukemia/lymphoma: A working print-score classification for epidemiological studies. *International Journal of Cancer* 59: 491–493.

Table 4 Main clinical features of ATLL

	Japan[a]	Caribbean[b]
Age at onset	58 years (range 27–82)	47 years
Sex ratio male/female	1.4	0.6
Lymphadenopathy	60%	70%
Hepatomegaly	26%	27%
Splenomegaly	22%	31%
Specific skin lesion	39%	41%
Hypercalcemia	32%	51%

[a]Based on a series of 187 Japanese patients and
[b]from a series of 57 patients, with 46 of Carribean origin, seen in the United Kingdom.

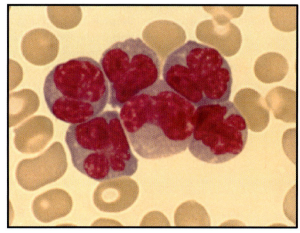

Figure 1 Peripheral blood smear from a Caribbean ATLL leukemic patient showing a cluster of atypical lymphoid cells with multilobulated nuclei (May-Grunwald-Giemsa staining).

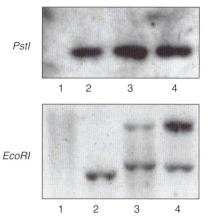

Figure 2 HTLV-1 proviral integration as detected by Southern blot analysis. High molecular weight DNA was extracted from (1) negative control peripheral blood mononuclear cells and (2, 3, 4) from peripheral leukemic cells obtained from three different HTLV-1 patients diagnosed with ATLL. Because *EcoRI* restriction sites are absent from HTLV-1 provirus sequence, the observation of two bands (lane 3 and 4) demonstrates the clonal integration of two proviruses in the DNA of the leukemic cells.

than Southern blot is currently used for diagnosis of the monoclonal integration of HTLV-1 provirus(es) in the tumoral cells. Using this technique, oligoclonal or polyclonal proliferation of HTLV-1 infected T cell can also be detected in the peripheral blood lymphocytes of most HAM/TSP patients as well as in healthy HTLV-1 seropositive carriers. This also allows the precise identification of integration sites, the identification of each infected clone, and the tracing of the kinetics of the infected cells *in vivo*. The detection of persistent HTLV-1 infected monoclonal cell population could therefore be used as a method for monitoring individuals at risk for developing ATLL.

Importantly, monoclonal expansion of HTLV-1 infected cells in carriers is directly associated with the onset of ATLL.

Specific Features and Complications

Hypercalcemia is very common among ATLL patients. It is detected in 20–30 % of the patients at admission and in more than 70 % of them during the entire clinical course, with or without lytic bone lesions. *In vitro*, ATLL cells induce the differentiation of hematopoietic precursor cells to osteoclasts through the expression of the RANKL (receptor activator nuclear factor kB) ligand on their surface, in cooperation with elevated M-CSF (mononuclear phagocyte colony-stimulating factor) serum levels. This accelerates bone resorption and ultimately causes hypercalcemia. The degree of hypercalcemia might also be linked to the expression of the parathyroid hormone-related protein (PTHrP), since PTHrP also induces RANK ligand expression on osteoblasts.

Skin infiltration of a clonal population of HTLV-1 tumor cells represents a frequent ATLL clinical feature. It is present in 20–40% of all ATLL patients and in more than 50% of smoldering ATLL patients. Various cutaneous lesions have been described including papules, nodules, erythroderma, plaques, tumors, and ulcerative lesions. ATLL cells densely infiltrate both dermis and epidermis forming Pautrier's micro-abscesses. When skin lesions dominate the clinical picture, the disease is referred to as 'cutaneous ATLL' (**Figure 3**). In such cases, establishing the difference between these lesions and other cutaneous T-cell lymphomas (CTCLs) is complicated. A possible role of CCR4 (CC chemokine receptor 4) in skin invasion by ATLL cells has been suggested. Other factors such as expression of cutaneous lymphocyte antigen (CLA) by leukemic cells and inflammatory responses in injured skin that would lead to upregulation of thymus and activation-regulated chemokine (TARC) and macrophage-derived chemokine (MDC) are also likely to contribute to skin involvement of ATLL.

ATLL is also known to frequently invade the gastrointestinal tract, even though the exact incidence has not been determined yet. CCR9 is known to be involved in T-cell homing to the gastrointestinal tract. A recent study demonstrated that CCR9 is expressed by HTLV-1 T cells and ATLL cells expressing Tax and suggested that it may play a role in the gastrointestinal involvement of ATLL.

The frequency of opportunistic infections is quite high in ATLL patients, indicating that T-cell-mediated immunity is severely impaired in such patients. Infestation by *Strongyloides stercoralis* (S.s) is frequent among HTLV-1 seropositive carriers and S.s infected individuals are often infected by HTLV-1 in highly endemic areas. Numerous studies showed the presence of S.S in acute or lymphoma ATLL, suggesting that infection by S.S might play a significant role as a candidate cofactor for HTLV-1 induced leukemogenesis.

Pulmonary complications are also frequent in ATLL patients. These are either leukemic infiltrate or opportunistic infections. Central nervous system localization rarely occurs in ATLL (10%).

ATLL Pathogenesis

The development of an ATLL has been clearly associated with an early acquired HTLV-1 infection. The risk of being infected for a child born from an HTLV-1 infected mother is positively correlated with prolonged (>6 months) breastfeeding. After a limited number of replication cycles using its reverse transcriptase during primary infection, HTLV-1 replicates and increases its copy number through the proliferation of infected cells, thus using the cellular DNA polymerase which possesses a proofreading activity. Such a model could explain, in part, the very high genetic stability of HTLV-1. The clonal proliferation of the HTLV-1-infected CD4+ lymphocytes is likely to be linked to the pleiotropic effects of the viral Tax protein. *In vitro*, Tax expression induces expression of several cellular cytokines such as IL-2, IL-15, but also alters the cell-cycle control machinery, inhibits apoptosis, and promotes genetic instability. As seen above, the fact that oligo/monoclonal-infected cell populations are present in the PBMCs of HTLV-1 carriers or of HAM/TSP patients suggests that neither

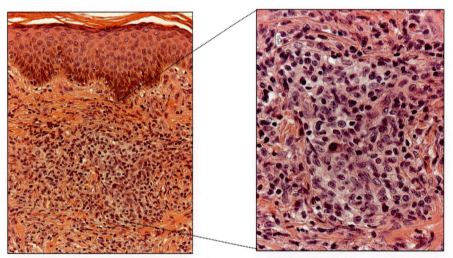

Figure 3 Histological analysis of skin invasion in an HTLV-1 patient from West Africa with a cutaneous ATLL showing an infiltration of the dermis by tumoral pleomorphic cells. These tumor cells are mature activated lymphoid T cells (CD2+, CD3+, CD4+, CD8–, and CD25+). Kindly provided by Dr. Michel Huerre, Institut Pasteur, Paris.

the monoclonal integration of the virus nor the CD4+ cell proliferation *per se* is sufficient to cause the disease. *In vivo*, the level of Tax expression is still debated. Several observations paradoxically suggested that HTLV-1 was transcriptionally silent *in vivo*, but it was shown lately that a high proportion (10%–80%) of naturally infected CD4+ peripheral blood mononuclear cells isolated from HAM/TSP patients or from HTLV-1 asymptomatic carriers are capable of expressing Tax *ex vivo*. It was also demonstrated that autologous $CD8^+$ T cells rapidly kill CD4+ cells that express Tax *in vitro* through a perforin-dependent mechanism. Altogether, these results suggest that virus-specific cytotoxic T lymphocytes (CTLs) participate in a highly efficient immune surveillance mechanism that tirelessly eradicates Tax-expressing HTLV-1-infected CD4+ T cells *in vivo*. Such CTL control of the HTLV-1-infected cell proliferation may also result in the prevention of ATLL. The role of the genetic background, especially HLA haplotypes, in such control could be crucial. A final step toward ATLL leukemogenesis consists in the accumulation of alterations in the host genome due to the pleiotropic effects of Tax (**Figure 4**).

Therapeutic Aspects of ATLL

The survival rate of ATLL patients, especially those who develop the acute leukemic or lymphoma forms, remains poor, and ATLL remains therefore one of the most severe lymphoproliferations. Treatment of ATLL patients using conventional chemotherapy (CHOP) has been and remains in most cases the standard first-line therapy. It has very limited benefit, since HTLV-1 cells are resistant to most apoptosis-inducing agents. A historical Japanese survey based on 818 ATLL cases reported a median survival time of 9 months, with a survival rate of only 27% and 10% at 2–4 years respectively. Survival is different when subtypes are considered into account, with a 4-year survival rate of 66% for the smoldering type, of 27% for the chronic type and of 5–6% for the lymphoma and acute types. Spontaneous regressions are observed in very few cases. The main prognosis factors associated with a poor response and a poor survival rate in ATLL patients are: a high LDH (lactate dehydrogenase) value, high leukemic counts (both of them reflecting a high tumor burden), hypercalcemia, and a poor clinical performance. Furthermore, the main obstacles to an efficient response to treatment are infectious complications (*pneumocystis carinii*, *cryptococcus meningitis*), disseminated herpes zoster), hypercalcemia, and liver or kidney dysfunction.

Various strategies other than CHOP have been used during clinical trials. This includes combination chemotherapy (CHOP plus methotrexate, CHOP in combination with etoposide, vindesine, ranimustine, mitoxantrone, or adriamycin), granulocyte macrophage-colony-stimulating factor (GM-CSF), supported combination chemotherapy, and other chemotherapy treatments that are highly toxic for bone marrow. Zidovudine (AZT) has been shown to inhibit HTLV-1 transmission *in vitro*. Several studies with AZT and interferon-alpha (IFN-α) have been conducted, both in the USA and in Europe (France and UK). They all demonstrated that response and survival in patients is more efficient when the drugs are used as first-line therapy. Until recently, the mechanism of action of AZT was a matter of debate. A recent study convincingly demonstrated that AZT can act as an inhibitor of the telomerase functions. More importantly, it was also shown that the p53 status

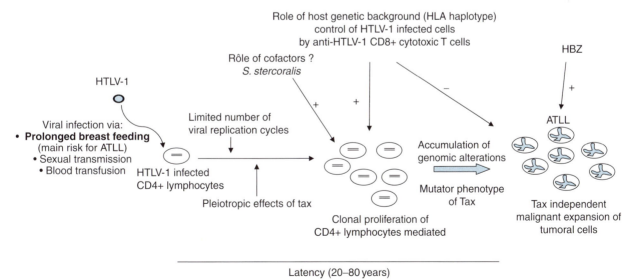

Figure 4 Pathogenesis of ATLL. Natural course from HTLV-1 primary infection by prolonged breast-feeding to the onset of the malignant proliferation of ATLL cells. HBZ: HTLV-1 bZIP factor. Adapted from Mahieux R and Gessain A (2003) HTLV-1 and associated adult T-cell leukemia/lymphoma. *Reviews in Clinical and Experimental Hematology* 7: 336–362.

(wild-type or mutated) is predictive of the response to AZT. Nuclear factor kappa B (NF-κB) inhibitor (Bay 11–7082) was found to be efficient for preventing the tumor growth in NOD-SCID (non-obese diabetic-severe combined immunodeficiency/gamma (null)) animals previously inoculated with HTLV-1 infected cells. Proteasome inhibitor PS-341 (Bortezomib) alone or combined with anti-CD25 (anti-Tac) therapy was partially successful in treating NOD/SCID animals or Tax transgenic mice. Allogenic bone marrow transplantation as well as allogenic hematopoietic stem cells transplantation have been performed on a limited number of patients. The estimated overall survival seems to compare favorably with historical data on chemotherapy, but definitive assessments are difficult to establish.

Tropical Spastic Paraparesis/HTLV-1 Associated Myelopathy

Spastic paraparesis without evidence of spinal cord compression has been described from various tropical and intertropical areas such as the Caribbean region, South and Central Africa, and India. Apart from cassava-related myelopathy and/or malnutrition in several African regions, lathyrism in India, and some infectious processes, the etiology of numerous spastic meylopathies remained elusive until 1985. The demonstration by our laboratory of the association between HTLV-1 and a form of tropical spastic paraparesis (TSP) of Unknown etiology, frequent in Martinique (French West Indies), led to the evaluation of the role of this oncoretrovirus, in this disease. Such an association was rapidly confirmed in Colombia and Jamaica. A year later, the association of HTLV-1 with a chronic spastic myelopathy of unknown etiology was documented in southern Japan, and this clinical entity was named HTLV-1 associated myelopathy (HAM). Soon after, it was recognized that HTLV-1-associated TSP and HAM were the same disease, the hybrid term HAM/TSP was adopted and WHO diagnostic criteria were established.

Hundreds of HAM/TSP patients have now been described in several HTLV-1 endemic areas, including Japan, several of the Caribbean islands (especially Jamaica, Trinidad, Martinique, Haiti), numerous countries of South and Central America (Brazil, Peru, Colombia), of Central and South Africa, as well as in Iran. Sporadic cases of HAM/TSP have also been described in Melanesia, in West African countries as well as in immigrants from high HTLV-1 endemic areas living in Europe and the USA. In Japan, the life-time risk among HTLV-1 carriers is estimated to be less than 2%, that is, lower than ATLL. HAM/TSP mainly occurs in adults, with a mean age at onset of 40–50 years, but some rare cases of HAM/TSP in children have been reported. In Japan, in contrast to ATLL, HAM/TSP is more common in women than in men at all ages, with a sex ratio (M/F) of around 1:2,3.

Diagnostic Criteria

Initially, HAM/TSP diagnostic criteria included: (1) chronic spastic paraparesis, which usually slowly progresses, with signs of bilateral pyramidal tract lesions manifested by increased knee reflexes, ankle clonus, and extensor plantar responses; (2) minor sensory signs of involvement of the posterior columns and spinothalamic tract; (3) a history of insidious onset with gait disturbance without an episode of complete remission; (4) no evidence of spinal cord compression or swelling on magnetic resonance imaging, myelography, or computed/tomographic scan; and (5) presence of specific anti-HTLV-1 antibodies in the serum and the cerebrospinal fluid (CSF).

In more than 90% of the cases, the neurological features of HAM/TSP involve: spasticity and/or hyperreflexia of the lower extremities, urinary bladder disturbance, lower extremity muscle weakness, and in around 50% of the cases, sensory disturbances with low back pain. Impotence is also frequent. Central functions and cranial nerves are usually spared. The evolution is generally chronic and progressive, and after 10 years of evolution, roughly 50% of the patients are wheel-chaired. The incubation period (between primary infection which occurs mainly in adults) to onset of the myelopathy signs ranges usually from years to decades, but HAM/TSP also developed within 3.3 years in 50% of the post-transfusion-associated cases.

Biologically, besides high level of antibodies directed against HTLV-1 antigens both in blood and CSF, ATLL-like cells can sometimes be detected on the blood smear and in the CSF. Furthermore, a high HTLV-1 proviral load is frequently observed in the peripheral blood lymphocytes from HAM/TSP patients. A mild to moderate increase of proteins may be present in the CSF. However, intrathecal production of specific antibody provides additional data to support the diagnosis of HAM/TSP and also contributes to eliminate other differential diagnoses. Multiple spotty high intensities in deep and subcortical areas on T2-weighted images are the most frequent findings in brain magnetic resonance imaging. A mild atrophy of the thoracic spinal cord can also be observed in few cases. From a pathological point of view, this neurological syndrome is characterized by a chronic inflammation with perivascular lymphocytic cuffing and mild parenchymal lymphocytic infiltrates (**Figure 5(a)**). The cells are mostly CD4+ in early disease and mostly CD8+ in latter disease. Pyramidal tract damage with myelin and axonal loss, mainly in the lower thoracic spinal cord are observed.

HAM/TSP can be associated with other HTLV-1-associated symptoms such as uveitis, myositis, pulmonary alveolitis, and arthritis. The coincidence of ATLL and HAM/TSP has been rarely reported.

Recently, a large group of neurologists established a proposal for a modification of the diagnostic criteria of

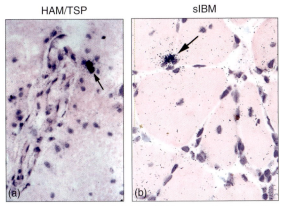

Figure 5 Detection of HTLV-1 *mRNA* expression by *in situ* hybridization with a ^{33}P *Tax* antisense riboprobe, on (a) frozen section from thoracic cord of a HAM/TSP patient, and (b) from a muscle section of a patient with an HTLV-1-associated sporadic inclusion body myositis (sIBM). Adapted from Ozden S, Seilhean D, Gessain A, Hauw J-J, and Gout O (2002) Severe demyelinating myelopathy with low human t cell lymphotropic virus type 1 expression after transfusion in an immunosuppressed patient. *Clinical Infectious Diseases* 34: 855–860 and Ozden S, Cochet M, Mikol J, *et al*. (2004) Direct evidence for a chronic CD8+-T-cell-mediated immune reaction to tax within the muscle of a human T-cell leukemia/lymphoma virus type 1-infected patient with sporadic inclusion body myositis. *Journal of Virology* 78: 10320–10327.

HAM/TSP. This was due to the fact that in some patients, especially those with early disease, HAM/TSP can be suspected but the complete WHO criteria are not met. These authors suggest that a HAM/TSP definite case corresponds to a nonremitting progressive spastic paraparesis with sufficiently impaired gait to be perceived by the patient. Sensory symptoms or signs may or may not be present. When present, they remain subtle and without a clear-cut sensory level. Urinary and anal sphincter signs or symptoms may or may not be present. Biologically, presence of HTLV-1 antibodies in serum and CSF confirmed by Western blot and/or a positive polymerase chain reaction (PCR) for HTLV-1 in blood and/or CSF should be demonstrated. Lastly, exclusion of other disorders that can resemble HAM/TSP is necessary.

HAM/TSP Pathogenesis

The pathogenesis of HAM/TSP is still poorly understood and viral and host factors as the proviral load and the immune response are considered to play a major role in disease progression. It remains unknown if the myelin and axonal loss, observed in HAM/TSP lesions, is a primary or secondary process and if it results from a direct viral effect or from an immune mediated process, as suggested by the marked lymphocytic infiltrates. Viral infection as demonstrated by *in situ* hybridization against *tax* mRNA has been clearly identified in some infiltrates by CD4+ lymphocytes by several teams, while the presence of HTLV-1 in neurons or oligodendrocytes remains controversial but unlikely.

At least three mechanisms have been proposed to explain the role of HTLV-1 in HAM/TSP development. The neurological damage is, in the first model, a direct consequence of the destruction by an antiviral attack mediated by cytotoxic T lymphocytes. These CTLs, frequent in HAM/TSP patients, both in the peripheral blood and CSF, are directed against HTLV-1 antigens, especially some immuno-dominant Tax epitopes expressed by HTLV-1-infected cells. In a second model, an autoimmune response could be due to either a chronic peripheral activation of autoreactive T cells by HTLV-1 infection, with infiltration of T lymphocytes into the CNS, or to a molecular mimicry, which could impair the immunological tolerance to the myelin antigens. A third possibility would be the bystander damage hypothesis. In this model, activated CD4+ T cells infected by HTLV-1, activated microglia, and CD8+ T cells could release into the spinal cord some myelinotoxic cytokines such as TNF-α or could directly damage the glial cells.

Therapeutic aspects of HAM/TSP

The long-term prognosis of HAM/TSP remains severe with a chronic evolution of a progressive disabling disorder without remission. Furthermore, its secondary complications may eventually lead to death, in some cases, after several years of evolution. Many small studies and cohort data have been performed using, among other drugs, corticosteroid therapy, danazol, vitamin C, interferon alpha, zidovudine, with very limited success. Recently, a randomized trial with zidovudine plus lamivudine has also been performed. Globally, a few short-term benefits have been observed, especially in early disease, but no treatment has been successful in chronic advanced disease. Failure to detect any clinical improvement is believed to be due to irreversible central nervous system damage in such patients. New controlled studies of both antiviral and anti-inflammatory agents are urgently required.

HAM/TSP is probably not the only neurological disorder associated with HTLV-1 infection. Indeed, an amyotrophic lateral sclerosis-like syndrome, as well as some peripheral neuropathies, and mild cognitive deficits have been described in some patients with HTLV-1 infection. It remains however difficult to prove a true association between such diseases and HTLV-1 infection, especially in high HTLV-1 endemic areas.

Infective Dermatitis

Infective dermatitis (ID) is a rare dermatological condition that was originally described in Jamaican children by Sweet in 1966. Subsequently, in 1990, La Grenade *et al.*

linked ID to HTLV-1 infection. A large series of patients with infective dermatitis have been described in Jamaica and more recently in Brazil. However sporadic cases of this clinical entity have also been reported in many other HTLV-1 endemic areas including Trinidad, Colombia, French West-Indies and French Guyana, Japan (where the disease seems very rare), and recently in Senegal (West Africa). In the great majority of cases, ID occurs in children from low socioeconomic backgrounds.

Infective dermatitis is a unique clinical entity characterized by a chronic and severe exudative dermatitis involving mainly the scalp, external ear and retroauricular areas, eyelid margins, paranasal skin, neck, axillae, and groin (**Figure 6**). Other symptoms include a chronic watery nasal discharge without other signs of rhinitis and/or crusting of anterior nares. A generalized fine papular rash is common in most of the severe cases. Clinically, the two differential diagnoses are atopic dermatitis and seborrheic dermatitis. In infective dermatitis, positive cultures for *Staphylococcus aureus* and/or beta-haemolytic streptococci are frequent from the anterior nares and skin samples. The evolution is typically chronic with relapse and several flares of superinfected lesions. Infective dermatitis responds well to antibiotic treatment, especially co-trimoxazole. However, relapse is very common if antibiotics are withdrawn. Infective dermatitis occurs mainly in young children (range 1–12 years), with an average age of onset of 2–6 years depending on the studies. Around 60% of the cases occur in girls. Anemia is frequent with a raised erythrocyte sedimentation rate and a hyper-immunoglobulinemia (IgD and IgE). The CD4 count, as well as the CD4/CD8 ratio, are elevated. Presence of rare ATLL-like cells is common in the peripheral blood. Pathological examination revealed an inflammatory lymphocytic infiltrate within the skin lesions. Epidemiological studies with long-term follow-up of ID patients have indicated that such disease may be associated with the later development of ATLL or of HAM/TSP. In a recent series from Brazil, neurological examination diagnosed six HAM/TSP cases among 20 children and adolescents with ID.

Mothers of children with ID are nearly always infected by HTLV-1. Furthermore, they have breastfed their child for a long period of time (mean 20 months). Thus, nearly all ID children were infected by their mothers. This implies that reduction of HTLV-1 transmission from mother to child would be likely to prevent the occurrence of such disease as well as of ATLL. Such preventive medicine, currently performed in several areas of high HTLV-1 endemicity (Japan, West Indies), is based on HTLV-1 screening programs during pregnancy followed by adequate counseling that is adapted to the socioeconomic situation of each infected mother. The pathogenesis of ID is unknown but environmental, as well as host genetic factors are very likely involved in its occurrence.

Myositis

Myositis is an inflammatory myopathy that is represented by a heterogeneous group of muscle disorders. It is characterized by an acquired muscle weakness and inflammatory infiltrates of the muscle tissues. Depending on the clinical and histological features, myositis can be classified into dermatomyositis, polymyositis, and inclusion-body myositis. In 1989, Morgan *et al.* reported for the first time that 11 out of 13 Jamaican patients with idiopathic adult polymyositis were seropositive for HTLV-1. Later on, several epidemiological studies conducted in Jamaica, Martinique, and Japan, and based on larger series of patients, also indicated a higher prevalence of HTLV-1 antibodies in polymyositis patients as compared to that of the general population. Such data suggested a possible link between HTLV-1 and polymyositis. Lately, several sporadic cases of polymyositis, occurring in HTLV-1 infected patients of various origins, have been reported.

HTLV-1-associated polymyositis is a chronic disease that occurs mainly in adults 30–50 years of age. Patients complain mainly of musculo-skeletal symptoms including myalgia, joint and back pain, and proximal muscle weakness with muscle atrophia and falls. The four limbs can be affected. A patient's clinical pattern typically displays an elevation of muscle enzymes, especially creatine phosphokinase (CPK), and significant electromyographic abnormalities. Histological findings indicate an inflammatory myopathy with an increased variation in the fiber size and a mononuclear infiltrate located both between and within muscle cells, which appear often degenerate and necrotic. Muscle regeneration may also be present. This infiltration

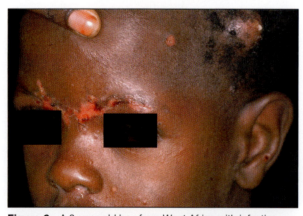

Figure 6 A 3-year-old boy from West Africa with infective dermatitis lesions: erosive dermatitis of the eyebrows, abscess of the scalp, and nasal discharge. Adapted from Mahe A, Meertens L, Ly F, *et al.* (2004) Human T-cell leukemia/lymphoma virus type 1-associated infective dermatitis in Africa: A report of five cases from Senegal. *British Journal of Dermatology* 150: 958–965.

is mainly due to T lymphocytes (both CD4+ and CD8+), but macrophages are also present. Expression of HTLV-1 *mRNAs* as well as of the viral proteins can be sporadically detected within the infiltrating T lymphocytes but not within muscle fibers. T-cell clones that are specifically directed against HTLV-1 Tax epitopes proliferate within the muscle lesions and may thus contribute to the pathogenesis. Patients respond poorly to corticosteroid therapy in most cases, and there is usually no sustained effect. Of note, HTLV-1 associated polymyositis is frequently associated with HAM/TSP.

A few cases of sporadic inclusion-myositis (sIBM) have also been reported in HTLV-1-infected patients. Such a disorder is a distinct form of chronic inflammatory myopathy for which a viral etiology has often been suspected. The lesions are characterized by vacuolated fibers that contain paired helical filaments similar to those of Alzheimer's diseases (**Figure 5(b)**). Immunohistochemical staining for ubiquitin may help to differentiate sIBM from polymyositis. sIBM can be associated with HAM/TSP.

Uveitis

Uveitis is an intraocular sight-threatening inflammatory disorder that can be caused by various infections, including tuberculosis, syphilis, toxoplasmosis, and cytomegalovirus. Uveitis, which does not have an infectious etiology, can occur in Bechet's diseases, Vogt–Koyanagi–Harada disease, and sarcoidosis. In roughly 40%, no cause can be identified and such disease is classified as idiopathic or unexplained uveitis. In the early 1990s, epidemiological studies demonstrated a higher prevalence of such idiopathic uveitis in HTLV-1 endemic areas from southern Japan than in areas where HTLV-1 has a lower prevalence. At the same time, some cases of uveitis were also observed in HAM/TSP patients. A large seroepidemiological study led by Mochizuki *et al.*, finally demonstrated without ambiguity that HTLV-1 infection was associated with idiopathic uveitis.

This clinico-virological entity is more frequent in females (around 60%) and the average onset age of the disease is 45 years, with few cases occurring in children. The major symptoms at initial presentation include sudden onset of occular floaters, foggy, and/or blurred vision. Unilateral disease is more common than bilateral (60% vs. 40%). The ocular signs consist mainly of iritis, vitreous opacities, and retinal vasculitis. Retinal exudates and hemorrhages are less frequent. Intermediate uveitis is more frequent than anterior or posterior lesions. Interestingly, co-morbid conditions include Graves disease and HAM/TSP. HTLV-1 uveitis responds well to topical and/or systemic corticosteroids and the intra-occular inflammation is markedly improved after steroid therapy. The visual prognosis is good in most cases. However, the inflammation tends to recur in about half of the cases. A specific opthalmological and systemic evaluation is necessary to eliminate other causes of uveitis in a given HTLV-1 infected individual. HTLV-1 antibodies are present in the aqueous humor at a level similar to that found in the plasma. HTLV-1-infected lymphocytes can be detected in the anterior chamber of the eye. Cytokines such as IL-6 and TNF-α produced by infiltrating lymphocytes are considered to play a major role in HTLV-1-associated uveitis pathogenesis. Infra-clinical uveitis can be detected through systematic examination in some HTLV-1 infected persons without any ocular symptoms.

Other Ocular Lesions Associated with HTLV-1 Infection

Kerato-conjonctivitis/sicca as well as interstitial chronic keratitis, are observed in HTLV-1 infected individuals, especially in HAM/TSP patients. The latter keratitis is also frequently associated with uveitis.

See also: Human T-Cell Leukemia Viruses: General Features.

Further Reading

Araujo AQ and Silva MT (2006) The HTLV-1 neurological complex. *Lancet Neurology* 5: 1068–1076.

Bazarbachi A, Ghez D, Lepelletier Y, et al. (2004) New therapeutic approaches for adult T-cell leukaemia. *Lancet Oncology* 5: 664–672.

Cleghorn F and Manns A (1994) Adult T-Cell leukemia/lymphoma: A working point-score classification for epidemiological studies. *International Journal of Cancer* 59: 491–493.

De Castro-Costa CM, Araujo AQ, Barreto MM, et al. (2006) Proposal for diagnostic criteria of tropical spastic paraparesis/HTLV-I-associated myelopathy (HAM/TSP). *AIDS Research and Human Retroviruses* 22: 931–935.

Grassmann R, Aboud M, and Jeang KT (2005) Molecular mechanisms of cellular transformation by HTLV-1 Tax. *Oncogene* 24: 5976–5985.

La Grenade L (1996) HTLV-I-associated infective dermatitis: Past, present, and future. *Journal of Acquired Immune Deficiency Syndromes and Human Retrovirology* 13(supplement 1): S46–S49.

La Grenade L, Manns A, Fletcher V, et al. (1998) Clinical, pathologic, and immunologic features of human T-lymphotrophic virus type I-associated infective dermatitis in children. *Archives of Dermatology* 134: 439–444.

Leon-Monzon M, Illa I, and Dalakas MC (1994) Polymyositis in patients infected with human T-cell leukemia virus type I: The role of the virus in the cause of the disease. *Annals of Neurology* 36: 643–649.

Levine PH, Cleghorn F, Manns A, et al. (1994) Adult T-cell leukemi/lymphoma: A working pring-score classification for epidemiological studies. *International Journal of Cancer* 59: 491–493.

Mahe A, Meertens L, Ly F, et al. (2004) Human T-cell leukaemia/lymphoma virus type 1-associated infective dermatitis in Africa: A report of five cases from Senegal. *British Journal of Dermatology* 150: 958–965.

Mahieux R and Gessain A (2003) A. HTLV-1 and associated adult T-cell leukemia/lymphoma. *Reviews in Clinical and Experimental Hematology* 7: 336–362.

Manns A, Hisada M, and La Grenade L (1999) Human T-lymphotropic virus type I infection. *Lancet* 353: 1951–1958.

Matsuoka M (2005) Human T-cell leukemia virus type I (HTLV-I) infection and the onset of adult T-cell leukemia (ATL). *Retrovirology* 2: 27.

Mochizuki M, Ono A, Ikeda E, *et al.* (1996) HTLV-I uveitis. *Journal of Acquired Immune Deficiency Syndromes and Human Retrovirology* 13(supplement 1): S50–S56.

Ozden S, Cochet M, Mikol J, *et al.* (2004) Direct evidence for a chronic CD8+-T-cell-mediated immune reaction to tax within the muscle of a human T-cell leukemia/lymphoma virus type 1-infected patient with sporadic inclusion body myositis. *Journal of Virology* 78: 10320–10327.

Ozden S, Mouly V, Prevost MC, Gessain A, Butler-Browne G, and Ceccaldi PE (2005) Muscle wasting induced by HTLV-1 tax-1 protein: An *in vitro* and *in vivo* study. *American Journal of Pathology* 167: 1609–1619.

Ozden S, Seilhean D, Gessain A, Hauw J-J, and Gout O (2002) Severe demyelinating myelopathy with low human t cell lymphotropic virus type 1 expression after transfusion in an immunosuppressed patient. *Clinical Infectious Diseases* 34: 855–860.

Pinheiro SR, Martins-Filho OA, Ribas JG, *et al.* (2006) Immunologic markers, uveitis, and keratoconjunctivitis sicca associated with human T-cell lymphotropic virus type 1. *American Journal of Ophthalmology* 142: 811–815.

Proietti FA, Carneiro-Proietti AB, Catalan-Soares BC, and Murphy EL (2005) Global epidemiology of HTLV-I infection and associated diseases. *Oncogene* 24: 6058–6068.

Shimoyama M and The Lymphoma Study Group (1991) Diagnostic criteria and classification of clinical subtypes of adult T-cell leukemia-lymphoma. *British Journal of Haematology* 79: 428–437.

Taylor GP and Matsuoka M (2005) Natural history of adult T-cell leukemia/lymphoma and approaches to therapy. *Oncogene* 24: 6047–6057.

Hypovirulence

N K Van Alfen and P Kazmierczak, University of California, Davis, CA, USA

© 2008 Elsevier Ltd. All rights reserved.

Glossary

Ascomycete A division of fungi whose members produce spores in a saclike structure called the ascus.

Canker A dead section of bark on the branches or main trunks of trees.

Hyphae Vegetative growth structures of the fungus.

Hypovirulence A reduction of virulence.

Fungal Viruses

Hypovirulence is a phenomenon that can occur in virus-infected fungi. Although fungal viruses are now known to be common and widespread within all groups of fungi, their discovery was fairly recent. It was not until the early 1960s that viruses were found in fungi, the first discovery being of viral particles that were isolated from the edible mushroom *Agaricus bisporus*. Fungal viruses became more widely known after it was discovered that strains of *Penicillium* spp., which were strong inducers of interferon, were found to be infected with a double-stranded RNA (dsRNA) virus.

Most descriptions of fungal viruses suggest that they are not normally associated with clearly defined symptoms. The lack of known infectivity cycles of fungal viruses is a major barrier to the study of these viruses since it limits the ability to demonstrate Koch's postulates. Transmission of fungal viruses is known to naturally occur only during reproduction by the formation of spores and other such structures and by cytoplasmic fusion between hyphae of different strains of the same fungus. In this respect, transmission of fungal viruses is more similar to the transmission of plasmids than it is to the transmission of other viruses, that is, the infectious agent never leaves the cytoplasmic environment. With a few exceptions, the genomic material of known fungal viruses is dsRNA, and in viruses where sequences are known, they contain a virus-associated RNA-dependent RNA polymerase (RdRp). Because fungal virus genomes are clearly viral in origin and affinity, the apparent lack of a viral-like infectivity cycle has led researchers to postulate an ancient origin for fungal viruses.

The most extensively studied fungal viruses cause symptoms that can generally be classified into three general groups of symptoms: (1) killer phenotype, (2) hypovirulence, and (3) debilitation. The killer systems consist of a helper totivirus and associated satellite dsRNAs, which encode a protein toxin that is secreted and is lethal to strains of the same fungus lacking the virus. The best studied examples of killer systems are those found in the common baker's yeast, *Saccharomyces cerevisiae*. Viruses that reduce the virulence of plant pathogenic fungi either directly, or by debilitation, are the subject of this article.

Hypovirulence

A reduction of the virulence of a plant pathogenic fungus is known as hypovirulence. Generally, anything that reduces the ability of a pathogenic fungus to grow normally will concomitantly reduce its virulence. So, by this general definition, viruses that debilitate a pathogen cause hypovirulence. It is useful, however, to distinguish this type of debilitating hypovirulence from a symptom of virus infection that is more directly associated with reducing virulence expression and/or other developmental processes of a fungus without significant effects on fungal growth. Virus-caused hypovirulence of both types has been described in diverse groups of plant pathogenic fungi. These mycoviruses that have hypovirulence as an associated symptom may have either ssRNA or dsRNA genomes and are found in the virus families *Chrysoviridae*, *Hypoviridae*, *Narnaviridae*, *Partitiviridae*, *Reoviridae*, and *Totiviridae*.

Perturbation of Virulence/Development

Other than the viruses associated with the killer phenotype of yeast, the most extensively studied fungal virus is cryphonectria hypovirus 1 (CHV1) that infects the ascomycete *Cryphonectria parasitica*. This fungus is the causal agent of chestnut blight and is the prime example of a virus that is able to cause the hypovirulence symptom without associated debilitation. Chestnut blight was first reported in North America during the summer of 1904 in the New York City Zoological Park. It was probably imported on nursery stock of Chinese or Japanese chestnuts, *Castanea mollissima* and *Castanea crenata*, respectively. The disease rapidly spread to the American chestnut, *Castanea dentata*, growing in surrounding woodlands and forests. Within 50 years, the disease spread throughout the range of the highly susceptible native tree. Few American chestnuts now survive in their natural range because of this disease. Prior to the blight, the American chestnut was the most common tree of the Eastern deciduous forest and was found on over 200 million acres across eastern North America. This is one of the most devastating diseases of recorded history.

Chestnut blight was also accidentally introduced into Europe, but during its spread in Italy it was noticed that chestnut trees in some orchards were recovering from the disease, that is, the cankers caused by the fungus stopped spreading. When the bark was peeled back, the fungal growth was found to be confined to the uppermost layers of the bark with no penetration occurring into the living tissues of the tree. The isolates of the fungus recovered from these cankers were of low virulence in assays, so were called hypovirulent by the French plant pathologist, J. Grente. Using genetically marked virulent and hypovirulent strains of the fungus, Van Alfen and colleagues demonstrated that a cytoplasmic element found in the hypovirulent strain was responsible for the symptoms and could be transferred cytoplasmically to the virulent strain, converting it into a hypovirulent strain. This cytoplasmic element was later found to be a fungal virus that is a member of the type species of the family *Hypoviridae*. **Figure 1** shows the phenotypic differences between virulent and hypovirulent isolates of *C. parasitica* on both the chestnut tree and in culture.

C. parasitica is also host to viruses classified in several other families of viruses: these include the *Reoviridae*, *Narnaviridae*, and *Chrysoviridae*. A hypovirulent phenotype of *C. parasitica* has also been described in strains infected by the virus NB631, a mitochondrial virus classified in the family *Narnaviridae*. This virus reduces virulence and growth of the fungus somewhat, but sporulation is normal. Two reoviruses of *C. parasitica*, C18 and 9B21 have been isolated. The best characterized of the two, 9B21, reduces virulence of the fungus significantly, but sporulation and pigmentation are not significantly affected. A very

Figure 1 Phenotypic changes in *C. parasitica* caused by the hypovirus CHV1. The photograph on the right is a tree that is infected with a virulent strain of the fungus. The photograph below shows this strain in culture. In addition to producing orange/brown pigments, virulent strains produce numerous conidia in asexual fruiting bodies, both in culture and on trees. The tree on the left is infected with a fungus containing CHV1. Strains of the fungus infected with this virus typically produce superficial healing cankers that do not kill the tree. Virus-infected strains of the fungus, as shown growing in culture, are not pigmented; they remain white and produce very few asexual conidia in culture. Canker photograph courtesy of Linda Haugen, University of Georgia, http://www.forestryimages.org. Culture photograph reproduced from McCabe PM, Pfeiffer PL, and Van Alfen NK (1999) The influence of dsRNA viruses on the biology of plant pathogenic fungi. *Trends in Microbiology* 7: 377–381, with permission from Elsevier.

unusual feature of 9B21 is that purified particles of the virus are able to infect fungal protoplasts.

Hypovirulence symptoms caused by viruses have been reported in many plant pathogenic fungi. Generally these viruses remain uncharacterized and in most cases the virulence of the fungus is reduced in association with a general debilitation of the infected fungus.

Structure and Classification of CHV1

Once it was demonstrated that a cytoplasmic element caused transmissible hypovirulence of the chestnut blight fungus there was a search for a fungal virus as the potential cause. This search culminated in the report of dsRNA associated with hypovirulent strains of the fungus. Further studies of the dsRNA showed that it was associated with fungal vesicles and that no detectable capsid was associated with this dsRNA, although RdRp activity was associated with the vesicles. The dsRNA of this 12.7 kb virus was cloned, sequenced, and an infectious clone of the virus was transformed into *C. parasitica*, confirming that this virus is responsible for transmissible hypovirulence. Based on the symptoms caused by the virus it has formally been named CHV1. This virus is the type member of the family *Hypoviridae* and was the first virus family described whose members have no capsid. Comparing the genome of CHV1 with other known viruses suggested that it is most similar to plant potyviruses and other picorna-like viruses.

Further study of the nature of the RdRp associated with fungal vesicles supported the hypothesis that the genome of this virus is derived from the replicative form of an ssRNA virus that lost its capsid during evolution. The coding strand of CHV1 contains two open reading frames: ORF A encodes a polyprotein that is processed into two polypeptides, p29 and p40 and ORF B which encodes a polyprotein from which only a single 48 kDa polypeptide p48 has thus far been shown to be autocatalytically processed. The sequences contained within ORF B indicate that it also contains RNA polymerase and helicase domains. **Figure 2** shows a schematic diagram of the genomic organization and expression strategy of CHV1.

Transmission of CHV1

There is no evidence that a typical infection cycle exists in the relationship of CHV1 with its fungal host. In order for infection of new hosts to occur the virus is transmitted through cytoplasmic exchange processes of the fungus. This cytoplasmic exchange occurs during fusion of hyphae (vegetative growth structures of the fungus) that happens frequently in fungal colonies and can occur between related strains of the fungus. In fungi, this hyphal fusion is controlled by vegetative incompatibility (*vic*) genes of the fusing strains; fusions most easily occur when alleles of the *vic* genes are the same in the two fungi. In *C. parasitica*, there are six known *vic* genes that control hyphal anastomosis. The more differences that occur in the *vic* loci, the less likely that transfer of CHV1 will occur. This vegetative incompatibility reaction is a defense in *C. parasitica* and other fungi against the spread of viruses. CHV1 spreads within a population with greater success if that population has few differences in alleles at the *vic* loci. Since strains of the fungus that contain CHV1 are incapable of sexual mating, infection by the virus suppresses allelic recombination at the *vic* loci and, as a result, may increase the effectiveness of virus spread within a population over time.

Very little is known about movement of the virus within the host fungus. However, the rate of movement of dsRNA within the fungus has been measured, and is rapid. The average rate of movement of the dsRNA within a fungal colony was found to be approximately 16 mm d^{-1}. This is 3–4 times faster than the colony growth rate during the same time period. Transmission electron microscope studies on the ultrastructure of hyphal anastomosis in *C. parasitica* reported that the dsRNA containing vesicles move from hypovirulent to virulent strains within 4–6 h after anastomosis. The incompatibility reaction controlled by the *vic* genes is an imperfect defense, since virus transfer can occur despite allelic differences in *vic* genes and virus transfer has been reported between different species of *C. parasitica*. There is clearly much left to be learned about virus movement within and between fungi.

The only other means by which the virus is known to spread is through asexual spores of the fungus. Although the

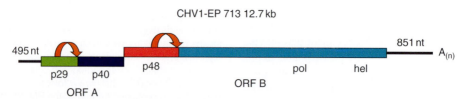

Figure 2 The genomic organization and expression strategy of CHV1–EP713, the prototypic strain of CHV1. Redrawn from Hillman BI and Suzuki N (2004) Viruses of the chestnut blight fungus, *Cryphonectria parasitica*. In: Maramorosch K and Shatkin AJ (eds.) *Advances in Virus Research*, vol. 63, pp. 423–472. San Diego, CA: Academic Press.

virus suppresses developmental processes of the fungus, it does not completely eliminate asexual sporulation. Many fewer asexual spores are produced by CHV1-infected strains of the fungus, but those asexual spores that are produced may contain the virus.

Hypovirulence as Developmental Perturbation

Hypovirulence caused by CHV1 is of interest not only because it provides a means of biological control for an important plant disease, but also because it offers the potential to understand critical control point(s) for development in a filamentous fungal pathogen. The perturbation of development, without fungal growth being affected, suggests that the virus interferes with development in specific ways that leave normal vegetative growth functions undisturbed. Understanding how the virus acts to perturb development in this fungus may provide insights of value on how to reduce the economic impact of filamentous fungi.

The symptoms of CHV1 infection of *C. parasitica* are a reduction of asexual sporulation, reduced pigmentation, female sterility, and hypovirulence (**Figure 1**). Careful growth measurements show little or no adverse effects of viral infection on fungal vegetative growth in culture. While there have been some claims that CHV1 reduces vegetative growth, these growth effects are likely an artifact of prolonged growth of the fungus in culture. We have found similar reduced growth rates in culture if some strains of the fungus, particularly EP713, are kept in culture too long. Such slow growth mutations are common in cultured fungi, and in our experience, are independent of virus infection. To prevent the accumulation of such growth mutations in prolonged culture, we routinely single-spore EP155 to select for a normal growth strain, and then reintroduce CHV1 by anastomosis to recreate EP713 with normal vegetative growth characteristics. Routine single-spore isolation is necessary to prevent the slower growth mutant strains from developing in culture.

Ultrastructural examination of virus-infected strains of the fungus show no significant cytopathology in CHV1-containing cells. The only differences observed between infected and uninfected cells are unique vesicles that are described as being located in association with unique Golgi in areas devoid of ribosomes and other cellular components. Recent cytological fractionation studies have isolated these small vesicles that accumulate in CHV1-infected strains and have shown that they are associated with the *trans*-Golgi network. These vesicles contain the virus and its replication-associated enzymes. Figure 3 shows the accumulation of these vesicles in the CHV1-infected strain, EP802.

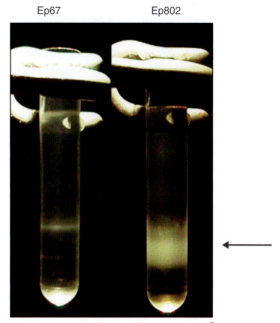

Figure 3 Subcellular fractionation on a Ficoll/^{2}H$_2$O gradient showing vesicle accumulation in CHV1-infected strain EP802 on the right and the isogenic virus free strain EP67 on the left. Gradient fractions were analyzed by Western and kex2 activity to show peak kex2 activity, AP-1μ, dsRNA, and viral helicase all cofractionate within this band. The endoprotease kex2 and the subunit for the AP-1 adaptor protein are markers for the *trans*-Golgi network.

Two general experimental approaches have been used to understand the molecular basis of how CHV1 causes symptoms in *C. parasitica*: (1) seek evidence that the virus perturbs normal fungal signal transduction pathways, and (2) identify commonalities in genes and proteins differentially expressed between normal and virus-infected strains of the fungus. A significant body of literature has been generated regarding how internal signal transduction pathways are affected in virus-infected strains, but, to date, there is no evidence of direct virus effects on signal transduction. As expected, signal transduction differs temporally and quantitatively between virus-infected and noninfected strains of the fungus. The most reasonable explanation of the published observations is that perturbation of signal transduction is the consequence of virus downregulation of development, rather than the cause. There is in fact no evidence that the symptoms are caused by direct virus perturbation of signal transduction and no testable hypothesis has emerged from these studies as to how the virus may be affecting signal transduction to cause the observed symptoms.

The approach that sought commonalities between differentially expressed proteins and genes associated with virus infection is leading to an understanding of how CHV1 may be causing symptoms. Given that the symptom of virus infection is a lack of development in infected strains, significant differences in expression between the two strains were expected and observed. Early two-dimensional gel and differential hybridization studies identified a number of protein and gene expression differences between the infected and uninfected strains. A relatively small number of these genes/proteins that were highly expressed in uninfected but not in infected strains were characterized to seek commonalities between them.

Three of the genes highly expressed in uninfected strains, and not expressed in CHV1-infected strains, encoded sex pheromones. Two of the genes contain an 83 bp ORF that encodes an identical 23-amino-acid peptide. This small peptide is similar to fungal lipopeptide sex pheromones: a C-terminal CAAX box with an asparagine residue 8–11 amino acids upstream of the box. CAAX is a prenylation signal, and farnesyl groups have been detected on the pheromones in other fungi. It was found that these peptides were expressed only in mating type 2 of uninfected strains of the fungus. Based on this similarity and loss of sexual mating upon deletion of the genes, the genes were named *mf2–1* and *mf2–2*. Deletion of *mf2–2* also resulted in a significant reduction in production of asexual fruiting bodies in culture and thus reduced numbers of asexual spores. Transcription run-on studies confirmed that these genes were transcriptionally downregulated in CHV1-infected cells.

Because *mf2–1* and *mf2–2* were so similar in structure to yeast pheromones, investigations were carried out to isolate the pheromone precursor gene of the opposite mating type. The gene isolated was shown to be similar in structure to the *S. cerevisiae* α-factor pheromone. There is a single copy of this gene expressed only by mating type 1 strains; it was named *mf1–1*. When the gene was deleted from *C. parasitica*, conidia or mycelial fragments normally used as spermatia in mating were sterile. Expression of this gene is significantly downregulated in CHV1-infected strains.

Differential protein expression studies identified two secreted proteins that are highly expressed by uninfected strains, but are downregulated in virus-infected cells. One of these is an extracellular laccase and the other an extracellular structural protein that was named cryparin. This later protein is a hydrophobin, a class of cell-surface proteins widely found in fungi and lichens. All hydrophobins are located on hyphal surfaces and confer hydrophobic properties to the surfaces, but they appear to have evolved different functions in various fungi. In *C. parasitica*, cryparin accumulates in fruiting bodies and is necessary for the eruption of the fungal fruiting bodies through the bark of the host tree. **Figure 4** shows the results of the deletion of cryparin.

The role of the laccase enzyme downregulated in CHV1-infected strains is not known. Knockout mutants did not exhibit any detectable phenotype, but did reveal the existence of multiple laccase enzymes expressed by the fungus. Laccases have been implicated in a number of

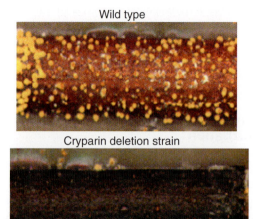

Figure 4 Stromal pustule (asexual fruiting bodies) eruption on chestnut wood. Sterile stem pieces of chestnut wood were inoculated with EP67 (upper panel) and the cryparin deletion strain (lower panel). Adapted from Kazmierczak P, Kim DH, Turina M, and Van Alfen NK (2005) A hydrophobin of the chestnut blight fungus, Cryphonectria parasitica, is required for stromal pustule eruption. *Eukaryotic Cell* 4: 931–936 with permission from American Society for Microbiology.

roles related to fungal development and virulence including degradation of lignin, formation of fruiting bodies, and pigment production.

The isolation and characterization of a number of the gene products downregulated in virus-infected cells was done to determine if there were any commonalities among them that could point to how the virus acts to cause symptoms. All of the gene products characterized are secreted into the extracellular environment and three of the proteins have sequences that suggest they are post-translationally processed by the same secretion pathway. Each protein is processed by an endoprotease after the signal peptide is cleaved, and the recognition signals suggest that they are processed by the same enzyme. **Figure 5** shows the preproproteins of Mf1–1, the fungal mating pheromone of mating type 1 strains, cryparin, and laccase. These three proteins have a signal peptide that presumably is cleaved early in the secretion process, followed by a propeptide region that terminates with a recognition sequence for processing by a kex2-like serine protease. These processing signals suggest that these three proteins are secreted via a kex2-like protein processing and secretion pathway. Kex2 processing is a post-Golgi function and the kex2 enzyme is a standard marker for *trans*-Golgi vesicles.

The processing of some secreted proteins by the kex2 pathway is highly conserved in eukaryotes. A similar pathway has been found in plants and animals and is utilized for the processing of specific proteins. It is generally not utilized for the secretion of most proteins. The specific role of this processing pathway in secretion appears to vary

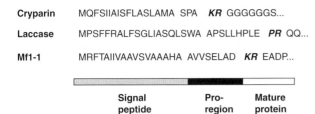

Figure 5 Amino acid N-terminal sequences derived from cloned mRNAs of three developmentally regulated host genes downregulated by CHV1. The signal peptide directs the protein to the ER for secretion and into the Golgi network for processing of the proprotein by the kex2 endoprotease at the residues lysine/arginine or proline/arginine which results in the mature form of the secreted protein.

between organisms. It was first discovered in yeast because it is involved in the processing and secretion of the viral encoded killer protein. This pathway was later found to be necessary for the secretion of the yeast alpha sex pheromone. The fact that three randomly isolated developmentally regulated host proteins, which are downregulated by the virus, are all processed by the same enzyme during secretion is an important clue in understanding how the virus perturbs development. Research has shown that not all secreted proteins of *C. parasitica* are affected by the virus; the secreted rennin-like endoprotease, endothiapepsin, for instance does not have a kex2 processing signal sequence nor it is downregulated by the virus.

The possibility that the virus directly affects secretion of some proteins, and thereby causes at least some of the symptoms associated with infection, is supported by evidence that the virus replicates using *trans*-Golgi vesicles. The vesicle fraction that contains the dsRNA genome of the virus and the viral encoded RNA polymerase and helicase co-purifies with kex2 enzymatic activity. This vesicle fraction has also been shown to contain cryparin, one of the proteins downregulated by the virus. The vesicle fraction in CHV1-infected cells that contains the virus dsRNA, viral encoded proteins, and kex2 activity are present in much greater amounts in virus-containing cells than in uninfected cells. This proliferation of vesicles in *C. parasitica* is typical of the effect of RNA viruses on their hosts, that is, the use of host membrane systems for replication by many RNA viruses cause a proliferation of these membranes within the host cells. We have estimated that in CHV1-infected cells there is at least a fivefold increase in this specific vesicle fraction when compared with noninfected cells. As previously mentioned, evidence from microscopic and cell fractionation observations suggest that CHV1 effects on membranes is specific to these *trans*-Golgi vesicles.

Recent research from our laboratory has shown that the CHV1 encoded protease p29 specifically integrates into the membranes of this *trans*-Golgi vesicle fraction. This protein is a close relative of the plant potyviral protease HC-Pro. HC-Pro is a papain-like protease that is autoprotelytic and which facilitates aphid transmission and promotes potyviral genome amplification. P29 is encoded on ORF A of CHV1 and is translated as part of the polyprotein p69 from which it is autocatalytically cleaved. Transformation of p29 into *C. parasitica* causes a loss of pigmentation, reduction in asexual sporulation, and suppression of host laccase expression, and there is evidence that it may be involved in suppression of RNA silencing by the host. Enhancement of both dsRNA accumulation and transmission of the virus through asexual spores are also functions attributed to CHV1 p29.

Recent studies show that p29 integrates into the *trans*-Golgi vesicle membranes of the host and that no other viral elements are required for this integration. While within the membrane p29 is fully susceptible to proteolytic digestion suggesting that it is primarily on the cytoplasmic side of the membrane. Deletion analysis of p29 showed that the C-terminal sequences of p29 mediate the membrane association. These transformation studies also confirmed the previous studies that showed p29 to be responsible for causing the reduced asexual sporulation and loss of pigment symptoms of virus infection. Further studies of the effect of p29 membrane integration on virus-caused host membrane proliferation and symptom production are in progress.

Based on the evidence available to date, CHV1 utilizes *trans*-Golgi vesicles of its host for replication and perhaps movement. *trans*-Golgi vesicles are critical for the final processing and secretion of proteins through the cytoplasmic membrane of the fungus, and for transport of proteins to certain cellular compartments. Studies of the viral protein p29 clearly link this protein to these vesicles and to causing some of the virus infection symptoms. The simplest hypothesis for how the virus is able to cause these and perhaps additional symptoms is that use of *trans*-Golgi vesicles for virus replication disrupts normal function of the vesicles and interferes with protein secretion or compartmentalization. The proprotein signal sequences directing kex2 endoprotease processing in a number of viral downregulated host proteins, such as the sex pheromone, cryparin, and laccase, also point to the *trans*-Golgi network as being a key to understanding the cause of virus symptom induction. The role of kex2 processing, a *trans*-Golgi vesicle function, in fungal development and its perturbation by the virus are currently under investigation.

There are precedents for RNA virus disruption of vesicle trafficking. Protein 3A of poliovirus specifically inhibits endoplasmic reticulum (ER)-to-Golgi traffic in mammalian cells, causing proteins otherwise destined for export to accumulate in ER-derived membranes. Many other viruses have been shown to have dramatic effects on the host membrane systems and a number of nonstructural virus proteins have been shown to interact with cellular factors responsible for the targeting and docking of membranes within the various organelles and membrane systems of virus hosts. Although no vesicle membrane proteins have been isolated yet from *C. parasitica* that are involved in vesicle function, the viral protein p29 which has an extensive cytoplasmic domain upon integration into vesicle membranes is a likely candidate for interaction with any host factors that direct cellular membrane traffic. An understanding of how the virus specifically targets these *trans*-Golgi membrane vesicles, and how this targeting affects normal function of these vesicles, will likely lead to a better understanding of virus symptom production, and provide new insights into critical host developmental controls.

See also: Fungal Viruses; Hypoviruses.

Further Reading

Boland GJ (2004) Fungal viruses, hypovirulence, and biological control of *Sclerotinia* species. *Canadian Journal of Plant Pathology* 26: 6–18.

Choe SS, Dodd DA, and Kirkegaard K (2005) Inhibition of cellular protein secretion by picornaviral 3A proteins. *Virology* 337: 18–29.

Ghabrial SA (1994) New developments in fungal virology. *Advances in Virus Research* 43: 303–388.

Ghabrial SA (1998) Origin, adaptation and evolutionary pathways of fungal viruses. *Virus Genes* 16: 119–131.

Hillman BI and Suzuki N (2004) Viruses of the chestnut blight fungus, *Cryphonectria parasitica*. In: Maramorosch K and Shatkin AJ (eds.) *Advances in Virus Research, vol.* 63, pp. 423–472. San Diego, CA: Academic Press.

Jacob-Wilk D, Turina M, and Van Alfen NK (2006) Mycovirus cryphonectria hypovirus 1 elements cofractionate with *trans*-Golgi network membranes of the fungal host *Cryphonectria parasitica*. *Journal of Virology* 80: 6588–6596.

Kazmierczak P, Kim DH, Turina M, and Van Alfen NK (2005) A hydrophobin of the chestnut blight fungus, *Cryphonectria parasitica*, is required for stromal pustule eruption. *Eukaryotic Cell* 4: 931–936.

Marsh M (2005) *Current Topics in Microbiology and Immunology, Vol. 285: Membrane Trafficking in Viral Replication*. Heidelberg: Springer.

McCabe PM, Pfeiffer PL, and Van Alfen NK (1999) The influence of dsRNA viruses on the biology of plant pathogenic fungi. *Trends in Microbiology* 7: 377–381.

McCabe PM and Van Alfen NK (2002) Molecular basis of symptom expression by the *Cryphonectria* hypovirus. In: Tavantzis SM (ed.) *dsRNA Genetic Elements*, pp. 125–144. Washington, DC: CRC Press.

Milgroom MG and Cortesi P (2004) Biological control of chestnut blight with hypovirulence: A critical analysis. *Annual Review of Phytopathology* 42: 311–338.

Nuss DL (2005) Hypovirulence: Mycoviruses at the fungal–plant interface. *Nature Reviews Microbiology* 3: 632–642.

Turina M, Zhang L, and Van Alfen NK (2006) Effect of *Cryphonectria hypovirus 1* (CHV1) infection on Cpkk1, a mitogen-activated protein kinase kinase of the filamentous *Cryphonectria parasitica*. *Fungal Genetics and Biology* 43: 764–774.

Villanueva RA, Rouille Y, and Dubuisson J (2005) Interactions between virus proteins and host cell membranes during the viral life cycle. *International Review of Cytology – A Survey of Cell Biology* 245: 171–244.

Hypoviruses

D L Nuss, University of Maryland Biotechnology Institute, Rockville, MD, USA

© 2008 Elsevier Ltd. All rights reserved.

Glossary

Anastomosis The fusion of fungal hyphae resulting in exchange of cytoplasmic material and hypovirus transmission.

Hypovirulence Virus-mediated attenuation of fungal virulence.

Vegetative incompatibility A system controlled by at lease six genetic loci that determines the ability of two fungal strains to undergo anastomosis.

Introduction

The discovery of hypoviruses, a group of RNA viruses that reduce the virulence (hypovirulence) of the chestnut blight fungus *Cryphonecria parasitica*, has stimulated intensive research into the potential of using fungal viruses for biological control of fungal diseases. Documented examples of virus-mediated hypovirulence have been reported for fungal diseases of plants that range from trees to turfgrass and involve mycoviruses, which include representatives from the *Totiviridae, Chrysoviridae, Reoviridae, Narnaviridae,* and *Hypoviridae* (the hypoviruses). However, the hypoviruses remain the most thoroughly studied of the hypovirulence-associated viruses, primarily due to several significant advances in hypovirus molecular biology.

A classic example of the havoc that can result from the introduction of an exotic organism, the North American chestnut blight epidemic, first reported in 1905, caused the destruction of millions of mature chestnut trees by 1950. *Cryphonecria parasitica* was subsequently introduced into Italy during the 1930s, threatening European chestnut forests and orchards. However, for reasons still to be completely understood, the European chestnut blight epidemic was much less severe than that witnessed in North America. One clear contribution to this reduced severity was the prevalence of *C. parasitica* strains exhibiting a reduced virulence phenotype (hypovirulence), first described by an Italian forest pathologist in the 1950s. French investigators subsequently showed that the hypovirulence phenotype was transmissible following anastomosis (fusion of hyphae) between vegetatively compatible *C. parasitica* strains, implicating a cytoplasmic genetic element as the causative agent of the phenotype. The observation that a hypovirulent strain could produce a curative effect when inoculated onto existing cankers on diseased trees stimulated a successful government-sponsored biological control program using hypovirulent *C. parasitica* strains for management of chestnut blight in French chestnut plantations. Recent studies in Switzerland also support the view that natural hypovirulent strains retard disease progression in European forest ecosystems.

In 1977, Peter Day and co-workers at the Connecticut Agricultural Experiment Station reported that hypovirulent *C. parasitica* strains harbor double-stranded (ds) RNAs, providing the first indication of the nature of cytoplasmic elements responsible for the phenotype. Subsequent surveys of dsRNAs associated with different North American and European hypovirulent strains revealed considerable variations in concentration, number, and size of dsRNA components. By the late 1980s, it was clear that a detailed molecular analysis of the dsRNAs associated with a single hypovirulent strain was required to bring some measure of order to the mounting confusion generated by such surveys. This resulted in the cloning and complete sequence determination of the prototypic hypovirus, now designated CHV1-EP713, in 1991. This milestone was followed in 1992 by the construction of an infectious full-length (12 712 bp) cDNA clone of CHV1-EP713 RNA by Choi and Nuss. This development furnished direct evidence that hypoviruses are indeed the causative agents responsible for transmissible hypovirulence and provided the means for facile manipulation of the hypovirus genome. Current interest in hypoviruses extends past biological control potential to their utility as unique experimental tools for probing fundamental processes underlying fungal pathogenesis and mycovirus–fungal host interactions.

Taxonomy and Genetic Organization

Hypoviruses are classified within the family *Hypoviridae*, consisting of the genus *Hypovirus* and four species designated *C. parasitica* hypovirus 1–4 (CHV1–CHV4). Hypovirus taxonomy is not based on virus structure, since this group of viruses does not encode a coat protein, but on genome organization, sequence similarity and symptom expression. Hypovirus genetic information is found predominantly as dsRNA associated with membrane vesicles ranging in diameter from 50 to 80 nm. As observed for fungal viruses generally, hypoviruses exhibit no extracellular phase in their life cycle. Infections cannot be initiated by inoculation with an infected cell extract or enriched fractions. Instead, these viruses are transmitted by cytoplasmic mixing as a result of fusion (anastomosis) between vegetatively compatible strains or to a variable degree in asexual spores (conidia).

The hypovirus species designations CHV1–CHV4 were assigned in the order in which their genome sequence was completed. The primary nucleotide sequence for the coding strand of the prototypic member of the genus *Hypovirus*, CHV1-EP713, specifies two large open reading frames (ORFs) designated ORF A and ORF B (**Figure 1**). A second closely related member, CHV1-Euro7, has the same organization and shares approximately 90% identity at the nucleotide level with CHV1-EP713. The type member of the CHV2 species, CHV2-NB58, shares only about 60% nucleotide sequence identity with CHV1-EP713 and lacks a portion of ORF A that, for the CHV1 species, encodes a functional *cis*-acting cysteine protease. Sequenced members of the CHV3 and CHV4 species, CHV4-GH2 and CHV4-SR2, respectively, are several kbp shorter than the CHV1 and CHV2 species, ~9.2–9.8 kbp versus 12.5–12.7 kbp, contain a single large ORF rather than two ORFs and are both more distantly related phylogenetically to species CHV1 than CHV1 is to CHV2.

Hypovirus Gene Expression Strategy

Although hypovirus genetic information is readily recovered from infected cultures as linear dsRNA, the absence of a discrete virus particle and an extracellular infection phase presents some difficulties in precisely defining the hypovirus genome. Synthetic copies of the coding strand are infectious by electroporation into fungal spheroplasts and phylogenetic analyses suggest a common ancestry with the positive strand RNA plant potyviruses. Thus, one could consider hypoviruses as having a positive strand RNA genome and the dsRNA as representing accumulated replicative form RNA. Irrespective of this complication, direct analysis of hypovirus dsRNAs, cDNA cloning studies, and *in vitro* translational analyses has provided the following view of genetic organization and expression strategies for the prototypic hypovirus CHV1-EP713 shown in **Figure 2**. One strand contains a 3′-poly A tail while the complementary strand contains a 5′-poly U tract. All of the CHV1-EP713 coding information appears to reside within two contiguous ORFs on the 12 712 nt long polyadenylated strand. ORF A encodes two polypeptides, p29 and p40, that are released from a polyprotein precursor, p69, by an autoproteolytic event between Gly-248 and Gly-249, mediated by a

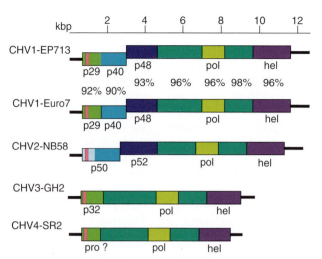

Figure 1 Genetic organization of sequenced hypovirus genomes representing the four species that comprise the genus *Hypovirus*, family *Hypoviridae*. Amino acid identity levels for coding regions of the two sequenced members of the CHV1 species, CHV1-EP713 and CHV1-Euro7, are indicated between representations of the two viral genomes. Protein coding regions homologous to CHV1-EP713 encoded p29, p40, p48, polymerase and helicase are color coded. The magenta regions represent a short conserved cysteine rich domain. Note that the genomes of CHV3-GH2 and CHV4-SR2 contain a single ORF. Modified from Dawe AL and Nuss DL (2001) Hypoviruses and chestnut blight: Exploiting viruses to understand and modulate fungal pathogenesis. *Annual Review of Genetics* 35: 1–29, with permission from Annual Reviews.

cysteine-like protease catalytic domain located within p29. ORF B has the capacity to encode a polyprotein of 3165 amino acids and contains unmistakable RNA-dependent RNA polymerase and helicase motifs. Proteolytic processing of only a portion of the ORF B polyprotein has been elucidated in the form of the autoproteolytic release of a 48 kDa protein, p48, from the N-terminus. This cleavage event occurs between Gly-418 and Ala-419 and is catalyzed by essential residues Cys-341 and His-388 within p48. The junction between ORF A and ORF B is defined by the sequence 5′-UAAUG-3′, in which the UAA portion clearly serves as the termination codon for ORF A and the AUG portion is thought to serve as the initiation codon for ORF B. While the mechanism involved in ribosome transition through the junction is not known, this unusual pentanucleotide sequence is found at the ORF A/ORF B junction for all confirmed CHV1 species. There is clearly a need for additional fine detailed mapping of the processing cascades for hypovirus-encoded polyproteins.

The genome of CHV2-NB58, type strain of species CHV2, also consists of a two ORF configuration with a UAAUG junction (**Figure 2**). However, ORF A lacks the p29 papain-like catalytic or cleavage sites and directs the translation of a 50 kDa protein product. ORF B of CHV2-NB58 does contain a p48 homolog, p52. The N-terminal portion of the single ORF of the CHV3-type strain CHV3-GH2 contains a protease, p32, with similarity to p29. A putative protease domain has been identified at the N-terminal portion of the CHV4-type strain CHV4-SR2, but protease cleavage has not been demonstrated.

In most hypovirus-infected *C. parasitica* isolates, the full-length viral dsRNA is accompanied by a constellation of shorter dsRNA species. These ancillary dsRNAs appear to be generated by internal deletion events, are replicated only in the presence of the full-length viral RNA, and are not associated with any function or phenotypic effect.

Hypovirus–Host Interactions

The phenotypic changes that are associated with hypovirus infection are not limited to hypovirulence. Additional hypovirus-mediated phenotypic changes can include altered colony morphology, female infertility, reduced asexual sprroulation (conidiation), and reduced pigmentation. The pleiotropic nature of these changes suggests that hypoviruses might perturb one or several cellular signaling pathways. Consistent with this view, alterations in G-protein, cyclic AMP-mediated-, mitogen-activated protein kinase-, and calcium/calmodulin/inositol trisphosphate-dependent-signaling pathways have been reported for hypovirus-infected fungal strains.

The construction of an ordered expressed sequence tag (EST) library representing approximately 2200 *C. parasitica* genes and development of a corresponding *C. parasitica* cDNA microarray platform has provided a view of host transcriptional responses to hypovirus infection. The modulation of transcript accumulation for approximately 13.4% of the 2200 cDNAs was shown following CHV1-EP713 infection. These transcriptional profiling studies also resulted in the identification of a

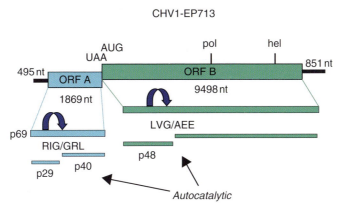

Figure 2 Expression strategy for prototypic hypovirus CHV1-EP713. The CHV1-EP713 coding strand consists of 12 712 nucleotides excluding a poly(A) tail, and contains two major coding domains designated ORF A and ORF B. Details are discussed in the text. Modified from Dawe AL and Nuss DL (2001) Hypoviruses and chestnut blight: Exploiting viruses to understand and modulate fungal pathogenesis. *Annual Review of Genetics* 35: 1–29 with permission from Annual Reviews.

subset of hypovirus responsive genes regulated through the G-protein signaling pathway and revealed a linkage between viral and mitochondrial hypovirulence.

Hypovirus-encoded symptom determinants and important replication elements have been mapped (**Figure 3**) by a combination of approaches that include (1) the construction of recombinant chimeras from hypoviruses that differ in their influence on host phenotype, (2) mutagenesis of a hypovirus infectious cDNA clone, and (3) cellular expression of viral coding domains independent of virus infection.

In analogy with plant viruses, CHV1-EP713 and CHV1-Euro7 can be viewed as severe and mild hypovirus isolates, respectively. Although these two CHV1 isolates cause quite different phenotypic changes in their fungal host, they share a high level of sequence similarity that has allowed the construction of viable chimeric viruses to begin mapping the determinants responsible for the differences in phenotypic changes.

Differences in colony morphologies were found to map to a region extending from a position just downstream of the p48 coding domain (map position 3575) to map position 9879, with clear indications of multiple discrete determinants. More specifically, the region extending from position 3575 to 5310 was able to confer a CHV1-EP713-like colony morphology when inserted into a CHV1-Euro7 genetic background. The CHV1-EP713 p48 coding region was found to be a dominant determinant contributing to suppression of asexual spore formation on the canker face.

The chimeric hypoviruses also proved to be very useful reagents when coupled with a pathway specific promoter/reporter system to map viral determinants responsible for altering G-protein/cAMP-mediated signaling. A common undesired side effect of hypovirus-mediated virulence attenuation is a significantly reduced ability of the fungal host to colonize and produce spores on the corresponding plant host. This reduces the ability of hypovirulent fungal strains to persist and spread through the ecosystem. Thus, from a practical perspective, a better understanding of the nature of viral symptom determinants and their relative effects on specific regulatory pathways and expression of gene clusters provides the means for a more rational approach for engineering hypoviruses that exhibit a desired balance between virulence attenuation and ecological fitness.

Additional insights into the functional role of viral coding regions were indirectly provided during efforts to develop hypoviruses as gene expression vectors. The nucleotide sequence corresponding to the first 24 codons of p29 was found to be required for viral replication, while the remaining 598 codons of ORF A, including all of the p40 coding region, was found to be dispensable. Substantial alterations were also tolerated in the pentanucleotide UAAUG that contains the ORF A termination codon and the overlapping putative ORF B initiation codon. For example, replication competence was maintained following either a frameshift mutation that caused a two-codon extension of ORF A or a modification that produced a single-ORF genomic organization. Further charactereization of p40 revealed a role as an accessory function in viral RNA amplification with a functional domain extending from Thr(288) to Asn(313).

Expression of the CHV1-EP713 encoded papain-like protease, p29, in the absence of virus infection was shown to cause a subset of phenotypic changes exhibited by CHV1-EP713-infected strains, for example, a white phenotype (reduction in orange pigmentation), reduced asexual sporulation and a slight reduction in the production of fungal laccase activity. By deleting all but the first 24 N-terminal codons of p29 in the context of the CHV1-EP713 infectious cDNA clone (mutant virus Δp29), it was also possible to show that the p29 protein is dispensable

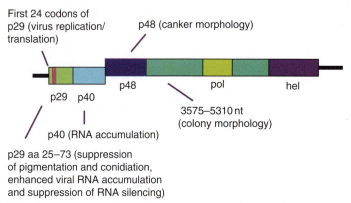

Figure 3 Emerging map of CHV1-EP713 symptom determinants and essential/dispensable replication elements as described in the text. Adapted from Dawe AL and Nuss DL (2001) Hypoviruses and chestnut blight: Exploiting viruses to understand and modulate fungal pathogenesis. *Annual Review of Genetics* 35: 1–29, with permission from Annual Reviews.

for viral replication and to demonstrate a near restoration of orange pigment production and a moderate increase in conidation levels relative to wild-type CHV1-EP713-infected fungal colonies. Deletion of p29 had no effect on virus-mediated virulence attenuation. A gain-of-function analysis involving progressive repair of the Δp29 mutant was also devised to map the p29 symptom determinant domain to a region extending from Phe-25 to Gln-73. When expressed from a chromosomally integrated cDNA copy, p29 elevated RNA accumulation and vertical transmission (through conidia) of the Δp29 mutant virus to levels observed for wild-type CHV1-EP713. Additional mutational studies indicated a linkage between p29-mediated changes in host phenotype in the absence of virus infection and p29-mediated in *trans*-enhancement of viral RNA accumulation and transmission.

The multifunctional nature of p29 was recently extended to include suppressor of RNA silencing both in the natural fungal host and in a heterologous plant system. This activity was predicted based on similarities between p29 and the well-characterized potyvirus-encoded suppressor of RNA silencing HC-Pro and has provided the first circumstantial evidence that RNA silencing in fungi may serve as an antiviral defense mechanism, a well-established function in plants.

Prospects for Biological Control

Chestnut blight cankers on American chestnut trees can be controlled by the direct application of hypovirus-infected hypovirulent *C. parasitica* strains to the canker margin, provided that the hypovirulent and resident virulent strains are vegetatively compatible. Anastomosis between the two strains results in hypovirus transmission and conversion of the canker to the hypovirulent phenotype, thereby preventing additional expansion of the canker and promoting formation of new callus tissue. However, since treated trees do not become resistant to new infections and North American *C. parasitica* populations generally exhibit a very diverse vegetative compatibility structure, this form of treatment is labor intensive and has not proved to be practical for effective control of chestnut blight in a North American forest ecosystem.

High levels of VC diversity and the severe phenotypic characteristics of the hypoviruses that have been introduced for biocontrol are thought to be two of the main contributing factors to the inability of introduced hypoviruses to spread through North America *C. parasitica* populations. In this regard, transgenic hypovirulent strains that contain a nuclear copy of the hypovirus cDNA exhibit novel hypovirus transmission properties that have been predicted to reduce these limitations. Hypovirus RNA is not transmitted to ascospores during mating. However, the hypovirus cDNA present in transgenic hypovirulent strains is inherited by a portion of the ascospore progeny, followed by the production of cDNA-derived cytoplasmically replicating viral RNA. Additionally, since the progeny represent a spectrum of vegetative compatibility groups, launching of the cytoplasmic viral RNA into these new vegetative compatibility groups is predicted to expand vegetative dissemination of hypovirus genetic information.

Three field trials with CHV1-EP713 transgenic strains have confirmed hypovirus transmission to ascospore progeny in a forest setting and also exposed the requirement for significant improvements in formulation and delivery methods. These trials also confirmed predictions that the CHV1-EP713 hypovirus performs poorly as a biological control agent because it severely reduces the ability of the fungal host to colonize, expand, and produce asexual spores on chestnut tissue. In this regard, the availability of an infectious cDNA clone for the mild hypovirus CHV1-Euro7 has provided the means to construct a transgenic hypovirulent strain that combines the properties of enhanced colonization and spore production with a novel mode of hypovirus transmission to ascospore progeny.

The infectious hypovirus cDNA clones have also been used to expand hypovirus host range to include several other pathogenic fungi that include the *Eucalyptus* canker pathogen *Cryphonectria cubensis* and the fruit tree pathogens *Phomopsis G-type* and *Valsa ceratosperma*. Hypovirus-based management of these economically important fungal diseases in controlled agricultural settings may not face the same problems of formulation and application that has been encountered for control of chestnut blight in a complex forest ecosystem.

See also: Fungal Viruses; Hypovirulence.

Further Reading

Dawe AL and Nuss DL (2001) Hypoviruses and chestnut blight: Exploiting viruses to understand and modulate fungal pathogenesis. *Annual Review of Genetics* 35: 1–29.

Hillman BI and Suzuki N (2004) Viruses of the chestnut blight fungus, *Cryphonectria parasitica*. *Advances in Virus Research* 63: 423.

Nuss DL (2005) Hypovirulence: Mycoviruses at the fungal–plant interface. *Nature Reviews Microbiology* 3: 632.

Segers GC, van Wezel R, Zhang X, Hong Y, and Nuss DL (2006) Hypovirus papain-like protease p29 suppresses RNA silencing in the natural fungal host and in a heterologous plant system. *Eukaryotic Cell* 5: 896.

Set
978-0-12-373935-3

Volume 2 of 5
978-0-12-373937-7